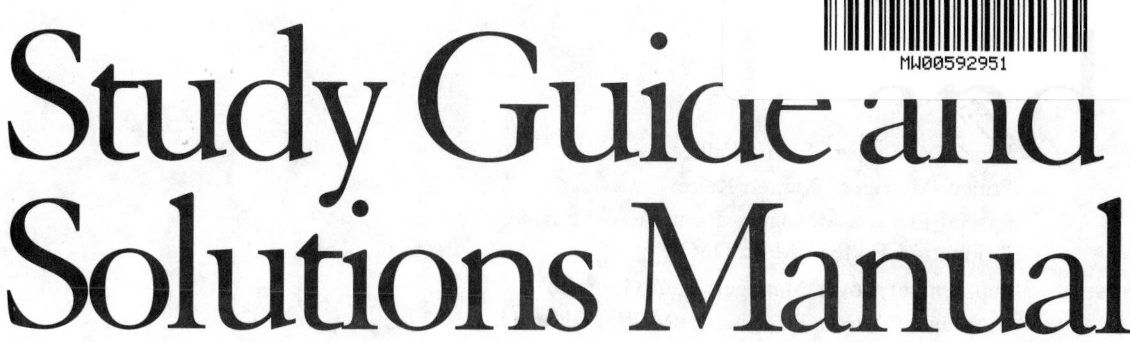

Study Guide and Solutions Manual

Organic Chemistry

PAULA YURKANIS BRUICE

THIRD EDITION

Upper Saddle River, NJ 074548

Executive Editor: John Challice
Project Manager: Kristen Kaiser
Special Projects Manager: Barbara A. Murray
Production Editor: Marc DeCarlo
Supplement Cover Manager: Paul Gourhan
Supplement Cover Designer: PM Workshop Inc.
Manufacturing Manager: Trudy Pisciotti

Printed in the United States of America

10 9 8 7 6 5 4 3 2 1

ISBN 0-13-017859-4

Prentice-Hall International (UK) Limited, London
Prentice-Hall of Australia Pty. Limited, Sydney
Prentice-Hall Canada, Inc., Toronto
Prentice-Hall Hispanoamericana, S.A., Mexico
Prentice-Hall of India Private Limited, New Delhi
Pearson Education Asia Pte. Ltd., Singapore
Prentice-Hall of Japan, Inc., Tokyo
Editora Prentice-Hall do Brazil, Ltda., Rio de Janeiro

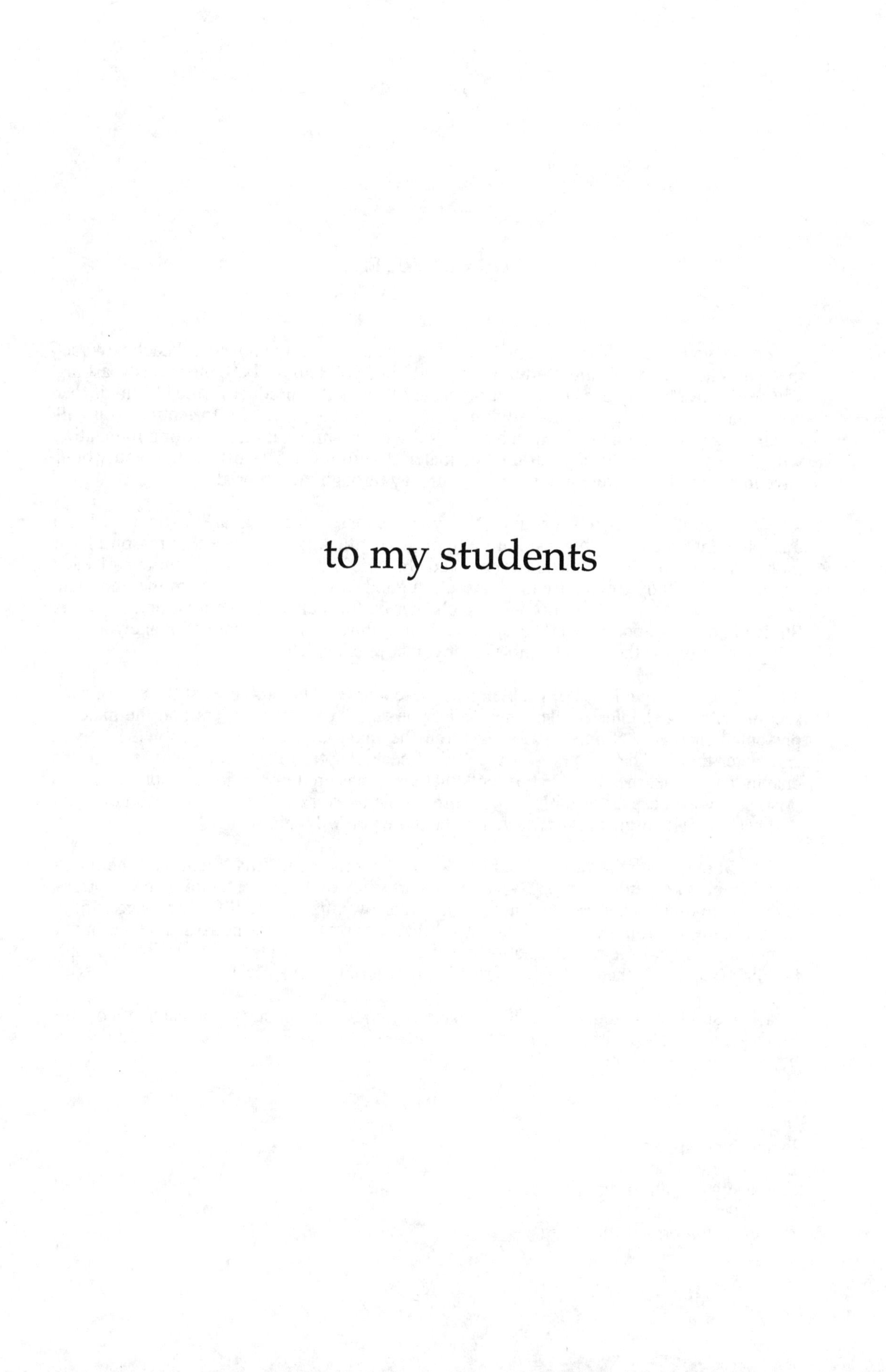

to my students

TO THE STUDENT

As you study organic chemistry, make certain you fully understand each new fact that you encounter. While studying the material, you should be continuously asking yourself "why?": Why does that reaction occur? Why is that product formed? Why is that compound more stable? If you truly understand each new piece of information, you will be creating a foundation upon which to lay subsequent information. A strong foundation will allow you to master a great deal of material with much less effort than you would have to put out if you were memorizing your way through the material.

Often it is the new vocabulary that you encounter when you are first exposed to a discipline that can be the biggest hurdle to mastering the material. For that reason I have included a list of the important terms and their definitions at the beginning of each chapter in the Study Guide. Reading these is a good way to review some of the important aspects of the chapter. I did not include chapter outlines either in the textbook or in this Study Guide/Solutions Manual because the best outline is the one that you make yourself, using the book together with the notes that you have taken in class.

There are two kinds of problems in the textbook. The problems at the end of each section within a chapter are designed to let you see if you have understood the material presented in that section, and to reinforce the material. Do these problems as you encounter them. The problems at the end of each chapter integrate the concepts in the chapter and sometimes include concepts that were mastered in previous chapters. Try to do as many of these as possible. The more problems you do, the more comfortable you will become with organic chemistry, and the more you will enjoy it.

Organic chemists use curved arrows to show the bonds that break and the bonds that form in an organic reaction. When you start studying organic reactions in Chapter 3, take time to do the exercise on drawing curved arrows that you will find on page 115. It only takes a few minutes but I have found that it makes an enormous difference in my students being able to write reaction mechanisms. There is also an exercise in model building (page 143) that will help you with the material in Chapter 4.

Good luck in your study. If you have any comments or suggestions about how the Study Guide could be improved for those students who will follow you, I would be very happy to hear from you.

Paula Yurkanis Bruice
Department of Chemistry
University of California
Santa Barbara, CA 93106
pybruice@bioorganic.ucsb.edu

TABLE OF CONTENTS

CHAPTER 1
Electronic Structure and Bonding. Acids and Bases

Important Terms

acid	a substance that donates a proton.
acid-base reaction	a reaction in which an acid donates a proton to a base.
acid dissociation constant	a measure of the degree to which an acid dissociates.
acidity	is a measure of how easily a compound gives up a proton.
antibonding molecular orbital	a molecular orbital that results when two atomic orbitals with opposite signs interact. Electrons in an antibonding orbital decrease bond strength.
atomic number	tells how many protons (or electrons) the neutral atom has.
atomic orbital	an orbital associated with an atom.
atomic weight	the average mass of the atoms in the naturally occurring element.
aufbau principle	states that an electron will always go into the available orbital with the lowest energy.
base	a substance that accepts a proton.
basicity	describes the tendency of a compound to share its electrons with a proton.
bond dissociation energy	the amount of energy required to break a bond (or the amount of energy released when a bond is formed).
bonding molecular orbital	a molecular orbital that results when two atomic orbitals with the same sign interact. Electrons in a bonding orbital increase bond strength.
bond length	the internuclear distance between two atoms at minimum energy (maximum stability).
Brønsted acid	a substance that donates a proton.
Brønsted base	a substance that accepts a proton.
buffer solution	solution of an acid and its conjugate base.
carbanion	a compound containing a negatively charged carbon.
carbocation	a compound containing a positively charged carbon.
condensed structure	a structure that does not show some (or all) of the covalent bonds.
conjugate acid	the compound formed when a base accepts a proton.
conjugate base	the compound formed when an acid loses a proton.
core electrons	electrons in filled shells.

covalent bond a bond created as a result of sharing electrons.

degenerate orbitals orbitals that have the same energy.

delocalized electrons electrons that do not belong to a single atom nor are they shared in a bond between two atoms.

dipole a positive end and a negative end.

dipole moment (μ) a measure of the separation of charge in a bond or in a molecule.

double bond a sigma bond and a pi bond.

electron affinity the energy given off when an atom acquires an electron.

electronegative describes an element that readily acquires an electron.

electronegativity the tendency of an atom to pull electrons toward itself.

electropositive describes an element that readily loses an electron.

electrostatic attraction an attractive force between opposite charges.

electrostatic potential map a map that allows you to see how electrons are distributed in a molecule.

equilibrium constant the ratio of products to reactants at equilibrium (or the ratio of the rate constants for the forward and reverse reactions).

excited-state electronic configuration the electronic configuration that results when an electron in the ground state has been moved to a higher energy orbital.

formal charge the number of valence electrons − (the number of nonbonding electrons + 1/2 the number of bonding electrons).

free radical (radical) a species with an unpaired electron.

ground-state electronic configuration a description of which orbitals the electrons of an atom occupy when they are all in their lowest energy orbitals.

Heisenberg uncertainty principle states that both the precise location and the momentum of an atomic particle cannot be simultaneously determined.

Henderson-Hasselbalch equation $pK_a = pH + \log[HA]/[A^-]$

Hund's rule states that when there are degenerate orbitals, an electron will occupy an empty orbital before it will pair up with another electron.

hybrid orbital an orbital formed by hybridizing (mixing) atomic orbitals.

hydride ion a negatively charged hydrogen.

hydrogen ion (proton) a positively charged hydrogen.

inductive electron withdrawal	the tendency of an atom or a group of atoms to pull electrons toward itself.
ionic bond	a bond formed through the attraction of two ions of opposite charges.
ionic compound	a compound composed of a positive ion and negative ion.
ionization energy	the energy required to remove an electron from an atom.
isotopes	atoms with the same number of protons but a different number of neutrons.
Kekulé structure	a model that represents the bonds between atoms as lines.
Lewis acid	a substance that accepts an electron pair.
Lewis base	a substance that donates an electron pair.
Lewis structure	a model that represents the bonds between atoms as lines or dots and the nonbonding electrons as dots.
lone pair electrons	valence electrons not used in bonding.
mass number	the number of protons plus the number of neutrons in an atom.
molecular orbital	an orbital associated with a molecule.
molecular orbital (MO) theory	describes a model in which the electrons occupy orbitals as they do in atoms but the orbitals extend over the entire molecule.
node	a region within an orbital where there is zero probability of finding an electron.
nonbonding electrons	valence electrons not used in bonding.
nonpolar covalent bond	a bond formed between two atoms that share the bonding electrons equally.
octet rule	states that an atom will give up, accept, or share electrons in order to achieve a filled shell. Because a filled second shell contains eight electrons, this is known as the octet rule.
orbital	the volume of space around the nucleus where an electron is most likely to be found.
orbital hybridization	mixing of atomic orbitals.
organic compound	a compound that contains carbon.
Pauli exclusion principle	states that no more than two electrons can occupy an orbital and that the two electrons must have opposite spin.
pH	the pH scale is used to describe the acidity of a solution ($pH = -\log[H^+]$).
pi (π) bond	a bond formed as a result of side-to-side overlap of *p* orbitals.

pK_a describes the tendency of a compound to lose a proton (pK_a = -log K_a, where K_a is the acid dissociation constant).

polar covalent bond a bond formed between two atoms that do not share the bonding electrons equally.

potential map (electrostatic potential map) a map that allows you to see how electrons are distributed in a molecule.

proton a positively charged hydrogen; a positively charged atomic particle.

proton transfer reaction a reaction in which a proton is transferred from an acid to a base.

quantum mechanics the use of mathematical equations to describe the behavior of electrons in atoms or molecules.

radical (free radical) a species with an unpaired electron.

resonance a compound with delocalized electrons is said to have resonance.

resonance contributors structures with localized electrons that approximate the true structure of a compound with delocalized electrons.

resonance hybrid the actual structure of a compound with delocalized electrons.

sigma (σ) bond a bond with a symmetrical distribution of electrons about the internuclear axis.

single bond a single pair of electrons shared between two atoms.

tetrahedral bond angle the bond angle (109.5°) formed by any two bonds of an sp^3 hybridized carbon.

tetrahedral carbon an sp^3 hybridized carbon; a carbon that forms covalent bonds using four sp^3 hybrid orbitals.

trigonal planar carbon an sp^2 hybridized carbon.

triple bond a sigma bond plus two pi bonds.

valence electrons an electron in an outermost shell.

valence shell electron pair repulsion (VSEPR) model a model that combines the concepts of atomic orbitals and shared electron pairs with minimization of electron repulsions.

wave equation an equation that describes the behavior of each electron in an atom or a molecule.

wave functions a series of solutions of a wave equation.

Solutions to Problems

1. The atomic number = the number of protons.
The mass number = the number of protons + neutrons.

Therefore:

The isomer of oxygen with a mass number = 16 has 8 protons and 8 neutrons.
The isomer of oxygen with a mass number = 17 has 8 protons and 9 neutrons.
The isomer of oxygen with a mass number = 18 has 8 protons and 10 neutrons.

2. Using the aufbau principle and the Pauli exclusion principle, and remembering that the first shell has one *s* orbital and the second shell has one *s* orbital and three *p* orbitals, one can see that potassium's 19th electron is in a 4*s* orbital.

$1s\ (2e^-),\ 2s\ (2e^-),\ 2p\ (6e^-),\ 3s\ (2e^-),\ 3p\ (6e^-),\ 4s\ (1e^-)$

3. Cl $1s^2\ 2s^2\ 2p^6\ 3s^2\ 3p^5$
Br $1s^2\ 2s^2\ 2p^6\ 3s^2\ 3p^6\ 4s^2\ 3d^{10}\ 4p^5$
I $1s^2\ 2s^2\ 2p^6\ 3s^2\ 3p^6\ 4s^2\ 3d^{10}\ 4p^6\ 5s^2\ 4d^{10}\ 5p^5$

Notice that because the three atoms are in the same vertical row of the periodic table, they all have the same number of valence electrons and the valence electrons are in similar orbitals (2 are in an *s* orbital and 5 are in *p* orbitals).

4. **a.** KCl has the most polar bond. From Table 1.3 you can calculate that the difference in the electronegativities of the atoms sharing the bonding electrons is 2.2 (3.0 - 0.8), whereas it is 1.8 for LiBr, 1.6 for NaI, and 0 for Cl_2.

b. Cl_2 has the least polar bond because the two chlorine atoms share the bonding electrons equally.

5. **a.** Lithium's hydrogen is red, each hydrogen of H_2 is green, and HF has a blue hydrogen. Therefore, of the three molecules, HF has the most positively charged (blue) hydrogen.

b. Because the hydrogen of LiH is red, LiH has the most negatively charged hydrogen.

c. The electrostatic potential map of H_2 does not have blue (+) and red (-) areas. Therefore, we know that it has the most nonpolar bond.

6. Solved in the text.

7. To answer this question, compare the electronegativities of the two atoms sharing the bonding electrons using Table 1.3 on page 10 of the text. (Note that if the atoms being compared are in the same row of the periodic chart, the atom on the right is the more electronegative; if the atoms being compared are in the same column, the one closer to the top of the column is the more electronegative.)

a. $\overset{\delta+}{H_3C}-\overset{\delta-}{Cl}$

b. $\overset{\delta-}{F}-\overset{\delta+}{Br}$

c. $\overset{\delta+}{H_3C}-\overset{\delta-}{NH_2}$

d. $\overset{\delta+}{H_3C}-\overset{\delta-}{OH}$

e. $\overset{\delta-}{HO}-\overset{\delta+}{Br}$

f. $\overset{\delta-}{H_3C}-\overset{\delta+}{MgBr}$

g. $\overset{\delta+}{I}-\overset{\delta-}{Cl}$

h. $\overset{\delta+}{H_2N}-\overset{\delta-}{OH}$

8. By answering this question you will see that a formal charge is a book-keeping device. It does *not necessarily* tell you which atom has the greatest electron density or is the most electron deficient.

a. oxygen **b.** oxygen (it is the most red)

In this example, the atom with the formal negative charge **is** the atom with the greatest electron density.

c. oxygen **a.** hydrogen (it is the most blue)

In this example, the atom with the formal positive charge **is not** the most electron deficient atom.

9.

a. $^{-}$:Ö:N:Ö:$^{-}$ (with :Ö: double-bonded above N; + on N)

b. :Ö::N^{+}::Ö:

c. :Ö::N̈:Ö:$^{-}$

d. :Ö::C::Ö:

e. $^{-}$:Ö:C:Ö:H (with :Ö: double-bonded above C)

f. :N:::N:

g. H:C:N:H^{+} (H H above; H H below)

h. H:C:C^{+}:H (H above each C; H H below)

i. H:C̈$^{-}$:H (H below)

j. Na^{+} $^{-}$:Ö:H

k. H:N:H^{+} (H above, H below) :C̈l:$^{-}$

l. Na^{+} $^{-}$:Ö:C:Ö:$^{-}$ Na^{+} (with :Ö: double-bonded above C)

10.

```
    H H                        H H H                      H
    ·· ··                      ·· ·· ··                   ··
a.  H:C:C:Ö:H            b.  H:C:C:C:Ö:H                H:C:H
    ·· ··  ··                  ·· ·· ··  ··               ··
    H H                        H H H                    H: C : Ö:H
                                                            ··  ··
    H    H                     H H   H                  H:C:H
    ··   ··                    ·· ·· ·· ··                ··
  H:C:Ö:C:H                 H:C:C:Ö:C:H                   H
    ·· ·· ··                   ·· ··  ·· ··
    H    H                     H H    H
```

11.

```
a.  H     H H        b.  H     H             c.  H     H   Ö:
    |  ··  | |           |     |   ··            |     |   ||
  H-C-N-C-C-H          H-C-----C-Cl:           H-C-----C---C-H
    |  |  | |            |     |   ··            |     |
    H  H  H H            H   H-C-H               H   H-C-H
                               |                       |
                               H                       H

                 H
  d.             |
       H  H-C-H  H H H      H     H
       |    |    | | |      |     |
     H-C----C----C-C-C------C-----C-H
       |    |    | | |      |     |
       H  H-C-H  H H H    H-C-H   H
            |               |
            H               H
```

12. He_2^+ has three electrons. Two are in a bonding orbital and one is in an antibonding orbital. Because there are more electrons in the bonding orbital than in the antibonding orbital, He_2^+ exists.

13. Bonding electrons in shells farther from the nucleus form **longer** and **weaker** bonds due to poorer overlap of the bonding orbitals. Therefore:

a. **relative lengths** of the bonds in the halogens are: $\mathbf{Br_2 > Cl_2}$.
relative strengths of the bonds are: $\mathbf{Cl_2 > Br_2}$.

b. **relative lengths:** HBr > HCl > HF
relative strengths: HF > HCl > HBr

14. The carbon-carbon bonds form as a result of sp^3—sp^3 overlap.
The carbon-hydrogen bonds form as a result of sp^3—s overlap.

15. The σ bond has more effective orbital-orbital overlap because it is a stronger bond than a π bond. We know the σ bond is stronger because the σ bond in ethane has a bond dissociation energy of 88 kcal/mol, whereas the bond dissociation energy of the double bond (σ + π) in ethene is 152 kcal/mole, which is less than twice as strong.

16. We know from the bond angles of H_2O that the bond angles would be 104.5° if H_3O^+ had *two* nonbonded pairs of electrons, and we know from the bond angles of CH_4 that the bond angles would be 109.5° if it H_3O^+ had *no* nonbonded pairs. Because H_3O^+ has *one* pair of nonbonded electrons, its bond angles will be greater than 104.5° and less than 109.5°.

17. The hydrogens of the ammonium ion are the atoms with the least electron density—they are the most blue so they have the most positive (least negative) electrostatic potential.

18. Water is the most polar—it has a deep red area and the most intense blue area. Methane is the least polar—it is all the same color with no red or blue areas.

19. The carbon-carbon sigma bond formed by sp^2—sp^2 overlap is stronger because an sp^2 orbital has 33.3% *s* character while an sp^3 orbital has 25% *s* character.
Since electrons in an *s* orbital are closer to the nucleus than those in a *p* orbital, greater *s* character in interacting orbitals results in a stronger (and shorter) bond.

20.

a. 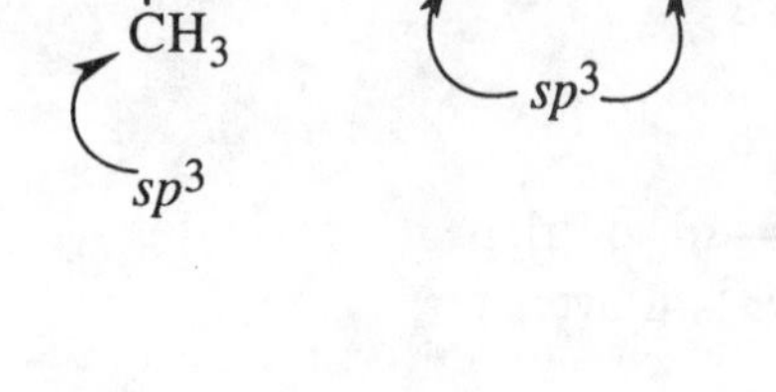 $CH_3CH(CH_3)CH{=}CHCH_2C{\equiv}CCH_3$

sp^3 sp^2 sp sp^3 sp^3

b. 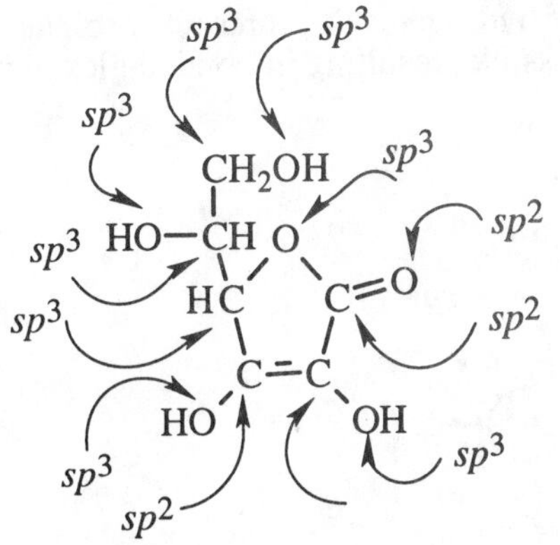 [structures of ascorbic acid and caffeine with each atom labeled sp^3 or sp^2]

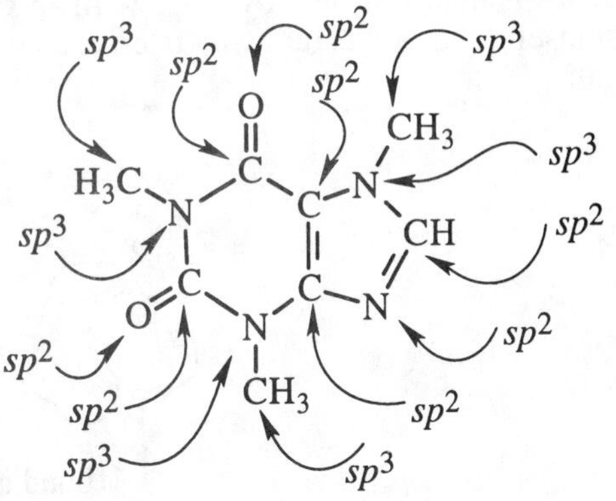

21. a. Because beryllium does not have any unpaired electrons in its ground state, it cannot form any bonds unless it promotes an electron. After promotion, hybridization of the two orbitals (an *s* and a *p*) that contain unpaired electrons results in two *sp* hybrid orbitals.

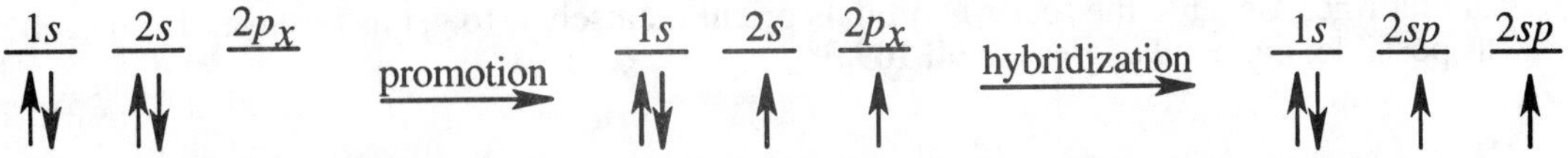

1*s*	2*s*	2p_x		1*s*	2*s*	2p_x		1*s*	2*sp*	2*sp*
↑↓	↑↓		promotion →	↑↓	↑	↑	hybridization →	↑↓	↑	↑

Each *sp* orbital of beryllium overlaps with an *s* orbital of hydrogen. The two *sp* orbitals orient themselves to get as far away from each other as possible, resulting in a bond angle in BeH_2 of **180°**.

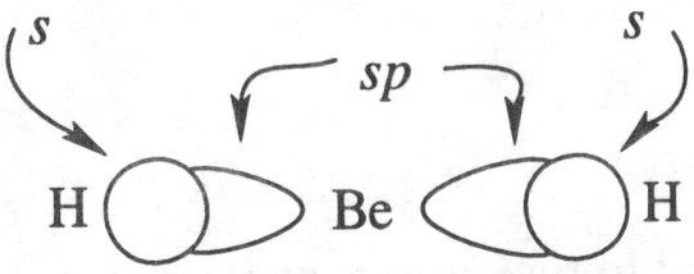

bond angle = 180°

Notice that because beryllium does not have an electron in a *p* orbital, it cannot form a π bond.

b. Without promotion, boron could form only one bond because it has only one unpaired electron. Promotion gives it three unpaired electrons. When the three orbitals (one *s* and two *p*'s) containing the unpaired electrons are hybridized, three sp^2 orbitals result.

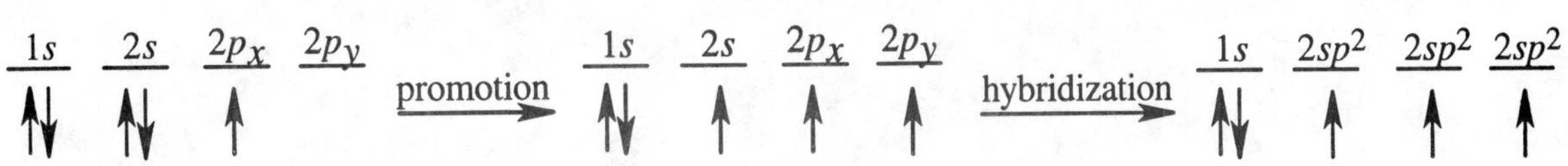

1*s*	2*s*	2p_x	2p_y		1*s*	2*s*	2p_x	2p_y		1*s*	2sp^2	2sp^2	2sp^2
↑↓	↑↓	↑		promotion →	↑↓	↑	↑	↑	hybridization →	↑↓	↑	↑	↑

Each sp^2 hybrid orbital overlaps with an *s* orbital of hydrogen. The three sp^2 orbitals orient themselves to get as far away from each other as possible, resulting in bond angles in BH_3 of **120°**.

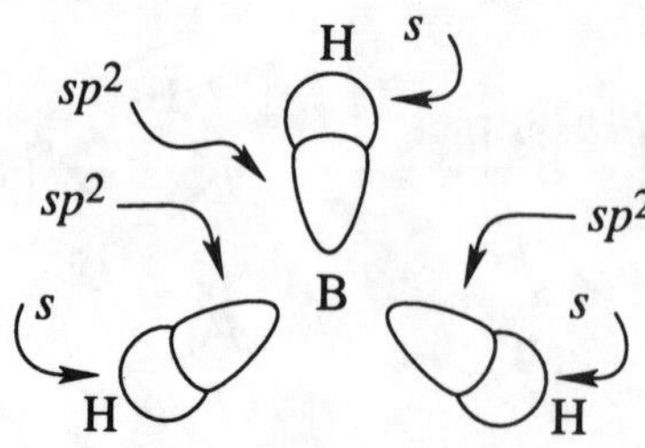

bond angle = 120°

c. The carbon in CCl_4 is bonded to four atoms, so it uses four sp^3 hybrid orbitals. Each carbon-chlorine bond is formed by the overlap of an sp^3 orbital of carbon with a *p* orbital of chlorine. Because the four sp^3 orbitals orient themselves to get as far away from each other as possible, the bond angles are all **109.5°**.

Cl
Cl—C—Cl
Cl

bond angle = 109.5°

d. The carbon in CO_2 is bonded to two atoms, so it uses two *sp* hybrid orbitals. Each carbon-oxygen bond is a double bond. One of the bonds of each double bond is formed by the overlap of an *sp* orbital of carbon with an sp^2 orbital of oxygen. The second bond of the double bond is formed as a result of side-to-side overlap of a *p* orbital of carbon with a *p* orbital of oxygen. Because the two *sp* orbitals orient themselves to get as far away from each other as possible, the bond angle in CO_2 is **180°**.

O=C=O

bond angle = 180°

e. The double-bonded carbon and the double-bonded oxygen in HCOOH use sp^2 hybrid orbitals.

The single-bonded oxygen uses sp^3 hybrid orbitals and each hydrogen uses an *s* orbital.

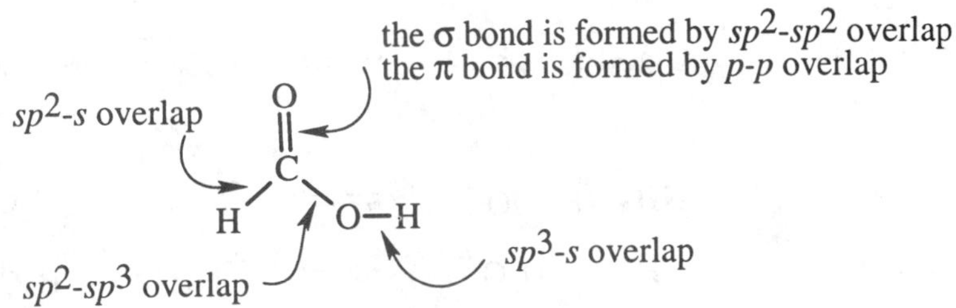

f. The triple bond consists of one σ bond and two π bonds. Each nitrogen uses an *sp* hybrid orbital to form the σ bond and two *p* orbitals to form the two π bonds.

N≡N the σ bond is formed by *sp*-*sp* overlap
each π bond is formed by *p*-*p* overlap

22. The electrostatic potential map of ammonia is not symmetrical in shape and is not symmetrical in the distribution of the charge—the nitrogen end is more electron rich that the three hydrogens. The electrostatic potential map of the ammonium ion is symmetrical in shape and is symmetrical the distribution of the charge because the nitrogen atoms forms a bond with each of the four hydrogens and the four bond angles are all the same.

23. **a, d, e, h**

24. **a.** **(1)** $^+NH_4$ **(2)** HCl **(3)** H_2O **(4)** H_3O^+

b. **(1)** $^-NH_2$ **(2)** Br^- **(3)** NO_3^- **(4)** HO^-

25.

<u>if the nonbonding electrons are not shown:</u>

a. CH_3OH as an acid $\quad CH_3OH + HO^- \rightleftharpoons CH_3O^- + H_2O$

CH_3OH as a base $\quad CH_3OH + H_3O^+ \rightleftharpoons CH_3\overset{+}{O}H(H) + H_2O$

b. NH_3 as an acid $\quad NH_3 + HO^- \rightleftharpoons {}^-NH_2 + H_2O$

NH_3 as a base $\quad NH_3 + H_3O^+ \rightleftharpoons {}^+NH_4 + H_2O$

<u>if the nonbonding electrons are shown:</u>

a. CH_3OH as an acid $\quad CH_3\ddot{O}H + H\ddot{O}:^- \rightleftharpoons CH_3\ddot{O}:^- + H_2\ddot{O}:$

CH_3OH as a base $\quad CH_3\ddot{O}H + H_3\ddot{O}^+ \rightleftharpoons CH_3\overset{..+}{O}H(H) + H_2\ddot{O}:$

b. NH_3 as an acid $\quad \ddot{N}H_3 + H\ddot{O}:^- \rightleftharpoons {}^-:\ddot{N}H_2 + H_2\ddot{O}:$

NH_3 as a base $\quad \ddot{N}H_3 + H_3\ddot{O}^+ \rightleftharpoons {}^+NH_4 + H_2\ddot{O}:$

26. **a.** The lower the pK_a, the stronger the acid, so the compound with $pK_a = 5.2$ is the stronger acid.

b. The greater the dissociation constant, the stronger the acid, so the compound with dissociation constant = 3.4×10^{-3} is the stronger acid.

27.

$$K_a = K_{eq}[H_2O]$$

$$K_{eq} = \frac{K_a}{[H_2O]} = \frac{4.53 \times 10^{-6}}{55.5} = 8.16 \times 10^{-8}$$

28. Using the following reaction as an example:

$$CH_3OH + NH_3 \rightleftharpoons CH_3O^- + {}^+NH_4$$

$$K_{eq} = \frac{[\text{products}]}{[\text{reactants}]}$$

$$K_{eq} = \frac{[CH_3O^-][\overset{+}{N}H_4]}{[CH_3OH][NH_3]}$$

$$K_a \text{ reactant acid} = \frac{[H^+][CH_3O^-]}{[CH_3OH]}$$

$$K_a \text{ product acid} = \frac{[H^+][NH_3]}{[\overset{+}{N}H_4]}$$

$$K_{eq} = \frac{K_a \text{ reactant acid}}{K_a \text{ product acid}} = \frac{\dfrac{[H^+][CH_3O^-]}{[CH_3OH]}}{\dfrac{[H^+][NH_3]}{[\overset{+}{N}H_4]}}$$

$$= \frac{\dfrac{[CH_3O^-]}{[CH_3OH]}}{\dfrac{[NH_3]}{[\overset{+}{N}H_4]}}$$

$$= \frac{[CH_3O^-][\overset{+}{N}H_4]}{[NH_3][CH_3OH]}$$

29. **a.** CH_3COO^- is the stronger base.
Because HCOOH is the stronger acid, it has the weaker conjugate base.

b. $^-NH_2$ is the stronger base.
Because H_2O is the stronger acid, it has the weaker conjugate base.

c. H_2O is the stronger base.
Because $CH_3OH_2^+$ is the stronger acid, it has the weaker conjugate base.

30. Notice that in each case, the equilibrium goes away from the strong acid and toward the weak acid.

a. $HCl + H_2O \rightleftharpoons Cl^- + H_3O^+$

$pK_a = -7$; $pK_a = -1.7$

CH_3COH (C=O) $+ H_2O \rightleftharpoons CH_3CO^-$ (C=O) $+ H_3O^+$

$pK_a = 4.8$; $pK_a = -1.7$

CH_3COH (C=O) $+ HO^- \rightleftharpoons CH_3CO^-$ (C=O) $+ H_2O$

$pK_a = 4.8$; $pK_a = 15.7$

CH_3COH (C=O) $+ H_3O^+ \rightleftharpoons CH_3COH$ (C=$\overset{+}{O}$H) $+ H_2O$

$pK_a = -1.7$; $pK_a = -6.1$

$CH_3NH_2 + HO^- \rightleftharpoons CH_3\bar{N}H + H_2O$

$pK_a = 40$; $pK_a = 15.7$

$CH_3NH_2 + H_3O^+ \rightleftharpoons CH_3\overset{+}{N}H_3 + H_2O$

$pK_a = -1.7$; $pK_a = 10.7$

b. $HCl + H_2O \rightleftharpoons H_3O^+ + Cl^-$

$pK_a = -7$; $pK_a = -1.7$

$NH_3 + H_2O \rightleftharpoons \overset{+}{N}H_4 + HO^-$

$pK_a = 15.7$; $pK_a = 9.4$

31. The conjugate acids have the following relative strengths:

$CH_3\overset{+}{O}H_2 > CH_3COH$ (C=O) $> CH_3\overset{+}{N}H_3 > CH_3OH > CH_3NH_2$

The bases, therefore, have the following relative strengths:

$CH_3NH^- > CH_3O^- > CH_3NH_2 > CH_3CO^-$ (C=O) $> CH_3OH$

32.

$$K_{eq} = \frac{K_a \text{ of the reactant acid}}{K_a \text{ of the product acid}}$$

For **a** the reactant acid is HCl and the product acid is H_3O^+.

For **b** the reactant acid is CH_3COOH and the product acid is H_3O^+

For **c** the reactant acid is H_2O and the product acid is $CH_3\overset{+}{N}H_3$.

For **d** the reactant acid is $CH_3\overset{+}{N}H_3$ and the product acid is H_3O^+.

a. $K_{eq} = \dfrac{10^7}{10^{1.7}} = 10^{7 - 1.7} = 10^{5.3} = 2.0 \times 10^5$

b. $K_{eq} = \dfrac{10^{-4.8}}{10^{1.7}} = 10^{-4.8 - 1.7} = 10^{-6.5} = 3.2 \times 10^{-7}$

c. $K_{eq} = \dfrac{10^{-15.7}}{10^{-10.7}} = 10^{-15.7 + 10.7} = 10^{-5} = 1.0 \times 10^{-5}$

d. $K_{eq} = \dfrac{10^{-10.7}}{10^{1.7}} = 10^{-10.7 - 1.7} = 10^{-12.4} = 4.0 \times 10^{-13}$

33.

a. CH_3OCH_2OH because of the electron-withdrawing CH_3O group

b. $CH_3CH_2CH_2\overset{+}{O}H_2$ because oxygen is more electron-withdrawing than nitrogen

c. $CH_3CH_2OCH_2OH$ because the electron-withdrawing oxygen is closer to the OH group

d. $CH_3CH_2\overset{O}{\overset{\|}{C}}OH$ because the electron-withdrawing C=O is closer to the OH group

34.

$CH_3\underset{F}{\underset{|}{C}}HOH$ > $\underset{F}{\underset{|}{C}}H_2CH_2OH$ > $\underset{Cl}{\underset{|}{C}}H_2CH_2OH$ > CH_3CH_2OH

The first listed compound is the most acidic because the electron-withdrawing fluorine is closest to the bond that holds the acidic hydrogen. The next two compounds have the electron-withdrawing substituents equally far from the O-H bond: since fluorine is more electronegative than chlorine, the fluoro-substituted compound is the stronger acid. The last-listed compound does not have an electonegative substituent, so it is the least acidic.

35. Solved in the text.

36. **a.** Because HF is the weakest acid, F^- is the strongest base.
b. Because HI is the strongest acid, I^- is the weakest base.

37. **a.** oxygen **b.** H_2S **c.** CH_3SH

As you saw in Problem 35, the size of an atom is more important than its electronegativity in determining stability. So even though oxygen is more electronegative than sulfur, a thiol is a stronger acid than an alcohol because the sulfur atom is larger, causing it to be less electron dense and, therefore, a more stable base. The more stable the base, the stronger its conjugate acid.

38.

a. $CH_3C{\equiv}\overset{+}{N}H$

b. CH_3CH_3

c. $F_3C(=O)OH$ (F_3CCOH with C=O)

d. an sp^2 hybridized oxygen

e. $sp > sp^2 > sp^3$

f. $sp > sp^2 > sp^3$

g. HNO_3 is more acidic than HNO_2.
When the structure of each molecule is drawn out, you can see that HNO_3 has a positively charged nitrogen and HNO_2 does not. The positive charge causes greater electron withdrawal from the O-H bond, making it easier to break.

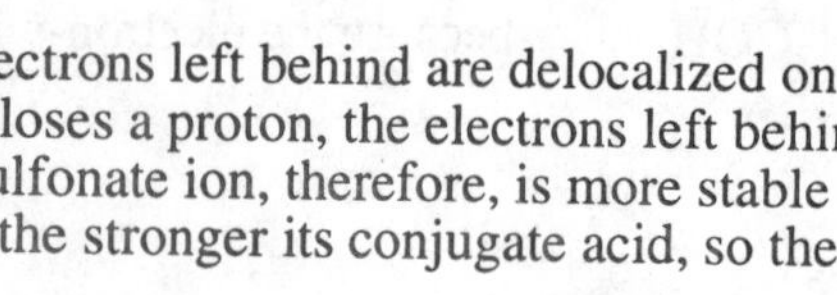

H—O—$\overset{+}{N}$(=O)—O^- H—O—N=O

39. When a sulfonic acid loses a proton, the electrons left behind are delocalized onto three oxygen atoms. In contrast, when a carboxylic acid loses a proton, the electrons left behind are delocalized onto two oxygen atoms. The sulfonate ion, therefore, is more stable than the carboxylate ion. The more stable the base, the stronger its conjugate acid, so the sulfonic acid is a stronger acid than the carboxylic acid.

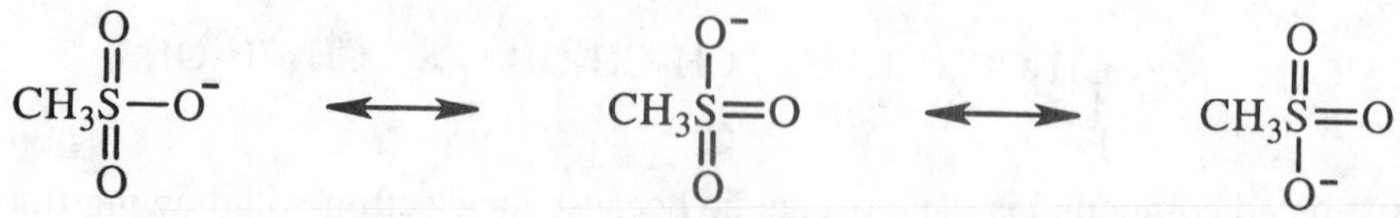

$CH_3S(=O)_2$—O^- ⟷ $CH_3S(=O)_2(O^-)$ ⟷ $CH_3S(=O)_2(O^-)$

a sulfonate ion

$CH_3C(=O)$—O^- ⟷ $CH_3C(O^-)$=O

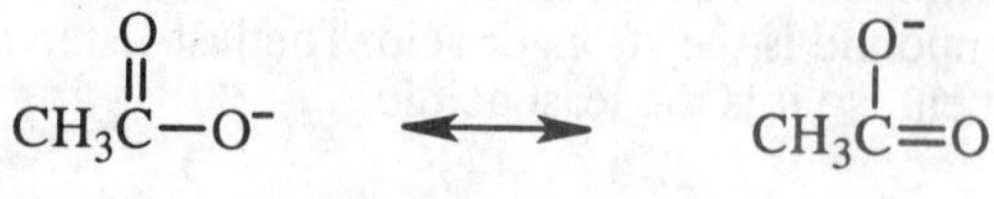

a carboxylate ion

40. greater than pH 10.4

As long as the pH is greater than the pK_a of the compound, the majority of the compound will be in its basic form.

41. **a.** 10.4 (two log units more basic than the pK_a)

b. 2.7 (one log unit more acidic than the pK_a)

c. 4.9 (If 10% is in its basic form, 90% will be in its acidic form, so you need to be one log unit more acidic than the pK_a.)

d. 7.3 (pH = pK_a)

e. 9.3 (If 1% is in its acidic form, 99% will be in its basic form, so you need to be two log units more basic than the pK_a.)

42. **a.**
1. neutral
2. neutral
3. equal amounts of both
4. charged
5. charged
6. charged
7. charged

b.
1. charged
2. charged
3. charged
4. charged
5. equal amounts of both
6. neutral
7. neutral

c.
1. neutral
2. neutral
3. neutral
4. neutral
5. neutral
6. neutral
7. neutral

43. **a.** **1.** pH = 4.9

2. pH = 10.7

b. **1.** pH > 6.9 Because the basic form is the form in which the compound is charged, you need to be more than two units more basic than the pK_a.

2. pH < 8.7 Because the acidic form is the form in which the compound is charged, you need to be more than two units more acidic than the pK_a.

44.

a. CH_3COO^- **c.** H_2O **e.** CH_3CH_2OH **g.** $HC{\equiv}N$ **i.** NO_3^-

b. $CH_3CH_2\overset{+}{N}H_3$ **d.** CH_3CH_2OH **f.** $\overset{+}{N}H_4$ **h.** NO_2^- **j.** Br^-

45.

a. $ZnCl_2 + CH_3OH \rightleftharpoons CH_3\overset{+}{O}H$ (with $^{-}ZnCl_2$ bonded to O)

b. $FeBr_3 + Br^- \rightleftharpoons Br{-}\overset{-}{Fe}Br_3$

c. $AlCl_3 + Cl^- \rightleftharpoons Cl{-}\overset{-}{Al}Cl_3$

d. $BF_3 + HCH{=}O \rightleftharpoons HCH{=}\overset{+}{O}{-}\overset{-}{B}F_3$

46. **a, b, c,** and **h** are Bronsted acids (protonating-donating acids). Therefore, they react with HO^- by donating a proton to it.
d, e, f, and **g** are Lewis acids and react with HO^- by accepting a pair of electrons from it.

a. $CH_3OH + HO^- \rightleftharpoons CH_3O^- + H_2O$

b. $\overset{+}{N}H_4 + HO^- \rightleftharpoons NH_3 + H_2O$

c. $CH_3\overset{+}{N}H_3 + HO^- \rightleftharpoons CH_3NH_2 + H_2O$

d. $BF_3 + HO^- \rightleftharpoons HO{-}\overset{-}{B}F_3$

e. $\overset{+}{C}H_3 + HO^- \rightleftharpoons CH_3OH$

f. $FeBr_3 + HO^- \rightleftharpoons HO{-}\overset{-}{Fe}Br_3$

g. $AlCl_3 + HO^- \rightleftharpoons HO{-}\overset{-}{Al}Cl_3$

h. $CH_3COOH + HO^- \rightleftharpoons CH_3COO^- + H_2O$

47.

a. $\text{H}:\ddot{\text{O}}:\overset{\ddot{\text{O}}:}{\ddot{\text{C}}}:\ddot{\text{O}}:\text{H}$

b. $^{-}:\ddot{\text{O}}:\overset{\ddot{\text{O}}:}{\ddot{\text{C}}}:\ddot{\text{O}}:^{-}$

c. $\text{H}:\overset{\ddot{\text{O}}:}{\ddot{\text{C}}}:\text{H}$

d. $\text{H}:\underset{\text{H}}{\ddot{\text{N}}}:\underset{\text{H}}{\ddot{\text{N}}}:\text{H}$

e. $\text{H}:\underset{\text{H}}{\overset{\text{H}}{\text{C}}}:\underset{\text{H}}{\ddot{\text{N}}}:\text{H}$

f. $\text{H}:\underset{\text{H}}{\overset{\text{H}}{\text{C}}}:\overset{+}{\text{N}}::\text{N}:$

g. $\ddot{\text{O}}=\text{C}=\ddot{\text{O}}$

h. $:\text{N}:::\overset{+}{\text{O}}:$

i. $\text{H}:\underset{\text{H}}{\ddot{\text{N}}}:\ddot{\text{O}}:^{-}$

48.

a. *sp*^{3}, tetrahedral
b. *sp*^{2}, trigonal
c. *sp*^{3}, tetrahedral
d. *sp*^{2}, trigonal
e. *sp*^{3}, tetrahedral
f. *sp*^{2}, trigonal
g. *sp*, linear
h. *sp*^{3}, tetrahedral
i. *sp*^{3}, tetrahedral

49.

a. $CH_3CH_2CH_3$ b. $CH_3CH{=}CH_2$ c. $CH_3C{\equiv}CCH_3$ or $CH_3CH_2C{\equiv}CH$

50.

a. 107.3 ° c. 109.5° e. 104.5° g. 109.5°

b. 107.3° d. 104.5°* f. 120° h. 180°

*104.5° is the correct prediction based on the bond angle in water, but the bond angle is actually somewhat larger (111.7°) because the bond opens up to minimize the nonbonding interactions between the electron clouds of the relatively bulky CH_3 groups.

51. formal charge = the number of valence electrons − (the number of nonbonding electrons + 1/2 the number of bonding electrons)

a. formal charge = 6 − (6 + 1) = 6 − 7 = − 1

$\text{H}:\ddot{\underset{\cdot\cdot}{\text{O}}}:^{-}$

b. formal charge = 6 − (5 + 1) = 6 − 6 = 0

$\text{H}:\ddot{\underset{\cdot\cdot}{\text{O}}}\cdot$

c. formal charge = 5 - (4) = +1

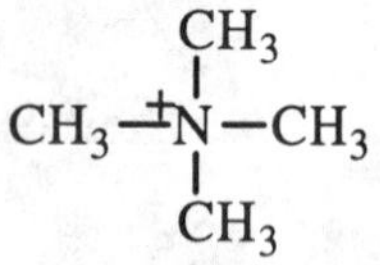

$$CH_3-\overset{CH_3}{\underset{CH_3}{\overset{|}{\underset{|}{N^+}}}}-CH_3$$

d. formal charge = 5 - (2 + 2) = 5 - 4 = +1

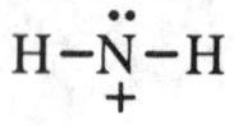

$$H-\ddot{N}^{+}-H$$

e. formal charge = 4 - (2 + 2) = 4 - 4 = 0

$$H-\ddot{C}-H$$

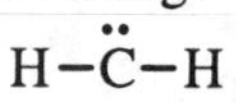

f. N has a formal charge = 5 - (4) = +1

B has a formal charge = 3 - (4) = - 1

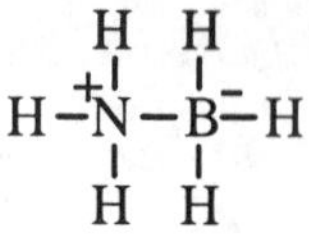

$$H-\overset{H}{\underset{H}{\overset{|}{\underset{|}{N^+}}}}-\overset{H}{\underset{H}{\overset{|}{\underset{|}{B^-}}}}-H$$

g. formal charge = 4 - (2 + 3) = 4 - 5 = - 1

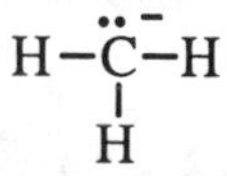

$$H-\underset{H}{\underset{|}{\ddot{C}^-}}-H$$

h. formal charge = 6 - (2 + 3) = 6 - 5 = +1

$$CH_3-\underset{H}{\underset{|}{\ddot{O}^+}}-CH_3$$

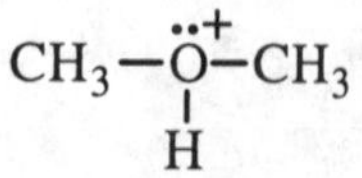

52.

	1s	2s	$2p_x$	$2p_y$	$2p_z$	3s	$3p_x$	$3p_y$	$3p_z$	4s
a. Ca	⇅	⇅	⇅	⇅	⇅	⇅	⇅	⇅	⇅	⇅
b. Ca^{2+}	⇅	⇅	⇅	⇅	⇅	⇅	⇅	⇅	⇅	
c. Ar	⇅	⇅	⇅	⇅	⇅	⇅	⇅	⇅	⇅	
d. Mg^{2+}	⇅	⇅	⇅	⇅	⇅					

53.

```
     H  O                          H
     |  ||                         |
a. H-C--C-H                  H     O     H
     |                       |     |     |
     H                  d. H-C-----C-----C-H
                             |     |     |
                             H   H-C-H   H
                                   |
                                   H

     H     H                 H  H  H
     |     |                 |  |  |
b. H-C--O--C-H          e. H-C--C--C--C≡N
     |     |                 |  |  |
     H     H                 H  O  H
                                |
                                H

                                       H           H
                                       |           |
     H  O                    H     H H-C-H H     H-C-H H
     |  ||                   |     |   |   |       |   |
c. H-C--C--O-H          f. H-C-----C---C---C-------C---C-H
     |                       |     |   |   |       |   |
     H                       H   H-C-H H   H     H-C-H H
                                   |                 |
                                   H                 H
```

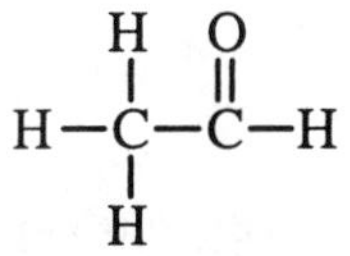

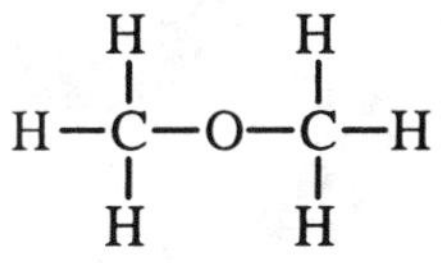

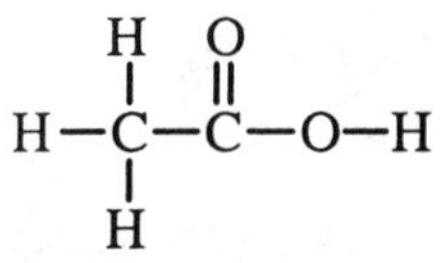

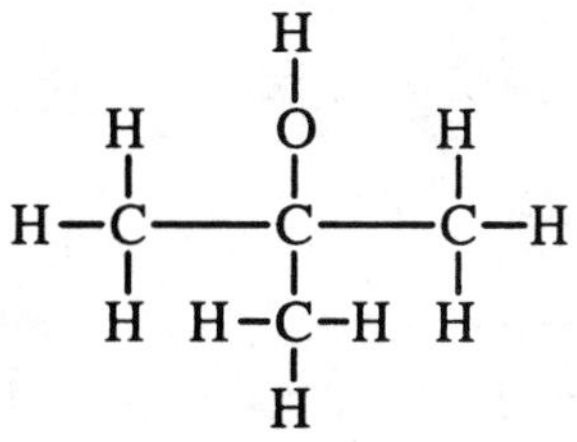

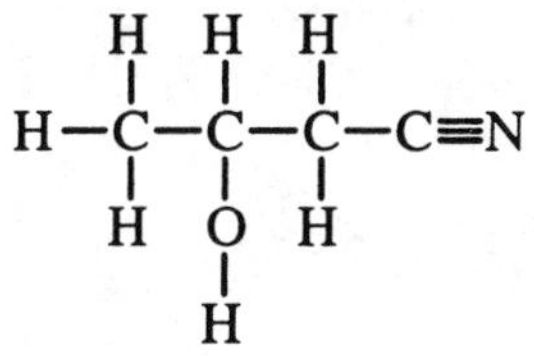

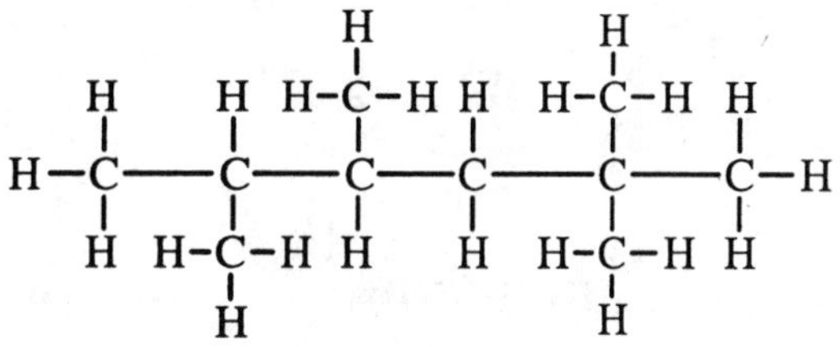

54.

a. $\overrightarrow{CH_3-Br}$

b. $\overleftarrow{CH_3-Li}$

c. $\overleftarrow{HO-NH_2}$

d. $\overrightarrow{I-Br}$

e. $\overrightarrow{CH_3-OH}$

f. $(CH_3)_2\overleftarrow{N-H}$

55.

a. $CH_3CH{=}CH_2$ (sp^2 ⇓ on the CH carbon)

b. $CH_3\overset{O}{\overset{\|}{C}}CH_3$ (O ⇐ sp^2)

c. CH_3CH_2OH (sp^3 ⇓ on O)

d. $CH_3C{\equiv}N$ (sp ⇓ on N)

e. $CH_3CH{=}NCH_3$ (sp^2 ⇓ on N)

f. $CH_3OCH_2CH_3$ (sp^3 ⇓ on O)

56. The open arrow points to the shorter of the two indicated bonds in each compound.

For **1**, **2**, and **3**: a triple bond is shorter than a double bond which is shorter than a single bond.

For **4** and **5**: Because an *s* orbital is closer to the nucleus than a *p* orbital, the more *s* character in the hybridized orbital, the shorter the bond. Therefore, a hydrogen bonded to an *sp* hybridized carbon is shorter than a hydrogen bonded to an sp^2 hybridized carbon, which is shorter than a hydrogen bonded to an sp^3 hybridized carbon. (See Table 1.7 on page 37 of the text.)

1. $CH_3CH{=}CHC{\equiv}CH$ (sp^3, sp^2, sp^2, sp, sp)

2. $CH_3C(=O)CH_2OH$ (O: sp^2; sp^3, sp^2, sp^3, sp^3)

3. $CH_3NHCH_2CH_2N{=}CHCH_3$ (sp^3 sp^3 sp^3 sp^3 sp^2 sp^2 sp^3)

4. $H_2C{=}CHC{\equiv}C{-}H$ (sp^2, sp^2, sp, sp)

5. $H_2C{=}CHC{\equiv}C{-}C(CH_3)_2{-}H$ (sp^2, sp^2, sp, sp, sp^3; CH_3: sp^3)

57.

a. at pH = 3 CH_3COOH

at pH = 6 CH_3COO^-

at pH = 10 CH_3COO^-

at pH = 14 CH_3COO^-

b. at pH = 3 $CH_3CH_2\overset{+}{N}H_3$

at pH = 6 $CH_3CH_2\overset{+}{N}H_3$

at pH = 10 $CH_3CH_2\overset{+}{N}H_3$

at pH = 14 $CH_3CH_2NH_2$

c. at pH = 3 CF_3CH_2OH

at pH = 6 CF_3CH_2OH

at pH = 10 CF_3CH_2OH

at pH = 14 $CF_3CH_2O^-$

58.

(H)(Cl)C=C(H)(Cl), both Cl on the same side

The dipole moment is 2.95 because the two Cl's are withdrawing electrons in the same direction.

(H)(Cl)C=C(Cl)(H), Cl on opposite sides

The dipole moment is 0 because the two Cl's are withdrawing electrons in the same direction.

59. The atoms bonded to the sp^2 hybridized carbons lie in the same plane. If the two atoms are not both bonded to the sp^2 hybridized carbons, they will not lie in the same plane.

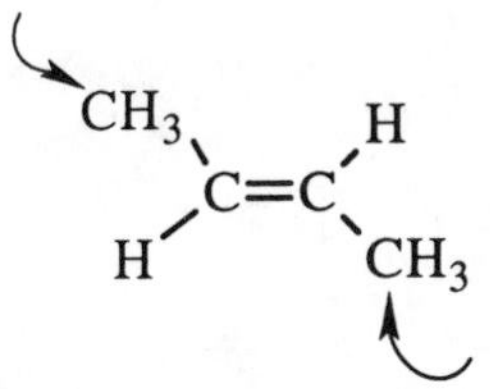

$CH_3CH=CHCH_3$ (arrows on the two CH_3 groups): yes

$CH_3CH=CHCH_2CH_3$ (arrows on CH_3 and the terminal CH_3 of CH_2CH_3): no

3-methylcyclohexene (CH_3 and ring carbon marked): no

1-methylcyclohexene (CH_3 and ring carbon marked): yes

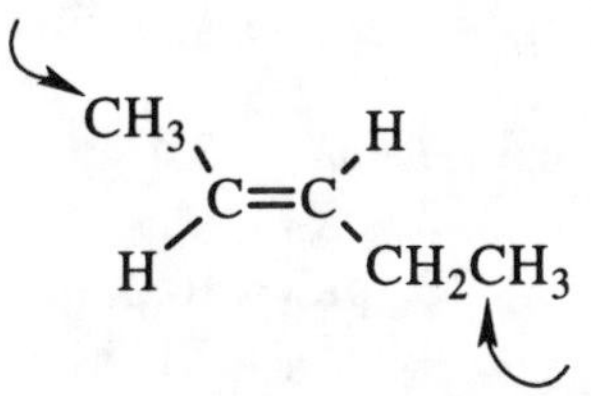

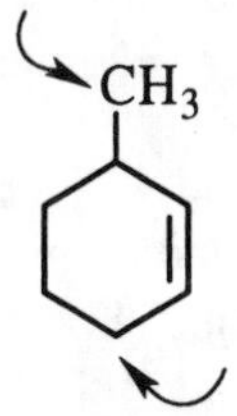

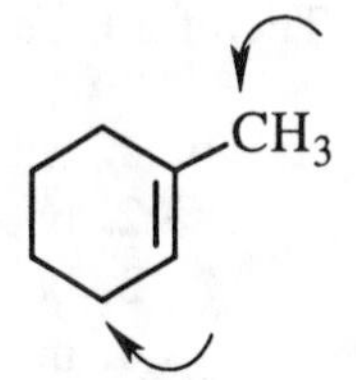

60.

a. $CH_3COOH + CH_3O^- \rightleftharpoons CH_3COO^- + CH_3OH$

b. $CH_3CH_2OH + {}^-NH_2 \rightleftharpoons CH_3CH_2O^- + NH_3$

c. $CH_3COOH + CH_3NH_2 \rightleftharpoons CH_3COO^- + CH_3\overset{+}{N}H_3$

d. $CH_3CH_2OH + HCl \rightleftharpoons CH_3CH_2\overset{+}{O}H_2 + Cl^-$

61.

a. $H{-}CH_2{-}C{\equiv}C{-}H$: sp^3 carbon (H–C–H 109.5°); sp, sp carbons (C–C≡C 180°)

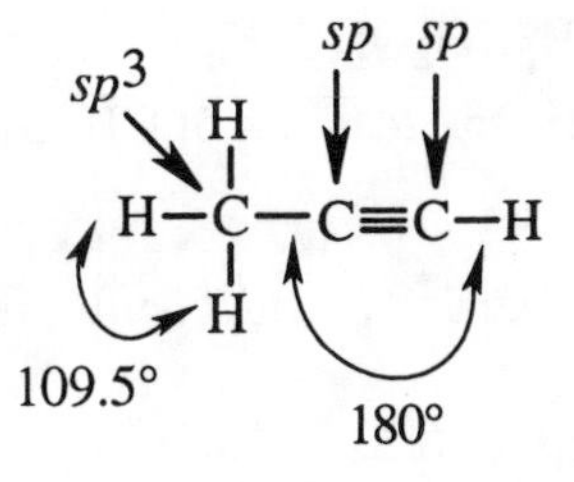

b. $CH_3{-}CH{=}CH_2$: sp^3 carbon (H–C–H 109.5°); sp^2, sp^2 carbons; bond angles 120°, 120°, 120°

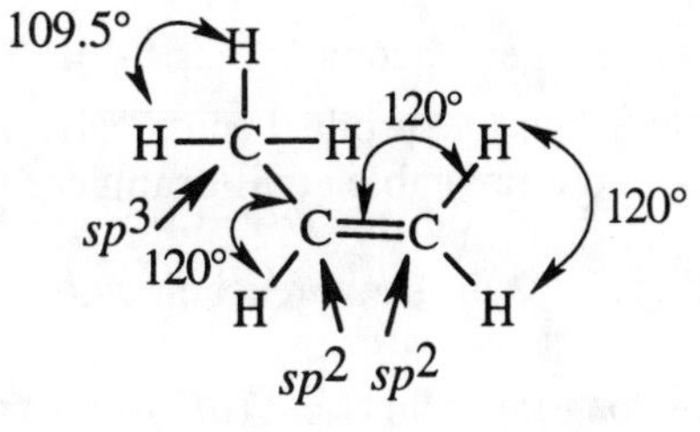

c. $H{-}CH_2{-}CH_2{-}CH_2{-}H$ (propane) the 3 carbons are all sp^3 hybridized
all the bond angles are 109.5°

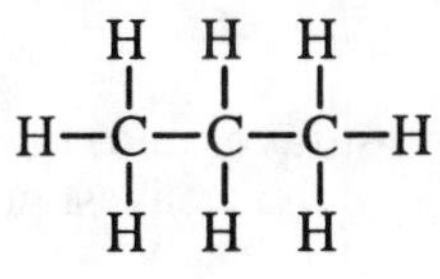

62. The log of $10^{-4} = -4$, log of $10^{-5} = -5$, log of $10^{-6} = -6$, etc.
Because the $pK_a = -\log K_a$, the pK_a of an acid with a K_a of 10^{-4} = is - (-4) = 4.
An acid with a K_a of 4.0×10^{-4} is a stronger acid than one with a K_a of 1.0×10^{-4}.
Therefore, the pK_a can be estimated as being between 3 and 4.

a. **1.** between 3 and 4
2. between -2 and -1
3. between 10 and 11
4. between 9 and 10
5. between 3 and 4

b. **1.** $pK_a = 3.4$
2. $pK_a = -1.3$
3. $pK_a = 10.2$
4. $pK_a = 9.1$
5. $pK_a = 3.7$

c. Because the lower the pK_a the stronger the acid, nitric acid is the strongest acid.

63.

a.

$CH_3CH_2\underset{\mid \atop Cl}{C}HCOOH > CH_3\underset{\mid \atop Cl}{C}HCH_2COOH > ClCH_2CH_2CH_2COOH > CH_3CH_2CH_2COOH$

b. The electron-withdrawing substituent makes the carboxylic acid more acidic, since the substituent pulls the bonding electrons in the O-H bond away from the hydrogen.

c. The closer the electron-withdrawing Cl is to the acidic proton, the more it increases the acidity of the compound.

64. The reaction with the most favorable equilibrium constant is the one with the strongest reactant acid reacting to give the weakest product acid.

a. **1.** CH_3OH is a stronger reactant acid ($pK_a = 15.5$) than CH_3CH_2OH ($pK_a = 15.9$), and both reactions have the same product acid ($^+NH_4$). Therefore, the reaction of CH_3OH with NH_3 has the more favorable equilibrium constant.

2. Both reactions have the same reactant acid (CH_3CH_2OH). The product acids are different:
$^+NH_4$ is a stronger product acid ($pK_a = 9.4$) than $CH_3NH_3^+$ ($pK_a = 10.7$). Therefore, the reaction of CH_3CH_2OH with CH_3NH_2 has the more favorable equilibrium constant.

b. From parts **1.** and **2.** we have seen that the reaction with the most favorable equilibrium constant is the one with the strongest reactant acid reacting to give the weakest product acid.

Now we have to compare "apples" and "oranges" because the reaction with the most favorable equilibrium constant in **1.** and the reaction with the most favorable equilibrium constant in **2.** do not have any species in common: **1.** has the stronger reactant acid, while **2.** has the weaker product acid. Therefore, the equilibrium constants have to be calculated.

The reaction of CH_3OH with NH_3 has an equilibrium constant = 7.9×10^{-7}
($K_{eq} = K^{-15.5}/K^{-9.4} = K^{-6.1} = 7.9 \times 10^{-7}$).

The reaction of CH_3CH_2OH with CH_3NH_2 has an equilibrium constant = 6.3×10^{-6}
($K_{eq} = K^{-15.9}/K^{-10.7} = K^{-5.2} = 6.3 \times 10^{-6}$).

Therefore, the reaction of CH_3CH_2OH with CH_3NH_2 has the greatest equilibrium constant of the four reactions.

65. From the following equilibria you can see that a carboxylic acid is neutral when it is in its acidic form and charged when it is in its basic form. An amine is charged when it is in its acidic form and neutral when it is in its basic form.

$$RCOOH \rightleftharpoons RCOO^- + H^+$$

$$R\overset{+}{N}H_3 \rightleftharpoons RNH_2 + H^+$$

Because charged species will dissolve in water and neutral species will dissolve in ether, you want all of the compound you want to separate in either the charged form or the neutral form. In other words, you want all of the compound in either the acidic form or the basic form.

From the Henderson-Hasselbalch equation it can be calculated that in order to obtain a 100:1 ratio of acidic form:basic form, the pH must be two pH units lower than the pK_a of the compound; and in order to obtain a 100:1 ratio of basic form:acidic form, the pH must be two pH units greater than the pK_a of the compound.

a. If both compounds are to dissolve in water, they both must be charged. Therefore, the carboxylic acid must be in its basic form, and the amine must be in its acidic form. To accomplish this, the pH will have to be at least two pH units greater than the pK_a of the carboxylic acid and at least two pH units less than the pK_a of the ammonium ion. In other words, it must be between pH 6.8 and pH 8.7.

b. For the carboxylic acid to dissolve in water, it must be charged (in its basic form), so the pH will have to be greater than 6.8. For the amine to dissolve in ether, it will have to be neutral (in its basic form), so the pH will have to be greater than 12.7 to have essentially all of it in the neutral form. Therefore, the pH of the water layer must be greater than 12.7.

c. To dissolve in ether, the carboxylic acid will have to be neutral, so the pH will have to be less than 2.8 to have essentially all the carboxylic acid in the acidic (neutral) form. To dissolve in water, the amine will have to be charged, so the pH will have to be less than 8.7 to have essentially all the amine in the acidic form. Therefore, the pH of the water layer must be less than 2.8.

66. Charged compounds will dissolve in water and uncharged compounds will dissolve in ether. The acidic forms of carboxylic acids and alcohols are neutral and the basic forms are charged. The acidic forms of amines are charged and the basic forms are neutral.

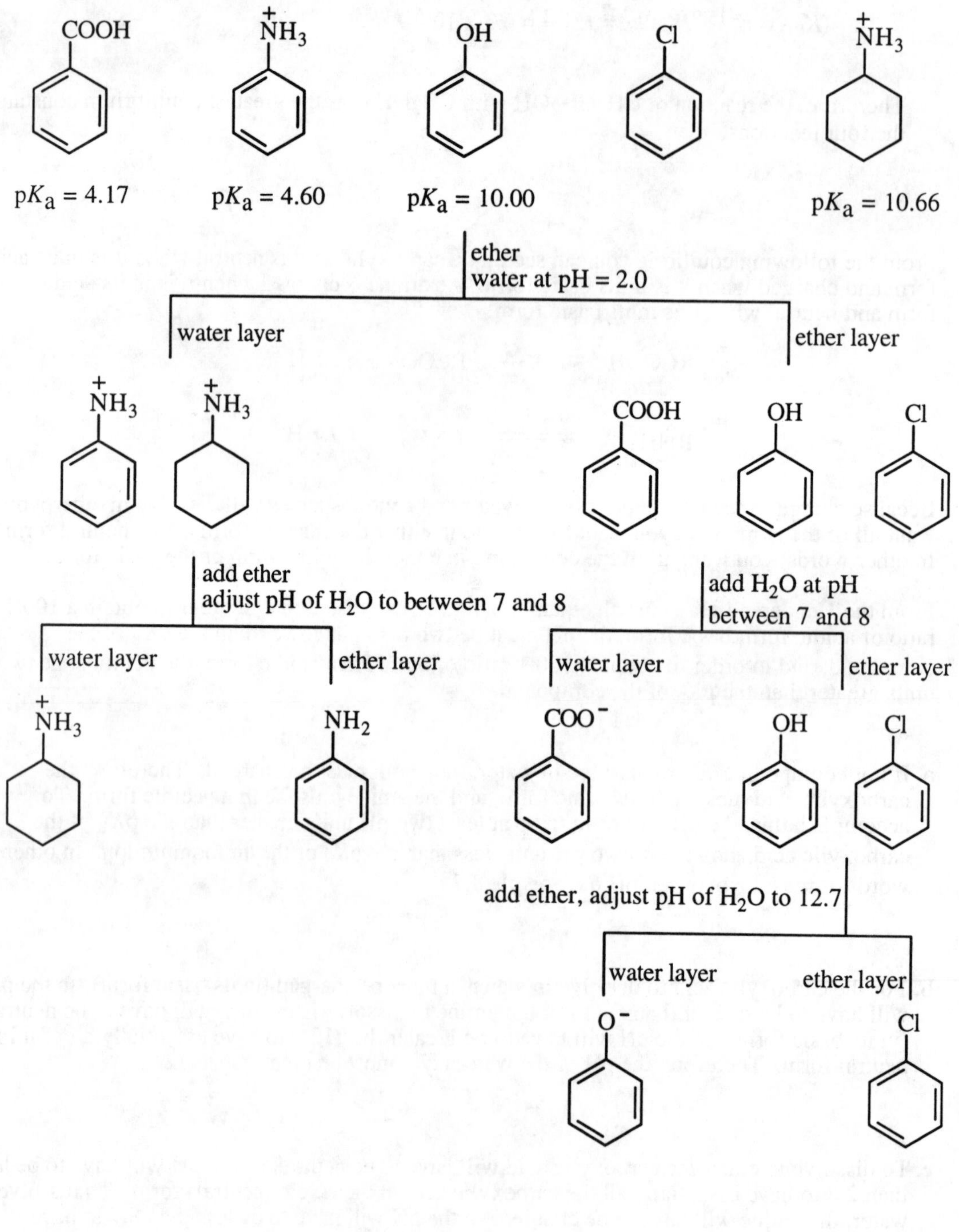

COOH (benzoic acid) $pK_a = 4.17$; $\overset{+}{N}H_3$ (anilinium) $pK_a = 4.60$; OH (phenol) $pK_a = 10.00$; Cl (chlorobenzene); $\overset{+}{N}H_3$ (cyclohexylammonium) $pK_a = 10.66$

ether
water at pH = 2.0

water layer: $\overset{+}{N}H_3$ (anilinium), $\overset{+}{N}H_3$ (cyclohexylammonium)

ether layer: COOH, OH, Cl

(water layer) add ether
adjust pH of H_2O to between 7 and 8

water layer: $\overset{+}{N}H_3$ (cyclohexylammonium)

ether layer: NH_2 (aniline)

(ether layer) add H_2O at pH
between 7 and 8

water layer: COO^-

ether layer: OH, Cl

add ether, adjust pH of H_2O to 12.7

water layer: O^-

ether layer: Cl

67.

$$pK_a = pH + \log \frac{[HA]}{[A^-]}$$

The above equation, called the Henderson-Hasselbalch equation, shows that:

1. When the value of the pH is equal to the value of the pK_a, the concentration of buffer in the acidic form equals the concentration of buffer in the basic form.
2. When the solution is more acidic than the pK_a, more buffer species is in the acidic form than in the basic form
3. When the solution is more basic than the pK_a, more buffer species is in the basic form than in the acidic form.

$$pK_a = pH + \log \frac{[HA]}{[A^-]}$$

Because the pH of the blood (7.4) is greater than the pK_a of the buffer (6.1), more buffer species is in the basic form than in the acidic form. Therefore, the buffer is better at neutralizing excess acid.

68. For a discussion of how to do problems such as Problems 68 and 69, see **Special Topic I** in the Solutions Manual (pH, pK_a, and Buffers).

a.

$$\text{fraction present in the acidic form} = \frac{\text{amount in the acidic form}}{\text{amount in the acidic form} + \text{amount in the basic form}}$$

$$= \frac{[HA]}{[HA] + [A^-]}$$

Because there are two unknowns, we must define one in terms of the other.

By using the definition of the acid dissociation constant, we can determine $[A^-]$ in terms of [HA].

$$K_a = \frac{[H^+][A^-]}{[HA]}$$

$$[A^-] = \frac{K_a [HA]}{[H^+]}$$

Substituting the value of $[A^-]$ into the equation gives the fraction present in the acidic form:

$$\frac{[HA]}{[HA] + [A^-]} = \frac{[HA]}{[HA] + \frac{K_a [HA]}{[H^+]}} = \frac{1}{1 + \frac{K_a}{[H^+]}} = \frac{[H^+]}{[H^+] + K_a}$$

Therefore, the percent present in the acidic form is given by:

$$\frac{[H^+]}{[H^+] + K_a} \times 100$$

Because the pK_a of the acid is given as 5.3, we know that K_a is 5.0×10^{-6} (pK_a = -log K_a). Because the pH of the solution is given as 5.7, we know that $[H^+]$ is 2.0×10^{-6} (pH = -log$[H^+]$).

Substituting into the above equation gives:

$$\frac{2.0 \times 10^{-6}}{2.0 \times 10^{-6} + 5.0 \times 10^{-6}} \times 100$$

$$\frac{2.0 \times 10^{-6}}{7.0 \times 10^{-6}} \times 100 = 29\%$$

b. percent present in the acidic form = $\frac{[H^+]}{[H^+] + K_a} = .80$

$$[H^+] = .80\left([H^+] + K_a\right)$$

$$[H^+] = .80\,[H^+] + .80\,K_a$$

$$.20\,[H^+] = .80\,K_a$$

$$[H^+] = 4\,K_a$$

$$[H^+] = 4 \times 5.0 \times 10^{-6}$$

$$[H^+] = 20 \times 10^{-6}$$

$$\text{pH} = 4.7$$

69.

a.

$$K_a = \frac{[H^+][A^-]}{[HA]}$$

$$1.74 \times 10^{-5} = \frac{x^2}{1.0 - x}$$

$$1.74 \times 10^{-5} = x^2$$

$$x = 4.16 \times 10^{-3}$$

$$\text{pH} = 2.38$$

b.

$$K_a = \frac{[H^+][A^-]}{[HA]}$$

$$2.00 \times 10^{-11} = \frac{x^2}{0.1 - x}$$

$$2.00 \times 10^{-12} = x^2$$

$$x = 1.41 \times 10^{-6}$$

$$\text{pH} = 5.85$$

c. This question can be answered by plugging the numbers into the Henderson-Hasselbalch equation.

$$pK_a = pH + \log \frac{[acid]}{[base]}$$

$$3.76 = pH + \frac{0.3}{0.1}$$

$$3.76 = pH + \log 3$$

$$3.76 = pH + 0.48$$

$$pH = 3.76 - 0.48 = 3.28$$

Chapter 1 Practice Test

1. Answer the following:

   a. Which is a stronger acid, HCl or HBr?

   b. Which is a stronger base, NH_3 or H_2O?

   c. Which bond has a greater dipole moment, a carbon-oxygen bond or a carbon-fluorine bond?

   d. Which has a dipole moment of zero, $CHCl_3$ or CCl_4?

2. What is the hybridization of the carbon atom in each of the following compounds?

   $\overset{+}{C}H_3$ $\quad$ $\overset{-}{C}H_3$ $\quad$ $\overset{\bullet}{C}H_3$

3. Draw the Lewis structure for HCO_3^-.

4. The following compounds are drawn in their acidic forms, and their pK_a's are given. Draw the form in which each compound would predominantly exist at pH = 8.

CH_3COOH	CH_3CH_2OH	$CH_3\overset{+}{O}H_2$	$CH_3CH_2\overset{+}{N}H_3$
pK_a = 4.8	pK_a = 15.9	pK_a = -2.5	pK_a = 11.2

5. Which compound has greater bond angles, H_3O^+ or $^+NH_4$?

6. What is the conjugate base of NH_3?

7. Give the structure of a compound that contains five carbons, two of which are sp^2 hybridized and three of which are sp^3 hybridized.

8. a. What products would be formed from the following reaction?

   $CH_3OH \; + \; \overset{+}{N}H_4 \; \rightleftharpoons$

   b. Does the reaction favor reactants or products?

9. a. What orbitals do carbon's electrons occupy before promotion?

b. What orbitals do carbon's electrons occupy after promotion?

10. Which of the following compounds is a stronger acid?

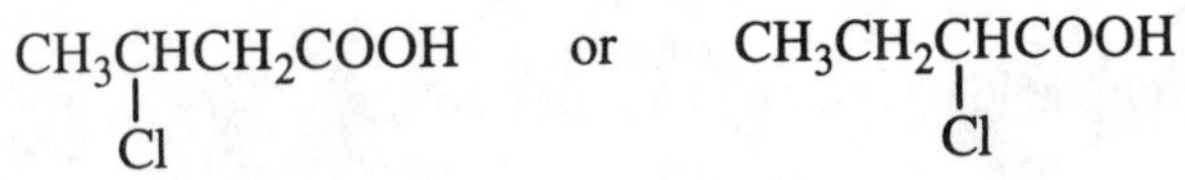

$CH_3CH(Cl)CH_2COOH$ or $CH_3CH_2CH(Cl)COOH$

11. For each of the following compounds indicate the hybridization of the atom to which the arrow is pointing.

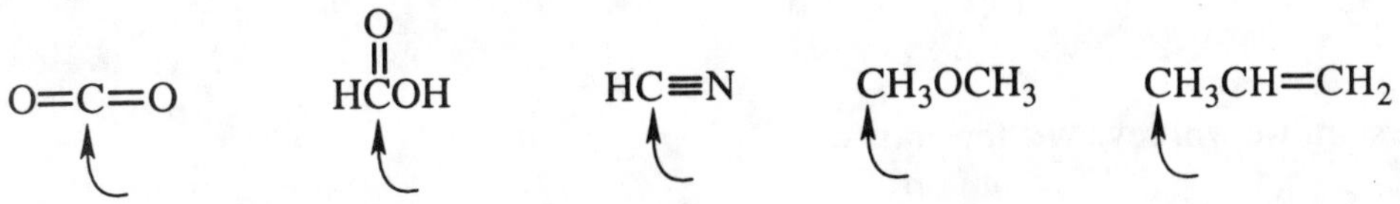

O=C=O (arrow to C) HC(=O)OH (arrow to C) HC≡N (arrow to C) CH_3OCH_3 (arrow to C) $CH_3CH{=}CH_2$ (arrow to CH)

12. Indicate whether each of the following statements is true or false.

a. A pi bond is stronger than a sigma bond. T F

b. A triple bond is shorter than a double bond. T F

c. The oxygen-hydrogen bonds in water are formed by the overlap of an sp^2 orbital of oxygen with an *s* orbital of hydrogen. T F

d. HO^- is a stronger base than $^-NH_2$. T F

e. A double bond is stronger than a single bond. T F

f. A tetrahedral carbon has bond angles of 107.5°. T F

g. A Lewis acid is a compound that accepts a share in a pair of electrons. T F

ANSWERS TO ALL THE PRACTICE TESTS CAN BE FOUND AT THE END OF THE SOLUTIONS MANUAL

SPECIAL TOPIC I

pH, pK_a , and Buffers

This is a continuation of the discussion on acids and bases found in Section 1.16 - 1.21 on pages 40 - 56 of the text. Now we will see how the pH of solutions of acids and bases can be calculated. We will look at three different kinds of solutions.

1. A solution made by dissolving a strong acid or a strong base in water.

2. A solution made by dissolving a weak acid or a weak base in water.

3. A solution made by dissolving a weak acid and its conjugate base in water. Such a solution is known as a **buffer solution.**

Before we start, we will review a few terms.

An acid is a compound that donates a proton, and a base is a compound that accepts a proton.

The degree to which an acid (HA) dissociates is described by its acid dissociation constant (K_a).

$$HA \rightleftharpoons H^+ + A^-$$

$$K_a = \frac{[H^+][A^-]}{[HA]}$$

The strength of an acid can be indicated by its acid dissociation constant or by its pK_a value.

$$\mathbf{pK_a = -\log K_a}$$

The stronger the acid, the **larger** its dissociation constant and the **smaller** its pK_a value. For example, an acid with a dissociation constant of 1×10^{-2} (pK_a = 2) is stronger than an acid with a dissociation constant of 1×10^{-4} (pK_a = 4).

While the pK_a scale is used to describe the strength of an acid, the pH scale is used to describe the acidity of a solution. In other words, the pH scale describes the concentration of hydrogen ions in a solution.

$$\mathbf{pH = -\log [H^+]}$$

The smaller the pH, the more acidic the solution. Acidic solutions have pH values < 7; a neutral solution has a pH = 7; basic solutions have pH values > 7.

A solution with a pH = 2 is more acidic than a solution with a pH = 4.
A solution with a pH = 12 is more basic than a solution with a pH = 8.

Determining the pH of a Solution

To determine the pH of a solution, the concentration of hydrogen ion $[H^+]$ in the solution must be determined.

Strong Acids

A strong acid is one that dissociates completely in solution. Strong acids have pK_a values < 1.

Because a strong acid dissociates completely, the concentration of hydrogen ions is the same as the concentration of the acid: a 1.0 M HCl solution contains 1.0 M $[H^+]$; a 1.5 M HCl solution contains 1.5 M $[H^+]$. Therefore, to determine the pH of a strong acid, the $[H^+]$ value does not have to be calculated; it is the same as the molarity of the strong acid.

solution	$[H^+]$	pH
1.0 M HCl	1.0 M	0
1.0×10^{-2} M HCl	1.0×10^{-2} M	2.0
6.4×10^{-4} M HCl	6.4×10^{-4} M	3.2

Strong Bases

Strong bases are compounds such as NaOH or KOH that dissociate completely in water.

Because they dissociate completely, the $[HO^-]$ is the same as the molarity of the strong base.

The pOH scale describes the basicity of a solution. The larger the pOH, the more basic the solution.

$$\mathbf{pOH = -\log [HO^-]}$$

$[HO^-]$ and $[H^+]$ are related by the water ionization constant (K_w).

$$\mathbf{K_w = [H^+][HO^-] = 10^{-14}}$$

$$\mathbf{pH + pOH = 14}$$

solution	$[HO^-]$	pOH	pH
1.0 M NaOH	1.0 M	0	14.0 – 0 = 14.0
1.0×10^{-4} M NaOH	1.0×10^{-4} M	4.0	14.0 – 4.0 = 10.0
7.8×10^{-2} M NaOH	7.8×10^{-2} M	1.1	14.0 – 1.1 = 12.9

Weak Acids

A weak acid does not dissociate completely in solution. This means that [H+] must be calculated before the pH can be determined.

Acetic acid (CH_3COOH) is an example of a weak acid. It has an acid dissociation constant of 1.74×10^{-5} ($pK_a = 4.76$). The pH of a 1.00 M solution of acetic acid can be calculated as follows:

$$CH_3COOH \rightleftharpoons H^+ + CH_3COO^-$$

$$K_a = \frac{[H^+][CH_3COO^-]}{[CH_3COOH]}$$

Each molecule of acetic acid that dissociates forms one proton and one molecule of acetate ion. Thus the concentration of protons in solution equals the concentration of acetate ions. Each has a concentration that can be represented by *x*. The concentration of acetic acid therefore is whatever we started with minus *x*.

$$1.74 \times 10^{-5} = \frac{(x)(x)}{1.00 - x}$$

The denominator (1.00 - *x*) can be simplified to 1.00 because 1.00 is much greater than *x*. (When we finally calculate the value of *x*, we see that it is 0.004. And 1.00 - 0.004 = 1.00.)

$$1.74 \times 10^{-5} = \frac{x^2}{1.00}$$

$$x = 4.17 \times 10^{-3}$$

$$pH = -\log 4.17 \times 10^{-3}$$

$$pH = 2.38$$

Formic acid (HCOOH) has a pK_a value of 3.75. The pH of a 1.50 M solution of formic acid can be calculated as follows:

A compound with a $pK_a = 3.75$ has an acid dissociation constant of 1.78×10^{-4}.

$$HCOOH \rightleftharpoons H^+ + HCOO^-$$

$$K_a = \frac{[H^+][HCOO^-]}{[HCOOH]}$$

$$1.78 \times 10^{-4} = \frac{(x)(x)}{1.50 - x} = \frac{x^2}{1.50}$$

$$x^2 = 2.67 \times 10^{-4}$$

$$x = 1.63 \times 10^{-2}$$

$$\text{pH} = -\log 1.63 \times 10^{-2}$$

$$\text{pH} = 1.79$$

Weak Bases

When a weak base is dissolved in water, it accepts a proton from water, creating hydroxide ion.

Determining the concentration of hydroxide allows the pOH to be determined, and this in turn allows the pH to be determined.

The pH of a 1.20 M solution of sodium acetate can be calculated as follows:

$$CH_3COO^- + H_2O \rightleftharpoons CH_3COOH + HO^-$$

$$\frac{K_w}{K_a} = \frac{[HO^-][CH_3COOH]}{[CH_3COO^-]}$$

$$\frac{1.00 \times 10^{-14}}{1.74 \times 10^{-5}} = \frac{(x)(x)}{1.20 - x}$$

$$5.75 \times 10^{-10} = \frac{x^2}{1.20}$$

$$x^2 = 6.86 \times 10^{-10}$$

$$x = 2.62 \times 10^{-5}$$

$$\text{pOH} = -\log 2.62 \times 10^{-5}$$

$$\text{pOH} = 4.58$$

$$\text{pH} = 14.00 - 4.58$$

$$\text{pH} = 9.42$$

Notice that by setting up the equation equal to K_w/K_a, we can avoid the introduction of a new term K_b.

Buffer Solutions

A buffer solution is a solution that maintains nearly constant pH despite the addition of small amounts of H^+ or HO^-. That is because a buffer solution contains both a weak acid and its conjugate base. The weak acid can donate a proton to any HO^- added to the solution, and the conjugate base can accept any H^+ that is added to the solution, so the addition of HO^- or H^+ does not significantly change the pH of the solution.

(In order to maintain approximately constant pH, the amount of H^+ or HO^- added to the solution cannot exceed the concentration of the conjugate acid or base in the solution.)

A buffer can maintain nearly constant pH in a range of one pH unit on either side of the pK_a of the conjugate acid. For example, an acetic acid/sodium acetate mixture can be used as a buffer in the pH range 3.76 - 5.76 because acetic acid has a $pK_a = 4.76$; methylammonium ion/methylamine can be used as a buffer in the pH range 9.7 - 11.7 because the methylammonium ion has a $pK_a = 10.7$.

The pH of a buffer solution can be determined from the Henderson-Hasselbalch equation. This equation comes directly from the expression defining the acid dissociation constant. Its derivation is found on page 53 of the text.

Henderson-Hasselbalch equation

$$pK_a = pH + \log\frac{[HA]}{[A^-]}$$

The pH of an acetic acid/sodium acetate buffer solution (pK_a of acetic acid = 4.76) that is 1.00 M in acetic acid and 0.50 M in sodium acetate is calculated as follows:

$$\begin{aligned} pK_a &= pH + \log\frac{[HA]}{[A^-]} \\ 4.76 &= pH + \log\frac{1.00}{0.50} \\ 4.76 &= pH + \log 2 \\ 4.76 &= pH + 0.30 \\ pH &= 4.46 \end{aligned}$$

Remember from Section 1.19 that compounds exist primarily in their acidic forms in solutions that are more acidic than their pK_a's and primarily in their basic forms in solutions that are more basic than their pK_a's. Therefore, it could have been predicted that the above solution will have a pH less than the pK_a of acetic acid because there is more conjugate acid than conjugate base present in the solution.

There are three ways a buffer solution can be prepared:

1. Weak Acid and Weak Base

A buffer solution can be prepared by mixing a solution of a weak acid with a solution of its conjugate base.

The pH of a formic acid/sodium formate buffer (pK_a of formic acid = 3.75) solution prepared by mixing 25 ml of 0.10 M formic acid and 15 ml of 0.20 M sodium formate is calculated as follows:

$$\text{molarity} = \frac{\text{moles}}{\text{liters}} = \frac{\text{millimoles}}{\text{milliliters}}$$

The number of millimoles (mmol) of each of the buffer components can be determined by multiplying the number of milliliters (ml) by the molarity (M) .

$$25 \text{ ml x } 0.10 \text{ M} = 2.5 \text{ mmol formic acid}$$

$$15 \text{ ml x } 0.20 \text{ M} = 3.0 \text{ mmol sodium formate}$$

$$pK_a = \text{pH} + \log \frac{[\text{HA}]}{[\text{A}^-]}$$

$$3.75 = \text{pH} + \log \frac{2.5}{3.0}$$

$$3.75 = \text{pH} + \log 0.83$$

$$3.75 = \text{pH} - 0.08$$

$$\text{pH} = 3.83$$

It could have been predicted that the above solution would have a pH greater than the pK_a of formic acid because there is more conjugate base than conjugate acid present in the solution.

2. Weak Acid and Strong Base

A buffer solution can be prepared by mixing a solution of a weak acid with a strong base such as NaOH. The NaOH reacts completely with the weak acid, thereby creating the conjugate base. For example, if 20 mmol of a weak acid and 5 mmol of a strong base are added to a solution, the 5 mmol of strong base will react with 5 mmol of weak acid, creating 5 mmol weak base and leaving behind 15 mmol of weak acid.

The pH of a solution prepared by mixing 10 ml of a 2.0 M solution of a weak acid with a pK_a of 5.86 with 5.0 ml of a 1.0 M solution of sodium hydroxide can be calculated as follows:

When the 20 mmol of HA and the 5.0 mmol of HO^- are mixed, the 5.0 mmol of strong base will react with 5.0 mmol of HA, with the result that 5.0 mmol of A^- will be formed and 15 mmol (20 mmol - 5.0 mmol) of HA will be left unreacted.

$$10 \text{ ml} \times 2.0 \text{ M} = 20 \text{ mmol HA} \longrightarrow 15 \text{ mmol HA}$$

$$5.0 \text{ ml} \times 1.0 \text{ M} = 5.0 \text{ mmol A}^- \longrightarrow 5.0 \text{ mmol A}^-$$

$$\begin{aligned} pK_a &= pH + \log \frac{[HA]}{[A^-]} \\ 5.86 &= pH + \log \frac{15}{5} \\ 5.86 &= pH + \log 3 \\ 5.86 &= pH + 0.48 \\ pH &= 5.38 \end{aligned}$$

3. Weak Base and Strong Acid

A buffer solution can be prepared by mixing a solution of a weak base with a strong acid such as HCl. The strong acid will react completely with the weak base, thereby forming the conjugate acid.

The pH of an ethylammonium ion/ethylamine buffer (pK_a of $CH_3CH_2NH_3^+ = 11.2$) prepared by mixing 30 ml of 0.20 M ethylamine with 40 ml of 0.10 M HCl can be calculated as follows:

$$30 \text{ ml} \times 0.20 \text{ M} = 6.0 \text{ mmol RNH}_2 \longrightarrow 2.0 \text{ mmol RNH}_2$$

$$40 \text{ ml} \times 0.10 \text{ M} = 4.0 \text{ mmol H}^+ \longrightarrow 4.0 \text{ mmol RNH}_3^+$$

$$\begin{aligned} pK_a &= pH + \log \frac{[HA]}{[A^-]} \\ 11.2 &= pH + \log \frac{4.0}{2.0} \\ 11.2 &= pH + \log 2.0 \\ 11.2 &= pH + 0.30 \\ pH &= 10.9 \end{aligned}$$

Fraction Present in Acidic or Basic Form

A common question to ask is what fraction of a buffer will be in a particular form; either what fraction will be in the acidic form or what fraction will be in the basic form. This is an easy question to answer if you remember the following formulas:

$$\textbf{fraction present in the acidic form} = \frac{[\mathrm{H}^+]}{K_a + [\mathrm{H}^+]}$$

$$\textbf{fraction present in the basic form} = \frac{K_a}{K_a + [\mathrm{H}^+]}$$

What fraction of an acetic acid/sodium acetate buffer (pK_a of acetic acid = 4.76) is present in the acidic form at pH = 5.20?

$$\frac{[\mathrm{H}^+]}{K_a + [\mathrm{H}^+]} = \frac{6.31 \times 10^{-6}}{1.74 \times 10^{-5} + 6.31 \times 10^{-6}}$$

$$= \frac{6.31 \times 10^{-6}}{17.4 \times 10^{-6} + 6.31 \times 10^{-6}}$$

$$= \frac{6.31 \times 10^{-6}}{23.7 \times 10^{-6}} = \frac{6.31}{23.7}$$

$$= 0.26$$

What fraction of a formic acid/sodium formate buffer (pK_a formic acid = 3.75) is present in the basic form at pH = 3.90?

$$\frac{K_a}{K_a + [\mathrm{H}^+]} = \frac{1.78 \times 10^{-4}}{1.78 \times 10^{-4} + 1.26 \times 10^{-4}}$$

$$= \frac{1.78 \times 10^{-4}}{3.04 \times 10^{-4}} = \frac{1.78}{3.04}$$

$$= 0.586$$

$$= 0.59$$

The formulas describing the fraction present in the acidic or basic form are obtained from the definition of the acid dissociation constant.

$$\text{fraction present in the acidic form} = \frac{[\mathrm{HA}]}{[\mathrm{HA}]+[\mathrm{A}^-]}$$

$$K_a = \frac{[\mathrm{H}^+][\mathrm{A}^-]}{[\mathrm{HA}]}$$

$$[\mathrm{A}^-] = \frac{K_a[\mathrm{HA}]}{[\mathrm{H}^+]}$$

$$\text{fraction present in the acidic form} = \frac{[\mathrm{HA}]}{[\mathrm{HA}]+[\mathrm{A}^-]} = \frac{[\mathrm{HA}]}{[\mathrm{HA}]+\dfrac{K_a[\mathrm{HA}]}{[\mathrm{H}^+]}} = \frac{1}{1+\dfrac{K_a}{[\mathrm{H}^+]}}$$

$$= \frac{[\mathrm{H}^+]}{K_a+[\mathrm{H}^+]}$$

$$\text{fraction present in the basic form} = \frac{[\mathrm{A}^-]}{[\mathrm{HA}]+[\mathrm{A}^-]}$$

$$K_a = \frac{[\mathrm{H}^+][\mathrm{A}^-]}{[\mathrm{HA}]}$$

$$[\mathrm{HA}] = \frac{[\mathrm{H}^+][\mathrm{A}^-]}{K_a}$$

$$\text{fraction present in the basic form} = \frac{[\mathrm{A}^-]}{[\mathrm{HA}]+[\mathrm{A}^-]} = \frac{[\mathrm{A}^-]}{[\mathrm{A}^-]+\dfrac{[\mathrm{H}^+][\mathrm{A}^-]}{K_a}}$$

$$= \frac{1}{1+\dfrac{[\mathrm{H}^+]}{K_a}}$$

$$= \frac{K_a}{K_a+[\mathrm{H}^+]}$$

Preparing Buffer Solutions

The type of calculations discussed on pages 36-38 can be used to determine how to make a buffer solution.

For example, how can 100 ml of a 1.00 M buffer solution of pH = 4.24 be prepared if you have available to you 1.50 M solutions of acetic acid, sodium acetate, HCl, and NaOH?

acetic acid has a pK_a = 4.76

fraction present in the acidic form at pH = 4.24 $= \dfrac{[H^+]}{K_a + [H^+]}$

$$\frac{[H^+]}{K_a + [H^+]} = \frac{5.75 \times 10^{-5}}{1.74 \times 10^{-5} + 5.75 \times 10^{-5}}$$

$$= \frac{5.75 \times 10^{-5}}{7.49 \times 10^{-5}}$$

$$= 0.77$$

$$= \frac{5.75 \times 10^{-5}}{7.49 \times 10^{-5}}$$

$$= 0.77$$

If a 1.00 M buffer solution is desired, the buffer must be 0.77 M in acetic acid and 0.23 M in sodium acetate.

Recalling that

$$M = \frac{\text{moles}}{\text{liter}}$$

$$M = \frac{\text{millimoles}}{\text{millimeter}} = \frac{\text{mmol}}{\text{ml}}$$

There are three ways such a buffer solution can be prepared:

1. By mixing the appropriate amounts of acetic acid and sodium acetate in water, and adding water to obtain a final volume of 100 ml.

The amount of acetic acid needed:

$$[CH_3COOH] = 0.77 \text{ M}$$

$$\frac{x \text{ mmol}}{100 \text{ ml}} = 0.77 \text{ M}$$

$$x = 77 \text{ mmol}$$

Therefore, we need to have 77 mmol of acetic acid in the final solution.

To obtain 77 mmol of acetic acid from a 1.50 M solution of acetic acid:

$$\frac{77 \text{ mmol}}{y \text{ ml}} = 1.50 \text{ M}$$

$$y = 51.3 \text{ ml}$$

Notice that the formula M = mmol/ml was used twice. The first time it was used to determine the number of mmol of acetic acid that was needed in the final solution. The second time it was used to determine how that number of mmol could be obtained from an acetic acid solution of a known concentration.

The amount of sodium acetate needed:

$$[CH_3COO^-] = 0.23 \text{ M}$$

$$\frac{x \text{ mmol}}{100 \text{ ml}} = 0.23$$

$$x = 23 \text{ mmol}$$

To obtain 23 mmol of sodium acetate from a 1.50 M solution of sodium acetate:

$$\frac{23 \text{ mmol}}{y \text{ ml}} = 1.50 \text{ M}$$

$$y = 15.3 \text{ ml}$$

The desired buffer solution can be prepared using:

51.3 ml 1.50 M acetic acid

15.3 ml 1.50 M sodium acetate

33.4 ml H_20

2. By mixing the appropriate amounts of acetic acid and sodium hydroxide, and adding water to obtain a final volume of 100 ml.

Sodium hydroxide is used to convert some of the acetic acid into sodium acetate. This means that acetic acid will be the source of both acetic acid and sodium acetate.

The concentrations needed are:

$$[CH_3COOH] = 1.00 \text{ M}$$

$$[NaOH] = 0.23 \text{ M}$$

The amount of acetic acid needed:

$$[CH_3COOH] = 1.00 \text{ M}$$

$$\frac{x \text{ mmol}}{100 \text{ ml}} = 1.00 \text{ M}$$

$$x = 100 \text{ mmol}$$

To obtain 100 mmol of acetic acid from a 1.50 M solution of acetic acid:

$$\frac{100 \text{ mmol}}{y \text{ ml}} = 1.50 \text{ M}$$

$$y = 66.7 \text{ ml}$$

The amount of sodium hydroxide needed:

$$[NaOH] = 0.23 \text{ M}$$

$$\frac{x \text{ mmol}}{100 \text{ ml}} = 0.23 \text{ M}$$

$$x = 23 \text{ mmol}$$

To obtain 23 mmol of sodium hydroxide from a 1.50 M solution of NaOH:

$$\frac{23 \text{ mmol}}{y \text{ ml}} = 1.50 \text{ M}$$

$$y = 15.3 \text{ ml}$$

The desired buffer solution can be prepared using: 66.7 ml 1.5 M acetic acid
15.3 ml 1.5 M NaOH
18.0 ml H_2O

3. By mixing the appropriate amounts of sodium acetate and hydrochloric acid, and adding water to obtain a final volume of 100 ml.

Hydrochloric acid is used to convert some of the sodium acetate into acetic acid. This means that sodium acetate will be the source of both acetic acid and sodium acetate.

The concentrations needed are:

$$[CH_3COONa] = 1.00 \text{ M}$$
$$[HCl] = 0.77 \text{ M}$$

The amount of sodium acetate needed:

$$[CH_3COONa] = 1.00 \text{ M}$$
$$\frac{x \text{ mmol}}{100 \text{ ml}} = 1.00 \text{ M}$$
$$x = 100 \text{ mmol}$$

To obtain 100 mm of sodium acetate from a 1.50 M solution of sodium acetate:

$$\frac{100 \text{ mmol}}{y \text{ ml}} = 1.5 \text{ M}$$
$$y = 66.7 \text{ ml}$$

The amount of hydrochloric acid needed:

$$[HCl] = 0.77 \text{ M}$$
$$\frac{x \text{ mmol}}{100 \text{ ml}} = 0.77 \text{ M}$$
$$x = 77 \text{ mmol}$$

To obtain 77 mm of hydrochloric acid from a 1.50 M solution of HCl:

$$\frac{77 \text{ mmol}}{y \text{ ml}} = 1.50 \text{ M}$$
$$y = 51.3 \text{ ml}$$

100 ml of a 1.00 M acetic acid/acetate buffer cannot be made from these reagents, because the volumes needed (66.7 + 51.3) add up to more than 100. To make this buffer using sodium acetate and hydrochloric acid, you would need to use a more concentrated solution of sodium acetate or a more concentrated solution of HCl.

Problems on pH, pK_a, and Buffers

1. Calculate the pH of each of the following solutions.

   a. 1×10^{-3} M HCl

   b. 0.60 M HCl

   c. 1.40×10^{-2} M HCl

   d. 1×10^{-3} M KOH

   e. 3.70×10^{-4} M NaOH

   f. 1.20 M solution of an acid with a pK_a = 4.23

   g. 1.60×10^{-2} M sodium acetate (pK_a of acetic acid = 4.76)

2. Calculate the pH of each of the following buffer solutions:

   a. A buffer prepared by mixing 20 ml of 0.10 M formic acid and 15 ml of 0.50 M sodium formate (pK_a of formic acid = 3.75).

   b. A buffer prepared by mixing 10 ml of 0.50 M aniline and 15 ml of 0.10 M HCl (pK_a of the anilinium ion = 4.60).

   c. A buffer prepared by mixing 15 ml of 1.00 M acetic acid and 10 ml of 0.50 M NaOH (pK_a of acetic acid = 4.76).

3. What fraction of a carboxylic acid with pK_a = 5.23 would be ionized at pH = 4.98?

4. What would be the concentration of formic acid and sodium formate in a 1.00 M buffer solution with a pH = 3.12?

5. You have found a bottle labeled 1.00 M RCOOH. You want to determine what carboxylic acid it is, so you decide to determine its pK_a. How would you do this?

6. a. How would you prepare 100 ml of a buffer solution that is 0.30 M in acetic acid and 0.20 M in sodium acetate using a 1.00 M acetic acid solution and a 2.00 M sodium acetate solution?

   b. The pK_a of acetic acid is 4.76. Would the pH of the above solution be greater or less than 4.76?

7. You have 100 ml of a 1.50 M acetic acid/sodium acetate buffer solution that has a pH = 4.90. How could you change the pH of the solution to 4.50?

8. You have 100 ml of a 1.00 M solution of an acid with a $pK_a = 5.62$ to which you add 10 ml of 1.00 M sodium hydroxide. What fraction of the acid will be in the acidic form? How much more sodium hydroxide will you need to add in order to have 40% of the acid in the acidic form?

9. Describe three ways to make a 1.00 M acetic acid/sodium acetate buffer solution with a pH = 4.00.

10. You have available to you 1.50 M solutions of acetic acid, sodium acetate, potassium hydroxide, and hydrochloric acid. How would you make 50 ml of each of the buffers described in the preceding problem?

11. How would you make a 1.0 M buffer solution with a pH = 3.30?

12. You are planning to carry out a reaction that will produce protons. In order for the reaction to take place at constant pH, it will be carried in a solution buffered at pH = 4.2. Would it be better to use a formic acid/formate buffer or an acetic acid/acetate buffer?

Answers to Problems on pH, pK_a, and Buffers

1.

a. $pH = -\log 1 \times 10^{-3}$

$pH = 3$

b. $pH = -\log 0.60$

$pH = 0.22$

c. $pH = -\log 1.40 \times 10^{-2}$

$pH = 1.85$

d. $pOH = -\log 1 \times 10^{-3}$

$pOH = 3$
$pH = 14 - 3 = 11$

e. $pOH = -\log 3.70 \times 10^{-4}$

$pOH = 3.43$
$pH = 10.57$

f. $pK_a = 4.23$, $K_a = 5.89 \times 10^{-5}$

$$K_a = \frac{[H^+][A^-]}{[HA]}$$

$$5.89 \times 10^{-5} = \frac{x^2}{1.20}$$

$$x^2 = 7.07 \times 10^{-5}$$

$$x = 8.41 \times 10^{-3}$$

$$pH = 2.08$$

g.

$$\frac{K_w}{K_a} = \frac{[HO^-][HA]}{[A^-]}$$

$$\frac{1.0 \times 10^{-14}}{1.74 \times 10^{-5}} = \frac{x^2}{1.60 \times 10^{-2}}$$

$$5.75 \times 10^{-10} = \frac{x^2}{1.60 \times 10^{-2}}$$

$$x^2 = 9.20 \times 10^{-12}$$

$$x = 3.03 \times 10^{-6}$$

$$pOH = 5.52$$

$$pH = 14.00 - 5.52 = 8.48$$

2. **a.** formic acid: 20 ml x 0.10 M = 2.0 mmol
sodium formate: 15 ml x 0.50 M = 7.5 mmol

$$pK_a = pH + \log \frac{[HA]}{[A^-]}$$
$$3.75 = pH + \log \frac{2.0}{7.5}$$
$$3.75 = pH + \log 0.27$$
$$3.75 = pH + (-0.57)$$
$$pH = 4.32$$

b. aniline: 10 ml x 0.50 = 5.0 mmol ⟶ 3.5 mmol NH_2
HCl: 15 ml x 0.10 = 1.5 mmol ⟶ 1.5 mmol $^+NH_3$

$$pK_a = pH + \log \frac{[HA]}{[A^-]}$$
$$4.60 = pH + \log \frac{1.5}{3.5}$$
$$4.60 = pH + \log 0.43$$
$$4.60 = pH + (-0.37)$$
$$pH = 4.97$$

c. acetic acid: 15 ml x 1.00 = 15 mmol ⟶ 10 mmol acetic acid
NaOH: 10 ml x 0.50 = 5.0 mmol ⟶ 5.0 mmol acetic acid

$$pK_a = pH + \log \frac{[HA]}{[A^-]}$$
$$4.76 = pH + \log \frac{10}{5.0}$$
$$4.76 = pH + \log 2$$
$$4.76 = pH + 0.30$$
$$pH = 4.46$$

3. The ionized form is the basic form.

$$\frac{K_a}{K_a + [H^+]} = \frac{5.89 \times 10^{-6}}{5.89 \times 10^{-6} + 10.47 \times 10^{-6}} = \frac{5.89 \times 10^{-6}}{16.36 \times 10^{-6}}$$

$$= 0.36$$

4.

$$pK_a = pH + \log\frac{[HA]}{[A^-]}$$

$$3.75 = 3.12 + \log\frac{[HA]}{[A^-]}$$

$$0.63 = \log\frac{[HA]}{[A^-]}$$

$$4.27 = \frac{[HA]}{[A^-]}$$

$$[HA] = 4.27[A^-]$$

$$[HA]+[A^-] = 1.0\text{ M}$$

$$4.27[A^-]+[A^-] = 1.0\text{ M}$$

$$5.27[A^-] = 1.0\text{ M}$$

$$[A^-] = 0.19\text{ M}$$

$$[\text{sodium formate}] = 0.19\text{ M}$$

$$[\text{formic acid}] = 0.81\text{ M}$$

5.

$$pK_a = pH + \log\frac{[HA]}{[A^-]}$$

$$\text{when } [HA] = [A^-],$$

$$pK_a = pH$$

Preparing a solution of x mmol of RCOOH and 1/2 x mmol NaOH will give a solution in which $[RCOOH] = [RCOO^-]$.

For example: 20 ml of 1.00 M RCOOH = 20 mmol
10 ml of 1.00 M NaOH = 10 mmol

This will give a solution that has 10 mmol RCOOH and 10 mmol $RCOO^-$.

The pH of this solution is the pK_a of RCOOH.

6. **a.**

$$\frac{x \text{ mmol}}{100 \text{ ml}} = 0.30 \text{ M} \qquad \frac{x \text{ mmol}}{100 \text{ ml}} = 0.20 \text{ M}$$

$$x = 30 \text{ mmol of acetic acid} \qquad x = 20 \text{ mmol of sodium acetate}$$

$$\frac{30 \text{ mmol}}{y \text{ ml}} = 1.00 \text{ M} \qquad \frac{20 \text{ mmol}}{y \text{ ml}} = 2.00 \text{ M}$$

$$y = 30 \text{ ml of 1.00 M acetic acid} \qquad y = 10 \text{ ml of 2.00 M acetic acid}$$

The buffer solution could be prepared by mixing: 30 ml of 1.00 M acetic acid
10 ml of 2.00 M sodium acetate
60 ml of water

b. Because the concentration of buffer in the acidic form (0.30 M) is greater than the concentration of buffer in the basic form (0.20 M), the pH of the solution will be less than 4.76.

7.

$$pK_a = \text{pH} + \log\frac{[\text{HA}]}{[\text{A}^-]}$$

original solution

$$4.76 = 4.90 + \log\frac{[\text{HA}]}{[\text{A}^-]}$$

$$-0.14 = \log\frac{[\text{HA}]}{[\text{A}^-]}$$

$$0.72 = \frac{[\text{HA}]}{[\text{A}^-]}$$

$$[\text{HA}] = 0.72[\text{A}^-]$$

$$[\text{HA}] + [\text{A}^-] = 1.50 \text{ M}$$

$$0.72[\text{A}^-] + [\text{A}^-] = 1.50 \text{ M}$$

$$1.72[\text{A}^-] = 1.50 \text{ M}$$

$$[\text{A}^-] = 0.87 \text{ M}$$

$$[\text{HA}] = 0.63 \text{ M}$$

desired solution

$$4.76 = 4.50 + \log\frac{[HA]}{[A^-]}$$

$$0.26 = \log\frac{[HA]}{[A^-]}$$

$$1.82 = \frac{[HA]}{[A^-]}$$

$$[HA] = 1.82[A^-]$$

$$[HA]+[A^-] = 1.50\ M$$

$$1.82[A^-]+[A^-] = 1.50\ M$$

$$2.82[A^-] = 1.50\ M$$

$$[A^-] = 0.53\ M$$

$$[HA] = 0.97\ M$$

The original solution contains 87 mmol of A^- (100 ml x 0.87 M).
The desired solution with a pH = 4.50 must contain 53 mmol of A^-.
Therefore, 34 mmol of A^- (87 - 53 = 34) must be converted to HA.
This can be done by adding 34 mmol of HCl to the original solution.

If you have a 1.00 M HCl solution, you will need to add 34 ml to the original solution in order to change its pH from 4.90 to 4.50.

$$\frac{34\ \text{mmol}}{x\ \text{ml}} = 1.00\ M$$

$$x = 34\ \text{ml}$$

Note that after adding HCl to the original solution, it will no longer be a 1.50 M buffer; it will be more dilute (154 mm/134 ml = 1.12 M).

The change in the concentration of the buffer solution will be less if a more concentrated solution of HCl is used to change the pH. If you have a 2.00 M HCl solution:

$$\frac{34\ \text{mmol}}{x\ \text{ml}} = 2.00\ M$$

$$x = 17\ \text{ml}$$

You will need to add 17 ml to the original solution, and the concentration of buffer species will be 1.28 M (154 mm/117 ml = 1.28 M).

8.

acid: 100 ml x 1.00 = 100 mmol ⟶ 90 mmol HA

NaOH: 10 ml x 1.00 = 10 mmol ⟶ 10 mmol A^-

Therefore, 90% is in the acidic form.

For 40% to be in the acidic form you need:

40 mmol HA

60 mmol A^-

You need to have 60 mmol rather than 10 mmol in the basic form. To get the additional 50 mmol in the basic form, you would need to add 50 ml of 1.0 M NaOH.

9.

$$pK_a = pH + \log\frac{[HA]}{[A^-]}$$

$$4.76 = 4.00 + \log\frac{[HA]}{[A^-]}$$

$$0.76 = \log\frac{[HA]}{[A^-]}$$

$$5.75 = \frac{[HA]}{[A^-]}$$

$$[HA] = 5.75[A^-]$$

$$[HA] + [A^-] = 1.0\text{ M}$$

$$5.75[A^-] + [A^-] = 1.0\text{ M}$$

$$6.75[A^-] = 1.0\text{ M}$$

$$[A^-] = 0.15\text{ M}$$

$$[HA] = 0.85\text{ M}$$

a.

[acetic acid] = 0.85 M

[sodium acetate] = 0.15 M

b.

[acetic acid] = 1.00 M

[NaOH] = 0.15 M

c.

[sodium acetate] = 1.00 M

[HCl] = 0.85 M

10. a.

$$\frac{x \text{ mmol}}{50 \text{ ml}} = 0.85 \text{ M}$$

$$x = 42.5 \text{ mmol of acetic acid}$$

$$\frac{42.4 \text{ mmol}}{y \text{ ml}} = 1.50 \text{ M}$$

$$y = 28.3 \text{ ml of 1.50 M acetic acid}$$

$$\frac{x \text{ mmol}}{50 \text{ ml}} = 0.15 \text{ M}$$

$$x = 7.5 \text{ mmol of sodium acetate}$$

$$\frac{7.5 \text{ mmol}}{y \text{ ml}} = 1.50 \text{ M}$$

$$y = 5.0 \text{ ml of 1.50 M sodium acetate}$$

28.3 ml of 1.50 M acetic acid
5.0 ml of 1.50 M sodium acetate
16.7 ml of H_2O

b.

$$\frac{x \text{ mmol}}{50 \text{ ml}} = 1.00 \text{ M}$$

$$x = 50 \text{ mmol of acetic acid}$$

$$\frac{50 \text{ mmol}}{y \text{ ml}} = 1.50 \text{ M}$$

$$y = 33.3 \text{ ml of 1.50 M acetic acid}$$

$$\frac{x \text{ mmol}}{50 \text{ ml}} = 0.85 \text{ M}$$

$$x = 42.5 \text{ mmol of HCl}$$

$$\frac{42.5 \text{ mmol}}{y \text{ ml}} = 1.50 \text{ M}$$

$$y = 28.3 \text{ ml of 1.50 M HCl}$$

33.3 ml of 1.50 M acetic acid
5.0 ml of 1.50 M NaOH
11.7 ml of H_2O

c.

$$\frac{x \text{ mmol}}{50 \text{ ml}} = 1.00 \text{ M}$$

$$x = 50 \text{ mmol of sodium acetate}$$

$$\frac{50 \text{ mmol}}{y \text{ ml}} = 1.50 \text{ M}$$

$$y = 33.3 \text{ ml of } 1.50 \text{ M sodium acetate}$$

$$\frac{x \text{ mm}}{50 \text{ ml}} = 0.85 \text{ M}$$

$$x = 42.5 \text{ mm}$$

$$\frac{42.5 \text{ mm}}{y \text{ ml}} = 1.5 \text{ M}$$

$$y = 28.3 \text{ ml of } 1.5 \text{ M HCl}$$

We cannot make the required buffer with these solutions, because 33.3 ml + 28.3 ml > 50 ml.

11. Because formic acid has a $pK_a = 3.75$, a formic acid/formate buffer can be a buffer at pH = 3.30.

$$pK_a = \text{pH} + \log\frac{[\text{HA}]}{[\text{A}^-]}$$

$$3.75 = 3.30 + \log\frac{[\text{HA}]}{[\text{A}^-]}$$

$$0.45 = \log\frac{[\text{HA}]}{[\text{A}^-]}$$

$$2.82 = \frac{[\text{HA}]}{[\text{A}^-]}$$

$$[\text{HA}] = 2.82[\text{A}^-]$$

$$[\text{HA}] + [\text{A}^-] = 1.0 \text{ M}$$

$$2.82[\text{A}^-] + [\text{A}^-] = 1.0 \text{ M}$$

$$3.82[\text{A}^-] = 1.0 \text{ M}$$

$$[A^-] = 0.26 \text{ M}$$
$$[HA] = 0.74 \text{ M}$$

The solution must have [formic acid] = 0.74 M and [sodium formate] = 0.26 M.

12. At pH = 4.20, 74% of the formate buffer will be in the basic form.

pH = 4.20, $[H^+] = 6.31 \times 10^{-5}$ Formic acid has a $pK_a = 3.75$, $K_a = 1.78 \times 10^{-4}$.

$$\frac{K_a}{K_a + [H^+]} = \frac{1.78 \times 10^{-4}}{1.78 \times 10^{-4} + 6.31 \times 10^{-5}} = \frac{1.78 \times 10^{-4}}{1.78 \times 10^{-4} + 0.63 \times 10^{-4}}$$
$$= \frac{1.78 \times 10^{-4}}{2.41 \times 10^{-4}}$$
$$= 0.74$$

At pH = 4.20, 22% of the acetate buffer will be in the basic form.

pH = 4.20, $[H^+] = 6.31 \times 10^{-5}$ Acetic acid has a $pK_a = 4.76$, $K_a = 1.74 \times 10^{-5}$.

$$\frac{K_a}{K_a + [H^+]} = \frac{1.74 \times 10^{-5}}{1.74 \times 10^{-5} + 6.31 \times 10^{-5}}$$
$$= \frac{1.78 \times 10^{-5}}{8.05 \times 10^{-5}}$$
$$= 0.22$$

The reaction to be carried out will generate protons that will react with the basic form of the buffer in order to keep the pH constant. Therefore, the formate buffer is preferred because it has a greater percentage of the buffer in the basic form.

CHAPTER 2
An Introduction to Organic Compounds: Nomenclature, Physical Properties, and Representation of Structure

Important Terms

alcohol — a compound with an OH group in place of one of the hydrogens of an alkane (ROH).

alkane — a hydrocarbon that contains only single bonds.

alkyl halide — a compound with a halogen in place of one of the hydrogens of an alkane.

alkyl substituent — a substituent formed by removing a hydrogen from an alkane.

amine — a compound in which one or more of the hydrogens of NH_3 is replaced by an alkyl substituent (RNH_2, R_2NH, R_3N).

angle strain — the strain introduced into a molecule as a result of its bond angles being distorted from their ideal values.

anti conformer — the staggered conformer in which the largest substituents bonded to the two carbons are opposite each other. It is the most stable of the staggered conformers.

asymmetrical ether — an ether with two different substituents bonded to the oxygen.

axial bond — a bond of the chair form of cyclohexane that is perpendicular to the plane in which the chair is drawn (an up-down bond).

banana bonds — the bonds in small rings that are slightly bent as a result of orbitals overlapping at an angle rather than overlapping head-on.

boat conformation — the conformation of cyclohexane that roughly resembles a boat.

boiling point — the temperature at which a liquid vaporizes.

chair conformation — the conformation of cyclohexane that roughly resembles a chair. It is the most stable conformation of cyclohexane.

cis fused — two rings fused together in such a way that if the second ring were considered to be two substituents of the first ring, the two substituents would be on the same side of the first ring.

cis isomer — the isomer with both hydrogens on the same side of the double bond.

cis-trans stereoisomers (geometric isomers) — geometric (or *E*, *Z*) isomers.

common name — nonsystematic nomenclature.

conformation — the three-dimensional shape of a molecule at a given instant.

conformational analysis — the investigation of various conformations of a compound and their relative stabilities.

conformational isomers (conformers) different conformations of a molecule.

constitutional isomers (structural isomers) molecules that have the same molecular formula but differ in the way the atoms are connected.

cycloalkane an alkane with its carbon chain arranged in a closed ring.

1,3-diaxial interaction the interaction between an axial substituent and the other two axial substituents on the same side of the cyclohexane ring.

dipole-dipole interaction an interaction between the dipole of one molecule and the dipole of another.

eclipsed conformation a conformation in which the bonds on adjacent carbons are parallel to each other as viewed looking down the carbon-carbon bond.

equatorial bond a bond of the chair form of cyclohexane that juts out from the ring in approximately the same plane that contains the chair.

ether a compound in which an oxygen is bonded to two alkyl groups (ROR).

flagpole hydrogens the two hydrogens in the boat conformation of cyclohexane that are closest to each other.

functional group the center of reactivity of a molecule.

gauche conformer a staggered conformer in which the largest substituents bonded to the two carbons are gauche to each other.

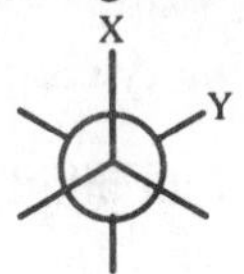

The substituents are gauche to each other.

gauche interaction the interaction between two atoms or groups that are gauche to each other.

geometric isomers (cis-trans stereoisomers) cis-trans (or *E*, *Z*) isomers.

half-chair conformer the least stable conformation of cyclohexane.

homolog a member of a homologous series.

homologous series a family of compounds in which each member differs from the next by one methylene group.

hydrocarbon a compound that contains only carbon and hydrogen.

hydrogen bond an unusually strong dipole-dipole attraction (5 kcal/mol) between a hydrogen bonded to O, N, or F and the nonbonding electrons of a different O, N, or F.

induced dipole-induced dipole interaction an interaction between a temporary dipole in one molecule and the dipole that the temporary dipole induces in another molecule.

IUPAC nomenclature	systematic nomenclature.
London forces (van der Waals forces)	induced dipole-induced dipole interactions.
melting point	the temperature at which a solid becomes a liquid.
methylene group	a CH_2 group.
Newman projection	a way to represent the three-dimensional spatial relationships of atoms by looking down the length of a particular carbon-carbon bond.
packing	the property that determines how well individual molecules fit into a crystal lattice.
parent hydrocarbon	the longest continuous carbon chain in a molecule.
perspective formula	a way to represent the three-dimensional spatial relationships of atoms using two adjacent solid lines, one solid wedge and one hatched wedge.
polarizability	the ease with which an electron cloud of an atom can be distorted.
primary alcohol	an alcohol in which the OH group is bonded to a primary carbon.
primary alkyl halide	an alkyl halide in which the halogen is bonded to a primary carbon.
primary amine	an amine with one alkyl group bonded to the nitrogen.
primary carbon	a carbon bonded to only one other carbon.
primary hydrogen	a hydrogen bonded to a primary carbon.
quaternary ammonium salt	a nitrogen compound with four alkyl groups bonded to the nitrogen
ring-flip (chair-chair interconversion)	the conversion of a chair conformer of cyclohexane into the other chair conformer. Bonds that are axial in one chair conformer are equatorial in the other chair conformer.
sawhorse projection	a way to represent the three-dimensional spatial relationships of atoms by looking at the carbon-carbon bond from an oblique angle.
secondary alcohol	an alcohol in which the OH group is bonded to a secondary carbon.
secondary alkyl halide	an alkyl halide in which the halogen is bonded to a secondary carbon.
secondary amine	an amine with two alkyl groups bonded to the nitrogen.
secondary carbon	a carbon bonded to two other carbons.
secondary hydrogen	a hydrogen bonded to a secondary carbon.
skeletal structure	a structure that shows the carbon-carbon bonds as lines and does not show the carbon-hydrogen bonds.

skew-boat conformer one of the conformations of a cyclohexane ring.

solubility the extent to which a compound dissolves in a solvent.

solvation the interaction between a solvent and another molecule (or ion).

staggered conformation a conformation in which the bonds on one carbon bisect the bond angle on the adjacent carbon when viewed looking down the carbon-carbon bond.

steric hindrance hindrance due to groups occupying a volume of space.

steric strain the repulsion between the electron cloud of an atom or group of atoms and the electron cloud of another atom or group of atoms.

straight-chain alkane an alkane in which the carbons form a continuous chain with no branches.

structural isomers (constitutional isomers) molecules that have the same molecular formula but differ in the way the atoms are connected.

symmetrical ether an ether with two identical substituents bonded to the oxygen.

systematic nomenclature IUPAC nomenclature.

tertiary alcohol an alcohol in which the OH group is bonded to a tertiary carbon.

tertiary alkyl halide an alkyl halide in which the halogen is bonded to a tertiary carbon.

tertiary amine an amine with three alkyl groups bonded to the nitrogen.

tertiary carbon a carbon bonded to three other carbons.

tertiary hydrogen a hydrogen bonded to a tertiary carbon.

torsional strain the repulsion felt by the bonding electrons of one substituent as they pass close to the bonding electrons of another substituent.

trans-fused two rings fused together in such a way that if the second ring were considered to be two substituents of the first ring, the two substituents would be on opposite sides of the first ring.

trans isomer the isomer with identical substituents on opposite sides of the double bond.

twist-boat conformer one of the conformations of a cyclohexane ring.

van der Waals forces (London forces) induced dipole-induced dipole interactions.

Solutions to Problems

1.

a. CH_3CHOH
$\quad\quad\ |$
$\quad\ \ CH_3$

b. $CH_3CHCH_2CH_2F$
$\quad\quad\ |$
$\quad\ \ CH_3$

c. CH_3CH_2CHI
$\quad\quad\quad\quad\ |$
$\quad\quad\quad\ \ CH_3$

d.
$\quad\quad\ CH_3$
$\quad\quad\ \ |$
CH_3CCH_2Cl
$\quad\quad\ \ |$
$\quad\quad\ CH_3$

e.
$\quad\quad\ CH_3$
$\quad\quad\ \ |$
CH_3CNH_2
$\quad\quad\ \ |$
$\quad\quad\ CH_3$

f. $CH_3CHCH_2CH_2CH_2CH_2CH_2Br$
$\quad\quad\ |$
$\quad\ \ CH_3$

2.

a.
$\quad\quad\quad\ CH_3$
$\quad\quad\quad\ \ |$
$CH_3CHCHCH_2CH_2CH_3$
$\quad\quad\ |$
$\quad\ \ CH_3$

b.
$\quad\ CH_3\quad\ \ CH_3\ CH_3$
$\quad\quad |\quad\quad\quad |\quad\ \ |$
$CH_3CHCH_2C—CHCH_2CH_3$
$\quad\quad\quad\quad\quad |$
$\quad\quad\quad\quad CHCH_3$
$\quad\quad\quad\quad\ \ |$
$\quad\quad\quad\quad CH_3$

c.
$\quad\quad\quad\quad\quad\ \ CH_2CH_3$
$\quad\quad\quad\quad\quad\quad |$
$CH_3CH_2CH_2CCH_2CH_2CH_2CH_2CH_2CH_3$
$\quad\quad\quad\quad\quad\quad |$
$\quad\quad\quad\quad\quad\ \ CH_2CH_3$

d.
$\quad\quad\ CH_3$
$\quad\quad\ \ |$
$CH_3CCH_2CHCH_2CH_2CH_2CH_3$
$\quad\quad\ \ |\quad\quad\ |$
$\quad\quad\ CH_3\ CH_2$
$\quad\quad\quad\quad\quad |$
$\quad\quad\quad\quad\ CH_2$
$\quad\quad\quad\quad\quad |$
$\quad\quad\quad\quad\ CH_3$

e.
$\quad\ CH_3\quad\quad\quad\quad CH_3$
$\quad\quad |\quad\quad\quad\quad\quad |$
$CH_3CHCH_2CHCHCH_2CH_2CH_3$
$\quad\quad\quad\quad\quad\ |$
$\quad\quad\quad\quad\ CH_2$
$\quad\quad\quad\quad\quad |$
$\quad\quad\quad\quad CHCH_3$
$\quad\quad\quad\quad\quad |$
$\quad\quad\quad\quad\ CH_3$

f. $CH_3CH_2CH_2CHCH_2CH_2CH_2CH_3$
$\quad\quad\quad\quad\quad\quad\quad\ |$
$\quad\quad\quad\quad\quad\quad CH_3CCH_3$
$\quad\quad\quad\quad\quad\quad\quad\ |$
$\quad\quad\quad\quad\quad\quad\quad CH_3$

3. a.

#1 $CH_3CH_2CH_2CH_2CH_2CH_2CH_2CH_3$

octane

#2 $CH_3CHCH_2CH_2CH_2CH_2CH_3$
CH_3 (on C-2)

2-methylheptane

#3 $CH_3CH_2CHCH_2CH_2CH_2CH_3$
CH_3 (on C-3)

3-methylheptane

#4 $CH_3CH_2CH_2CHCH_2CH_2CH_3$
CH_3 (on C-4)

4-methylheptane

#5 CH_3 (above C-2)
$CH_3CCH_2CH_2CH_2CH_3$
CH_3 (below C-2)

2,2-dimethylhexane

#6 CH_3 (above C-3)
$CH_3CH_2CCH_2CH_2CH_3$
CH_3 (below C-3)

3,3-dimethylhexane

#7 CH_3 CH_3
$CH_3CH—CHCH_2CH_2CH_3$

2,3-dimethylhexane

#8 CH_3 CH_3
$CH_3CHCH_2CHCH_2CH_3$

2, 4-dimethylhexane

#9 CH_3 CH_3
$CH_3CHCH_2CH_2CHCH_3$

2,5-dimethylhexane

#10 CH_3 CH_3
$CH_3CH_2CH-CHCH_2CH_3$

3,4-dimethylhexane

#11 CH_3 CH_3
$CH_3C—CHCH_2CH_3$
CH_3

2,2,3-trimethylpentane

#12 CH_3 CH_3
$CH_3CCH_2CHCH_3$
CH_3

2,2,4-trimethylpentane

#13 CH_3 CH_3
$CH_3CH—CCH_2CH_3$
CH_3

2,3,3-trimethylpentane

#14 CH_3 CH_3 CH_3
$CH_3CH—CH—CHCH_3$

2,3,4-trimethylpentane

#15 CH_3CH_3
$CH_3C—CCH_3$
CH_3CH_3

2,2,3,3-tetramethylbutane

#16 $CH_3CH_2CHCH_2CH_2CH_3$
CH_2CH_3

3-ethylhexane

#17 CH_3
$CH_3CH_2CHCHCH_3$
CH_2CH_3

3-ethyl-2-methylpentane

#18 CH_3
$CH_3CH_2CCH_2CH_3$
CH_2CH_3

3-ethyl-3-methylpentane

b. IUPAC names are under the compound.

c. Only #1 (octane) and #2 (isooctane) have common names.

d. #2, #7, #8, #9, #12, #13, #14, #17

e. #3, #8, #10, #11

f. #5, #11, #12, #15

4.

a. 2,2,4-trimethylhexane

b. 2,2-dimethylbutane

c. 2,5-dimethylheptane

d. 3,3-diethylhexane

e. 3,3-diethyl-4-methyl-5-propyloctane

f. 3-methyl-4-propylheptane

g. 5-ethyl-4,4-dimethyloctane

h. 4-isopropyloctane

5.

a.  OH

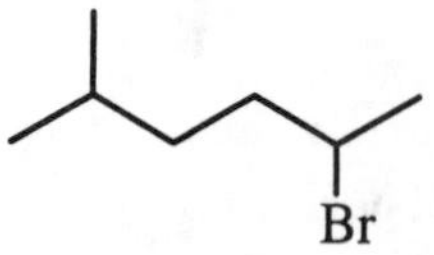

b.

c. Br

d. 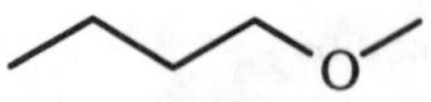 O

6.

a. 1-ethyl-2-methylcyclopentane

b. ethylcyclobutane

c. 4-ethyl-1,2-dimethylcyclohexane

d. 3,6-dimethyldecane

e. 2-cyclopropylpentane

f. 1-ethyl-3-isobutylcyclohexane

g. 5-isopropylnonane

h. 1-*sec*-butyl-4-isopropylcyclohexane

7.

a. *sec*-butyl chloride
2-chlorobutane
secondary

b. isoheptyl chloride
1-chloro-5-methylhexane
primary

c. cyclohexyl bromide
bromocyclohexane
secondary

d. isopropyl fluoride
2-fluoropropane
secondary

8.

a. CH_2Cl (on cyclohexane ring)

Note that the name of a $-CH_2Cl$ substituent is "chloromethyl", since a chloro is in place of one of the hydrogens of a methyl substituent.

chloromethylcyclohexane

b.

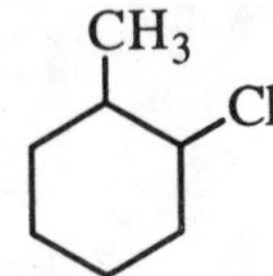

CH_3 / Cl (1,3 on cyclohexane ring)

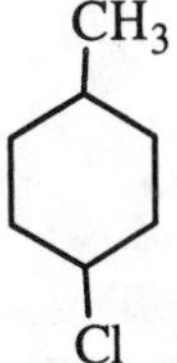

1-chloro-2-methylcyclohexane 1-chloro-3-methylcyclohexane 1-chloro-4-methylcyclohexane

c. Cl, CH_3 (both on same ring carbon)

1-chloro-1-methylcyclohexane

9.

a. $CH_3CH_2CH_2CH_2CH_3$ b. $CH_3C(CH_3)_2CH_3$ c. $CH_3CH(CH_3)CH_2CH_3$

pentane dimethylpropane methylbutane

10. a.

1. methoxyethane
2. ethoxyethane
3. 4-methoxyoctane
4. 1-propoxybutane
5. 2-isopropoxypentane
6. 1-isopropoxy-3-methylbutane

b. No.

c.

1. ethyl methyl ether
2. diethyl ether
3. no common name
4. butyl propyl ether
5. no common name
6. isopentyl isopropyl ether

11.

CH_3OH methyl alcohol
methanol

CH_3CH_2OH ethyl alcohol
ethanol

$CH_3CH_2CH_2OH$ propyl alcohol
propanol

$CH_3CH_2CH_2CH_2OH$ butyl alcohol
butanol

$CH_3CH_2CH_2CH_2CH_2OH$ pentyl alcohol
pentanol

$CH_3CH_2CH_2CH_2CH_2CH_2OH$ hexyl alcohol
hexanol

12.

a. 1-pentanol
primary

b. 4-methylcyclohexanol
secondary

c. 5-chloro-2-methyl-2-pentanol
tertiary

d. 5-methyl-3-hexanol
secondary

e. 2,6-dimethyl-4-octanol
secondary

f. 4-chloro-3-ethylcyclohexanol
secondary

13.

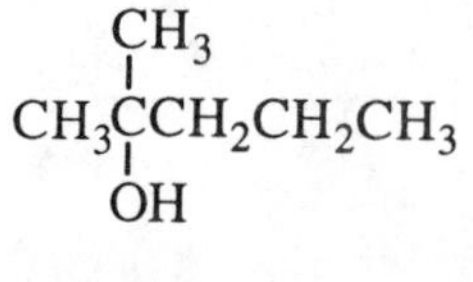

$CH_3C(CH_3)(OH)CH_2CH_2CH_3$ (structure: CH_3 above and OH below the second carbon)
2-methyl-2-pentanol

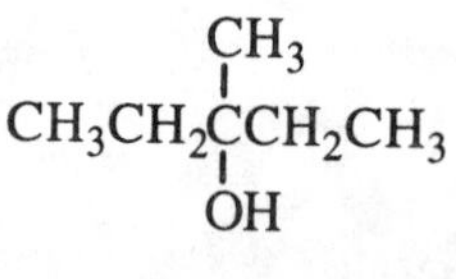

$CH_3CH_2C(CH_3)(OH)CH_2CH_3$ (structure: CH_3 above and OH below the third carbon)
3-methyl-3-pentanol

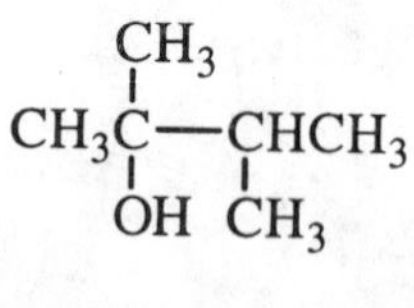

$CH_3C(CH_3)(OH)—CH(CH_3)CH_3$ (structure: CH_3 above and OH below C; CH_3 below CH)
2,3-dimethyl-2-butanol

14.

a. hexylamine
1-hexanamine

b. butylpropylamine
N-propyl-1-butanamine

c. *sec*-butylisobutylamine
N-isobutyl-2-butanamine (notice that the longest continuous chain has 4 carbons)

d. diethylproplyamine
N,N-diethyl-1-propanamine

e. cyclohexylamine
cyclohexanamine

15.

a. 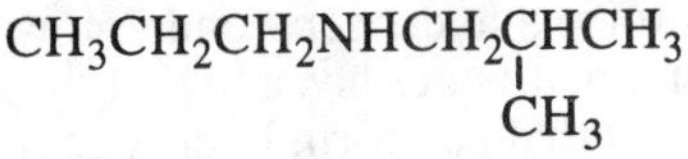 $CH_3CH_2CH_2NHCH_2CHCH_3$ (with CH_3 on the CH)

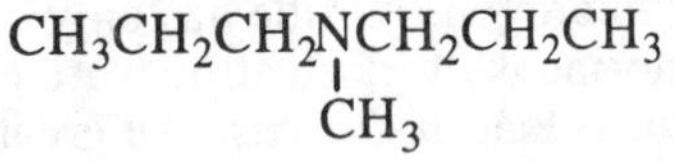

b. $CH_3CH_2NHCH_2CH_3$

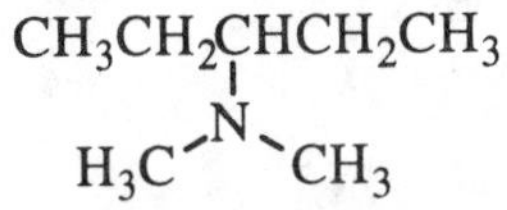

c. 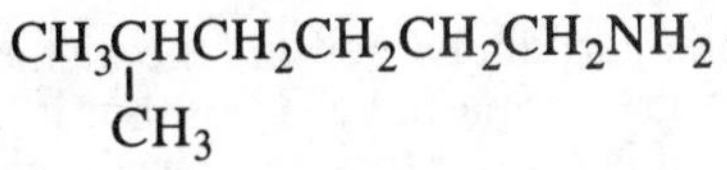 $CH_3CHCH_2CH_2CH_2CH_2NH_2$ (with CH_3 on the CH)

d. $CH_3CH_2CH_2NCH_2CH_2CH_3$ (with CH_3 on the N)

e. $CH_3CH_2CHCH_2CH_3$ (with $N(CH_3)_2$, shown as $H_3C-N-CH_3$, on the CH)

f. 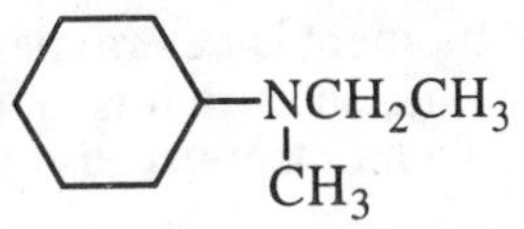 cyclohexane ring$-NCH_2CH_3$ (with CH_3 on the N)

16.

a. 6-methyl-1-heptanamine
isooctylamine
primary

b. 3-methyl-*N*-propyl-1-butanamine
isopentylpropylamine
secondary

c. *N*-ethyl-*N*-methylethanamine
diethylmethylamine
tertiary

d. 2,5-dimethylcyclohexanamine
no common name
primary

17.

a. The bond angle is predicted to be similar to the bond angle in water (104.5°)
b. The bond angle is predicted to be similar to the bond angle in ammonia (107.3°)
c. The bond angle is predicted to be similar to the bond angle in water (104.5°)
d. The bond angle is predicted to be similar to the bond angle in the ammonium ion (109.5°)

18.

a. 1, 4, and 5

b. 1, 2, 4, 5, and 6

19.

a. Each water molecule has two hydrogens that can form hydrogen bonds, while each alcohol molecule has only one hydrogen that can form a hydrogen bond. Therefore, there are more hydrogen bonds between water molecules than between alcohol molecules.

b. Each water molecule has two hydrogens that can form hydrogen bonds, while each ammonia has three hydrogens that can form hydrogen bonds. However, oxygen is more electronegative than nitrogen, so the hydrogen bonds between water molecules are stronger than the hydrogen bonds between ammonia molecules. The fact that water has a higher boiling point than ammonia indicates that the difference in electronegativity is more important than the number of hydrogens that can form hydrogen bonds in determining the difference in the boiling points.

c. Each water molecule has two hydrogens that can form hydrogen bonds while each molecule of hydrogen fluoride has only one hydrogen that can form a hydrogen bond. However, fluorine is more electronegative than oxygen. The fact that water has a higher boiling point means that *in this case* the greater number of hydrogens that can form hydrogen bonds is more important than the difference in electronegativity in determining the difference in the boiling points.

20. The dipole moment is determined by both the size of the partial charges and the distance between them. Fluorine, because it is more electronegative, has greater partial charges, but it has a smaller dipole moment because the carbon-fluorine bond is shorter than the carbon-chlorine bond.

21.

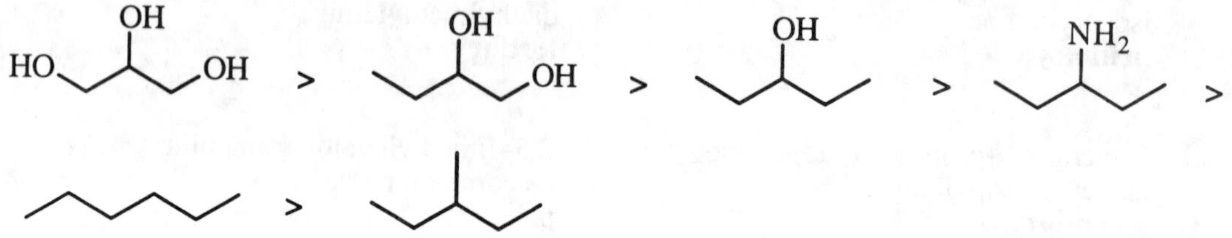

HO, OH, OH > OH, OH > OH > NH_2 >

\>

22.

a. $CH_3CH_2CH_2CH_2CH_2CH_2Br$ > $CH_3CH_2CH_2CH_2CH_2Br$ > $CH_3CH_2CH_2CH_2Br$

b. $CH_3CH_2CH_2CH_2CH_2CH_2CH_2CH_3$ > $CH_3\underset{CH_3}{CH}CH_2CH_2CH_2CH_2CH_3$ > $CH_3\overset{H_3C}{\underset{H_3C}{C}}-\overset{CH_3}{\underset{CH_3}{C}}CH_3$

c. $CH_3CH_2CH_2CH_2CH_2OH$ > $CH_3CH_2CH_2CH_2OH$ > $CH_3CH_2CH_2CH_2Cl$ >
$CH_3CH_2CH_2CH_2CH_3$

23.

a. $HOCH_2CH_2CH_2OH > CH_3CH_2CH_2OH > CH_3CH_2CH_2CH_2OH > CH_3CH_2CH_2CH_2Cl$

b. NH_2 > OH > CH_3

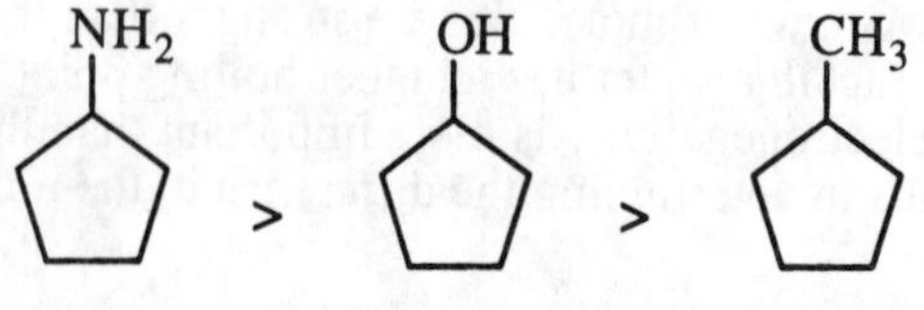

The amine is more soluble in water than the alcohol, because the amine has two hydrogens that can form hydrogen bonds with water.

24. Because cyclohexane is a nonpolar compound it will have the lowest solubility in the most polar solvent, which, of the solvents given, is ethanol.

$CH_3CH_2CH_2CH_2CH_2OH$ 1-pentanol

$CH_3CH_2OCH_2CH_3$ diethyl ether

CH_3CH_2OH ethanol

O tetrahydrofuran

25.

a.

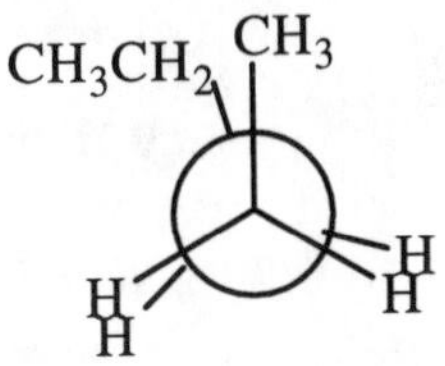

A: CH_3CH_2, CH_3, H, H, H, H

B: CH_3, H, CH_2CH_3, H, H, H

C: H CH_3, H, H, CH_2CH_3, H

D: CH_3, H, H, H, H, CH_2CH_3

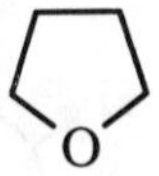

E: H CH_3, H, CH_3CH_2, H, H

F: CH_3, CH_3CH_2, H, H, H, H

b.

A C E A B D F

0 60 120 180 240 300 360

Degrees of Rotation

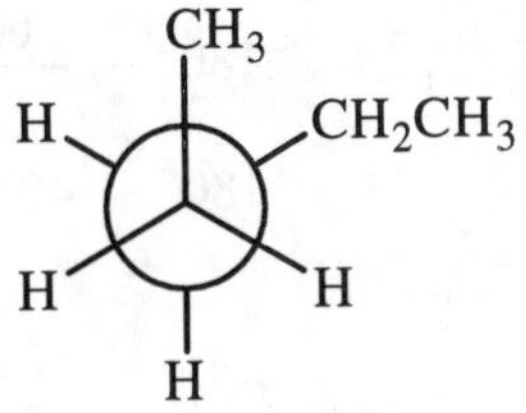

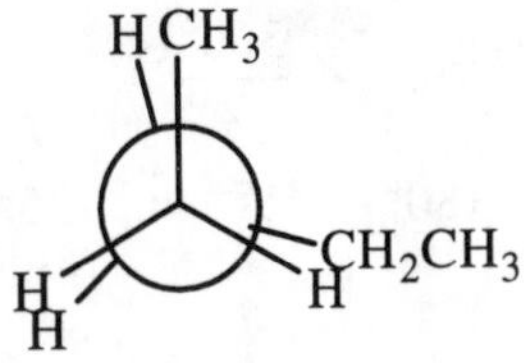

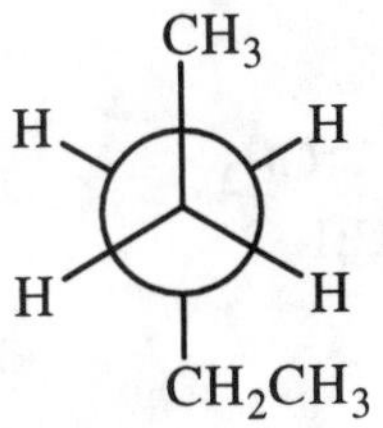

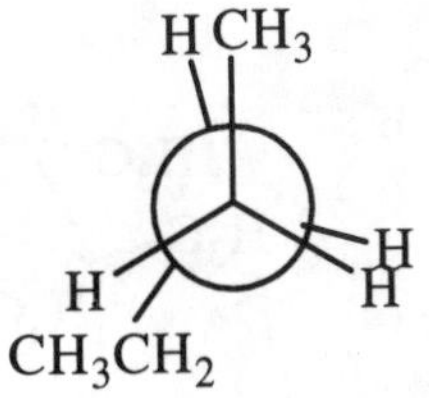

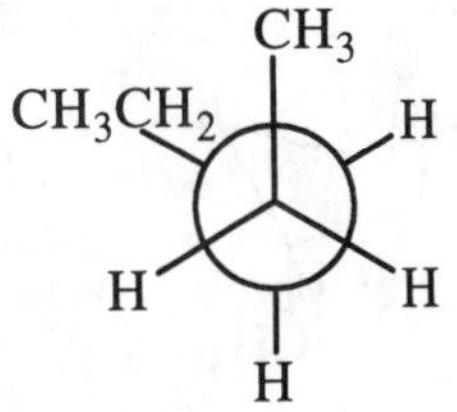

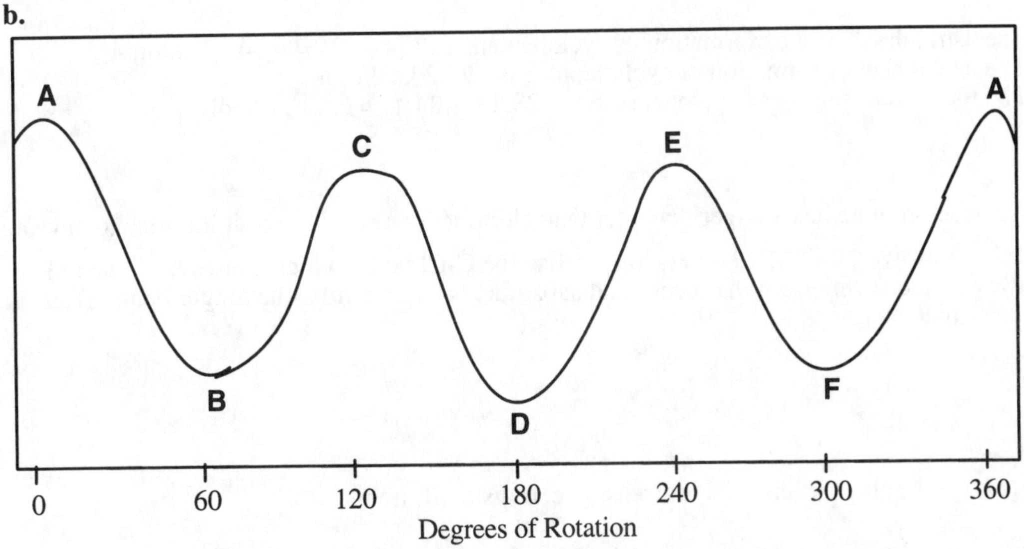

26.

a. b. 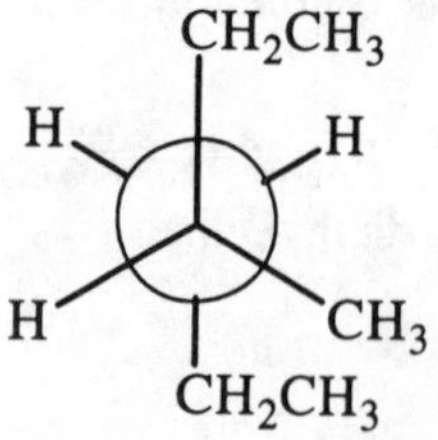 c. (Newman projections)

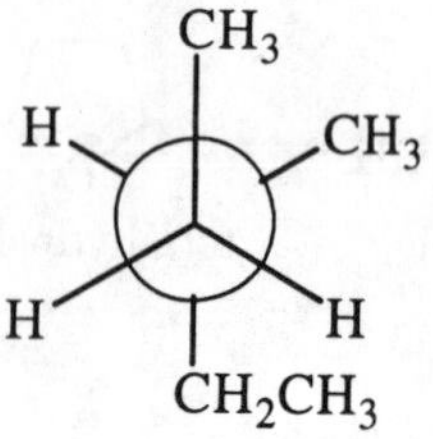

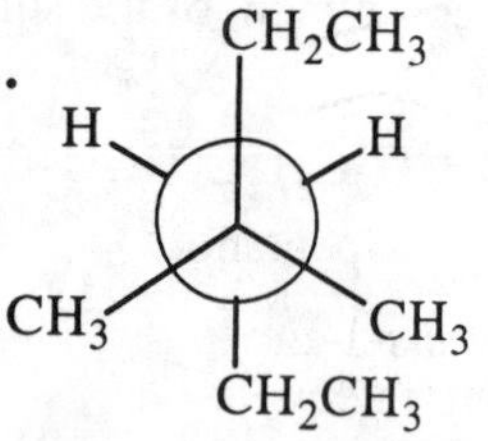

27.

a. $180° - \frac{360°}{8}$

$180° - 45° = 135°$

b. $180° - \frac{360°}{9}$

$180° - 40° = 140°$

28.

a. b. (chair conformations)

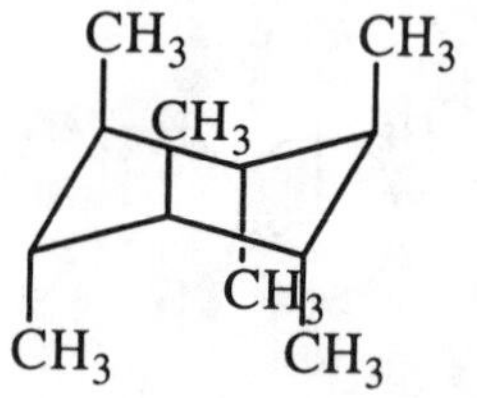

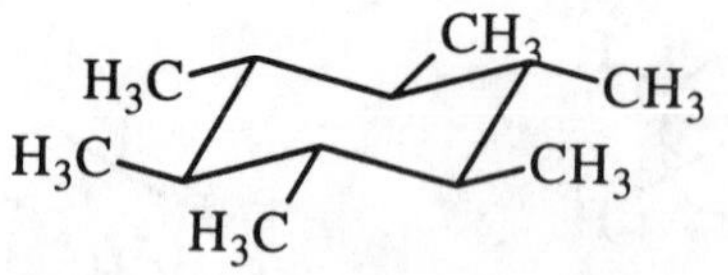

29. The "strainless" heat of formation of cycloheptane is 7 (- 4.92) = - 34.4 kcal/mol
The actual heat of formation of cycloheptane is - 28.2 kcal/mol
The total strain energy of cycloheptane is - 28.2 - (- 34.4) = 6.2 kcal/mol

30. Because bromine has a larger diameter than chlorine, one would expect bromine to have a larger ΔG°. However, the C-Br bond is longer than the C-Cl bond, which causes bromine to be farther away than chlorine from the other axial substituents. Apparently, the longer bond offsets the larger diameter.

31. **a.** cis **b.** cis **c.** cis **d.** trans **e.** trans **f.** trans

32.

a. CH_2CH_3, CH_3 (chair)

b. CH_2CH_3, CH_3 (chair)

c. *trans*-1-Ethyl-2-methylcyclohexane is more stable because both substituents are in equatorial positions.

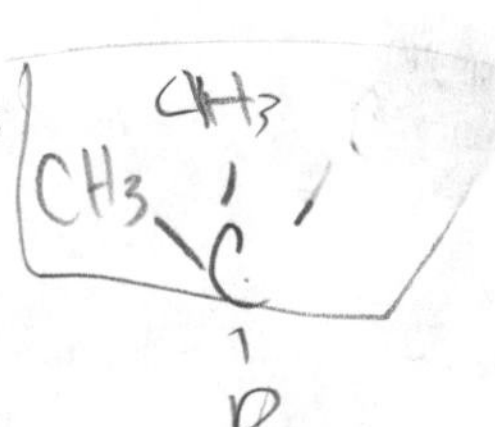

33.

a. one equatorial and one axial
b. both equatorial and both axial
c. both equatorial and both axial
d. one equatorial and one axial
e. one equatorial and one axial
f. both equatorial and both axial

34.

a. $\begin{array}{c} \qquad\qquad\quad CH_3 \\ CH_3CH_2CHOCCH_3 \\ \qquad\quad CH_3\;CH_3 \end{array}$

b. $\begin{array}{l} CH_3CHCH_2CH_2CH_2CH_2OH \\ \qquad\; CH_3 \end{array}$

c. $\begin{array}{l} CH_3CH_2CHNH_2 \\ \qquad\qquad CH_3 \end{array}$

d. $\begin{array}{l} \qquad\; CH_3 \\ CH_3CCH_2Br \\ \qquad\; CH_3 \end{array}$

e. CH_3, CH_3 (cyclohexane)

f. $\begin{array}{c} CH_3 \\ CHCH_3 \\ CH_3CH_2CH_2CHCHCH_2CH_2CH_2CH_3 \\ CHCH_3 \\ CH_3 \end{array}$

g. $\begin{array}{l} \qquad\quad CH_2CH_3 \\ CH_3CH_2N \\ \qquad\quad CH_2CH_3 \end{array}$

h. (ring structure)

i. $\begin{array}{c} CH_3CH_2CH_2CHCH_2CH_2CH_3 \\ CH_3CCH_3 \\ CH_3 \end{array}$

j. $\begin{array}{l} \qquad\qquad\qquad\qquad Br \\ CH_3CHCH_2CH_2CCH_2CH_2CH_3 \\ \qquad CH_3 \qquad\quad\; Br \end{array}$

k. CH_3, OH (cyclopentane)

l. $\begin{array}{l} \qquad\; CH_3 \\ CH_3CHCHCH_2CH_2CH_3 \\ \qquad\quad OCH_2CH_3 \end{array}$

m. 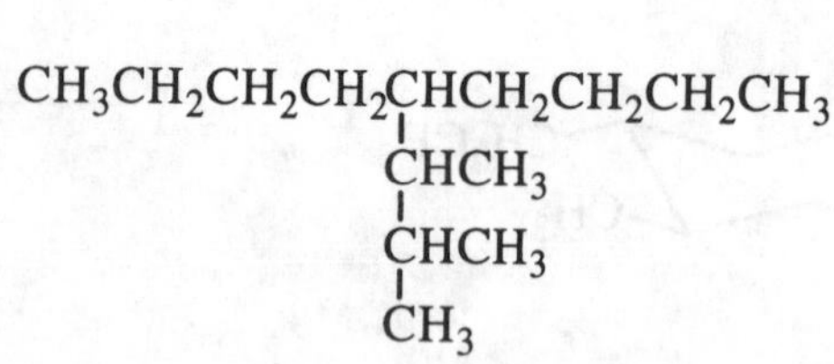 $CH_3CH_2CH_2CH_2CHCH_2CH_2CH_2CH_3$ with $CHCH_3$–$CHCH_3$–CH_3 branch on C5

$$\begin{array}{l} CH_3CH_2CH_2CH_2\underset{\textstyle |}{C}HCH_2CH_2CH_2CH_3 \\ \qquad\qquad\qquad\quad CHCH_3 \\ \qquad\qquad\qquad\quad\ | \\ \qquad\qquad\qquad\quad CHCH_3 \\ \qquad\qquad\qquad\quad\ | \\ \qquad\qquad\qquad\quad CH_3 \end{array}$$

n.

$$\begin{array}{c} \qquad\qquad CH_3 \\ \qquad\qquad | \\ CH_3CH_2CH\text{—}CHCH_2CH_2CH_2CH_3 \\ \quad\ | \\ \quad CH_3 \end{array}$$

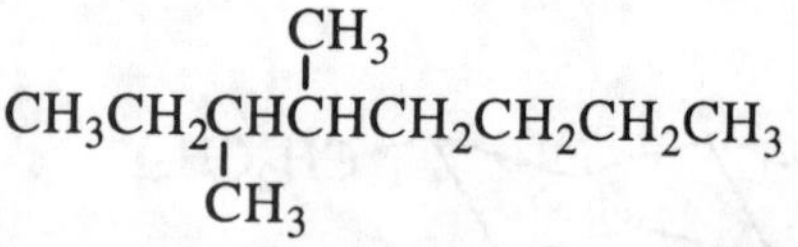

35. **a.** 5-bromo-2-methyloctane

b. 2,2,6-trimethylheptane

c. 2,3,5-trimethylhexane

d. 3,3-diethylpentane

e. 5-methyl-3-hexanol

f. 3-ethoxyheptane

g. 1-bromo-4-methylcyclohexane

h. *N,N*-dimethylcyclohexanamine

i. 3-ethylcyclohexanol

j. 1,3-dimethoxypropane

36. **a.** 3 **b.** 6 **c.** 3

37.

a.  (cyclohexane ring)

b.

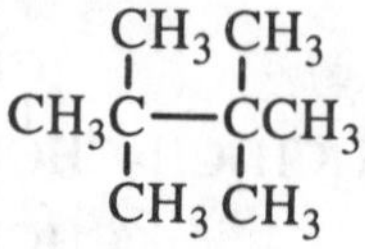

$$\begin{array}{c} CH_3\ CH_3 \\ |\quad\ \ | \\ CH_3C\text{—}CCH_3 \\ |\quad\ \ | \\ CH_3\ CH_3 \end{array}$$

c.

$$\begin{array}{c} CH_3CHCH_2CHCH_3 \\ |\qquad\ \ | \\ CH_3\quad\ CH_3 \end{array}$$

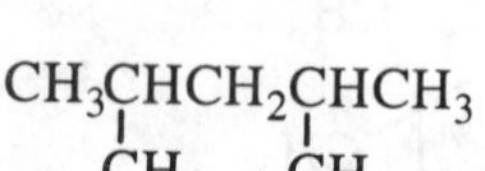

38.

a. 1-ethoxypropane
ethyl propyl ether

b. 4-methyl-1-pentanol
isohexyl alcohol

c. 2-butanamine
sec-butylamine

d. 2-chlorobutane
sec-butyl chloride

e. 2-methylpentane
isohexane

f. 2-bromo-2-methylbutane
tert-pentyl bromide

g. cyclohexanol
cyclohexyl alcohol

h. bromocyclopentane
cyclopentyl bromide

i. 2-propanamine
isopropylamine

39. a. 1-bromohexane
b. pentyl chloride
c. 1-butanol
d. 1-hexanol
e. hexane
f. 1-pentanol
g. 1-bromopentane
h. butyl alcohol
i. octane
j. isopentyl alcohol
(The alcohol has the higher boiling point because it forms stronger hydrogen bonds. If you were asked to compare their solubilities in water, the amine is more soluble because it forms more hydrogen bonds)
k. hexylamine

40. a. correct
b. 4-ethyl-2,2-dimethylheptane
c. 3-methylcyclohexanol
d. 2,2-dimethylcyclohexanol
e. 5-(2-methylpropyl)nonane
f. 1-bromo-3-methylbutane
g. correct
h. 2,5-dimethylheptane
i. 5-bromo-2-pentanol
j. 3-ethyl-2-methyloctane
k. 2,3,3-trimethyloctane
l. 5-methyl-*N,N*-dimethyl-3-hexanamine

41. The only one is 2,2,3-trimethylbutane.

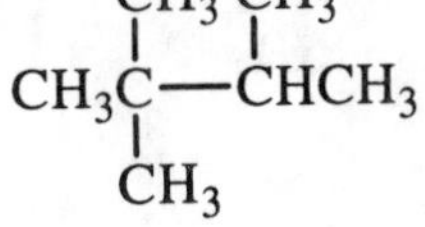

$$CH_3C(CH_3)_2-CH(CH_3)CH_3$$

42.

a.

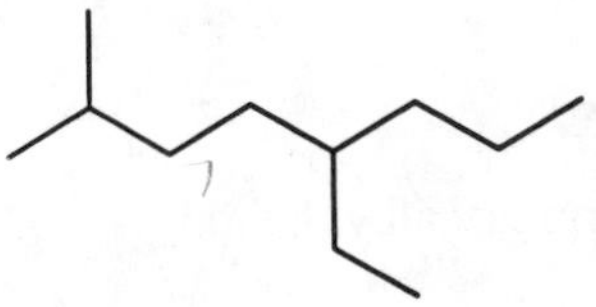

b.

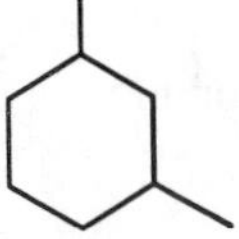

c.

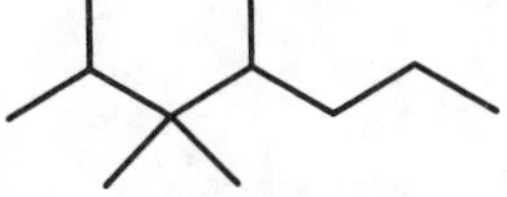

d.

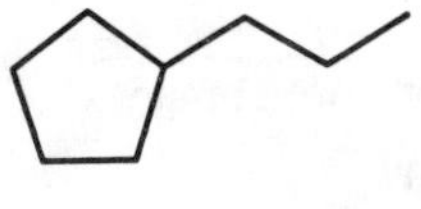

e.

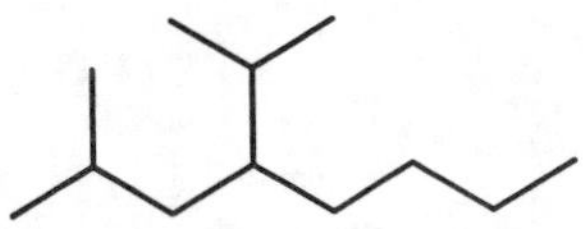

f.

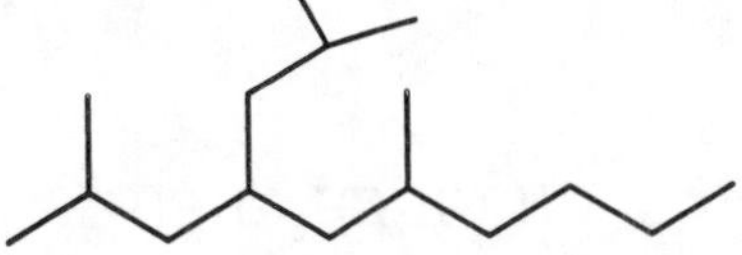

43.

a. 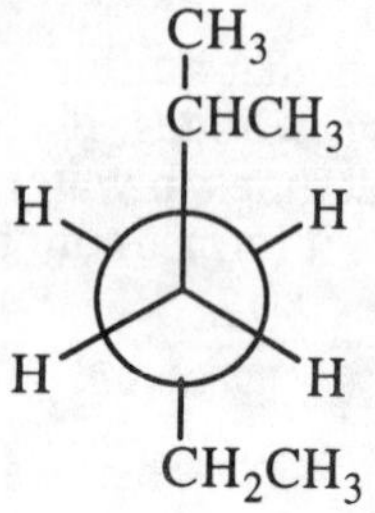 CH_3 / $CHCH_3$ (front), H, H, H, H, CH_2CH_3

most stable

b. 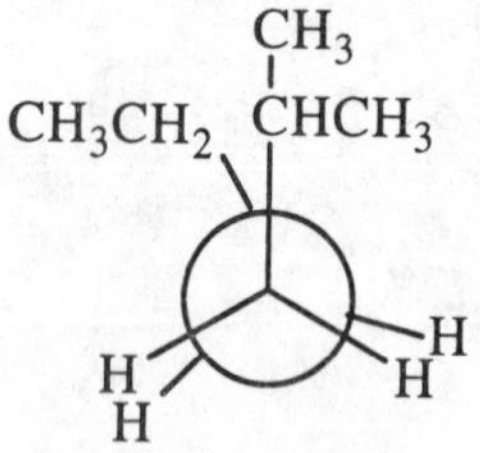 CH_3 / CH_3CH_2 $CHCH_3$, H, H, H, H, H

least stable

c. There are six carbon-carbon bonds in the compound and they all rotate, so there are five other carbon-carbon bonds, in addition to the C_3-C_4 bond, that can rotate.

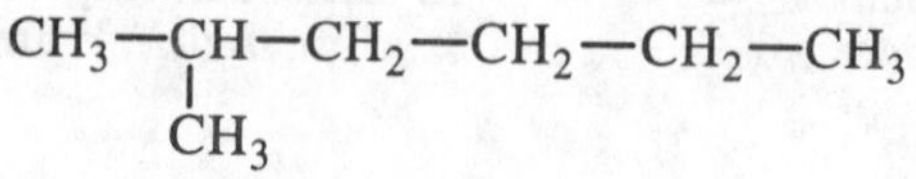

$$CH_3-\underset{\displaystyle CH_3}{\underset{|}{CH}}-CH_2-CH_2-CH_2-CH_3$$

d. Three of the carbon-carbon bonds have staggered conformers that are equally stable because each is bonded to a carbon with three identical substituents.

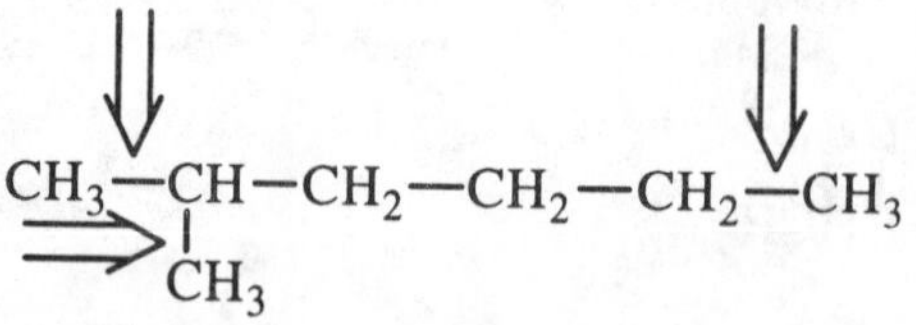

$$CH_3-\underset{\displaystyle CH_3}{\underset{|}{CH}}-CH_2-CH_2-CH_2-CH_3$$

44.

$CH_3CH_2CH_2CH_2CH_2Br$ **a.** 1-bromopentane primary alkyl halide
b. pentyl bromide

$CH_3CH_2CH_2\underset{\displaystyle Br}{\underset{|}{C}}HCH_3$ **a.** 2-bromopentane secondary alkyl halide
b. none

$CH_3CH_2\underset{\displaystyle Br}{\underset{|}{C}}HCH_2CH_3$ **a.** 3-bromopentane secondary alkyl halide
b. none

$$\begin{array}{c} CH_3 \\ | \\ CH_3CHCH_2CH_2Br \end{array}$$

a. 1-bromo-3-methylbutane primary alkyl halide
b. isopentyl bromide

$$\begin{array}{c} CH_3 \\ | \\ CH_3CH_2CHCH_2Br \end{array}$$

a. 1-bromo-2-methylbutane primary alkyl halide
b. none

$$\begin{array}{c} Br \\ | \\ CH_3CH_2CCH_3 \\ | \\ CH_3 \end{array}$$

a. 2-bromo-2-methylbutane tertiary alkyl halide
b. *tert*-pentyl bromide

$$\begin{array}{c} Br \\ | \\ CH_3CHCHCH_3 \\ | \\ CH_3 \end{array}$$

a. 2-bromo-3-methylbutane secondary alkyl halide
b. none

$$\begin{array}{c} CH_3 \\ | \\ CH_3CCH_2Br \\ | \\ CH_3 \end{array}$$

a. 1-bromo-2,2-dimethylpropane primary alkyl halide
b. neopentyl bromide

c. Four isomers do not have common names.

d. Four isomers are primary alkyl halides.

e. Three isomers are secondary alkyl halides.

f. One isomer is a tertiary alkyl halide.

45.
a. methoxyethane
b. 1-propanol
c. 4-propyl-1-nonanol
d. 5-isopropyl-2-methyloctane or 2-methyl-5-(1-methylethyl)octane
e. 6-chloro-4-ethyl-3-methyloctane
f. 5-methyl-3-propyl-1-hexanol
g. 6-isobutyl-2,3-dimethyldecane

46.

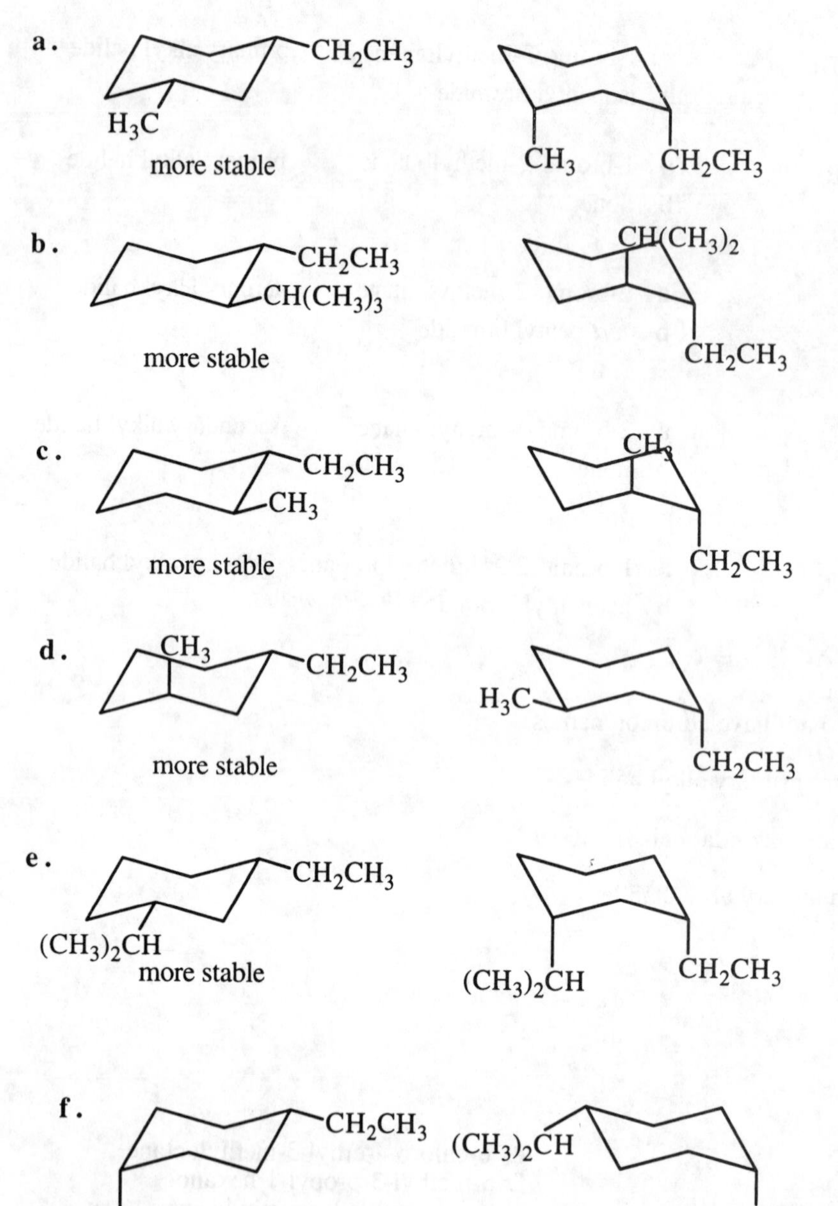

47. Alcohols with low molecular weights are more water soluble than alcohols with high molecular weights because, as a result of having fewer carbons, they have a smaller nonpolar component that has to be dragged into water.

48.

a.

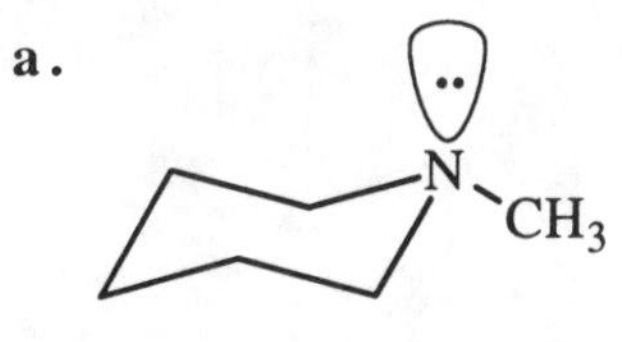

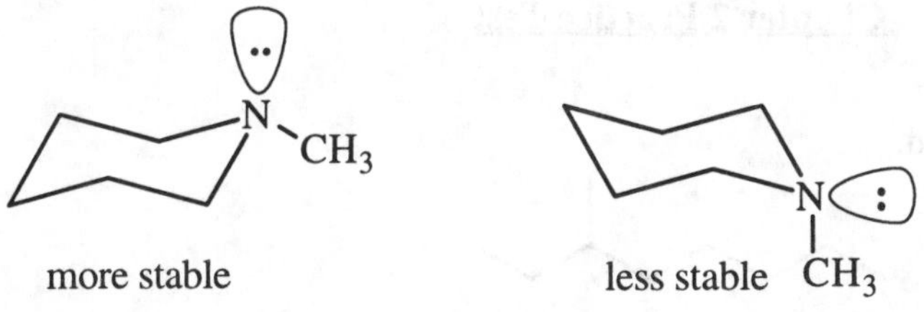

more stable

N : CH$_3$

less stable

b. There is more room for a substituent in the equatorial position. Because we have been told that the more stable conformer has the methyl group in the equatorial position, we can conclude that the methyl group takes up more room than the nonbonded pair of electrons.

49. Six ethers have molecular formula = $C_5H_{12}O$.

$CH_3OCH_2CH_2CH_2CH_3$
1-methoxybutane
butyl methyl ether

$CH_3CHCH_2CH_3$ with OCH_3 on C-2
2-methoxybutane
sec-butyl methyl ether

$CH_3CH_2OCH_2CH_2CH_3$
1-ethoxypropane
ethyl propyl ether

CH_3CHCH_3 with OCH_2CH_3 on C-2
2-ethoxypropane
ethyl isopropyl ether

CH_3COCH_3 with CH_3 above and CH_3 below on C
2-methoxy-2-methylpropane
tert-butyl methyl ether

$CH_3CHCH_2OCH_3$ with CH_3 on C-2
1-methoxy-2-methylpropane
isobutyl methyl ether

50. **a.** 6-methyl-*N*-methyl-3-heptanamine

b. 3-ethyl-2,5-dimethylheptane

c. 1,4-dichloro-5-methylheptane

d. 5-(1,1-dimethylpropyl)nonane or 5-neopentylnonane

e. 5-(2-ethylbutyl)-3,3-dimethyldecane

51.

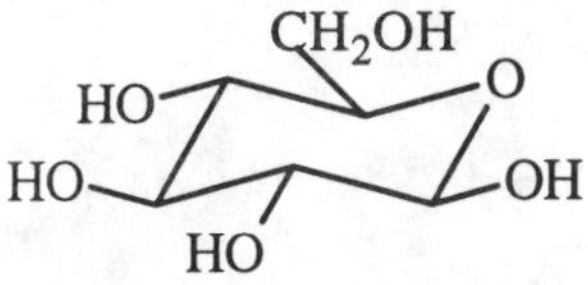

CH_2OH
HO
O
HO
OH
HO

52. **a.** 1-Hexanol because its 6 carbons can better engage in van der Waals interactions. The OH group of 3-hexanol will make it harder for is 6 carbons to lie close to the 6 carbons of another molecule.

b. The floppy ethyl groups in diethyl ether interfere with the ability of the oxygen atom to engage in hydrogen bonding with water.

Chapter 2 Practice Test

1. Name the following compound.

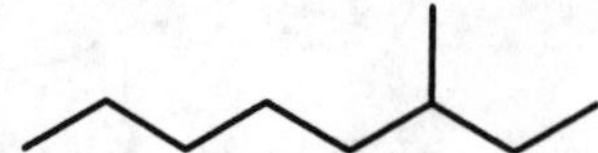

2. Draw the following conformers of hexane considering rotation about the C_3—C_4 bond.

a. the most stable of all the conformers

b. the least stable of all the conformers

c. a gauche conformer

3. Give two names for each of the following compounds.

a. $CH_3CH_2CHCH_3$ (Cl on C-3)

b. $CH_3CHCH_2CH_2CH_2OH$ (CH_3 on C-2)

c. cyclopentane—Br

4. Label the three compounds in each set in order of decreasing boiling point. (Label the highest boiling compound #1, the next #2, and the lowest boiling #3.)

a. $CH_3CH_2CH_2CH_2CH_2Br$ $\quad$ $CH_3CH_2CH_2Br$ $\quad$ $CH_3CH_2CH_2CH_2Br$

b. $CH_3CH_2CH_2CH_2CH_3$ $\quad$ $CH_3CH_2CH_2CH_2OH$ $\quad$ $CH_3CH_2CH_2CH_2Cl$

c. CH_3C—CCH_3 (each central C bearing CH_3 above and CH_3 below) $\quad$ $CH_3CH_2CH_2CH_2CH_2CH_2CH_2CH_3$ $\quad$ $CH_3CHCH_2CH_2CH_2CH_2CH_3$ (CH_3 on C-2)

5. Give the systematic name for each of the following compounds.

a. $CH_3CHCH_2CH_2CHCH_2CH_3$ (CH_3 on C-2, OH on C-5)

b. $CH_3CH_2CHOCH_2CH_3$ ($CH_2CH_2CH_2CH_3$ on C-3)

c. Br—cyclopentane—CH_3

d. $CH_3CHCHCH_2CH_2CH_2Cl$ (Cl on C-3, CH_2CH_3 on C-2)

6. Draw the other chair conformer.

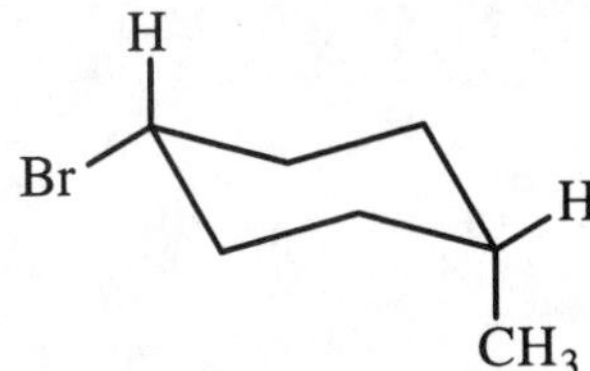

7. Draw the most stable conformer of trans-1-isopropyl-3-methylcyclohexane

8. Which of the following has:

a. the higher boiling point: diethyl ether or butyl alcohol?

b. the greater solubility in water: 1-butanol or 1-pentanol?

c. the higher boiling point: hexane or isohexane?

d. the higher boiling point: pentylamine or ethylmethylamine?

e. the greater solubility in water: ethyl alcohol or ethyl chloride?

9. Give two names for each of the following compounds.

a. $CH_3CH(CH_3)CH_2CH_2Br$

b. $CH_3CH(CH_3)CH_2CH_2OH$

c. $CH_3CH(CH_3)CH_2CH_2NH_2$

10. Which is more stable:

a. A staggered conformer or an eclipsed conformer?

b. The chair conformer of methylcyclohexane with the methyl group in the axial position or the chair conformer of methylcyclohexane with the methyl group in the equatorial position?

c. Cyclohexane or cyclobutane?

11. Give the structure of the following.

a. a secondary alkyl bromide that has three carbons

b. a secondary amine that has three carbons

c. an alkane with no secondary hydrogens

d. a constitutional isomer of butane

e. three compounds with molecular formula C_3H_8O

CHAPTER 3
Reactions of Alkenes. Thermodynamics and Kinetics

Important Terms

acid-catalyzed reaction a reaction catalyzed by an acid.

acyclic noncyclic.

addition reaction a reaction in which atoms or groups are added to the reactant.

alkene a hydrocarbon that contains a double bond.

alkoxymercuration-demercuration addition of alcohol to an alkene using a mercuric acetate followed by sodium borohydride and sodium hydroxide.

allyl group

$CH_2{=}CHCH_2{-}$

allylic carbon an sp^3 carbon adjacent to a vinyl carbon.

Arrhenius equation relates the rate constant of a reaction to the energy of activation and to the temperature at which the reaction is carried out ($k = Ae^{-E_a/RT}$).

carbene a carbon with a nonbonding pair of electrons and an energy empty orbital.

carbocation rearrangement the rearrangement of a carbocation to a more stable carbocation.

catalyst a compound that increases the rate at which a reaction occurs without being consumed in the reaction. Because it does not change the equilibrium constant of the reaction, it doesn't change the amount of product that is formed.

catalytic hydrogenation the addition of hydrogen to a double or a triple bond with the aid of a metal catalyst.

cis isomer the isomer with the hydrogens on the same side of the double bond.

cis-trans stereoisomers (geometric isomers) geometric (or *E*, *Z*) isomers.

concerted reaction a reaction in which all the bond-making and bond-breaking processes take place in a single step.

constitutional isomers (structural isomers) molecules that have the same molecular formula but differ in the way the atoms are connected.

dimer a molecule formed by the joining together of two identical molecules.

***E* isomer** the isomer with the high-priority groups on opposite sides of the double bond.

electrophile an electron deficient atom or molecule.

electrophilic addition reaction	an addition reaction in which the first species that adds to the reactant is an electrophile.
endergonic reaction	a reaction with a positive $\Delta G°$.
endothermic reaction	a reaction with a positive $\Delta H°$.
enthalpy	the heat given off ($-\Delta H°$) or the heat absorbed ($+\Delta H°$) during the course of a reaction.
entropy	a measure of the freedom of motion in a system.
exergonic reaction	a reaction with a negative $\Delta G°$.
exothermic reaction	a reaction with a negative $\Delta H°$.
experimental energy of activation ($E_a = \Delta H^\dagger - RT$)	a measure of the approximate energy barrier to a reaction. (It is approximate because it does not contain an entropy component.)
first-order rate constant	the rate constant of a first-order reaction.
first-order reaction (unimolecular reaction)	a reaction whose rate is dependent on the concentration of one reactant.
free energy of activation ($\Delta G^\dagger$)	the true energy barrier to a reaction.
free radical	an atom or molecule with an unpaired electron.
functional group	the center of reactivity of a molecule.
geometric isomers (cis-trans stereoisomers)	cis-trans (or *E*, *Z*) isomers.
Gibbs standard free energy change	the difference between the free energy content of the products and the free energy content of the reactants at equilibrium under standard conditions (1M, 25 °C, 1 atm).
halohydrin	an organic molecule that contains a halogen atom and an OH group.
Hammond postulate	states that the transition state will be more similar in structure to the species (reactants or products) that it is closer to energetically.
heat of hydrogenation	the heat ($\Delta H°$) released in a hydrogenation reaction.
heterogeneous catalyst	a catalyst that is insoluble in the reaction mixture.
heterolytic bond cleavage (heterolysis)	breaking a bond with the result that both bonding electrons stay with one of the atoms.
homogeneous catalyst	a catalyst that is soluble in the reaction mixture.

homolytic bond cleavage (homolysis) breaking a bond with the result that each of the atoms gets one of the bonding electrons.

hormone a compound that controls growth and other changes in tissues.

hydration addition of water to a compound.

1,2-hydride shift the movement of a hydride ion from one carbon to an adjacent carbon.

hydroboration-oxidation the addition of borane to an alkene (or to an alkyne) followed by reaction with hydrogen peroxide and hydroxide ion.

hydrogenation addition of hydrogen.

hyperconjugation delocalization of electrons by overlap of carbon-hydrogen or carbon-carbon σ bonds with an empty p orbital.

initiation step the step in which radicals are created, or the step in which the radical needed for the first propagation step is created.

intermediate a species formed during a reaction that is not the final product of the reaction.

kinetics the field of chemistry that deals with the rates of chemical reactions.

kinetic stability is indicated by $\Delta G^\dagger$. If $\Delta G^\dagger$ is large, the compound is kinetically stable (not very reactive). If $\Delta G^\dagger$ is small, the compound is kinetically unstable (is very reactive).

Markovnikov's rule the actual rule is: "When a hydrogen halide adds to an asymmetrical alkene, the addition occurs such that the halogen attaches itself to the carbon atom of the alkene bearing the least number of hydrogen atoms." A more useful version is: The electrophile adds to the sp^2 carbon that is bonded to the greater number of hydrogens.

mechanism of the reaction a description of the step-by-step process by which reactants are changed into products.

1,2-methyl shift the movement of a methyl group with its bonding electrons from one carbon to an adjacent carbon.

nucleophile an electron-rich atom or molecule.

olefin an older term for an alkene.

oxymercuration-demercuration addition of water using a mercuric salt of a carboxylic acid as a catalyst, followed by reaction with sodium borohydride.

pericyclic reaction a concerted reaction that takes place as the result of a cyclic rearrangement of electrons.

peroxide effect the addition of HBr to a π bond in the reverse of the usual order, caused by the presence of peroxide in the reaction mixture.

pheromone a chemical substance used for the purpose of communication.

primary alkyl radical an alkyl radical with the unpaired electron on a primary carbon.

primary carbocation a carbocation with the positive charge on a primary carbon.

propagation step in the first of a pair of propagation steps, a radical (or an electrophile or a nucleophile) reacts to produce another radical (or an electrophile or a nucleophile) that reacts in the second propagation step to produce the radical (or the electrophile or the nucleophile) that was the reactant in the first propagation step.

radical (often called a free radical) an atom or molecule with an unpaired electron.

radical addition reaction an addition reaction in which the first species that adds is a radical.

radical chain reaction a reaction in which radicals are formed and react in repeating propagating steps.

radical inhibitor a compound that traps radicals.

radical initiator a compound that creates radicals.

rate constant a measure of how easy it is to reach the transition state of a reaction (to get over the energy barrier of the reaction).

rate-determining step or rate-limiting step the step in a reaction that has the transition state with the highest energy.

reaction coordinate diagram describes the energy changes that take place during the course of a reaction.

regioselective reaction a reaction that leads to the preferential formation of one constitutional isomer over another.

saturated hydrocarbon a hydrocarbon that is completely saturated with hydrogen (contains no double or triple bonds).

secondary alkyl radical an alkyl radical with the unpaired electron on a secondary carbon.

secondary carbocation a carbocation with the positive charge on a secondary carbon.

second-order rate constant the rate constant of a second-order reaction.

second-order reaction (bimolecular reaction) a reaction whose rate is dependent on the concentration of two reactants, or on the square of the concentration of a single reactant.

solvation the interaction between a solvent and another molecule (or ion).

steric effects an effect due to the space occupied by a substituent.

steric hindrance refers to bulky groups at the site of a reaction that make it difficult for the reactants to approach one another.

termination step two radicals combine to produce a molecule in which all the electrons are paired.

tertiary alkyl radical an alkyl radical with the unpaired electron on a tertiary carbon.

tertiary carbocation a carbocation with the positive charge on a tertiary carbon.

thermodynamics the field of chemistry that describes the properties of a system at equilibrium.

thermodynamic stability is indicated by $\Delta G°$. If $\Delta G°$ is negative, the products are more stable than the reactants. If $\Delta G°$ is positive, the reactants are more stable than the products.

trans isomer the isomer with identical substituents on opposite sides of the double bond.

transition state the highest point on a hill in a reaction coordinate diagram. In the transition state, bonds in the reactant that will break are partially broken and bonds in the product that will form are partially formed.

unsaturated hydrocarbon a hydrocarbon that contains one or more double or triple bonds.

vicinal describes a compound with groups bonded to adjacent carbons.

vinyl group

$CH_2{=}CH{-}$

vinylic carbon a carbon doubly bonded to another carbon.

***Z* isomer** the isomer with the high-priority groups on the same side of the double bond.

Solutions to Problems

1. **a.** C_5H_8 **b.** C_4H_6 **c.** $C_{10}H_{16}$ **d.** C_8H_{10}

2. **a.** 3 **b.** 4 **c.** 1 **d.** 3

3.

a. $CH_3CH{=}CH_2$

b. $CH_3C{\equiv}CH$

$CH_2{=}C{=}CH_2$

c. $HC{\equiv}CCH_2CH_3$

$CH_3C{\equiv}CCH_3$

$CH_2{=}CHCH{=}CH_2$

$CH_2{=}C{=}CHCH_3$

4.

a. CH_3, CH_3

c. $CH_3CH_2OCH{=}CH_2$

b. $CH_3C{=}C(CH_3)CH_2CH_2CH_2Br$ (with CH_3 on C-2)

d. $CH_2{=}CHCH_2OH$

5. **a.** 4-methyl-2-pentene
b. 2-chloro-3,4-dimethyl-3-hexene
c. 1-bromocyclopentene
d. 1-bromo-4-methyl-3-hexene
e. 1,5-dimethylcyclohexene
f. 1-butoxy-1-propene

6. **a.** 1 and 3

b.

1. H_3C, CH_2CH_3 / H, H — $C{=}C$ — cis; H_3C, H / H, CH_2CH_3 — $C{=}C$ — trans

3. H_3C, CH_3 / H, H — $C{=}C$ — cis; H_3C, H / H, CH_3 — $C{=}C$ — trans

7.

a.

1. $CH_3CH_2(H)C{=}C(CH_3)(H)$ — **Z**; $CH_3CH_2(H)C{=}C(H)(CH_3)$ — **E**

2. $CH_3CH_2(Cl)C{=}C(CH_2CH_3)(H)$ — **E**; $CH_3CH_2(Cl)C{=}C(H)(CH_2CH_3)$ — **Z**

3. $CH_3CH_2CH_2CH_2(CH_3CH_2)C{=}C(CH_2Cl)(CH(CH_3)CH_3)$ — **Z**; $CH_3CH_2CH_2CH_2(CH_3CH_2)C{=}C(CH(CH_3)CH_3)(CH_2Cl)$ — **E**

4. $HOCH_2CH_2(O{=}CH)C{=}C(C(CH_3)_3)(C{\equiv}CH)$ — **Z**; $HOCH_2CH_2(O{=}CH)C{=}C(C{\equiv}CH)(C(CH_3)_3)$ — **E**

b. $H_3C(H)C{=}C(CH(CH_3)CH_3)(CH_2CH_2CH_2CH_3)$

8.

nucleophiles: H^- CH_3O^- $CH_3C{\equiv}CH$ NH_3

electrophiles: $AlCl_3$ $CH_3\overset{+}{C}HCH_3$

9.

a. $CH_3\overset{O}{\overset{\|}{C}}-O-H$ + $H\ddot{O}:^-$ ⟶ $CH_3\overset{O}{\overset{\|}{C}}-O^-$ + H_2O

b. cyclohexene + Br^+ ⟶ 2-bromocyclohexyl cation (Br, +)

c. $CH_3\overset{\ddot{O}:}{\overset{\|}{C}}-O-H$ + $H-\overset{+}{O}-H$ (H) ⟶ $CH_3\overset{+\ddot{O}H}{\overset{\|}{C}}-O-H$ + H_2O

d. $(CH_3)_3C-Cl$ ⟶ $(CH_3)_3C^+$ + Cl^-

10. a. Because the equilibrium constants for all the monosubstituted cyclohexanes in Table 2.10 (on page 101) of the text are positive, they all have negative $\Delta G°$'s.

b. the *tert*-butyl substituent

c. the *tert*-butyl substituent

d. $\Delta G° = -2.303\ RT \times \log K_{eq}$
$\Delta G° = -1.36 \times \log 18$
$\Delta G° = -1.36 \times 1.26$
$\Delta G° = -1.71$ kcal/mol

11. a. Solved in the text.

b. $\Delta G° = -2.303\ RT \log K_{eq}$
$-2.1 = -2.303 \times 1.986 \times 10^{-3} \times 298 \times \log K_{eq}$
$\log K_{eq} = 1.54$
$K_{eq} = 35$

$$K_{eq} = \frac{[\text{isopropylcyclohexane}]_{\text{equatorial}}}{[\text{isopropylcyclohexane}]_{\text{axial}}} = \frac{35}{1}$$

$$\frac{[\text{isopropylcyclohexane}]_{\text{equatorial}}}{[\text{isopropylcyclohexane}]_{\text{equatorial}} + [\text{isopropylcyclohexane}]_{\text{axial}}} = \frac{35}{35+1}$$

$$= \frac{35}{36}$$

$$= 0.97 = 97\%$$

c. Because the isopropyl substituent is a larger substituent than the fluoro substituent.

12. $\Delta S°$ is more significant in reactions in which the number of reactant molecules and the number of product molecules are not the same.

a. A + B ⇌ C

b. A + B ⇌ C

13. **a.**

$\Delta G° = \Delta H° - T\Delta S°$

(recall that T = °C + 273)

$\Delta G° = -12 - (303)(.01)$

$\Delta G° = -12 - 3 = -15$ kcal/mol

$\Delta G° = -(2.303)\ R\ T \log K_{eq}$

$\Delta G° = -(2.303)\ (1.986 \times 10^{-3})\ (303) \log K_{eq}$

$-15 = -1.39 \log K_{eq}$

$\log K_{eq} = 10.79$

$K_{eq} = 6.2 \times 10^{10}$

b.

$\Delta G° = \Delta H° - T\Delta S°$

$\Delta G° = -12 - (423)(.01)$

$\Delta G° = -12 - 4 = -16$ kcal/mol

$\Delta G° = -(2.303)\ (1.986 \times 10^{-3})\ (423) \log K_{eq}$

$-16 = -1.93 \log K_{eq}$

$\log K_{eq} = 8.29$

$K_{eq} = 1.9 \times 10^{8}$

c. For this reaction: the greater the temperature, the more negative (or less positive) the $\Delta G°$.

d. For this reaction: the greater the temperature, the smaller the K_{eq}, because $\log K_{eq} = \Delta G°/-2.303\ RT$.

14. **a.**

bonds broken		bonds formed	
π bond of ethene	61	C—H	101
H—Cl	103	C—Cl	82
	164		183

ΔH° = 164 - 183 = - 19 kcal/mol

b.

bonds broken		bonds formed	
π bond of ethene	61	C—H	101
H—H	104	C—H	101
	165		202

ΔH° = 165 - 202 = - 37 kcal/mol

c. Both are exothermic.

d. Both reactions have sufficiently negative ΔH°'s to expect that they would be exergonic as well.

15. **a.** a and b **b.** b **c.** c

16. A thermodynamically **unstable** product is one that is less stable than the reactant.

A kinetically **unstable** product is one that has a large rate constant (reforms reactant rapidly), whereas a kinetically **stable** product is one that has a small rate constant (reforms reactant slowly).

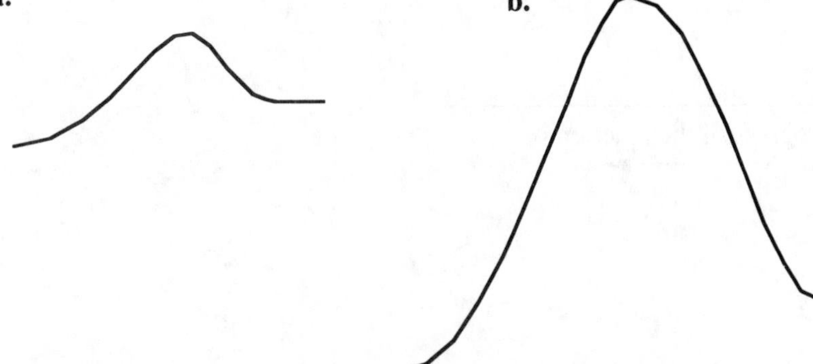

17. **a.** Because the experimental activation energy is in the numerator of a negative exponent, increasing the experimental activation energy will decrease the rate constant of a reaction.

b. Because temperature is in the denominator of a negative exponent, increasing the temperature will increase the rate constant of a reaction

18. The rate constant for a reaction can be increased by **decreasing** the stability of the reactant or by **increasing** the stability of the transition state.

19. **a.** rate = k[methyl chloride] [HO^-]
rate = $1 \times 10^{-5}\ M^{-1}s^{-1}(0.1\ M)\ (0.1\ M)$
rate = $1 \times 10^{-7}\ M\ s^{-1}$

b. Decreasing the concentration of methyl chloride will decrease the rate of the reaction to $1 \times 10^{-8}\ M\ s^{-1}$.

c. Changing the concentration will not affect the rate constant (k) of the reaction.

20.

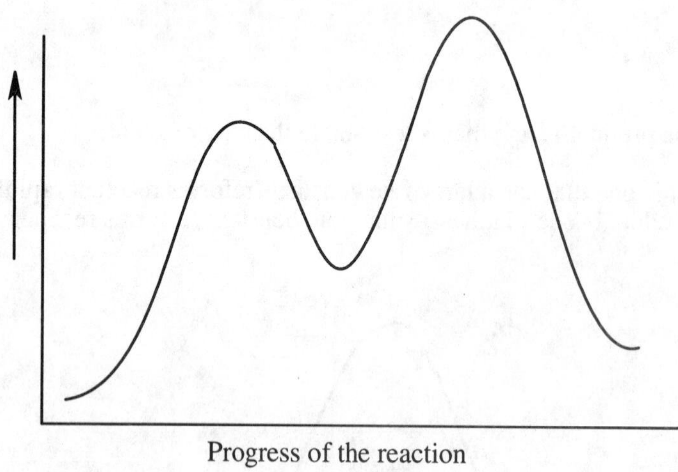

21. The second step is the rate-determining step because it has the transition state with the highest energy.
Notice that the second step is rate-determining even though the first step has the greatest energy of activation (steepest hill to climb). That is because it is easier for the intermediate that is formed in the first step to go back to starting material than to undergo the second step of the reaction. So, the second step is the rate-limiting step.

22.

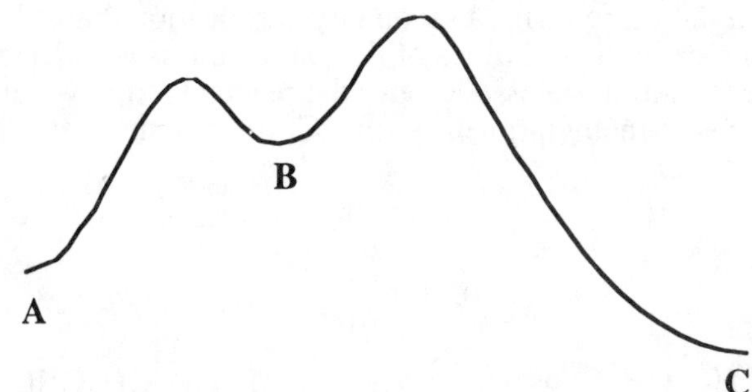

a. one

b. two

c. k_2 (In this particular diagram, $k_2 > k_1$: if you had made the transition state for the second step higher, you could have had a diagram in which $k_1 > k_2$.)

d. k_{-1}

e. k_{-1}

f. B to C

g. C to B

23.

a. $CH_3CH_2\overset{\overset{CH_3}{|}}{\underset{+}{C}}CH_3 > CH_3CH_2\underset{+}{C}HCH_3 > CH_3\overset{\overset{CH_3}{|}}{C}HCH_2\underset{+}{C}H_2$

b. The chlorine atom decreases the stability of the carbocation because, since it is more electronegative than a hydrogen, it is more effective than a hydrogen at withdrawing electrons away from the positively charged carbon. This increases the concentration of positive charge on the carbocation which makes it less stable. And the closer the chlorine atom is to the positively charged carbon, the more effective it is at withdrawing electrons away from that carbon.

$CH_3\underset{\underset{CH_3}{|}}{C}HCH_2\overset{+}{C}H_2 > CH_3\underset{\underset{Cl}{|}}{C}HCH_2\overset{+}{C}H_2 > CH_3CH_2\underset{\underset{Cl}{|}}{C}H\overset{+}{C}H_2$

24. a. The bond orbitals of the carbon that is adjacent to the positively charged carbon are the bond orbitals that are available for overlap with the vacant *p* orbital. Because the methyl cation does not have a carbon adjacent to the positively charged carbon, there are no bond orbitals available for overlap with the vacant *p* orbital.

b. An ethyl cation is more stable because it has three carbon-hydrogen bond orbitals available for overlap with the vacant *p* orbital, while a methyl cation does not have any carbon-hydrogen bond orbitals available for overlap with the vacant *p* orbital.

25. 2-Methylpropene will react faster. 2-Methylpropene forms a tertiary carbocation in the rate-limiting step of the reaction, while 2-butene forms a secondary carbocation that is less stable than a tertiary carbocation. The difference in the stabilities of the carbocations is reflected in the difference in the stabilities of the transition states. Because the reactants (the two alkenes) have the same stability and the reaction of 2-methylpropene with HCl forms a more stable transition state, 2-methylpropene will react faster.

26.

a. $CH_3CH_2CHCH_3$ with Br on C2 (2-bromobutane)

b. $CH_3CH_2CCH_3$ with CH_3 and Br on C2

c. cyclopentane ring bearing CH_3 and Br on the same carbon

d. $CH_3CH_2CHCH_3$ with Br on C2

27.

a. $CH_2{=}CCH_3$ with CH_3 on C2

b. cyclohexane–$CH_2CH{=}CH_2$

c. cyclohexane–$C({=}CH_2)CH_3$

d. cyclohexane$=CHCH_3$ or cyclohexene–CH_2CH_3

28.

CH_2 (methylenecyclohexane) This compound is more highly regioselective because the choice is between forming a tertiary carbocation or a primary carbocation. In the other compound, the choice is between forming a tertiary carbocation or a secondary carbocation, so the difference in the stability of the two possible carbocations is not as great.

29. As long as the pH is less than 16.0 because when the pH = pK_a, half the compound is in its neutral acid form and half is in its negatively charged basic form. As the pH is decreased below 16.0 more and more of the compound is in its acid form.

30.

a. $CH_3CH_2CH_2CHCH_3$ with OH on C2

c. $CH_3CH_2CH_2CH_2CHCH_3$ with OH and $CH_3CH_2CH_2CHCH_2CH_3$ with OH

b. cyclohexane–OH

d. cyclohexane with CH_3 and OH on the same carbon

31.

$$CH_3\underset{\underset{CH_3}{|}}{C}{=}CH_2 \xrightarrow{H^+} CH_3\overset{+}{\underset{\underset{CH_3}{|}}{C}}CH_3 \xrightarrow{CH_3\underset{\underset{CH_3}{|}}{C}HCH_2CH_2\ddot{O}H} CH_3\underset{\underset{CH_3}{|}}{C}HCH_2CH_2\overset{+}{\underset{\underset{H}{|}}{O}}{-}\overset{\overset{CH_3}{|}}{\underset{\underset{CH_3}{|}}{C}}CH_3$$

$$\xrightarrow{-H^+} CH_3\underset{\underset{CH_3}{|}}{C}HCH_2CH_2O\overset{\overset{CH_3}{|}}{\underset{\underset{CH_3}{|}}{C}}CH_3$$

32. **a.**

1. $CH_3\overset{\overset{CH_3}{|}}{\underset{\underset{Cl}{|}}{C}}CH_3$ 2. $CH_3\overset{\overset{CH_3}{|}}{\underset{\underset{Br}{|}}{C}}CH_3$ 3. $CH_3\overset{\overset{CH_3}{|}}{\underset{\underset{OH}{|}}{C}}CH_3$ 4. $CH_3\overset{\overset{CH_3}{|}}{\underset{\underset{OCH_3}{|}}{C}}CH_3$

b. The first step in all the reactions is addition of an electrophilic proton (H^+) to the carbon of the CH_2 group.
The *tert*-butyl carbocation is formed as an intermediate in each of the reactions.

c. The nucleophile that adds to the *tert*-butyl carbocation is different in each reaction.
In answers #3 and #4, a proton is lost from the group that was the nucleophile in the second step of the reaction.

3.

$$CH_3\overset{\overset{CH_3}{|}}{\underset{\underset{+\underset{H}{OH}}{|}}{C}}CH_3 \xrightarrow{-H^+} CH_3\overset{\overset{CH_3}{|}}{\underset{\underset{OH}{|}}{C}}CH_3$$

4.

$$CH_3\overset{\overset{CH_3}{|}}{\underset{\underset{+\underset{H}{O}CH_3}{|}}{C}}CH_3 \xrightarrow{-H^+} CH_3\overset{\overset{CH_3}{|}}{\underset{\underset{OCH_3}{|}}{C}}CH_3$$

33.

a. cyclohexene + $CH_3OH \xrightarrow{H^+}$ cyclohexyl$-OCH_3$

b. $CH_2{=}\overset{\overset{CH_3}{|}}{C}CH_3 + CH_3OH \xrightarrow{H^+} CH_3O\overset{\overset{CH_3}{|}}{\underset{\underset{CH_3}{|}}{C}}CH_3$

c. $CH_3CH{=}CHCH_3 + CH_3CH_2OH \xrightarrow{H^+} CH_3CH_2CH(OCH_2CH_3)CH_3$

or

$CH_2{=}CH_2 + CH_3CH_2CH(OH)CH_3 \xrightarrow{H^+} CH_3CH_2CH(OCH_2CH_3)CH_3$

d. $CH_3CH{=}CHCH_3 + H_2O \xrightarrow{H^+} CH_3CH(OH)CH_2CH_3$

e. cyclopentene $+ H_2O \xrightarrow{H^+}$ cyclopentanol (OH)

f. $CH_3CH_2CH{=}CHCH_2CH_3 + H_2O \xrightarrow{H^+} CH_3CH_2CH(OH)CH_2CH_2CH_3$

34. Solved in the text.

35.

a. $CH_3CH(CH_3)CH{=}CH_2 \xrightarrow{H^+} CH_3CH(CH_3)\overset{+}{C}HCH_3 \longrightarrow CH_3\overset{+}{C}(CH_3)CH_2CH_3 \xrightarrow{Br^-} CH_3CBr(CH_3)CH_2CH_3$

secondary → **tertiary**

b. methylenecyclohexane (CH_2) $\xrightarrow{H^+}$ 1-methylcyclohexyl cation (CH_3, +) $\xrightarrow{Br^-}$ 1-bromo-1-methylcyclohexane (Br, CH_3)

c. 1-methylcyclohexene (CH_3) $\xrightarrow{H^+}$ 1-methylcyclohexyl cation (CH_3, +) $\xrightarrow{Br^-}$ 1-bromo-1-methylcyclohexane (Br, CH_3)

d.

1-Bromo-1-methylcyclohexane and 3-bromo-1-methylcyclohexane will be obtained in approximately equal amounts because in each case the initially formed carbocation is secondary.

e.

3-Bromo-1-methylcyclohexane and 4-bromo-1-methylcyclohexane will be obtained in approximately equal amounts because in each case the initially formed carbocation is secondary.

f. $CH_2{=}CHC(CH_3)_2CH_3 \xrightarrow{H^+} CH_3\overset{+}{C}HC(CH_3)_2CH_3 \longrightarrow CH_3CH(CH_3)\overset{+}{C}(CH_3)CH_3 \xrightarrow{Br^-} CH_3CH(CH_3){-}CBr(CH_3)CH_3$

secondary   tertiary

36. a. The first step in the reaction of ethene with Br_2 forms a cyclic bromonium ion, while the first step in the reaction of ethene with HBr forms a carbocation.

b. If the bromide ion were to attack the positively charged bromine atom, a highly unstable compound (with a negative charge on carbon and a positive charge on bromine) would be formed.

$H_2C\text{–}\overset{+}{Br}\text{–}CH_2 \text{ (cyclic)} + :\ddot{Br}:^- \longrightarrow \overset{-}{C}H_2{-}CH_2{-}\overset{+}{Br}{-}Br$

37. The nucleophile that is present in greater concentration is more apt to collide with the carbocation intermediate. Therefore, if the solvent is a nucleophile, the major product will come from reaction of the solvent with the carbocation intermediate, because the concentration of the solvent is much greater than the concentration of the other nucleophile. (For example, in "a" the concentration of CH_3OH is much greater than the concentration of Cl^-.)

a. $ClCH_2C(CH_3)(OCH_3)CH_3$ **major** and $ClCH_2C(CH_3)(Cl)CH_3$

b. $CH_3CH(I)CH_3$ **major** and $CH_3CH(Br)CH_3$

c. $CH_3CH_2CH(OH)CH_3$ **major** and $CH_3CH_2CH(Cl)CH_3$

d. $CH_3CH_2CH(OCH_3)CH_3$ **major** and $CH_3CH_2CH(Br)CH_3$

38. As elements, sodium and potassium achieve an outer shell of eight electrons by losing the single electron they have in the 3*s* (in the case of Na) or 4*s* (in the case of K) orbital, thereby becoming Na^+ and K^+. In order to form a covalent bond, they would have to regain electrons in these orbitals, thus losing the stability associated with having an outer shell of eight electrons and no extra electrons.

39. Because chlorine is more electronegative than iodine, iodine will be the electrophile and it will add to the sp^2 carbon bonded to the greater number of hydrogens.

$$CH_3CH_2CH{=}CH_2 + I{-}Cl \longrightarrow CH_3CH_2\overset{+}{C}HCH_2I \xrightarrow{:\ddot{Cl}:^-} CH_3CH_2CH(Cl)CH_2I$$

40.

a. $CH_2(Br)CH(Br)CH_2CH_3$ **b.** $CH_2(Br)CH(OH)CH_2CH_3$ **c.** $CH_2(Br)CH(OCH_2CH_3)CH_2CH_3$ **d.** $CH_2(Br)CH(OCH_3)CH_2CH_3$

41. Notice that the addition of water or the addition of an alcohol to an alkene can be carried out using an acid catalyst (Section 3.13) or by mercuration/demercuration (Section 3.16).

a. cyclopentene $\xrightarrow[H_2O]{H^+}$ cyclopentanol (OH)

or

$\xrightarrow[2.\ NaBH_4]{1.\ Hg(OAc)_2,\ H_2O,\ THF}$

c. $CH_2{=}C(CH_3)CH_2CH_3 \xrightarrow[H_2O]{H^+} CH_3C(CH_3)(OH)CH_2CH_3$

or

$\xrightarrow[2.\ NaBH_4]{1.\ Hg(OAc)_2,\ H_2O,\ THF}$

b. cyclopentene $\xrightarrow[CH_3CH_2OH]{H^+}$ ethoxycyclopentane (OCH_2CH_3)

or

$\xrightarrow[2.\ NaBH_4]{1.\ Hg(O_2CCF_3)_2,\ CH_3CH_2OH}$

d. $CH_2{=}\overset{CH_3}{\overset{|}{C}}CH_2CH_3 \xrightarrow[CH_3OH]{H^+} CH_3\underset{OCH_3}{\underset{|}{\overset{CH_3}{\overset{|}{C}}}}CH_2CH_3$

or

$\xrightarrow[2.\ NaBH_4]{1.\ Hg(O_2CCF_3)_2,\ CH_3OH}$

42. Because one mole of BH_3 reacts with three moles of an alkene, you need one third of a mole of BH_3 to react with each mole of alkene. Therefore, you need two thirds of a mole of BH_3 to react with two moles of an alkene (in this case, 1-pentene).

43.

a. $CH_3\overset{CH_3}{\overset{|}{C}}{=}CHCH_3 \xrightarrow[2.\ HO^-,\ H_2O_2,\ H_2O]{1.\ BH_3} CH_3\overset{CH_3}{\overset{|}{C}}H\underset{OH}{\underset{|}{C}}HCH_3$

b. 1-methylcyclohexene (CH_3) $\xrightarrow[2.\ HO^-,\ H_2O_2,\ H_2O]{1.\ BH_3}$ 2-methylcyclohexanol (CH_3, OH)

44.

1-methylcyclohexene (CH_3) + Br• ⟶ 2-bromo-1-methylcyclohexyl radical (CH_3, •, Br)

2-bromo-1-methylcyclohexyl radical (CH_3, •, Br) + HBr ⟶ 1-bromo-2-methylcyclohexane (CH_3, Br) + Br•

45. Because alkene A has the smaller (less negative) heat of hydrogenation, it is more stable.

46. Solved in the text.

47. The reaction of 3-methylcyclohexene with HBr and peroxide would form approximately equal amounts of 1-bromo-2-methylcyclohexane (the desired product) and 1-bromo-3-methylcyclohexane because the bromine radical could add to either the 1-position or the 2-position of the alkene since in both cases a secondary radical would be formed. Thus, only half as much of the desired product would be formed from 3-methylcyclohexene than from 1-methylcyclohexene.

CH_3 Br• CH_3 Br + CH_3 Br

HBr HBr

CH_3 Br + CH_3 Br

1-bromo-2-methylcyclohexane 1-bromo-3-methylcyclohexane

If 3-methylcyclohexene were treated with HBr in the absence of peroxide, little 1-bromo- 2-methylcyclohexane (the desired product) would be formed because the secondary carbocation with a positive charge at the 2-position would rearrange to a more stable tertiary carbocation.

CH_3 H^+ CH_3 + + CH_3 + CH_3 +

Br^- Br^- Br^-

CH_3 Br CH_3 Br Br CH_3

1-bromo-3-methylcyclohexane 1-bromo-2-methylcyclohexane **minor** 1-bromo-1-methylcyclohexane

48.

a. $CH_3CH{=}CH_2 \xrightarrow[CH_3OH]{H^+} CH_3\underset{\displaystyle OCH_3}{\underset{|}{C}}HCH_3$

b. $CH_3CH_2CH{=}CHCH_3 + Br_2 \longrightarrow CH_3CH_2\underset{\displaystyle Br}{\underset{|}{C}}H\overset{\displaystyle Br}{\overset{|}{C}}HCH_3$

c. methylenecyclohexane (CH_2) $\xrightarrow[2.\ HO^-,\ H_2O_2,\ H_2O]{1.\ BH_3}$ cyclohexylmethanol (CH_2OH)

d. methylenecyclohexane (CH_2) or 1-methylcyclohexene (CH_3) $\xrightarrow[H_2O]{H^+}$ 1-methylcyclohexanol (HO, CH_3)

e. methylenecyclohexane (CH_2) or 1-methylcyclohexene (CH_3) $\xrightarrow{HBr}$ 1-bromo-1-methylcyclohexane (Br, CH_3)

f. cyclohexene + $CH_3CH_2OH \xrightarrow{H^+}$ ethoxycyclohexane (OCH_2CH_3)

49. **a.** 3,8-dibromo-4-nonene
b. (*Z*)-4-ethyl-3,7-dimethyl-3-octene
c. 1,5-dimethylcyclopentene
d. 3-ethyl-2-methyl-2-heptene

50.

a. $CH_2{=}CHCH_2CH_2CH_2CH_3$

b. $CH_2{=}CH\overset{\displaystyle CH_3}{\overset{|}{C}}HCH_2CH_3$

c. $CH_2{=}CH\underset{\displaystyle CH_3}{\underset{|}{\overset{\displaystyle CH_3}{\overset{|}{C}}}}CH_3$

51.

a. 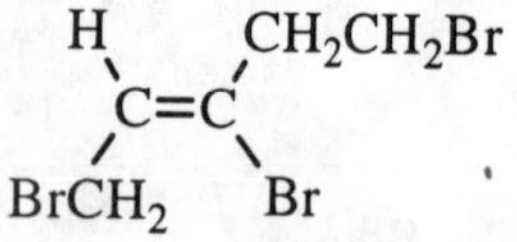 $BrCH_2(H)C=C(CH_2CH_2Br)Br$ — H and CH_2CH_2Br on top; $BrCH_2$ and Br on bottom

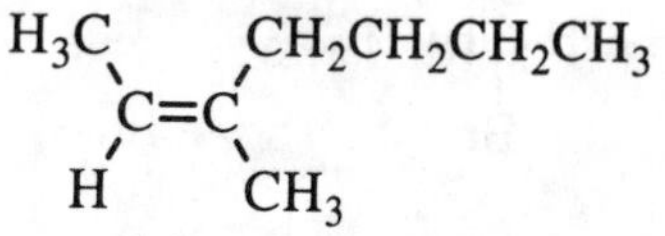

b. $H_3C(H)C=C(CH_2CH_2CH_2CH_3)CH_3$ — H_3C and $CH_2CH_2CH_2CH_3$ on top; H and CH_3 on bottom

c. $BrCH_2(Br)C=C(CH(CH_3)CH_3)CH_2CH_2CH_3$ — $BrCH_2$ and $CHCH_3$ (with CH_3) on top; Br and $CH_2CH_2CH_3$ on bottom

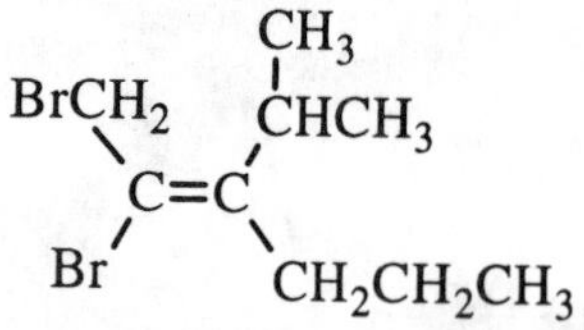

d. $CH_2{=}CHBr$

e. 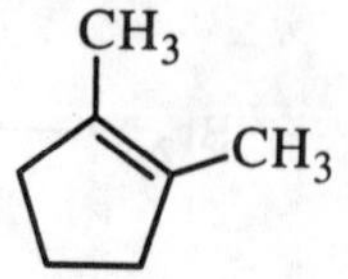 1,2-dimethylcyclopentene (CH_3, CH_3 on the double-bond carbons of a cyclopentene ring)

f. $CH_2{=}CHCH_2NHCH_2CH{=}CH_2$

52.

$CH_2{=}CHCH_2CH_2CH_2CH_3$
1-hexene

$CH_3CH{=}CHCH_2CH_2CH_3$
2-hexene

$CH_3CH_2CH{=}CHCH_2CH_3$
3-hexene

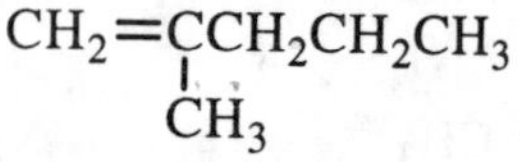

$CH_2{=}C(CH_3)CH_2CH_2CH_3$
2-methyl-1-pentene

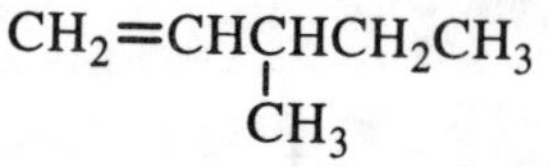

$CH_2{=}CHCH(CH_3)CH_2CH_3$
3-methyl-1-pentene

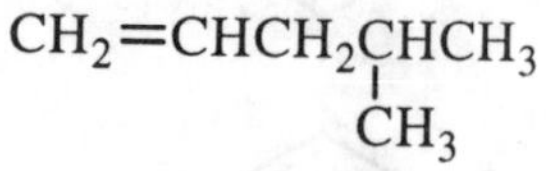

$CH_2{=}CHCH_2CH(CH_3)CH_3$
4-methyl-1-pentene

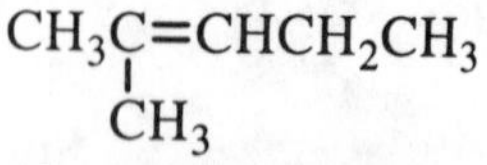

$CH_3C(CH_3){=}CHCH_2CH_3$
2-methyl-2-pentene

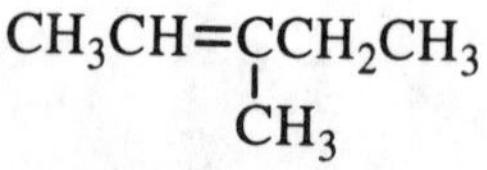

$CH_3CH{=}C(CH_3)CH_2CH_3$
3-methyl-2-pentene

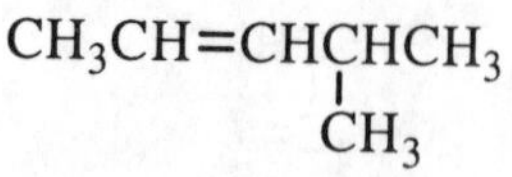

$CH_3CH{=}CHCH(CH_3)CH_3$
4-methyl-2-pentene

$CH_2{=}C(CH_3)CH(CH_3)CH_3$
2,3-dimethyl-1-butene

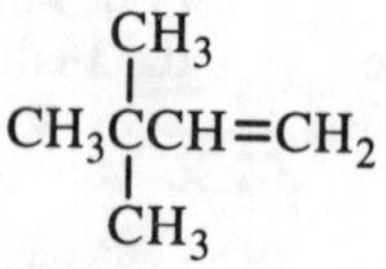

$CH_3C(CH_3)_2CH{=}CH_2$
3,3-dimethyl-1-butene

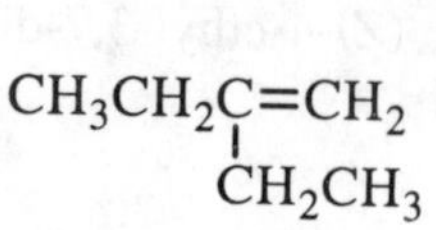

$CH_3CH_2C(CH_2CH_3){=}CH_2$
2-ethyl-1-butene

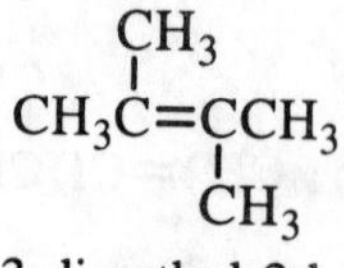

$CH_3C(CH_3){=}C(CH_3)CH_3$
2,3-dimethyl-2-butene

Of the compounds shown above, the following have *E* and *Z* isomers:

2-hexene, 3-hexene, 3-methyl-2-pentene, 4-methyl-2-pentene

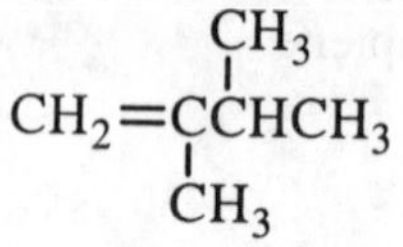

53.

a. $CH_3\overset{CH_3}{\overset{|}{C}}=CHCH_3$ + HBr ⟶ $CH_3\overset{CH_3}{\overset{|}{\underset{\underset{Br}{|}}{C}}}-CH_2CH_3$

b. $CH_3\overset{CH_3}{\overset{|}{C}}=CHCH_3$ + HBr $\xrightarrow{\text{peroxide}}$ $CH_3\overset{CH_3}{\overset{|}{C}}H-\underset{\underset{Br}{|}}{C}HCH_3$

c. $CH_3\overset{CH_3}{\overset{|}{C}}=CHCH_3$ + HI ⟶ $CH_3\overset{CH_3}{\overset{|}{\underset{\underset{I}{|}}{C}}}-CH_2CH_3$

d. $CH_3\overset{CH_3}{\overset{|}{C}}=CHCH_3$ + HI $\xrightarrow{\text{peroxide}}$ $CH_3\overset{CH_3}{\overset{|}{\underset{\underset{I}{|}}{C}}}-CH_2CH_3$

e. $CH_3\overset{CH_3}{\overset{|}{C}}=CHCH_3$ + ICl ⟶ $CH_3\overset{CH_3}{\overset{|}{\underset{\underset{Cl}{|}}{C}}}-\underset{\underset{I}{|}}{C}HCH_3$

f. $CH_3\overset{CH_3}{\overset{|}{C}}=CHCH_3$ $\xrightarrow[\text{Pd/C}]{H_2}$ $CH_3\overset{CH_3}{\overset{|}{C}}H-CH_2CH_3$

g. $CH_3\overset{CH_3}{\overset{|}{C}}=CHCH_3$ + HBr + excess NaCl ⟶ $CH_3\overset{CH_3}{\overset{|}{\underset{\underset{Cl}{|}}{C}}}-CH_2CH_3$

h. $CH_3\overset{CH_3}{\overset{|}{C}}=CHCH_3$ $\xrightarrow[\text{2. NaBH}_4]{\text{1. Hg(OAc)}_2\text{, H}_2\text{O}}$ $CH_3\overset{CH_3}{\overset{|}{\underset{\underset{OH}{|}}{C}}}-CH_2CH_3$

i. $CH_3\overset{CH_3}{\overset{|}{C}}=CHCH_3$ + H_2O $\xrightarrow{\text{trace HCl}}$ $CH_3\overset{CH_3}{\overset{|}{\underset{\underset{OH}{|}}{C}}}-CH_2CH_3$

j. $CH_3\overset{CH_3}{\overset{|}{C}}=CHCH_3$ + Br_2 $\xrightarrow[CH_2Cl_2]{}$ $CH_3\overset{CH_3}{\overset{|}{\underset{\underset{Br}{|}}{C}}}-\underset{\underset{Br}{|}}{C}HCH_3$

k. $CH_3\overset{CH_3}{\overset{|}{C}}=CHCH_3$ + Br_2 $\xrightarrow{H_2O}$ $CH_3\overset{CH_3}{\overset{|}{\underset{\underset{HO}{|}}{C}}}-\underset{\underset{Br}{|}}{C}HCH_3$

l. $CH_3C(CH_3){=}CHCH_3 + Br_2 \xrightarrow{CH_3OH} CH_3C(CH_3)(OCH_3){-}CH(Br)CH_3$

(product: CH_3 on the quaternary carbon; CH_3O and Br below)

j. $CH_3C(CH_3){=}CHCH_3 + Br_2 \xrightarrow[CH_2Cl_2]{} CH_3C(CH_3)(Br){-}CH(Br)CH_3$

k. $CH_3C(CH_3){=}CHCH_3 + Br_2 \xrightarrow{H_2O} CH_3C(CH_3)(OH){-}CH(Br)CH_3$

(product: HO and Br below)

l. $CH_3C(CH_3){=}CHCH_3 + Br_2 \xrightarrow{CH_3OH} CH_3C(CH_3)(OCH_3){-}CH(Br)CH_3$

(product: CH_3O and Br below)

m. $CH_3C(CH_3){=}CHCH_3 \xrightarrow[2.\ H_2O_2,\ HO^-]{1.\ BH_3} CH_3CH(CH_3){-}CH(OH)CH_3$

n. $CH_3C(CH_3){=}CHCH_3 \xrightarrow[2.\ NaBH_4]{1.\ Hg(O_2CCF_3)_2,\ CH_3OH} CH_3C(CH_3)(OCH_3){-}CH_2CH_3$

54.
a. (*E*)-3-methyl-3-hexene
b. *trans*-8-methyl-4-nonene or (*E*)-8-methyl-4-nonene
c. *trans*-9-bromo-2-nonene or (*E*)-9-bromo-2-nonene
d. 2,4-dimethyl-1-pentene
e. 2-ethyl-1-pentene
f. *cis*-2-pentene or (*Z*)-2-pentene

55.

$H{-}\ddot{\underset{..}{O}}{:}^- + H_3C{-}CH_2Br \longrightarrow H_2C{=}CH_2 + H_2O + Br^-$

(Curved arrows: hydroxide lone pair attacks an H on the carbon bearing three H's; the C–H bond electrons move to form the C=C bond; the C–Br bond electrons move onto Br.)

56.

$CH_3CHCH{=}CH_2$ (with CH_3 on C-2) + H–Br ⟶ $CH_3\overset{+}{C}H$–... $CH_3C{-}\overset{+}{C}HCH_3$ (H and CH_3 on C-2) ⟶ $CH_3\overset{+}{C}CH_2CH_3$ (with CH_3)

:Br:⁻ ↓ :Br:⁻ ↓

$CH_3CHCHCH_3$ (Br on C-3, CH_3 on C-2) $CH_3CCH_2CH_3$ (Br and CH_3 on C-2)

57. $\Delta G° = -RT\,2.303 \log K_{eq}$

$\log K_{eq} = -\Delta G°/RT\,2.303$

$\log K_{eq} = -\Delta G°/1.36$ kcal/mol

a. $\log K_{eq} = -2.72/1.36$ kcal/mol
$\log K_{eq} = -2$
$K_{eq} = [B]/[A] = 0.01$

b. $\log K_{eq} = -0.65/1.36$ kcal/mol
$\log K_{eq} = -0.48$
$K_{eq} = [B]/[A] = 0.33$

c. $\log K_{eq} = 2.72/1.36$ kcal/mol
$\log K_{eq} = 2$
$K_{eq} = [B]/[A] = 100$

d. $\log K_{eq} = 0.65/1.36$ kcal/mol
$\log K_{eq} = 0.48$
$K_{eq} = [B]/[A] = 3.0$

58.

Starting material: cyclohexyl–$CH_2CH{=}CH_2$

- Br_2, CH_2Cl_2 ⟶ cyclohexyl–CH_2CHCH_2Br (Br on C-2)
- 1. BH_3 2. H_2O_2, HO^- ⟶ cyclohexyl–$CH_2CH_2CH_2OH$
- H_2/Pt ⟶ cyclohexyl–$CH_2CH_2CH_3$
- H^+, H_2O ⟶ cyclohexyl–CH_2CHCH_3 (OH on C-2)
- HBr ⟶ cyclohexyl–CH_2CHCH_3 (Br on C-2)
- H^+, CH_3OH ⟶ cyclohexyl–CH_2CHCH_3 (OCH_3 on C-2)
- HBr, peroxide ⟶ cyclohexyl–$CH_2CH_2CH_2Br$
- Br_2, H_2O ⟶ cyclohexyl–CH_2CHCH_2Br (OH on C-2)

59. If the number of carbons is 40, $C_nH_{2n+2} = C_{40}H_{82}$. Therefore, a compound with molecular formula $C_{40}H_{56}$ is missing 26 hydrogens. This means that it has a total of 13 rings and π bonds, since each ring and π bond causes the compound to have two fewer hydrogens. Knowing that it has two rings and no triple bonds, one can conclude that it has 11 double bonds.

60.

a. $CH_3{-}Cl$ — Cl uses a 3*p* orbital in bond formation, while Br uses a 4*p* orbital. Cl, therefore, forms a shorter and stronger bond with carbon.

b. $CH_3CH_2CH_2{-}H$ — A primary radical is less stable than a secondary radical, so it is harder to break a bond that results in the formation of a primary radical than it is to break a bond that results in the formation of a secondary radical.

c. $CH_3{-}CH_3$ — A methyl radical is less stable than a primary radical, so it is harder to break a bond that results in the formation of two methyl radicals than it is to break a bond that results in the formation of a methyl radical and a primary radical.

d. $Br{-}Br$ — I forms a weaker bond than Br because I uses a 5*p* orbital in bond formation, while Br uses a 4*p* orbital.

61.

a. cyclohexene $\xrightarrow[\text{Pd/C}]{H_2}$

b. $CH_3CH_2CH_2CH{=}CH_2 \xrightarrow{HCl}$

c. methylenecyclohexane ($=CH_2$) $\xrightarrow[\text{peroxide}]{HBr}$

d. cyclohexane with $CH_2CH{=}CH_2$ substituent $\xrightarrow[H_2O]{H^+}$

e. $CH_3CH_2CH{=}CHCH_2CH_3 \xrightarrow[H_2O]{Br_2}$

f. $CH_3CH_2CH{=}CHCH_2CH_3 \xrightarrow[\text{NaBr excess}]{Cl_2}$

or

$CH_3CH_2CH{=}CHCH_2CH_3 \xrightarrow[\text{NaCl excess}]{Br_2}$

62.

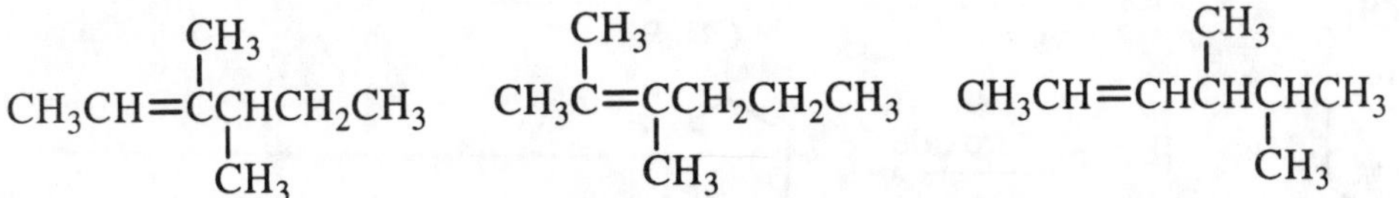

CH_3 $CH_3CH{=}CCHCH_2CH_3$ CH_3	CH_3 $CH_3C{=}CCH_2CH_2CH_3$ CH_3	CH_3 $CH_3CH{=}CHCHCHCH_3$ CH_3
3,4-dimethyl-2-hexene	2,3-dimethyl-2-hexene	4,5-dimethyl-2-hexene

2,3-Dimethyl-2-hexene is the most stable because it has the most alkyl substituents bonded to the sp^2 carbons. Because it is the most stable, it has the smallest heat of hydrogenation.

4,5-Dimethyl-2-hexene has the fewest alkyl substituents bonded to the sp^2 carbons. making it the least stable and thereby giving it the greatest heat of hydrogenation.

63. **a.** *Z* **b.** *E* **c.** *E* **d.** *Z*

64. If the number of carbons is 30, $C_nH_{2n+2} = C_{30}H_{62}$. A compound with molecular formula $C_{30}H_{50}$ is missing 12 hydrogens. Because it has no rings, squalene has 6 π bonds (12/2 = 6).

65.

a. -I > -Br > -OH > $-CH_3$

b. -OH > $-CH_2Cl$ > $-CH{=}CH_2$ > $-CH_2CH_2OH$

c. $-CH{=}CH_2$ > $-CH(CH_3)_2$ > $-CH_2CH_2CH_3$ > $-CH_3$

d. -OH > $-NH_2$ > $-CH_2OH$ > $-CH_2NH_2$

e. -Cl > $-COCH_3$ > -CN > $-CH{=}CH_2$

66.

a. 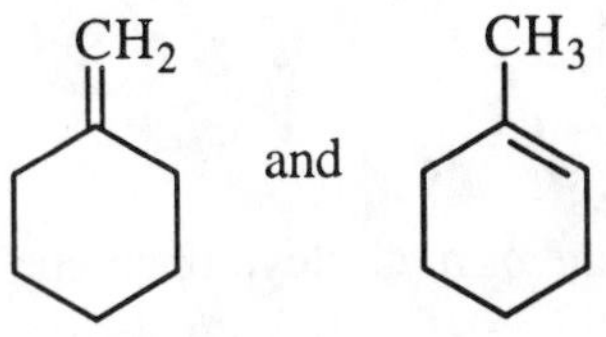 CH_2 and CH_3

b. No.

$$\text{methylenecyclohexane } (=CH_2) + HBr \xrightarrow{\text{peroxide}} \text{cyclohexyl-}CH_2Br$$

$$\text{1-methylcyclohexene } (CH_3) + HBr \xrightarrow{\text{peroxide}} \text{1-bromo-2-methylcyclohexane}$$

c. Yes.

$$\text{methylenecyclohexane } (=CH_2) + HCl \longrightarrow \text{1-chloro-1-methylcyclohexane } (H_3C,\ Cl)$$

$$\text{1-methylcyclohexene } (CH_3) + HCl \longrightarrow \text{1-chloro-1-methylcyclohexane } (H_3C,\ Cl)$$

d. Peroxide has no effect on the addition of HCl, so both compounds will give the same product (the product shown in **c**).

67.

a. $CH_3\overset{+}{C}(CH_3)CH_3$ (CH₃ on the central C, + below)

b. $CH_3\dot{C}HCH_2CH_3$

c. $CH_3\overset{+}{C}HCH_3$

d. $CH_3CH_2CH_2\dot{C}HCH_3$

e. $CH_3C(CH_3){=}CHCH_2CH_3$

f. 1-methylcyclohexene (CH_3 on the ring double bond)

g. $CH_3\overset{+}{C}HCH_2CH_2Cl$ (Cl on the terminal CH_2)

68.

a. 2-pentene
b. correct
c. 3-methyl-1-hexene
d. 3-heptene
e. 4-ethylcyclohexene
f. 2-chloro-3-hexene
g. correct
h. 2-methyl-1-hexene
i. 1-methylcyclopentene or methylcyclopentene
j. 3-methyl-2-pentene

69. **a.** 2 **b.** 3 **c.** C to D **d.** D **e.** B **f.** endergonic **g.** exergonic

70. **a.** To determine their relative rates of hydration, the rate constant of each alkene is divided by the smallest rate constant of the series (3.51×10^{-8}).

propene = $4.95 \times 10^{-8}/3.51 \times 10^{-8} = 1.41$

cis-2-butene = $8.32 \times 10^{-8}/3.51 \times 10^{-8} = 2.37$

trans-2-butene = $3.51 \times 10^{-8}/3.51 \times 10^{-8} = 1$

2-methyl-2-butene = $2.15 \times 10^{-4}/3.51 \times 10^{-8} = 6.12 \times 10^{3}$

2,3-dimethyl-2-butene = $3.42 \times 10^{-4}/3.51 \times 10^{-8} = 9.74 \times 10^{3}$

b. Both compounds form the same carbocation, but since (*Z*)-2-butene is less stable than (*E*)-2-butene, (*Z*)-2-butene has a smaller free energy of activation.

c. 2-Methyl-2-butene reacts faster because it forms a tertiary carbocation in the rate-limiting step, while *cis*-2-butene forms a less stable secondary carbocation.

d. Both compounds form tertiary carbocation intermediates. However, 2,3-dimethyl-2-butene has two sp^2 carbons that can react with a proton to form the tertiary carbocation while only one sp^2 carbon of 2-methyl-2-butene can react with a proton to form the tertiary carbocation. This means there will be more collisions with the proper orientation to lead to a productive reaction in the case of 2,3-dimethyl-2-butene.

71.

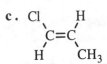

a. H H / C=C / H Cl

b. Cl H / C=C / H CH_3

c. Cl H / C=C / H CH_3

72.

a.

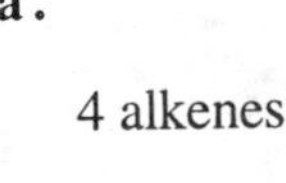

4 alkenes

CH_2 CH_3 CH_3 CH_3

b. 1-Methylcyclopentene is the most stable.

c. Because 1-methylcyclopentene is the most stable, it would have the least negative heat of hydrogenation.

73. Only a symmetrical alkene will form the same alkyl halide when it reacts with HBr in the presence of peroxides that it forms when it reacts with HBr in the absence of peroxides.

a.

b. There are many more possibilities for a symmetrical alkene for an alkene with an even number of carbon atoms.

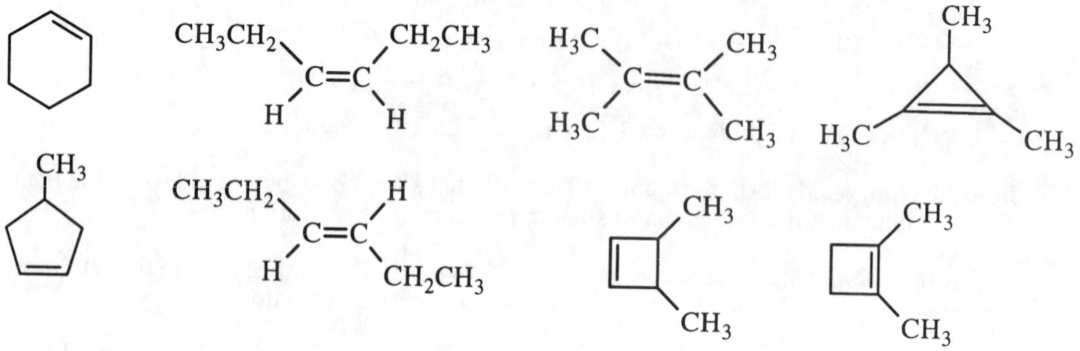

CH_3CH_2 CH_2CH_3 C=C H H

H_3C CH_3 C=C H_3C CH_3

CH_3 H_3C CH_3

CH_3

CH_3CH_2 H C=C H CH_2CH_3

CH_3 CH_3

CH_3 CH_3

74. **a.** 2-Methylpropene is more stable than 1-butene, and it has a more stable transition state for reaction with HCl. Therefore, to predict whether 2-methylpropene or 1-butene will react faster, you need to know whether the difference in the transition state stabilities is greater or less than the difference in the stabilities of the alkenes.

b. Knowing that 2-methylpropene reacts faster tells you that the difference in the stabilities of the transition states is greater than the difference in the stabilities of the alkenes.

1. Therefore, propene will react faster than ethene because propene forms a secondary carbocation, while ethene forms a primary carbocation.

2. 2-Methyl-1-butene will react faster because it forms a tertiary carbocation, while 3-methyl-1-butene forms a secondary carbocation.

75. $\Delta G^\circ = \Delta H^\circ - T\Delta S^\circ$

a. $\Delta G^\circ = -20 - (298)(-25 \times 10^{-3}) = -20 + 7.5$

$\Delta G^\circ = -12.5$ kcal/mol

$\Delta G^\circ = -2.303\ RT \log K_{eq}$

$-12.5 = -2.303 \times 1.986 \times 10^{-3} \times 298 \log K_{eq}$

$-12.5 = -1.36 \log K_{eq}$

$9.19 = \log K_{eq}$

$1.6 \times 10^9 = K_{eq}$

b. $\Delta G° = -20 - (398)(-25 \times 10^{-3})$
$\Delta G° = -10.0$ kcal/mol

$-10.0 = -2.303 \times 1.986 \times 10^{-3} \times 398 \log K_{eq}$
$-10.0 = -1.82 \log K_{eq}$
$5.49 = \log K_{eq}$
$3.12 \times 10^5 = K_{eq}$

76. No, he should not follow the student's advice. Markovnikov's rule would indicate that the secondary carbocation is more stable than the primary carbocation. But because of the electron-withdrawing fluorines in this compound, the primary carbocation will be more stable than the secondary carbocation, since in the latter the positive charge is closer to the fluorines. So the major product will be 1,1,1-trifluoro-3-iodopropane, not 1,1,1-trifluoro-2-iodopropane that would be predicted to be the major product by Markovnikov's rule.

77.

a. $\Delta G° = -2.303\, RT \log K_{eq}$
$\Delta G° = -2.303\, RT \log 10$
$\Delta G° = -2.303 \times 1.986 \times 10^{-3} \times 298 \times 1$
$\Delta G° = -1.36$ kcal/mol

b. $\Delta G° = \Delta H° - 0$
$-1.36 = \Delta H° - 0$
$\Delta H° = -1.36$ kcal/mol

c. $\Delta G° = 0 - T\Delta S°$
$-1.36 = 0 - 298\Delta S°$
$\Delta S° = 1.36/298 = 4.56 \times 10^{-3}$ kcal/(mol deg) = 4.56 eu

78.

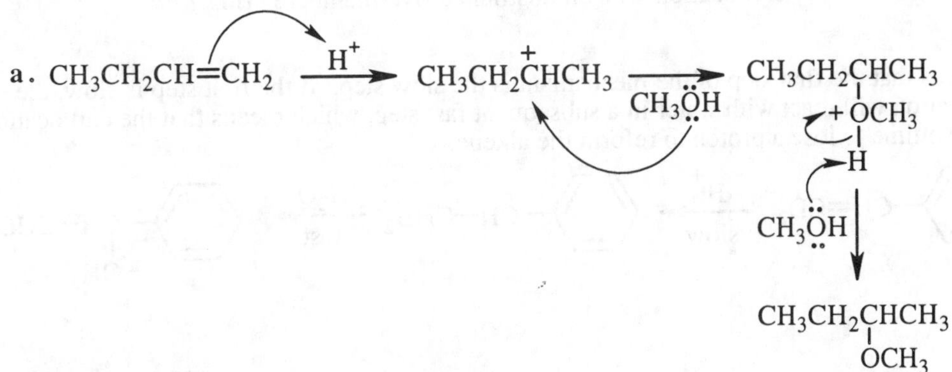

b. the first step
c. H^+
d. 1-butene
e. the *sec*-butyl cation
f. methanol

79. **a.** Both 1-butene and 2-butene react with HCl to form 2-chlorobutane.

b. Both alkenes form the same carbocation, but because 2-butene is more stable than 1-butene, 2-butene has the greater free energy of activation.

c. Because 1-butene has the smaller free energy of activation, it will react more rapidly with HCl.

d. Both compounds form the same carbocation, but since (*Z*)-2-butene is less stable, it will react more rapidly with HCl.

80.

a. [reaction scheme: cyclobutyl–C(CH_3)=CH_2 + H–Br → cyclobutyl–$\overset{+}{C}$(CH_3)–CH_3 → ring expansion to 2,2-dimethylcyclopentyl cation (**secondary**) → with :Br:⁻ → 1-bromo-2,2-dimethylcyclopentane; the secondary cation rearranges (methyl shift) to the 1,2-dimethylcyclopentyl cation (**tertiary**) → with :Br:⁻ → 1-bromo-1,2-dimethylcyclopentane]

secondary

tertiary

b. The initially formed carbocation is tertiary.

c. The rearranged carbocation is secondary, which then undergoes another rearrangement to a more stable tertiary carbocation.

d. The initially formed carbocation rearranges in order to release the strain in the four-membered ring. (A tertiary carbocation with a strained four-membered ring is less stable than a secondary carbocation with an unstrained five-membered ring.)

81. It tells us that the first step of the mechanism is the slow step. If the first step is slow, the carbocation will react with water in a subsequent fast step, which means that the carbocation will not have time to lose a proton to reform the alkene.

$$C_6H_5-CH{=}CD_2 \xrightarrow[\text{slow}]{H^+} C_6H_5-\overset{+}{C}H-CHD_2 \xrightarrow[\text{fast}]{H_2O} C_6H_5-\underset{+OH_2}{\underset{|}{C}H}-CHD_2 \xrightarrow{-H^+} C_6H_5-\underset{OH}{\underset{|}{C}H}-CHD_2$$

If the first step were not the slow step, an equilibrium would be set up between the alkene and the carbocation, and because the carbocation could lose either H^+ or D^+ when it reformed the alkene, all the deuterium (D) would not be retained in the alkene.

$$C_6H_5-CH=CD_2 \underset{-H^+}{\overset{H^+}{\rightleftharpoons}} C_6H_5-\overset{+}{C}H-CHD_2 \xrightarrow{-D^+} C_6H_5-CH=CHD$$

82. **a.** Addition of a proton to the double bond, following Markovnikov's rule, forms a carbocation intermediate. The alcohol group in the same molecule is the nucleophile that reacts with the carbocation.

$$CH_2=C(CH_3)CH_2CH(CH_2OH)CH_3 \xrightarrow{H^+} CH_3-\overset{+}{C}(CH_3)CH_2CH(CH_2\ddot{O}H)CH_3 \longrightarrow \text{protonated 2,2,4-trimethyltetrahydrofuran} \xrightarrow{-H^+} \text{2,2,4-trimethyltetrahydrofuran}$$

b. The weak oxygen-oxygen bond breaks homolytically, forming a radical that abstracts a hydrogen atom from $HCCl_3$. The radical that is formed adds to the double bond. A π electron of the other double bond pairs with the unpaired electron, forming a bond that divides the ring. The new radical abstracts a hydrogen atom from $HCCl_3$, and the propagation steps are repeated.

$$RO-OR \longrightarrow 2\ RO\cdot$$

$$RO\cdot + H-CCl_3 \longrightarrow ROH + \cdot CCl_3$$

$$\text{1,5-cyclooctadiene} + \cdot CCl_3 \longrightarrow \text{(CCl}_3\text{-substituted cyclooctenyl radical)} \longrightarrow \text{(CCl}_3\text{-substituted bicyclo[3.3.0]octyl radical)}$$

$$\text{(CCl}_3\text{-substituted bicyclo[3.3.0]octyl radical)} + HCCl_3 \longrightarrow \text{(CCl}_3\text{-substituted bicyclo[3.3.0]octane)} + \cdot CCl_3$$

c. A proton adds to the alkene, forming a secondary carbocation, which undergoes a ring-expansion rearrangement to form a more stable tertiary carbocation.

H_3C $CH{=}CH_2$ $\xrightarrow{H^+}$ H_3C $\overset{+}{C}HCH_3$ $\longrightarrow$ H_3C CH_3 $\xrightarrow{H_2\ddot{O}:}$ H_3C HO^+ H CH_3 $\xrightarrow{-H^+}$ HO CH_3 H_3C

d. At first glance this appears to be a difficult mechanism, but examination of the reaction shows that the only electrophile available to the alkene adds to the sp^2 carbon bonded to the most hydrogens (the one that results in the formation of the most stable carbocation. This is followed by the addition of the nucleophilic nitrogen to the other sp^2 carbon.

$CH_3CH_2\underset{CH_3}{C}{=}CH_2 + CH_2{=}\overset{+}{N}{=}\overset{-}{N} \longrightarrow CH_3CH_2\underset{CH_3}{\overset{+}{C}}CH_2CH_2\ddot{N}{=}\ddot{N}^- \longrightarrow$ N=N, H_3C, CH_2CH_3

83. A step-by-step description of how to solve this problem is given in the box entitled "Calculating Kinetic Parameters" on page 135 of the text.

E_a can be determined from the Arrhenius equation ($\log k = -E_a/2.303\,RT$), because a plot of $\log k$ versus $1/T$ gives a slope $= -E_a/2.303\,R$

$\log 2.11 \times 10^{-5} = -4.68$	$T = 304$	$1/T = 3.29 \times 10^{-3}$
$\log 4.44 \times 10^{-5} = -4.35$	$T = 313$	$1/T = 3.19 \times 10^{-3}$
$\log 1.16 \times 10^{-4} = -3.94$	$T = 324.5$	$1/T = 3.08 \times 10^{-3}$
$\log 2.10 \times 10^{-4} = -3.68$	$T = 332.8$	$1/T = 3.00 \times 10^{-3}$
$\log 4.34 \times 10^{-4} = -3.36$	$T = 342.2$	$1/T = 2.92 \times 10^{-3}$

slope = - 3600

$E_a = -(\text{slope}) \times 2.303 \times R$

$E_a = -(-3600) \times 2.303 \times 1.98 \times 10^{-3}$ kcal

$E_a = 16.4$ kcal

To find $\Delta G°$:

$-\Delta G° = RT\, 2.303 \log kh/Tk_b$

From the graph used to determine E_a, one can find the rate constant (k) at 30°. (It is $1.84 \times 10^{-5}\ s^{-1}$.)

$-\Delta G° = 1.98 \times 10^{-3} \times 303 \times 2.303 \log (1.84 \times 10^{-5} \times 6.625 \times 10^{-27})/(303 \times 1.3805 \times 10^{-10})$

$-\Delta G° = 1.98 \times 10^{-3} \times 303 \times 2.303 \times -23.5$

$\Delta G° = 32.5$ kcal

To find $\Delta H°$:

$\Delta H° = E_a - RT$

$\Delta H° = 16.4 - 1.98 \times 10^{-3} \times 303$

$\Delta H° = 16.4 - 0.6$

$\Delta H° = 15.8$

To find $\Delta S°$:

$\Delta S° = (\Delta H° - \Delta G°)/T$

$\Delta S° = (15.8 - 32.5)/303$

$\Delta S° = (-16.7)/303$

$\Delta S° = 0.055$ kcal/(mol deg) = 55 eu

84.

a. $CCl_3H + HO^- \longrightarrow Cl_3C:^- \longrightarrow Cl_2C: + Cl^-$

$+ H_2O$

b. $Cl_3C{-}CO_2^-\ Na^+ \xrightarrow{\Delta} Cl_3C:^- \longrightarrow Cl_2C: + Cl^-$

$+ CO_2$

$+ Na^+$

Chapter 3 Practice Test

1. For each pair, indicate the more stable.

a. 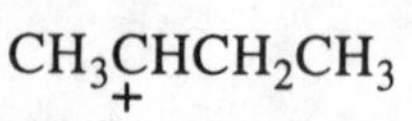 $CH_3\overset{+}{C}HCH_2CH_3$ or $CH_3\overset{+}{C}(CH_3)CH_3$

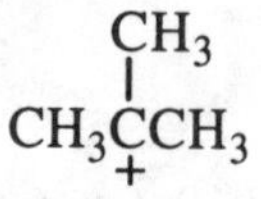

b. $CH_3CH_2CH_2\dot{C}H_2$ or $CH_3CH_2\dot{C}HCH_3$

c. $CH_3CH_2\overset{+}{C}H_2$ or $CH_3CH(Cl)\overset{+}{C}H_2$

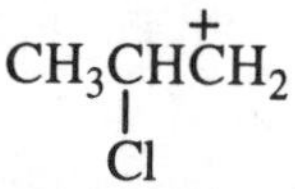

d. 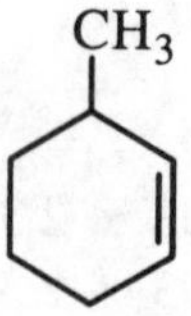 3-methylcyclohexene (CH₃ on ring carbon adjacent to the double bond) or 1-methylcyclohexene (CH₃ on the double-bond carbon)

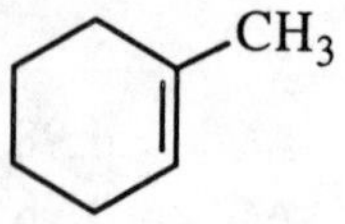

e. 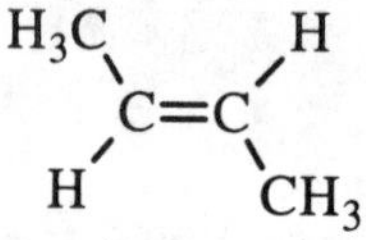 $H_3C(H)C{=}C(H)CH_3$ (trans: H_3C and CH_3 on opposite sides) or $H_3C(H)C{=}C(CH_3)H$ (cis: H_3C and CH_3 on the same side)

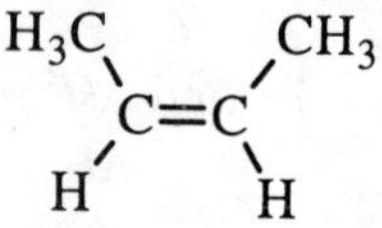

2. Which would be a better compound to use as a starting material for the synthesis of 2-bromopentane?

$CH_3CH_2CH_2CH{=}CH_2$ or $CH_3CH_2CH{=}CHCH_3$

3. The addition of H_2 in the presence of Pd/C to alkenes **A** and **B** results in the formation of the same alkane. The addition of H_2 to alkene **A** has a heat of hydrogenation ($-\Delta H°$) of 29.7 kcal/mol, while the addition of H_2 to alkene **B** has a heat of hydrogenation of 27.3 kcal/mol. Which is the more stable alkene, **A** or **B**?

4. What is the major product of each of the following reactions?

a. 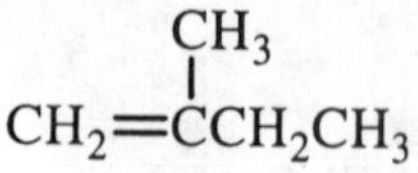 $CH_2{=}C(CH_3)CH_2CH_3 + HBr \longrightarrow$

b. 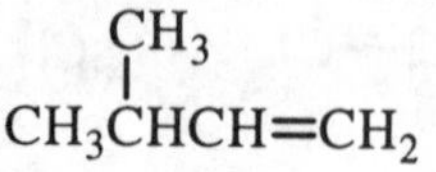 $CH_3CH(CH_3)CH{=}CH_2 + HCl \longrightarrow$

c. $CH_3CH_2CH{=}CH_2 + Cl_2 \xrightarrow{H_2O}$

d. 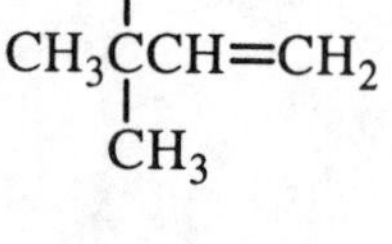 $CH_3C(CH_3)_2CH{=}CH_2 + HBr \longrightarrow$

5. Write out the propagation steps that occur in the addition of HBr to 1-methylcyclohexene in the presence of peroxides.

6. Label the following substituents in order of decreasing priority in the *E, Z* system of nomenclature. Label the highest priority #1.

$-\overset{\displaystyle O}{\overset{\|}{C}}CH_3$ $\quad -CH{=}CH_2 \quad -Cl \quad -C{\equiv}N$

7. Correct the incorrect names.

a. 3-pentene **c.** 2-ethyl-2-butene

b. 2-vinylpentane **d.** 2-methylcyclohexene

8. Indicate how each of the following compounds could be synthesized using an alkene as one of the starting materials.

a. $\longrightarrow$ $CH_3C(CH_3)_2CH_2CH_3$

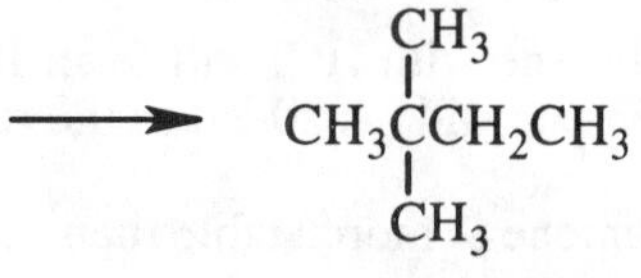

b. $\longrightarrow$ (cyclopentane ring bearing OH and adjacent Cl)

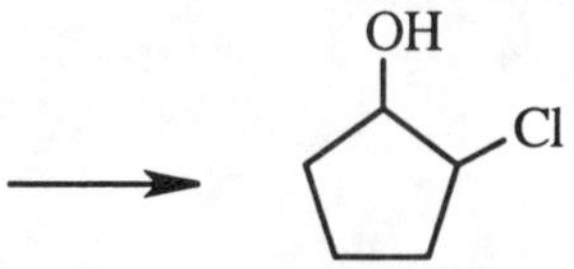

c. $\longrightarrow$ (cyclohexane ring bearing $CH_2CH_2CH_2Br$)

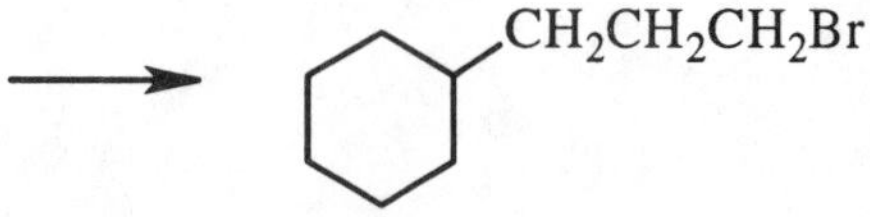

d. $\longrightarrow$ (cyclohexane ring bearing CH_3 and adjacent OH)

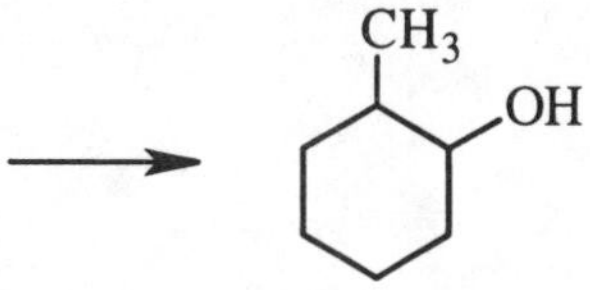

9. Indicate the carbocations that you would expect to rearrange to give a more stable carbocation.

$CH_3CH_2\overset{+}{C}HCH(CH_3)CH_3$ $\quad CH_3CH_2\overset{+}{C}HCH_3$ $\quad$ (cyclohexyl cation with CH_3 on the cationic carbon) $\quad$ (cyclohexyl cation with CH_3 on the adjacent carbon)

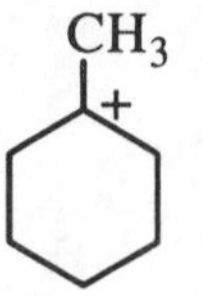

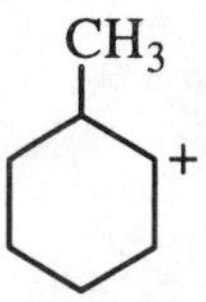

10. What product would be obtained from the hydroboration-oxidation of the following alkenes?

a. 2-methyl-2-butene **b.** 1-ethylcyclopentene

11. Indicate whether each of the following statements is true or false.

a. The addition of Br_2 to 2-butene to form 1,2-dibromobutane is a concerted reaction. T F

b. Decreasing the entropy of the products compared to the entropy of the reactants makes the equilibrium constant more favorable. T F

c. An exergonic reaction is one with a $-\Delta G°$. T F

d. An alkene is an electrophile. T F

e. The higher the energy of activation, the more slowly the reaction will take place. T F

f. Another name for *trans*-2-butene is *Z*-2-butene. T F

g. The reaction of 1-butene with HCl will form 1-chlorobutane as the major product if hydrogen peroxide is added to the reaction mixture. T F

h. 2,3-Dimethyl-2-pentene is more stable than 3,4-dimethyl-2-pentene. T F

SPECIAL TOPIC II

An Exercise in Drawing Curved Arrows "Pushing Electrons"

This is an extension of what you learned about drawing curved arrows in Section 3.6 on pages 125 of the text. Working through these problems will take just a little of your time. It will, however, be time well spent, because curved arrows will be used throughout the course and it is important that you are comfortable with this notation. (You will not actually encounter the reaction steps shown in this exercise for weeks or even months, so don't worry about the actual chemical changes in the compounds.)

Chemists use curved arrows to show how electrons move as covalent bonds break and/or new covalent bonds form. The tail of the arrow is positioned at the point where the electrons are in the reactant, and the head of the arrow points to where these same electrons end up in the product.

In the following reaction step, the bond between bromine and a carbon of the cyclohexane ring breaks and both electrons in the bond end up with bromine in the product. Thus, **the arrow starts at the electrons that carbon and bromine share in the reactant,** and **the head of the arrow points at the bromine atom** because this is where the two electrons end up in the product. The carbon of the cyclohexane ring is positively charged in the product because it has lost the two electrons it was sharing with bromine. The bromine is negatively charged in the product because it has gained the electrons that it shared with carbon in the reactant. The fact that two electrons moved in this example is indicated by the two-sided arrowhead.

$$C_6H_{11}\text{—}\ddot{B}r: \longrightarrow C_6H_{11}^{+} + :\ddot{B}r:^{-}$$

Notice that the arrow always starts at a bond or at a nonbonding pair of electrons. It does not start at a negative charge.

$$CH_3\overset{+}{C}HCH_3 + :\ddot{C}l:^{-} \longrightarrow CH_3CH(:\ddot{C}l:)CH_3$$

In the following reaction step a bond is being formed between the oxygen of water and a carbon of the other reactant. The arrow starts at the nonbonding electrons of the oxygen and points at the atom (the carbon) that will share the electrons. The oxygen in the product is positively charged because the electrons that oxygen had to itself in the reactant are now being shared with carbon. The carbon that was positively charged in the reactant is not charged in the product, because it has gained a share in a pair of electrons.

$$CH_3\overset{+}{C}HCH_3 + H_2\ddot{O}: \longrightarrow CH_3CH(\overset{+}{:}OH_2)CH_3$$

In the next examples, a bond breaks and a bond forms in the same step. As in the previous examples, the arrow starts at the point where the electrons are in the reactant, and the head of the arrow points to where these same electrons end up in the product (at the carbon atom in the first example, and between the carbons in the next example). Notice that the atom that loses a share in a pair of electrons (C in the first example, H in the second) ends up with a positive charge.

$$CH_3C(CH_3)_2\!-\!\overset{+}{C}HCH_3 \longrightarrow CH_3\overset{+}{C}(CH_3)\!-\!CH(CH_3)CH_3$$

$$CH_2(H)\!-\!\overset{+}{C}HCH_3 \longrightarrow CH_2{=}CHCH_3 + H^+$$

In the examples that follow, a nucleophile attacks an atom, causing a bond to break, with the result that the bonding electrons move to a positively charged atom.

$$CH_3\!-\!C(={\overset{+}{O}H})\!-\!CH_3 + H_2\ddot{O}: \longrightarrow CH_3\!-\!C(:\ddot{O}H)(\overset{+}{:}OH_2)\!-\!CH_3$$

$$CH_3CH_2\!-\!\overset{+}{\ddot{O}}H_2 + H_2\ddot{O}: \longrightarrow CH_3CH_2\!-\!\ddot{O}H + H_3\overset{+}{O}:$$

$$H_2C\!-\!\overset{+}{Br}\!-\!CH_2 \text{ (bromonium ion)} + H\ddot{O}:^- \longrightarrow H_2C(:\ddot{O}H)\!-\!CH_2(:\ddot{B}r:)$$

Problem 1. Draw curved arrows to show the movement of the electrons in the following reaction steps. (You will find the answers immediately after this exercise.)

a. $CH_3CH_2C(CH_3)_2\ddot{B}r: \longrightarrow CH_3CH_2C^+(CH_3)_2 + :\ddot{B}r:^-$

b. cyclopentyl–$\ddot{C}l: \longrightarrow$ cyclopentyl cation (+) $+ :\ddot{C}l:^-$

c. cyclohexyl–$\ddot{O}H$ (with $^+$H on O) $\longrightarrow$ cyclohexyl cation (+) $+ H_2\ddot{O}:$

d. $CH_3CH_2{-}MgBr \longrightarrow CH_3\ddot{C}H_2^- + {}^+MgBr$

e. $CH_3CH_2\overset{+}{C}HCH_3 + :\ddot{B}r:^- \longrightarrow CH_3CH_2CH(:\ddot{B}r:)CH_3$

Frequently chemists do not show the nonbonding electrons when they write reactions. The following are the same reaction steps you just saw except that the nonbonding electrons are not shown.

Problem 2. Draw curved arrows to show the movement of the electrons.

a. $CH_3CH_2C(CH_3)_2{-}Br \longrightarrow CH_3CH_2C^+(CH_3)_2 + Br^-$

b. cyclopentyl–Cl $\longrightarrow$ cyclopentyl cation (+) $+ Cl^-$

c. cyclohexyl–OH (with $^+$H on O) $\longrightarrow$ cyclohexyl cation (+) $+ H_2O$

d. $CH_3CH_2{-}MgBr \longrightarrow CH_3\bar{C}H_2 + {}^+MgBr$

The nonbonding electrons on Br^- in example **e** have to be shown in the reactant because an arrow can start only at a bond or at a pair of nonbonding electrons. Bromine's nonbonding electrons do not have to be shown in the product.

e. $CH_3CH_2\overset{+}{C}HCH_3$ + $:\ddot{Br}:^-$ ⟶ $CH_3CH_2CH(Br)CH_3$

In the examples in Problems 1 and 2 that show a reaction step in which a bond breaks, both of the bonding electrons end up on one of the atoms that shared the electrons in the bond. Because two electrons move, a two-sided arrowhead is used to show their movement.

In the examples in Problems 1 and 2 that show a reaction step in which a bond is formed, the two electrons that form the bond come from the same atom. Because two electrons move, a two-sided arrowhead is used to show their movement.

Sometimes a bond breaks in such a way that each of the bonded atoms gets one of the bonding electrons. A one-sided arrowhead represents the movement of one electron.

cyclohexyl–$\ddot{Br}:$ ⟶ cyclohexyl• + $\cdot\ddot{Br}:$

Sometimes a bond is formed using one electron from one atom and one electron from the other atom that forms the bond. Because one electron moves, a one-sided arrowhead is used to show its movement.

$CH_3\dot{C}HCH_3$ + $\cdot\ddot{Br}:$ ⟶ $CH_3CH(Br)CH_3$

Problem 3. Draw curved arrows to show the movement of the electrons.

a. $CH_3CH_2-\ddot{Cl}:$ ⟶ $CH_3\dot{C}H_2$ + $\cdot\ddot{Cl}:$

b. cyclopentyl–$\ddot{Br}:$ ⟶ cyclopentyl• + $\cdot\ddot{Br}:$

c. $CH_3\ddot{O}-\ddot{O}CH_3$ ⟶ $CH_3\ddot{O}\cdot$ + $\cdot\ddot{O}CH_3$

d. $(CH_3)_3C-\ddot{N}=\ddot{N}-C(CH_3)_3$ ⟶ $(CH_3)_3C\cdot$ + $:N\equiv N:$ + $\cdot C(CH_3)_3$

The following are the same reaction steps you just saw except that the nonbonding electrons are not shown.

Problem 4. Draw curved arrows to show the movement of the electrons.

a. $CH_3CH_2{-}Cl \longrightarrow CH_3\dot{C}H_2 + \cdot Cl$

b. (cyclopentyl)—Br $\longrightarrow$ (cyclopentyl radical)$\cdot$ + $\cdot Br$

c. $CH_3O{-}OCH_3 \longrightarrow CH_3O\cdot + \cdot OCH_3$

d. $(CH_3)_2C(CH_3){-}N{=}N{-}C(CH_3)_2CH_3 \longrightarrow CH_3\dot{C}(CH_3)_2 \quad N{\equiv}N \quad \cdot C(CH_3)_2CH_3$

In many reaction steps, two pairs of electrons move simultaneously. In each of the examples, follow the arrows to see how the electrons move. Notice how the movement of the electrons allows you to determine the structure of the products and the charges on the products.

$CH_3CH{=}CH_2 + H{-}\ddot{B}r{:} \longrightarrow CH_3\overset{+}{C}HCH_3 + {:}\ddot{B}r{:}^-$

$H\ddot{O}{:}^- + CH_3CH_2{-}\ddot{B}r{:} \longrightarrow CH_3CH_2{-}\ddot{O}H + {:}\ddot{B}r{:}^-$

$CH_3{-}C({:}\ddot{O}H)(CH_3){-}\ddot{C}l{:} \longrightarrow CH_3{-}C({=}\overset{+}{\ddot{O}}H){-}CH_3 + {:}\ddot{C}l{:}^-$

(2-bromocyclohexane with H) $\longrightarrow$ (cyclohexene) $+ Br^- + H^+$

Problem 5. Draw curved arrows to show the movement of the electrons that result in the formation of the given product.

a. $CH_3{-}\underset{CH_3}{\overset{:\ddot{O}H}{|}}C{-}\overset{+}{\underset{H}{\ddot{O}H}} \longrightarrow CH_3{-}\overset{+\ddot{O}H}{\overset{\|}{C}}{-}CH_3 + H_2O$

b. $CH_3CH_2CH{=}CH_2 + H{-}Cl \longrightarrow CH_3CH_2\overset{+}{C}H{-}CH_3 + Cl^-$

c. $CH_3CH_2{-}Br + \ddot{N}H_3 \longrightarrow CH_3CH_2{-}\overset{+}{N}H_3 + Br^-$

Problem 6. Draw curved arrows to show the movement of the electrons.

a. $CH_3CH{=}CHCH_3 + H{-}\overset{+}{\underset{H}{\ddot{O}}}{-}H \longrightarrow CH_3\overset{+}{C}H{-}CH_2CH_3 + H_2\ddot{O}:$

b. $CH_3CH_2CH_2CH_2{-}\ddot{Cl}: + \overset{-}{:C}{\equiv}N \longrightarrow CH_3CH_2CH_2CH_2{-}C{\equiv}N + :\ddot{Cl}:^-$

c. $CH_3{-}\underset{OH}{\overset{:\ddot{O}H}{|}}C{-}\overset{+}{\underset{H}{\ddot{O}CH_3}} \longrightarrow CH_3{-}\overset{+\ddot{O}H}{\overset{\|}{C}}{-}OH + CH_3\ddot{O}H$

d. $CH_3{-}\overset{\ddot{O}:}{\overset{\|}{C}}{-}H + CH_3{-}MgBr \longrightarrow CH_3{-}\underset{CH_3}{\overset{:\ddot{O}:^-}{|}}C{-}H + {}^{+}MgBr$

Problem 7. Draw curved arrows to show the movement of the electrons.

a. $CH_3-\overset{O}{\overset{\|}{C}}-CH_3$ + CH_3CH_2-MgBr ⟶ $CH_3-\underset{CH_2CH_3}{\underset{|}{\overset{O^-}{\overset{|}{C}}}}-CH_3$ + ^+MgBr

b. $CH_3CH_2CH_2-Br$ + $CH_3\ddot{\underset{..}{O}}:^-$ ⟶ $CH_3CH_2CH_2-OCH_3$ + Br^-

c. $CH_3CH=CH_2$ + $\cdot\ddot{\underset{..}{Br}}:$ ⟶ $CH_3\dot{C}H-CH_2-Br$

d. $CH_3-\underset{CH_3}{\underset{|}{\overset{:\ddot{O}:^-}{\overset{|}{C}}}}-OCH_2CH_3$ ⟶ $CH_3-\overset{O}{\overset{\|}{C}}-CH_3$ + $CH_3CH_2O^-$

Problem 8. Draw curved arrows to show the movement of the electrons.

a. $H\ddot{\underset{..}{O}}:^-$ + $CH_3\underset{H}{\underset{|}{C}}H-\overset{Br}{\overset{|}{C}}HCH_3$ ⟶ $CH_3CH=CHCH_3$ + H_2O + Br^-

b. $CH_3CH_2C\equiv C-H$ + $\overset{-}{\ddot{N}}H_2$ ⟶ $CH_3CH_2C\equiv C^-$ + NH_3

c. $CH_3\underset{CH_3}{\underset{|}{\overset{CH_3}{\overset{|}{C}}}}-\overset{+}{C}HCH_2CH_3$ ⟶ $CH_3\underset{+}{\overset{CH_3}{\overset{|}{C}}}-\underset{CH_3}{\underset{|}{C}}HCH_2CH_3$

d. $\underset{H}{\underset{|}{CH_2}}-\underset{+}{\overset{CH_3}{\overset{|}{C}}}CH_3$ ⟶ $CH_2=\overset{CH_3}{\overset{|}{C}}CH_3$ + H^+

Donation of a proton to an electronegative atom such as oxygen or nitrogen, or removal of a proton from an electronegative atom, can be shown in two different ways. One way is to show it using curved arrows.

$$CH_3CH_2\ddot{\underset{\cdot\cdot}{O}}H + H-\overset{\cdot\cdot}{\underset{H}{\overset{|+}{O}}}-H \rightleftharpoons CH_3CH_2\overset{H}{\underset{\cdot\cdot +}{O}}H + H_2\ddot{O}:$$

$$CH_3\overset{+}{\underset{H}{N}}H_2 + H_2\underset{\cdot\cdot}{O}: \rightleftharpoons CH_3NH_2 + H_3\overset{+}{O}:$$

Alternatively, a shortcut commonly used for such proton transfers is to show the proton that is being donated as H+ and the proton being removed as - H+.

$$CH_3CH_2\ddot{\underset{\cdot\cdot}{O}}H \underset{-H^+}{\overset{H^+}{\rightleftharpoons}} CH_3CH_2\overset{H}{\underset{\cdot\cdot +}{O}}H$$

$$CH_3\overset{+}{N}H_3 \underset{H^+}{\overset{-H^+}{\rightleftharpoons}} CH_3NH_2$$

Problem 9. Show the proton transfers in the following sequence using the shortcut method (H+ and - H+).

$$CH_3-\underset{+:\underset{H}{O}H}{\overset{OH}{\overset{|}{\underset{|}{C}}}}-\ddot{N}H_2 \rightleftharpoons CH_3-\underset{:\underset{\cdot\cdot}{O}H}{\overset{OH}{\overset{|}{\underset{|}{C}}}}-\ddot{N}H_2 \rightleftharpoons CH_3-\underset{:\underset{\cdot\cdot}{O}H}{\overset{OH}{\overset{|}{\underset{|}{C}}}}-\overset{+}{N}H_3$$

Some reactions occur in one step; others require two or more steps. Curved arrows are used to show the movement of electrons in each step of the reaction.

Problem 10. Draw curved arrows to show the movement of the electrons in each step of the following reaction sequences.

a. $CH_3CH{=}CH_2 + H{-}\ddot{\underset{..}{Br}}: \longrightarrow CH_3\overset{+}{C}H{-}CH_3 + :\ddot{\underset{..}{Br}}:^-$

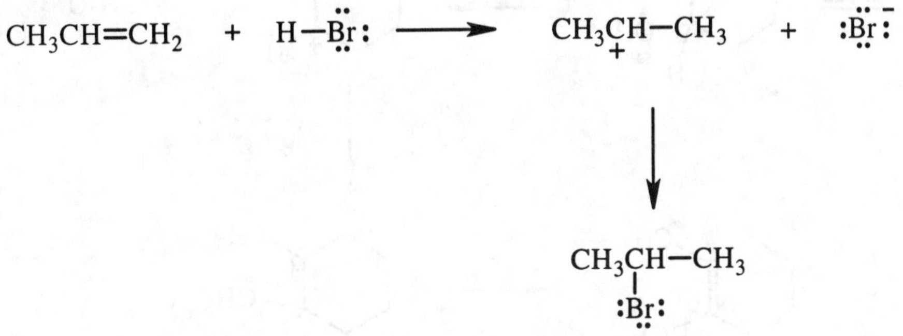

$\downarrow$

$CH_3CH(:\underset{..}{Br}:){-}CH_3$

b. $CH_3C(CH_3)_2{-}Cl \longrightarrow CH_3C^+(CH_3)_2 + Cl^-$

$\downarrow \ddot{N}H_3$

$CH_3C(CH_3)_2{-}\overset{+}{N}H_3 \rightleftharpoons CH_3C(CH_3)_2{-}NH_2$

c. $CH_3{-}\overset{\ddot{O}:}{\overset{\|}{C}}{-}Cl + H\ddot{\underset{..}{O}}:^- \longrightarrow CH_3{-}C(:\ddot{O}:^-)(:\underset{..}{O}H){-}Cl \longrightarrow CH_3{-}\overset{\ddot{O}:}{\overset{\|}{C}}{-}OH + Cl^-$

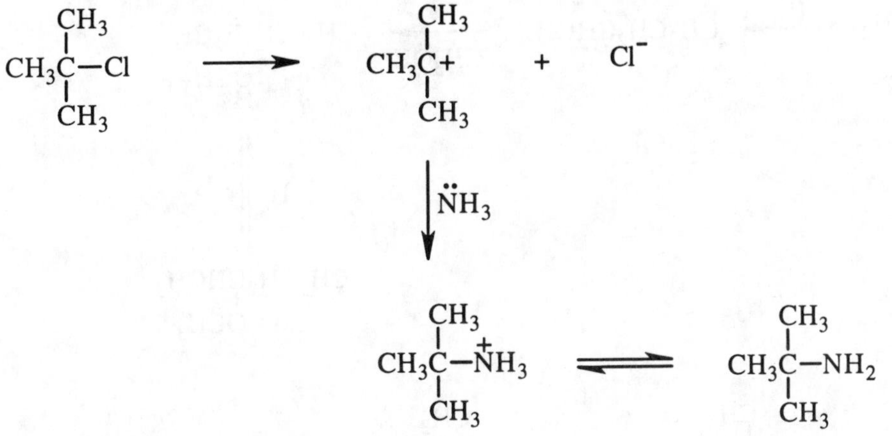

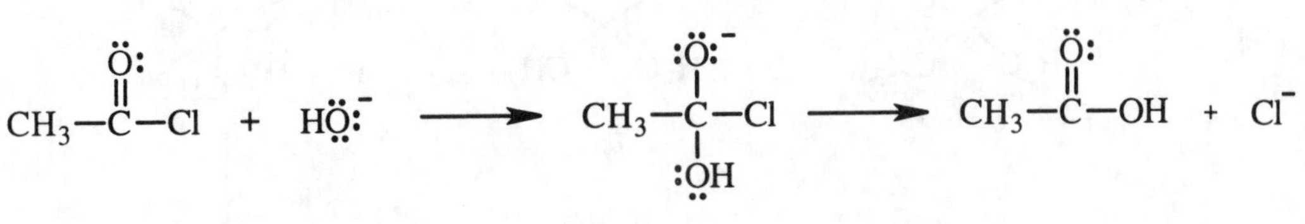

Problem 11. Draw curved arrows to show the movement of the electrons in each step of the following reactions. Show any proton transfers to and from oxygen or nitrogen using the shortcut method (H^+ and - H^+).

a. [reaction scheme: 1,2-dimethylcyclohexanol ⇌ protonated alcohol → carbocation + H_2O → rearranged carbocation → 1,2-dimethylcyclohexene + H^+]

CH3 CH3 OH H ⇌ CH3 CH3 $\overset{+}{O}H$ H H → CH3 CH3 + H + H_2O

H^+ + CH3 CH3 ← CH3 + CH3 H

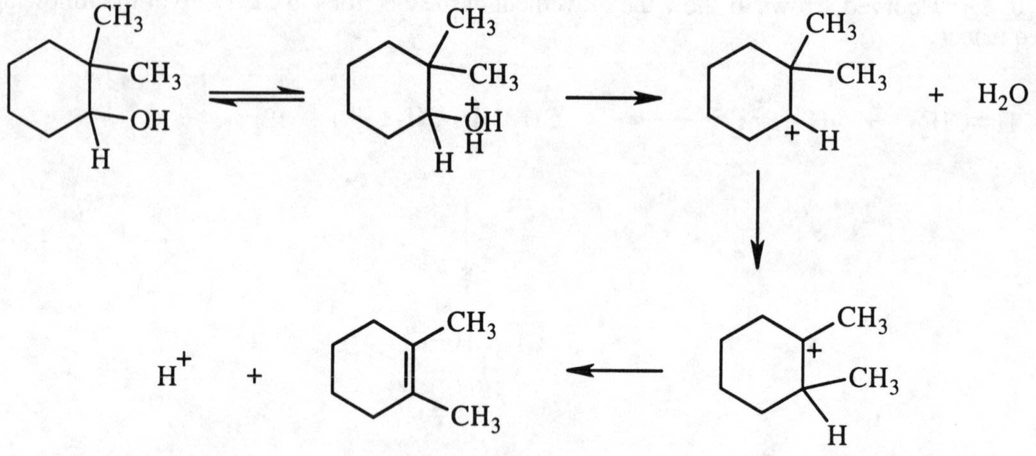

b. $CH_3CH_2CH{=}CH_2 \xrightarrow{H^+} CH_3CH_2\overset{+}{C}HCH_3 \xrightarrow{CH_3\ddot{O}H} CH_3CH_2CHCH_3(\overset{+}{:O}CH_3H)$

H^+ ⇅ - H^+

$CH_3CH_2CHCH_3(OCH_3)$

c. [reaction scheme: 4,4-dimethyl-1-methylenecyclohexa-2,5-diene → (H^+) cation ⟷ resonance form → methyl-shifted cation → 1,2,4-trimethylbenzene + H^+]

CH2 H3C CH3 $\xrightarrow{H^+}$ CH3 + H3C CH3 ⟷ CH3 + H3C CH3 → CH3 H3C + H CH3

CH3 H3C CH3 + H^+

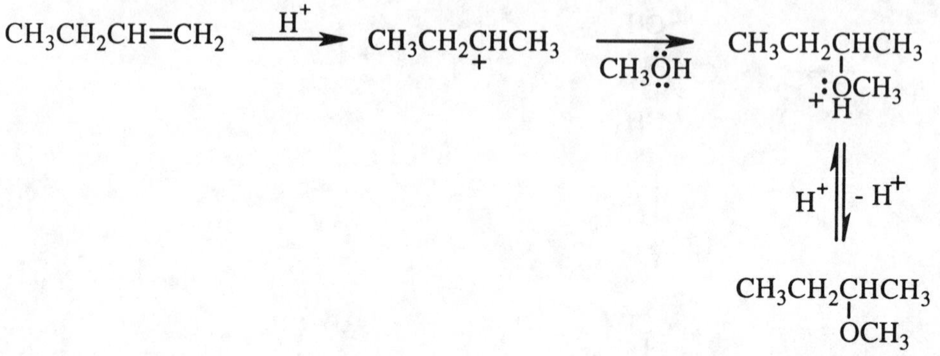

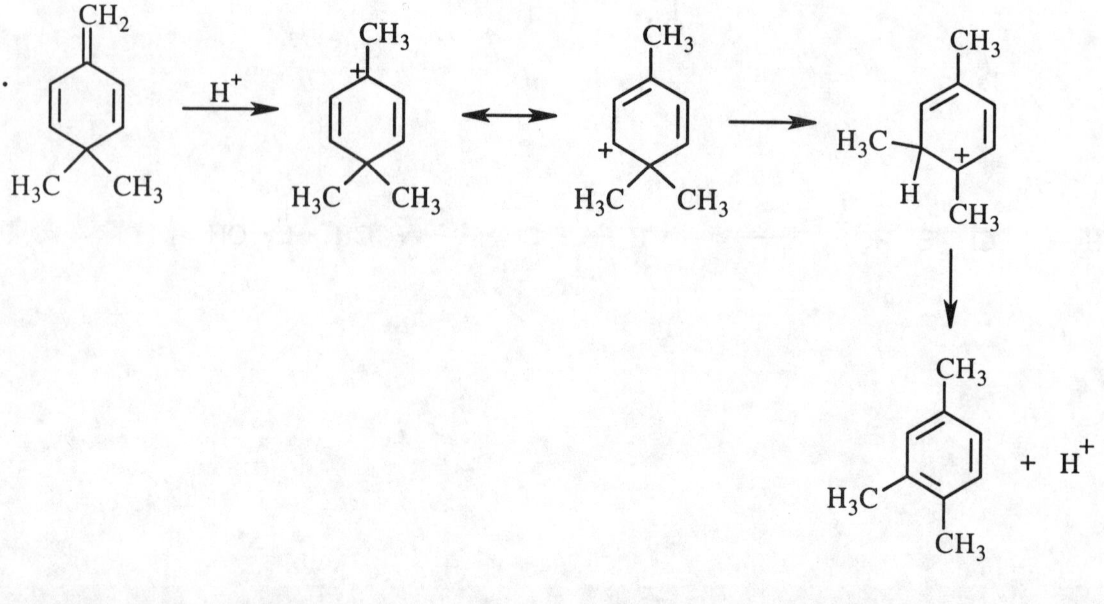

Problem 12. Use what the curved arrows tell you about electron movement to determine the product of each reaction step.

a. $CH_3CH_2\ddot{\underset{..}{O}}:^-$ + $CH_3-\ddot{\underset{..}{Br}}:$ $\longrightarrow$

b. $CH_3-\overset{+\ddot{O}H}{\overset{\|}{C}}-OCH_3$ + $H_2\ddot{\underset{..}{O}}:$ $\longrightarrow$

c. $H\ddot{\underset{..}{O}}:^-$ + $CH_3CH_2\underset{H}{\underset{|}{C}H}-CH_2-Br$ $\longrightarrow$

d. $CH_3CH_2-\overset{:\ddot{O}:^-}{\overset{|}{\underset{OH}{\underset{|}{C}}}}-NH_2$ $\longrightarrow$

e. $CH_3CH_2-\overset{O}{\overset{\|}{C}}-H$ + CH_3-MgBr $\longrightarrow$

f. $CH_3-\overset{CH_3}{\overset{|}{\underset{CH_3}{\underset{|}{C}}}}-\overset{+}{\underset{H}{O}}H$ $\longrightarrow$

g. $CH_3-\overset{:\ddot{O}H}{\overset{|}{\underset{OH}{\underset{|}{C}}}}-\overset{+}{\underset{H}{\ddot{O}}}CH_3$ $\longrightarrow$

Answers to Electron Pushing Problems

Problem 1

a. $CH_3CH_2C(CH_3)_2{-}\ddot{Br}:$ ⟶ $CH_3CH_2C^+(CH_3)_2$ + $:\ddot{Br}:^-$

b. cyclopentyl$-\ddot{Cl}:$ ⟶ cyclopentyl$^+$ + $:\ddot{Cl}:^-$

c. cyclohexyl$-\ddot{O}^+H(H)$ ⟶ cyclohexyl$^+$ + $H_2\ddot{O}:$

d. $CH_3CH_2{-}MgBr$ ⟶ $CH_3\ddot{C}^-H_2$ + ^+MgBr

e. $CH_3CH_2\overset{+}{C}HCH_3$ + $:\ddot{Br}:^-$ ⟶ $CH_3CH_2CH(:\ddot{Br}:)CH_3$

Problem 2

a. $CH_3CH_2C(CH_3)_2{-}Br$ ⟶ $CH_3CH_2C^+(CH_3)_2$ + Br^-

b. cyclopentyl$-Cl$ ⟶ cyclopentyl$^+$ + Cl^-

c. cyclohexyl$-O^+H(H)$ ⟶ cyclohexyl$^+$ + H_2O

d. $CH_3CH_2{-}MgBr$ ⟶ $CH_3\bar{C}H_2$ + ^+MgBr

Problem 3

a. $CH_3CH_2-\ddot{\underset{..}{Cl}}: \longrightarrow CH_3\dot{C}H_2 + \cdot\ddot{\underset{..}{Cl}}:$

b. $C_5H_9-\ddot{\underset{..}{Br}}: \longrightarrow C_5H_9\cdot + \cdot\ddot{\underset{..}{Br}}:$

c. $CH_3\ddot{\underset{..}{O}}-\ddot{\underset{..}{O}}CH_3 \longrightarrow CH_3\ddot{\underset{..}{O}}\cdot + \cdot\ddot{\underset{..}{O}}CH_3$

d. $(CH_3)_3C-\ddot{N}=\ddot{N}-C(CH_3)_3 \longrightarrow (CH_3)_3C\cdot \quad \ddot{N}\equiv\ddot{N} \quad \cdot C(CH_3)_3$

Problem 4

a. $CH_3CH_2-Cl \longrightarrow CH_3\dot{C}H_2 + \cdot Cl$

b. $C_5H_9-Br \longrightarrow C_5H_9\cdot + \cdot Br$

c. $CH_3O-OCH_3 \longrightarrow CH_3O\cdot + \cdot OCH_3$

d. $(CH_3)_3C-N=N-C(CH_3)_3 \longrightarrow (CH_3)_3C\cdot \quad N\equiv N \quad \cdot C(CH_3)_3$

Problem 5

a. $CH_3-C(:\ddot{O}H)(CH_3)-\overset{+}{O}H_2 \longrightarrow CH_3-C(=\overset{+}{\ddot{O}}H)-CH_3 + H_2O$

b. $CH_3CH_2CH{=}CH_2 + H{-}Cl \longrightarrow CH_3CH_2\overset{+}{C}H-CH_3 + Cl^-$

c. $CH_3CH_2{-}Br + \ddot{N}H_3 \longrightarrow CH_3CH_2-\overset{+}{N}H_3 + Br^-$

Problem 6

a. $CH_3CH{=}CHCH_3 + H{-}\overset{+}{\ddot{O}}(H){-}H \longrightarrow CH_3\overset{+}{C}H-CH_2CH_3 + H_2\ddot{O}:$

b. $CH_3CH_2CH_2CH_2{-}\ddot{Cl}: + {}^-:C{\equiv}N \longrightarrow CH_3CH_2CH_2CH_2-C{\equiv}N + :\ddot{Cl}:^-$

c. $CH_3-C(:\ddot{O}H)(OH)-\overset{+}{\ddot{O}}(H)CH_3 \longrightarrow CH_3-C(=\overset{+}{\ddot{O}}H)-OH + CH_3\ddot{O}H$

d. $CH_3-C(=\ddot{O}:)-H + CH_3{-}MgBr \longrightarrow CH_3-C(:\ddot{O}:^-)(CH_3)-H + {}^+MgBr$

Problem 7

a. $CH_3-\overset{\overset{O}{\|}}{C}-CH_3 + CH_3CH_2-MgBr \longrightarrow CH_3-\overset{\overset{O^-}{|}}{\underset{\underset{CH_2CH_3}{|}}{C}}-CH_3 + {}^{+}MgBr$

b. $CH_3CH_2CH_2-Br + CH_3\ddot{O}:^- \longrightarrow CH_3CH_2CH_2-OCH_3 + Br^-$

c. $CH_3CH=CH_2 + \cdot\ddot{Br}: \longrightarrow CH_3\dot{C}H-CH_2-Br$

d. $CH_3-\overset{\overset{:\ddot{O}:^-}{|}}{\underset{\underset{CH_3}{|}}{C}}-OCH_2CH_3 \longrightarrow CH_3-\overset{\overset{O}{\|}}{C}-CH_3 + CH_3CH_2O^-$

Problem 8

a. $H\ddot{O}:^- + CH_3\underset{\underset{H}{|}}{C}H-\overset{\overset{Br}{|}}{C}HCH_3 \longrightarrow CH_3CH=CHCH_3 + H_2O + Br^-$

b. $CH_3CH_2C\equiv C-H + \ddot{N}H_2^- \longrightarrow CH_3CH_2C\equiv C^- + NH_3$

c. $CH_3\overset{\overset{CH_3}{|}}{\underset{\underset{CH_3}{|}}{C}}-\overset{+}{C}HCH_2CH_3 \longrightarrow CH_3\overset{\overset{CH_3}{|}}{\underset{+}{C}}-\underset{\underset{CH_3}{|}}{C}HCH_2CH_3$

d. $\underset{\underset{H}{|}}{C}H_2-\overset{\overset{CH_3}{|}}{\underset{+}{C}}CH_3 \longrightarrow CH_2=\overset{\overset{CH_3}{|}}{C}CH_3 + H^+$

Problem 9

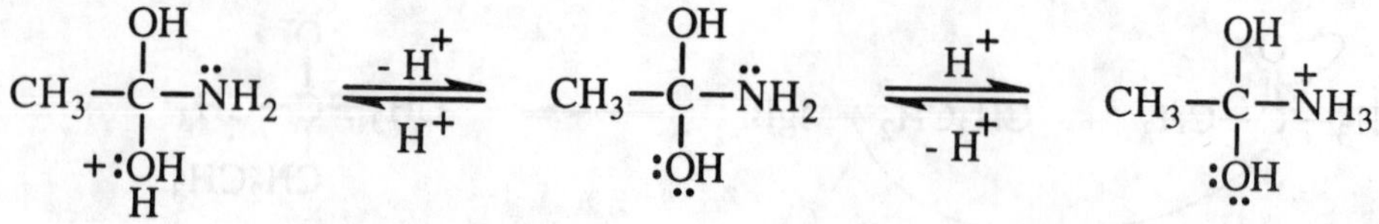

$$CH_3-\underset{+:\ddot{O}H_2}{\overset{OH}{C}}-\ddot{N}H_2 \underset{H^+}{\overset{-H^+}{\rightleftharpoons}} CH_3-\underset{:\ddot{O}H}{\overset{OH}{C}}-\ddot{N}H_2 \underset{-H^+}{\overset{H^+}{\rightleftharpoons}} CH_3-\underset{:\ddot{O}H}{\overset{OH}{C}}-\overset{+}{N}H_3$$

Problem 10

a. $CH_3CH{=}CH_2 + H{-}\ddot{B}r: \longrightarrow CH_3\underset{+}{C}H-CH_3 + :\ddot{B}r:^-$

$\downarrow$

$CH_3\underset{:\ddot{B}r:}{C}H-CH_3$

b. $CH_3\underset{CH_3}{\overset{CH_3}{C}}-Cl \longrightarrow CH_3\underset{CH_3}{\overset{CH_3}{C}}+ \quad + \quad Cl^-$

$\downarrow$ $\ddot{N}H_3$

$CH_3\underset{CH_3}{\overset{CH_3}{C}}-\overset{+}{N}H_3 \underset{H^+}{\overset{-H^+}{\rightleftharpoons}} CH_3\underset{CH_3}{\overset{CH_3}{C}}-NH_2$

c. $CH_3-\overset{\ddot{O}:}{\overset{\|}{C}}-Cl + H\ddot{O}:^- \longrightarrow CH_3-\underset{:\ddot{O}H}{\overset{:\ddot{O}:^-}{C}}-Cl \longrightarrow CH_3-\overset{\ddot{O}:}{\overset{\|}{C}}-OH + Cl^-$

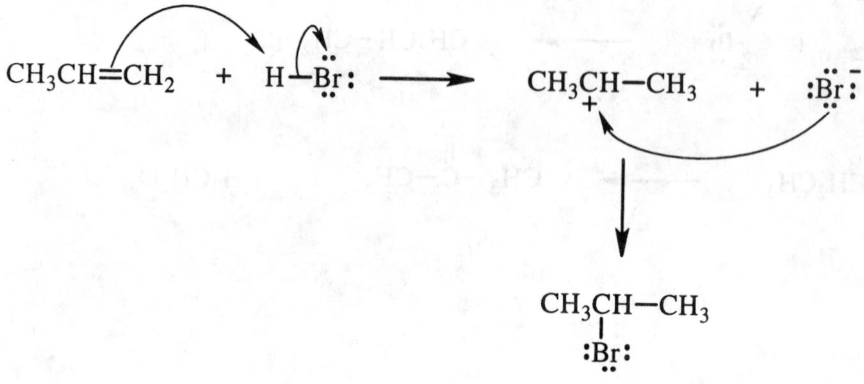

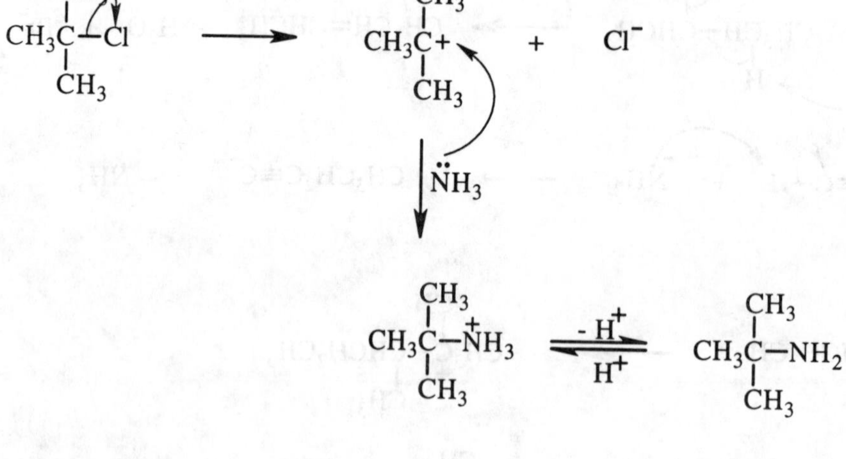

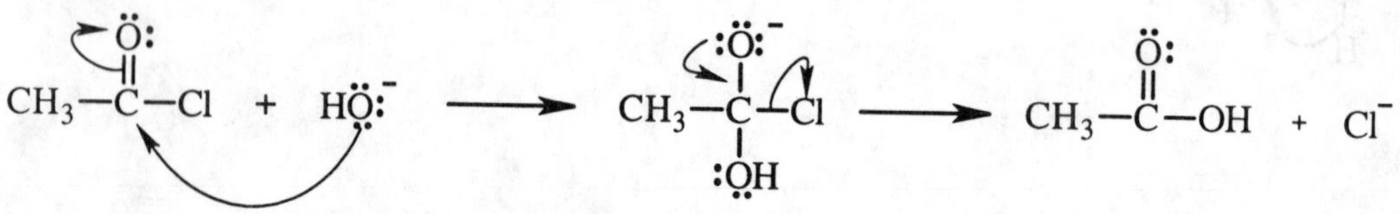

Problem 11

a.

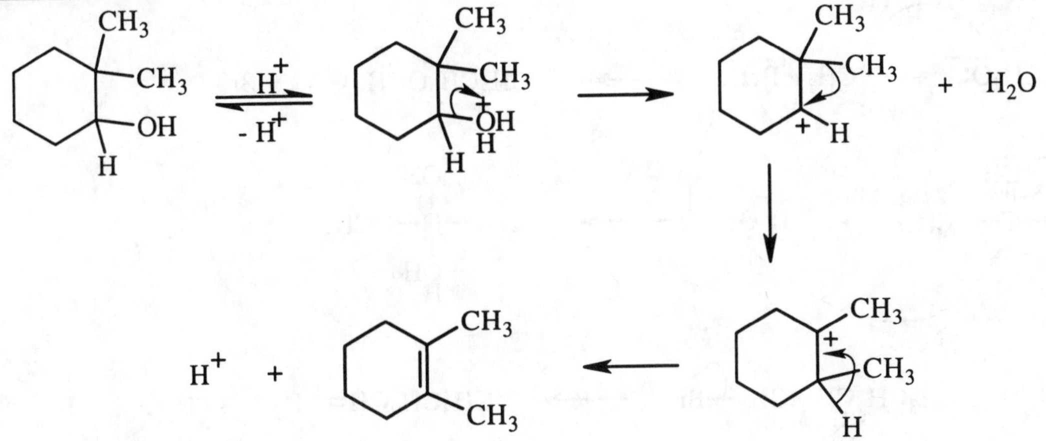

CH_3 CH_3 OH H $\underset{-\,H^+}{\overset{H^+}{\rightleftharpoons}}$ CH_3 CH_3 $\overset{+}{O}H$ H H $\longrightarrow$ CH_3 CH_3 + H + H_2O

$\downarrow$

H^+ + CH_3 CH_3 $\longleftarrow$ CH_3 + CH_3 H

b. $CH_3CH_2CH{=}CH_2 \xrightarrow{H^+} CH_3CH_2\overset{+}{C}HCH_3 \xrightarrow{CH_3\ddot{O}H} CH_3CH_2CH(\overset{+}{O}HCH_3)CH_3$

$\underset{-\,H^+}{\overset{H^+}{\rightleftharpoons}}$

$CH_3CH_2CH(OCH_3)CH_3$

c.

CH_2 H_3C CH_3 $\xrightarrow{H^+}$ CH_3 + H_3C CH_3 $\longleftrightarrow$ CH_3 + H_3C CH_3 $\longrightarrow$ CH_3 H_3C + H CH_3

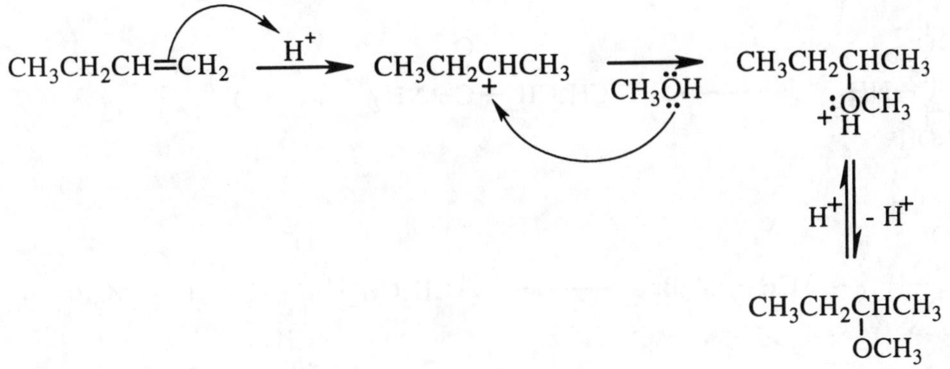

$\downarrow$

CH_3 H_3C CH_3 + H^+

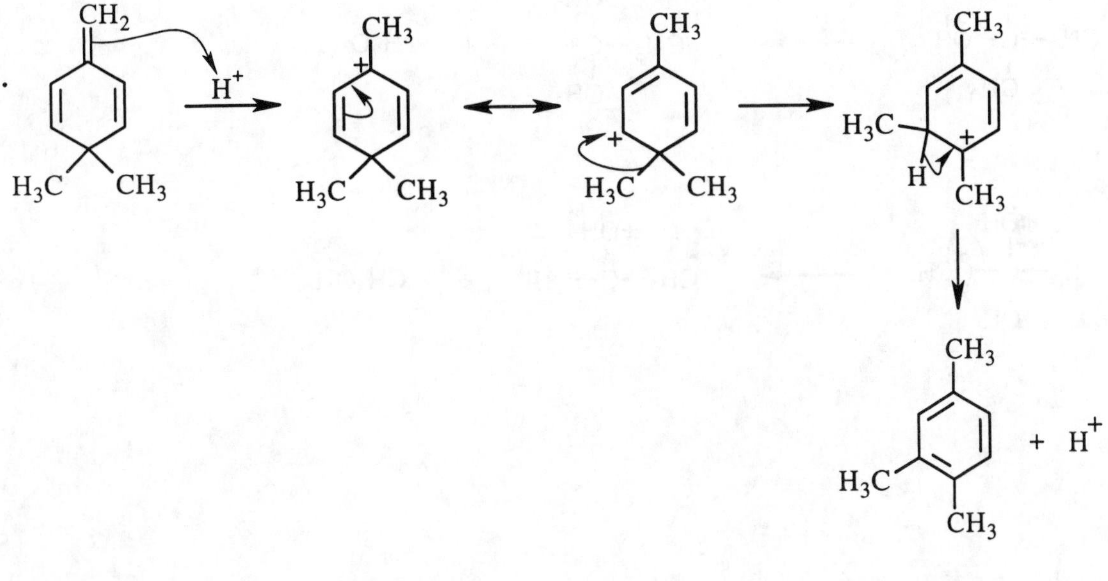

Problem 12

a. $CH_3CH_2\ddot{O}:^-$ + $CH_3-\ddot{Br}:$ ⟶ $CH_3CH_2OCH_3$ + Br^-

b. $CH_3-C(=\overset{+}{O}H)-OCH_3$ + $H_2\ddot{O}:$ ⟶ $CH_3-C(OH)(\overset{+}{O}H_2)-OCH_3$

c. $H\ddot{O}:^-$ + $CH_3CH_2CH(H)-CH_2-Br$ ⟶ $CH_3CH_2CH{=}CH_2$ + H_2O + Br^-

d. $CH_3CH_2-C(-\ddot{O}:^-)(OH)-NH_2$ ⟶ $CH_3CH_2-C(=O)-NH_2$ + HO^-

e. $CH_3CH_2-C(=O)-H$ + CH_3-MgBr ⟶ $CH_3CH_2-C(O^-)(CH_3)-H$ + ^+MgBr

f. $CH_3-C(CH_3)_2-\overset{+}{O}H_2$ ⟶ $CH_3-C^+(CH_3)_2$ + H_2O

g. $CH_3-C(:\ddot{O}H)(OH)-\overset{+}{O}(H)CH_3$ ⟶ $CH_3-C(=\overset{+}{O}H)-OH$ + CH_3OH

SPECIAL TOPIC III

Kinetics

This is a continuation of the discussion on kinetics found in Section 3.7 of the text. Note that the rate laws are derived in Appendix III.

1. How long would it take for the reactant of a first-order reaction to decrease to one-half its initial concentration if the rate constant is 4.5×10^{-3} sec^{-1} and the initial concentration of the reactant is: **a.** 1.0 M? **b.** 0.50 M?

2. How long would it take for the reactants of a second-order reaction to decrease to one-half their initial concentration if the rate constant is 2.3×10^{-2} M^{-1} sec^{-1} and the initial concentration of both reactants is: **a.** 1.0 M? **b.** 0.50 M?

3. How many half-lives are required for a first-order reaction to reach > 99% completion?

4. The initial concentration of a reactant undergoing a first-order reaction is 0.40 M. After 5 minutes, the concentration of the reactant is 0.27 M; after an additional 5 minutes, the concentration of the reactant is 0.18 M; and after an additional 5 minutes, the concentration of the reactant is 0.12 M.

a. What is the average rate of the reaction during each five-minute interval?
b. What is the rate constant of the reaction?

5. What percentage of a compound, undergoing a first-order reaction with a rate constant of 2.7×10^{-5} sec^{-1}, would have reacted at the end of two hours?

6. How long would it take for a first-order reaction with a rate constant of 5.3×10^{-4} sec^{-1} to reach 70% completion?

7. The following data were obtained in a study of the rate of inversion of sucrose at 25°C. The initial concentration of sucrose was 1.00 M.

Time (minutes)	0	30	60	90	130	180
Sucrose inverted (M)	0	0.100	0.195	0.277	0.373	0.468

a. What is the order of the reaction?
b. What is the rate constant of the reaction?

8. Calculate the activation energy of a first-order reaction that is 20% complete in 15 minutes at 40°C or 20% complete in 3 minutes at 60°C.

9. Analysis of an aqueous solution of sucrose shows that 80 grams of the original 100 grams of sucrose remain after 10 hours. At this rate, how much sucrose would be left after 24 hours?

Solutions to Problems in Special Topic III

1. a.

$$\text{half-life of a first-order reaction} = t_{1/2} = \frac{0.693}{k_1}$$

$$= \frac{0.693}{4.5 \times 10^{-3}\ \text{sec}^{-1}}$$

$$= 154 \text{ seconds}$$

$$= 2 \text{ minutes, 34 seconds}$$

b. The half-life of a first-order reaction is independent of concentration, so the answer is the same as for **a** (15.4 seconds).

2.

a. $$\text{half-life of a second-order reaction} = t_{1/2} = \frac{1}{k_2\, a}$$

$$= \frac{1}{2.3 \times 10^{-2}\ \text{M}^{-1}\ \text{sec}^{-1}\ (1.00\ \text{M})}$$

$$= 43 \text{ seconds}$$

b. $$t_{1/2} = \frac{1}{2.3 \times 10^{-2}\ \text{M}^{-1}\ \text{sec}^{-1}\ (0.50\ \text{M})}$$

$$= \frac{1}{1.15 \times 10^{-2}}$$

$$= 87 \text{ seconds}$$

3. One-half of the compound reacts during the first half-life. One half of what is left after the first half-life reacts during the second half-life. One half of what is left after the second half-life reacts during the third half-life, etc.

number of half-lives	percentage completion	
1	0.50 (100) = 50	50
2	0.50 (50) = 25	75
3	0.50 (25) = 12.5	87.5
4	0.50 (12.5) = 6.25	93.8
5	0.50 (6.25) = 3.125	96.9
6	0.50 (3.125) = 1.5625	98.4
7	0.50 (1.5625) = 0.78125	99.2

Seven half-lives are required.

4. a.

$$\text{rate} = \frac{\text{change in concentration}}{\text{change in time}}$$

1st interval: rate $= \frac{0.40\text{ M} - 0.27\text{ M}}{5\text{ min}} = \frac{0.13\text{ M}}{5\text{ min}} = 2.60 \times 10^{-2}\text{ M min}^{-1}$

2nd interval: rate $= \frac{0.27\text{ M} - 0.18\text{ M}}{5\text{ min}} = \frac{0.09\text{ M}}{5\text{ min}} = 1.80 \times 10^{-2}\text{ M min}^{-1}$

3rd interval: rate $= \frac{0.18\text{ M} - 0.12\text{ M}}{5\text{ min}} = \frac{0.06\text{ M}}{5\text{ min}} = 1.20 \times 10^{-2}\text{ M min}^{-1}$

Thus, the **rate** of the reaction decreases with decreasing concentration.

b. rate = k [reactant], using the average concentration of the reactant during the 5 minute interval.

1st interval: $2.60 \times 10^{-2}\text{ M min}^{-1} = k\,(0.335\text{ M})$

$\boldsymbol{k} = 7.8 \times 10^{-2}\text{ min}^{-1}$

2nd interval: $1.80 \times 10^{-2}\text{ M min}^{-1} = k\,(0.225\text{ M})$

$\boldsymbol{k} = 8.0 \times 10^{-2}\text{ min}^{-1}$

3rd interval: $1.20 \times 10^{-2}\text{ M min}^{-1} = k\,(0.15\text{ M})$

$\boldsymbol{k} = 8.0 \times 10^{-2}\text{ min}^{-1}$

The **rate constant**, as its name indicates, is constant during the course of the reaction.

5. $\boldsymbol{a}$ = the initial concentration
$\boldsymbol{x}$ = the concentration that has reacted at time = $\boldsymbol{t}$

$$2.303 \log \frac{a}{a-x} = k_1 t$$

$$2.303 \log \frac{100}{100-x} = 2.70 \times 10^{-5}\text{ sec}^{-1}\ (2\text{ hours})$$

$$\log \frac{100}{100-x} = \frac{2.70 \times 10^{-5}\text{ sec}^{-1}\ (7200\text{ sec})}{2.303}$$

$$\log \frac{100}{100-x} = 0.0844$$

$$\frac{100}{100-x} = 1.21$$

$$100 = 121 - 1.21x$$

$$1.21x = 21$$

$$x = 17.4\%$$

6.

$$2.303 \log \frac{a}{a-x} = k_1 t$$

$$2.303 \log \frac{100}{100-70} = 5.30 \times 10^{-4}\ \text{sec}^{-1}\ t$$

$$t = \frac{2.303 \log 3.33}{5.30 \times 10^{-4}\ \text{sec}^{-1}}$$

$$t = 2272 \text{ seconds}$$

$$= 37.9 \text{ minutes}$$

7.

1st order

$$k_1 = \frac{-2.303 \log \frac{a-x}{a}}{t}$$

$$k_1 = \frac{-2.303 \log \frac{1.000-0.100}{1.000}}{30}$$

$$= \mathbf{3.51 \times 10^{-3}}$$

$$k_1 = \frac{-2.303 \log \frac{1.000-0.195}{1.000}}{60}$$

$$= \mathbf{3.62 \times 10^{-3}}$$

$$k_1 = \frac{-2.303 \log \frac{1.000-0.277}{1.000}}{90}$$

$$= \mathbf{3.60 \times 10^{-3}}$$

$$k_1 = \frac{-2.303 \log \frac{1.000-0.373}{1.000}}{130}$$

$$= \mathbf{3.59 \times 10^{-3}}$$

2nd order

$$k_2 = \frac{\frac{1}{a-x} - \frac{1}{a}}{t}$$

$$k_2 = \frac{\frac{1}{1.000-0.100} - \frac{1}{1.000}}{30}$$

$$= \frac{1.111-1.000}{30}$$

$$= \mathbf{3.70 \times 10^{-3}}$$

$$k_2 = \frac{\frac{1}{1.000-0.195} - 1.000}{60}$$

$$= \mathbf{4.03 \times 10^{-3}}$$

$$k_2 = \frac{\frac{1}{1.000-0.277} - 1.000}{90}$$

$$= \mathbf{4.26 \times 10^{-3}}$$

$$k_2 = \frac{\frac{1}{1.000-0.373} - 1.000}{130}$$

$$= \mathbf{4.57 \times 10^{-3}}$$

$$k_1 = \frac{-2.303 \log \dfrac{1.000 - 0.468}{1.000}}{180} \qquad k_2 = \frac{\dfrac{1}{1.000 - 0.468} - 1.000}{180}$$

$$= \mathbf{3.51 \times 10^{-3}} \qquad = \mathbf{4.89 \times 10^{-3}}$$

a. Because the calculated rate constants are relatively constant when the data are plugged into a first-order equation and vary considerably when the data are plugged into a second-order equation, one can conclude that the reaction is first-order.

b. 3.59×10^{-3} min^{-1}

8.

$$k \text{ at } 40°\text{C} = \frac{2.303 \log \dfrac{1.00}{1.00 - 0.20}}{15 \text{ min}} = 1.49 \times 10^{-2} \text{ min}^{-1}$$

$$k \text{ at } 60°\text{C} = \frac{2.303 \log \dfrac{1.00}{1.00 - 0.20}}{3 \text{ min}} = 7.44 \times 10^{-2} \text{ min}^{-1}$$

$$\log k_2 - \log k_1 = \frac{-E_a}{2.303\,R}\left(\frac{1}{T_2} - \frac{1}{T_1}\right)$$

$$\log 7.44 \times 10^{-2} - \log 1.49 \times 10^{-2} = \frac{-E_a}{2.303 \times 1.986 \times 10^{-3} \text{ kcal}}\left(\frac{1}{333} - \frac{1}{313}\right)$$

$$-1.13 - (-1.83) = \frac{-E_a}{4.57 \times 10^{-3}}(0.00300 - 0.00319)$$

$$0.70 = \frac{-E_a}{4.57 \times 10^{-3}}(-0.00019)$$

$$E_a = \frac{0.70 \times 4.57 \times 10^{-3}}{0.00019}$$

$$E_a = 16.8 \text{ kcal/mol}$$

9. Sucrose is hydrolyzed to form a mixture of glucose and fructose. Because there is an excess water (it is the solvent), the reaction is a pseudo first-order reaction.

First, the rate constant of the reaction must be determined:

$$2.303 \log \frac{a}{a - x} = k_1 t$$

$$2.303 \log \frac{100}{80} = k_1 \text{ x } 10 \text{ hours}$$

$$k_1 = 2.23 \text{ x } 10^{-2} \text{ hr}^{-1}$$

$$2.303 \log \frac{a}{a - x} = k_1 t$$

$$2.303 \log \frac{100}{100 - x} = 2.23 \text{ x } 10^{-2} \text{ (24 hr)}$$

$$\log \frac{100}{100 - x} = \frac{0.535}{2.303}$$

$$\frac{100}{100 - x} = 1.71$$

$$100 = 171 - 1.71x$$

$$1.71x = 71$$

$$x = 41.5 \text{ g have reacted}$$

Therefore, 58.5 g would be left.

CHAPTER 4
Stereochemistry: The Arrangement of Atoms in Space; The Stereochemistry of Addition Reactions

Important Terms

absolute configuration — the three-dimensional structure of a chiral compound. The configuration is designated by *R* or *S*. The configuration is called absolute to distinguish it from a relative configuration.

achiral (optically inactive) — a molecule or object that contains a plane of symmetry.

amine inversion — results when a compound containing an sp^3 hybridized nitrogen with a nonbonding pair of electrons rapidly turns inside out.

anti addition — an addition reaction in which the two added substituents add to opposite sides of the molecule.

biochemistry — the chemistry associated with living organisms.

chiral (optically active) — a chiral molecule has a nonsuperimposable mirror image.

chirality center — a carbon bonded to four different substituents.

cis isomer — the isomer with substituents on the same side of a cyclic structure, or the isomer with the hydrogens on the same side of a double bond.

cis-trans isomers — geometric (or *E*, *Z*) isomers.

chromatography — a separation technique in which the mixture to be separated is dissolved in a solvent and the solvent is passed through a column packed with an adsorbent stationary phase.

configuration — the three-dimensional structure of a chiral compound. The configuration is designated by *R* or *S*.

configurational isomers — stereoisomers that cannot interconvert unless a covalent bond is broken. Cis-trans isomers and optical isomers are configurational isomers.

conformational isomers or conformers — stereoisomers that can interconvert readily at room temperature. Isomers resulting from rotation about carbon-carbon single bonds and from amine inversion are conformational isomers.

constitutional isomers (structural isomers) — molecules that have the same molecular formula but differ in the way the atoms are connected.

dextrorotatory — the enantiomer that rotates polarized light in a clockwise direction.

diastereomer — a configurational isomer that is not an enantiomer.

diastereotopic hydrogens — two hydrogens bonded to the same carbon that will result in a pair of diastereomers when each of them is replaced in turn with deuterium.

enantiomers nonsuperimposable mirror-image molecules.

enantiomerically pure only one enantiomer is present in an enantiomerically pure sample.

enantiomeric excess how much excess of one enantiomer is present in a mixture of a pair of enantiomers.

enantiotopic hydrogens two hydrogens bonded to a carbon that is bonded to two other groups that are nonidentical.

enzyme a protein that catalyzes a biological reaction.

erythro enantiomers the pair of enantiomers with similar groups on the same side (in the case of stereoisomers with adjacent chirality centers) when drawn in a Fischer projection.

Fischer projection a method of representing the spatial arrangement of groups bonded to a chirality center. The chirality center is the point of intersection of two perpendicular lines; the horizontal lines represent bonds that project out of the plane of the paper toward the viewer, and the vertical lines represent bonds that project back from the plane of the paper away from the viewer.

homotopic hydrogens two hydrogens bonded to a carbon that is bonded to two other groups that are identical.

isomers nonidentical compounds with the same molecular formula.

isomers that contain chirality centers these can be enantiomers, diastereomers, and meso compounds.

levorotatory the enantiomer that rotates polarized light in a counterclockwise direction.

meso compound a compound that possesses chirality centers and a plane of symmetry.

observed rotation the amount of rotation observed in a polarimeter.

optical purity how much excess of one enantiomer is present in a mixture of a pair of enantiomers.

optically active (chiral) rotates the plane of polarized light.

optically inactive (achiral) does not rotate the plane of polarized light.

pair of enantiomers a pair of nonsuperimposable mirror image molecules.

perspective formula a method of representing the spatial arrangement of groups bonded to a chirality center. Two bonds are drawn in the plane of the paper; a solid wedge is used to depict a bond that projects out of the plane of the paper toward the viewer, and a hatched wedge is used to represent a bond that projects back from the paper away from the viewer.

plane of symmetry an imaginary plane that bisects a molecule into a pair of mirror images.

plane-polarized light light that oscillates in a single plane passing through the diretion the light travels.

polarimeter an instrument that measures the rotation of polarized light.

polarized light light that oscillates in only one plane.

prochirality center a carbon (bonded to two hydrogens) that will become a chirality center if one of the hydrogens is replaced by deuterium.

pro-*R*-hydrogen replacing this hydrogen with deuterium creates a chirality center with the *R* configuration.

pro-*S*-hydrogen replacing this hydrogen with deuterium creates a chirality center with the *S* configuration.

racemic mixture (racemic modification, racemate) a mixture of equal amounts of a pair of enantiomers.

***R* configuration** after assigning relative priorities to the four groups bonded to a chirality center, if the lowest-priority group is on a vertical axis in a Fischer projection (or pointing away from the viewer in a perspective formula), an arrow drawn from the highest-priority group to the next highest-priority group goes in a clockwise direction.

regioselective describes a reaction that leads to the preferential formation of one constitutional isomer over another.

relative configuration the configuration of a compound relative to the configuration of another compound.

resolution of a racemic mixture separation of a racemic mixture into the individual enantiomers.

***S* configuration** after assigning relative priorities to the four groups bonded to a chirality center, if the lowest-priority group is on a vertical axis in a Fischer projection (or pointing away from the viewer in a perspective formula), an arrow drawn from the highest-priority group to the next highest-priority group goes in a counterclockwise direction.

specific rotation the amount of rotation that will be caused by a compound with a concentration of 1.0 g/ml in a sample tube 1.0 decimeter long.

stereochemistry the field of chemistry that deals with the structure of molecules in three dimensions.

stereoisomers isomers that differ in the way the atoms are arranged in space.

stereoselective describes a reaction that leads to the preferential formation of one stereoisomer over another.

stereospecific describes a reaction in which the reactant can exist as stereoisomers and each stereoisomeric reactant leads to a different stereoisomeric product.

structural isomers (constitutional isomers)	molecules that have the same molecular formula but differ in the way the atoms are connected.
syn addition	an addition reaction in which the two added substituents add to the same side of the molecule.
threo enantiomers	the pair of enantiomers with similar groups on opposite sides (in the case of stereoisomers with adjacent chirality centers) when drawn in a Fischer projection.
trans isomer	the isomer with substituents on the opposite sides of a cyclic structure, or the isomer with the hydrogens on the opposite sides of a double bond.

Exercise in Model Building

Do the following exercises using molecular models.

1. Build the enantiomers of 2-bromobutane.
   a. Try to superimpose them.
   b. Show that they are mirror images.
   c. Which one is (*R*)-2-bromobutane?

2. Build the erythro enantiomers of 3-bromo-2-butanol.
   a. Where are the Br and OH substituents (relative to each other) in the Fischer projection? (Recall that in a Fischer projection, the horizontal lines represent bonds that point out of the plane of the paper toward the viewer, and the dashed lines represent bonds that point back from the plane of the paper away from the viewer.)
   b. Where are the Br and OH substituents (relative to each other) in the most stable conformer considering rotation about the C-2—C-3 bond?

3. **a.** Build the two compounds shown on the top of page 202 of the text.
   b. Show that they are superimposable.

4. Build the three stereoisomers of 2,3-dibromobutane.

5. Build the four stereoisomers of 2,3-dibromopentane.
   Why does 2,3-dibromopentane have four stereoisomers, while 2,3-dibromobutane has only three stereoisomers?

6. Build (*S*)-2-pentanol.

7. Build (2*S*,3*R*)-3-bromo-2-butanol. Rotate your model so it is in a Fischer projection. Compare its structure with the structure of (2*S*,3*R*)-3-bromo-2-butanol shown on page 206 of the text.

8. Build the compounds shown in Problem 32 on page 209 of the text. Name the compounds.

9. Build (*S*)-1-bromo-2-methylbutane. Substitute the Br^- with an HO^- to form 2-methyl-1-butanol. What is the configuration of your model of 2-methyl-1-butanol?

10. Build ethanol. Which of the hydrogens is the pro-*R*-hydrogen?

11. Build two models of *trans*-2-pentene. Add Br_2 to opposite sides of the double bond, forming the two enantiomers shown on page 227 of the text. Rotate them so they are Fischer projections. Are they erythro or threo enantiomers? Compare your answer with that given in the text.

12. Build models of the molecules shown in Problem 80 on page 238 of the text. What is the configuration of the chirality center in each of the molecules?

Solutions to Problems

1.

a. 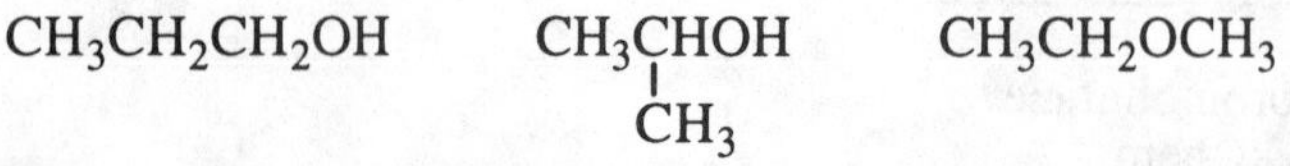 $CH_3CH_2CH_2OH$ $CH_3CH(CH_3)OH$ $CH_3CH_2OCH_3$

b. There are seven constitutional isomers with molecular formula $C_4H_{10}O$.

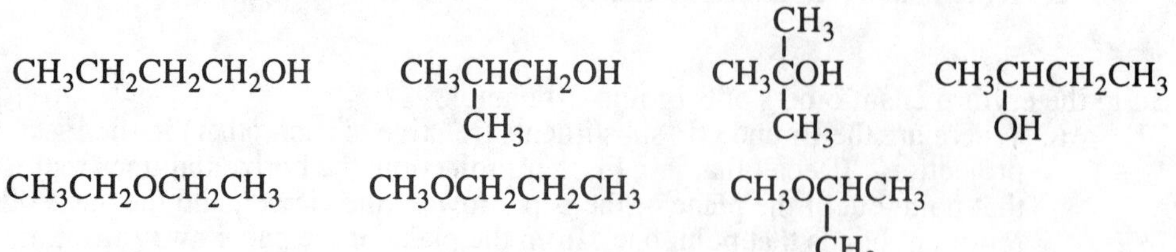

$CH_3CH_2CH_2CH_2OH$ $CH_3CH(CH_3)CH_2OH$ $CH_3C(CH_3)_2OH$ $CH_3CH(OH)CH_2CH_3$

$CH_3CH_2OCH_2CH_3$ $CH_3OCH_2CH_2CH_3$ $CH_3OCH(CH_3)CH_3$

2.

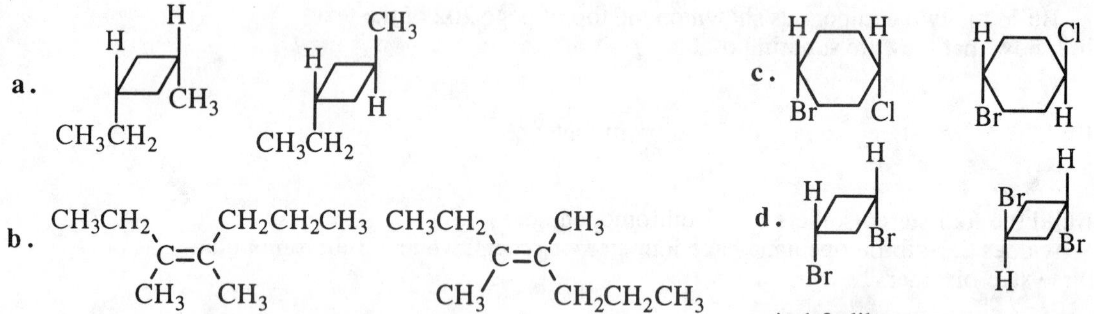

a. [cyclobutane structures with H, CH_3, CH_3CH_2 substituents]

b. $CH_3CH_2(CH_3)C{=}C(CH_3)CH_2CH_2CH_3$ $CH_3CH_2(CH_3)C{=}C(CH_3)CH_2CH_2CH_3$

c. [cyclohexane structures with H, Br, Cl substituents]

d. *cis*-1,3-dibromocyclobutane *trans*-1,3-dibromocyclobutane

3. a, c, and f have chirality centers.

4. a, c, and f, because in order to be able to exist as a pair of enantiomers, the compound must have a chirality center (except in the case with compounds with unusual structures. See Problem 81.)

5. Solved in the text.

6. a. F, G, J, L, N, P, Q, R, S, Z

b. A, C, D, H, I, M, O, T, U, V, W, X, Y

Whether B, E, and K are chiral or achiral depends on how they are drawn. For example, if B is drawn with two equal loops, it is achiral; if the loops differ in size, it is chiral.

7. Draw the first enantiomer with the groups in any order you want. Then draw the second enantiomer by drawing the mirror image of the first enantiomer.

a. 1. CH_3 / Br–C(H)–CH_2OH CH_3 / $HOCH_2$–C(H)–Br

2. CH_2CH_2Cl / CH_3CH_2–C(H)–CH_3 CH_2CH_2Cl / H_3C–C(H)–CH_2CH_3

3. OH / $(CH_3)_2CH$–C(H)–CH_3 OH / H_3C–C(H)–$CH(CH_3)_2$

b. 1. CH_3 / Br–C–H / CH_2OH CH_3 / H–C–Br / CH_2OH

2. CH_2CH_2Cl / H–C–CH_3 / CH_2CH_3 CH_2CH_2Cl / CH_3–C–H / CH_2CH_3

3. CH_3 / H–C–OH / $CH(CH_3)_2$ CH_3 / HO–C–H / $CH(CH_3)_2$

8. **a.** *R* **b.** *R* **c.** *R* **d.** *R*

9. The easiest way to determine if two compounds are identical or enantiomers is to determine their configurations. If both are *R* (or both are *S*), they are identical; if one is *R* and the other is *S*, they are enantiomers.

a. enantiomers **b.** enantiomers **c.** enantiomers **d.** enantiomers

10.

a. CH_2OH (1) CH_3 (3) CH_2CH_2OH (2)   H (4)

b. CH=O (2)   OH (1) CH_3 (4) CH_2OH (3)

c. $CH(CH_3)_2$ (2) CH_2CH_2Br (3)   Cl (1) $CH_2CH_2CH_2Br$ (4)

d. $CH=CH_2$ (2) CH_2CH_3 (3)   phenyl (1) CH_3 (4)

C attached to 2 C's (for $CH=CH_2$)

C attached to 3 C's (for phenyl)

11. **a.** *S* **b.** *R* **c.** *S* **d.** *S*

12.

$$[\alpha] = \frac{\alpha}{\text{concentration} \times \text{length}} = \frac{+13.4°}{\frac{2}{10} \times 2.5 \text{ dm}} = \frac{+13.4°}{0.5} = +268°$$

13. **a.** levorotatory **b.** dextrorotatory

14. **a.** - 24° **b.** 0°

15.

a. 50% of the mixture is excess (+)-mandelic acid.

$$\text{optical purity} = 0.50 = \frac{\text{observed specific rotation}}{\text{specific rotation of the pure enantiomer}}$$

$$0.50 = \frac{\text{observed specific rotation}}{+158°}$$

$$\text{observed specific rotation} = +79°$$

b. 0° (It is a racemic mixture.)

c. 50% of the mixture is excess (-)-mandelic acid.

observed specific rotation = - 79°

16.

$$\text{optical purity} = \frac{+1.4°}{+8.7°} = .16 = 16\% \text{ excess } R \text{ enantiomer}$$

100% - 16% = 84% is a racemic mixture

R enantiomer = 1/2 (84%) + 16% = 42% + 16% = 58%

17. Solved in the text.

18. **a.** enantiomers

b. identical compounds (Therefore, they are not isomers.)

c. diastereomers

19. Solved in the text.

20. **a.** First find the sp^3 carbons that are bonded to four different substituents; these are the chirality centers. Cholesterol has eight chirality centers. They are indicated by arrows.

b. $2^8 = 256$

c. Only one of the stereoisomers is found in nature.

21. As a result of the double bond, the compound has a cis isomer and a trans isomer. Because the compound also has a chirality center, the cis isomer can exist as a pair of enantiomers and the trans isomer can exist as a pair of enantiomers.

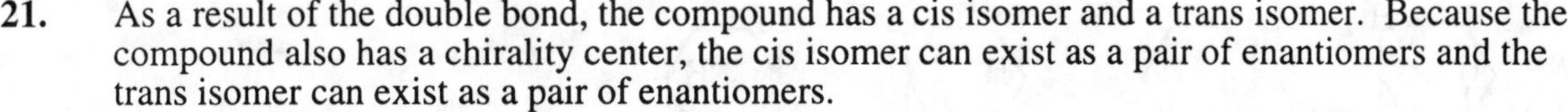

cis enantiomers

trans enantiomers

22.

a.

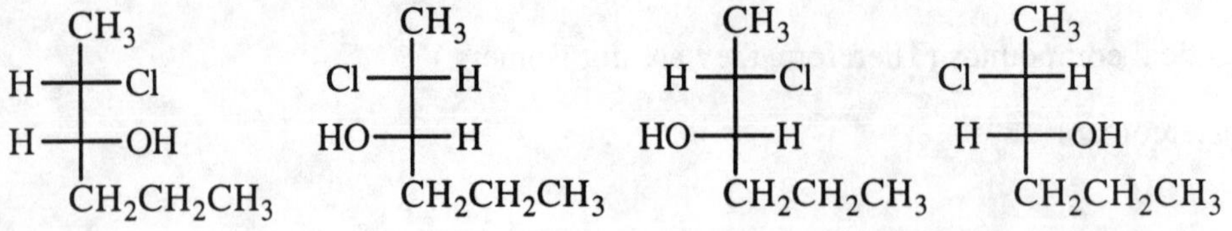

CH_3; H—C—Cl; H—C—OH; $CH_2CH_2CH_3$

CH_3; Cl—C—H; HO—C—H; $CH_2CH_2CH_3$

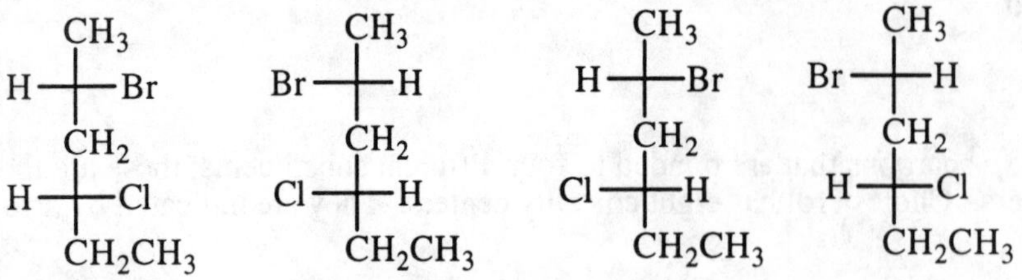

CH_3; H—C—Cl; HO—C—H; $CH_2CH_2CH_3$

CH_3; Cl—C—H; H—C—OH; $CH_2CH_2CH_3$

b.

CH_3; H—C—Br; CH_2; H—C—Cl; CH_2CH_3

CH_3; Br—C—H; CH_2; Cl—C—H; CH_2CH_3

CH_3; H—C—Br; CH_2; Cl—C—H; CH_2CH_3

CH_3; Br—C—H; CH_2; H—C—Cl; CH_2CH_3

c.

CH_3; H—C—Cl; H—C—Cl; CH_2CH_3

CH_3; Cl—C—H; Cl—C—H; CH_2CH_3

CH_3; H—C—Cl; Cl—C—H; CH_2CH_3

CH_3; Cl—C—H; H—C—Cl; CH_2CH_3

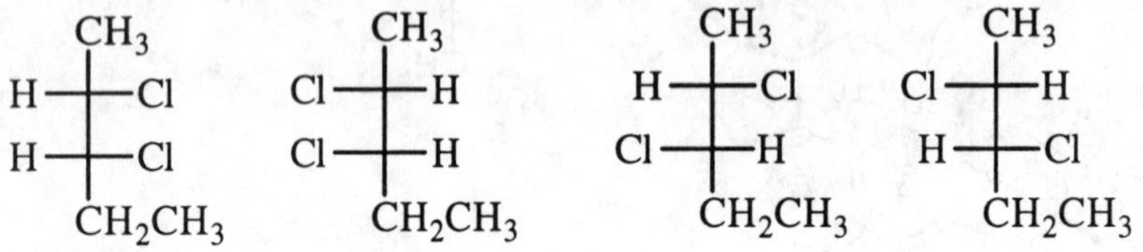

d. 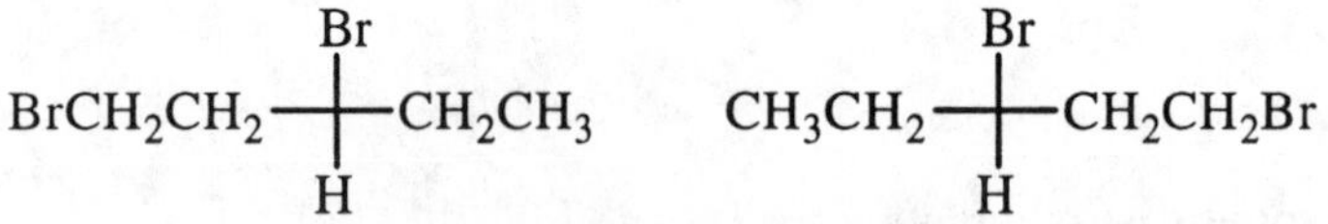 $BrCH_2CH_2$—C(Br)(H)—CH_2CH_3 CH_3CH_2—C(Br)(H)—CH_2CH_2Br

23.

H, H, Cl, Br   H, H, Cl, Br   Cl, H, H, Br   Cl, H, H, Br

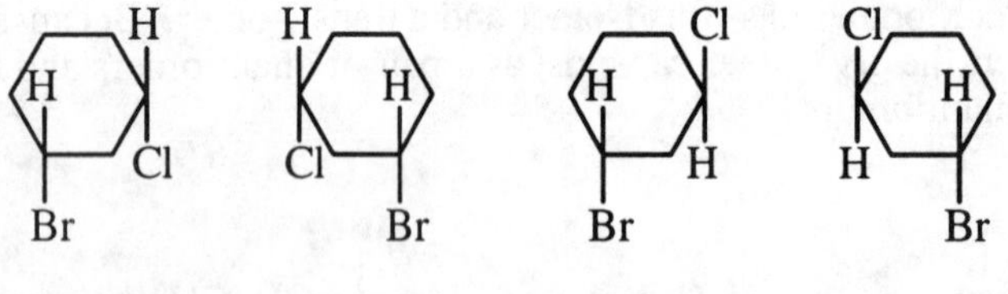

24.

The trans compound exists as a pair of enantiomers.

H, CH_3, $C(CH_3)_3$, H   H, CH_3, $(CH_3)_3C$, H

As a result of ring-flip, each enantiomer has two chair conformations.
In each case, the more stable conformation is the one with the larger group (the *tert*-butyl group) in the equatorial position.

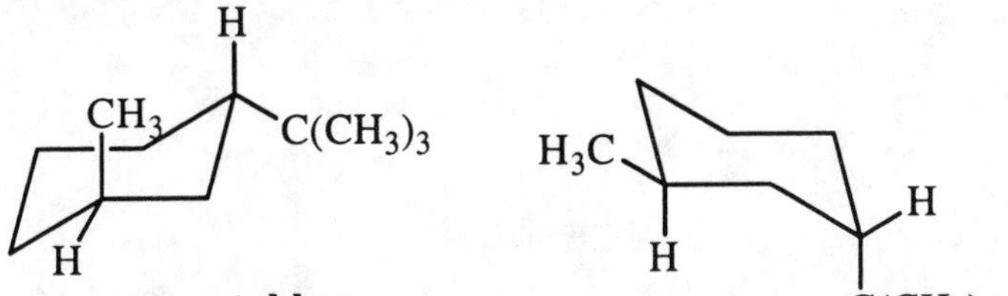

H, CH_3, $C(CH_3)_3$, H — **more stable**; H_3C, H, H, $C(CH_3)_3$

H, $(CH_3)_3C$, CH_3, H; H, $(CH_3)_3C$, CH_3, H — **more stable**

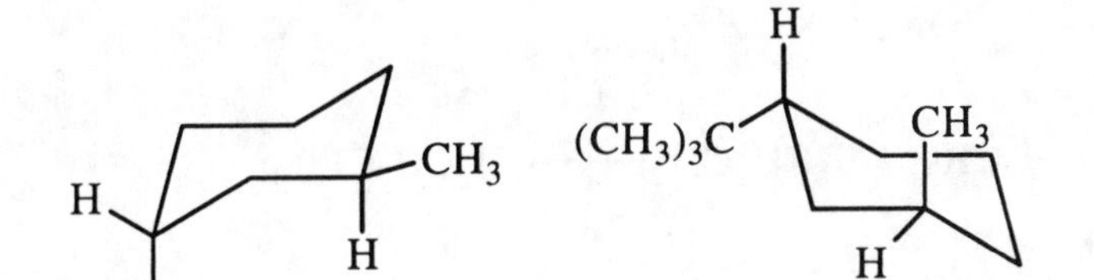

25.

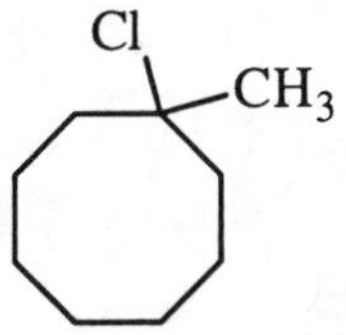

Cl, CH_3 — 1-chloro-1-methylcyclooctane

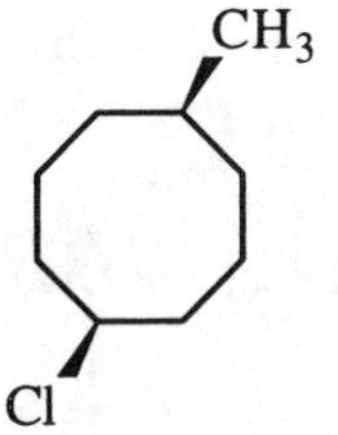

CH_3, Cl — *cis*-1-chloro-5-methylcyclooctane

CH_3, Cl — *trans*-1-chloro-5-methylcyclooctane

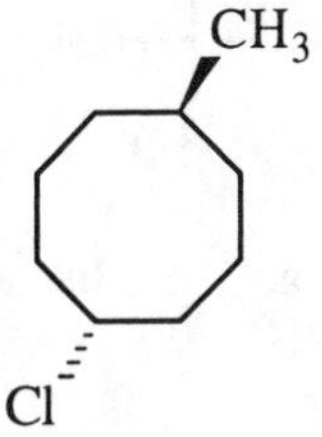

26. There is more than one diastereomer for **a, b,** and **d**.
c has only one diastereomer.

To draw a diastereomer of **a, b,** or **d**, switch any one pair of substituents bonded to one of the chirality centers. Because any one pair can be switched, your diastereomer won't be the same as the one drawn here unless you happened to switch the same pair.

a. CH_3; H—OH; HO—H; CH_3

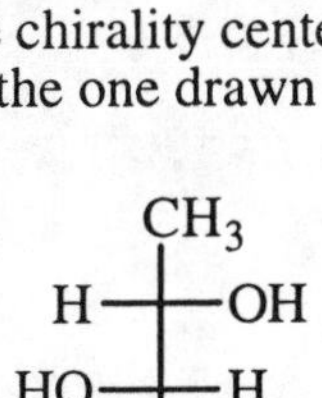

c. CH_3, H, C=C, H, CH_3

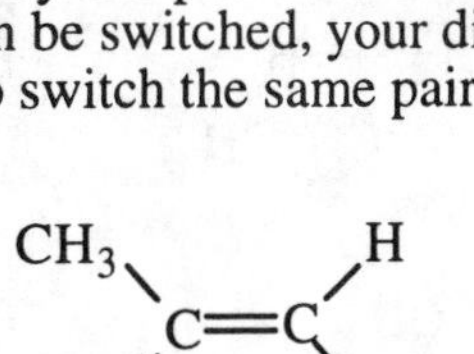

b. Cl, H, C—C, H, Cl, CH_3CH_2, CH_3

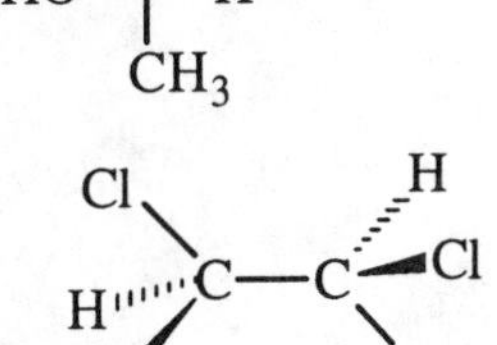

d. H, CH_3, HO, H

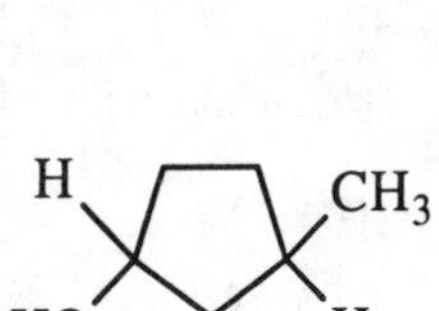

27. **b**, **d**, and **f**

(**c** and **e** do not have a stereoisomer that is a meso compound, because they do not have chirality centers.)

28.

a. CH_3CH_2–C(CH$_3$)(H)–CH_2Br $BrCH_2$–C(CH$_3$)(H)–CH_2CH_3

```
              CH3                  CH3
a. CH3CH2 ────┼──── CH2Br   BrCH2 ──┼── CH2CH3
              H                     H
```

b.

```
                CH3                      CH3
b. ClCH2CH2 ────┼──── CH2CH3   CH3CH2 ───┼─── CH2CH2Cl
                H                        H
```

c. no stereoisomers

d.

```
              Br                    Br
d. CH3CH2 ────┼──── CH2OH   HOCH2 ──┼── CH2CH3
              H                     H
```

e. no stereoisomers because the compound does not have a chirality center

f.

```
     CH3           CH3           CH3           CH3
  H ──┼── OH   HO ──┼── H    H ──┼── OH   HO ──┼── H
  H ──┼── Br   Br ──┼── H   Br ──┼── H     H ──┼── Br
     CH3           CH3           CH3           CH3
```

g.

```
    CH2CH3        CH2CH3        CH2CH3
  H ──┼── Cl    H ──┼── Cl   Cl ──┼── H
  H ──┼── Cl   Cl ──┼── H     H ──┼── Cl
    CH2CH3        CH2CH3        CH2CH3
```

h.

```
     CH3           CH3           CH3
  H ──┼── Cl    H ──┼── Cl   Cl ──┼── H
     CH2           CH2           CH2
  H ──┼── Cl   Cl ──┼── H     H ──┼── Cl
     CH3           CH3           CH3
```

i.

j.

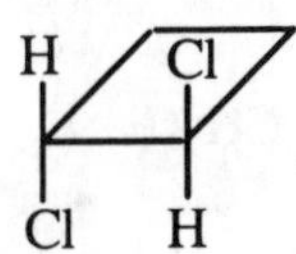
H
Cl
Cl
H

k.

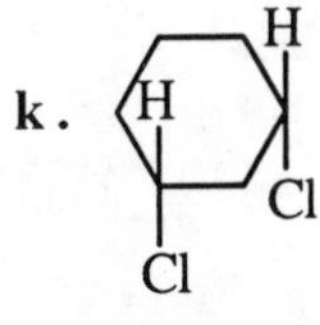
H
H
Cl
Cl

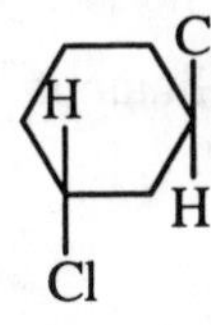
Cl
H
H
Cl

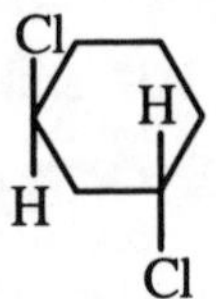
Cl
H
H
Cl

l.

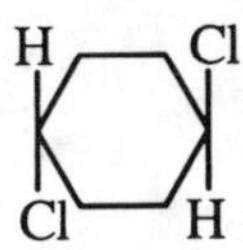
H
Cl
Cl
H

m.

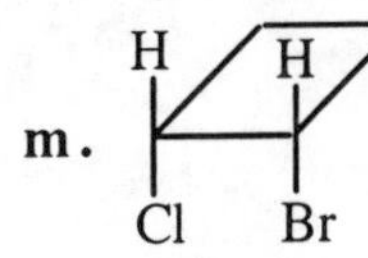
H
H
Cl
Br

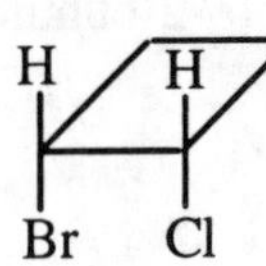
H
H
Br
Cl

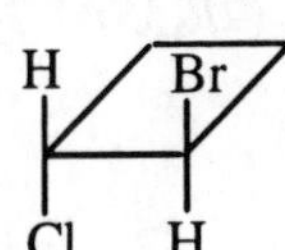
H
Br
Cl
H

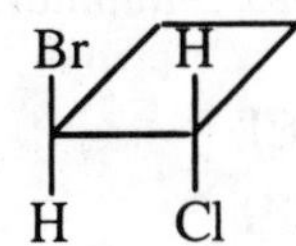
Br
H
H
Cl

n.

29.

a.

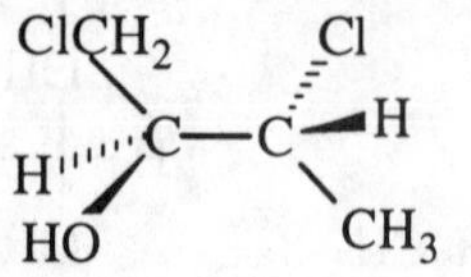

(2*S*,3*R*)-1,3-dichloro-2-butanol

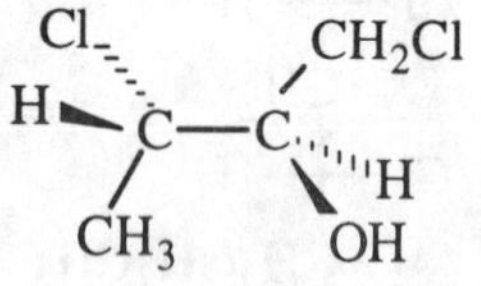

(2*R*,3*S*)-1,3-dichloro-2-butanol

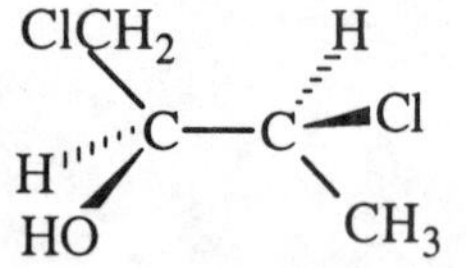

(2*S*,3*S*)-1,3-dichloro-2-butanol

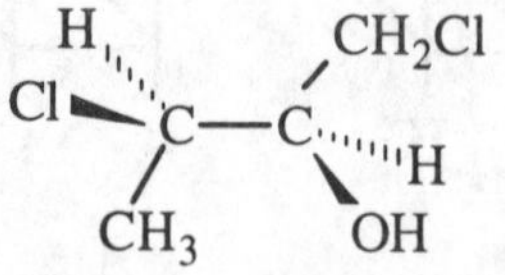

(2*R*,3*R*)-1,3-dichloro-2-butanol

b.

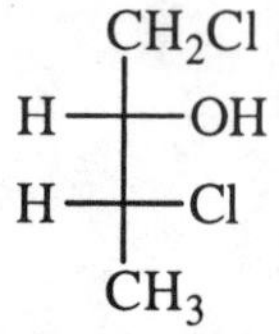

(2*S*,3*R*)-1,3-dichloro-2-butanol

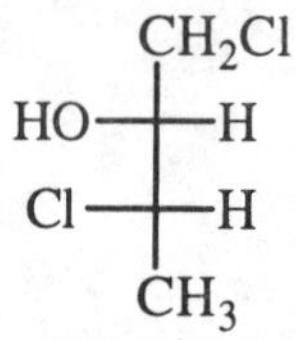

(2*R*,3*S*)-1,3-dichloro-2-butanol

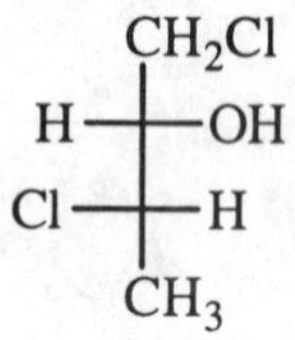

(2*S*,3*S*)-1,3-dichloro-2-butanol

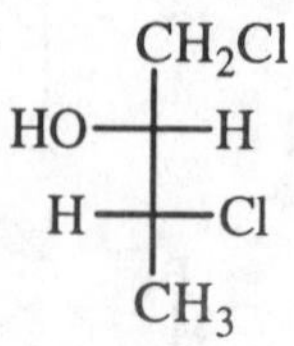

(2*R*,3*R*)-1,3-dichloro-2-butanol

30. Your answer might be correct yet not look like the answers shown here. If you can get the answer shown here by interchanging **two** pairs of groups bonded to a chiralty center, then your answer is correct. If you get the answer shown here by interchanging **one** pair of groups bonded to a chiralty center, then your answer is not correct.

a.

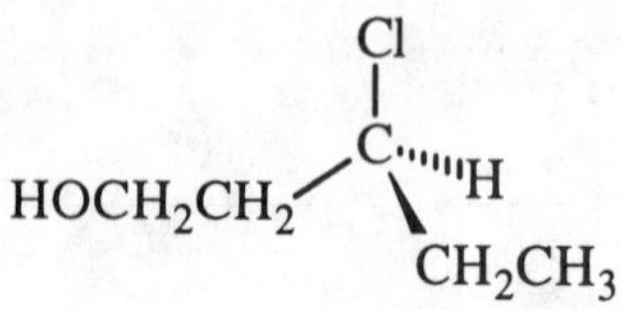

c.

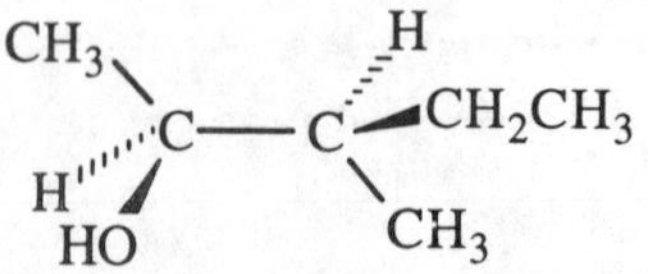

b.

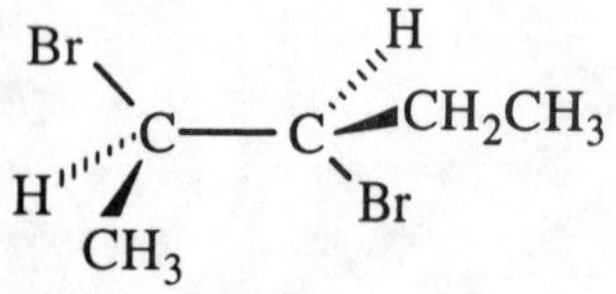

d.

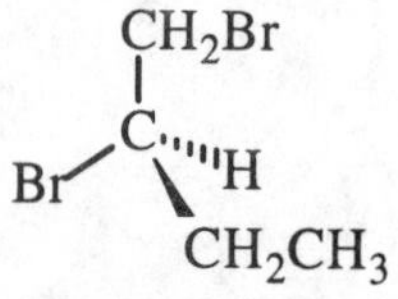

31. **a.** Because there are two chirality centers, there are four possible stereoisomers.

b.

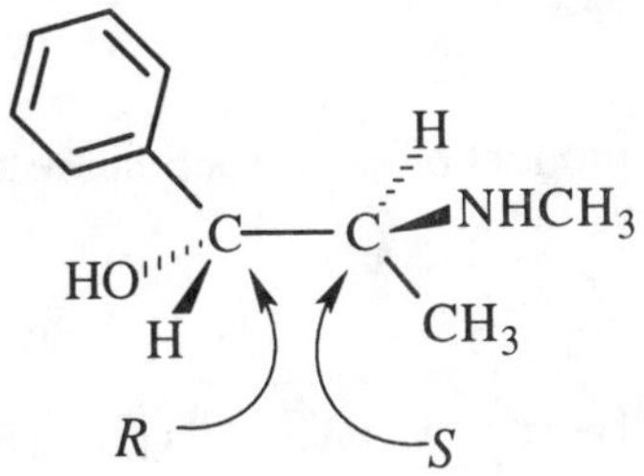

(Structure: phenyl–C(HO, H)–C(H, NHCH$_3$)–CH$_3$; arrows labeled *R* and *S*)

32. **a.** (2*R*,3*R*)-2,3-dichloropentane

b. (2*R*,3*R*)-2-bromo-3-chloropentane

c. (1*R*,3*S*)-1,3-cyclopentanediol

d. (3*R*,4*S*)-3-chloro-4-methylhexane

33. Solved in the text.

34. We see that the *S*-alkyl halide reacts with HO^- to form the *S*-alcohol. We were told that the product (the *S*-alcohol) is (+). We can, therefore, conclude that the (-) alcohol has the *R* configuration.

35. From the structures given on page 214 of the text, you can determine the configuration about the chirality center in each compound.

a. *R* **b.** *R* **c.** *S*

36. **a.** enantiotopic

b. homotopic

c. diastereotopic

d. homotopic

e. diastereotopic

f. diastereotopic (remember that cis-trans isomers are diastereomers)

37. Compound I has two stereoisomers because it has a chirality center.

Compound II has only one stereoisomer because it does not have a chirality center.

Compound III has a chirality center but, because of the nonbonding electrons, the two enantiomers rapidly interconvert, so it exists as a single compound.

38. **a.** no **b.** no **c.** no **d.** yes **e.** no **f.** no

39. Only the stereoisomers of the major product of each reaction are shown.

a. 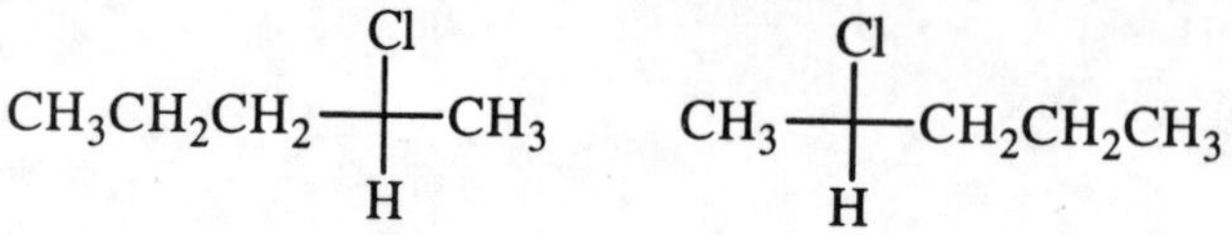 Fischer projections: $CH_3CH_2CH_2$–C(Cl)(H)–CH_3 and CH_3–C(Cl)(H)–$CH_2CH_2CH_3$ (Cl on top, H on bottom)

b. Fischer projections: $CH_3CH_2CH_2$–C(OH)(H)–CH_2CH_3 and CH_3CH_2–C(OH)(H)–$CH_2CH_2CH_3$ (OH on top, H on bottom)

c. 1-bromo-1-methylcyclopentane (Br, CH_3 on the same ring carbon) This compound has no stereoisomers.

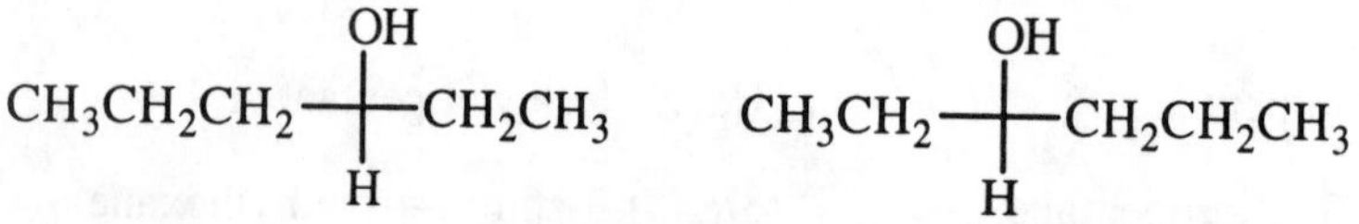

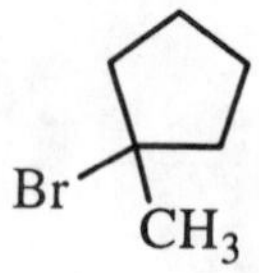

d. $CH_3C(CH_3)(Br)CH_2CH_3$ This compound has no stereoisomers.

e. 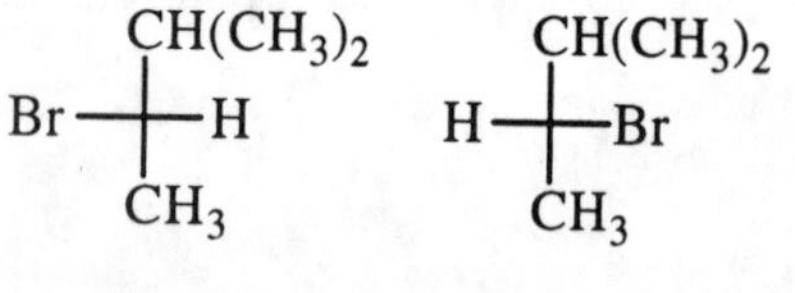 Fischer projections: top $CH(CH_3)_2$, bottom CH_3; Br–C–H and H–C–Br

f. 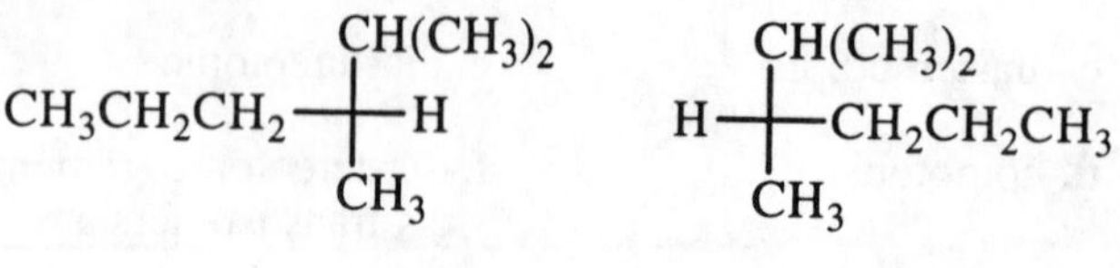 Fischer projections: top $CH(CH_3)_2$, bottom CH_3; $CH_3CH_2CH_2$–C–H and H–C–$CH_2CH_2CH_3$

40.

a. 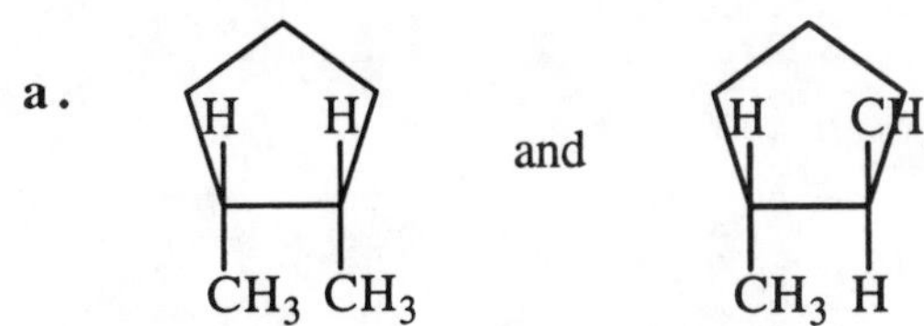

b. 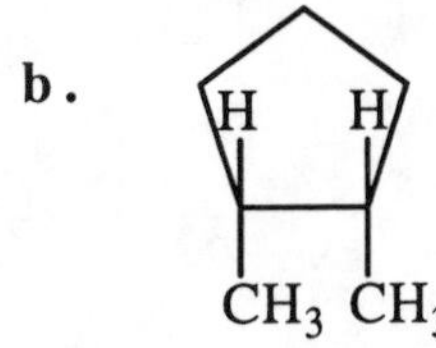

Hydrogen will be more likely to add from above the plane of the double bond (where the hydrogen substituent is) than from below the plane of the double bond (where the methyl substituent is) because hydrogen provides less steric hindrance than a methyl group.

41.

a. OH

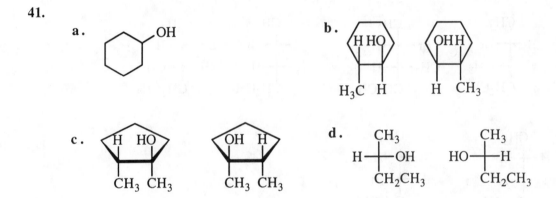

b. H HO / H_3C H   OH H / H CH_3

c. H HO / CH_3 CH_3 OH H / CH_3 CH_3

d. CH_3 / H—OH / CH_2CH_3 CH_3 / HO—H / CH_2CH_3

42.

$$CH_2{=}C(CH_2CH_3)CH_2CH_2CH_3 \xrightarrow{Br_2} BrCH_2\overset{*}{C}(CH_2CH_3)(Br)CH_2CH_2CH_3$$

$$CH_2{=}C(CH_2CH_3)CH_2CH_2CH_3 \xrightarrow[Pt]{H_2} CH_3\overset{*}{C}H(CH_2CH_3)CH_2CH_2CH_3$$

$$CH_2{=}C(CH_2CH_3)CH_2CH_2CH_3 \xrightarrow[2.\,HO^-,\ H_2O_2]{1.\,BH_3} HOCH_2\overset{*}{C}H(CH_2CH_3)CH_2CH_2CH_3$$

(* indicates a chirality center)

Each of the reactions forms a compound with one chirality center from a compound with no chirality centers. Therefore, each of the products is a racemic mixture.

43. If the reaction formed a carbocation intermediate, both the erythro pair of enantiomers and the threo pair of enantiomers would be formed because both syn and anti addition could occur.

44.

a.

CH_3
H—Br
CH_2CH_3

CH_3
Br—H
CH_2CH_3

b.

CH_2CH_3
H—Br
$CH_2CH_2CH_3$

CH_2CH_3
Br—H
$CH_2CH_2CH_3$

c.

CH_3
$CH_3CH_2CCH_2CH_3$
Br

d.

CH_3
H—Br
H—CH_3
CH_2CH_3

CH_3
Br—H
CH_3—H
CH_2CH_3

CH_3
H—Br
CH_3—H
CH_2CH_3

CH_3
Br—H
H—CH_3
CH_2CH_3

e.

CH_3
H—Br
Br—H
CH_2CH_3

CH_3
Br—H
H—Br
CH_2CH_3

f.

CH_2Br
H—Br
$CH_2CH_2CH_2CH_3$

CH_2Br
Br—H
$CH_2CH_2CH_2CH_3$

45. Two different bromonium ions are formed because Br^+ can add to the double bond either from the top of the plane or from the bottom of the plane defined by the alkene, and the two bromonium ions are formed in equal amounts. Attacking the less hindered carbon of one bromonium ion forms one stereoisomer, while attacking the less hindered carbon of the other bromonium ion forms the other stereoisomer.

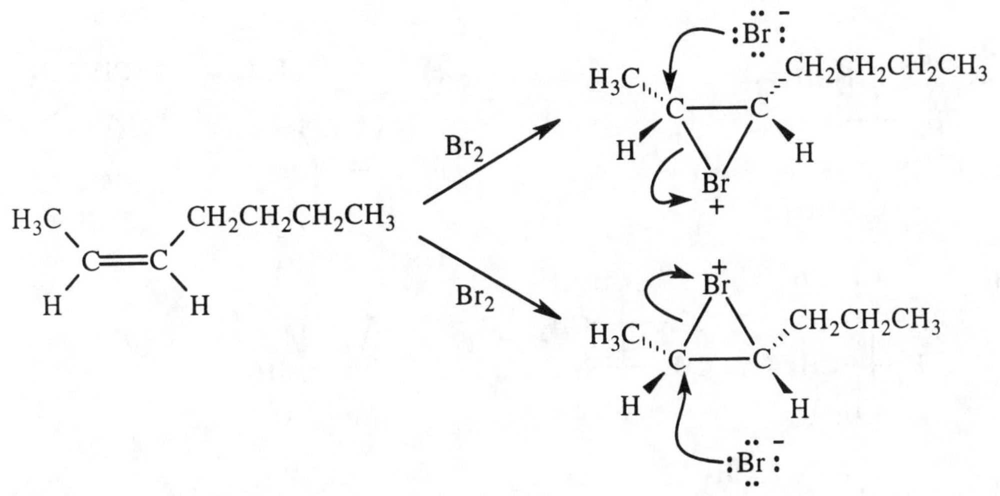

46.

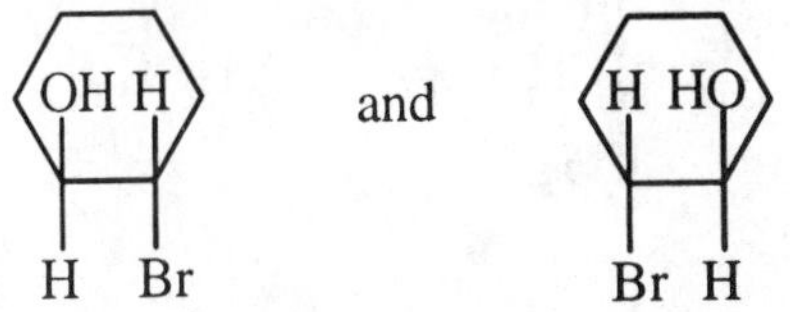

mechanism of the reaction

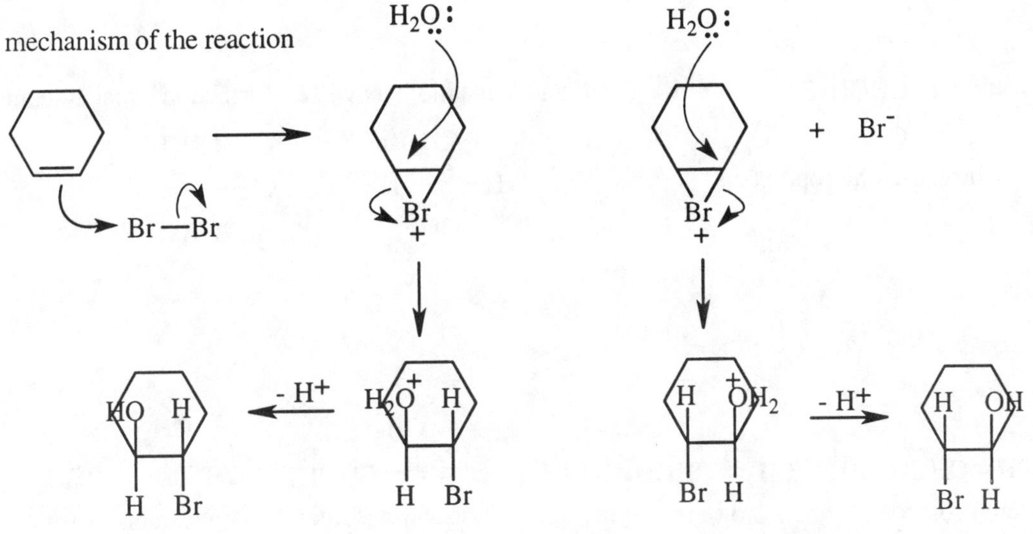

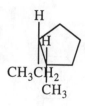

47.

a. Fischer projection: CH_2CH_3 (top); CH_3—C—Br; CH_3—C—Br; CH_2CH_3 (bottom)

b. CH_2CH_3; CH_3—C—H; H—C—CH_3; CH_2CH_3 and CH_2CH_3; H—C—CH_3; CH_3—C—H; CH_2CH_3

c. Cyclopentane structures with CH_3, Br, Br, CH_3 substituents

d. Cyclopentane structures with CH_2CH_3, Br, Br, CH_3 substituents

e. Cyclopentane structure with H, H, CH_3, CH_3 substituents

f. Cyclopentane structures with H, H, CH_2CH_3, CH_3 substituents (H, H, CH_3CH_2, CH_3)

48.

a. CH_3CHCH_2Br with Cl on C-2

1-bromo-2-chloropropane

b. The *R* and *S* enantiomers will be formed in equal amounts.

CH_2Br; H—C—Cl; CH_3 and CH_2Br; Cl—C—H; CH_3

49.

$CH_3CH{=}CHCH_2CH_3$
2 stereoisomers
[cis and trans]

$CH_2{=}CHCH_2CH_2CH_3$
no stereoisomers

$CH_3C(CH_3){=}CHCH_3$
no stereoisomers

$CH_3CH(CH_3)CH{=}CH_2$
no stereoisomers

$CH_3CH_2C(CH_3){=}CH_2$
no stereoisomers

cyclopentane
no stereoisomers

methylcyclobutane (CH_3)
no stereoisomers

1,1-dimethylcyclopropane (CH_3, CH_3)
no stereoisomers

1,2-dimethylcyclopropane (H_3C, CH_3)
3 stereoisomers
[Cis is a meso compound.]
[Trans is a pair of enantiomers.]

50.

a. 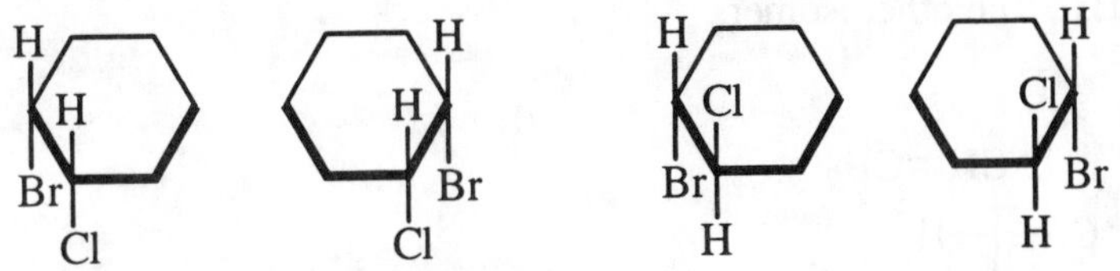 H, H, Br, Cl   H, H, Br, Cl   H, Cl, Br, H   H, Cl, Br, H

b.

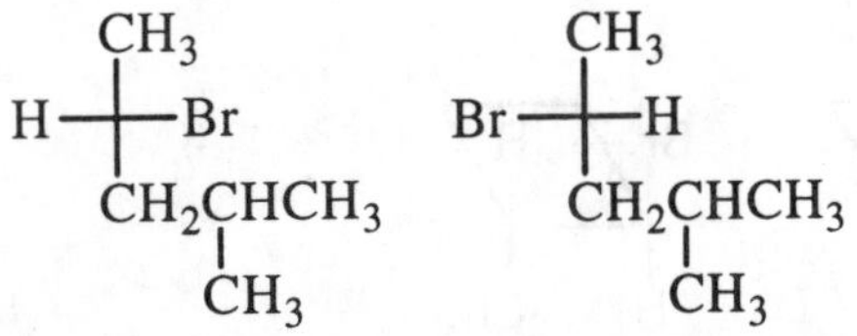

CH_3
H—C—Br
CH_2CHCH_3
CH_3

CH_3
Br—C—H
CH_2CHCH_3
CH_3

c. H, H, Cl, Cl   Cl, H, H, Cl   Cl, H, H, Cl

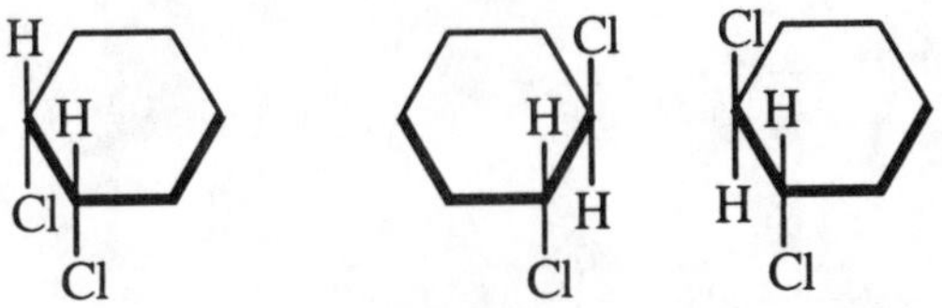

d.

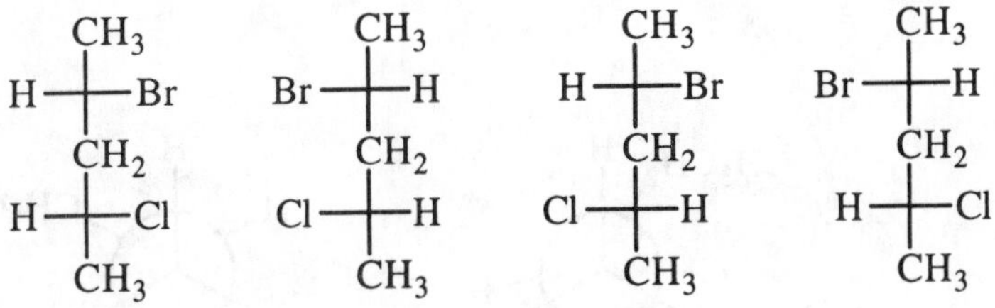

CH_3
H—C—Br
CH_2
H—C—Cl
CH_3

CH_3
Br—C—H
CH_2
Cl—C—H
CH_3

CH_3
H—C—Br
CH_2
Cl—C—H
CH_3

CH_3
Br—C—H
CH_2
H—C—Cl
CH_3

e. 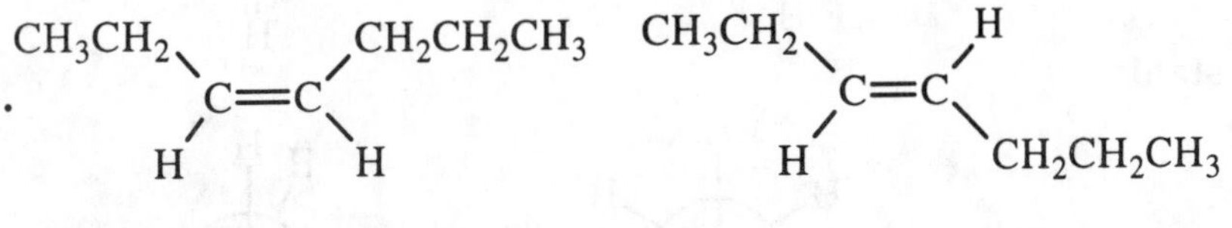 CH_3CH_2(H)C=C(H)$CH_2CH_2CH_3$ CH_3CH_2(H)C=C(H)$CH_2CH_2CH_3$

f. 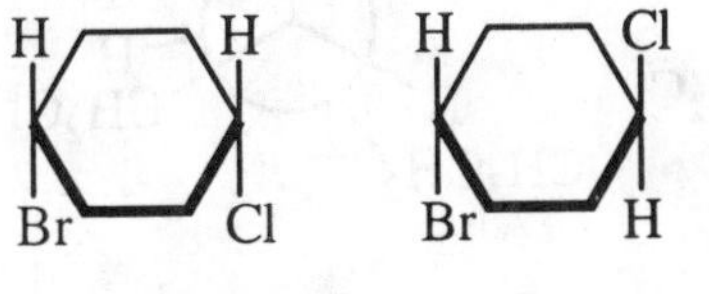 H, H, Br, Cl   H, Cl, Br, H

g. H, H, CH_3, CH_3 H, CH_3, CH_3, H CH_3, H, H, CH_3

h. CH_3, CH_3, C, H, Br, C=C, H, H CH_3, H, C, Br, CH_3, C=C, H, H CH_3, H, C, H, Br, C=C, CH_3, H CH_3, H, C, Br, H, C=C, H, CH_3

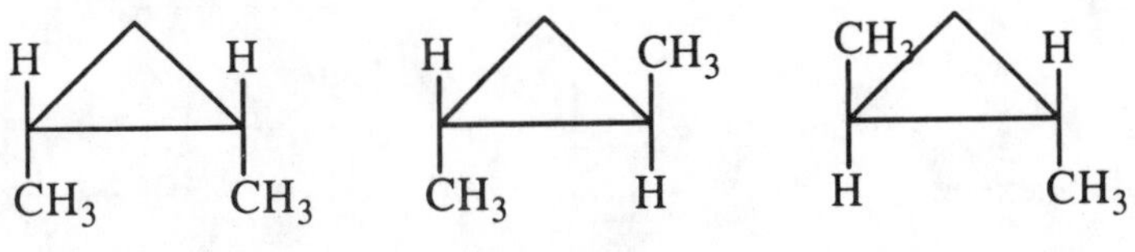

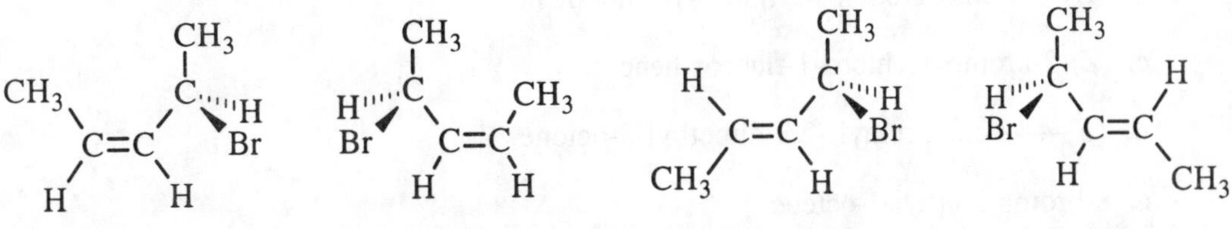

i. 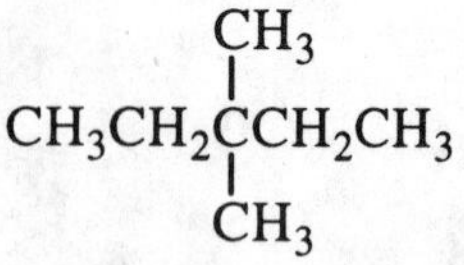 $CH_3CH_2C(CH_3)_2CH_2CH_3$ no other isomers

j. 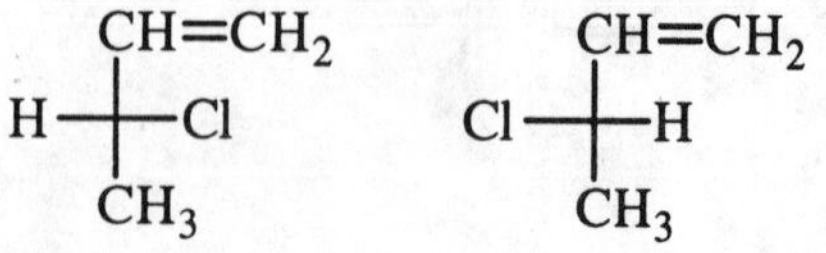 (two Fischer projections: $CH=CH_2$ top, CH_3 bottom; H and Cl on the horizontal, swapped between the two)

k.

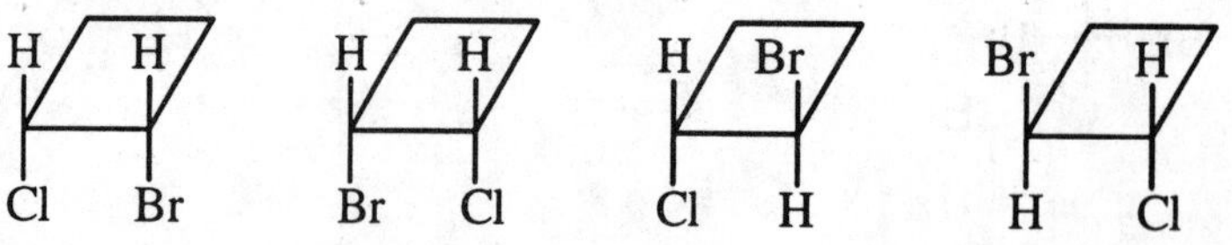

l.

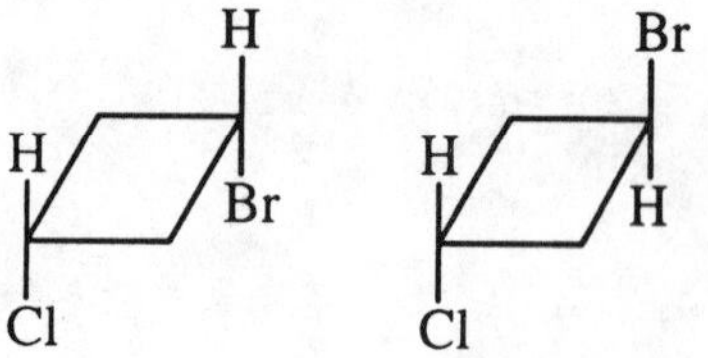

51.

most stable

least stable

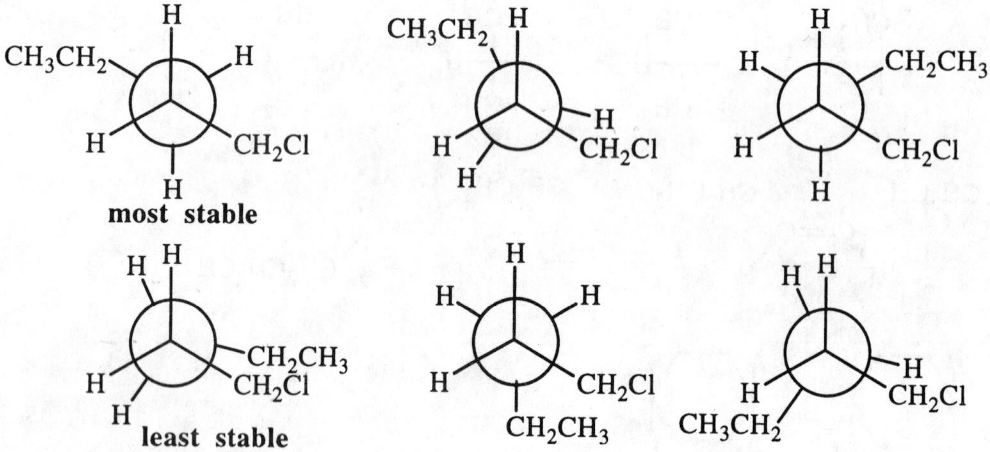

52. **a.** (*S*)-2-ethyl-4-methyl-1-hexene

b. (*E*)-1-bromo-2-chloro-2-fluoro-1-iodoethene

c. (*Z*)-2-bromo-1-chloro-1-fluoroethene

d. (*E*)-4-(2-chloroethyl)-2,3-dimethyl-3-octene

e. 8-bromo-2-ethyl-1-octene

f. (*E*)-1,3-dibromo-4,7-dimethyl-3-octene

g. (*S*)-2-methyl-1,2,5-pentanetriol

h. (2*R*,3*R*)-3-chloro-2-pentanol

53. Mevacor has eight chirality centers.

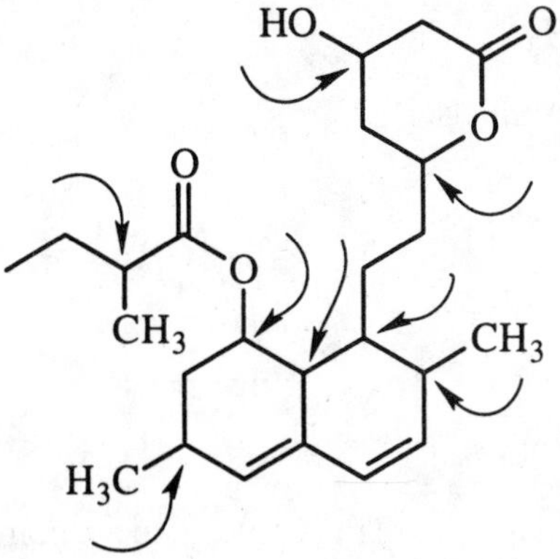

54. **a.** enantiomers (By naming them, you see that one is *S* and the other is *R*.)

b. constitutional isomers (One is 3-chloro-2-pentanol and the other is 2-chloro-3-pentanol.)

c. enantiomers (They have chirality centers and are mirror images.)

d. constitutional isomers

e. diastereomers

f. enantiomers

g. identical

h. enantiomers

i. constitutional isomers

j. enantiomers

k. identical

l. diastereomers

m. identical

n. identical

o. diastereomers

p. constitutional isomers

55. **a.**

1.

$$\begin{array}{c}CH_3\\ H-\!\!+\!\!-Cl\\ CH_2CH_3\end{array}\qquad\begin{array}{c}CH_3\\ Cl-\!\!+\!\!-H\\ CH_2CH_3\end{array}$$

Both cis and trans give these products.

2.

$$\begin{array}{c}CH_3\\ H-\!\!+\!\!-OH\\ CH_2CH_3\end{array}\qquad\begin{array}{c}CH_3\\ HO-\!\!+\!\!-H\\ CH_2CH_3\end{array}$$

Both cis and trans give these products.

3.

$$\begin{array}{c}CH_3\\ H-\!\!+\!\!-Br\\ CH_2CH_3\end{array}\qquad\begin{array}{c}CH_3\\ Br-\!\!+\!\!-H\\ CH_2CH_3\end{array}$$

Both cis and trans give these products.

4.

$$\begin{array}{c}CH_3\\ H-\!\!+\!\!-Br\\ Br-\!\!+\!\!-H\\ CH_3\end{array}\qquad\begin{array}{c}CH_3\\ Br-\!\!+\!\!-H\\ H-\!\!+\!\!-Br\\ CH_3\end{array}\qquad\qquad\begin{array}{c}CH_3\\ H-\!\!+\!\!-Br\\ H-\!\!+\!\!-Br\\ CH_3\end{array}$$

Cis gives the threo pair. Trans gives a meso compound.

5.

$$\begin{array}{c}CH_3\\ H-\!\!+\!\!-OH\\ Br-\!\!+\!\!-H\\ CH_3\end{array}\qquad\begin{array}{c}CH_3\\ HO-\!\!+\!\!-H\\ H-\!\!+\!\!-Br\\ CH_3\end{array}\qquad\qquad\begin{array}{c}CH_3\\ H-\!\!+\!\!-OH\\ H-\!\!+\!\!-Br\\ CH_3\end{array}\qquad\begin{array}{c}CH_3\\ HO-\!\!+\!\!-H\\ Br-\!\!+\!\!-H\\ CH_3\end{array}$$

Cis gives the threo pair. Trans gives the erythro pair.

6. $CH_3CH_2CH_2CH_3$

Both cis and trans give this product.

7.

$$\begin{array}{c}CH_3\\ H-\!\!+\!\!-OH\\ CH_2CH_3\end{array}\qquad\begin{array}{c}CH_3\\ HO-\!\!+\!\!-H\\ CH_2CH_3\end{array}$$

Both cis and trans give these products.

8.

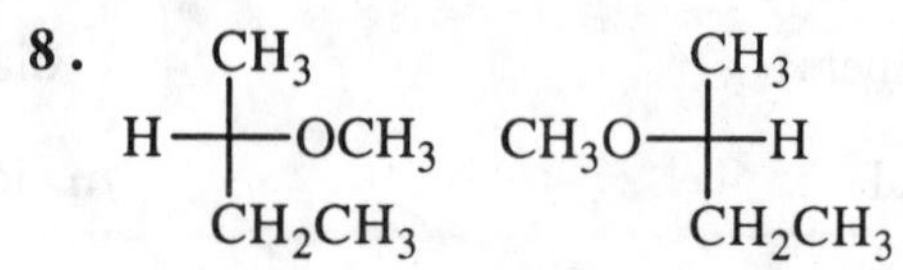

Both cis and trans give these products.

b. For the cis and trans isomers to form different products, the reaction must form two new chirality centers in the product.
Therefore, the cis and trans alkenes form different products when they react with Br_2 in CH_2Cl_2 and when they react with Br_2 in H_2O.

56. c — does not have stereoisomers; it is, therefore, achiral.

e and **h** — have 2 stereoisomers (cis and trans); both are achiral.

a, d, f, i, and **j** — have 3 stereoisomers; 2 are chiral and 1 is achiral.

b and **g** — have 4 stereoisomers, all of which are chiral.

57. **a.** (*R*)-citric acid (^{14}C has a higher priority than ^{12}C)

b. The reaction is catalyzed by an enzyme. Only one stereoisomer is formed in an enzyme-catalyzed reaction because an enzyme has a chiral binding sitewhich allows reagents to be delived to only one side of the functional group of the compound.

c. The product of the reaction will be achiral because if it doesn't have a ^{14}C, the two CH_2COOH groups will be identical so it will not have a chirality center.

58.

a. 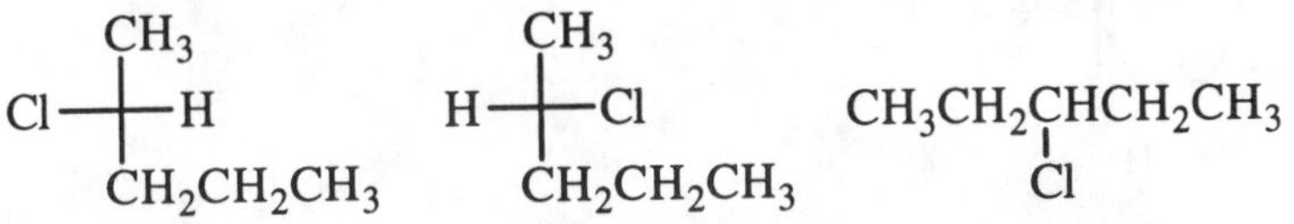

b. 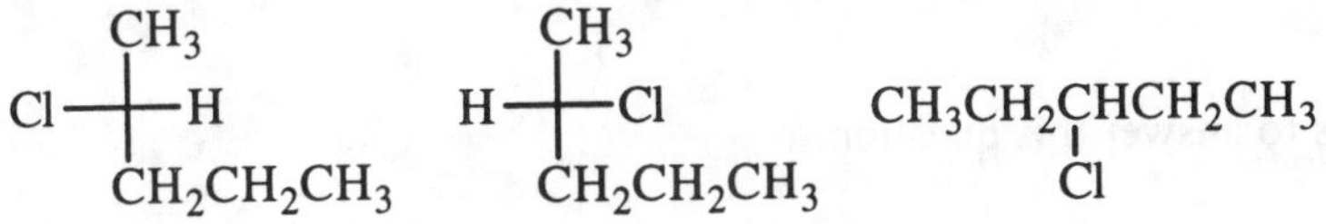

c. 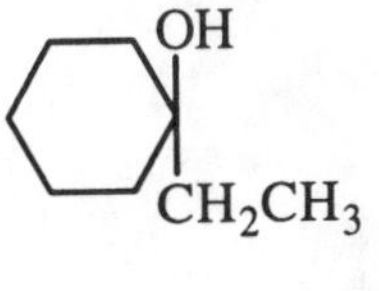

d. 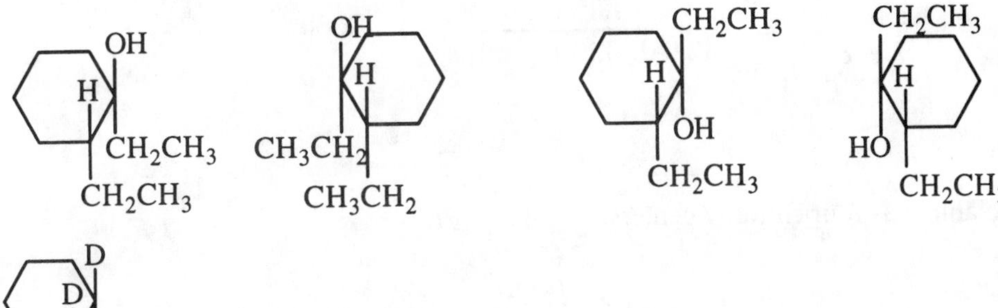

e. 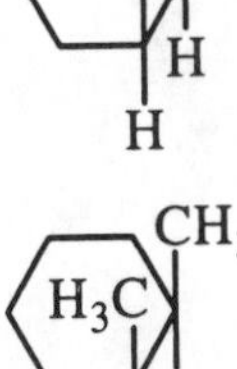

f. 

g.

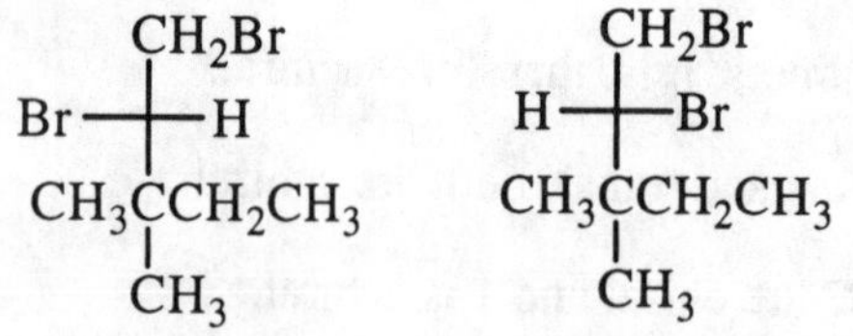

```
      CH2Br                 CH2Br
  Br──┼──H              H──┼──Br
   CH3CCH2CH3           CH3CCH2CH3
      │                    │
      CH3                  CH3
```

h.

```
        CH2CH3                   CH2CH3
   H ───┼─── CH3          CH3 ───┼─── H
 CH3 ───┼─── H              H ───┼─── CH3
        CH2CH2CH3                CH2CH2CH3
```

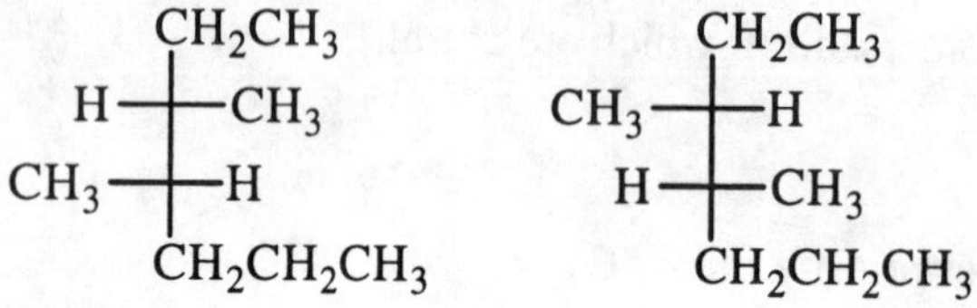

i.

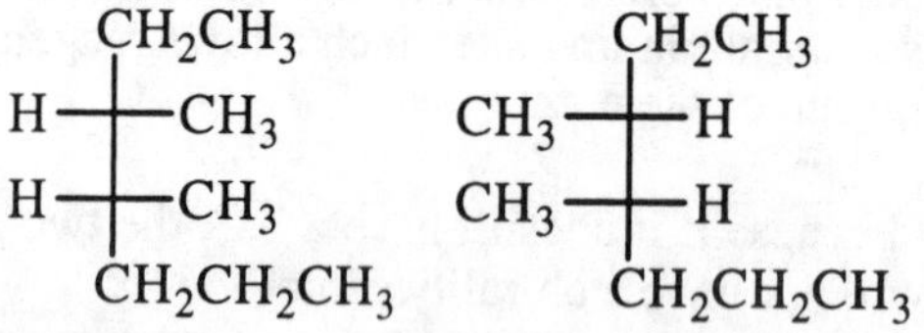

```
      CH2CH3                 CH2CH3
  H ──┼── CH3          CH3 ──┼── H
  H ──┼── CH3          CH3 ──┼── H
      CH2CH2CH3              CH2CH2CH3
```

j. 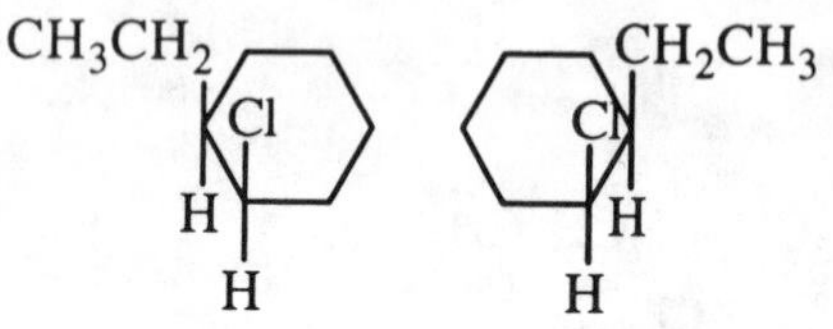 CH_3CH_2 Cl H H CH_2CH_3 Cl H H

59. Use molecular models to answer this question.

a. *R* **b.** *S* **c.** *S* **d.** *R* **e.** *R* **f.** *S*

60.

$$[\alpha] = \frac{\alpha}{l \times c} = \frac{-1.8°}{[2.0 \text{ dm}]\,[0.15 \text{g/ml}]} = -6.0°$$

61. Butaclamol has four chirality centers.

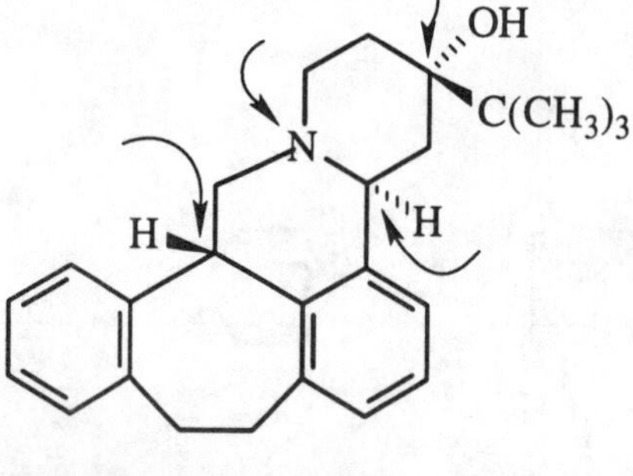

OH
$C(CH_3)_3$
N
H
H

62. **a**, **b**, **c**, and **g** are chiral. If a plane were to bisect **d**, **e**, or **f**, the left-hand side would be the mirror image of the right-hand side. These, therefore, are achiral.

63. *R* and *S* are related to (+) and (-) in that if one configuration (say, *R*) is (+), the other one is (-).

Because some compounds with the *R* configuration are (+) and some are (-), there is no way to determine whether a particular *R* enantiomer is (+) or (-) without putting the compound in a polarimeter or going to the library to see if someone else has previously determined how it rotates light.

64.

a.

CH_3		CH_3
H—+—Br		Br—+—H
Br—+—H		H—+—Br
CH_2CH_3		CH_2CH_3

b.

CH_3		CH_3
H—+—Br		Br—+—H
H—+—Br		Br—+—H
CH_2CH_3		CH_2CH_3

c.

CH_3		CH_3
H—+—Cl		Cl—+—H
CH_2CH_3		CH_2CH_3

d. $BrCH_2CH_2CH_2CH_3$

e.

CH_2CH_3
H—+—Br
H—+—Br
CH_2CH_3

f.

CH_2CH_3		CH_2CH_3
H—+—Br		Br—+—H
Br—+—H		H—+—Br
CH_2CH_3		CH_2CH_3

g.

First structure:
CH_2CH_3
Br — C — CH_3
$CHCH_3$
CH_3

Second structure:
CH_2CH_3
CH_3 — C — Br
$CHCH_3$
CH_3

h.

First structure:
CH_3
H — C — Br
CH_2CH_3

Second structure:
CH_3
Br — C — H
CH_2CH_3

i.

CH_3
H — C — Cl
H — C — Cl
CH_3

j.

First structure:
CH_3
H — C — Cl
Cl — C — H
CH_3

Second structure:
CH_3
Cl — C — H
H — C — Cl
CH_3

k.

CH_2CH_3
H — C — CH_3
H — C — CH_3
CH_2CH_3

l.

First structure:
CH_2CH_3
H — C — CH_3
CH_3 — C — H
CH_2CH_3

Second structure:
CH_2CH_3
CH_3 — C — H
H — C — CH_3
CH_2CH_3

65. **a.** The compound has two chirality centers and, therefore, a maximum of four stereoisomers, since $2^2 = 4$.

Isomer 1:
CH_2OH
H — C — OH
H — C — OH
H — C — OH
CH_2OH

Isomer 2:
CH_2OH
H — C — OH
HO — C — H
H — C — OH
CH_2OH

Isomer 3:
CH_2OH
H — C — OH
H — C — OH
HO — C — H
CH_2OH

Isomer 4:
CH_2OH
HO — C — H
HO — C — H
H — C — OH
CH_2OH

b. The first two isomers are optically inactive because they are meso compounds.

66.

a. R

CH_2CH_2Br

$CH_3CH_2CH_2$—C—H

Br

b. S

HC=O

H—C—OH

HO—C—H S

Br

c. S

CH_3

H—C—Br

H—C—Br R

CH_2CH_3

67.

a. and **b.**

H

CH_2CH_3

ethylcyclobutane

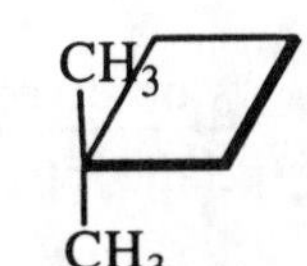

1,1-dimethylcyclobutane

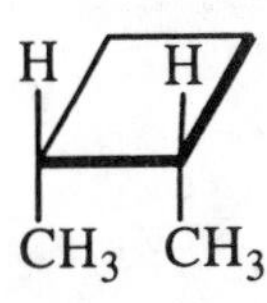

cis-1,2-dimethylcyclobutane

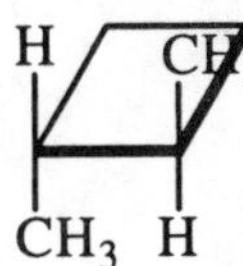

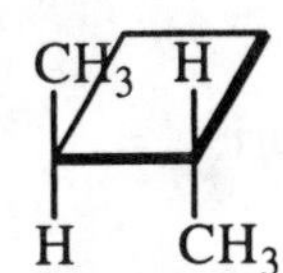

trans-1,2-dimethylcyclobutane

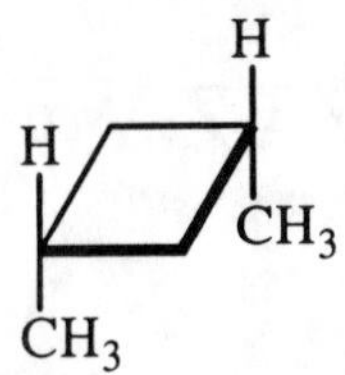

cis-1,3-dimethylcyclobutane

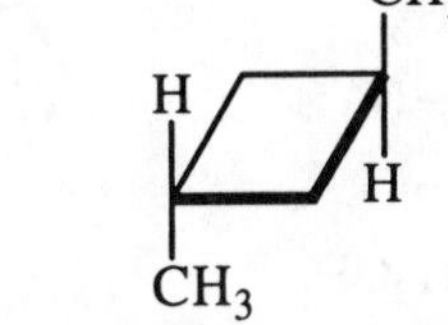

trans-1,3-dimethylcyclobutane

c. **1.** ethylcyclobutane
1,1-dimethylcyclobutane
1,2-dimethylcyclobutane
1,3-dimethylcyclobutane

2. **a.** the three isomers of 1,2-dimethylcyclobutane
b. the two isomers of 1,3-dimethylcyclobutane

3. **a.** 1,2-dimethylcyclobutane
b. 1,3-dimethylcyclobutane

4. the two trans stereoisomers of 1,2-dimethylcyclobutane

5. all the isomers except the two trans stereoisomers of 1,2-dimethylcyclobutane

6. *cis*-1,2-dimethylcyclobutane
(Note: *cis*-1,3-dimethylcyclobutane is not a meso compound, because it does not have chirality centers.)

7. the two trans stereoisomers of 1,2-dimethylcyclobutane

8. a. *cis*-1,3-dimethylcyclobutane and *trans*-1,3-dimethylcyclobutane
b. *cis*-1,2-dimethylcyclobutane and either of the enantiomers of *trans*-1,2-dimethylcyclobutane

68.

$$\text{observed specific rotation} = \frac{\text{observed rotation}}{\text{concentration x length}} = \frac{-6.52°}{0.187 \times 1} = -34.9°$$

$$\%\ \text{optical purity} = \frac{\text{observed specific rotation}}{\text{specific rotation of the pure enantiomer}} \times 100$$

$$\%\ \text{optical purity} = \frac{-34.9°}{-39.0°} \times 100$$

$$\%\ \text{optical purity} = 89.5\%$$

69.

a. Br / H—C—Cl / $CH_2CH_2CH_3$

b. CH_3 / Cl—C—H / H—C—Cl / CH_2CH_3

c. H H / H_3C CH_3

d. H Br / Br H or Br H / H Br

e. CH_2CH_3 / H—C—CH_3 / H—C—CH_3 / H—C—CH_3 / CH_2CH_3 CH_2CH_3 / CH_3—C—H / H—C—CH_3 / CH_3—C—H / CH_2CH_3

70. In the transition state for amine inversion, the nitrogen atom is sp^2 hybridized, which means it has bond angles of 120°. A nitrogen atom in a three-membered ring cannot achieve a 120° bond angle, so the amine inversion that would interconvert the enantiomers cannot occur. Therefore, the eneantiomers can be separated.

71. The compound shown below would not be obtained, because none of the bonds attached to C-2 were broken during the reaction. Therefore, the configuration at C-2 cannot change.

$$\begin{array}{c} CH_3 \\ Br-\!\!\!+\!\!\!-H \\ H-\!\!\!+\!\!\!-Br \\ CH_2CH_3 \end{array}$$

72. The fact that the optical purity is 72% means that there is 72% enantiomeric excess of the *S* isomer and 28% racemic mixture. Therefore, the actual amount of the *S* isomer in the sample is 72% + 1/2(28%) = 86%. So the amount of the *R* isomer in the sample is 100% - 86% = 14%.

73.

$$\xrightarrow[\text{peroxide}]{HBr} \begin{array}{c} CH_3 \\ H-\!\!\!+\!\!\!-Br \\ H-\!\!\!+\!\!\!-CH_3 \\ CH(CH_3)_2 \end{array} \quad \begin{array}{c} CH_3 \\ Br-\!\!\!+\!\!\!-H \\ CH_3-\!\!\!+\!\!\!-H \\ CH(CH_3)_2 \end{array} \quad \begin{array}{c} CH_3 \\ H-\!\!\!+\!\!\!-Br \\ CH_3-\!\!\!+\!\!\!-H \\ CH(CH_3)_2 \end{array} \quad \begin{array}{c} CH_3 \\ Br-\!\!\!+\!\!\!-H \\ H-\!\!\!+\!\!\!-CH_3 \\ CH(CH_3)_2 \end{array}$$

$$\xrightarrow[CH_3OH]{Br_2} \begin{array}{c} H \\ Br-\!\!\!+\!\!\!-CH_3 \\ CH_3O-\!\!\!+\!\!\!-CH_3 \\ CH(CH_3)_2 \end{array} \quad \begin{array}{c} H \\ CH_3-\!\!\!+\!\!\!-Br \\ CH_3-\!\!\!+\!\!\!-OCH_3 \\ CH(CH_3)_2 \end{array}$$

$$\xrightarrow[H_2O]{H^+} \begin{array}{c} OH \\ CH_3CH_2-\!\!\!+\!\!\!-CH(CH_3)_2 \\ CH_3 \end{array} \quad \begin{array}{c} OH \\ (CH_3)_2CH-\!\!\!+\!\!\!-CH_2CH_3 \\ CH_3 \end{array}$$

$$\xrightarrow[Pt]{H_2} \begin{array}{c} H \\ CH_3CH_2-\!\!\!+\!\!\!-CH(CH_3)_2 \\ CH_3 \end{array} \quad \begin{array}{c} H \\ (CH_3)_2CH-\!\!\!+\!\!\!-CH_2CH_3 \\ CH_3 \end{array}$$

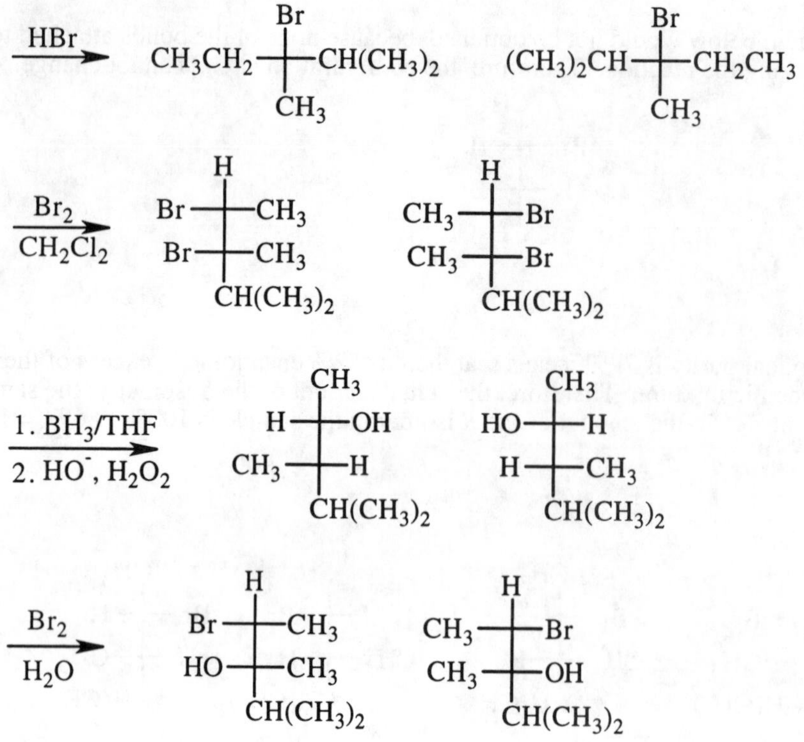

74.

a.

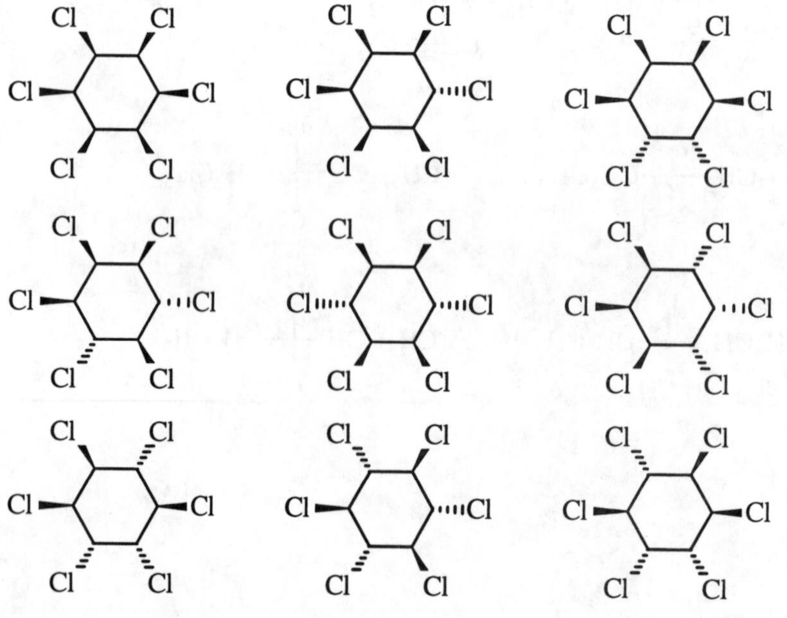

b. 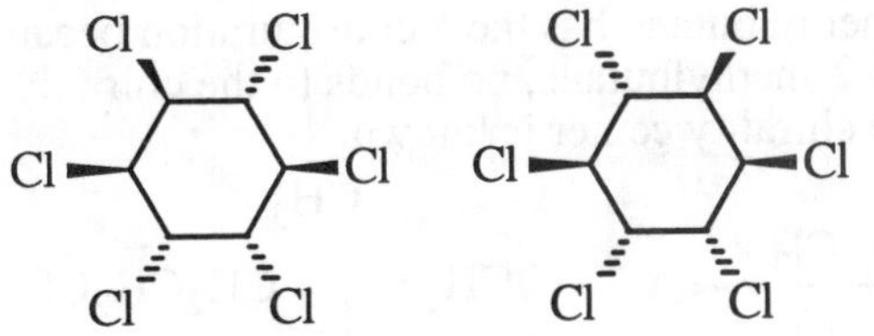 This is a pair of enantiomers because they are nonsuperimposable mirror images.

c. 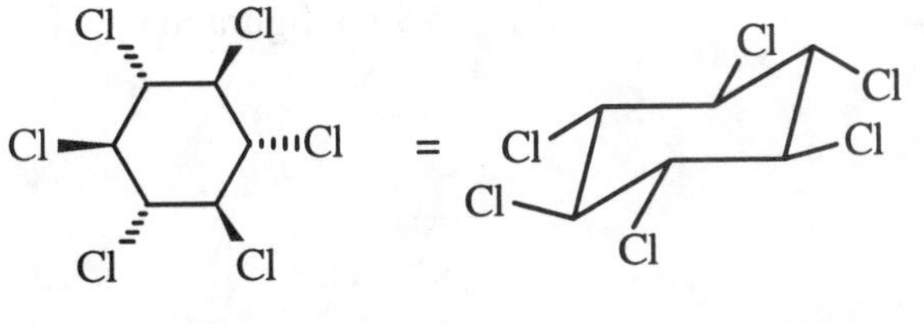 This is the most stable isomer because, since the chloro substituents are all trans to each other, they can all be in the more stable equatorial position. (There is more room for a substituent in the equatorial position.)

75. Yes.

76. A six-membered ring is too small to accommodate a trans double bond, but a ten-membered ring is large enough to accommodate a trans double bond.

77. Because fumarate is the trans isomer and it forms an erythro product, the enzyme must catalyze an anti addition of D_2O.

78.

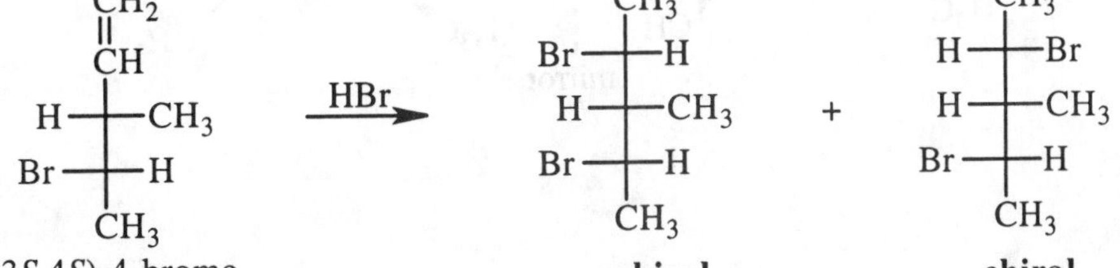

$\xrightarrow{HBr}$

(3*S*,4*S*)-4-bromo-3-methyl-1-pentene

achiral + **chiral**

The configurations at C-3 and C-4 do not change, because no bonds to these carbons are broken. The new chirality center at C-2 can have either the *R* or the *S* configuration.
The product with C-2 in the *R* configuration has a plane of symmetry; therefore, it is achiral.
The product with C-2 in the *S* configuration does not have a plane of symmetry; therefore, it is chiral.
(Notice that the secondary carbocation that is formed by the addition of a proton to the double bond does not rearrange to a tertiary carbocation. That is because the positively charged carbon of the tertiary carbocation that would be formed is adjacent to a carbon that is bonded to an electron-withdrawing bromine atom. This decreases the stability of the tertiary carbocation, making it less stable than the originally formed secondary carvocation)

79. We can determine that (-)-1,4-dichloro-2-methylbutane has the *S* configuration because in the process of forming it from (*S*)-(+)-1-chloro-2-methylbutane, no bonds to the chirality center are broken. Therefore, the configuration of the chirality center is known.

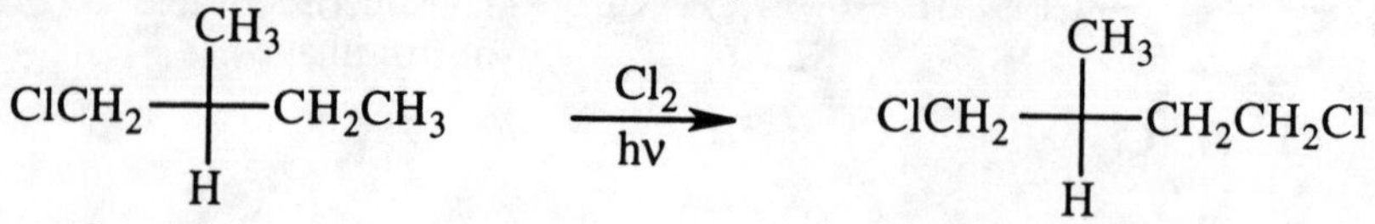

$$ClCH_2-C(CH_3)(H)-CH_2CH_3 \xrightarrow[h\nu]{Cl_2} ClCH_2-C(CH_3)(H)-CH_2CH_2Cl$$

S-configuration *S*-configuration

80.

a. 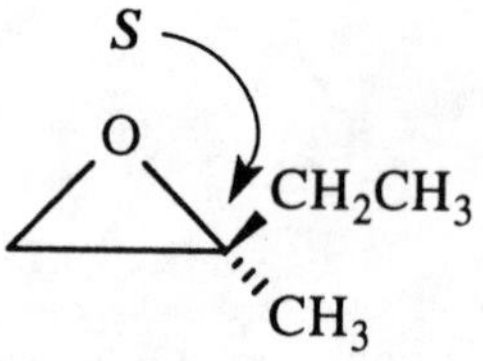 *S* — CH_2CH_3, CH_3, O

b. 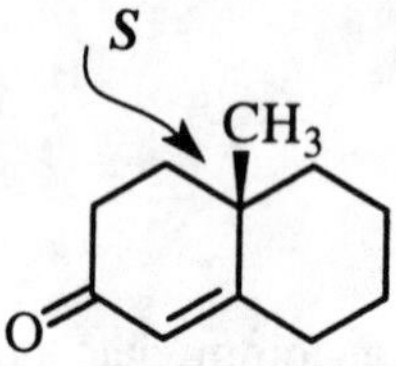 *S* — CH_3, O

c. 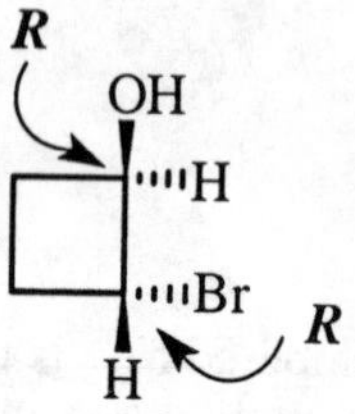 *R* — OH, H; *R* — Br, H

81. a. The compounds do not have any chirality centers.

b. Yes, it is chiral. Because of its unusual geometry, it is a chiral molecule, even though it has no chirality centers, since it cannot be superimposed on its mirror image. This may be easier to understand if you build models.

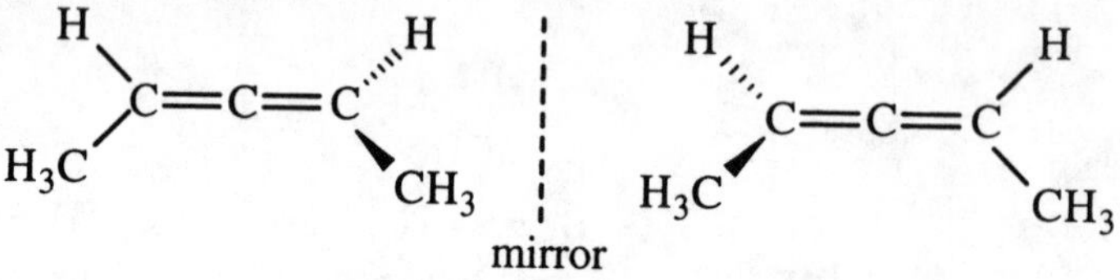

H, H_3C, C=C=C, H, CH_3 | H, H_3C, C=C=C, H, CH_3

mirror

Chapter 4 Practice Test

1. Are the following compounds identical or a pair of enantiomers?

H — top; CH_3CH_2 — left; CH_2OH — right; CH_3 — bottom

CH_3 — top; H — left; CH_2CH_3 — right; CH_2OH — bottom

2. 30 ml of a solution containing 2.4 g of a compound rotates the plane of polarized light -0.48° in a polarimeter with a 2 decimeter sample tube. What is the specific rotation of the compound?

3. Which are meso compounds?

CH_3	CH_3	CH_3	CH_3	CH_3
H—C—Cl	H—C—Cl	Br—C—Cl	H—C—Cl	Cl—C—H
H—C—Cl	Cl—C—H	Br—C—Cl	H—C—Cl	H—C—Cl
CH_2CH_3	CH_2CH_3	CH_2Cl	CH_3	CH_3

4. Draw the constitutional isomers of C_4H_9Cl.

5. Give the stereochemistry of the products that would be obtained from each of the following reactions.

a. 1-butene + HCl

b. 2-pentene + HBr

c. *trans*-3-hexene + Br_2

d. *trans*-3-heptene + Br_2

6. *S*-(-)-2-methyl-1-butanol can be oxidized to (+)-2-methylbutanoic acid without breaking any of the bonds to the chirality center. What is the configuration of (-)-2-methylbutanoic acid?

CH_3 — top; H — left; CH_2OH — right; CH_2CH_3 — bottom

S-(-)-2-methyl-1-butanol

CH_3 — top; H — left; COOH — right; CH_2CH_3 — bottom

(+)-2-methylbutanoic acid

7. What stereoisomers would be obtained from each of the following reactions?

a. 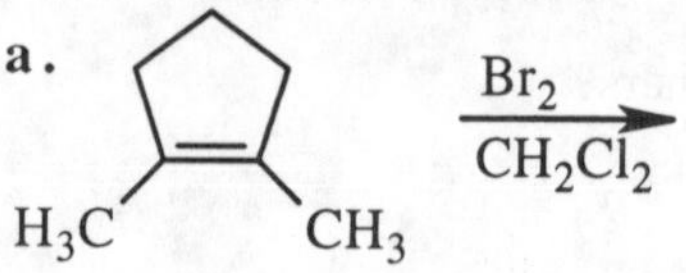 [1,2-dimethylcyclopentene, with H_3C and CH_3 on the double-bond carbons] $\xrightarrow[CH_2Cl_2]{Br_2}$

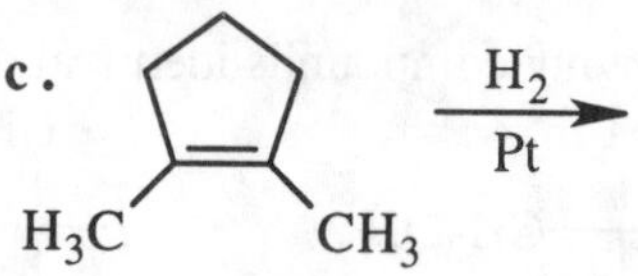

b. 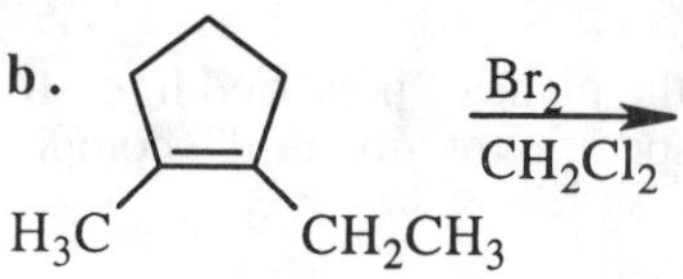 [1-ethyl-2-methylcyclopentene, with H_3C and CH_2CH_3 on the double-bond carbons] $\xrightarrow[CH_2Cl_2]{Br_2}$

c. [1,2-dimethylcyclopentene, with H_3C and CH_3 on the double-bond carbons] $\xrightarrow[Pt]{H_2}$

d. 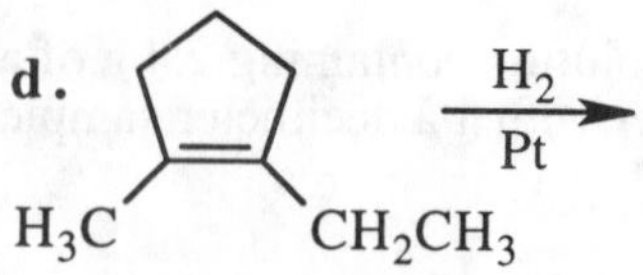 [1-ethyl-2-methylcyclopentene, with H_3C and CH_2CH_3 on the double-bond carbons] $\xrightarrow[Pt]{H_2}$

8. Which of the following have the *R* configuration?

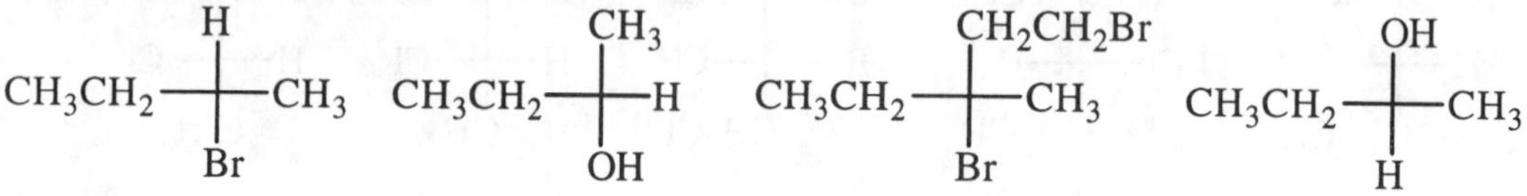

[Fischer projection 1: top H; left CH_3CH_2; right CH_3; bottom Br]

[Fischer projection 2: top CH_3; left CH_3CH_2; right H; bottom OH]

[Fischer projection 3: top CH_2CH_2Br; left CH_3CH_2; right CH_3; bottom Br]

[Fischer projection 4: top OH; left CH_3CH_2; right CH_3; bottom H]

9. Draw a diastereomer of the following compound.

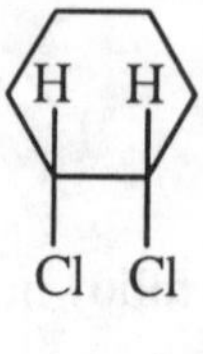

[Structure: H, H; Cl, Cl]

10. Indicate whether each of the following statements is true or false.

a. Diastereomers have the same melting points. T F

b. The addition of HBr to 3-methyl-2-pentene is a stereospecific reaction. T F

c. The addition of HBr to 3-methyl-2-pentene is a stereoselective reaction. T F

d. The addition of HBr to 3-methyl-2-pentene is a regioselective reaction. T F

e. Meso compounds do not rotate polarized light. T F

f. 2,3-Dichloropentane has a stereoisomer that is a meso compound. T F

g. In the most stable conformation of *cis*-1-ethyl-2-methylcyclohexane, both substituents are in equatorial positions. T F

h. A compound with three chirality centers can have a maximum of nine stereoisomers. T F

CHAPTER 5
Reactions of Alkynes. Introduction to Multistep Synthesis

Important Terms

alkylation reaction a reaction that adds an alkyl group to a reactant.

alkyne a hydrocarbon that contains a triple bond.

geminal dihalide a compound with two halogen atoms bonded to the same carbon.

internal alkyne an alkyne with the triple bond not at the end of the carbon chain.

keto-enol tautomers a ketone and its isomeric α,β-unsaturated alcohol.

$$\underset{\text{keto}}{RCH_2\overset{\overset{\displaystyle O}{\|}}{C}R} \rightleftharpoons \underset{\text{enol}}{RCH{=}\overset{\overset{\displaystyle OH}{|}}{C}R}$$

pi-complex a complex formed between an electrophile and a triple bond.

radical anion a species with a negative charge and an unpaired electron.

retrosynthesis (retrosynthetic analysis) working backward (on paper) from a target molecule to available starting materials.

tautomerization interconversion of tautomers.

tautomers isomers that differ in the location of a double bond and a hydrogen.

terminal alkyne an alkyne with the triple bond at the end of the carbon chain.

thermal cracking using heat to break a molecule apart.

vinylic cation a compound with a positive charge on a vinylic carbon.

vinylic radical a compound with an unpaired electron on a vinylic carbon.

Solutions to Problems

1. A noncyclic hydrocarbon with 14 carbons and no π bonds would have a molecular formula = $C_{14}H_{30}$ (C_nH_{2n+2}). Because a compound has two fewer hydrogens for every ring and π bond, a compound with one ring and 4 π bonds would have 10 fewer hydrogens than the C_nH_{2n+2} formula. Thus, the molecular formula is $\mathbf{C_{14}H_{20}}$.

2.

a. $ClCH_2CH_2C{\equiv}CCH_2CH_3$

b. 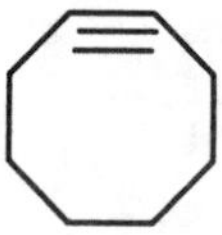 (cyclooctene ring structure)

c. $CH_3CHC{\equiv}CH$ with CH_3 on C-2

d. $HC{\equiv}CCH_2Cl$

e. $HC{\equiv}CCH_2CCH_3$ with two CH_3 groups on C-4

f. $CH_3C{\equiv}CCH_3$

3. **a.** 5-bromo-2-pentyne
b. 6-bromo-2-chloro-4-octyne
c. 1-methoxy-2-pentyne
d. 3-ethyl-1-hexyne

4.

$HC{\equiv}CCH_2CH_2CH_2CH_3$
1-hexyne
butylacetylene

$CH_3C{\equiv}CCH_2CH_2CH_3$
2-hexyne
methylpropylacetylene

$CH_3CH_2C{\equiv}CCH_2CH_3$
3-hexyne
diethylacetylene

$CH_3CH_2CHC{\equiv}CH$ with CH_3 on C-3
3-methyl-1-pentyne
sec-butylacetylene

$CH_3CHCH_2C{\equiv}CH$ with CH_3 on C-4
4-methyl-1-pentyne
isobutylacetylene

$CH_3CHC{\equiv}CCH_3$ with CH_3 on C-4
4-methyl-2-pentyne
isopropylmethylacetylene

$CH_3CC{\equiv}CH$ with two CH_3 groups on C-3
3,3-dimethyl-1-butyne
tert-butylacetylene

5. An internal alkyne is more stable because it has two substituents bonded to the *sp* carbons, while an internal alkyne has only one substituent bonded to the *sp* carbon.
Alkyl substituents bonded to *sp* carbons stabilize the compound by hyperconjugation—just like alkyl substituents bonded to sp^2 carbons stabilize alkenes (Section 3.19).

6. The less stable reactant will be the more reactive reactant, if the less stable reactant has the more stable transition state, or if the less stable reactant has the less stable transition state **and** the difference in the stabilities of the reactants is greater than the difference in the stabilities of the transition states.

7.

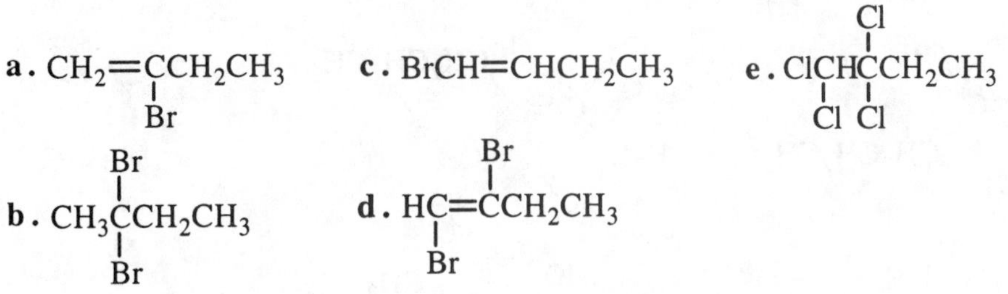

a. $CH_2{=}C(Br)CH_2CH_3$

b. $CH_3CBr_2CH_2CH_3$

c. $BrCH{=}CHCH_2CH_3$

d. $HC(Br){=}C(Br)CH_2CH_3$

e. $ClCH(Cl)CCl_2CH_2CH_3$

8.

a. 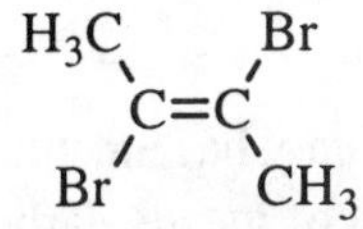 $H_3C(Br)C{=}C(Br)CH_3$ (H₃C and Br on one carbon; Br and CH₃ on the other, anti)

Only anti addition occurs, because the intermediate is a cyclic bromonium ion.

b. 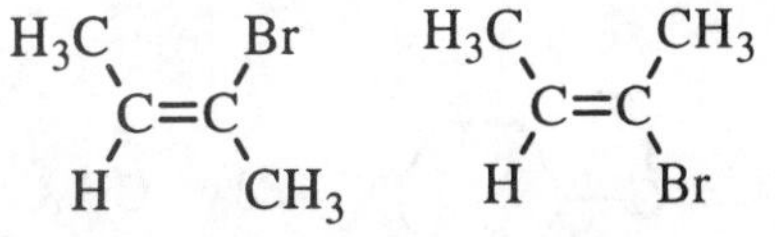 $H_3C(H)C{=}C(Br)CH_3$ and $H_3C(H)C{=}C(CH_3)Br$

Both anti and syn addition can occur, because the intermediate is a radical.

9.

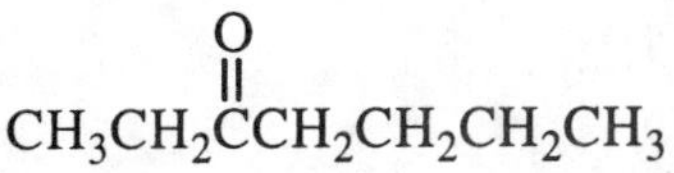

$CH_3CH_2C({=}O)CH_2CH_2CH_2CH_3$ and $CH_3CH_2CH_2C({=}O)CH_2CH_2CH_3$

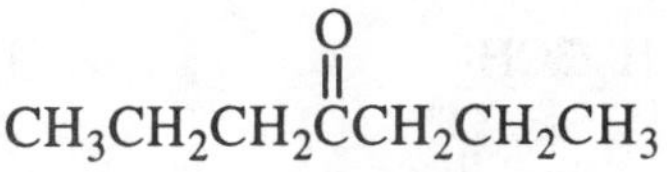

10.

a. $CH_3C{\equiv}CH$ b. $CH_3CH_2C{\equiv}CCH_2CH_3$ c. 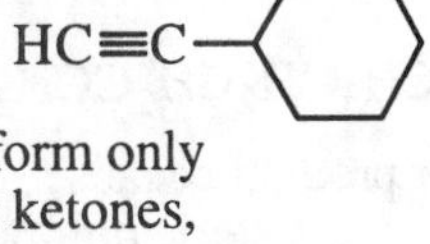 $HC{\equiv}C$–cyclohexyl

The best answer for "**b**" is 3-hexyne, because it would form only the desired ketone. 2-Hexyne would form two different ketones, so only half of the product would be the desired ketone.

11.

a. $CH_2{=}C(OH)CH_3$ Because the ketone has identical substituents bonded to the carbonyl carbon, it has only one enol tautomer.

b. $CH_3CH{=}C(OH)CH_2CH_2CH_3$ and $CH_3CH_2C(OH){=}CHCH_2CH_3$

(H)(H₃C)C=C(OH)(CH₂CH₂CH₃) — *E*; (H)(H₃C)C=C(CH₂CH₂CH₃)(OH) — *Z*

(HO)(CH₃CH₂)C=C(H)(CH₂CH₃) — *E*; (HO)(CH₃CH₂)C=C(CH₂CH₃)(H) — *Z*

c. $CH_2{=}C(OH)$–cyclohexyl and $CH_3C(OH){=}$cyclohexylidene

Because each enol has indentical groups on one of its sp^2 carbons, *E* and *Z* isomers are not possible for either one.

12.

a. (1) $CH_3CH_2COCH_3$ (2) $CH_3CH_2CH_2CHO$

b. (1) $CH_3CH_2COCH_3$ (2) $CH_3CH_2COCH_3$

c. (1) $CH_3CH_2CH_2COCH_3$ and $CH_3CH_2COCH_2CH_3$

(2) $CH_3CH_2CH_2COCH_3$ and $CH_3CH_2COCH_2CH_3$

major product because there is less steric hindrance to addition of BH_3 at the 2-position

13. Ethyne (acetylene)

An alkyne can form an aldehyde only if the OH group adds to a terminal *sp* carbon. When water adds to a terminal alkyne, the proton adds to the terminal *sp* carbon. Therefore, the only way the OH group can add to a terminal *sp* carbon is if there are two terminal *sp* carbons in the alkyne. In other words, the alkyne must be ethyne.

14.

a. $CH_3CH_2CH_2C{\equiv}CH$ or $CH_3CH_2C{\equiv}CCH_3$ $\xrightarrow[Pt]{H_2}$ $CH_3CH_2CH_2CH_2CH_3$

b. $CH_3C{\equiv}CCH_3$ $\xrightarrow[\text{Lindlar's catalyst}]{H_2}$ (cis) $H_3C(H)C{=}C(H)CH_3$

c. $CH_3CH_2C{\equiv}CCH_3$ $\xrightarrow[NH_3\ \text{liq}]{Na}$ (trans) $CH_3CH_2(H)C{=}C(H)CH_3$

d. $CH_3CH_2CH_2CH_2C{\equiv}CH$ $\xrightarrow[\text{Lindlar's catalyst}]{H_2}$ or $\xrightarrow[NH_3]{Na}$ $CH_3CH_2CH_2CH_2CH{=}CH_2$

15. The base used to remove a proton must be stronger than the base that is formed as a result of proton removal. A terminal alkyne has a pK_a ~ 25. Therefore, the base used to remove a proton from a terminal alkyne must be a stronger base than the terminal alkyne. In other words, any base whose conjugate acid has a pK_a greater than 25 can be used.

16. The electronegativities of carbon atoms decrease in the order: $sp > sp^2 > sp^3$.
The more electronegative the carbon atom, the less stable it will be with a positive charge.

a. $CH_3\overset{+}{C}H_2$ b. $H_2C{=}\overset{+}{C}H$

17. Solved in the text.

18.

a. $HC{\equiv}CH \xrightarrow[\text{2. } CH_3CH_2CH_2Br]{\text{1. } ^-NH_2} CH_3CH_2CH_2C{\equiv}CH$

b. $HC{\equiv}CH \xrightarrow[\text{2. } CH_3Br]{\text{1. } ^-NH_2} CH_3C{\equiv}CH \xrightarrow[\text{Na/NH}_3\text{ liq}]{H_2\text{/Lindlar's catalyst or}} CH_3CH{=}CH_2$

c. product of **a** $\xrightarrow[\text{2. } HO^-,\ H_2O_2,\ H_2O]{\text{1. disiamylborane}} CH_3CH_2CH_2CH_2\overset{O}{\overset{\|}{C}}H$

d. $HC{\equiv}CH \xrightarrow[\text{2. } CH_3Br]{\text{1. } ^-NH_2} CH_3C{\equiv}CH \xrightarrow{\text{excess HCl}} CH_3\underset{Cl}{\overset{Cl}{C}}CH_3$

e. product of **b** $\xrightarrow{HBr} CH_3\underset{Br}{C}HCH_3$

f. $HC{\equiv}CH \xrightarrow[\text{2. } CH_3Br]{\text{1. } ^-NH_2} CH_3C{\equiv}CH \xrightarrow[\text{2. } CH_3Br]{\text{1. } ^-NH_2} CH_3C{\equiv}CCH_3 \xrightarrow[\text{Lindlar's catalyst}]{H_2}$ $\overset{H_3C}{\underset{H}{}}C{=}C\overset{CH_3}{\underset{H}{}}$ (cis)

19.

a. $CH_3C{\equiv}CCH_2CH_2CH_3$

b. $CH_3CH_2C{\equiv}C\underset{CH_2CH_3}{C}HCH_2CH_2CH_3$

c. $CH_3C{\equiv}CH$

d. $CH_2{=}CHC{\equiv}CH$

e. $CH_3OC{\equiv}CH$

f. $CH_3\underset{CH_3}{\overset{CH_3}{C}}C{\equiv}C\overset{CH_3}{C}HCH_2CH_3$

g. $BrC{\equiv}CCH_2CH_2CH_3$

h. $HC{\equiv}CCH_2Br$

i. $CH_3CH_2C{\equiv}CCH_2CH_3$

j. $CH_3\underset{CH_3}{\overset{CH_3}{C}}C{\equiv}C\underset{CH_3}{\overset{CH_3}{C}}CH_3$

k. cyclopentyl$-C{\equiv}CH$

l. $CH_3C{\equiv}CCH_2\overset{CH_3}{C}H\underset{CH_3}{C}HCH_3$

20. **a.** 5-bromo-2-hexyne **c.** 5,5-dimethyl-2-hexyne

b. 5-methyl-2-octyne **d.** 6-chloro-2-methyl-3-heptyne

21.

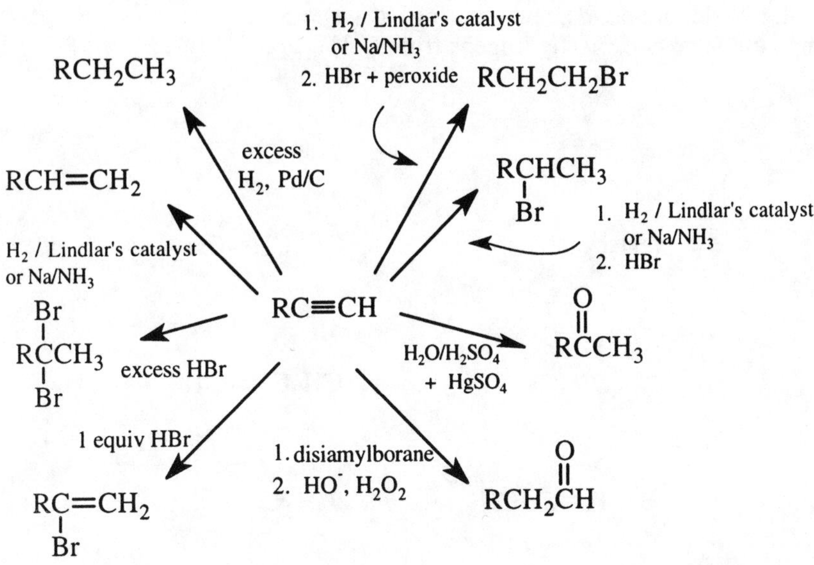

22.

a. $CH_3CH_2CH_2C{\equiv}CH \xrightarrow[\text{2. HO}^-,\ H_2O_2]{\text{1. disiamylborane}} CH_3CH_2CH_2CH_2\overset{O}{\overset{\|}{C}}H$

b. $CH_3CH_2CH{=}CH_2 \xrightarrow[\text{2. HO}^-,\ H_2O_2]{\text{1. BH}_3} CH_3CH_2CH_2CH_2OH$

c. $CH_3CH_2CH_2C{\equiv}CCH_2CH_2CH_3 + H_2O \xrightarrow{H_2SO_4} CH_3CH_2CH_2\overset{O}{\overset{\|}{C}}CH_2CH_2CH_2CH_3$

This symmetrical alkyne will give the greatest yield of the desired ketone. Because the reactant is not a terminal alkyne, the reaction can take place without the mercuric ion catalyst.

d. $CH_3CH_2CH_2CH{=}CH_2 + HBr \xrightarrow{\text{peroxide}} CH_3CH_2CH_2CH_2CH_2Br$

23. **a.** H_2/Lindlar's catalyst **b.** Na, NH_3 **c.** excess H_2/Pt

24. The molecular formula of the hydrocarbon is $C_{32}H_{56}$.

$C_nH_{2n+2} = C_{32}H_{66}$

With one triple bond, two double bonds, and one ring, $(\pi + r) = 5$.
Therefore, the compound is missing 10 hydrogens from C_nH_{2n+2}.

25.

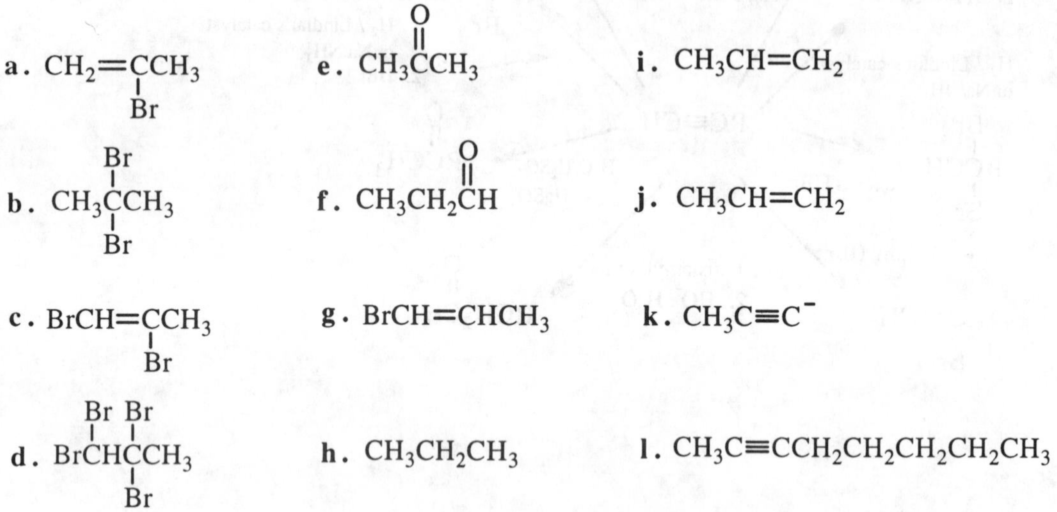

a. $CH_2{=}C(Br)CH_3$

b. $CH_3CBr_2CH_3$

c. $BrCH{=}C(Br)CH_3$

d. $BrCH(Br)CBr_2CH_3$

e. $CH_3C(=O)CH_3$

f. $CH_3CH_2CH{=}O$

g. $BrCH{=}CHCH_3$

h. $CH_3CH_2CH_3$

i. $CH_3CH{=}CH_2$

j. $CH_3CH{=}CH_2$

k. $CH_3C{\equiv}C^-$

l. $CH_3C{\equiv}CCH_2CH_2CH_2CH_2CH_3$

26.

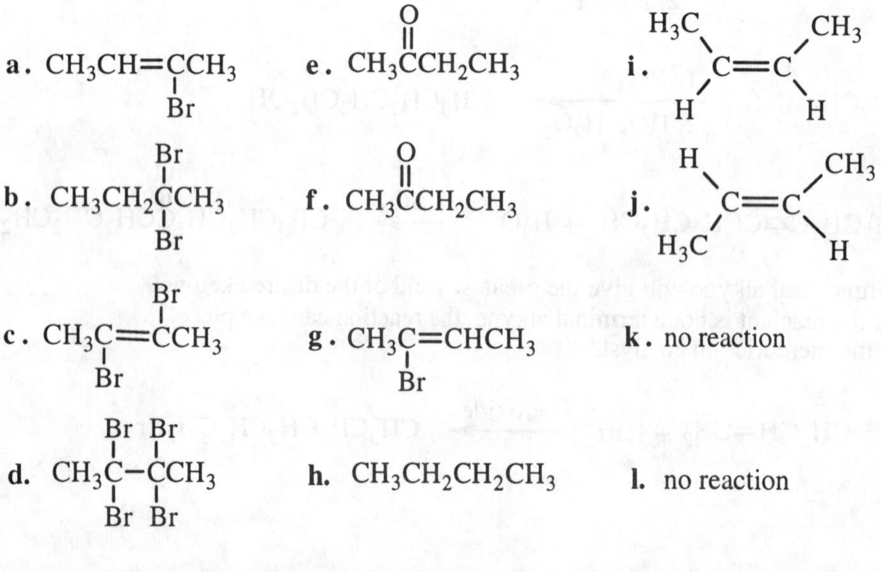

a. $CH_3CH{=}C(Br)CH_3$

b. $CH_3CH_2CBr_2CH_3$

c. $CH_3C(Br){=}C(Br)CH_3$

d. $CH_3CBr_2{-}CBr_2CH_3$

e. $CH_3C(=O)CH_2CH_3$

f. $CH_3C(=O)CH_2CH_3$

g. $CH_3C(Br){=}CHCH_3$

h. $CH_3CH_2CH_2CH_3$

i. $(H_3C)(H)C{=}C(CH_3)(H)$ (cis: H_3C and CH_3 on the same side)

j. $(H)(H_3C)C{=}C(CH_3)(H)$ (trans: H_3C and CH_3 on opposite sides)

k. no reaction

l. no reaction

27.

a. 1. $CH_3CHC{\equiv}CH$ (with CH_3 on C2) $\xrightarrow[\text{2. } CH_3Br]{\text{1. } NaNH_2}$ $CH_3CHC{\equiv}CCH_3$ (with CH_3 on C2) $\xrightarrow[\text{Lindlar's catalyst}]{H_2}$ $CH_3CHCH{=}CHCH_3$ (with CH_3 on C2)

$\downarrow$ H_2O, H_2SO_4

$CH_3\overset{+}{C}HCHCH_2CH_3$ ... (hydride shift: $CH_3CCHCH_2CH_3$ with H on C2, CH_3 on C2, + on C3) $\longrightarrow$ $CH_3\overset{+}{C}CH_2CH_2CH_3$ (with CH_3 on C2) $\xrightarrow{H_2O}$ $CH_3CCH_2CH_2CH_3$ (with OH and CH_3 on C2)

2. $CH_3CHC{\equiv}CH$ (with CH_3 on C2) $\xrightarrow[\text{2. } CH_3Br]{\text{1. } NaNH_2}$ $CH_3CHC{\equiv}CCH_3$ (with CH_3 on C2) $\xrightarrow[\text{Lindlar's catalyst}]{H_2}$ $CH_3CHCH{=}CHCH_3$ (with CH_3 on C2)

$\downarrow$ 1. BH_3 2. H_2O_2, HO^-

$CH_3CHCH_2CHCH_3$ (with CH_3 on C2 and OH on C4)

b. 4-Methyl-2-pentanol also will be obtained from **1.** because in the third step of the synthesis, the proton can add to either of the sp^2 hybridized carbons.

Some 2-methyl-3-pentanol will be obtained if the nucleophile reacts with the carbocation before rearrangement to the more stable tertiary carbocation takes place.

$CH_3CHCH_2CHCH_3$ (with CH_3 on C2 and OH on C4)
4-methyl-2-pentanol

$CH_3CHCHCH_2CH_3$ (with CH_3 on C2 and OH on C3)
2-methyl-3-pentanol

4-Methyl-3-pentanol also will be obtained from **2.** because in the third step of the synthesis, boron can add to either of the sp^2 hybridized carbons.

Some 2-methyl-3-pentanol will be obtained if the nucleophile reacts with the carbocation before rearrangement takes place.

$CH_3CHCHCH_2CH_3$ (with CH_3 on C2 and OH on C3)
2-methyl-3-pentanol

28. **a.** 3-heptyne **c.** correct **e.** correct

b. 5-methyl-3-heptyne **d.** 6,7-dimethyl-3-octyne **f.** correct

29. **c** and **e** are keto-enol tautomers.

30.

a. $HC\equiv CH \xrightarrow[\text{2. }CH_3CH_2Br]{\text{1. }NaNH_2} CH_3CH_2C\equiv CH \xrightarrow[\text{Lindlar's catalyst}]{H_2} CH_3CH_2CH{=}CH_2 \xrightarrow{Br_2,\ CCl_4} CH_3CH_2CHBrCH_2Br$

b. $HC\equiv CH \xrightarrow[HgSO_4]{H_2O,\ H_2SO_4} \underset{\text{enol}}{CH_2{=}CHOH} \rightleftharpoons CH_3CH{=}O$

c. $HC\equiv CH \xrightarrow[\text{2. }CH_3Br]{\text{1. }NaNH_2} HC\equiv CCH_3 \xrightarrow[HgSO_4]{H_2O,\ H_2SO_4} CH_3C({=}O)CH_3$

d. $HC\equiv CH \xrightarrow[\text{2. }CH_3Br]{\text{1. }NaNH_2} HC\equiv CCH_3 \xrightarrow[\text{2. }CH_3CH_2Br]{\text{1. }NaNH_2} CH_3C\equiv CCH_2CH_3 \xrightarrow{Na,\ NH_3}$ (trans-2-pentene)

e. $HC\equiv CH \xrightarrow[\text{2. }CH_3Br]{\text{1. }NaNH_2} HC\equiv CCH_3 \xrightarrow[\text{2. }CH_3CH_2Br]{\text{1. }NaNH_2} CH_3C\equiv CCH_2CH_3 \xrightarrow[\text{Lindlar's catalyst}]{H_2}$ (cis-2-pentene)

f. $HC\equiv CH \xrightarrow[\text{2. }CH_3Br]{\text{1. }NaNH_2} HC\equiv CCH_3 \xrightarrow[\text{2. }CH_3CH_2Br]{\text{1. }NaNH_2} CH_3C\equiv CCH_2CH_3 \xrightarrow{H_2,\ Pd/C}$ (pentane)

31. **a.** Syn addition of H_2 forms *cis*-2-butene. And when Br_2 adds to *cis*-2-butene, the threo pair of enantiomers is formed.

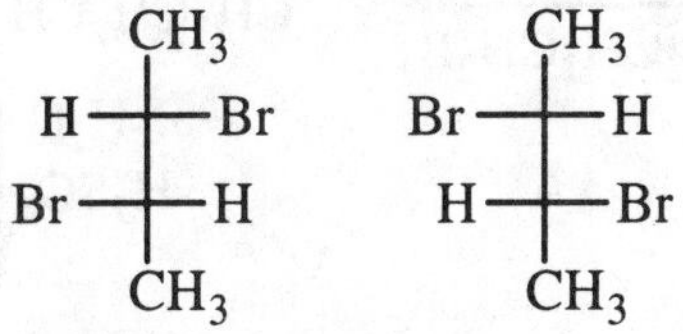

CH_3 / H—C—Br / Br—C—H / CH_3 and CH_3 / Br—C—H / H—C—Br / CH_3

b. Reaction with sodium and liquid ammonia forms *trans*-2-butene. And when Br_2 adds to *trans*-2-butene, a meso compound is formed.

CH_3 / H—C—Br / H—C—Br / CH_3

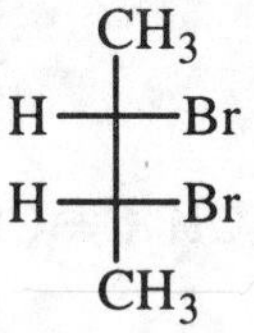

c. Anti addition of Cl_2 forms *trans*-2,3-dichloro-2-butene. And when Br_2 adds to *trans*-2,3-dichloro-2-butene, a meso compound is formed.

CH_3 / Cl—C—Br / Cl—C—Br / CH_3

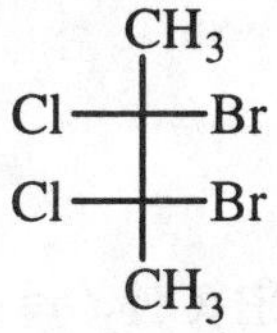

32.

a. $CH_3CH_2\overset{O}{\overset{\|}{C}}CH_3$ **b.** $CH_3CH_2CH_2\overset{O}{\overset{\|}{C}}CH_3$ **c.** 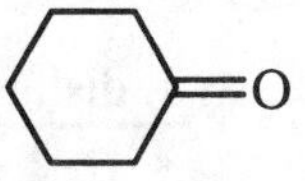 =O **d.** 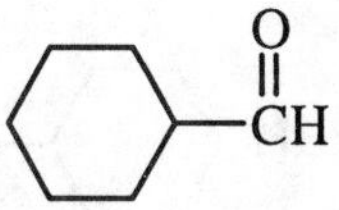 $-\overset{O}{\overset{\|}{C}}H$

33.

a. $HC \equiv CH \xrightarrow[\text{2. } CH_3CH_2CH_2CH_2Br]{\text{1. } NaNH_2} CH_3CH_2CH_2CH_2C \equiv CH \xrightarrow[H_2SO_4]{H_2O,\ HgSO_4} CH_3CH_2CH_2CH_2\overset{O}{\overset{\|}{C}}CH_3$

b. $HC \equiv CH \xrightarrow[\text{2. } CH_3CH_2Br]{\text{1. } NaNH_2} CH_3CH_2C \equiv CH \xrightarrow[\text{or } Na/NH_3 \text{ liq}]{H_2 \text{ Lindlar's catalyst}} CH_3CH_2CH{=}CH_2 \xrightarrow{HBr} CH_3CH_2\underset{Br}{\underset{|}{C}}HCH_3$

c. $HC \equiv CH \xrightarrow[\text{2. } CH_3CH_2CH_2Br]{\text{1. } NaNH_2} CH_3CH_2CH_2C \equiv CH \xrightarrow[\text{or } Na/NH_3 \text{ liq}]{H_2 \text{ Lindlar's catalyst}} CH_3CH_2CH_2CH{=}CH_2 \xrightarrow[H_2SO_4]{H_2O} CH_3CH_2CH_2\underset{OH}{\underset{|}{C}}HCH_3$

d. cyclohexyl$-C \equiv CH \xrightarrow[\text{2. } HO^-,\ H_2O_2]{\text{1. disiamylborane}}$ cyclohexyl$-CH_2\overset{O}{\overset{\|}{C}}H$

e. cyclohexyl$-C \equiv CH \xrightarrow[HgSO_4]{H_2O,\ H_2SO_4}$ cyclohexyl$-\overset{O}{\overset{\|}{C}}-CH_3$

f. phenyl$-C \equiv CCH_3 \xrightarrow[\text{Lindlar's catalyst}]{H_2}$ (cis) phenyl$(H)C{=}C(H)CH_3$

34. She can make 3-octyne by using 1-hexyne instead of 1-butyne. She would then need to use ethyl bromide (instead of butyl bromide) for the alkylation step:

$$CH_3CH_2CH_2CH_2C{\equiv}CH \xrightarrow[\text{2. } CH_3CH_2Br]{\text{1. } NaNH_2} CH_3CH_2CH_2CH_2C{\equiv}CCH_2CH_3$$

Or she could make the 1-butyne she needed by alkylating ethyne:

$$HC{\equiv}CH \xrightarrow[\text{2. } CH_3CH_2Br]{\text{1. } NaNH_2} CH_3CH_2C{\equiv}CH$$

$$CH_3CH_2C{\equiv}CH \xrightarrow{\text{1. } NaNH_2 \quad \text{2. } CH_3CH_2CH_2CH_2Br} CH_3CH_2CH_2CH_2C{\equiv}CCH_2CH_3$$

35.

a. $$HO(CH_2)_8C{\equiv}CH \xrightarrow{\text{excess } {}^{-}NH_2} {}^{-}O(CH_2)_8C{\equiv}C^{-} \xrightarrow{CH_3(CH_2)_3Br} {}^{-}O(CH_2)_8C{\equiv}C(CH_2)_3CH_3$$

$$\xrightarrow{EtOH} HO(CH_2)_8C{\equiv}C(CH_2)_3CH_3 \xrightarrow[\text{Lindlar's catalyst}]{H_2}$$

$$HOCH_2CH_2CH_2CH_2CH_2CH_2CH_2CH_2(H)C{=}C(H)CH_2CH_2CH_2CH_3$$

(cis; both H on the same side)

spodoptol

b. The most easily removed hydrogen in the starting material is the one bonded to oxygen. Therefore, to remove the hydrogen bonded to the *sp* carbon, two equivalents of base are required. Notice that when the dianion reacts with the alkyl bromide, the negatively charged carbon atom is more reactive than the negatively charged oxygen atom because the carbon is the strongest base and, therefore, the best nucleophile.

36. **a.** Only one product is obtained from the hydroboration-oxidation of 2-butyne because the alkyne is symmetrical. Two different products can be obtained from hydroboration-oxidation of asymmetrical 2-pentyne.

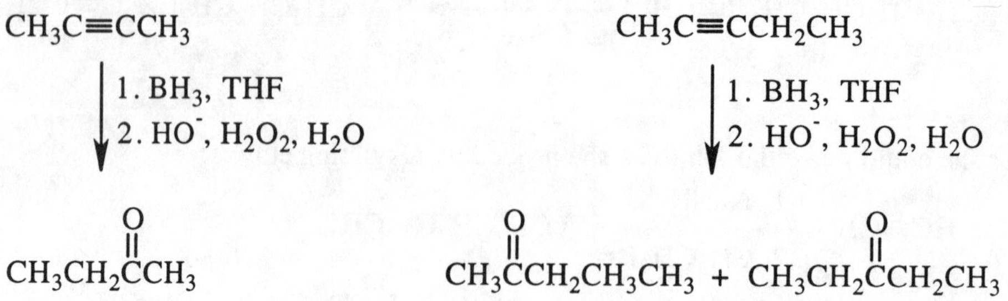

$$CH_3C{\equiv}CCH_3 \xrightarrow[\text{2. } HO^-,\ H_2O_2,\ H_2O]{\text{1. } BH_3,\ THF} CH_3CH_2\overset{O}{\overset{\|}{C}}CH_3$$

$$CH_3C{\equiv}CCH_2CH_3 \xrightarrow[\text{2. } HO^-,\ H_2O_2,\ H_2O]{\text{1. } BH_3,\ THF} CH_3\overset{O}{\overset{\|}{C}}CH_2CH_3CH_3 + CH_3CH_2\overset{O}{\overset{\|}{C}}CH_2CH_3$$

b. Only one product will be obtained from hydroboration-oxidation of any symmetrical alkyne, such as ethyne, 3-hexyne, and 4-octyne.

$HC{\equiv}CH$	$CH_3CH_2C{\equiv}CCH_2CH_3$	$CH_3CH_2CH_2C{\equiv}CCH_2CH_2CH_3$
ethyne	3-hexyne	4-octyne

37. There are two reasons that hyperconjugation is less effective in stabilizing a charge on a vinylic cation. First, an sp^2-s bond is stronger than an sp^3-s bond, so the sp^2 orbitals are less able to donate electron density to the adjacent positively charge carbon. Second, an sp^2 carbon has 120°bond angles compared to the 109.5° bond angles of an sp^3 carbon, so the orbitals of an sp^2 carbon are farther away, which makes them less able to interact with the positively charged carbon.

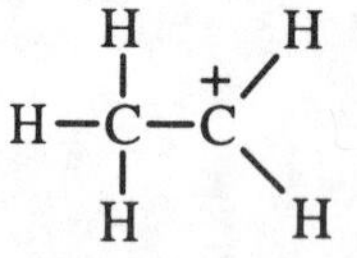

$$H_3C{-}\overset{+}{C}H_2$$

an alkyl cation

$$H_2C{=}\overset{+}{C}{-}H$$

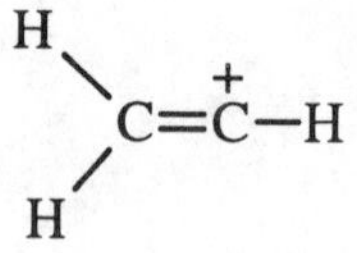

a vinylic cation

Chapter 5 Practice Test

1. What reagents could be used to convert the given starting material into the desired product?

a.  cyclohexyl–C≡CH ⟶ cyclohexyl–$CH_2\overset{O}{\overset{\|}{C}}H$

b. 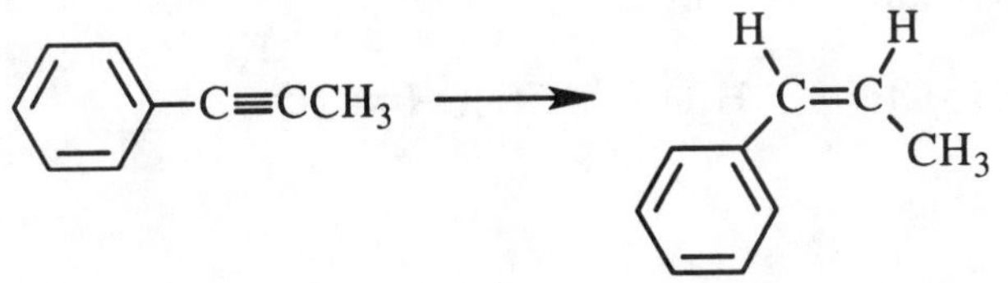 phenyl–$C{\equiv}CCH_3$ ⟶ (H)(phenyl)C=C(H)(CH_3)

2. Draw the enol tautomer of the following compound:

(cyclohexanone)

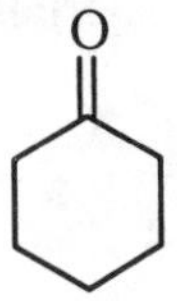

3. Give the structure of *sec*-butyl isobutyl acetylene.

4. Indicate whether each of the following statements is true or false:

a. A terminal alkyne is more stable than an internal alkyne. T F

b. Propyne is more reactive than propene toward reaction with HBr. T F

c. 1-Butyne is more acidic than 1-butene. T F

d. An sp^2 hybridized carbon is more electronegative than an sp^3 hybridized carbon. T F

5. Give the systematic name for the following compound:

$CH_3\underset{\displaystyle CH_3}{\underset{|}{C}}HC{\equiv}CCH_2CH_2Br$

6. What alkyne would be the best reagent to use for the synthesis of the following ketone?

$$CH_3CH_2CH_2\overset{\overset{\displaystyle O}{\|}}{C}CH_3$$

7. Rank the following compounds in order of decreasing acidity:
(Label the most acidic compound #1.)

NH_3 $CH_3C{\equiv}CH$ CH_3CH_3 H_2O $CH_3CH{=}CH_2$

8. Give an example of a ketone that has two enol tautomers.

9. Show how the target molecules could be prepared from the given starting materials.

a. $CH_3CH_2C{\equiv}CH \longrightarrow CH_3CH_2CH_2CH_2CH_2CH_3$

b. $CH_3CH_2C{\equiv}CH \longrightarrow CH_3CH_2CH_2CH_2Br$

c. $CH_3CH_2C{\equiv}CH \longrightarrow CH_3CH_2\overset{\overset{\displaystyle O}{\|}}{C}CH_2CH_2CH_3$

CHAPTER 6
Electron Delocalization and Resonance

Important Terms

allylic carbon	a carbon adjacent to an sp^2 carbon of a carbon-carbon double bond.
allylic cation	a compound with a positive charge on an allylic carbon.
antibonding π molecular orbital	the molecular orbital formed by side-to-side overlapping of out-of-phase *p* orbitals.
aromatic compound	compounds that are unusually stable because of large delocalization energies (see Section 14.1).
benzylic carbon	a carbon, joined to other atoms by single bonds, that is bonded to a benzene ring.
benzylic cation	a compound with a positive charge on a benzylic carbon.
bonding π molecular orbital	the molecular orbital formed by side-to-side overlapping of in-phase *p* orbitals.
contributing resonance structure	a structure with localized electrons that approximates the true structure of a compound with delocalized electrons.
delocalization energy (resonance energy)	the extra stability associated with a compound as a result of having delocalized electrons.
delocalized electrons	electrons that are not localized on a single atom or between two atoms.
linear combination of molecular orbitals (LCAO)	the combination of atomic orbitals to produce molecular orbitals.
localized electrons	electrons that are restricted to a particular locality.
nonbonding molecular orbital	the *p* orbitals are too far apart to overlap significantly, so the molecular orbital that results neither favors nor disfavors bonding.
resonance	a compound with delocalized electrons is said to have resonance.
resonance contributor (resonance structure)	a structure with localized electrons that approximates the true structure of a compound with delocalized electrons.
resonance energy (delocalization energy)	the extra stability associated with a compound as a result of having delocalized electrons.
resonance hybrid	the actual structure of a compound with delocalized electrons; it is represented by two or more structures with localized electrons.
separated charges	a positive and a negative charge that can be neutralized by the movement of electrons.

Solutions to Problems

1. a. 1. If stereoisomers are not included, three different monosubstituted compounds can be formed. If stereoisomers are included, four different monosubstituted compounds can be formed because the second listed compound has a chirality center.

$BrC{\equiv}CC{\equiv}CCH_2CH_3$ $\quad$ $HC{\equiv}CC{\equiv}CCH(Br)CH_3$ $\quad$ $HC{\equiv}CC{\equiv}CCH_2CH_2Br$

2. If stereoisomers are not included, two different monosubstituted compounds can be formed. If stereoisomers are included, three different monosubstituted compounds can be formed because the first compound has a double bond that can have cis-trans isomers.

$BrCH{=}CHC{\equiv}CCH{=}CH_2$ $\quad$ $CH_2{=}C(Br)C{\equiv}CCH{=}CH_2$

b. 1. Ignoring stereoisomers, five different monosubstituted compounds can be formed.

$BrC{\equiv}CC{\equiv}CCH(Br)CH_3$ $\quad$ $BrC{\equiv}CC{\equiv}CCH_2CH_2Br$ $\quad$ $HC{\equiv}CC{\equiv}CCH(Br)CH_2Br$

$HC{\equiv}CC{\equiv}CC(Br)_2CH_3$ $\quad$ $HC{\equiv}CC{\equiv}CCH_2CHBr_2$

2. Ignoring stereoisomers, five different monosubstituted compounds can be formed.

$BrCH{=}C(Br)C{\equiv}CCH{=}CH_2$ $\quad$ $BrCH{=}CHC{\equiv}CC(Br){=}CH_2$ $\quad$ $BrCH{=}CHC{\equiv}CCH{=}CHBr$

$CH_2{=}C(Br)C{\equiv}CC(Br){=}CH_2$ $\quad$ $Br_2C{=}CHC{\equiv}CCH{=}CH_2$

2. Ladenburg benzene is a better proposal. It would form 1 monosubstituted compound, three disubstituted compounds, and would not add Br_2, all in accordance with what early chemists knew about the structure of benzene.

Dewar benzene is not in accordance with what early chemists knew about the structure of benzene, because it would form two monosubstituted compounds, 6 disubstituted compounds, and it would add Br_2.

3. a. From examination of the contributing resonance structures in Figure 6.4 on page 274 of the text, one can conclude that all the carbon-oxygen bonds in the carbonate ion should be the same length.

b. Because the two negative charges are shared equally by three oxygens, each oxygen should have two thirds of a negative charge.

4. **a.** **2**, **3**, **5**, and **6**

b. **2.**

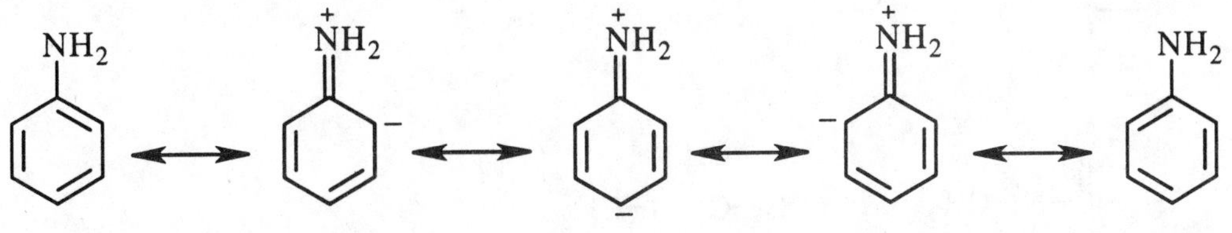

NH$_2$ ⟷ $\overset{+}{N}H_2$ ⟷ $\overset{+}{N}H_2$ ⟷ $\overset{+}{N}H_2$ ⟷ NH$_2$

3. Because the substituent is not able to participate in resonance, the only resonance structures are the two resonance structures of the benzene ring.

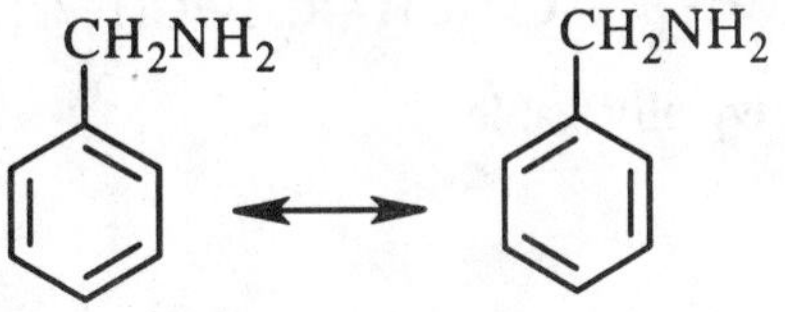

CH$_2$NH$_2$ ⟷ CH$_2$NH$_2$

5.

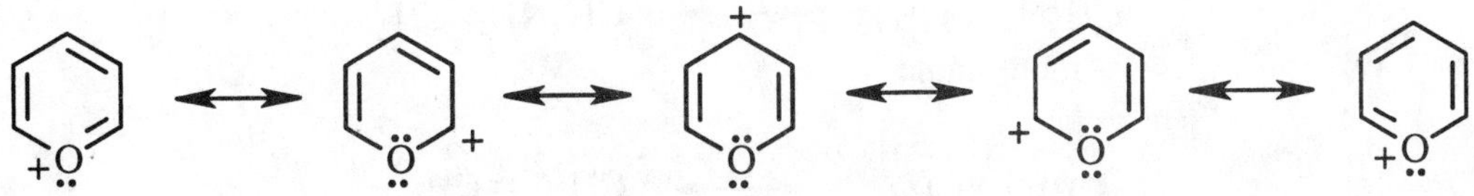

6.

$CH_3CH{=}CH{-}CH{=}CH{-}\overset{+}{C}H_2$ ⟷ $CH_3CH{=}CH{-}\overset{+}{C}H{-}CH{=}CH_2$ ⟷ $CH_3\overset{+}{C}H{-}CH{=}CH{-}CH{=}CH_2$

5. The resonance contributor that makes the greater contribution to the hybrid is labelled "**A**".

a. Solved in the text.

b.  $CH_3{-}\overset{O}{\overset{\|}{C}}{-}OCH_3$ $CH_3{-}\overset{O^-}{\overset{|}{C}}{=}\overset{+}{O}CH_3$

A   **B**

c. 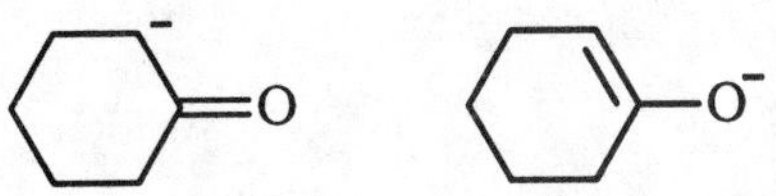 =O   –O$^-$

B   **A**

d. (structures A and B: cyclohexenone and its resonance form with $-O^-$ and a positive charge)

A B

e. $CH_3-\overset{+OH}{\overset{\|}{C}}-\underset{CH_3}{\underset{|}{N}}-CH_3$ $\quad$ $CH_3-\overset{OH}{\overset{|}{C}}=\underset{CH_3}{\underset{|}{\overset{+}{N}}}-CH_3$

B A

f. $CH_3-\overset{+}{C}H-CH=CH-CH_3$ $\quad$ $CH_3CH=CH-\overset{+}{C}H-CH_3$

equally stable

6. a. Because nitrogen is less electronegative than oxygen, it is more able to stabilize the positive charge by resonance.

$$CH_3\ddot{N}H-\overset{+}{C}H_2 \longleftrightarrow CH_3\overset{+}{N}H=CH_2$$

more stable

$$CH_3\ddot{\underset{..}{O}}-\overset{+}{C}H_2 \longleftrightarrow CH_3\overset{+}{\underset{..}{O}}=CH_2$$

b. In order for resonance to occur, the atoms involved in resonance must be in the same plane. The two *tert*-butyl groups prevent the positively charged carbon and the benzene ring from being in the same plane, which means that the carbocation cannot be stabilized by resonance.

more stable

c. The compound with delocalized electrons is more stable than the compound in which all the electrons are localized.

$$CH_3\ddot{\underset{..}{O}}-\overset{+}{C}H_2 \longleftrightarrow CH_3\overset{+}{\underset{..}{O}}=CH_2$$

more stable

$$CH_3\ddot{\underset{..}{O}}-CH_2\overset{+}{C}H_2$$

all the electrons are localized

d. The compound with delocalized electrons is more stable than the compound in which all the electrons are localized.

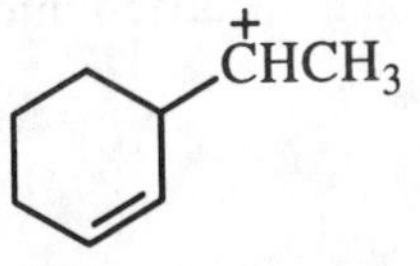

$\overset{+}{C}HCH_3$ ⟷ $\overset{+}{C}HCH_3$

$\overset{+}{C}HCH_3$

all the electrons are localized

7. **a.** Solved in the text.

b. The contributing resonance structures show that there are four sites that could react with HO^-.

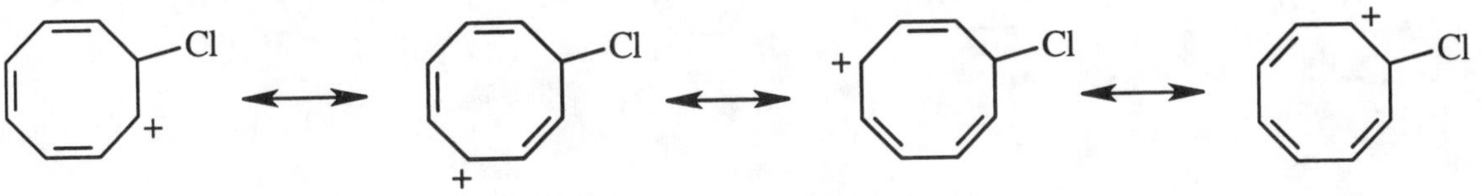

Cl + ⟷ Cl + ⟷ + Cl ⟷ + Cl

c. The contributing resonance structures show that there are three sites that could react with a bromine radical.

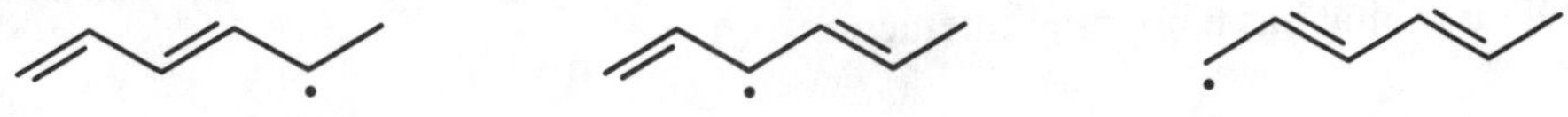

d. The contributing resonance structures show that there are two sites that could be protonated.

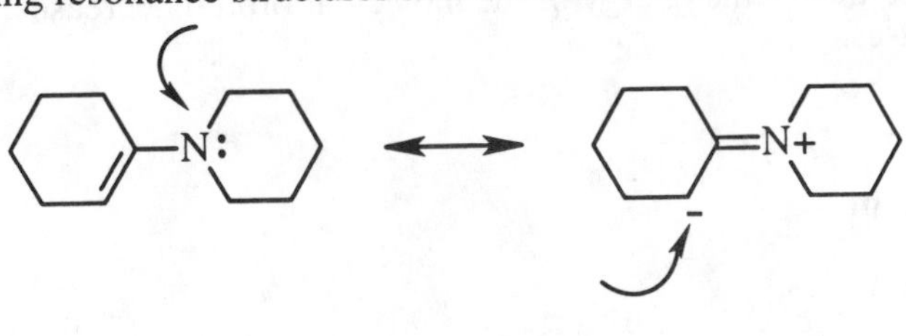

N: ⟷ =N+

8. Solved in the text.

9. The compound shown is the stronger acid in both **a** and **b** because the negative charge that results when it loses a proton can be delocalized by resonance. No resonance is possible for the other compound in either **a** or **b**.

$$CH_3CH{=}CHOH \xrightarrow{-H^+} CH_3CH{=}CHO^- \longleftrightarrow CH_3\bar{C}HCH{=}O$$

10. **a.** Ethylamine is a stronger base because when the nonbonding electrons on the nitrogen in aniline are protonated, they can no longer be delocalized into the benzene ring.

b. Ethoxide ion is a stronger base because a negatively charged oxygen is a stronger base than a neutral nitrogen.

c. Ethoxide ion is a stronger base because when the phenolate ion is protonated, the pair of electrons that is protonated can no longer be delocalized into the benzene ring.

11. We learned in Chapter 1 of the text that a carboxylic acid is a stronger acid than a protonated amine, which is a stronger acid than an alcohol. So the only question here is the relative acidity of an aromatic alcohol: it is slightly more acidic than a protonated amine.

$C_6H_5-COOH > C_6H_5-OH > C_6H_5-CH_2\overset{+}{N}H_3 > C_6H_5-CH_2OH$

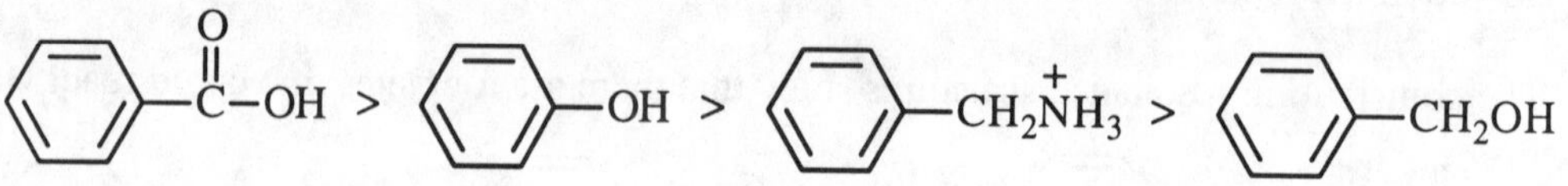

12. **a.** The ψ_1 molecular orbital of 1,3-butadiene has 3 bonding interactions and the ψ_2 molecular orbital has 2 bonding interactions.
b. The ψ_1 molecular orbital of 1,3,5,7-octatetraene has 7 bonding interactions and the ψ_2 molecular orbital has 6 bonding interactions.

Notice that the ψ_1 molecular orbital has one bonding orbital between each of the overlapping p orbitals. Notice also that as the energy of the molecular orbital decreases, the number of bonding interactions decreases.

13. **a, d, f, g, i, j, l, m**

14. **a.** and **b.**

1. $\overset{-}{\ddot{C}}H_2-\overset{+}{N}\equiv N:$ ⟷ $CH_2=\overset{+}{N}=\overset{-}{\ddot{N}}:$

More stable, because the negative charge is on nitrogen rather than on carbon.

2. $:N\equiv\overset{+}{N}-\overset{-}{\ddot{O}}:$ ⟷ $\overset{-}{\ddot{N}}=\overset{+}{N}=\ddot{O}:$

More stable, because the negative charge is on oxygen rather than on nitrogen.

3.  $^{-}$:Ö–N̈=Ö: ⟷ :O=N̈–Ö:$^{-}$

Both are equally stable.

Additional resonance structures could be drawn for each of these three species, but they are relatively unstable because they have an incomplete octet.

15.

a. 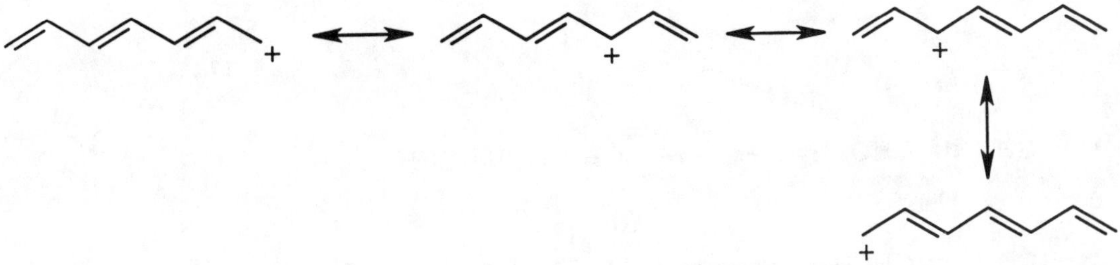 + ⟷ + ⟷ + ⟷ +

b. 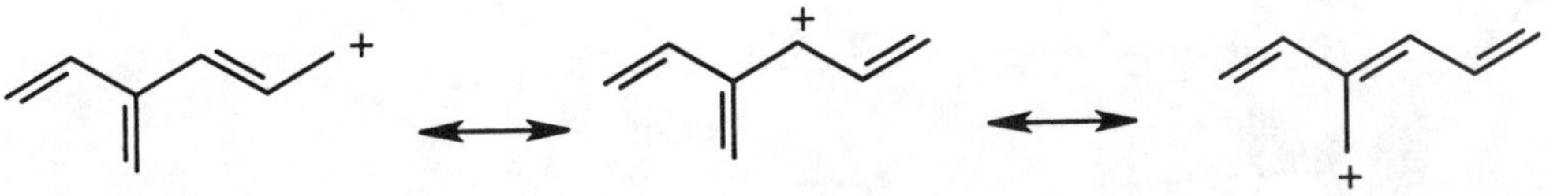 + ⟷ + ⟷ +

c. 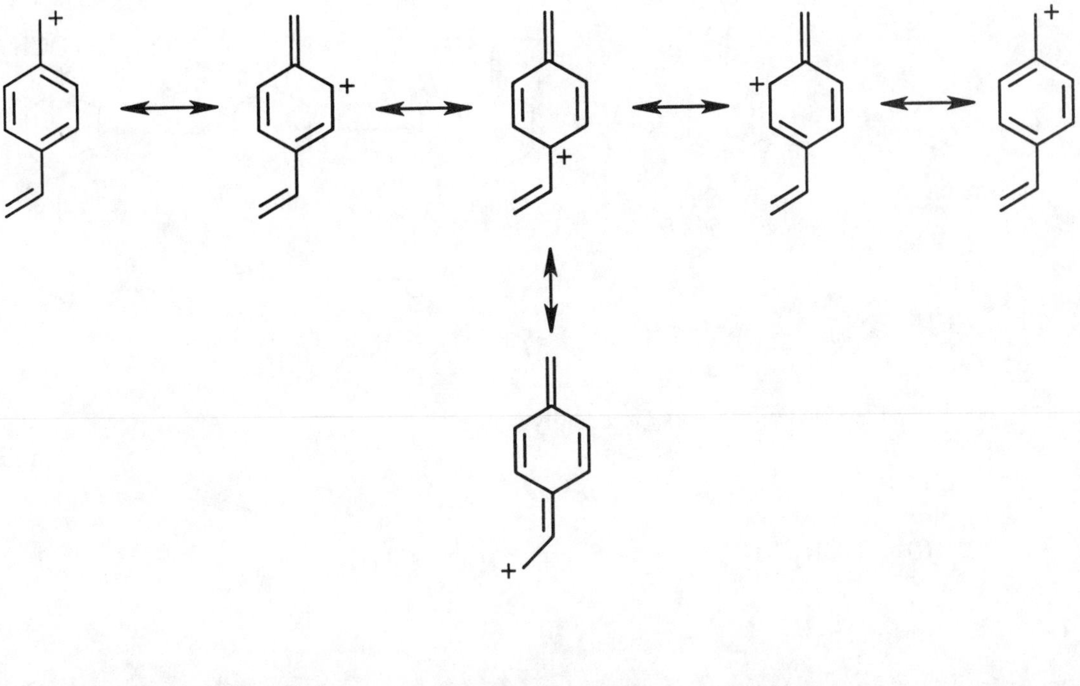 + ⟷ + ⟷ + ⟷ + ⟷ + ⟷ +

d.

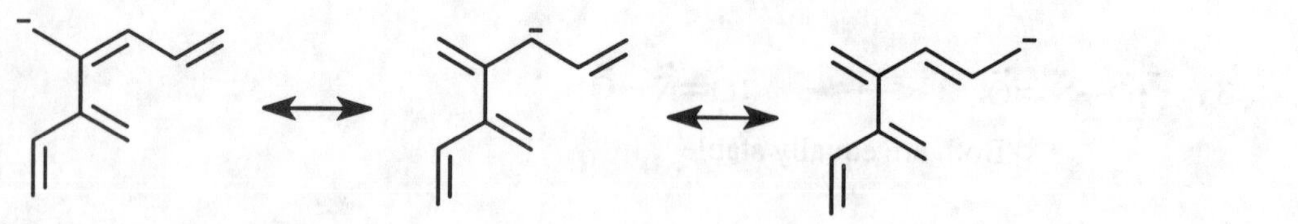

16. **a.** different compounds
b. resonance contributors
c. different compounds
d. resonance contributors
e. different compounds

17. **a.**

1. $CH_3CH{=}CHOCH_3 \longleftrightarrow CH_3\overset{-}{C}HCH{=}\overset{+}{O}CH_3$
   major — minor

2. $CH_3CH{=}CH\overset{+}{C}H_2 \longleftrightarrow CH_3\overset{+}{C}HCH{=}CH_2$
   minor — major

3. $CH_3\overset{-}{C}HC{\equiv}N \longleftrightarrow CH_3CH{=}C{=}\overset{-}{N}$
   minor — major

4. 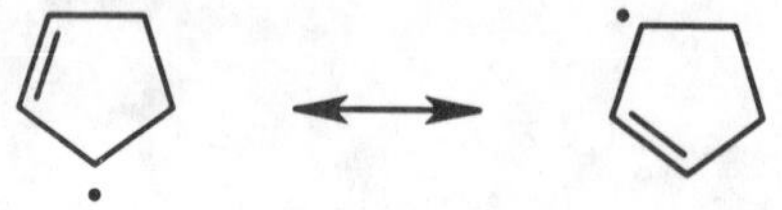 The two contributors are equally stable.

5. 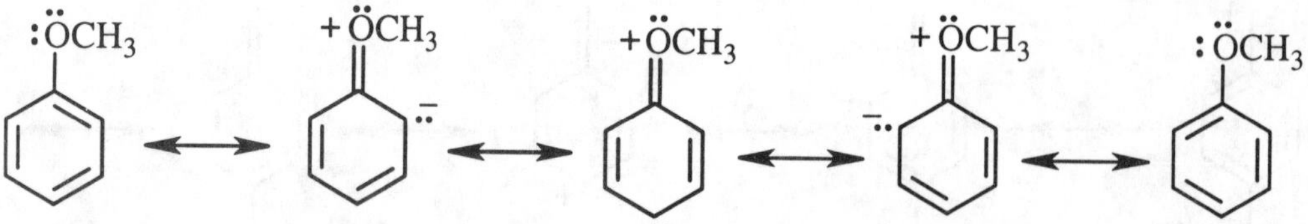 major — minor — minor — minor — major

6. 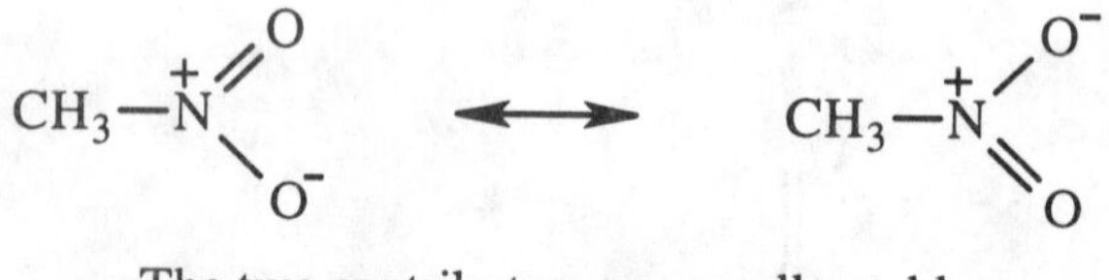 $CH_3{-}\overset{+}{N}(=O)O^- \longleftrightarrow CH_3{-}\overset{+}{N}(O^-)=O$

   The two contributors are equally stable.

7. $CH_3CH_2\overset{O}{\overset{\|}{C}}OCH_2CH_3 \longleftrightarrow CH_3CH_2\overset{O^-}{\overset{|}{C}}{=}\overset{+}{O}CH_2CH_3$
   major — minor

8. $CH_3CH{=}CHCH{=}CH\dot{C}H_2$ (minor) ⟷ $CH_3CH{=}CH\dot{C}HCH{=}CH_2$ (major) ⟷ $CH_3\dot{C}HCH{=}CHCH{=}CH_2$ (major)

9. $HC(=O)NHCH_3$ (major) ⟷ $HC(O^-){=}\overset{+}{N}HCH_3$ (minor)

10. $CH_2{=}CH\bar{C}H_2$ ⟷ $\bar{C}H_2CH{=}CH_2$

The two contributors are equally stable.

11. 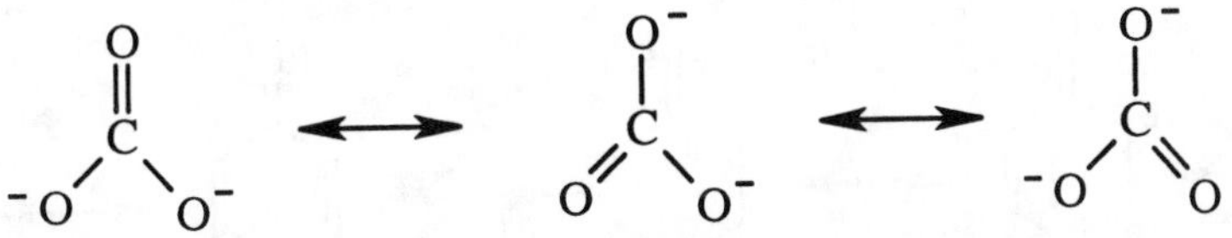 $O{=}C(O^-)O^-$ ⟷ $^-O{-}C(=O)O^-$ ⟷ $^-O{-}C(O^-){=}O$

The three contributors are equally stable.

12. $HC(=O)CH{=}CH\bar{C}H_2$ (minor) ⟷ $HC(=O)\bar{C}HCH{=}CH_2$ (minor) ⟷ $HC(O^-){=}CHCH{=}CH_2$ (major)

13. 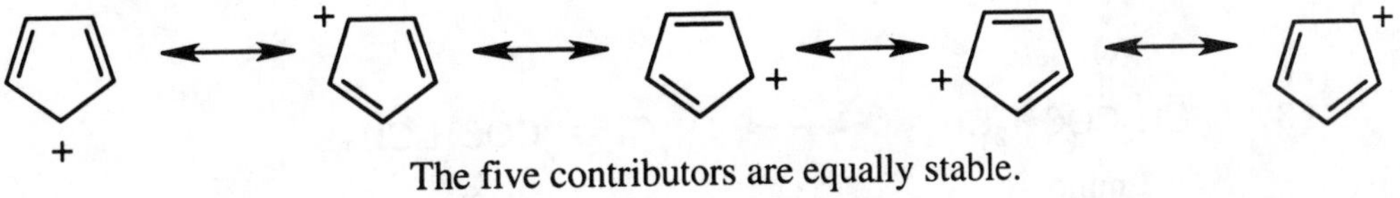 Five cyclopentadienyl cation contributors, with the + charge on each of the five ring carbons in turn.

The five contributors are equally stable.

14. $CH_3\bar{C}H{-}\overset{+}{N}(=O)O^-$ (minor) ⟷ $CH_3\bar{C}H{-}\overset{+}{N}(O^-){=}O$ (minor) ⟷ $CH_3CH{=}\overset{+}{N}(O^-)O^-$ (major)

15. The electrons can move in two different direction. They can move out of the benzene ring toward the alkyl group; they can move into the benzene ring away from the alkyl group.

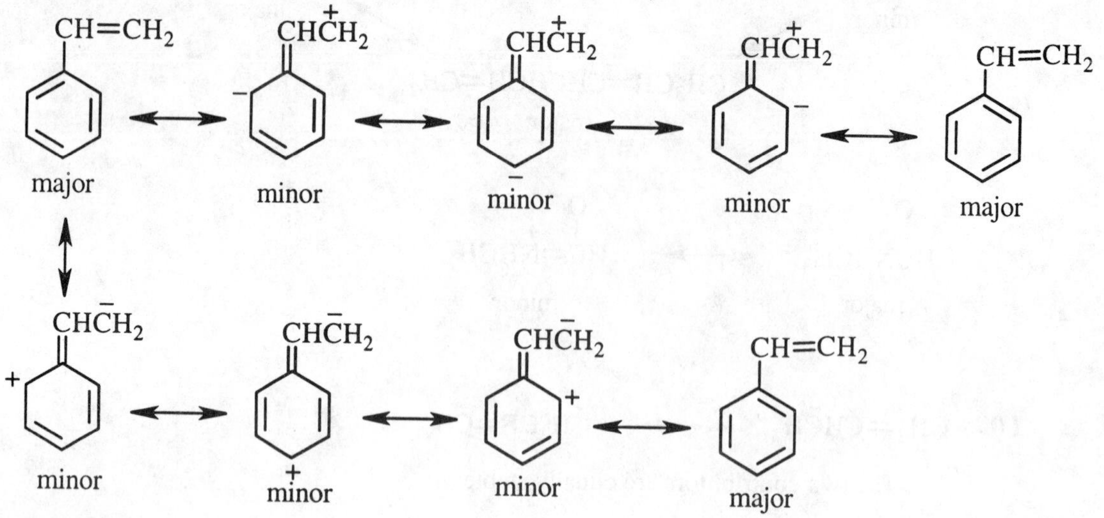

$CH{=}CH_2$ (major) ⟷ $\overset{+}{C}HCH_2$ (minor) ⟷ $\overset{+}{C}HCH_2$ (minor) ⟷ $\overset{+}{C}HCH_2$ (minor) ⟷ $CH{=}CH_2$ (major)

⟷

$\overset{-}{C}HCH_2$ (minor) ⟷ $\overset{-}{C}HCH_2$ (minor) ⟷ $\overset{-}{C}HCH_2$ (minor) ⟷ $CH{=}CH_2$ (major)

16. Cl (major) ⟷ +Cl (minor) ⟷ +Cl (minor) ⟷ +Cl (minor) ⟷ Cl (major)

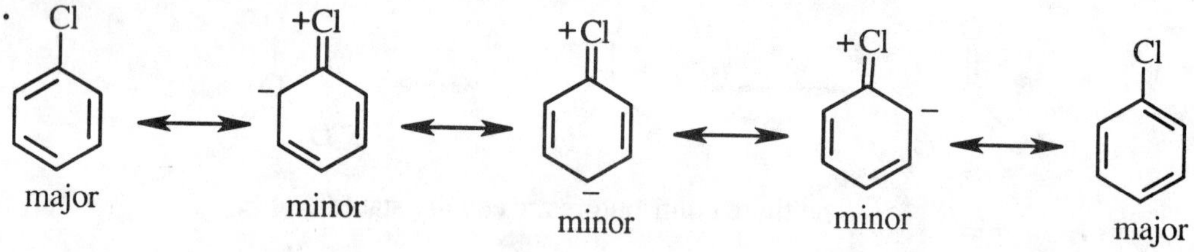

17. $CH_3\overset{O}{\overset{\|}{C}}-\overset{-}{C}H-\overset{O}{\overset{\|}{C}}CH_3$ ⟷ $CH_3\overset{O}{\overset{\|}{C}}-CH{=}\overset{O^-}{\overset{|}{C}}CH_3$ ⟷ $CH_3\overset{O^-}{\overset{|}{C}}{=}CH-\overset{O}{\overset{\|}{C}}CH_3$

minor major major

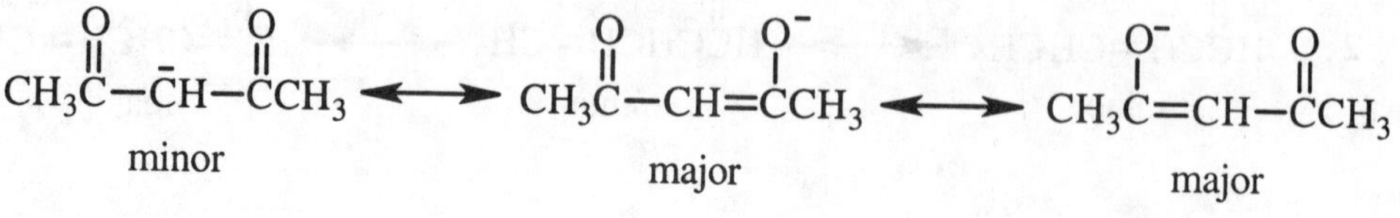

18. $\overset{-}{C}H_2\overset{O}{\overset{\|}{C}}OCH_2CH_3$ ⟷ $CH_2{=}\overset{O^-}{\overset{|}{C}}OCH_2CH_2$

minor major

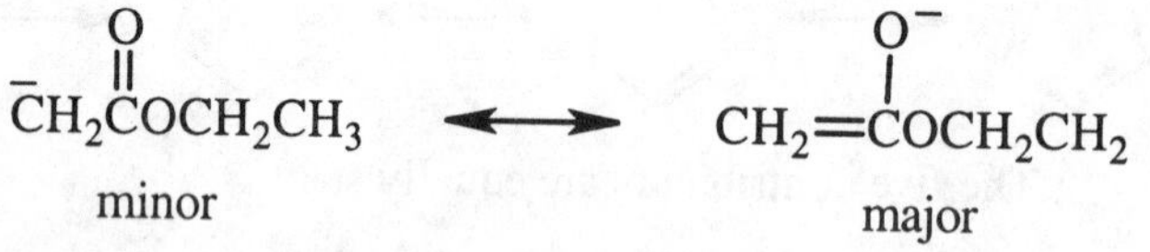

b. 4, 6, 10, 11, and 13

18.

a. $CH_3\overset{+}{C}HCH{=}CH_2$

This makes the greater contribution because the positive charge is on a secondary carbon.

b. CH_3 (on + carbon of cyclopentene)

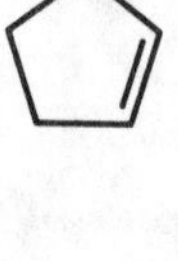

This makes the greater contribution because the positive charge is on a tertiary carbon.

c. O^- (on benzene ring)

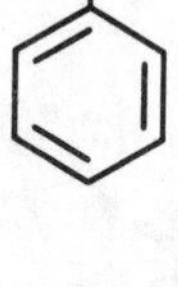

This makes the greater contribution because the negative charge is on an oxygen atom.

19. **a.** The resonance contributors show that the carbonyl oxygen has the greater electron density.

$$CH_3\overset{\overset{\ddot{O}:}{\|}}{C}-\ddot{O}CH_3 \longleftrightarrow CH_3\overset{\overset{:\ddot{O}:^-}{|}}{C}=\overset{+}{O}CH_3$$

b. The compound on the right has the greater amount of electron density on its nitrogen because the compound on the left has a resonance contributor with a positive charge on the nitrogen.

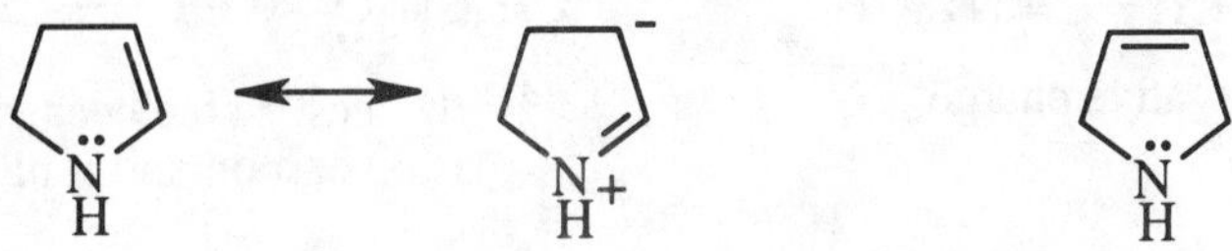

c. The compound with the cyclohexane ring has the greater amount of electron density on its oxygen because the nonbonding electrons on the nitrogen can be delocalized onto the oxygen. There is less delocalization onto oxygen by the nonbonding electrons in the compound with the benzene ring because the nonbonding electrons can also be delocalized away from the oxygen into the benzene ring.

20. The methyl group on benzene can lose a proton easier than the methyl group on cyclohexane because the electrons left behind in the former can be delocalized by resonance into the benzene ring.

21. The carbocation is stable because the positive charge is shared by 10 carbon atoms (the central carbon and 3 carbons of each of the 3 benzene rings).

22.

$$CH_3CH_2\ddot{O}-\overset{:\ddot{O}:^-}{\overset{|}{C}}=CH-C\equiv N:$$

#1 most stable because the negative charge is on oxygen

$$CH_3CH_2\ddot{O}-\overset{\ddot{O}:}{\overset{\|}{C}}-CH=C=\ddot{N}:^-$$

#2 because the negative charge is on nitrogen

$$CH_3CH_2\ddot{O}-\overset{\ddot{O}:}{\overset{\|}{C}}-\underset{-}{\ddot{C}H}-C\equiv N:$$

#3 because the negative charge is on carbon

$$CH_3CH_2\overset{+}{\ddot{O}}=\overset{:\ddot{O}:^-}{\overset{|}{C}}-\underset{-}{\ddot{C}H}-C\equiv N:$$

#4 the least stable because the negative charge is on carbon and it has separated charges

23. The more the electrons that are left behind when the proton is removed can be delocalized, the easier it is to remove the proton (i.e., the more acidic the compound). The negative charge that would be generated in the first compound can be delocalized onto two other carbons; the negative charge that would be generated in the second compound can be delocalized onto one other carbon; the negative charge that would be generated in the last compound cannot be delocalized.

24.

a. $CH_3\overset{O}{\overset{\|}{C}}O^-$ the negative charge is shared by 2 oxygens

b. $CH_3\overset{O}{\overset{\|}{C}}\underset{-}{C}H\overset{O}{\overset{\|}{C}}CH_3$ the negative charge is shared by a carbon and 2 oxygens

c. $CH_3CH_2\underset{-}{C}H\overset{O}{\overset{\|}{C}}CH_3$ the negative charge is shared by a carbon and an oxygen

d. $CH_3\overset{NH}{\overset{\|}{C}}NH_2$ this compound has delocalized electrons

e. $CH_3\underset{CH_3}{\overset{-}{C}}-\overset{O}{\overset{\|}{C}}H$ the negative charge is shared by a carbon and an oxygen

f. (cyclic imide anion with two C=O, N^-) the negative charge is shared by a nitrogen and 2 oxygens

25.

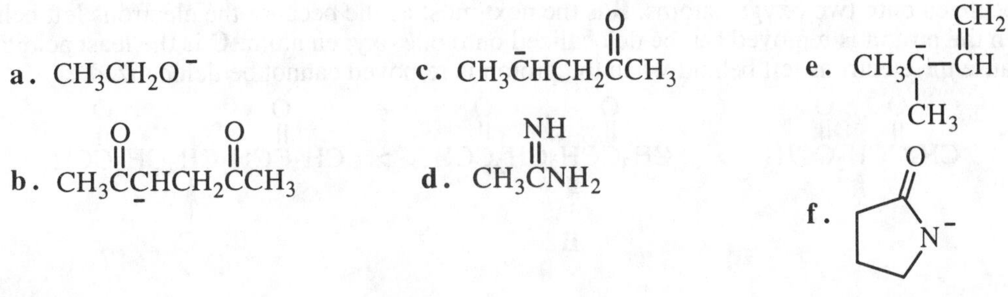

a. $CH_3CH_2O^-$

b. $CH_3\overset{O}{\overset{\|}{C}}\underset{-}{C}HCH_2\overset{O}{\overset{\|}{C}}CH_3$

c. $CH_3\overset{-}{C}HCH_2\overset{O}{\overset{\|}{C}}CH_3$

d. $CH_3\overset{NH}{\overset{\|}{C}}NH_2$

e. $CH_3\overset{-}{C}(=CH_2)CH(CH_3)$

f. (pyrrolidinone with N^-)

26. The carbocation leading to 1,1-dichloroethane is more stable than the carbocation leading to 1,2-dichloroethane because the positive charge on the intermediate leading to 1,1-dichloroethane is shared by carbon and chlorine. Since the more stable carbocation is the one that is easier to form, the final product of the reaction is 1,1-dichloroethane.

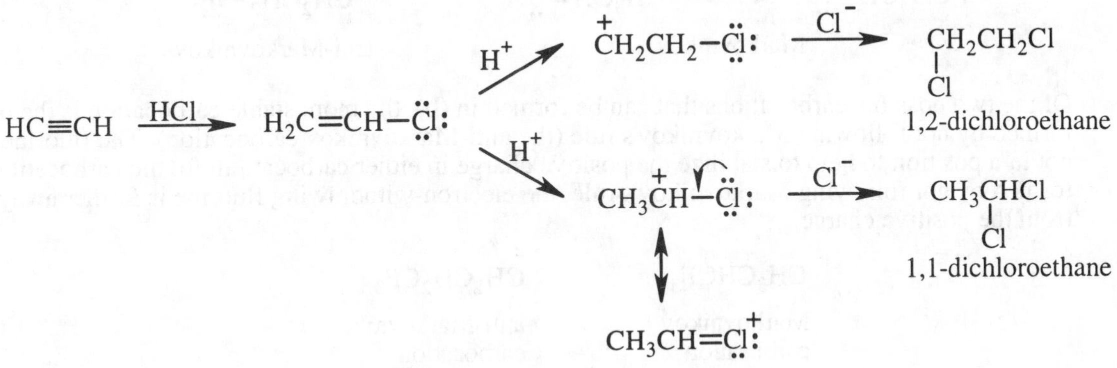

$$HC{\equiv}CH \xrightarrow{HCl} H_2C{=}CH{-}Cl \xrightarrow{H^+} \overset{+}{C}H_2CH_2{-}Cl \xrightarrow{Cl^-} CH_2CH_2Cl \text{ (1,2-dichloroethane)}$$

$$H_2C{=}CH{-}Cl \xrightarrow{H^+} CH_3\overset{+}{C}H{-}Cl \longleftrightarrow CH_3CH{=}\overset{+}{Cl} \qquad CH_3\overset{+}{C}H{-}Cl \xrightarrow{Cl^-} CH_3CHCl_2 \text{ (1,1-dichloroethane)}$$

27. The resonance contributors of pyrrole are more stable because the positive charge is on nitrogen. In furan, the positive charge is on oxygen, which being more electronegative, is less stable with a positive charge.

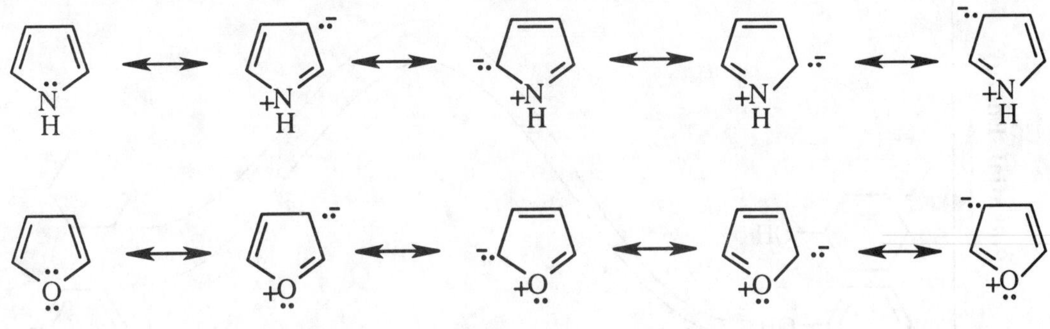

(resonance contributors of pyrrole: neutral pyrrole ⟷ four contributors with $\overset{+}{N}$–H and a negative charge on a ring carbon)

(resonance contributors of furan: neutral furan ⟷ four contributors with $\overset{+}{O}$ and a negative charge on a ring carbon)

28. **A** is the most acidic because the electrons left behind when the proton is removed can be delocalized onto two oxygen atoms. **B** is the next most acidic because the electrons left behind when the proton is removed can be delocalized onto one oxygen atom. **C** is the least acidic because the electrons left behind when the proton is removed cannot be delocalized.

$$\underset{\uparrow \atop \textbf{A}}{CH_3\overset{O}{\overset{\|}{C}}CH_2\overset{O}{\overset{\|}{C}}CH_3} > \underset{\uparrow \atop \textbf{B}}{CH_3\overset{O}{\overset{\|}{C}}CH_2CH_2\overset{O}{\overset{\|}{C}}CH_3} > \underset{\uparrow \atop \textbf{C}}{CH_3\overset{O}{\overset{\|}{C}}CH_2CH_2CH_2\overset{O}{\overset{\|}{C}}CH_3}$$

29. Of the two possible carbocations that can be formed in **(a)**, the more stable carbocation is the one formed by following Markovnikov's rule. It is more stable because the positive charge is shared by carbon and fluorine.

$$CH_3\overset{+}{C}H-\ddot{\underset{\cdot\cdot}{F}}: \longleftrightarrow CH_3CH=\overset{+}{\underset{\cdot\cdot}{F}}: \qquad \overset{+}{C}H_2CH_2-\ddot{\underset{\cdot\cdot}{F}}:$$

Markovnikov anti-Markovnikov

Of the two possible carbocations that can be formed in **(b)**, the more stable carbocation is the one formed by not following Markovnikov's rule (the anti-Markovnikov carbocation). The fluorine is not in a position to help to stabilize the positive charge in either carbocation. In the carbocation formed by not following Markovnikov's rule, the electron-withdrawing fluorine is farther away from the positive charge.

$$CH_3\overset{+}{C}HCF_3 \qquad \overset{+}{C}H_2CH_2CF_3$$

Markovnikov carbocation anti-Markovnikov carbocation

30. **a.** and **d.**

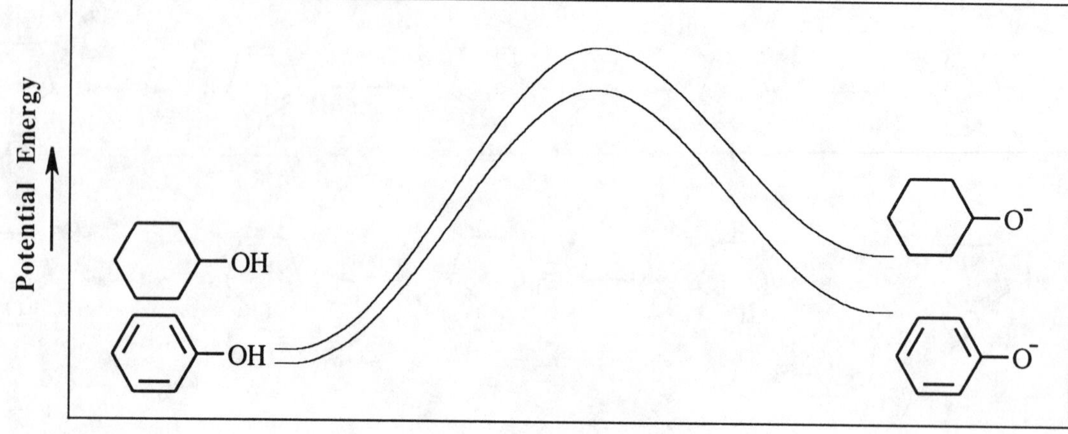

b. :ÖH ⟷ +ÖH ⟷ +ÖH ⟷ +ÖH ⟷ :ÖH

c. :Ö:⁻ ⟷ Ö: ⟷ Ö: ⟷ Ö: ⟷ :Ö:⁻

d. The resonance contributors in "**c**" are more stable than the resonance contributors in "**b**". Therefore, the phenolate ion has greater resonance stabilization than phenol. Thus, in the energy diagram, the difference in energy between the phenolate ion and the cyclohexoxide ion is greater than the difference in energy between phenol and cyclohexanol.

e. Because of the greater resonance stabilization of the phenolate ion compared to phenol, phenol has a larger K_a than cyclohexanol.

f. Because it has a larger K_a, phenol is a stronger acid.

31.

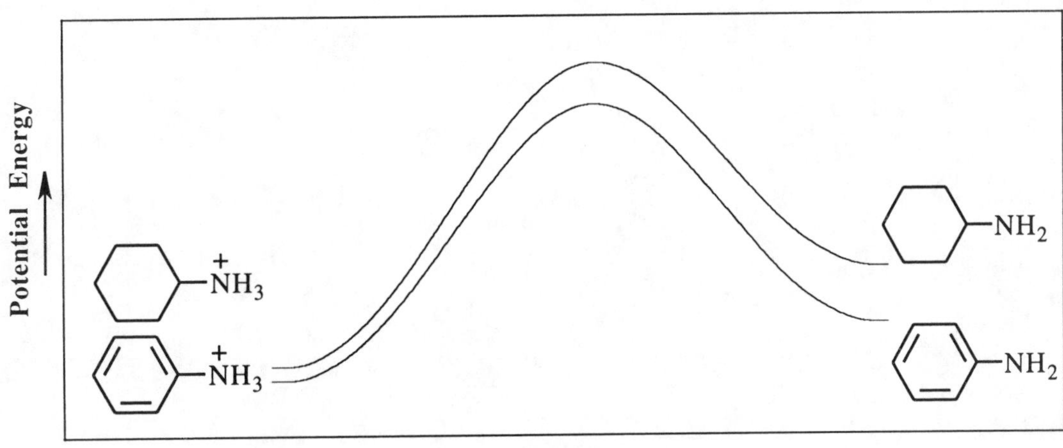

b. $+NH_3$ ⟷ $+NH_3$

c. :NH_2 ⟷ +NH_2 ⟷ +NH_2 ⟷ +NH_2 ⟷ :NH_2

d. Aniline has greater resonance stabilization than the anilinium ion. Thus in the energy diagram, the difference in energy between aniline and cyclohexylamine is greater than the difference in energy between the anilinium ion and the cyclohexylammonium ion.

e. Because of the greater resonance stabilization of aniline compared to the anilinium ion, the anilinium ion has a larger K_a than the cyclohexylammonium ion.

f. Because it has a larger K_a, the anilinium ion is a stronger acid than the cyclohexylammonium ion. Therefore, cyclohexylamine is a stronger base than aniline. (The stronger the acid, the weaker its conjugate base.)

Chapter 6 Practice Test

1. For each of the following pairs of compounds indicate the one that is the more stable.

   a. 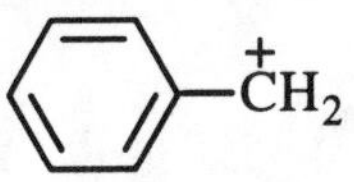 $-\overset{+}{C}H_2$ or $-\overset{+}{C}H_2$

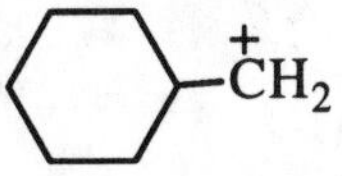

   b. $CH_3\overset{-}{C}HCH_3$ or $CH_3\overset{-}{C}HC{\equiv}CH$

   c. $CH_3\overset{-}{C}HCH_2\overset{O}{\overset{\|}{C}}CH_3$ or $CH_3\overset{-}{C}H\overset{O}{\overset{\|}{C}}CH_3$

   d. $CH_2{=}CH\overset{+}{C}H_2$ or $CH_2{=}CHCH_2\overset{+}{C}H_2$

   e. 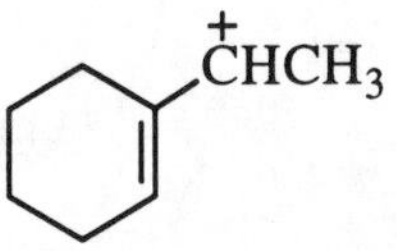 $\overset{+}{C}HCH_3$ or $\overset{+}{C}HCH_3$

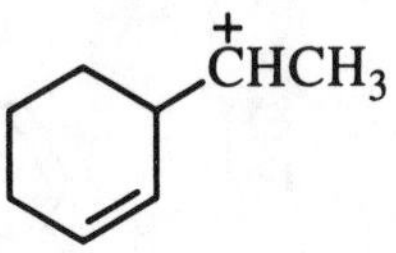

2. Draw resonance contributors for the following.

   a. $CH_3CH{=}CH{-}\ddot{\underset{..}{O}}CH_3$

   b. $CH_3CH{=}CH{-}CH{=}CH{-}\overset{+}{C}H_2$

   c. $\overset{-}{C}H_2{-}CH{=}CH{-}\overset{O}{\overset{\|}{C}}H$

3. Which compounds do not have delocalized electrons?

   $CH_3CH_2NHCH{=}CHCH_3$ $\quad$ $CH_3\underset{+}{\overset{CH_3}{\overset{|}{C}}}CH_2CH{=}CH_2$ $\quad$ $CH_2{=}CHCH_2CH{=}CH_2$

   $CH_2{=}CH\overset{O}{\overset{\|}{C}}CH_3$ $\quad$ $CH_3CH_2NHCH_2CH{=}CHCH_3$

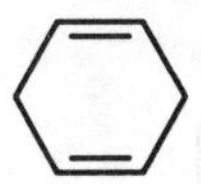

4. Which of the following pairs are resonance contributors?

a. CH_3CH_2OH and CH_3OCH_3

b. $CH_3\overset{O}{\overset{\|}{C}}OH$ and $CH_3\overset{O}{\overset{\|}{C}}O^-$

c. $CH_3\overset{O}{\overset{\|}{C}}OH$ and $CH_3\overset{O^-}{\overset{|}{C}}=\overset{+}{O}H$

d. $CH_3CH_2\overset{O}{\overset{\|}{C}}H$ and $CH_3CH=\overset{OH}{\overset{|}{C}}H$

5. Which is a stronger base?

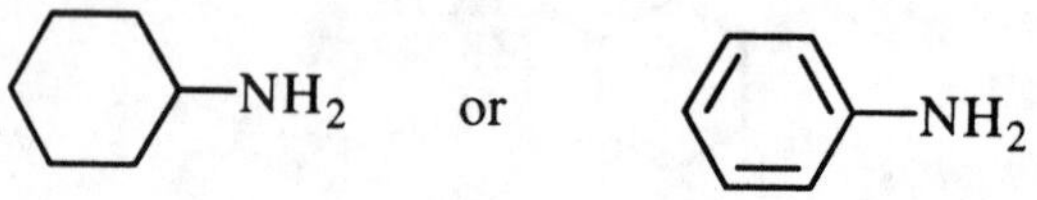

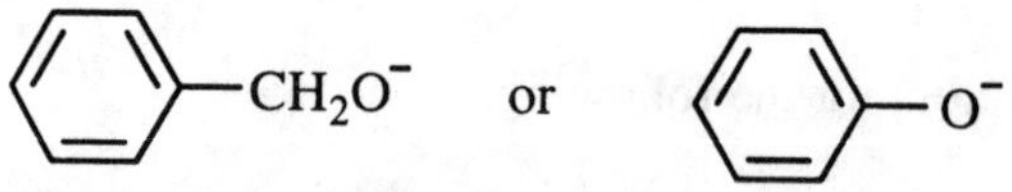

cyclohexyl$-NH_2$ or phenyl$-NH_2$

phenyl$-CH_2O^-$ or phenyl$-O^-$

6. Draw resonance contributors for the following.

a. phenyl$-\ddot{N}H_2$

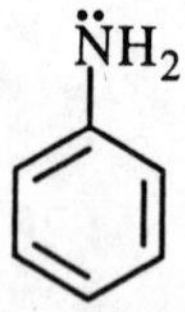

b. phenyl$-\overset{+}{N}H_3$

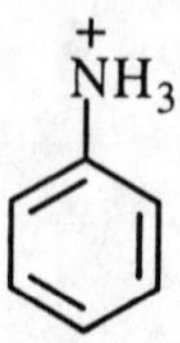

c. phenyl$-:\ddot{O}:^-$

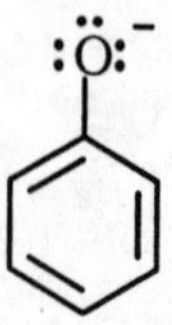

7. For each of the following pairs of compounds indicate the one that is the more acidic.

a. 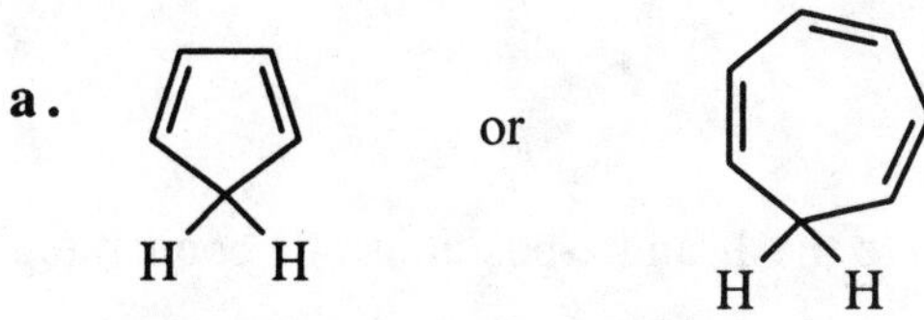

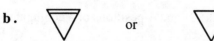

b. (cyclopropene) or (cyclopropane)

c. $C_6H_5-CH_2\overset{+}{N}H_3$ or $C_6H_5-\overset{+}{N}H_3$

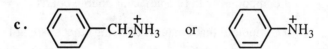

8. Which resonance contributor makes a greater contribution to the resonance hybrid?

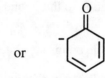

a. O^- or O

b. CH_3 or CH_3

CHAPTER 7
Reactions of Dienes

Important Terms

1,2-addition (direct addition)	addition to the 1- and 2-positions of a conjugated system.
1,4-addition (conjugate addition)	addition to the 1- and 4-positions of a conjugated system.
allene	a compound with two adjacent double bonds.
bicyclic compound	a compound that contains two rings that share at least one carbon.
bridged bicyclic compound	a bicyclic compound in which the rings share two nonadjacent carbons.
common intermediate	an intermediate that two compounds have in common.
conjugate addition (1,4-addition)	addition to the 1- and 4-positions of a conjugated system.
conjugated double bonds	double bonds separated by one single bond.
cumulated double bonds	double bonds that are adjacent to one another.
cycloaddition reaction	a reaction in which two π-electron-containing reactants combine to form a cyclic product.
[4 + 2] cycloaddition reaction	a cycloaddition reaction in which four π electrons come from one reactant and two π electrons come from the other reactant.
Diels-Alder reaction	a [4 + 2] cycloaddition reaction.
diene	a hydrocarbon with two double bonds.
dienophile	an alkene that reacts with a diene in a Diels-Alder reaction.
direct addition (1,2-addition)	addition to the 1- and 2-positions of a conjugated system.
endo	a substituent is endo if it and the bridge are on opposite sides of the bicyclic compound.
equilibrium control	thermodynamic control.
exo	a substituent is exo if it and the bridge are on the same side of the bicyclic compound.
fused bicyclic compound	a bicyclic compound in which the rings share two adjacent carbons.
isolated double bonds	double bonds separated by more than one single bond.
kinetic control	when a reaction is under kinetic control, the relative amounts of the products depend on the rates at which they are formed.

kinetic product	the product that is formed the fastest.
pericyclic reaction	a reaction that takes place in one step as a result of a cyclic reorganization of electrons.
polyene	a compound that has several double bonds.
secondary orbital overlap	the interaction of the *p* orbitals of the dienophile with the *p* orbitals of the new double bond formed in the diene.
***s*-cis-conformation**	the conformation in which two double bonds of a conjugated diene are on the same side of a connecting single bond.
***s*-trans-conformation**	the conformation in which two double bonds of a conjugated diene are on opposite sides of a connecting single bond.
spirocyclic compound	a bicyclic compound in which the rings share one carbon.
tetraene	a compound with four double bonds.
thermodynamic control	when a reaction is under thermodynamic control, the relative amounts of the products depend on their stabilities.
thermodynamic product	the most stable product.
triene	a compound with three double bonds

Solutions to Problems

1. **a.** 1,5-cyclooctadiene
   b. 1-hepten-4-yne
   c. 4-methyl-1,4-hexadiene
   d. 2,4-dimethyl-4-hexen-1-ol
   e. 1,6-dimethyl-1,3-cyclohexadiene
   f. 3-butyn-1-ol
   g. 1,3,5-heptatriene
   h. 5-vinyl-5-octen-1-yne

2. **a** and **c** have only two stereoisomers because in each case there are two identical substituents bonded to one of the sp^2 carbons.

a.

$(CH_3)_2C{=}CH{-}CH{=}CH{-}CH_3$

(*E*)-2-methyl-2,4-hexadiene

(*Z*)-2-methyl-2,4-hexadiene

b.

$CH_3CH{=}CH{-}CH{=}CH{-}CH_2CH_3$

(2*E*,4*E*)-2,4-pentadiene

(2*E*,4*Z*)-2,4-pentadiene

(2*Z*,4*E*)-2,4-pentadiene

(2*Z*,4*Z*)-2,4-pentadiene

c.

$H_2C{=}CH{-}CH{=}CH{-}CH_3$

(*E*)-1,3-pentadiene

(*Z*)-1,3-pentadiene

3.

$$\underset{\text{2,5-dimethyl-2,4-hexadiene}}{\mathrm{CH_3\overset{\overset{\displaystyle CH_3}{|}}{C}{=}CHCH{=}\overset{\overset{\displaystyle CH_3}{|}}{C}CH_3}} > \underset{\text{2,4-hexadiene}}{\mathrm{CH_3CH{=}CHCH{=}CHCH_3}} > \underset{\text{1,3-pentadiene}}{\mathrm{CH_3CH{=}CHCH{=}CH_2}} > \underset{\text{1,4-pentadiene}}{\mathrm{CH_2{=}CHCH_2CH{=}CH_2}}$$

4. ψ_3 has three nodes (two vertical and one horizontal).
ψ_4 has four nodes (three vertical and one horizontal).

5.

Fischer projections (top to bottom: CH_3; H–C–Br; CH_2; CH_2; H–C–Br; CH_3):

- CH_3 / H—C—Br / CH_2 / CH_2 / H—C—Br / CH_3
- CH_3 / H—C—Br / CH_2 / CH_2 / Br—C—H / CH_3
- CH_3 / Br—C—H / CH_2 / CH_2 / H—C—Br / CH_3

$$\mathrm{CH_2{=}CHCH_2CH_2\overset{\overset{\displaystyle CH_3}{|}}{\underset{\underset{\displaystyle Cl}{|}}{C}}CH_3}$$

no other isomers are possible

6.

a. 1-chloro-1-methylcyclohept-4-ene (ring structure with CH_3 and Cl on the same carbon)

The intermediate is a tertiary carbocation.

b. $\mathrm{CH_2{=}CHCH_2CH_2CH_2\overset{\overset{\displaystyle CH_3}{|}}{\underset{\underset{\displaystyle Br}{|}}{C}}CH_3}$

The intermediate is a tertiary carbocation

c. $\mathrm{CH_2{=}CHCH_2CH_2{-}\overset{\overset{\displaystyle CH_3}{|}}{\underset{\underset{\displaystyle H}{|}}{C}}{-}CH_2Br}$ $\quad$ $\mathrm{BrCH_2{-}\overset{\overset{\displaystyle CH_3}{|}}{\underset{\underset{\displaystyle H}{|}}{C}}{-}CH_2CH_2CH{=}CH_2}$

The intermediate is a tertiary radical.

d. $\mathrm{HC{\equiv}CCH_2CH_2{-}\overset{\overset{\displaystyle Cl}{|}}{\underset{\underset{\displaystyle H}{|}}{C}}{-}CH_2Cl}$ $\quad$ $\mathrm{ClCH_2{-}\overset{\overset{\displaystyle Cl}{|}}{\underset{\underset{\displaystyle H}{|}}{C}}{-}CH_2CH_2C{\equiv}CH}$

An alkene is more reactive than an alkyne.

7. The indicated double bond is the most reactive in an electrophilic substitution reaction because addition of an electrophile to this double bond forms the most stable carbocation (a tertiary allylic carbocation).

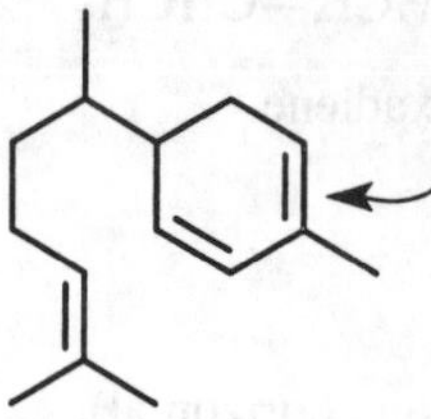

8.

$$CH_2{=}CH{-}CH{=}CH{-}CH{=}CH_2 \xrightarrow{HBr} CH_3{-}\underset{Br}{CH}{-}CH{=}CH{-}CH{=}CH_2$$

$$CH_3{-}CH{=}CH{-}\underset{Br}{CH}{-}CH{=}CH_2$$

$$CH_3{-}CH{=}CH{-}CH{=}CH{-}\underset{Br}{CH_2}$$

9.

a. $CH_3\underset{Cl}{C}H\underset{Cl}{C}HCH{=}CHCH_3$ + $CH_3\underset{Cl}{C}HCH{=}CH\underset{Cl}{C}HCH_3$

1,2-addition product — 1,4-addition product

b. $CH_3CH_2\overset{Br}{\underset{CH_3}{C}}{-}\underset{CH_3}{C}{=}CHCH_3$ + $CH_3CH_2\underset{CH_3}{C}{=}\overset{Br}{\underset{CH_3}{C}}CHCH_3$

1,2-addition product — 1,4-addition product

c. $CH_3CH{=}CH\overset{Br}{\underset{CH_3}{C}}CH_2CH_3$ + $CH_3\underset{Br}{C}HCH{=}\underset{CH_3}{C}CH_2CH_3$

1,2-addition product — 1,4-addition product

d. 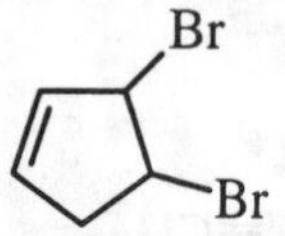 3,4-dibromocyclopentene (Br, Br) + 3,5-dibromocyclopentene (Br, Br)

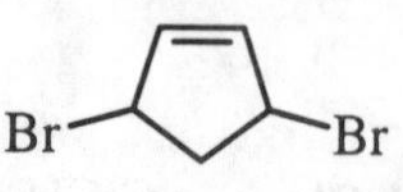

1,2-addition product — 1,4-addition product

10. **a.** The initial addition of H^+ to C-1 forms **A** (the 1,2-product) and **B** (the 1,4-product). Notice that in this case A and B are the same, so only one product is actually formed.
The initial addition of H^+ to C-4 forms **C** (the 1,2-product) and **D** (the 1,4-product).

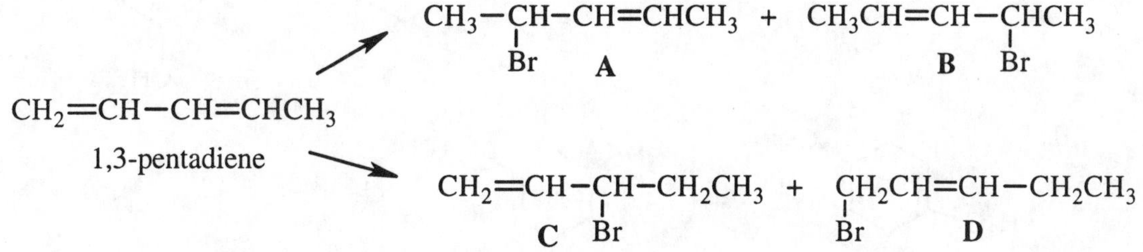

$CH_2{=}CH{-}CH{=}CHCH_3$ (1,3-pentadiene) → $CH_3{-}CH(Br){-}CH{=}CHCH_3$ (**A**) + $CH_3CH{=}CH{-}CH(Br)CH_3$ (**B**)

$CH_2{=}CH{-}CH{=}CHCH_3$ (1,3-pentadiene) → $CH_2{=}CH{-}CH(Br){-}CH_2CH_3$ (**C**) + $CH_2(Br)CH{=}CH{-}CH_2CH_3$ (**D**)

b. The major product obtained from addition of HBr to 1,3-pentadiene is 4-bromo-2-pentene (**A** or **B**) for two reasons.

1. The carbocation leading to **A** and **B** is more stable (the positive charge is shared by two secondary allylic carbons) than the carbocation leading to **C** and **D** (the positive charge is shared by a secondary allylic and a primary allylic carbon).

2. **A** and **B** are the same compound, so there are two routes to the formation of 4-bromo-2-pentene from 1,3-pentadiene; in contrast, **C** and **D** are different compounds.

11. **a.** Solved in the text.

b. $CH_3CH{=}CHC(CH_3)(Cl)CH_3$ — kinetic product; $CH_3CH(Cl)CH{=}C(CH_3)CH_3$ — thermodynamic product

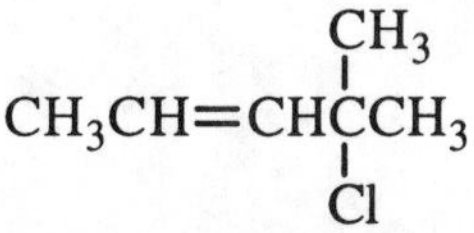

c. kinetic product: 3-chloro-3-methylcyclohexene (CH₃ and Cl on the same ring carbon, adjacent to the double bond); thermodynamic product: 5-chloro-1-methylcyclohexene (CH₃ on the double-bond carbon, Cl on ring)

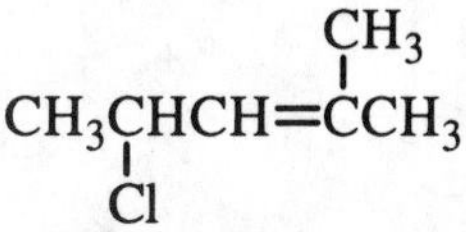

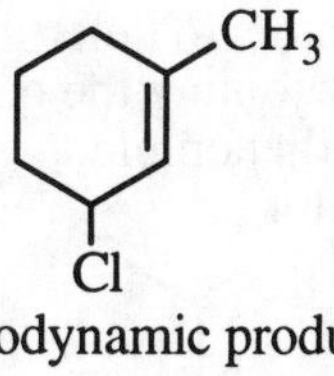

d. kinetic product: 1-chloro-1-(CH=CHCH₃)cyclohexane (Cl and $CH{=}CHCH_3$ on the same ring carbon); thermodynamic product: cyclohexylidene=$CHCH(Cl)CH_3$

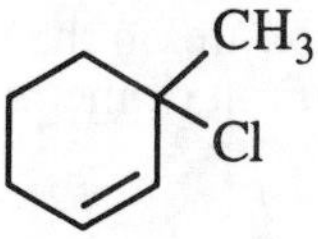

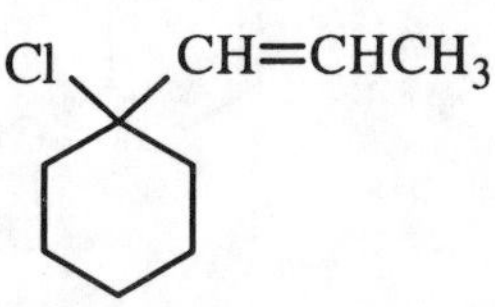

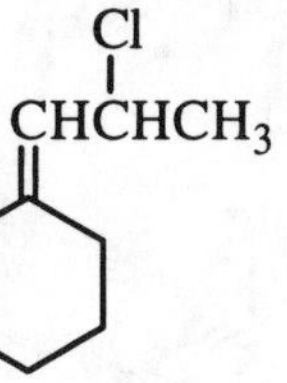

12.

a.

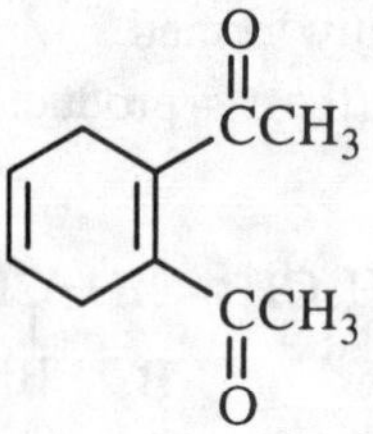

b.

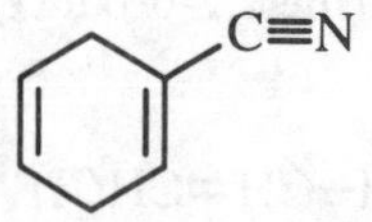

c.

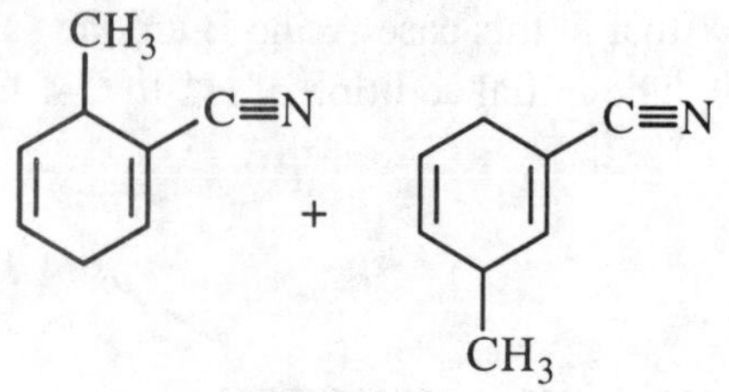

d.

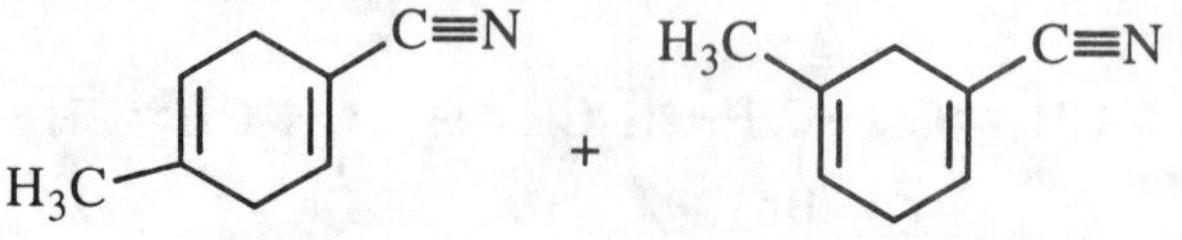

13. **a.** It is not optically active because it is a meso compound. (It has 2 chirality centers and a plane of symmetry.)

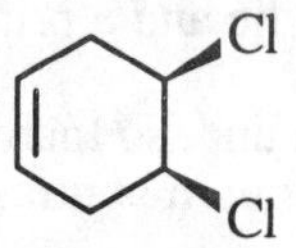

b. It is not optically active because it is a racemic mixture. (Identical amounts of the enantiomers will be obtained.)

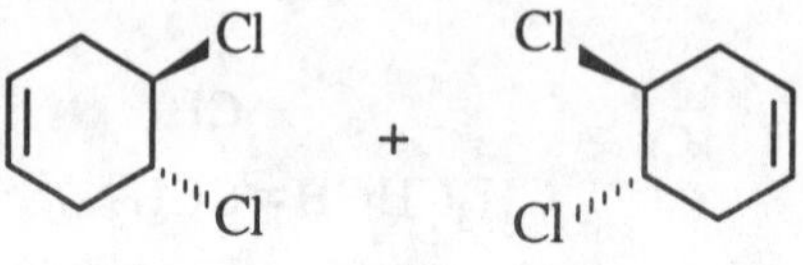

14. First draw the resonance contributors to determine where the charges are on the reactants. The major product is obtained by joining the negatively charged carbon of the diene with the positively charged carbon of the dienophile.

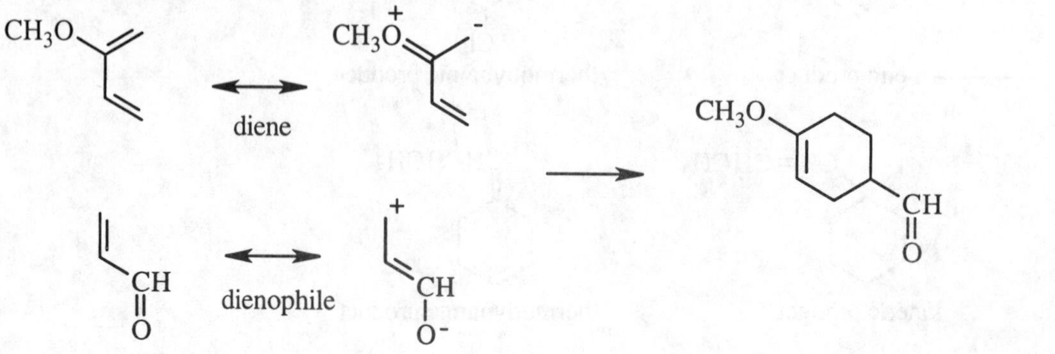

diene

dienophile

15. **a** and **d** will not react, because they are both locked in an *s*-trans conformation.

c and **e** will react, because they are both locked in an *s*-cis conformation.

b and **f** will react, because they can rotate into an *s*-cis conformation.

16. Solved in the text.

17.

a. 1,3-butadiene + $H(C{\equiv}N)C{=}CH_2$ (acrylonitrile)

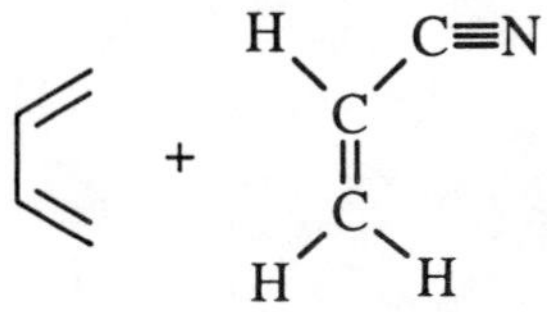

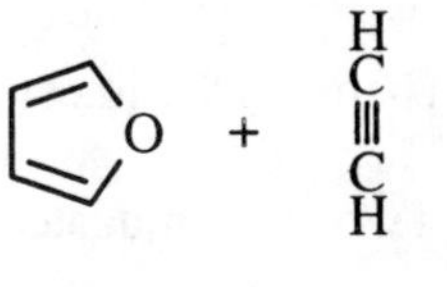

b. 1,3-butadiene + $HC(=O)CH{=}CHC(=O)H$

c. cyclopentadiene + $CH_3OC(=O)C{\equiv}CC(=O)OCH_3$

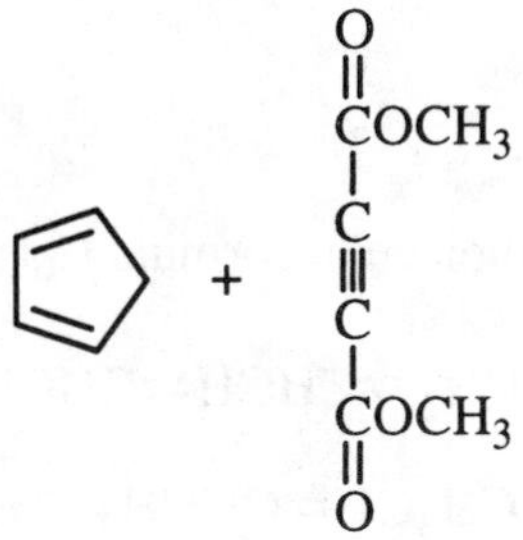

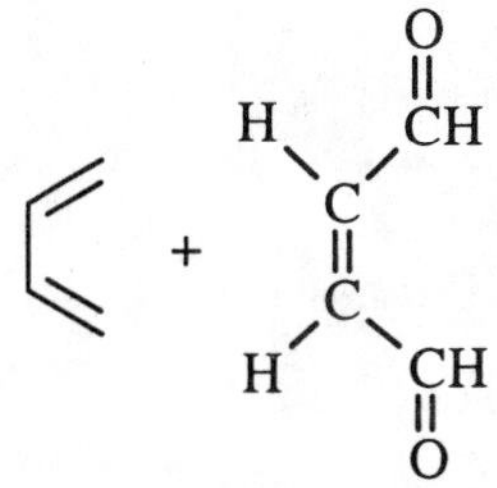

d. 1,3-butadiene + 2-cyclohexenone

e. furan + $HC{\equiv}CH$

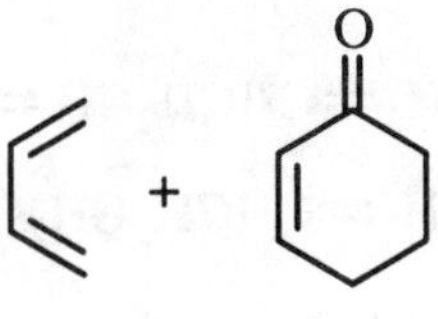

f. 1,3-butadiene + $HOC(=O)CH{=}CHC(=O)OH$

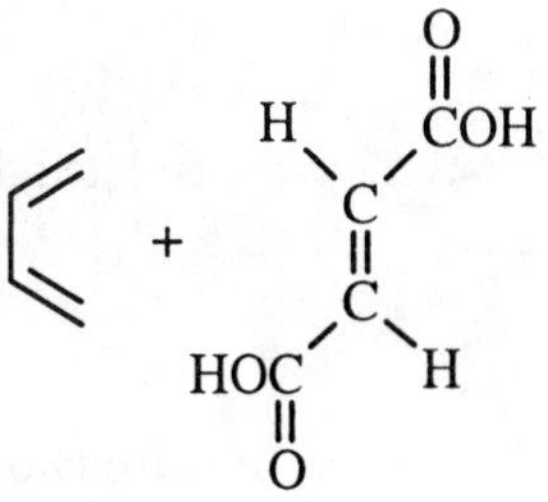

18. **a.** *exo*-2-chlorobicyclo[2.2.1]heptane

b. 3-methylbicyclo[4.4.0]decane

c. bicyclo[2.2.2]octane

d. *cis*-6,7-dimethylbicyclo[3.2.0]heptane

19. **a.** 1-octen-6-yne

b. (*Z*)-3-hexen-1-ol

c. 1,5-octadiyne

d. 5-chloro-1,3-cyclohexadiene

e. 1-methyl-1,3,5-cycloheptatriene

20.

a. $ClCH_2$, CH_3 / C=C / H, H / C=C / H, CH_3

c. H, CH_3 / C=C / CH_3CH_2, $CH_2CH_2CH_3$ / C=C / CH_3, H

b. H, CH_3 / C=C / CH_3CH_2, H / C=C / H, $CH_2CH_2CH_3$

d. Br / CH_3CH, H / C=C / H, CH_2CH_3 / C=C / Br, H

21. a. There are six linear dienes with molecular formula C_6H_{10}.

b. Two are conjugated dienes. $CH_2{=}CHCH{=}CHCH_2CH_3$

$CH_3CH{=}CHCH{=}CHCH_3$

c. Two are isolated dienes. $CH_2{=}CHCH_2CH{=}CHCH_3$

$CH_2{=}CHCH_2CH_2CH{=}CH_2$

There are also two cumulated dienes. $CH_2{=}C{=}CHCH_2CH_2CH_3$

$CH_3CH{=}C{=}CHCH_2CH_3$

22. Both compounds form the same product when they are hydrogenated, so the difference in heats of hydrogenation will depend only on the difference in the stabilities of the reactants. Because 1,2-pentadiene has cumulated double bonds and 1,4-pentadiene has isolated double bonds, 1,2-pentadiene is less stable and, therefore, will have a more negative heat of hydrogenation.

$$\underset{\text{1,2-pentadiene}}{CH_2{=}C{=}CHCH_2CH_3} \xrightarrow[\text{Pd/C}]{H_2} CH_3CH_2CH_2CH_2CH_3$$

$$\underset{\text{1,4-pentadiene}}{CH_2{=}CHCH_2CH{=}CH_2} \xrightarrow[\text{Pd/C}]{H_2} CH_3CH_2CH_2CH_2CH_3$$

23. (3*E*,6*E*)-3,7,11-trimethyl-1,3,6,10-dodecatetraene

The configuration of the double bond at the #1 and #10 positions is not specified because isomers are not possible at those positions since there are 2 hydrogens on C-1 and two methyl groups on C-11.

24. The reaction of 1,3-cyclohexadiene with Br_2 forms 3,4-dibromocyclohexene as the 1,2-addition product and 3,6-dibromocyclohexene as the 1,4-addition product. (Recall that in naming the compounds, the double bond is at the 1,2-position.) The reaction of 1,3-cyclohexadiene with HBr forms 3-bromocyclohexene as both the 1,2-addition product and the 1,4-addition product.

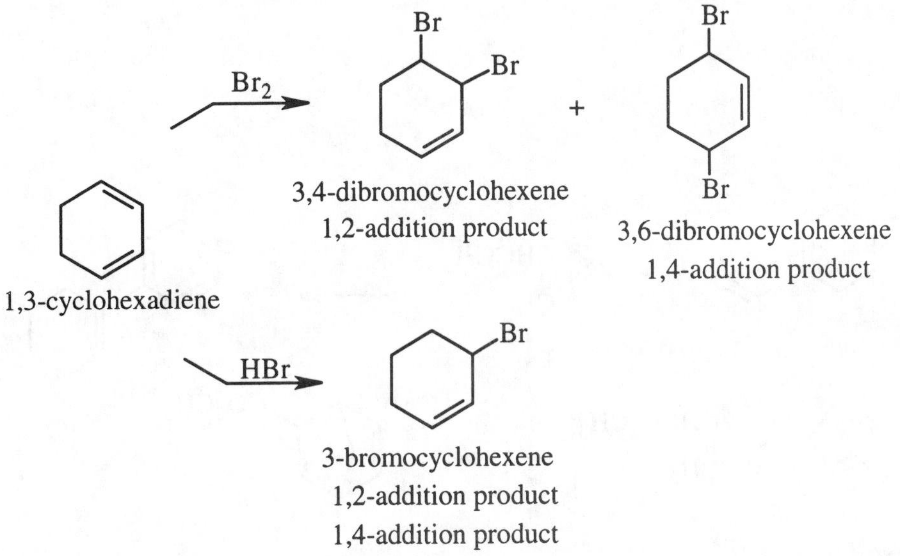

Br_2 →

Br Br

3,4-dibromocyclohexene
1,2-addition product

\+

Br Br

3,6-dibromocyclohexene
1,4-addition product

1,3-cyclohexadiene

HBr →

Br

3-bromocyclohexene
1,2-addition product
1,4-addition product

25.

a. 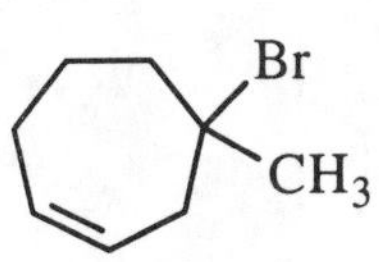 Br CH_3 b. 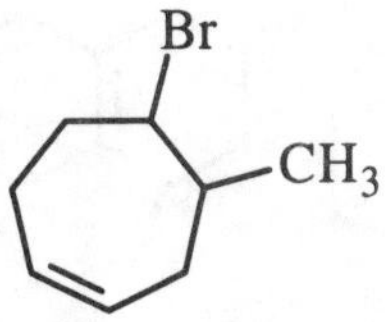 Br CH_3 c. 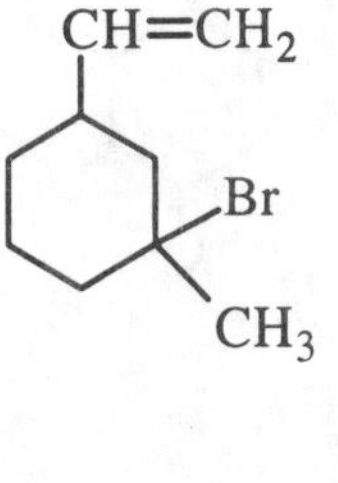 CH=CH_2 Br CH_3 d. 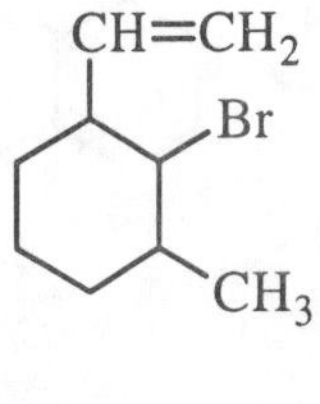 CH=CH_2 Br CH_3

26.

a. $CH_2{=}CHCH{=}CHCH{=}CH_2 \xrightarrow{HBr} \underset{Br}{CH_3CHCH{=}CHCH{=}CH_2}$ (**A**) +
1,3,5-hexatriene

$CH_3CH{=}CH\underset{Br}{CH}CH{=}CH_2$ (**B**) + $CH_3CH{=}CHCH{=}CHCH_2Br$ (**C**)

b. **A** and **B** predominate if the reaction is under kinetic control because the secondary allylic carbocations are more stable than the primary allylic carbocation.

c. **C** predominates if the reaction is under thermodynamic control because it is the most stable diene. (It is the most substituted conjugated diene.)

27.

a. 2-phenyl-1,3-butadiene + $CH_2{=}CHCCH_3$ (with C=O) → 1-(3-phenylcyclohex-3-enyl) ketone bearing $\overset{O}{\|}CCH_3$

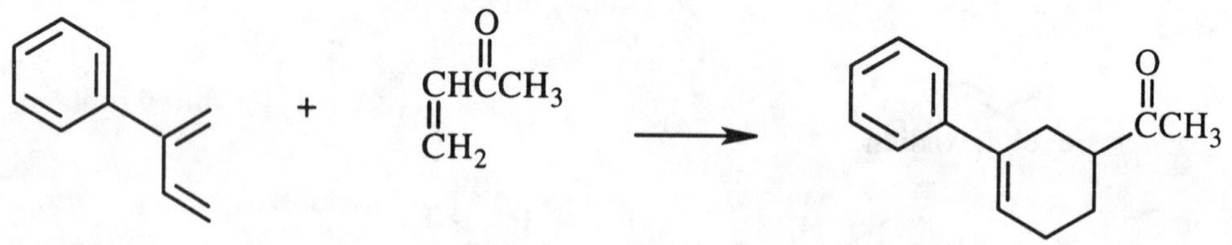

b. cyclopentadiene + $CH_2{=}CHCO_2CH_3$ → bicyclic norbornene product bearing CO_2CH_3

c. 1,3-butadiene + $CH_2{=}CHCCH_3$ (with C=O) → cyclohex-3-enyl methyl ketone

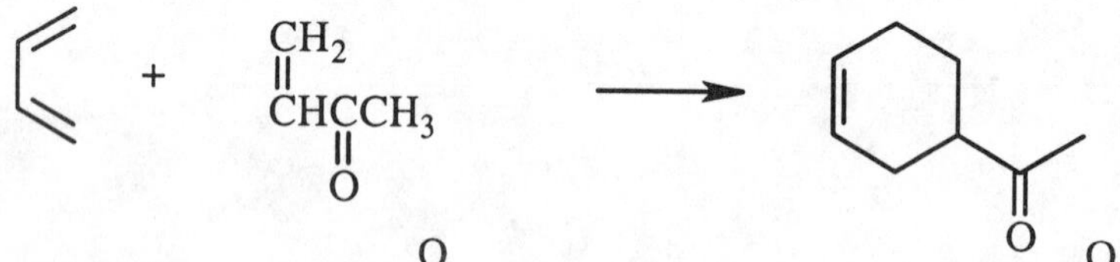

d. 1,2-dimethylenecyclohexane + 1,4-benzoquinone → tricyclic diketone product

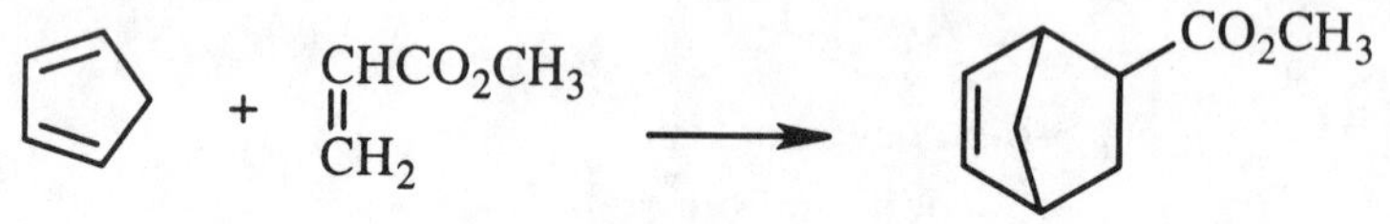

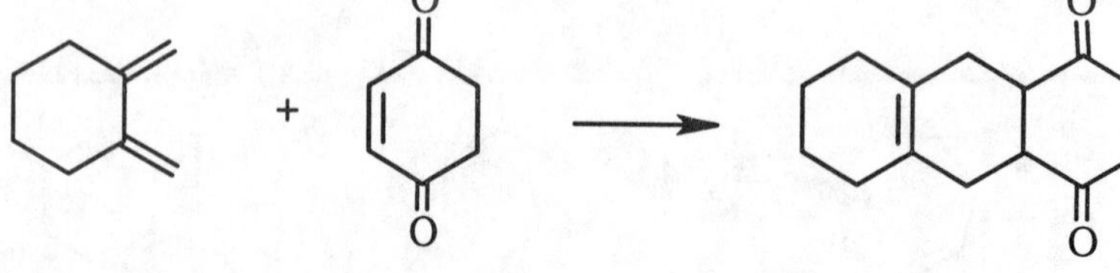

28.

a. 1. [cyclohexa-1,3-diene] + Br_2 $\xrightarrow{H_2O}$ [product: OH, Br]

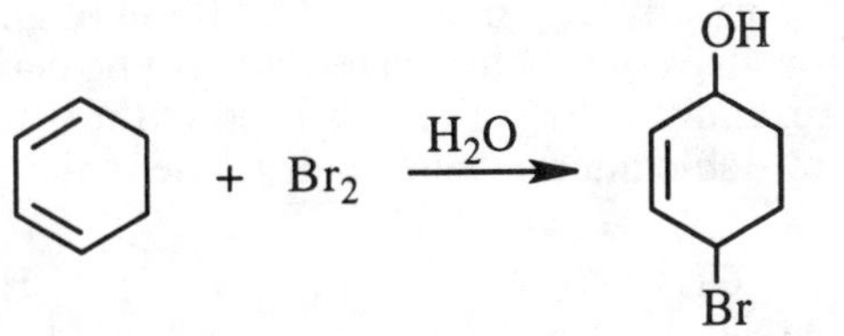

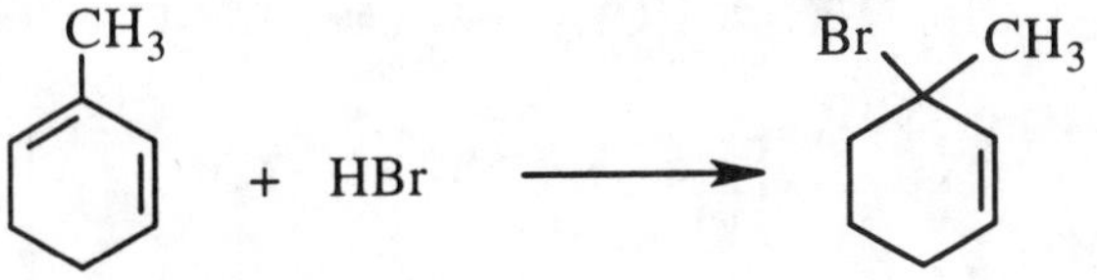

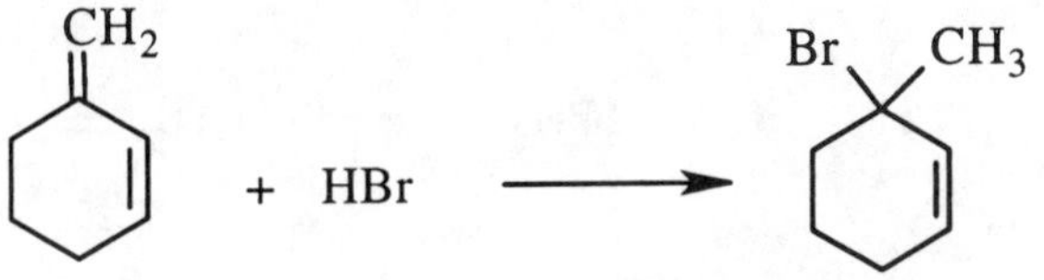

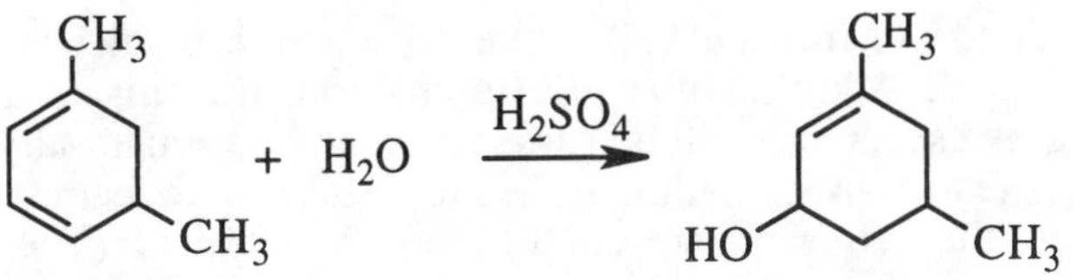

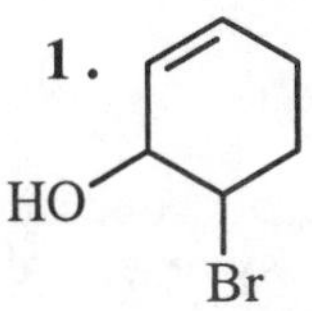

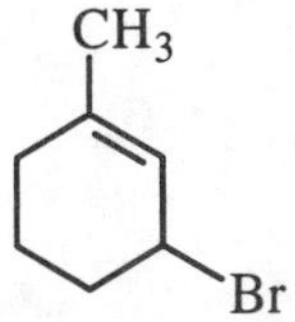

2. [CH_3-substituted cyclohexadiene] + HBr ⟶ [product: Br, CH_3]

or [CH_2=cyclohexene] + HBr ⟶ [product: Br, CH_3]

3. [CH_3, CH_3 cyclohexadiene] + H_2O $\xrightarrow{H_2SO_4}$ [product: CH_3, HO, CH_3]

or [CH_3, CH_3 cyclohexadiene] + H_2O $\xrightarrow{H_2SO_4}$ [product: CH_3, HO, CH_3]

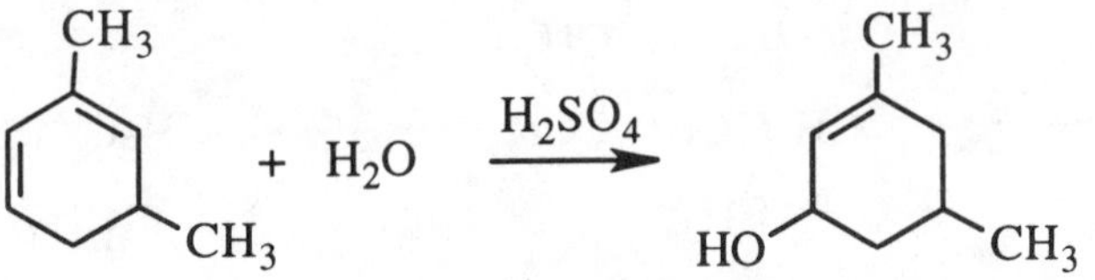

b.

1. [HO, Br cyclohexene] 2. [CH_3, Br cyclohexene] 3. [HO, CH_3, CH_3 cyclohexene]

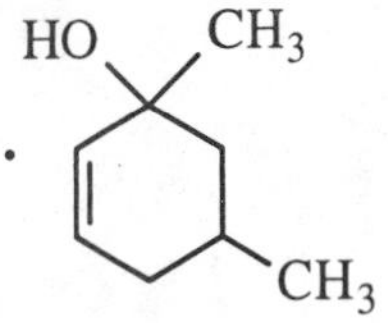

29. The diene is the nucleophile, and the dienophile is the electrophile in a Diels-Alder reaction.

a. An electron-donating substituent in the diene would increase the rate of the reaction.

b. An electron-donating substituent in the dienophile would decrease the rate of the reaction.

c. An electron-withdrawing substituent in the dienophile would increase the rate of the reaction.

30. **a.** Addition of an electrophile to C-1 forms a carbocation with two resonance contributors, a tertiary allylic carbocation and a secondary allylic carbocation. Addition of an electrophile to C-4 forms a carbocation with two resonance contributors, a tertiary allylic carbocation and a primary allylic carbocation. Therefore, addition to C-1 results in formation of the more stable carbocation intermediate, and the more stable intermediate leads to the major products.

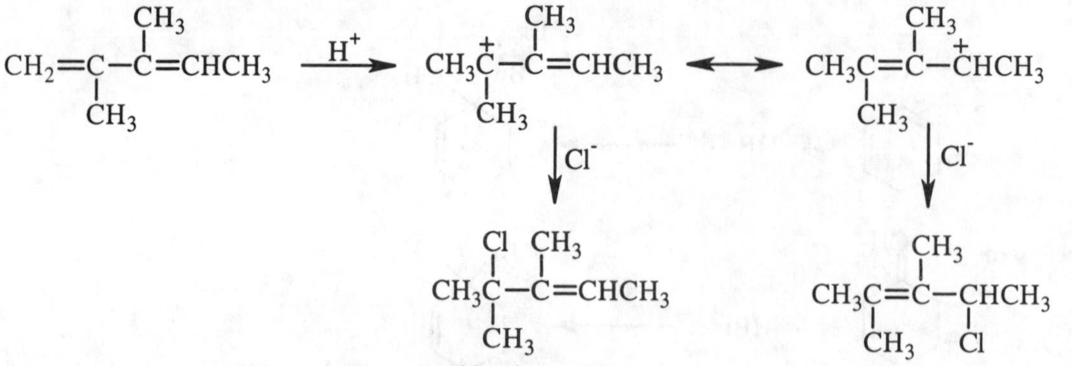

$$CH_2{=}C(CH_3){-}C(CH_3){=}CHCH_3 \xrightarrow{H^+} CH_3\overset{+}{C}(CH_3){-}C(CH_3){=}CHCH_3 \longleftrightarrow CH_3C(CH_3){=}C(CH_3){-}\overset{+}{C}HCH_3$$

$$\downarrow Cl^- \qquad\qquad \downarrow Cl^-$$

$$CH_3C(Cl)(CH_3){-}C(CH_3){=}CHCH_3 \qquad CH_3C(CH_3){=}C(CH_3){-}CH(Cl)CH_3$$

kinetic product — thermodynamic product

b. Addition of an electrophile to C-1 forms a carbocation with two resonance contributors; both are tertiary allylic carbocations. Addition of an electrophile to C-4 forms a carbocation with two resonance contributors, a secondary allylic carbocation and a primary allylic carbocation. Therefore, addition to C-1 results in formation of the more stable carbocation. Only one product is formed, because the two resonance contributors are identical. (The carbocation is symmetrical.)

$$CH_2{=}C(CH_3){-}CH{=}C(CH_3)CH_3 \xrightarrow{H^+} CH_3\overset{+}{C}(CH_3)CH{=}C(CH_3)CH_3 \longleftrightarrow CH_3C(CH_3){=}CH\overset{+}{C}(CH_3)CH_3$$

$$\searrow Cl^- \qquad \swarrow Cl^-$$

$$CH_3C(Cl)(CH_3)CH{=}C(CH_3)CH_3$$

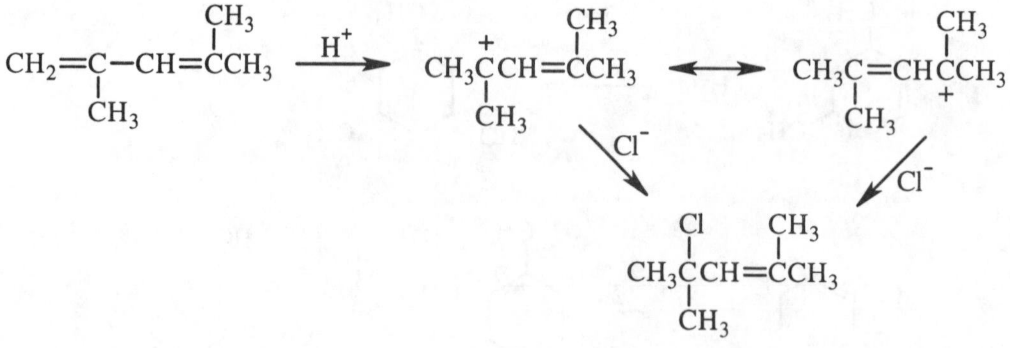

This is the only product because the carbocation is symmetrical.

31.

a.

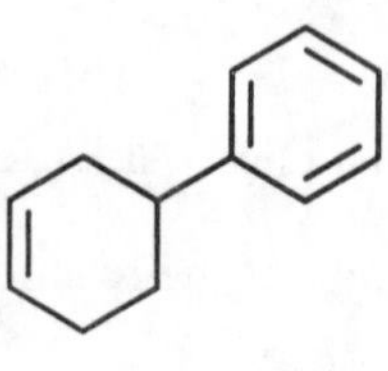

b.

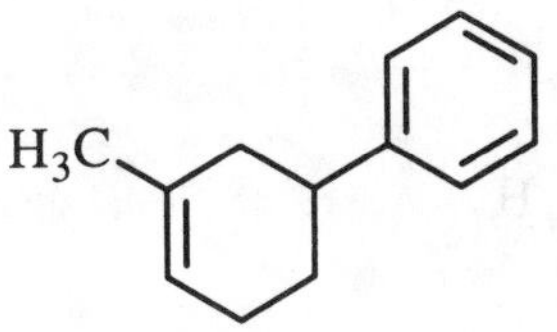

H$_3$C and H$_3$C

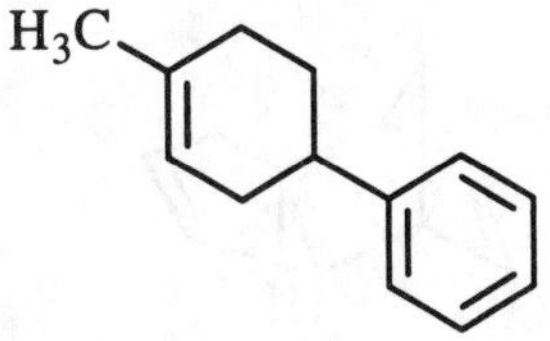

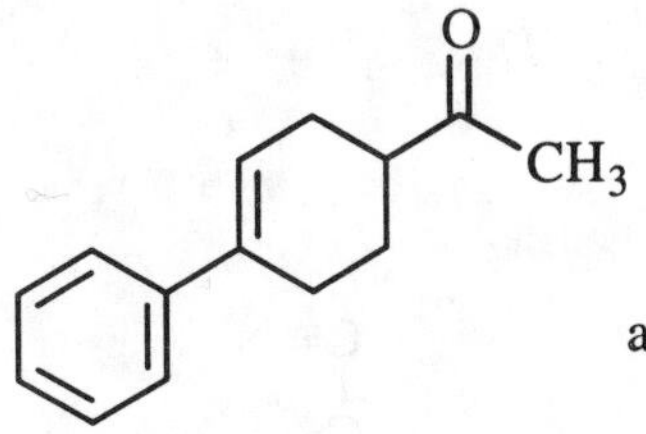

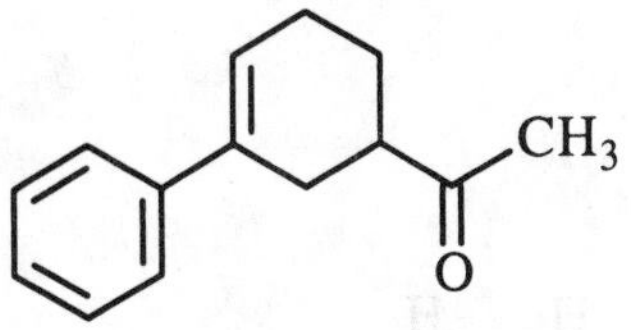

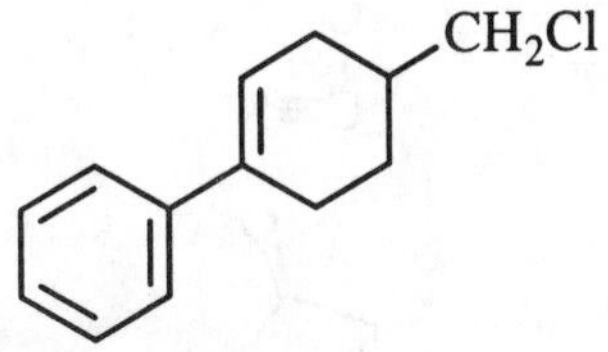

c.

O CH$_3$ and CH$_3$ O

d.

CH$_2$Cl and CH$_2$Cl

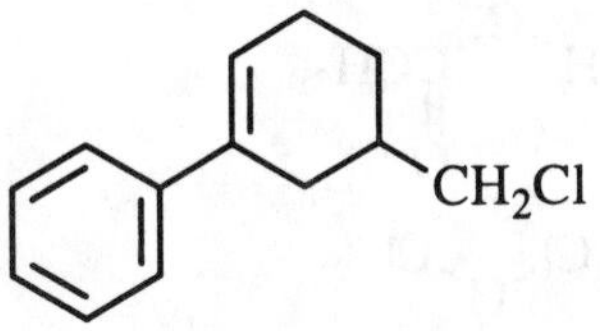

32. The first pair is the preferred set of reagents because it has the more nucleophilic diene and the more electrophilic dienophile.

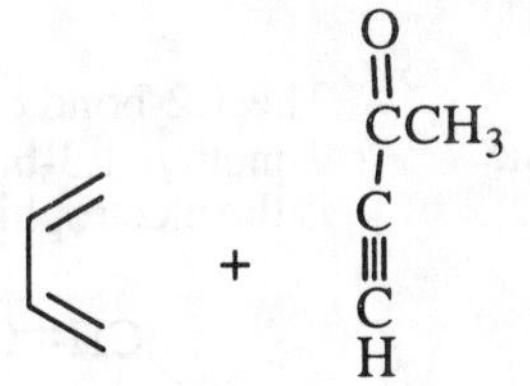

O ‖ CCH$_3$ | C ≡ C | H + or HC≡CH + O ‖ CCH$_3$

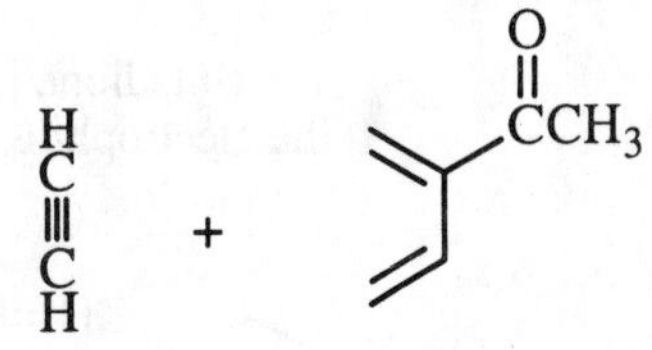

33. A Diels-Alder reaction is a reaction between a nucleophilic diene and an electrophilic dienophile.

a. The compound shown below is more reactive in both **1** and **2**, because electron delocalization increases the electrophilicity of the dienophile.

$$CH_2{=}CH\overset{O}{\overset{\|}{C}}H \longleftrightarrow \overset{+}{C}H_2CH{=}\overset{O^-}{\overset{|}{C}}H$$

b. The compound shown below is more reactive, because electron delocalization increases the nucleophilicity of the diene.

$$CH_2{=}CHCH{=}CHOCH_3 \longleftrightarrow \overset{-}{C}H_2CH{=}CHCH{=}\overset{+}{O}CH_3$$

34.

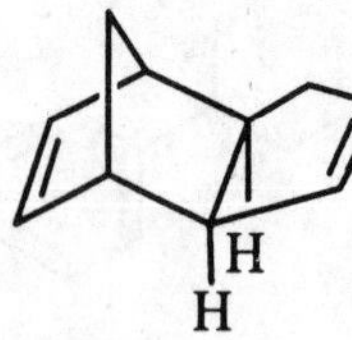

exo 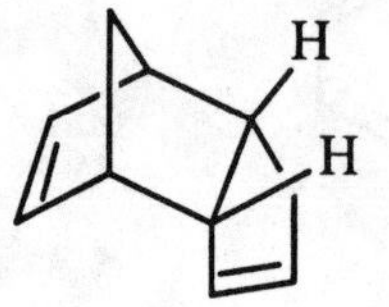 endo

35.

a. 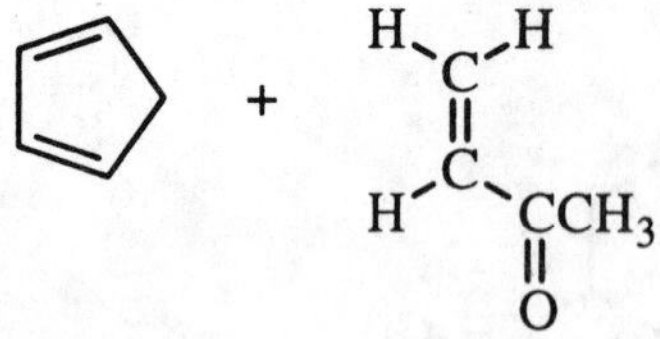 +

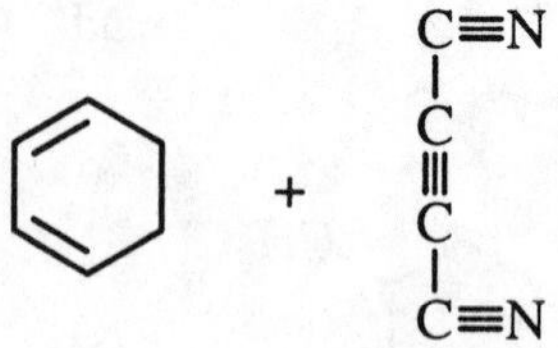

b. 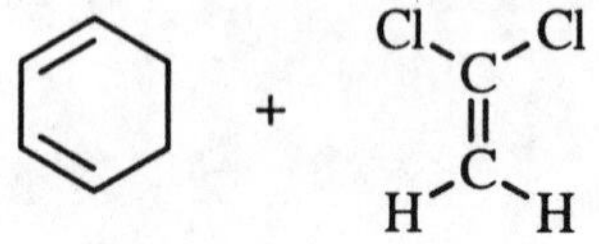 +

c. +

d. 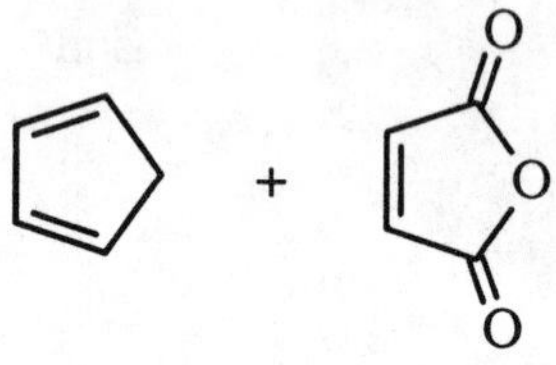 +

36.

	1,3-Butadiene is the electrophile.	The 3,4-bond of 2-methyl-1,3-butadiene is the electrophile.	The 1,2-bond of 2-methyl-1,3-butadiene is the electrophile.
1,3-Butadiene is the nucleophile.	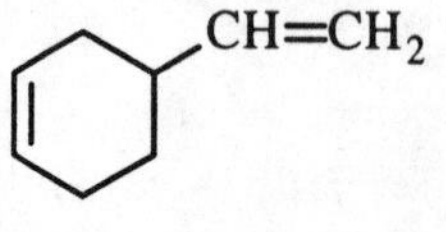 CH=CH$_2$	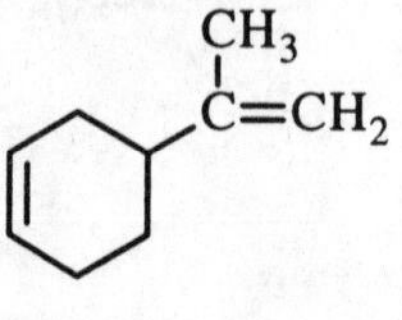 CH$_3$, C=CH$_2$	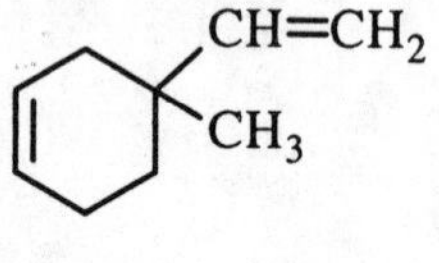 CH=CH$_2$, CH$_3$
2-Methyl-1,3-butadiene is the nucleophile. (the 1-position is on top)	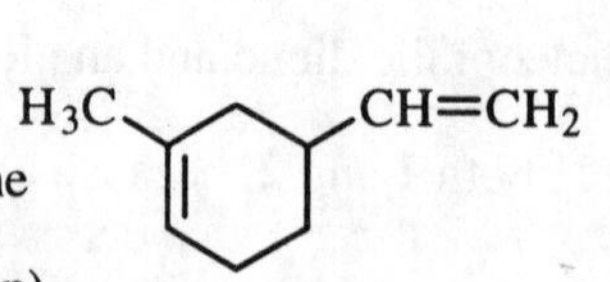 H$_3$C, CH=CH$_2$	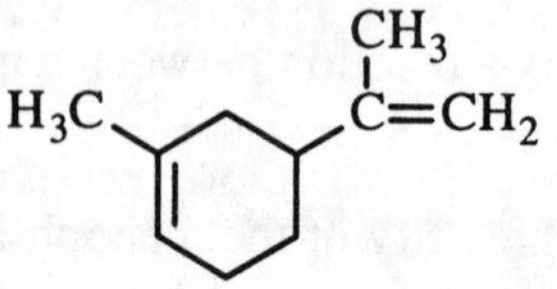 H$_3$C, CH$_3$, C=CH$_2$	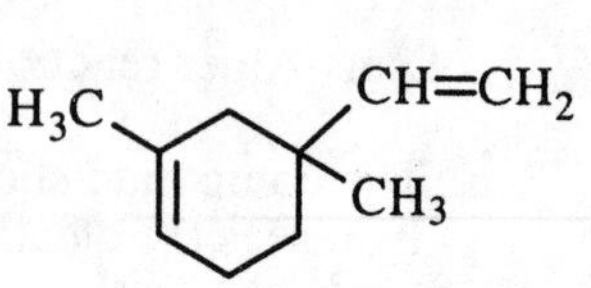 H$_3$C, CH=CH$_2$, CH$_3$
2-Methyl-1,3-butadiene is the nucleophile. (the 4-position is on top)	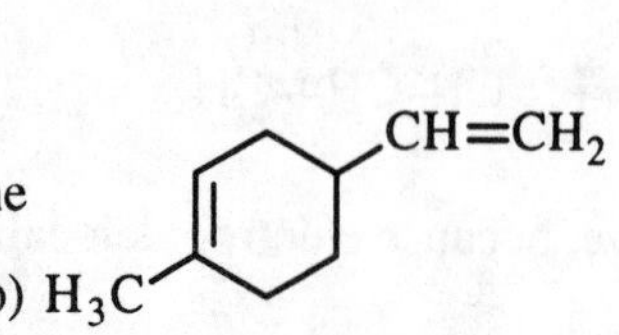 H$_3$C, CH=CH$_2$	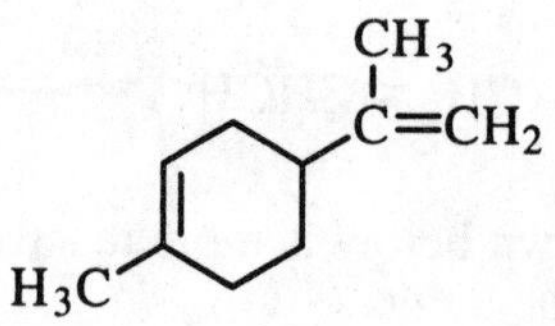 H$_3$C, CH$_3$, C=CH$_2$	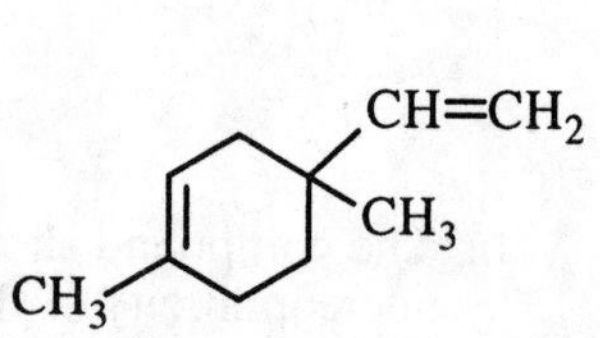 CH$_3$, CH=CH$_2$, CH$_3$

37.

a. $\xrightarrow{Br_2}$ **A** + **B** + **C**

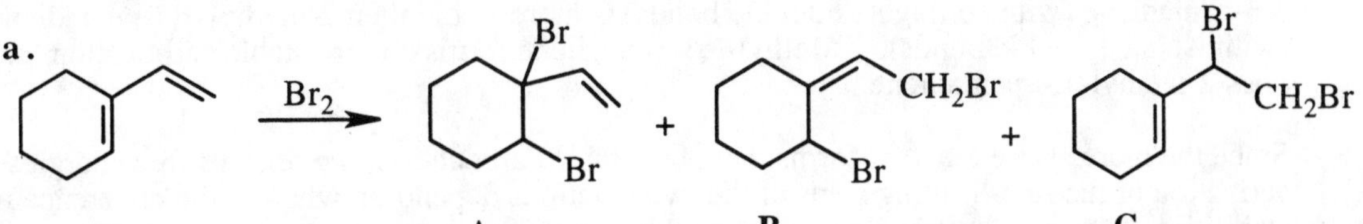

b. **A** has two chirality centers but only two stereoisomers are obtained because addition of Br_2 can occur only in an anti fashion.

A

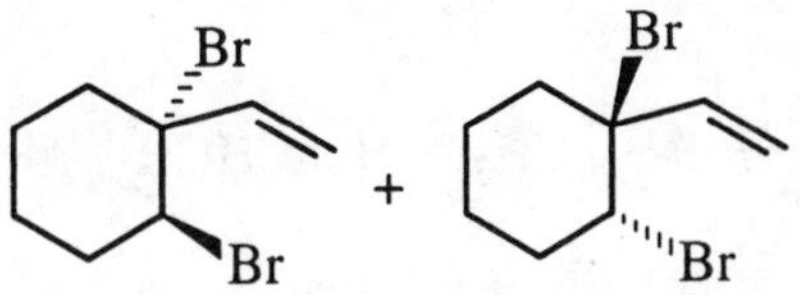

Four stereoisomers can be obtained for **B** because it has a chirality center and a double bond that can be in either the *E* or *Z* configuration.

B

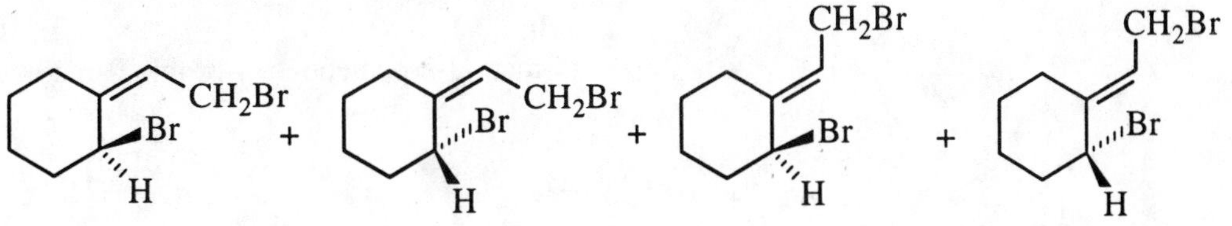

C has two stereoisomers because it has one chirality center.

C

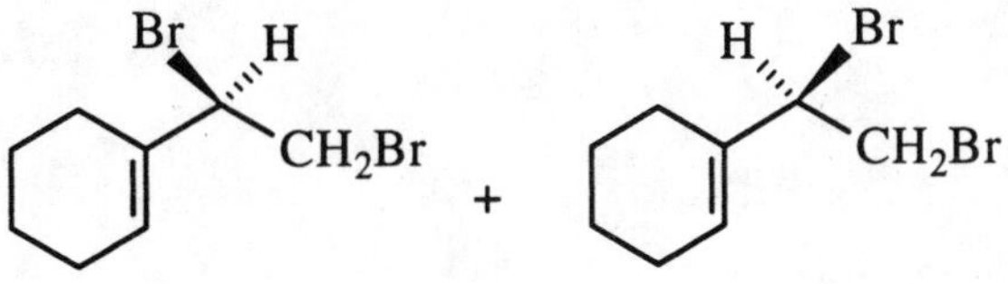

38. The rate-limiting step of the reaction is the formation of the carbocation intermediate. 2-Methyl-1,3-pentadiene (with conjugated double bonds) is more stable than 2-methyl-1,4-pentadiene (with isolated double bonds). 2-Methyl-1,3-pentadiene forms a more stable carbocation than does 2-methyl-1,4-pentadiene.

Since the more stable reactant forms the more stable carbocation, the relative free energies of activation of the rate-limiting steps of the two reactions depend on whether the difference in the stabilities of the reactants is greater or less than the difference in the stabilities of the transition states (which depend on the difference in stabilities of the carbocations). Because the difference in the stabilities of the reactants is less than the difference in the stabilities of the transition states, the rate of reaction of HBr with 2-methyl-1,3-pentadiene is the faster reaction. (If the difference in the stabilities of the reactants had been greater than the difference in the stabilities of the transition states, the rate of reaction of HBr with 2-methyl-1,4-pentadiene would have been the faster reaction.)

$$CH_2{=}\overset{\overset{CH_3}{|}}{C}CH{=}CHCH_3 \xrightarrow{H^+} CH_3\underset{+}{\overset{\overset{CH_3}{|}}{C}}CH{=}CHCH_3 \longleftrightarrow CH_3\overset{\overset{CH_3}{|}}{C}{=}CH\overset{+}{C}HCH_3$$

2-methyl-1,3-pentadiene

$\downarrow Br^-$ (each carbocation)

$$CH_3\underset{\underset{Br}{|}}{\overset{\overset{CH_3}{|}}{C}}CH{=}CHCH_3 \qquad CH_3\overset{\overset{CH_3}{|}}{C}{=}CH\underset{\underset{Br}{|}}{C}HCH_3$$

4-bromo-4-methyl-2-pentene 4-bromo-2-methyl-2-pentene

$$CH_2{=}\overset{\overset{CH_3}{|}}{C}CH_2CH{=}CH_2 \xrightarrow{H^+} CH_3\underset{+}{\overset{\overset{CH_3}{|}}{C}}CH_2CH{=}CH_2 \xrightarrow{Br^-} CH_3\underset{\underset{Br}{|}}{\overset{\overset{CH_3}{|}}{C}}CH_2CH{=}CH_2$$

2-methyl-1,4-pentadiene 4-bromo-4-methyl-1-pentene

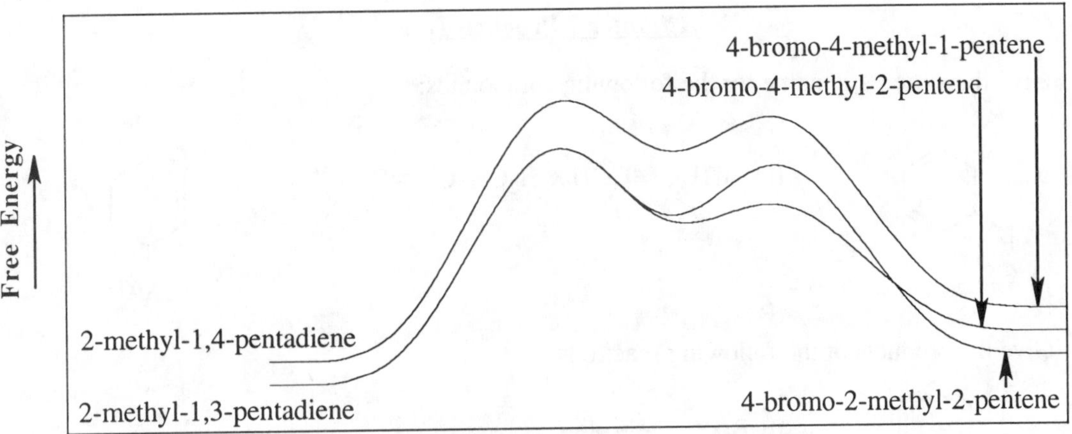

Progress of the Reaction

39. His recrystallization was not successful. Because maleic anhydride is a dienophile, it reacts with cyclopentadiene in a Diels-Alder reaction.

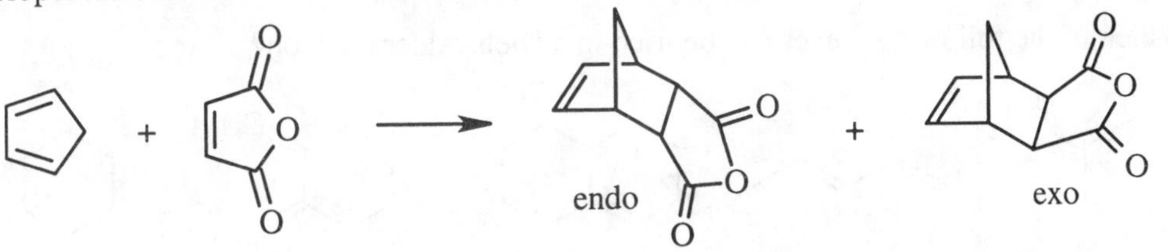

40. Maleic anhydride reacts with cyclopentadiene as in the above problem. The function of maleic acid in this reaction is to remove the cyclopentadiene, since removal of a product drives the equilibrium toward products. (See Le Châtelier's principle on page 369 of the text.)

41. The bridgehead carbon cannot have the 120° bond angle required for the sp^2 hybridized carbon of a double bond. With a 120° bond angle, the compound would be too strained to exist.

Chapter 7 Practice Test

1. Give the systematic name for the following compounds:

a. 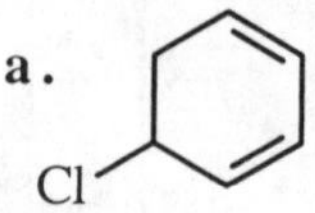 [structure: chlorocyclohexadiene]

b. $CH_3C{\equiv}CCH_2CH_2CH_2CH{=}CH_2$

c. 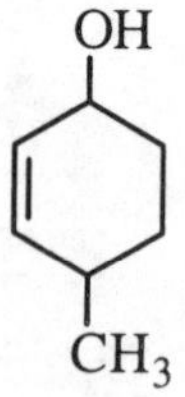 [structure: cyclohexene ring with OH and CH_3]

2. Give the products of the following reactions:

a. 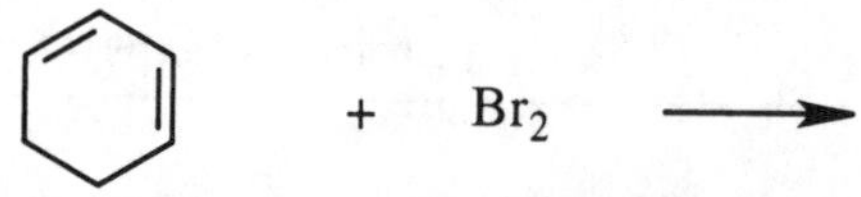 [cyclohexene] + Br_2 ⟶

b. 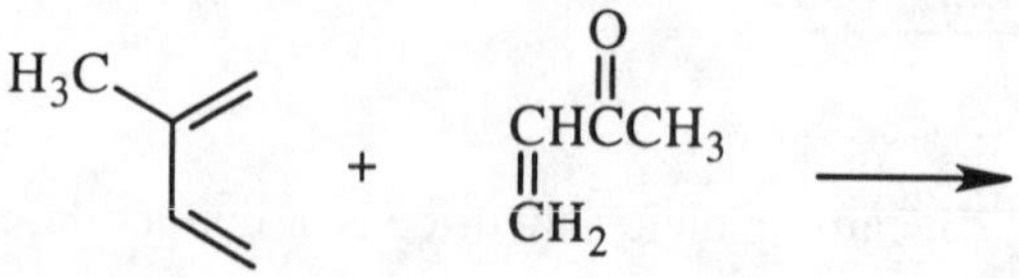 H_3C–diene + $CH_2{=}CHCCH_3$ (C=O) ⟶

3. Which of the following dienes can be used in a Diels-Alder reaction?

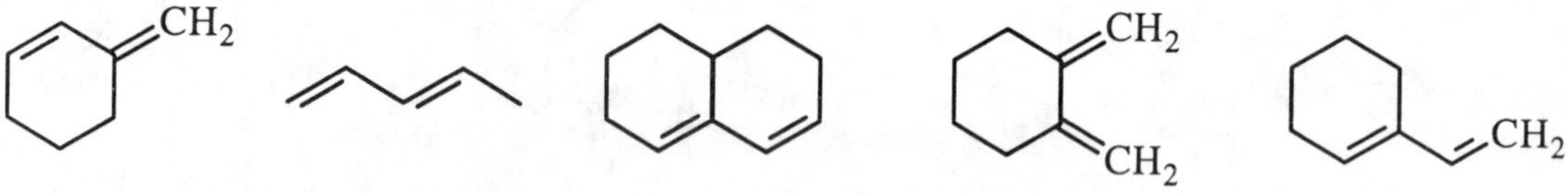

[structures: CH_2; ; ; CH_2, CH_2; CH_2]

4. Indicate whether each of the following statements is true or false.

a. A conjugated diene is more stable than an isomeric isolated diene. T F

b. A single bond formed by an sp^2—sp^2 overlap is longer than a single bond formed by an sp^2—sp^3 overlap. T F

c. The thermodynamically controlled product is the major product obtained when the reaction is carried out under mild conditions. T F

d. 1,3-Hexadiene is more stable than 1,4-hexadiene. T F

5. Give the four products that would be obtained from the following reaction, ignoring their configurations.

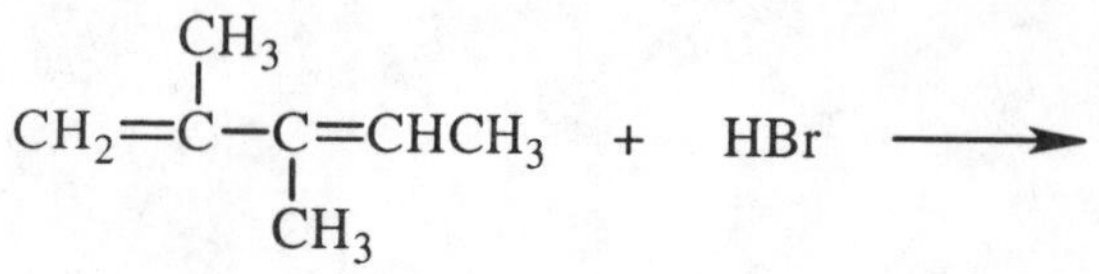

$$CH_2{=}\underset{}{\overset{CH_3}{\overset{|}{C}}}-\underset{\underset{CH_3}{|}}{C}{=}CHCH_3 \quad + \quad HBr \longrightarrow$$

6. What reactants are necessary for the synthesis of the following compound via a Diels-Alder reaction?

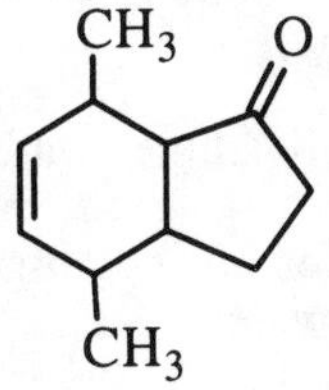

CH_3 O

CH_3

7. Two 1,2-products and two 1,4-products can be obtained from the following reaction:

CH_3

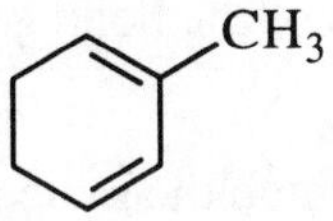

+ HCl ⟶

a. Which are the predominant 1,2- and 1,4-products?

b. Which of the above two products is the product of kinetic control?

8. What reagents could be used to convert the given starting material into the given product?

⟶ HO— —Br

9. Give the product of the following reaction, showing its configuration:

$$CH_2{=}CHCH{=}CH_2 \quad + \quad \begin{matrix} H_3C & & CH_3 \\ & C{=}C & \\ H & & H \end{matrix} \quad \xrightarrow{\Delta}$$

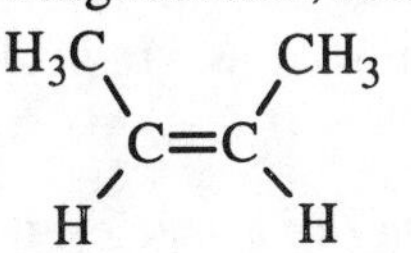

10. For each of the following reactions give the major 1,2- and 1,4-products. Label the product of kinetic control and the product of thermodynamic control.

a. $CH_2{=}CH{-}\overset{CH_3}{\overset{|}{C}}{=}CH_2$ + HCl ⟶

b. 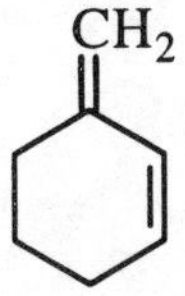 CH_2 + HBr ⟶

CHAPTER 8
Reactions of Alkanes. Radicals

Important Terms

alkane — a hydrocarbon that contains only single bonds.

combustion — a reaction with oxygen that takes place at high temperatures and converts alkanes to carbon dioxide and water

free radical — an atom or a molecule with an unpaired electron.

homolytic bond cleavage — breaking a bond with the result that each of the atoms of the bond gets one of thebonding electrons.

initiation step — the step in which radicals are created, or the step in which the radical needed for the first propagating step is created.

paraffin — an older word for alkanes, which means "little affinity"

propagation step — in the first of a pair of propagation steps, a radical (or an electrophile or a nucleophile) reacts to produce another radical (or an electrophile or a nucleophile) that reacts in the second propagation step to produce the radical (or the electrophile or the nucleophile) that was the reactant in the first propagation step.

radical (often called a free radical) — an atom or a molecule with an unpaired electron.

radical chain reaction — a reaction in which radicals are formed and react in repeating propagating steps.

radical inhibitor — a compound that traps radicals.

radical substitution reaction — a substitution reaction that has a radical intermediate.

reactivity-selectivity principle — the greater the reactivity of a species, the less selective it will be.

saturated hydrocarbon — a hydrocarbon that contains only single bonds (is saturated with hydrogen).

termination step — two radicals combine to produce a molecule in which all the electrons are paired.

Solutions to Problems

1.

Initiation:

$Cl_2 \xrightarrow{h\nu} 2\ Cl\cdot$

Propagation:

$Cl\cdot$ + cyclohexane → cyclohexyl radical + HCl

cyclohexyl radical + Cl_2 → chlorocyclohexane + $Cl\cdot$

Termination:

$Cl\cdot + Cl\cdot \longrightarrow Cl_2$

$Cl\cdot$ + cyclohexyl radical → chlorocyclohexane

cyclohexyl radical + cyclohexyl radical → bicyclohexyl

2.

$Cl_2 \xrightarrow{h\nu} 2\ Cl\cdot$

$Cl\cdot + CH_4 \longrightarrow \cdot CH_3 + HCl$

$\cdot CH_3 + Cl_2 \longrightarrow CH_3Cl + Cl\cdot$

$Cl\cdot + CH_3Cl \longrightarrow \cdot CH_2Cl + HCl$

$\cdot CH_2Cl + Cl_2 \longrightarrow CH_2Cl_2 + Cl\cdot$

$Cl\cdot + CH_2Cl_2 \longrightarrow \cdot CHCl_2 + HCl$

$\cdot CHCl_2 + Cl_2 \longrightarrow CHCl_3 + Cl\cdot$

$Cl\cdot + CHCl_3 \longrightarrow \cdot CCl_3 + HCl$

$\cdot CCl_3 + Cl_2 \longrightarrow CCl_4 + Cl\cdot$

$Cl\cdot + \cdot CCl_3 \longrightarrow CCl_4$

3. Solved in the text.

4.

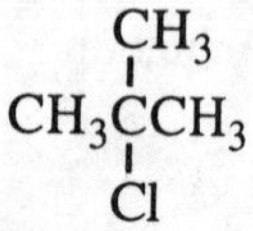

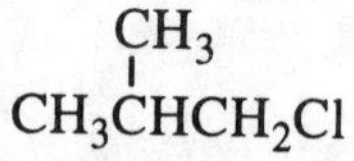

$$\underset{\displaystyle 1 \times x = x}{CH_3\underset{Cl}{\overset{CH_3}{C}}CH_3} \qquad \underset{\displaystyle 9 \times 1 = 9}{CH_3\overset{CH_3}{C}HCH_2Cl}$$

$$\%\ \text{1-chloro-2-methylpropane} = \frac{9}{9+x} = 0.64$$

$$9 = 0.64\,(9 + x)$$

$$9 = 5.76 + 0.64x$$

$$3.24 = 0.64x$$

$$x = 5$$

It is 5 times easier for a chloride radical to remove a hydrogen atom from a tertiary carbon than from a primary carbon.

5. **a.** 3 **b.** 3 **c.** 1 **d.** 5 **e.** 2 **f.** 1 **g.** 5 **h.** 4

6. Note: the denominator used in each of these problems is obtained by adding the numerators.

a. $CH_3CH_2CH_2CH_2CH_2Cl$ $\quad CH_3CH_2CH_2\underset{Cl}{C}HCH_3$ $\quad CH_3CH_2\underset{Cl}{C}HCH_2CH_3$

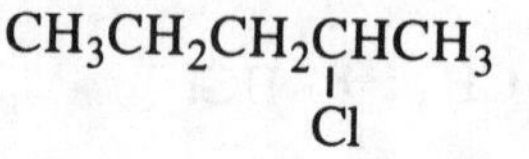

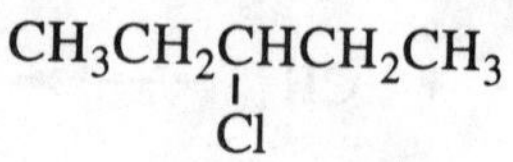

$6 \times 1 = 6$ $\qquad 4 \times 3.8 = 15.2$ $\qquad 2 \times 3.8 = 7.6$

$\frac{6}{28.8} = 21\%$ $\qquad \frac{15.2}{28.8} = 53\%$ $\qquad \frac{7.6}{28.8} = 26\%$

b. $ClCH_2\overset{CH_3}{C}HCH_2CH_2\overset{CH_3}{C}HCH_3$ $\quad CH_3\underset{Cl}{\overset{CH_3}{C}}CH_2CH_2\overset{CH_3}{C}HCH_3$ $\quad CH_3\overset{CH_3}{C}H\underset{Cl}{C}HCH_2\overset{CH_3}{C}HCH_3$

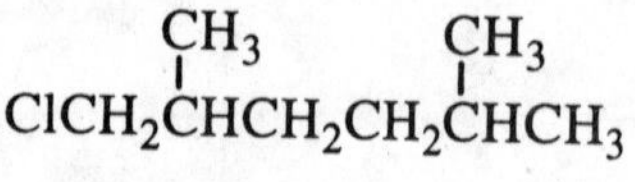

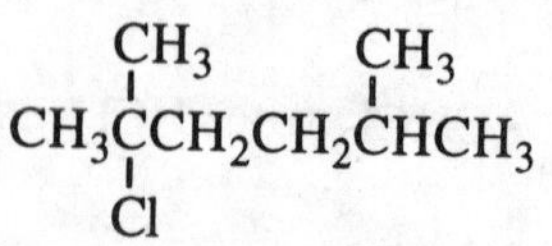

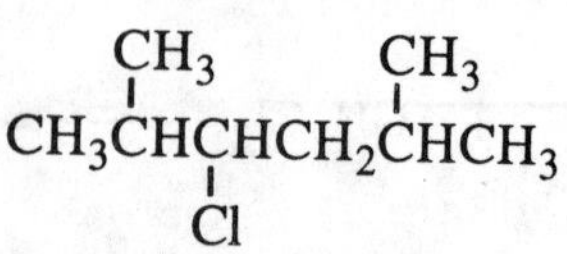

$12 \times 1 = 12$ $\qquad 2 \times 5 = 10$ $\qquad 4 \times 3.8 = 15.2$

$\frac{12}{37.2} = 32\%$ $\qquad \frac{10}{37.2} = 27\%$ $\qquad \frac{15.2}{37.2} = 41\%$

g. 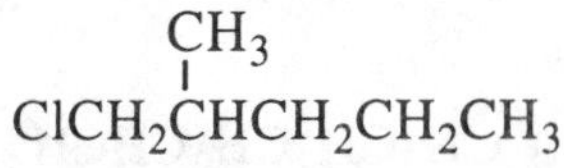 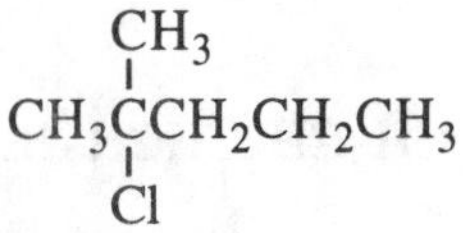 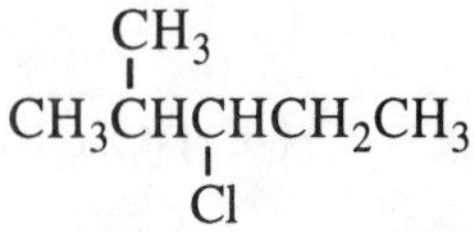 $ClCH_2CH(CH_3)CH_2CH_2CH_3$ | $CH_3C(CH_3)(Cl)CH_2CH_2CH_3$ | $CH_3CH(CH_3)CH(Cl)CH_2CH_3$

$6 \times 1 = 6$ $\qquad \frac{6}{29.2} = 21\%$

$1 \times 5 = 5$ $\qquad \frac{5}{29.2} = 17\%$

$2 \times 3.8 = 7.6$ $\qquad \frac{7.6}{29.2} = 26\%$

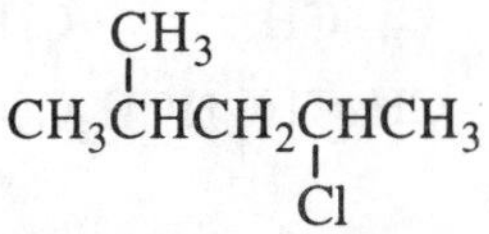

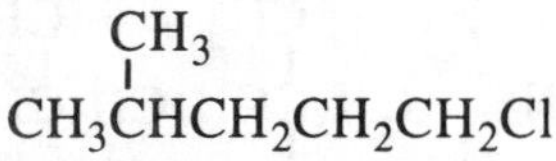

$CH_3CH(CH_3)CH_2CH(Cl)CH_3$ | $CH_3CH(CH_3)CH_2CH_2CH_2Cl$

$2 \times 3.8 = 7.6$ $\qquad \frac{7.6}{29.2} = 26\%$

$3 \times 1 = 3$ $\qquad \frac{3}{29.2} = 10\%$

7.

a. 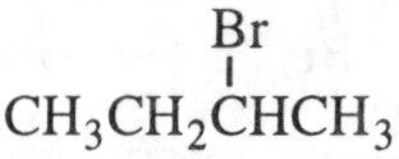 $CH_3CH_2CH(Br)CH_3$ $\qquad CH_3CH_2CH_2CH_2Br$

$4 \times 82 = 328$ $\qquad 6 \times 1 = 6$

$$\frac{328}{328 + 6} = \frac{328}{334} = 98\%$$

b. 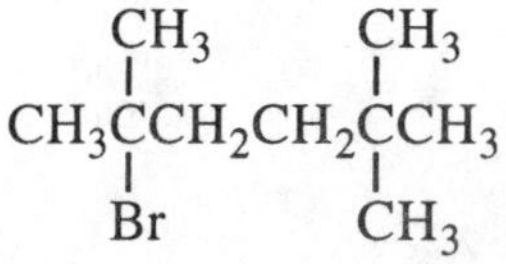 $CH_3C(CH_3)(Br)CH_2CH_2C(CH_3)_2CH_3$

$1 \times 1600 = 1600$

$\frac{1600}{1943} = 82\%$

other products:

$$\begin{array}{r} 1600 \\ 9 \times 1 = 9 \\ 2 \times 82 = 164 \\ 2 \times 82 = 164 \\ 6 \times 1 = 6 \\ \hline 1943 \end{array}$$

8.

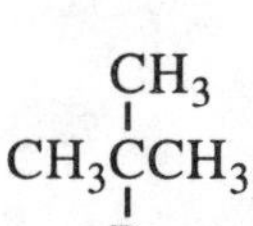

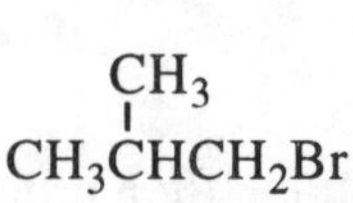

$CH_3C(CH_3)(Br)CH_3$ $\qquad CH_3CH(CH_3)CH_2Br$

$1 \times 1600 = 1600$ $\qquad 9 \times 1 = 9$

$\frac{1600}{1609} = 99.4\%$ for bromination
versus 36% for chlorination

9.

a. $CH_3CH_2CH_2CH_2CH_2Br$ $\quad$ $CH_3CH_2CH(Br)CH_2CH_3$ $\quad$ $CH_3CH_2CH_2CH(Br)CH_3$

$\frac{6}{498} = 1\%$ $\quad$ $\frac{164}{498} = 33\%$ $\quad$ $\frac{328}{498} = 66\%$

b. $BrCH_2CH(CH_3)CH_2CH_2CH(CH_3)CH_3$ $\quad$ $CH_3C(CH_3)(Br)CH_2CH_2CH(CH_3)CH_3$ $\quad$ $CH_3CH(CH_3)CH(Br)CH_2CH(CH_3)CH_3$

$\frac{12}{3540} = 0.3\%$ $\quad$ $\frac{3200}{3540} = 90.4\%$ $\quad$ $\frac{328}{3540} = 9.3\%$

g. $BrCH_2CH(CH_3)CH_2CH_2CH_3$ $\quad$ $CH_3C(CH_3)(Br)CH_2CH_2CH_3$ $\quad$ $CH_3CH(CH_3)CH(Br)CH_2CH_3$

$\frac{6}{1937} = 0.3\%$ $\quad$ $\frac{1600}{1937} = 82.6\%$ $\quad$ $\frac{164}{1937} = 8.5\%$

$CH_3CH(CH_3)CH_2CH(Br)CH_3$ $\quad$ $CH_3CH(CH_3)CH_2CH_2CH_2Br$

$\frac{164}{1937} = 8.5\%$ $\quad$ $\frac{3}{1937} = 0.2\%$

10. a. Chlorination, because the halogen is substituting for a primary hydrogen

b. Bromination, because the halogen is substituting for a tertiary hydrogen.

c. Since the molecule has only one kind of hydrogen, only one monohalogenated product will be obtained with bromination and with chlorination.

11.

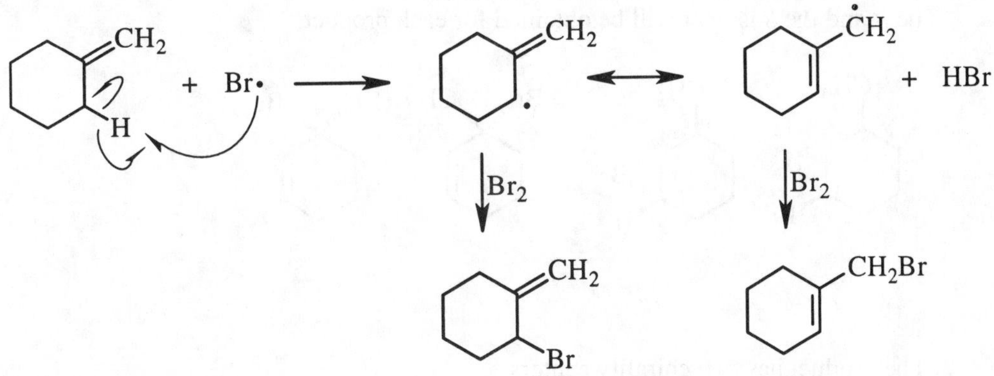

CH_2 + Br• ⟶ CH_2 ⟷ $\dot{C}H_2$ + HBr

Br_2 ↓ ↓ Br_2

CH_2 Br CH_2Br

12. Solved in the text.

13.

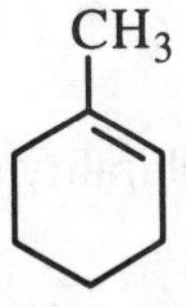

CH_3

1-methylcyclohexene

a. 1. There are two sets of allylic hydrogens. The major products are obtained by removing an allylic hydrogen from the carbon that results in the unpaired electron being shared by a secondary carbon and and tertiary carbon.

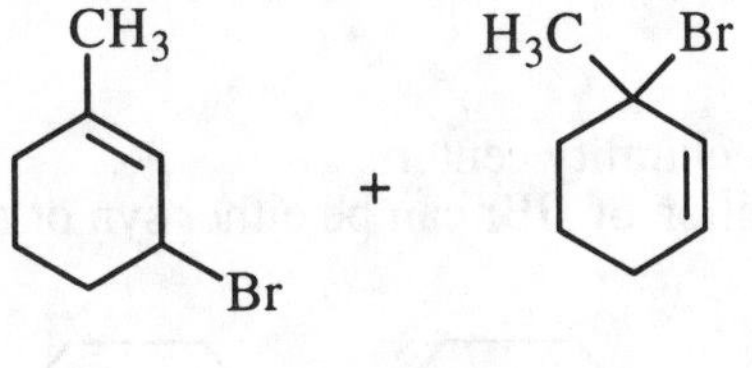

CH_3 Br + H_3C Br

2. H_3C Br Br 3. H_3C Br 4. CH_3 Br

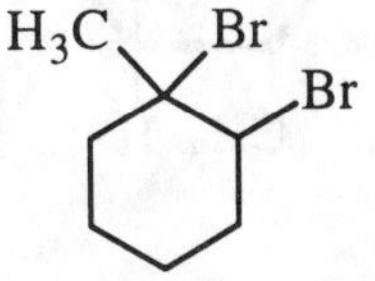

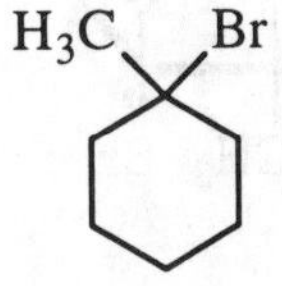

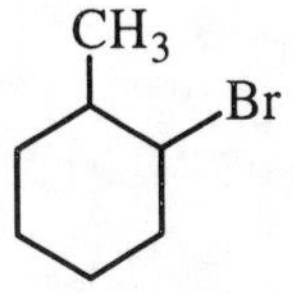

b. **1.** Each of the products has one chirality center.
The *R* and the *S* isomer will be obtained for each product.

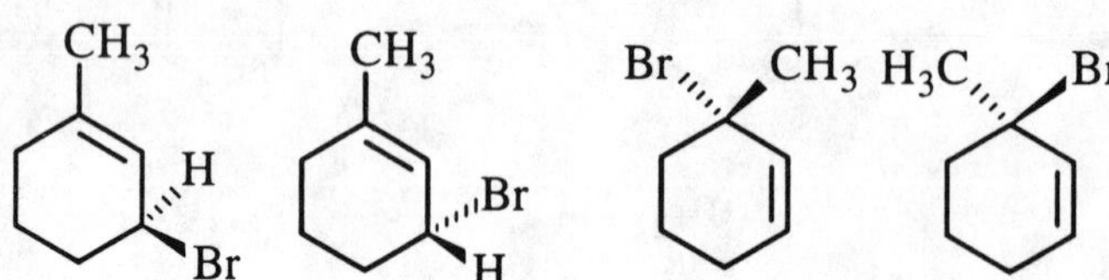

2. The product has two chirality centers.
Because addition of Br_2 is anti, only the products with the bromine atoms on opposite sides of the ring are obtained.

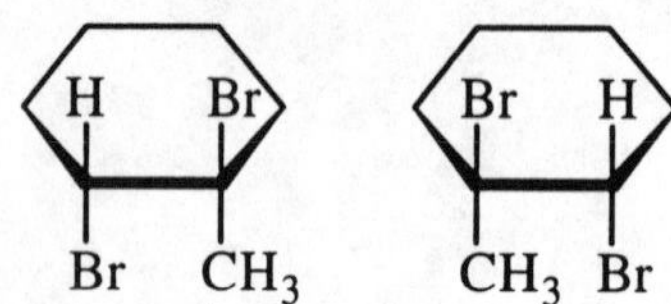

3. The products does not have a chirality center, so it does not have stereoisomers.

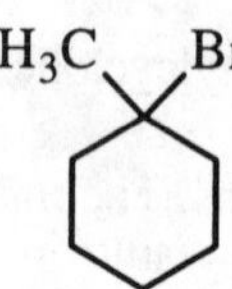

4. The product has two chirality centers.
Because radical addition of HBr can be either syn or anti, four products are obtained.

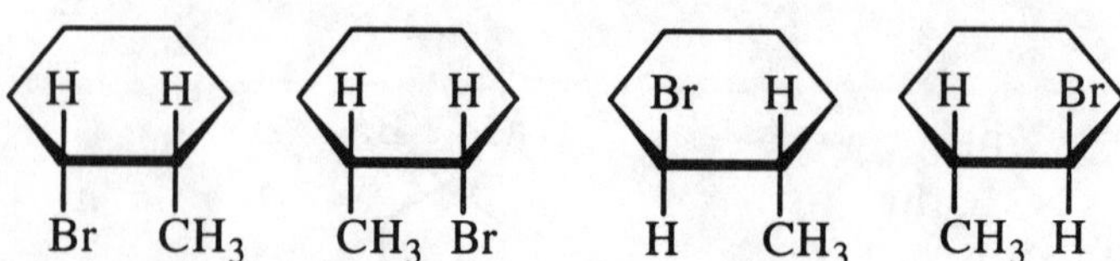

14.

a. $CH_2{=}CHCH(Br)CH_2CH_3$ + $BrCH_2CH{=}CHCH_2CH_3$

b. $BrCH_2C(CH_3){=}CHCH_3$ + $CH_2{=}C(CH_3)CH(Br)CH_3$

major products

c. $CH_3CH_2CH_2Cl$

d. $CH_3CH_2C(CH_3)(Br)CH_2CH_2CH_3$ major product

e. $BrCH_2CH_2CH_2Br$

f. 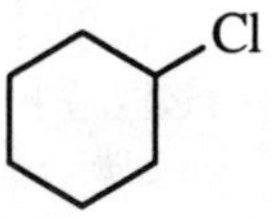 chlorocyclohexane (cyclohexane ring with Cl)

g. no reaction

h. 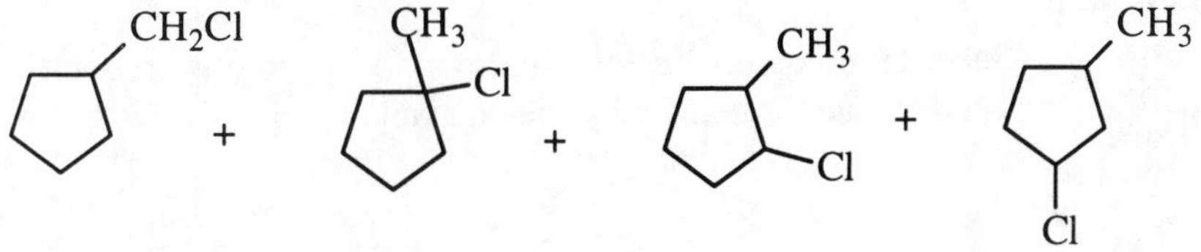 (chloromethyl)cyclopentane (CH_2Cl on cyclopentane) + 1-chloro-1-methylcyclopentane (CH_3, Cl) + 1-chloro-2-methylcyclopentane (CH_3, Cl) + 1-chloro-3-methylcyclopentane (CH_3, Cl)

15.

a. $CH_3C(CH_3)_2CH_3$

dimethylpropane

b. $CH_3CH_2CH(CH_3)CH_2CH_2CH_3$

3-methylhexane

16.

$CH_3CHCH_2CH_3$ (with CH_3 on C-2)

2-methylbutane

a primary alkyl halide

$CH_2CHCH_2CH_3$ (Cl on C-1, CH_3 on C-2)

Substitution of any one of 6 hydrogens leads to this product.
percentage formed per hydrogen available = 29/6 = 4.8

$CH_3CHCH_2CH_2$ (CH_3 on C-2, Cl on C-4)

Substitution of any one of 3 hydrogens leads to this product.
percentage formed per hydrogen available = 14/3 = 4.7

a secondary alkyl halide

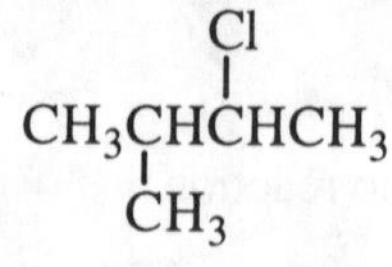

Substitution of any one of 2 hydrogens leads to this product.
percentage formed per hydrogen available = 33/2 = 16.5

a tertiary alkyl halide

$CH_3CCH_2CH_3$ (Cl and CH_3 on C-2)

Substitution of any one of 1 hydrogen leads to this product.
percentage formed per hydrogen available = 24/1 = 24

Because 16.5/4.8 = 3.4, a secondary hydrogen is 3.4 times easier to remove than a primary hydrogen.

Because 24/4.8 = 5, a tertiary hydrogen is 5 times easier to remove than a primary hydrogen.

17. The chlorine radical is more reactive at the higher temperature (600°C), so it is less selective.

18. Abstraction of a hydrogen atom from ethane by an iodine radical is a highly endothermic reaction ($\Delta H°$ = 98 - 71 = 27 kcal/mol), so the iodine radicals will reform I_2 rather than abstract a hydrogen atom.

19.

a. Theroetically five monobromination products are possible, but some will be obtained in very small amounts.

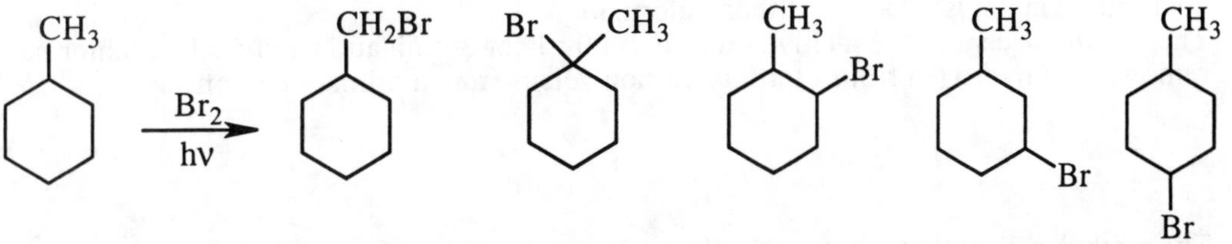

CH_3 $\xrightarrow[h\nu]{Br_2}$ CH_2Br Br CH_3 CH_3 Br CH_3 Br CH_3 Br

b. A bromine radical is very selective, much preferring to remove a hydrogen atom from a tertiary carbon than from a secondary or primary carbon. 1-Bromo-1-methylcyclohexane would be obtained in greatest yield because it is the only one formed by removing a hydrogen atom from a tertiary carbon.

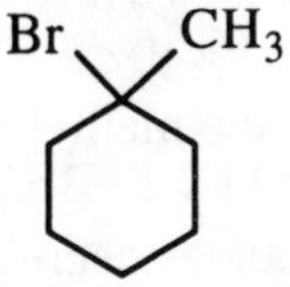

Br CH_3

20.

a. 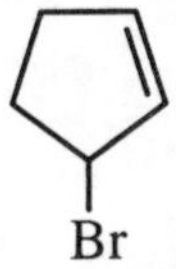 Br

b. $CH_2{=}CHCHCH_2CH_3$ (Br on C-3)

or

$BrCH_2CH{=}CHCH_2CH_3$

c. 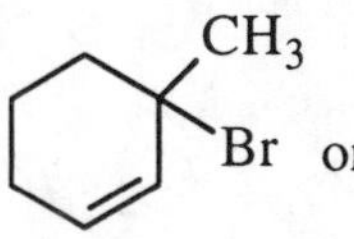 CH_3 Br or CH_3 Br

d. CH_3CHCH_2Cl (with CH_3 on C-2)

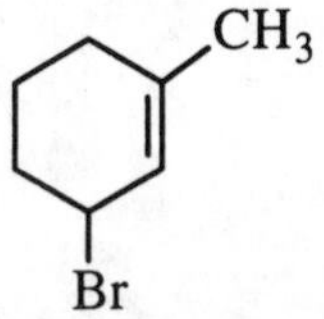

e. CH_3CCH_3 (with CH_3 and Br on C-2)

f. Br $CHCH_3$ (on benzene ring)

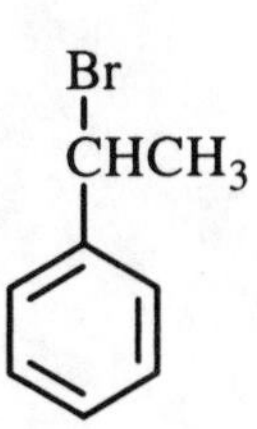

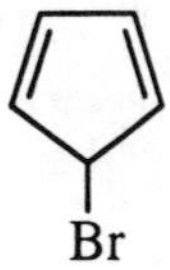

g. Br

h. CH_3 Br CH_3

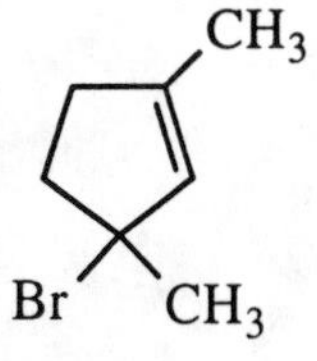

In **b** and **c**, because one product is under kinetic control and the other is under thermodynamic control, the major product will depend on the conditions under which the reaction is carried out.

21. It is harder to break a C-H bond than a C-D bond. (Deuterium is an isotope of hydrogen; it has an additional neutron.)
Because a bromine radical is less reactive than a chlorine radical, a bromine radical has a greater preference for the more easily broken C-H bond. Bromination, therefore, would have a greater deuterium kinetic isotope effect than chlorination.
This is reminiscent of the realtive rates of 1600 : 1 for bromination and 5 : 1 for chlorination for removal of a hydrogen from a tertiary carbon versus from a primary carbon.

22. The methyl radical that is created in the first propagation step reacts with Br_2, forming bromomethane.

$$Br\cdot + CH_4 \longrightarrow \cdot CH_3 + HBr$$

$$\cdot CH_3 + Br_2 \longrightarrow CH_3Br + \cdot Br$$

If HBr is added to the reaction mixture, the methyl radical that is created in the first propagation step can react with Br_2 or with the added HBr. Because reaction with HBr reforms methane, the overall rate of formation of bromomethane is decreased.

$$Br\cdot + CH_4 \longrightarrow \cdot CH_3 + HBr$$

$$\cdot CH_3 + Br_2 \longrightarrow CH_3Br + \cdot Br$$

$$\cdot CH_3 + HBr \longrightarrow CH_4 + \cdot Br$$

23. **a.** One difference between this reaction and the monochlorination of ethane is the source of the chlorine radical. This reaction has two sets of propagation steps because two different radicals are generated in the initiation step. This reaction also has several more termination steps because of the two different radicals generated in the initiation step.

initiation

$$CH_3\underset{CH_3}{\overset{CH_3}{C}}OCl \longrightarrow CH_3\underset{CH_3}{\overset{CH_3}{C}}O^{\bullet} + Cl\cdot$$

propagation

$$CH_3\underset{CH_3}{\overset{CH_3}{C}}O\cdot + CH_3CH_3 \longrightarrow CH_3\underset{CH_3}{\overset{CH_3}{C}}OH + CH_3\dot{C}H_2$$

$$CH_3\dot{C}H_2 + CH_3\underset{CH_3}{\overset{CH_3}{C}}OCl \longrightarrow CH_3CH_2Cl + CH_3\underset{CH_3}{\overset{CH_3}{C}}O\cdot$$

propagation

$$Cl\cdot + CH_3CH_3 \longrightarrow HCl + CH_3\dot{C}H_2$$

$$CH_3\dot{C}H_2 + CH_3\underset{CH_3}{\overset{CH_3}{C}}OCl \longrightarrow CH_3CH_2Cl + CH_3\underset{CH_3}{\overset{CH_3}{C}}O\cdot$$

termination

$$CH_3\dot{C}H_2 + Cl\cdot \longrightarrow CH_3CH_2Cl$$

$$CH_3\underset{CH_3}{\overset{CH_3}{C}}O\cdot + CH_3\underset{CH_3}{\overset{CH_3}{C}}O\cdot \longrightarrow CH_3\underset{CH_3}{\overset{CH_3}{C}}O{-}O\underset{CH_3}{\overset{CH_3}{C}}CH_3$$

$$CH_3\underset{CH_3}{\overset{CH_3}{C}}O\cdot + Cl\cdot \longrightarrow CH_3\underset{CH_3}{\overset{CH_3}{C}}OCl$$

$$CH_3\underset{CH_3}{\overset{CH_3}{C}}O\cdot + CH_3\dot{C}H_2 \longrightarrow CH_3\underset{CH_3}{\overset{CH_3}{C}}OCH_2CH_3$$

$$Cl\cdot + Cl\cdot \longrightarrow Cl_2$$

$$CH_3\dot{C}H_2 + CH_3\dot{C}H_2 \longrightarrow CH_3CH_2CH_2CH_3$$

24.

a. $\Delta H° =$ bonds broken - bonds formed

$\Delta H°$ = [105 kcal/mol + 58 kcal/mol] - [84 kcal/mol + 103 kcal/mol]

$\Delta H°$ = 163 - 187 = -24 kcal/mol

b.

	$\Delta H°$
$CH_3{-}H + {}^{\bullet}Cl \longrightarrow {}^{\bullet}CH_3 + H{-}Cl$	105 - 103 = 2 kcal/mol
${}^{\bullet}CH_3 + Cl{-}Cl \longrightarrow CH_3{-}Cl + {}^{\bullet}Cl$	58 - 84 = - 26 kcal/mol
	- 24 kcal/mol

c. If you cancel the things that are the same on opposite sides of the equations and then add the two equations, you are left with the equation in "**a**".

$$CH_3{-}H + \cancel{{}^{\bullet}Cl} \longrightarrow \cancel{{}^{\bullet}CH_3} + H{-}Cl$$

$$\cancel{{}^{\bullet}CH_3} + Cl{-}Cl \longrightarrow CH_3{-}Cl + \cancel{{}^{\bullet}Cl}$$

$$CH_4 + Cl_2 \longrightarrow CH_3Cl + HCl$$

25. The first propagation step is very endothermic, so it would not be able to compete with the first propagation step shown in Problem 24b.

	$\Delta H°$
$CH_3{-}H + {}^{\bullet}Cl \longrightarrow CH_3{-}Cl + {}^{\bullet}H$	105 - 84 = 21 kcal/mol
${}^{\bullet}H + Cl{-}Cl \longrightarrow H{-}Cl + {}^{\bullet}Cl$	58 - 103 = - 45 kcal/mol
	- 24 kcal/mol

Chapter 8 Practice Test

1. How many monochlorinated products would be obtained from the following reaction? (Ignore stereoisomers.)

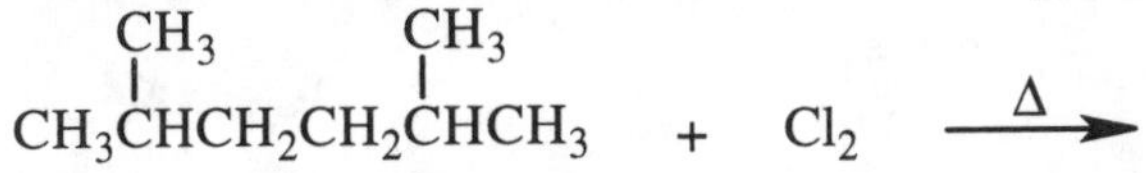

$$CH_3\overset{\overset{\displaystyle CH_3}{|}}{C}HCH_2CH_2\overset{\overset{\displaystyle CH_3}{|}}{C}HCH_3 \quad + \quad Cl_2 \quad \xrightarrow{\Delta}$$

2. Give the first propagation step in the monochlorination of ethane.

3. When (*S*-)-2-bromopentane is brominated, 2,3-dibromopentanes are formed. Which of the following compounds are **not** formed?

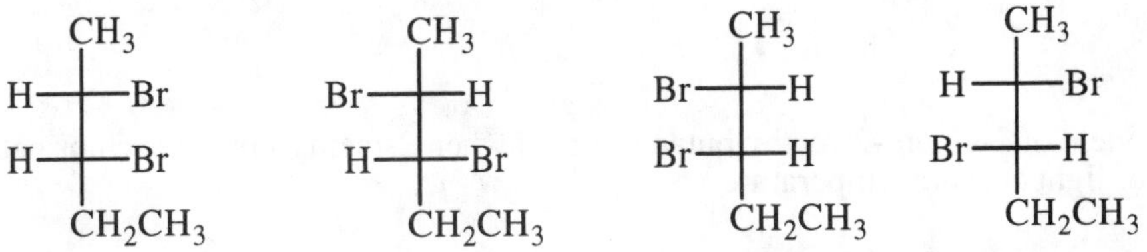

CH_3	CH_3	CH_3	CH_3
H—┼—Br	Br—┼—H	Br—┼—H	H—┼—Br
H—┼—Br	H—┼—Br	Br—┼—H	Br—┼—H
CH_2CH_3	CH_2CH_3	CH_2CH_3	CH_2CH_3

4. Determine the $\Delta H°$ of the two propagation steps in the monochlorination of ethane, using Table 3.1 on page 130 of the text.

5. Label the radicals in order of decreasing stability. (Label the most stable #1.)

$\dot{C}H_2CH_2CH_2CH{=}CH_2$ $\qquad$ $CH_3CH_2\dot{C}HCH{=}CH_2$

$CH_3CH_2CH_2CH{=}\dot{C}H$ $\qquad$ $CH_3\dot{C}HCH_2CH{=}CH_2$

6. Give the product(s) of the following reactions, ignoring stereoisomers.

a. 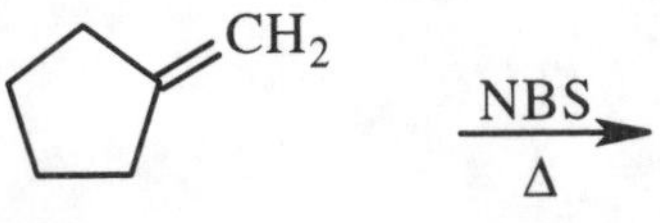 (cyclopentane)$=CH_2$ $\xrightarrow[\Delta]{NBS}$

b. $CH_3CH_2CH{=}CH_2 \xrightarrow[\Delta]{NBS}$

7. a. Give the products that will be obtained from monochlorination of the following alkane at room temperature. (Ignore stereoisomers.)

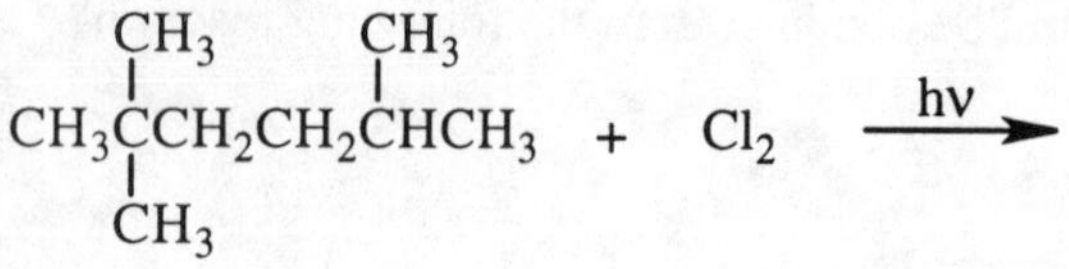

$$CH_3C(CH_3)_2CH_2CH_2CH(CH_3)CH_3 + Cl_2 \xrightarrow{h\nu}$$

b. What product will be obtained in greatest yield?

c. What product would be obtained in greatest yield if the alkane were brominated instead of chlorinated?

8. Calculate the yield of 2-chloro-2-methylbutane formed when 2-methylbutane is chlorinated in the presence of light at room temperature.

9. Give the products of the following reactions.

a. 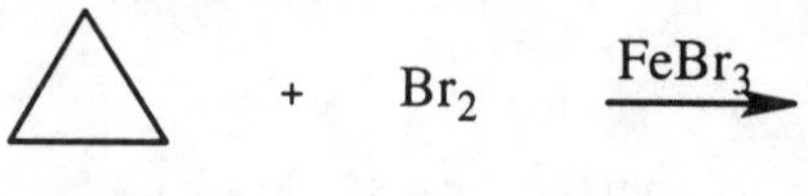 △ + Br_2 $\xrightarrow{FeBr_3}$

b. 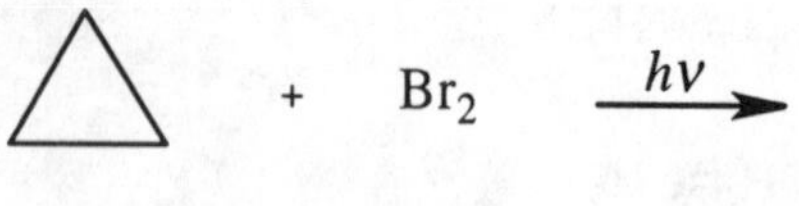 △ + Br_2 $\xrightarrow{h\nu}$

c. 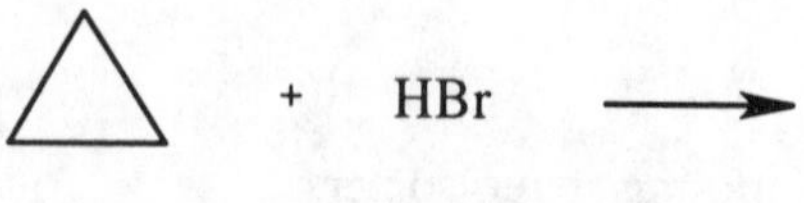 △ + HBr $\longrightarrow$

CHAPTER 9
Reactions at an *sp*^{3} Hybridized Carbon I: Substitution Reactions of Alkyl Halides

Important Terms

aprotic solvent — a solvent that does not have a hydrogen bonded to an oxygen or to a nitrogen.

backside attack — nucleophilic attack on the side of the carbon opposite to the side bonded to the leaving group.

base — a substance that accepts a proton.

basicity — describes the tendency of a compound to share its electrons with a proton.

bimolecular — involving two molecules.

complete racemization — formation of a pair of enantiomers in equal amounts.

dielectric constant — a measure of how well a solvent can insulate opposite charges from one another.

direct displacement mechanism — a reaction in which a nucleophile displaces the leaving group in a single step.

elimination reaction — a reaction that removes atoms or groups from the reactant to form a π bond.

first-order reaction — a reaction whose rate is dependent on the concentration of one reactant.

intimate ion pair — results when the covalent bond that joined the cation and anion has broken, but the cation and anion are still next to each other.

inversion of configuration — turning the carbon inside out like an umbrella so that the resulting product has a configuration opposite to that of the reactant.

ion-dipole interaction — the interaction between an ion and the dipole of a molecule.

kinetics — the field of chemistry that deals with the rates of chemical reactions.

leaving group — the group that is displaced in a nucleophilic substitution reaction.

Le Châtelier's principle — states that if an equilibrium is disturbed, the components of the equilibrium will adjust in a way that will offset the disturbance.

nucleophile — an electron-rich atom or molecule.

nucleophilicity — a measure of how readily an atom or molecule with a pair of nonbonded electrons attacks an atom.

nucleophilic substitution reaction — a reaction in which a nucleophile substitutes for an atom or group.

partial racemization — formation of a pair of enantiomers in unequal amounts.

protic solvent	a solvent that has a hydrogen bonded to an oxygen or to a nitrogen.
rate law	the equation that shows the relationship between the rate of a reaction and the concentration of the reactants.
second-order reaction	a reaction whose rate is dependent on the concentration of two reactants.
S_N1 reaction	a first-order nucleophilic substitution reaction.
S_N2 reaction	a second-order nucleophilic substitution reaction.
solvent-separated ion pair	results when the cation and anion are separated by one or more solvent molecules.
solvolysis	reaction with the solvent.
steric effects	effects due to the fact that groups occupy a certain volume of space.
steric hindrance	refers to bulky groups at the site of a reaction that make it difficult for the reactants to approach each other.
substitution reaction	a reaction that changes one substituent of a reactant for another.
unimolecular	involving one molecule.

Solutions to Problems

1. It decreases the magnitude of the rate constant which causes the reaction to be slower.

2.

$$\mathrm{CH_3CH_2CH_2CH_2CH_2Br} > \mathrm{CH_3\overset{\overset{CH_3}{|}}{C}HCH_2CH_2Br} > \mathrm{CH_3CH_2\overset{\overset{CH_3}{|}}{C}HCH_2Br} > \mathrm{CH_3CH_2\underset{\underset{CH_3}{|}}{\overset{\overset{CH_3}{|}}{C}}Br}$$

3.

a. 2-Butanol will be formed from the S_N2 reaction between 2-bromobutane and hydroxide ion. The configuration of the product cannot be specified, because the configuration of the reactant is not specified.

b. (*S*)-2-Butanol will be formed from the S_N2 reaction between (*R*)-2-bromobutane and hydroxide ion.

c. (*R*)-3-Hexanol will be formed from the S_N2 reaction between (*S*)-3-chlorohexane and hydroxide ion.

d. 3-Pentanol (it does not have a chirality center) will be formed from the S_N2 reaction between 3-iodopentane (it does not have a chirality center) and hydroxide ion.

4.

a. RO^- because ROH is a weaker acid than RSH.

b. RS^- because it is less well solvated by water.

5. Solved in the text.

6. **a.** aprotic **b.** aprotic **c.** protic **d.** aprotic

7.

a. CH_3CH_2Br + HO^- — HO^- is a stronger nucelophile than H_2O.

b. $CH_3\underset{\underset{CH_3}{|}}{C}HCH_2Br$ + HO^- — less steric hindrance

c. CH_3CH_2Cl + CH_3S^- — CH_3S^- is a stronger nucleophile than CH_3O^- in a solvent that can form hydrogen bonds.

d. CH_3Br + I^- — Br^- is a weaker base than Cl^-, so Br^- is a better leaving group.

e. $BrCH_2CH_2CH_2CH_2NH_2$ — Because of strain a four-membered ring is less stable than a five-membered ring, so the more stable five-membered ring is easier to form.

8. Solved in the text.

9. Reaction of an alkyl halide with ammonia gives a low yield of primary amine because as soon as the primary amine is formed, it can react with another molecule of alkyl halide to form a secondary amine.
The alkyl azide is treated with hydrogen after all the alkyl halide has reacted with azide ion. Therefore, when the primary amine is formed, there is no alkyl halide for it to react with to form a secondary amine.

10. In each reaction, the conjugate acid of the leaving group is stronger than the conjugate acid of the nucleophile. That means the leaving group is the weaker base (better leaving group).

H_2O pK_a = 15.7	HCl pK_a = -7.0	RC≡CH pK_a = 25	HBr pK_a = -9.0
H_2S pK_a = 7.0	HBr pK_a = -9.0	HC≡N pK_a = 9.1	HI pK_a = -10.0
ROH pK_a = 15.5	HI pK_a = -10.0	NH_4 pK_a = 9.4	HBr pK_a = -9.0
RSH pK_a = 10.5	HBr pK_a = -9.0	$CH_3CH_2NH_2CH_3$ pK_a = 10.8	HI pK_a = -10.0
NH_3 pK_a = 36	HCl pK_a = -7.0		

11.

a. $CH_3CH_2OCH_3$ b. $CH_3CH_2N_3$ c. $CH_3CH_2\overset{+}{N}(CH_3)_3\ Br^-$

12.

$$CH_3CH_2Cl + K^+I^- \xrightleftharpoons{\text{acetone}} CH_3CH_2I + K^+Cl^-$$

K^+Cl^- precipitates out in acetone, which drives the reaction to the right.

13.

$$(CH_3)_3CBr > CH_3CH(CH_3)Br > CH_3CH_2CH_2Br > CH_3Br$$

14.

$$CH_3CH_2C(CH_3)(Br)CH_2CH_3 > CH_3CH(Br)CH_2CH_2CH_3 > CH_3CH(Cl)CH_2CH_2CH_3 > ClCH_2CH_2CH_2CH_2CH_3$$

15. a. The secondary carbocation that is formed initially will undergo a 1,2-hydride shift to form a tertiary carbocation.

$$CH_3CH(CH_3)\overset{+}{C}HCH_2CH_3 \longrightarrow CH_3\overset{+}{C}(CH_3)CH_2CH_3$$

b. The secondary carbocation that is formed initially will undergo a 1,2-hydride shift to form a tertiary carbocation.

(2-methylcyclohexyl cation with + on ring carbon adjacent to the CH_3-bearing carbon) $\longrightarrow$ (1-methylcyclohexyl cation, + on the CH_3-bearing carbon)

c. The secondary carbocation that is formed initially will undergo a 1,2-methyl shift to form a tertiary carbocation.

$$CH_3CH_2C(CH_3)_2{-}\overset{+}{C}HCH_3 \longrightarrow CH_3CH_2\overset{+}{C}(CH_3){-}CH(CH_3)CH_3$$

d. The same substitution product will be obtained from S_N1 and S_N2 reactions. Notice that because the configuration of the chirality center is not specified, the configuration of the product cannot be specified. If the configuration at C-4 were specified, one stereoisomer would be obtained from an S_N2 reaction (the product with the inverted configuration compared to the configuration of the reactant), and two products would be obtained (one with the inverted configuration and one with the retained configuration) from an S_N1 reaction.

16.

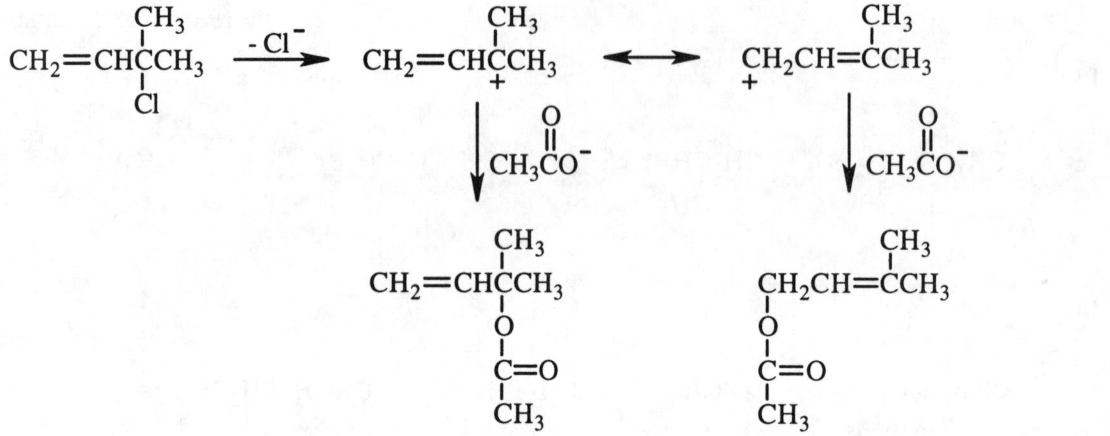

$CH_2{=}CHC(CH_3)(Cl)CH_3 \xrightarrow{-Cl^-} CH_2{=}CH\overset{+}{C}(CH_3)CH_3 \longleftrightarrow \overset{+}{C}H_2CH{=}C(CH_3)CH_3$

$\downarrow CH_3CO_2^-$ (each carbocation)

$CH_2{=}CHC(CH_3)(OC(=O)CH_3)CH_3$

kinetic product
(tertiary carbocation versus pirmary carbocation))

$CH_3C(=O)OCH_2CH{=}C(CH_3)CH_3$

thermodynamic product
(more substituted double bond)

17. The trans isomer will be formed in greater yield because the departing bromide ion can block the approach of the incoming nucleophile to the side of the carbocation vacated by the bromide ion.

18.

a. 1. (cyclohexane: H, Br; CH_3, H) $\xrightarrow[S_N2]{CH_3O^-}$ (cyclohexane: H, H; CH_3, OCH_3)

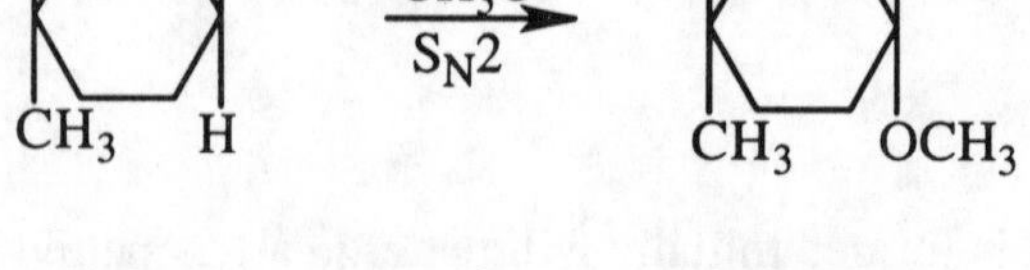

2. (chair: H, H; CH_3, Cl) $\xrightarrow[S_N2]{HO^-}$ (chair: H, OH; CH_3, H)

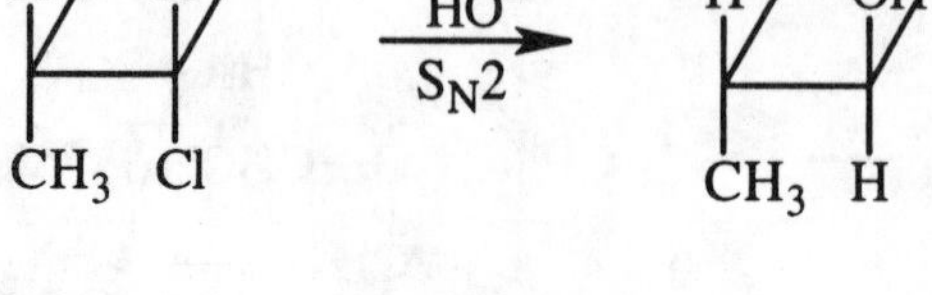

b. 1. 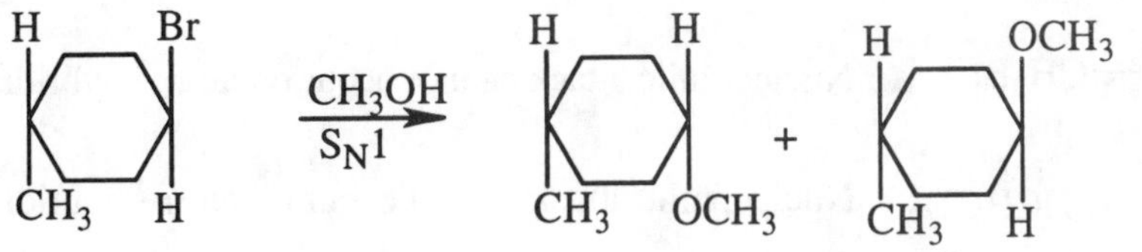 [structures: cyclohexane ring with H, Br, CH_3, H] $\xrightarrow[S_N1]{CH_3OH}$ [cyclohexane ring with H, H, CH_3, OCH_3] + [cyclohexane ring with H, OCH_3, CH_3, H]

2. 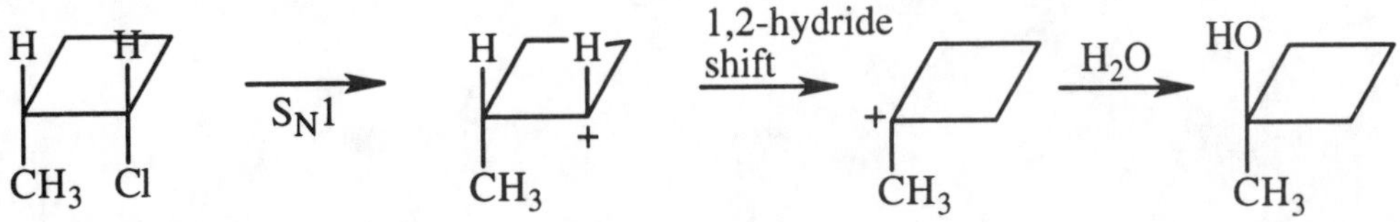 [structure: H, H, CH_3, Cl] $\xrightarrow{S_N1}$ [cation: H, H, CH_3, +] $\xrightarrow{\text{1,2-hydride shift}}$ [cation: +, CH_3] $\xrightarrow{H_2O}$ [HO, CH_3]

19. Because the rate of an S_N1 reaction is not affected by increasing the concentration of the nucleophile, while the rate of an S_N2 reaction is increased when the concentration of the nucleophile is increased, you first have to determine whether the reactions are S_N1 or S_N2 reactions.

a is an S_N2 reaction because the configuration of the product is inverted compared with the reactant.

b is an S_N2 reaction because the reactant is a primary alkyl halide.

c is an S_N1 reaction because the reactant is a tertiary alkyl halide.

Because they are S_N2 reactions, **a** and **b** will go faster if the concentration of the nucleophile is increased.

Because it is an S_N1 reaction, the rate of **c** will not change if the concentration of the nucleophile is increased.

20.

a. $CH_3CH_2CH_2I$ — I^- is a weaker base than Br^-, so I^- is a better leaving group.

b. CH_3OCH_2Cl — The electron-withdrawing methoxy group increases the electrophilicity of the carbon bonded to the leaving group.

c. $CH_3CH_2CH(CH_3)Br$ — less steric hindrance

d. $CH_3CH_2CH(CH_3)CH_2Br$ — A primary carbon is less sterically hindered than a secondary carbon.

e. 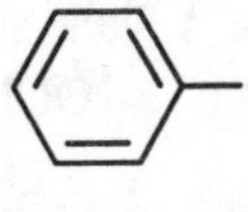 $C_6H_5CH_2CH_2Br$ (benzene ring–CH_2CH_2Br) — A primary carbon is less sterically hindered than a secondary carbon.

f. 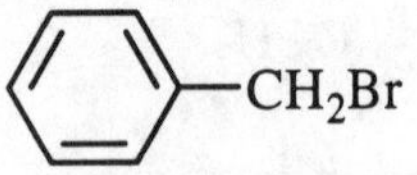 C_6H_5–CH_2Br Nucleophilic attack cannot occur on an sp^2 hybridized carbon.

g. $CH_3CH{=}CHCH(Br)CH_3$ Nucleophilic attack cannot occur on an sp^2 hybridized carbon.

21.

a. $CH_3CH_2CH_2I$ I^- is a weaker base than Br^-, so I^- is a better leaving group.

b. CH_3OCH_2Cl The methoxy group stabilizes the carbocation by resonance electron donation.

c. The two compounds are equally reactive.

d. $CH_3CH_2CH_2CH(CH_3)Br$ A secondary carbocation is more stable than a primary carbocation.

e. 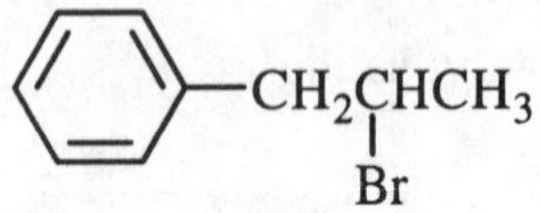 C_6H_5–$CH_2CH(Br)CH_3$ A secondary carbocation is more stable than a primary carbocation.

f. 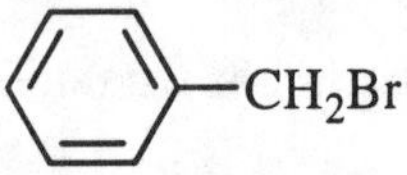 C_6H_5–CH_2Br A benzyl cation is more stable than an aryl cation.

g. $CH_3CH{=}CHCH(Br)CH_3$ An allyl cation is more stable than a vinyl cation.

22.

a. $$CH_3CH{=}CHCH_2Br \xrightarrow{CH_3O^-} CH_3CH{=}CHCH_2OCH_3$$

b. $$CH_3CH{=}CHCH_2Br \longrightarrow CH_3CH{=}CH\overset{+}{C}H_2 \longleftrightarrow CH_3\overset{+}{C}HCH{=}CH_2$$

$$\downarrow CH_3O^- \qquad\qquad \downarrow CH_3O^-$$

$$CH_3CH{=}CHCH_2OCH_3 \quad + \quad CH_3CH(OCH_3)CH{=}CH_2$$

23.

a. An S_N2 reaction, because of the high concentration of a good nucleophile.

CH_2CH_3, H, CH_3, $CH_3CH_2CH_2O$

b. An S_N1 reaction, because a poor nucleophile is used.

$CH_2CH_2CH_3$, CH_3, H, $+NH_3$ $CH_2CH_2CH_3$, H, CH_3, H_3N+

c. An S_N2 reaction, because of the high concentration of a good nucleophile.

H, CH_3, H, OCH_3

d. An S_N1 reaction, because a poor nucleophile is used. Notice that the initially formed secondary carbocation rearranges to a more stable tertiary carbocation.

H, CH_3, + → CH_3, + $\xrightarrow{CH_3O^-}$ OCH_3, CH_3

The product does not have a chirality center.

e. An S_N2 reaction, because of the high concentration of a good nucleophile.

CH_2CH_3, CH_3, H, OCH_3

f. An S_N1 reaction, because a poor nucleophile is used.

CH_2CH_3, H, CH_3, CH_3O CH_2CH_3, CH_3, H, OCH_3

24. **a.** Solved in the text.

b. $$\frac{S_N2}{S_N2 + S_N1} = \frac{3.2 \times 10^{-5} \times 1 \times 10^{-3}}{3.2 \times 10^{-8} + 1.5 \times 10^{-6}} = \frac{3.2 \times 10^{-8}}{3.2 \times 10^{-8} + 150 \times 10^{-8}} = \frac{3.2}{153} = 2\%$$

25. **a.** Increasing the polarity will decrease the rate of the reaction because the concentration of charge on the reactants is greater than the concentration of charge on the transition state.

b. Increasing the polarity will decrease the rate of the reaction because the concentration of charge on the reactants is greater than the concentration of charge on the transition state.

c. Increasing the polarity will increase the rate of the reaction because the concentration of charge on the reactants is less than the concentration of charge on the transition state.

26.

a. $CH_3Br + HO^- \longrightarrow CH_3OH + Br^-$

HO^- is a better nucleophile than H_2O.

b. $CH_3I + HO^- \longrightarrow CH_3OH + I^-$

I^- is a better leaving group than Cl^-.

c. $CH_3Br + NH_3 \longrightarrow CH_3\overset{+}{N}H_3 + Br^-$

NH_3 is a better nucleophile than H_2O.

d. $CH_3Br + HO^- \xrightarrow{\text{DMSO}} CH_3OH + Br^-$

DMSO will not stabilize the nucleophile by hydrogen bonding.

e. $CH_3Br + NH_3 \xrightarrow{\text{EtOH}} CH_3\overset{+}{N}H_3 + Br^-$

The more polar solvent will be more able to stabilize the transition state.

27. Solved in the text.

28. Acetate ion will be a more reactive nucleophile in dimethyl sulfoxide because dimethyl sulfoxide will not stabilize the nucleophile by hydrogen bonding, while methanol will stabilize it by hydrogen bonding.

29. Only an S_N1 reaction will give the product with retention of configuration. Since the S_N1 reaction is favored by a polar solvent, a greater percentage of the reaction will take place by an S_N1 pathway in 50% water/50% ethanol, the more polar of the two solvents.

30. a.
1. 3° > 2° > 1°
2. An S_N1 reaction is not affected by the strength of the nucleophile, but a weak nucleophile favors an S_N1 by disfavoring an S_N2 reaction.
3. An S_N1 reaction is not affected by the concentration of the nucleophile, but a low concentration of a nucleophile favors an S_N1 by disfavoring an S_N2 reaction.
4. An aprotic polar solvent favors an S_N1 reaction if the reactant is charged. A protic polar solvent favors an S_N1 reaction if the reactant is not charged.

b.
1. 1° > 2° > 3°
2. A strong nucleophile favors an S_N2 reaction.
3. A high concentration of a nucleophile favors an S_N2 reaction.
4. An aprotic polar solvent favors an S_N2 reaction if either of the reactants is charged. A protic polar solvent favors an S_N2 reaction if neither of the reactants is charged.

31. If the atoms are in the same row, the stronger base is the stronger nucleophile. If the atoms are in the same column, the larger atom is the stronger nucleophile because the solvent will form stronger hydrogen bonds with the smaller atom.

a. HO^- c. H_2S e. I^-

b. $^-NH_2$ d. HS^- f. Br^-

32. The weaker base is the better leaving group.

a. H_2O c. H_2S e. I^-

b. NH_3 d. HS^- f. Br^-

33.

a. HO^- c. HS^- e. CH_3NH_2 g. $CH_3\overset{O}{\overset{\|}{C}}O^-$

b. CH_3O^- d. $CH_3CH_2S^-$ f. $^-C{\equiv}N$ h. $CH_3C{\equiv}C^-$

34.

a. $CH_3CH_2S^- > CH_3CH_2O^- > CH_3\overset{\overset{O}{\|}}{C}O^-$

b. 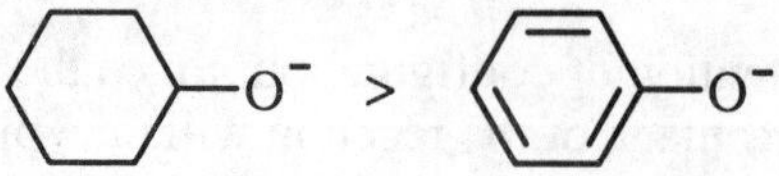 (cyclohexyl)$-O^-$ > (phenyl)$-O^-$

c. $NH_3 > H_2O$

d. $I^- > Br^- > Cl^-$

35. The pK_a would increase (it would be a weaker acid) because of a decreased tendency to form a charged species in a less polar solvent.

36.

a. 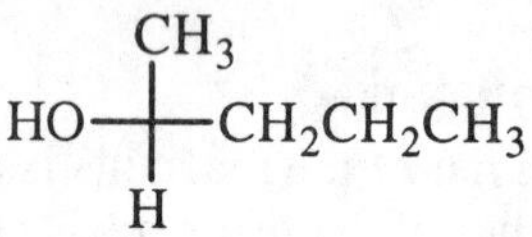 HO–C(CH$_3$)(H)–$CH_2CH_2CH_3$

b. $CH_3CH_2CH_2$–C(CH$_3$)(H)–OCH_3 + CH_3O–C(CH$_3$)(H)–$CH_2CH_2CH_3$

c. (cyclohexane ring: H, H; H_3C, OCH_3) + (cyclohexane ring: H, H; CH_3O, CH_3)

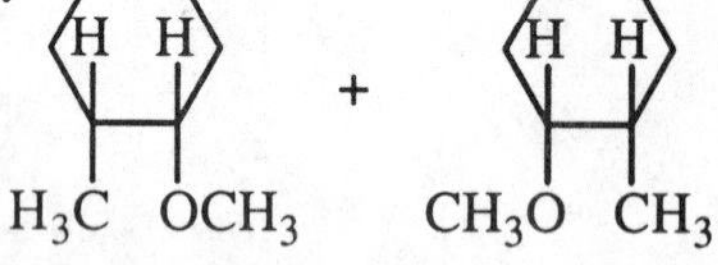

d. 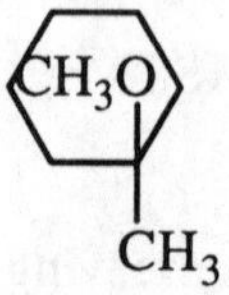 (cyclohexane ring with CH_3O and CH_3 on the same carbon)

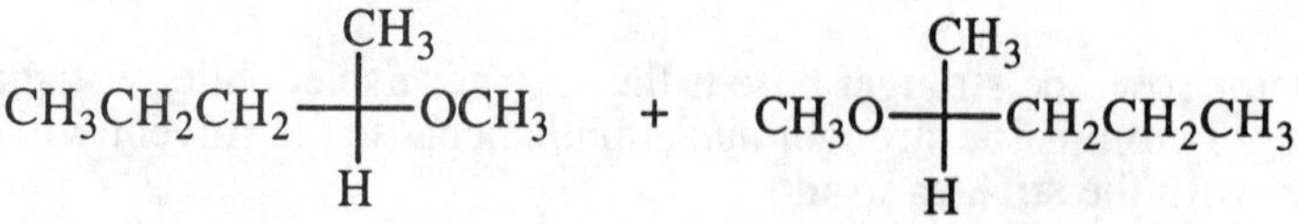

e. 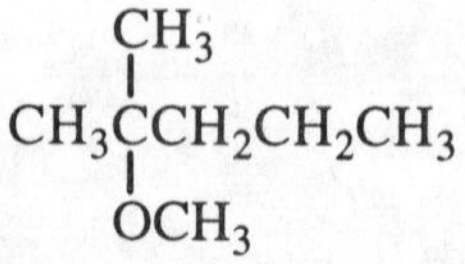 $CH_3\overset{\overset{CH_3}{|}}{\underset{\underset{OCH_3}{|}}{C}}CH_2CH_2CH_3$

f. 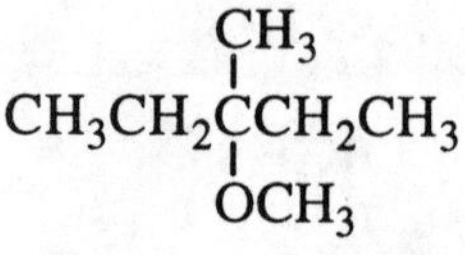 $CH_3CH_2\overset{\overset{CH_3}{|}}{\underset{\underset{OCH_3}{|}}{C}}CH_2CH_3$

37. Methoxide ion will be a stronger nucleophile in DMSO because DMSO cannot stabilize the anion by hydrogen bonding.

38.

a. $(CH_3)_3CCH_2Cl \xrightarrow{CH_3S^-} (CH_3)_3CCH_2SCH_3 + Cl^-$

The nucleophile is less sterically hindered.

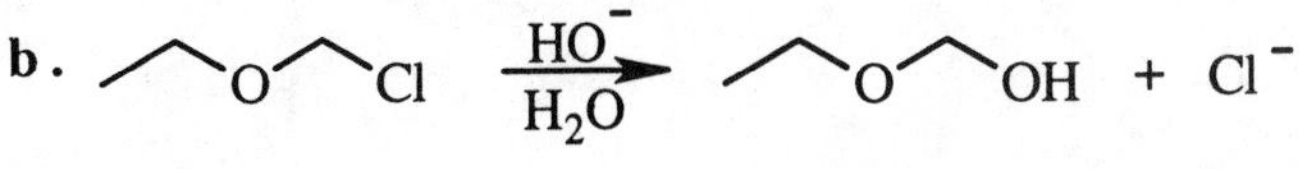

b. $CH_3CH_2OCH_2Cl \xrightarrow[H_2O]{HO^-} CH_3CH_2OCH_2OH + Cl^-$

The electron-withdrawing oxygen increases the electrophilicity of the carbon that the nucleophile attacks.

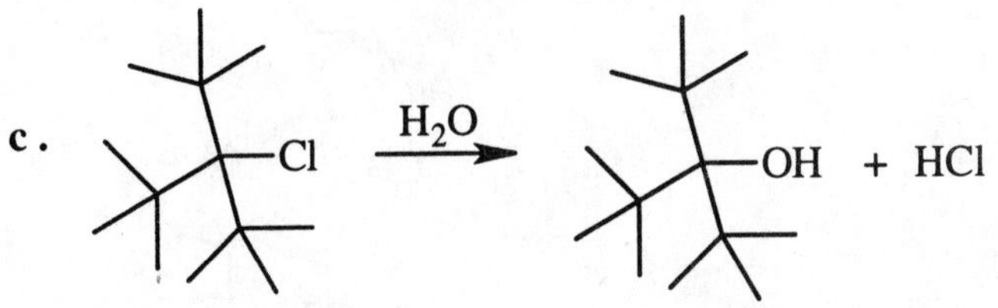

c. $\xrightarrow{H_2O}$ —OH + HCl

Steric strain is decreased when Cl^- dissociates to form the carbocation in the rate-limiting step since the hybridization of the carbon atom changes from sp^3 to sp^2, allowing the bond angle between the bulky groups to increase from 109.5° to 120°.

d. $(CH_3)_3CBr \xrightarrow{H_2O} (CH_3)_3COH + HBr$

Because the reactants are neutral, the reaction will be faster in the more polar solvent.

39. The all-cis isomer is more reactive in an S_N2 reaction.

In an S_N2 reaction, the leaving group must be in an axial position in order to allow backside attack to occur without steric hindrance from the cyclohexane ring. When the bromine is in the axial position in the all-cis isomer, both methyl substituents are in equatorial positions, so the reaction takes place via the most stable conformer. In the other isomer, the conformer with bromine in the axial position (the reactive conformer) has both methyl substituents in less stable axial positions.

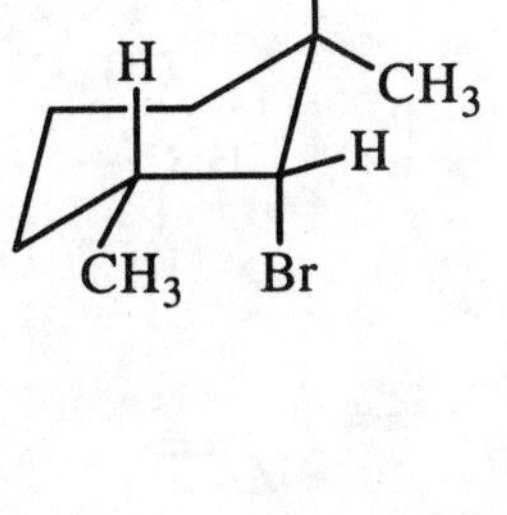

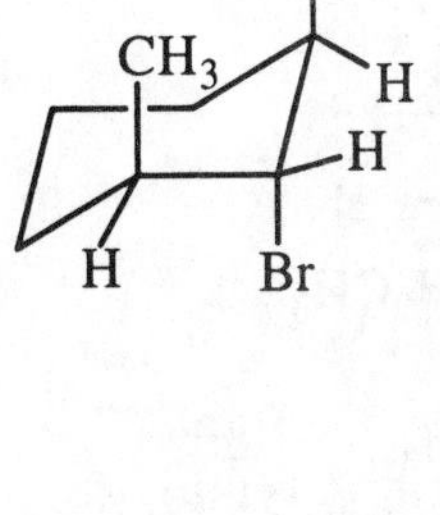

H, H, CH_3, H, CH_3, Br

CH_3, CH_3, H, H, H, Br

40.

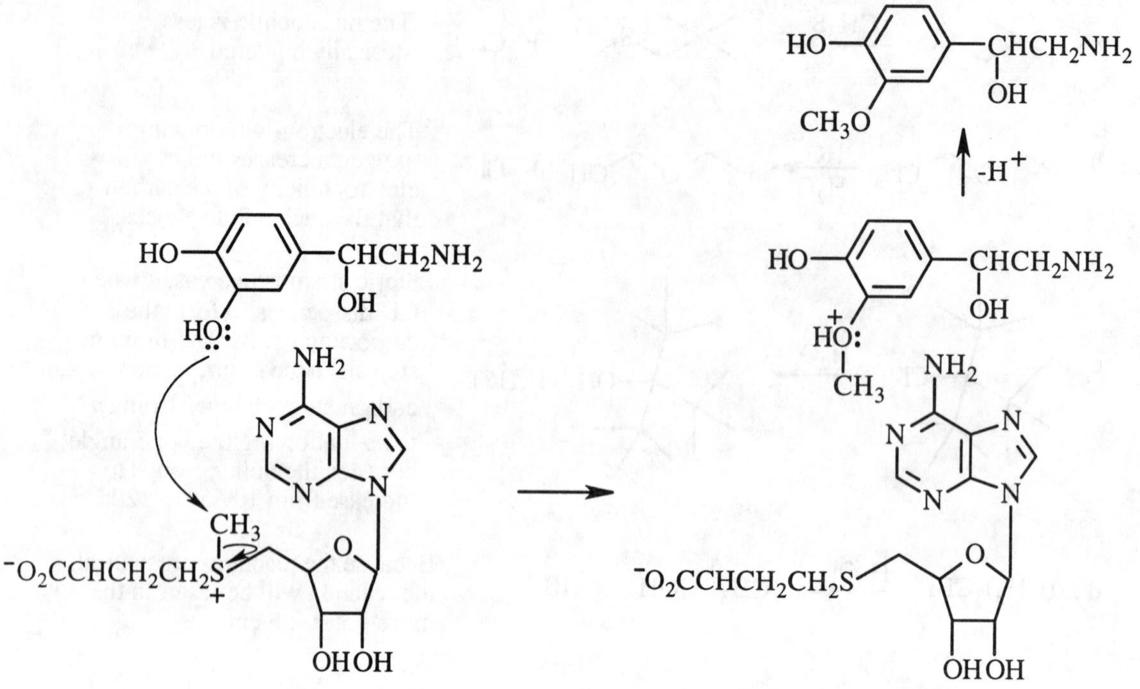

HO, HO: —CHCH$_2$NH$_2$ / OH; NH$_2$; N; N; N; N; O; CH$_3$; $^-O_2CCHCH_2CH_2\overset{+}{S}$; OH OH

→

HO, $\overset{+}{H}O$: / CH$_3$ —CHCH$_2$NH$_2$ / OH; NH$_2$; N; N; N; N; O; $^-O_2CCHCH_2CH_2S$; OH OH

↑ -H$^+$

HO, CH$_3$O —CHCH$_2$NH$_2$ / OH

41.

a. CH$_3$; CH$_3$O—|—H; H$_3$C—|—H; CH$_2$CH$_3$

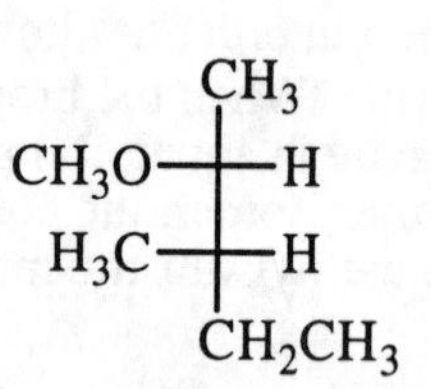

b. CH$_3$; CH$_3$O—|—H; H—|—CH$_3$; CH$_2$CH$_3$

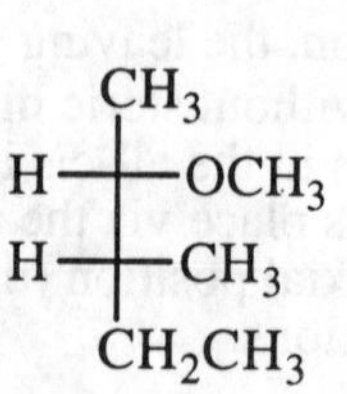

c. CH$_3$; H—|—OCH$_3$; H$_3$C—|—H; CH$_2$CH$_3$

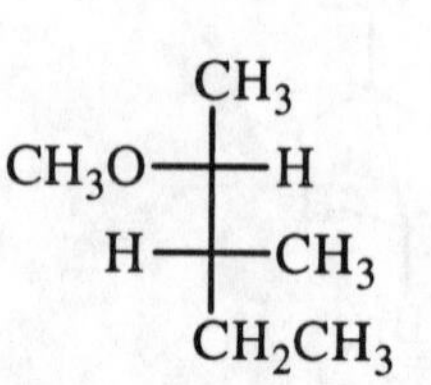

d. CH$_3$; H—|—OCH$_3$; H—|—CH$_3$; CH$_2$CH$_3$

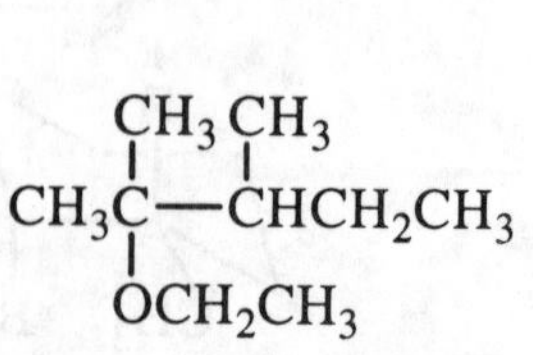

e. CH$_3$C(CH$_3$)(OCH$_2$CH$_3$)—CH(CH$_3$)CH$_2$CH$_3$

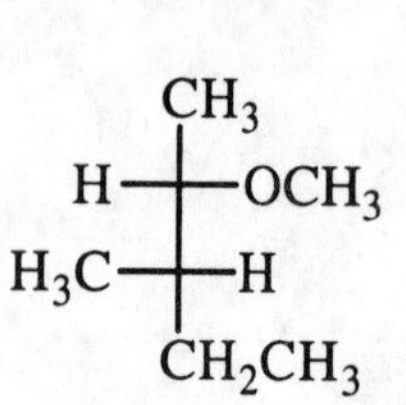

f. C$_6$H$_5$—CH$_2$OCH$_2$CH$_3$

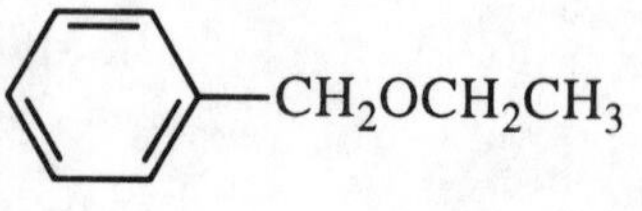

42. Sodium acetate is a poor nucleophile, so we can assume the reaction takes place primarily by an S_N1 pathway.

$CH_2{=}CHC(CH_3)(Br)CH_3 \longrightarrow CH_2{=}CH\overset{+}{C}(CH_3)CH_3 \longleftrightarrow \overset{+}{C}H_2CH{=}C(CH_3)CH_3$

$\downarrow CH_3CO_2^-$ (both cations)

$CH_2{=}CHC(CH_3)(OCOCH_3)CH_3$ — kinetic product

$CH_3COOCH_2CH{=}C(CH_3)CH_3$ — thermodynamic product

43. **a.** The reaction with quinuclidine had the larger rate constant because quinuclidine is less sterically hindered as a result of the substituents on the nitrogen being pulled back into a ring structure.

b. The reaction with quinuclidine had the larger rate constant for the same reason given in **a**.

c. Isopropyl iodide had the larger ratio because, since it is more sterically hindered than methyl iodide, it is more affected by differences in the amount of steric hindrance in the nucleophile.

44. Because methanol is a poor nucleophile, the reaction will take place predominately via an S_N1 pathway. A secondary benzylic carbocation is more stable and, therefore, is easier to form than a secondary carbocation, so the product will be the one shown.

(Structure: tetrahydronaphthalene with OCH_3 at the benzylic carbon and Br on the ring)

45.

a. cyclohexane $\xrightarrow[h\nu]{Br_2}$ cyclohexyl–Br

b. cyclohexane $\xrightarrow[h\nu]{Br_2}$ cyclohexyl–Br $\xrightarrow{CH_3O^-}$ cyclohexyl–OCH_3

c. cyclohexane $\xrightarrow[h\nu]{Br_2}$ cyclohexyl–Br $\xrightarrow{HO^-}$ cyclohexyl–OH

46.

a. Fischer projection: top CH_2CH_3; CH_3O—C—H; CH_3—C—H; bottom CH_2CH_3 (3*R*,4*S*)

b. Fischer projection: top CH_2CH_3; CH_3O—C—H; H—C—CH_3; bottom CH_2CH_3 (3*R*,4*R*)

c. Fischer projection: top CH_2CH_3; H—C—OCH_3; H—C—CH_3; bottom CH_2CH_3 (3*S*,4*R*)

d. Fischer projection: top CH_2CH_3; H—C—OCH_3; CH_3—C—H; bottom CH_2CH_3 (3*S*,4*S*)

47. Tetrahydrofuran can solvate a charge better because the floppy ethyl substituents of diethyl ether provide steric hindrance, making it difficult for the nonbonding electrons of the oxygen to approach the compound to be solvated.

48.

a. CH_3, C, Br, CH_3 → CH_3, C+, CH_3 + Br^- → CH_3, CH_3 $\xrightarrow{H_2\ddot{O}:}$ H, +OH, CH_3, CH_3 $\xrightarrow{-H^+}$ OH, CH_3, CH_3

b. H_3C, $CHCH_3$, Br → H_3C, $\overset{+}{C}HCH_3$ → H, +, CH_3, CH_3 $\xrightarrow{H_2O}$ H, H, HO^+, CH_3, CH_3 + H, H_3C, CH_3, +OH, H $\xrightarrow{-H^+}$ H, HO, CH_3, CH_3 + H, H_3C, CH_3, OH

49. The cis isomer would be expected to react faster in an S_N1 reaction.
The rate-limiting step in an S_N1 reaction is formation of the carbocation intermediate. Both isomers form the same carbocation. The cis isomer (with one substituent in the equatorial position and one in the axial position) is less stable than the trans isomer (with both substituents in the equatorial position). Since the cis isomer is less stable, it has the smaller energy of activation and, therefore, the faster reaction rate.

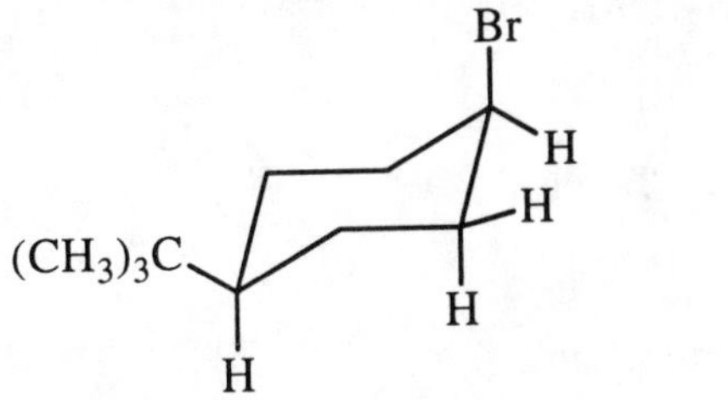

Br H H $(CH_3)_3C$ H H — *cis*-1-bromo-4-*tert*-butylcyclohexane

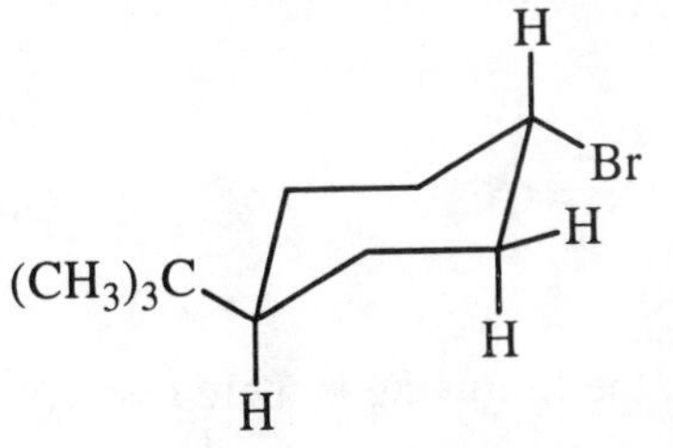

H Br H $(CH_3)_3C$ H H — *trans*-1-bromo-4-*tert*-butylcyclohexane

50. A Diels-Alder reaction between hexachlorocyclopentadiene and 3,4-dichlorocyclopentene forms Chlordane.

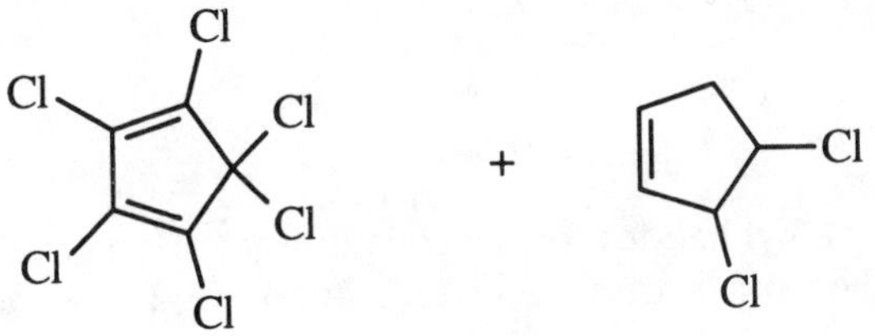

Cl Cl Cl Cl Cl Cl + Cl Cl

51. It will not undergo an S_N2 reaction, because of steric hindrance to backside attack.

It will not undergo an S_N1 reaction, because the carbocation that would be formed is unstable, since the ring structure prevents it from achieving the 120° bond angles required for an sp^2 hybridized carbon.

Chapter 9 Practice Test

1. Which of the following is more reactive in an S_N1 reaction?

   a. $CH_3CH_2CH_2CH_2CH(CH_3)Br$ or $CH_3CH_2CH_2CH(CH_3)CH_2Br$

   b. $CH_3CH{=}C(Br)CH_3$ or $CH_3CH{=}CHCH(Br)CH_3$

2. Which of the following is more reactive in an S_N2 reaction?

   a. $CH_3CH_2CH(CH_3)Br$ or $CH_3CH_2CH(CH_2CH_3)Br$

   b. C_6H_5–Br or C_6H_5–CH_2Br

3. Which of the following alkyl halides form a constitutional isomer as a result of an S_N1 reaction that is different from the isomer formed as a result of an S_N2 reaction?

   $CH_3CH(CH_3)CH(Br)CH_3$ $\quad$ $CH_3C(CH_3)(Br)CH_2CH_3$ $\quad$ (cyclopentane with H_3C and Br on the same carbon) $\quad$ (cyclopentane with CH_3 and Br on adjacent carbons)

4. Indicate whether each of the following statements is true or false.

   a. Increasing the concentration of the nucleophile favors an S_N1 reaction over an S_N2 reaction. T F

   b. Ethyl iodide is more reactive than ethyl chloride in an S_N2 reaction. T F

   c. In an S_N1 reaction, the product with the retained configuration is obtained in greater yield. T F

   d. The rate of a substitution reaction in which none of the reactants is charged will increase if the polarity of the solvent is increased. T F

   e. An S_N2 reaction is a two-step reaction. T F

   f. The pK_a of a carboxylic acid is greater in water than it is in a mixture of dioxane and water. T F

5. Answer the following:

a. Which is a stronger base, CH_3O^- or CH_3S^-?

b. Which is a better nucleophile in an aqueous solution, CH_3O^- or CH_3S^-?

6. For each of the following pairs of S_N2 reactions, indicate the one that occurs with the greater rate constant.

a. $CH_3CH_2CH_2Cl + HO^-$ or $CH_3CH(Cl)CH_3 + HO^-$

b. $CH_3CH_2CH_2Cl + HO^-$ or $CH_3CH_2CH_2I + HO^-$

c. $CH_3CH_2CH_2Br + HO^-$ or $CH_3CH_2CH_2Br + H_2O$

d. $CH_3CH(Br)CH_3 \xrightarrow[H_2O/CH_3OH]{CH_3O^-}$ or $CH_3CH(Br)CH_3 \xrightarrow[CH_3OH]{CH_3O^-}$

e. $BrCH_2CH_2CH_2CH_2NHCH_3$ or $BrCH_2CH_2CH_2NHCH_3$

7. Circle the aprotic solvents.

a. dimethyl sulfoxide b. diethyl ether c. ethanol d. hexane

8. How would increasing the polarity of the solvent affect the following?

a. the rate of the S_N2 reaction of methyl amine with 2-bromobutane

b. the rate of the S_N1 reaction of methyl amine with 2-bromobutane

c. the rate of the S_N2 reaction of methoxide ion with 2-bromobutane

d. the pK_a of acetic acid

e. the pK_a of phenol

CHAPTER 10

Reactions at an *sp*^{3} Hybridized Carbon II: Elimination Reactions of Alkyl Halides; Competition Between Substitution and Elimination

Important Terms

anti elimination — an elimination reaction in which the substituents being eliminated are removed from opposite sides of the molecule.

anti-periplanar — parallel substituents on opposite sides of a molecule.

bifunctional molecule — a molecule with two functional groups.

dehydrohalogenation — elimination of a proton and a halide ion.

deuterium kinetic isotope effect — ratio of the rate constant obtained for a compound containing hydrogen and the rate constant obtained for an identical compound in which one or more of the hydrogens have been replaced by deuterium.

elimination reaction — a reaction that removes atoms or groups from the reactant to form a π bond.

β-elimination reaction or 1,2-elimination reaction — an elimination reaction where the groups being eliminated are bonded to adjacent carbons.

E1 reaction — a first-order elimination reaction.

E2 reaction — a second-order elimination reaction.

intermolecular reaction — a reaction that takes place between two molecules.

intramolecular reaction — a reaction that takes place within a molecule.

kinetic isotope effect — a comparison of the rate of reaction of a compound with the rate of reaction of a compound in which one of the atoms has been replaced by an isotope.

syn elimination — an elimination reaction in which substituents being eliminated are removed from the same side of the molecule.

syn-periplanar — parallel substituents on the same side of a molecule.

target molecule — the desired end product of a synthesis.

Williamson ether synthesis — formation of an ether from the reaction of an alkoxide ion with an alkyl halide.

Zaitsev's rule — the more stable alkene product is obtained by removing a proton from the β-carbon that is bonded to the fewest hydrogens.

Solutions to Problems

1. The reaction of 2-bromo-2,3-dimethylbutane with sodium *tert*-butoxide forms two alkene elimination products. Because of steric hindrance, the least stable alkene (2,3-dimethyl-1-butene) is the one that is easier to make.

$$\underset{\text{2-bromo-2,3-dimethylbutane}}{\mathrm{CH_3C(Br)(CH_3)-CH(CH_3)CH_3}} \xrightarrow{(CH_3)_3CO^-} \underset{\text{2,3-dimethyl-1-butene}}{\mathrm{CH_2{=}C(CH_3)-CH(CH_3)CH_3}} + \underset{\text{2,3-dimethyl-2-butene}}{\mathrm{CH_3C(CH_3){=}C(CH_3)CH_3}}$$

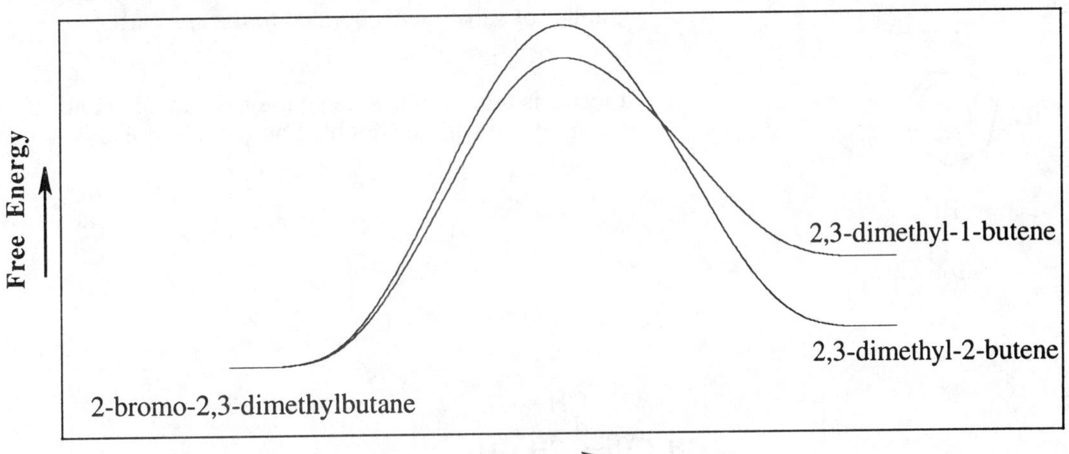

2.

a. $CH_3CH{=}CHCH_3$ Removal of a β-hydrogen from the most substituted carbon forms the most stable "alkene-like" transition state.

b. $CH_2{=}CHCH_2CH_3$ Removal of a β-hydrogen from the least substituted carbon forms the most stable "carbanion-like" transition state.

c. $CH_3C(CH_3){=}CHCH_2CH_3$ Removal of a β-hydrogen from the most substituted carbon forms the most stable "alkene-like" transition state.

d. $CH_3CH(CH_3)CH{=}CHCH_3$ Removal of a β-hydrogen from the least substituted carbon forms the most stable "carbanion-like" transition state.

e. (1,3-cyclohexadiene) The β-hydrogen is removed that will lead to a conjugated alkene.

f. $CH_3CH{=}CHCH{=}CH_2$ The β-hydrogen is removed that will lead to a conjugated alkene.

3.

a. $CH_3CH_2CHCH_2CH_3$ (Br on C-3)

It has four hydrogens that can be removed to form an alkene with two substituents on the sp^2 carbons so it has a greater probability of an effective collision with the nucleophile than the other alkyl halide that has only two such hydrogens.

b. C_6H_5–$CH_2CHCH_2CH_3$ (Br on C-2)

It forms the more stable alkene; the new double bond is conjugated with the phenyl substituent.

c. $CH_3CHCHCH_2CH_3$ (CH_3 on C-2, Br on C-3)

It forms the more stable alkene; the alkene has a greater number of substituents bonded to the sp^2 carbons.

d. (bromocycloheptene)

It forms is more stable alkene; the new double bond is conjugated with the double bond that is already there.

4.

CH_3CH–CCH_2CH_3 (CH_3 on C-2; CH_3 and Br on C-3)

3-bromo-2,3-dimethylpentane

$CH_3C{=}CCH_2CH_3$ (CH_3 on each sp^2 carbon) > $CH_3CHC{=}CHCH_3$ (CH_3 on C-2 and C-3) > 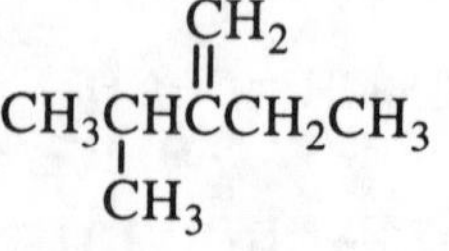

Four alkyl substituents are bonded to the sp^2 carbons.

Three alkyl substituents are bonded to the sp^2 carbons.

Two alkyl substituents are bonded to the sp^2 carbons.

5. The major product is the one predicted by Zaitsev's rule, because the fluoride ion dissociates off in the first step, forming a carbocation. Loss of a proton from the carbocation will follow Zaitsev's rule, as it does in all E1 reactions.

6.

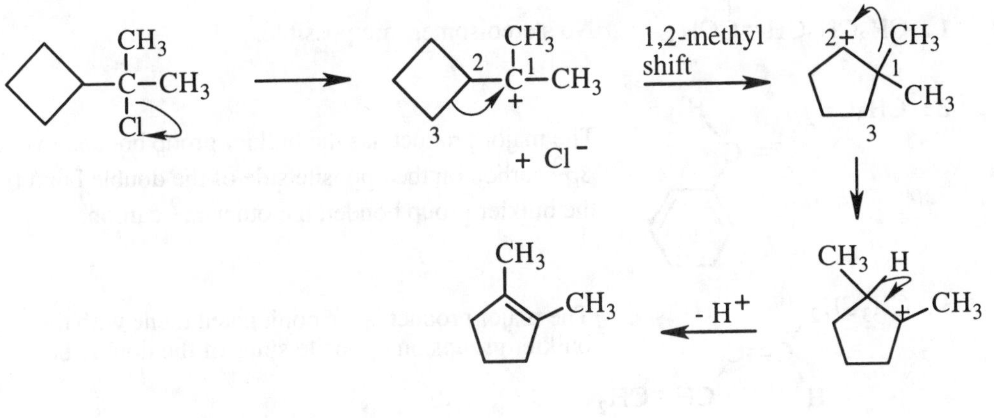

Cyclobutyl–$C(CH_3)_2Cl$ → cyclobutyl–$\overset{+}{C}(CH_3)_2$ + Cl^- —1,2-methyl shift→ 2,2-dimethylcyclopentyl cation → (hydride shift) 1,2-dimethylcyclopentyl cation —$-H^+$→ 1,2-dimethylcyclopentene

7.

a. E2 beause a strong base is used.
$CH_3CH{=}CHCH_3$

b. E1 because a weak base is used.
$CH_3CH{=}CHCH_3$

c. E1 because a weak base is used.
$CH_3C(CH_3){=}CH_2$

d. E2 because a strong base is used.
$CH_3C(CH_3){=}CH_2$

e. E1 because a weak base is used.
$CH_3C(CH_3){=}C(CH_3)CH_3$

f. E2 because a strong base is used.
$CH_3C(CH_3)_2CH{=}CH_2$

8.

a.

$$\frac{E2}{E2 + E1} = \frac{7.1 \times 10^{-5} \times 5.0}{7.1 \times 10^{-5} \times 5.0 + 1.5 \times 10^{-5}} = \frac{35.5 \times 10^{-5}}{35.5 \times 10^{-5} + 1.5 \times 10^{-5}} = \frac{35.5}{37} = 96\%$$

b.

$$\frac{E2}{E2 + E1} = \frac{7.1 \times 10^{-5} \times 2.5 \times 10^{-3}}{7.1 \times 10^{-5} \times 2.5 \times 10^{-3} + 150 \times 10^{-7}} = \frac{1.78 \times 10^{-7}}{1.78 \times 10^{-7} + 150 \times 10^{-7}} = \frac{1.78}{152} = 1.2\%$$

9.

a. 1. $CH_3CH_2CH{=}C(CH_3)CH_3$ No stereoisomers are possible.

2. 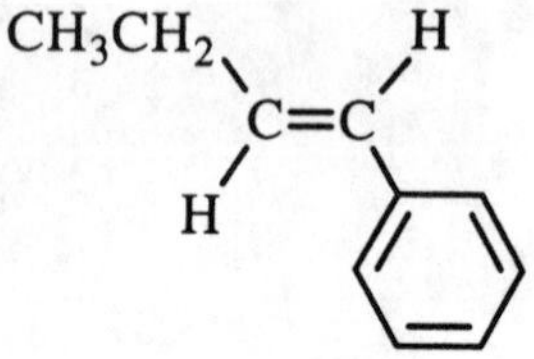 $CH_3CH_2(H)C{=}C(H)C_6H_5$

The major product has the bulkier group bonded to one sp^2 carbon on the opposite side of the double bond from the bulkier group bonded the other sp^2 carbon.

3. 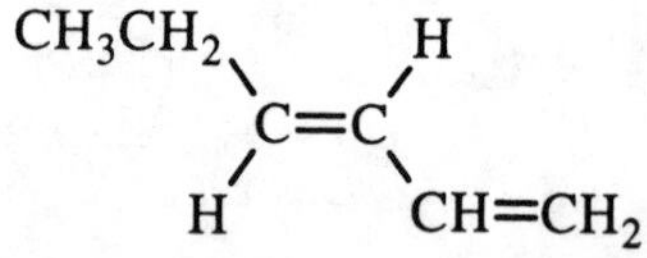 $CH_3CH_2(H)C{=}C(H)CH{=}CH_2$

The major product is the conjugated diene with the bulkier groups on opposite sides of the double bond.

b. In none of the reactions is the major product dependent on whether you started with the *R* or *S* enantiomer.

10.

a. 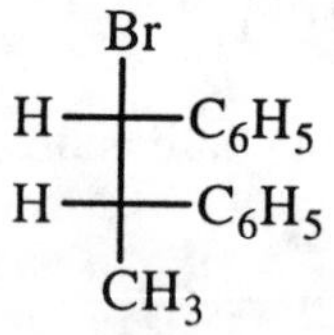 Br, H, C_6H_5, H, C_6H_5, CH_3

1*S*,2*S*

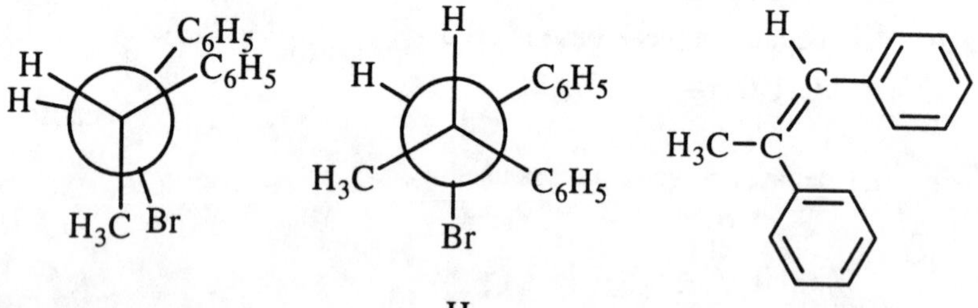

C_6H_5, C_6H_5, H, H, H_3C, Br

H, H, C_6H_5, H_3C, C_6H_5, Br

H, C, $H_3C{-}C$

b. 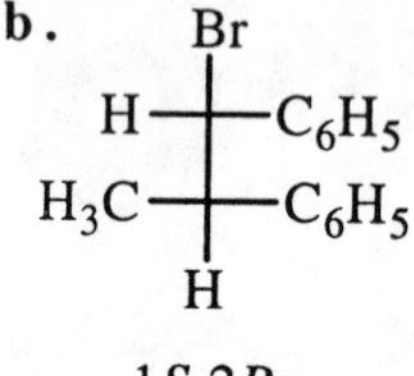 Br, H, C_6H_5, H_3C, C_6H_5, H

1*S*,2*R*

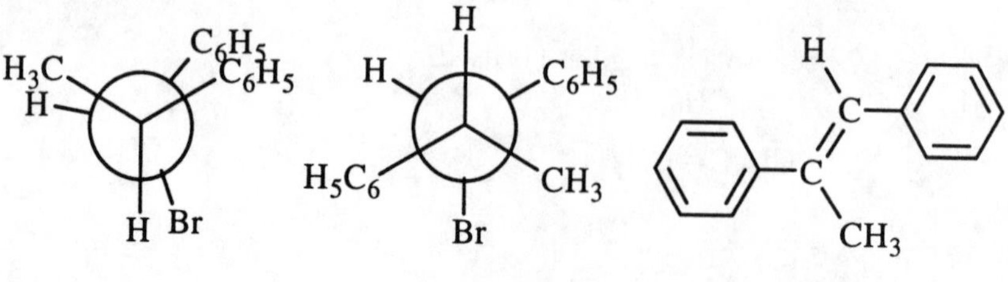

C_6H_5, C_6H_5, H_3C, H, H, Br

H, H, C_6H_5, H_5C_6, CH_3, Br

H, C, C, CH_3

11.

a. 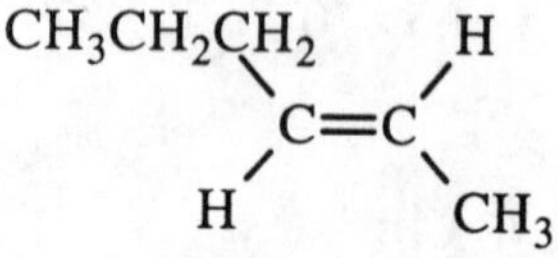 $CH_3CH_2CH_2(H)C{=}C(H)CH_3$

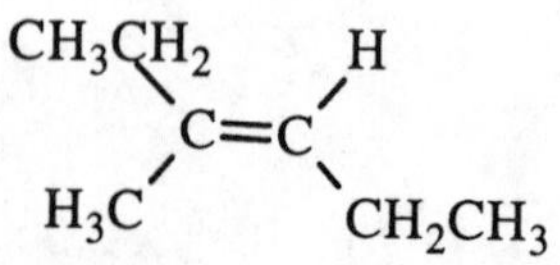

b. $CH_3CH_2CH{=}C(CH_3)CH_3$

c. $CH_3CH_2(H_3C)C{=}C(H)CH_2CH_3$

d. 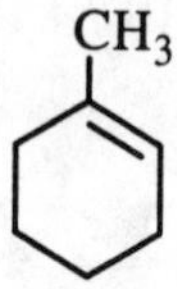 1-methylcyclohexene (CH_3)

12. Elimination occurs when the H and Br to be eliminated are in axial positions.

When Br is in an axial position in the cis isomer, it has an axial hydrogen on each of the adjacent carbons. The one bonded to the same carbon as the ethyl group will be more apt to be the one eliminated with Br because the product formed is more stable and, therefore, more easily formed than the product formed when the other H is eliminated with Br.

When Br is in an axial position in the trans isomer, it has an axial hydrogen on only one adjacent carbon, and it is not the carbon that is bonded to the ethyl group. Therefore, a different product is formed.

cis-1-bromo-2-ethylcyclohexane

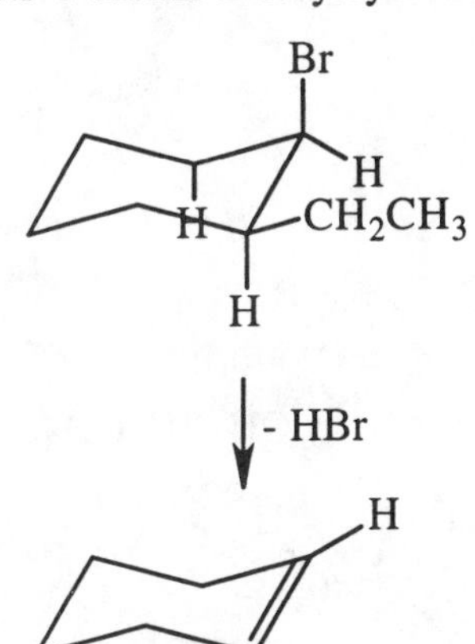

CH_2CH_3

1-ethylcyclohexene

trans-1-bromo-2-ethylcyclohexane

Br

H

H

H

CH_2CH_3

- HBr

H

H

CH_2CH_3

3-ethylcyclohexene

13. In order to undergo an E2 reaction, the substituents that are to be eliminated must both be in axial positions.

When bromine and an adjacent hydrogen are in axial positions, the large *tert*-butyl substituent is in an equatorial position in the cis isomer and in an axial position in the trans isomer.

Because a large substituent is more stable in an equatorial position than in an axial position, elimination of the cis isomer occurs through its more stable chair conformer, while elimination of the trans isomer has to occur through its less stable chair conformer. The cis isomer, therefore, reacts more rapidly in an E2 reaction.

more stable of the chair conformers of *cis*-1-bromo-4-*tert*-butylcyclohexane

less stable of the chair conformers of *trans*-1-bromo-4-*tert*-butylcyclohexane

14.

a. $CH_3CH(Cl)(CH_2)_3CH_3 \xrightarrow[S_N2/E2]{CH_3O^-} CH_3CH(OCH_3)(CH_2)_3CH_3$ + *trans*-$CH_3CH{=}CHCH_2CH_2CH_3$ + *cis*-$CH_3CH{=}CHCH_2CH_2CH_3$ (minor) + $CH_2{=}CHCH_2CH_2CH_2CH_3$ (minor)

b. $CH_3CH(Cl)(CH_2)_3CH_3 \xrightarrow[S_N1/E1]{CH_3OH} CH_3CH(OCH_3)(CH_2)_3CH_3$ + *trans*-$CH_3CH{=}CHCH_2CH_2CH_3$ + $CH_3(CH_2)_3CH(OCH_3)CH_3$ + *cis*-$CH_3CH{=}CHCH_2CH_2CH_3$ (minor) + $CH_2{=}CHCH_2CH_2CH_2CH_3$ (minor)

c.

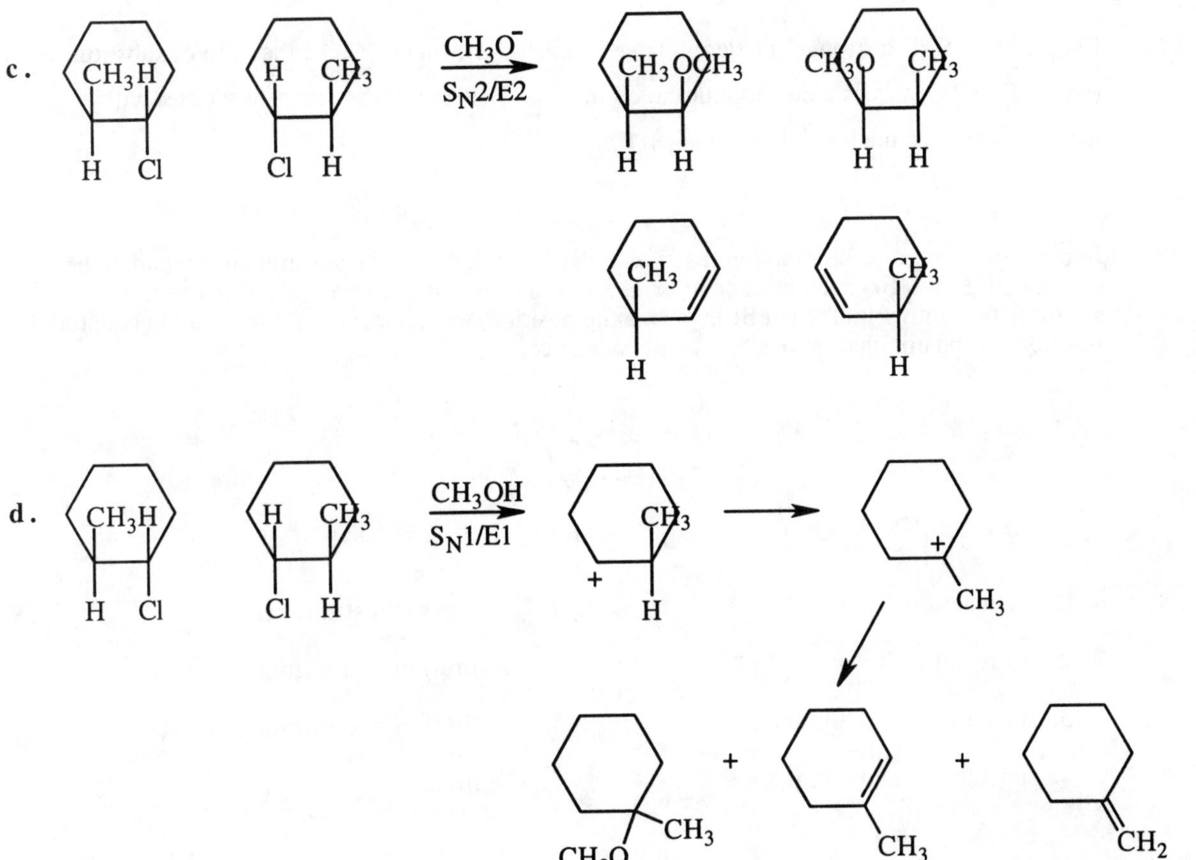

e.

CH₃
H'''''C
CH₃O CH₂CH₃

f.

CH₃
H'''''C
CH₃O CH₂CH₃
+
CH₃
C'''''H
CH₃CH₂ OCH₃

15. The rate-limiting step in an E1 reaction is carbocation formation. Because the proton is removed in a subsequent fast step, the difference in the rate of removal of an H^+ versus a D^+ would not be reflected in the rate constant. Therefore, the deuterium kinetic isotope effect would be close to 1.

16. Because CH_3S^- is a stronger nucleophile and weaker base than CH_3O^-, the ratio of substitution (where CH_3S^- reacts as a nucleophile) to elimination (where CH_3S^- reacts as a base) will increase when the nucleophile is changed from CH_3O^- to CH_3S^-.

17. In order to undergo an elimination reaction under E2 conditions, the substituents that are to be eliminated (H and Br) must both be in axial positions. Drawing the compound in the chair conformation shows that when Br is in an axial position, the adjacent hydrogens are in equatorial positions, so an elimination reaction can't take place.

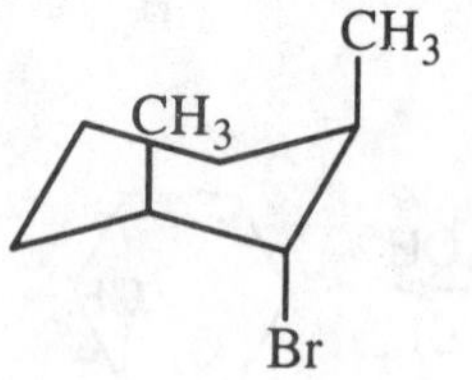

CH_3

CH_3

Br

18. **a. 1.** no reaction

2. no reaction

3. substitution and elimination

4. substitution and elimination

b. 1. primarily substitution

2. substitution and elimination

3. substitution and elimination

4. elimination

19.

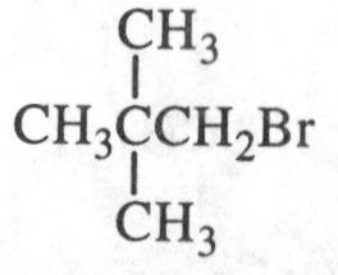

$$\begin{array}{c} CH_3 \\ | \\ CH_3CCH_2Br \\ | \\ CH_3 \end{array}$$

1-bromo-2,2-dimethylpropane

a. The bulky *tert*-butyl substituent blocks the backside of the carbon bonded to the bromine to nucleophilic attack, making an S_N2 reaction difficult. An S_N1 reaction is difficult because the carbocation formed when the bromide ion departs is an unstable primary carbocation.

b. It cannot undergo an E2 reaction, because the β-carbon is not bonded to a hydrogen. It cannot undergo an E1 reaction, because that would require formation of a primary carbocation.

20. Because a strong base is used in the Williamson ether synthesis, the reaction is an S_N2 reaction, so a competing E2 reaction can also occur.

a. $CH_3CH_2CH_2Br + CH_3CH_2CH_2CH_2O^- \longrightarrow CH_3CH_2CH_2CH_2OCH_2CH_2CH_3$ (butyl propyl ether) $+ CH_3CH{=}CH_2$ (propene)

b. $CH_3CH_2CH_2CH_2Br + CH_3CH_2CH_2O^- \longrightarrow CH_3CH_2CH_2CH_2OCH_2CH_2CH_3$ (butyl propyl ether) $+ CH_3CH_2CH{=}CH_2$ (1-butene)

21.

a. $CH_3CH_2CH(CH_3)OH \xrightarrow{Na} CH_3CH_2CH(CH_3)O^- \xrightarrow{CH_3CH_2CH_2Br} CH_3CH_2CH(CH_3)OCH_2CH_2CH_3$

b. $CH_3C(CH_3)_2OH \xrightarrow{Na} CH_3C(CH_3)_2O^- \xrightarrow{\text{cyclohexyl-Br}} \text{cyclohexyl-}OC(CH_3)_2CH_3$

c. $C_6H_5OH \xrightarrow{Na} C_6H_5O^- \xrightarrow{C_6H_5CH_2Br} C_6H_5CH_2O{-}C_6H_5$

d. $CH_3CH_2CH_2CH_2CH_2OH \xrightarrow{Na} CH_3CH_2CH_2CH_2CH_2O^- \xrightarrow{CH_3CH_2Br} CH_3CH_2CH_2CH_2CH_2OCH_2CH_3$

or

$CH_3CH_2OH \xrightarrow{Na} CH_3CH_2O^- \xrightarrow{CH_3CH_2CH_2CH_2CH_2Br} CH_3CH_2CH_2CH_2CH_2OCH_2CH_3$

Both methods can be used to synthesize the target molecule. The first method is preferred because there is less steric hindrance in the S_N2 reaction.

22.

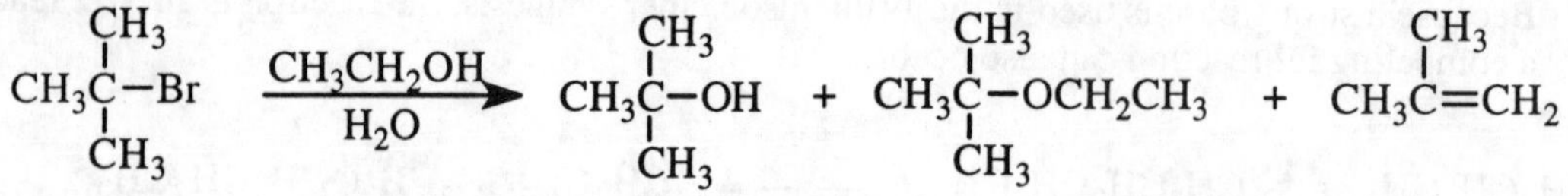

$$CH_3C(CH_3)_2{-}Br \xrightarrow[H_2O]{CH_3CH_2OH} CH_3C(CH_3)_2{-}OH + CH_3C(CH_3)_2{-}OCH_2CH_3 + CH_3C(CH_3){=}CH_2$$

23.

a.

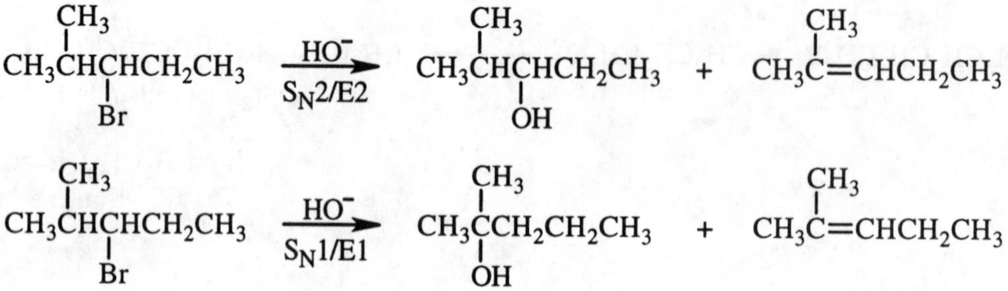

$$CH_3CH(CH_3)CH(Br)CH_2CH_3 \xrightarrow[S_N2/E2]{HO^-} CH_3CH(CH_3)CH(OH)CH_2CH_3 + CH_3C(CH_3){=}CHCH_2CH_3$$

$$CH_3CH(CH_3)CH(Br)CH_2CH_3 \xrightarrow[S_N1/E1]{HO^-} CH_3C(CH_3)(OH)CH_2CH_2CH_3 + CH_3C(CH_3){=}CHCH_2CH_3$$

b.

$$CH_3CH_2C(CH_3)(Br)CH_2CH_3 \xrightarrow[S_N2/E2]{HO^-} (H_3C)(H)C{=}C(CH_3)(CH_2CH_3) + (H_3C)(H)C{=}C(CH_2CH_3)(CH_3)$$

The first alkene has H_3C and CH_3 on the same side (E); the second alkene has H_3C and CH_2CH_3 on the same side (Z), labelled **minor**.

$$CH_3CH_2C(CH_3)(Br)CH_2CH_3 \xrightarrow[S_N1/E1]{HO^-} CH_3CH_2C(CH_3)(OH)CH_2CH_3 + (H_3C)(H)C{=}C(CH_3)(CH_2CH_3)$$

$$(H_3C)(H)C{=}C(CH_2CH_3)(CH_3)$$

minor

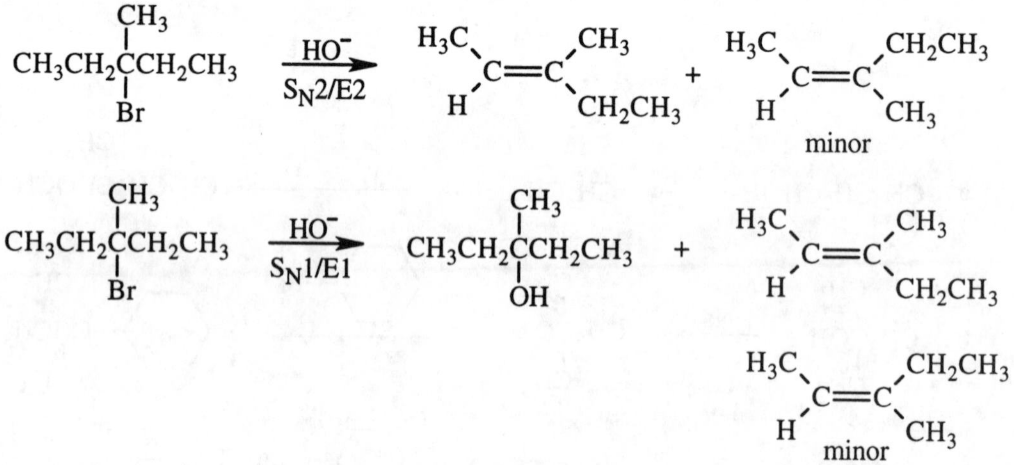

24. Because an allene is less stable than an alkyne, it is harder to make.

25.

a. HO〜〜〜Br because it forms a six-membered ring, while the other compound would form a seven-membered ring. A seven-membered ring is more strained than a six-membered ring so the six-membered ring is formed more easily (see Table 2.9 on p. 97 of the text).

b. HO〜〜Br because it forms a five-membered ring, while the other compound would form a four-membered ring. A four-membered ring is more strained than a five-membered ring so the five-membered ring is formed more easily.

c. HO〜〜〜〜Br because it forms a seven-membered ring, while the other compound would form an eight-membered ring. An eight-membered ring is more strained than a seven-membered ring so the seven-membered ring is formed more easily; also, the Br and OH are less likely to be in the proper position realtive to one anotherfor reaction because there are more bonds around which rotation to an unfovorable conformation can occur in the compound that leads to the eight-membered ring (Section 22.6).

26. No. You would get the desired product but the rate of its formation would be slower because of steric hindrance since it is a secondary alkyl halide.

$HOCH_2CH_2CH_2CH(Br)CH_3 \xrightarrow{Na}$ 2-methyltetrahydrofuran (ring O, CH_3)

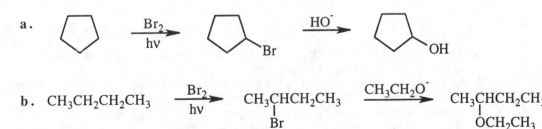

27.

a. cyclopentane $\xrightarrow[h\nu]{Br_2}$ bromocyclopentane (Br) $\xrightarrow{HO^-}$ cyclopentanol (OH)

b. $CH_3CH_2CH_2CH_3 \xrightarrow[h\nu]{Br_2} CH_3CH(Br)CH_2CH_3 \xrightarrow{CH_3CH_2O^-} CH_3CH(OCH_2CH_3)CH_2CH_3$

c. (cyclohexyl)$CH=CH_2$ $\xrightarrow[CH_2Cl_2]{Br_2}$ (cyclohexyl)$CH(Br)CH_2Br$ $\xrightarrow[\text{excess}]{^-NH_2}$ (cyclohexyl)$C\equiv CH$ $\xrightarrow[2.\ HO^-,\ H_2O_2,\ H_2O]{1.\ \text{disiamylborane}}$ (cyclohexyl)$CH_2\overset{O}{\overset{\|}{C}}H$

d. $CH_3CH_2CH_2CH_2Br \xrightarrow{HO^-} CH_3CH_2CH=CH_2 \xrightarrow{Br_2} CH_3CH_2\underset{Br}{\underset{|}{C}}HCH_2Br \xrightarrow[\text{excess}]{^-NH_2} CH_3CH_2C\equiv CH \xrightarrow[2.CH_2CH_2Br]{1.\ ^-NH_2} CH_3CH_2C\equiv CCH_2CH_3 \xrightarrow[H_2SO_4]{H_2O} CH_3CH_2\overset{O}{\overset{\|}{C}}CH_2CH_2CH_3$

e. $BrCH_2CH_2CH_2CH_2Br \xrightarrow[\text{excess}]{HO^-} CH_2=CHCH=CH_2 \xrightarrow{\Delta}$ (4-vinylcyclohexene, $CH=CH_2$) $\xrightarrow{H_2,\ Pd/C}$ (cyclohexyl)CH_2CH_3

28. a.

1. 3° > 2° > 1°
2. An E1 reaction is not affected by the strength of the base, but a weak base favors an E1 reaction by disfavoring an E2 reaction.
3. An E1 reaction is not affected by the concentration of the base, but a low concentration of a base favors an E1 reaction by disfavoring an E2 reaction.
4. An aprotic polar solvent favors an E1 reaction if the reactant is charged.
   A protic polar solvent favors an E1 reaction if the reactant is not charged.

b.

1. $3° > 2° > 1°$
2. A strong base favors an E2 reaction.
3. A high concentration of a base favors an E2 reaction.
4. An aprotic polar solvent favors an E2 reaction if either of the reactants is charged. A protic polar solvent favors an E2 reaction if neither of the reactants is charged.

29. The predominant product is the elimination product because tertiary alkyl halides react with nucleophiles to form an elimination product and little substitution product.

$$CH_3C(CH_3)(Cl)CH_3 + CH_3CH_2O^- \longrightarrow CH_3C(CH_3){=}CH_2$$

predominant product

Rather than a tertiary alkyl halide and a primary alkoxide ion, he should have used a primary alkyl halide and a tertiary alkoxide ion.

$$CH_3CH_2Cl + CH_3C(CH_3)_2O^- \longrightarrow CH_3C(CH_3)_2OCH_2CH_3 + Cl^-$$

30.

a. $$(CH_3)_3CI \xrightarrow[H_2O]{HO^-} (H_3C)_2C{=}CH_2 + I^-$$

Because I^- is a better leaving group than Cl^-.

b.

(chair structure: H, H, CH_3, H, CH_3, Br)

This compound is the only one that can undergo an E2 elimination reaction because the other compound does not have a H trans to the Br.

(chair structure: CH_3, CH_3, H, H, H, Br)

This compound will undergo an elimination reaction more slowly because it can react only by an E1pathway.

31. The minor products that would be obtained from anti-Zaitsev elimination are not shown.

a. 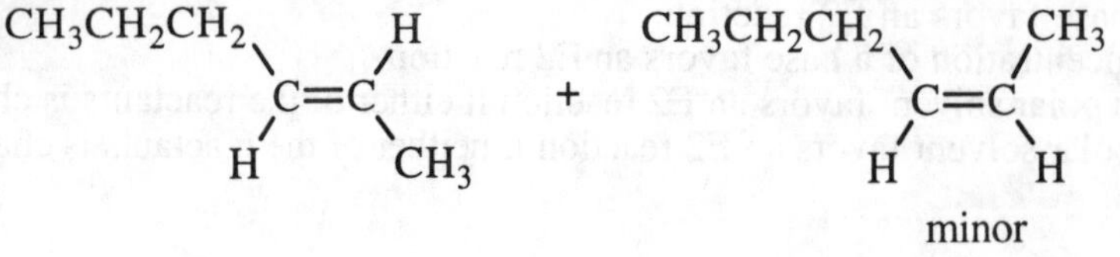 $CH_3CH_2CH_2$ H $CH_3CH_2CH_2$ CH_3
C=C + C=C
H CH_3 H H
minor

b. 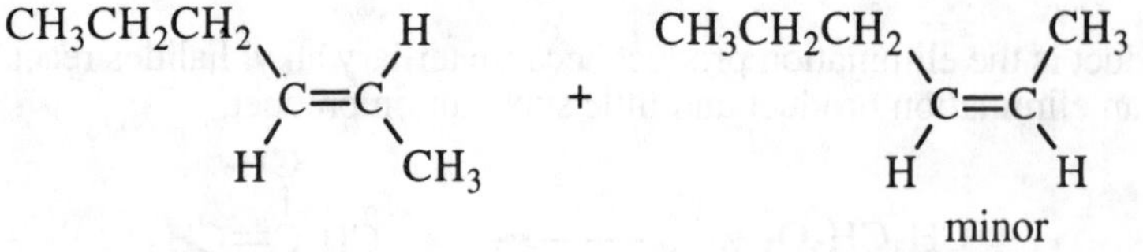 $CH_3CH_2CH_2$ H $CH_3CH_2CH_2$ CH_3
C=C + C=C
H CH_3 H H
minor

c. H + H
CH_3 CH_3

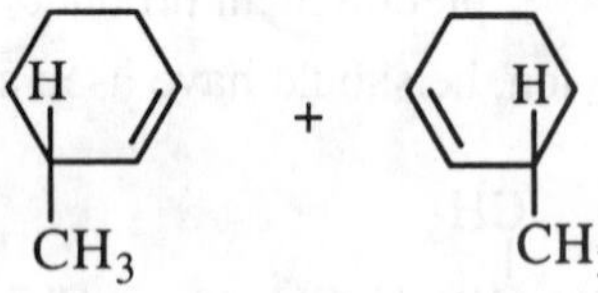

d. 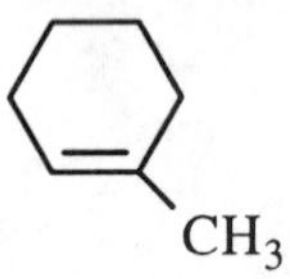 CH_3

e. 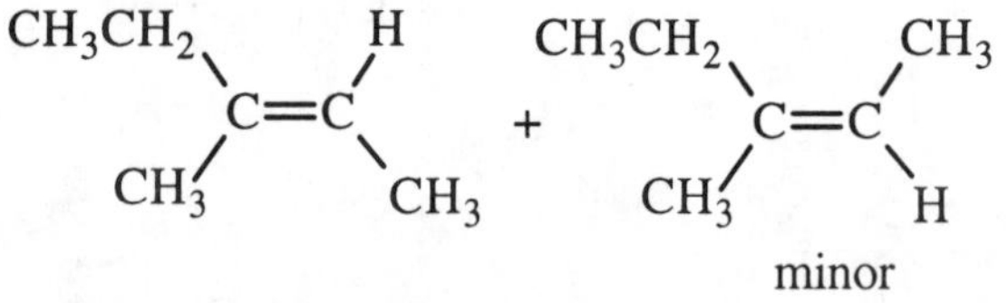 CH_3CH_2 H CH_3CH_2 CH_3
C=C + C=C
CH_3 CH_3 CH_3 H
minor

f. 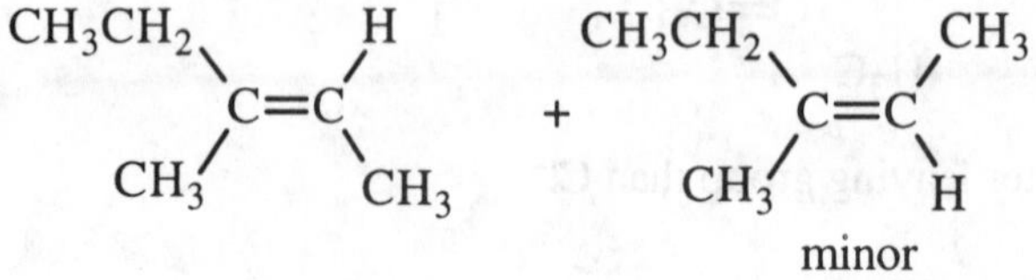 CH_3CH_2 H CH_3CH_2 CH_3
C=C + C=C
CH_3 CH_3 CH_3 H
minor

32. a. ethoxide ion because elimination is favored by the bulkier base, and *tert*-butoxide ion is bulkier than ethoxide ion

b. ^-SCN because elimination is favored by the stronger base, and ^-OCN is a stronger base than ^-SCN

c. Br^- because elimination is favored by the stronger base, and Cl^- is a stronger base than Br^-

d. CH_3S^- because elimination is favored by the stronger base, and CH_3O^- is a stronger base than CH_3S^-

33. The first compound has two axial hydrogens adjacent to Br, the second has one axial hydrogen adjacent to Br but it cannot form the more substituted (more stable) alkene that can be formed by the first compound. The last compound cannot undergo an E2 reaction because it does not have an axial hydrogen adjacent to Br.

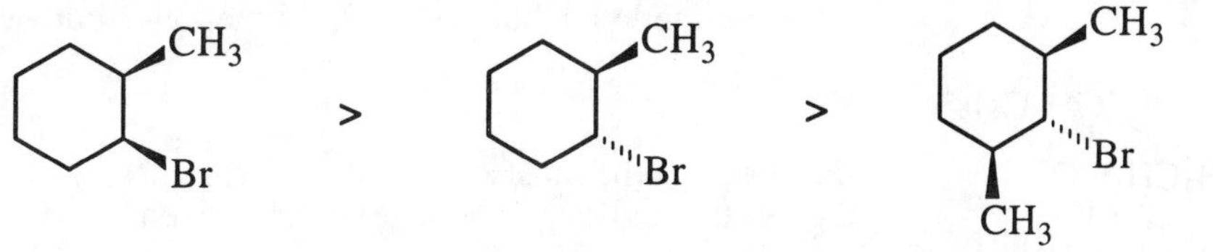

CH_3 Br > CH_3 Br > CH_3 Br CH_3

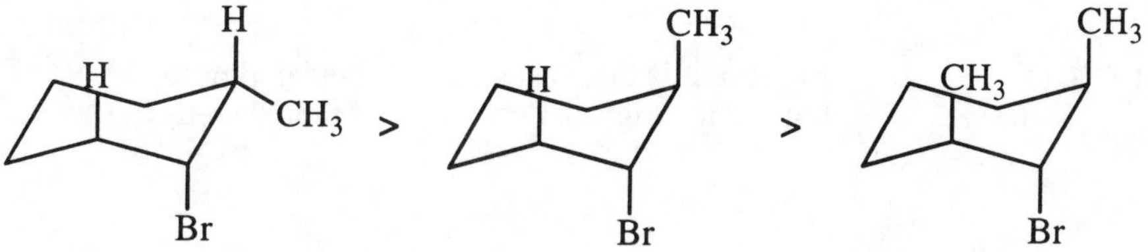

H H CH_3 Br > CH_3 H Br > CH_3 CH_3 Br

34.

a. 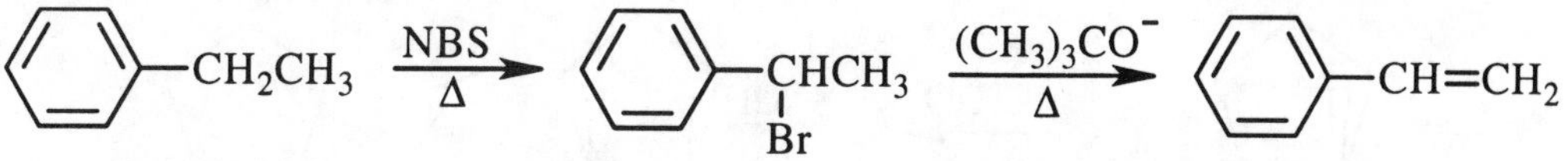 $C_6H_5CH_2CH_3 \xrightarrow[\Delta]{NBS} C_6H_5CHBrCH_3 \xrightarrow[\Delta]{(CH_3)_3CO^-} C_6H_5CH{=}CH_2$

Notice that a bulky base and heat are used in the last step to encourage elimination over substitution.

b. $CH_3CH_2CH{=}CH_2 \xrightarrow[\text{peroxide}]{HBr} CH_3CH_2CH_2CH_2Br \xrightarrow[2.\ HO^-]{1.\ NH_3} CH_3CH_2CH_2CH_2NH_2$

After the S_N2 reaction with NH_3, the solution is made basic so the final amine product is in its basic form.

c. 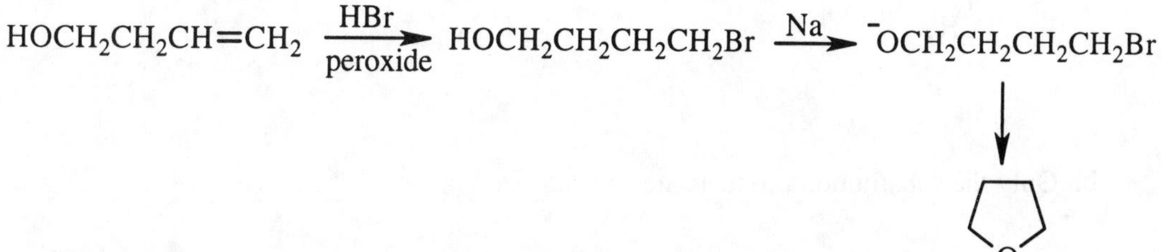 $HOCH_2CH_2CH{=}CH_2 \xrightarrow[\text{peroxide}]{HBr} HOCH_2CH_2CH_2CH_2Br \xrightarrow{Na} {}^-OCH_2CH_2CH_2CH_2Br \rightarrow$ O (tetrahydrofuran)

d. 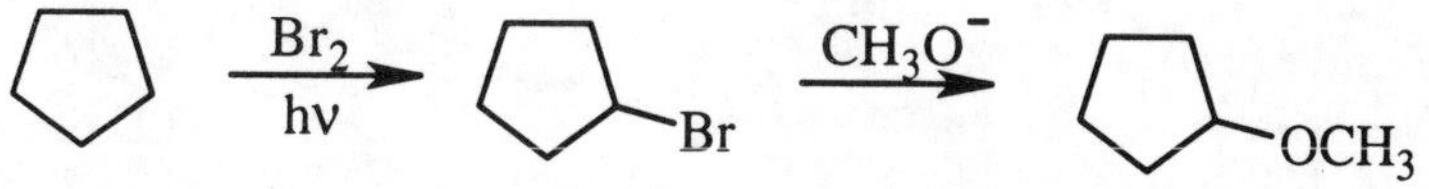 cyclopentane $\xrightarrow[h\nu]{Br_2}$ cyclopentyl–Br $\xrightarrow{CH_3O^-}$ cyclopentyl–OCH_3

e. $CH_3CH_2CH{=}CH_2 \xrightarrow[\Delta]{NBS} CH_3CHBrCH{=}CH_2 \xrightarrow[\Delta]{(CH_3)_3CO^-} CH_2{=}CHCH{=}CH_2$

35.

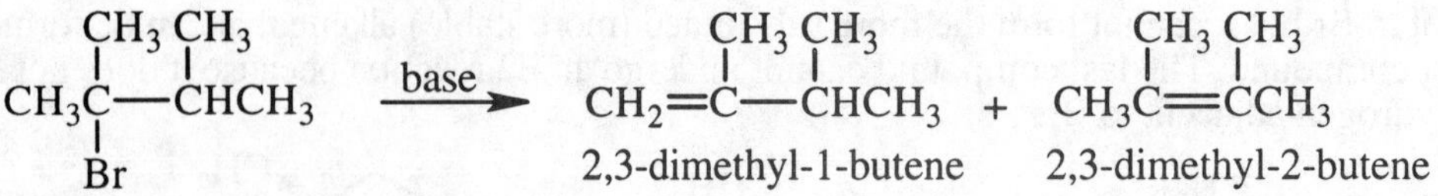

$$CH_3C(CH_3)(Br)-CH(CH_3)CH_3 \xrightarrow{\text{base}} CH_2{=}C(CH_3)-CHCH_3(CH_3) + CH_3C(CH_3){=}C(CH_3)CH_3$$

2,3-dimethyl-1-butene 2,3-dimethyl-2-butene

a. 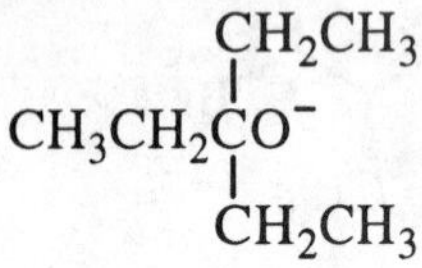 $CH_3CH_2C(CH_2CH_3)(CH_2CH_3)O^-$ Because it is the most sterically hindered base, it gives the highest percentage of the 1-alkene.

b. $CH_3CH_2O^-$ Because it is the least sterically hindered base, it gives the highest percentage of the 2-alkene.

36.

a. These are S_N2/E2 reactions.

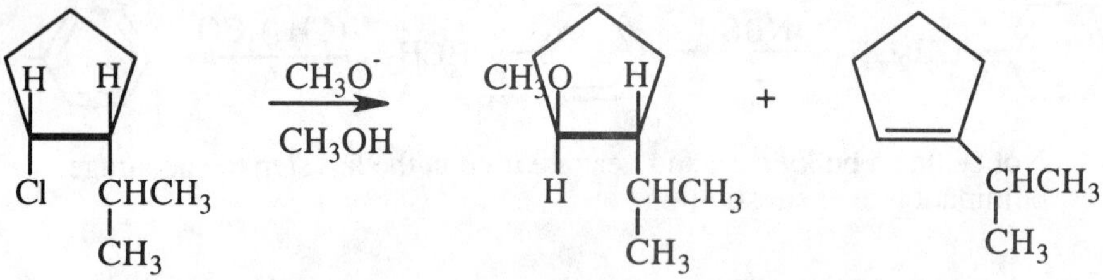

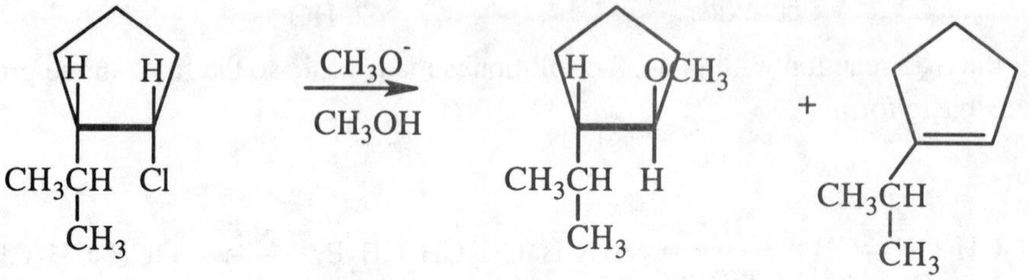

(cyclopentane with H, H; Cl, CHCH$_3$(CH$_3$)) $\xrightarrow[CH_3OH]{CH_3O^-}$ (CH$_3$O, H; H, CHCH$_3$(CH$_3$)) + (cyclopentene with CHCH$_3$(CH$_3$))

(cyclopentane with H, H; CH$_3$CH(CH$_3$), Cl) $\xrightarrow[CH_3OH]{CH_3O^-}$ (H, OCH$_3$; CH$_3$CH(CH$_3$), H) + (cyclopentene with CH$_3$CH(CH$_3$))

b. Only the substitution products are optically active.

c.

$$\text{(cis-1-chloro-2-isopropylcyclopentane)} \xrightarrow[CH_3OH]{CH_3O^-} \text{(trans-1-methoxy-2-isopropylcyclopentane)} + \text{3-isopropylcyclopentene}$$

$$\text{(cis-1-chloro-2-isopropylcyclopentane, enantiomer)} \xrightarrow[CH_3OH]{CH_3O^-} \text{(trans-1-methoxy-2-isopropylcyclopentane, enantiomer)} + \text{3-isopropylcyclopentene}$$

Only the substitution products are optically active.

d. The cis enantiomers form the substitution products more rapidly because there is less steric hindrance from the adjacent substituent.

e. The cis enantiomers form the elimination products more rapidly because the alkenes formed from the cis enantiomers are more substituted and, therefore, more stable. The more stable the alkene, the lower the energy of the transition state leading to its formation.

37.

a. cyclohexane $\xrightarrow[h\nu]{Br_2}$ bromocyclohexane $\xrightarrow{HO^-}$ cyclohexene $\xrightarrow[CH_2Cl_2]{Cl_2}$ trans-1,2-dichlorocyclohexane + enantiomer

b. cyclohexane $\xrightarrow[h\nu]{Br_2}$ bromocyclohexane $\xrightarrow{HO^-}$ cyclohexene $\xrightarrow[\Delta]{NBS}$ 3-bromocyclohexene $\xrightarrow{H_2O}$ 2-cyclohexen-1-ol

38. In an E2 reaction, both groups to be eliminated must be in axial positions. When the bromine is in the axial position in the cis isomer, the *tert*-butyl substituent is in the more stable equatorial position. When the bromine is in the axial position in the trans isomer, the *tert*-butyl substituent is in the less stable axial position. So elimination takes place via the most stable conformer in the cis isomer and via a less stable chair conformer in the trans isomer, so the cis isomer undergoes elimination more rapidly.

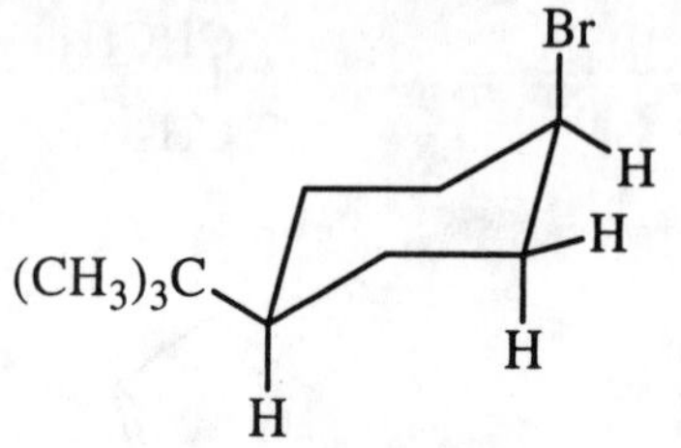

cis-1-bromo-4-*tert*-butylcyclohexane

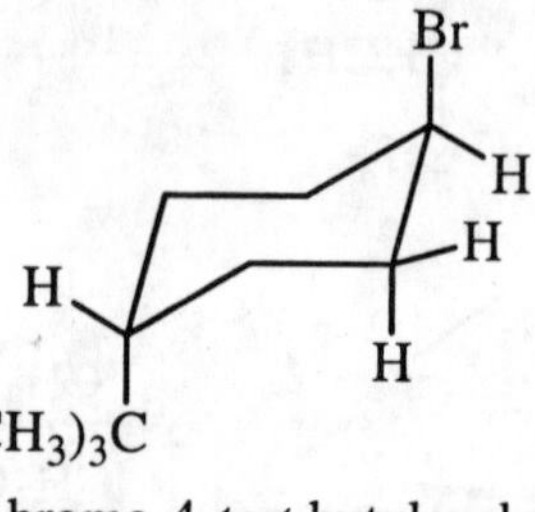

trans-1-bromo-4-*tert*-butylcyclohexane

39. Minor anti-Zaitsev products aren't shown.

"e" is the only compound for which both *E* and *Z* stereoisomers will be formed because "e" is the only compound with two hydrogens bonded to the β-carbon.

a. $CH_3CH_2(CH_3)C{=}CH(CH_3)$

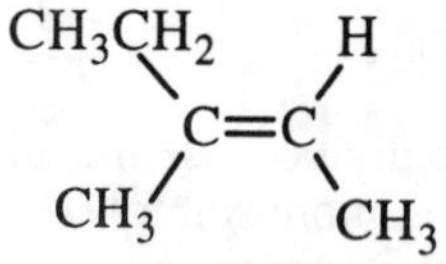

b. $CH_3CH_2(CH_3)C{=}C(CH_3)H$

c. $CH_3CH_2(CH_3)C{=}C(CH_3)H$

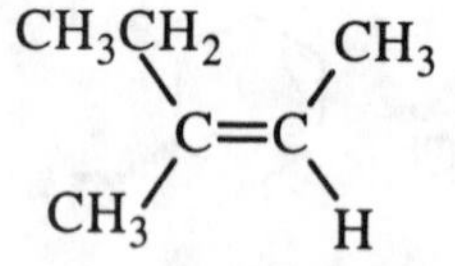

d. $CH_3CH_2(CH_3)C{=}CH(CH_3)$

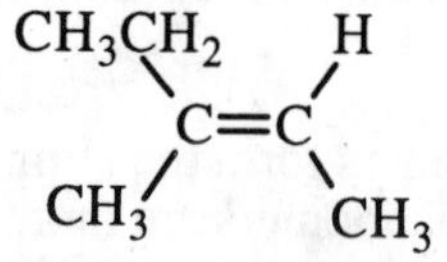

e. $(CH_3)_3C(CH_3CH_2)C{=}CH(CH_3)$
major

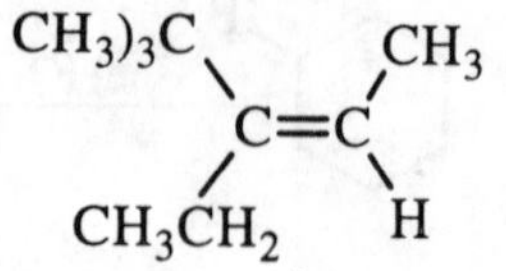

$(CH_3)_3C(CH_3CH_2)C{=}C(CH_3)H$
minor

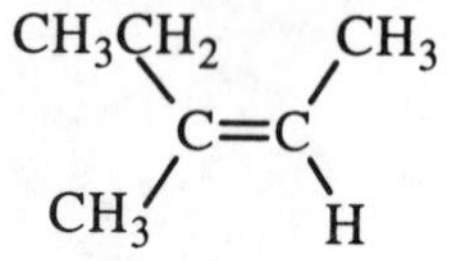

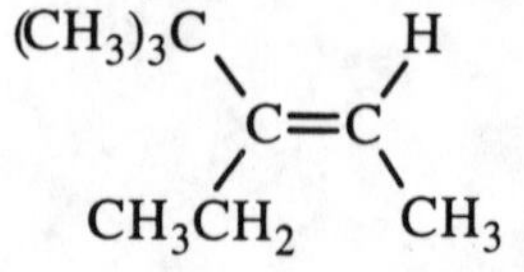

40. In order to undergo an E2 reaction, a chlorine and a hydrogen on an adjacent carbon must be trans to one another so they can both be in the required axial positions. Every Cl in the following compound has a Cl trans to it so no Cl has a H trans to it.

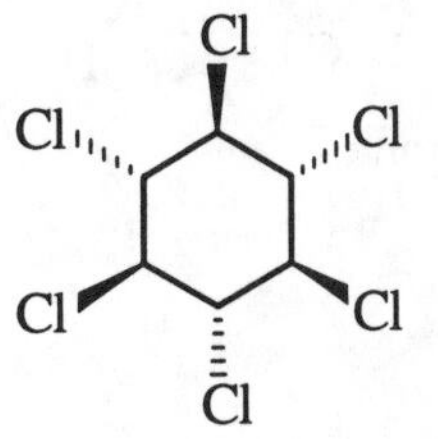

Cl Cl Cl Cl Cl Cl

41. The silver ion increases the ease of departure of the halogen atom.

$$CH_3CH{=}CHCH_2{-}\ddot{Br}: + Ag^+ \longrightarrow CH_3CH{=}CHCH_2{-}\overset{+}{\ddot{Br}}{:}Ag \longrightarrow CH_3CH{=}CH\overset{+}{C}H_2 + AgBr$$

42.

a. 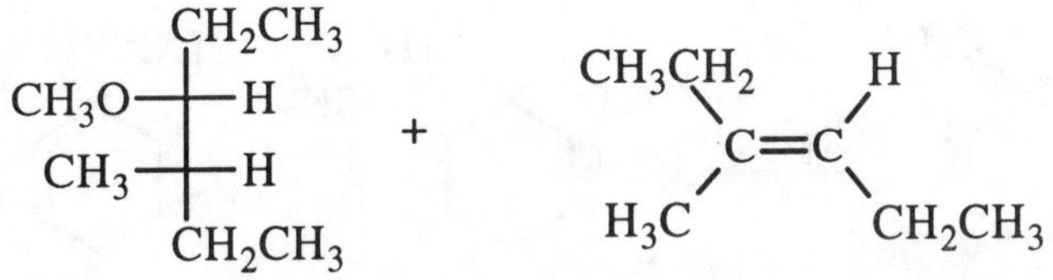 CH_3O—C—H, CH_3—C—H (top CH_2CH_3, bottom CH_2CH_3) + CH_3CH_2(H)C=C(H)(CH_2CH_3) with H_3C

(3*R*,4*S*)-3-methoxy-4-methylhexane

b. 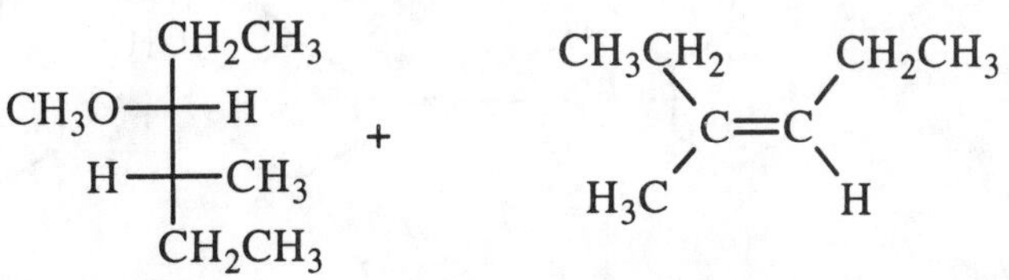 CH_3O—C—H, H—C—CH_3 (top CH_2CH_3, bottom CH_2CH_3) + CH_3CH_2, H_3C C=C CH_2CH_3, H

(3*R*,4*R*)-3-methoxy-4-methylhexane

c. 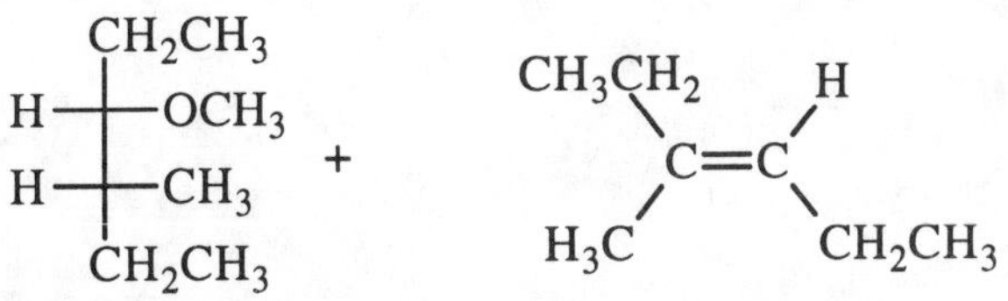 H—C—OCH_3, H—C—CH_3 (top CH_2CH_3, bottom CH_2CH_3) + CH_3CH_2, H_3C C=C H, CH_2CH_3

(3*S*,4*R*)-3-methoxy-4-methylhexane

d. 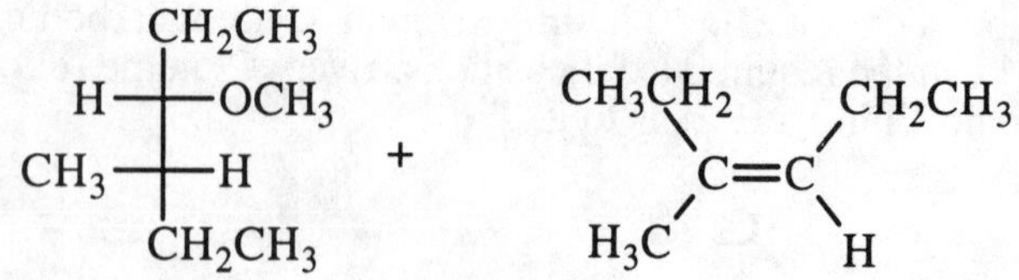

(3*S*,4*S*)-3-methoxy-4-methylhexane

43.

a. $CH_3CH_2CD{=}CH_2$ and $CH_3CH_2CH{=}CH_2$

b. The deuterium-containing compound results from elimination of HBr, while the non-deuterium-containing compound results from elimination of DBr. The deuterium-containing compound will be obtained in greater yield because a carbon-hydrogen bond is easier to break than a carbon-deuterium bond.

44.

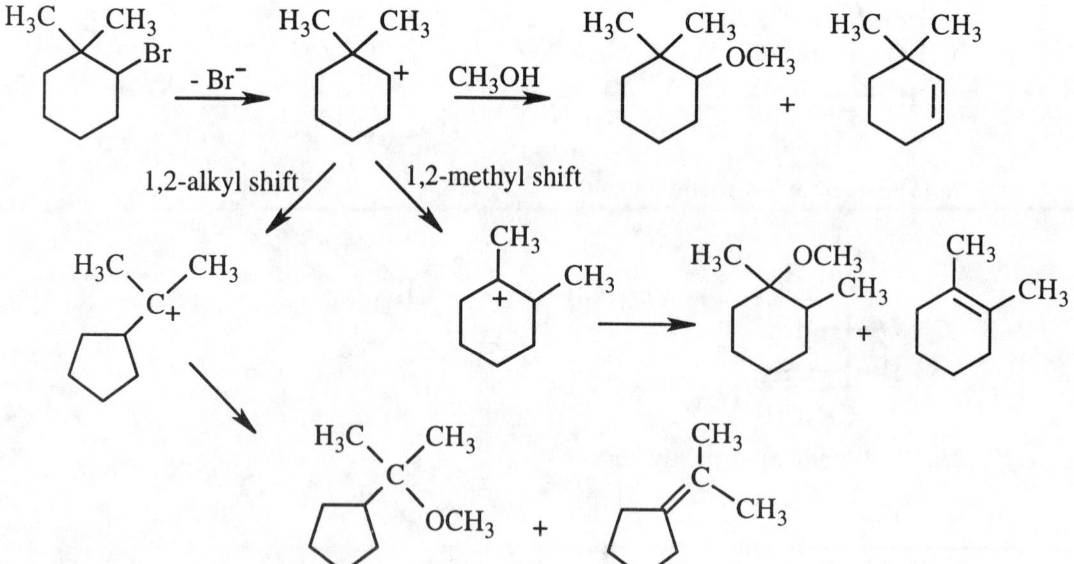

45.

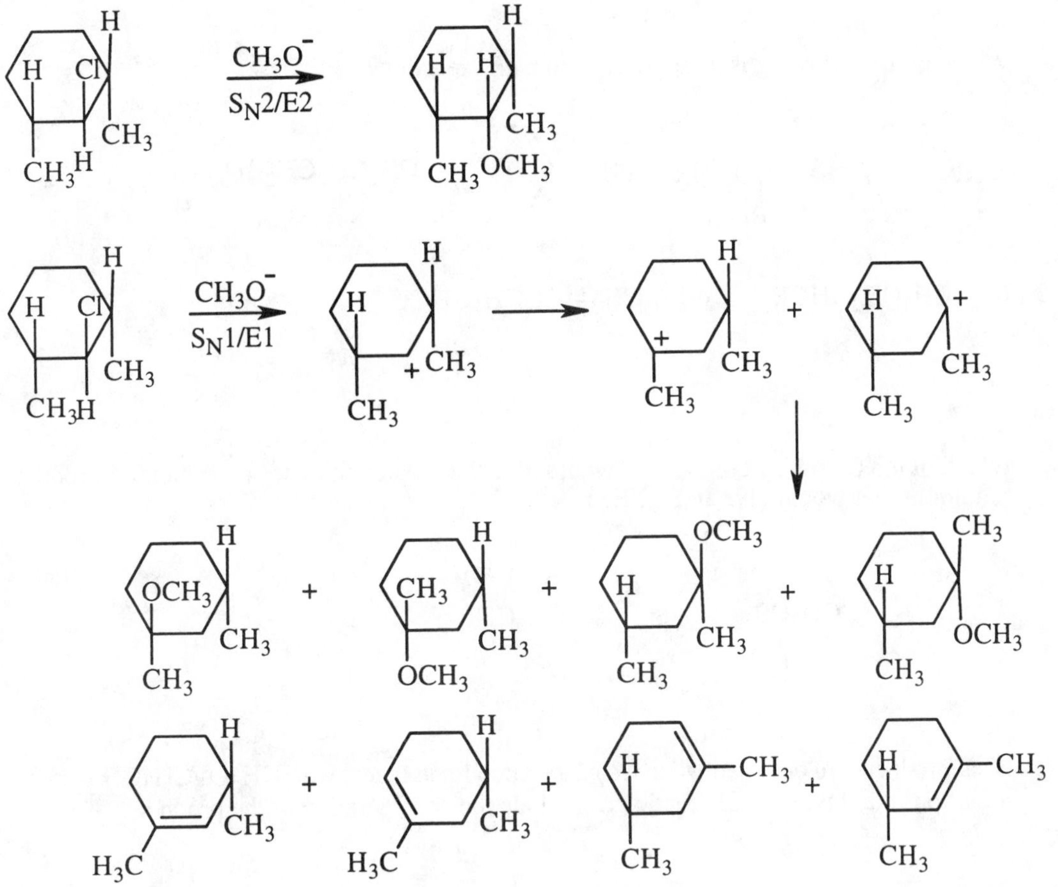

46. The number of atoms in the ring is given by n. Three and four membered rings have strain, so they are harder to make than 5 and 6 membered rings. The three-membered ring is formed faster than the four-membered ring because the compound leading to the three-membered ring has one less carbon-carbon single bond that can rotate to give a conformer in which the reacting groups are positioned too far from one another for reaction.

Even though the compound that forms the five-membered ring has one more bond that can rotate to give a conformer in which the reacting groups are positioned too far from one another for reaction than the compound that forms the four-membered ring, the five-membered ring is formed faster because it is relatively strain free. So lack of strain more than makes up for the lower probability of having the reacting groups in the proper position for reaction.

Now the rate of the ring-forming reaction gets slower as the size of the ring being formed gets larger because the reactant has more bonds that can rotate to give conformers in which the reacting groups are positioned too far from one another for reaction.

Chapter 10 Practice Test

1. Which of the following is more reactive in an E2 reaction?

a. 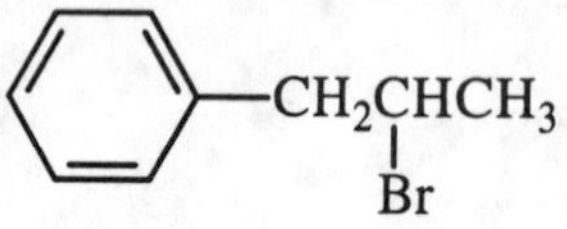 C_6H_5–CH_2CHCH_3 (with Br on CH) or C_6H_5–$CH_2CH_2CH_2Br$

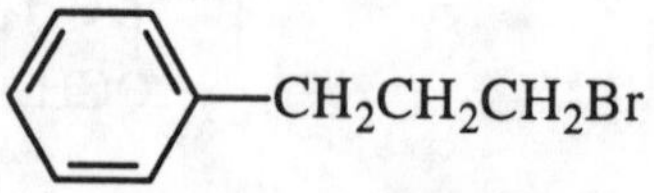

b. $CH_3CH_2CHCH_3$ (with Br on CH) or $CH_2{=}CHCH_2CHCH_3$ (with Br on CH)

2. Which of the following compounds would give the greater amount of substitution product under conditions that would give an S_N2/E2 reaction?

CH_3CBr (with two CH_3 on C) or CH_3CHBr (with CH_3 on C)

3. What products are obtained when (*R*)-2-bromobutane reacts with CH_3O^-/CH_3OH under conditions that favor S1/E1 reactions? Include the configuration of the products.

4. What alkoxide ion and what alcohol should be used to synthesize the following ethers?

a. $CH_3CH_2COCH_2CH_2CH_3$ (with two CH_3 on C)

b. cyclohexyl–O–phenyl

c. cyclohexyl–O–CH_3

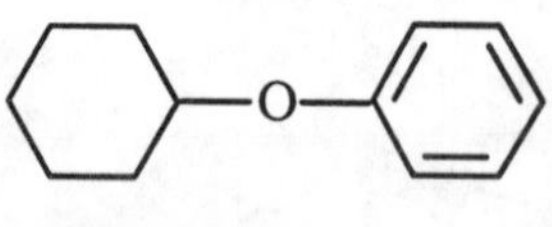

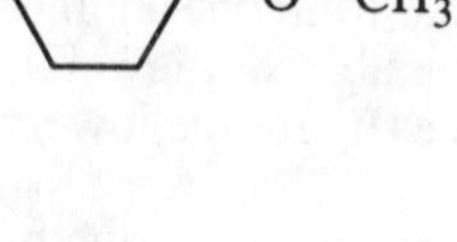

5. For each of the following pairs of E2 reactions, indicate the one that occurs with the greater rate constant.

a. $CH_3CH_2CH_2Cl + HO^-$ or $\underset{Cl}{CH_3\underset{|}{C}HCH_3} + HO^-$

b. $CH_3CH_2CH_2Cl + HO^-$ or $CH_3CH_2CH_2I + HO^-$

c. $CH_3CH_2CH_2Br + HO^-$ or $CH_3CH_2CH_2Br + H_2O$

d. $\underset{Br}{CH_3\underset{|}{C}HCH_3} \xrightarrow[H_2O/CH_3OH]{CH_3O^-}$ or $\underset{Br}{CH_3\underset{|}{C}HCH_3} \xrightarrow[CH_3OH]{CH_3O^-}$

e. $\underset{Br}{CH_3\underset{|}{C}HCH_3} + HO^-$ or $CH_3\overset{CH_3}{\underset{Br}{C}}CH_3 + HO^-$

6. Give the major product of an E2 reaction of each of the following compounds with hydroxide ion.

a. 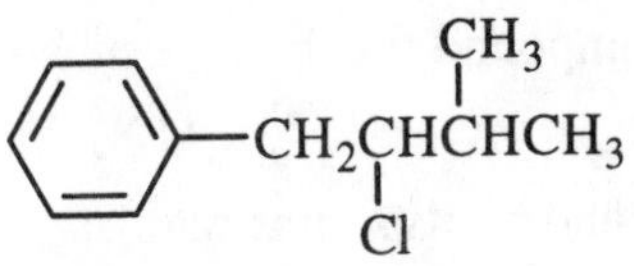 $C_6H_5{-}CH_2\underset{Cl}{CH}\overset{CH_3}{CH}CH_3$

b. (cyclopentene ring bearing Br)

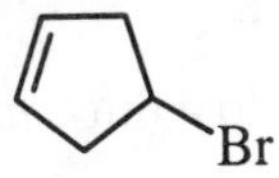

c. $CH_3\underset{Br}{CH}\overset{CH_3}{CH}CH_3$

d. $CH_3\underset{F}{CH}\overset{CH_3}{CH}CH_3$

7. a. Which would be more reactive in an E2 reaction, *cis*-1-bromo-2-methylcyclohexane or *trans*-1-bromo-2-methylcyclohexane?

b. Which would be more reactive in an E1 reaction, *cis*-1-bromo-2-methylcyclohexane or *trans*-1-bromo-2-methylcyclohexane?

CHAPTER 11

Reactions at an sp^3 Hybridized Carbon III: Substitution and Elimination Reactions of Compounds with Leaving Groups Other than Halogen. Organometallic Compounds

Important Terms

alcohol — an organic compound with an OH functional group (ROH).

alkyl tosylate — an ester of *para*-toluenesulfonic acid.

antibiotic — a compound that interferes with the growth of a microorganism.

anti-Zaitsev elimination (Hofmann elimination) — a hydrogen is removed from the β-carbon bonded to the most hydrogens.

arene oxide — an aromatic compound that has had one of its double bonds converted to an epoxide.

Cope elimination reaction — elimination from an amine oxide.

coupling reaction — a reaction that joins two alkyl groups.

crown ether — a cyclic molecule that possesses several ether linkages.

crown-guest complex — the complex formed when a crown ether binds a substrate.

cryptand — a three-dimensional polycyclic compound that binds a substrate by encompassing it.

cryptate — the complex formed when a cryptand binds a substrate.

dehydration — loss of water.

epoxide (oxirane) — an ether in which the oxygen is incorporated into a three-membered ring.

ether — a compound containing an oxygen bonded to two carbons (ROR).

exhaustive methylation — reaction of an amine with excess methyl iodide, resulting in the formation of a quaternary ammonium iodide.

Gilman reagent — a dialkylcopper reagent used to replace a halogen with an alkyl group.

Grignard reagent — the compound that results when magnesium is inserted between the carbon and halogen of an alkyl halide (RMgBr, RMgCl).

Hofmann elimination (anti-Zaitsev elimination) — a hydrogen is removed from the β-carbon bonded to the most hydrogens.

Hofmann elimination reaction — elimination of a proton and a tertiary amine from a quaternary ammonium hydroxide.

inclusion compound a compound that specifically binds a metal ion or an organic molecule.

ionophore a compound that transports metal ions by binding them tightly.

mercapto group an -SH group.

molecular recognition the recognition of one molecule by another as a result of specific interactions.

organocuprates $(R)_2CuLi$; prepared by treating an organolithium reagent with cuprous iodide.

organolithium compound RLi; prepared by adding lithium to an alkyl halide.

organomagnesium compound RMgBr; prepared by adding an alkyl halide to magnesium shavings.

organometallic compound a compound with a carbon-metal bond.

oxirane an ether in which the oxygen is incorporated into a three-membered ring.

phase transfer catalysis catalysis of a reaction by providing a way to bring a polar reagent into a nonpolar phase so that the reaction between a polar and a nonpolar compound can occur.

phase transfer catalyst a compound that carries a polar reagent into a nonpolar phase.

pinacol rearrangement rearrangement of a vicinal diol.

quaternary ammonium ion a cation containing a nitrogen bonded to four alkyl groups (R_4N^+).

ring-expansion rearrangement rearrangement of a carbocation in which the positively charged carbon is bonded to a cyclic compound, and as a result of rearrangement, the size of the ring increases by one carbon.

sulfonate ester the ester of a sulfonic acid (RSO_2OR).

sulfonium salt $(R)_3S^+$

thioether (sulfide) the sulfur analog of an ether (RSR).

thiol (mercaptan) the sulfur analog of an alcohol (RSH).

transmetallation metal exchange.

vicinal diol (vicinal glycol) a compound with OH groups bonded to adjacent carbons.

Solutions to Problems

1. The relative reactivity would be: tertiary > primary > secondary.

If secondary alcohols reacted by an S_N2 mechanism, they would be less reactive than primary alcohols because they are more sterically hindered than primary alcohols.

2. All four alcohols react by an S_N1 mechanism because they are either secondary or tertiary alcohols.

a. $CH_3CH_2CH(OH)CH_3 \xrightarrow{H^+} CH_3CH_2CH(\overset{+}{O}H_2)CH_3 \xrightarrow{-H_2O} CH_3CH_2\overset{+}{C}HCH_3 \xrightarrow{Br^-} CH_3CH_2CH(Br)CH_3$

b. 1-methylcyclopentanol $\xrightarrow{H^+}$ protonated alcohol ($\overset{+}{O}H_2$) $\xrightarrow{-H_2O}$ 1-methylcyclopentyl cation $\xrightarrow{Cl^-}$ 1-chloro-1-methylcyclopentane

The carbocation that is initially formed in **c** and the carbocation that is initially formed in **d** rearrange in order to form more stable carbocations.

c. $CH_3C(CH_3)_2CH(OH)CH_3 \xrightarrow{H^+} CH_3C(CH_3)_2CH(\overset{+}{O}H_2)CH_3 \xrightarrow{-H_2O} CH_3C(CH_3)_2\overset{+}{C}HCH_3$ (secondary carbocation) $\rightarrow CH_3\overset{+}{C}(CH_3)CH(CH_3)CH_3$ (tertiary carbocation) $\xrightarrow{Br^-} CH_3C(CH_3)(Br)CH(CH_3)CH_3$

d. (reaction scheme: 1-(2-methylcyclohexyl)ethanol $\xrightarrow{H^+}$ protonated alcohol $\xrightarrow{-H_2O}$ secondary carbocation $\rightarrow$ tertiary carbocation $\xrightarrow{Cl^-}$ 1-chloro-1-ethyl-2-methylcyclohexane)

secondary carbocation

tertiary carbocation

3. Activating an alcohol for reaction with a nucleophile by converting the alcohol into a sulfonate ester results in a product with an inverted configuration when compared with the configuration of the alcohol because one S_N2 reaction takes place.

a. $CH_3CH_2CH(OH)CH_3$ $\xrightarrow{TsCl}$ **W** (OTs) $\xrightarrow[S_N2]{NaC\equiv N}$ **X** (C≡N)

W

X
inversion

Activating an alcohol for reaction with a nucleophile by converting the alcohol into an alkyl halide results in a product with the same configuration as the alcohol because two successive S_N2 reactions take place.

b. $CH_3CH_2CH(OH)CH_3$ $\xrightarrow[\text{pyridine, } S_N2]{PBr_3}$ **Y** (Br) $\xrightarrow[S_N2]{NaC\equiv N}$ **Z** (C≡N)

Y
inversion

Z
retention

4. All the syntheses below are shown to be accomplished by converting the alcohol into a sulfonate ester and then treating the sulfonate ester with the desired nucleophile:

a could also have been carried out by converting the alcohol directly into the alkyl iodide with HI;

b-f could have been carried out by converting the alcohol into an alkyl halide and then adding the desired nucleophile.

a. $CH_3CH_2CH_2CH_2OH \xrightarrow[\text{2. KBr}]{\text{1. TsCl}} CH_3CH_2CH_2CH_2Br$

b. $CH_3CH_2CH_2CH_2OH \xrightarrow[\text{2. } CH_3CH_2\overset{\displaystyle O}{\overset{\|}{C}}O^-]{\text{1. TsCl}} CH_3CH_2CH_2CH_2O\overset{\displaystyle O}{\overset{\|}{C}}CH_2CH_3$

c. $CH_3CH_2CH_2CH_2OH \xrightarrow[\text{2. } CH_3O^-]{\text{1. TsCl}} CH_3CH_2CH_2CH_2OCH_3$

d. $CH_3CH_2CH_2CH_2OH \xrightarrow[\text{2. } CH_3CH_2NH_2]{\text{1. TsCl}} CH_3CH_2CH_2CH_2\overset{+}{N}H_2CH_2CH_3 \xrightarrow{HO^-} CH_3CH_2CH_2CH_2NHCH_2CH_3$

Notice that in "**d**" a neutral amine (CH_3NH_2) is used instead of CH_3NH^-. That is because the weaker base (CH_3NH_2) favors the desired substitution reaction, while the stronger base (CH_3NH^-) would favor elimination at the expense of substitution (Section 10.8).

e. $CH_3CH_2CH_2CH_2OH \xrightarrow[\text{2. } H_2S]{\text{1. TsCl}} CH_3CH_2CH_2CH_2\overset{+}{S}H_2 \xrightarrow{HO^-} CH_3CH_2CH_2CH_2SH$

f. $CH_3CH_2CH_2CH_2OH \xrightarrow[\text{2. } ^-C{\equiv}N]{\text{1. TsCl}} CH_3CH_2CH_2CH_2C{\equiv}N$

5. The product of each reaction is an alkene.

$$CH_3CH_2\underset{OH}{CHCH_3} \xrightarrow[\Delta]{H^+} CH_3CH_2\underset{\overset{+}{O}H_2}{CHCH_3} \xrightarrow{-H_2O} CH_3CH_2\overset{+}{C}HCH_3 \longrightarrow CH_3CH{=}CHCH_3 + H^+$$

$$CH_3CH_2\underset{Br}{CHCH_3} \xrightarrow{HO^-} CH_3CH{=}CHCH_3 + H_2O + Br^-$$

Recall that alkenes undergo electrophilic addition reactions, and the first step in an electrophilic addition reaction is addition of an electrophile to the alkene. The acid-catalyzed dehydration reaction is reversible because the electrophile (H^+) needed for the first step of the reverse electrophilic addition reaction is available.

Base-promoted elimination reaction of a hydrogen halide is not reversible because an electrophile is not available to react with the alkene.

6. 1-Bromo-1-methylcyclopentane would be formed because Br^- would react with the carbocation. If the carbocation were able to lose a proton before it reacts with Br^-, HBr could add to the resulting alkene, thereby forming the same alkyl halide as would be formed by addition of Br^- to the carbocation.

CH$_3$ + (2-methylcyclopentyl cation) → CH$_3$ + (1-methylcyclopentyl cation) $\xrightarrow{Br^-}$ Br, CH$_3$ (1-bromo-1-methylcyclopentane)

1-methylcyclopentyl cation $\xrightarrow{-H^+}$ CH$_3$ (1-methylcyclopentene) $\xrightarrow{HBr}$ 1-bromo-1-methylcyclopentane

7. **a.** In order to synthesize an asymmetrical ether by this method, two different alcohols would have to be heated with sulfuric acid. Therefore, three different ethers would be obtained as products. Consequently, the desired ether would account for considerably less than half of the total amount of ether that is synthesized.

$$ROH + R'OH \xrightarrow[\Delta]{H_2SO_4} ROR + ROR' + R'OR'$$

b. It could be synthesized by means of the Williamson ether synthesis. (See Section 10.9 on page 418 of the text.)

$$CH_3CH_2CH_2OH \xrightarrow{Na} CH_3CH_2CH_2O^- \xrightarrow{CH_3CH_2Br} CH_3CH_2CH_2OCH_2CH_3 + Br^-$$

8.

a. CH_3 CH_3 $\ddot{O}H$ $\xrightarrow{H^+}$ CH_3 CH_3 $\overset{+}{O}H$ H $\xrightarrow{-H_2O}$ CH_3 CH_3 + → + CH_3 H CH_3 $\xrightarrow{-H^+}$ CH_3 CH_3

b. Because of the difficulty of forming a primary carbocation, loss of water and the hydride shift are concerted. The carbocation that is formed is stabilized by electron delcoalization. In other words, the positive charge is shared by carbon and oxygen.

$\ddot{O}$ $CH_2\ddot{O}H$ $\xrightarrow{H^+}$ $\ddot{O}$ H $CH_2-\overset{+}{O}H$ H → $\ddot{O}$ + CH_3 ⟷ H $\overset{+}{O}$ CH_3 → $\ddot{O}$ CH_3

c. [Reaction scheme: cyclopropyl–C(CH$_3$)$_2$–ÖH $\xrightarrow{H^+}$ cyclopropyl–C(CH$_3$)$_2$–$\overset{+}{O}H_2$ $\xrightarrow{-H_2O}$ cyclopropyl–$\overset{+}{C}$(CH$_3$)$_2$ → 2,2-dimethylcyclobutyl cation → 1,2-dimethylcyclobutyl cation; :Br:$^-$ → 1-bromo-1,2-dimethylcyclobutane]

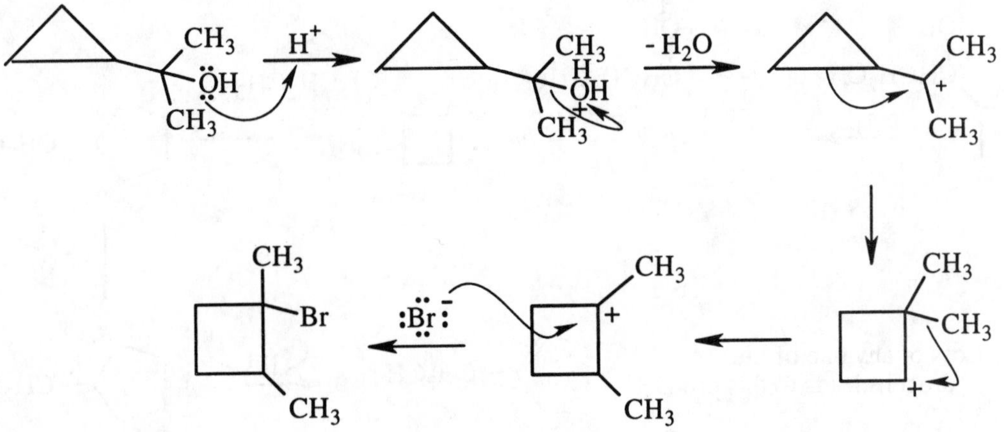

9.

a. $CH_3CH_2C(CH_3)(OH)-CH(CH_3)CH_3$ $\xrightarrow{H^+}$ $CH_3CH_2C(CH_3)(\overset{+}{O}H_2)-CH(CH_3)CH_3$ $\xrightarrow{-H_2O}$ $CH_3CH_2\overset{+}{C}(CH_3)-CH(CH_3)CH_3$ $\xrightarrow{-H^+}$ $(CH_3)(CH_3CH_2)C=C(CH_3)_2$

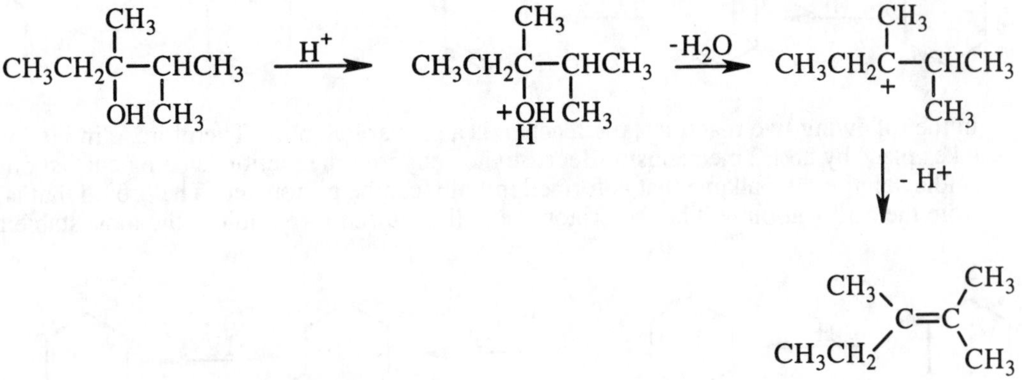

b. $CH_3CH_2CH_2CH(OH)-C(CH_3)_2CH_3$ $\xrightarrow{H^+}$ $CH_3CH_2CH_2CH(\overset{+}{O}H_2)-C(CH_3)_2CH_3$ $\xrightarrow{-H_2O}$ $CH_3CH_2CH_2\overset{+}{C}H-C(CH_3)_2CH_3$

carbocation rearrangement

$CH_3CH_2CH_2CH(CH_3)-\overset{+}{C}(CH_3)CH_3$ $\xrightarrow{-H^+}$ $(CH_3)(CH_3CH_2CH_2)C=C(CH_3)_2$

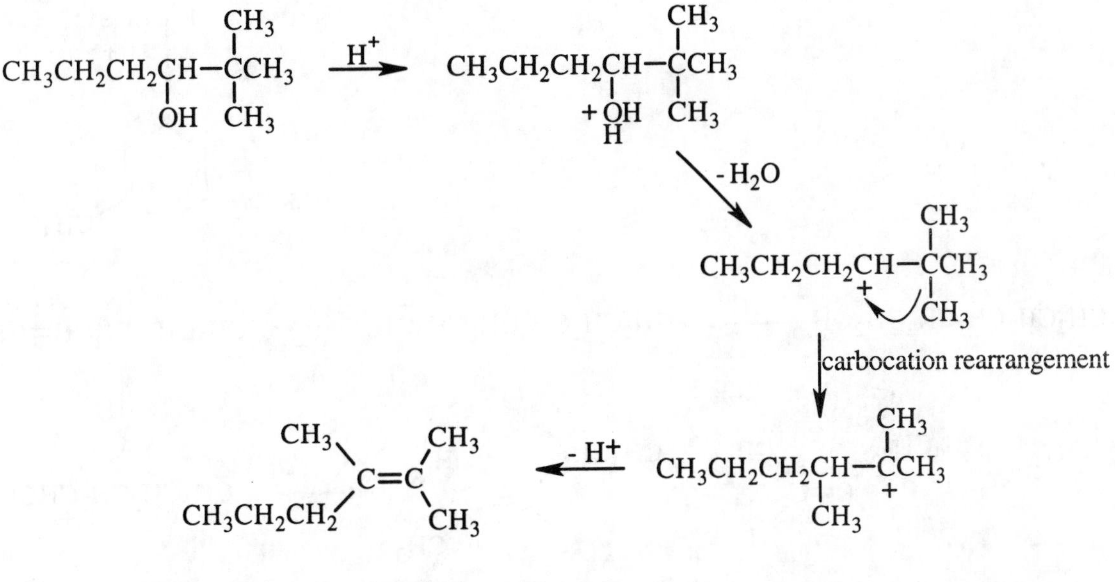

c.

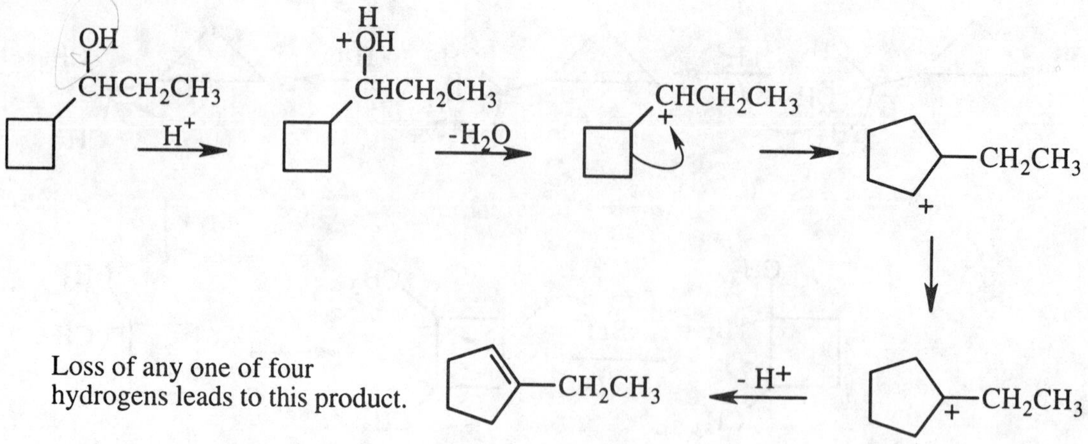

Loss of any one of four hydrogens leads to this product.

d.

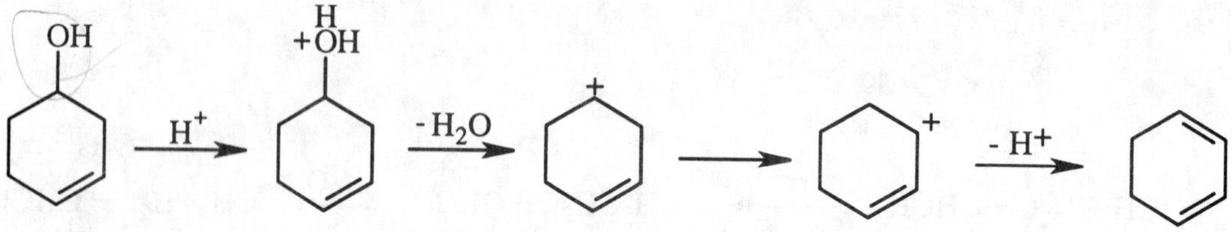

In the following two reactions, the reactant is a primary alcohol. Therefore, elimination of water takes place by an E2 mechanism. Because the dehydration reaction is being carried out in an acidic solution, the alkene that is formed initially can be protonated. The proton that is then lost from the carbocation will be the proton that will result in formation of the most stable alkene.

e.

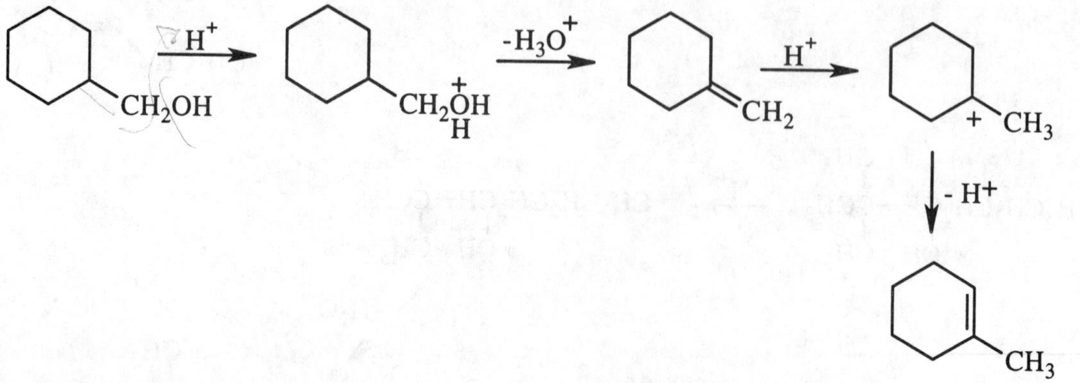

f. $CH_2CH_2CH_2CH_2CH_2OH \xrightarrow{H^+} CH_2CH_2CH_2CH_2CH_2\overset{+}{O}H_2 \xrightarrow[-H^+]{-H_2O} CH_3CH_2CH_2CH{=}CH_2$

$\xrightarrow{H^+} CH_2CH_2CH_2\overset{+}{C}HCH_3 \xrightarrow{-H^+}$ (E)-$CH_3CH_2CH{=}CHCH_3$ + (Z)-$CH_3CH_2CH{=}CHCH_3$

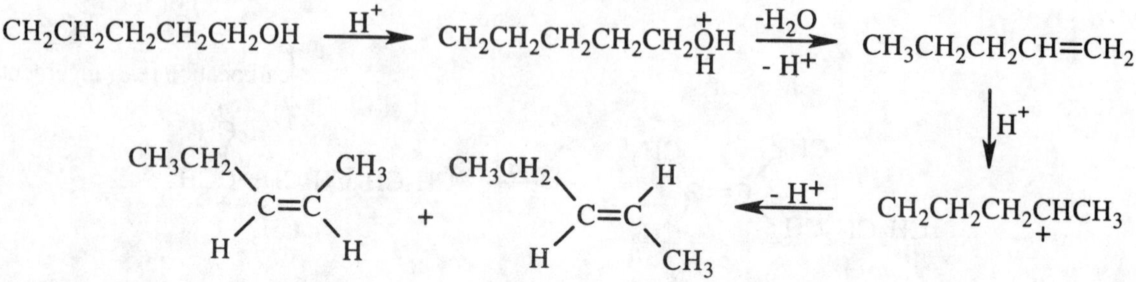

10. We saw that HCl does not cleave ethers, because Cl^- is not a strong enough nucleophile. F^- is an even weaker nucleophile, so HF cannot cleave ethers. Therefore, ethers can be cleaved only with HBr or HI.

11. Notice that if there is excess concentrated HI, the initially formed alcohol can be converted to an alkyl iodide in c, d, and f.

a. Solved in the text.

b. Cleavage will occur by an S_N1 pathway because the carbocation that is formed is relatively stable (benzyl); I^- will attack the benzyl carbocation.

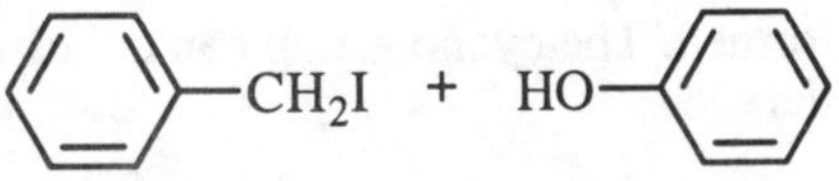

$$C_6H_5CH_2I \quad + \quad HO-C_6H_5$$

c. Cleavage will occur by an S_N1 pathway because the carbocation that is formed is relatively stable (tertiary); I^- will attack the tertiary carbocation.

$$CH_3CH_2CH_2OH \quad + \quad CH_3CH_2C(CH_3)(I)CH_3$$

$$\downarrow I^-$$

$$CH_3CH_2CH_2I$$

d. $HOCH_2CH_2CH_2CH_2CH_2I \xrightarrow{I^-} ICH_2CH_2CH_2CH_2CH_2I$

e. CH_3I + (cyclohexenol) $\underset{\text{of the enol}}{\overset{\text{tautomerization}}{\rightleftharpoons}}$ (cyclohexanone)

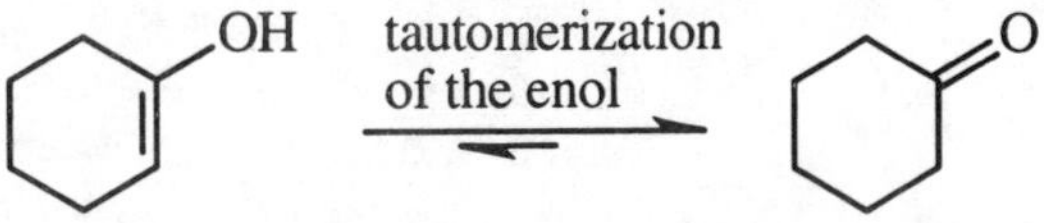

enol ketone

f. Cleavage will occur by an S_N1 pathway because the carbocation that is formed is relatively stable (tertiary); I^- will attack the tertiary carbocation.

$$HOCH_2CH_2CH_2C(CH_3)(I)CH_3 \xrightarrow{I^-} ICH_2CH_2CH_2C(CH_3)(I)CH_3$$

12.

a. 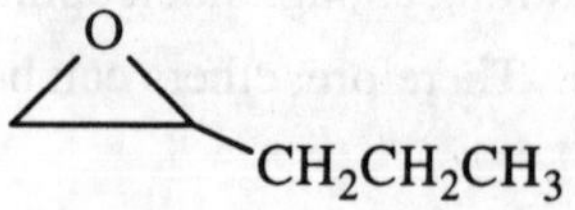 (epoxide with $CH_2CH_2CH_3$ substituent)

b. 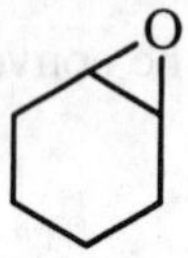 (cyclohexene oxide)

c. 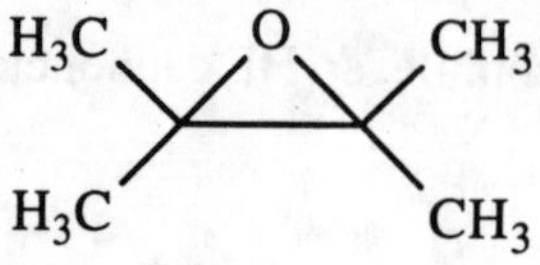 (tetramethyl epoxide: H_3C, H_3C, CH_3, CH_3)

d. 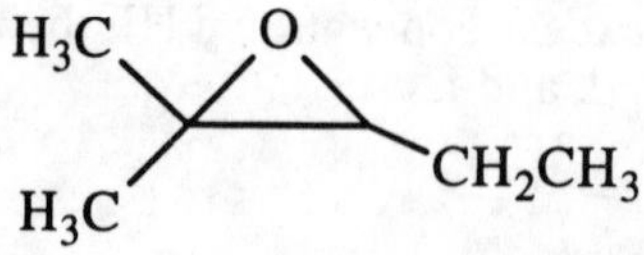 (epoxide with H_3C, H_3C, CH_2CH_3)

13. Two stereoisomers will be formed. The cyano group can attack either of the carbons bonded to the oxygen from the backside.

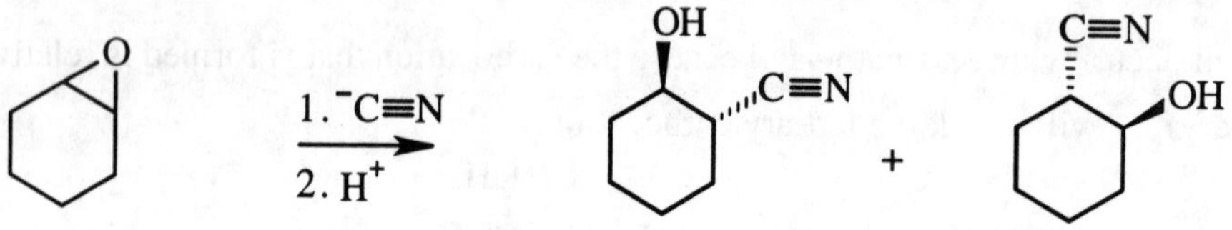

cyclohexene oxide $\xrightarrow[2.\ H^+]{1.\ ^-C\equiv N}$ trans-2-hydroxycyclohexanecarbonitrile (OH, C≡N) + trans-2-hydroxycyclohexanecarbonitrile (C≡N, OH)

14.

a. 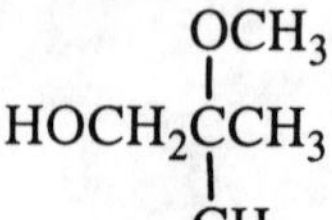 $HOCH_2C(OCH_3)(CH_3)CH_3$

b. 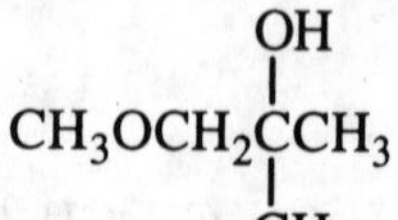 $CH_3OCH_2C(OH)(CH_3)CH_3$

c. 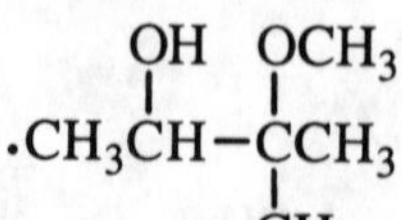 $CH_3CH(OH)-C(OCH_3)(CH_3)CH_3$

d. 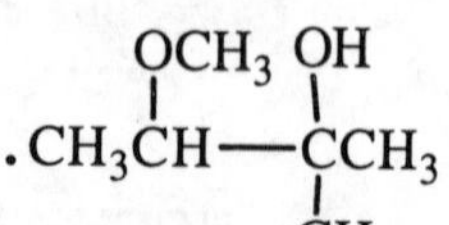 $CH_3CH(OCH_3)-C(OH)(CH_3)CH_3$

15. The reactivity of tetrahydrofuran is more similar to a noncyclic ether because the five-membered ring does not have the strain that makes the epoxide reactive.

16. The carbocation leading to 1-naphthol can be stabilized without destroying the aromaticity of the intact benzene ring. The carbocation leading to 2-naphthol can be stabilized only by destroying the aromaticity of the intact benzene ring. therefore, the carbocation leading to 1-naphthol is more stable.

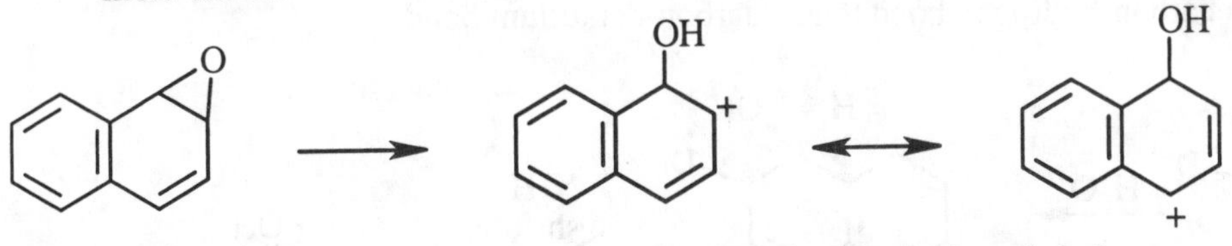

OH OH O + +

carbocation leading to 1-naphthol

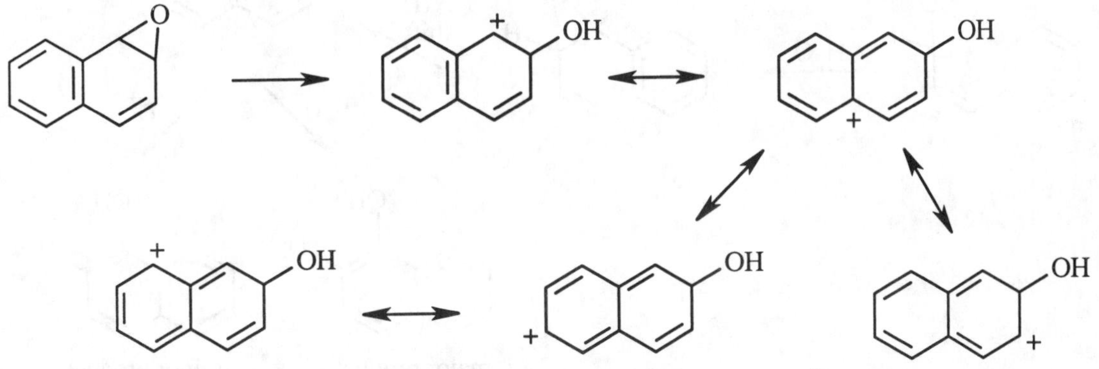

O + OH OH +

+ OH OH OH + +

carbocation leading to 2-naphthol

17.

without an NIH shift

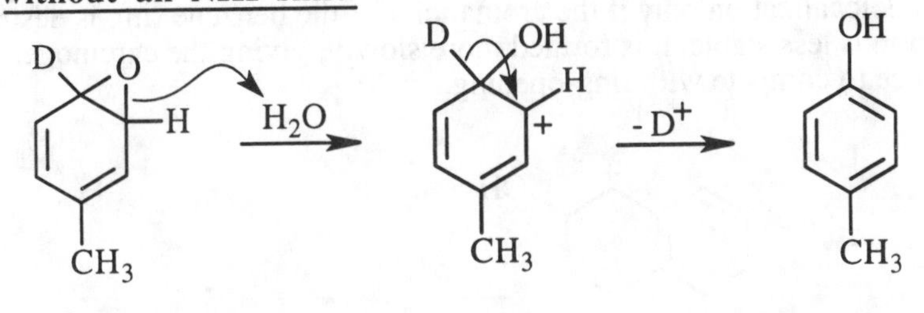

D O H H_2O D OH H + $- D^+$ OH

CH_3 CH_3 CH_3

with an NIH shift

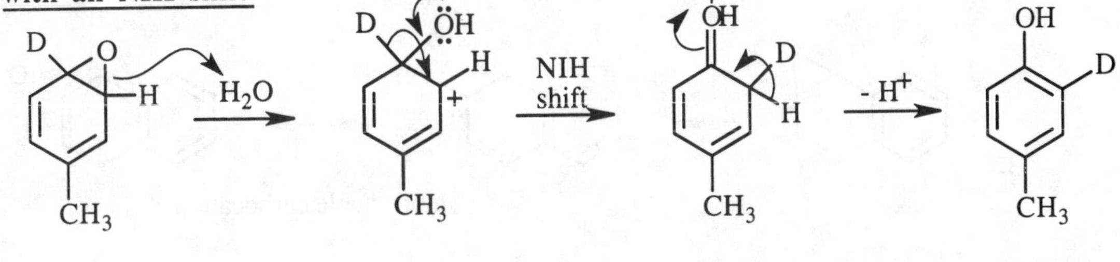

D O H H_2O D ÖH H + NIH shift $\overset{+}{O}H$ D H $- H^+$ OH D

CH_3 CH_3 CH_3 CH_3

18. The epoxide opens in the direction that will give the most stable carbocation. The carbocation undergoes an NIH shift and, as a result of the NIH shift, both reactants form the same ketone intermediate. Because they form the same intermediate, they form the same products. The deuterium-containing product is the major product because in the last step of the reaction it is easier to break a carbon-hydrogen bond than a carbon-deuterium bond.

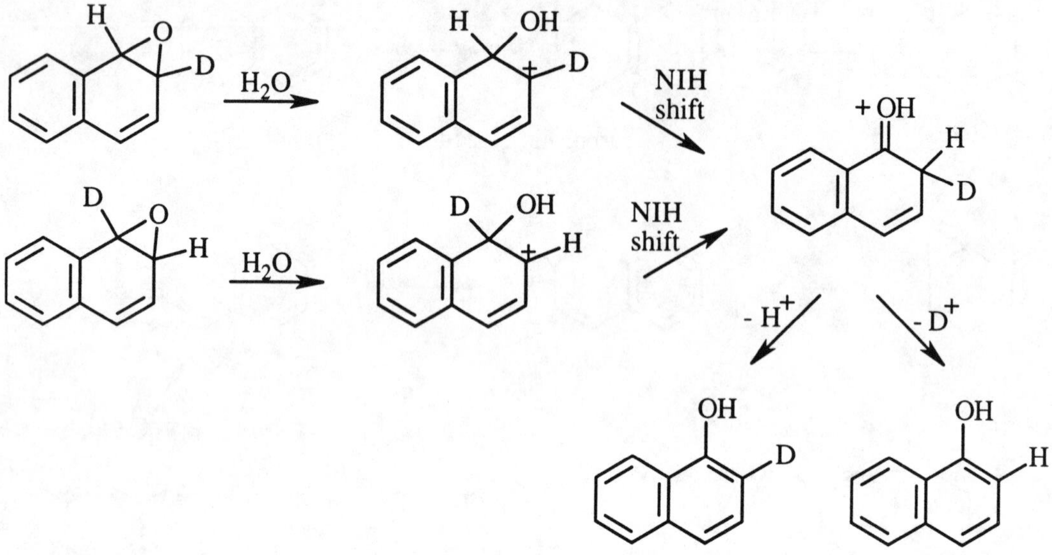

H O D $\xrightarrow{H_2O}$ H OH + D — NIH shift → +OH H D

D O H $\xrightarrow{H_2O}$ D OH + H — NIH shift ↗

$-H^+$ ↙ ↘ $-D^+$

OH D — major product

OH H — minor product

19. a. Solved in the text.

b. The compound without the double bond in the second ring is more apt to be carcinogenic. It opens to form a less stable carbocation than the other compound because it can be stabilized by electron delocalization only if the aromaticity of the benzene ring is destroyed. Because the carbocation is less stable, it is formed more slowly, giving the carcinogenic pathway a better chance to compete with ring-opening.

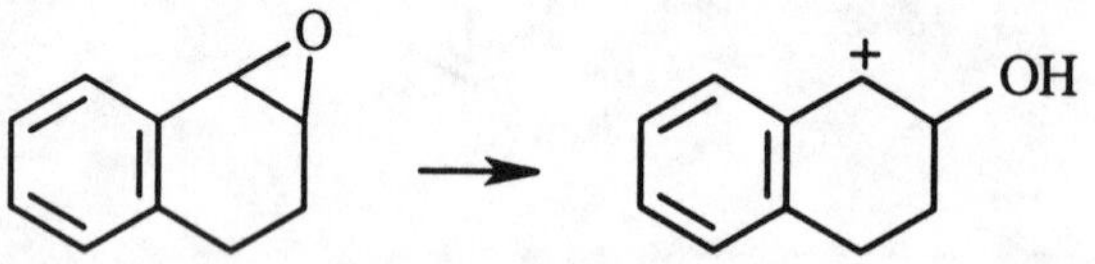

O → + OH

less stable carbocation

O → OH + ⟷ OH +

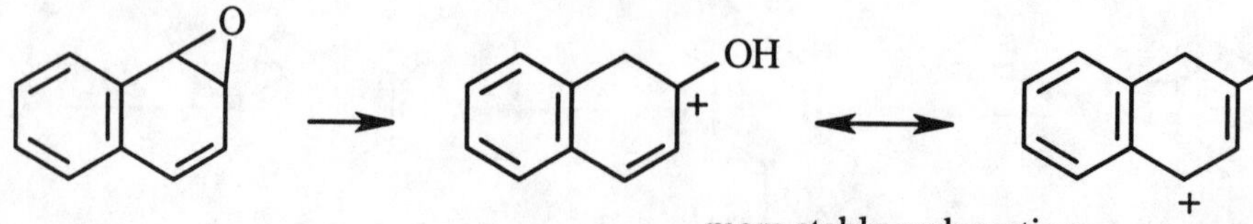

more stable carbocation

The other arene oxide opens to give a more stable carbocation. Consequently, the carcinogenic pathway will be less able to compete with the ring-opening reaction.

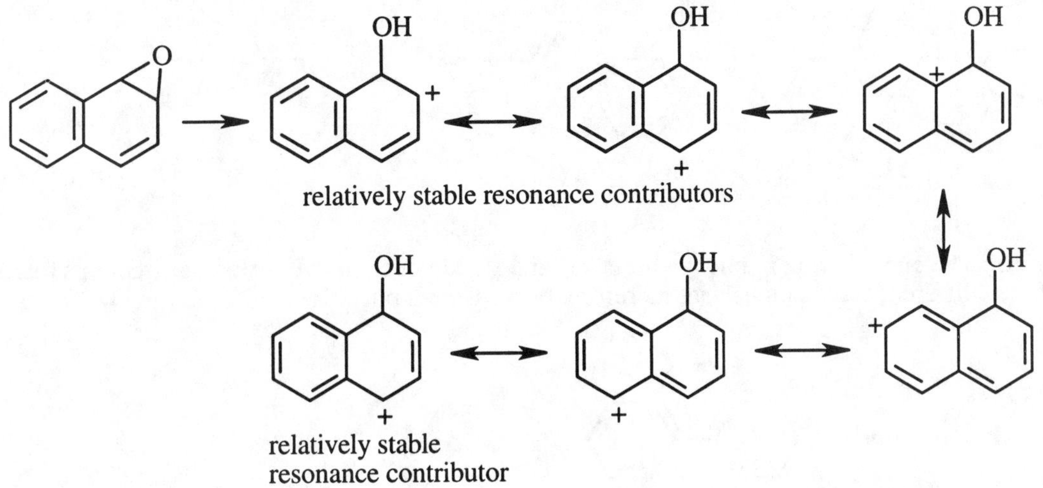

relatively stable resonance contributors

relatively stable resonance contributor

20. Each arene oxide will open in the direction that forms the most stable carbocation. Thus, the arene oxide opens so the positive charge can be stabilized by electron delocalization from the methoxide group.

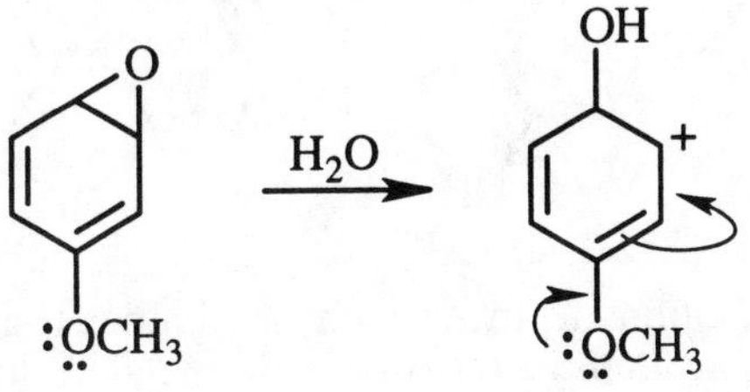

H_2O

:ÖCH$_3$:ÖCH$_3$

The arene oxide opens to form the most stable carbocation intermediate, which is the one where the positive charge is father away from the electron-withdrawing NO_2 group.

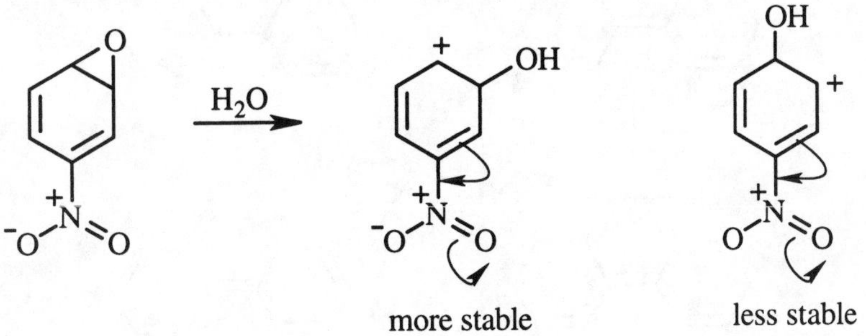

H_2O

more stable   less stable

21. **a.** Note that a bond joining two rings cannot be epoxidized.

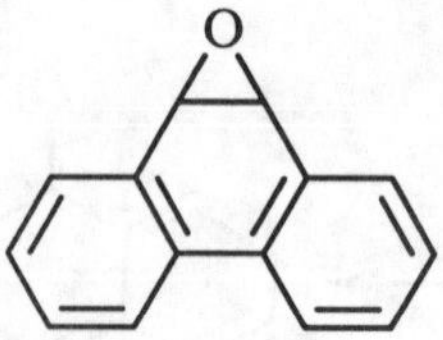

O

I II O III O

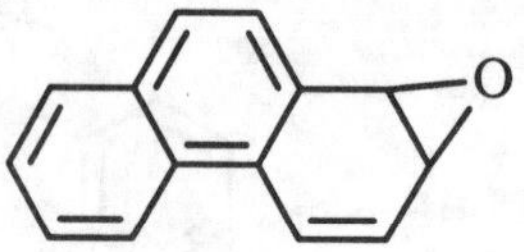

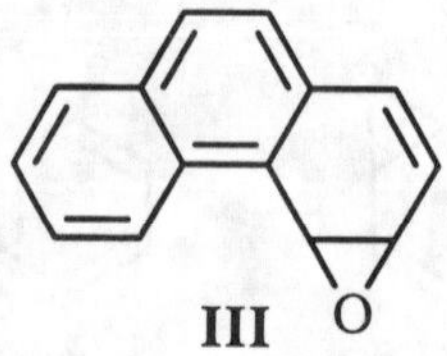

b. The epoxide ring in phenanthrenes II and III can open in two different directions to give two different carbocations and, therefore, two different phenols.

I ⟶ OH

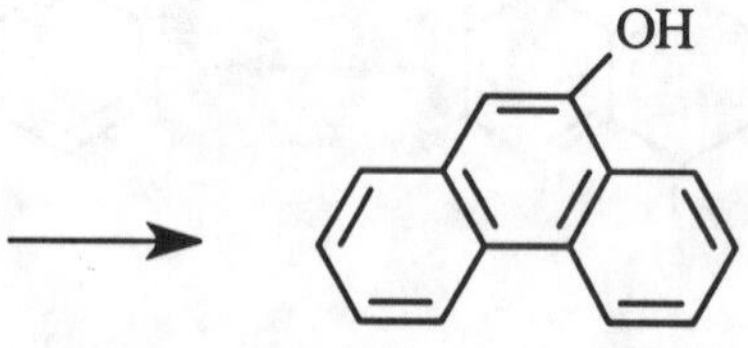

II ⟶ OH OH

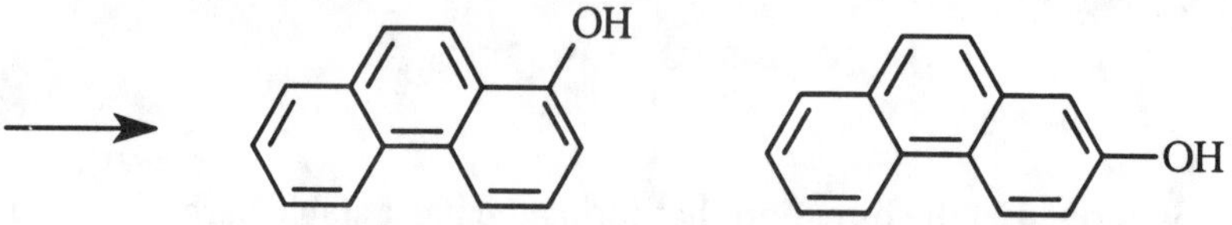

III ⟶ OH HO

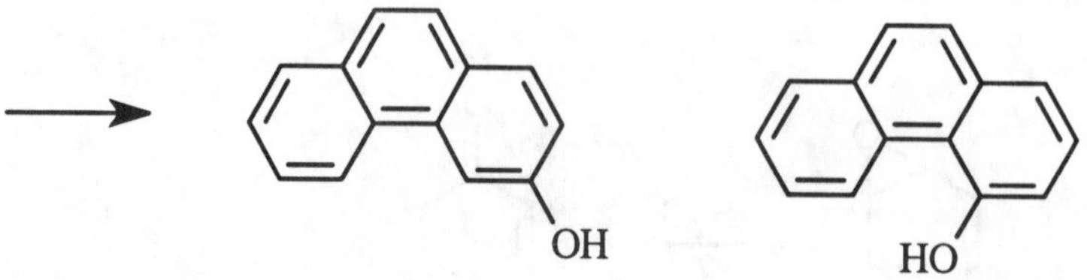

c. The two different carbocations formed by phenanthrenes II and III differ in stability. One carbocation is more stable than the other because it can be stabilized by resonance without disrupting the aromaticity of the adjacent ring. The more stable carbocation leads to the major product.

II ⟶ OH + + OH

OH OH

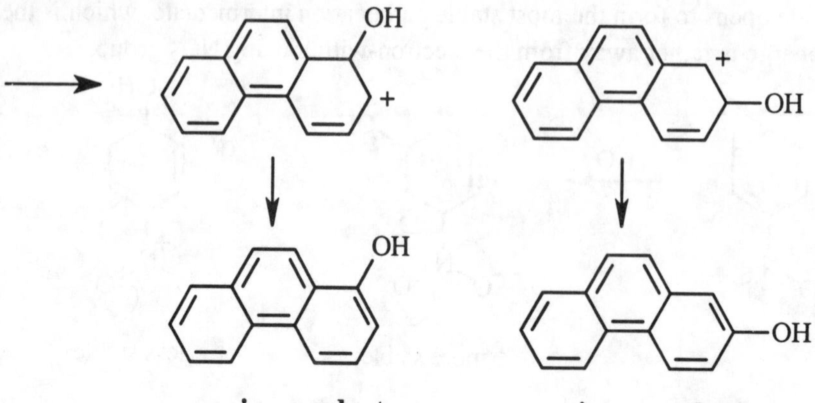

major product **minor product**

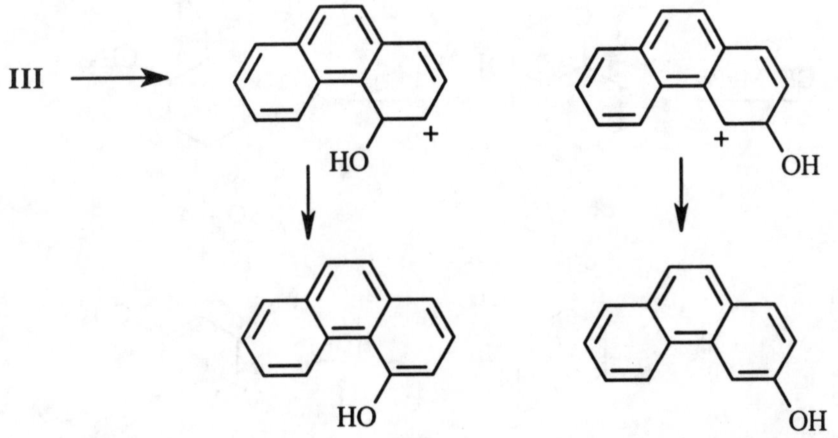

III →

HO (+) → major product

(+) OH → minor product

HO
major product

OH
minor product

d. Phenanthrene oxide I is the most carcinogenic because it is the only one that opens to form a carbocation that cannot be stabilized without disrupting the aromaticity of the other ring(s).

22.

a. $CH_3CH_2CH_2CH_2CH_2OH$ **b.** 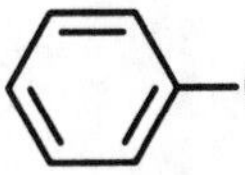 $C_6H_5-CH_2CH_2CH_2OH$ **c.** 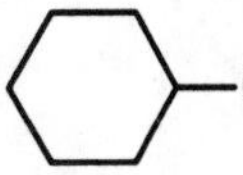 $C_6H_{11}-CH_2CH_2OH$

23.

a. 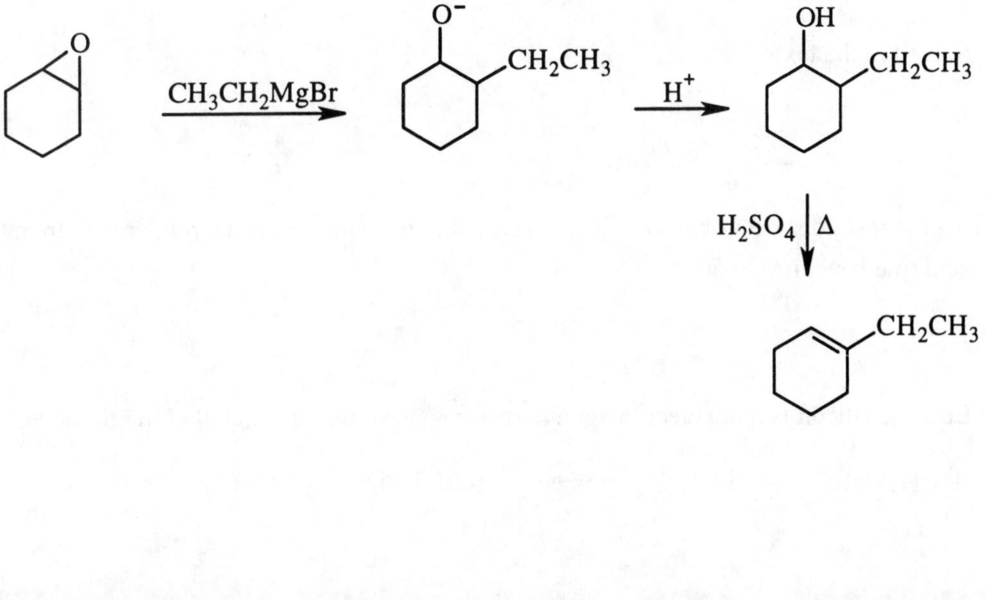 cyclohexene oxide $\xrightarrow{CH_3CH_2MgBr}$ 2-ethylcyclohexan-1-olate (O^-, CH_2CH_3) $\xrightarrow{H^+}$ 2-ethylcyclohexanol (OH, CH_2CH_3)

H_2SO_4 / Δ ↓

1-ethylcyclohexene (CH_2CH_3)

b. [epoxycyclohexane] $\xrightarrow{CH_3MgBr}$ [2-methylcyclohexanolate, O^-, CH_3] $\xrightarrow{H^+}$ [2-methylcyclohexanol, OH, CH_3]

H_2SO_4 / Δ ↓

[1-bromo-1-methylcyclohexane, CH_3, Br] $\xleftarrow{HBr}$ [1-methylcyclohexene, CH_3]

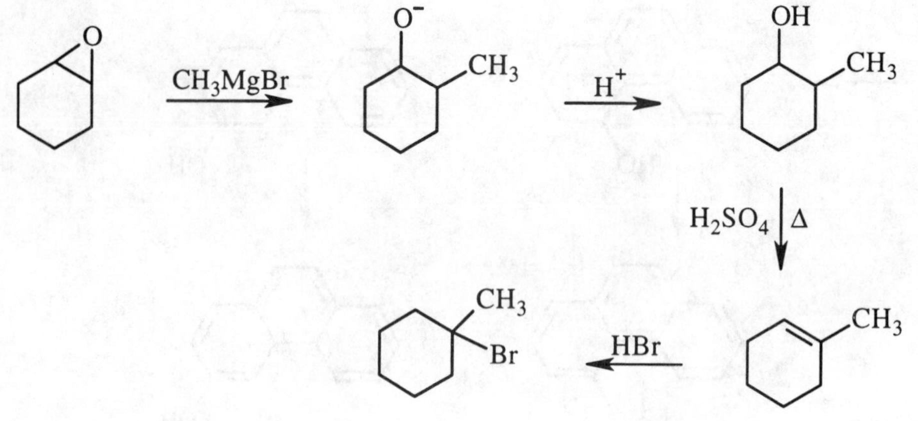

c. [epoxycyclohexane] $\xrightarrow{\text{phenyllithium (Li)}}$ [2-phenylcyclohexanolate, O^-] $\xrightarrow{H^+}$ [2-phenylcyclohexanol, OH]

H_2SO_4 / Δ ↓

[cyclohexylbenzene] $\xleftarrow[Pd/C]{H_2}$ [1-phenylcyclohexene]

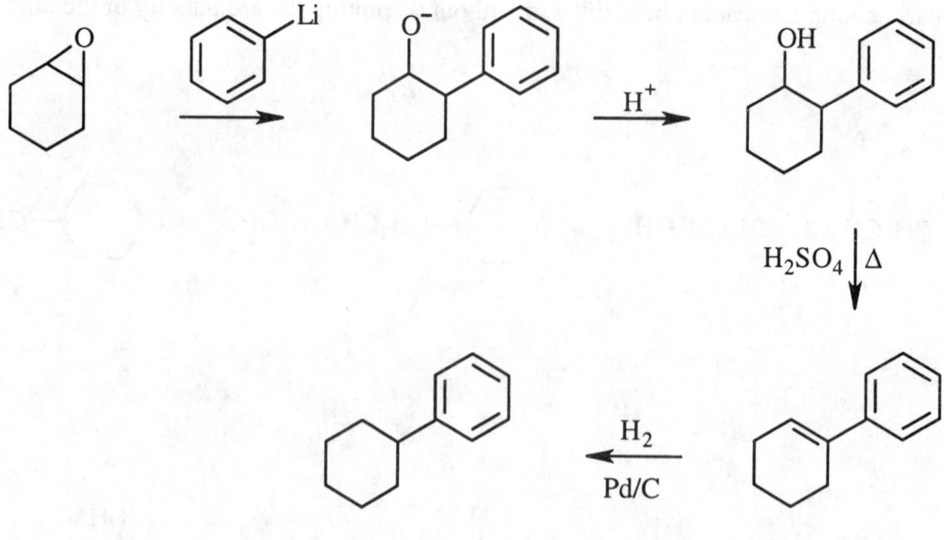

24. Solved in the text.

25. All the reactions occur because, in each case, the reactant acid is a stronger acid than the product acid (methane, $pK_a = 50$).

26. Because silicon is more electronegative than magnesium, transmetallation will occur.

$$4\ CH_3MgCl + SiCl_4 \longrightarrow (CH_3)_4Si + 4\ MgCl_2$$

27.

$$CH_3(CH_2)_3CH_2Br \xrightarrow[\text{hexane}]{\text{Li}} CH_3(CH_2)_3CH_2Li \xrightarrow{\text{CuI}} (CH_3CH_2CH_2CH_2CH_2)_2CuLi$$

$$\downarrow \; CH_3(CH_2)_7CH{=}CH(CH_2)_7CH_2Br \; (cis, \text{H}, \text{H})$$

$$CH_3(CH_2)_7CH{=}CH(CH_2)_{12}CH_3 \; (cis, \text{H}, \text{H})$$

28.

a. $CH_3CH_2SH \xrightarrow{HO^-} CH_3CH_2S^- \xrightarrow{CH_3CH_2Br} CH_3CH_2SCH_2CH_3 + Br^-$

b. The reaction cannot be done with methanethiol and *tert*-butyl bromide, because a tertiary alkyl halide would form an elimination product rather than a substitution product.

$$(CH_3)_3CSH \xrightarrow{HO^-} (CH_3)_3CS^- \xrightarrow{CH_3Br} (CH_3)_3CSCH_3 + Br^-$$

c. The highest yield is obtained by having the less substituted of the two R groups of the thioether be the alkyl halide and the more substituted be the thiol.

$$CH_2{=}CHCH(CH_3)SH \xrightarrow{HO^-} CH_2{=}CHCH(CH_3)S^- \xrightarrow{CH_3CH_2Br} CH_2{=}CHCH(CH_3)SCH_2CH_3 + Br^-$$

d. The synthesis must be done this way because an sp^2 carbon cannot undergo backside attack.

$$C_6H_5\text{–}SH \xrightarrow{HO^-} C_6H_5\text{–}S^- \xrightarrow{C_6H_5\text{–}CH_2Br} C_6H_5\text{–}SCH_2\text{–}C_6H_5 + Br^-$$

29. It is difficult for a weaker base to displace a stronger base.

Br^- (pK_a of HBr = -9) is a weaker base by about 7 pK_a units than the base it is displacing when it displaces water (pK_a $CH_3OH_2^+$ = -1.7).

$$Br^- + CH_3\overset{H}{\underset{+}{O}}H \xrightarrow{\Delta} CH_3Br + H_2O$$

Br^- is a weaker base by about 18 pK_a units than the base it is displacing when it displaces ammonia (pK_a $CH_3NH_3^+$ = 9.4). Thus, ammonia is far too basic to be displaced by Br^-.

$$Br^- + CH_3\overset{+}{N}H_3 \xrightarrow{\Delta} \text{no reaction}$$

30. The only difference is the leaving group.

$$H\!-\!CH_2\!-\!CH(CH_3)\!-\!\overset{+}{N}(CH_3)_3 \quad + \quad HO^- \qquad\qquad H\!-\!CH_2\!-\!CH(CH_3)\!-\!Br \quad + \quad HO^-$$

31.

$$CH_3\underset{CH_3}{C}{=}CH_2 \quad + \quad \overset{CH_3}{\underset{CH_3}{N}}CH_2CH_2CH_3$$

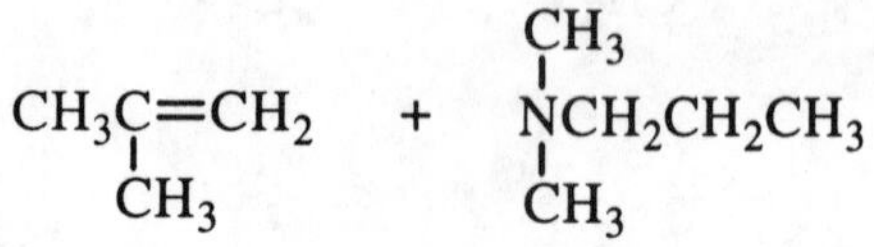

32. In each case, a proton is removed from the β-carbon that is bonded to the greater number of hydrogens.

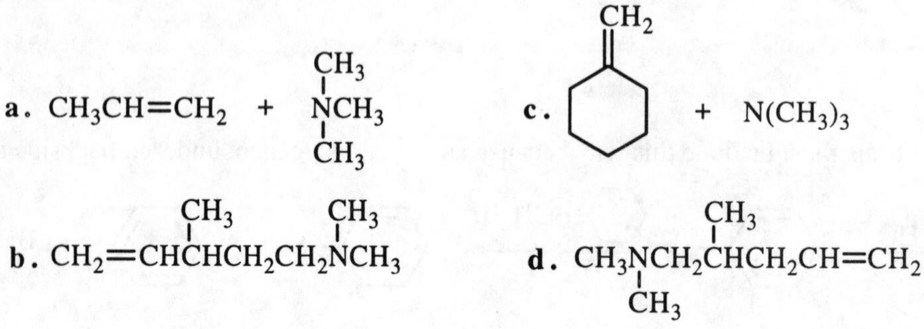

a. $CH_3CH{=}CH_2 \; + \; \overset{CH_3}{\underset{CH_3}{N}}CH_3$

b. $CH_2{=}CH\overset{CH_3}{C}HCH_2CH_2\overset{CH_3}{N}CH_3$

c. methylenecyclohexane (CH_2 exocyclic) + $N(CH_3)_3$

d. $CH_3\underset{CH_3}{N}CH_2\overset{CH_3}{C}HCH_2CH{=}CH_2$

33.

$$\text{piperidine (N–H)} + CH_3\text{–}I \longrightarrow \text{piperidinium } (H_3C,\ H)\ N^+\ I^- \underset{H^+}{\overset{-H^+}{\rightleftharpoons}} \text{N–}CH_3\text{ piperidine} \xrightarrow{CH_3\text{–}I} \text{piperidinium } (H_3C,\ CH_3)\ N^+\ I^-$$

34. **a.** Solved in the text.

b.

$$CH_3CH_2CH_2CH(Br)CH_3 \xrightarrow{(CH_3)_3N} CH_3CH_2CH_2CH(CH_3)\text{–}\overset{+}{N}(CH_3)_3\ Br^-$$

$$\xrightarrow{Ag_2O,\ H_2O} CH_3CH_2CH_2CH(CH_3)\text{–}\overset{+}{N}(CH_3)_3\ HO^-$$

$$\xrightarrow{\Delta} (CH_3)_3N + CH_3CH_2CH_2CH{=}CH_2$$

c.

$$\text{pyrrolidine (N–H)} \xrightarrow[K_2CO_3]{CH_3I\ \text{excess}} \text{N,N-dimethylpyrrolidinium } (H_3C,\ CH_3)\ I^- \xrightarrow[H_2O]{Ag_2O} \text{N,N-dimethylpyrrolidinium } (H_3C,\ CH_3)\ HO^- \xrightarrow{\Delta} CH_2{=}CHCH_2CH_2N(CH_3)_2$$

$$\xrightarrow{CH_3I} CH_2{=}CHCH_2CH_2\overset{+}{N}(CH_3)_3\ I^- \xrightarrow[H_2O]{Ag_2O} CH_2{=}CHCH_2CH_2\overset{+}{N}(CH_3)_3\ HO^- \xrightarrow{\Delta} CH_2{=}CHCH{=}CH_2$$

35. Because the major product is the one obtained by removing a proton from the β-carbon bonded to the most hydrogens, we can conclude that the Cope elimination has a "carbanion-like" transition state.

36.

a. 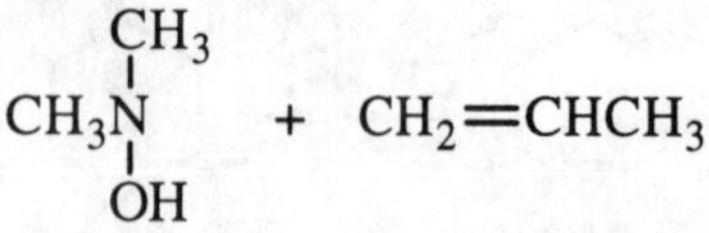 $CH_3\overset{CH_3}{\underset{OH}{N}} \quad + \quad CH_2{=}CHCH_3$

c. 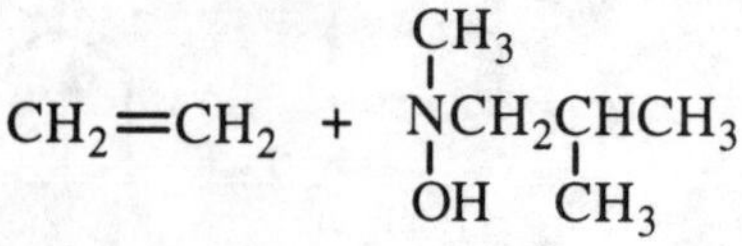 $CH_2{=}CH_2 \quad + \quad \overset{CH_3}{\underset{OH}{N}}CH_2\underset{CH_3}{C}HCH_3$

b. 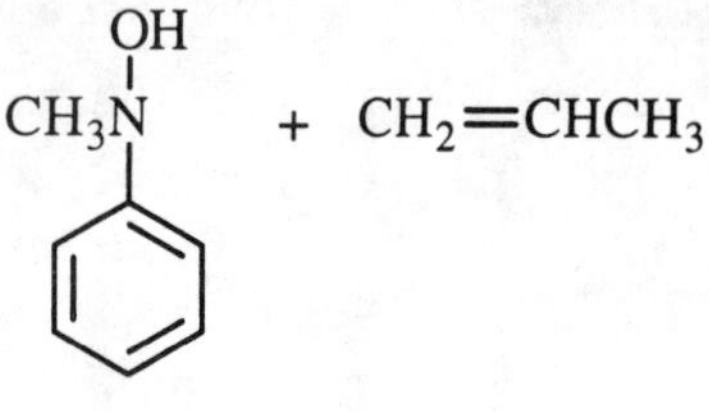 $CH_3\overset{OH}{\underset{C_6H_5}{N}} \quad + \quad CH_2{=}CHCH_3$

d. 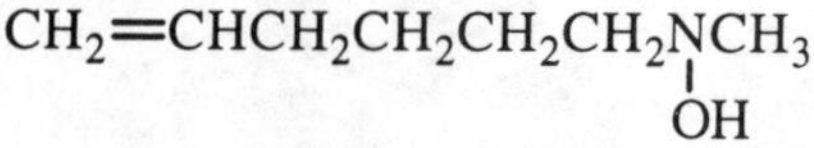 $CH_2{=}CHCH_2CH_2CH_2CH_2\underset{OH}{N}CH_3$

37.

a. 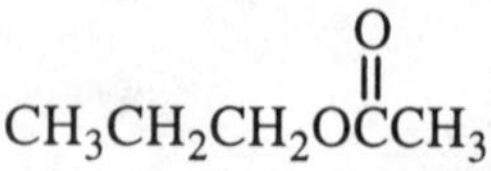 $CH_3CH_2CH_2O\overset{O}{\overset{\|}{C}}CH_3$

h. 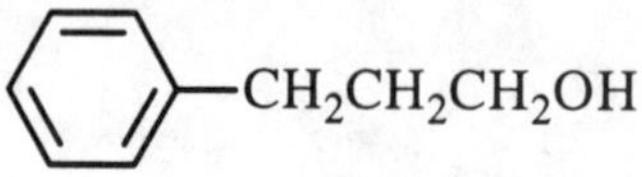 $C_6H_5{-}CH_2CH_2CH_2OH$

b. $CH_3CH_2CH_2CH_2Br$

i. 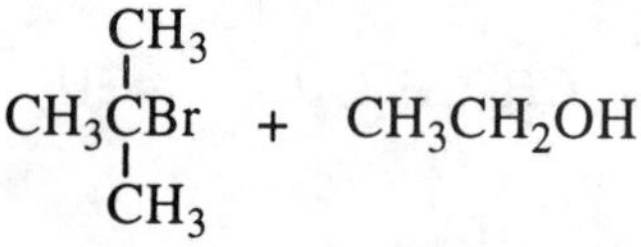 $CH_3\overset{CH_3}{\underset{CH_3}{C}}Br \quad + \quad CH_3CH_2OH$

c. $CH_3\underset{CH_3}{C}HCH_2CH_2O{-}C_6H_5$

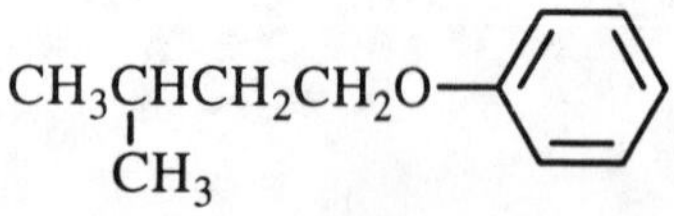

j. 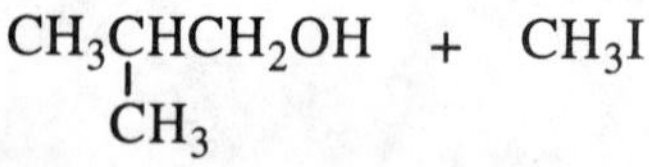 $CH_3\underset{CH_3}{C}HCH_2OH \quad + \quad CH_3I$

d. 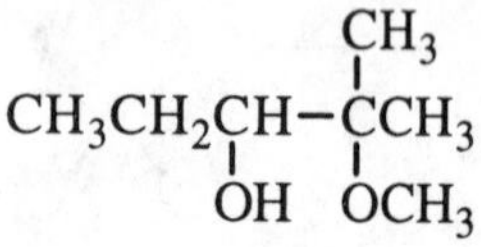 $CH_3CH_2\underset{OH}{C}H{-}\overset{CH_3}{\underset{OCH_3}{C}}CH_3$

k. 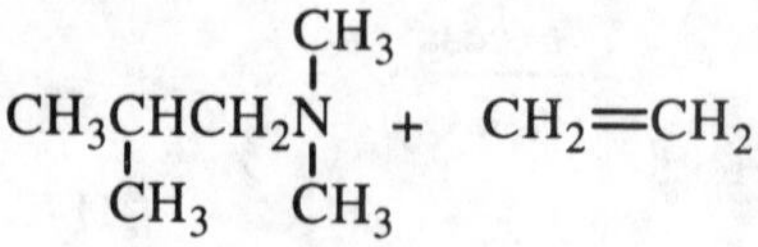 $CH_3\underset{CH_3}{C}HCH_2\overset{CH_3}{\underset{CH_3}{N}} \quad + \quad CH_2{=}CH_2$

e. 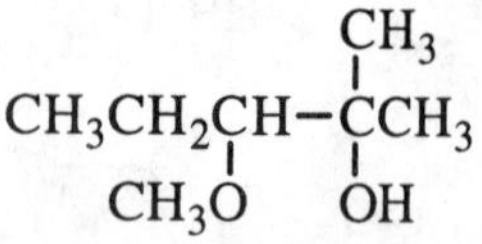 $CH_3CH_2\underset{CH_3O}{C}H{-}\overset{CH_3}{\underset{OH}{C}}CH_3$

l. 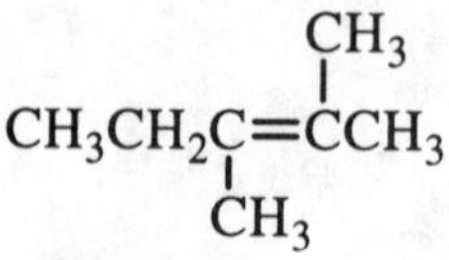 $CH_3CH_2\underset{CH_3}{C}{=}\overset{CH_3}{C}CH_3$

m. 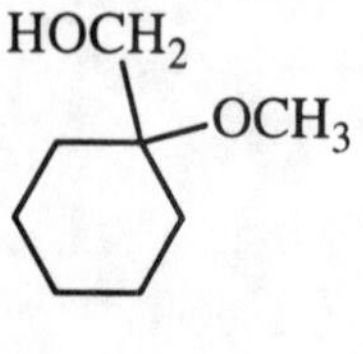 $HOCH_2$, OCH_3

f. $CH_3CH{=}CHCH_3$

n. 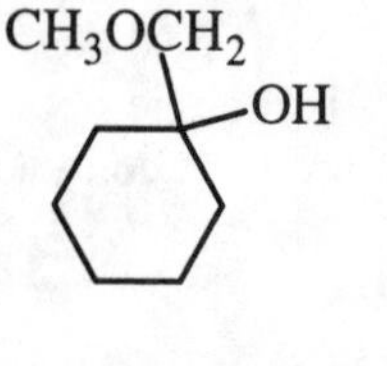 CH_3OCH_2, OH

g. 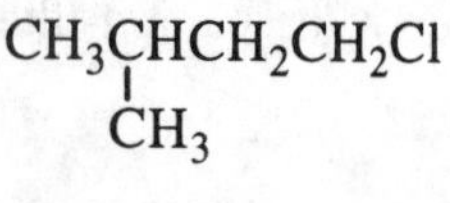 $CH_3\underset{CH_3}{C}HCH_2CH_2Cl$

38.

a. $C_6H_5CH_2CH_2OH$

The other alcohol cannot undergo dehydration because its β-carbon is not bonded to a hydrogen.

b. OH (cyclohex-2-en-1-ol)

A secondary allylic carbocation is more stable than a secondary carbocation.

c. H_3C OH (1-methylcyclohexanol)

A tertiary carbocation is more stable than a secondary carbocation.

d. $C_6H_5CH(OH)CH_3$

The rate-limiting step in both dehydrations is carbocation formation: a secondary benzylic carbocation is more stable than a secondary carbocation and, therefore, easier to form.

e. $C_6H_5CH(OH)CH_3$

Secondary alcohols undergo dehydration faster than primary alcohols.

f. $CH_3C(CH_3)(OH)CH_2CH_3$

A tertiary carbocation is more stable than a secondary carbocation.

39. **c** is the only one that can be used to form a Grignard reagent.

The Grignard reagent formed from **a** will be destroyed by reacting with the proton of the alcohol group.

The Grignard reagent formed from **b** will be destroyed by reacting with the proton of the carboxylic acid group.

The Grignard reagent formed from **d** will be destroyed by reacting with the proton of the amino group.

40. Only one S_N2 reaction takes place in reactions "a" and "b", so the product has the inverted configuration compared to the configuration of the reactant.

Two S_N2 reactions take place in reaction "c", so the product has the same configuration compared to the configuration of the reactant.

a. $HO-C(CH_2CH_3)(H)-D$ + $R-SO_2-Cl$ ⟶ $RSO_2O-C(CH_2CH_3)(H)-D$ $\xrightarrow[S_N2]{HO^-}$ (inversion) $D-C(CH_2CH_3)(H)-OH$

(*R*)-1-deuterio-1-propanol ⟶ (*S*)-1-deuterio-1-propanol

b. $HO-C(CH_2CH_3)(H)-D$ + $R-SO_2-Cl$ ⟶ $RSO_2O-C(CH_2CH_3)(H)-D$ $\xrightarrow[S_N2]{CH_3O^-}$ (inversion) $D-C(CH_2CH_3)(H)-OCH_3$

(*R*)-1-deuterio-1-propanol ⟶ (*S*)-1-deuterio-1-methoxypropane

c. $HO-C(CH_2CH_3)(H)-D$ $\xrightarrow[S_N2]{PBr_3}$ (inversion) $D-C(CH_2CH_3)(H)-Br$ $\xrightarrow[S_N2]{CH_3O^-}$ (inversion) $CH_3O-C(CH_2CH_3)(H)-D$

(*R*)-1-deuterio-1-propanol ⟶ (*R*)-1-deuterio-1-methoxypropane

41. 2-hexene and 3-hexene

$$CH_3CH_2CH_2CH_2CH_2CH_2OH \xrightarrow[\Delta]{H_2SO_4} CH_3CH_2CH_2CH_2CH{=}CH_2 \longrightarrow CH_3CH_2CH_2CH_2\overset{+}{C}HCH_3$$

$$CH_3CH_2CH_2CH_2\overset{+}{C}HCH_3 \xrightarrow{-H^+} CH_3CH_2CH_2CH{=}CHCH_3 \text{ (2-hexene)}$$

$$CH_3CH_2CH_2CH{=}CHCH_3 \xrightarrow{+H^+} CH_3CH_2CH_2\overset{+}{C}HCH_2CH_3$$

$$CH_3CH_2CH_2\overset{+}{C}HCH_2CH_3 \xrightarrow{-H^+} CH_3CH_2CH{=}CHCH_2CH_3 + CH_3CH_2CH_2CH{=}CHCH_3$$

3-hexene (cis and trans) + 2-hexene (cis and trans)

42. 2,3-Dimethyl-2-butanol will dehydrate faster because it is a tertiary alcohol, while 3,3-dimethyl-2-butanol is a secondary alcohol.

$CH_3CH(CH_3)-C(CH_3)(OH)CH_3$ 2,3-dimethyl-2-butanol **a tertiary alcohol**

$CH_3CH(OH)-C(CH_3)_2CH_3$ 3,3-dimethyl-2-butanol **a secondary alcohol**

43.

a. N,N-dimethylcyclohexylamine $\xrightarrow{H_2O_2}$ $H_3C-N^{+}(CH_3)(O^-)$-cyclohexyl $\xrightarrow{\Delta}$ cyclohexene + $CH_3N(OH)CH_3$

b. (cyclohexane: H, H, CH_3, OH) $\xrightarrow[\text{pyridine}]{TsCl}$ (H, H, CH_3, OTs) $\xrightarrow{^-C{\equiv}N}$ (H, $C{\equiv}N$, CH_3, H)

c. 1-chlorocyclohexene $\xrightarrow{(CH_3CH_2CH_2)_2CuLi}$ 1-propylcyclohexene ($CH_2CH_2CH_3$)

d. $CH_3CH(CH_3)-C(CH_3)(OH)CH_3$ $\xrightarrow[\Delta]{H_2SO_4}$ $(H_3C)_2C{=}C(CH_3)_2$

44.

a. cyclohexanol (OH) $\xrightarrow[\Delta]{H_2SO_4}$ cyclohexene $\xrightarrow[Pd/C]{H_2}$ cyclohexane

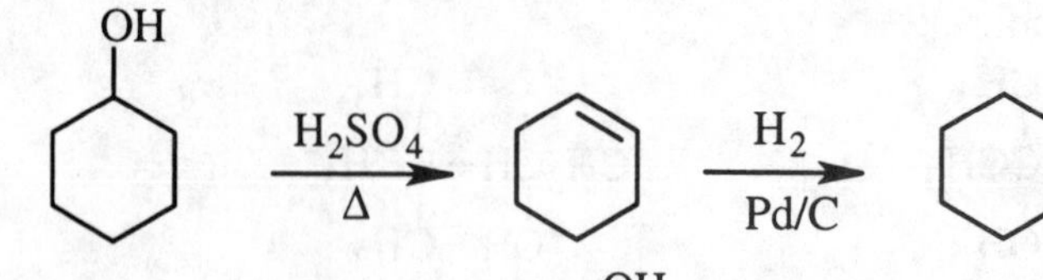

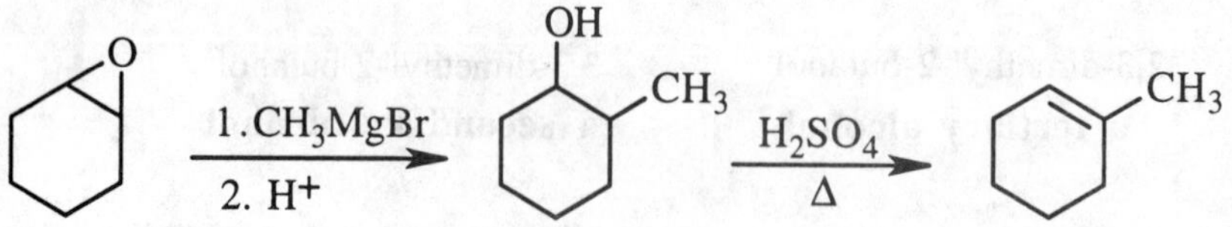

b. cyclohexene oxide (O) $\xrightarrow[2.\ H^+]{1.\ CH_3MgBr}$ 2-methylcyclohexanol (OH, CH_3) $\xrightarrow[\Delta]{H_2SO_4}$ 1-methylcyclohexene (CH_3)

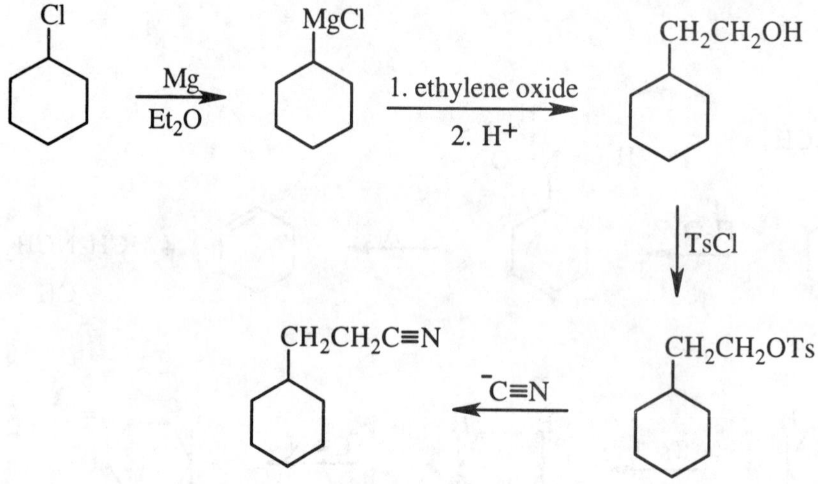

c. chlorocyclohexane (Cl) $\xrightarrow[Et_2O]{Mg}$ cyclohexylmagnesium chloride (MgCl) $\xrightarrow[2.\ H^+]{1.\ \text{ethylene oxide}}$ cyclohexyl–CH_2CH_2OH

$\downarrow$ TsCl

cyclohexyl–CH_2CH_2OTs $\xrightarrow{^-C{\equiv}N}$ cyclohexyl–$CH_2CH_2C{\equiv}N$

d. $CH_3CH_2C{\equiv}CH \xrightarrow{^-NH_2} CH_3CH_2C{\equiv}C^-$

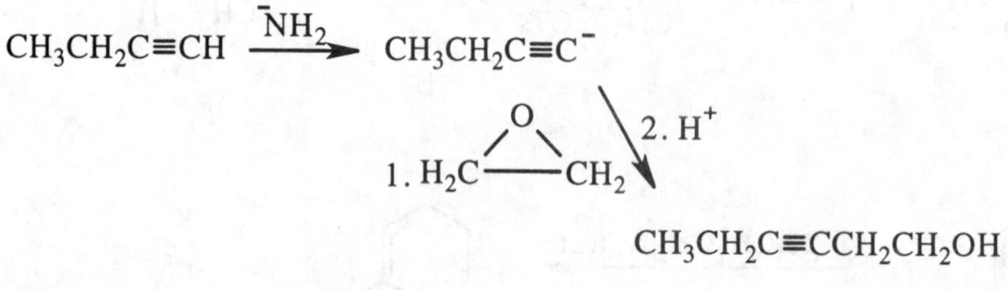

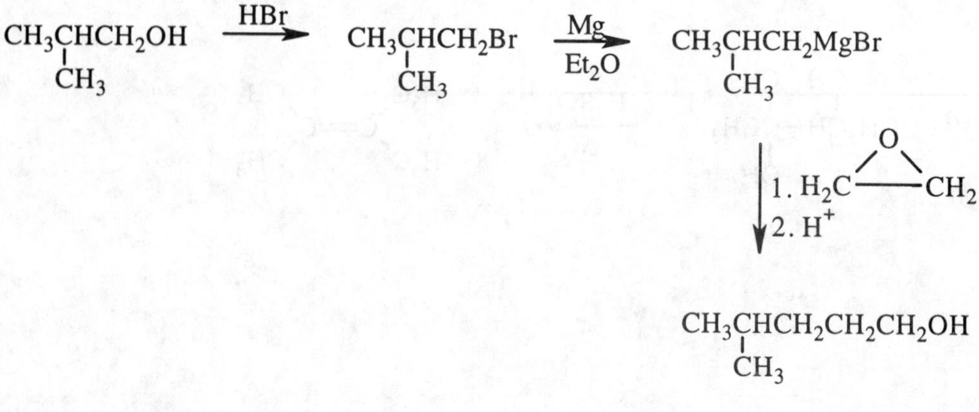

1. H_2C—CH_2 (O) (ethylene oxide)
2. H^+

$CH_3CH_2C{\equiv}CCH_2CH_2OH$

e. $CH_3\underset{\displaystyle CH_3}{CH}CH_2OH \xrightarrow{HBr} CH_3\underset{\displaystyle CH_3}{CH}CH_2Br \xrightarrow[Et_2O]{Mg} CH_3\underset{\displaystyle CH_3}{CH}CH_2MgBr$

1. H_2C—CH_2 (O) (ethylene oxide)
2. H^+

$CH_3\underset{\displaystyle CH_3}{CH}CH_2CH_2CH_2OH$

45.

$CH_3CHCH{-}CH_2$ (epoxide, Cl on C2) + $CH_3\ddot{O}:^-$ ⟶ $CH_3CHCH{-}CH_2OCH_3$ (with $:\ddot{O}:^-$ and Cl) ⟶ $CH_3CH{-}CHCH_2OCH_3$ (epoxide) + Cl^-

46. a. If an NIH shift occurs, both carbocations will form the same keto-intermediate. Because it is about four times easier to break a carbon-hydrogen bond (k_3) compared with a carbon-deuterium bond (k_4), about 80% of the deuterium will be retained.

H OH D + k_1 H_2O H O D k_2 H_2O H OH D +

NIH shift D O H NIH shift

D OH k_3 k_4 H OH

b. If an NIH shift does not occur, 50% of the deuterium will be retained because the epoxide can open equally easily in either direction, and k_1 is equal to k_2.

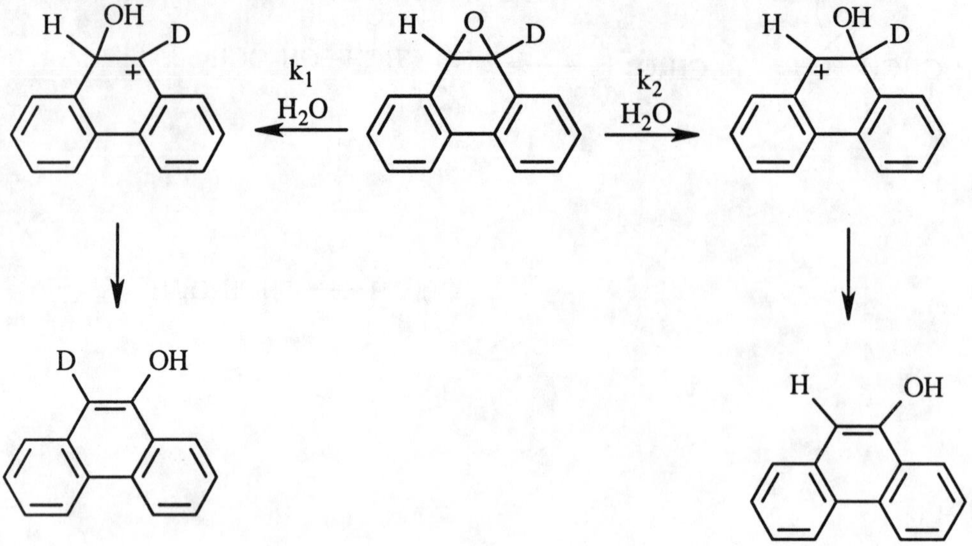

H OH D + $\xleftarrow[H_2O]{k_1}$ H O D $\xrightarrow[H_2O]{k_2}$ H OH D +

↓ ↓

D OH H OH

47. 3-Methyl-2-butanol is a secondary alcohol and, therefore, will undergo an S_N1 reaction. The carbocation intermediate that is formed rearranges to a more stable tertiary carbocation.

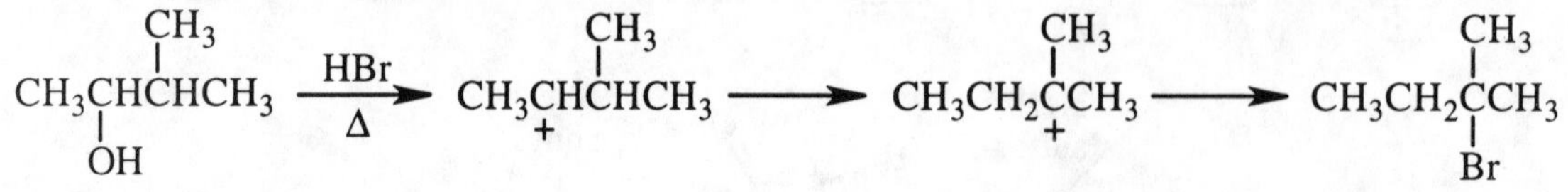

$$\underset{\text{OH}}{\overset{\text{CH}_3}{\text{CH}_3\text{CHCHCH}_3}} \xrightarrow[\Delta]{\text{HBr}} \overset{\text{CH}_3}{\underset{+}{\text{CH}_3\text{CHCHCH}_3}} \longrightarrow \overset{\text{CH}_3}{\underset{+}{\text{CH}_3\text{CH}_2\text{CCH}_3}} \longrightarrow \underset{\text{Br}}{\overset{\text{CH}_3}{\text{CH}_3\text{CH}_2\text{CCH}_3}}$$

3-methyl-2-butanol

2-Methyl-1-propanol is a primary alcohol and therefore will undergo an S_N2 reaction. Because carbocations are not formed in S_N2 reactions, carbocation rearrangement cannot occur.

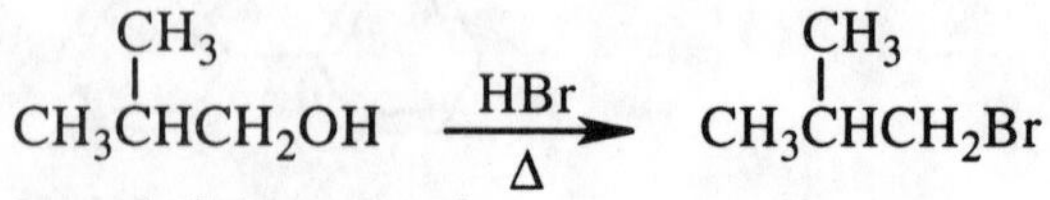

$$\overset{\text{CH}_3}{\text{CH}_3\text{CHCH}_2\text{OH}} \xrightarrow[\Delta]{\text{HBr}} \overset{\text{CH}_3}{\text{CH}_3\text{CHCH}_2\text{Br}}$$

2-methyl-1-propanol

48.

H_3C, CH_3, CH_3, OH (cycloheptane) $\xrightarrow[\Delta]{H_2SO_4}$ H_3C, CH_3, CH_3, + $\longrightarrow$ CH_3, +, CH_3, CH_3 $\longrightarrow$ CH_3, CH_3, CH_3, +

$\downarrow$ $-H^+$ $\quad$ $\downarrow$ $-H^+$ $\quad$ $\downarrow$ $-H^+$

H_3C, CH_3, CH_3 (cycloheptene) $\quad$ CH_3, $=CH_2$, CH_3 $\quad$ CH_3, CH_3, CH_3 (cyclohexene)

49.

$$CH_3\underset{CH_3}{\underset{|}{C}}HOH \xrightarrow{Na} CH_3\underset{CH_3}{\underset{|}{C}}HO^-$$

$$CH_3\underset{CH_3}{\underset{|}{C}}HOH \xrightarrow[\Delta]{H_2SO_4} CH_2CH{=}CH_2 \xrightarrow[\text{peroxide}]{HBr} CH_3CH_2CH_2Br \xrightarrow{CH_3\underset{CH_3}{\underset{|}{C}}HO^-} CH_3\underset{CH_3}{\underset{|}{C}}HOCH_2CH_2CH_3$$

isopropyl propyl ether

50. *N*-Methylpiperidine forms 1,4-pentadiene.

N-methylpiperidine (N–CH_3) $\xrightarrow[2.\ Ag_2O]{1.\ CH_3I}$ $\overset{+}{N}$ (CH_3, CH_3) HO^- $\xrightarrow{\Delta}$ $CH_2{=}CHCH_2CH_2CH_2\underset{}{\overset{CH_3}{\overset{|}{N}}}CH_3$

$\downarrow$ 1. CH_3I 2. Ag_2O 3. Δ

$$CH_2{=}CHCH_2CH{=}CH_2$$

2-Methylpiperidine forms 1,5-hexadiene.

1. CH_3I excess; 2. Ag_2O → (N,N-dimethyl-2-methylpiperidinium HO^-) $\xrightarrow{\Delta}$ $CH_2{=}CHCH_2CH_2CH_2CH_2N(CH_3)CH_3$

β-carbon bonded to the greatest number of hydrogens

1. CH_3I 2. Ag_2O 3. Δ

$CH_2{=}CHCH_2CH_2CH{=}CH_2$

3-Methylpiperidine forms 2-methyl-1,4-pentadiene.

1. CH_3I excess; 2. Ag_2O → (N,N-dimethyl-3-methylpiperidinium HO^-) $\xrightarrow{\Delta}$ $CH_2{=}CHCH_2CH(CH_3)CH_2N(CH_3)CH_3$

1. CH_3I 2. Ag_2O 3. Δ

$CH_2{=}CHCH_2C(CH_3){=}CH_2$

4-Methylpiperidine forms 3-methyl-1,4-pentadiene.

1. CH_3I excess; 2. Ag_2O → (N,N-dimethyl-4-methylpiperidinium HO^-) $\xrightarrow{\Delta}$ $CH_2{=}CHCH(CH_3)CH_2CH_2N(CH_3)CH_3$

1. CH_3I 2. Ag_2O 3. Δ

$CH_2{=}CHCH(CH_3)CH{=}CH_2$

51. Cyclopropane does not react with HO^-, because cyclopropane does not contain a leaving group; a carbanion is far too basic to serve as a leaving group. Ethylene oxide reacts with HO^- because ethylene oxide contains an RO^- leaving group.

52. **Diethyl ether** is the ether that would be obtained in greatest yield, because it is a symmetrical ether. Since it is symmetrical, only one alcohol is used in its synthesis. Therefore, it is the only ether that would be formed.

The synthesis of an asymmetrical ether requires two different alcohols. Therefore, the asymmetrical ether is one of three different ethers that can be formed.

53.

a. $HOCH_2CH_2CH_2CH_2OH$ ⇌ (H^+ / $-H^+$) $HO^+(H)-CH_2CH_2CH_2CH_2\ddot{O}H$ → protonated tetrahydrofuran ($\overset{+}{O}H$) + H_2O ⇌ ($-H^+$ / H^+) tetrahydrofuran

b. tetrahydropyran + H—Br →(Δ) protonated tetrahydropyran ($\overset{+}{O}H$) + $:\ddot{Br}:^-$ → $H\ddot{O}CH_2CH_2CH_2CH_2CH_2Br$ →(H—Br, Δ) $HO^+(H)-CH_2CH_2CH_2CH_2CH_2Br$ + $:\ddot{Br}:^-$ → $BrCH_2CH_2CH_2CH_2CH_2Br$

54.

a. cyclohexene $\xrightarrow[\Delta]{NBS}$ 3-bromocyclohexene (Br) $\xrightarrow[Et_2O]{(CH_3)_2CuLi}$ 3-methylcyclohexene (CH_3)

b. cyclohexanol (OH) $\xrightarrow{PBr_3}$ bromocyclohexane (Br) $\xrightarrow[Et_2O]{Mg}$ cyclohexylmagnesium bromide (MgBr) $\xrightarrow[2.\ H^+]{1.\ \text{ethylene oxide}}$ cyclohexyl-CH_2CH_2OH

c. $CH_3C(CH_3)(OH)CH_2CH_2CH_3 \xrightarrow[\Delta]{H_2SO_4} CH_3C(CH_3){=}CHCH_2CH_3 \xrightarrow[\text{peroxide}]{HBr} CH_3CH(CH_3)CH(Br)CH_2CH_3$

d. 2-butanol (OH) $\xrightarrow{PBr_3}$ 2-bromobutane (Br) $\xrightarrow[Et_2O]{Mg}$ (MgBr) $\xrightarrow{D_2O}$ (D)

e. $CH_3CH_2CH{=}CH_2 \xrightarrow[\text{peroxide}]{HBr} CH_3CH_2CH_2CH_2Br \xrightarrow{Li} CH_3CH_2CH_2CH_2Li \xrightarrow[Et_2O]{CuI} (CH_3CH_2CH_2CH_2)_2CuLi$

use some of the product of the fisrt step as the reagent for this step

$(CH_3CH_2CH_2CH_2)_2CuLi \xrightarrow{CH_3CH_2CH_2CH_2Br} CH_3CH_2CH_2CH_2CH_2CH_2CH_2CH_3$

55.

$$H_2C\overset{O}{\frown}CH_2 + HO^- \longrightarrow {}^-OCH_2CH_2OH$$

$$^-OCH_2CH_2OH + H_2C\overset{O}{\frown}CH_2 \longrightarrow HOCH_2CH_2OCH_2CH_2O^-$$

$$HOCH_2CH_2OCH_2CH_2O^- + H_2C\overset{O}{\frown}CH_2 \longrightarrow HOCH_2CH_2OCH_2CH_2OCH_2CH_2O^-$$

$$HOCH_2CH_2OCH_2CH_2OCH_2CH_2O^- \xrightarrow{H_2O} HOCH_2CH_2OCH_2CH_2OCH_2CH_2OH + HO^-$$

56.

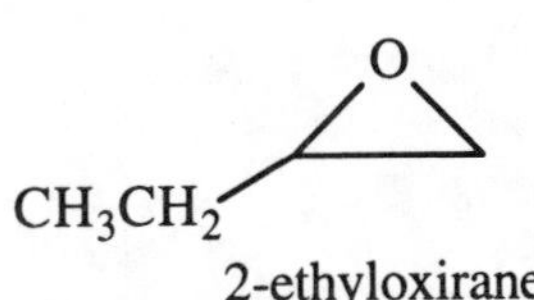

2-ethyloxirane

a. $CH_3CH_2CH(OH)CH_2OH$

0.1 M HCl is a dilute solution of HCl in water

b. $CH_3CH_2CH(OCH_3)CH_2OH$

c. $CH_3CH_2CH(OH)CH_2CH_2CH_3$

d. $CH_3CH_2CH(OH)CH_2OH$

e. $CH_3CH_2CH(OH)CH_2OCH_3$

57. Ethyl alcohol is not obtained as a product, because it reacts with the excess HI and forms ethyl iodide.

$$CH_3CH_2OCH_2CH_3 \xrightarrow[\Delta]{HI} CH_3CH_2I + CH_3CH_2OH$$

$$CH_3CH_2OH \xrightarrow{HI,\ \Delta} CH_3CH_2I + H_2O$$

58.

a.

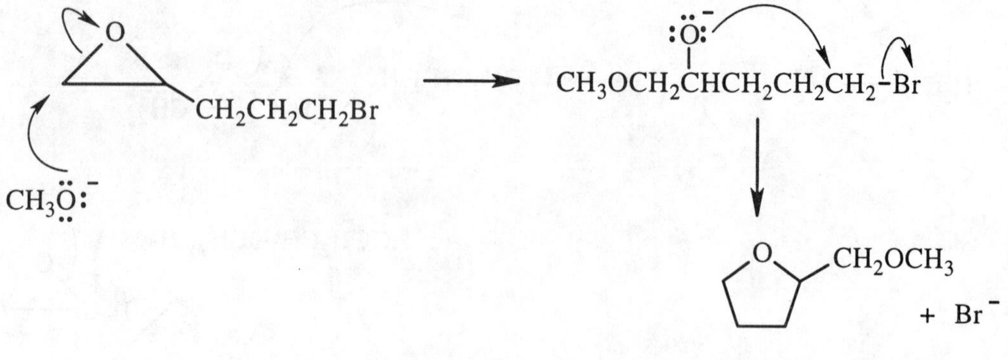

$CH_2CH_2CH_2Br$ (epoxide) + $CH_3\ddot{O}:^-$ → $CH_3OCH_2CH(O^-)CH_2CH_2CH_2-Br$ → (tetrahydrofuran ring) CH_2OCH_3 + Br^-

b.

$CH_2CH_2CH_2Br$ (epoxide) + $CH_3\ddot{O}:^-$ → $^-:\ddot{O}CH_2CH(OCH_3)CH_2CH_2CH_2-Br$ → (tetrahydropyran ring) OCH_3 + Br^-

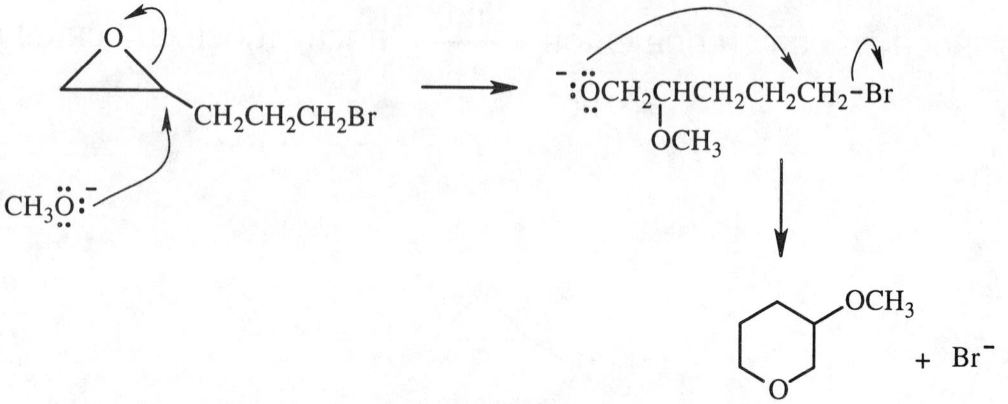

c. The six-membered ring is formed by attack on the more sterically hindered carbon of the epoxide. Attack on the less sterically hindered carbon is preferred.

59.

A = Mg	**D** = ethylene oxide	**G** = ethylene oxide
B = diethyl ether	**E** = H^+	**H** = H^+
C = CH_3MgBr	**F** = Na	

60. Greg did not get any of the expected product, because the Grignard reagent reacted with the hydrogen of the alcohol group. Addition of HCl/H_2O protonated the alkoxide ion and opened the epoxide ring.

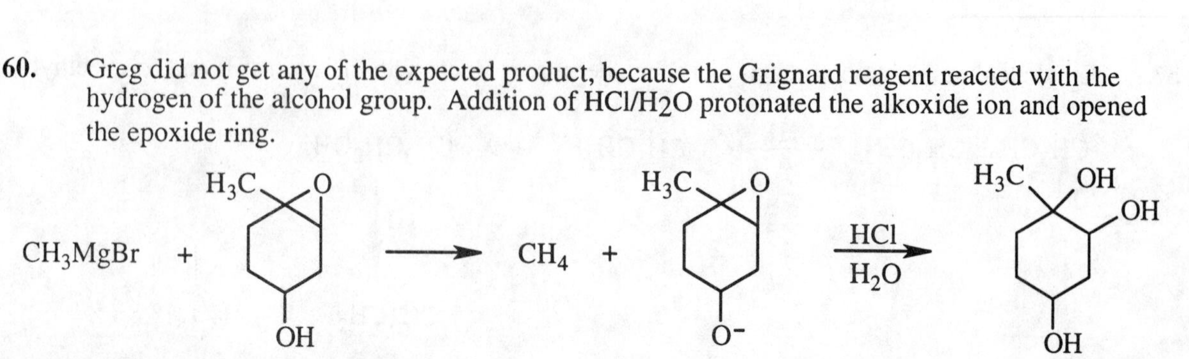

CH_3MgBr + (H_3C, O epoxide; OH) → CH_4 + (H_3C, O epoxide; O^-) $\xrightarrow[H_2O]{HCl}$ (H_3C, OH; OH; OH)

61. **A** is the substitution product that is formed by methoxide ion attacking a carbon of the three-membered ring and eliminating the amino group, thereby opening the ring.

B is the product of a Hofmann elimination reaction: methoxide ion removes a proton from a methyl group bonded to a carbon, eliminating the amino group. The red color disappears when Br_2 is added to **B**, because Br_2 adds to the double bond.

When the aziridinium ion reacts with methanol, only the substitution product is formed.

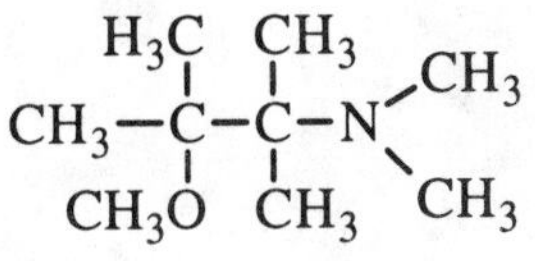

$CH_3-C(CH_3)(OCH_3)-C(CH_3)_2-N(CH_3)_2$

A

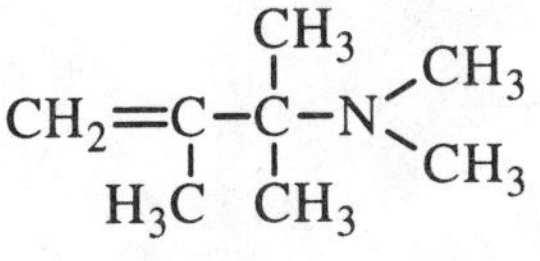

$CH_2=C(CH_3)-C(CH_3)_2-N(CH_3)_2$

B

62.

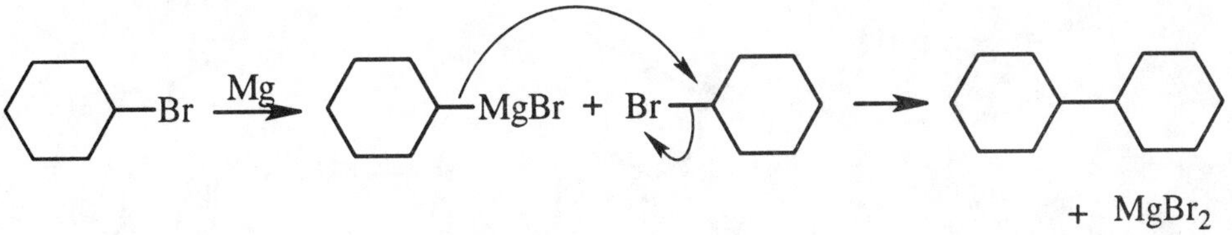

Br, Mg, MgBr + Br, + $MgBr_2$

63.

a. H^+, $-H^+$, H, O+, OH, +, HO, H_2O, H, OH, +, $-H^+$, H^+, OH, HO

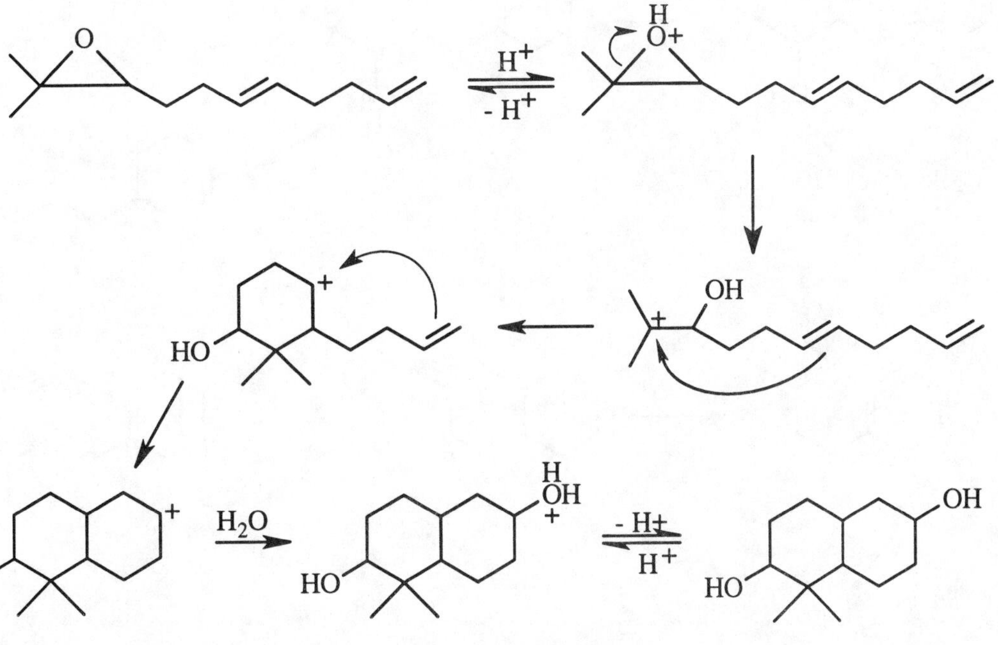

b. :ÖH → H^+ → +OH H → + + H_2O → + ↔ + H H → H^+ +

c. OH → H^+ → +OH H → + + H_2O → + → + → - H^+

64.

a. Br—Br; $H_2\ddot{O}$:; + Br^-; $\overset{+}{O}H_2$; HO^-; + H_2O; HO^-; + Br^-; + H_2O

b. Only the trans isomers will be formed because the epoxide undergoes backside attack by methoxide ion.

OH, OCH_3; CH_3O, OH

65. **c** is faster than **b,** and **b** is faster than **a.**

In order to form the epoxide, the alkoxide ion must attack the backside of the carbon that is bonded to Br. This means that the OH and Br substituents must both be in axial positions. To be diaxial, they must be trans to each other.

a does not form an epoxide, because the OH and Br substituents are cis to each other.

b and **c** can form epoxides because the OH and Br substituents are trans to each other. The rate of the reaction is given by $k'K_{eq}$, where k' is the rate constant for the elimination reaction, and K_{eq} is the equilibrium constant for the equatorial/axial conformers.

When the OH and Br substituents are in the required diaxial position, the large *tert*-butyl substituent is in the axial position in **b** and in the equatorial position in **c**.

Because the more stable conformer is the one with the large *tert*-butyl group in the equatorial position, the more stable conformer of **c** has the OH and Br substituents in the required diaxial position (K_{eq} is large), while those substituents in **b** are in the required diaxial position in its less stable conformer (K_{eq} is small). Therefore, **c** reacts faster than **b**.

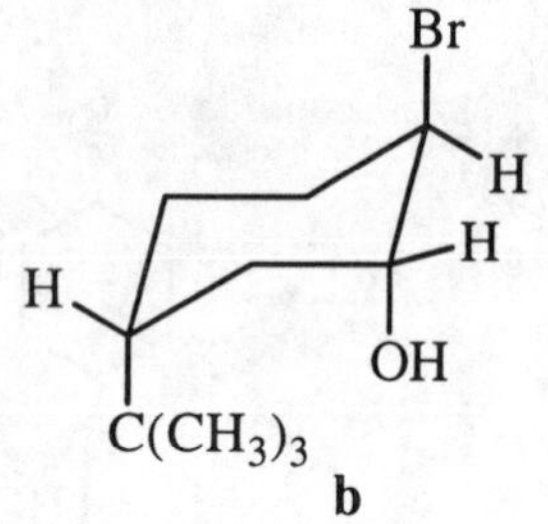

Br, H, H, OH, H, $C(CH_3)_3$ — **b**

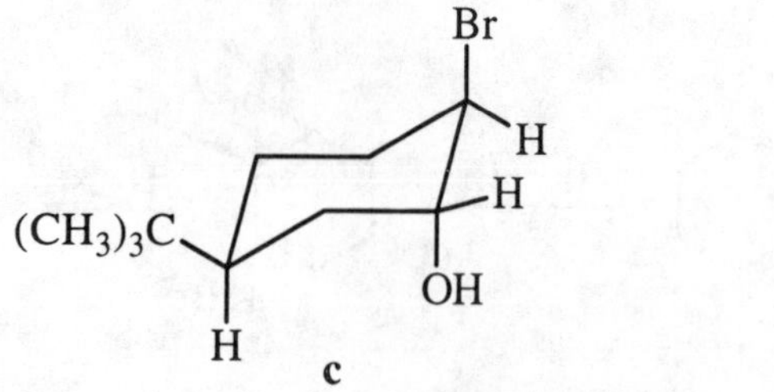

Br, H, H, $(CH_3)_3C$, OH, H — **c**

66.

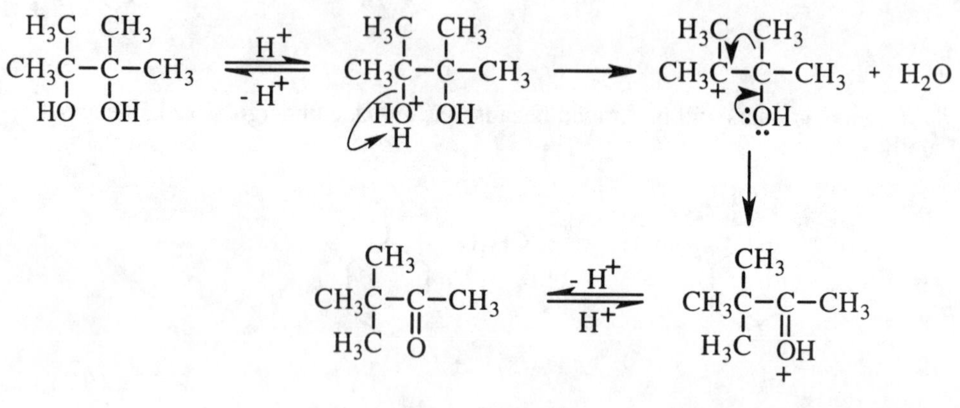

$$CH_3C(CH_3)(OH)-C(CH_3)(OH)-CH_3 \underset{-H^+}{\overset{H^+}{\rightleftharpoons}} CH_3C(CH_3)(\overset{+}{O}H_2)-C(CH_3)(OH)-CH_3 \longrightarrow CH_3\overset{+}{C}(CH_3)-C(CH_3)(\ddot{O}H)-CH_3 + H_2O$$

$$\downarrow$$

$$CH_3C(CH_3)(CH_3)-C(=O)-CH_3 \underset{H^+}{\overset{-H^+}{\rightleftharpoons}} CH_3C(CH_3)(CH_3)-C(=\overset{+}{O}H)-CH_3$$

67.

a. 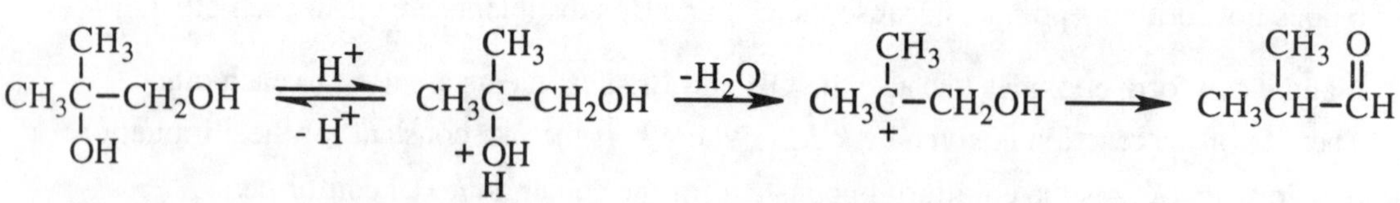 $$CH_3C(CH_3)(OH)-CH_2OH \underset{-H^+}{\overset{H^+}{\rightleftharpoons}} CH_3C(CH_3)(\overset{+}{O}H_2)-CH_2OH \xrightarrow{-H_2O} CH_3\overset{+}{C}(CH_3)-CH_2OH \longrightarrow CH_3CH(CH_3)-CH{=}O$$

b. Dehydration of the primary alcohol group cannot occur because it cannot lose water via an E1 pathway, because a primary carbocation cannot be formed. It cannot lose water via an E2 pathway, because the β-carbon is not bonded to a hydrogen.

68.

OH, OH; $\underset{-H^+}{\overset{H^+}{\rightleftharpoons}}$ OH, $^+OH_2$; $\xrightarrow{\Delta}$:ÖH, +; + H_2O; +OH; $\underset{H^+}{\overset{-H^+}{\rightleftharpoons}}$ O

69.

a. HO OH, H_3C, CH_3 $\xrightarrow{H^+}$ HO $^+OH_2$, H_3C, CH_3 $\longrightarrow$ CH_3, HÖ, CH_3, + or CH_3, HÖ, +, CH_3

$\downarrow$ HO^+, CH_3, CH_3 $\underset{H^+}{\overset{-H^+}{\rightleftharpoons}}$ O, CH_3, CH_3

$\downarrow$ HO^+=C, CH_3, CH_3 $\underset{H^+}{\overset{-H^+}{\rightleftharpoons}}$ O=C, CH_3, CH_3

b. OH, CH=CH_2 $\xrightarrow{H^+}$ ÖH, CH—CH_3 (+) $\longrightarrow$ ^+OH, CH_3 $\underset{H^+}{\overset{-H^+}{\rightleftharpoons}}$ O, CH_3

Chapter 11 Practice Test

1. Which of the following reagents is the best one to use in order to convert methyl alcohol into methyl bromide?

   Br^- HBr Br_2 NaBr Br^+

2. Which of the following reagents is the best one to use in order to convert methyl alcohol into methyl chloride?

   Cl_2/CH_2Cl_2 $Cl_2/h\nu$ Cl^- $SOCl_2$ NaCl

3. Indicate how the following compounds could be prepared using the given starting material.

   a. $(H)(H)C{=}C(CH_3)(Br) \longrightarrow (H)(H)C{=}C(CH_3)(CH_2CH_3)$

   b. cyclopentyl–$CH_2OH \longrightarrow$ cyclopentyl–$CH_2CH_2CH_2OH$

4. Give two names for the following compound.

   (epoxide ring with one carbon bearing CH_2CH_3 and CH_2CH_3)

5. a. What would be the major product obtained from the reaction of the epoxide in the above problem in methanol containing O.1 M HCl?

   b. What would be the major product obtained from the reaction of the epoxide in the above problem in methanol containing O.1 M $NaOCH_3$?

6. Give the major product that is obtained when each of the following alcohols is heated in the presence of H_2SO_4.

   a. $CH_3CH_2C(CH_3)(OH){-}CH(CH_3)CH_3$

   b. $CH_3CH_2CH_2CH(OH){-}C(CH_3)_2CH_3$

   c. $CH_3CH_2CH_2CH_2CH_2OH$

   d. (cyclohexene ring with OH substituent)

7. Indicate whether the following statements are true or false.

a. Tertiary alcohols are easier to dehydrate than secondary alcohols. T F

b. Alcohols are more acidic than thiols. T F

c. Alcohols have higher boiling points than thiols. T F

8. What products would be obtained from heating the following ethers with one equivalent of HI?

a. 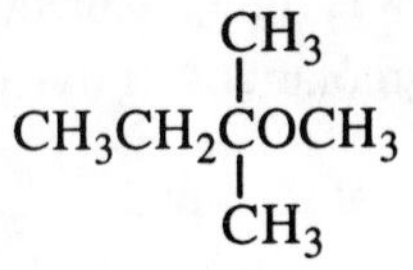 $CH_3CH_2C(CH_3)_2OCH_3$

b. 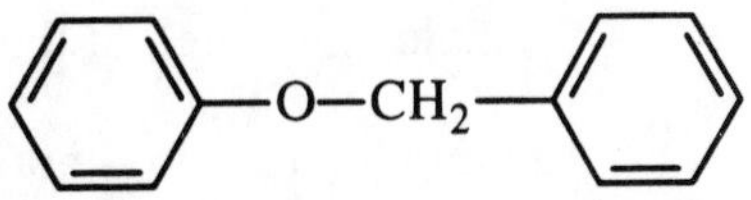 C_6H_5–O–CH_2–C_6H_5

9. What alcohols would be formed from the reaction of ethylene oxide with the following Grignard reagents?

a. $CH_3CH_2CH_2MgBr$

b. 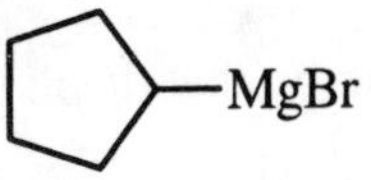 cyclopentyl–MgBr

10. Give the major product of each of the following reactions.

a. 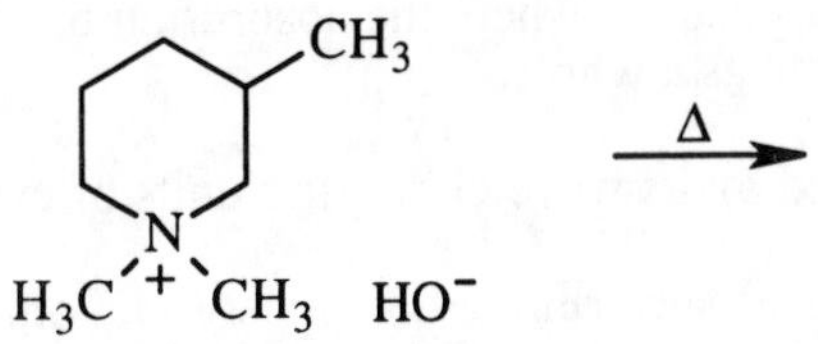 3-methyl-N,N-dimethylpiperidinium (H_3C, CH_3 on N^+) HO^- $\xrightarrow{\Delta}$

b. 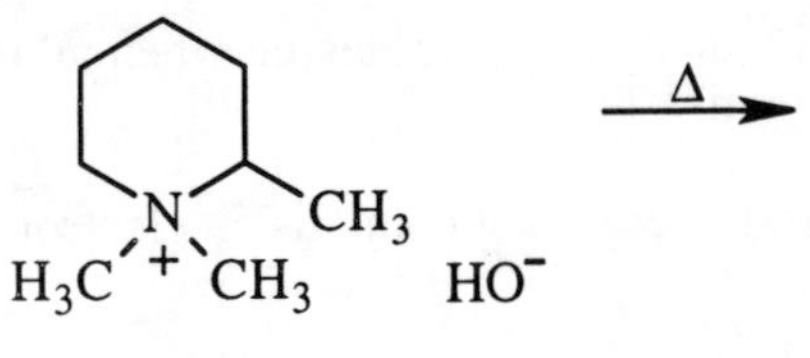 2-methyl-N,N-dimethylpiperidinium (H_3C, CH_3 on N^+) HO^- $\xrightarrow{\Delta}$

c. 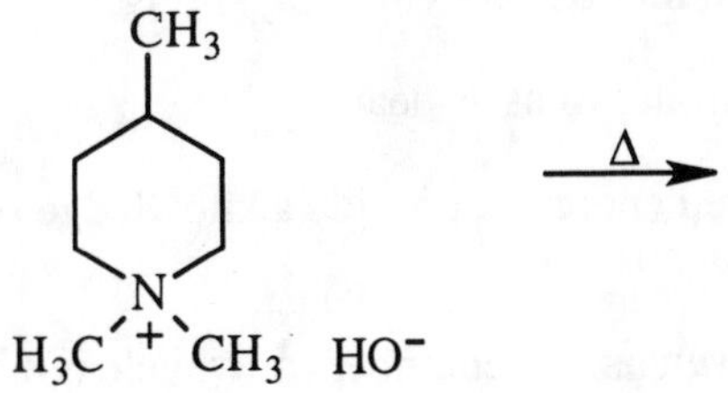 4-methyl-N,N-dimethylpiperidinium (H_3C, CH_3 on N^+) HO^- $\xrightarrow{\Delta}$

CHAPTER 12
Mass Spectrometry, Infrared Spectroscopy, and Ultraviolet/Visible Spectroscopy

Important Terms

absorption band a peak in a spectrum that occurs as a result of absorption of energy.

auxochrome a substituent that, when attached to a chromophore, alters the λ_{max} and the intensity of absorption of UV/Vis radiation.

base peak the peak in a mass spectrum with the greatest intensity.

Beer-Lambert law the relationship between absorbance of UV/Vis light, concentration of the sample, the length of the light path, and the molar absorptivity ($A = cl\varepsilon$).

bending vibration a vibration that does not occur along the line of the bond.

blue shift a shift to a shorter wavelength.

chromophore the part of a molecule responsible for a UV or visible spectrum.

α-cleavage homolytic cleavage of an alpha substituent.

electromagnetic radiation radiant energy that displays wave properties.

electronic transition promotion of an electron from its HOMO to its LUMO.

excited state the electronic configuration of an atom or a molecule that results when an electron in the ground-state electronic configuration has been moved to a higher-energy orbital.

fingerprint region the right-hand third of an IR spectrum where the absorption bands are characteristic of the compound as a whole.

frequency the velocity of a wave divided by its wavelength (has units of cycles/s).

functional group region the left-hand two thirds of an IR spectrum, where most functional groups show absorption bands.

ground state the electronic configuration of an atom or a molecule when all the electrons are in their lowest-energy orbitals.

highest occupied molecular orbital (HOMO) the molecular orbital of highest energy that contains an electron.

Hooke's law an equation that describes the motion of a vibrating spring.

infrared radiation electromagnetic radiation familiar to us as heat.

infrared spectroscopy spectroscopy that uses infrared energy to provide a knowledge of the functional groups in a compound.

infrared spectrum (IR spectrum) a plot of relative absorption versus wavenumber (or wavelength) of absorbed infrared radiation.

λ_{max}	the wavelength at which there is maximum UV/Vis absorbance.
lowest unoccupied molecular orbital (LUMO)	the molecular orbital of lowest energy that does not contain an electron.
mass spectrometry	provides a knowledge of the molecular weight and certain structural features of a compound.
mass spectrum	a plot of the relative abundance of the positively charged fragments produced in a mass spectrometer versus their *m/z* values.
McLafferty rearrangement	rearrangement of the molecular ion of a ketone that contains a γ-hydrogen; the bond between the α- and β-carbons breaks, and a γ-hydrogen migrates to the oxygen.
molar absorptivity	the absorbance obtained from a 1.00 M solution in a cell with a 1.00 cm light path.
molecular ion (M)	the peak in a mass spectrum with the largest *m/z*.
nominal molecular mass	mass to the nearest whole number.
radical cation	a species with a positive charge and an unpaired electron.
red shift	a shift to a longer wavelength.
spectroscopy	study of the interaction of matter and electromagnetic radiation.
stretching frequency	the frequency at which a stretching vibration occurs.
stretching vibration	a vibration occurring along the line of the bond.
ultraviolet light	electromagnetic radiation with wavelengths ranging from 180 to 400 nm.
UV/Vis spectroscopy	the absorption of electromagnetic radiation to determine information about conjugated systems.
visible light	electromagnetic radiation with wavelengths ranging from 400 to 780 nm.
wavelength	distance from any point on one wave to the corresponding point on the next wave.
wavenumber	the number of waves in one centimeter.

Solutions to Problems

1. The peak at $m/z = 57$ will be more intense for 2,2-dimethylpropane than for 2-methylbutane or for pentane. That is because the peak at $m/z = 57$ is due to loss of a methyl group: loss of a methyl group from 2,2-dimethylpropane results in the formation of a tertiary carbocation, while loss of a methyl group from 2-methylbutane and pentane results in the formation of a secondary and primary carbocation, respectively.

$$\left[CH_3\overset{CH_3}{\underset{CH_3}{C}}CH_3\right]^{+\cdot} \longrightarrow CH_3\overset{CH_3}{\underset{+}{C}}CH_3 \quad \dot{C}H_3$$

2,2-dimethylpropane → $m/z = 57$, a tertiary carbocation

$$\left[CH_3\overset{CH_3}{C}HCH_2CH_3\right]^{+\cdot} \longrightarrow CH_3\overset{+}{C}HCH_2CH_3 \quad \dot{C}H_3$$

2-methylbutane → $m/z = 57$, a secondary carbocation

$$\left[CH_3CH_2CH_2CH_2CH_3\right]^{+\cdot} \longrightarrow CH_3CH_2CH_2\overset{+}{C}H_2 \quad \dot{C}H_3$$

pentane → $m/z = 57$, a primary carbocation

Note that the mass spectrum of 2-methylbutane can be distinguished from those of the other isomers by the peak at $m/z = 43$. The peak at $m/z = 43$ will be most intense for 2-methylbutane because such a peak is due to loss of an ethyl group, which results in the formation of a secondary carbocation. Pentane gives a less intense peak at $m/z = 43$ because loss of an ethyl group from pentane forms a primary carbocation. 2,2-Dimethylpropane cannot form a peak at $m/z = 43$, because it does not have an ethyl group.

$$\left[CH_3\overset{CH_3}{C}HCH_2CH_3\right]^{+\cdot} \longrightarrow CH_3\overset{CH_3}{\underset{+}{C}}H \quad CH_3\dot{C}H_2$$

$m/z = 43$

$$\left[CH_3CH_2CH_2CH_2CH_3\right]^{+\cdot} \longrightarrow CH_3CH_2\overset{+}{C}H_2 \quad CH_3\dot{C}H_2$$

$m/z = 43$

2. Peaks should occur at $m/z = 57$ for loss of an ethyl group (86-29), and at $m/z = 71$ for loss of a methyl group (86-15).

$$\left[CH_3CH_2\overset{CH_3}{C}HCH_2CH_3\right]^{+\cdot} \longrightarrow CH_3CH_2\overset{CH_3}{\underset{+}{C}}H \quad \dot{C}H_2CH_3$$

$m/z = 57$

$$\left[CH_3CH_2\overset{CH_3}{C}HCH_2CH_3\right]^{+\cdot} \longrightarrow CH_3CH_2\overset{+}{C}HCH_2CH_3 \quad \dot{C}H_3$$

$m/z = 71$

A secondary carbocation is formed in both cases. Because an ethyl radical is more stable than a methyl radical, the base peak will most likely be at $m/z = 57$.

3. Solved in the text.

4. **a.** A alkane has an even-mass molecular ion. If a CH_2 group (14) of an alkane is replaced by an NH group (15), or if a CH_3 group (15) of an alkane is replaced by an NH_2 group (16), the molecular ion will have an odd mass. A second nitrogen in the molecular ion will cause it to have an even mass. Thus, for a molecular ion to have an odd mass, an odd number of nitrogens must be present.

b. An even-mass molecular ion either has no nitrogens or has an even number of nitrogens.

5.

$$\text{number of carbon atoms} = \frac{\text{relative intensity of M + 1 peak}}{0.011 \text{ x relative intensity of M peak}}$$

$$= \frac{3.81}{0.011 \text{ x } 43.27} = \frac{3.81}{0.467} = 8 \text{ carbon atoms}$$

6. The calculated exact masses show that only C_6H_{14} has an exact mass of 86.10955.

C_6H_{14}

$$\begin{aligned} 6(12.00000) &= 72.00000 \\ 14(1.007825) &= \underline{14.10955} \\ & 86.10955 \end{aligned}$$

$C_4H_{10}N_2$

$$\begin{aligned} 4(12.00000) &= 48.00000 \\ 10(1.007825) &= 10.07825 \\ 2(14.0031) &= \underline{28.0064} \\ & 86.08465 \end{aligned}$$

$C_4H_6O_2$

$$\begin{aligned} 4(12.00000) &= 48.00000 \\ 6(1.007825) &= 6.04695 \\ 2(15.9949) &= \underline{31.9898} \\ & 86.03675 \end{aligned}$$

7. Because the compound contains chlorine, the M + 2 peak is one third the size of the M peak. Loss of a chlorine atom from either the M + 2 peak (80 - 37) or the M peak (78 - 35) gives a peak with *m/z* = 43.

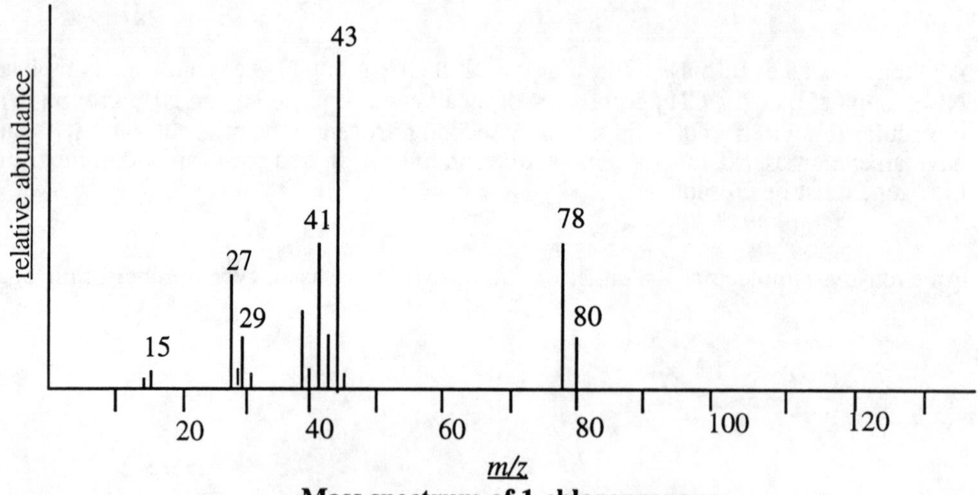

Mass spectrum of 1-chloropropane

8. The base peak at *m/z* = 73, due to loss of a methyl group (88-15), indicates that Figure 12.7a is the mass spectrum of **2-methoxy-2-methylpropane.**

The base peak at *m/z* = 59 (88-29), due to loss of an ethyl group, indicates that Figure 12.7b is the mass spectrum of **2-methoxybutane.**

The base peak at *m/z* = 45 (88-43), due to loss of a propyl group, indicates that Figure 12.7c is the mass spectrum of **1-methoxybutane.**

$CH_3C(CH_3)(OCH_3)CH_3$ — 2-methoxy-2-methylpropane

$CH_3CH_2CH(OCH_3)CH_3$ — 2-methoxybutane

$CH_3CH_2CH_2CH_2OCH_3$ — 1-methoxybutane

9.

$$R{-}CH_2{-}\ddot{O}H \xrightarrow{-e^-} R{-}CH_2{-}\dot{O}^{+}H \longrightarrow R\cdot + CH_2{=}\overset{+}{O}H$$

m/z = 31

10. Both ketones have molecular ions with $m/z = 86$. Therefore, you can determine that both ketones have five carbon atoms and no rings or double bonds other than the double bond of the carbonyl group. Possible ketones are 2-pentanone, 3-pentanone, and 3-methyl-2-butanone.

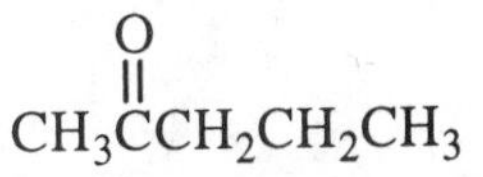

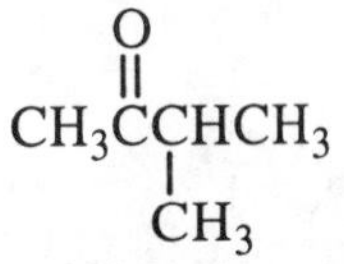

$CH_3C(=O)CH_2CH_2CH_3$ — 2-pentanone

$CH_3CH_2C(=O)CH_2CH_3$ — 3-pentanone

$CH_3C(=O)CH(CH_3)CH_3$ — 3-methyl-2-butanone

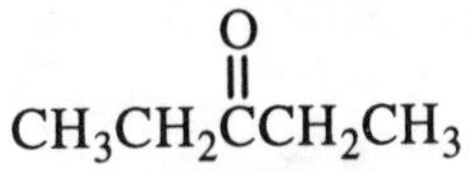

The fact that Figure 12.9a shows a peak at $m/z = 43$ for loss of a propyl group (86-43) indicates that it is the mass spectrum of either 2-pentanone or 3-methyl-2-butanone, since each of these has a propyl group. The fact that the spectrum has a peak at $m/z = 58$, indicating loss of ethene, indicates that the compound has a γ-hydrogen that enables it to undergo a McLafferty rearrangement. Therefore, it must be 2-pentanone, since 3-methyl-2-butanone does not have a γ-hydrogen.

The fact that Figure 12.9b shows a peak at $m/z = 57$ for loss of an ethyl group (86-29), indicates that it is the mass spectrum of 3-pentanone.

11.

a. $CH_3CH_2CH_2CH_2CH_2\ddot{O}H$

$\downarrow -e^-$

$CH_3CH_2CH_2CH_2{-}CH_2{-}\dot{O}^{+}H$ ($m/z = 88$) $\xrightarrow{\alpha\text{-cleavage}}$ $CH_3CH_2CH_2\dot{C}H_2$ + $CH_2{=}\overset{+}{\ddot{O}}H$ ($m/z = 31$)

$CH_3CH_2CH(H)CH_2CH_2{-}\dot{O}^{+}H$ $\longrightarrow$ $CH_3CH_2\dot{C}HCH_2\overset{+}{C}H_2$ ($m/z = 70$) + H_2O

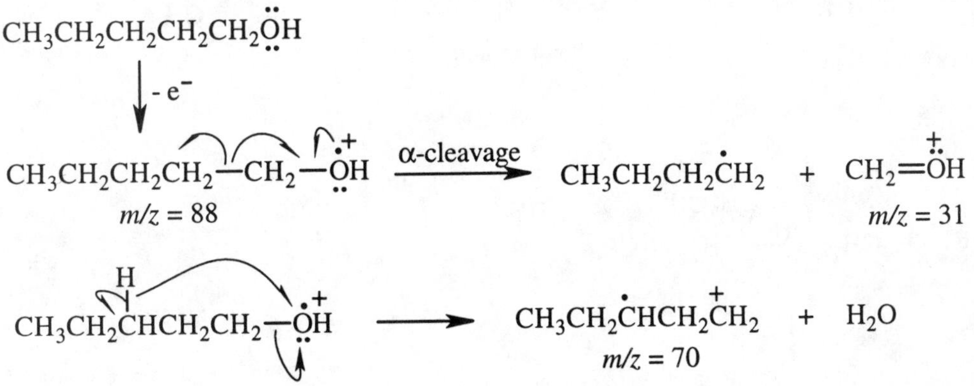

b. $CH_3\overset{\overset{\large\ddot{O}:}{\|}}{C}CH_2CH_2CH_2CH_3$

$\downarrow$ - e^-

$CH_3\text{—}\overset{\overset{\cdot +}{\ddot{O}:}}{\overset{\|}{C}}CH_2CH_2CH_2CH_3 \xrightarrow{\alpha\text{-cleavage}} CH_3CH_2CH_2CH_2C{\equiv}\overset{+}{\ddot{O}} + \dot{C}H_3$

$m/z = 100$ $\quad$ $m/z = 85$

$CH_3\overset{\overset{+\cdot}{:O}}{\overset{\|}{C}}\text{—}CH_2CH_2CH_2CH_2CH_3 \xrightarrow{\alpha\text{-cleavage}} CH_3C{\equiv}\overset{+}{\ddot{O}} + CH_3CH_2CH_2\dot{C}H_2$

$m/z = 43$

$CH_3\overset{\overset{+\cdot}{:O}}{\overset{\|}{C}}CH_2\text{—}CH_2\text{—}\overset{H}{\overset{|}{C}}HCH_3 \xrightarrow[\text{rearrangement}]{\text{McLafferty}} CH_3\overset{\overset{+}{:OH}}{\overset{\|}{C}}\dot{C}H_2 + CH_2{=}CHCH_3$

$m/z = 58$

c. $CH_3\overset{CH_3}{\underset{CH_3}{C}}\text{—}\ddot{\underset{\cdot\cdot}{Br}}: \xrightarrow{-\,e^-} CH_3\overset{CH_3}{\underset{CH_3}{C}}\text{—}\overset{\cdot +}{\underset{\cdot\cdot}{Br}}: \longrightarrow CH_3\overset{CH_3}{\underset{CH_3}{C}}+ \quad + \quad :\dot{\underset{\cdot\cdot}{Br}}:$

$m/z = 136$ and 138 $\quad$ $m/z = 57$

d. $CH_3CH_2\text{—}\ddot{\underset{\cdot\cdot}{O}}\text{—}\overset{CH_2CH_3}{\underset{CH_3}{C}}CH_2CH_2CH_3$

$\downarrow$ -e^-

$CH_3CH_2\text{—}\overset{+\cdot}{\underset{\cdot\cdot}{O}}\text{—}\overset{CH_2CH_3}{\underset{CH_3}{C}}CH_2CH_2CH_3 \longrightarrow CH_3CH_2\text{—}\dot{\underset{\cdot\cdot}{O}}: + \; {}^{+}\overset{CH_2CH_3}{\underset{CH_3}{C}}CH_2CH_2CH_3$

$m/z = 144$ $\quad$ $m/z = 99$

$CH_3CH_2\text{—}\overset{+\cdot}{\underset{\cdot\cdot}{O}}\text{—}\overset{CH_2CH_3}{\underset{CH_3}{C}}CH_2CH_2CH_3 \xrightarrow{\alpha\text{-cleavage}} CH_3CH_2\text{—}\overset{+}{\underset{\cdot\cdot}{O}}{=}\overset{CH_2CH_3}{C}CH_2CH_2CH_3 + \dot{C}H_3$

$m/z = 129$

$$CH_3CH_2-\overset{+\bullet}{\ddot{O}}-\underset{CH_3}{\overset{CH_2CH_3}{C}}CH_2CH_2CH_3 \xrightarrow{\alpha\text{-cleavage}} CH_3CH_2-\overset{+}{\ddot{O}}=\underset{CH_3}{C}CH_2CH_2CH_3 + CH_3\dot{C}H_2$$

$m/z = 115$

$$CH_3CH_2-\overset{+\bullet}{\ddot{O}}-\underset{CH_3}{\overset{CH_2CH_3}{C}}-CH_2CH_2CH_3 \xrightarrow{\alpha\text{-cleavage}} CH_3CH_2-\overset{+}{\ddot{O}}=\underset{CH_3}{C}CH_2CH_3 + CH_3CH_2\dot{C}H_2$$

$m/z = 101$

$$CH_3-CH_2-\overset{+\bullet}{\ddot{O}}-\underset{CH_3}{\overset{CH_2CH_3}{C}}-CH_2CH_2CH_3 \xrightarrow{\alpha\text{-cleavage}} CH_2=\overset{+}{\ddot{O}}-\underset{CH_3}{\overset{CH_2CH_3}{C}}CH_2CH_2CH_3 + \dot{C}H_3$$

$m/z = 129$

e. $$CH_3CH_2\underset{CH_3}{C}H\ddot{\underset{\cdot\cdot}{Cl}}: \xrightarrow{-e^-} CH_3CH_2\underset{CH_3}{C}H-\overset{+\bullet}{Cl}: \longrightarrow CH_3CH_2\underset{CH_3}{\overset{+}{C}}H + :\ddot{\underset{\cdot\cdot}{Cl}}:$$

$m/z = 92$ and 94 $\qquad$ $m/z = 57$

$$CH_3CH_2\underset{CH_3}{C}H-\overset{+\bullet}{Cl}: \xrightarrow{\alpha\text{-cleavage}} CH_3CH_2CH=\overset{+}{Cl}: + \dot{C}H_3$$

$m/z = 77$ and 79

$$CH_3CH_2-\underset{CH_3}{C}H-\overset{+\bullet}{Cl}: \xrightarrow{\alpha\text{-cleavage}} CH_3CH=\overset{+}{Cl}: + CH_3\dot{C}H_2$$

$m/z = 63$ and 65

f. $CH_3CH_2CHCH_2CH_2CH_2CH_3$ (with :ÖH on C3)

$\downarrow$ $-e^-$

$CH_3CH_2CHCH_2CH_2CH_2CH_3$ (with $\dot{O}H^{+}$) $\xrightarrow{\alpha\text{-cleavage}}$ $CH_3CH_2CH_2CH_2CH{=}\overset{+}{O}H + CH_3\dot{C}H_2$

$m/z = 116$ $m/z = 87$

$CH_3CH_2CHCH_2CH_2CH_2CH_3$ (with $\dot{O}H^{+}$) $\xrightarrow{\alpha\text{-cleavage}}$ $CH_3CH_2CH{=}\overset{+}{O}H + CH_3CH_2CH_2\dot{C}H_2$

$m/z = 59$

$CH_2CH_2CHCH_2CH_2CH_2CH_3$ (H on C1, $\dot{O}H^{+}$ on C3) $\longrightarrow$ $\dot{C}H_2CH_2\overset{+}{C}HCH_2CH_2CH_2CH_3 + H_2O$

$m/z = 98$

$CH_3CH_2CHCH_2CHCH_2CH_3$ ($\dot{O}H^{+}$ on C3, H on C5) $\longrightarrow$ $CH_3CH_2\overset{+}{C}HCH_2\dot{C}HCH_2CH_3 + H_2O$

$m/z = 98$

12. When (*Z*)-2-pentene reacts with water and an acid catalyst, 3-pentanol and 2-pentanol are formed. Both alcohols have a molecular weight of 88. (Notice that the first step in solving this problem is to use chemical knowledge to identify the products.) The fact that there is no molecular ion peak is consistent with the fact that the compounds are alcohols.

(*Z*)-2-pentene ($H_3C(H)C{=}C(H)CH_2CH_3$) $+ H_2O \xrightarrow{H_2SO_4}$ $CH_3CH_2CH(OH)CH_2CH_3$ (3-pentanol) $+$ $CH_3CH(OH)CH_2CH_2CH_3$ (2-pentanol)

Figure 12.10a shows a base peak at $m/z = 59$ due to loss of an ethyl group (88-29), indicating that it is the spectrum of 3-pentanol.

Figure 12.10b shows a base peak at $m/z = 45$ due to loss of a propyl group (88-43), indicating that it is the spectrum of 2-pentanol.

13. **a.** 2000 cm^{-1} (The greater the wavenumber, the higher the energy.)

b. 8 μm (The smaller the wavelength, the higher the energy.)

c. 2 μm, because 2 μm = 5000 cm^{-1} (5000 cm^{-1} is a greater wavenumber than 3000 cm^{-1})

14.

the equation used for these calculations: $\tilde{\nu}\ (\text{cm}^{-1}) = \dfrac{10{,}000}{\lambda\ (\mu\text{m})}$

a. $\tilde{\nu}\ (\text{cm}^{-1}) = \dfrac{10{,}000}{4\ (\mu\text{m})} = 2500\ \text{cm}^{-1}$

b. $200\ (\text{cm}^{-1}) = \dfrac{10{,}000}{\lambda\ (\mu\text{m})}$

$$\lambda\ (\mu\text{m}) = \frac{10{,}000}{200\ (\text{cm}^{-1})}$$

$$\lambda\ (\mu\text{m}) = 50\ \mu\text{m}$$

15.

a. **1.** C=O stretch — A double bond is stronger than a single bond, so it takes more energy to stretch a double bond.

2. C—H stretch — It requires more energy to stretch a bond than to bend it.

3. C≡C stretch — A triple bond is stronger than a double bond, so it takes more energy to stretch a triple bond.

b. **1.** C—O — Vibrations of lighter atoms occur at higher frequencies.

2. C—C — Vibrations of lighter atoms occur at higher frequencies.

16. **a.** The carbon-nitrogen stretch of an amide because it has partial double-bond character as a result of electron delocalization.

$$\mathrm{R{-}C({=}O){-}NH_2} \longleftrightarrow \mathrm{R{-}C(O^-){=}\overset{+}{N}H_2}$$

b. The carbon-oxygen stretch of phenol because it has partial double-bond character as a result of electron delocalization.

$$\mathrm{C_6H_5{-}OH} \longleftrightarrow \mathrm{{}^-C_6H_5{=}\overset{+}{O}H}$$

c. The carbon-oxygen double-bond stretch of an amide because it has more double-bond character. Because nitrogen is less electronegative than oxygen, it can better accommodate a positive charge. Therefore, the resonance contributor with the C=N bond makes a greater contribution to the hybrid Than the resonance contributor with the C=O bond.

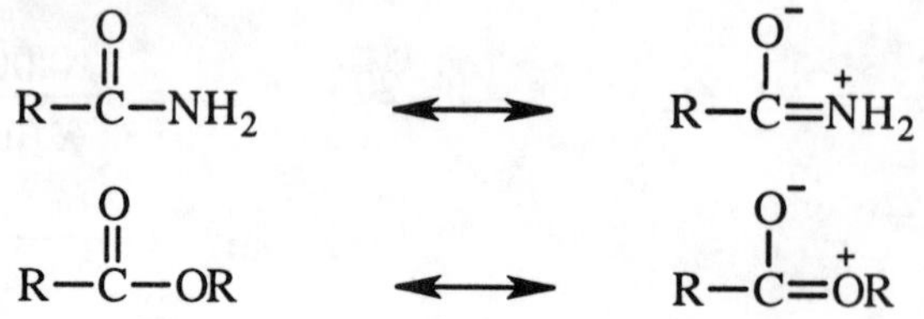

$$R{-}\overset{\overset{\displaystyle O}{\|}}{C}{-}NH_2 \longleftrightarrow R{-}\overset{\overset{\displaystyle O^-}{|}}{C}{=}\overset{+}{N}H_2$$

$$R{-}\overset{\overset{\displaystyle O}{\|}}{C}{-}OR \longleftrightarrow R{-}\overset{\overset{\displaystyle O^-}{|}}{C}{=}\overset{+}{O}R$$

d. The carbon-oxygen stretch, because a bond stretches at a higher frequency than it bends.

17. A carbonyl group bonded to an sp^3 hybridized carbon will exhibit an absorption band at a higher frequency because a carbonyl group bonded to an sp^2 hybridized carbon will have greater single-bond character as a result of electron delocalization.

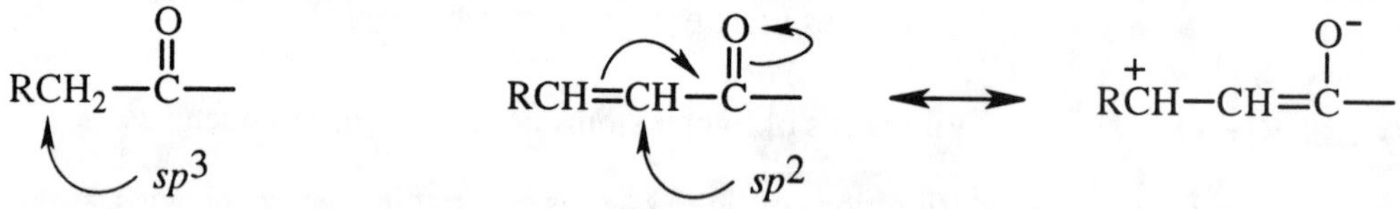

$$\underset{sp^3}{RCH_2}{-}\overset{\overset{\displaystyle O}{\|}}{C}{-} \qquad RCH{=}\underset{sp^2}{CH}{-}\overset{\overset{\displaystyle O}{\|}}{C}{-} \longleftrightarrow R\overset{+}{C}H{-}CH{=}\overset{\overset{\displaystyle O^-}{|}}{C}{-}$$

18. **a.** The C=O absorption band of an ester occurs at the highest frequency because an ester has the most double bond character since the predominant effect of the ester oxygen atom is inductive electron withdrawal.
The C=O absorption band of an amide occurs at the lowest frequency because an amide has the least double bond character since the predominant effect of the amide nitrogen atom is electron donation by resonance.

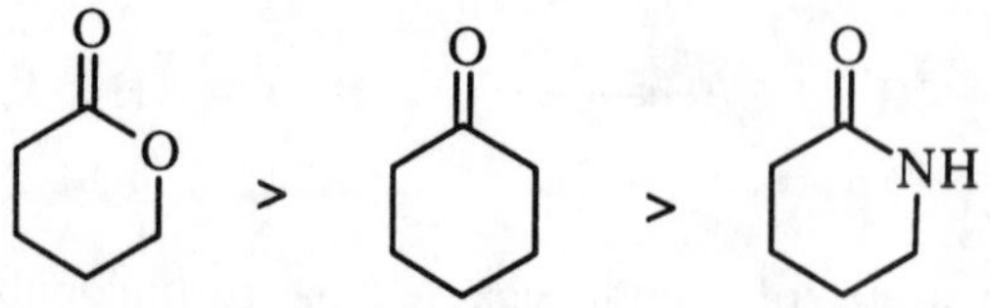

(six-membered lactone) > (cyclohexanone) > (six-membered lactam, NH)

b. The C=O absorption band of the three compounds decreases in the following order.

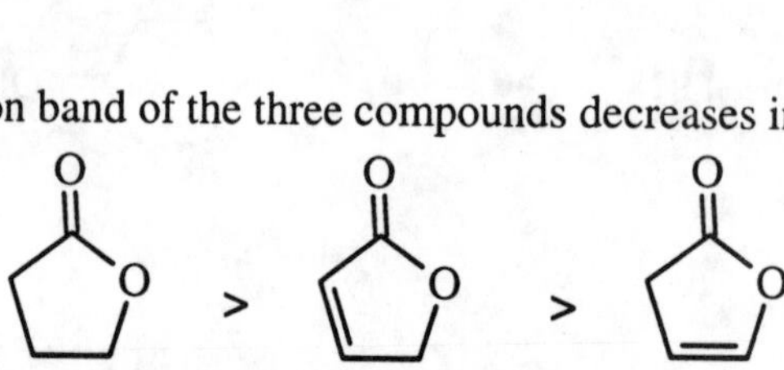

(five-membered lactone) > (five-membered lactone with C=C adjacent to C=O) > (five-membered lactone with C=C adjacent to ring O)

To explain the relative frequency of the C=O absorption, we need to look at the relative stabilities of the resonance contributors because the more stable the resonance contributor with a carbon-oxygen single bond, the lower the carbonyl absorption frequency will be. The resonance contributor without a conjugated double bond is the least stable, and the compound with a π cloud extending over five atoms is the most stable.

relative stabilities of the resonance contributors

O^- / O^+ < O^- / O^+ < O^- / O^+

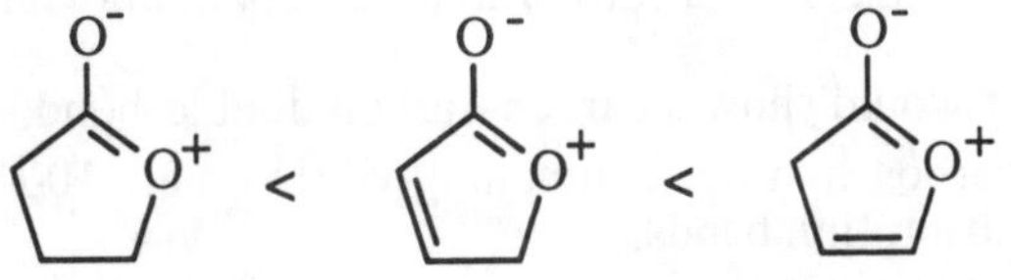

least stable **most stable**

c. Because alkyl groups are more electron-donating than a hydrogen, the absorption band for the carbonyl group bonded to two relatively electron-withdrawing hydrogens has the greatest frequency, while the absorption band for the carbonyl group bonded to two alkyl groups has the lowest frequency.

$$H-\overset{\overset{O}{\|}}{C}-H \quad > \quad CH_3-\overset{\overset{O}{\|}}{C}-H \quad > \quad CH_3-\overset{\overset{O}{\|}}{C}-CH_3$$

19. Ethanol dissolved in carbon disulfide will show the oxygen-hydrogen stretch at a greater wavenumber because there is extensive hydrogen bonding in the undiluted alcohol, and an oxygen-hydrogen bond is easier to stretch if it is hydrogen bonded.

20. **a.** Because oxygen is more electronegative than nitrogen, the O-H stretching vibration is associated with a greater change in dipole moment.

b. The O-H group of a carboxylic acid can form both intermolecular and intramolecular hydrogen bonds, while an alcohol can form only intermolecular hydrogen bonds. Therefore, the extent of hydrogen bonding is greater in a carboxylic acid, and hydrogen bonded OH groups have broader absorption bands.

21. The absorption band at 1700 cm^{-1} indicates that the compound has a carbonyl group.
The absence of an absorption band at 3300 cm^{-1} indicates that the compound is not a carboxylic acid.
The absence of an absorption band at 2700 cm^{-1} indicates that the compound is not an aldehyde.
The absence of an absorption band at 1100 cm^{-1} indicates that the compound is not an ester or an amide.
The compound, therefore, must be a **ketone**.

22. **a.** A primary amine would show a nitrogen-hydrogen stretch at 3500-3300 cm^{-1}, and a tertiary amine would not have this absorption band.

b. A straight-chain ketone would have a methyl substituent and therefore an absorption band at 1385-1365 cm^{-1} that a cyclic ketone would not have.

c. The cis isomer would show a carbon-hydrogen bending vibration at 730-675 cm^{-1}, while the trans isomer would show a carbon-hydrogen bending vibration at 960-980 cm^{-1}.

d. Cyclohexene would show a carbon-carbon double-bond stretching vibration at 1680-1600 cm^{-1} and a carbon-hydrogen stretching vibration at 3100-3020 cm^{-1}. Cyclohexane would not show these absorption bands.

23. **a.** An absorption band at 1150-1050 cm^{-1} would be present for the ether and absent for the alkane.

b. An absorption band at 1780-1650 cm^{-1} would be present for the carboxylic acid and absent for the alcohol.

c. An absorption band at 3300-2500 cm^{-1} would be present for the carboxylic acid and absent for the ester.

d. An absorption band 2960-2850 cm^{-1} would be present for cyclohexene and absent for benzene. (Also, an absorption band at 1500 cm^{-1} would be present for benzene and absent for cyclohexene.)

e. An absorption band at 1385-1365 cm^{-1} would be present for methylcyclohexane and absent for cyclohexane.

24. CO_2, 2-butyne, H_2, Cl_2, ethene

25. The absorption bands in the vicinity of 3000 cm^{-1} indicate that the compound has hydrogens attached to both sp^2 and sp^3 hybridized carbons. The lack of absorption at 1600 cm^{-1} and 1500 cm^{-1} indicates the compound does not have a benzene ring. The sp^2 hydrogens, therefore, must be those of an **alkene**.

The lack of significant absorption at 1600 cm^{-1} indicates it must be an alkene with a relatively small (if any) dipole moment change when the vibration occurs. The absorption band at 965 cm^{-1} indicates the compound is a ***trans*-alkene**.

The molecular ion with m/z = 84 suggests the compound has a molecular formula of C_6H_{12}. The base peak with m/z = 55 indicates the compound most easily loses an ethyl group (84 - 29 = 55), but it can also lose a methyl group (m/z = 69) or a propyl group (m/z = 41). Because it most easily loses an ethyl group, the ethyl group must be attached to an allylic carbon. The compound, therefore, is ***trans*-2-hexene**.

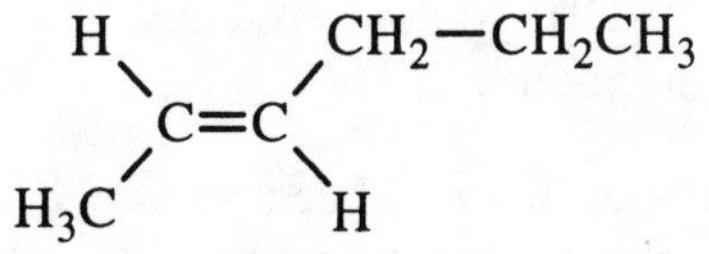

$$\underset{H_3C}{\overset{H}{}}\!\!>C=C<\!\!\underset{H}{\overset{CH_2-CH_2CH_3}{}}$$

trans-2-hexene

26. The absorption band at ~1700 cm^{-1} indicates the compound has a carbonyl group, and the absorption band at ~1600 cm^{-1} indicates the compound has a carbon-carbon double bond. The absorption bands in the vicinity of 3000 cm^{-1} indicate that the compound has hydrogens attached to both sp^2 and sp^3 hybridized carbons. The absorption band at ~1380 cm^{-1} indicates the compound has a methyl.
Because the compound has only four carbon and one oxygen , it must be **methyl vinyl ketone**.

$$CH_3\overset{O}{\overset{\|}{C}}CH{=}CH_2$$

27. a. UV light b. visible light

(See in Figure 12.11 on page 495 of the text.)

28.

$$A = c\ l\ \varepsilon$$

$$c = \frac{A}{l\varepsilon}$$

$$c = \frac{0.52}{12{,}600} = 4.1 \times 10^{-5}\ M$$

29.

a. [cyclohexadienone] > [3-methylcyclohex-2-enone] > [2-methylcyclohex-3-enone]

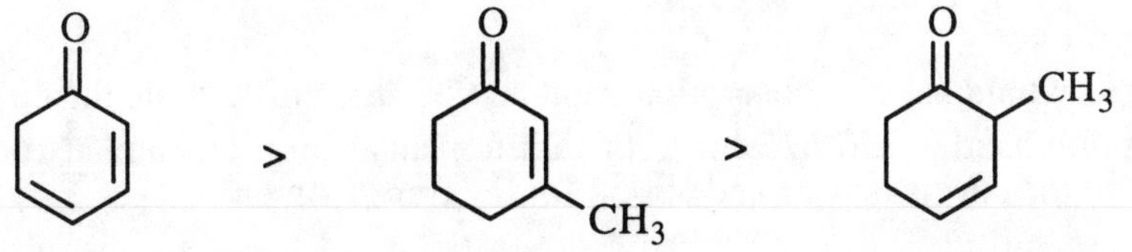

3 conjugated double bonds | 2 conjugated double bonds | isolated double bonds

b. C_6H_5–CH=CH–C_6H_5 > C_6H_5–C_6H_5 > C_6H_5–CH=CH$_2$ > C_6H_6

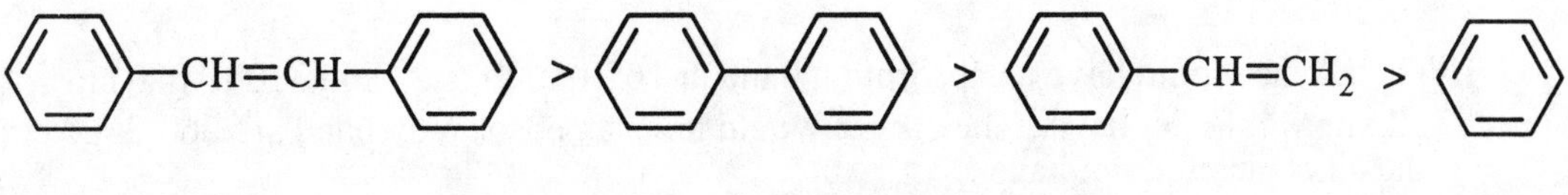

30. **a.** Blue is the result of absorption of longer wavelength light than purple. The compound on the right has an auxochrome ($N(CH_3)_2$) that will cause it to absorb at a longer wavelength. Therefore, it is the blue compound.

b. They will be the same color at pH = 3 because the $N(CH_3)_2$ group will be protonated and, therefore, will not possess the nonbonding pair of electrons that causes the compound to absorb light of a longer wavelength.

31. NADH is formed as a product, and it absorbs light at 340 nm. Therefore, the rate of the oxidation reaction can be determined by monitoring the increase in absorbance at 340 nm as a function of time.

$$CH_3CH_2OH + NAD^+ \xrightarrow{\text{alcohol dehydrogenase}} CH_3\overset{O}{\overset{\|}{C}}H + NADH$$

32. **a.** An absorption band at ~2700 cm^{-1} would be present for the aldehyde and absent for the ketone.

b. An absorption band at 720 cm^{-1} would be present for heptane and absent for methyl-cyclohexane.

c. Absorption bands at 990 cm^{-1} and 910 cm^{-1} would be present for the terminal alkene and absent for the internal alkene.

d. An absorption band at 3500-3300 cm^{-1} would be present for the amide and absent for the ester.

e. Absorption bands at 1600 cm^{-1} and 1500 cm^{-1} and at 3100-3020 cm^{-1} would be present for the compound with the benzene ring and absent for the compound with the cyclohexane ring. An absorption band at 2960-2850 cm^{-1} would be present for the compound with the cyclohexane ring and absent for the compound with the benzene ring.

f. An absorption band at 3650-3200 cm^{-1} would be present for the alcohol and absent for the ether.

g. The trans isomer would have an absorption band at 980-960 cm^{-1}, while the cis isomer would have an absorption band at 730-675 cm^{-1}. In addition, an absorption band at 1680-1600 cm^{-1} would be present for the cis isomer and absent for the trans isomer.

h. An absorption band at ~1250 cm^{-1} would be present for the ester and absent for the ketone.

i. The secondary alcohol would have an absorption band at 1385-1365 cm^{-1} for the methyl group. The primary alcohol does not have a methyl group, so it would not have this absorption band.

j. The alkene would have an absorption band at 1680-1600 cm^{-1} and 3100-3020 cm^{-1} that the alkyne would not have. The alkyne would have an absorption band at 2260-2100 cm^{-1} that the alkene would not have.

33. The molecular ion peak for these compounds is $m/z = 86$; the peak at $m/z = 57$ is due to loss of an ethyl group (86-29), and the peak at $m/z = 71$ is due to loss of a methyl group (86-15).

a. 3-Methylpentane will be more apt to lose an ethyl group (forming a secondary carbocation and a primary radical) than a methyl group (forming a secondary carbocation and a methyl radical), so the peak at $m/z = 57$ would be more intense than the peak at $m/z = 71$.

$$\begin{array}{c} CH_3CH_2\underset{\displaystyle CH_3}{\underset{|}{C}}HCH_2CH_3 \end{array}$$

3-methylpentane

b. 2-Methylpentane has two pathways to lose a methyl group (forming a secondary carbocation and a methyl radical in each pathway), and it cannot form a secondary carbocation by losing an ethyl group. (Loss of an ethyl group would form a primary carbocation and a primary radical.) Therefore, it will be more apt to lose a methyl group than an ethyl group, so the peak at $m/z = 71$ would be more intense than the peak at $m/z = 57$.

$$\begin{array}{c} CH_3\underset{\displaystyle CH_3}{\underset{|}{C}}HCH_2CH_2CH_3 \end{array}$$

2-methylpentane

34. 1. the change in the dipole moment when the bond stretches or bends
2. the number of bonds that give rise to the absorption band
3. the concentration of the sample

35. **a.** The absorption at 236 nm is due to a $\pi \rightarrow \pi^*$ transition, and the absorption at 314 nm is due to an $n \rightarrow \pi^*$ transition.

b. the one at 236 nm (the $\pi \rightarrow \pi^*$ transition)

36. If the reaction had occurred, the intensity of the absorption bands at ~1700 cm^{-1} (due to the carbonyl group) and ~2700 cm^{-1} (due to the aldehyde hydrogen) shown by the reactant would have decreased. If all the aldehyde had reacted, these absorption bands would have disappeared.

37. The compound would have an M and an M + 4 peak of equal intensity, and there would be an M + 2 peak of twice the intensity of the M peak.

38. **a.** At pH < 4, the nonbonding electrons of the nitrogen atom of the $N(CH_3)_2$ group are protonated, so they cannot interfere with the fully conjugated system.

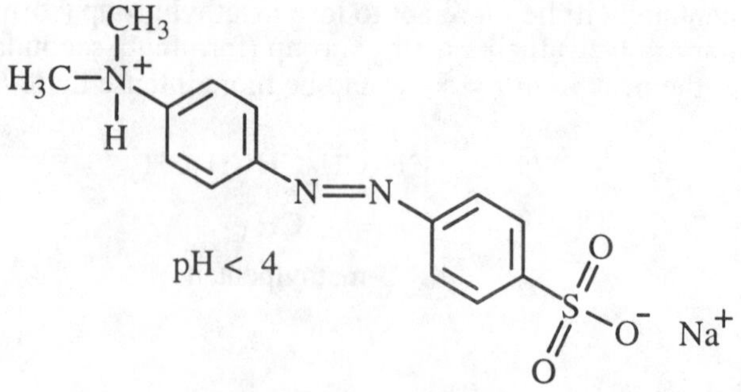

pH < 4

At pH > 4, the nitrogen of the $N(CH_3)_2$ group is not protonated and the nonbonding pair of electrons can be delocalized into the benzene ring. This decreases the conjugation and, therefore, light of shorter wavelengths will be absorbed; hence, the transition from red to yellow.

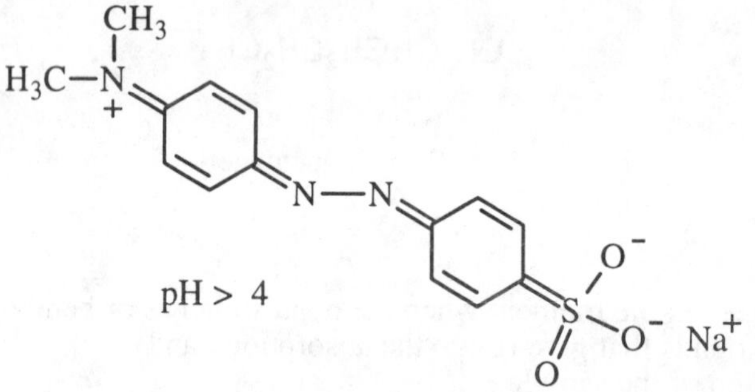

pH > 4

b. In acidic solutions, the three benzene rings are isolated from one another.

In basic solutions, as a result of loss of the proton from the OH groups, there is greater degree of conjugation.

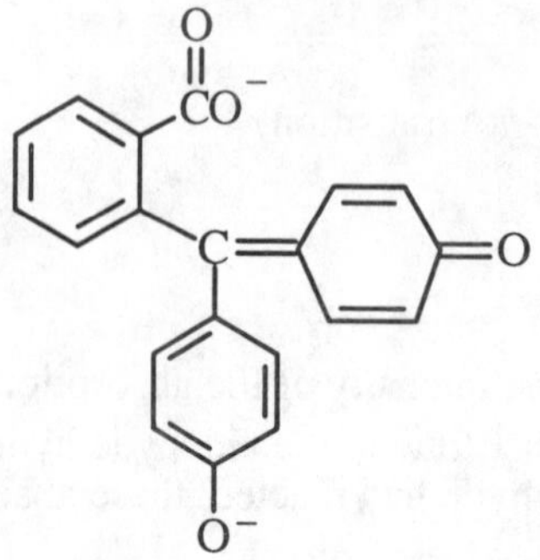

39. If the force constants are approximately the same, the lighter atoms absorb at higher frequencies.

C—C > C—N > C—O

40. The molecular weight of each of the alcohols is 158. The peak at $m/z = 140$ is due to loss of water (158-18); each of the alcohols could show such a peak. The peaks at $m/z = 87$, 115, and 143 are due to loss of a group with five carbons, a group with three carbons, and a methyl group, respectively. Only 2,2,4-trimethyl-4-heptanol would lose all three groups.

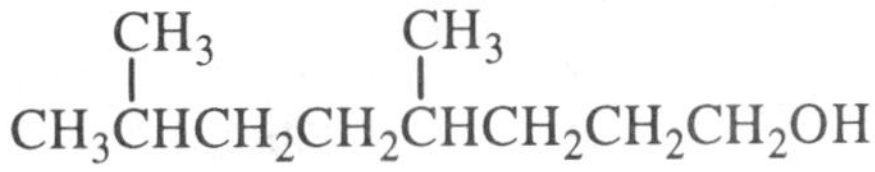

4,7-dimethyl-1-octanol

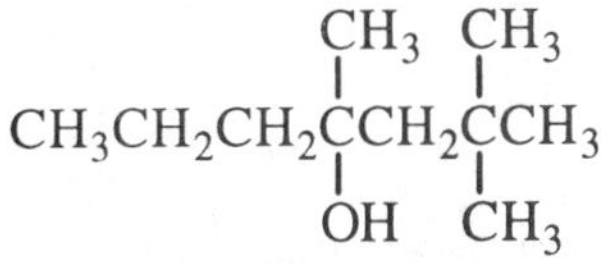

2,2,4-trimethyl-4-heptanol

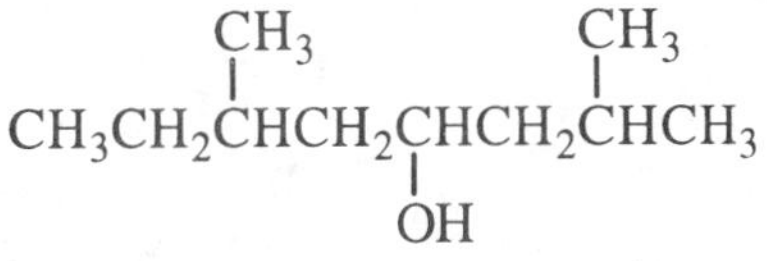

2,6-dimethyl-4-octanol

41.

$CH_2{=}CHCH_2CH_2CH{=}CH_2$

1,5-hexadiene

$CH_3CH{=}CHCH{=}CHCH_3$

2,4-hexadiene

The easiest way to distinguish the two compounds is by the presence or absence of an absorption band at ~1370 cm^{-1} due to the methyl group that 2,4-hexadiene has but that 1,5-hexadiene does not have. In addition, 2,4-hexadiene has conjugated double bonds, and therefore its double bonds have some single-bond character due to resonance. Consequently, they are easier to stretch than the isolated double bonds of 1,5-hexadiene. Thus, the carbon-carbon double bond stretch of 2,4-hexadiene will be at a lower wavenumber than the carbon-carbon double-bond stretch of 1,5-hexadiene.

42.

a. 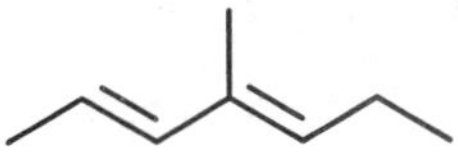 and 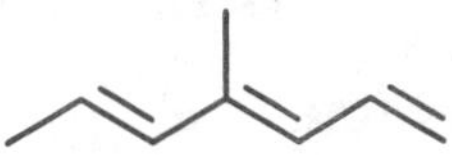

The λ_{max} will be at a longer wavelength.

b. $CH_2{=}CHCH{=}CHCH{=}CH_2$

It will show 1 absorption band.

a $\pi \rightarrow \pi^*$ transition

and

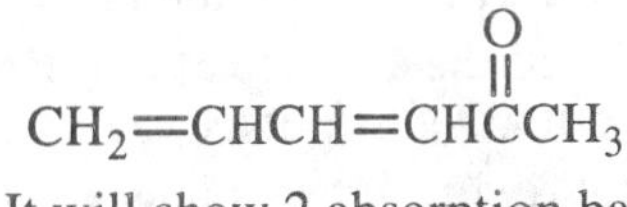

It will show 2 absorption bands.

a $\pi \rightarrow \pi^*$ transition

and

a $n \rightarrow \pi^*$ transition

c. 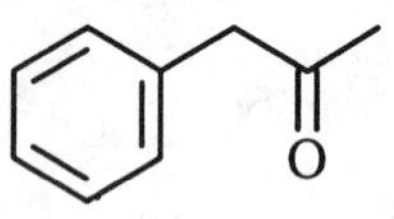 and

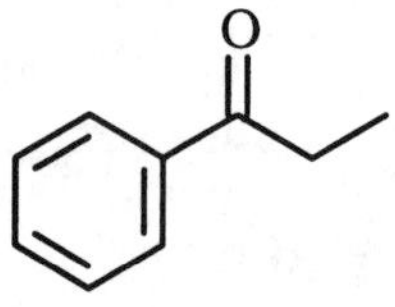

The λ_{max} will be at a longer wavelength because the carbonyl group is conjugated with the benzene ring.

d. 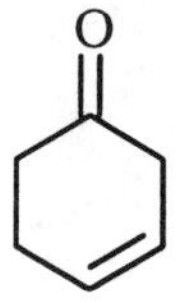 and

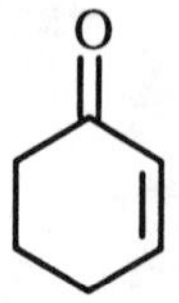

The λ_{max} will be at a longer wavelength because the double bonds are conjugated.

e. 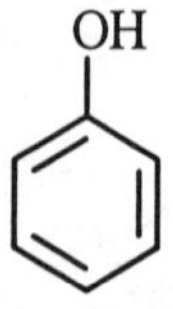 OH and OCH_3

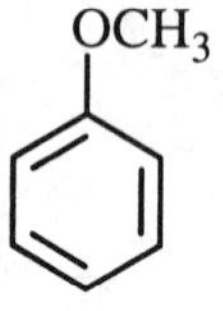

Since phenol has a $pK_a = 10$, the λ_{max} will be at a longer wavelength at pH = 11 than at pH = 7. (The λ_{max} of the phenolate ion is at a longer wavelength than the λ_{max} of phenol.)

The λ_{max} is pH independent.

43. **a.** The absorption band at ~2100 cm^{-1} indicates a carbon-carbon triple bond, and the absorption band at ~3300 cm^{-1} indicates a hydrogen bonded to an *sp* hybridized carbon.

$$CH_3CH_2CH_2CH_2C{\equiv}CH$$

b. The absence of an absorption band at ~2700 cm^{-1} indicates that the compound is not an aldehyde, and the absence of a broad absorption band in the vicinity of 3000 cm^{-1} indicates that the compound is not a carboxylic acid. The ester and the ketone can be distinguished by the absorption band that is present at ~1200 cm^{-1} that indicates the carbon-oxygen single bond of an ester.

$$CH_3CH_2\overset{O}{\overset{\|}{C}}OCH_2CH_3$$

c. The absorption band at ~1360 cm^{-1} indicates the presence of a methyl group.

$C(CH_3)_3$ (attached to benzene ring)

44. A hydrocarbon with eight carbons (C_nH_{2n+2}) has a molecular formula of C_8H_{18} and a molecular weight of 114. Since the molecular ion is 112, the compound must be a hydrocarbon with eight carbons and one double bond or one ring. Possible compounds are cyclooctane, methylcycloheptane, all the dimethylcyclohexanes, ethylcyclohexane, propylcyclopentane, isopropylcyclopentane, all the ethylmethyl cyclopentanes, all the trimethylcyclopentanes, all octenes, all methylheptenes, all ethylhexenes, etc.

45.

3600 cm^{-1} — 3000	3000 — 1800	1800 — 1400	1400 — 1000
OH 3300-3000	*sp*^{3} CH 2950	C=O 1700	C–O 1250-1050
NH 3600-3200	RC(=O)H 2700	C=C 1600	C–N 1230-1030
sp CH 3300	C≡C 2100	benzene ring 1500	
sp^2 CH 3050	C≡N 2250		

46. 4-Methyl-2-pentanone would show peaks at $m/z = 85$ (loss of a methyl group) and at $m/z = 43$ (loss of an isobutyl group) and a peak at $m/z = 58$ due to a McLafferty rearrangement.

$CH_3C(=O)CH_2CH(CH_3)CH_3$

4-methyl-2-pentanone

2-Methyl-3-pentanone would show peaks at $m/z = 71$ (loss of an ethyl group) and at $m/z = 57$ (loss of an isopropyl group). Because it doesn't have a γ-hydrogen, it cannot undergo a McLafferty rearrangement.

$$CH_3CH(CH_3)C(=O)CH_2CH_3$$

2-methyl-3-pentanone

47. The absorption bands at ~2700 cm^{-1} for the aldehydic hydrogen and at ~1380 cm^{-1} for the methyl group would distinguish the compounds.

A would have the band at ~2700 cm^{-1} but not the one at ~1380 cm^{-1}.
B would have the band at ~1380 cm^{-1} but not the one at ~2700 cm^{-1}.
C would have the both the band at ~2700 cm^{-1} and the one at ~1380 cm^{-1}.

48. 1-Hexyne will show absorption bands at ~3300 cm^{-1} for a hydrogen bonded to an *sp* hybridized carbon and at ~2100 cm^{-1} for the triple bond.

$$CH_3CH_2CH_2CH_2C{\equiv}CH$$

1-hexyne

2-Hexyne will show the absorption band at ~2100 cm^{-1} but not the one at ~3300 cm^{-1}.

$$CH_3CH_2CH_2C{\equiv}CCH_3$$

2-hexyne

3-Hexyne will show neither the absorption band at ~3300 cm^{-1} nor the one at ~2100 cm^{-1} (no change in dipole moment).

$$CH_2CH_2C{\equiv}CCH_2CH_3$$

3-hexyne

49. **a.** The broad absorption band at ~3300 cm^{-1} is characteristic of the oxygen-hydrogen stretch of an alcohol, and the absence of absorption bands at ~1600 cm^{-1} and ~3100 cm^{-1} indicates that it is not the alcohol with a carbon-carbon double bond.

$$CH_3CH_2CH_2CH_2OH$$

b. The absorption band at ~1685 cm^{-1} indicates a carbon-oxygen double bond. The absence of a strong and broad absorption band at ~3000 cm^{-1} rules out the carboxylic acid, and the absence of an absorption band at ~2700 cm^{-1} rules out the aldehyde. Thus, it must be the ketone.

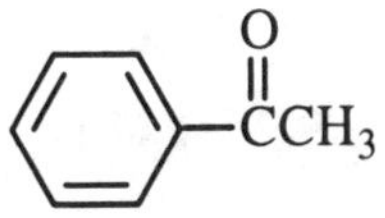

O
||
—CCH$_3$

c. The absorption band at ~1700 cm^{-1} indicates a carbon-oxygen double bond. The absence of an absorption band at ~1600 cm^{-1} rules out the ketones with the benzene or cyclohexene rings. The absence of absorption bands at ~2100 cm^{-1} and ~3300 cm^{-1} rules out the ketone with the carbon-carbon triple bond. Thus, it must be 4-ethylcyclohexanone.

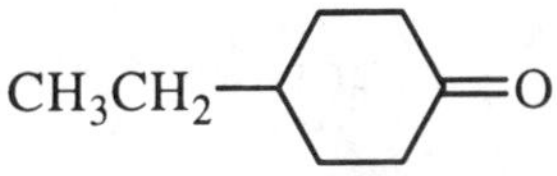

CH_3CH_2— —=O

50. Before the addition of acid, the compound is colorless because the benzene rings are not conjugated with each other, since they are separated from each other by two single bonds. In the presence of acid, a carbocation is formed in which the three benzene rings are conjugated with each other. The conjugated carbocation is highly colored.

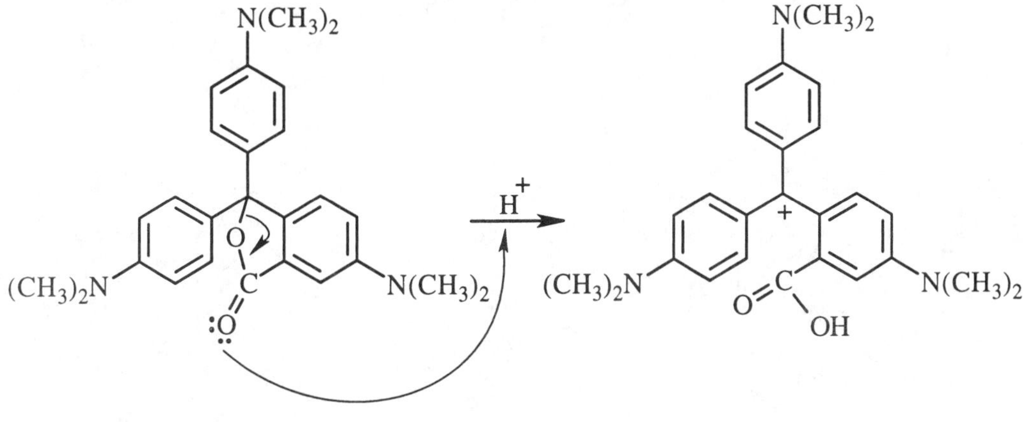

$N(CH_3)_2$ $N(CH_3)_2$

H^+ →

$(CH_3)_2N$ O C $N(CH_3)_2$ $(CH_3)_2N$ + O=C OH $N(CH_3)_2$

:O:

51. **a.** The absorption band at ~2700 cm^{-1} indicates that the compound is an aldehyde (carbon-hydrogen stretch of an aldehyde hydrogen). The absence of an absorption band at ~1600 cm^{-1} rules out the aldehyde with the benzene ring. Thus, it must be the other aldehyde.

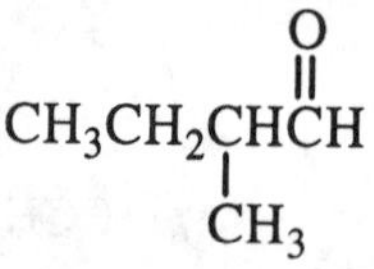

$$CH_3CH_2\underset{\displaystyle CH_3}{\underset{|}{C}}H\overset{\displaystyle O}{\overset{\|}{C}}H$$

b. The absorption bands at ~3350 cm^{-1} and ~3200 cm^{-1} indicate that the compound is an amide (nitrogen-hydrogen stretch). The absence of an absorption band at ~3050 cm^{-1} indicates that the compound does not have hydrogens bonded to sp^2 carbons. Therefore, it is not the amide that has a benzene ring.

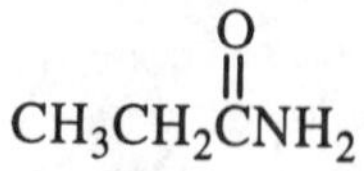

$$CH_3CH_2\overset{\displaystyle O}{\overset{\|}{C}}NH_2$$

c. The absence of absorption bands at ~1600 cm^{-1} and ~1500 cm^{-1} indicates that the compound does not have a benzene ring. Thus, it must be the ketone. This is confirmed by the absence of an absorption band at ~1380 cm^{-1}, indicating that the compound does not have a methyl group. (Notice the first fundamental of the 1700 cm^{-1} absorption band at 3400 cm^{-1}.)

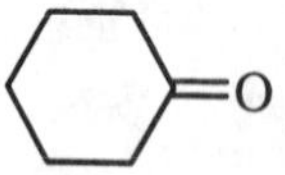

(cyclohexanone structure: cyclohexane ring =O)

52.

a.
$R\overset{O}{\overset{\|}{C}}{-}H$ 2700
$\overset{O}{\overset{\|}{C}}$ 1700
C=C 1600

b.
$\overset{O}{\overset{\|}{C}}$ 1700
(benzene ring) 1500, 1600
C−O ~1250, ~1050

c.
$\overset{O}{\overset{\|}{C}}$ 1700
C−N ~1030
N−H 3500-3300

d.
C−O ~1050
O−H 3600-3200 (broad)

e.
sp CH 3300 (narrow)
C≡C 2100

f.
$\overset{O}{\overset{\|}{C}}$ 1700
$\overset{O}{\overset{\|}{C}}OH$ ~3000 (broad)
C−O ~1250

53. The absorption band at ~1700 cm^{-1} indicates that the compound has a carbonyl group, and the absence of an absorption band at ~1380 cm^{-1} indicates that it has no methyl groups. The absence of an absorption band at ~1600 cm^{-1} indicates the compound does not have a carbon-carbon double bond, and the absence of an absorption band at ~3050 cm^{-1} indicates the compound does not have hydrogens bonded to sp^2 carbons.

From the molecular formula you can deduce that the compound is **cyclopentanone**.

54. Since only benzene absorbs light of 260 nm, the concentration of benzene can be determined by measuring the amount of absorbance at 260 nm, using the Beer-Lambert law, since the length of the light path of the cell is known and the molar absorptivity of benzene at 260 nm can be looked up (or can be determined from the absorbance at 260 nm given by a solution of benzene with a known concentration).

the Beer-Lambert law: $A = c\,l\,\varepsilon$

55. The broad absorption band at ~3300 cm^{-1} indicates that the compound is an OH group. Therefore, the compound is **benzyl alcohol**.

56. The broad absorption band at ~3300 cm^{-1} indicates that the compound has an OH group. The absence of absorption at ~2950 cm^{-1} indicates the compound does not have any hydrogens bonded to sp^3 hybridized carbons. Therefore, the compound is phenol.

57. The M peak at 100 indicates that the compound has a molecular weight of 100, and the relative intensities of the M and M + 1 peaks indicate that the compound has seven carbons. When $n = 7$, $C_nH_{2n+2} = C_7H_{16}$. A compound with that molecular formula has a molecular weight of 100.

$$\text{number of carbon atoms} = \frac{M+1}{0.011 \times M} = \frac{2.10}{0.011 \times 27.32} = \frac{2.10}{0.301} = 7$$

58.

$$\tilde{\nu} = \frac{1}{2 \times 3.1416 \times 3 \times 10^{10}\ \text{cm s}^{-1}} \sqrt{\frac{10 \times 10^{5}\ \text{g s}^{-2} \left(\frac{12}{6.02} + \frac{12}{6.02}\right) \times 10^{-23}\ \text{g}}{\frac{12}{6.02} \times 10^{-23}\ \text{g} \times \frac{12}{6.02} \times 10^{-23}\ \text{g}}}$$

$$\tilde{\nu} = \frac{1}{18.85 \times 10^{10}} \sqrt{10.0 \times 10^{28}}$$

$$\tilde{\nu} = \frac{1}{18.85 \times 10^{10}} \times 3.16 \times 10^{14}$$

$$\tilde{\nu} = 1676\ \text{cm}^{-1}$$

59. **a.** The IR spectrum indicates that the compound is an aliphatic ketone with at least one methyl group. The M peak at m/z = 100 indicates that the ketone is a hexanone. The peak at 43 (100-57) for loss of a butyl group and the peak at 85 for loss of a methyl group (100-15) suggest that the compound is **2-hexanone**.

$CH_3C(=O)CH_2CH_2CH_2CH_3$

2-hexanone

This is confirmed by the peak at 58 for loss of propene (100-42) as a result of a McLafferty rearrangement.

$CH_3C(=\overset{+\bullet}{O}:)CH_2-CH_2-CH(H)CH_3$ $\xrightarrow[\text{rearrangement}]{\text{McLafferty}}$ $CH_3C(=\overset{+}{O}H)\dot{C}H_2$ + $CH_2{=}CHCH_3$

$m/z = 58$

b. The absorption bands at ~1700 cm^{-1} and ~2700 cm^{-1} indicate that the compound is an aldehyde. The molecular ion peak at m/z = 72 indicates that the aldehyde contains four carbons. The peak at m/z = 44 for loss of a group with molecular weight 28 indicates that ethene has been lost as a result of a McLafferty rearrangement.

$H-CH_2-CH_2-CH_2CH(=\overset{\bullet+}{O}:)$ $\xrightarrow[\text{rearrangement}]{\text{McLafferty}}$ $\dot{C}H_2CH(=\overset{+}{O}H)$ + $CH_2{=}CH_2$

$m/z = 44$

A McLafferty rearrangement can occur only if the aldehyde has a γ-hydrogen. The only four-carbon aldehyde that has a γ-hydrogen is **butanal**.

$$CH_3CH_2CH_2\overset{\overset{\displaystyle O}{\|}}{C}H$$

butanal

c. The equal heights of the M and M + 2 peaks at 162 and 164 indicate that the compound contains bromine. The peak at m/z = 83 (162 - 79) is for the carbocation that is formed when the bromine atom is eliminated. The IR spectrum does not indicate the presence of any functional groups, and it shows that there are no methyl groups present. The m/z peak = 83 indicates a carbocation with a molecular formula of C_6H_{11}. The fact that the compound does not contain a methyl group indicates that the compound is **bromocyclohexane**.

bromocyclohexane

60. Calculating the term in Hooke's law that depends on the masses of the atoms joined by a bond for a carbon-hydrogen bond and for a carbon-carbon bond shows why smaller atoms give rise to greater wavenumbers.

<u>for a carbon-hydrogen bond</u>

$$\frac{m_1 + m_2}{m_1 \times m_2} = \frac{12 + 1}{12 \times 1} = \frac{13}{12} = 1.08$$

<u>for a carbon-carbon bond</u>

$$\frac{m_1 + m_2}{m_1 \times m_2} = \frac{12 + 12}{12 \times 12} = \frac{24}{144} = 0.17$$

Chapter 12 Practice Test

1. Give one IR absorption band that could be used to distinguish each of the following pairs of compounds. Indicate the compound for which the band would be present.

a. $CH_3CH_2CH_2\overset{O}{\overset{\|}{C}}H$ and $CH_3CH_2CH_2\overset{O}{\overset{\|}{C}}CH_3$

b. $CH_3CH_2\overset{O}{\overset{\|}{C}}NH_2$ and $CH_3CH_2\overset{O}{\overset{\|}{C}}OCH_3$

c. $CH_3CH_2CH_2CH_2OH$ and $CH_3CH_2CH_2OCH_3$

d. 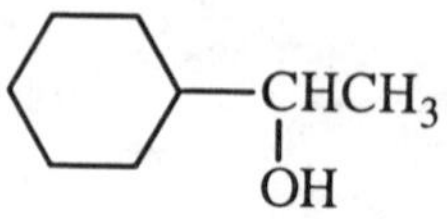 (cyclohexyl)$-\underset{OH}{\underset{|}{C}}HCH_3$ and (cyclohexyl)$-CH_2CH_2OH$

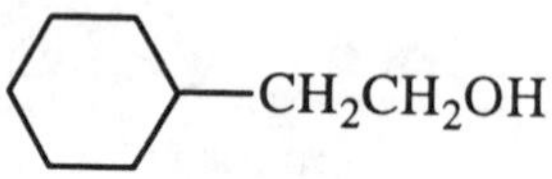

e. $CH_3CH_2CH_2\overset{O}{\overset{\|}{C}}OCH_3$ and $CH_3CH_2CH_2\overset{O}{\overset{\|}{C}}CH_3$

f. $CH_3CH_2CH{=}CHCH_3$ and $CH_3CH_2C{\equiv}CCH_3$

g. $CH_3CH_2C{\equiv}CH$ and $CH_3CH_2C{\equiv}CCH_3$

2. Which compound has the greater λ_{max}? (300 nm is a greater λ_{max} than 250 nm.)

a.  (1,4-cyclohexadiene) or (1,3-cyclohexadiene)

b. 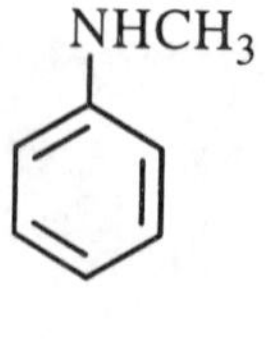 (benzene ring with $NHCH_3$) or (benzene ring with CH_2NH_2)

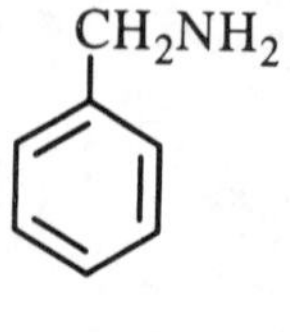

c. 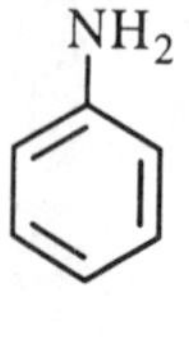 (benzene ring with NH_2) or (benzene ring with $+NH_3$)

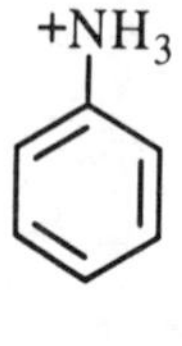

3. A 0.038 M solution of cyclohexanone shows an absorbance of 0.75 at 280 nm in a 1.00 cm cell. What is the molar absorptivity of cyclohexanone at 280 nm?

4. The major peaks shown in the mass spectrum of a tertiary alcohol are at m/z = 73, 87, 98, and 101. Identify the alcohol.

5. How could you distinguish between the IR spectra of the following compounds?

a. $C_6H_{11}CH_2OH$ (cyclohexylmethanol) and $C_6H_5CH_2OH$ (benzyl alcohol)

b. C_6H_5CHO and $C_6H_5COCH_3$

c. C_6H_5COOH and $C_6H_5COOCH_3$

d. $C_6H_5COCH_3$ and $C_6H_5CH_2OCH_3$

e. $C_6H_5CH{=}CH_2$ and $C_6H_5CH_2CH_3$

6. Indicate whether each of the following is true or false.

   a. The O-H stretch of a concentrated solution of an alcohol occurs at a higher frequency than the O-H stretch of a dilute solution. T F

   b. Light of 2 μm is of higher energy than light of 3 μm. T F

   c. It takes more energy for a bending vibration than for a stretching vibration. T F

   d. Propyne will not have an absorption band at 3100 cm^{-1} because there is no change in the dipole moment. T F

   e. Light of 8 μm has the same energy as light of 1250 cm^{-1}. T F

   f. Ultraviolet radiation is higher in energy than infrared radiation. T F

   g. The M + 2 peak of an alkyl chloride is half the height of the M peak. T F

   h. A chromophore exhibiting both $n \rightarrow \pi^*$ and $\pi \rightarrow \pi^*$ transitions will have the $n \rightarrow \pi^*$ transition at a longer wavelength. T F

CHAPTER 13
NMR Spectroscopy

Important Terms

applied magnetic field — the externally applied magnetic field.

chemically equivalent protons — protons with the same connectivity relationship to the rest of the molecule.

chemical shift — location of a signal occurring in an NMR spectrum. It is measured downfield from a reference compound (most often TMS).

^{13}C NMR — nuclear magnetic resonance that shows carbon (^{13}C) nuclei.

COSY spectrum — a 2-D NMR spectrum showing ^{1}H-^{1}H correlations.

coupled protons — protons that split each other. Coupled protons have the same coupling constant.

coupling constant — the distance (in hertz) between two adjacent peaks of a split NMR signal.

DEPT ^{13}C NMR spectrum — a group of four ^{13}C NMR spectra that distinguish CH_3, CH_2, and CH groups.

diamagnetic anisotropy — the term used to describe the greater freedom of π electrons to move in response to a magnetic field as a consequence of their greater polarizability compared with σ electrons.

diamagnetic shielding — shielding by the local magnetic field that opposes the applied magnetic field.

2-D NMR — two-dimensional nuclear magnetic resonance.

doublet — an NMR signal split into two peaks.

doublet of doublets — an NMR signal split into four peaks of approximately equal height. Caused by splitting a signal into a doublet by one hydrogen and into another doublet by another (nonequivalent) hydrogen.

downfield — father to the left-hand side of the spectrum.

effective magnetic field — the magnetic field that a nucleus "senses" through the surrounding cloud of electrons.

Fourier transform NMR (FT-NMR) — a technique in which all the nuclei are excited simultaneously by an rf pulse, their relaxation monitored, and the data mathematically converted to a spectrum.

geminal coupling — the mutual splitting of two nonidentical protons bonded to the same carbon.

gyromagnetic ratio — the ratio of the magnetic moment of a rotating charged particle to its angular momentum.

HETCOR spectrum a 2-D NMR spectrum showing ^{13}C-^{1}H correlations.

^{1}H NMR nuclear magnetic resonance that shows hydrogen nuclei.

high-resolution NMR spectroscopy NMR spectroscopy that uses a spectrometer with a high operating frequency.

long-range coupling splitting of a proton by a proton more than 3 σ bonds away.

magnetic resonance imaging (MRI) NMR used in medicine. The difference in the way water is bound in different tissues produces the signal variation between organs as well as between healthy and diseased states.

methine hydrogen a tertiary hydrogen.

MRI scanner an NMR spectrometer used in medicine for whole-body NMR.

multiplet an NMR signal split by two non-equivalent sets of protons.

multiplicity the number of peaks in an NMR signal.

$N + 1$ rule an ^{1}H NMR signal for a hydrogen with N equivalent hydrogens bonded to an adjacent carbon is split into $N + 1$ peaks. A ^{13}C NMR signal for a carbon bonded to N hydrogens is split into $N + 1$ peaks.

NMR spectroscopy the absorption of electromagnetic radiation to determine the structural features of an organic compound. In the case of ^{1}H NMR spectroscopy, it determines the carbon-hydrogen framework.

operating frequency the frequency at which an NMR spectrometer operates.

proton exchange the transfer of a proton from one molecule to another.

quartet an NMR signal split into four peaks.

reference compound a compound added to the sample whose NMR spectrum is to be taken. The position of the signals in the NMR spectrum are measured from the position of the signal given by the reference compound.

rf radiation radiation in the radiofrequency region of the electromagnetic spectrum.

shielding caused by electron donation to the environment of a proton. The electrons shield the proton from the full effect of the applied magnetic field. The more a proton is shielded, the farther to the right its signal appears in an NMR spectrum.

singlet an unsplit NMR signal.

spin-coupled ^{13}C NMR spectrum a ^{13}C NMR spectrum in which each signal for a carbon is split by the hydrogens bonded to that carbon.

spin-coupling coupling between the carbon nucleus and certain nearby atoms.

spin-spin coupling the splitting of a signal in an NMR spectrum described by the $N + 1$ rule.

α-spin state nuclei in this spin state have their magnetic moments oriented in the same direction as the applied magnetic field.

β-spin state nuclei in this spin state have their magnetic moments oriented opposite to the direction of the applied magnetic field.

splitting diagram (splitting tree) a diagram that describes the splitting of a set of protons.

triplet an NMR signal split into three peaks.

upfield father to the right-hand side of the spectrum.

Solutions to Problems

1.

$$\nu = \frac{\gamma}{2\pi} B_o$$

$$= \frac{26.75 \times 10^7 \text{ rad T}^{-1} \text{ sec}^{-1} \times 1.0 \text{ T}}{2(3.1416) \text{ rad}}$$

$$= 43 \times 10^6 \text{ sec}^{-1}$$

$$= 43 \times 10^6 \text{ Hz} = 43 \text{ MHz}$$

2. **a.**

$$\nu = \frac{\gamma}{2\pi} B_o$$

$$B_o = \frac{\nu \times 2\pi}{\gamma}$$

$$B_o = \frac{360 \times 10^6 \times 2(3.1416)}{26.75 \times 10^7}$$

$$B_o = \frac{226.2}{26.75}$$

$$B_o = 8.46 \text{ T}$$

b.

$$B_o = \frac{\nu \times 2\pi}{\gamma}$$

$$B_o = \frac{500 \times 10^6 \times 2(3.1416)}{26.75 \times 10^7}$$

$$B_o = \frac{314.2}{26.75}$$

$$B_o = 11.75 \text{ T}$$

From these calculations, you can see that the greater the operating frequency of the instrument (360 MHz versus 500 MHz), the more powerful the magnet (8.46 T versus 11.75 T) required to operate it.

3.

a. 4
b. 4
c. 3
d. 3
e. 2
f. 3
g. 1
h. 5
i. 3
j. 3
k. 3
l. 4
m. 2
n. 1
o. 3

4. **a** would give two signals, **b** would give one signal, and **c** would give three signals.

5.

#1

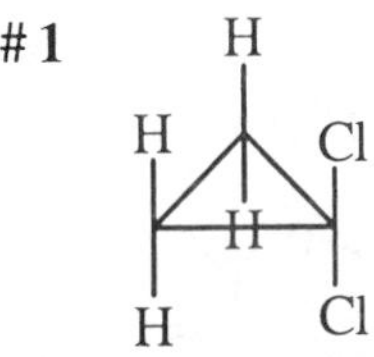

All the H's are equivalent.

#2

H
H Cl
H
Cl H

The H's attached to the front of the molecule are equivalent, and the methylene H's are equivalent.

#3

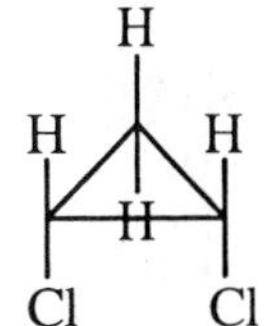

The H's attached to the front of the molecule are equivalent, and the methylene H's are not equivalent.

6.

a. $\frac{120 \text{ Hz}}{300 \text{ MHz}} = 0.4 \text{ ppm}$

b. The answer would still be 0.4 ppm because the chemical shift is independent of the operating frequency of the spectrometer.

c. $\frac{x \text{ Hz}}{100 \text{ MHz}} = 0.4 \text{ ppm}$

$x = 40$

40 Hz downfield from TMS

7. a. The chemical shift is independent of the operating frequency. Therefore, if the two signals differ by **1.5 ppm** in a 300-MHz spectrometer, they will still differ by **1.5 ppm** in a 100-MHz spectrometer.

b. $\frac{\text{Hz}}{\text{MHz}} = \text{ppm}$

$\frac{90 \text{ Hz}}{300 \text{ MHz}} = 0.3 \text{ ppm}$

$\frac{\text{x Hz}}{100 \text{ MHz}} = 0.3 \text{ ppm}$

$\text{x} = 30 \text{ Hz}$

8. Magnesium is less electronegative than silicon. (See Table 11.3 on page 454 of the text.) Therefore, the peak for $(CH_3)_2Mg$ would be upfield from the TMS peak.

9.

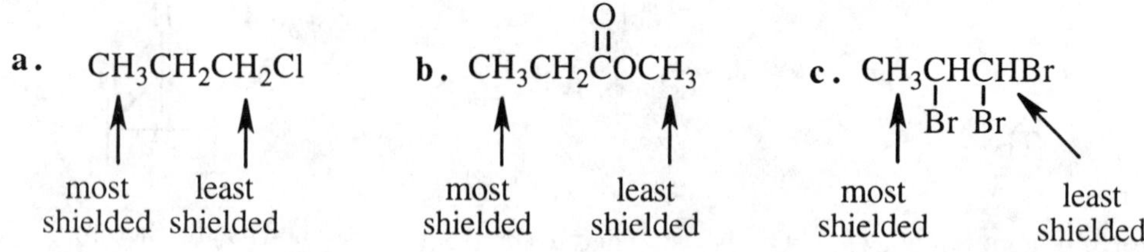

10. The greater the chemical shift (the larger numerical value for the chemical shift), the farther downfield the signal is from the TMS signal.

a. $CH_3CH_2C\underline{H}(Cl)CH_3$

b. $CH_3C\underline{H_2}CH(Cl)CH_3$

c. $CH_3CH_2C\underline{H_2}Cl$

d. $CH_3CH_2CH{=}C\underline{H_2}$

e. $CH_3CH_2C(=O)\underline{H}$

f. $CH_3CH(CH_3)OC\underline{H_3}$

g. $CH_3C\underline{H}(CH_3)C(=O)CH_2CH_3$

h. $CH_3CH(Br)C\underline{H}(Br)Br$

11.

a. $\overset{a}{CH_3}\overset{b}{CH_2}\overset{c}{CH_2}Cl$

b. $\overset{a}{CH_3}CH(\overset{a}{CH_3})\overset{c}{C}H(Cl)\overset{d}{CH}\overset{b}{CH_3}$

c. $\overset{a}{CH_3}\overset{b}{CH_2}\overset{c}{CH_2}C(=O)O\overset{d}{CH_3}$

d. $\overset{a}{CH_3}\overset{b}{CH}(Br)\overset{c}{CH}Br_2$

e. $\overset{a}{CH_3}\overset{b}{CH}(CH_3)\overset{d}{CH_2}O\overset{c}{CH_3}$

f. $\overset{a}{CH_3}\overset{b}{CH_2}C(=O)\overset{c}{H}$

g. $\overset{a}{CH_3}\overset{b}{CH_2}\overset{d}{CH_2}C(=O)\overset{c}{CH_3}$

h. $Cl\overset{b}{CH_2}\overset{a}{CH_2}CH_2Cl$

i. $\overset{a}{CH_3}\overset{b}{CH_2}\overset{d}{CH}(O\overset{c}{CH_3})CH_2CH_3$

j. $\overset{a}{CH_3}\overset{c}{CH_2}\overset{d}{CH_2}O\overset{e}{CH}(CH_3)\overset{b}{CH_3}$

12. Each of the compounds would show two signals, but the ratio of the integrals for the two signals would be different for each of the compounds. The ratio of the integrals for the signals given by the first compound would be 1 : 4.5, the ratio of the integrals for the signals given by the second compound would be 1 : 3, and the ratio of the integrals for the signals given by the third compound would be 1 : 2.

13. Solved in the text.

14. The heights of the integrals for the signals in the spectrum shown in Figure 13.7 are about 3.5 and 5.2. The ratio of the integrals, therefore, is 5.2/3.5 = 1.5. This matches the ratio of the integrals calculated for **1,4-dimethyl benzene**. (Later we will see that a signal at ~7 ppm is characteristic of a benzene ring.)

$CH_3-C_6H_4-CH_3$	$CH_3C{\equiv}CH$	CH_3COCH_3 (C=O)	$CH_3C(CH_3)_2OCH_3$
$\frac{6}{4} = 1.5$	$\frac{3}{1} = 3$	$\frac{3}{3} = 1$	$\frac{9}{3} = 3$

15. From the direction of the electron flow around the benzene ring pictured in Figure 13.8 on page 548 of the text, you can see that the magnetic field induced in the region of the hydrogens that protrude out from the compound shown in this problem is in the same direction as the applied magnetic field, while the magnetic field induced in the region of the hydrogens that protrude into the center of the compound is in the opposite direction of the applied magnetic field.

Thus, the signal at 9.25 ppm is for the hydrogens that protrude out because they need a smaller applied magnetic field to come into resonance due to the fact that the induced and applied magnetic fields are in the same direction. The signal at -2.88 ppm is for the hydrogens that protrude inward because a greater applied field is necessary to make up for the magnetic field that is induced in the opposite direction.

16. **a.** A triplet is caused by two equivalent protons on an adjacent carbon. The two protons can be aligned in three different ways: both with the field, one with the field and one is against the field, or both against the field. That is why the signal is a triplet. There is only one way to align two protons that are with the field or two protons that are against the field. However, there are two ways to align two protons if one is with and one against the field: with and against or against and with. Consequently, the peaks in a triplet have relative intensities of 1 : 2: 1.

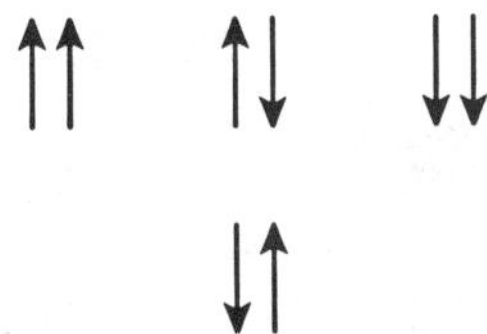

b. A quintet is caused by four equivalent protons on adjacent carbons. The four protons can be aligned in five different ways. The following possible arrangements for the alignment of four protons explain why the relative intensities of a quintet are 1: 4 : 6 : 4 : 1.

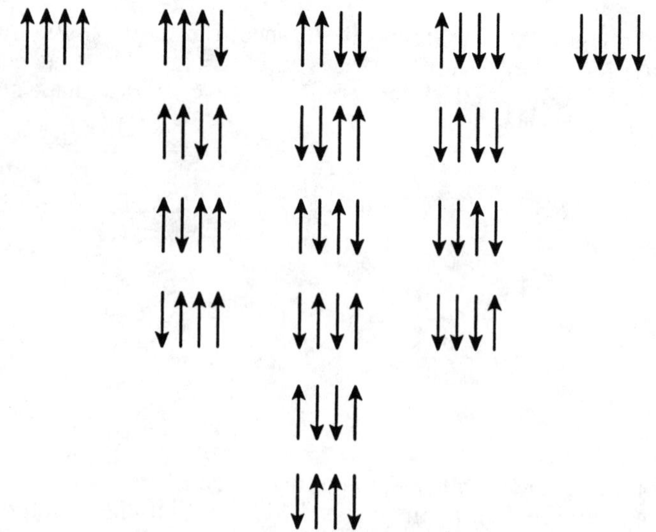

17. The signal farthest downfield in both spectra is the signal for the hydrogens bonded to the carbon that is also bonded to the halogen. Because chlorine is more electronegative than iodine, the farthest downfield signal should be farther downfield in the ^{1}H NMR spectrum for 1-chloropropane than in the ^{1}H NMR spectrum for 1-iodopropane.
Therefore, the **first spectrum** in Figure 13.13 is the ^{1}H NMR spectrum for **1-iodopropane**, and the **second spectrum** in Figure 13.13 is ^{1}H the NMR spectrum for **1-chloropropane**.

18.

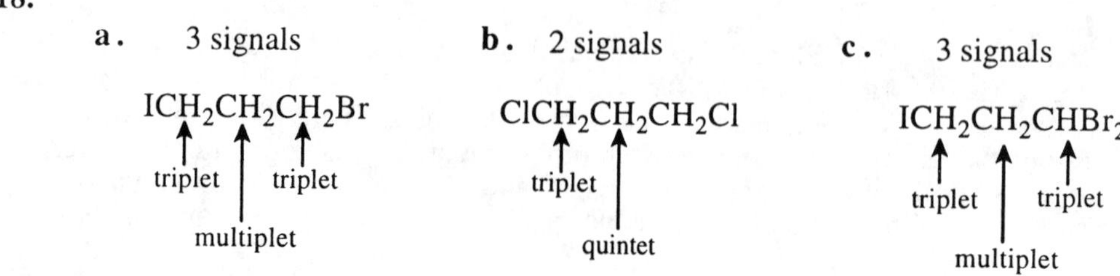

19.

$CH_3C(Br)_2CH_3$	$BrCH_2CH_2CH_2Br$	$CH_3CH_2CHBr_2$	$CH_3CHBrCH_2Br$
1 signal	2 signals	3 signals 2 triplets and a multiplet	3 signals 2 doublets and a multiplet

20. **a.** The signal at ~7.2 ppm indicates the presence of a benzene ring. If the ring has a single substituent, it has a molecular formula of C_6H_5. Subtracting C_6H_5 from the molecular formula of the compound (C_9H_{12} - C_6H_5) gives a substituent with a molecular formula of C_3H_7. The triplet at ~0.9 ppm indicates a methyl group adjacent to a methylene group. The identical integration under the signals at 1.6 ppm and 2.6 ppm indicate two methylene groups. Thus, the compound is propyl benzene.

$C_6H_5-CH_2CH_2CH_3$

b. The triplet and quartet indicate a CH_3CH_2 group bonded to an atom that is not bonded to any hydrogens. The molecular formula of $C_5H_{10}O$ indicates that the compound is diethyl ketone.

$CH_3CH_2C(=O)CH_2CH_3$

c. The signals between ~7-8 ppm indicate the presence of a benzene ring. The observation that there are 3 benzene rings signals indicates the compound has a mono-substituted benzene ring or a 1,3-disubstituted benzene ring. Since none of the 3 benzene ring signals is a singlet, we know that the compound has a mono-substituted benzene ring. Subtracting C_6H_5 from the molecular formula of the compound ($C_9H_{10}O_2$ - C_6H_5) shows that the substituent has a molecular formula of $C_3H_5O_2$.

The triplet and quartet suggest a CH_3CH_2 group. Therefore, we can conclude that the compound is one of the following.

$C_6H_5-C(=O)OCH_2CH_3$ or $C_6H_5-OC(=O)CH_2CH_3$

A **B**

Compound **A** withdraws electrons from the benzene ring, while compound **B** donates electrons into the benzene ring. Because one of the peaks of the benzene ring signal is at a higher frequency (farther downfield) than usual (>8 ppm) suggests that an electron-withdrawing substituent is present on the benzene ring.

This is confirmed by the signal for the CH_2 group at 3.8 ppm, indicating that the methylene group is adjacent to the electron-withdrawing oxygen. Thus, the compound is compound **A**.

$C_6H_5-C(=O)OCH_2CH_3$

21.

a. 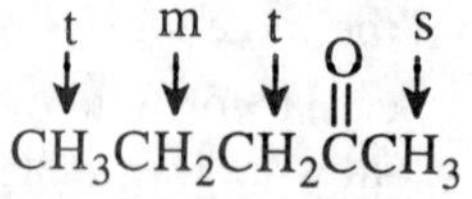 $CH_3CH_2CH_2\overset{O}{\overset{\|}{C}}CH_3$ — t, m, t, s

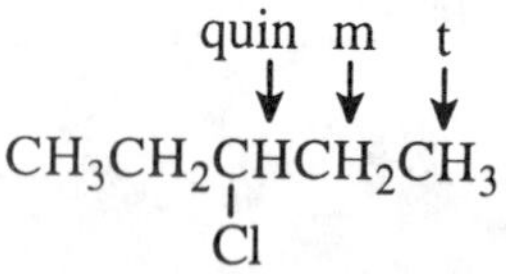

b. 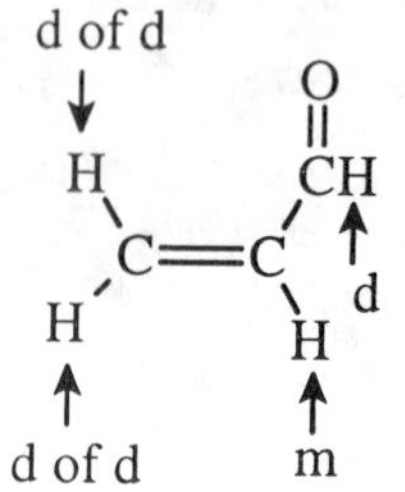 $H_2C{=}CH{-}CH{=}O$ — H (d of d), H (d of d), H (m), CH (d)

c. $CH_3CH_2\underset{Cl}{CH}CH_2CH_3$ — quin, m, t

d. $H_2C{=}CHCl$ — H (d of d), H (d of d), H (d of d)

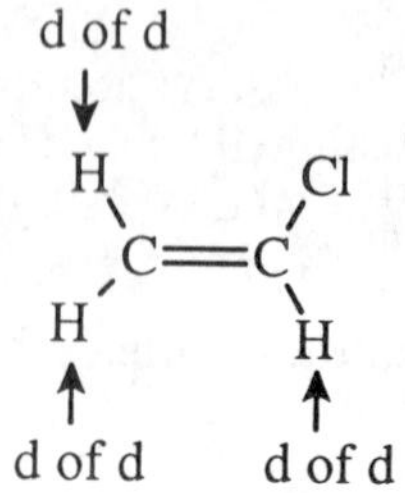

e. 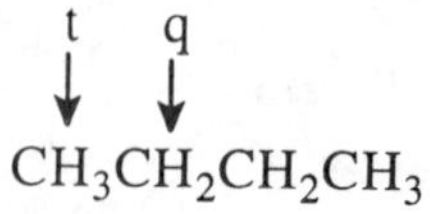 $CH_3CH_2CH_2CH_3$ — t, q

f. 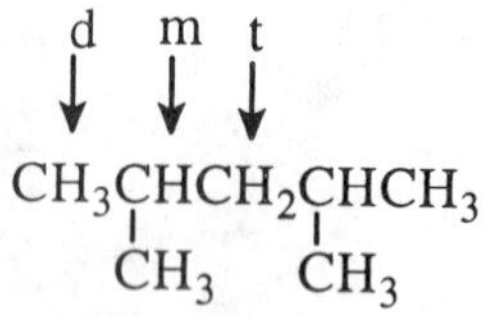 $CH_3\underset{CH_3}{CH}CH_2\underset{CH_3}{CH}CH_3$ — d, m, t

g. 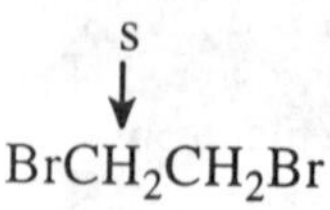 $BrCH_2CH_2Br$ — s

h. 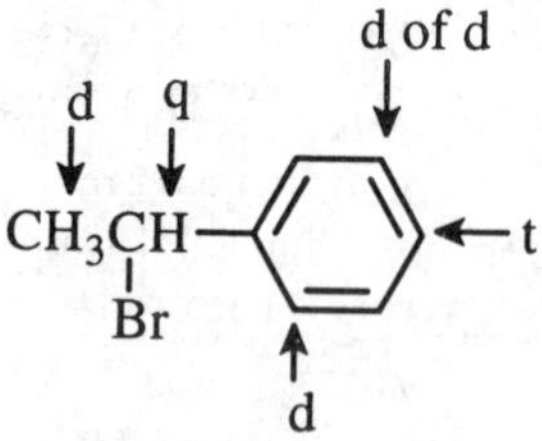 $CH_3\underset{Br}{CH}{-}C_6H_5$ — d, q; ring: d of d, t, d

i. 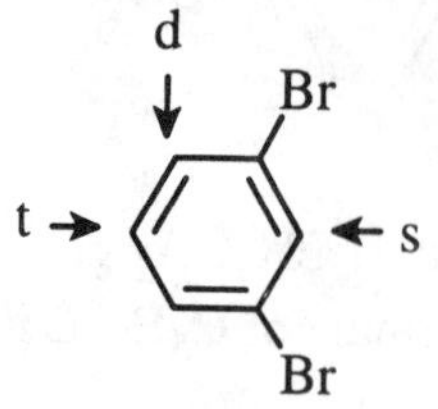 1,3-dibromobenzene (Br, Br) — d, t, s

j. 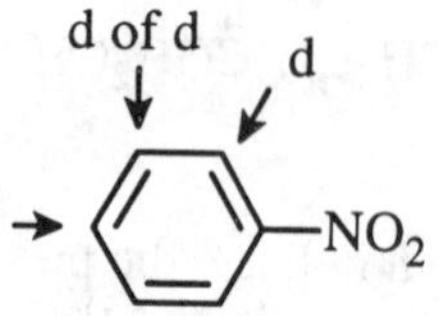 nitrobenzene ($-NO_2$) — d of d, d, t

k. 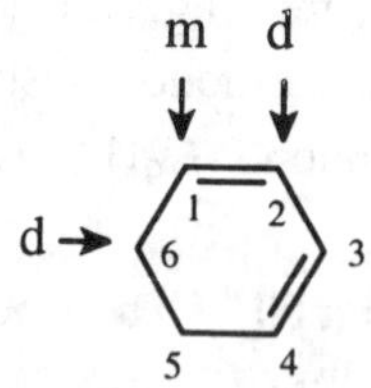 d, m, d (ring positions 1–6)

The H on C-2 will not be split by the H on C-3, since they are equivalent.

The H's on C-6 will not be split by the H's on C-5, since the H's on C-5 and C-6 are equivalent.

l. 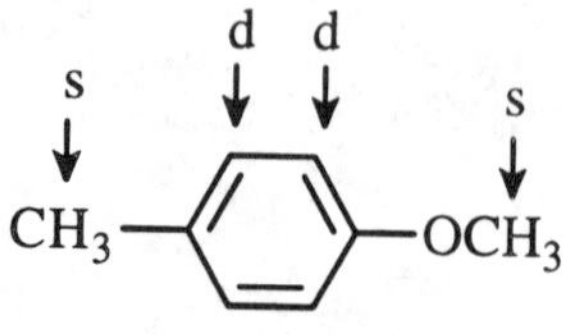 $CH_3{-}C_6H_4{-}OCH_3$ — s, d, d, s

m. 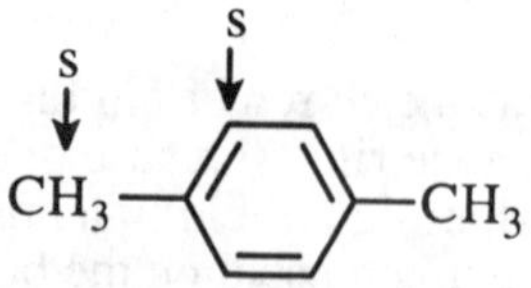 $CH_3{-}C_6H_4{-}CH_3$ — s, s

n. 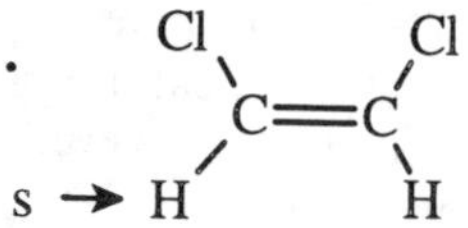 $ClHC{=}CHCl$ — s (H)

o. 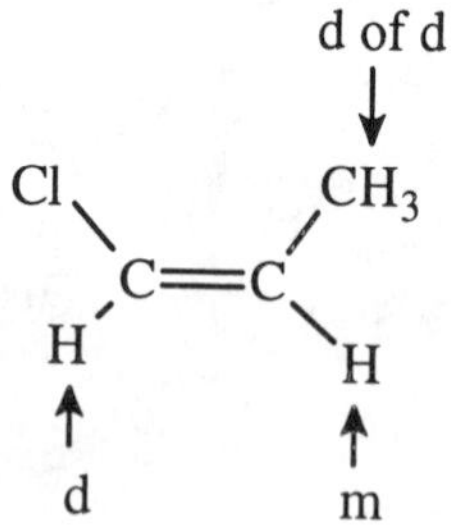 $ClHC{=}CHCH_3$ — H (d), H (m), CH_3 (d of d)

22. The contributing resonance structures show that the nitro group's ability to withdraw electrons by resonance causes the C-2 and C-4 positions of the resonance hybrid (the actual structure of the compound) to have a partial positive charge. The nitro group's ability to withdraw electrons inductively (through the sigma bonds) causes the partial positive charge to be greater at C-2 than at C-4. Therefore, the H_c proton has the least electron dense (less shielded) environment, so its signal appears at the highest frequency. The C-3 position does not have a partial positive charge.

Therefore, the H_a proton has the most electron dense (most shielded) environment, so its signal appears at the lowest frequency.

23.

a. $CH_3OCH_2CH_2OCH_3$

b. $CH_3\overset{O}{\overset{\|}{C}}CH_2CH_2\overset{O}{\overset{\|}{C}}CH_3$

c. $H\overset{O}{\overset{\|}{C}}-C_6H_4-\overset{O}{\overset{\|}{C}}H$

$CH_3OCH_2C{\equiv}CCH_2OCH_3$

24. Each spectrum is described going from left to right.

a. triplet quintet

b. two triplets close to each other singlet multiplet
(Appendix VI on page A-19 of the text indicates that a methylene adjacent to an RO and a methylene adjacent to a Br appear at about the same place.)

c. quartet triplet

d. singlet quartet triplet

e. singlet

(Equivalent H's do not split each other.)

f. quartet singlet triplet

g. three singlets

h. doublet multiplet doublet

i. triplet quintet

j. triplet doublet of doublets multiplet doublet

k. singlet (Equivalent H's do not split each other.)

l. singlet (Equivalent H's do not split each other.)

m. three doublets of doublets

n. triplet singlet quintet

25. The IR spectrum indicates a benzene ring (1600 cm^{-1} and 1500^{-1}) with hydrogens on sp^2 carbons, and no carbonyl group. The absorption bands in the 1250^{-1}-1000^{-1} region suggests there are two C-O single bonds, one with no double bond character and one with some double bond character. The two singlets in the ^1H NMR spectrum with about the same integration suggest two methyl groups, one of which is adjacent to an oxygen. That the benzene ring protons (6.7-7.1 ppm) consist of two doublets indicates a 1,4-disubstituted benzene ring.

CH_3O—⟨benzene ring⟩—CH_3

26. Solved in the text.

27.

a.

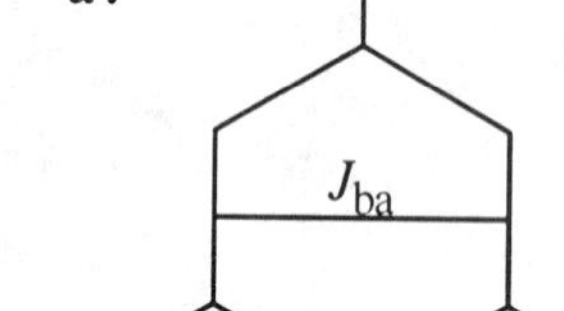

b.

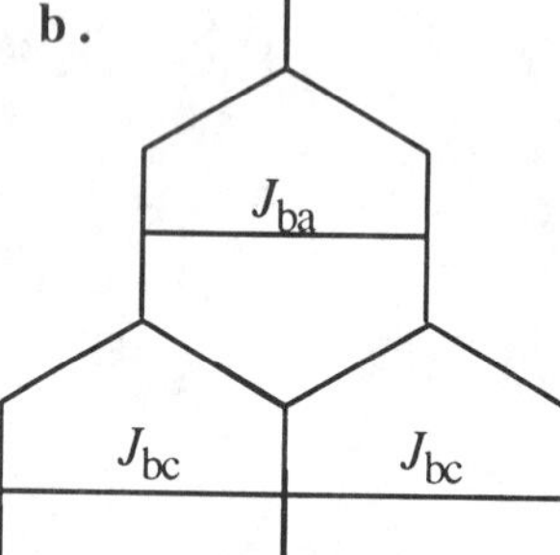

28. The greater the extent of hydrogen bonding, the greater the chemical shift. Therefore, the ^1H NMR spectrum of pure ethanol would have the signal for the OH proton at a greater chemical shift because it would be hydrogen bonded to a greater extent.

29.

30. Figure 13.28 is the ^{1}H NMR spectrum of propanamide. Notice that the signals for the N-H protons are unusually broad. Because of the partial double bond character of the C-N bond, the two N-H protons are not chemically equivalent. The quartet and triplet are characteristic of an ethyl group.

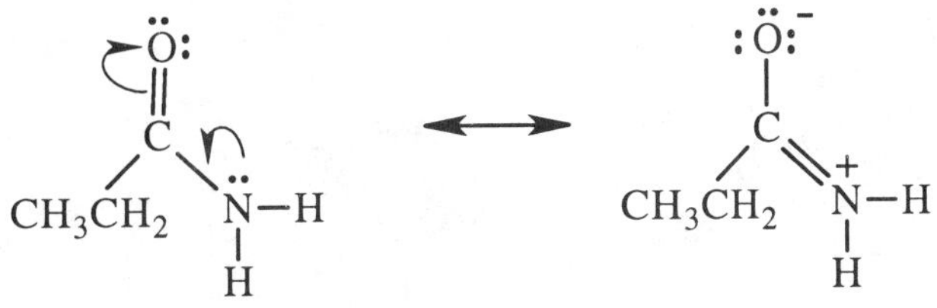

$CH_3CH_2C(=O)NH_2 \longleftrightarrow CH_3CH_2C(O^-)=N^+H_2$

31. a. 1. 3 2. 3 3. 2 4. 3 5. 4 6. 3 7. 2 8. 3 9. 4 10. 3

b. 1. $CH_3CH_2CH_2Br$

2. $CH_3CH_2OCH_3$

3. CH_3CHCH_3 (Br on the central carbon)

4. $CH_3C(CH_3)_2OCH_3$

5. $CH_3CH_2COCH_3$ (C=O)

6. $(H_3C)_2C{=}CH_2$

7. HC(H)=C(Br)H (H, H on one carbon; Br, H on the other)

8. $CH_3CH(CH_3)CHO$

9. chlorobenzene (C_6H_5Cl)

10. $CH_3COCH_2CH_2COCH_3$

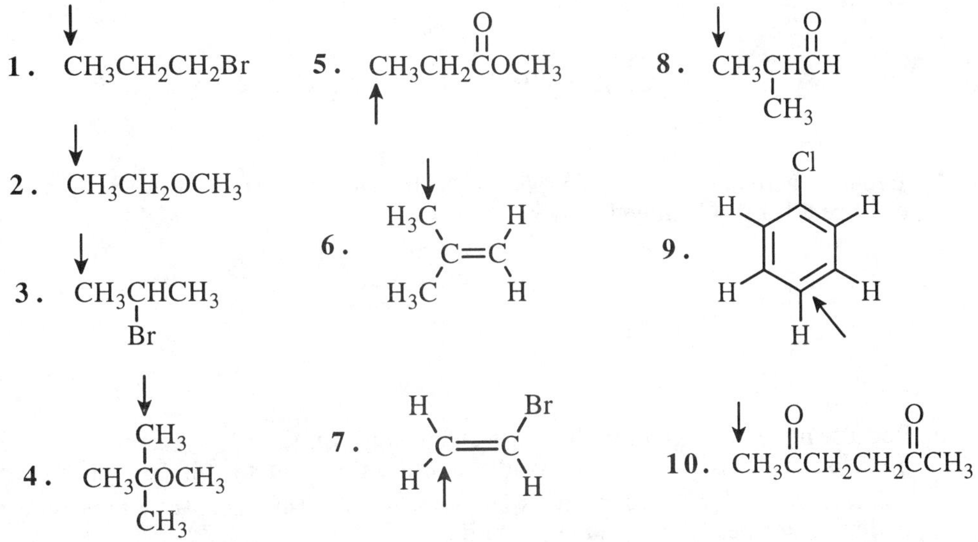

32. Each spectrum is described going from left to right.

1. triplet triplet quartet

2. triplet quartet quartet

3. doublet quartet

4. singlet quartet quartet

5. singlet quartet triplet quartet

33.

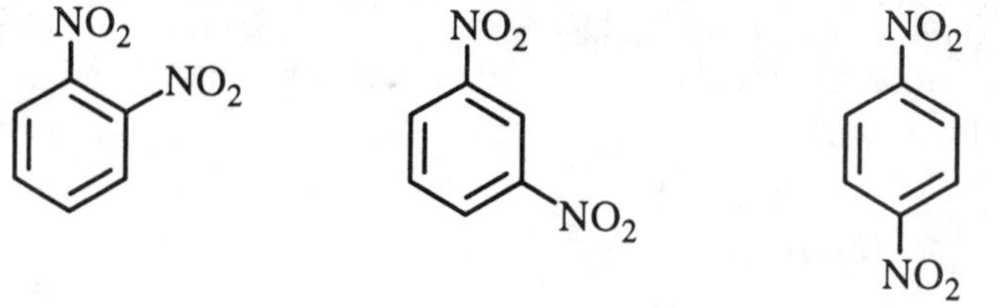

	NO_2 / NO_2 (1,2-dinitrobenzene)	NO_2 / NO_2 (1,3-dinitrobenzene)	NO_2 / NO_2 (1,4-dinitrobenzene)
a. ^{1}H NMR	2 signals	3 signals	1 signal
b. ^{13}C NMR	3 signals	4 signals	2 signals

34. **a.** The signal at 210 is for a carbonyl carbon. There are ten other carbons in the compound and five other signals. That suggests the compound is a ketone with identical five-carbon alkyl groups.

$$CH_3CH_2CH_2CH_2CH_2\overset{\overset{\displaystyle O}{\|}}{C}CH_2CH_2CH_2CH_2CH_3$$

b. Because there are only four signals for the six carbons of the benzene ring, the compound must be a 1,4-disubstituted benzene.

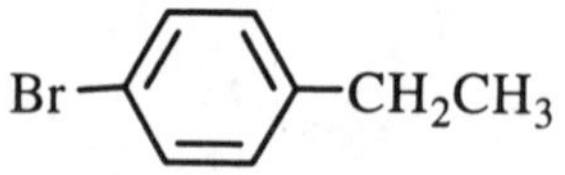

$$Br-C_6H_4-CH_2CH_3$$

c. Because none of the three carbons are chemically equivalent, one carbon must be bonded to no chlorine atoms, one must be bonded to one chlorine atom, and one must be bonded to two chlorine atoms. We would need a spin-coupled ^{13}C NMR spectrum or an ^{1}H NMR spectrum to distinguish between the three possibilities.

$$CH_3\underset{\underset{\displaystyle Cl}{|}}{C}HCHCl_2 \quad \text{or} \quad CH_3\overset{\overset{\displaystyle Cl}{|}}{\underset{\underset{\displaystyle Cl}{|}}{C}}CH_2Cl \quad \text{or} \quad Cl\underset{\underset{\displaystyle Cl}{|}}{C}HCH_2CH_2Cl$$

d. The molecular formula indicates the compound has one double bond. Each of the two sp^2 carbons must be bonded to the same groups because the six carbons only exhibit three signals. Whether the compound is *cis*-3-hexene or *trans*-3-hexene can't be determined from the spectrum. The coupling constants in an ^{1}H NMR spectrum could distinguish the compounds because the coupling constant for trans protons is greater than the coupling constant for cis protons.

$$CH_3CH_2CH=CHCH_2CH_3$$

35.

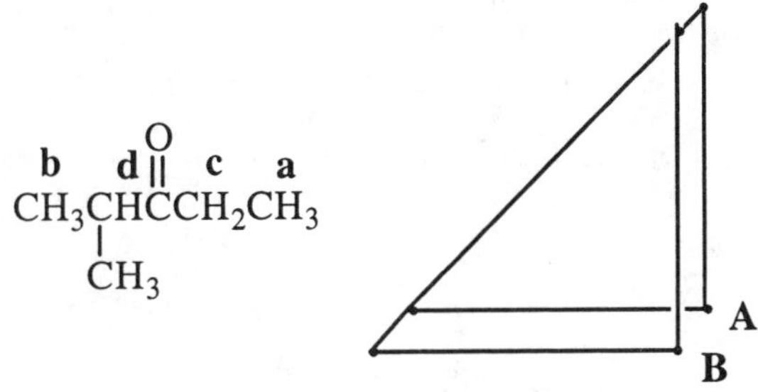

Point **A** shows that the "**a**" protons are split by the "**c**" protons.
Point **B** shows that the "**b**" protons are split by the "**d**" protons.

36. a. **1.** 5 **2.** 5 **3.** 4 **4.** 2 **5.** 3 **6.** 2

b. **1.** 7 **2.** 7 **3.** 5 **4.** 2 **5.** 2 **6.** 3

37.

a.

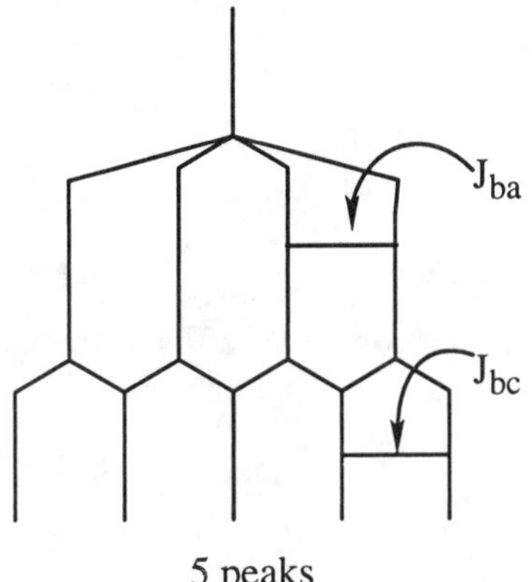

5 peaks

b.

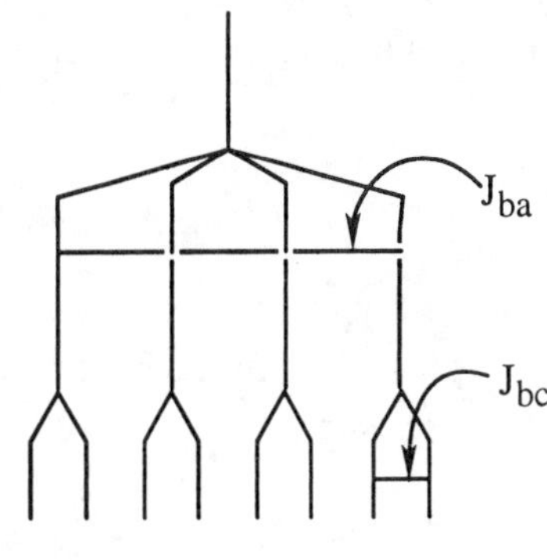

8 peaks

38.

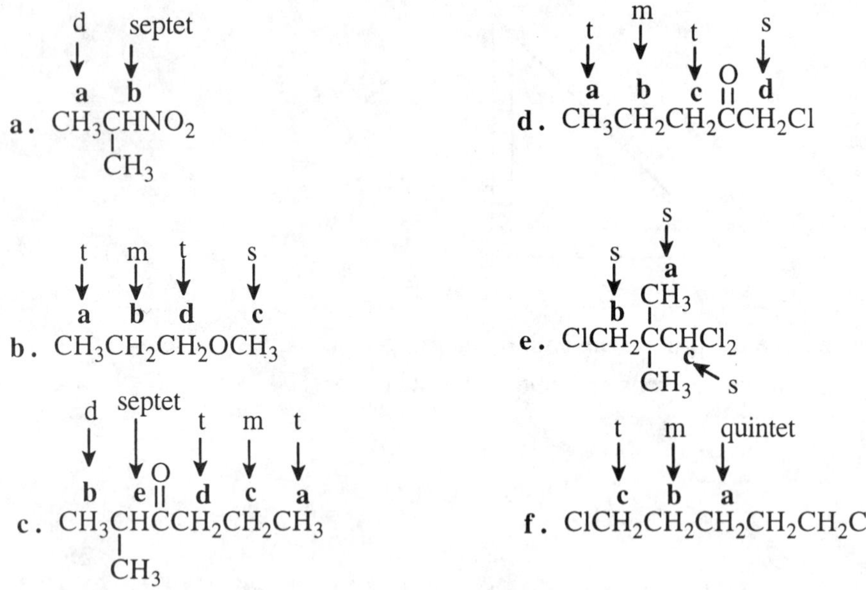

39. **a.** The spectrum must be that of **2-bromopropane** because the NMR spectrum has two signals and the signal farthest upfield is a doublet.

b. The spectrum must be that of **1-nitropropane** because the NMR spectrum has three signals and the signals farthest upfield and farthest downfield are triplets.

c. The spectrum must be that of **ethyl methyl ketone** because the NMR spectrum has three signals and the signals are a triplet, a singlet, and a quartet.

40. By dividing the value of the integration steps by the smallest one, the ratios of the hydrogens are found to be 3 : 2 : 1 : 9.

$$\frac{40.5}{13} = 3.1 \qquad \frac{27}{13} = 2.1 \qquad \frac{13}{13} = 1 \qquad \frac{118}{13} = 9.1$$

Assuming that the ratios are given in the downfield to upfield direction, a possible compound could be the following ester.

```
      CH3    O
      |      ||
  CH3C—CHCOCH3
      |   |
    CH3 CH2Cl
```

41.

a. $CH_3CH_2CH_2OCH_3$ (4 signals) and $CH_3CH_2OCH_2CH_3$ (2 signals)

b. $BrCH_2CH_2CH_2Br$ (2 signals) and $BrCH_2CH_2CH_2NO_2$ (3 signals)

c. $CH_3CH(CH_3)-CH(CH_3)CH_3$ (2 signals) and $CH_3C(CH_3)_2CH_2CH_3$ (3 signals)

d. $CH_3C(CH_3)_2-COCH_3$ (C=O) (2 signals with integration 3 : 1) and $CH_3C(OCH_3)_2CH_3$ (2 signals with integration 1 : 1)

e. $CH_3-C_6H_4-C(CH_3)_3$ (para) (3 signals with integration 9 : 4 : 3) and $C_6H_5-CH_2C(CH_3)_3$ (3 signals with integration 9 : 5 : 2)

f.

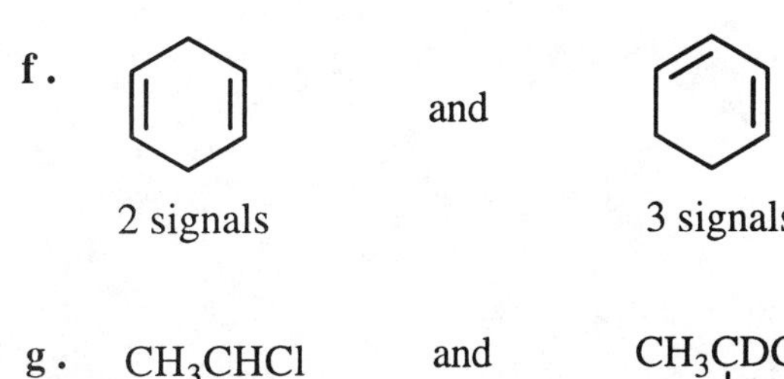

2 signals and 3 signals

g. $CH_3CHCl(CH_3)$ (2 signals) and $CH_3CDCl(CH_3)$ (1 signal)

h. Fischer projections: H–C(Cl)–CH_3 / D–C(Cl)–H and D–C(Cl)–CH_3 / H–C(Cl)–H

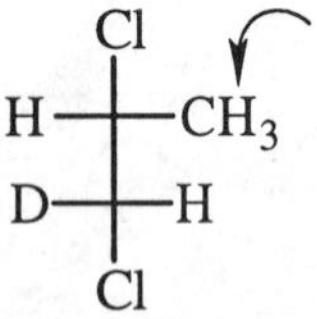

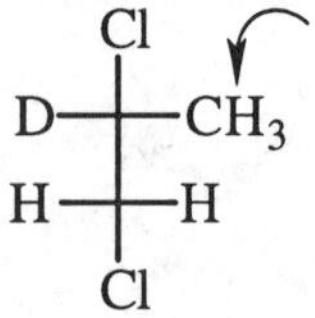

signal farthest upfield = doublet

signal farthest upfield = singlet

i. 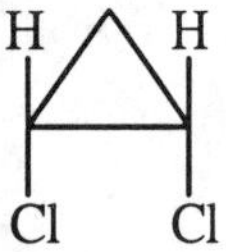 (cyclopropane structures) and

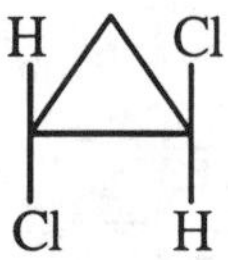

3 signals

2 signals

42. **a.** Chemical shift in ppm is independent of the operating frequency.

b. Chemical shift in hertz is proportional to the operating frequency.

c. The coupling constant is independent of the operating frequency.

d. The frequency required for NMR is lower than that required for IR, which is lower than that required for UV/Vis.

43.

a. 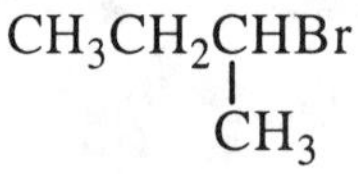 CH_3CH_2CHBr with CH_3 on the CHBr carbon

b. $CH_3CH_2CH_2CH_2Br$

c. 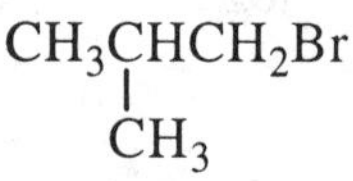 CH_3CHCH_2Br with CH_3 on the CH carbon

44.

a. 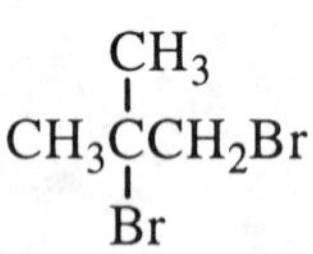 CH_3CCH_2Br with CH_3 and Br on the central carbon

b. $C_6H_5CHCH_3$ with Br on the CH carbon

c. $CH_3CH_2COCH_2CH_3$ (C=O)

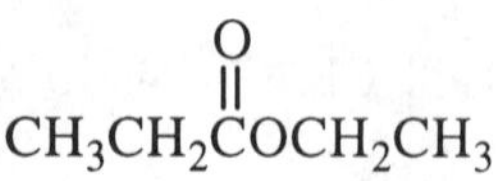

45.

CH_3CCH_3 with CH_3 and Cl on the central carbon

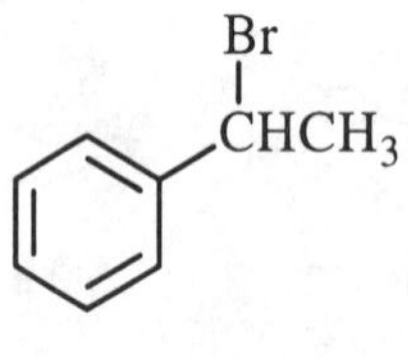

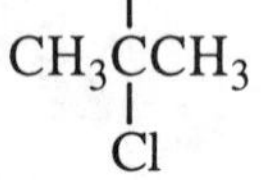

tert-butyl chloride
Compound A

$CH_3CHCH_2CH_3$ with Cl on the CH carbon

sec-butyl chloride
Compound B

46.

a. $CH_3\overset{O}{\overset{\|}{C}}CH_2\underset{CH_3}{\overset{CH_3}{C}}CH_3$

b. $CH_3\underset{CH_3}{CH}\overset{O}{\overset{\|}{C}}CH_2CH_2CH_3$

c. $CH_3\underset{CH_3}{CH}\overset{O}{\overset{\|}{C}}\underset{CH_3}{CH}CH_3$

47.

$CH_3CH_2CH{=}CH_2$
1-butene

$\overset{H_3C}{\underset{H}{}}C{=}C\overset{CH_3}{\underset{H}{}}$
cis-2-butene

$CH_3\overset{CH_3}{C}{=}CH_2$
2-methylpropene

It would be better to use ^{13}C NMR because you would only have to look at the number of signals in each spectrum: 1-butene will show four signals, *cis*-2-butene will show two signals, and 2-methylpropene will show three signals. (In the ^1H NMR spectrum, 1-butene will show four signals, and *cis*-2-butene and 2-methylpropene will both show two signals.)

48.

a. $CH_3\overset{O}{\overset{\|}{C}}CH_3$

b. $CH_3CH_2CH_2CH_2OCH_3$

c. $CH_3\underset{CH_3}{\overset{CH_3}{C}}{-}\overset{O}{\overset{\|}{C}}OCH_3$

d. $CH_3\underset{Cl}{CH}\overset{O}{\overset{\|}{C}}OCH_3$

e. $CH_3CH_2CH_2\overset{O}{\overset{\|}{C}}OH$

49.

$CH_3CH_2\overset{O}{\overset{\|}{C}}OCH_3$
A
3 signals
singlet, quartet, triplet
(singlet farthest downfield)

$CH_3\overset{O}{\overset{\|}{C}}OCH_2CH_3$
B
3 signals
singlet, quartet, triplet
(quartet farthest downfield)

$H\overset{O}{\overset{\|}{C}}OCH_2CH_2CH_3$
C
4 signals
singlet, triplet
multiplet, quartet

$H\overset{O}{\overset{\|}{C}}O\overset{CH_3}{CH}CH_3$
D
3 signals
singlet, septet
doublet

C can be distinguished from **A**, **B**, and **D** because **C** has 4 signals and the others have 3 signals.

D can be distinguished from **A** and **B** because the 3 signals of **D** are a singlet, a septet, and a doublet, while the 3 signals of **A** and **B** are a singlet, a quartet, and a triplet.

A and **B** can be distinguished because the signal farthest downfield in **A** is a singlet, while the signal farthest downfield in **B** is a quartet.

50. The protons at 5.1 ppm and 5.3 ppm can be distinguished by their coupling constants, since the coupling constant for trans protons is greater than the coupling constant for cis protons. (See Figure 13.20 on page 560 of the text).

~ 5.1 ppm → H (=CH₂ proton, trans to CH₂OH); ~5.3 ppm → H (=CH₂ proton, cis to CH₂OH); ~6.0 ppm → H on C bearing CH_2OH; ~4.1 ppm → CH_2; ~2.9 ppm → OH

$$H_2C{=}CH{-}CH_2OH$$

51. If addition of HBr to propene follows Markovnikov's rule, the product of the reaction will give an NMR spectrum with two signals (a doublet and a septet). If addition of HBr does not follow Markovnikov's rule, the product will give an NMR spectrum with three signals (two triplets and a multiplet).

$CH_3CH{=}CH_2$ + HBr ⟶ $CH_3CH(Br)CH_3$ (Markovnikov product, 2 signals) $CH_3CH_2CH_2Br$ (anti-Markovnikov product, 3 signals)

52. **a.** The signals at ~7.2 ppm indicate that the compounds whose spectra are shown in **a** and **b** both contain a benzene ring. From the molecular formula, you know that there are five additional carbons in **a**, and the hydrogens on these carbons must all be accounted for by two singlets with integral ratios of ~ 1 : 4.5, indicating that the compound is **2,2-dimethyl-1-phenylpropane.**

$C_6H_5{-}CH_2C(CH_3)_3$

2,2-dimethyl-1-phenylpropane

b. The two singlets have integral ratios of ~ 1 : 3, indicating a methyl substituent and a *tert*-butyl substituent. The signals at ~7.1 and ~7.3 ppm indicate that the benzene ring has two kinds of hydrogens. Thus, the compound is **4-methyl-1-*tert*-butylbenzene.**

$H_3C{-}C_6H_4{-}C(CH_3)_3$

4-methyl-1-*tert*-butylbenzene

53.

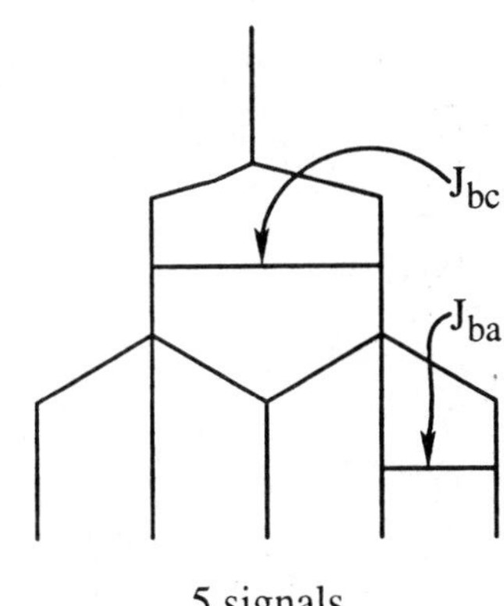

5 signals

54.

a.

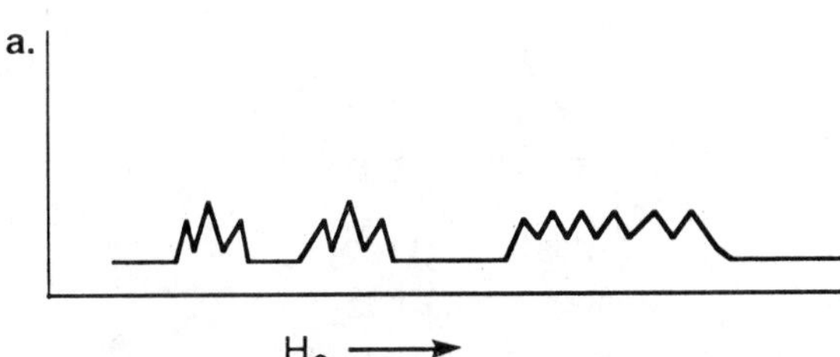

b.

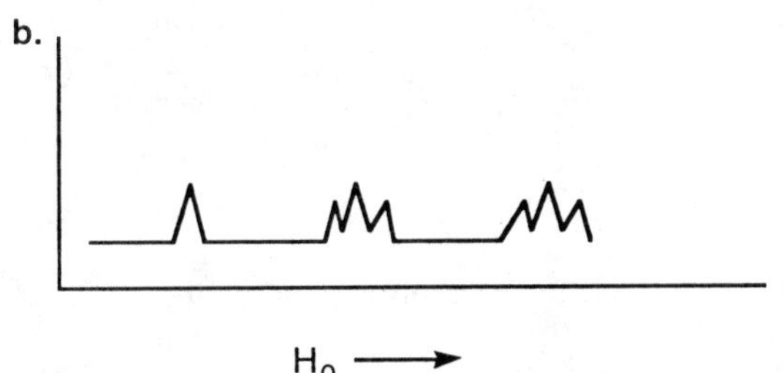

c.

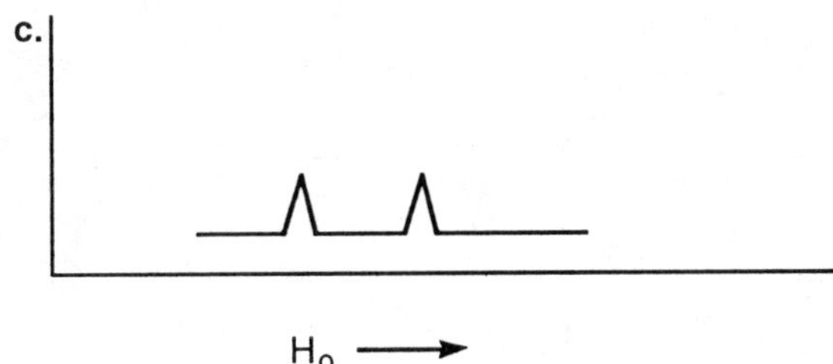

d.

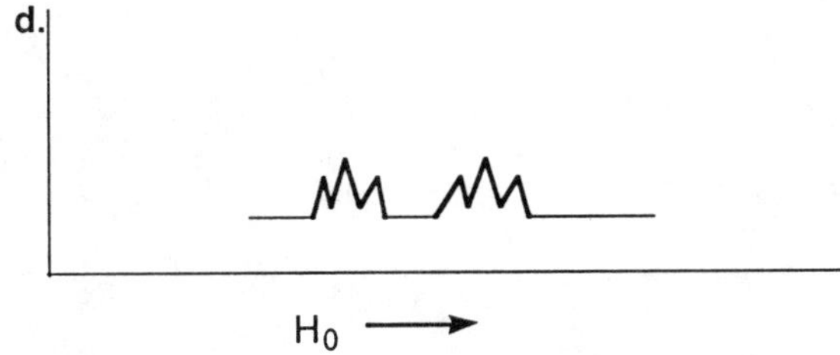

e.

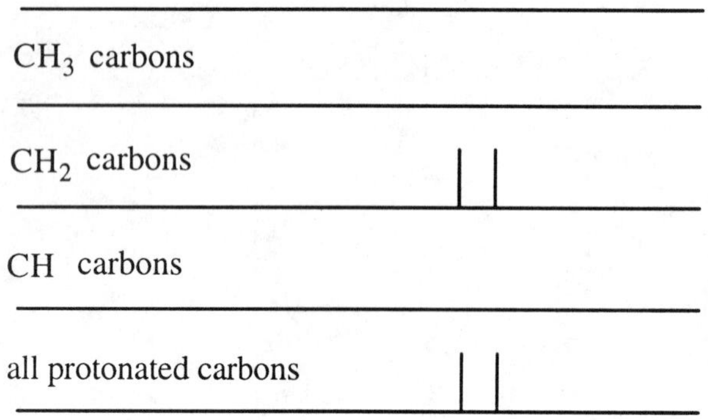

CH_3 carbons

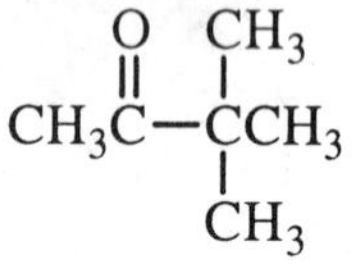

CH_2 carbons

CH carbons

all protonated carbons

55.

a. 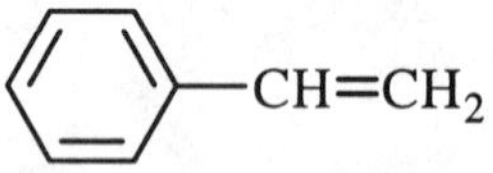 $C_6H_5-CH{=}CH_2$

b. $CH_3\overset{O}{\overset{\|}{C}}-C(CH_3)_2CH_3$

c. $CH_3C(CH_3)_2-\overset{O}{\overset{\|}{C}}-C(CH_3)_2CH_3$

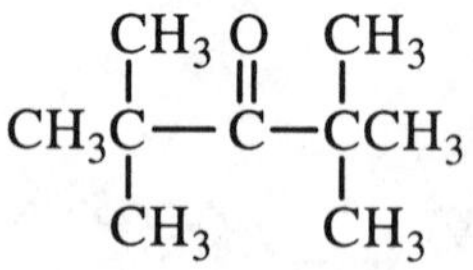

d. $CH_2{=}CHOCH_2CH_3$

56. Methyl bromide gives a signal at 2.7 ppm for its three methyl protons, and 2-bromo-2-methylpropane gives a signal at 1.8 ppm for its nine methyl protons. If equal amounts of each were present in solution, the ratio of the hydrogens (and, therefore, the ratio of the relative integrals) would be 3 : 1. Because the relative integrals are 1 : 6, there must be an 18-fold greater concentration of methyl bromide in the mixture.

$$\frac{(CH_3)_3CBr}{CH_3Br} = \frac{3}{1} \times \frac{1}{18} = \frac{1}{6}$$

57.

$$\Delta E = h\nu$$

$$\Delta E = \frac{9.54 \times 10^{-11}\ \text{cal}}{\text{mol}} \times \frac{1\ \text{mol}}{6 \times 10^{23}\ \text{atoms}} \times 60 \times 10^{6}\ \text{Hz}$$

$$\Delta E = 9.54 \times 10^{-27}\ \text{cal}$$

58. All four spectra show a singlet at ~2.0 ppm suggesting they are all esters with a methyl group attached to the carbonyl group. So now the problem becomes determining the nature of the group attached to the oxygen.

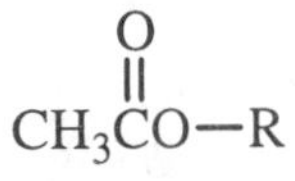

$$CH_3\overset{\overset{O}{\|}}{C}O{-}R$$

The signal farthest downfield in the first spectrum is a triplet, indicating that the carbon attached to the oxygen is bonded to a CH_2 group. The signal farthest upfield is also a triplet, indicating that it too is attached to a CH_2 group. The presence of two multiplets confirms the structure.

$$CH_3\overset{\overset{O}{\|}}{C}OCH_2CH_2CH_2CH_3$$

The signal farthest downfield in the second spectrum is a multiplet, indicating that the carbon attached to the oxygen is attached to two non-equivalent carbons bonded to hydrogens. The signal farthest upfield is a triplet, indicating that it is attached to a CH_2 group. The doublet at ~1.2 ppm is a methyl group attached to a carbon bonded to a one hydrogen.

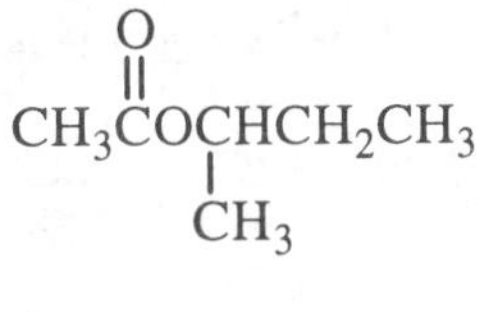

$$CH_3\overset{\overset{O}{\|}}{C}O\underset{\underset{CH_3}{|}}{C}HCH_2CH_3$$

The signal farthest downfield and the signal farthest upfield in the third spectrum are doublets, indicating that the carbon attached to the oxygen and the terminal carbon of the group are both attached to a CH group.

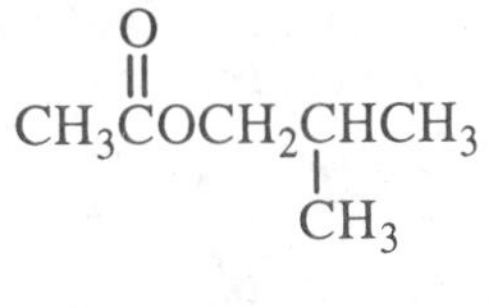

$$CH_3\overset{\overset{O}{\|}}{C}OCH_2\underset{\underset{CH_3}{|}}{C}HCH_3$$

The group attached to the oxygen in the fourth spectrum has only one kind of hydrogen.

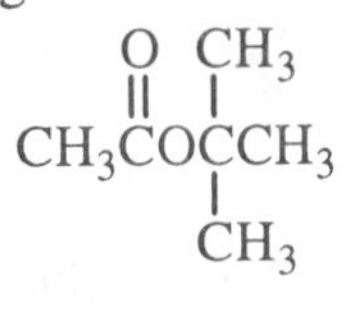

$$CH_3\overset{\overset{O}{\|}}{C}O\overset{\overset{CH_3}{|}}{\underset{\underset{CH_3}{|}}{C}}CH_3$$

59.

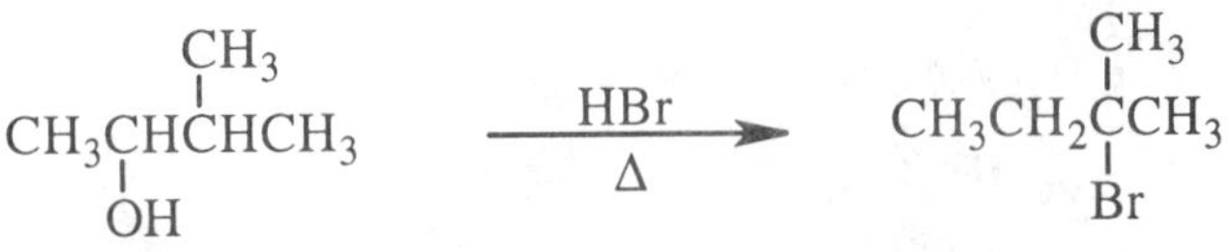

$$\underset{\text{compound A}}{CH_3\underset{\underset{OH}{|}}{C}H\overset{\overset{CH_3}{|}}{C}HCH_3} \xrightarrow[\Delta]{HBr} \underset{\text{compound B}}{CH_3CH_2\overset{\overset{CH_3}{|}}{\underset{\underset{Br}{|}}{C}}CH_3}$$

60. **a.** The IR spectrum indicates that the compound is a ketone. The doublet in the NMR spectrum at ~0.9 ppm suggests an isopropyl group. There is a singlet at ~2.1 ppm on top of a multiplet, and a doublet at ~2.2 ppm. Knowing that the compound contains 6 carbons helps in the identification.

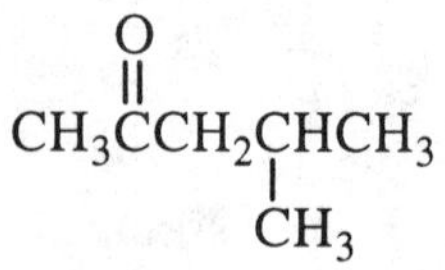

$$\overset{\displaystyle O}{\overset{\|}{CH_3CCH_2}}\underset{\displaystyle CH_3}{\underset{|}{CH}}CH_3$$

b. The IR spectrum indicates that this oxygen-containing compound is not a carbonyl compound or an alcohol; the absorption band at ~1000 cm^{-1} suggests that it is an ether. The doublet at ~1.1 ppm suggests an isopropyl group. Since there is only one other signal in the NMR spectrum, the compound must be a symmetrical ether. The compound is diisopropyl ether.

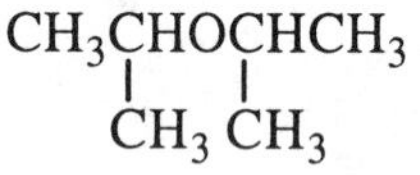

$$CH_3\underset{\displaystyle CH_3}{\underset{|}{CH}}O\underset{\displaystyle CH_3}{\underset{|}{CH}}CH_3$$

c. The absorption band at ~3400 cm^{-1} indicates an alcohol. The two signals at 7.3 and 8.1 ppm indicate a 1,4-substituted benzene ring with a strongly electron-withdrawing substituent. The two triplets indicate that the four-carbon substituent is not branched.

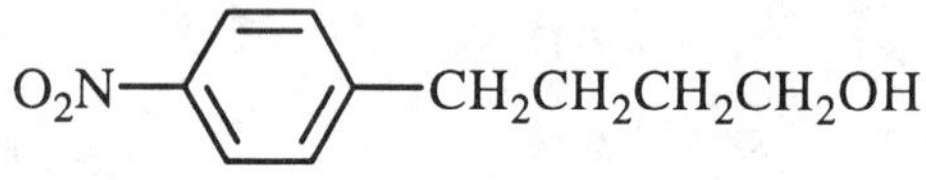

$$O_2N-C_6H_4-CH_2CH_2CH_2CH_2OH$$

d. The absorption bands at ~1700 cm^{-1} and 2700 cm^{-1} indicate the compound is an aldehyde. The two doublets at ~7.0 and 7.8 ppm indicate a 1,4-disubstituted benzene ring. That none of the NMR signals is a doublet suggests that the aldehyde group is attached directly to the benzene ring. The two triplets and two multiplets indicate an unbranched substituent. The triplet at ~ 4.0 ppm indicates that the group giving this signal is next to an electron-withdrawing group.

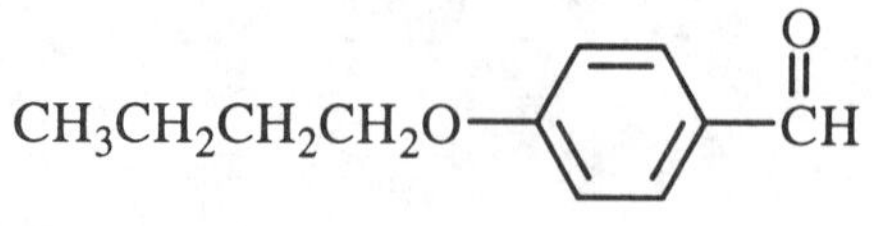

$$CH_3CH_2CH_2CH_2O-C_6H_4-\overset{\displaystyle O}{\overset{\|}{C}}H$$

61. The singlet at 210 indicates a carbonyl group. The splitting of the other two signals indicates an isopropyl group. The molecular formula indicates that it must have two isopropyl groups.

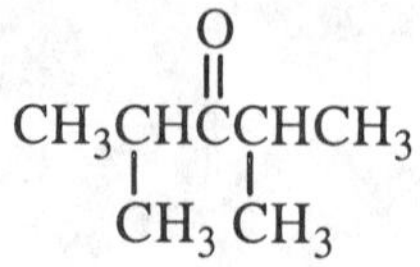

$$CH_3\underset{\displaystyle CH_3}{\underset{|}{CH}}\overset{\displaystyle O}{\overset{\|}{C}}\underset{\displaystyle CH_3}{\underset{|}{CH}}CH_3$$

62. **a.** The IR spectrum indicates it is a ketone; the mass spectrum tells you it is a ketone with 7 carbons. The fact that there are only 3 signals in the NMR spectrum suggests it is a symmetrical ketone. The splitting pattern confirms the structure.

$$CH_3CH_2CH_2\overset{\overset{\displaystyle O}{\|}}{C}CH_2CH_2CH_3$$

b. The M + 2 peak tells you the compound contains chlorine; the IR spectrum indicates it is a ketone; the NMR spectrum indicates it has a monosubstituted benzene ring. The singlet at ~4.7 indicates that the group giving this signal is in a strongly electron-withdrawing environment. The integration tells you that the two kinds of hydrogens are in a 2.5 : 1 ratio.

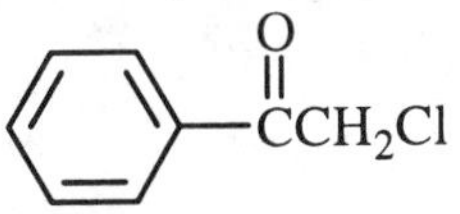

$$C_6H_5\text{—}\overset{\overset{\displaystyle O}{\|}}{C}CH_2Cl$$

63. The DEPT ^{13}C NMR spectrum indicates that the compound has five carbons bonded to hydrogens. The molecular formula tells us that the compound has six carbons and an oxygen. The sixth carbon is not bonded to a hydrogen, suggesting a ketone. The following ketone has the single CH_3 group, the single CH group, and the three CH_2 groups indicated by the DEPT spectrum.

$$CH_3\overset{\overset{\displaystyle O}{\|}}{C}CH_2CH_2CH{=}CH_2$$

Chapter 13 Practice Test

1. How many signals would you expect to see in the ^{1}H NMR spectrum of each of the following compounds?

$CH_3CH_2CH_2C(=O)CH_3$ $\quad$ $CH_3CH_2CH(Cl)CH_2CH_3$ $\quad$ $CH_2{=}CHC(=O)H$

$C_6H_5{-}NO_2$ (nitrobenzene) $\quad$ 1,3-dichlorobenzene (Cl, Cl) $\quad$ $CH_3CH(CH_3)CH_2CH(CH_3)CH_3$

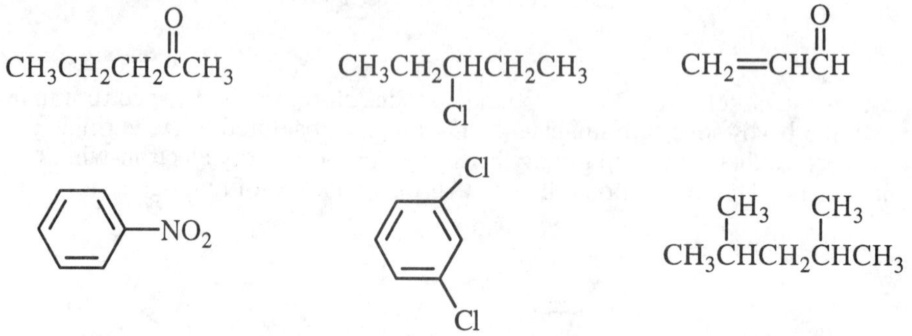

2. Indicate the multiplicity of each of the indicated sets of protons. (For example, is it a singlet, doublet, triplet, quartet, quintet, multiplet, or doublet of doublets?)

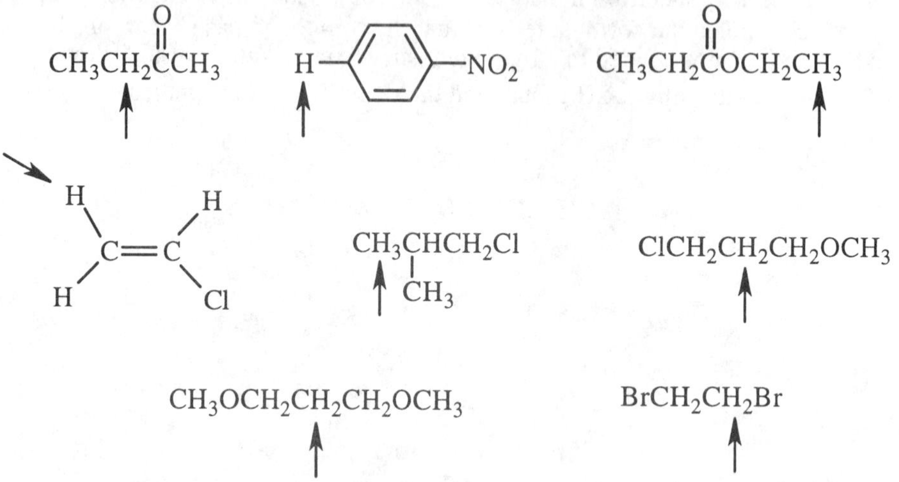

$CH_3CH_2C(=O)CH_3$ $\quad$ $H{-}C_6H_4{-}NO_2$ $\quad$ $CH_3CH_2C(=O)OCH_2CH_3$

$H_2C{=}CHCl$ $\quad$ $CH_3CH(CH_3)CH_2Cl$ $\quad$ $ClCH_2CH_2CH_2OCH_3$

$CH_3OCH_2CH_2CH_2OCH_3$ $\quad$ $BrCH_2CH_2Br$

3. How could you distinguish the following compounds using ^{1}H NMR spectroscopy?

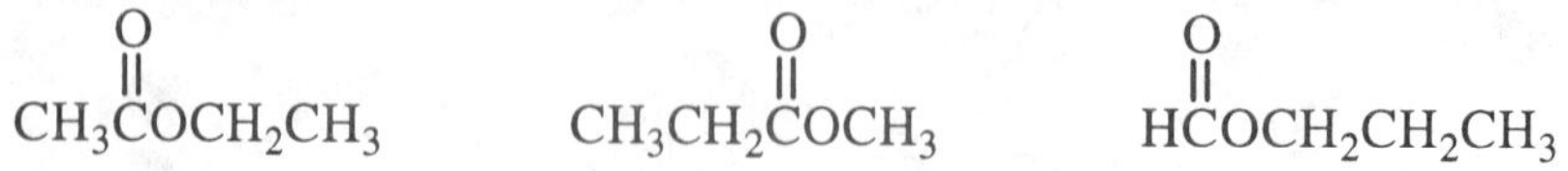

$CH_3C(=O)OCH_2CH_3$ $\quad$ $CH_3CH_2C(=O)OCH_3$ $\quad$ $HC(=O)OCH_2CH_2CH_3$

4. Indicate whether each of the following statements is true or false.

a. The peaks on the right of an NMR spectrum are deshielded compared to the peaks on the left. T F

b. Dimethyl ketone has the same number of signals in its ^{1}H NMR spectrum as in its ^{13}C NMR spectrum. T F

c. In the ^{1}H NMR spectrum of the compound shown below, the lowest frequency signal (the one farthest upfield is a singlet and the signal farthest downfield is a doublet. T F

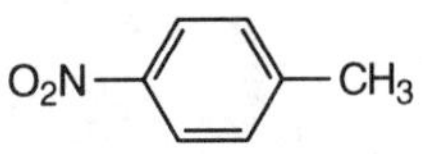

O_2N–(benzene ring)–CH_3

5. For each compound:

1. indicate the number of signals you would expect to see in its proton-decoupled ^{13}C NMR spectrum.

2. indicate the carbon that would give the highest frequency (farthest downfield) signal.

3. indicate the multiplicity of that signal in a spin-coupled ^{13}C NMR spectrum.

a. $CH_3CH_2CH_2Cl$ **b.** $CH_3CH_2\overset{O}{\overset{\|}{C}}OCH_3$ **c.** $CH_3\underset{Br}{\underset{|}{C}}HCH_3$

CHAPTER 14
Aromaticity. Reactions of Benzene

Important Terms

aliphatic compound	a nonaromatic organic compound.
annulene	a monocyclic hydrocarbon with alternating double and single bonds.
antiaromatic	a cyclic and planar compound with an uninterrupted cloud of electrons containing an even number of pairs of π electrons.
aromatic	a cyclic and planar compound with an uninterrupted cloud of electrons containing an odd number of pairs of π electrons.
aromatic compound	a cyclic and planar compound with an uninterrupted cloud of electrons containing an odd number of pairs of π electrons.
Clemmensen reduction	a reaction that reduces the carbonyl group of a ketone to a methylene group using Zn(Hg)/HCl.
electrophilic aromatic substitution reaction	a reaction in which an electrophile substitutes for a hydrogen of an aromatic ring.
Friedel-Crafts acylation	an electrophilic substitution reaction that puts an acyl group on a benzene ring.
Friedel-Crafts alkylation	an electrophilic substitution reaction that puts an alkyl group on a benzene ring.
halogenation	reaction with halogen (Br_2, Cl_2).
heteroatom	an atom other than a carbon atom or a hydrogen atom.
heterocyclic compound (heterocycle)	a cyclic compound in which one or more of the atoms of the ring are heteroatoms.
Hückel's rule or the 4*n* + 2 rule	the number of π electrons in a cyclic uninterrupted π cloud that cyclic and planar compound must have in order to be aromatic.
nitration	substitution of a nitro group (NO_2) for a hydrogen of a benzene ring.
principle of microscopic reversibility	states that the mechanism for a reaction in the forward direction has the same intermediates and the same transition states as the mechanism for the reaction in the reverse direction.
sulfonation	substitution of a hydrogen of a benzene ring with a sulfonic acid group (SO_3H).
Wolff-Kishner reduction	a reaction that reduces the carbonyl group of a ketone to a methylene group using NH_2NH_2/HO^-.

<u>**Solutions to Problems**</u>

1. **a.** In the case of 9 pairs of π electrons, there are 18 electrons.
Therefore, $4n + 2 = 18$ and $n = 4$.

b. Because it has an odd number of pairs of π electrons, it will be aromatic if it is cyclic, planar, and if every atom in the ring has a *p* orbital.

2. **b, c, e,** and **g** are aromatic.

b is aromatic, because it is cyclic, planar, every atom in the ring has a *p* orbital, and it has seven pairs of π electrons.

c is aromatic, because it is cyclic, planar, every atom in the ring has a *p* orbital, and it has three pairs of π electrons.

e is aromatic, because it is cyclic, planar, every atom in the ring has a *p* orbital, and it has nine pairs of π electrons.

g is aromatic, because it is cyclic, planar, every atom in the ring has a *p* orbital, and it has five pairs of π electrons.

a is not aromatic, because it has two pairs of π electrons.

d is not aromatic, because it has two pairs of π electrons and every atom in the ring does not have a *p* orbital

f is not aromatic, because it has two pairs of π electrons and every atom in the ring does not have a *p* orbital

h is not aromatic, because it is not cyclic.

3. **a.** Solved in the text.

b. There are five monobromophenanthrenes.

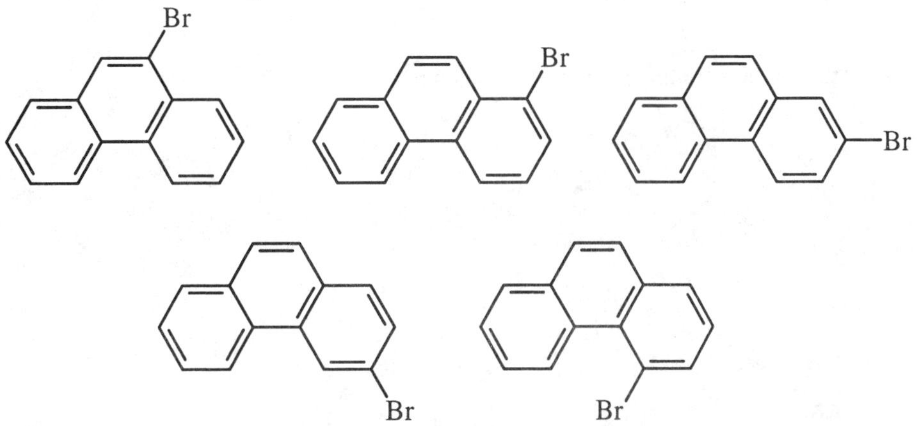

Br Br Br

Br Br

4. To be aromatic, a compound must be cyclic and planar, every atom in the ring must have a *p* orbital, and the π cloud must contain an odd number of pairs of π electrons.

[10]-Annulene is cyclic, every atom in the ring has a *p* orbital, and it has the correct number of π electrons to be aromatic (five pairs). Knowing it is not aromatic, we can conclude that it is not planar.

[12]-Annulene has an even number of pairs of π electrons (six pairs) in its π cloud, so it cannot be aromatic.

5.

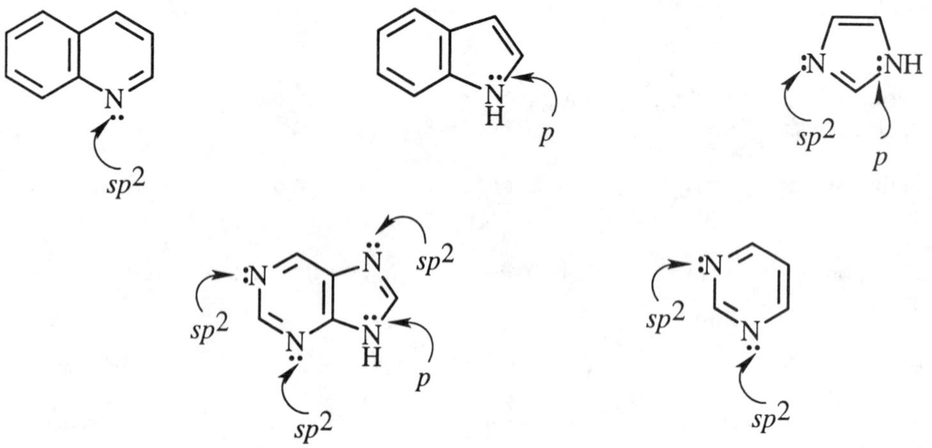

N
sp^2

N H
p

:N :NH
sp^2 *p*

:N N N N H
sp^2 sp^2 sp^2 *p*

:N N
sp^2 sp^2

6.

a. 1. Notice that each resonance contributor has a charge of -1.

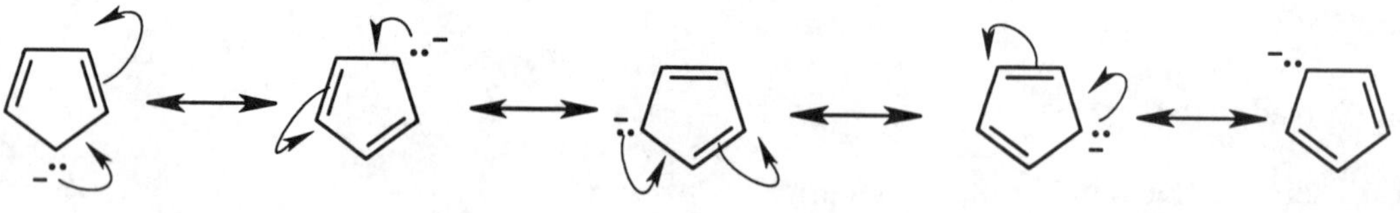

2. Notice that each resonance contributor has a net charge of 0.

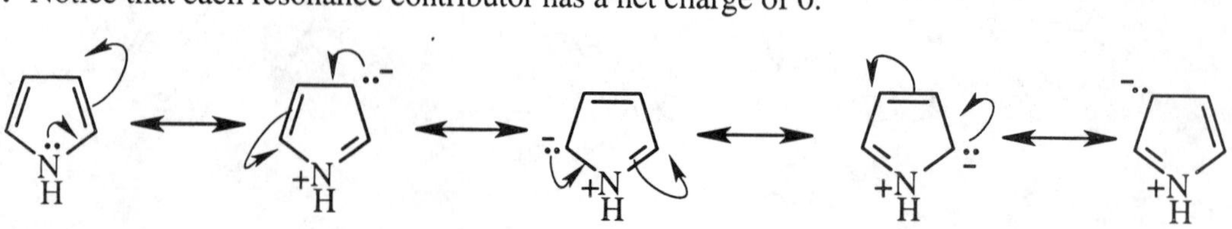

N H +N H +N H +N H +N H

b. 1. five ring atoms

2. four ring atoms

7. Cyclopentadiene has a lower pK_a. When cyclopentadiene loses a proton, a relatively stable aromatic compound is formed. When cycloheptatriene loses a proton, an unstable antiaromatic compound is formed. It is, therefore, easier to lose a proton from cyclopentadiene.

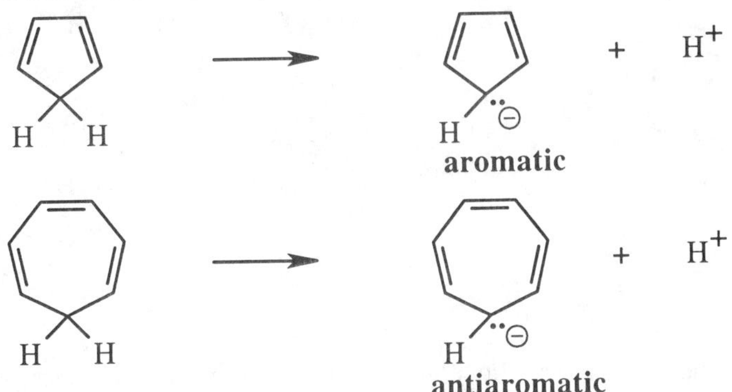

8. **a.** The resonance contributors marked with an X are the least stable because in these contributors the two negative charges are on adjacent carbons.

b. Because these contributors are the least stable, they make the smallest contribution to the hybrid.

9. **a.** Cyclopropane has a lower pK_a because an antiaromatic compound is formed when cyclopropene loses a proton.

antiaromatic

b. 3-Bromocyclopropene is more soluble in water, because it is more apt to ionize since heterolytic cleavage of its carbon-bromine bond forms an aromatic compound.

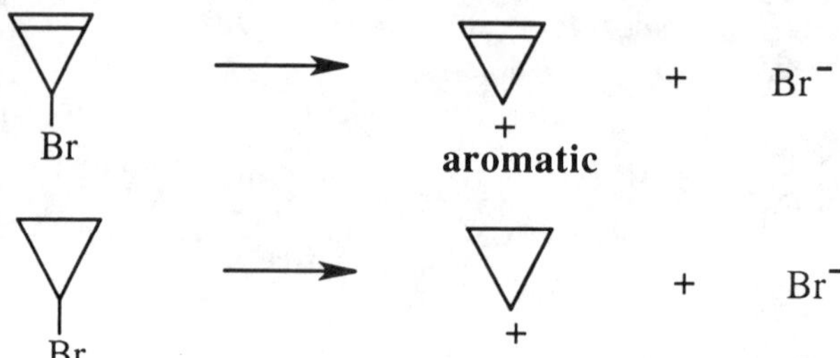

10. **a** is antiaromatic because it is cyclic, planar, every atom in the ring has a *p* orbital, and it has two pairs of π electrons.

When you did Problem 2, you found that **b, c, e,** and **g** are aromatic.

d, f, and **h** are neither aromatic nor antiaromatic. **d** and **f** have ring atoms that do not have *p* orbitals, and **h** is not cyclic

11. The compounds with completely filled bonding orbitals and electrons in no other orbitals are aromatic (the cycloheptatrienyl cation and the cylopropenyl cation).

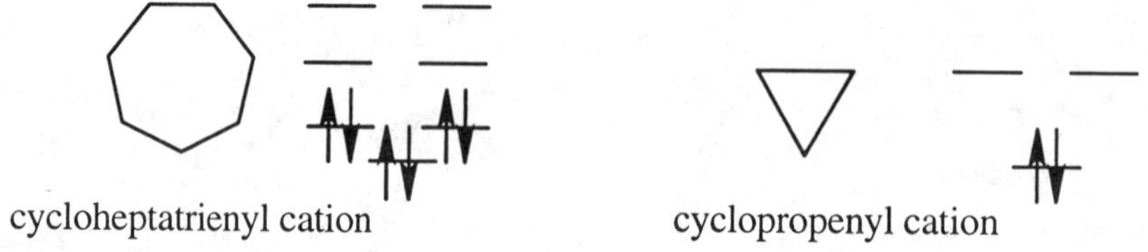

The compound with unpaired electrons in degenerate orbitals is antiaromatic (the cycloheptatrienyl anion).

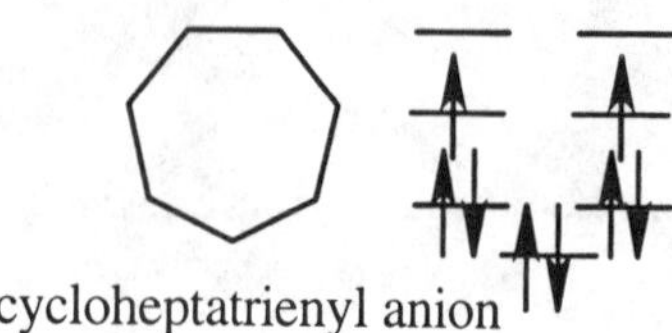

12. Cyclobutadiene has 1 bonding orbital, 2 nonbonding orbitals, and 1 antibonding orbital.

Because each of its four atoms has a *p* orbital, it has 4 π electrons; 2 electrons are in the bonding π molecular orbital, and each of the two nonbonding molecular orbitals contains one electron.

13. Because a cyclic compound has an odd number of bonding molecular orbitals, all the bonding molecular orbitals will be filled if the compound has an odd number of pairs of π electrons.

If the compound has an even number of pairs of π electrons, electrons will be left over after all the bonding molecular orbitals are filled. These electrons will have to go into degenerate nonbonding or antibonding molecular orbitals and will be unpaired.

14. Because benzene is an aromatic molecule, it is particularly stable. Therefore, electrophilic addition to benzene is an endergonic reaction because benzene is more stable than the non-aromatic addition product. (See Figure 14.3 on page 612 of the text.)

An alkene is much less stable than benzene because it does not have any delocalized electrons. Electrophilic addition to an alkene is an exergonic reaction because an alkene is less stable than the addition product. (The overall reaction trades a σ bond and a π bond into two σ bonds; a σ bond is stronger and, therefore, of lower energy than a π bond.)

15. Hydrated ferric bromide cannot activate Br_2 for nucleophilic attack by accepting a pair of electrons from it, because it has already accepted a pair of electrons from water.

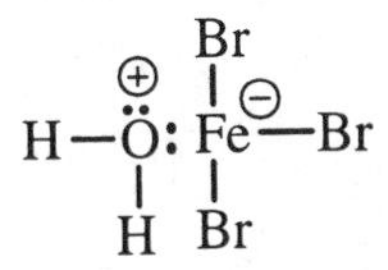

```
      ⊕  Br
         | ⊖
  H—Ö: Fe—Br
     |   |
     H   Br
```

16. In sulfonation, **A** to **B** has a smaller rate constant (a higher energy hill and, therefore, a slower reaction) than **B** to **C**; **A** to **B** is the rate-determining step.

In desulfonation (the reverse reaction), **C** to **B** has a smaller rate constant than **B** to **A**, but **B** to **A** is the rate-determining step. That is because once **B** is formed, it is easier for it to go back to **C** than to proceed to **A**, so **B** to **A** is the bottleneck (or rate-determining) step.

This example shows that the rate-determining step of a reaction is not necessarily the step with the smallest rate constant; it is the step that has the transition state with the highest energy on the reaction coordinate.

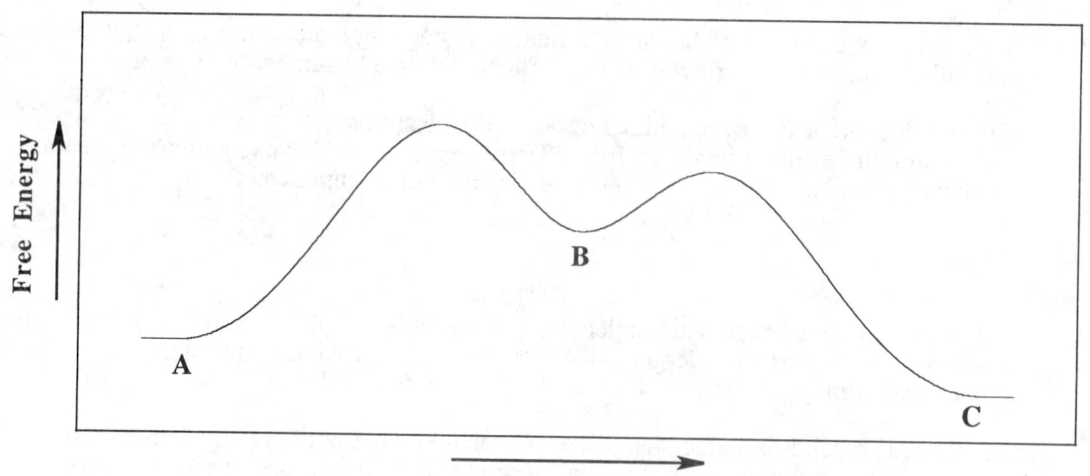

17.

$$R-\overset{\overset{\displaystyle :\ddot{O}}{\|}}{C}-O-\overset{\overset{\displaystyle O}{\|}}{C}-R + AlCl_3 \longrightarrow \left[R-C\equiv\overset{+}{O}: \longleftrightarrow R-\overset{+}{C}=\ddot{O}:\right] + R-\overset{\overset{\displaystyle O}{\|}}{C}-O\bar{A}lCl_3$$

acylium ion

18.

a. CH_2CH_3 (on a benzene ring)

c. $CH_3CH_2CHCH_3$ (benzene ring on the CH)

e. $CH_3\overset{\overset{\displaystyle CH_3}{|}}{C}CH_3$ (benzene ring on the central C)

b. CH_3CHCH_3 (benzene ring on the CH)

d. $CH_3\overset{\overset{\displaystyle CH_3}{|}}{C}CH_2CH_3$ (benzene ring on the central C)

f. $CH_2CH{=}CH_2$ (on a benzene ring)

19.

a. Benzene $\xrightarrow[\text{AlCl}_3]{\text{CH}_3\text{CH(Cl)CH}_2\text{CH}_2\text{CH}_3}$ C_6H_5–$CH(CH_3)CH_2CH_2CH_3$ ($CH_3\underset{|}{C}HCH_2CH_2CH_3$ attached to benzene)

b. Benzene $\xrightarrow[\text{2. H}_2\text{O}]{\text{1. CH}_3\text{CH}_2\text{CH}_2\text{CH}_2\text{CCl(=O)} + \text{AlCl}_3}$ C_6H_5–$C(=O)CH_2CH_2CH_2CH_3$

$\downarrow$ Zn(Hg)/HCl/Δ or $NH_2NH_2/HO^-/\Delta$

C_6H_5–$CH_2CH_2CH_2CH_2CH_3$

20.

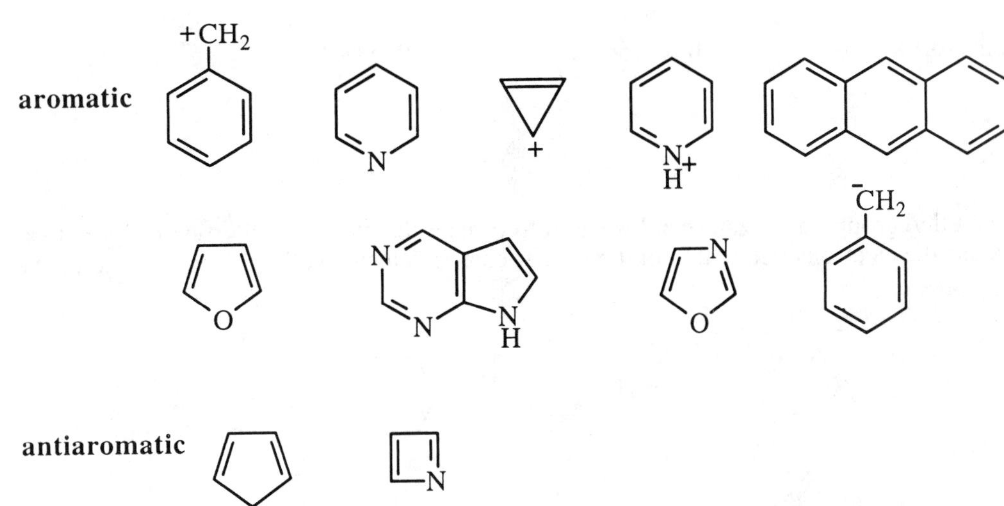

By not being flat, cyclooctatetraene avoids being antiaromatic.

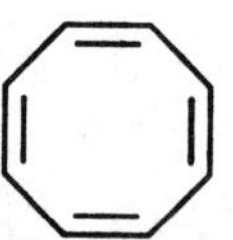

Notice that the nitrogen atoms in compounds such as those to the right are sp^3 hybridized, because if they were sp^2 hybridized, the compounds would be unstable antiaromatic compounds.

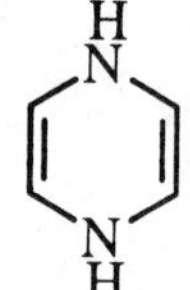

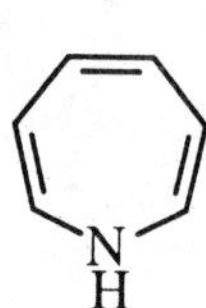

21.

a. $CH_3\overset{CH_3}{C}CH_3$ (attached to a benzene ring)

b. CH_3CHCH_3 (attached to a benzene ring)

c. $CH_3\overset{CH_3}{C}CH_2CH_3$ (attached to a benzene ring)

d. CH_2Cl (attached to a benzene ring)

22.

a. 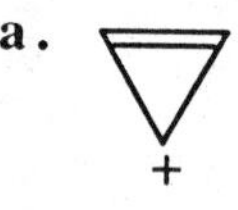

It is aromatic.

b. +

It is aromatic.

c. 

It is aromatic.

23. The methyl group on benzene can lose a proton easier than the methyl group on cyclohexane because the electrons left behind in the former can be delocalized by resonance into the benzene ring.

$-CH_2-H$ $\qquad$ $-CH_2-H$

24. Cl^- is formed because the central three-membered ring in the compound is aromatic when it has a positive charge. If Cl^+ were formed, the central three-membered ring in the compound would have a negative charge and, therefore, would be antiaromatic.

25. The ^{1}H NMR spectrum is the spectrum of 1-phenylbutane: the benzene ring protons show a signal at ~ 7.2 ppm. The two triplets and two multiplets indicate a butyl substituent.

$C_6H_5-CH_2CH_2CH_2CH_3$

Therefore, the acyl chloride has a straight chain propyl group and a carbonyl group that will be reduced to a methylene group.

$CH_3CH_2CH_2\overset{O}{\overset{\|}{C}}Cl$

26. **a.** In fulvene, the electrons move toward the five-membered ring because the resonance contributor that has a negative charge on a ring carbon is aromatic.

fulvene

b. In calicene, the electrons move toward the five-membered ring because both rings are aromatic in the resonance contributor that has a negative charge on a carbon of the five-membered ring and a positive charge on a carbon of the three-membered ring.

calicene

27. **a.** Pyridine is a stronger base because its nonbonding electrons are not part of the π cloud. Thus, when it is protonated, it is still aromatic.

The nonbonding electrons of pyrrole are part of the π electron cloud, so when pyrrole is protonated, it loses its aromaticity.

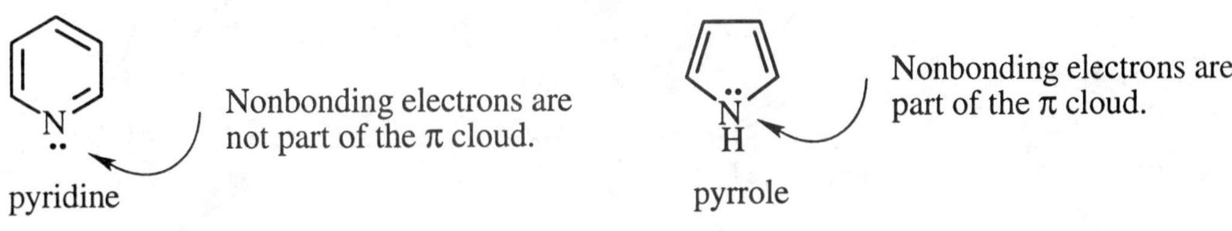
N
Nonbonding electrons are not part of the π cloud.
pyridine
N
H
Nonbonding electrons are part of the π cloud.
pyrrole

b. The compound with the carbon-nitrogen double bond is a stronger base because it has increased electron density on one of the nitrogens as a result of resonance.

$$CH_3\overset{\overset{\ddot{N}H}{\|}}{C}-\ddot{N}H_2 \longleftrightarrow CH_3\overset{\overset{:\ddot{N}H^-}{|}}{C}=\overset{+}{N}H_2$$

28.

a. $C_6H_5CH_2CH_2CH(CH_3)CH=CH_2 \xrightarrow{H^+} C_6H_5CH_2CH_2CH(CH_3)\overset{+}{C}HCH_3 \xrightarrow{\text{1,2-hydride shift}} C_6H_5CH_2CH_2\overset{+}{C}(CH_3)CH_2CH_3 \longrightarrow$ (cyclized cation, CH_3, CH_2CH_3, H) $\xrightarrow{-H^+}$ (indane product with CH_3 and CH_2CH_3)

b. $C_6H_5CH=CH_2 \xrightarrow{H^+} C_6H_5\overset{+}{C}H-CH_3$ + $C_6H_5CH=CH_2 \longrightarrow C_6H_5CH(CH_3)-CH_2-\overset{+}{C}H-C_6H_5 \longrightarrow$ (cyclized cation, CH_3, H) $\xrightarrow{-H^+}$ (1-methyl-3-phenylindane, CH_3)

29.

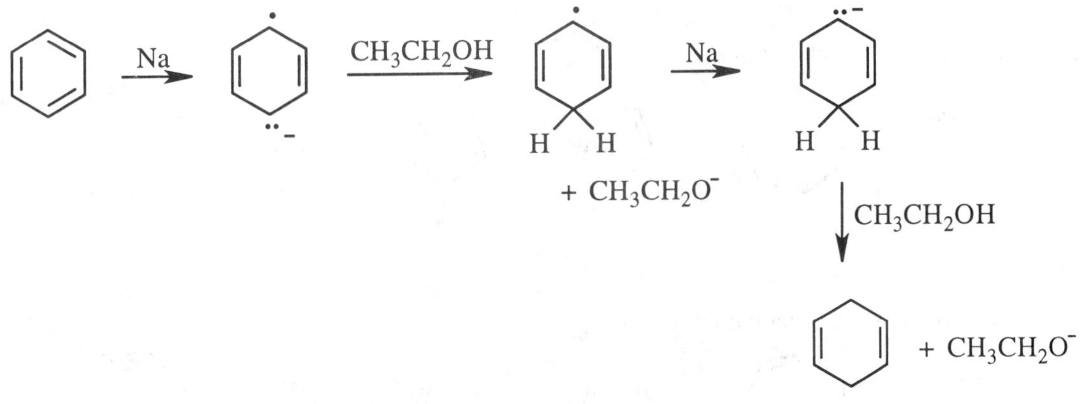

Na → ; CH_3CH_2OH → H H + $CH_3CH_2O^-$; Na → H H ; CH_3CH_2OH → + $CH_3CH_2O^-$

30. From the three resonance structures for the cyclohexadienyl anion, the bond orders in the resonance hybrid can be calculated. (For example, the C1-C2 bond is represented by a single bond in one resonance structure and by a double bond in two resonance structures, which gives it a bond order 1+ 2/3 in the hybrid.)

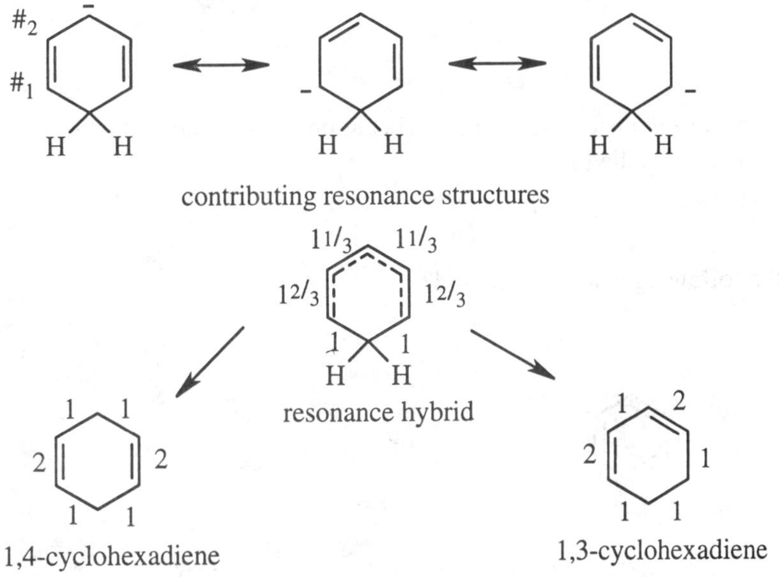

#2 #1 H H ⟷ H H ⟷ H H

contributing resonance structures

1 1/3 1 1/3 1 2/3 1 2/3 1 1 H H

resonance hybrid

1 1 2 2 1 1 — 1,4-cyclohexadiene

1 2 2 1 1 1 — 1,3-cyclohexadiene

Comparing the bond order difference in each bond in the anion and in the corresponding bond in the possible products, there is a (1/3 + 1/3 + 1/3 + 1/3 = 4/3) 4/3 difference in bond order when 1,4-cyclohexadiene is formed and a (1/3 + 1/3 + 2/3 + 2/3 = 6/3) 6/3 difference in bond order when 1,3-cyclohexadiene is formed. Thus, the principle of least motion predicts that 1,4-cyclohexadiene will be the major product, since its formation from the anion involves less of a change in electronic configuration.

Chapter 14 Practice Test

1. Which are aromatic compounds?

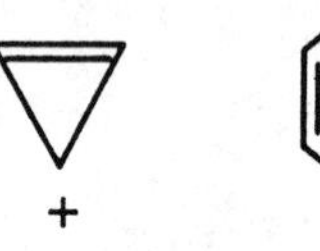

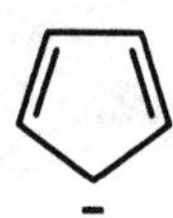

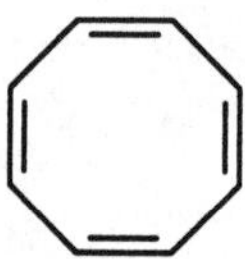

2. Which are antiaromatic compounds?

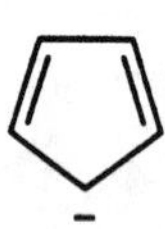

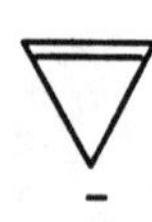

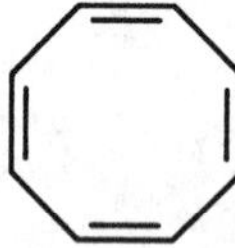

3. Which compound has the greater resonance energy?

or
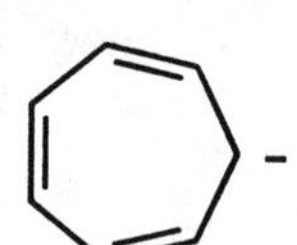

4. Draw the resonance contributors for the carbocation itermediate that is formed when benzene reacts with an electrophile (Y^+).

5. Which of the following is a stronger acid?

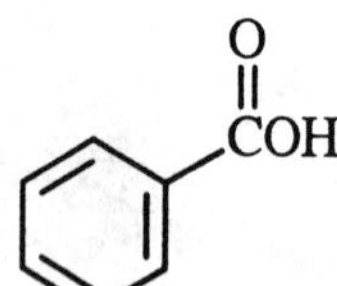

or
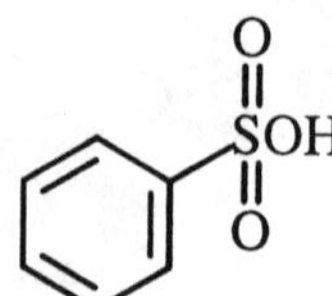

or
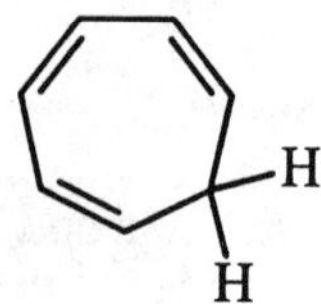

6. Give two alkyl halides, two alkenes, and two alcohols that could be used in a reaction with benzene to form 2-phenylbutane.

7. What acid anhydride would you use in a synthesis of 1-phenylpropane?

8. Which in each of the following pairs is more stable?

a. $\overset{+}{C}HCH_3$ or $\overset{+}{C}HCH_3$

b. $CH_3\overset{+}{C}HCH_2CH_3$ or $CH_3\overset{+}{C}HCH{=}CH_2$

c. $CH_3\overset{-}{C}HCH_2\overset{\overset{O}{\|}}{C}CH_3$ or $CH_3\overset{-}{C}H\overset{\overset{O}{\|}}{C}CH_3$

9. Give the mechanism for formation of the nitronium ion from nitric acid and sulfuric acid.

CHAPTER 15
Reactions of Substituted Benzenes

Important Terms

activating substituent — a substituent that increases the reactivity of an aromatic ring. Electron-donating substituents activate aromatic rings toward electrophilic attack, and electron-withdrawing substituents activate aromatic rings toward nucleophilic attack.

arenediazonium salt

$$\mathrm{Ar}\overset{+}{\mathrm{N}}\equiv\mathrm{N}\quad \mathrm{X}^{-}$$

azo linkage — an -N=N- bond.

benzyl group

C_6H_5–CH_2—

benzyne intermediate — a compound with a triple bond in place of one of the double bonds of benzene.

cine substitution — substitution at the carbon adjacent to the carbon that was bonded to the leaving group.

deactivating substituent — a substituent that decreases the reactivity of an aromatic ring. Electron-withdrawing substituents deactivate aromatic rings toward electrophilic attack, and electron-donating substituents deactivate aromatic rings toward nucleophilic attack.

direct substitution — substitution at the carbon that was bonded to the leaving group.

donate electrons by resonance — donation of electrons through *p* orbital overlap with neighboring π bonds.

fused rings — rings that share two adjacent carbons.

inductive electron donation — donation of electrons through a σ bond.

inductive electron withdrawal — withdrawal of electrons through a σ bond.

meta director — a substituent that directs an incoming substituent meta to an existing substituent.

nitrosamine — an amine with a nitroso (-N=O) substituent bonded to its nitrogen atom

***N*-nitroso compound** — an amine with a nitroso (-N=O) substituent bonded to its nitrogen atom

nucleophilic aromatic substitution a reaction in which a nucleophile substitutes for an atom or group bonded to a benzene ring.

ortho/para director a substituent that directs an incoming substituent ortho and para to an existing substituent.

phenyl group

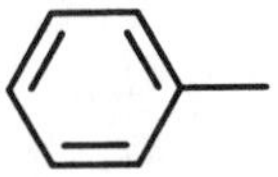

C_6H_5—

resonance electron donation donation of electrons through *p* orbital overlap with neighboring π bonds.

resonance electron withdrawal withdrawal of electrons through *p* orbital overlap with neighboring π bonds.

Sandmeyer reaction the reaction of an arenediazonium ion with a cuprous salt.

Schiemann reaction the reaction of an arenediazonium ion with HBF_4.

S_NAr reaction a nucleophilic aromatic substitution reaction.

withdraw electrons by resonance withdrawal of electrons through *p* orbital overlap with neighboring π bonds.

Solutions to Problems

1.

a. CH_3 / NH_2

b. CH_3 / OH

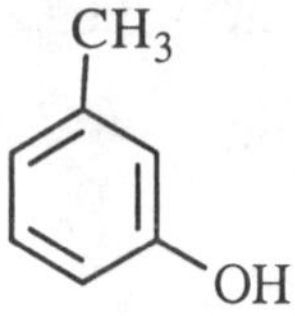

c. CH_3 / CH_3

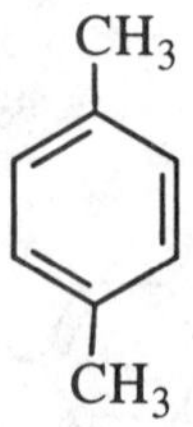

2.

a. Cl / Cl

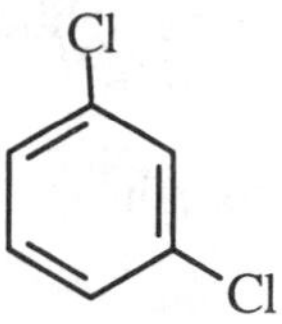

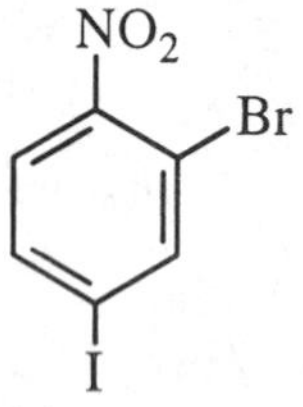

b. OH / Br

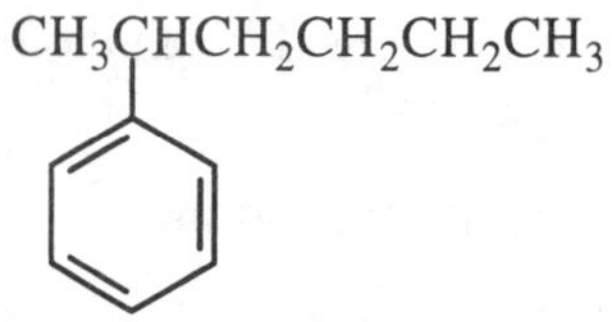

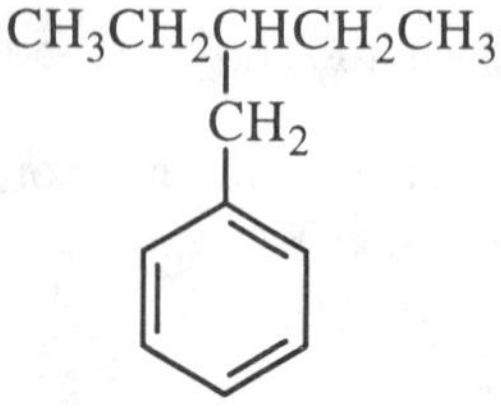

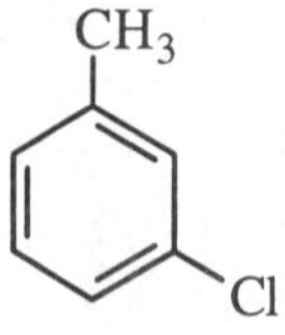

c. NH_2 / NO_2

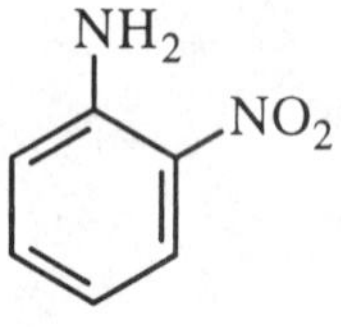

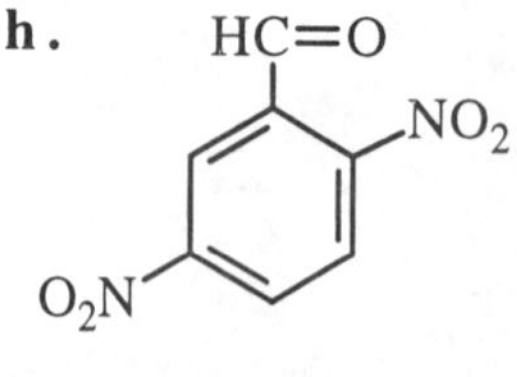

d. NO_2 / Br / I

e. $CH_3CHCH_2CH_2CH_2CH_3$

f. $CH_3CH_2CHCH_2CH_3$ / CH_2

g. CH_3 / Cl

h. HC=O / NO_2 / O_2N

i. CH_3 / CH_3

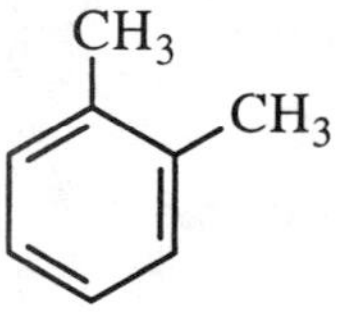

j. C≡N / Cl

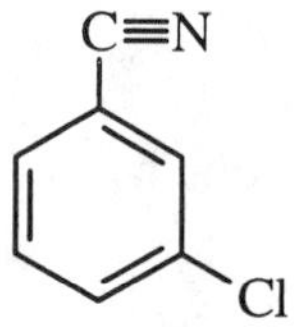

3. **a.** 1,3,5-tribromobenzene

b. *meta*-nitrophenol or 3-nitrophenol

c. *para*-bromotoluene or 4-bromotoluene

d. *ortho*-dichlorobenzene or 1,2-dichlorobenzene

4.

a. 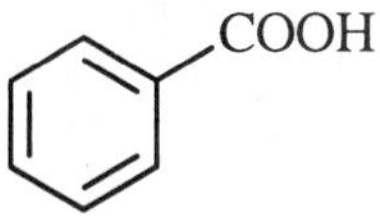 C_6H_5COOH (benzene ring with COOH)

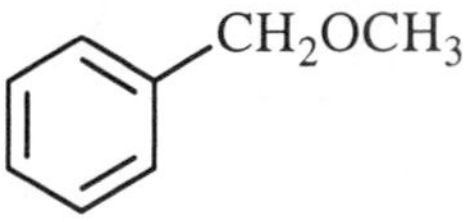

c. benzene ring with COOH and COOH (1,4-positions)

b. $C_6H_5CH_2OCH_3$

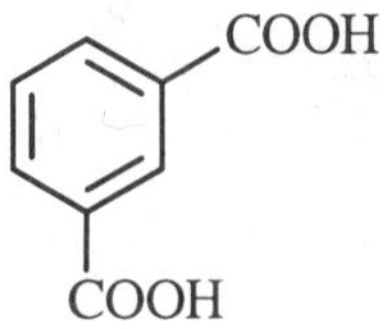

d. 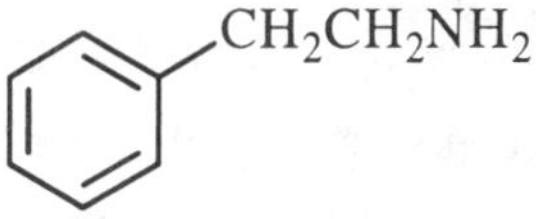 $C_6H_5CH_2CH_2NH_2$

5. **a.** Solved in the text.

b. 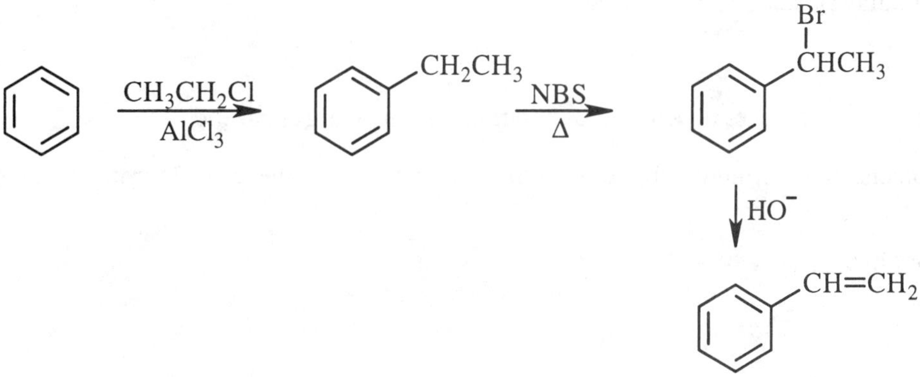 benzene $\xrightarrow[AlCl_3]{CH_3CH_2Cl}$ $C_6H_5CH_2CH_3$ $\xrightarrow[\Delta]{NBS}$ $C_6H_5CH(Br)CH_3$ $\xrightarrow{HO^-}$ $C_6H_5CH{=}CH_2$

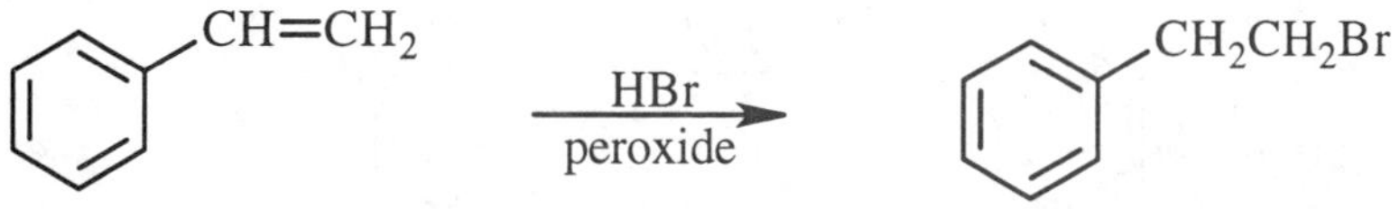

c. $C_6H_5CH{=}CH_2$ (product of **b**) $\xrightarrow[peroxide]{HBr}$ $C_6H_5CH_2CH_2Br$

d. 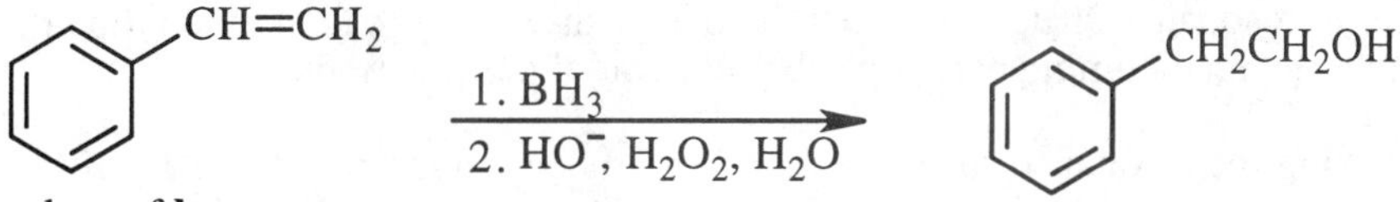 $C_6H_5CH{=}CH_2$ (product of **b**) $\xrightarrow[2.\ HO^-,\ H_2O_2,\ H_2O]{1.\ BH_3}$ $C_6H_5CH_2CH_2OH$

e. 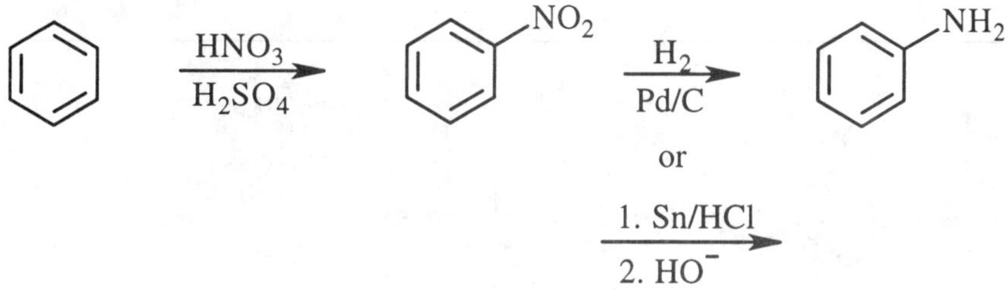 benzene $\xrightarrow[H_2SO_4]{HNO_3}$ $C_6H_5NO_2$ $\xrightarrow[Pd/C]{H_2}$ $C_6H_5NH_2$

or

$\xrightarrow[2.\ HO^-]{1.\ Sn/HCl}$

f. 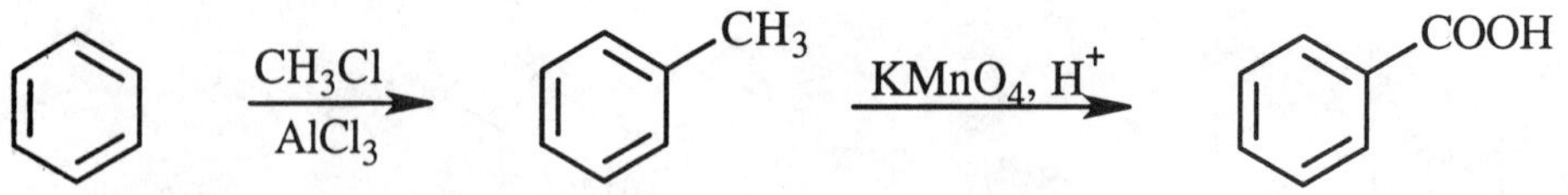 benzene $\xrightarrow[AlCl_3]{CH_3Cl}$ toluene (CH_3) $\xrightarrow{KMnO_4,\ H^+}$ benzoic acid (COOH)

6. **a.** donates electrons by resonance and withdraws electrons inductively
b. donates electrons inductively
c. withdraws electrons by resonance and withdraws electrons inductively
d. donates electrons by resonance and withdraws electrons inductively
e. donates electrons by resonance and withdraws electrons inductively
f. withdraws electrons inductively

7. **a.** phenol > toluene > benzene > bromobenzene > nitrobenzene
b. toluene > chloromethylbenzene > dichloromethylbenzene > difluoromethylbenzene

8. Solved in the text.

9. **a.**

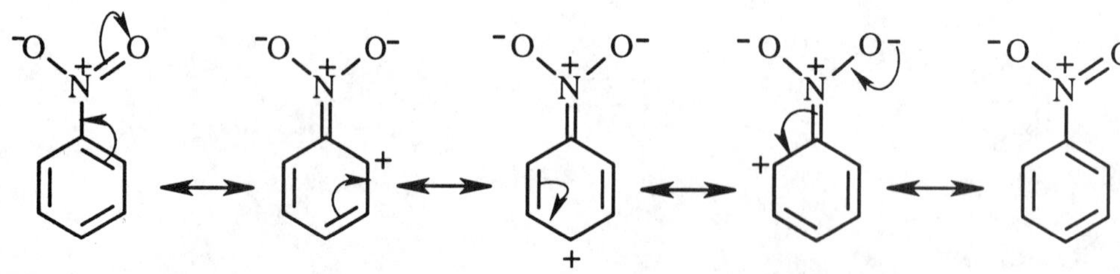

The above resonance structures show there is positive charge density on the ortho and para positions, so an incoming electrophile will avoid those positions.

By drawing the resonance structures for the carbocation intermediates formed by putting a substituent at various positions, you will see that the most stable carbocation is obtained when the substituent is placed in the meta position.

b.

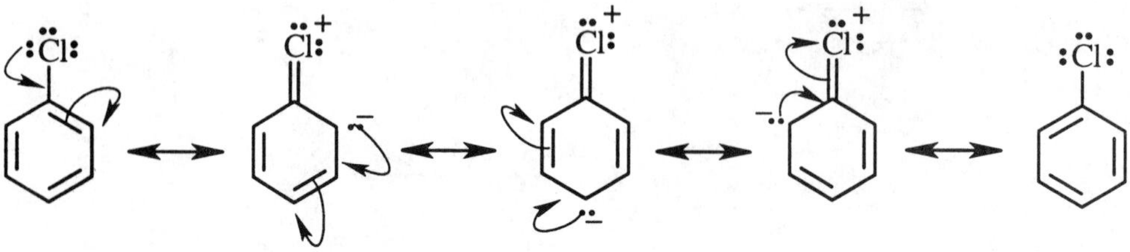

Resonance electron donation increases the electron density at the ortho and para positions, so an incoming nucleophile will be attracted to those positions.

By drawing the resonance structures for the carbocation intermediates formed by putting a substituent at various positions, you will see that the most stable carbocation is obtained when the substituent is placed in the ortho or para position.

10.

a. $CH_2CH_2CH_3$ + $CH_2CH_2CH_3$ NO_2 ; NO_2

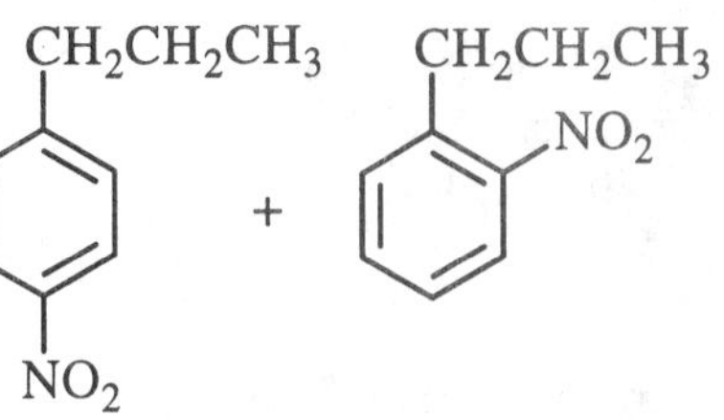

c. O ‖ CH ; NO_2

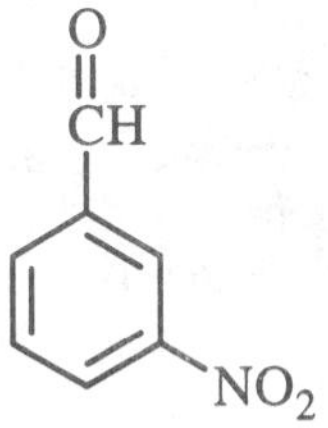

d. SO_3H ; NO_2

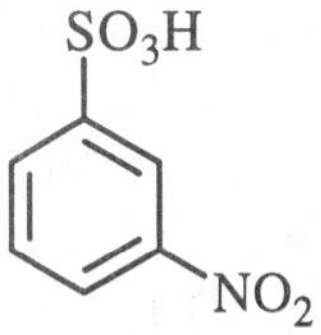

b. Br + Br NO_2 ; NO_2

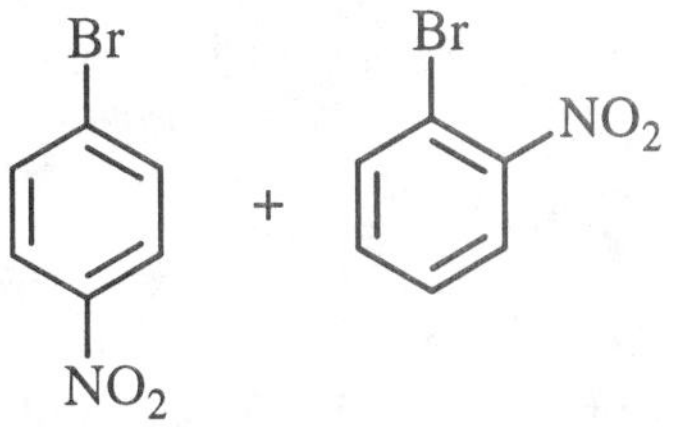

f. C≡N ; NO_2

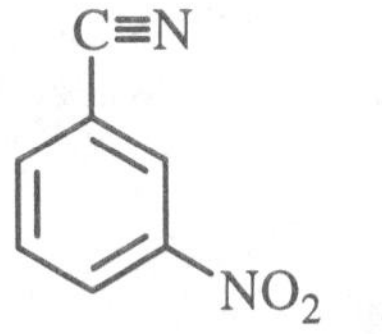

e. NO_2 + NO_2

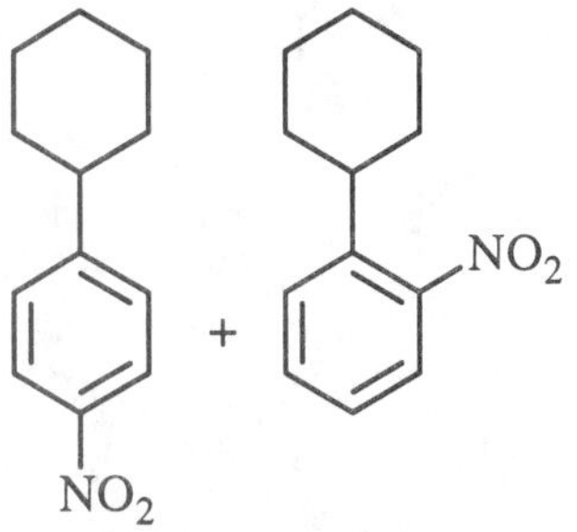

11. They are all meta directors:

a. CF_3 withdraws electrons inductively from the ring.

b. N=O withdraws electrons inductively and withdraws electrons by resonance. You could draw resonance contributors for electron donation into the ring by resonance. However, the most stable resonance contributors are obtained by electron flow out of the benzene ring toward oxygen, the most electronegative atom in the compound.

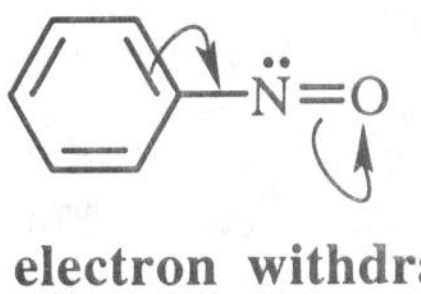

N̈=O N̈=O

resonance electron withdrawal out of the ring 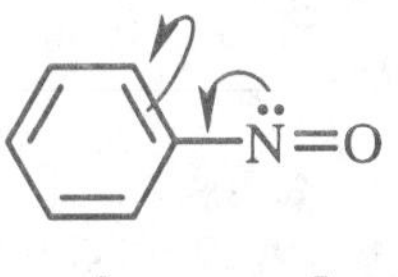 **resonance electron donation into of the ring**

c. NO_2 withdraws electrons inductively and withdraws electrons by resonance.

12.

a. $\mathrm{ClCH_2\overset{\overset{O}{\|}}{C}OH}$

b. $\mathrm{O_2NCH_2\overset{\overset{O}{\|}}{C}OH}$

c. $\mathrm{H_3\overset{+}{N}CH_2\overset{\overset{O}{\|}}{C}OH}$

d. $\mathrm{HO\overset{\overset{O}{\|}}{C}CH_2\overset{\overset{O}{\|}}{C}OH}$

The negatively charged compound has a greater amount of electron donation by resonance.

e. $\mathrm{H\overset{\overset{O}{\|}}{C}OH}$

A hydrogen is more electron-withdrawing than a methyl group.

f.

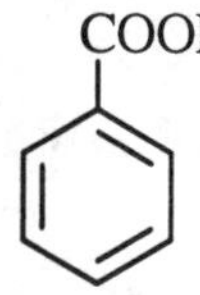

g. $\mathrm{FCH_2\overset{\overset{O}{\|}}{C}OH}$

Fluorine is more electronegative than chlorine.

h.

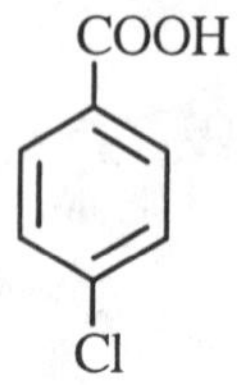

We know that F withdraws electrons inductively more than Cl, because F is more electronegative than Cl.
We know that F donates electrons by resonance better than Cl, because F donation involves 2*p*-2*p* overlap, while Cl donation involves 3*p*-2*p* overlap. Because Table 15.1 on page 639 of the text shows that Cl is more deactivating than F, we know that overall Cl withdraws electrons better than F.

13. When *para*-nitrophenol loses a proton, the electrons that held the proton can be delocalized by resonance onto the nitro substituent. Therefore, the *para*-nitro substituent decreases the pK_a by resonance electron withdrawal and by inductive electron withdrawal.

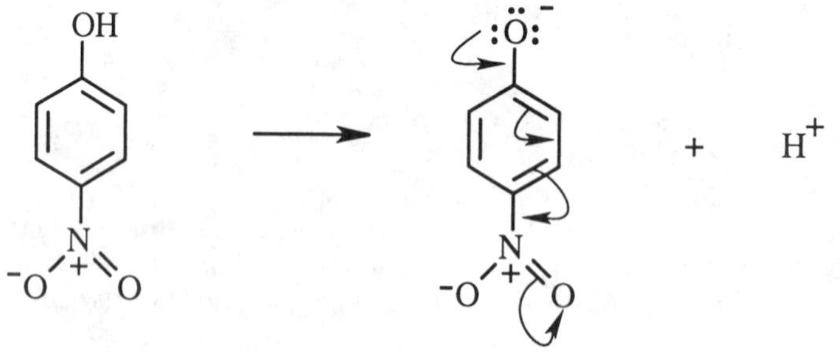

When *meta*-nitrophenol loses a proton, the electrons that held the proton cannot be delocalized by resonance onto the nitro substituent. Therefore, the *meta*-nitro substituent can only decrease the pK_a by inductive electron withdrawal. Therefore, the para isomer has a lower pK_a.

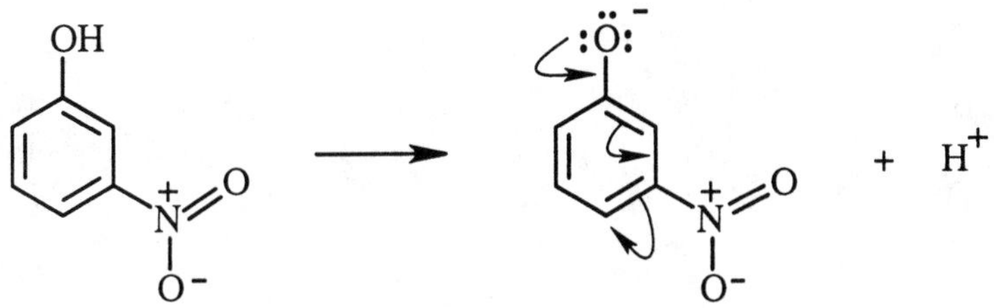

OH

$:\ddot{O}:^-$

+ H^+

14. In acid solution, some of the amine is protonated, and the protonated substituent is a meta director. Because a greater percentage of the compound is protonated at pH = 3.5 than at pH = 4.5, more of the meta isomer is formed at pH = 3.5.

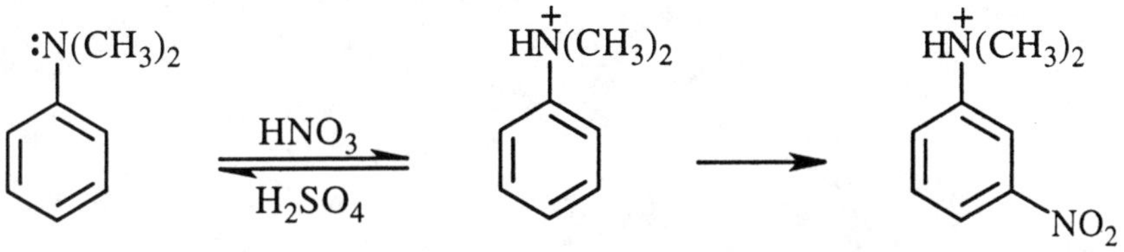

$:N(CH_3)_2$ $\xrightleftharpoons[H_2SO_4]{HNO_3}$ $H\overset{+}{N}(CH_3)_2$ ⟶ $H\overset{+}{N}(CH_3)_2$, NO_2

15. No reaction will occur in **a** and **c**, because a Friedel-Crafts reaction cannot be carried out on a ring that possesses a meta director.

a. no reaction

c. no reaction

b. NH_2, Br, Br, Br

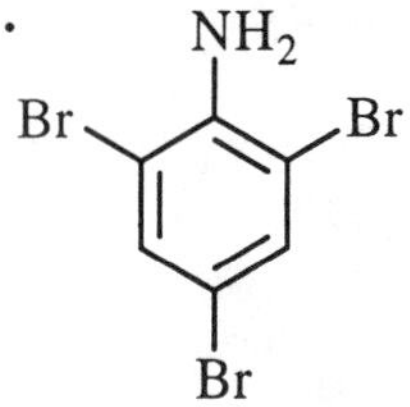

d. CH_3, CH_3 + CH_3, CH_3

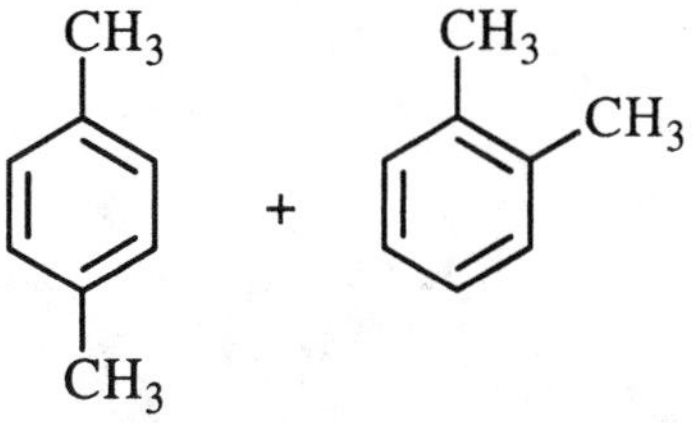

16.

a. $\xrightarrow[FeCl_3]{Cl_2}$ Cl $\xrightarrow[H_2SO_4]{HNO_3}$ Cl, NO_2 $\xrightarrow[2.\ HO^-]{1.\ Sn/HCl}$ Cl, NH_2

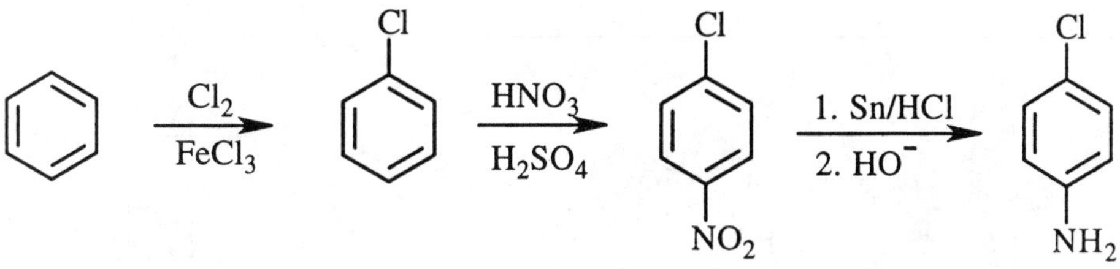

b. benzene $\xrightarrow[AlCl_3]{CH_3Cl}$ toluene (CH_3) $\xrightarrow[H^+,\ \Delta]{KMnO_4}$ benzoic acid (COOH) $\xrightarrow[H_2SO_4]{HNO_3}$ m-nitrobenzoic acid (COOH, NO_2)

c. benzene $\xrightarrow[AlCl_3]{CH_3Cl}$ toluene (CH_3) $\xrightarrow[H_2SO_4]{HNO_3}$ p-nitrotoluene (CH_3, NO_2) $\xrightarrow[H^+,\ \Delta]{KMnO_4}$ p-nitrobenzoic acid (COOH, NO_2)

d. benzene $\xrightarrow[2.\ H_2O]{1.\ CH_3CH_2\overset{O}{\overset{\|}{C}}Cl\ +\ AlCl_3}$ $CH_3CH_2\overset{O}{\overset{\|}{C}}$–$C_6H_5$ $\xrightarrow[FeBr_3]{Br_2}$ $CH_3CH_2\overset{O}{\overset{\|}{C}}$–$C_6H_4$–Br (meta)

$\downarrow$ NH_2NH_2 | HO^-, Δ

$CH_3CH_2CH_2$–C_6H_4–Br (meta)

e. benzene $\xrightarrow[2.\ H_2O]{1.\ CH_3CH_2\overset{O}{\overset{\|}{C}}Cl\ +\ AlCl_3}$ $CH_3CH_2\overset{O}{\overset{\|}{C}}$–$C_6H_5$ $\xrightarrow[HCl\ \Delta]{Zn(Hg)}$ $CH_3CH_2CH_2$–C_6H_5

$\downarrow$ Br_2 | $FeBr_3$

$CH_3CH_2CH_2$–C_6H_4–Br (ortho)

f. Benzene $\xrightarrow[\text{2. } H_2O]{\text{1. } CH_3CH_2COCl \text{ + } AlCl_3}$ $C_6H_5CCH_2CH_3$ (O above C) $\xrightarrow[\text{HCl } \Delta]{\text{Zn(Hg)}}$ $C_6H_5CH_2CH_2CH_3$

$\xrightarrow{\text{NBS } \Delta}$ C_6H_5BrCHCH$_2$CH$_3$ $\xrightarrow{HO^-}$ $C_6H_5CH{=}CHCH_3$ $\xrightarrow[\text{2. } H_2O_2, \, HO^-, \, H_2O]{\text{1. } BH_3/THF}$ $C_6H_5CH_2CH(OH)CH_3$

g. Benzene $\xrightarrow[H_2SO_4]{HNO_3}$ nitrobenzene (NO_2) $\xrightarrow[\text{2. } HO^-]{\text{1. Sn, HCl}}$ aniline (NH_2) $\xrightarrow[0°]{NaNO_2, \, HCl}$ benzenediazonium chloride ($\overset{+}{N}{\equiv}N$ Cl^-)

$\xrightarrow{H_2O}$ phenol (OH) $\xrightarrow[H_2SO_4]{HNO_3}$ o-nitrophenol (OH, NO_2)

h. Benzene $\xrightarrow[H_2SO_4]{HNO_3}$ nitrobenzene (NO_2) $\xrightarrow[H_2SO_4]{HNO_3}$ m-dinitrobenzene (NO_2, NO_2) $\xrightarrow{(NH_4)_2S}$ m-nitroaniline (NH_2, NO_2)

$\xrightarrow[0°]{NaNO_2, \, HCl}$ m-nitrobenzenediazonium chloride ($\overset{+}{N}{\equiv}N$ Cl^-, NO_2) $\xrightarrow{H_2O}$ m-nitrophenol (OH, NO_2)

i. benzene $\xrightarrow[AlCl_3]{CH_3CHClCH_3}$ $C_6H_5CH(CH_3)_2$ $\xrightarrow[\Delta]{NBS}$ $C_6H_5CBr(CH_3)_2$ $\xrightarrow{HO^-}$ $C_6H_5C(CH_3)=CH_2$

(Structures: CH_3CHCH_3 on benzene; CH_3CCH_3 with Br on benzene; $CH_3C=CH_2$ on benzene)

j. benzene $\xrightarrow[H_2SO_4]{HNO_3}$ nitrobenzene (NO_2) $\xrightarrow[FeCl_3]{Cl_2}$ m-chloronitrobenzene (NO_2, Cl) $\xrightarrow[2.\ HO^-]{1.\ Sn/HCl}$ m-chloroaniline (NH_2, Cl)

17.

a. p-fluoroanisole (OCH_3, F) $\xrightarrow[H_2SO_4]{HNO_3}$ 4-fluoro-2-nitroanisole (OCH_3, NO_2, F)

Methoxy is strongly activating, while fluorine is deactivating, so methoxy will do the directing.

b. phthalic acid (COOH, COOH) $\xrightarrow[FeCl_3]{Cl_2}$ 4-chlorophthalic acid (COOH, COOH, Cl) + 3-chlorophthalic acid (COOH, COOH, Cl)

COOH directs to the meta position. The same product will be obtained regardless of which COOH is chosen to be the director.

Less of this compound will be obtained because of steric hindrance.

c. p-chlorobenzoic acid (COOH, Cl) $\xrightarrow[FeBr_3]{Br_2}$ 3-bromo-4-chlorobenzoic acid (COOH, Br, Cl)

COOH directs to its meta position and Cl directs to its ortho position, so they both direct to the same position on the ring.

18. Solved in the text.

19. More of the desired bromo-substituted compound will be obtained if the ions in solution are Cu^+ and Br^- than if they are Cu^+ and Br^- and Cl^-.

20.

benzene $\xrightarrow[H_2SO_4]{HNO_3}$ nitrobenzene (NO_2) $\xrightarrow[2.\ HO^-]{1.\ Sn/HCl}$ or $\xrightarrow{H_2,\ Pd/C}$ aniline (NH_2) $\xrightarrow[HCl, 0°]{NaNO_2}$ benzenediazonium chloride ($\overset{+}{N}{\equiv}N$ Cl^-)

21. **a.** The synthesis was not successful.

b. As benzenediazonium chloride warms up, nitrogen is lost, which means that the compound will react with whatever nucleophile is available. The nucleophile in greatest concentration is water, so the major product of the reaction is phenol.

22. **a.** Solved in the text.

b. benzene $\xrightarrow[AlCl_3]{CH_3Cl}$ toluene (CH_3) $\xrightarrow[H_2SO_4]{HNO_3}$ p-nitrotoluene (CH_3, NO_2) $\xrightarrow[2.\ HO^-]{1.\ Sn/HCl}$ or $\xrightarrow{H_2,\ Pd/C}$ p-toluidine (CH_3, NH_2)

$\xrightarrow[HCl]{NaNO_2,\ 0°}$ p-methylbenzenediazonium chloride (CH_3, $+N{\equiv}N$ Cl^-) $\xrightarrow{CuC{\equiv}N}$ p-methylbenzonitrile (CH_3, $C{\equiv}N$)

c. The reaction steps are the same as those in **a** (answered on page 656 of the text) except for the last step. In the last step the *meta*-bromo-substituted ion should undergo reaction with water rather than with CuBr.

$\overset{+}{N}\equiv N\ Cl^-$ Br $\xrightarrow{H_2O}$ OH Br

d. $\xrightarrow[H_2SO_4]{HNO_3}$ NO_2 $\xrightarrow[\text{or } H_2,\ Pd/C]{1.\ Sn/HCl\ \ 2.\ HO^-}$ NH_2 $\xrightarrow[HCl\ 0°]{NaNO_2}$ $\overset{+}{N}\equiv N\ Cl^-$ $\xrightarrow{H_2O}$ OH $\xrightarrow[FeCl_3]{Cl_2}$ OH Cl

or

$\xrightarrow[FeCl_3]{Cl_2}$ Cl $\xrightarrow[H_2SO_4]{HNO_3}$ Cl NO_2 $\xrightarrow[\text{or } H_2,\ Pd/C]{1.\ Sn/HCl\ \ 2.\ HO^-}$ Cl NH_2 $\xrightarrow{NaNO_2,\ HCl,\ 0°}$ Cl $\overset{+}{N}\equiv N\ Cl^-$ $\xrightarrow{H_2O}$ Cl OH

e. The nitro group cannot be placed on the benzene ring first, because a Friedel-Crafts reaction cannot be carried out on a ring with a meta director. Because formyl chloride and formic anhydride do not exist, the only way a single carbon can be put on the ring is by methylation. Therefore, the methyl group has to be converted into a meta director and then converted back.

CH_3Cl, $AlCl_3$ → CH_3 ; NBS, Δ → CH_2Br ; HO^- → CH_2OH ; MnO_2, Δ → $\overset{O}{\overset{\|}{C}}H$; HNO_3, H_2SO_4 → $\overset{O}{\overset{\|}{C}}H$, NO_2 ; NH_2NH_2, HO^- Δ → CH_3, NO_2

f. CH_3Cl, $AlCl_3$ → CH_3 ; NBS, Δ → CH_2Br ; HO^- → CH_2OH ; MnO_2, Δ → $\overset{O}{\overset{\|}{C}}H$; Cl_2, $FeCl_3$ → $\overset{O}{\overset{\|}{C}}H$, Cl

23. You can see why nucleophilic attack occurs on the neutral nitrogen if you compare the products of nucleophilic attack on the two nitrogens. Nucleophilic attack on the neutral nitrogen forms a stable product, while nucleophilic attack on the positively charged nitrogen would form an unstable compound with two charged nitrogen atoms.

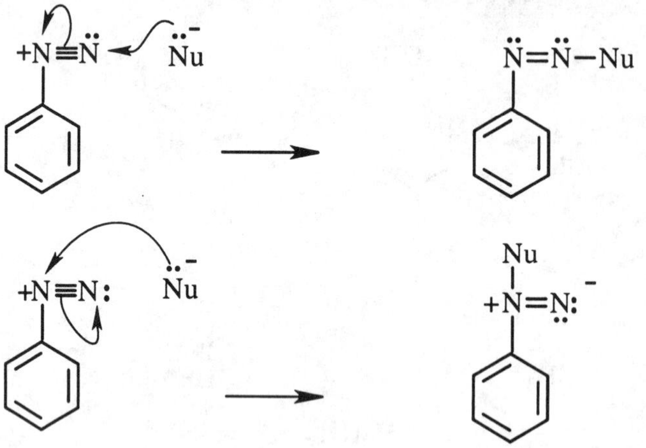

+N≡N̈ ⟵ N̈u⁻ ⟶ N̈=N̈−Nu

+N≡N: N̈u⁻ ⟶ Nu–+N=N̈:⁻

The terminal nitrogen is electrophilic because of electron withdrawal by the positively charged nitrogen. If you draw the resonance contributors, you can see that the "neutral" nitrogen is electron deficient.

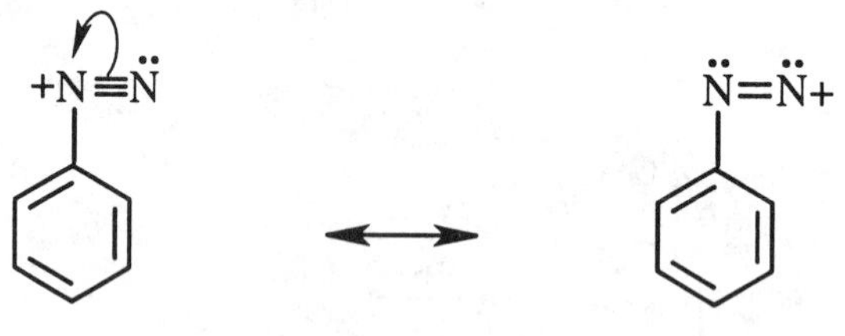

+N≡N̈ ⟷ N̈=N̈+

24. Because a diazonium ion is electron withdrawing, it deactivates the benzene ring. A deactivated benzene ring would be too unreactive to undergo an electrophilic substitution reaction at the cold temperature necessary to keep the benzenediazonium from decomposing.

25.

a. 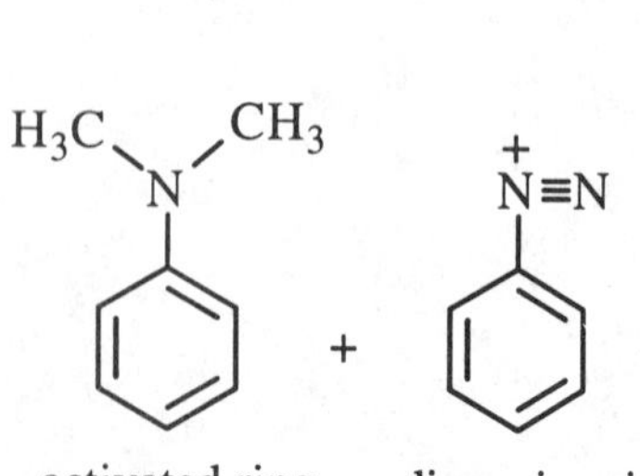 H_3C–N–CH_3 (on benzene) + $\overset{+}{N}$≡N (on benzene)

activated ring diazonium ion

b. 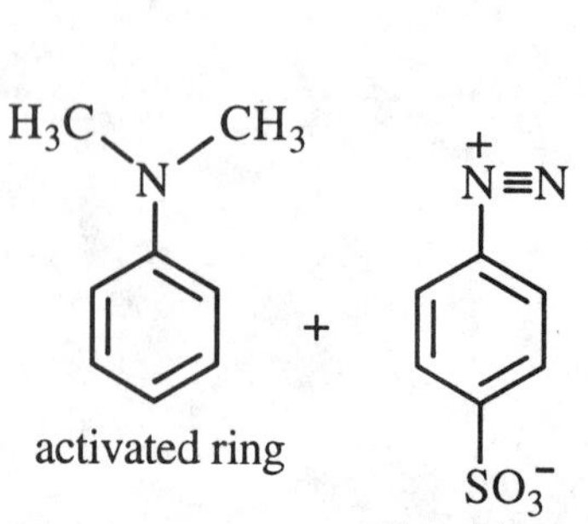 H_3C–N–CH_3 (on benzene) + $\overset{+}{N}$≡N (on benzene with SO_3^-)

activated ring diazonium ion

26. The reaction of an amine with sodium nitrite and HCl converts the amino group into an excellent leaving group. Substitution and elimination reactions can then occur by both S_N1/E1 and S_N2/E2 pathways. The same products are obtained by both pathways. Because the reaction is carried out in an aqueous solution, the nucleophiles are Cl^- and H_2O.

$$CH_3\underset{\displaystyle CH_3}{\underset{|}{C}}HNH_2 \xrightarrow[HCl]{NaNO_2} CH_3\underset{\displaystyle CH_3}{\underset{|}{C}}H\overset{+}{N}\equiv N$$

isopropylamine

E2/S_N2 →

$$CH_3\underset{\displaystyle Cl}{\underset{|}{C}}HCH_3 + CH_3\underset{\displaystyle OH}{\underset{|}{C}}HCH_3$$

$$CH_3CH{=}CH_2 + N_2$$

S_N1/E1 →

$$CH_3\underset{\displaystyle Cl}{\underset{|}{C}}HCH_3 + CH_3\underset{\displaystyle OH}{\underset{|}{C}}HCH_3$$

$$CH_3CH{=}CH_2 + N_2$$

27. The first step in the reaction is formation of the methyldiazonium ion as a result of abstraction of a proton from the carboxylic acid by diazomethane. Diazomethane is both explosive and toxic, so it should be synthesized only in small amounts by experienced laboratory workers. In the second step of the reaction, the carboxylate ion displaces nitrogen from the methyldiazonium ion. High yields are obtained, since the only side product is N_2 gas.

$$R-\overset{\displaystyle O}{\overset{\|}{C}}-OH + \overset{-}{\ddot{C}}H_2-\overset{+}{N}\equiv N \rightleftharpoons R-\overset{\displaystyle O}{\overset{\|}{C}}-\ddot{\underset{\cdot\cdot}{O}}:^- + CH_3-\overset{+}{N}\equiv N$$

diazomethane — methyldiazonium ion

$$\downarrow$$

$$R-\overset{\displaystyle O}{\overset{\|}{C}}-OCH_3 + N_2$$

28. From the resonance contributors, you can see that the reason that *meta*-chloronitrobenzene does not react with hydroxide ion is because the negative charge that is generated on the benzene ring cannot be delocalized onto the nitro substituent. Electron delocalization onto the nitro substituent can occur only if the nitro substituent is ortho or para to the site of nucleophilic attack.

Cl, NO_2 (benzene) + $H\ddot{\underset{\cdot\cdot}{O}}:^-$ ⟶ Cl, OH, NO_2 ⟷ Cl, OH, NO_2 ⟷ Cl, OH, NO_2

29.

a. 1-chloro-2,4-dinitrobenzene > *p*-chloronitrobenzene > chlorobenzene

b. chlorobenzene > *p*-chloronitrobenzene > 1-chloro-2,4-dinitrobenzene

30.

a. benzene $\xrightarrow[FeBr_3]{Br_2}$ bromobenzene $\xrightarrow[H_2SO_4]{HNO_3}$ o-bromonitrobenzene $\xrightarrow[\Delta]{^-NH_2}$ o-nitroaniline

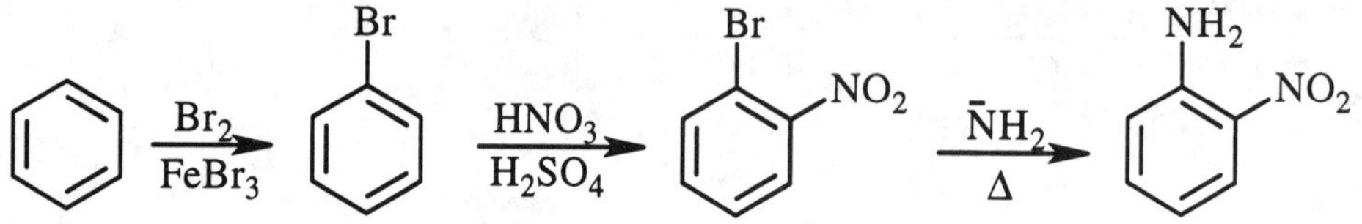

b. benzene $\xrightarrow[FeBr_3]{Br_2}$ bromobenzene $\xrightarrow[H_2SO_4]{HNO_3}$ p-bromonitrobenzene $\xrightarrow[\Delta]{^-NH_2}$ p-nitroaniline $\xrightarrow{CH_3I}$ $NHCH_3$ / NO_2 (para)

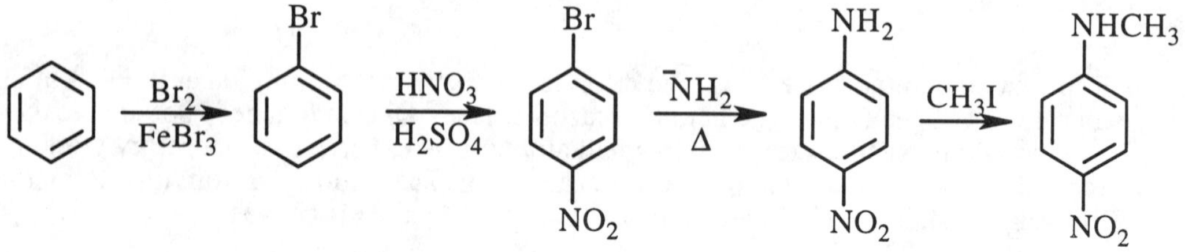

c. benzene $\xrightarrow[FeBr_3]{Br_2}$ bromobenzene $\xrightarrow[H_2SO_4]{HNO_3}$ p-bromonitrobenzene $\xrightarrow[\Delta]{CH_3O^-}$ OCH_3 / NO_2 (para) $\xrightarrow{1.\ Sn/HCl\quad 2.\ HO^-}$ OCH_3 / NH_2 (para) $\xrightarrow[HCl,\ 0°]{NaNO_2}$ OCH_3 / $+N{\equiv}N\ Cl^-$ (para) $\xrightarrow{CuBr}$ OCH_3 / Br (para)

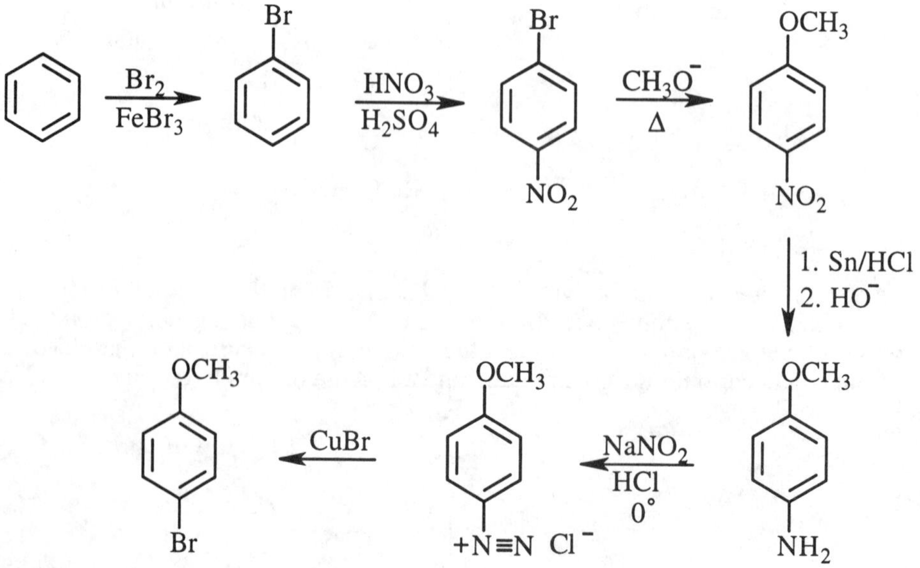

31.

a. 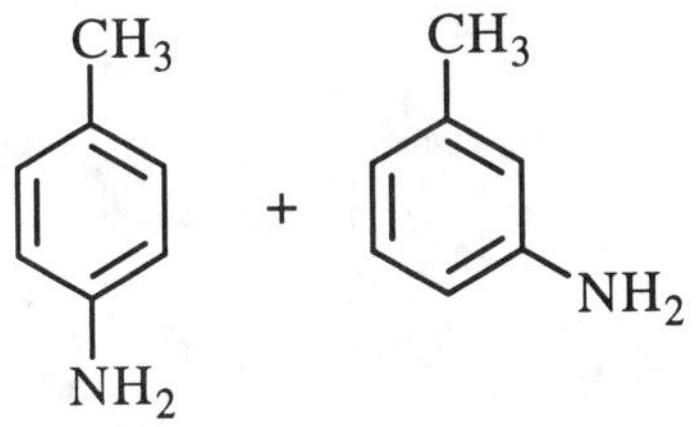 CH_3 / NH_2 + CH_3 / NH_2

b. 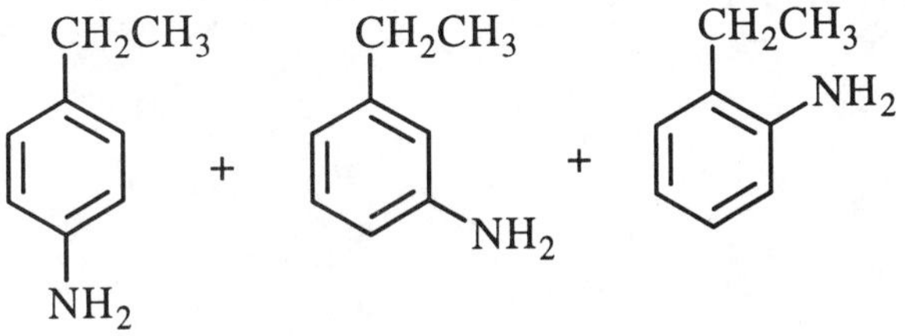 CH_2CH_3 / NH_2 + CH_2CH_3 / NH_2 + CH_2CH_3 / NH_2

c. 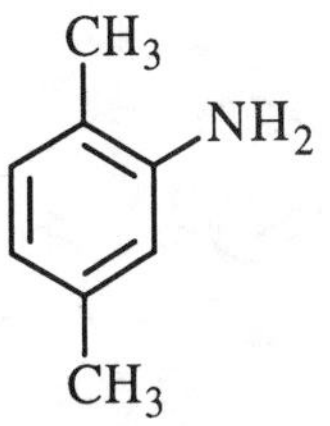 CH_3 / NH_2 / CH_3

32. From the resonance contributors, you can see that some bonds are approximated by two double bonds and one single bond (for example, the C1-C2 bond), while some are approximated by two single bonds and one double bond (for example, the C2-C3 bond). Therefore, we can conclude that all the carbon-carbon bonds in naphthalene are not the same length.

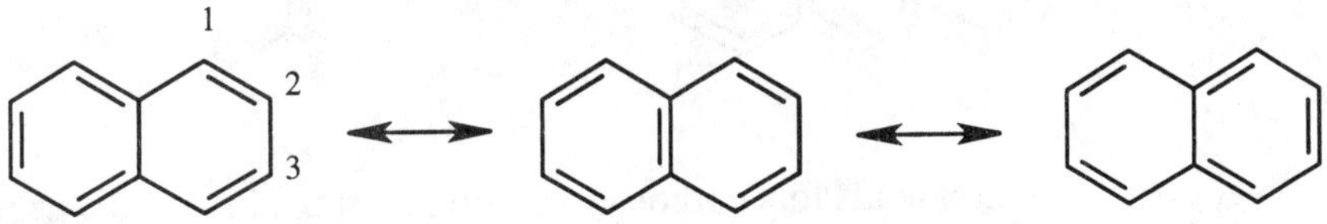

1 2 3

33.

Y
+
Y
+
Y
+
Y
+
Y⁺ #1 position
Y
+
Y
+
Y
+

Y⁺ #2 position
+ Y
Y
+
Y
+
+ Y
+ Y
Y
+

Substitution at the #1 position leads to a carbocation with seven resonance contributors, and substitution at the #2 position leads to a carbocation with six resonance contributors.

Of the seven resonance contributors obtained from substitution at the #1 position, four are more stable than the others because they have an intact benzene ring.

Of the six resonance contributors obtained from substitution at the #2 position, only two have an intact benzene ring. Substitution, therefore, is favored at the #1 position because the intermediate carbocation is more stable.

34.

a. OCH_3 Cl

c. Cl CH_3

b. Cl Br + Br Cl

d. Cl NO_2 + NO_2 Cl

35.

a. CH_2CH_3 OH

c. H_3C CH_2CH_3 C=C H

e.

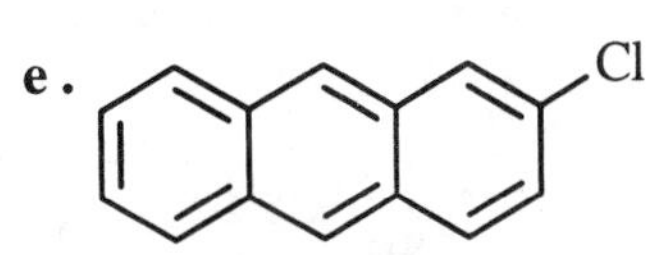

g. OCH_3 NO_2

b. SO_3H NO_2

d. NH_2 Br

f. CH=CH_2 Cl

h. CH_3 Cl Cl

36.

a. bromomethylbenzene or benzyl bromide
b. dibromomethylbenzene
c. 2,6-dimethylphenol
d. *m*-bromobenzoic acid
e. *o*-bromotoluene
f. *m*-ethylanisole
g. 3,5-dichlorobenzenesulfonic acid
h. 2-chloro-4-ethylazobenzene
i. *p*-nitrostyrene
j. 1,2,4-tribromobenzene
k. *p*-cyclohexyltoluene

37.

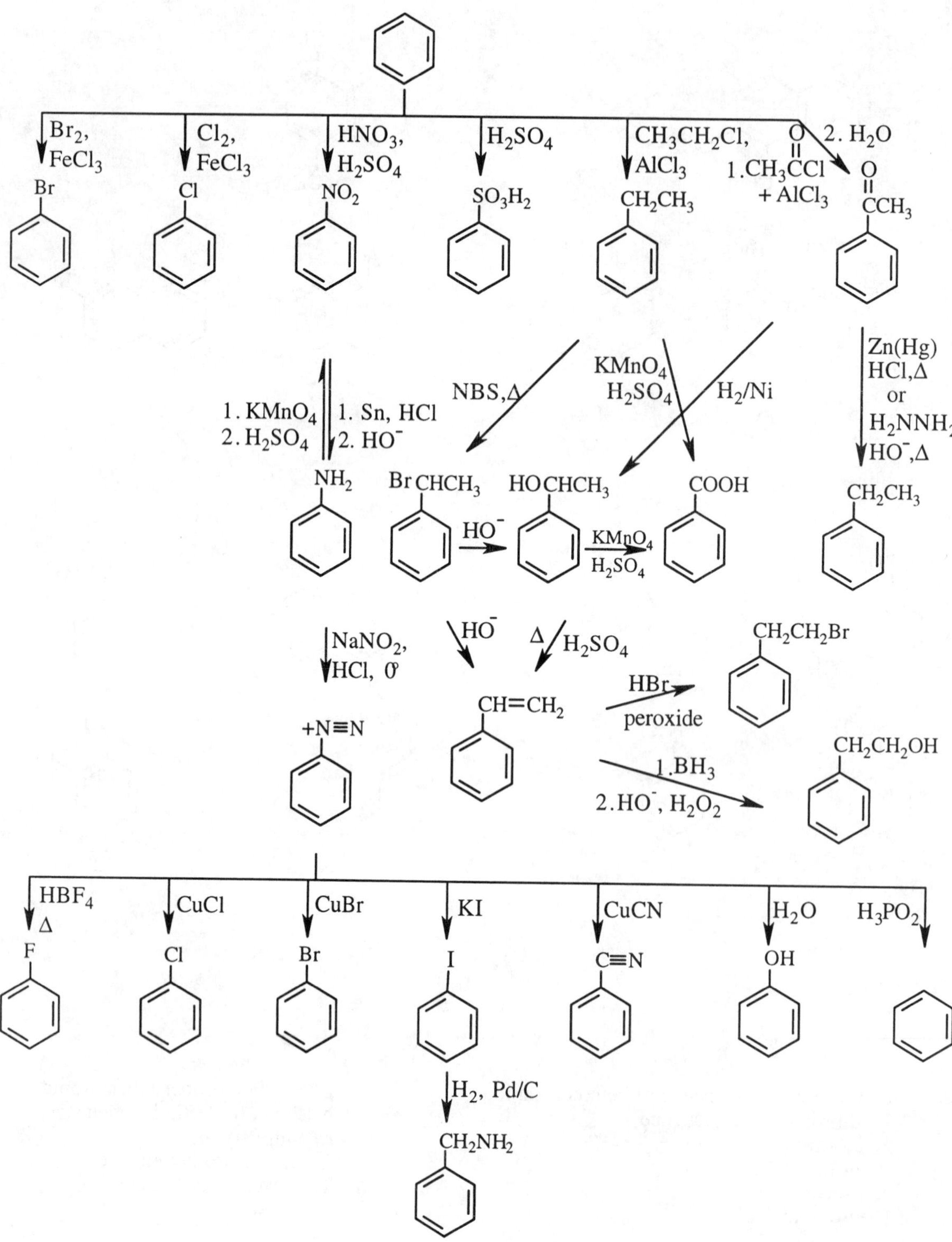
Br$_2$, FeCl$_3$
Br
Cl$_2$, FeCl$_3$
Cl
HNO$_3$, H$_2$SO$_4$
NO$_2$
H$_2$SO$_4$
SO$_3$H$_2$
CH$_3$CH$_2$Cl, AlCl$_3$
CH$_2$CH$_3$
1. CH$_3$CCl + AlCl$_3$
2. H$_2$O
O
CCH$_3$
Zn(Hg) HCl,Δ or H$_2$NNH$_2$ HO$^-$,Δ
CH$_2$CH$_3$
NBS,Δ
KMnO$_4$ H$_2$SO$_4$
H$_2$/Ni
1. KMnO$_4$ 2. H$_2$SO$_4$
1. Sn, HCl 2. HO$^-$
NH$_2$
Br CHCH$_3$
HO$^-$
HO CHCH$_3$
KMnO$_4$ H$_2$SO$_4$
COOH
NaNO$_2$, HCl, 0°
HO$^-$
Δ H$_2$SO$_4$
CH=CH$_2$
HBr peroxide
CH$_2$CH$_2$Br
1. BH$_3$ 2. HO$^-$, H$_2$O$_2$
CH$_2$CH$_2$OH
+N≡N
HBF$_4$ Δ
F
CuCl
Cl
CuBr
Br
KI
I
CuCN
C≡N
H$_2$O
OH
H$_3$PO$_2$
H$_2$, Pd/C
CH$_2$NH$_2$

38. The chloro substituent primarily withdraws electrons inductively. (It only minimally donates electrons by resonance.) The closer it is to the COOH group, the more it can withdraw electrons from the OH bond and the stronger the acid.

The nitro substituent withdraws electrons inductively. It also withdraws electrons by resonance if it is ortho or para to the COOH group. Therefore, the ortho and para isomers are the strongest acids, and the ortho isomer is a stronger acid than the para isomer because of the greater inductive electron withdrawal from the closer position.

The amino substituent primarily donates electrons by resonance, but it can donate electrons by resonance to the COOH group only if it is ortho or para to it. From the pK_a's it is apparent that resonance electron donation to the COOH group is more efficient from the ortho position.

39.

a. COOH, NO_2

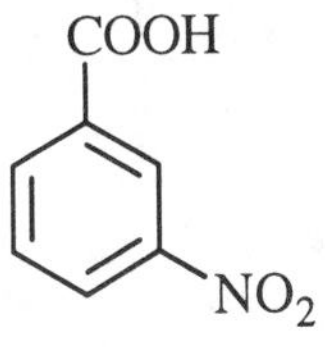

b. CH_3CHCH_3

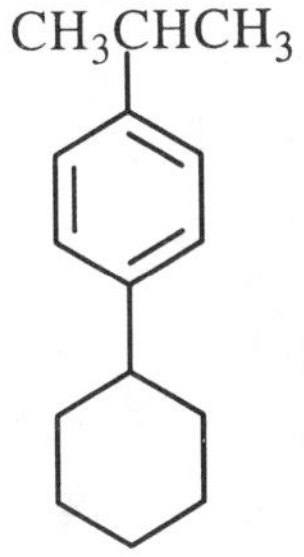

c. O, ‖, CCH_3

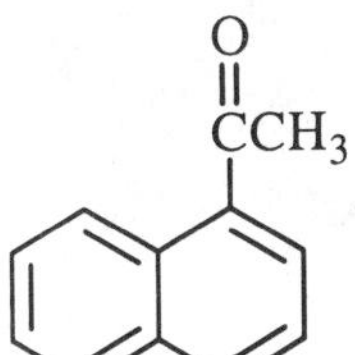

d. NH_2, CH_3, N=N

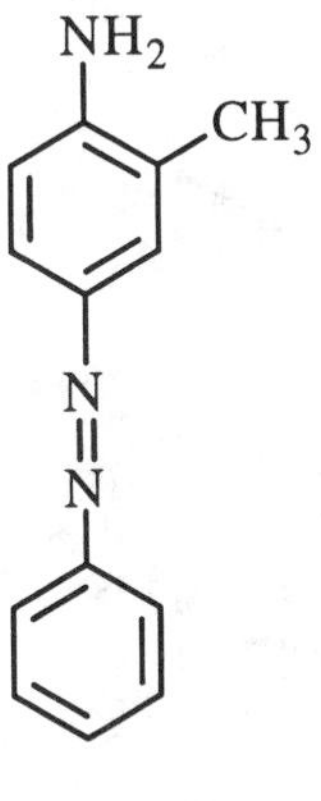

e. Br, O; Br, O

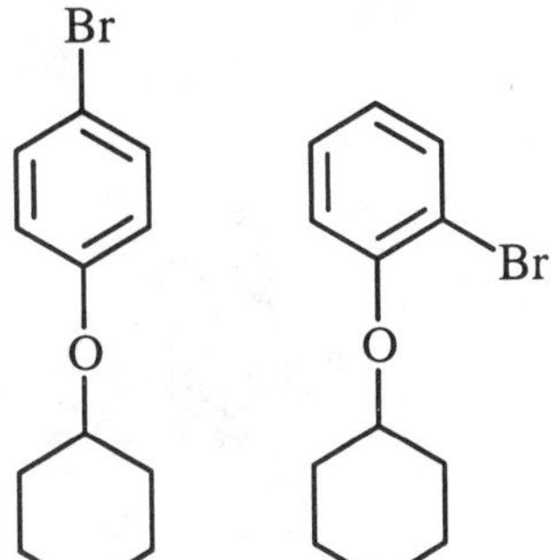

f. OH, SO_3H; OH, SO_3H

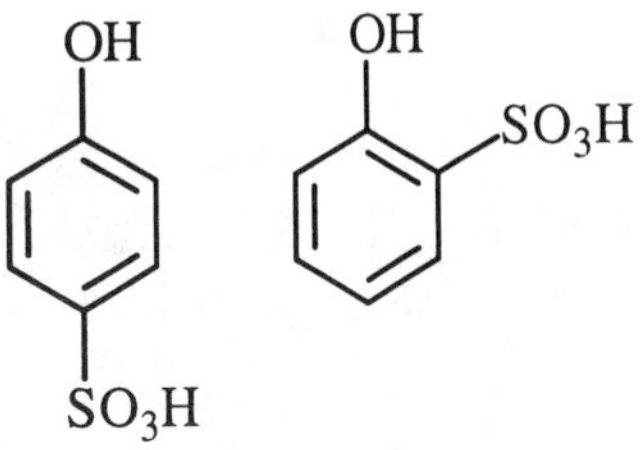

g. CH_2CH_3, Br + CH_2CH_3, Br

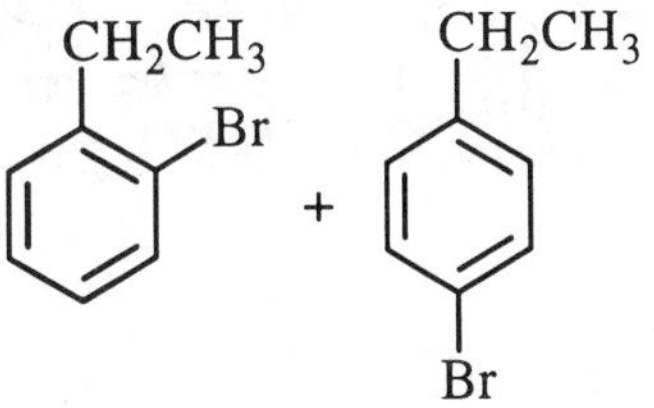

h. COOH, COOH

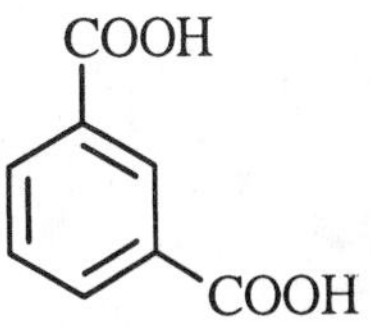

40.

a. benzene + H_2SO_4 ⟶ benzenesulfonic acid (SO_3H) $\xrightarrow[FeCl_3]{Cl_2}$ m-chlorobenzenesulfonic acid (SO_3H, Cl)

b. benzene $\xrightarrow[2.\ H_2O]{1.\ CH_3\overset{O}{\overset{\|}{C}}Cl + AlCl_3}$ $C_6H_5\overset{O}{\overset{\|}{C}}CH_3$ $\xrightarrow[FeCl_3]{Cl_2}$ m-Cl-$C_6H_4\overset{O}{\overset{\|}{C}}CH_3$ $\xrightarrow[HCl\ \Delta]{Zn(Hg)}$ m-Cl-$C_6H_4CH_2CH_3$

c. benzene $\xrightarrow[AlCl_3]{CH_3Cl}$ $C_6H_5CH_3$ $\xrightarrow[\Delta]{NBS}$ $C_6H_5CH_2Br$ $\xrightarrow{HO^-}$ $C_6H_5CH_2OH$

d. benzene $\xrightarrow[H_2SO_4]{HNO_3}$ $C_6H_5NO_2$ $\xrightarrow[FeBr_3]{Br_2}$ m-Br-$C_6H_4NO_2$ $\xrightarrow[2.\ HO^-]{1.\ Sn/HCl}$ m-Br-$C_6H_4NH_2$ $\xrightarrow[0°]{NaNO_2,\ HCl}$ m-Br-$C_6H_4\overset{+}{N}\equiv N$ $\xrightarrow{CuC\equiv N}$ m-Br-$C_6H_4C\equiv N$

e. benzene $\xrightarrow[H_2SO_4]{HNO_3}$ $C_6H_5NO_2$ $\xrightarrow[\text{or } H_2,\ Pd/C]{1.\ Sn/HCl\quad 2.\ HO^-}$ $C_6H_5NH_2$ $\xrightarrow[K_2CO_3]{CH_3I\ \text{excess}}$ $C_6H_5\overset{+}{N}(CH_3)_3\ I^-$

f. benzene $\xrightarrow[AlCl_3]{CH_3Cl}$ $C_6H_5CH_3$ $\xrightarrow[\Delta]{NBS}$ $C_6H_5CH_2Br$ $\xrightarrow{CH_3O^-}$ $C_6H_5CH_2OCH_3$

g. benzene $\xrightarrow[AlCl_3]{CH_2Cl_2}$ $C_6H_5CH_2Cl$ $\xrightarrow[AlCl_3]{\text{benzene}}$ $C_6H_5CH_2C_6H_5$ $\xrightarrow[FeCl_3]{Cl_2}$ $C_6H_5CH_2C_6H_4Cl$ (Cl para)

h. benzene $\xrightarrow[2.\ H_2O]{1.\ CH_3CH_2CH_2CH_2CCl(=O)\ +\ AlCl_3}$ $C_6H_5CCH_2CH_2CH_2CH_3$ (C=O)

$\downarrow$ Zn(Hg), HCl Δ

$C_6H_5CH_2CH_2CH_2CH_2CH_3$

i. benzene $\xrightarrow[AlCl_3]{CH_3Cl}$ $C_6H_5CH_3$ $\xrightarrow[H^+\ \Delta]{KMnO_4}$ C_6H_5COOH $\xrightarrow[H_2SO_4]{HNO_3}$ m-$NO_2C_6H_4COOH$

$\downarrow$ 1. Sn/HCl 2. HO^-

m-$NH_2C_6H_4COOH$ $\xrightarrow[HCl,\ 0°]{NaNO_2}$ m-$^+N\equiv NC_6H_4COOH$ $\xrightarrow{H_2O}$ m-HOC_6H_4COOH

j. benzene $\xrightarrow[AlCl_3]{CH_3Cl}$ toluene (CH_3) $\xrightarrow[H^+\ \Delta]{KMnO_4}$ benzoic acid (COOH) $\xrightarrow[FeBr_3]{Br_2}$ m-bromobenzoic acid (COOH, Br)

k. benzene $\xrightarrow[AlCl_3]{CH_3Cl}$ toluene (CH_3) $\xrightarrow[H_2SO_4]{HNO_3}$ p-nitrotoluene (CH_3, NO_2) $\xrightarrow[Pd/C]{H_2}$ p-toluidine (CH_3, NH_2) $\xrightarrow[0^\circ]{NaNO_2,\ HCl}$ p-methylbenzenediazonium (CH_3, $\overset{+}{N}\equiv N$) $\xrightarrow{H_2O}$ p-cresol (CH_3, OH)

l. benzene $\xrightarrow[FeBr_3]{Br_2}$ bromobenzene (Br) $\xrightarrow[H_2SO_4]{HNO_3}$ p-bromonitrobenzene (Br, NO_2) $\xrightarrow[\Delta]{^-NH_2}$ p-nitroaniline (NH_2, NO_2)

m. benzene $\xrightarrow[H_2SO_4]{HNO_3}$ nitrobenzene (NO_2) $\xrightarrow[FeBr_3]{Br_2}$ m-bromonitrobenzene (Br, NO_2) $\xrightarrow[2.\ HO^-]{1.\ Sn/HCl}$ m-bromoaniline (Br, NH_2) $\xrightarrow[0^\circ]{NaNO_2,\ HCl}$ m-bromobenzenediazonium (Br, $\overset{+}{N}\equiv N$) $\xrightarrow{KI}$ m-bromoiodobenzene (Br, I)

n. Benzene $\xrightarrow[FeBr_3]{Br_2}$ bromobenzene $\xrightarrow[FeBr_3]{Br_2}$ p-dibromobenzene (Br, Br) $\xrightarrow[Et_2O]{Mg}$ (MgBr, MgBr) $\xrightarrow{D_2O}$ (D, D)

41.

1. $C_6H_5CH_3$ — most reactive; $C_6H_5CHF_2$; $C_6H_5CF_3$ — least reactive, highest % meta product

2. $C_6H_5N^+(CH_3)_3$ ($H_3C-N^+-CH_3$, CH_3) — least reactive, highest % meta product; $C_6H_5CH_2-N^+(CH_3)_3$; $C_6H_5CH_2CH_2-N^+(CH_3)_3$ — most reactive

3. $C_6H_5OCH_2CH_3$ — most reactive; $C_6H_5CH_2OCH_3$; $C_6H_5COCH_3$ (O=C, ester $COCH_3$) — least reactive, highest % meta product

42. a. anisole > ethylbenzene > benzene > chlorobenzene > nitrobenzene

b. 2,4-dinitrophenol > 2,4-dinitrotoluene > 1-chloro-2,4-dinitrobenzene

c. *p*-cresol > *p*-xylene > toluene > benzene

d. phenol > propylbenzene > benzene > benzoic acid

e. *p*-chlorotoluene > *p*-nitrotoluene > 2-chloro-4-nitrotoluene > 2,4-dinitrotoluene

f. fluorobenzene > chlorobenzene > bromobenzene > iodobenzene

43.

a. O_2N–C_6H_4–$OCCH_3$ (para) and C_6H_4(ortho-NO_2)–$OCCH_3$ (C=O on the acyl carbon)

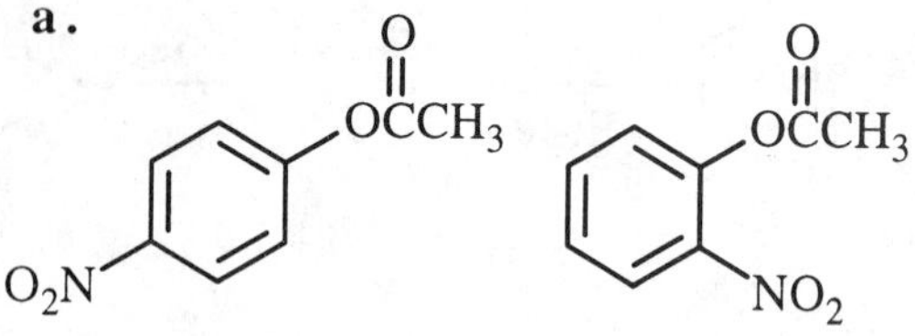

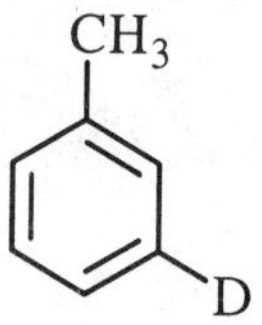

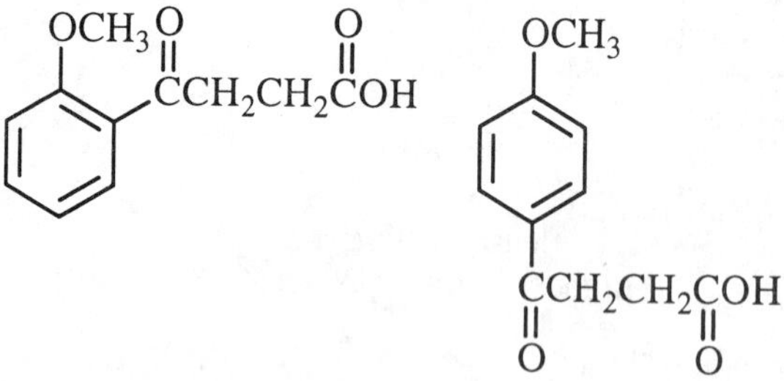

b. CH_3 and D (meta)

c. OCH_3 / CCH_2CH_2COH (ortho) and OCH_3 / CCH_2CH_2COH (para) (C=O on both carbonyl carbons)

d. N-piperidinyl with Br (para) and N-piperidinyl with Br (ortho)

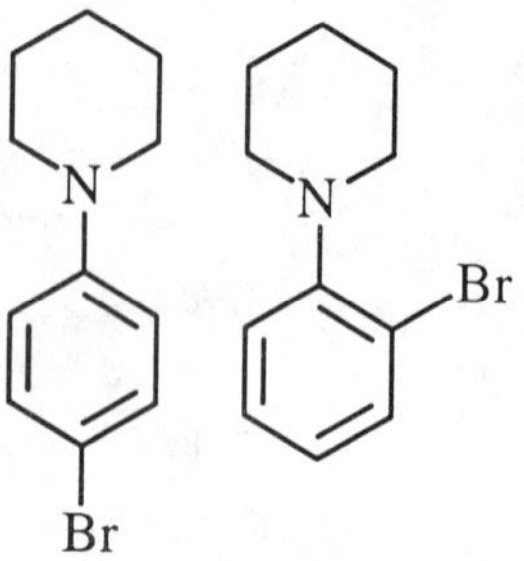

e. $CH_2CH_2CH_2OH$

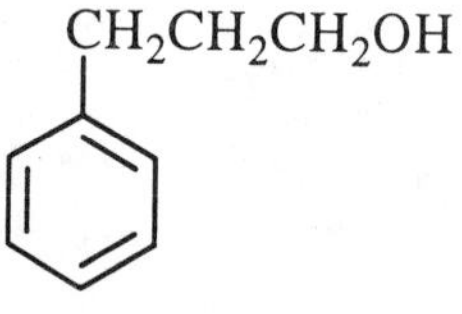

f. CF_3 and Cl (meta)

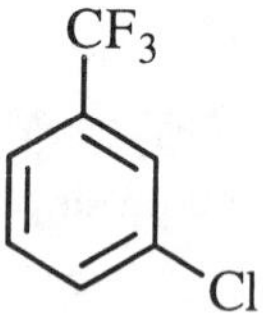

44. a. CH_2CH_3 donates electrons inductively but does not donate or withdraw electrons by resonance.

b. NO_2 withdraws electrons inductively and withdraws electrons by resonance.

c. Br deactivates the ring and directs ortho/para.

d. OH withdraws electrons inductively, donates electrons by resonance, and activates the ring.

e. $^+NH_3$ withdraws electrons inductively but does not donate or withdraw electrons by resonance.

45.

a. 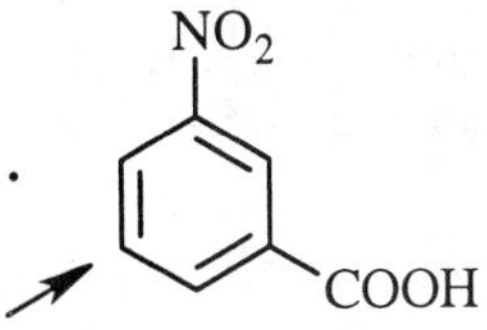 NO_2, COOH

d. 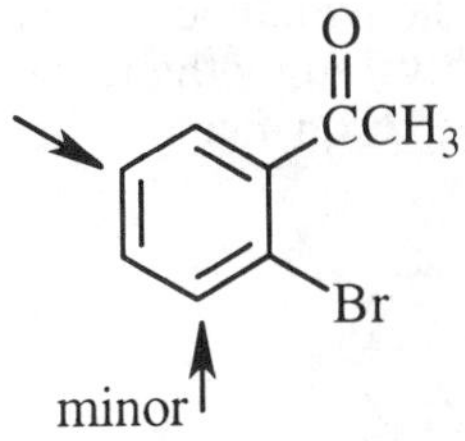 O, CCH_3, Br, minor

b. 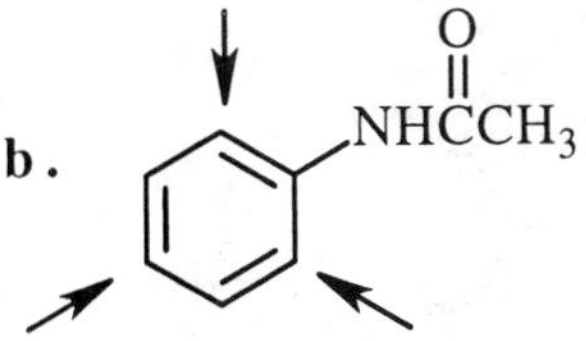 O, $NHCCH_3$

e. 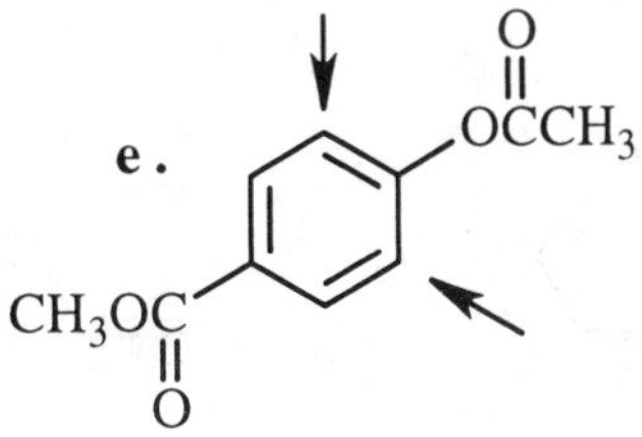 O, $OCCH_3$, CH_3OC, O

c. 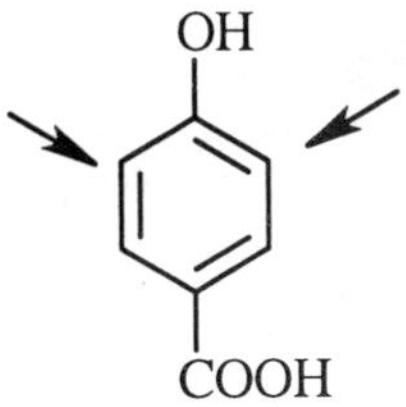 OH, COOH

f. 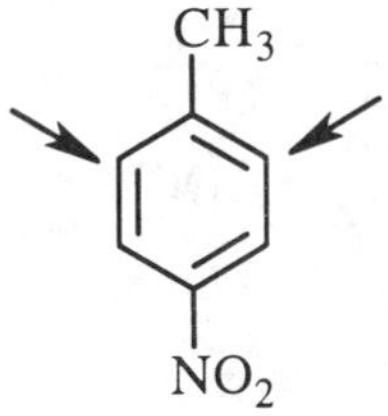 CH_3, NO_2

g. 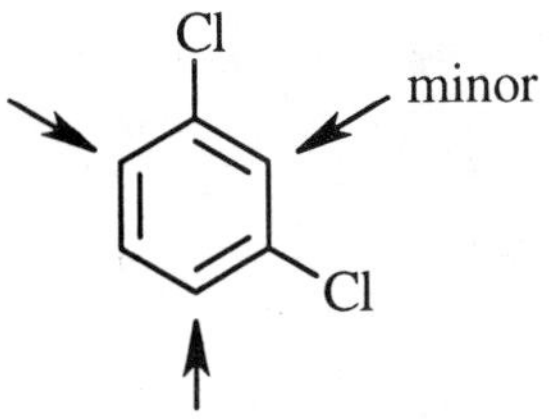 Cl, minor, Cl

j. 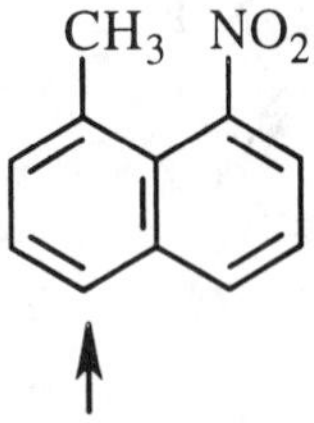 CH_3 NO_2

h. 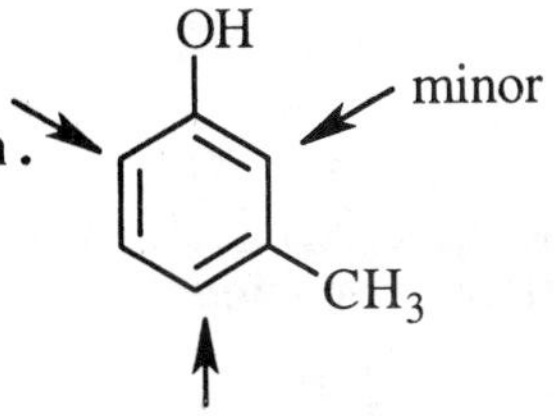 OH, minor, CH_3

k. 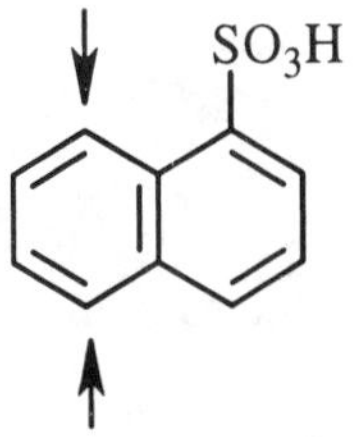 SO_3H

i. 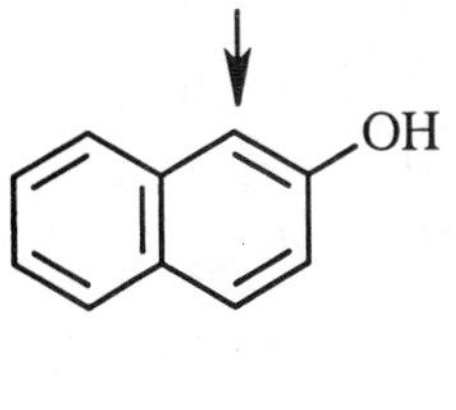 OH

l. 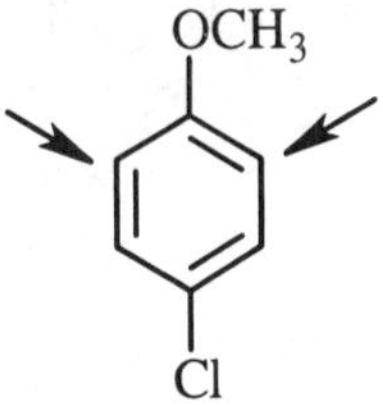 OCH_3, Cl

46. For each compound, determine which benzene ring is more highly activated. For example, in "**a**", the ring on the left is activated by the resonance electron-donating oxygen, while the ring on the right is deactivated by the resonance electron-withdrawing carbonyl group.

a. Br ... O—C(=O)— + Br— ... —O—C(=O)—

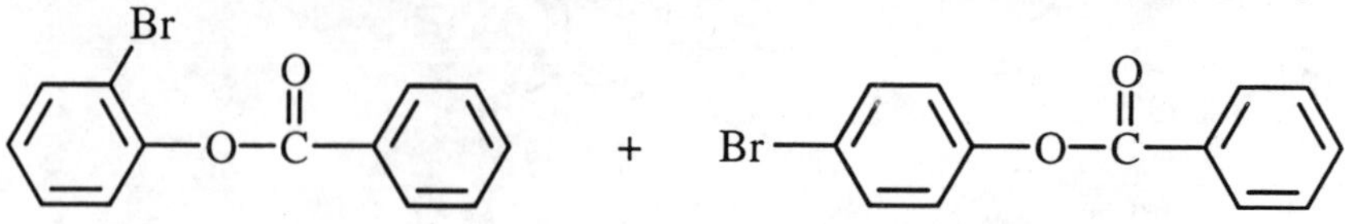

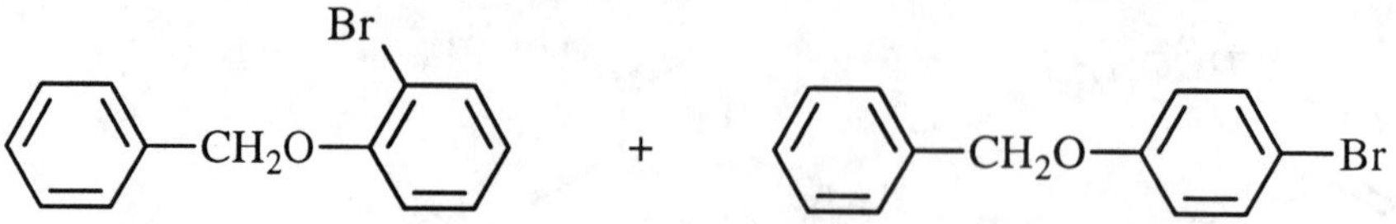

b. —CH_2O— (Br) + —CH_2O——Br

c. Br, H_3C— ... —C(=O)— ... —C(=O)OCH_3

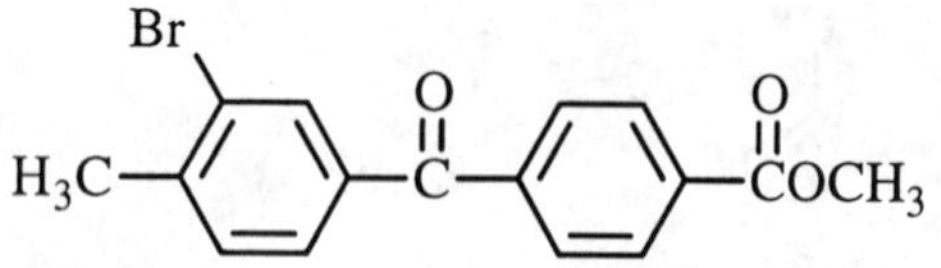

d. Br— (CH_3O) ... NO_2 + Br (CH_3O) ... NO_2

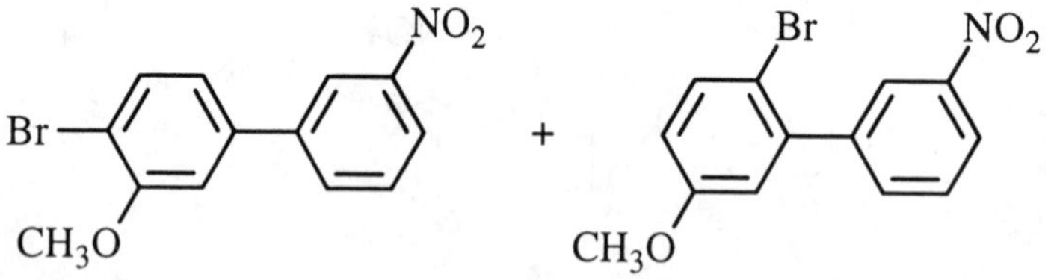

47. *Meta*-xylene will react more rapidly. In *meta*-xylene both methyl groups activate the same position, while in *para*-xylene each methyl group activates a different position.

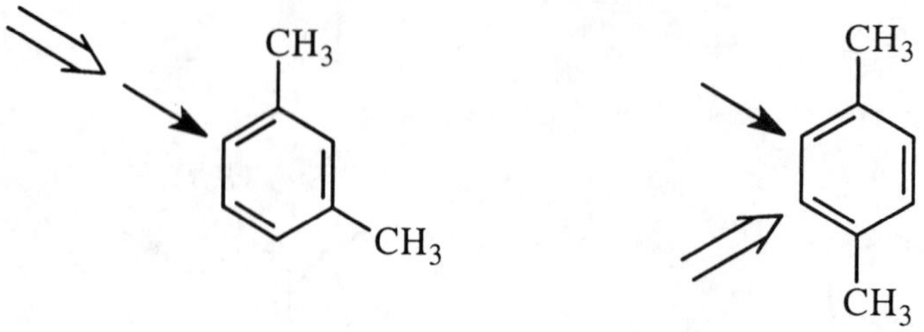

CH_3 CH_3 CH_3 CH_3

48.

a. 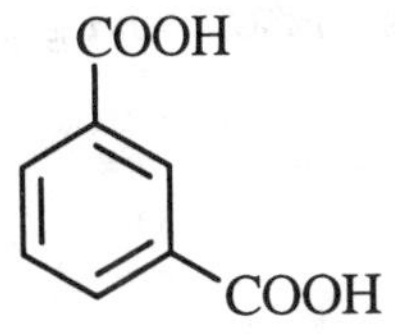 COOH / COOH (isophthalic acid structure)

b. 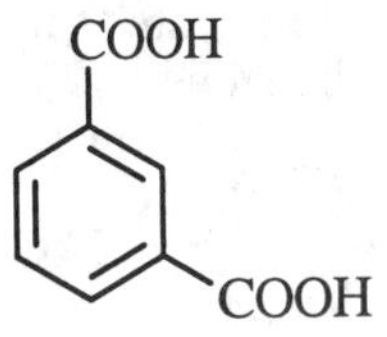 COOH / COOH

c. 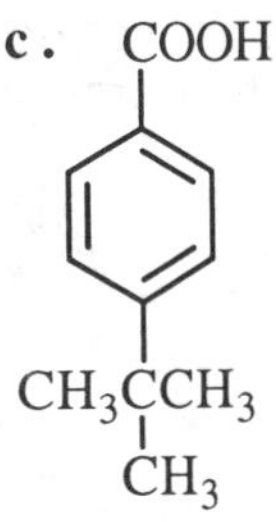 COOH / CH_3CCH_3 / CH_3

49. Yes, the advice is sound. The para isomer will form one product, the ortho isomer will form two products, and the meta isomer will form as many as four products, because the positions activatedby the ethyl substituent will be deactivated by the -CH=O substituent. One of the products will be formed in relatively low yield because of steric hindrance.

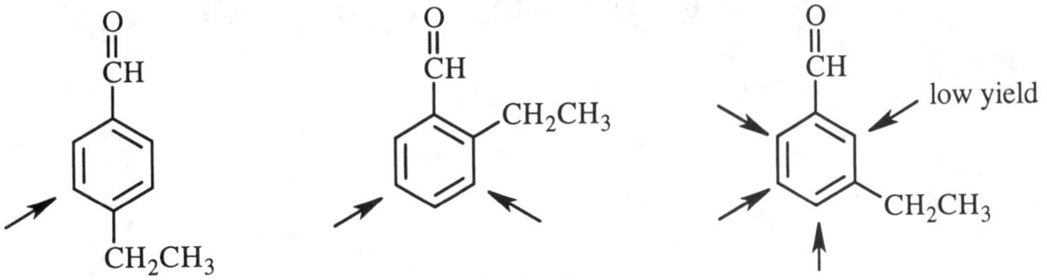

O || CH, CH_2CH_3 O || CH, CH_2CH_3 O || CH, low yield, CH_2CH_3

50. The carbocation formed by putting an electrophile at the ortho or para positions can be stabilized by resonance electron donation from the phenyl substituent.

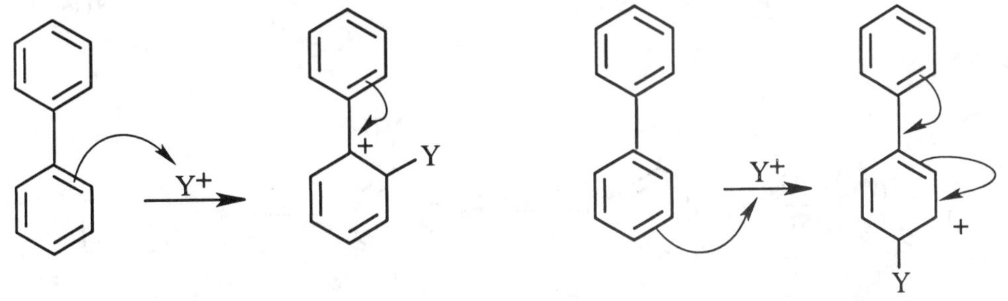

Y^+ → + Y Y^+ → + Y

The carbocation formed by putting an electrophile at the meta position cannot be stabilized by resonance electron donation from the phenyl substituent.

Y^+ → Y +

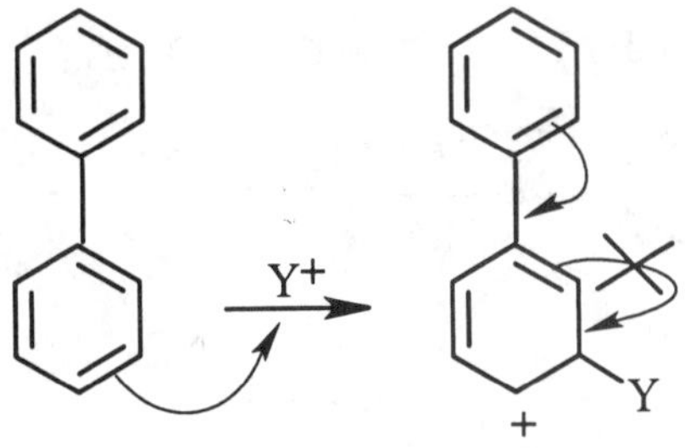

51. The spectrum indicates that the benzene ring has a substituent with two different kinds of hydrogens. The doublet and multiplet indicate that the substituent is an isopropyl group. Therefore, Compound A is isopropylbenzene.

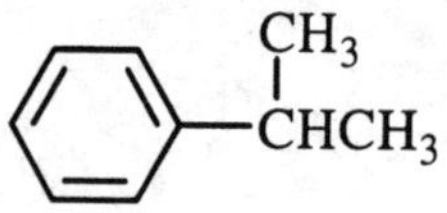

CH_3

C_6H_5–$CHCH_3$

isopropylbenzene

52.

1. Benzene $\xrightarrow[FeBr_3]{Br_2}$ bromobenzene (Br) $\xrightarrow[H_2SO_4]{HNO_3}$ p-bromonitrobenzene (Br, NO_2) $\xrightarrow[\Delta]{CH_3O^-}$ p-nitroanisole (OCH_3, NO_2) $\xrightarrow{1.\ Sn/HCl\quad 2.\ HO^-}$ p-methoxyaniline (OCH_3, NH_2)

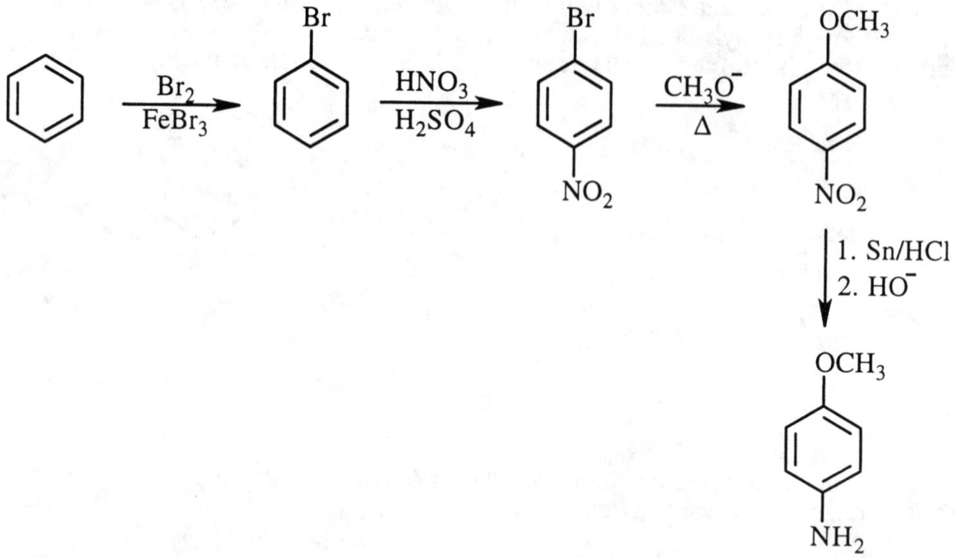

2. Benzene $\xrightarrow[H_2SO_4]{HNO_3}$ nitrobenzene (NO_2) $\xrightarrow[2.\ HO^-]{1.\ Sn/HCl}$ aniline (NH_2) $\xrightarrow[NaNO_2,\ 0°]{HCl}$ benzenediazonium ($\overset{+}{N}\equiv N$) $\xrightarrow{CH_3OH}$ anisole (OCH_3) $\xrightarrow[H_2SO_4]{HNO_3}$ p-nitroanisole (OCH_3, NO_2) $\xrightarrow[2.\ HO^-]{1.\ Sn/HCl}$ p-methoxyaniline (OCH_3, NH_2)

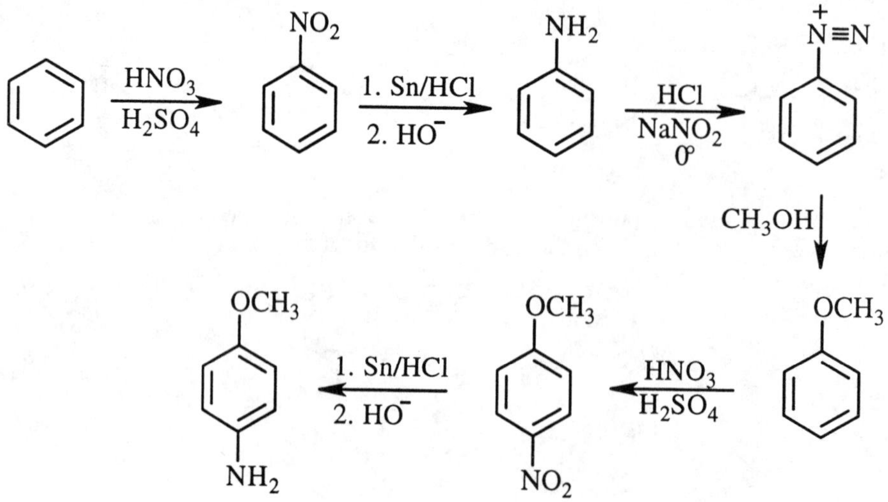

53. **a.** The hydroxy-substituted carbocation intermediate is more stable because the positive charge can be stabilized by resonance electron donation from the OH group.

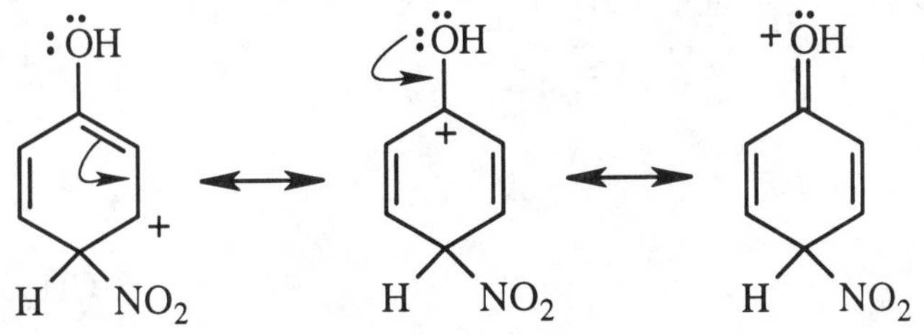

:ÖH　:ÖH　+ÖH

H NO$_2$　H NO$_2$　H NO$_2$

b. The carbanion with the negative charge meta to the nitro group is more stable because a negative charge in the meta position can be delocalized onto the nitro group but a negative charge in the ortho position cannot.

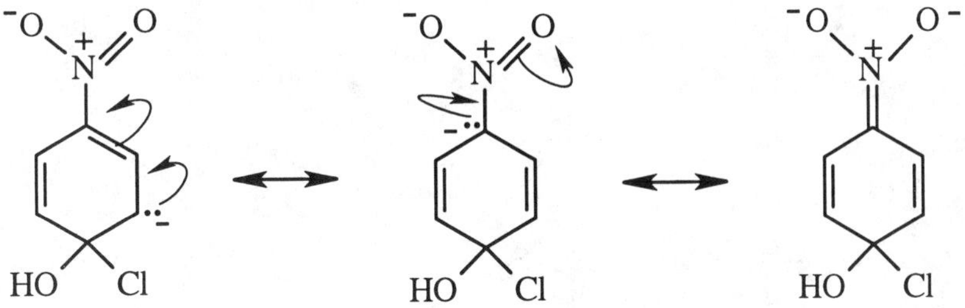

HO Cl　HO Cl　HO Cl

54.

a. CH_3Cl / $AlCl_3$ → CH_3; $CH_3CH(Cl)CH_3$ / $AlCl_3$ → CH_3, CH_3CHCH_3; NBS / Δ → CH_3, $CH_3C(Br)CH_3$; HO^- → CH_3, $CH_3C{=}CH_2$

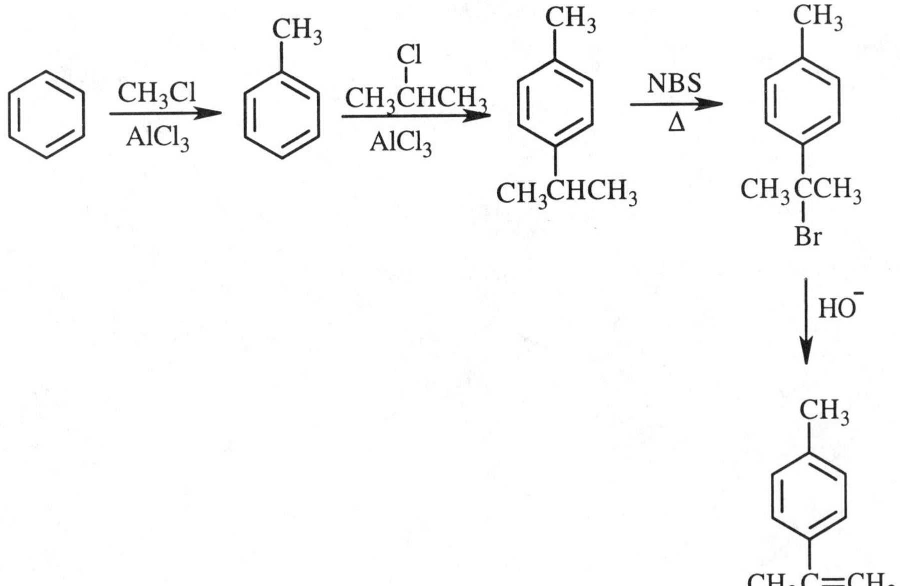

b. benzene $\xrightarrow[FeBr_3]{Br_2}$ bromobenzene (Br) $\xrightarrow[H_2SO_4]{HNO_3}$ 1-bromo-4-nitrobenzene (Br, NO_2) $\xrightarrow[FeBr_3]{Br_2}$ 1,2-dibromo-4-nitrobenzene (Br, Br, NO_2)

$\downarrow$ CH_3O^-, Δ

2-bromo-4-nitroanisole (OCH_3, Br, NO_2)

c. benzene $\xrightarrow[2.\ H_2O]{1.\ ClCOCH(CH_3)CH_3\ +\ AlCl_3}$ C_6H_5–$O{=}CCHCH_3$ (CH_3) $\xrightarrow[HCl\ \Delta]{Zn(Hg)}$ C_6H_5–CH_2CHCH_3 (CH_3)

$\downarrow$ 1. CH_3CCl (with C=O) + $AlCl_3$; 2. H_2O

para-($\overset{O}{\|}CCH_3$)-C_6H_4–CH_2CHCH_3 (CH_3) $\xrightarrow{H_2SO_4}$ product with $\overset{O}{\|}CCH_3$, SO_3H (ortho to the alkyl group), CH_2CHCH_3 (CH_3)

55. Aniline is a poor substrate because an NH_2 substituent shares its electrons better than an OH substituent. Therefore, an NH_2 substituent complexes with the Lewis acid, which often is the catalyst in an electrophilic substitution reaction. And because the NH_2 substituent is a stronger base than the OH substituent, it is more easily protonated (pK_a protonated OH group = -2; pK_a protonated NH_2 group = 5). The complexed (or protonated) NH_2 group is a deactivating meta director.

The problem can be circumvented by protecting the NH_2 group with an acetyl group. The acetyl group can be removed by hydrolysis (reaction with water) after the electrophilic substitution reaction is done.

NH_2 (benzene ring) $\xrightarrow[\text{pyridine}]{CH_3CCl\ (C{=}O)}$ $NHCCH_3$ (C=O) (benzene ring) $\xrightarrow{\text{electrophilic substitution}}$ $NHCCH_3$ (C=O), Y, Y (ring) $\xrightarrow[HO^-]{H_2O}$ NH_2, Y, Y (ring)

56. a. The first three compounds will not show a carbonyl stretch at 1700 cm^{-1} and the bottom four will show this absorption band. The first three can be differentiated from one another by the presence or absence of the indicated absorption bands.

CH_2OH (cyclohexane)	CH_2OH (benzene)	CH_2OCH_3 (benzene)
band at 3300 no band at 1600	band at 3300 band at 1600	no band at 3300 band at 1600

The last four compounds all have an absorption band at 1700 cm^{-1}.
They can be differentiated by the presence or absence of the indicated absorption bands.

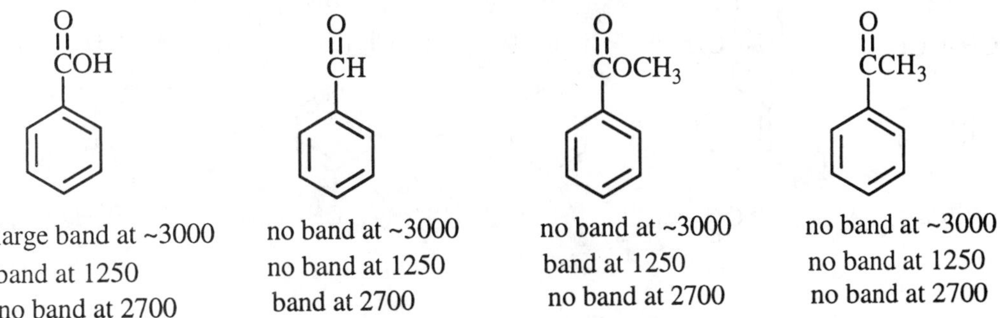

b. This is the only compound without the characteristic benzene ring hydrogens at ~ 7-8 ppm

CH_2OH

Only two compounds will have two signals other than the signals for the benzene ring hydrogens. They can be distinguished by integration (3:2 versus 2:1), or by the two sharp singlets for the ester, versus the somewhat broader singlet for the hydrogen bonded to oxygen.

CH_2OH CH_2OCH_3

The following four compound have only one signal (a singlet) in addition to the benzene ring hydrogens. The four can be identified by the positions of the signal.

$\overset{O}{\|\|}$ CCH_3	$\overset{O}{\|\|}$ $COCH_3$	$\overset{O}{\|\|}$ CH	$\overset{O}{\|\|}$ COH
~ 2 ppm	~ 3 ppm	~ 9-10 ppm	~ 10-12 ppm

57. The rate-determining step in the S_N1 reaction is the formation of the tertiary carbocation. An electron-donating substituent will stabilize the carbocation and cause it to be more easily formed. An electron-withdrawing substituent will destabilize the carbocation and cause it to be less easily formed.

Br / CH_3CCH_3 / OCH_2CH_3 > Br / CH_3CCH_3 / $\underset{\|\|\ O}{OCCH_3}$ > Br / CH_3CCH_3 / $CH_2CH_2CH_3$ > Br / CH_3CCH_3 / $CHClCH_3$ > Br / CH_3CCH_3 / SO_3H

58. A fluoro substituent is more electronegative than a chloro substituent. Therefore, nucleophilic attack on the carbon bearing the fluoro substituent will be easier than nucleophilic attack on the carbon bearing the chloro substituent.

A fluoro substituent is a stronger base than a chloro substituent, so elimination of the halogen in the second step of the reaction will be harder for a fluoro-substituted benzene than for a chloro-substituted benzene.

The fact that the fluoro-substituted compound is more reactive tells you that attack of the nucleophile on the aromatic ring is the rate-determining step of the reaction.

59. **a.** The alkyl diazonium ion is very unstable. Loss of N_2 and a 1,2-hydride shift forms a *tert*-butyl carbocation, which can undergo either substitution or elimination.

$CH_3CH(CH_3)CH_2NH_2 \xrightarrow[HCl]{NaNO_2} CH_3C(H)(CH_3)CH_2-\overset{+}{N}\equiv N \longrightarrow CH_3\overset{+}{C}(CH_3)CH_3$

$\xrightarrow{H_2O}$ $CH_3C(OH)(CH_3)CH_3$; $\xrightarrow{-H^+}$ $CH_3C(CH_3)=CH_2$

b. The cation formed from the diazonium ion will undergo a pinacol rearrangement.

$CH_3-C(OH)(CH_3)-C(NH_2)(CH_3)-CH_3 \xrightarrow[HCl]{NaNO_2} CH_3-C(OH)(CH_3)-C(\overset{+}{N}\equiv N)(CH_3)-CH_3 \xrightarrow{-N_2} CH_3-C(\ddot{O}H)(CH_3)-\overset{+}{C}(CH_3)-CH_3$

$\downarrow$

$CH_3-C(=\overset{+}{O}H)-C(CH_3)_2-CH_3 \xrightarrow{-H^+} CH_3-C(=O)-C(CH_3)_2-CH_3$

60.

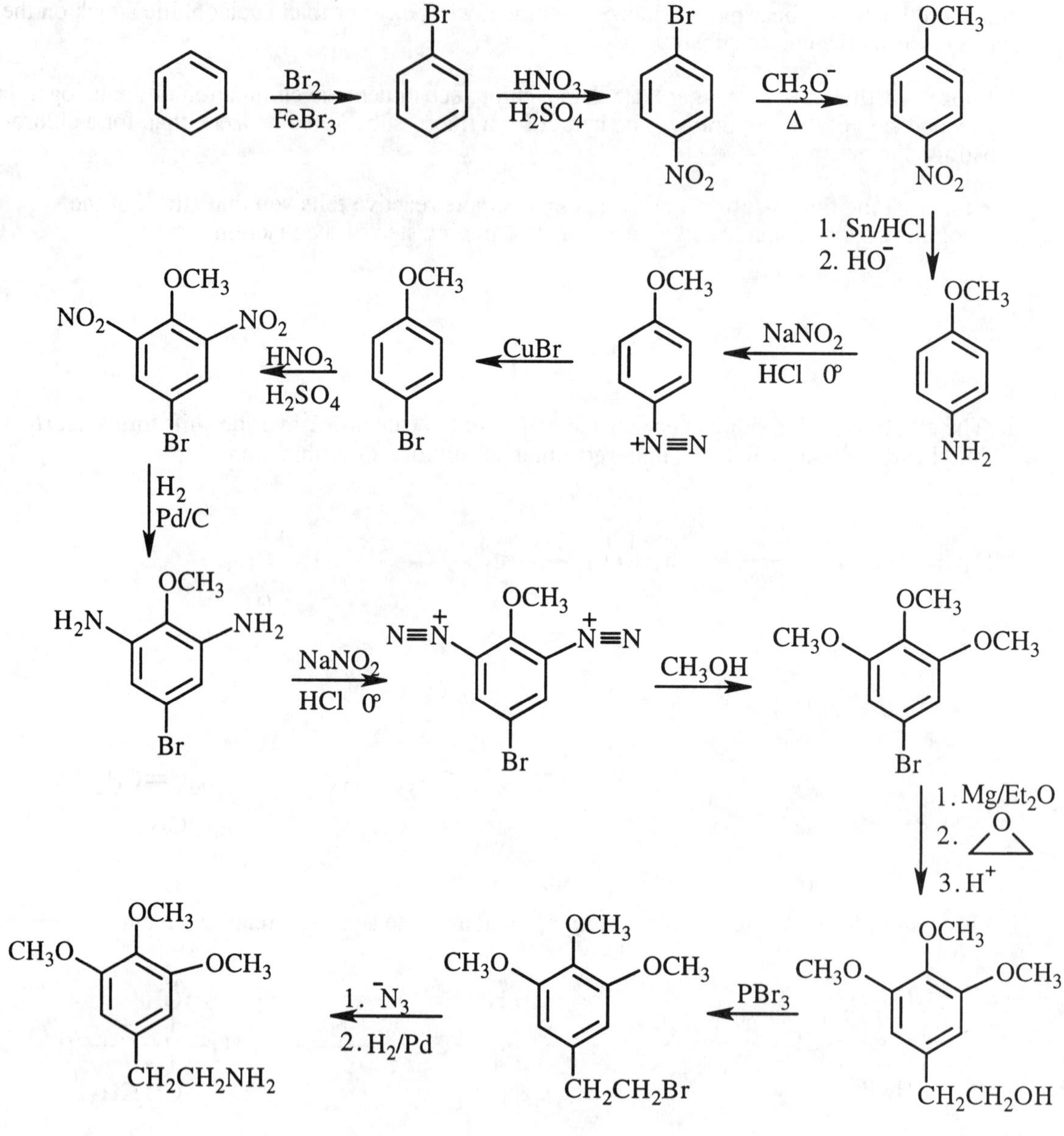

Benzene $\xrightarrow[FeBr_3]{Br_2}$ bromobenzene $\xrightarrow[H_2SO_4]{HNO_3}$ 1-bromo-4-nitrobenzene $\xrightarrow[\Delta]{CH_3O^-}$ 4-nitroanisole (OCH_3, NO_2) $\xrightarrow[2.\ HO^-]{1.\ Sn/HCl}$ 4-methoxyaniline (OCH_3, NH_2) $\xrightarrow[HCl\ 0°]{NaNO_2}$ 4-methoxybenzenediazonium (OCH_3, $+N{\equiv}N$) $\xrightarrow{CuBr}$ 4-bromoanisole (OCH_3, Br) $\xrightarrow[H_2SO_4]{HNO_3}$ 4-bromo-2,6-dinitroanisole (OCH_3, NO_2, NO_2, Br)

$\xrightarrow[Pd/C]{H_2}$ (OCH_3, H_2N, NH_2, Br) $\xrightarrow[HCl\ 0°]{NaNO_2}$ (OCH_3, $N{\equiv}\overset{+}{N}$, $\overset{+}{N}{\equiv}N$, Br) $\xrightarrow{CH_3OH}$ (OCH_3, CH_3O, OCH_3, Br)

$\xrightarrow{1.\ Mg/Et_2O\quad 2.\ \text{ethylene oxide}\quad 3.\ H^+}$ (OCH_3, CH_3O, OCH_3, CH_2CH_2OH) $\xrightarrow{PBr_3}$ (OCH_3, CH_3O, OCH_3, CH_2CH_2Br) $\xrightarrow[2.\ H_2/Pd]{1.\ ^-N_3}$ (OCH_3, CH_3O, OCH_3, $CH_2CH_2NH_2$)

61. The configuration of the chirality center in the reactant will be retained only if the chirality center undergoes two successive S_N2 reactions.

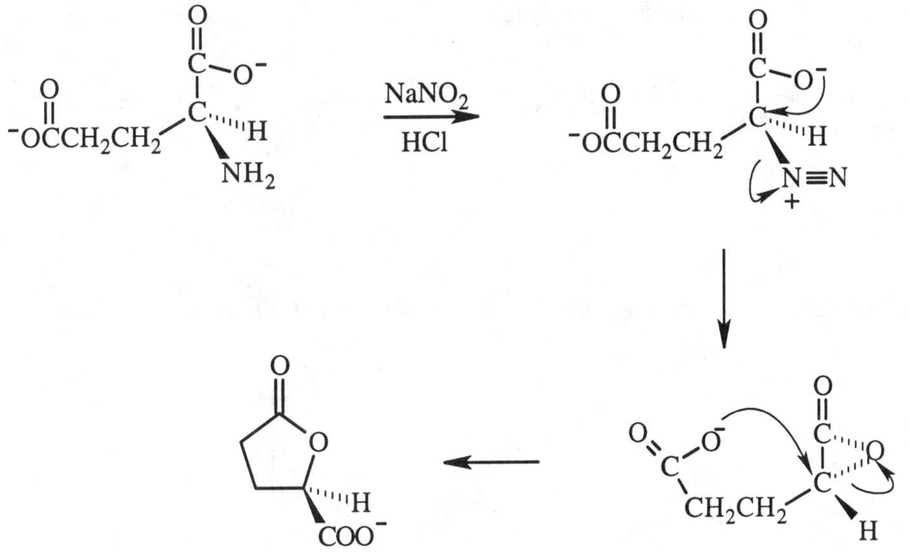

62. A chloro group is a better leaving group than the ammonium group, so the product is formed without hydroxide ion catalysis.

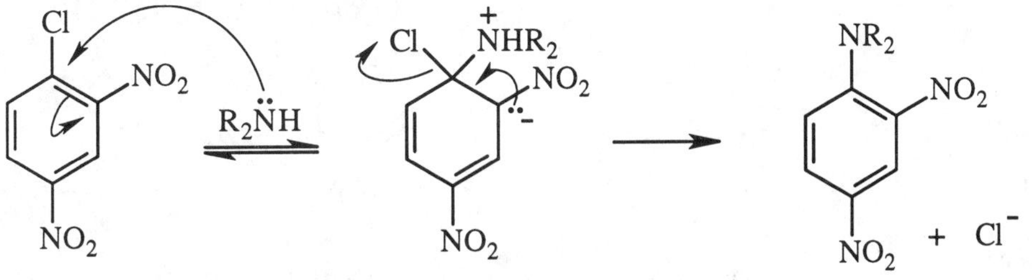

A methoxy group is a poorer leaving group than the ammonium group, so the ammonium group is eliminated, reforming starting materials. By removing a proton, hydroxide ion converts the ammonium group into an amino group. Since the amino group is a poorer leaving group than the methoxy group, the methoxy group is eliminated.

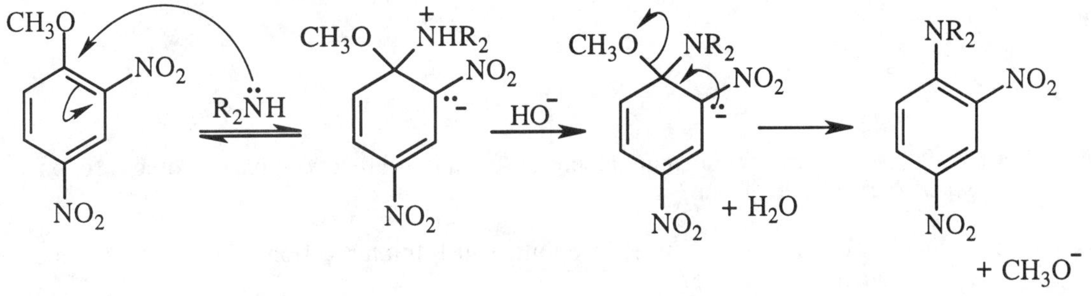

Chapter 15 Practice Test

1. Give one name for each of the following.

a. 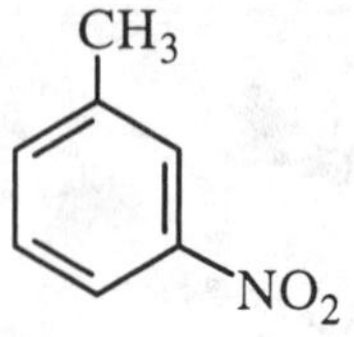 CH_3 … NO_2

b. 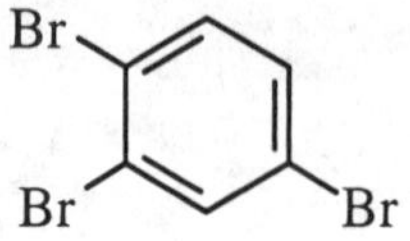 Br … Br … Br

c. 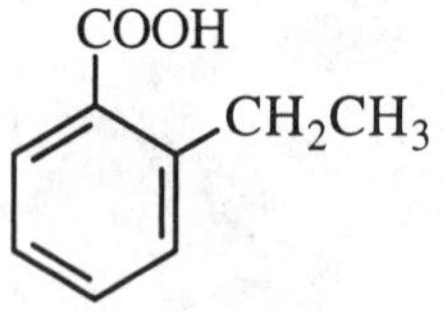 COOH … CH_2CH_3

d. 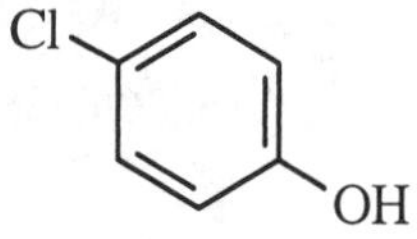 Cl … OH

2. Rank the following compounds in order of decreasing reactivity toward $Br_2/FeBr_3$.

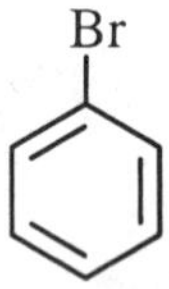

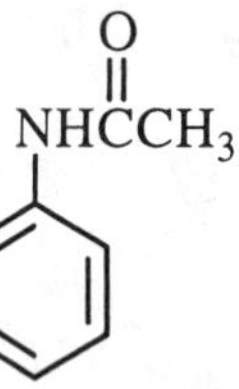

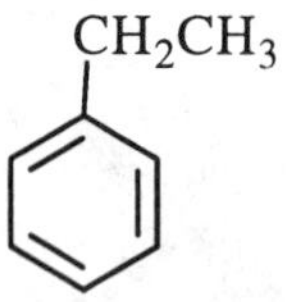

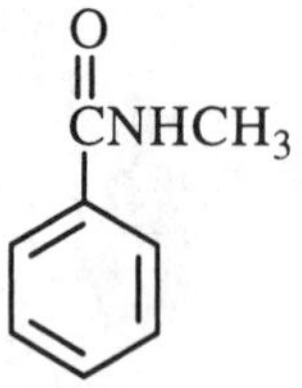

Br $NHCCH_3$ (C=O) CH_2CH_3 $CNHCH_3$ (C=O)

3. For each pair of compounds, indicate the one that is the stronger acid.

a. 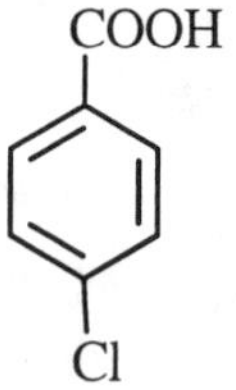 COOH / Cl or COOH / CH_3

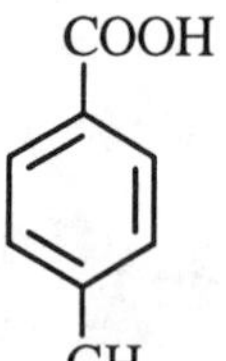

c. 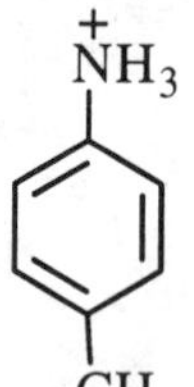 $\overset{+}{N}H_3$ / CH_3 or $\overset{+}{N}H_3$ / OCH_3

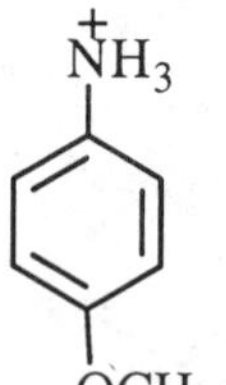

b.  OH / NO_2 or OH / Br

d.

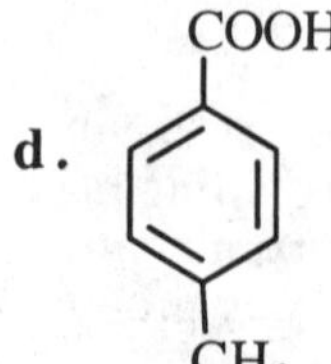

or

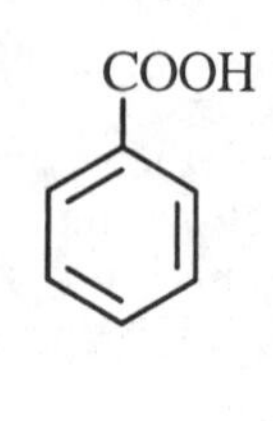

4. **a.** Which is more reactive in a nucleophilic substitution reaction, *para*-bromonitrobenzene or *para*-bromoethylbenzene?

b. Which is more reactive in an electrophilic substitution reaction, *para*-bromonitrobenzene or *para*-bromoethylbenzene?

5. Give the major product(s) of each of the following reactions.

a. 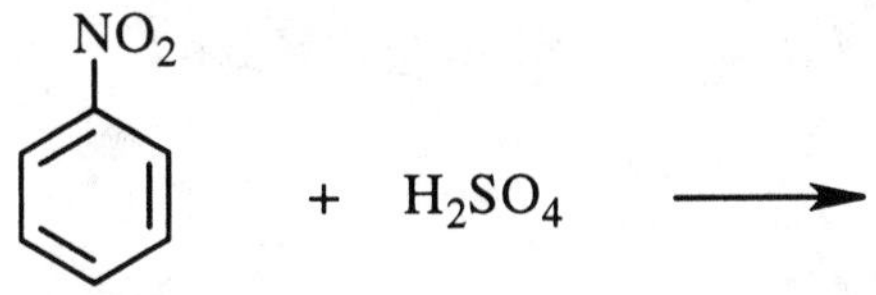 $C_6H_5NO_2$ + H_2SO_4 →

b. $C_6H_5OCH_3$ + CH_3Cl $\xrightarrow{AlCl_3}$

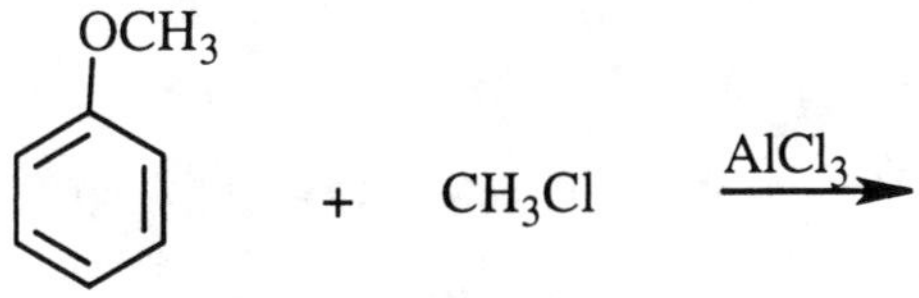

c. $C_6H_5CH_2CH_3$ $\xrightarrow[\Delta\ H^+]{Na_2Cr_2O_7}$

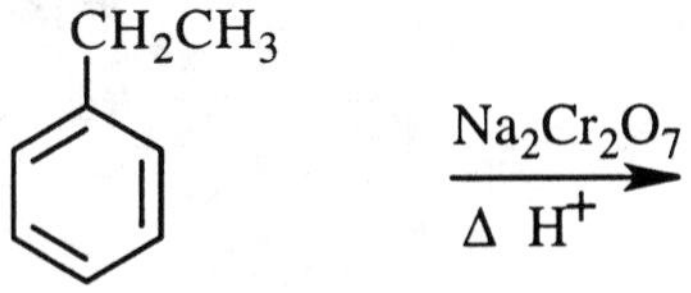

d. 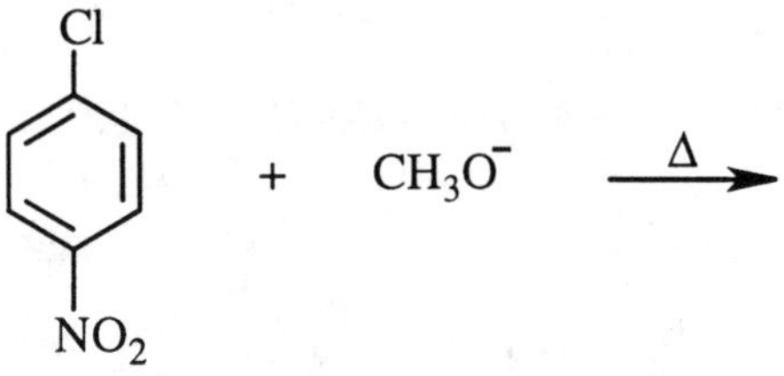 $p\text{-}ClC_6H_4NO_2$ + CH_3O^- $\xrightarrow{\Delta}$

e. 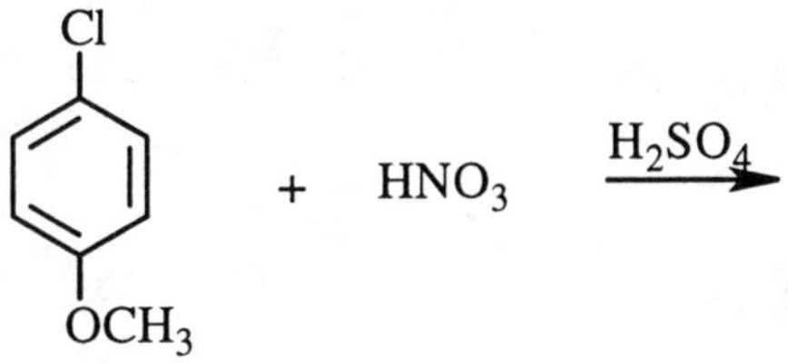 $p\text{-}ClC_6H_4OCH_3$ + HNO_3 $\xrightarrow{H_2SO_4}$

f. 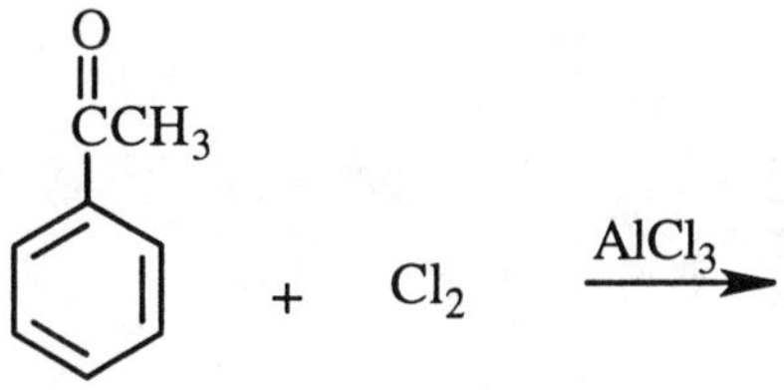 $C_6H_5COCH_3$ + Cl_2 $\xrightarrow{AlCl_3}$

6. Indicate whether each of the following reactions is true or false.

a. Benzoic acid is more reactive than benzene towards electrophilic substitution. T F

b. *para*-Chlorobenzoic acid is more acidic than *para*-methoxybenzoic acid. T F

c. A -CH=CH_2 group is a meta director. T F

d. *para*-Nitroaniline is more basic than *para*-chloroaniline. T F

CHAPTER 16
Carbonyl Compounds I:
Reactions of Carboxylic Acids and Their Derivatives with Oxygen and Nitrogen Nucleophiles

Important Terms

acid anhydride $R-\overset{\overset{O}{\|}}{C}-O-\overset{\overset{O}{\|}}{C}-R$

acyl adenylate a carboxylic acid derivative with AMP as the leaving group.

acyl group $R-\overset{\overset{O}{\|}}{C}-$

acyl halide $R-\overset{\overset{O}{\|}}{C}-Cl$ or $R-\overset{\overset{O}{\|}}{C}-Br$

acyl phosphate a carboxylic acid derivative with a phosphate leaving group.

acyl pyrophosphate a carboxylic acid derivative with a pyrophosphate leaving group.

alcoholysis reaction with an alcohol.

amide $R-\overset{\overset{O}{\|}}{C}-NH_2$ $R-\overset{\overset{O}{\|}}{C}-NHR$ $R-\overset{\overset{O}{\|}}{C}-NR_2$

amino acid an α-amino carboxylic acid.

aminolysis reaction with an amine.

biosynthesis synthesis in a biological system.

α-carbon a carbon adjacent to a carbonyl group.

carbonyl carbon the carbon of a carbonyl group.

carbonyl compound a compound that contains a carbonyl group.

carbonyl group a carbon doubly bonded to an oxygen.

carbonyl oxygen the oxygen of a carbonyl group.

carboxyl group $-\overset{\overset{O}{\|}}{C}-OH$ or $-COOH$

carboxylic acid $R-\overset{\overset{O}{\|}}{C}-OH$

carboxylic acid derivative a compound that is hydrolyzed to a carboxylic acid.

carboxyl oxygen the single-bonded oxygen of a carboxylic acid or ester.

detergent a salt of a sulfonic acid.

ester

$$R-\overset{\overset{\displaystyle O}{\|}}{C}-OR$$

fat a triester of glycerol that exists as a solid at room temperature.

fatty acid a long-chain carboxylic acid.

Fischer esterification reaction reaction of a carboxylic acid with excess alcohol and an acid catalyst.

Gabriel synthesis a method used to convert an alkyl halide into a primary amine.

hydrolysis reaction with water.

hydrophobic interactions the attractive forces of hydrocarbon chains in water.

imide a compound with two acyl groups bonded to a nitrogen.

lactam a cyclic amide.

lactone a cyclic ester.

micelle a spherical aggregation of molecules, each with a long hydrophobic tail and a polar head, arranged so the polar head points to the outside of the sphere.

mixed anhydride an acid anhydride with two different R groups.

neurotransmitter a compound that transmits nerve impulses across the synapses between nerve cells.

nitrile a compound that contains a carbon-nitrogen triple bond.

$$R-C\equiv N$$

nucleophilic acyl substitution reaction a reaction in which a group bonded to an acyl group is substituted by another group.

oil a triester of glycerol that exists as a liquid at room temperature.

saponification hydrolysis of a fat under basic conditions.

soap a sodium or potassium salt of a fatty acid.

symmetrical anhydride an acid anhydride with identical R groups.

tetrahedral intermediate the intermediate formed in a nucleophilic acyl substitution reaction.

thioester the sulfur analog of an ester.

$$R-\overset{\overset{\displaystyle O}{\|}}{C}-SR$$

transesterification reaction the reaction of an ester with an alcohol to form a different ester.

Solutions to Problems

1. The lactone would be a three-membered ring, which is too strained to form.

2. **a.** butanenitrile
propyl cyanide

b. ethanoic propanoic anhydride
acetic propionic anhydride

c. potassium butanoate
potassium butyrate

d. pentanoyl chloride
valeryl chloride

e. isobutyl butanoate
isobutyl butyrate

f. *N,N*-dimethylhexanamide

g. γ-butyrolactam

h. cyclopentanecarboxylic acid

3. The carbon-oxygen single bond in an alcohol is longer because, as a result of resonance, the carbon-oxygen single bond in a carboxylic acid has some double-bond character.

$RCH_2{-}OH$ (longer)

$RC(=O){-}OH$ (shorter) $\longleftrightarrow$ $RC(O^-){=}\overset{+}{O}H$

4. **a.** The bond between oxygen and the methyl group is the longest because it is a pure single bond, whereas the other two carbon-oxygen bonds have some double-bond character.

The bond between carbon and the carbonyl oxygen is the shortest because it has the most double bond character.

$CH_3{-}C(=O){-}O{-}CH_3$

(C=O: 3; C–O: 2; O–CH₃: 1)

1 = longest
3 = shortest

b. Notice that the longer the bond, the lower its IR stretching frequency.

$CH_3{-}C(=O){-}O{-}CH_3$

(C=O: 1; C–O: 2; O–CH₃: 3)

1 = highest frequency
3 = lowest frequency

5. The resonance contributors show that the carbonyl oxygen has a partial negative charge, while the carboxyl oxygen has a partial positive charge. Therefore, the carbonyl oxygen is more basic than the carboxyl oxygen.

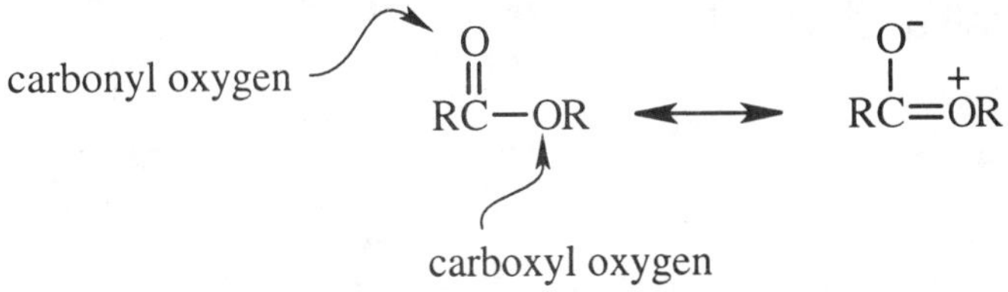

carbonyl oxygen

carboxyl oxygen

6. The more electronegative the base (Y) attached to the carbonyl carbon, the greater the double bond character of the C=O bond.

The weaker the base (Y) attached to the carbonyl carbon the less well it shares it electrons. When Y shares its electrons, it decreases the double bond character of the C=O bond, making it easier to stretch.

Therefore, the compounds have the indicated carbonyl IR absorption bands.

The carbonyl group of the acyl chloride stretches at the highest frequency because the chlorine atom is the most electronegative atom and the weakest base.

The carbonyl group of the amide stretches at the lowest frequency because the nitrogen atom is the least electronegative atom and the strongest base.

acyl chloride	~ 1800 cm^{-1}
acid anhydride	~ 1800 and 1750 cm^{-1}
ester	~ 1730 cm^{-1}
amide	~ 1640 cm^{-1}

The resonance contributors show why the acid anhydride has two carbonyl IR absorption bands.

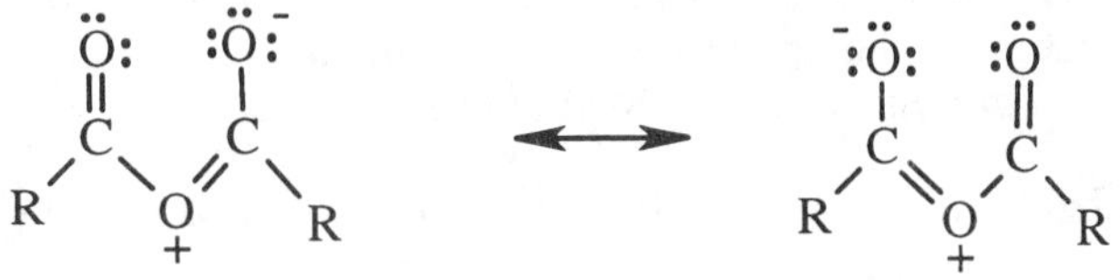

7.

a. $CH_3-\overset{\overset{\displaystyle O}{\|}}{C}-OCH_3 \ + \ NaCl \longrightarrow$ no reaction

b. $CH_3-\overset{\overset{\displaystyle O}{\|}}{C}-Cl \ + \ CH_3-\overset{\overset{\displaystyle O}{\|}}{C}-O^- \longrightarrow CH_3-\overset{\overset{\displaystyle O}{\|}}{C}-O-\overset{\overset{\displaystyle O}{\|}}{C}-CH_3 \ + \ Cl^-$

c. $CH_3-\overset{\overset{\displaystyle O}{\|}}{C}-O-\overset{\overset{\displaystyle O}{\|}}{C}-CH_3 \ + \ NaCl \longrightarrow$ no reaction

8. It is a true statement.

If the nucleophile is the stronger base, it will be harder to eliminate the nucleophile from the tetrahedral intermediate (**B**) than to eliminate the group attached to the acyl group in the reactant. In other words, the hill that has to be climbed from the intermediate (**B** to **A**) back to the reactants is higher than the hill that has to be climbed from the intermediate (**B** to **C**) to the products. Since the transition state with the highest energy is the transition state of the rate-limiting step, the first step is the rate-limiting step.

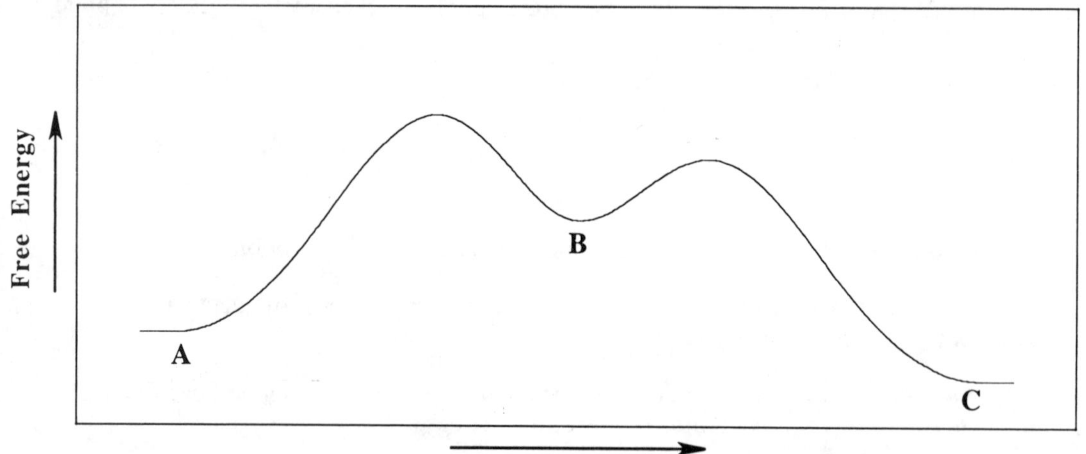

9. Solved in the text.

10. A protonated amine has a pK_a ~ 11. Therefore, the amine will be protonated by the acid that is produced in the reaction, and a protonated amine is not a nucleophile. Excess amine is used in order to have some unprotonated amine available to react as a nucleophile.

A protonated alcohol has a pK_a ~ -2. Therefore, the alcohol will not be protonated by the acid that is produced in the reaction.

11.

a. $CH_3-\overset{\overset{\ddot{O}:}{\|}}{C}-Cl + H_2\ddot{O}: \rightleftharpoons CH_3-\overset{:\ddot{O}:^-}{\underset{+\ddot{O}H_2}{C}}-Cl \underset{H^+}{\overset{-H^+}{\rightleftharpoons}} CH_3-\overset{:\ddot{O}:^-}{\underset{OH}{C}}-Cl \rightarrow CH_3-\overset{\overset{:\ddot{O}}{\|}}{C}-OH + Cl^-$

b. $CH_3-\overset{\overset{:\ddot{O}}{\|}}{C}-Br + CH_3\ddot{N}H_2 \rightleftharpoons CH_3-\overset{:\ddot{O}:^-}{\underset{+NH_2CH_3}{C}}-Br \underset{H^+}{\overset{-H^+}{\rightleftharpoons}} CH_3-\overset{:\ddot{O}:^-}{\underset{NHCH_3}{C}}-Br \rightarrow CH_3-\overset{\overset{:\ddot{O}}{\|}}{C}-NHCH_3 + Br^-$

12.

a. $CH_3CH_2CH_2OH$

b. $CH_3CH_2NH_2$

c. $(CH_3)_2NH$

d. $CH_3\overset{\overset{O}{\|}}{C}O^-$

e. HO—C₆H₄—NO₂ (para)

f. H_2O

13.

a. $CH_3\overset{\overset{O}{\|}}{C}-O-\overset{\overset{O}{\|}}{C}CH_3 + H_2\ddot{O}: \rightleftharpoons CH_3\overset{:\ddot{O}:^-}{\underset{+\ddot{O}H_2}{C}}-O-\overset{\overset{O}{\|}}{C}CH_3 \underset{H^+}{\overset{-H^+}{\rightleftharpoons}} CH_3\overset{:\ddot{O}:^-}{\underset{OH}{C}}-O-\overset{\overset{O}{\|}}{C}CH_3 \rightarrow CH_3\overset{\overset{O}{\|}}{C}OH + {}^-O\overset{\overset{O}{\|}}{C}CH_3$

b. The mechanisms are exactly the same.

$$CH_3\overset{O}{\overset{\|}{C}}-O-\overset{O}{\overset{\|}{C}}CH_3 + CH_3\ddot{O}H \rightleftharpoons CH_3\overset{:\ddot{O}:^-}{\underset{+\overset{|}{O}CH_3 \atop H}{\overset{|}{C}}}-O-\overset{O}{\overset{\|}{C}}CH_3 \underset{H^+}{\overset{-H^+}{\rightleftharpoons}} CH_3\overset{:\ddot{O}:^-}{\underset{\overset{|}{O}CH_3}{\overset{|}{C}}}-O-\overset{O}{\overset{\|}{C}}CH_3$$

$$\downarrow$$

$$CH_3\overset{O}{\overset{\|}{C}}OCH_3 + {}^-O\overset{O}{\overset{\|}{C}}CH_3$$

14. When an acid anhydride reacts with an amine, the tetrahedral intermediate does not have to lose a proton before it eliminates the carboxylate ion, because the carboxylate ion is a weaker base (pK_a of its conjugate acid is ~ 5) than an amine (pK_a of its conjugate acid is ~ 10).

$$CH_3\overset{O}{\overset{\|}{C}}-O-\overset{O}{\overset{\|}{C}}CH_3 + CH_3\ddot{N}H_2 \rightleftharpoons CH_3\overset{:\ddot{O}:^-}{\underset{+\overset{|}{N}H_2CH_3}{\overset{|}{C}}}-O-\overset{O}{\overset{\|}{C}}CH_3 .$$

15.

a. $$CH_3CH_2\overset{O}{\overset{\|}{C}}OCH_3 + H_2\ddot{O}: \rightleftharpoons CH_3CH_2\overset{:\ddot{O}:^-}{\underset{+\overset{|}{O}H \atop H}{\overset{|}{C}}}OCH_3$$

$$H^+ \Big\updownarrow -H^+$$

$$CH_3CH_2\overset{:\ddot{O}:^-}{\underset{\overset{|}{O}H}{\overset{|}{C}}}OCH_3$$

$$\swarrow\!\!\nearrow \qquad \nwarrow\!\!\searrow$$

$$CH_3CH_2\overset{O}{\overset{\|}{C}}OH + CH_3O^- \qquad CH_3CH_2\overset{O}{\overset{\|}{C}}OCH_3 + HO^-$$

$$\downarrow$$

$$CH_3CH_2\overset{O}{\overset{\|}{C}}O^- + CH_3OH$$

b. $HCO{-}O{-}C_6H_5$ + $CH_3\ddot{N}H_2$ ⇌ $HC(O^-)(O{-}C_6H_5)(\overset{+}{N}H_2CH_3)$

H^+ ⇅ $-H^+$

$HC(=O)NHCH_3$ + $^-O{-}C_6H_5$ ⟵ $HC(O^-)(O{-}C_6H_5)(NHCH_3)$

16. Solved in the text.

17. **1.** The carbonyl group of an ester is a weak electrophile.

2. Water is not a very strong nucleophile.

3. $^-OCH_3$ is a strong base and, therefore, a poor leaving group.

18. The mechanism for the acid-catalyzed reaction of acetic acid and methanol is the exact reverse of the mechanism for the acid-catalyzed hydrolysis of methyl acetate.

$CH_3{-}C(=O){-}OH \underset{-H^+}{\overset{H^+}{\rightleftharpoons}} CH_3{-}C(=\overset{+}{O}H){-}OH + CH_3\ddot{O}H \rightleftharpoons CH_3{-}C(OH)(OH){-}\overset{+}{O}(H)CH_3$

H^+ ⇅ $-H^+$

$CH_3{-}C(OH)(OH){-}OCH_3$

$-H^+$ ⇅ H^+

$CH_3{-}C(=O){-}OCH_3 \underset{H^+}{\overset{-H^+}{\rightleftharpoons}} CH_3{-}C(=\overset{+}{O}H){-}OCH_3 + H_2O \rightleftharpoons CH_3{-}C(\ddot{O}H)(\overset{+}{O}H_2)(OCH_3)$

19.

$$CH_3-\overset{O}{\overset{\|}{C}}-OCH_3 \underset{-H^+}{\overset{H^+}{\rightleftharpoons}} CH_3-\overset{\overset{+}{O}H}{\overset{\|}{C}}-OCH_3 + CH_3CH_2\ddot{O}H \rightleftharpoons CH_3-\underset{\underset{H}{+OCH_2CH_3}}{\overset{OH}{C}}-OCH_3$$

$$\underset{-H^+}{\overset{H^+}{\rightleftharpoons}}$$

$$CH_3-\underset{OCH_2CH_3}{\overset{OH}{C}}-OCH_3$$

$$\underset{H^+}{\overset{-H^+}{\rightleftharpoons}}$$

$$CH_3-\underset{OCH_2CH_3}{\overset{:\ddot{O}H}{C}}-\overset{+}{O}(H)CH_3 \rightleftharpoons CH_3-\overset{+OH}{\overset{\|}{C}}-OCH_2CH_3 + CH_3OH \underset{H^+}{\overset{-H^+}{\rightleftharpoons}} CH_3-\overset{O}{\overset{\|}{C}}-OCH_2CH_3$$

20. **a.** The conjugate base ($CH_3CH_2CH_2O^-$) of the reactant alcohol ($CH_3CH_2CH_2OH$) can be used to catalyze the reaction.

b. If H^+ is used as a catalyst, the amine will be protonated in the acidic solution and, therefore, will not be able to react as a nucleophile.

If HO^- is used as a catalyst, HO^- will be the strongest nucleophile present in solution, so it will attack the ester, and the product of the reaction will be a carboxylic acid rather than an amide.

If RO^- is used as a catalyst, RO^- will be the nucleophile, and the product of the reaction will be an ester rather than an amide.

21. **a.** The alcohol (CH_3CH_2OH) contained the ^{18}O label.

b. The carboxylic acid would have contained the ^{18}O label.

22. Solved in the text.

23. In the first step, protonation occurred to give the most stable carbocation (a tertiary carbocation).

$$CH_2{=}CHCH_2CH_2CH{=}C(CH_3)CH_3 \xrightarrow{H^+} CH_2{=}CHCH_2CH_2CH_2{-}\overset{+}{C}(CH_3)CH_3$$

$\xrightarrow{CH_3COOH}$ protonated 3,3-dimethylcyclohexyl acetate ($H\overset{+}{O}CCH_3$, with CH₃ and CH₃) $\xrightarrow{-H^+}$ 3,3-dimethylcyclohexyl acetate ($OCCH_3$, O, with CH_3 and CH_3)

24. The relative reactivities of the amides depend on the basicities of their leaving groups: the weaker the base, the more reactive the amide.

CH_3CONH–C$_6$H$_4$–NO_2 (para) > CH_3CONH–C$_6$H$_4$–NO_2 (meta) CH_3CONH–C$_6$H$_5$ > CH_3CONH–C$_6$H$_{11}$

A *para*-nitro-substituted aniline is less basic than a *meta*-substituted aniline because when the nitro group is in the para position, electrons can be delocalized onto the nitro group.

$CH_3C(=O){-}\ddot{N}H{-}C_6H_4{-}\overset{+}{N}(=O)O^-$

25. **a.** pentyl bromide **b.** isohexyl bromide **c.** benzyl bromide **d.** cyclohexyl bromide

26. The reaction of an alkyl halide ammonia leads to primary, secondary, and tertiary amines, and even quaternary ammonium ions. Thus, the yield of primary amine could be relatively low.

$$RBr \xrightarrow{NH_3} R\overset{+}{N}H_3 \xrightarrow{-H^+} RNH_2 \xrightarrow{RBr} R_2\overset{+}{N}H_2 \xrightarrow{-H^+} R_2NH$$

$$R_2NH \xrightarrow{RBr} R_3\overset{+}{N}H \xrightarrow{-H^+} R_3N \xrightarrow{RBr} R_4\overset{+}{N}\ Br^-$$

In contrast, the Gabriel synthesis forms only primary amines. The reaction of an alkyl halide with azide ion also forms only primary amines because the compound formed from the initial reaction of the two reagents is not nucleophilic, so polyalkylation does not occur.

27.

a. $CH_3CH_2CH_2Br$ b. CH_3CHCH_2Br (with CH_3 on C-2)   c. 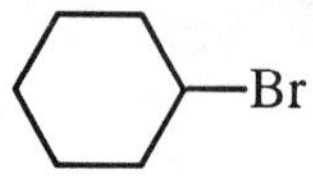 cyclohexyl–Br   d. $BrCH_2CH_2Br$

28. Solved in the text.

29.

a. $HOCCH_2CH_2COH$ (succinic acid) + CH_3COCCH_3 (acetic anhydride) ⟶ $CH_3C(O^-)(+OH–C(=O)CH_2CH_2COH)OCCH_3$

H^+ ⇅ $-H^+$

$CH_3C(O^-)(O–C(=O)CH_2CH_2COH)–OCCH_3$ ⟶ $CH_3C(=O)–O–C(=O)CH_2CH_2COH$ + CH_3COH ⟶ cyclic intermediate (^-O, $OCCH_3$, $\overset{+}{O}H$)

H^+ ⇅ $-H^+$

cyclic intermediate ($:\ddot{O}:^-$, $OCCH_3$, O) ⟶ succinic anhydride + CH_3COH

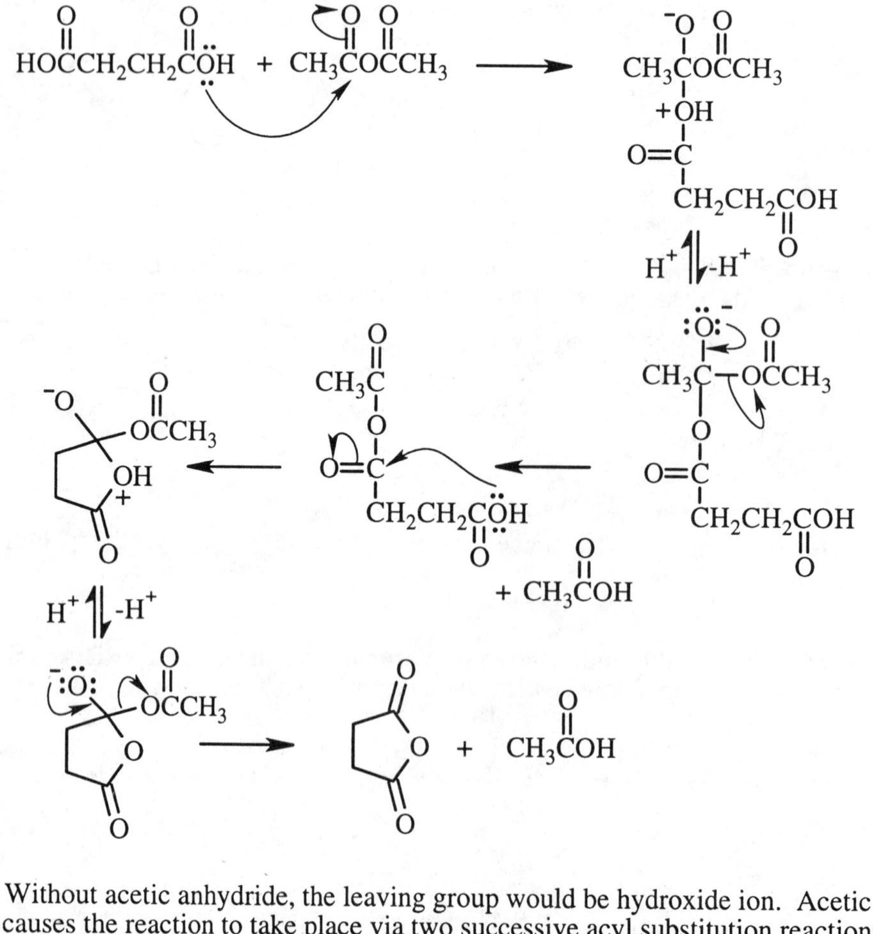

b. Without acetic anhydride, the leaving group would be hydroxide ion. Acetic anhydride causes the reaction to take place via two successive acyl substitution reactions. In both reactions the leaving group is acetate ion, which is much less basic than hydroxide ion.

30.

a. $(CH_3CH_2)_2N\overset{\overset{O}{\|}}{C}N(CH_2CH_3)_2$

b. $CH_3\overset{\overset{O}{\|}}{C}O\overset{\overset{O}{\|}}{C}CH_2\overset{\overset{O}{\|}}{C}O\overset{\overset{O}{\|}}{C}CH_3$

c. $H_2N\overset{\overset{O}{\|}}{C}NHCH_3$

d. no reaction

e. $^{+}NH_4 + HO\overset{\overset{O}{\|}}{C}OH \rightleftharpoons CO_2 + H_2O$

f. 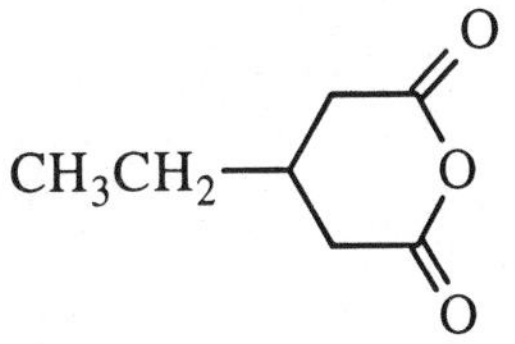

31.

a. $CH_3CH_2CH_2CH_2CH_2\overset{\overset{O}{\|}}{C}\underset{\underset{CH_3}{|}}{N}CH_3$

b. $CH_3CH_2CH_2\overset{\overset{CH_3}{|}}{\underset{\underset{CH_3}{|}}{C}}CH_2\overset{\overset{O}{\|}}{C}NH_2$

c. 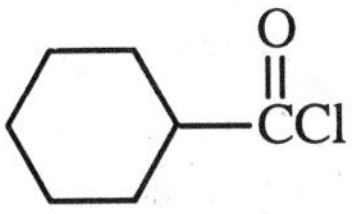

d. $CH_3CH_2C{\equiv}N$

e. $CH_3CH_2\overset{\overset{O}{\|}}{C}Br$

f. $CH_3\overset{\overset{O}{\|}}{C}O^{-}\ Na^{+}$

g. 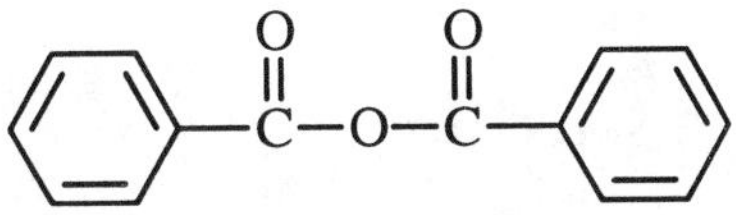

h. 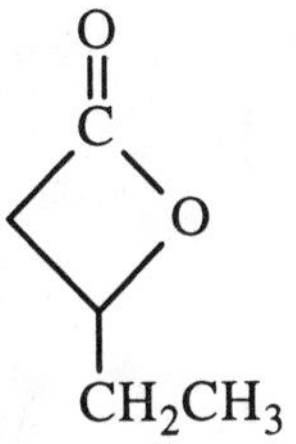

i. $CH_3\underset{\underset{CH_3}{|}}{C}HCH_2C{\equiv}N$

j. 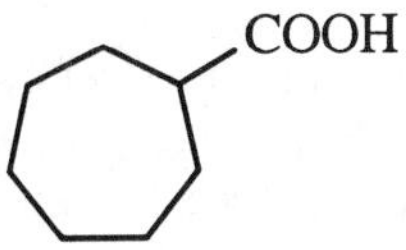

32.

a. 5-ethylheptanoic acid

b. propyl propanoate
propyl propionate

f. pentanoyl chloride
valeryl chloride

g. acetic benzoic anhydride

c. pentanenitrile
butyl cyanide

d. propanoic anhydride
propionic anhydride

e. *N,N*-dimethylbutanamide
N,N-dimethylbutyramide

h. *N*-methyl-3-butenamide

i. (*S*)-3-methylpentanoic acid
(*S*)-β-methylvaleric acid

j. (*R*)-3-methylhexanenitrile

33.

a. 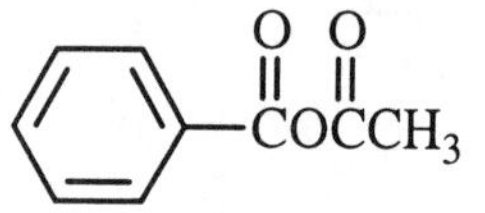 O O ‖ ‖ –COCCH$_3$

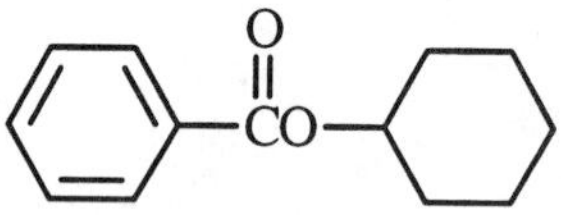

b. 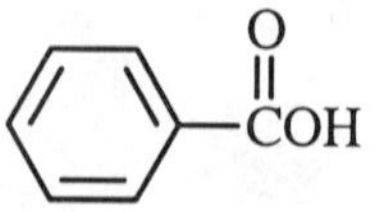 O ‖ –COH

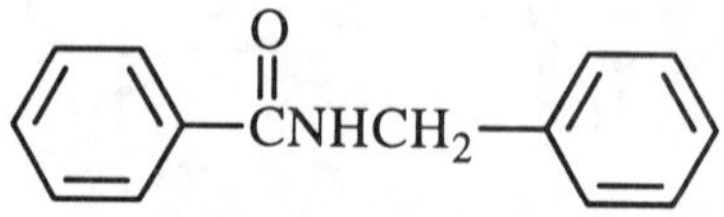

c. 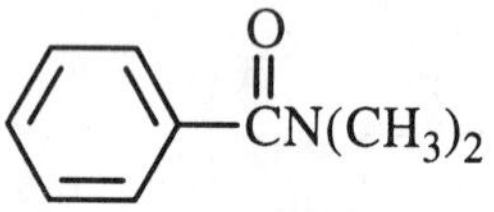 O ‖ –CN(CH$_3$)$_2$

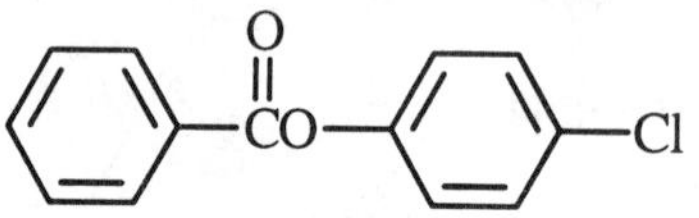

d. 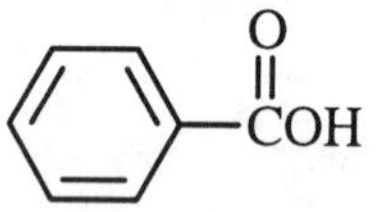 O ‖ –COH

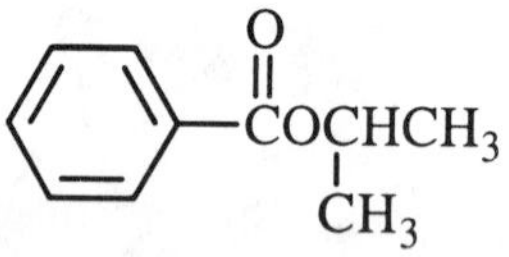

e. 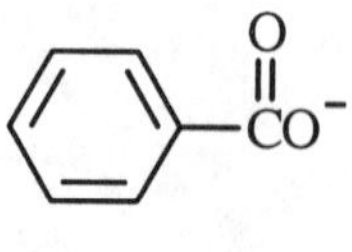 O ‖ –CO$^-$

f. O ‖ –CO–

g. O ‖ –CNHCH$_2$–

h. O ‖ –CO– –Cl

i. O ‖ –COCHCH$_3$ | CH$_3$

j. 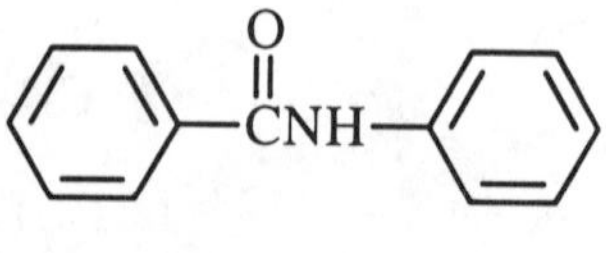 O ‖ –CNH–

34. **a.** The weaker the base attached to the acyl group, the easier it is to form the tetrahedral intermediate. (*para*-Chlorophenol is a stronger acid than phenol so the conjugate base of *para* chlorophenol is a weaker base than the conjugate base of phenol, etc.)

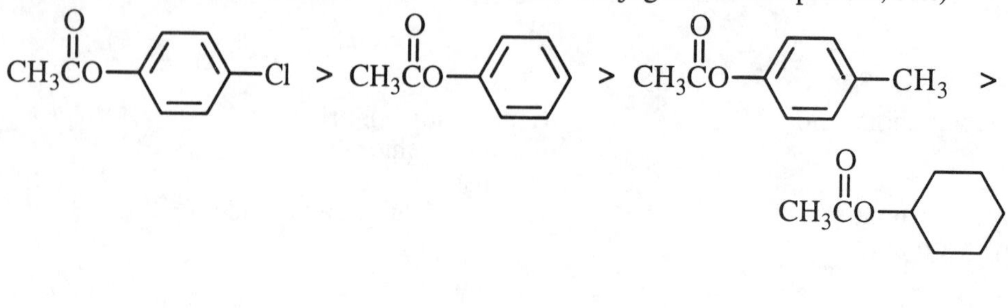

O ‖ CH$_3$CO– –Cl > O ‖ CH$_3$CO– > O ‖ CH$_3$CO– –CH$_3$ > O ‖ CH$_3$CO–

b. The tetrahedral intermediate collapses by eliminating the $^-$OR group of the tetrahedral intermediate. The weaker the basicity of the $^-$OR group, the easier it is to eliminate it. Thus, the rate of both formation of the tetrahedral intermediate and collapse of the tetrahedral intermediate is decreased by increasing the basicity of the $^-$OR group.

$CH_3C(=O)O\text{–}C_6H_4\text{–}Cl > CH_3C(=O)O\text{–}C_6H_5 > CH_3C(=O)O\text{–}C_6H_4\text{–}CH_3 > CH_3C(=O)O\text{–}C_6H_{11}$

35. a. Methyl acetate has a resonance contributor that butanone does not have, and this resonance contributor causes methyl acetate to be more polar than butanone. Because butanone is less polar, it has the lower dipole moment.

$$CH_3\text{–}\overset{O}{\overset{\|}{C}}\text{–}CH_2CH_3 \qquad CH_3\text{–}\overset{O}{\overset{\|}{C}}\text{–}\ddot{O}CH_3 \longleftrightarrow CH_3\text{–}\overset{O^-}{\overset{|}{C}}=\overset{+}{O}CH_3$$

b. Because it is more polar, the intermolecular forces holding methyl acetate molecules together are stronger, so we would expect methyl acetate to have a higher boiling point.

36. The best way to prepare an ester from a carboxylic acid is to first activate the carboxylic acid by converting it into an acyl chloride and then treating the acyl chloride with an alcohol. (The reaction of a carboxylic acid directly with an alcohol is a much slower reaction and has a much less favorable equilibrium constant.)

a. $C_6H_5\text{–}CH_2\overset{O}{\overset{\|}{C}}OH \xrightarrow{SOCl_2} C_6H_5\text{–}CH_2\overset{O}{\overset{\|}{C}}Cl \xrightarrow{CH_3CH_2OH} C_6H_5\text{–}CH_2\overset{O}{\overset{\|}{C}}OCH_2CH_3$

b. $CH_3CH_2CH_2\overset{O}{\overset{\|}{C}}OH \xrightarrow{SOCl_2} CH_3CH_2CH_2\overset{O}{\overset{\|}{C}}Cl \xrightarrow{CH_3CH_2OH} CH_3CH_2CH_2\overset{O}{\overset{\|}{C}}OCH_2CH_3$

c. 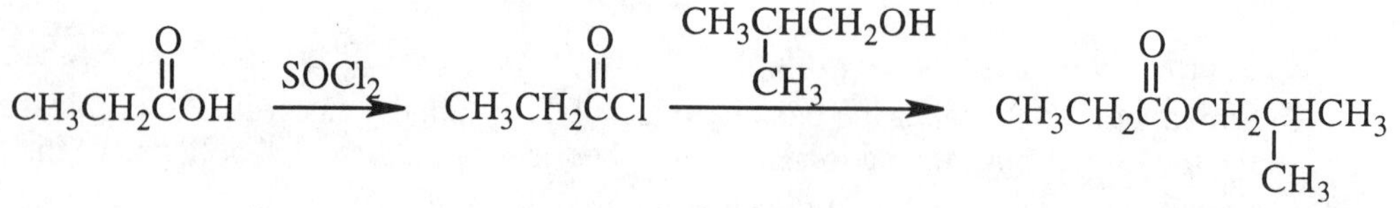 $CH_3CH_2\overset{O}{\overset{\|}{C}}OH \xrightarrow{SOCl_2} CH_3CH_2\overset{O}{\overset{\|}{C}}Cl \xrightarrow{CH_3CH(CH_3)CH_2OH} CH_3CH_2\overset{O}{\overset{\|}{C}}OCH_2CH(CH_3)CH_3$

d. 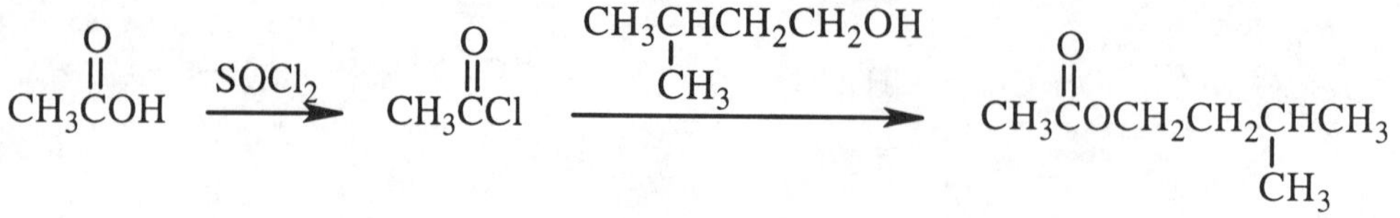 $CH_3\overset{O}{\overset{\|}{C}}OH \xrightarrow{SOCl_2} CH_3\overset{O}{\overset{\|}{C}}Cl \xrightarrow{CH_3CH(CH_3)CH_2CH_2OH} CH_3\overset{O}{\overset{\|}{C}}OCH_2CH_2CH(CH_3)CH_3$

37. Propyl formate is easy to distinguish because it is the only ester that will show four signals. The other three esters show three signals. Isopropyl formate can be distinguished by its unique splitting pattern: a singlet, a doublet, and a septet. The splitting patterns of the other two esters are the same: a singlet, a triplet, and a quartet. They can be distinguished because the peak farthest downfield in ethyl acetate is a quartet, while the peak farthest downfield in methyl propionate is a singlet.

$H\overset{O}{\overset{\|}{C}}OCH_2CH_2CH_3$

4 signals

$H\overset{O}{\overset{\|}{C}}OCH(CH_3)CH_3$	$CH_3\overset{O}{\overset{\|}{C}}OCH_2CH_3$	$CH_3CH_2\overset{O}{\overset{\|}{C}}OCH_3$
3 signals	3 signals	3 signals
singlet, doublet, septet	singlet, triplet, quartet	singlet, triplet, quartet
	The peak farthest downfield is a quartet.	The peak farthest downfield is a singlet.

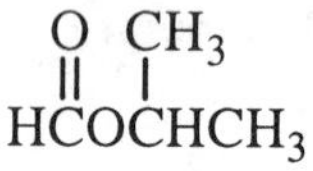

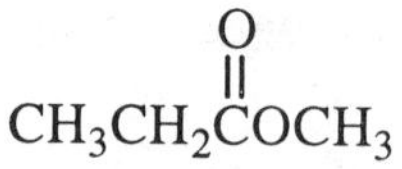

38. The reaction of methylamine with propionyl chloride generates a proton that will protonate unreacted amine, thereby destroying its nucleophilicity. If two equivalents of CH_3NH_2 are used, one equivalent will remain unprotonated and be able to react with propionyl chloride to form *N*-methylpropanamide.

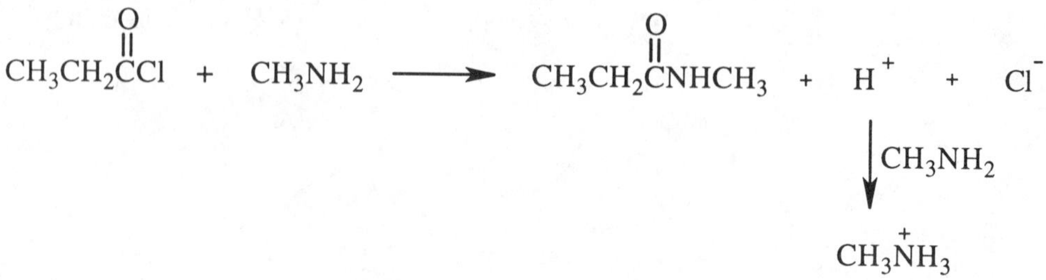

$$CH_3CH_2\overset{O}{\overset{\|}{C}}Cl + CH_3NH_2 \longrightarrow CH_3CH_2\overset{O}{\overset{\|}{C}}NHCH_3 + H^+ + Cl^-$$

$$H^+ \xrightarrow{CH_3NH_2} CH_3\overset{+}{N}H_3$$

39.

a. $CH_3-\overset{O}{\overset{||}{C}}-OH + H_2\overset{18}{O}: \rightleftharpoons CH_3-\underset{\underset{H}{\overset{+}{^{18}OH}}}{\overset{O^-}{\overset{|}{C}}}-OH \overset{-H^+}{\rightleftharpoons} CH_3-\underset{^{18}OH}{\overset{:\ddot{O}:^-}{\overset{|}{C}}}-OH$

$\rightleftharpoons CH_3-\overset{O}{\overset{||}{C}}-\overset{18}{OH} + H_2\overset{18}{\ddot{O}}: \rightleftharpoons CH_3-\underset{\underset{H}{\overset{+}{^{18}OH}}}{\overset{O^-}{\overset{|}{C}}}-\overset{18}{OH} \underset{H^+}{\overset{-H^+}{\rightleftharpoons}} CH_3-\underset{^{18}OH}{\overset{O^-}{\overset{|}{C}}}-\overset{18}{OH} \underset{-H^+}{\overset{H^+}{\rightleftharpoons}} CH_3-\underset{^{18}OH}{\overset{OH}{\overset{|}{C}}}-\overset{18}{\ddot{O}}:^- \rightleftharpoons CH_3-\overset{^{18}O}{\overset{||}{C}}-\overset{18}{OH}$

b. Only one isotopically labeled oxygen can be incorporated into the ester because the bond between the methyl group and the labeled oxygen does not break, so there is no way for the carbonyl oxygen to become labeled.

$CH_3-\overset{O}{\overset{||}{C}}-\overset{18}{O}-CH_3$

cannot become labeled

bond does not break

40. **a.** isopropyl alcohol and HCl

b. aqueous sodium hydroxide

c. ethylamine

d. water and HCl

41. The offset in the NMR spectrum shows that there is a signal at ~ 10 ppm, which is where the proton of a COOH group shows a signal. The two triplets and the multiplet are characteristic of a propyl group. The compound is butanoic acid.

$$CH_3CH_2CH_2\overset{O}{\overset{||}{C}}OH$$

The molecular formula shows that the unknown compound has one more carbon atom than butanoic acid. Since butanoic acid is formed from acid hydrolysis of the compound, the compound must be the methyl ester of butanoic acid.

$$CH_3CH_2CH_2\overset{O}{\overset{||}{C}}OCH_3$$

$$C_5H_{10}O_2$$

42. Aspartame has an amide group and an ester group that will be hydrolyzed in an aqueous solution of HCl. Because the hydrolysis is carried out in an acidic solution, the carboxylic acid groups and the amino groups will be in their acidic forms.

$$\mathrm{HO\overset{O}{\overset{\|}{C}}CH_2\underset{+NH_3}{\underset{|}{C}}H\overset{O}{\overset{\|}{C}}NH\underset{CH_2C_6H_5}{\underset{|}{C}}H\overset{O}{\overset{\|}{C}}OCH_3}$$

$$\downarrow \ \mathrm{H^+, H_2O}$$

$$\mathrm{HO\overset{O}{\overset{\|}{C}}CH_2\underset{+NH_3}{\underset{|}{C}}H\overset{O}{\overset{\|}{C}}OH} + \mathrm{H_3\overset{+}{N}\underset{CH_2C_6H_5}{\underset{|}{C}}H\overset{O}{\overset{\|}{C}}OH} + \mathrm{CH_3OH}$$

43. **a.** 1, 3, 4, 6, 7, 9 will not form the indicated products under the given conditions.

4 will form the product shown if it is heated to ~225 °C to dehydrate the salt that is formed at room temperature.

b. 9 will form the product shown in the presence of an acid catalyst.

44. The tertiary amine is a stronger nucleophile than the alcohol, so formation of the charged amide will be faster than formation of the new ester would have been. The charged amide is more reactive than an ester, so formation of the new ester by reaction of the alcohol with the charged amide will be faster than formation of the ester by reaction of the alcohol with the starting ester would have been. In other words, both reactions that occur in the presence of the tertiary amine are faster than the single reaction that occurs in the absence of the tertiary amine.

$$\mathrm{CH_3-\overset{O}{\overset{\|}{C}}-OCH_3} + \mathrm{:N\langle\ \rangle N:} \rightleftharpoons \mathrm{CH_3-\overset{O}{\overset{\|}{C}}-\overset{+}{N}\langle\ \rangle N:} \xrightleftharpoons{\mathrm{ROH}} \mathrm{CH_3-\overset{O}{\overset{\|}{C}}-OR}$$

$$+ \ \mathrm{CH_3OH} + \mathrm{:N\langle\ \rangle N:}$$

45. The amine is a stronger nucleophile than the alcohol, but since the acyl chloride is very reactive, it can react easily with both nucleophiles. Therefore, steric hindrance will be the most important factor in determining the products. The amino group is less sterically hindered than the alcohol group so that will be the group most easily acetylated.

CH₂OH / CH₂CH₃ piperidine (N–H) + $CH_3\overset{O}{\overset{\|}{C}}Cl$ ⟶

CH_2OH, CH_2CH_3, N–C(=O)–CH_3
major product

$CH_2O\overset{O}{\overset{\|}{C}}CH_3$, CH_2CH_3, N–C(=O)–CH_3
minor product

$CH_2O\overset{O}{\overset{\|}{C}}CH_3$, CH_2CH_3, N–H
minor product

46. Bromobutane undergoes an S_N2 reaction with NH_3 to form butylamine (**A**). Bromobutane can then react with butylamine to form dibutylamine (**B**). The amines each form an amide upon reaction with acetyl chloride. The IR spectrum of **C** exhibits an NH stretch at about 3300 cm^{-1}, while the IR spectrum of **D** does not exhibit this absorption band, because the nitrogen in **D** is not bonded to a hydrogen.

$$CH_3CH_2CH_2CH_2Br + NH_3 \longrightarrow \underset{\textbf{A}}{CH_3CH_2CH_2CH_2NH_2}$$

$$\downarrow NH_3$$

$$\underset{\textbf{B}}{(CH_3CH_2CH_2CH_2)_2NH}$$

$$CH_3\overset{O}{\overset{\|}{C}}Cl + \underset{\textbf{A}}{CH_3CH_2CH_2CH_2NH_2} \longrightarrow \underset{\textbf{C}}{CH_3\overset{O}{\overset{\|}{C}}NHCH_2CH_2CH_2CH_3}$$

$$CH_3\overset{O}{\overset{\|}{C}}Cl + \underset{\textbf{B}}{(CH_3CH_2CH_2CH_2)_2NH} \longrightarrow \underset{\textbf{D}}{CH_3\overset{O}{\overset{\|}{C}}N(CH_2CH_2CH_2CH_3)_2}$$

47.

a. 1. $ClCOCH_3$ (with C=O) 3. $CH_3CH_2CH_2NHCNHCH_2CH_2CH_3$ (with C=O)

2. CH_3OCOCH_3 (with C=O) 4. $CH_3CH_2OCNHCH_3$ (with C=O)

b. $ClCCl + 2\ H_2O \longrightarrow HOCOH + 2\ HCl$ (C=O in each)

$HOCOH \longrightarrow CO_2 + H_2O$

The HCl that is formed destroys lung tissue.

48.

$HOCCH_2CH_2COH + Cl{-}S({=}O){-}Cl \longrightarrow HOCCH_2CH_2C{-}O{-}S({=}O){-}Cl + H^+ + Cl^-$

$\downarrow$

O (protonated cyclic anhydride, O+ with H) $+ SOCl_2 + Cl^-$ $\underset{H^+}{\overset{-H^+}{\rightleftharpoons}}$ cyclic anhydride

49.

a. CH_3CF (with C=O)

b. $HOCCH_2CH_2CH_2\overset{+}{N}H_3\ Cl^-$ (with C=O)

c. $C_6H_5CNHCH_3$ (with C=O)

d. $HOCCH_2CH_2COH$ (with C=O groups)

e. (benzo-fused cyclic carbonate: O, O, =O)

f. $HOCCH_2CH_2CH_2OH$ (with C=O)

g. $CH_3CCH_2O^- + CH_3OCCH_3$ (with C=O)

h. (benzo-fused cyclic anhydride: O, =O, =O)

i. (benzene ring with $CONH_2$ and CO_2^-: O, NH_2, O^-, O)

j. (benzene ring with $COOCH_3$ and CH_2COOCH_3: O, OCH_3, O, OCH_3)

50.

$$\underset{\text{Compound A}}{\text{ClCH}_2\text{CH}_2\text{CH}_2\overset{\overset{\text{O}}{\parallel}}{\text{C}}\text{Cl}} + \text{CH}_3\text{OH} \longrightarrow \text{ClCH}_2\text{CH}_2\text{CH}_2\overset{\overset{\text{O}}{\parallel}}{\text{C}}\text{OCH}_3$$

51.

a. $$\text{CH}_3\overset{\overset{\text{O}}{\parallel}}{\text{C}}\text{NH}\underset{\underset{\text{CH}_3}{|}}{\text{C}}\text{HCH}_2\text{CH}_2\text{O}\overset{\overset{\text{O}}{\parallel}}{\text{C}}\text{CH}_3 + \text{CH}_3\overset{\overset{\text{O}}{\parallel}}{\text{C}}\text{OH}$$

b. $$\text{CH}_3\overset{\overset{\text{O}}{\parallel}}{\text{C}}\text{NH}\underset{\underset{\text{CH}_3}{|}}{\text{C}}\text{HCH}_2\text{CH}_2\text{OH}$$

Because the amino group is a stronger nucleophile than the OH group, the predominant product will be the amide if the reaction is stopped prematurely.

52. If the amine is tertiary, the nitrogen atom in the amide cannot get rid of its positive charge by losing a proton. An amide with a positively charged amino group is very reactive because the positively charged amino group is a weak base and, therefore, an excellent leaving group. Water will immediately react with the amide, and because the positively charged amine is a better leaving group than the OH group, the amine will be expelled and the product will be a carboxylic acid.

$$\text{CH}_3-\overset{\overset{\text{O}}{\parallel}}{\text{C}}-\text{Cl} \xrightarrow{\text{NR}_3} \text{CH}_3-\overset{\overset{\text{O}}{\parallel}}{\text{C}}-\overset{+}{\text{N}}\text{R}_3 \xrightarrow{\text{H}_2\text{O}} \text{CH}_3-\overset{\overset{\text{O}}{\parallel}}{\text{C}}-\text{O}^- + \text{H}\overset{+}{\text{N}}\text{R}_3$$

53. **a.** The steric hindrance provided by the methyl groups prevents methyl alcohol from attacking the carbonyl carbon.

b. No, because there would be no steric hindrance.

c. 2,4,6-trimethylbenzoic acid (COOH; H_3C, CH_3, CH_3) $\xrightarrow{\overset{-}{\text{C}}\text{H}_2\overset{+}{\text{N}}\equiv\text{N}}$ 2,4,6-trimethylbenzoate (COO^-; H_3C, CH_3, CH_3) $\xrightarrow{\text{CH}_3\overset{+}{\text{N}}\equiv\text{N}}$ methyl 2,4,6-trimethylbenzoate ($COOCH_3$; H_3C, CH_3, CH_3)

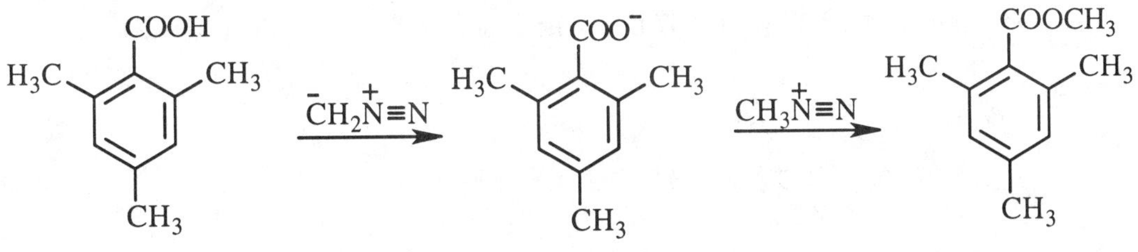

54. The more electronegative the base (Y) attached to the carbonyl carbon, the greater the double bond character of the C=O bond.

The weaker the base (Y) attached to the carbonyl carbon the less well it shares it electrons. when Y shares its electrons, it decreases the double bond character of the C=O bond, making it easier to stretch.

Therefore, the carbonyl IR absorption band decreases in the order:

$$CH_3\overset{\overset{O}{\|}}{C}Cl \;>\; CH_3\overset{\overset{O}{\|}}{C}OCH_3 \;>\; CH_3\overset{\overset{O}{\|}}{C}H \;>\; CH_3\overset{\overset{O}{\|}}{C}NH_2$$

The carbonyl group of the acyl chloride stretches at the highest frequency because the chlorine atom is the most electronegative atom and the weakest base.

The predominant effect of the oxygen of an ester is inductive electron withdrawal, so its carbonyl group stretches at a higher frequency than the carbonyl group of the aldehyde.

The carbonyl group of the amide stretches at the lowest frequency because the nitrogen atom is the strongest base, so it is best at sharing its electrons.

55. **a.**

$$4.02 = \frac{x^2}{(1-x)^2}$$

$$2.00 = \frac{x}{(1-x)}$$

$$2 - 2x = x$$

$$2 = 3x$$

$$x = 0.667$$

[ethyl acetate] = 0.667 times the concentration of acetic acid used

b.

$$4.02 = \frac{x^2}{(10-x)(1-x)}$$

$$x = 0.974$$

[ethyl acetate] = 0.974 times the concentration of acetic acid used

c.

$$4.02 = \frac{x^2}{(100 - x)(1 - x)}$$

$x = 0.997$

[ethyl acetate] = 0.997 times the concentration of acetic acid used

56. Each of the NMR spectra has signals between about 7-8 ppm, indicating that the compound has a benzene ring with one substituent (the benzene ring has hydrogens in three different environments) and one additional signal that is a singlet. From the molecular formulas it can be determined that the esters have the following structures. The singlet in each spectrum is due to a methyl group. Because the methyl group is farther downfield in the spectrum on the right, it is the spectrum of the compound on the right, since its methyl group is adjacent to an electron-withdrawing oxygen atom.

$CH_3C(=O)-O-C_6H_5$ $\qquad$ $C_6H_5-C(=O)-OCH_3$

The ester on the left will hydrolyze faster because phenol is a much weaker base than methyl alcohol.

57. It tells you that two hydroxide molecules have been used to build the transition state of the rate-limiting step.

$$RC(=O)-NH_2 \xrightarrow{HO^-} RC(O^-)(OH)-NH_2 \xrightarrow{HO^-} RC(O^-)(O^-)-NH_2 \xrightarrow{\text{rate-limiting step}} RC(=O)-O^- + {}^-NH_2$$

58.

a. $$CH_3CH_2CNH_2 \text{ (C=O)} \xrightarrow[H_2O,\ \Delta]{H^+} CH_3CH_2COH \text{ (C=O)} \xrightarrow{SOCl_2} CH_3CH_2CCl \text{ (C=O)}$$

b. $$CH_3CH_2CH_2CH_2OH \xrightarrow{SOCl_2} CH_3CH_2CH_2CH_2Cl \xrightarrow{^-C\equiv N} CH_3CH_2CH_2CH_2C\equiv N$$

$$\xrightarrow{H_2O,\ H^+,\ \Delta} \overset{+}{N}H_4 + CH_3CH_2CH_2CH_2COH \text{ (C=O)}$$

c. $CH_3(CH_2)_{10}COH$ (C=O) $\xrightarrow{SOCl_2}$ $CH_3(CH_2)_{10}CCl$ (C=O) $\xrightarrow[2.\ H_2O]{1.\ \text{benzene},\ AlCl_3}$ $CH_3(CH_2)_{10}C(=O)C_6H_5$

$\xrightarrow{Zn(Hg),\ HCl,\ \Delta \text{ or } NH_2NH_2,\ HO^-,\ \Delta}$ $CH_3(CH_2)_{11}C_6H_5$ $\xrightarrow[\Delta]{H_2SO_4}$ $CH_3(CH_2)_{11}C_6H_4SO_3H$ $\xrightarrow{NaOH}$ $CH_3(CH_2)_{11}C_6H_4SO_3^-\ Na^+$

d. $C_6H_5CH_3$ $\xrightarrow[\Delta]{KMnO_4}$ C_6H_5COH (C=O) $\xrightarrow{SOCl_2}$ C_6H_5CCl (C=O) $\xrightarrow{\text{excess } CH_3NH_2}$ $C_6H_5CNHCH_3$ (C=O)

e. $C_6H_5NH_2$ $\xrightarrow{CH_3CCl\ (C=O)}$ $C_6H_5NHCCH_3$ (C=O) $\xrightarrow[AlCl_3]{CH_3CCl\ (C=O)}$ $p\text{-}CH_3C(=O)C_6H_4NHCCH_3$ (C=O) $\xrightarrow[2.\ HO^-]{1.\ H_3O^+,\ \Delta}$ $p\text{-}CH_3C(=O)C_6H_4NH_2$

f. $C_6H_5CH_3$ $\xrightarrow[\Delta]{NBS}$ $C_6H_5CH_2Br$ $\xrightarrow{^-C{\equiv}N}$ $C_6H_5CH_2C{\equiv}N$ $\xrightarrow[H^+\ \Delta]{H_2O}$ $C_6H_5CH_2COOH$

59. The acid-catalyzed hydrolysis of acetamide forms acetic acid and ammonium ion. It is an irreversible reaction, because the pK_a of a acetic acid is less than the pK_a of the ammonium ion. Therefore, it is impossible to have the carboxylic acid in its reactive acidic form and ammonia in its reactive basic form.

If the solution is sufficiently acidic to have the carboxylic acid is in its acidic form, ammonia will also be in its acidic form so it will not be a nucleophile. If the pH of the solution is sufficiently basic to have ammonia in its nucleophilc basic form, the carboxylic acid will also be its basic form; a negatively charged carboxylic acid is not attacked by nucleophiles

CH_3COOH + $\overset{+}{N}H_4$ ⟶ no reaction
The ammonium ion is not nucleophilic.

CH_3CO^- + NH_3 ⟶ no reaction
The carboxylate ion is not attacked by nucleophiles.

60.

a. R—C—ÖH $\xrightarrow{H^+}$ R—C—OH(H)+ ⟶ R—C+ ⟶ (RC≡N) RC=N+—C—R ⟶ ($H_2\ddot{O}$:) RC=N—C—R (+OH H) $\underset{H^+}{\overset{-H^+}{\rightleftharpoons}}$ RC=N—C—R (OH) $\underset{-H^+}{\overset{H^+}{\rightleftharpoons}}$ RC=N(H)+—C—R (:OH) ⟶ RC(=OH+)—N(H)—R' $\underset{-H^+}{\overset{H^+}{\rightleftharpoons}}$ RC(=O)—N—R'

b. The Ritter reaction does not work with primary alcohols, because primary alcohols do not form carbocations.

c. The only difference in the two reactions is the electrophile that attaches to the nitrogen of the nitrile: it is a carbocation in the Ritter reaction and a proton in the acid-catalyzed hydrolysis of a nitrile.

61.

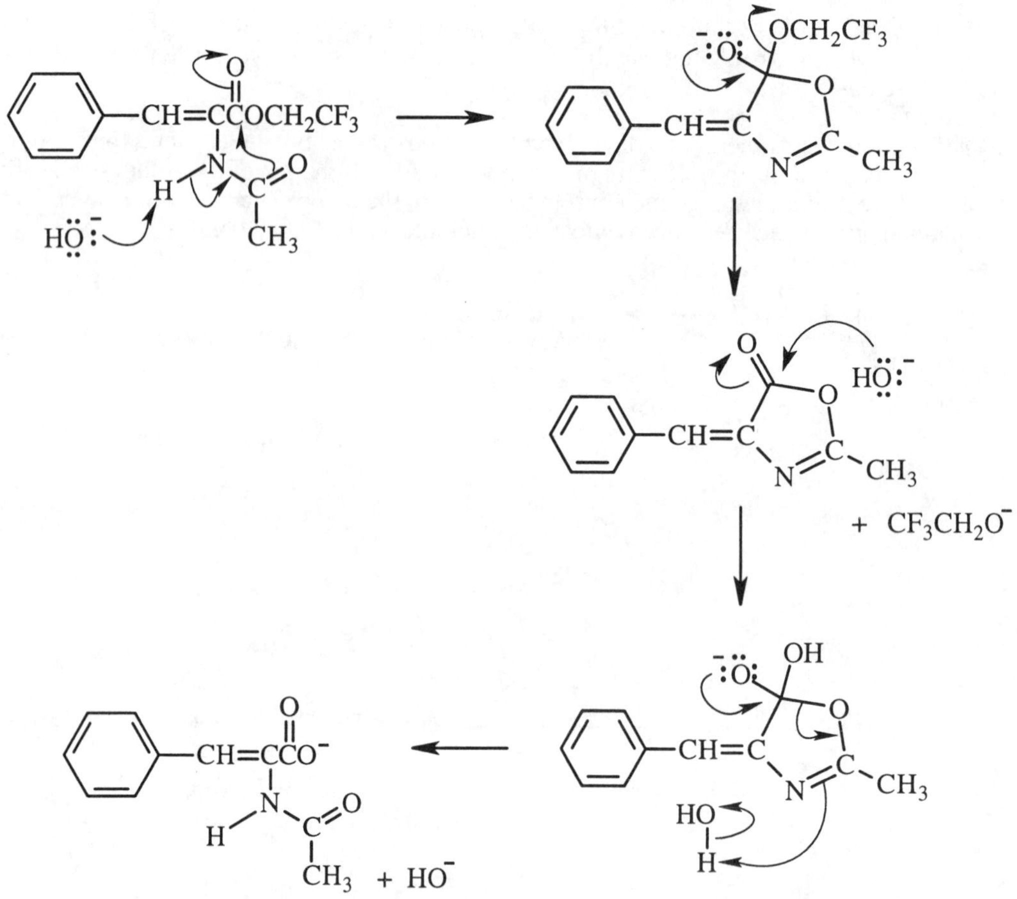

$C_6H_5CH{=}C(NHCOCH_3)COOCH_2CF_3$ + HO^- → (oxazolone tetrahedral intermediate with O^- and OCH_2CF_3) → (oxazolone, $C_6H_5CH{=}C$ ring with CH_3) + $CF_3CH_2O^-$ → (tetrahedral intermediate with O^- and OH, HO–H) → $C_6H_5CH{=}C(NHCOCH_3)CO_2^-$ + HO^-

62.

a. (six-membered lactone with CH_2CH_3)

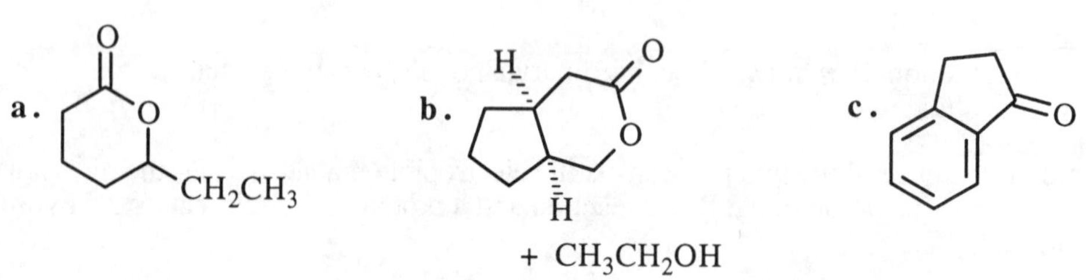

b. (bicyclic lactone with H, H) + CH_3CH_2OH

c. (indanone)

63.

Benzene $\xrightarrow[H_2SO_4]{HNO_3}$ $C_6H_5NO_2$ $\xrightarrow[Pd/C]{H_2}$ $C_6H_5NH_2$ $\xrightarrow{CH_3\overset{O}{\overset{\|}{C}}Cl}$ $C_6H_5NH\overset{O}{\overset{\|}{C}}CH_3$

$\xrightarrow{H_2SO_4}$ $HOSO_2-C_6H_4-NH\overset{O}{\overset{\|}{C}}CH_3$ $\xrightarrow{PCl_3}$ $ClSO_2-C_6H_4-NH\overset{O}{\overset{\|}{C}}CH_3$

$\xrightarrow{NH_3}$ $H_2NSO_2-C_6H_4-NH\overset{O}{\overset{\|}{C}}CH_3$ $\xrightarrow[2.\ HO^-]{1.\ H^+,\ H_2O}$ $H_2NSO_2-C_6H_4-NH_2$

64.

a. Benzene $\xrightarrow[AlCl_3]{CH_3Cl}$ toluene (CH_3) $\xrightarrow[H_2SO_4]{HNO_3}$ o-nitrotoluene (CH_3, NO_2) $\xrightarrow[\Delta\ H^+]{Na_2Cr_2O_7}$ o-nitrobenzoic acid ($COOH$, NO_2)

$\xrightarrow{H_2\ Pd/C}$ ($COOH$, NH_2) $\xrightarrow[0^\circ]{NaNO_2,\ HCl}$ ($COOH$, $\overset{+}{N}\equiv N$ Cl^-) $\xrightarrow{H_2O}$ ($COOH$, OH) $\xrightarrow{CH_3\overset{O}{\overset{\|}{C}}O\overset{O}{\overset{\|}{C}}CH_3}$ ($COOH$, $O\overset{O}{\overset{\|}{C}}CH_3$)

b. BenYene $\xrightarrow[\text{AlCl}_3]{\text{CH}_3\text{CH}_2\text{Br}}$ ethylbenzene (CH_3CH_2–C$_6$H$_5$) $\xrightarrow[\text{2. } H_2O]{\text{1. } (CH_3)_2CHCCl\text{ (O), AlCl}_3}$ CH_3CH_2–C$_6$H$_4$–$C(=O)CHCH_3$ (CH_3) $\xrightarrow[\Delta]{\text{NBS}}$ CH_3CHBr–C$_6$H$_4$–$O{=}CCHCH_3$ (CH_3) $\xrightarrow{^-C{\equiv}N}$ $CH_3CH\,C{\equiv}N$–C$_6$H$_4$–$O{=}CCHCH_3$ (CH_3) $\xrightarrow[H_2O,\ \Delta]{H^+}$ CH_3CHCOH (O)–C$_6$H$_4$–$O{=}CCHCH_3$ (CH_3) $\xrightarrow[\text{2. } H^+]{\text{excess, 1. } H_2NNH_2,\ HO^-,\ \Delta}$ CH_3CHCOH (O)–C$_6$H$_4$–CH_2CHCH_3 (CH_3)

65. A β-lactamase enzyme provides resistance to penicillin by reacting with it, opening the four-membered ring.

$-CH_2\ddot{O}H$ + β-lactam ring (C=O, N) → ring-opened ester (O, O–CH_2, NH)

The inhibitor of the β-lactamase enzyme has an excellent leaving group, so it is readily attacked by nucleophiles. When the enzyme attacks the inhibitor, a relatively stable phosphonyl-enzyme is formed, so its OH group is no longer available to react with penicillin.

$-CH_2\ddot{O}H$

$C_6H_5CH_2CNHCH_2P(=O)(O^-)-O-C_6H_4COO^- \rightarrow C_6H_5CH_2CNHCH_2P(=O)(O^-)-OCH_2-$

a phosphonyl-enzyme

$HO-C_6H_4-COO^-$

Hydroxylamine (H_2NOH) reactivates the enzyme by liberating the enzyme from the phosphonyl-enzyme.

$H_2N\ddot{O}H$

$C_6H_5CH_2CNHCH_2P(=O)(O^-)-OCH_2- \rightarrow C_6H_5CH_2CNHCH_2P(=O)(O^-)-ONH_2$

$-CH_2OH$

66. **a.**

HO, $NHCO_2CH_3$ $\underset{-H^+}{\overset{H^+}{\rightleftharpoons}}$ H, HO+, $\ddot{N}HCO_2CH_3$

+, $\ddot{N}HCO_2CH_3$

$N-CO_2CH_3$ $\underset{H^+}{\overset{-H^+}{\rightleftharpoons}}$ $\overset{+}{N}H-CO_2CH_3$

b. H_3C–(2-hydroxy-5-methylphenyl)–SO_2–(2,4-dinitrophenyl) + $CH_3\ddot{O}:^-$ → spiro intermediate (anionic Meisenheimer complex) ↔ resonance form with $N^+(O^-)_2$ → H_3C–(2-sulfinate)phenyl–O–(2,4-dinitrophenyl)

67.

$$C_6H_6 \xrightarrow[H_2SO_4]{HNO_3} C_6H_5NO_2 \xrightarrow[Pd/C]{H_2} C_6H_5NH_2 \xrightarrow{CH_3CCl(=O)} C_6H_5NHCCH_3(=O)$$

$$\xrightarrow[AlCl_3]{CH_3Cl} CH_3-C_6H_4-NHCCH_3(=O) \xrightarrow[\Delta]{Na_2Cr_2O_7,\ H_2SO_4} HOC(=O)-C_6H_4-NHCCH_3(=O) \xrightarrow{H^+,\ H_2O} HOC(=O)-C_6H_4-\overset{+}{N}H_3$$

$$(CH_3CH_2)_2NH + \text{ethylene oxide} \longrightarrow (CH_3CH_2)NCH_2CH_2OH$$

$$HOC(=O)-C_6H_4-\overset{+}{N}H_3 \xrightarrow{H^+,\ (CH_3CH_2)NCH_2CH_2OH} (CH_3CH_2)_2NCH_2CH_2OC(=O)-C_6H_4-\overset{+}{N}H_3 \xrightarrow{HO^-} (CH_3CH_2)_2NCH_2CH_2OC(=O)-C_6H_4-NH_2$$

68.

a.

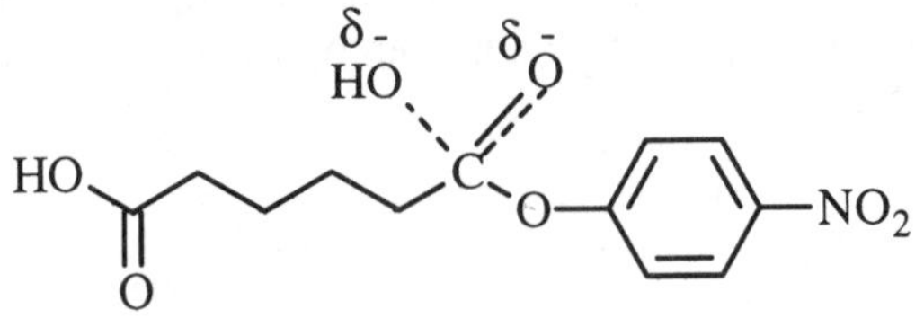

b.

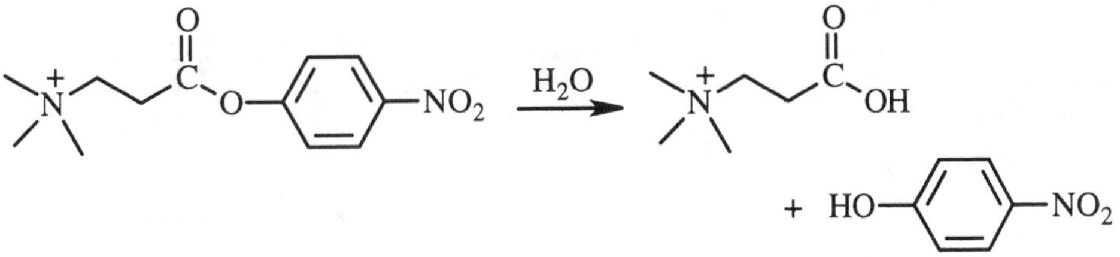

c.

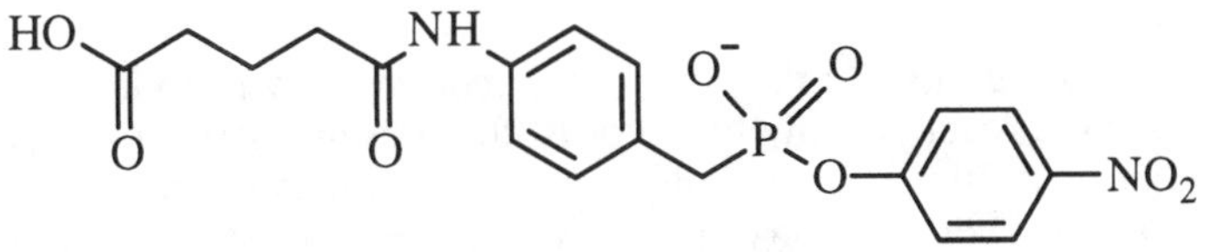

69. Because electron-withdrawing substituents have positive substituent constants and electron-donating substituents have negative substituent constants, a reaction with a positive ρ value is one in which compounds with electron-withdrawing substituents react more rapidly than compounds with electron-donating substituents, and a reaction with a negative ρ value is one in which compounds with electron-donating substituents react more rapidly than compounds with electron-withdrawing substituents.

electron withdrawal increases the rate of the reaction

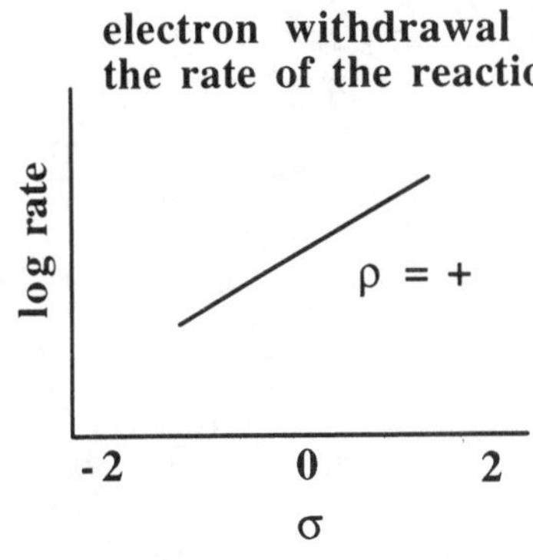

electron donation increases the rate of the reaction

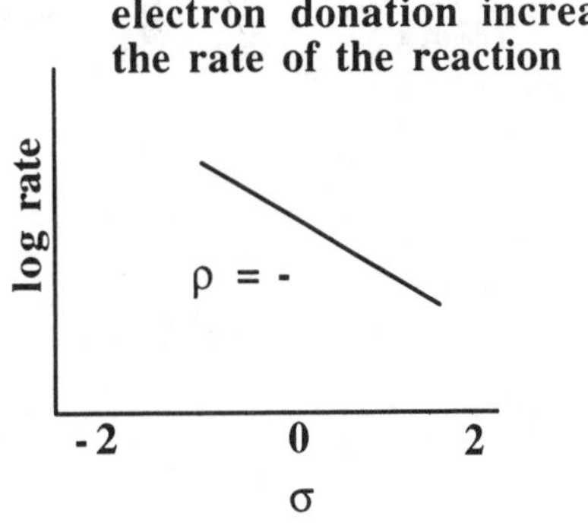

a. In the hydroxide-ion-promoted hydrolysis of a series of ethyl benzoates, electron-withdrawing substituents will increase the rate of the reaction by increasing the amount of positive charge on the carbonyl carbon, thereby making it more readily attacked by hydroxide ion. The ρ value for this reaction is, therefore, positive.

$$X{-}C_6H_4{-}\overset{O}{\overset{\|}{C}}{-}OCH_2CH_3 \;+\; H\ddot{\underset{\cdot\cdot}{O}}{:}^- \longrightarrow$$

In amide formation with a series of anilines, electron donation will increase the rate of the reaction by increasing the nucleophilicity of the aniline. The ρ value for this reaction is, therefore, negative.

$$C_6H_5{-}\overset{O}{\overset{\|}{C}}{-}Cl \;+\; H_2\underset{\cdot\cdot}{N}{-}C_6H_4{-}X \longrightarrow$$

b. Because ortho substituents are close to the site of the reaction. they introduce steric factors into the rate constant for the reaction. In other words, the presence of an ortho substituent can slow a reaction down, not because it can donate or withdraw electrons but because it can get in the way of the reactants. Therefore, any change in the rate is due to a combination of steric effects and the electron-donating or electron-withdrawing ability of the substituent. Because the change in rate cannot be attributed solely to the electron-donating or electron-withdrawing ability of the substituent, ortho-substituted compounds were not included in the study.

c. An electron-withdrawing substituent will make it easier for benzoic acid to lose a proton, so ionization will show a positive ρ value.

70.

$$C_6H_6 \xrightarrow[AlCl_3]{CH_3Cl} C_6H_5CH_3 \xrightarrow{H_2SO_4} o\text{-}CH_3C_6H_4SO_3H \xrightarrow[H_2SO_4,\ \Delta]{Na_2Cr_2O_7} o\text{-}HOOCC_6H_4SO_3H$$

$$\xrightarrow{\text{excess } SOCl_2} o\text{-}ClC(=O)C_6H_4S(=O)_2Cl \xrightarrow{NH_3} \text{saccharin (cyclic } C(=O){-}NH{-}S(=O)_2\text{)}$$

Chapter 16 Practice Test

1. For each of the following pairs of compounds indicate the one that is more reactive toward nucleophilic acyl substitution.

a. $CH_3\overset{O}{\overset{\|}{C}}OCH_3$ or $CH_3\overset{O}{\overset{\|}{C}}NHCH_3$

b. $CH_3\overset{O}{\overset{\|}{C}}OCH_3$ or $CH_3\overset{O}{\overset{\|}{C}}O-C_6H_5$

c. $CH_3\overset{O}{\overset{\|}{C}}O-C_6H_4-OCH_3$ or $CH_3\overset{O}{\overset{\|}{C}}O-C_6H_4-NO_2$

d. $CH_3\overset{O}{\overset{\|}{C}}O\overset{O}{\overset{\|}{C}}CH_3$ or $CH_3\overset{O}{\overset{\|}{C}}Cl$

2. Give the systematic name for each of the following:

a. $CH_3CH_2CH_2CH_2\overset{O}{\overset{\|}{C}}NHCH_2CH_3$

b. $CH_3CH_2\underset{CH_3}{\underset{|}{C}}HCH_2\overset{O}{\overset{\|}{C}}OH$

c. $C_6H_5-CH_2CH_2CH_2\overset{O}{\overset{\|}{C}}OCH_3$

d. $CH_3CH_2\overset{O}{\overset{\|}{C}}O\overset{O}{\overset{\|}{C}}CH_3$

3. Give an example of each of the following:

a. a symmetrical anhydride

b. a hydrolysis reaction

c. a transesterification reaction

4. What carbonyl compound would be obtained from collapse of each of the following tetrahedral intermediates?

a. $CH_3-C(OH)_2-\overset{+}{N}H_3$

b. $CH_3-C(OH)(\overset{+}{O}H_2)-OCH_3$

c. $CH_3-C(O^-)(OH)-NH_2$

d. $CH_3-C(OH)_2-\overset{+}{O}(H)CH_3$

5. Give the product of each of the following reactions.

a. $CH_3CH_2C{\equiv}N \quad + \quad H_2O \xrightarrow[\Delta]{H^+}$

b. $CH_3CH_2C{\equiv}N \quad + \quad H_2O \xrightarrow[\Delta]{HO^-}$

c. $CH_3CH_2CH_2\overset{O}{\overset{\|}{C}}O\overset{O}{\overset{\|}{C}}CH_2CH_3 \quad + \quad H_2O \longrightarrow$

d. $CH_3CH_2\overset{O}{\overset{\|}{C}}Cl \quad + \quad 2\ CH_3CH_2NH_2 \longrightarrow$

e. $C_6H_{11}-\overset{O}{\overset{\|}{C}}OCH_3 \quad + \quad \underset{\text{excess}}{CH_3CH_2OH} \xrightarrow{H^+}$ (cyclohexyl)

f. $CH_3CH_2CH_2\overset{O}{\overset{\|}{C}}OH \xrightarrow[\text{2. } CH_3CH_2CH_2OH]{\text{1. } SOCl_2}$

g. $CH_3CH_2\overset{O}{\overset{\|}{C}}NHCH_3 \quad + \quad H_2O \xrightarrow[\Delta]{H^+}$

h. $CH_3CH_2\overset{O}{\overset{\|}{C}}O-C_6H_5 \quad + \quad \underset{\text{excess}}{H_2O} \xrightarrow{H^+}$

CHAPTER 17
Carbonyl Compounds II:
Reactions of Carbonyl Compounds with Carbon and Hydrogen Nucleophiles;
Reactions of Aldehydes and Ketones with Oxygen and Nitrogen Nucleophiles;
Reactions of α, β-Unsaturated Carbonyl Compounds

Important Terms

acetal

$$R-\overset{\large OR}{\underset{\large OR}{C}}-H$$

aldehyde

$$R-\overset{\large O}{\overset{\|}{C}}-H$$

conjugate addition nucleophilic addition to the β-carbon of an α,β-unsaturated carbonyl compound.

cyanohydrin

$$R-\overset{\large OH}{\underset{\large C\equiv N}{C}}-R(H)$$

deoxygenation removal of an oxygen from a reactant.

***gem*-diol** a molecule with two OH groups on the same carbon.

direct addition nucleophilic addition to the carbonyl carbon.

enamine an α,β-unsaturated tertiary amine.

hemiacetal

$$R-\overset{\large OH}{\underset{\large OR}{C}}-H$$

hemiketal

$$R-\overset{\large OH}{\underset{\large OR}{C}}-R$$

hydrate (*gem*-diol) a compound with two OH groups on the same carbon.

hydrazone

$$\begin{matrix}R\diagdown\\ \quad C=N-NH_2\\ (H)R\diagup\end{matrix}$$

imine (Schiff base)

$$R_2C{=}N{-}R \quad \text{(one R may be H)}$$

ketal

$$R{-}C(OR)_2{-}R$$

ketone

$$R{-}\overset{O}{\overset{\|}{C}}{-}R$$

nucleophilic acyl substitution reaction — a reaction in which a group bonded to an acyl group is substituted by another group.

nucleophilic addition-elimination reaction — a nucleophilic addition reaction that is followed by an elimination reaction. Imine formation is an example: an amine adds to the carbonyl carbon, and water is eliminated.

nucleophilic addition reaction — a reaction that involves the addition of a nucleophile to a reagent.

oxime

$$R(H)R\,C{=}N{-}OH$$

phenylhydrazone

$$R(H)R\,C{=}N{-}NHC_6H_5$$

pH-rate profile — a plot of the rate constant of a reaction versus the pH of the reaction mixture.

prochiral carbonyl carbon — a carbonyl carbon that will become a chirality center if it is attacked by a group unlike any of the groups already bonded to it.

protecting group — a reagent that protects a functional group from a synthetic operation that it would otherwise not survive.

reduction reaction — a reaction in which a molecule gains electrons. In the case of an organic molecule, a reaction in which the number of C-H bonds is increased or the number of C-O, C-N, or C-X (X = halogen) bonds in decreased

reductive amination — the reaction of an aldehyde or a ketone with ammonia or with a primary amine in the presence of a reducing agent (H_2/Raney Ni).

Schiff base (imine)

$$\underset{(H)}{R}\!\!\!\begin{array}{l}R\diagdown\\ \quad C{=}N{-}R\\ R\diagup\end{array}$$

semicarbazone

$$\begin{array}{l}R\diagdown\qquad\quad O\\ \qquad\qquad\quad \|\\ \quad C{=}NNHCNH_2\\ R\diagup\\ (H)\end{array}$$

Wittig reaction — the reaction of an aldehyde or a ketone with a phosphonium ylide, resulting in formation of an alkene.

ylide — a compound with opposite charges on adjacent, covalently bonded atoms with complete octets.

Solutions to Problems

1. If the ketone functional group were anywhere else in these compounds, they would not be ketones and, therefore, would not have the "one" suffix.

2. **a.** 3-methylpentanal, β-methylvaleraldehyde
b. 4-heptanone, dipropyl ketone
c. 2-methyl-4-heptanone, isobutyl propyl ketone
d. 4-phenylbutanal, γ-phenylbutyraldehyde
e. 4-ethylhexanal, γ-ethylcaproaldehyde
f. 1-hepten-3-one, butyl vinyl ketone

3. **a.** 6-hydroxy-3-heptanone **b.** 2-oxocyclohexylmethanenitrile **c.** 3-formylpentanamide

4. **a.** 2-Heptanone is more reactive because it has less steric hindrance. There is little difference in the amount of steric hindrance provided at the carbonyl carbon (the site of nucleophilic attack) by a pentyl and a propyl group because they differ at a point somewhat removed from the site of nucleophilic attack. The difference in size between a methyl group and a propyl group is significant at the site of nucleophilic attack.

$CH_3C(=O)CH_2CH_2CH_2CH_2CH_3$ — 2-heptanone

$CH_3CH_2CH_2C(=O)CH_2CH_2CH_3$ — 4-heptanone

b. *para*-Nitroacetophenone is more reactive because the electron-withdrawing nitro group makes the carbonyl carbon more susceptible to nucleophilic attack compared with an electron-donating methoxy group.

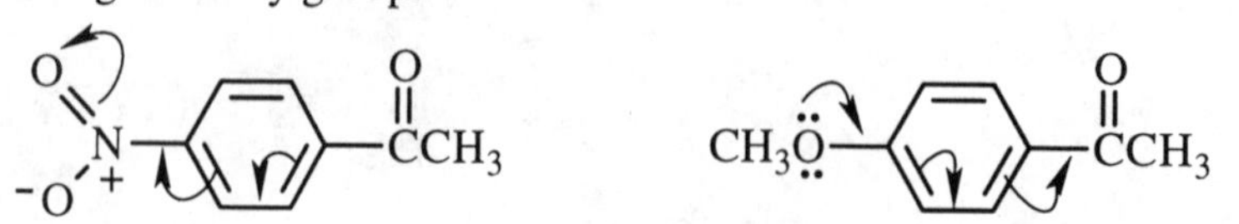

O_2N^+–C_6H_4–$C(=O)CH_3$ (with O^-; arrows showing electron withdrawal) $\qquad$ $CH_3\ddot{O}$–C_6H_4–$C(=O)CH_3$ (arrows showing electron donation)

5. **a.** Two isomers are obtained because the reaction creates chirality center in the product.

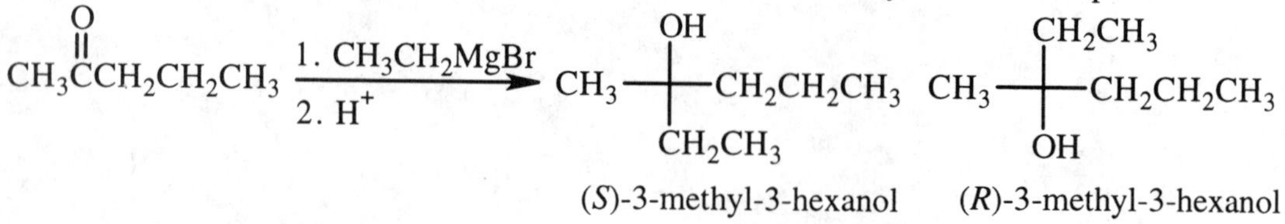

$$CH_3C(=O)CH_2CH_2CH_3 \xrightarrow[2.\ H^+]{1.\ CH_3CH_2MgBr} CH_3\text{–}C(OH)(CH_2CH_3)\text{–}CH_2CH_2CH_3 \quad CH_3\text{–}C(CH_2CH_3)(OH)\text{–}CH_2CH_2CH_3$$

(*S*)-3-methyl-3-hexanol $\qquad$ (*R*)-3-methyl-3-hexanol

b. Only one compound is obtained because the product does not have a chirality center.

$$CH_3\overset{O}{\overset{\|}{C}}CH_2CH_2CH_3 \xrightarrow[2.\ H^+]{1.\ CH_3MgBr} CH_3\underset{CH_3}{\underset{|}{\overset{OH}{\overset{|}{C}}}}CH_2CH_2CH_3$$

6.

$$CH_3\overset{O}{\overset{\|}{C}}CH_2CH_3 + CH_3CH_2CH_2MgBr$$

$$CH_3CH_2\overset{O}{\overset{\|}{C}}CH_2CH_2CH_3 + CH_3MgBr$$

7. The reaction of a Grignard reagent with formaldehyde forms a primary alcohol. (Grignard reagents react with all other aldehydes to form secondary alcohols.)

$$H\overset{O}{\overset{\|}{C}}H \xrightarrow[2.\ H^+]{1.\ RMgBr} RCH_2OH$$

8. If the compound is alkylated after nucleophilic attack, the alkylation on oxygen will compete with alkylation on carbon.

a. $HC\equiv CH \xrightarrow{^-NH_2} HC\equiv C^- \xrightarrow{CH_3CH_2\overset{O}{\overset{\|}{C}}H} HC\equiv C\overset{O^-}{\overset{|}{C}}HCH_2CH_3 \xrightarrow{H^+} HC\equiv C\overset{OH}{\overset{|}{C}}HCH_2CH_3$

b. $HC\equiv CH \xrightarrow{^-NH_2} HC\equiv C^- \xrightarrow{CH_3Br} HC\equiv CCH_3 \xrightarrow{^-NH_2} {}^-C\equiv CCH_3$

$\downarrow$ $C_6H_5\overset{O}{\overset{\|}{C}}H$

$C_6H_5\overset{OH}{\overset{|}{C}}HC\equiv CCH_3 \xleftarrow{H^+} C_6H_5\overset{O^-}{\overset{|}{C}}HC\equiv CCH_3$

c. 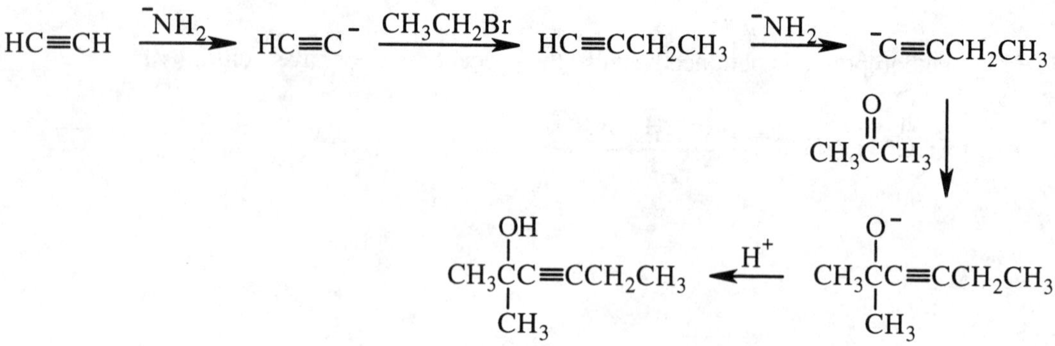 $HC{\equiv}CH \xrightarrow{^{-}NH_2} HC{\equiv}C^- \xrightarrow{CH_3CH_2Br} HC{\equiv}CCH_2CH_3 \xrightarrow{^{-}NH_2} {}^{-}C{\equiv}CCH_2CH_3$

$\xrightarrow{CH_3\overset{O}{\overset{\|}{C}}CH_3} CH_3\underset{CH_3}{\overset{O^-}{C}}C{\equiv}CCH_2CH_3 \xrightarrow{H^+} CH_3\underset{CH_3}{\overset{OH}{C}}C{\equiv}CCH_2CH_3$

9. No, an acid must be present in the reaction mixture in order to protonate the oxygen of the cyanohydrin. Otherwise the cyano group will be eliminated and the reactants will be reformed.

10. Strong acids like HCl and H_2SO_4 have very weak conjugate bases (Cl^- and HSO_4^-), which are excellent leaving groups. When these bases add to the carbonyl group, they are readily eliminated, reforming the starting materials. Cyanide ion is a strong enough base, so it is not eliminated unless the oxygen in the product is negatively charged.

11. Solved in the text.

12.

a. 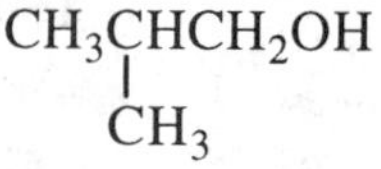 $CH_3\underset{CH_3}{C}HCH_2OH$

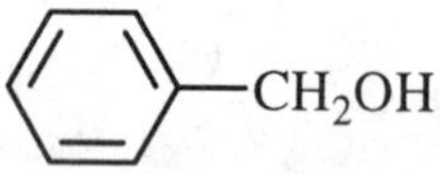

b. 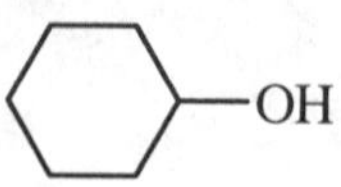 cyclohexyl–OH

c. phenyl–CH_2OH

d. 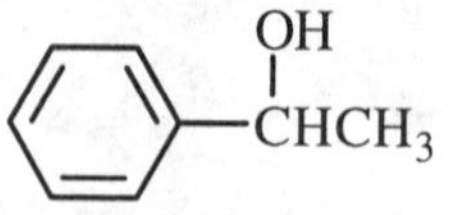 phenyl–$\overset{OH}{C}HCH_3$

13. The Grignard reagent will react with the proton from the carboxylic acid, forming an alkane and a carboxylate ion.

$$R\overset{O}{\overset{\|}{C}}OH + R'MgBr \longrightarrow R\overset{O}{\overset{\|}{C}}O^- + R'H + Mg^{2+} + Br^-$$

14. **a.** Solved in the text.

b.

1. $CH_3\overset{O}{\overset{\|}{C}}OR + 2\ CH_3MgBr$

2. Solved in the text.

4. $CH_3\overset{O}{\overset{\|}{C}}OR + 2\ CH_3CH_2MgBr$

6. 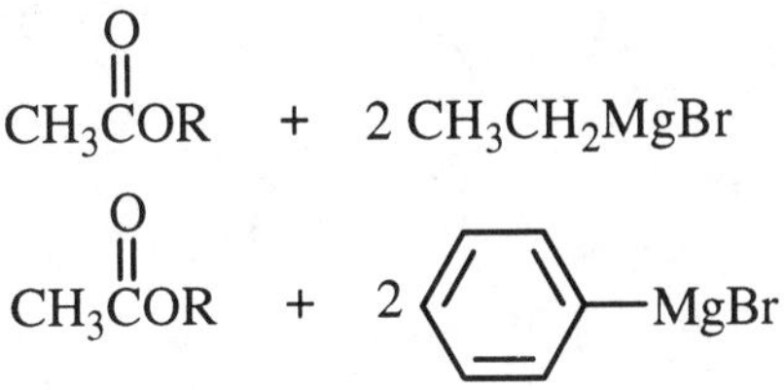 $CH_3\overset{O}{\overset{\|}{C}}OR + 2\ C_6H_5\text{—}MgBr$

15. If a secondary alcohol is formed from the reaction of a formate ester with excess Grignard reagent, the two alkyl substituents of the alcohol will be identical because they both come from the Grignard reagent. Therefore, only the following two alcohols can be prepared that way.

$CH_3\underset{OH}{\underset{|}{C}}HCH_3$ $\quad$ $CH_3CH_2\underset{OH}{\underset{|}{C}}HCH_2CH_3$

16.

a. $C_6H_5\text{—}\overset{O}{\overset{\|}{C}}NHCH_3$ **b.** $CH_3\overset{O}{\overset{\|}{C}}NH_2$ **c.** $CH_3\overset{O}{\overset{\|}{C}}NHCH_2CH_3$ **d.** $CH_3\overset{O}{\overset{\|}{C}}N(CH_2CH_3)_2$

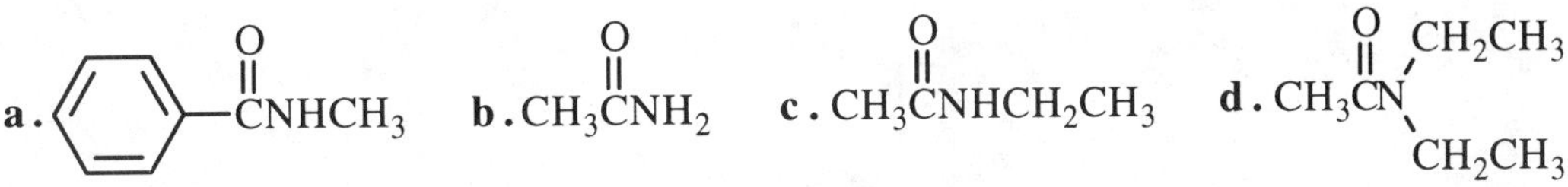

17.

a.	1. $LiAlH_4$	2. H_2O		
b.	H^+, H_2O, Δ			
c.	1. H^+, H_2O, Δ	2. $SOCl_2$	3. $LiAlH[OC(CH_3)_3]_3$, 80 °C	4. H_2O
d.	1. H^+, H_2O, Δ	2. $LiAlH_4$	3. H^+, H_2O	

18. The nonbonding electrons on the nitrogen that is attached to the carbonyl carbon are in resonance with the carbonyl group. This nitrogen, therefore, cannot act as a nucleophile since its nonbonding electrons are not available for nucleophilic attack.

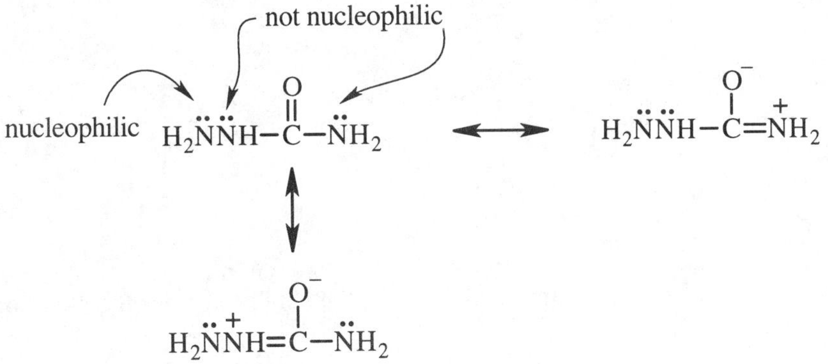

not nucleophilic

nucleophilic $H_2\ddot{N}\ddot{N}H\text{—}\overset{O}{\overset{\|}{C}}\text{—}\ddot{N}H_2 \longleftrightarrow H_2\ddot{N}\ddot{N}H\text{—}\overset{O^-}{\overset{|}{C}}=\overset{+}{N}H_2$

$\updownarrow$

$H_2\ddot{N}\overset{+}{N}H=\overset{O^-}{\overset{|}{C}}\text{—}\ddot{N}H_2$

19. **a.**

1. Imine hydrolysis mechanism (structures): cyclohexylidene=N–R + H^+ ⇌ cyclohexylidene=$\overset{+}{N}$(R)(H) + H_2O: ⇌ 1-NHR-1-$\overset{+}{O}H_2$-cyclohexane

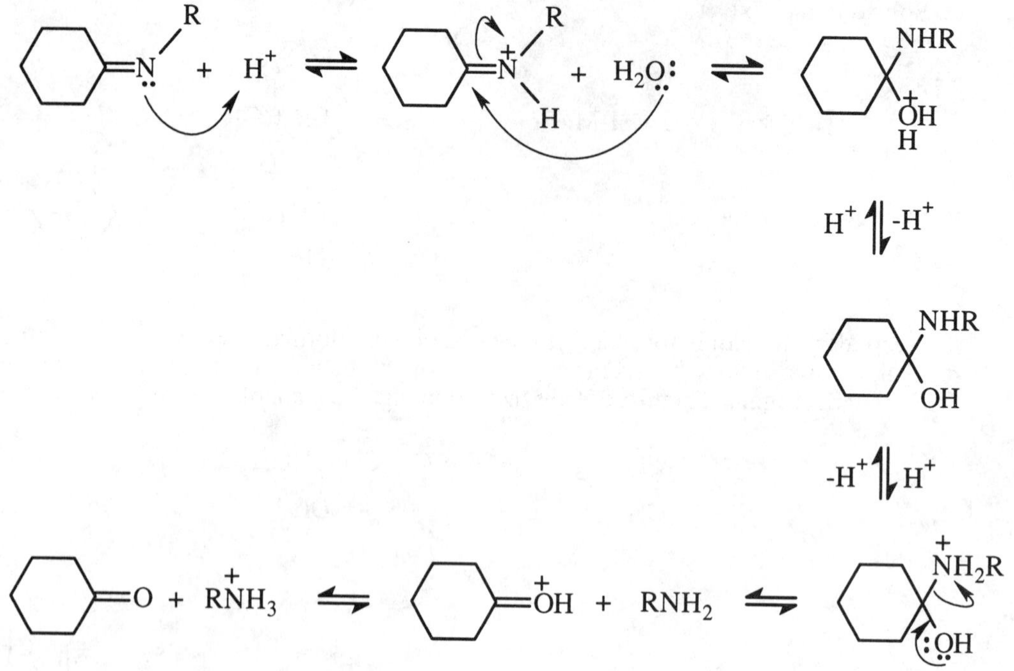

H^+ ⇅ $-H^+$

1-NHR-1-OH-cyclohexane

$-H^+$ ⇅ H^+

cyclohexanone (=O) + $R\overset{+}{N}H_3$ ⇌ cyclohexylidene=$\overset{+}{O}H$ + RNH_2 ⇌ 1-$\overset{+}{N}H_2R$-1-:OH-cyclohexane

2. Enamine hydrolysis mechanism (structures): 1-cyclohexenyl–N(R)R + H^+ ⇌ cyclohexylidene=$\overset{+}{N}$(R)R (with H, H on the ring carbon) + H_2O: ⇌ 1-NR_2-1-$\overset{+}{O}H_2$-cyclohexane

H^+ ⇅ $-H^+$

1-NR_2-1-OH-cyclohexane

$-H^+$ ⇅ H^+

cyclohexanone (=O) + $R_2\overset{+}{N}H_2$ ⇌ cyclohexylidene=$\overset{+}{O}H$ + R_2NH ⇌ 1-$H\overset{+}{N}R_2$-1-:OH-cyclohexane

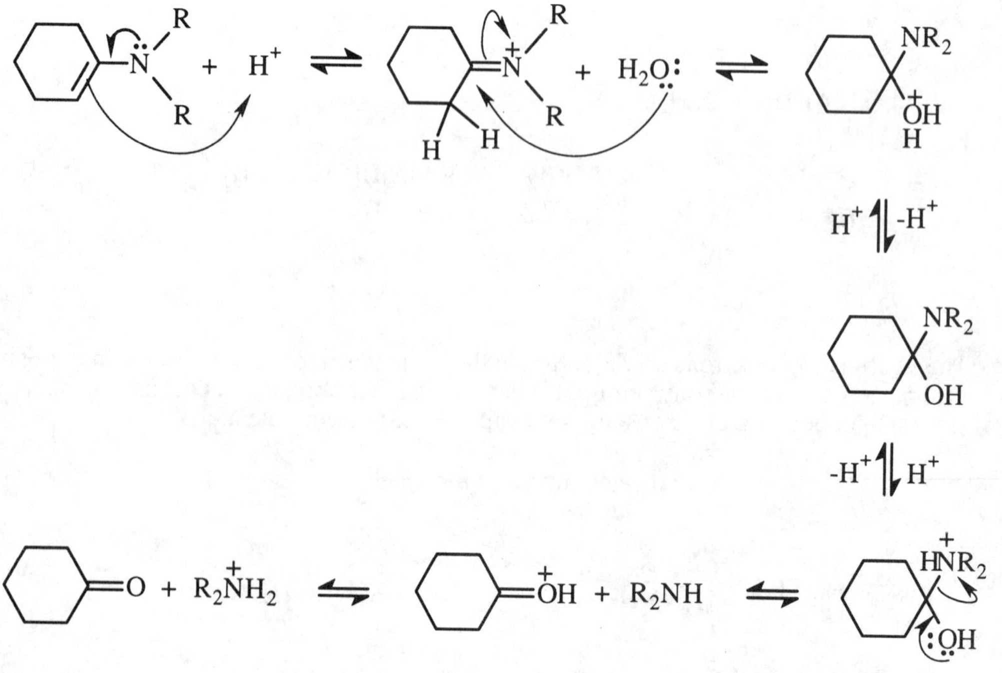

b. The only difference is in the first step of the mechanism: in imine hydrolysis the acid protonates the nitrogen; in enamine hydrolysis, the acid protonates the β-carbon.

20.

a. cyclopentanone $=O$ + $CH_3CH_2NH_2$ $\rightleftharpoons$ cyclopentylidene $=NCH_2CH_3$ + H_2O

b. cyclopentanone $=O$ + $CH_3CH_2NHCH_2CH_3$ $\rightleftharpoons$ cyclopentenyl$-N(CH_2CH_3)_2$ + H_2O

c. $C_6H_5-C(CH_3)=O$ + $CH_3(CH_2)_5NH_2$ $\rightleftharpoons$ $C_6H_5-C(CH_3)=N(CH_2)_5CH_3$ + H_2O

d. $C_6H_5-C(CH_3)=O$ + cyclohexyl$-NH_2$ $\rightleftharpoons$ $C_6H_5-C(CH_3)=N-$cyclohexyl + H_2O

21. A tertiary amine will be obtained because the primary amine synthesized in the first part of the reaction will react with the excess carbonyl compound, forming an imine that will be reduced to a secondary amine. The secondary amine will then react with the carbonyl compound, forming an enamine that will be reduced to a tertiary amine.

$$R_2C{=}O + NH_3 \xrightarrow[\text{Raney Ni}]{H_2} R_2CHNH_2 \xrightarrow[H_2,\ \text{Raney Ni}]{R_2C=O} R_2CHNHCHR_2 \xrightarrow[H_2,\ \text{Raney Ni}]{R_2C=O} (R_2CH)_3N$$

22.

$$R-\overset{O}{\overset{\|}{C}}-H + H\ddot{O}:^- \rightleftharpoons R-\overset{:\ddot{O}:^-}{\overset{|}{C}}(OH)-H \xrightarrow{H-O-H} \rightleftharpoons R-\overset{OH}{\overset{|}{C}}(OH)-H + H\ddot{O}:^-$$

23.

Electron-withdrawing groups decrease the stability of the aldehyde and increase the stability of the hydrate. Therefore, the three electron-withdrawing chlorines cause trichloroacetaldehyde to have a large equilibrium constant for hydrate formation.

$$K_{eq} = \frac{[\text{hydrate}]}{[\text{aldehyde}]\,[H_2O]}$$

24. Because an electron-withdrawing substituent decreases the stability of a ketone and increases the stability of a hydrate, the compound with the electron-withdrawing *para*-nitro substituents has the largest equilibrium constant for addition of water.

$$O_2N-C_6H_4-\overset{O}{\overset{\|}{C}}-C_6H_4-NO_2$$

25. **a.** Hemiacetals are unstable in basic solution because the base can remove a proton from an OH group, thereby providing an oxyanion that has sufficient driving force to expel the basic ^-OR group.

$$R-\underset{OR}{\underset{|}{\overset{:\ddot{O}-H}{\overset{|}{C}}}}-R \xrightleftharpoons{^-:\ddot{O}H} R-\underset{OR}{\underset{|}{\overset{:\ddot{O}:^-}{\overset{|}{C}}}}-R$$

b. In order for an acetal to form, the CH_3O group in the hemiacetal must eliminate an OH group. Hydroxide ion is too basic to be eliminated by a CH_3O group, but water can be eliminated by a CH_3O group. Since the OH group must be protonated before it can be eliminated, acetal formation must be carried out in an acidic solution.

$$R-\underset{OH}{\underset{|}{\overset{:\ddot{O}CH_3}{\overset{|}{C}}}}-H \not\longrightarrow$$

$$R-\underset{\overset{+}{O}H_2}{\underset{|}{\overset{:\ddot{O}CH_3}{\overset{|}{C}}}}-H \rightleftharpoons R-\overset{+\ddot{O}CH_3}{\overset{\|}{C}}-H + H_2O$$

c. Hydrate formation can be catalyzed by hydroxide ion because a group does not have to be eliminated after hydroxide ion attacks the aldehyde or ketone.

26.

$HOCH_2CH_2CH_2CH_2CH{=}O \underset{-H^+}{\overset{H^+}{\rightleftharpoons}} HOCH_2CH_2CH_2CH_2CH{=}\overset{+}{O}H \rightleftharpoons$ (cyclic protonated hemiacetal) $\underset{+H^+}{\overset{-H^+}{\rightleftharpoons}}$ 2-hydroxytetrahydropyran

$\underset{H^+}{\overset{-H^+}{\rightleftharpoons}}$ (protonated OH) $\rightleftharpoons$ oxocarbenium ion $\overset{CH_3OH}{\rightleftharpoons}$ (protonated methoxy) $\underset{H^+}{\overset{-H^+}{\rightleftharpoons}}$ 2-methoxytetrahydropyran

27.

a.

1. Solved in the text.

2. cyclohexanone $=O$ + $CH_3CH_2CH{=}P(C_6H_5)_3$

or cyclohexylidene $=P(C_6H_5)_3$ + $CH_3CH_2CH{=}O$

3. $C_6H_5CH{=}O$ + $CH_2{=}P(C_6H_5)_3$

or $C_6H_5CH{=}P(C_6H_5)_3$ + $CH_2{=}O$

4. $(C_6H_5)_2C{=}O$ + $CH_3CH{=}P(C_6H_5)_3$

or $(C_6H_5)_2C{=}P(C_6H_5)_3$ + $CH_3CH{=}O$

b.

1. Solved in the text.

2. $CH_3CH_2CH_2Br$ or cyclohexyl$-Br$

3. CH_3Br or $C_6H_5CH_2Br$

4. CH_3CH_2Br or $(C_6H_5)_2CHBr$

28. To review how to determine whether a chirality center has the *R* or *S* configuration, see Section 4.5.

a. $CH_3CH_2C(=O)C_6H_5 \xrightarrow{CH_3MgBr} CH_3CH_2C(OH)(CH_3)C_6H_5$

S

b. C_6H_5CHO (benzaldehyde) $\xrightarrow{CH_3MgBr}$ $C_6H_5CH(OH)CH_3$

R

c. $CH_3COCH_2CH_3$ $\xrightarrow{CH_3MgBr}$ $(CH_3)_2C(OH)CH_2CH_3$

No enantiomers are possible because the compound does not have a chirality center.

d. $CH_3CH_2COCH_2CH_2CH_3$ $\xrightarrow{CH_3MgBr}$ $CH_3CH_2C(OH)(CH_3)CH_2CH_2CH_3$

S

29.

a. $BrCH_2CH_2CH_2CH(OH)CH_2CH_3$ $\xrightarrow{Na}$ 2-ethyltetrahydrofuran (CH_2CH_3)

b. $C_6H_5CH_2CH_2CH_2CH(Cl)CH_3$ $\xrightarrow{AlCl_3}$ 1-methyltetralin (CH_3)

c. $C_6H_5CH_2CH_2CH_2CH_2Cl$ $\xrightarrow{AlCl_3}$ tetralin $\xrightarrow{NBS}$ 1-bromotetralin (Br) $\xrightarrow{KOH}$ 1,2-dihydronaphthalene

d. $BrCH_2CH_2CH_2CHCH_2CH{=}CH_2$ (with OH on the CH) $\xrightarrow{Na}$ [tetrahydrofuran ring]$-CH_2CH{=}CH_2$

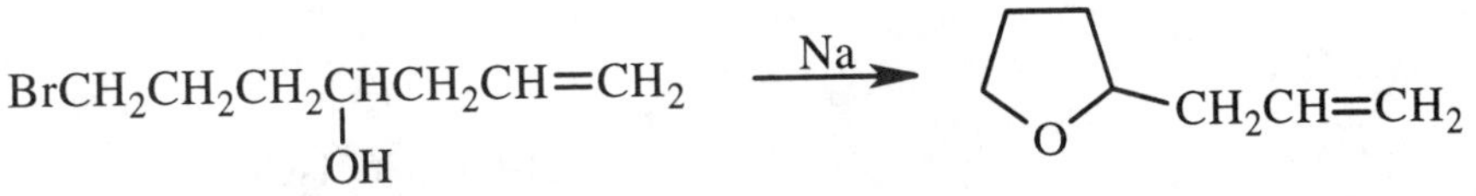

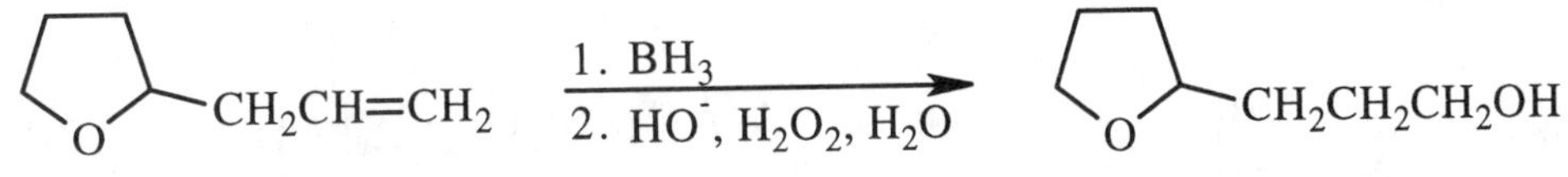

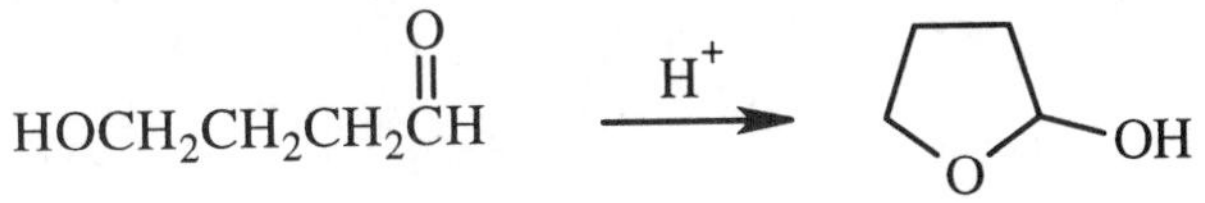

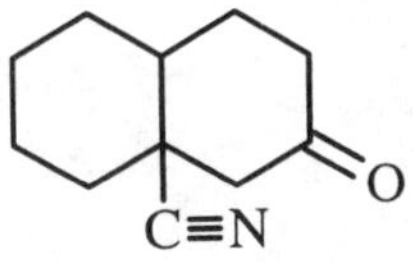

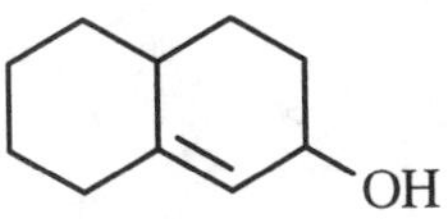

e. [tetrahydrofuran ring]$-CH_2CH{=}CH_2$ $\xrightarrow[2.\ HO^-,\ H_2O_2,\ H_2O]{1.\ BH_3}$ [tetrahydrofuran ring]$-CH_2CH_2CH_2OH$

product of **d**

f. $HOCH_2CH_2CH_2CH{=}O$ $\xrightarrow{H^+}$ [tetrahydrofuran ring]$-OH$

30.

a. [decalin ring system with C≡N at ring fusion carbon and =O on ring]

b. [octahydronaphthalene ring system with C=C and OH]

c. $CH_3C(CH_3){=}CHC(OH)(CH_3)CH_3$

d. $CH_3C(CH_3)_2-CH_2C(=O)CH_3$

31.

a. $CH_3CH(CH_3)CH{=}O$

b. $CH_3CH{=}CHCH_2CH_2CH{=}O$

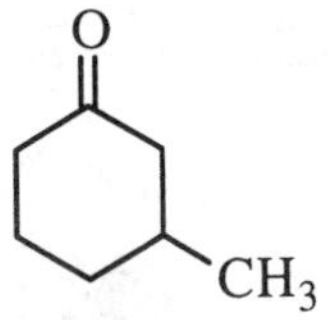

c. $CH_3CH(CH_3)CH_2CH_2C(=O)CH_2CH_2CH(CH_3)CH_3$

d. [cyclohexanone ring with CH_3 at position 3]

e. $CH_3C(=O)CH_2C(=O)CH_3$

f. $CH_3CH_2C(=O)CH(Br)CH_2CH_2CH_3$

g. $CH_3CH_2CH(Br)CH_2CH_2CH{=}O$

h. [cyclopentane ring with $CH{=}O$ and CH_2CH_3 on adjacent carbons]

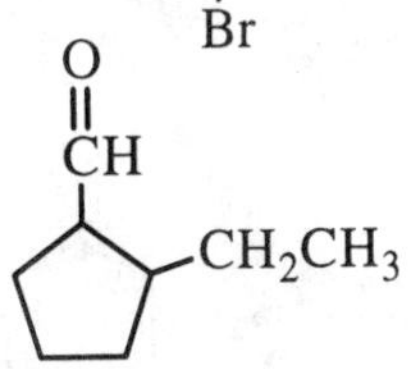

i. $CH_3C(=O)CH(CH_3)CH_2CH_2CH{=}O$

j. [benzene ring with $CH{=}O$ groups at 1 and 3 positions]

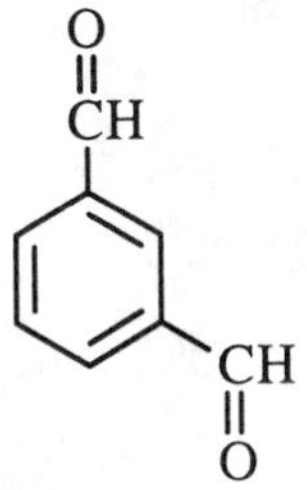

32.

a. $CH_3CH_2CH(OCH_2CH_3)_2$

OCH₂CH₃ above and OCH₂CH₃ below the CH carbon.

b. $C_6H_5C(=NNH_2)CH_2CH_3$

c. $C_6H_5CH_2CH_2CH_3$

d. $CH_3CH_2CH(OH)CH_3$

e. $CH_3CH_2C(OH)(C{\equiv}N)CH_2CH_3$

f. $CH_3CH_2CH_2CH_2OH$ + CH_3CH_2OH

g. $CH_3CH_2CH_2C(OCH_2CH_2O)CH_3$ (cyclic O–CH₂CH₂–O acetal on the C)

h. 3-cyano-3-methylcyclohexanone: ring C=O, with CH_3 and $C{\equiv}N$ on the same ring carbon

33.

CH_3CH_2CHO > $CH_3CH_2COCH_2CH_3$ > $CH_3CH(CH_3)CH_2COCH_2CH_3$ >

$CH_3CH_2CH(CH_3)COCH_2CH_3$ > $CH_3CH_2CH(CH_3)COCH(CH_3)CH_2CH_3$ > $CH_3CH_2CH(CH_3)COC(CH_3)_2CH_2CH_3$

34.

a. cyclohexanone $\xrightarrow[2.\ H^+,\ H_2O]{1.\ NaBH_4}$ cyclohexanol (OH)

b. cyclohexanone $\xrightarrow[2.\ H^+,\ H_2O]{1.\ NaBH_4}$ cyclohexanol (OH) $\xrightarrow[\Delta]{H_2SO_4}$ cyclohexene

c. cyclohexanone $\xrightarrow[\text{2. } H^+, H_2O]{\text{1. NaBH}_4}$ cyclohexanol (OH) $\xrightarrow{PBr_3}$ bromocyclohexane (Br)

d. cyclohexanone (O) + NH_3 $\xrightarrow[Pt]{H_2}$ cyclohexylamine (NH_2)

e. bromocyclohexane (Br), product of **c** $\xrightarrow{^-C{\equiv}N}$ cyclohexyl–$C{\equiv}N$ $\xrightarrow[Pt]{H_2}$ cyclohexyl–CH_2NH_2

product of **c**

f. cyclohexanone (O) + CH_3NHCH_3 ⟶ 1-(CH_3–N–CH_3)cyclohexene $\xrightarrow[Pt]{H_2}$ (CH_3–N–CH_3)cyclohexane

g. bromocyclohexane (Br) + $(CH_2{=}CH)CuLi$ $\xrightarrow{\text{ether}}$ cyclohexyl–$CH{=}CH_2$

product of **c**

or bromocyclohexane (Br) + $HC{\equiv}C^-$ ⟶ cyclohexyl–$C{\equiv}CH$ $\xrightarrow[\text{Lindlar's catalyst}]{H_2}$ cyclohexyl–$CH{=}CH_2$

product of **c**

h. cyclohexanone (O) $\xrightarrow[\text{Zn(Hg), HCl, }\Delta]{NH_2NH_2,\ HO^-,\ \Delta \text{ or}}$ cyclohexane

or

cyclohexanone (O) $\xrightarrow[\text{2. } H^+, H_2O]{\text{1. NaBH}_4}$ cyclohexanol (OH) $\xrightarrow[\Delta]{H_2SO_4}$ cyclohexene $\xrightarrow[Pt]{H_2}$ cyclohexane

i. cyclohexanone $\xrightarrow[\text{2. } H^+, H_2O]{\text{1. } CH_3CH_2MgBr}$ 1-ethylcyclohexanol (HO, CH_2CH_3) $\xrightarrow[\Delta]{H_2SO_4}$ 1-ethylcyclohexene (CH_2CH_3) $\xrightarrow{NaBH(O\overset{O}{\overset{\|}{C}}CH_3)_3}$ ethylcyclohexane (CH_2CH_3)

or

cyclohexanone + $CH_3CH{=}P(C_6H_5)_3$ $\longrightarrow$ ethylidenecyclohexane ($CHCH_3$) $\xrightarrow{NaBH(O\overset{O}{\overset{\|}{C}}CH_3)_3}$ ethylcyclohexane (CH_2CH_3)

35. The greater the electron-withdrawing ability of the para substituent, the greater the K_{eq} for hydrate formation.

p-nitroacetophenone ($\overset{O}{\overset{\|}{C}}CH_3$, NO_2) > p-chloroacetophenone ($\overset{O}{\overset{\|}{C}}CH_3$, Cl) > acetophenone ($\overset{O}{\overset{\|}{C}}CH_3$) > p-methoxyacetophenone ($\overset{O}{\overset{\|}{C}}CH_3$, OCH_3)

36.

a. $CH_3OH \xrightarrow{PBr_3} CH_3Br \xrightarrow[Et_2O]{Mg} CH_3MgBr \xrightarrow[\text{2. } H^+, H_2O]{\text{1. } H_2C{=}O} CH_3CH_2OH$

b. $CH_4 \xrightarrow[h\nu]{Br_2} CH_3Br \xrightarrow[Et_2O]{Mg} CH_3MgBr \xrightarrow[\text{2. } H^+, H_2O]{\text{1. ethylene oxide}} CH_3CH_2CH_2OH$

c. $CH_3CH_2CH_2OH \xrightarrow{PBr_3} CH_3CH_2CH_2Br \xrightarrow[Et_2O]{Mg} CH_3CH_2CH_2MgBr$

$\downarrow$ 1. $CH_3\overset{O}{\overset{\|}{C}}CH_3$ 2. H_3O^+

$CH_3CH_2CH{=}\underset{CH_3}{\underset{|}{C}}CH_3 \xleftarrow[\Delta]{H_2SO_4} CH_3CH_2CH_2\overset{OH}{\overset{|}{\underset{CH_3}{\underset{|}{C}}}}CH_3$

37.

a. 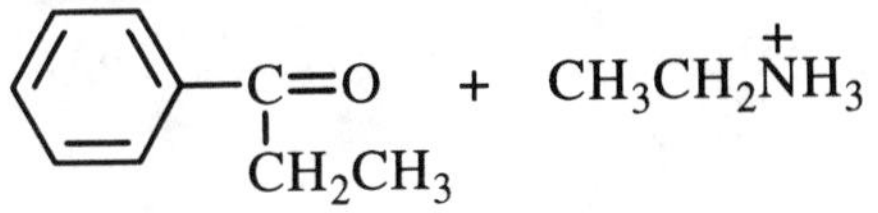 $C_6H_5C(=O)CH_2CH_3$ + $CH_3CH_2\overset{+}{N}H_3$

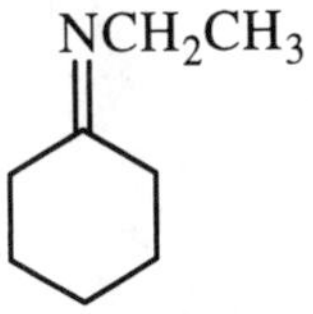

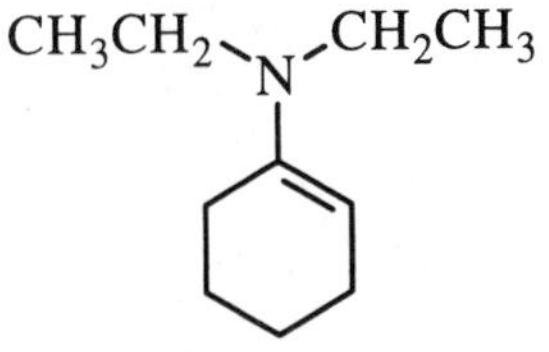

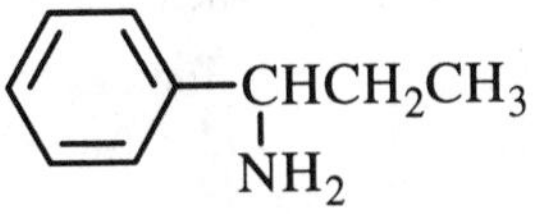

b. (cyclohexylidene)NCH_2CH_3

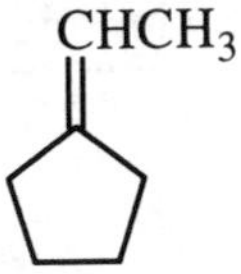

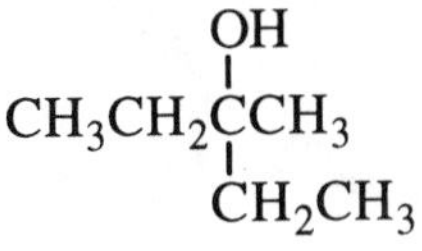

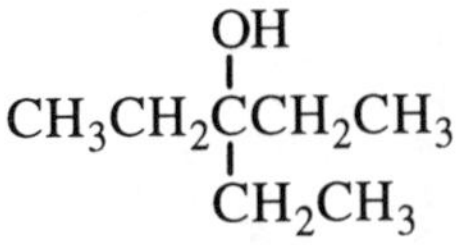

c. $(CH_3CH_2)_2N$–(cyclohexenyl)

d. $C_6H_5CH(NH_2)CH_2CH_3$

e. (cyclopentylidene)$CHCH_3$

f. $CH_3CH_2C(OH)(CH_2CH_3)CH_3$

g. $CH_3CH_2C(OH)(CH_2CH_3)CH_2CH_3$

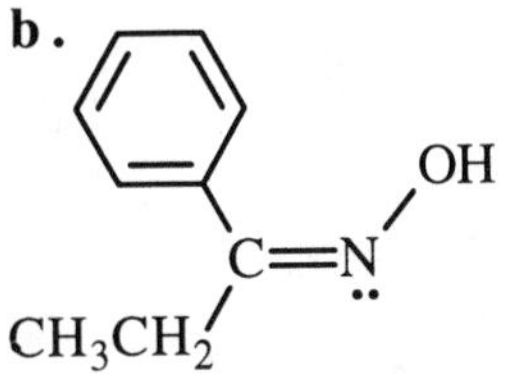

h. $CH_3C(CH_3)(Br)CH_2C(=O)CH_3$

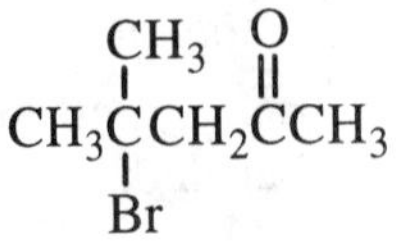

i. 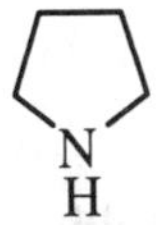 pyrrolidine (N–H)

38. a.

$H_2N-C(=S)-NH_2$ + CH_3CH_2-Br ⟶ $H_2N-C(=\overset{+}{S}-CH_2CH_3)-NH_2$ $\xrightarrow{HO^-}$ $H_2N-C(SCH_2CH_3)(OH)-NH_2$

⟶ $H_2N-C(=\overset{+}{O}H)-NH_2$ + $CH_3CH_2S^-$ ⟶ $H_2N-C(=O)-NH_2$ + CH_3CH_2SH

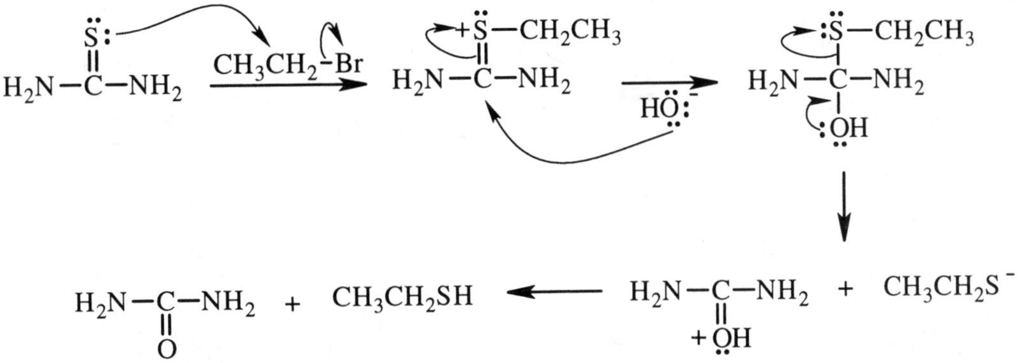

b. $CH_3CH_2CH_2CH_2CH_2SH$

39.

a. 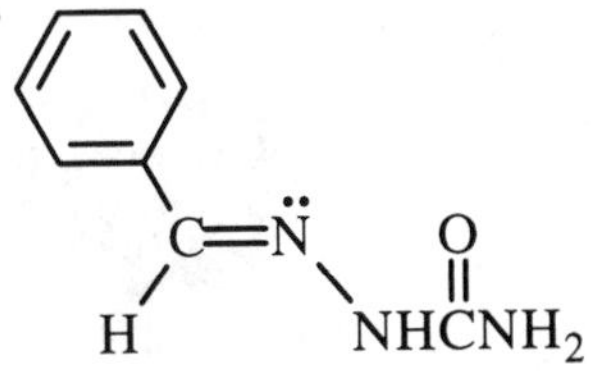 $C_6H_5(H)C=N-NHC(=O)NH_2$

b. $C_6H_5(CH_3CH_2)C=N-OH$

c. (cyclohexylidene)$=N-NH-$(3,5-dinitrophenyl)

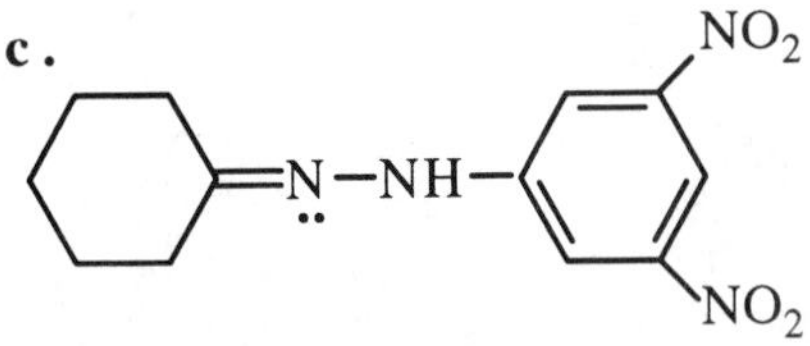

Because the ketone is symmetrical, *E* and *Z* isomers are not possible for this compound.

40. The offset shows there is a signal at ~11.8 ppm, indicating an aldehyde. The ^{1}H NMR spectrum is that of benzaldehyde. Phenylmagnesium bromide, therefore, must react with a compound with one carbon atom to form an alcohol that can be oxidized by the mild oxidizing agent (MnO_2) to benzaldehyde. Therefore, Compound Z must be formaldehyde.

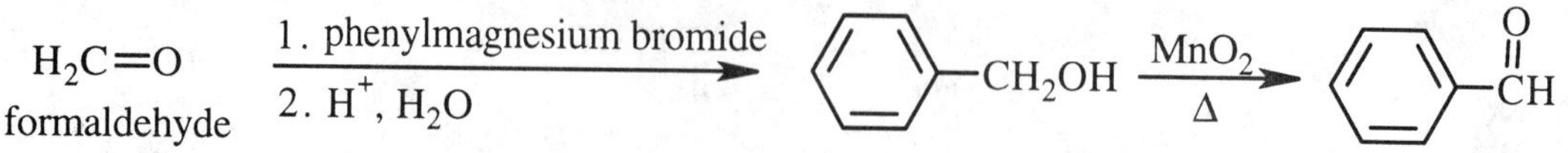

$H_2C{=}O$ (formaldehyde, **Compound Z**) $\xrightarrow[2.\ H^+,\ H_2O]{1.\ \text{phenylmagnesium bromide}}$ $C_6H_5CH_2OH$ $\xrightarrow[\Delta]{MnO_2}$ C_6H_5CHO

41.

a. [reaction mechanism scheme; labels: H^+, $-H^+$, H_2O:, HO$^+$ H, H, $-H^+$, H^+, HO H, $-H^+$ H^+, $CH_2CH_2CH_2\overset{+}{N}H_3$, $CH_2CH_2CH_2NH_2$, $\overset{+}{O}H$, O, :O:, N, H]

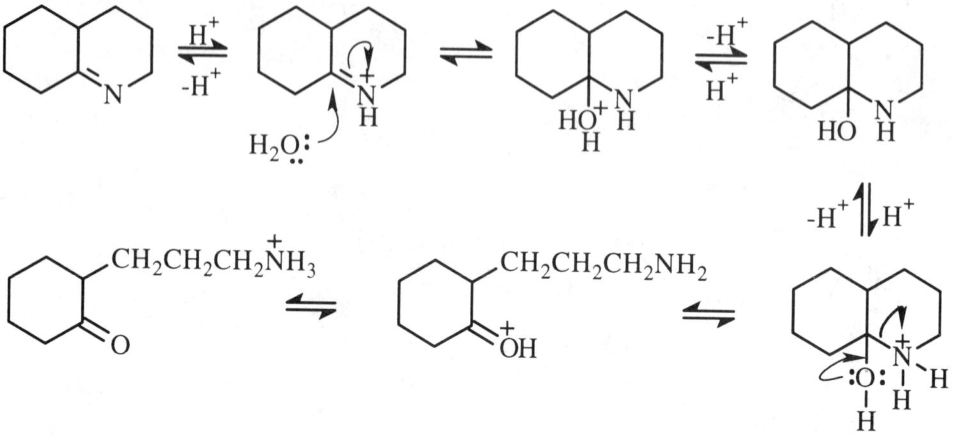

b. [reaction mechanism scheme; labels: :ÖCH$_3$, + H^+, +OCH$_3$, $H_2\ddot{O}$:, CH_3O, +ÖH, H, $-H^+$, H^+, OH, $-H^+$ H^+, CH_3O+, :ÖH, +OH, O, $-H^+$, $+H^+$]

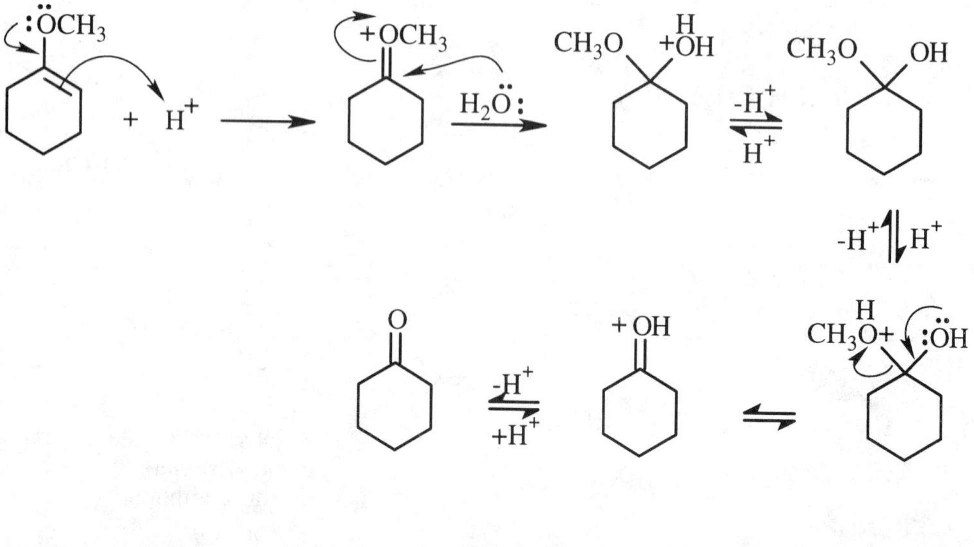

c. 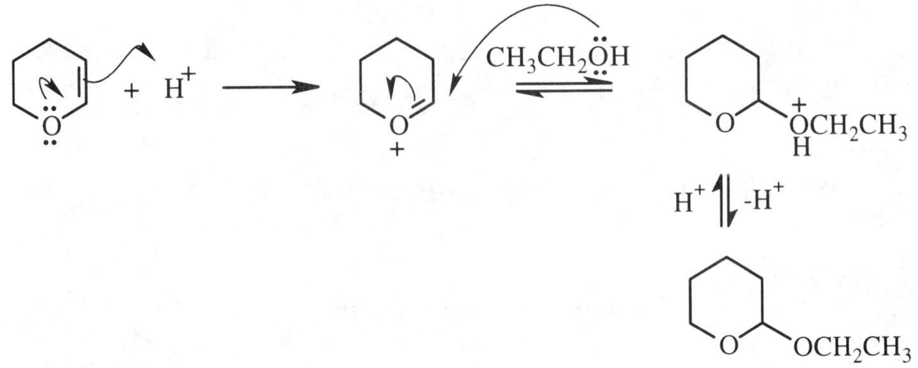 [reaction mechanism: dihydropyran + H^+ → oxocarbenium ion; + $CH_3CH_2\ddot{O}H$ ⇌ protonated acetal ($\overset{+}{O}CH_2CH_3$, H); H^+ ⇌ $-H^+$ → 2-ethoxytetrahydropyran (OCH_2CH_3)]

42. **a.** three signals in the ^{1}H NMR spectrum **b.** three signals in the ^{13}C NMR spectrum

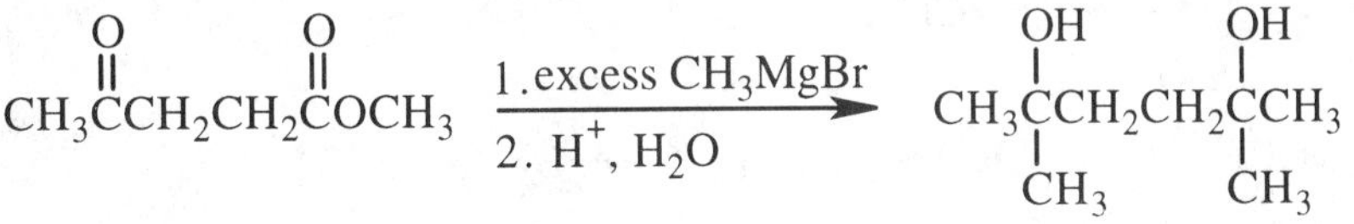

$$CH_3\overset{O}{\overset{\|}{C}}CH_2CH_2\overset{O}{\overset{\|}{C}}OCH_3 \xrightarrow[2.\ H^+,\ H_2O]{1.\text{excess } CH_3MgBr} CH_3\underset{CH_3}{\overset{OH}{C}}CH_2CH_2\underset{CH_3}{\overset{OH}{C}}CH_3$$

43.

a. 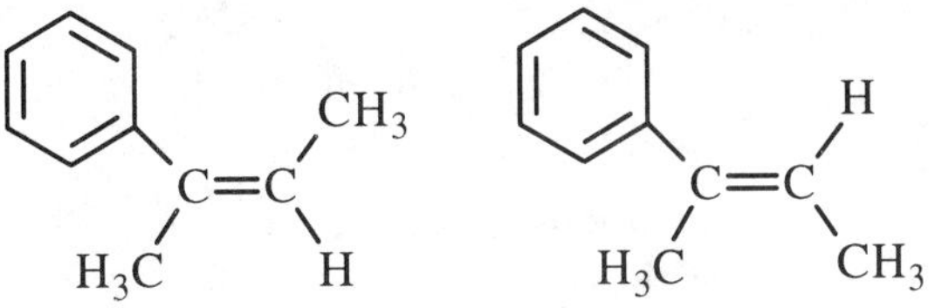 [structures: two stereoisomers of $C_6H_5C(CH_3)=CHCH_3$; labels CH_3, H_3C, H / H, H_3C, CH_3]

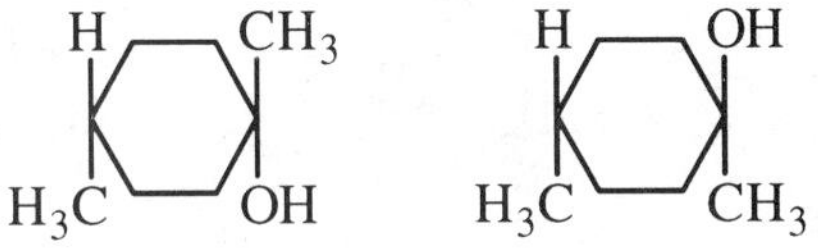

b. 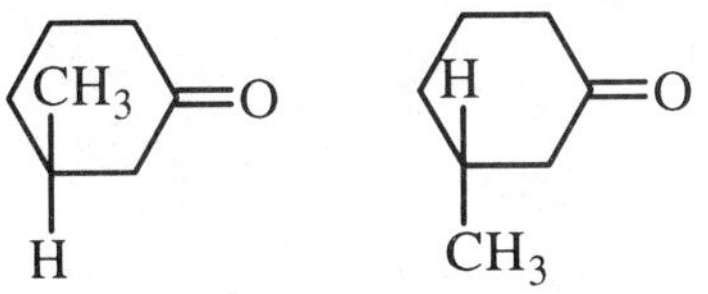 [structures: 3-methylcyclohexanone enantiomers; labels CH_3, =O, H / H, =O, CH_3]

c. [structures: cis/trans 1,4-dimethylcyclohexanol; labels H, CH_3, H_3C, OH / H, OH, H_3C, CH_3]

d. 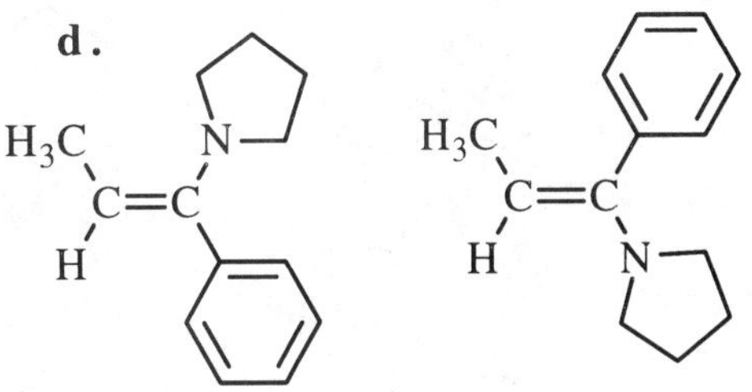 [structures: E/Z enamines; labels H_3C, N, C=C, H, phenyl / H_3C, C=C, H, N, phenyl]

e.

$$CH_3CH_2-\underset{H}{\overset{OH}{C}}-CH_2CH_2CH_2CH_3 \qquad CH_3CH_2CH_2CH_2-\underset{H}{\overset{OH}{C}}-CH_2CH_3$$

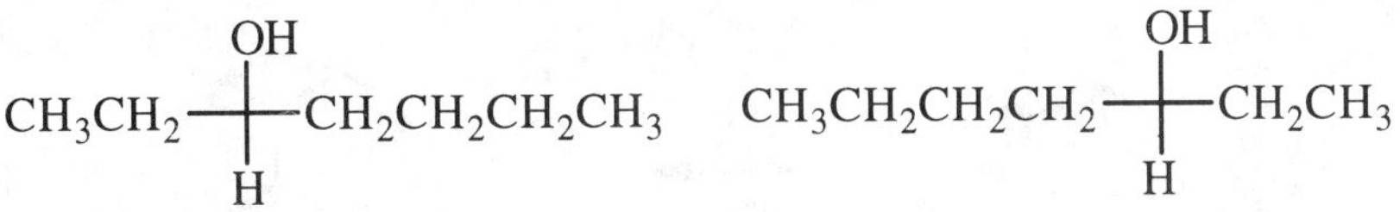

44.

a. 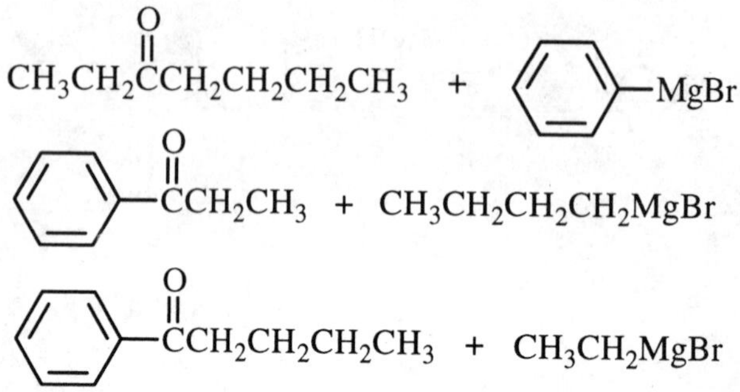 $CH_3CH_2C(=O)CH_2CH_2CH_2CH_3$ + C_6H_5MgBr

$C_6H_5C(=O)CH_2CH_3$ + $CH_3CH_2CH_2CH_2MgBr$

$C_6H_5C(=O)CH_2CH_2CH_2CH_3$ + CH_3CH_2MgBr

b. $CH_3CH_2C(=O)CH_2CH_2CH_3$ + CH_3CH_2MgBr

$CH_3CH_2C(=O)CH_2CH_3$ + $CH_3CH_2CH_2MgBr$

$CH_3CH_2CH_2C(=O)OCH_2CH_3$ + 2 CH_3CH_2MgBr

45.

a. 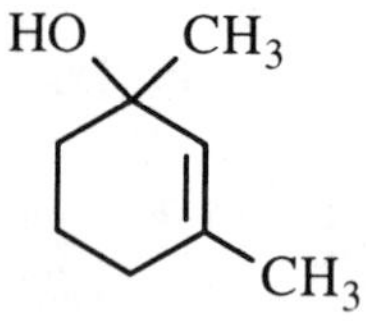 HO, CH_3; CH_3

b. 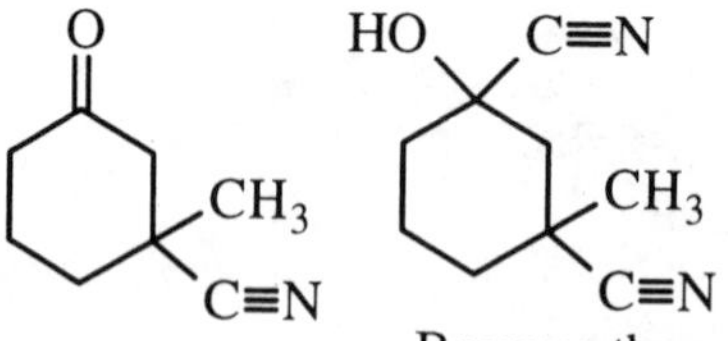 O; CH_3, C≡N → HO, C≡N; CH_3, C≡N

Because there is excess cyanide ion.

c. 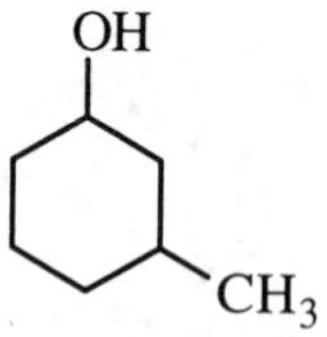 OH; CH_3

d. 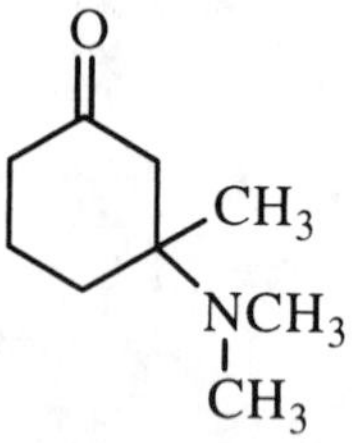 O; CH_3, NCH_3, CH_3

e. 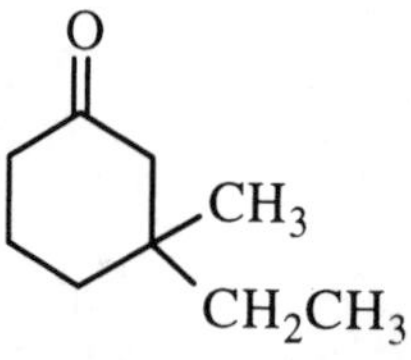 O; CH_3, CH_2CH_3

f. 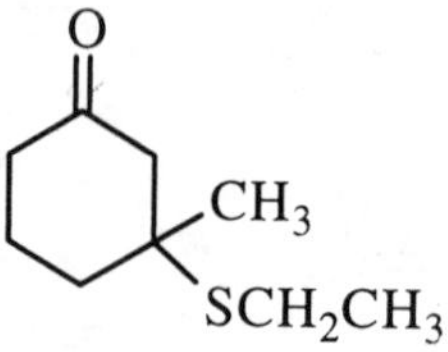 O; CH_3, SCH_2CH_3

46. Enovid would have its carbonyl stretch at a higher frequency. The carbonyl group in Norlutin has some single-bond character because of the conjugated double bonds. This causes the carbon-oxygen bond to be easier to stretch than the carbon-oxygen bond in Enovid that is not involved in resonance.

O ⟷ $^-$O, +

Norlutin

47. The absorption bands at 1600 cm^{-1}, 1500 cm^{-1}, and > 3000 cm^{-1} in the IR spectrum indicate the compound has a benzene ring. The absorption band 1720 cm^{-1} suggests it is a ketone with the carbonyl group not conjugated with the benzene ring.

The broad signal at 1.8 ppm in the ^1H NMR spectrum indicates an OH group. We also see there are three different kinds of hydrogens (other than the OH group and the benzene ring hydrogens). The doublet at 1.2 ppm indicates a methyl group adjacent to a carbon bonded to one hydrogen, and the doublet at 3.6 ppm are due to protons on a carbon that has the carbon bonded to one hydrogen on one side of it and the electron-withdrawing OH group on the other side.

The IR spectrum is the spectrum of 2-phenylpropanal and the ^1H NMR spectrum is the spectrum of 2-phenyl-1-propanol.

$$C_6H_5CH(CH_3)CHO \xrightarrow[2.\ H^+,\ H_2O]{1.\ NaBH_4} C_6H_5CH(CH_3)CH_2OH$$

48. The difference in the two reactions is a result of the difference in the leaving ability of a sulfonium group and a phosphonium group. Because the sulfonium group is a weaker base, it is a better leaving group. Therefore, it is eliminated by the oxyanion.

$$CH_3CH_2CH{=}O + (CH_3)_2\overset{+}{S}-\overset{-}{C}H_2 \longrightarrow CH_3CH_2CH(O^-)-CH_2-\overset{+}{S}(CH_3)_2 \longrightarrow CH_3CH_2\underset{\text{epoxide}}{CH-CH_2}(O) + (CH_3)_2S$$

$$CH_3CH_2CH{=}O + (C_6H_5)_3\overset{+}{P}-\overset{-}{C}H_2 \longrightarrow CH_3CH_2CH(O^-)-CH_2-\overset{+}{P}(C_6H_5)_3 \longrightarrow CH_3CH_2CH(-O-P(C_6H_5)_3-)CH_2$$

49.

a. $$C_6H_5COCH_3\ (\text{ester } C_6H_5C(=O)OCH_3) \xrightarrow[2.\ H^+,\ H_2O]{1.\ \text{excess } CH_3MgBr} C_6H_5C(OH)(CH_3)CH_3$$

b. 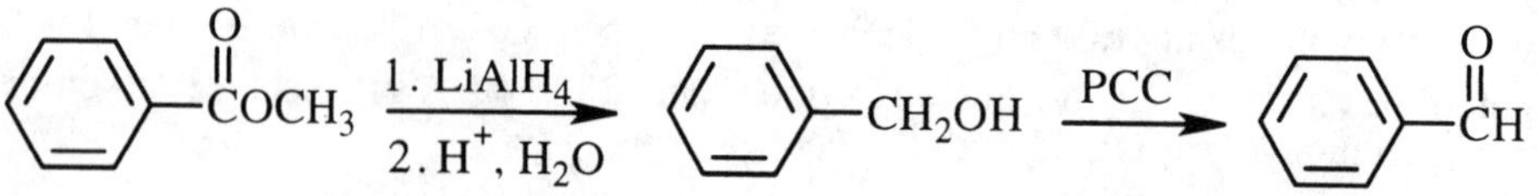

After you study Chapter 18, return to this problem and show how it can be done in one step.

c.

$$CH_3CH_2CH_2CH_2Br \xrightarrow[Et_2O]{Mg} CH_3CH_2CH_2CH_2MgBr \xrightarrow[2.\,H^+,\,H_2O]{1.\,CO_2} CH_3CH_2CH_2CH_2\overset{O}{\overset{\|}{C}}OH$$

or

$$CH_3CH_2CH_2CH_2Br \xrightarrow[HCl]{^-C\equiv N} CH_3CH_2CH_2CH_2C\equiv N \xrightarrow[\Delta]{H^+,\,H_2O} CH_3CH_2CH_2CH_2\overset{O}{\overset{\|}{C}}OH$$

d. 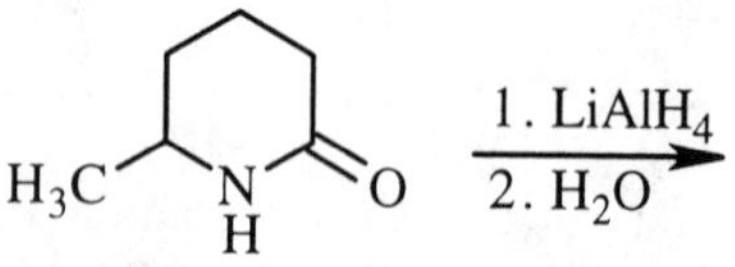

e. 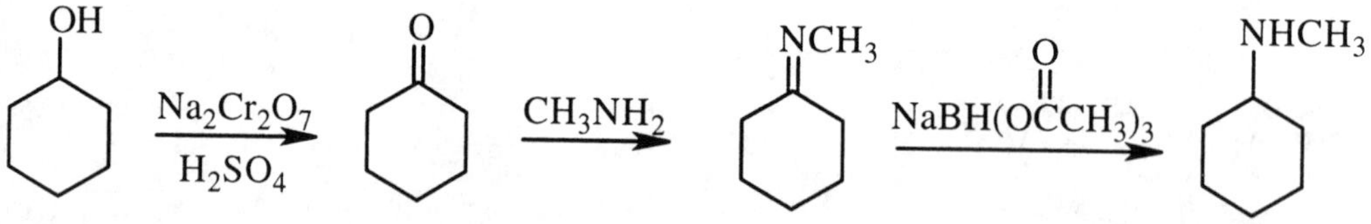

50.

a. 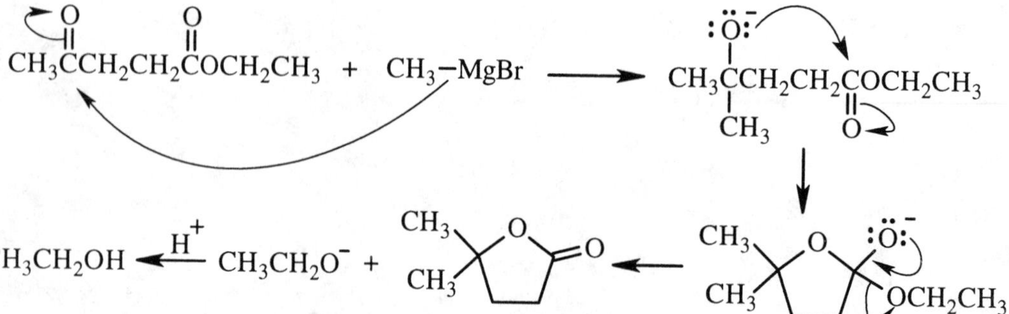

b.

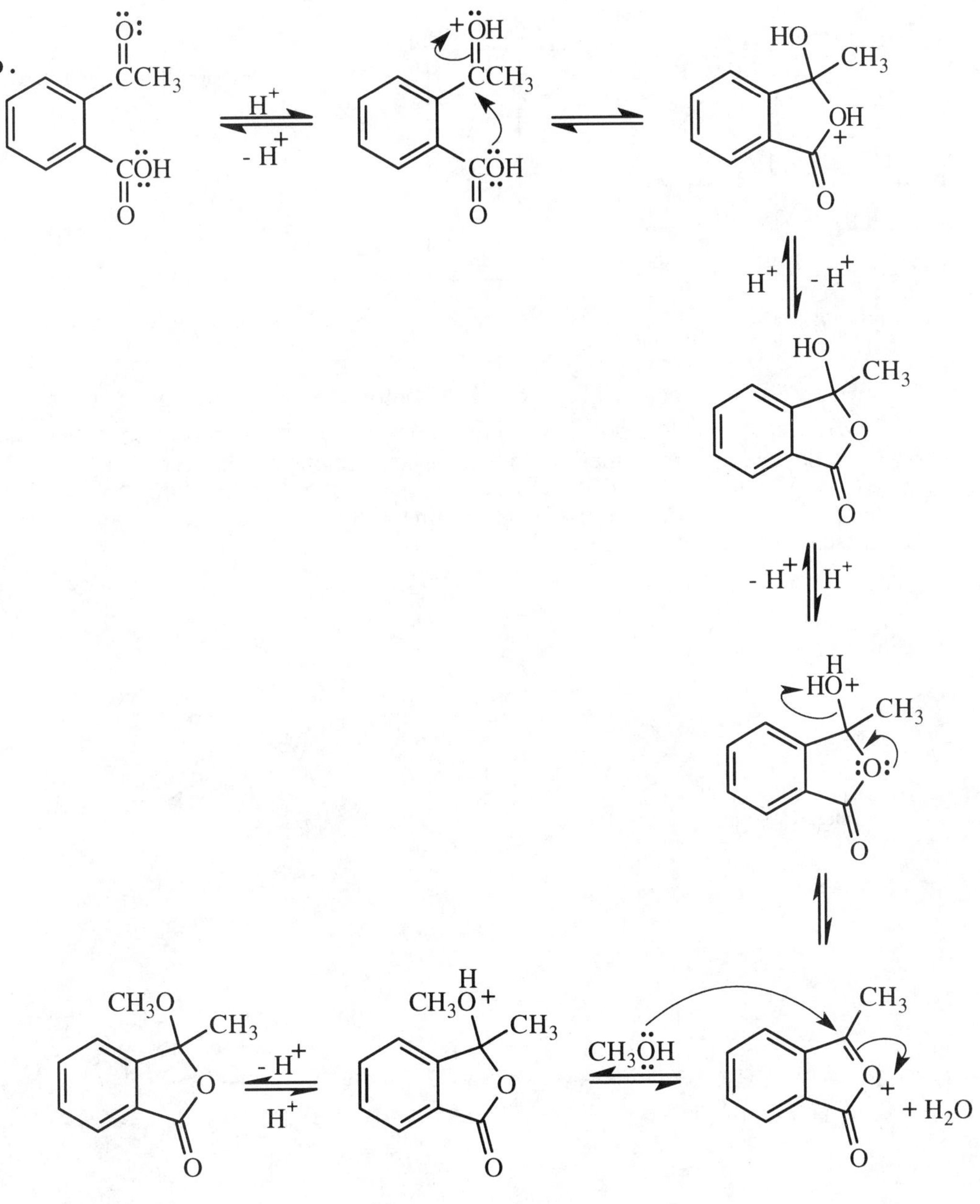

51. **a.** The OH group on C-5 of glucose reacts with the aldehyde group in an intramolecular reaction, forming a cyclic hemiacetal. Because the reaction creates a new chirality center, two cyclic hemiacetals can form, one with the *R* configuration at the new chirality center and one with the *S* configuration.

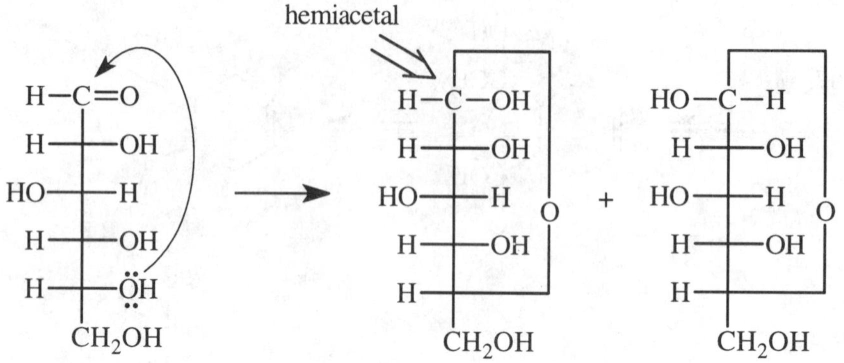

hemiacetal

H—C=O / H—OH / HO—H / H—OH / H—ÖH / CH_2OH → H—C—OH / H—OH / HO—H / H—OH / H— / CH_2OH (O ring) + HO—C—H / H—OH / HO—H / H—OH / H— / CH_2OH (O ring)

b. The two products can be drawn in their chair conformations by putting the largest group (CH_2OH) in the equatorial position and then putting the other groups in axial or equatorial positions depending on whether they are cis or trans to one another. The hemiacetal on the left has all but one of its OH groups in the more stable equatorial position, while the hemiacetal on the right has all its OH groups in the equatorial position. Therefore, the hemiacetal on the right is more stable.

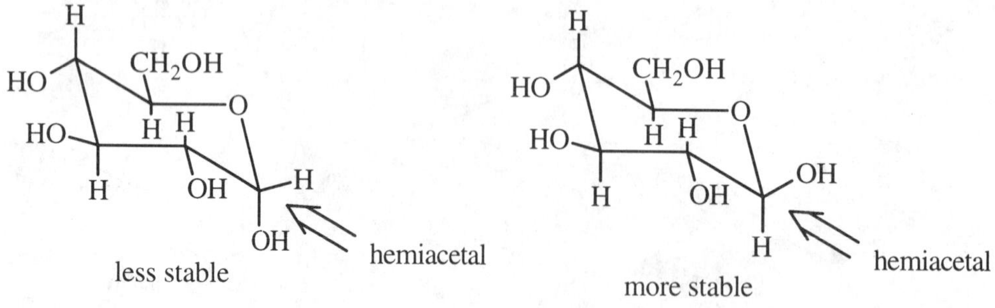

H, HO, CH_2OH, O, HO, H H, H, OH, H, OH — hemiacetal — less stable

H, HO, CH_2OH, O, HO, H H, OH, H, OH, H — hemiacetal — more stable

52.

CH_3CH=O $\underset{-H^+}{\overset{H^+}{\rightleftharpoons}}$ CH_3CH=$\overset{+}{O}H$ (+ CH_3CH=O) → $CH_3CH(OH)$–O–$\overset{+}{C}H$–CH_3 + $HCCH_3$=O → (ring: CH_3, $H\overset{+}{O}$, O, H_3C, O, CH_3) $\underset{-H^+}{\overset{H^+}{\rightleftharpoons}}$ (ring: CH_3, O, O, H_3C, O, CH_3)

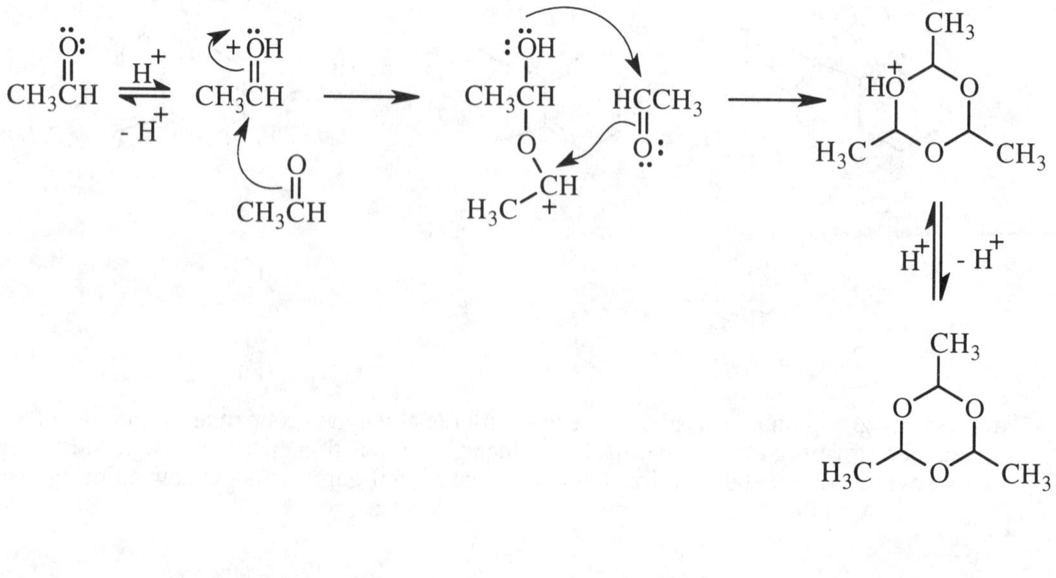

53. Converting the Fischer projection of (*R*)-mandelonitrile into a wedge-and-dash structure allows you to see that attack of cyanide ion occurred on the *si* face (decreasing priorities on the face closest to the observer are in a counterclockwise direction).

OH, H, $C{\equiv}N$ (*R*)-mandelonitrile = OH, H, $C{\equiv}N$ ⟹ 1 O, 2, CH 3, $^{-}\ddot{C}{\equiv}N$

54.

a. $C_6H_5CH{=}O$ + $(C_6H_5)_3P{=}CHCH_2CH_2CH_3$ **or** $C_6H_5CH{=}P(C_6H_5)_3$ + $O{=}CHCH_2CH_2CH_3$

b. cyclopentanone + $(C_6H_5)_3P{=}CHCH_2CH_3$ **or** $P(C_6H_5)_3$=cyclopentylidene + $O{=}CHCH_2CH_3$

c. $C_6H_5CH{=}O$ + $(C_6H_5)_3P{=}CH{-}C_6H_5$

d. 2-cyclohexenone + $(C_6H_5)_3P{=}CH_2$ **or** $P(C_6H_5)_3$=cyclohexenylidene + $O{=}CH_2$

55.

$CH_3C(CH_3){=}CHCH{=}O$ **A** $\xrightarrow[2.\ H^+,\ H_2O]{1.\ (CH_2{=}CH)_2CuLi}$ $CH_2{=}CHC(CH_3)_2CH_2CH{=}O$ **B**

B $\xrightarrow[2.\ H^+,\ H_2O]{1.\ CH_3Li}$ $CH_2{=}CHC(CH_3)_2CH_2CH(OH)CH_3$

56.

a.

CH_3SO_3H

$-H^+$

$-H^+$ H^+

$-H^+$

$+ H_2O$

b.

$-H^+$

57. The compound that gives the 1H NMR spectrum is 2-phenyl-2-butanol.

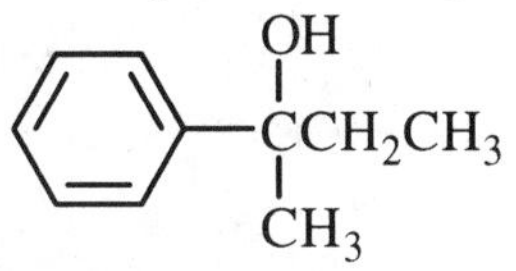

$$C_6H_5-C(OH)(CH_3)CH_2CH_3$$

Therefore, the compound that reacts with methylmagnesium bromide is 1-phenyl-1-propanone.

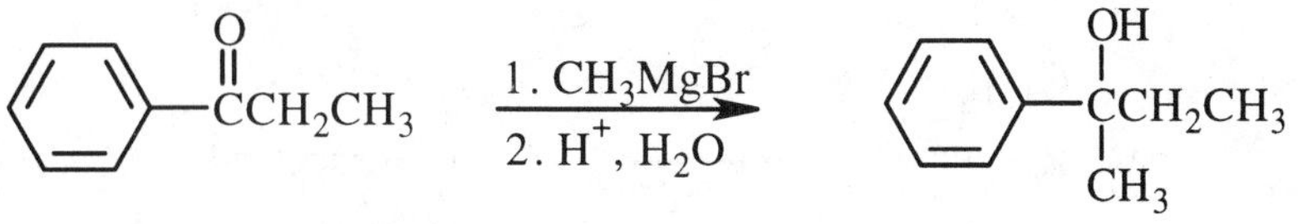

$$C_6H_5-\overset{O}{\overset{\|}{C}}CH_2CH_3 \xrightarrow[2.\ H^+,\ H_2O]{1.\ CH_3MgBr} C_6H_5-C(OH)(CH_3)CH_2CH_3$$

58.

$$\text{cyclohexanone} + \overset{-}{\ddot{C}}H_2-\overset{+}{N}\equiv N \longrightarrow \text{1-(}\overset{-}{:\ddot{O}:}\text{)-1-(}CH_2-\overset{+}{N}\equiv N\text{)cyclohexane} \longrightarrow \text{cycloheptanone} + N_2$$

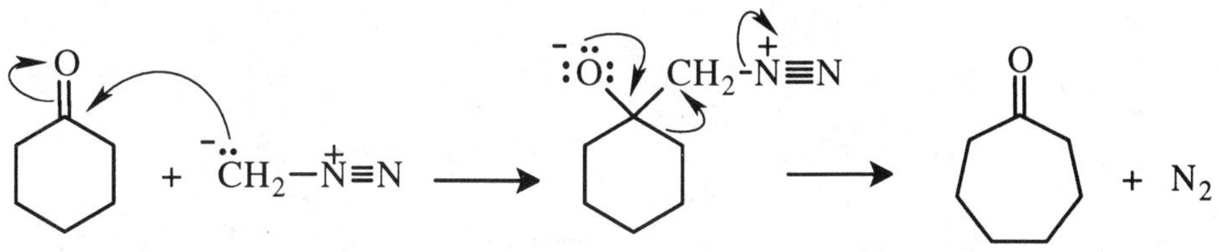

59. The data show that the amount of hydrate decreases with increasing pH until about pH = 6 and that increasing the pH beyond 6 has no effect on the amount of hydrate.

A hydrate is stabilized by electron-withdrawing groups. A COOH group is electron withdrawing, but a COO^- group is less so. In acidic solutions, where both carboxylic acid groups are in their acidic (COOH) forms, the compound exists as essentially all hydrate. As the pH of the solution increases and the COOH groups become COO^- groups, the amount of hydrate decreases. Above pH = 6, where both carboxyl groups are in their basic (COO^-)forms, there is only a small amount of hydrate.

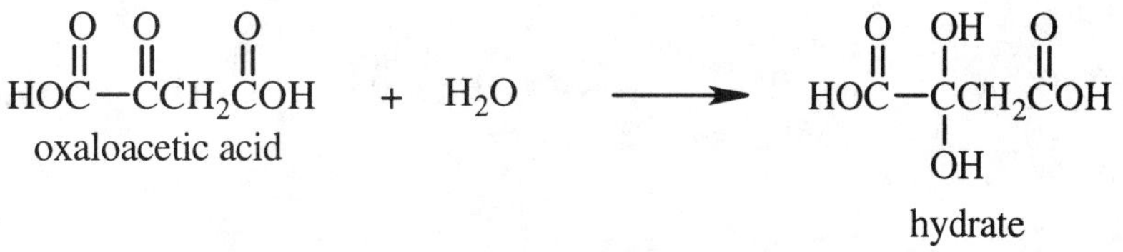

$$\underset{\text{oxaloacetic acid}}{HO\overset{O}{\overset{\|}{C}}-\overset{O}{\overset{\|}{C}}CH_2\overset{O}{\overset{\|}{C}}OH} + H_2O \longrightarrow \underset{\text{hydrate}}{HO\overset{O}{\overset{\|}{C}}-C(OH)_2CH_2\overset{O}{\overset{\|}{C}}OH}$$

60.

a. $(EtO)_3P\colon + CH_3CH(Br)CH_3 \longrightarrow CH_3CH(CH_3)\text{-}\overset{+}{P}(OEt)_2\text{-}O\text{-}CH_2CH_3 + \colon Br\colon^-$

$\downarrow$

$CH_3C(CH_3)(H)\text{-}P(=O)(OEt)_2 + CH_3CH_2Br \xrightarrow{B\colon^-} CH_3\overset{-}{C}(CH_3)\text{-}P(=O)(OEt)_2$

b. $R_2C{=}O + \colon\overset{-}{C}(CH_3)_2\text{-}P(=O)(OEt)_2 \longrightarrow {}^-\colon O\colon\text{-}CR_2\text{-}C(CH_3)_2\text{-}P(=O)(OEt)_2$

$\downarrow$

$R_2C{=}C(CH_3)_2 + (EtO)_2PO_2^- \longleftarrow$ (cyclic intermediate: $O\text{-}CR_2\text{-}C(CH_3)_2\text{-}P(O^-)(OEt)_2$, with O bonded to P)

c.

1. $(EtO)_3P\colon + BrCH_2\overset{O}{\overset{\|}{C}}CH_3 \longrightarrow (EtO)_2\overset{O}{\overset{\|}{P}}\text{-}CH_2\overset{O}{\overset{\|}{C}}CH_3 + CH_3CH_2Br$

$\downarrow B\colon^-$

$CH_3CH_2CH{=}CH\overset{O}{\overset{\|}{C}}CH_3 \xleftarrow{CH_3CH_2\overset{O}{\overset{\|}{C}}H} (EtO)_2\overset{O}{\overset{\|}{P}}\text{-}\overset{-}{C}H\overset{O}{\overset{\|}{C}}CH_3$

2. $(EtO)_3P\!:$ + $CH_3CH(Br)COCH_3$ (with C=O) ⟶ $(EtO)_2P(=O)-CH(CH_3)-C(=O)OCH_3$ + CH_3CH_2Br

↓ B:⁻

$(EtO)_2P(=O)-\overset{-}{C}(CH_3)-C(=O)OCH_3$ + cyclohexyl methyl ketone ⟶ methyl 3-cyclohexyl-2-methylbut-2-enoate (OCH_3 ester)

61.

morpholine-N: C=CHCH₃ (p-X-phenyl) + H−OH ⟶ morpholine-$\overset{+}{N}$=C−CH_2CH_3 (p-X-phenyl) + $H_2\ddot{O}$: ⟶ morpholine-N−C(−CH_2CH_3)(−$\overset{+}{O}H$−H) (p-X-phenyl)

a. The negative ρ value obtained when hydrolysis is carried out in a basic solution indicates that electron-donating substituents increase the rate of the reaction. This means that the rate-determining step must be protonation of the carbon (the first step), because the more electron-donating the substituent, the greater the negative charge on the carbon and the easier it will be to protonate.

b. The positive ρ value obtained when hydrolysis is carried out in an acidic solution indicates that electron-withdrawing substituents increase the rate of the reaction. This means that the rate-determining step must be attack of water on the imine carbon to form the tetrahedral intermediate, because electron withdrawal increases the electrophilicity of the imine carbon, making it more susceptible to nucleophilic attack.

62.

a.

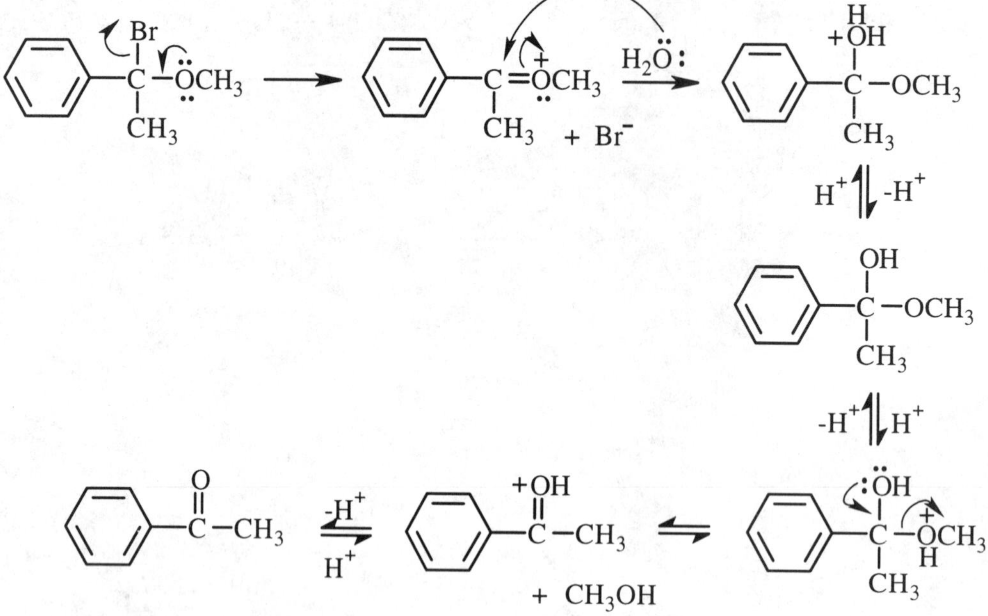

b.

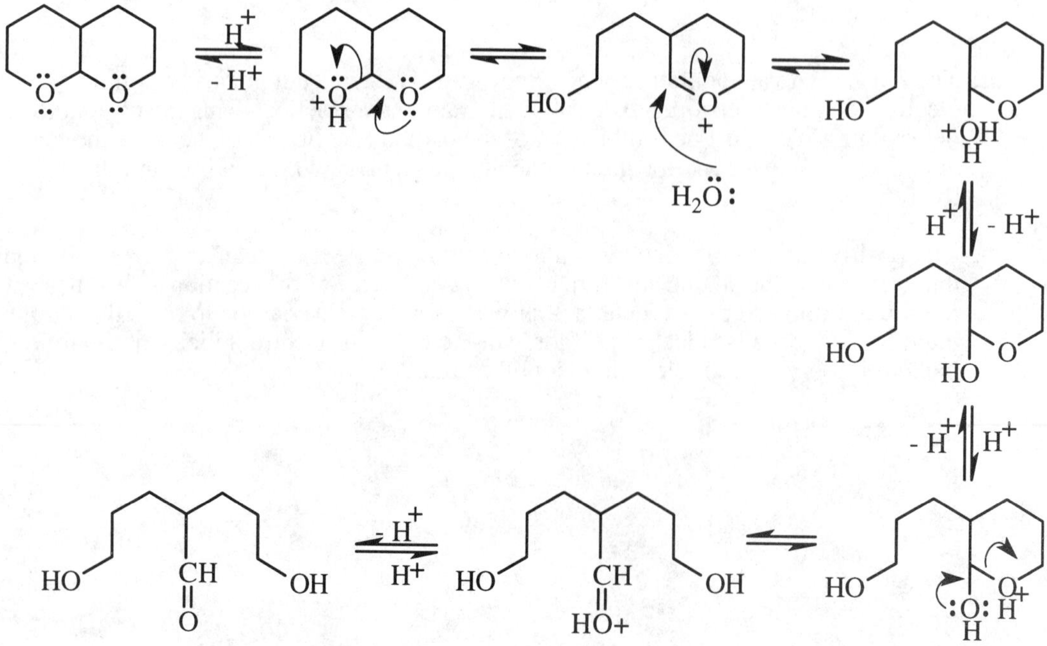

c.

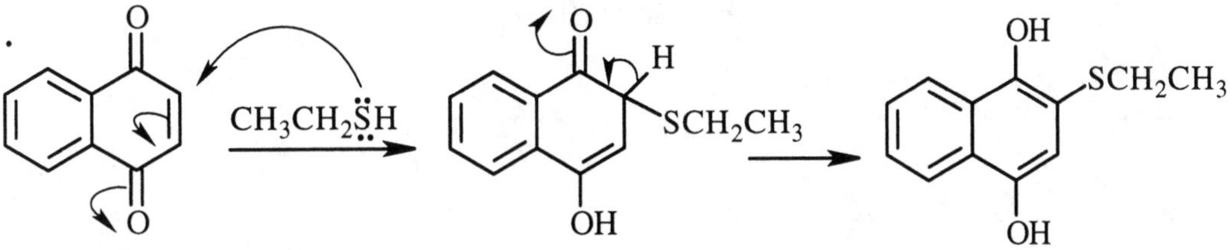

O, O, $CH_3CH_2\ddot{S}H$, O, H, SCH_2CH_3, OH, OH, SCH_2CH_3, OH

d.

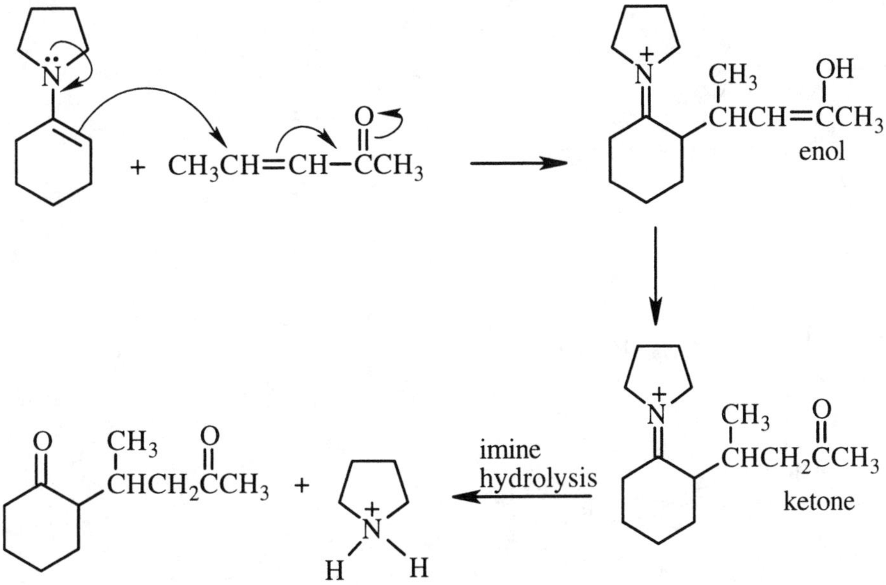

N, + $CH_3CH{=}CH{-}\overset{O}{\overset{\|}{C}}CH_3$

N+, CH_3, OH, $CHCH{=}CCH_3$, enol

N+, CH_3, O, $CHCH_2\overset{\|}{C}CH_3$, ketone

imine hydrolysis

O, CH_3, O, $CHCH_2\overset{\|}{C}CH_3$ + N+ H H

Chapter 17 Practice Test

1. Give the product of each of the following reactions.

a. C_6H_5–$\overset{O}{\overset{\|}{C}}CH_2CH_3$ + NH_2OH ⟶

b. cyclopentanone + piperidine (N–H) ⟶

c. CO_2 $\xrightarrow[\text{2. } H^+,\ H_2O]{\text{1. } CH_3CH_2CH_2MgBr}$

d. cyclohexanone + CH_3CH_2OH (excess) $\xrightarrow{H^+}$

e. $CH_3CH_2\overset{O}{\overset{\|}{C}}CH_2CH_3$ $\xrightarrow[\text{2. } H^+,\ H_2O]{\text{1. } CH_3MgBr}$

f. cyclohexanone + NH_3 $\xrightarrow[\text{Raney Ni}]{H_2}$

g. $CH_3CH_2\overset{O}{\overset{\|}{C}}CH_2CH_3$ $\xrightarrow[HCl]{^-C\equiv N}$

h. $CH_3CH{=}CH\overset{O}{\overset{\|}{C}}CH_3$ $\xrightarrow[\text{2. } H^+,\ H_2O]{\text{1. } LiAlH_4}$

i. C_6H_5–$\overset{O}{\overset{\|}{C}}OCH_2CH_3$ $\xrightarrow[\text{2. } H^+,\ H_2O]{\text{1. 2 } CH_3CH_2CH_2MgBr}$

2. Which of the following alcohols cannot be prepared by the reaction of an ester with excess Grignard reagent?

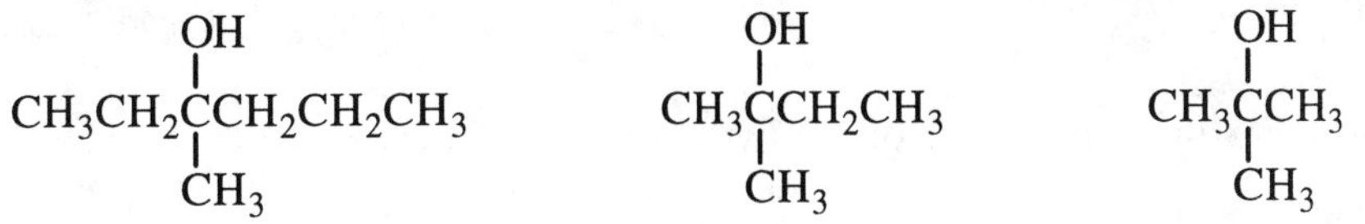

$CH_3CH_2C(OH)(CH_3)CH_2CH_2CH_3$ $\qquad$ $CH_3C(OH)(CH_3)CH_2CH_3$ $\qquad$ $CH_3C(OH)(CH_3)CH_3$

3. Which of the following ketones would form the greatest amount of hydrate in an aqueous solution?

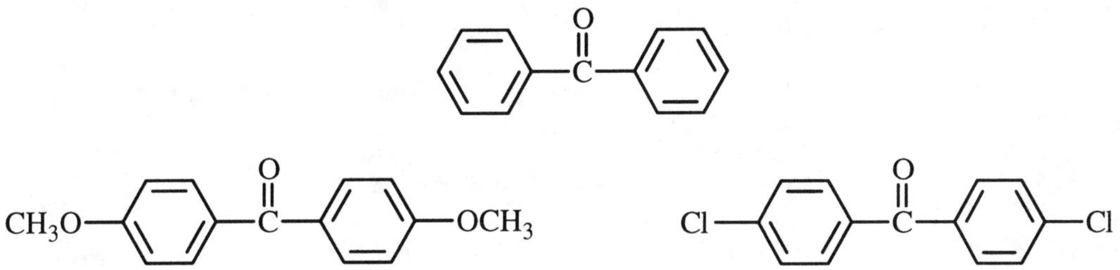

$C_6H_5COC_6H_5$

$CH_3O\text{-}C_6H_4\text{-}CO\text{-}C_6H_4\text{-}OCH_3$ $\qquad$ $Cl\text{-}C_6H_4\text{-}CO\text{-}C_6H_4\text{-}Cl$

4. Give an example of each the following:

a. an enamine

b. an acetal

c. an imine

d. a hemiacetal

e. a phenylhydrazone

5. Which is more reactive toward nucleophilic addition?

a. butanal or methyl propyl ketone

b. 4-heptanone or 3-pentanone

6. Indicate how the following compound could be prepared using the given starting material.

$$CH_3CH_2CH_2Br \longrightarrow CH_3CH_2CH_2\overset{O}{\overset{\|}{C}}OCH_2CH_3$$

CHAPTER 18
More About Oxidation-Reduction Reactions

Important Terms

Baeyer-Villiger oxidation	oxidation of aldehydes or ketones with H_2O_2 to form carboxylic acids or esters, respectively.
catalytic hydrogenation	addition of hydrogen to a double or triple bond in the presence of a metal catalyst.
dissolving metal reduction	a reduction using sodium or lithium metal dissolved in liquid ammonia.
epoxidation	formation of an epoxide.
functional group interconversion	a reaction that converts one functional group into another.
glycol	a compound containing two or more OH groups.
molozonide	an unstable intermediate containing a five-membered ring with three oxygens in a row that is formed from the reaction of an alkene with ozone.
oxidation	loss of electrons by an atom or molecule.
oxidation-reduction reaction	a reaction that involves the transfer of electrons from one atom or molecule to another.
oxidation state	the oxidation state of an carbon is given by the number of C-Z bonds, where Z = oxygen, nitrogen, or halogen
oxidative cleavage	an oxidation reaction that cleaves the reactant into two or more compounds.
oxidizing agent	the compound that is reduced in a redox reaction as it oxidizes the other compound
ozonide	a five-membered ring compound formed as a result of rearrangement of a molozonide.
ozonolysis	reaction of a carbon-carbon double or triple bond with ozone.
peroxyacid	a carboxylic acid with an OOH group instead of an OH group.
redox reaction	an oxidation-reduction reaction.
reducing agent	the compound that is oxdized in a redox reaction as it reduces the other compound
reduction	gain of electrons by an atom or molecule.
Rosenmund reduction	the reduction of an acyl chloride to an aldehyde using H_2/Lindlar's catalyst.
Swern oxidation	uses dimethyl sulfoxide, oxalyl chloride, and triethylamime to oxdize primary alcohols to aldehydes and secondary alcohols to ketones

Tollens test a test to determine the presence of an aldehyde. A positive result is indicated by the formation of a silver mirror.

vicinal diol (vicinal glycol) a 1,2-diol.

Solutions to Problems

1. **a.** reduction **c.** oxidation **e.** reduction

   b. neither **d.** oxidation **f.** neither

2. a. $CH_3CH_2CH_2CH_2CH_2OH$

   b. $CH_3CH_2CH_2CH_2NH_2$

   c. 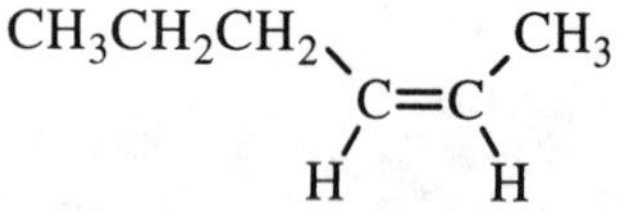 $CH_3CH_2CH_2(H)C{=}C(H)CH_3$ (cis)

   d. 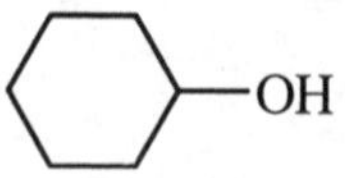 cyclohexyl–OH

   e. no reaction

   f. CH_3CH_2OH

   g. 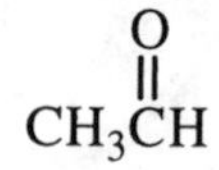 $CH_3CH{=}O$ (CH_3CH with C=O)

   h. 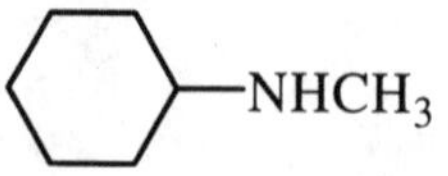 cyclohexyl–$NHCH_3$

3. A terminal alkyne cannot be reduced by Na, because Na donates an electron to the terminal *sp* carbon, causing a hydrogen atom to be removed.

$$2\ CH_3C\equiv CH + 2\,Na \longrightarrow 2\ CH_3C\equiv C^- + 2\,Na^+ + H_2$$

4. a. 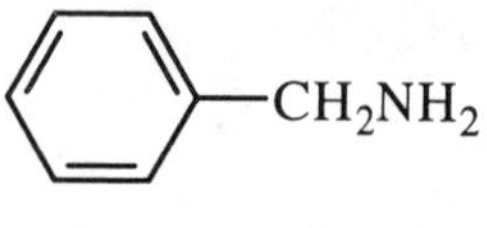 phenyl–CH_2NH_2

   b. 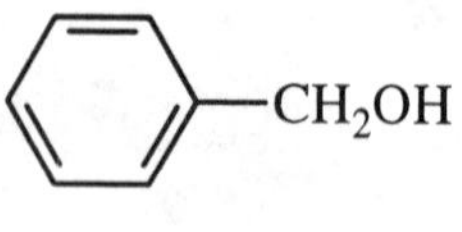 phenyl–CH_2OH

   c. 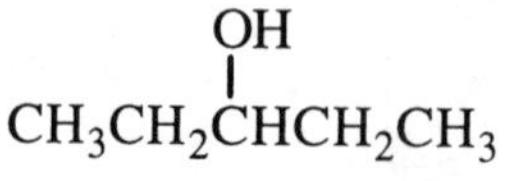 $CH_3CH_2CH(OH)CH_2CH_3$

   d. 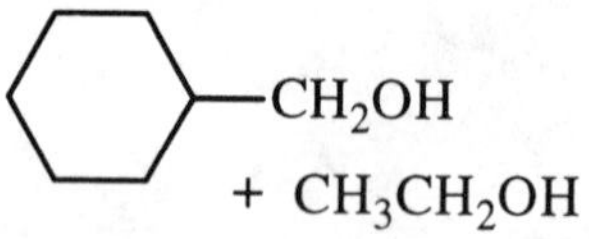 cyclohexyl–CH_2OH + CH_3CH_2OH

   e. $CH_2CH_2CH_2NHCH_2CH_3$

   f. $CH_3CH_2CH_2CH_2OH$

5. Carbon-nitrogen double and triple bonds can be reduced by $LiAlH_4$ because the bonds are polar. The hydride ion will be attracted to the partially positively charged carbon.

$\delta+$ $\delta-$ $>C{=}N-$ ← $:H^-$ $\delta+$ $\delta-$ $-C\equiv N$ ← $:H^-$

6.

a. 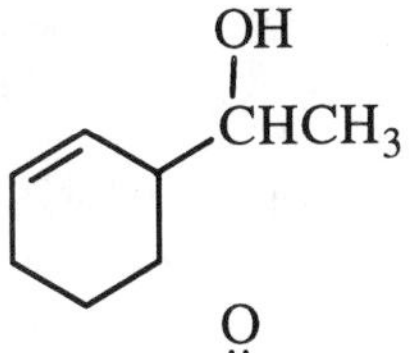 1-(cyclohex-2-enyl)ethanol structure: ring bearing $\overset{\text{OH}}{|}{\text{CHCH}_3}$

b. 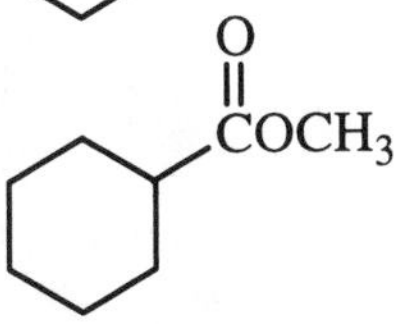 cyclohexyl ring bearing $\overset{\text{O}}{\|}{\text{COCH}_3}$

c. 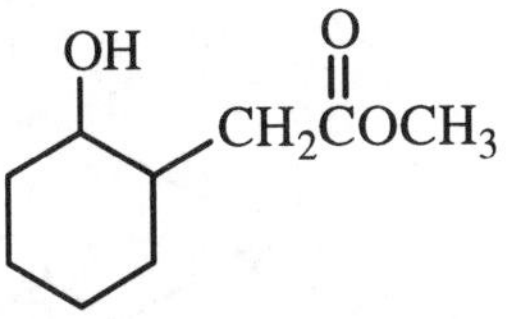 2-hydroxycyclohexyl ring (OH) bearing $CH_2\overset{\text{O}}{\|}{C}OCH_3$

d. 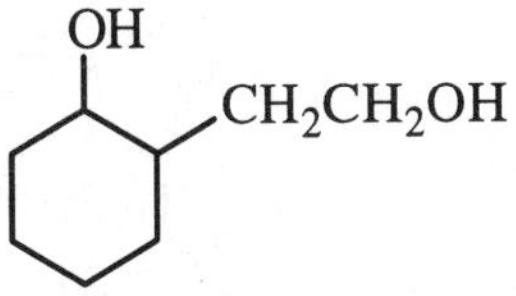 2-hydroxycyclohexyl ring (OH) bearing CH_2CH_2OH + CH_3OH

7.

a. 1. $CH_3CH_2\overset{O}{\overset{\|}{C}}CH_2CH_3$

3. A tertiary alcohol is not oxdized to a carbonyl group.

5. cyclohexanone (ring =O)

2. $CH_3CH_2CH_2CH_2\overset{O}{\overset{\|}{C}}OH$

4. $CH_3\overset{O}{\overset{\|}{C}}CH_2\overset{O}{\overset{\|}{C}}CH_2CH_3$

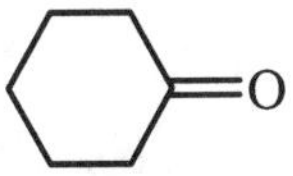

6. $HO\overset{O}{\overset{\|}{C}}CH_2CH_2\overset{O}{\overset{\|}{C}}OH$

b. 1. $CH_3CH_2\overset{O}{\overset{\|}{C}}CH_2CH_3$

3. A tertiary alcohol is not oxdized to a carbonyl group.

5. cyclohexanone (ring =O)

2. $CH_3CH_2CH_2CH_2\overset{O}{\overset{\|}{C}}H$

4. $CH_3\overset{O}{\overset{\|}{C}}CH_2\overset{O}{\overset{\|}{C}}CH_2CH_3$

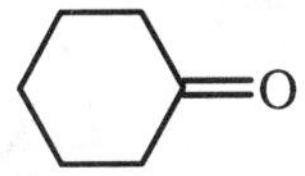

6. $H\overset{O}{\overset{\|}{C}}CH_2CH_2\overset{O}{\overset{\|}{C}}H$

8. Solved in the text.

9. Chromic acid needs to be protonated to convert a poor leaving group (^-OH) into a good leaving group (H_2O). The alcohols displaces water to form a protonated chromate ester, which loses a proton. The aldehyde is then formed in an elimination reaction.

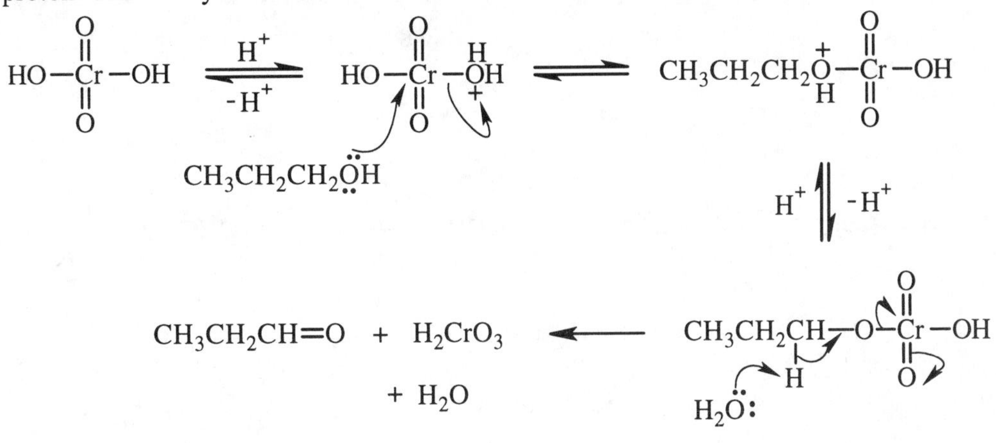

$$HO-\overset{O}{\overset{\|}{Cr}}(=O)-OH \underset{-H^+}{\overset{H^+}{\rightleftharpoons}} HO-\overset{O}{\overset{\|}{Cr}}(=O)-\overset{+}{O}H_2 \rightleftharpoons CH_3CH_2CH_2\overset{+}{O}(H)-\overset{O}{\overset{\|}{Cr}}(=O)-OH$$

(attacking nucleophile: $CH_3CH_2CH_2\ddot{O}H$)

$$\underset{-H^+}{\overset{H^+}{\rightleftharpoons}}$$

$$CH_3CH_2CH(H)-O-\overset{O}{\overset{\|}{Cr}}(=O)-OH \;(\text{base: } H_2\ddot{O}:) \longrightarrow CH_3CH_2CH{=}O + H_2CrO_3 + H_2O$$

10.

a. 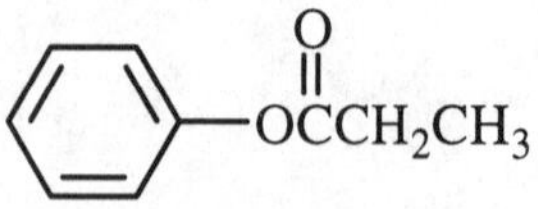

b. 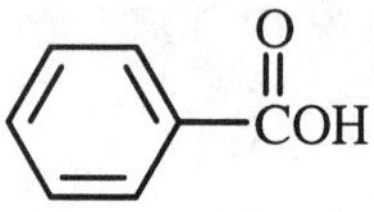

c. 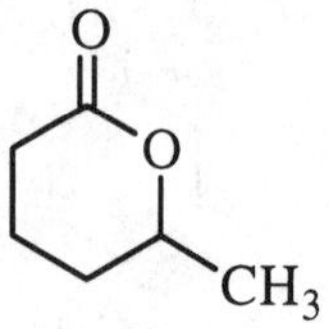

d. $\mathrm{CH_3\underset{CH_3}{CH}\overset{O}{\overset{\|}{C}}O\overset{CH_3}{\underset{CH_3}{C}}CH_3}$

e. $\mathrm{CH_3CH_2CH_2\overset{O}{\overset{\|}{C}}OH}$

f. no reaction

11. **a.** cyclohexene **b.** 1-butene **c.** *trans*-2-pentene **d.** *cis*-2-pentene

12.

a. 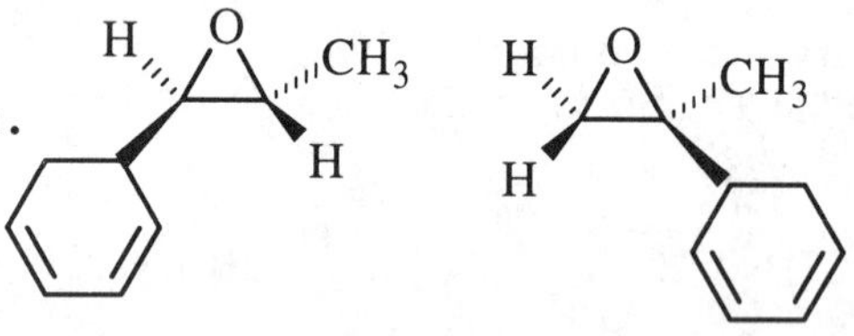

The reaction forms a product with two new chirality centers. Because only syn addition occurs, only two of the four possible stereoisomers are formed.

b. 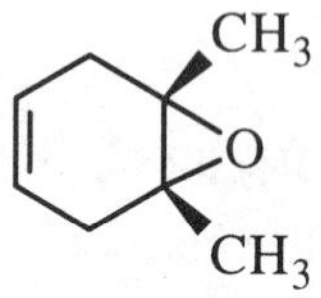

c. 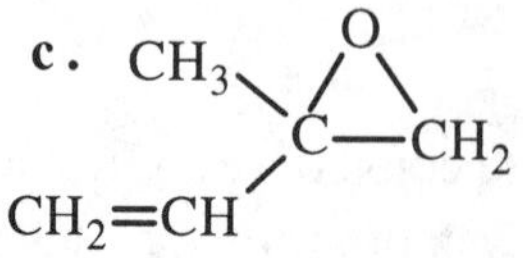

d. 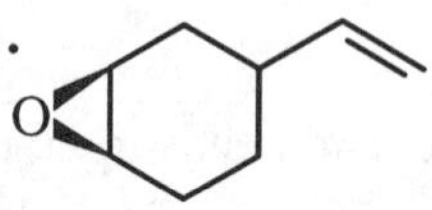

13.

a. 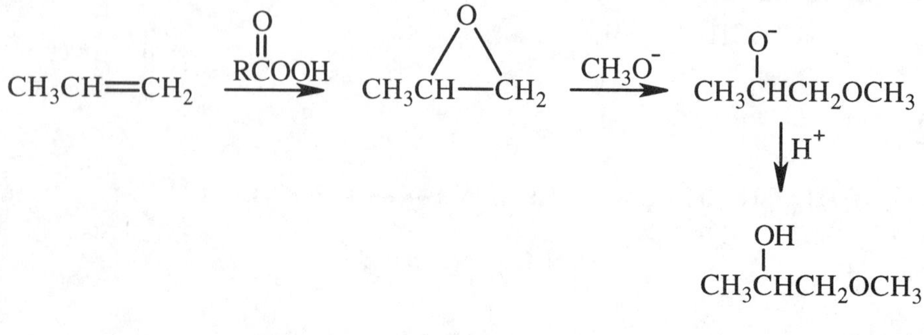

b. 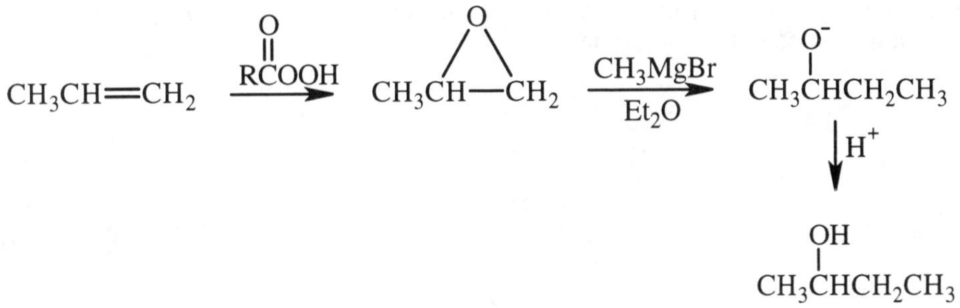 $CH_3CH{=}CH_2 \xrightarrow{RCOOH\ (peroxyacid,\ C{=}O)} CH_3CH{-}CH_2$ (epoxide, O bridging) $\xrightarrow[Et_2O]{CH_3MgBr} CH_3CH(O^-)CH_2CH_3 \xrightarrow{H^+} CH_3CH(OH)CH_2CH_3$

c. $CH_3CH(OH)CH_2CH_3 \xrightarrow{H_2CrO_4} CH_3C(=O)CH_2CH_3$
product of "**b**"

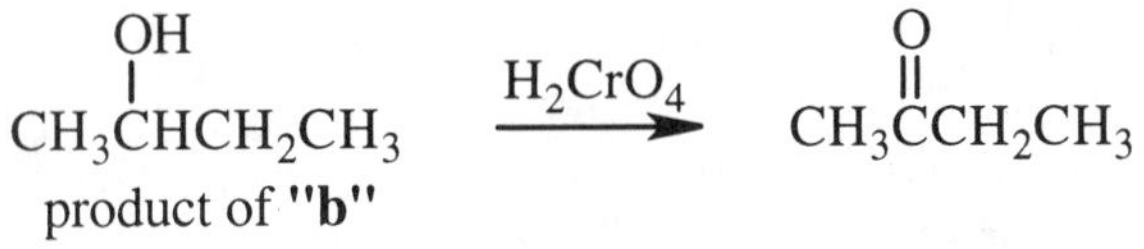

14. The bromonium ion is much less stable than an epoxide, because the C-Br bond of a bromonium ion is a much weaker bond than the C-O bond of an epoxide, and a weaker bond is easier to break.

The difference in the strength of the bonds can be understood by comparing the pK_a of HBr (-9) and the pK_a of an alcohol such as CH_3OH (15.5). The comparison shows that Br^- is a much weaker base than $^-OCH_3$, which means that it shares its electrons less well (is a better leaving group). The difference in basicity is even greater than this when comparing the leaving group in a bromonium ion and an epoxide, because the leaving group in a bromonium ion is a neutral bromine, which is a weaker base than Br^-.

15.

a. 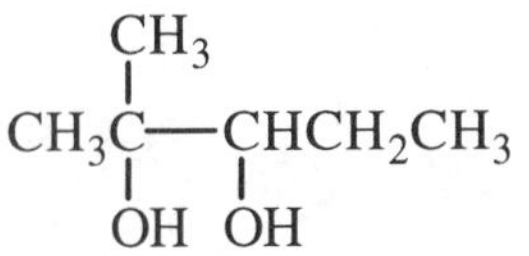 $CH_3C(CH_3)(OH){-}CH(OH)CH_2CH_3$

b. 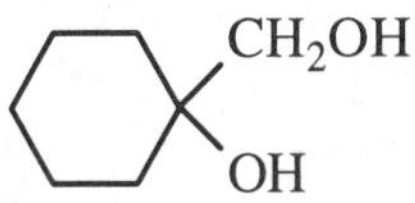 cyclohexane ring bearing CH_2OH and OH on the same carbon

16. a. Syn addition to the trans isomer forms the threo pair of enantiomers. (See Section 4.17)

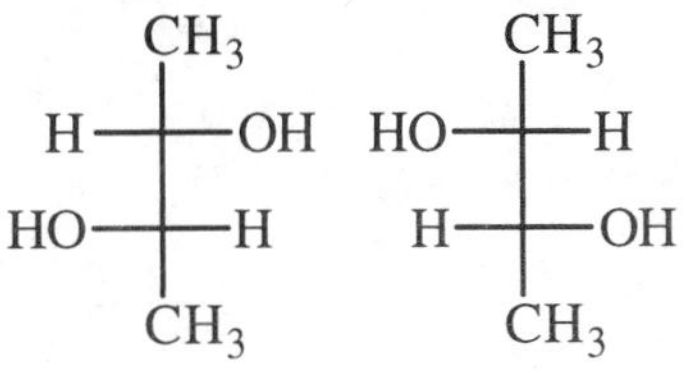

Fischer projections:

CH_3	CH_3
H—C—OH	HO—C—H
HO—C—H	H—C—OH
CH_3	CH_3

b. Syn addition to the cis isomer forms the erythro pair of enantiomers. In this case, the product is a meso compound, so only one stereoisomer is formed.

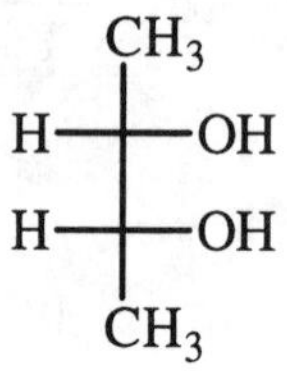

CH_3
H—|—OH
H—|—OH
CH_3

c. Syn addition to the cis isomer forms the erythro pair of enantiomers.

CH_3	CH_3
H—|—OH	HO—|—H
H—|—OH	HO—|—H
CH_2CH_3	CH_2CH_3

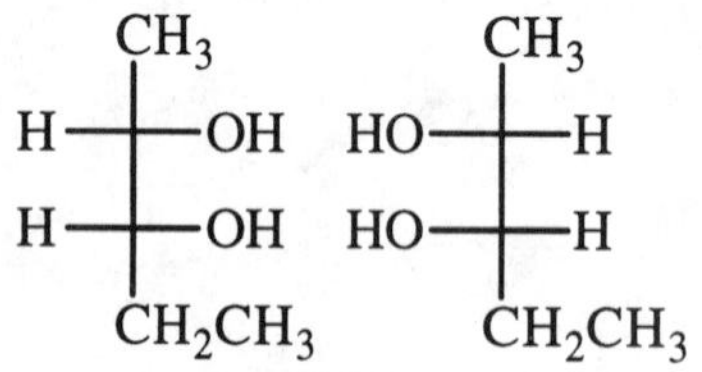

d. Syn addition to the trans isomer forms the threo pair of enantiomers.

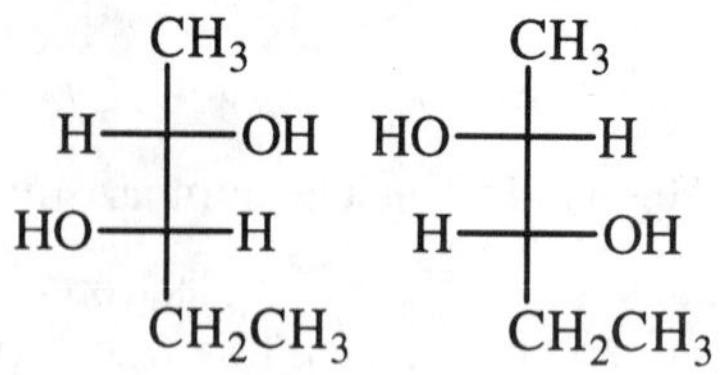

CH_3	CH_3
H—|—OH	HO—|—H
HO—|—H	H—|—OH
CH_2CH_3	CH_2CH_3

17. The fact that only one ketone is obtained means that the alkene must be symmetrical. The cyclic ketone that is obtained is

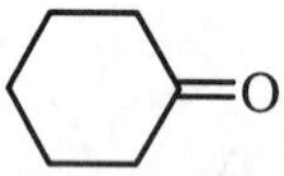

=O

Therefore, the alkene is

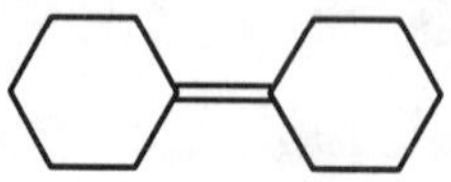

18.

a. The compound on the left is more reactive because when the bulky *tert*-butyl group is in the more stable equatorial position, the two OH groups are in equatorial positions. In the other compound, the two OH groups are in axial positions and, therefore, are too far away one another to form the cyclic intermediate.

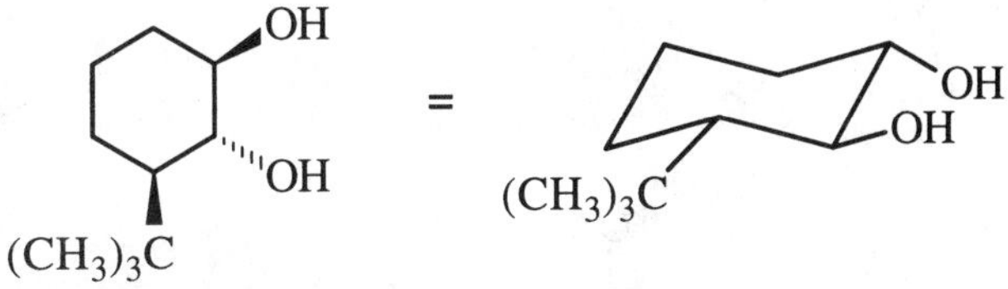

more reactive

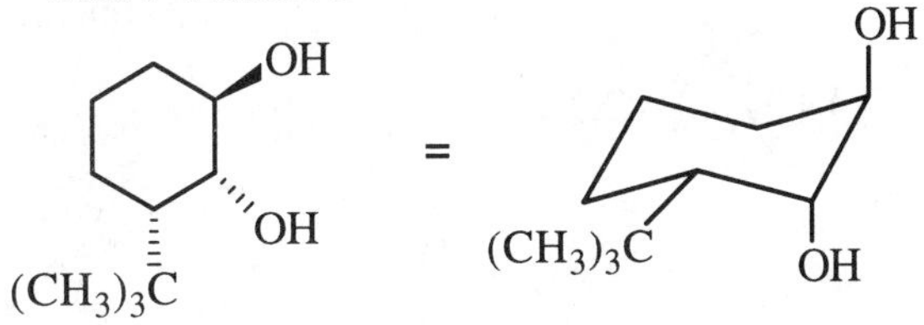

less reactive

b. The compound on the right is more reactive because when the bulky *tert*-butyl group is in the more stable equatorial position, the two OH groups are in equatorial positions. In the other compound, the two OH groups are in axial positions and, therefore, cannot form the cyclic intermediate.

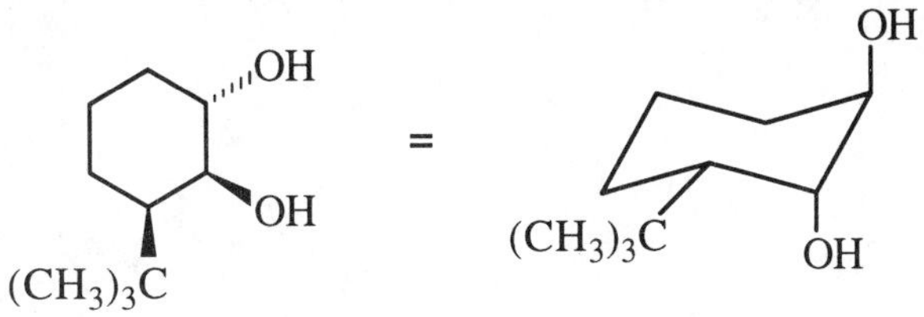

less reactive

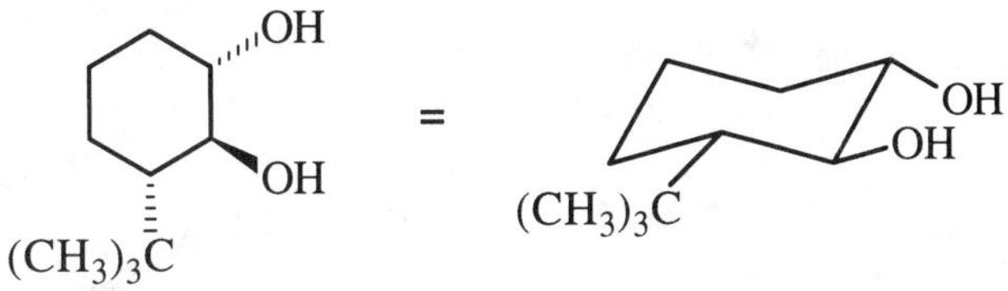

more reactive

19. Any alkene that forms two ketones when it is cleaved will form the same products when the ozonide is worked up under reducing conditions as it forms when it is worked up under oxidizing conditions. In other words, each sp^2 carbon must be bonded to two alkyl groups.

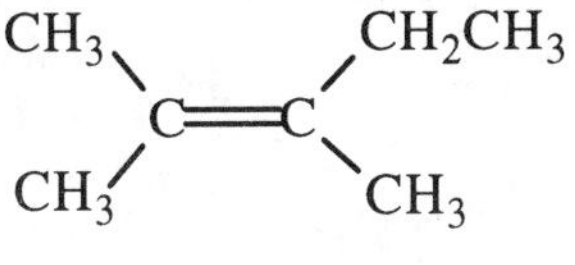

$(CH_3)_2C{=}C(CH_3)CH_2CH_3$

20.

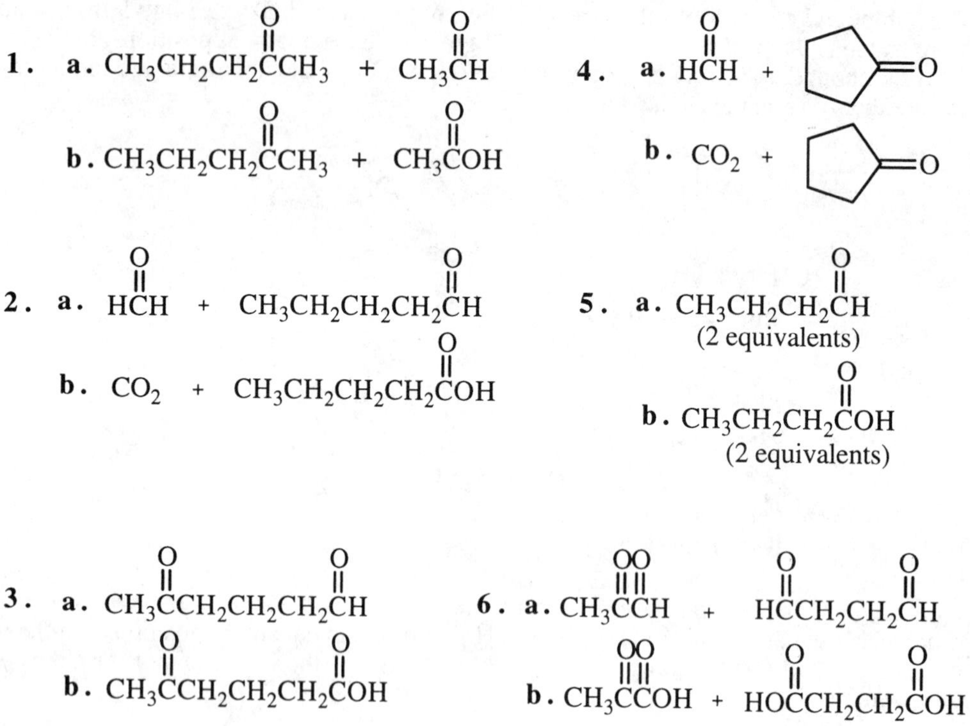

1. a. $CH_3CH_2CH_2\overset{O}{\overset{\|}{C}}CH_3$ + $CH_3\overset{O}{\overset{\|}{C}}H$

   b. $CH_3CH_2CH_2\overset{O}{\overset{\|}{C}}CH_3$ + $CH_3\overset{O}{\overset{\|}{C}}OH$

2. a. $H\overset{O}{\overset{\|}{C}}H$ + $CH_3CH_2CH_2CH_2\overset{O}{\overset{\|}{C}}H$

   b. CO_2 + $CH_3CH_2CH_2CH_2\overset{O}{\overset{\|}{C}}OH$

3. a. $CH_3\overset{O}{\overset{\|}{C}}CH_2CH_2CH_2\overset{O}{\overset{\|}{C}}H$

   b. $CH_3\overset{O}{\overset{\|}{C}}CH_2CH_2CH_2\overset{O}{\overset{\|}{C}}OH$

4. a. $H\overset{O}{\overset{\|}{C}}H$ + cyclopentanone (=O)

   b. CO_2 + cyclopentanone (=O)

5. a. $CH_3CH_2CH_2\overset{O}{\overset{\|}{C}}H$
   (2 equivalents)

   b. $CH_3CH_2CH_2\overset{O}{\overset{\|}{C}}OH$
   (2 equivalents)

6. a. $CH_3\overset{O}{\overset{\|}{C}}\overset{O}{\overset{\|}{C}}H$ + $H\overset{O}{\overset{\|}{C}}CH_2CH_2\overset{O}{\overset{\|}{C}}H$

   b. $CH_3\overset{O}{\overset{\|}{C}}\overset{O}{\overset{\|}{C}}OH$ + $HO\overset{O}{\overset{\|}{C}}CH_2CH_2\overset{O}{\overset{\|}{C}}OH$

21.

a. $(CH_3)_2C{=}C(CH_3)_2$

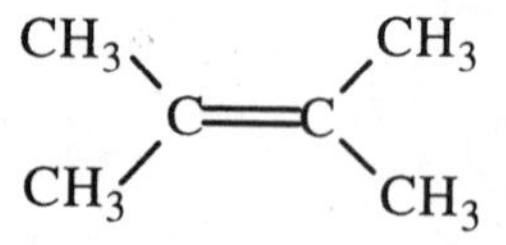

2,3-dimethyl-2-butene

b. $CH_3CH_2CH_2(H)C{=}C(H)CH_2CH_2CH_3$ (both propyl groups on the same side)

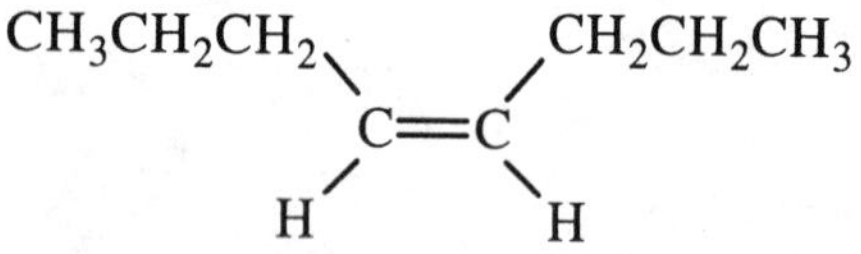

cis-4-octene

$CH_3CH_2CH_2(H)C{=}C(H)CH_2CH_2CH_3$ (propyl groups on opposite sides)

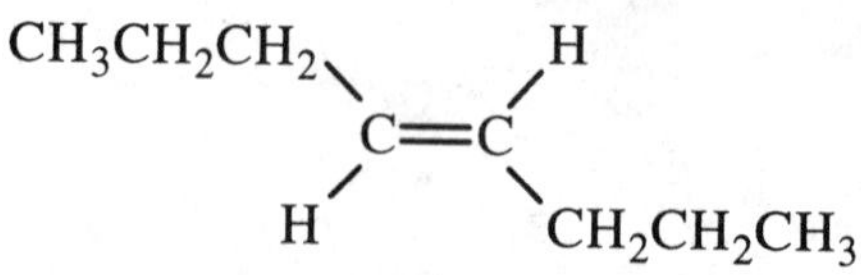

trans-4-octene

22. It does not tell you whether the double bond has the cis or the trans configuration.

23. Solved in the text.

24.

a. cyclohexyl–$C{\equiv}CH$

b. $CH_3CH_2C{\equiv}CCH_2CH_2CH_2C{\equiv}CCH_2CH_3$

25.

aldehyde
carboxylic acid
ester

acyl halide
acid anhydride
alkene

alkyl halide
ether
epoxide

26.

a. 2-bromopropane $\xrightarrow{HO^-}$ 2-propanol (OH) $\xrightarrow{H_2CrO_4}$ acetone (O)

acetone $\xrightarrow[2.\,H_2O]{1.\,NaBH_4}$ 2-propanol $\xrightarrow[\Delta]{HBr}$ 2-bromopropane

b. cyclohexene $\xrightarrow{KMnO_4,\ HO^-,\ H_2O,\ \text{cold or } 1.\ OsO_4\ 2.\ NaHSO_3,\ H_2O}$ cis-1,2-cyclohexanediol (OH, OH)

cyclohexene $\xrightarrow{1.\ RCOOH\ (\text{peroxyacid, } RC(=O)OOH)\quad 2.\ HO^-}$ trans-1,2-cyclohexanediol (OH, OH)

27.

a.

1. $CH_3CH_2Br \xrightarrow[Et_2O]{Mg} CH_3CH_2MgBr \xrightarrow{H_2C{=}O} CH_3CH_2CH_2O^- \xrightarrow{H^+} CH_3CH_2CH_2OH$

2. $CH_3CH_2Br \xrightarrow[Et_2O]{Mg} CH_3CH_2MgBr \xrightarrow{CO_2} CH_3CH_2\overset{O}{\overset{\|}{C}}O^- \xrightarrow[2.\,H^+,\ H_2O]{1.\,LiAlH_4} CH_3CH_2CH_2OH$

2. $CH_3CH_2Br \xrightarrow[HCl]{^-C{\equiv}N} CH_3CH_2C{\equiv}N \xrightarrow[\Delta]{H^+,\ H_2O} CH_3CH_2\overset{O}{\overset{\|}{C}}OH$

or

$\xrightarrow[\Delta]{^-OH,\ H_2O} CH_3CH_2\overset{O}{\overset{\|}{C}}O^-$

$\xrightarrow[2.\,H^+,\ H_2O]{1.\,LiAlH_4} CH_3CH_2CH_2OH$

b. $CH_3CH_2Br \xrightarrow{^-C{\equiv}N} CH_3CH_2C{\equiv}N \xrightarrow[Pd/C]{H_2} CH_3CH_2CH_2NH_2$

c. $CH_3CH_2CH_2Br \xrightarrow{HO^-} CH_3CH{=}CH_2 \xrightarrow[2.\ Zn,\ H_2O]{1.\ O_3} CH_3CH{=}O \xrightarrow[2.\ H_2,\ \text{Raney Nickel}]{1.\ NH_3} CH_3CH_2NH_2$

28.

a. (cyclohexyl)$CH_2Br \xrightarrow[Et_2O]{Mg}$ (cyclohexyl)$CH_2MgBr \xrightarrow[2.\ H^+]{1.\ CO_2}$ (cyclohexyl)CH_2COOH

or

(cyclohexyl)$CH_2Br \xrightarrow{^-C{\equiv}N}$ (cyclohexyl)$CH_2C{\equiv}N \xrightarrow[H_2O,\ \Delta]{H^+}$ (cyclohexyl)CH_2COOH

b. (cyclohexyl)$CH_2Br \xrightarrow[Et_2O]{Mg}$ (cyclohexyl)$CH_2MgBr \xrightarrow[2.\ H^+]{1.\ CH_3CHO}$ (cyclohexyl)$CH_2CH(OH)CH_3 \xrightarrow[H_2SO_4]{KMnO_4}$ (cyclohexyl)CH_2COCH_3

c. (cyclohexyl)$CH_2Br \xrightarrow{(CH_3)_3CO^-}$ methylenecyclohexane $\xrightarrow[2.\ Zn,\ H_2O]{1.\ O_3}$ cyclohexanone

A bulky base is used to encourage elimination.

d. (cyclohexyl)$CH_2Br \xrightarrow{HO^-}$ (cyclohexyl)$CH_2OH \xrightarrow[H_2SO_4]{Na_2Cr_2O_7}$ (cyclohexyl)$COOH$

29. **a.** oxidized **b.** reduced **c.** reduced **d.** oxidized **e.** oxidized **f.** oxidized **g.** reduced

30.

a. $CH_3CH_2CH_2CH_2\overset{O}{\overset{\|}{C}}OH$

b. $C_6H_5\overset{O}{\overset{\|}{C}}O^-$ + CO_2

c. $CH_3CH_2CH_2CH_2OH$

d. $CH_3CH_2CH_2CH_2OH$

e. $CH_3CH_2\overset{O}{\overset{\|}{C}}H$

f. $CH_3CH_2CH_2CH_2NHCH_3$

g. $C_6H_5\overset{O}{\overset{\|}{C}}OH$

h. $C_6H_5CH_2OH$ + $CH_3\overset{CH_3}{\overset{|}{C}}HOH$

i. (H₃C, H / O / H, CH₃ epoxide) + (H, H₃C / O / CH₃, H epoxide)

j. $C_6H_5CH_2OH$

k. $CH_3CH_2CH_2(H)C{=}C(H)CH_3$

l. $CH_3CH_2CH_2\overset{O}{\overset{\|}{C}}OH$ + $CH_3\overset{O}{\overset{\|}{C}}OH$

m. $C_6H_5CH_2CH_2CH_3$

n. cyclopentanone + $H\overset{O}{\overset{\|}{C}}H$

o. (cyclohexane: OH, H / H, CH_3) (cyclohexane: H, HO / H_3C, H)

p. (cyclohexane: H, H / OH, OH)

q. $^-O\overset{O}{\overset{\|}{C}}CH_2CH_2CH_2CH_2\overset{O}{\overset{\|}{C}}O^-$

r. $HO\overset{O}{\overset{\|}{C}}CH_2\overset{O}{\overset{\|}{C}}OH$

31.

a. $CH_3CH_2CH_2\overset{O}{\overset{\|}{C}}H \xrightarrow{H_2CrO_4} CH_3CH_2CH_2\overset{O}{\overset{\|}{C}}OH$

b. $CH_3CH_2CH_2CH_2OH \xrightarrow{H_2CrO_4} CH_3CH_2CH_2\overset{O}{\overset{\|}{C}}OH$

c. $CH_3CH_2CH_2CH_2Br \xrightarrow{HO^-} CH_3CH_2CH_2CH_2OH \xrightarrow{H_2CrO_4} CH_3CH_2CH_2\overset{O}{\overset{\|}{C}}OH$

d. $CH_3CH_2CH{=}CH_2 \xrightarrow[2.\ HO^-,\ H_2O_2]{1.\ BH_3} CH_3CH_2CH_2CH_2OH \xrightarrow{H_2CrO_4} CH_3CH_2CH_2\overset{O}{\overset{\|}{C}}OH$

32.

a. $CH_3CH_2CH_2CH{=}\underset{CH_3}{\underset{|}{C}}CH_3$

b. 1-methyl-2-ethylcyclohexene (CH_3, CH_2CH_3 on the ring double bond)

c. dimethylcycloheptadiene structure

or

dimethylcycloheptadiene structure

d. cyclohexylidene $=CHCH_2CH_3$

e. $C_6H_5{-}\underset{CH_3}{\underset{|}{C}}{=}CH_2$

f. bicyclic diene (octahydronaphthalene) structure

33.

a. $CH_3CH_2CH{=}CH_2 \xrightarrow[2.\ H_2O_2,\ HO^-]{1.\ BH_3} CH_3CH_2CH_2CH_2OH$

$CH_3CH_2CH_2CH_2OH \underset{}{\overset{H_2SO_4,\ \Delta}{\rightleftharpoons}} CH_3CH{=}CHCH_3 + H_2O$

$CH_3CH{=}CHCH_3 \xrightarrow[H_2SO_4]{KMnO_4} 2\ CH_3\overset{O}{\overset{\|}{C}}OH$

b. $CH_3CH_2Br \xrightarrow[Et_2O]{Mg} CH_3CH_2MgBr \xrightarrow[2.\ H_3O^+]{1.\ \text{ethylene oxide}} CH_3CH_2CH_2CH_2OH \xrightarrow{PCC} CH_3CH_2CH_2\overset{O}{\overset{\|}{C}}H$

c. cyclohexane $\xrightarrow[h\nu]{Br_2}$ bromocyclohexane $\xrightarrow{HO^-}$ cyclohexanol (substitution product) + cyclohexene (elimination product)

cyclohexene $\xrightarrow{KMnO_4,\ H_2SO_4}$ cyclohexanone + $HOCCH_2CH_2CH_2CH_2COH$ (with C=O at both terminal carbons)

34.

a. $HC\equiv CCH_2CH_3 \xrightarrow[2.\ CH_3Br]{1.\ ^-NH_2} CH_3C\equiv CCH_2CH_3 \xrightarrow[\text{Lindlar's catalyst}]{H_2}$ cis alkene (H, H on same side; CH_3, CH_2CH_3)

$\xrightarrow{RCOOH \text{ (peroxyacid, } RC(=O)OOH)}$ cis epoxide (H, H; CH_3, CH_2CH_3)

b. $HC\equiv CCH_2CH_3 \xrightarrow[2.\ CH_3Br]{1.\ ^-NH_2} CH_3C\equiv CCH_2CH_3 \xrightarrow[NH_3]{Na}$ trans alkene (H, CH_2CH_3; CH_3, H)

$\xrightarrow{RCOOH}$ trans epoxide (H, CH_2CH_3; CH_3, H)

35.

a. 1. $CH_3\overset{O}{\overset{\|}{C}}CH_2CH_2CH_2CH_2CH_2\overset{O}{\overset{\|}{C}}OH$

2. $HO\overset{O}{\overset{\|}{C}}\underset{CH_3}{\underset{|}{C}H}CH_2CH_2CH_2CH_2\overset{O}{\overset{\|}{C}}OH$

3. $CO_2 + HO\overset{O}{C}\underset{O}{C}CH_2CH_2CH_2CH_2\overset{O}{C}OH$

4. $CO_2 + HO\overset{O}{C}\underset{O}{C}CH_2\overset{O}{C}OH + HO\overset{O}{C}CH_2\overset{O}{C}OH$

5. $CO_2 + HO\overset{O}{C}CH_2\overset{O}{C}CH_2\overset{O}{C}OH + HO\overset{O}{C}-\overset{O}{C}OH$

b.

36.

a. $\xrightarrow[2.\ H_2O]{1.\ KMnO_4,\ HO^-,\ cold}$

b. $\xrightarrow{R\overset{O}{C}OOH}$ $\xrightarrow[H_2O]{H^+}$

c. $\xrightarrow{R\overset{O}{C}OOH}$ $\xrightarrow[2.\ H^+]{1.\ CH_3MgBr}$

d. $\xrightarrow[2.\ Zn,\ H_2O]{1.\ O_3}$

e. $\xrightarrow[2.\ Zn,\ H_2O]{1.\ O_3}$ $H\overset{O}{C}CH_2CH_2CH_2CH_2\overset{O}{C}H$ $\xrightarrow[2.\ H^+,\ H_2O]{1.\ NaBH_4}$

f. $\xrightarrow[H_2O]{H^+}$ OH $\xrightarrow[H_2SO_4]{Na_2Cr_2O_7}$ O $\xrightarrow{R\overset{O}{C}OOH}$

g. $\xrightarrow[h\nu]{Br_2}$ Br $\xrightarrow{HO^-}$ OH $\xrightarrow[H_2SO_4]{Na_2Cr_2O_7}$

h. cyclohexene $\xrightarrow[\Delta]{NBS}$ 3-bromocyclohexene (Br) $\xrightarrow{HO^-}$ 1,3-cyclohexadiene $\xrightarrow[2.\ H_2O_2]{1.\ O_3}$

37. Because the compound is produced as a result of ozonolysis of an alkene under oxidizing conditions, it must be a ketone or a carboxylic acid. Since there is no signal for an OH group, the compound must be a ketone. The NMR spectrum shows two signals with splitting that is characteristic of an ethyl group. Therefore, the compound produced as a result of ozonolysis must be 3-pentanone. The alkene that underwent ozonolysis, therefore, must be 3,4-diethyl-3-hexene.

$$(CH_3CH_2)_2C{=}C(CH_2CH_3)_2$$

3,4-diethyl-3-hexene

38.

benzene $\xrightarrow[AlCl_3]{CH_3\overset{O}{\overset{\|}{C}}Cl}$ **A** ($C_6H_5\overset{O}{\overset{\|}{C}}CH_3$) $\xrightarrow{R\overset{O}{\overset{\|}{C}}OOH}$ **B** ($C_6H_5O\overset{O}{\overset{\|}{C}}CH_3$) $\xrightarrow[H_2O]{H^+}$ **C** ($CH_3\overset{O}{\overset{\|}{C}}OH$) + **D** (phenol, OH)

C $\xrightarrow{SOCl_2}$ **G** ($CH_3\overset{O}{\overset{\|}{C}}Cl$) $\xrightarrow{NH_3}$ **H** ($CH_3\overset{O}{\overset{\|}{C}}NH_2$) $\xrightarrow[2.\ H_2O]{1.\ LiAlH_4}$ **I** ($CH_3CH_2NH_2$)

D $\xrightarrow{Br_2,\ FeBr_3}$ **E** (4-bromophenol: OH, Br) + **F** (2-bromophenol: OH, Br)

I $\xrightarrow[K_2CO_3]{CH_3I\ \text{excess}}$ **J** ($CH_3CH_2\overset{+}{N}(CH_3)_3\ I^-$) $\xrightarrow{Ag_2O}$ **K** ($CH_3CH_2\overset{+}{N}(CH_3)_3\ HO^-$) $\xrightarrow{\Delta}$ **L** ($CH_3N(CH_3)_2$) + **M** ($CH_2{=}CH_2$)

M $\xrightarrow{1.\ O_3\quad 2.\ H_2O_2}$ **N** (CO_2)

39. The rate-determining step in the chromic acid oxidation of an alcohol is the E2 elimination reaction of the chromate ester. 2-Propanol is oxidized more rapidly than 2-deuterio-2-propanol because it is easier to break the carbon-hydrogen bond than the carbon-deuterium bond.

$CH_3C(CH_3)(H)$–O–CrO_2OH with $H_2\ddot{O}$:

$CH_3C(CH_3)(D)$–O–CrO_2OH with $H_2\ddot{O}$:

40.

a. CH_3CH_2CHO $\xrightarrow{H_2CrO_4}$ CH_3CH_2COOH $\xrightarrow{SOCl_2}$ CH_3CH_2COCl

CH_3CH_2CHO $\xrightarrow{1.\ NaBH_4\quad 2.\ H_2O}$ $CH_3CH_2CH_2OH$

CH_3CH_2COCl + $CH_3CH_2CH_2OH$ → $CH_3CH_2COCH_2CH_2CH_3$ (ester, C=O)

b. $CH_3CH_2CH_2CH_2OH$ $\xrightarrow{PCC}$ $CH_3CH_2CH_2CHO$ $\xrightarrow{1.\ CH_3CH_2MgBr\quad 2.\ H_2O}$ $CH_3CH_2CH_2CH(OH)CH_2CH_3$ $\xrightarrow{H_2CrO_4}$ $CH_3CH_2CH_2C(=O)CH_2CH_3$

c. cyclohexanol (OH) $\xrightarrow{KMnO_4,\ H_2SO_4}$ cyclohexanone (O) $\xrightarrow{CH_3NH_2}$ cyclohexylidene NCH_3 $\xrightarrow{H_2,\ Pt}$ cyclohexyl $NHCH_3$

or

cyclohexanol (OH) $\xrightarrow{PBr_3}$ cyclohexyl Br $\xrightarrow{CH_3NH_2}$ cyclohexyl $NHCH_3$

d. cyclohexanone $\xrightarrow[2.\ H_2O]{1.\ NaBH_4}$ cyclohexanol $\xrightarrow[\Delta]{H_2SO_4}$ cyclohexene $\xrightarrow[2.\ H_2O_2]{1.\ O_3}$ $HO\overset{O}{\overset{\|}{C}}CH_2CH_2CH_2CH_2\overset{O}{\overset{\|}{C}}OH$

e. cyclohexanone $\xrightarrow[2.\ H^+]{1.\ CH_3MgBr}$ 1-methylcyclohexanol (CH_3, OH) $\xrightarrow[\Delta]{H_2SO_4}$ 1-methylcyclohexene (CH_3) $\xrightarrow[2.\ H_2O_2]{1.\ O_3}$ $HO\overset{O}{\overset{\|}{C}}CH_2CH_2CH_2CH_2\overset{O}{\overset{\|}{C}}CH_3$

41.

$$CH_3CH_2\underset{|\atop CH_3}{C}{=}CH_2$$

42. A cyclic intermediate is formed when HIO_4 cleaves a 1,2-diol. It is easier to form a cyclic intermediate if the two OH groups are on the same side of the molecule. Therefore, **A** is cleaved more easily because it is a cis diol, while **B** is a trans diol.

43.

a. cyclohexyl–$C{\equiv}CH$ $\xrightarrow[2.\ H_2O]{1.\ O_3}$ cyclohexyl–$\overset{O}{\overset{\|}{C}}OH$ + CO_2

b. cyclohexyl–$C{\equiv}CH$ $\xrightarrow[2.\ HO^-,\ H_2O_2]{1.\ \text{disiamylborane}}$ cyclohexyl–$CH_2\overset{O}{\overset{\|}{C}}H$ $\xrightarrow[H_2SO_4]{Na_2Cr_2O_7}$ cyclohexyl–$CH_2\overset{O}{\overset{\|}{C}}OH$

44. The compound is either **4-octyne** or **2,5-dimethyl-3-hexyne.**

$CH_3CH_2CH_2C{\equiv}CCH_2CH_2CH_3$ — 4-octyne

$CH_3\overset{CH_3}{\overset{|}{C}}HC{\equiv}C\overset{CH_3}{\overset{|}{C}}HCH_3$ — 2,5-dimethyl-3-hexyne

Because oxidative cleavage formed only one four-carbon carboxylic acid, you know the compound is either a symmetrical eight-carbon alkene or a symmetrical eight-carbon alkyne. Therefore, the compound has a molecular formula of C_8H_{14} or C_8H_{16} and, consequently, a molecular weight of 110 or 112.

0.5 g of the hydrocarbon was hydrogenated; this corresponds to 0.0045 mol (0.5/112 = 0.0045, 0.5/110 = 0.0045). The key here is to remember from general chemistry that 1 mol of a gas has a volume of 22.4 liters at standard temperature and pressure. Knowing this, it can be calculated that 0.0045 mol of H_2 = 100 mL H_2. Since about 200 mL of H_2 were consumed, the compound must have two π bonds. Thus, the unknown compound is an alkyne.

45. **a.** Only **4.** was successful.

b. **1.** $CH_3CH_2\overset{O}{\overset{\|}{C}}CH_3 + CH_3\overset{O}{\overset{\|}{C}}OH$

2. no reaction

3. 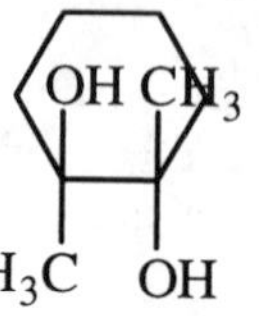 OH CH$_3$ / H$_3$C OH (cyclohexane ring with OH and CH$_3$ on each of two adjacent carbons)

c. **1.** $\xrightarrow[2.\ H_2O]{1.\ KMnO_4,\ HO^-,\ cold}$

2. $\xrightarrow[2.\ H^+,\ H_2O]{1.\ LiAlH_4}$

4. $\xrightarrow[2.\ H_2O]{1.\ KMnO_4,\ HO^-,\ cold}$

or

$\xrightarrow[2.\ NaHSO_3,\ H_2O]{1.\ OsO_4}$

46. The absorption bands at 1600 cm^{-1}, 1500 cm^{-1}, and > 3000 cm^{-1} in the IR spectrum indicate the compound has a benzene ring. The absorption band 2250 cm^{-1} indicates the compound has a triple bond, and the absence of an absorption band 3200 cm^{-1} tells you that it is not a terminal alkyne.

The two triplets and the multiplet in the ^{1}H NMR spectrum indicates a propyl group.

Compound **A** is 1-phenyl-1-propyne and compound **B** is propylbenzene.

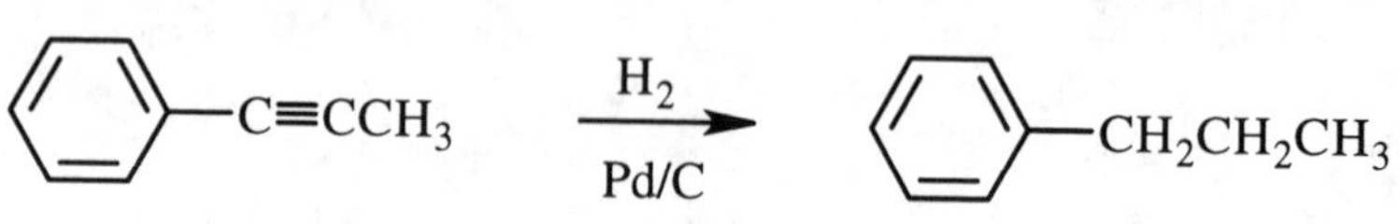

$C_6H_5-C{\equiv}CCH_3 \xrightarrow[Pd/C]{H_2} C_6H_5-CH_2CH_2CH_3$

A **B**

47. Diane has enough information to identify diols **A**, **C**, **E**, and **F**, but not enough information to distinguish between diols **B** and **D**.

Only the 4th compound will form two products upon cleavage with periodic acid. So it must be **F**.

The 1st and 5th compounds will not react with periodic acid. Because the 1st compound is optically inactive and the 5th compound is optically active, the 1st compound must be **C** and the 5th compound must be **E**.

The 3rd compound is the only optically active compound that forms one compound with periodic acid. Therefore, it must be **A**.

Diane could distinguish between **B** and **D** if she analyzed the products obtained from the reaction of **B** and **D** with periodic acid. **B** will form acetaldehyde (a 2-carbon aldehyde), while **D** will form propanal (a 3-carbon aldehyde).

1: CH_3; H—C—OH; CH_2; H—C—OH; CH_3 — **C**

2: CH_3; H—C—OH; H—C—OH; CH_3 — **B or D**

3: CH_3; H—C—OH; HO—C—H; CH_3 — **A**

4: CH_3; H—C—OH; HO—C—H; CH_2CH_3 — **F**

5: CH_3; H—C—OH; CH_2; HO—C—H; CH_3 — **E**

6: CH_2CH_3; H—C—OH; H—C—OH; CH_2CH_3 — **B or D**

48.

$$CH_2{=}CHCH_2OH \xrightarrow[Pt]{H_2} CH_3CH_2CH_2OH \xrightarrow[H_2SO_4]{KMnO_4} CH_3CH_2\overset{O}{\overset{\|}{C}}OH$$

$$CH_3CH_2\overset{O}{\overset{\|}{C}}OH \xrightarrow{SOCl_2} CH_3CH_2\overset{O}{\overset{\|}{C}}Cl$$

$$CH_3CH_2\overset{O}{\overset{\|}{C}}Cl \xrightarrow{CH_3CH_2CH_2OH} CH_3CH_2\overset{O}{\overset{\|}{C}}OCH_2CH_2CH_3$$

(The $CH_3CH_2CH_2OH$ used in the last step comes from the H_2/Pt reduction.)

49.

$$\underset{\textbf{A}}{CH_3\overset{OH}{\overset{|}{C}}H\underset{CH_3}{\underset{|}{C}}HCH_3} \quad \underset{\textbf{B}}{CH_3\overset{O}{\overset{\|}{C}}\underset{CH_3}{\underset{|}{C}}HCH_3} \quad \underset{\textbf{C}}{CH_2{=}\underset{CH_3}{\underset{|}{C}}CH_2CH_3} \quad \underset{\textbf{D}}{CH_3\underset{CH_3}{\underset{|}{C}}{=}CHCH_3}$$

$$\underset{\textbf{E}}{CH_3\overset{O}{\overset{\|}{C}}CH_2CH_3} \quad \underset{\textbf{F}}{CH_3\overset{O}{\overset{\|}{C}}CH_3} \quad \underset{\textbf{G}}{CH_3\overset{O}{\overset{\|}{C}}OH}$$

50.

H_3C — cyclopropene — CH_3

51. **a.** (2*R*,3*S*)-Tartaric acid is a meso compound. It will be formed by syn addition to the cis isomer. Syn addition of two OH groups can be carried out using a cold, basic solution of potassium permanganate or using osmium tetroxide followed by an aqueous solution of sodium bisulfite.

HOOC(H)C=C(H)COOH (cis) $\xrightarrow{KMnO_4,\ HO^-,\ H_2O}$ COOH / H—C—OH / H—C—OH / COOH

b. (2*R*,3*S*)-Tartaric acid can also be formed by anti addition to the trans isomer. Anti addition of two OH groups can be carried out by first forming an epoxide and then treating the epoxide with an aqueous solution of hydroxide.

HOOC(H)C=C(H)COOH (trans) $\xrightarrow[2.\ HO^-,\ H_2O]{1.\ RCOOH\ (peroxyacid,\ R\overset{O}{\overset{\|}{C}}OOH)}$ COOH / H—C—OH / H—C—OH / COOH

c. (2*R*,3*R*)-Tartaric acid and (2*S*,3*S*)-tartaric acid are the threo isomers. They will be formed by anti addition to the cis isomer.

HOOC(H)C=C(H)COOH (cis) $\xrightarrow{KMnO_4,\ HO^-,\ H_2O}$ COOH / H—C—OH / HO—C—H / COOH + COOH / HO—C—H / H—C—OH / COOH

d. The threo isomers can also be formed by syn addition to the trans isomer.

HOOC, H, C=C, H, COOH; 1. RCOOH (O above C); 2. HO^-, H_2O → COOH / H—OH / HO—H / COOH + COOH / HO—H / H—OH / COOH

52.

Methylcyclohexane, CH_3 —Br_2, hv→ **A** (CH_3, Br) —HO^-→ **B** (CH_3) —Br_2→ **C** (CH_3, Br, Br) + **D** (CH_3, Br, Br)

B —1. BH_3/THF, 2. H_2O_2, HO^-→ **E** (CH_3, OH)

B —1. O_3, 2. Zn, H_2O→ **I** $CH_3\overset{O}{\overset{\|}{C}}CH_2CH_2CH_2CH_2\overset{O}{\overset{\|}{C}}H$

B —H^+, H_2O→ **M** (CH_3, OH) —HBr→ **N** (CH_3, Br)

N —1. Mg, Et_2O, 2. ethylene oxide→ **O** (CH_3, CH_2CH_2OH)

E —H_2CrO_4→ **F** (CH_3, O)

F —CH_3CH_2OH excess, H^+→ **G** (CH_3, OCH_2CH_3, OCH_2CH_3)

F —$CH_3CH_2NH_2$→ **H** (CH_3, NCH_2CH_3)

I —H_2CrO_4→ **J** $CH_3\overset{O}{\overset{\|}{C}}CH_2CH_2CH_2CH_2\overset{O}{\overset{\|}{C}}OH$

J —$SOCl_2$→ **K** $CH_3\overset{O}{\overset{\|}{C}}CH_2CH_2CH_2CH_2\overset{O}{\overset{\|}{C}}Cl$

K —CH_3OH→ **L** $CH_3\overset{O}{\overset{\|}{C}}CH_2CH_2CH_2CH_2\overset{O}{\overset{\|}{C}}OCH_3$

53.

a. $CH_3CH(CH_3)CH_3 \xrightarrow[h\nu]{Br_2} CH_3CBr(CH_3)CH_3 \xrightarrow{HO^-} CH_3C(CH_3){=}CH_2 \xrightarrow[2.\ H_2O_2]{1.\ O_3} CH_3\overset{O}{\overset{\|}{C}}CH_3$

b. Note that 2-bromopropane (obtained from the first step) is divided into two portions; one portion is used to form propene, and the other is used to form the Grignard reagent.

$CH_3CH_2CH_3 \xrightarrow[h\nu]{Br_2} CH_3CHBrCH_3 \xrightarrow{HO^-} CH_3CH{=}CH_2$

$CH_3CH{=}CH_2 \xrightarrow[2.\ Zn,\ H_2O]{1.\ O_3} CH_3\overset{O}{\overset{\|}{C}}H$

$CH_3CHBrCH_3 \xrightarrow{Mg,\ Et_2O} CH_3CH(MgBr)CH_3$

$CH_3CH(MgBr)CH_3 + CH_3\overset{O}{\overset{\|}{C}}H \longrightarrow CH_3CH(O^-)CH(CH_3)CH_3 \xrightarrow{H_2SO_4,\ \Delta} CH_3CH{=}C(CH_3)CH_3$

c. $CH_3CH_2CH_3 \xrightarrow[h\nu]{Br_2} CH_3CHBrCH_3 \xrightarrow[Et_2O]{Mg} CH_3CH(MgBr)CH_3 \xrightarrow[2.\ H^+]{1.\ CH_3\overset{O}{\overset{\|}{C}}H} CH_3CH(OH)CH(CH_3)CH_3 \xrightarrow[H_2SO_4]{Na_2Cr_2O_7} CH_3CH(CH_3)\overset{O}{\overset{\|}{C}}CH_3$

d. $CH_3CH_3 \xrightarrow[h\nu]{Br_2} CH_3CH_2Br \xrightarrow[Et_2O]{Mg} CH_3CH_2MgBr \xrightarrow[2.\ H^+]{1.\ \text{ethylene oxide}} CH_3CH_2CH_2CH_2OH \xrightarrow{PBr_3} CH_3CH_2CH_2CH_2Br \xrightarrow{HO^-} CH_3CH_2CH{=}CH_2 \xrightarrow[2.\ Zn,\ H_2O]{1.\ O_3} CH_3CH_2\overset{O}{\overset{\|}{C}}H + H\overset{O}{\overset{\|}{C}}H$

The ethylene oxide needed for the reaction can be obtained from some of the bromoethane prepared in the first step.

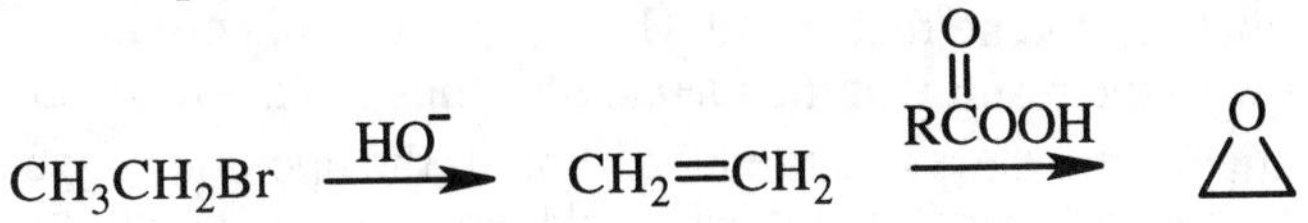

$$CH_3CH_2Br \xrightarrow{HO^-} CH_2{=}CH_2 \xrightarrow{RCOOH\ (\text{C=O})} \text{ethylene oxide}$$

54.

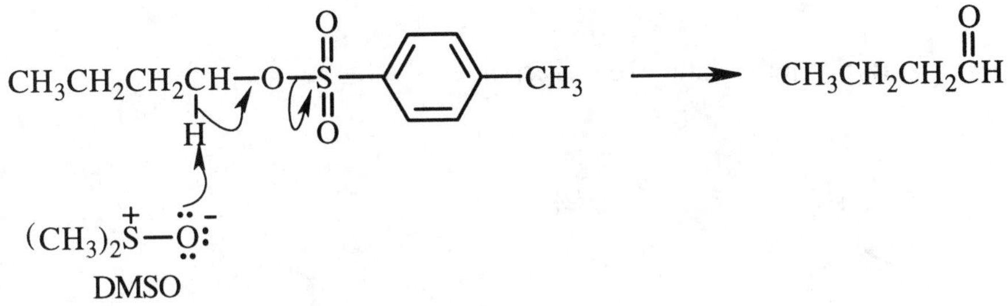

$$CH_3CH_2CH_2CH(H){-}O{-}SO_2{-}C_6H_4{-}CH_3 \longrightarrow CH_3CH_2CH_2\overset{O}{\overset{\|}{C}}H$$

$(CH_3)_2\overset{+}{S}{-}\ddot{O}:^-$

DMSO

55.

a.

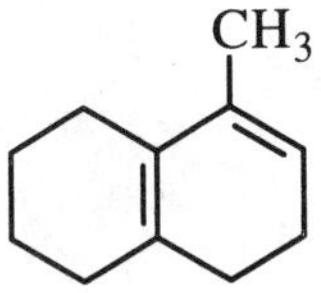

CH_3

b.

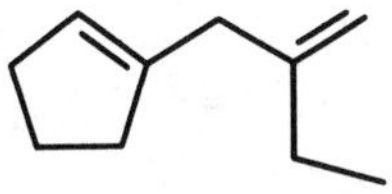

56.

$$Cl{-}\overset{O}{\overset{\|}{C}}{-}\overset{O}{\overset{\|}{C}}{-}Cl + {}^-{:}\ddot{O}{-}\overset{+}{S}(CH_3)_2 \longrightarrow Cl{-}C(O^-)(O{-}\overset{+}{S}(CH_3)_2){-}\overset{O}{\overset{\|}{C}}{-}Cl \longrightarrow O{=}C(O{-}\overset{+}{S}(CH_3)_2){-}\overset{O}{\overset{\|}{C}}{-}Cl + :\ddot{Cl}:^-$$

$$\downarrow$$

$$CH_3{-}\overset{+}{S}(Cl){-}CH_3 + O{=}C{=}O + C{\equiv}O + Cl^-$$

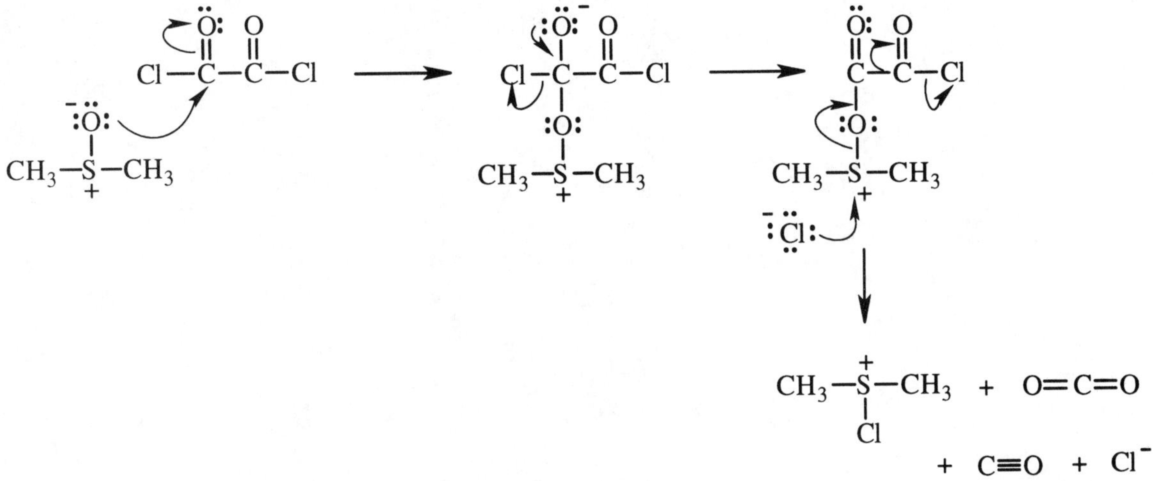

57. The fact that the HO substituent ends up on the least substituted sp^2 carbon of the double bond indicates that oxygen is the first species to add to the double bond. That H_2O_2 is one of the reagents suggests that the alkene reacts with H_2O_2, forming an epoxide. Br^- attacks the epoxide from the backside with the result that the OH and Br substituents form an anti addition product. The fact that Br^- attacks the most substituted carbon of the epoxide suggests that there is an acid catalyst at the surface of the enzyme that causes the epoxide to open to give the more stable partial carbocation.

58. Three structures (**A**, **B**, **C**) fit the data given. Terpineol is actually **C**.

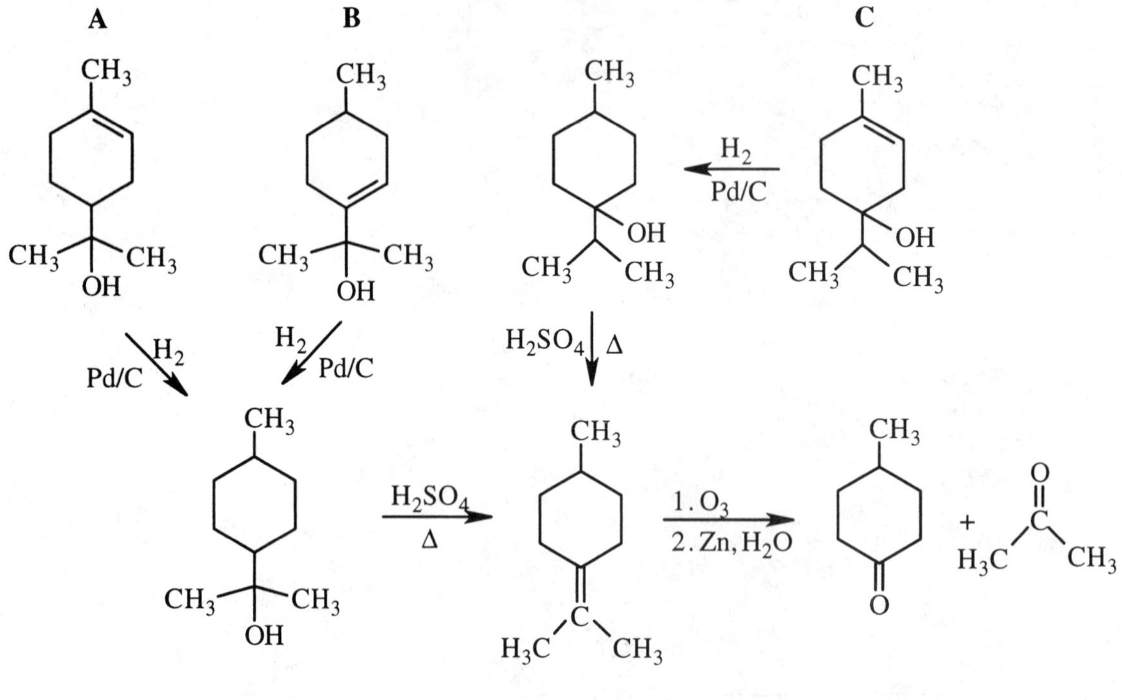

A **B** **C**

CH_3 CH_3 CH_3 CH_3

H_2 / Pd/C

OH OH

CH_3 CH_3 CH_3 CH_3 CH_3 CH_3 CH_3 CH_3

OH OH

H_2 / Pd/C H_2 / Pd/C H_2SO_4 / Δ

CH_3 CH_3 CH_3

H_2SO_4 / Δ

1. O_3 / 2. Zn, H_2O

O

+ C

H_3C CH_3

CH_3 CH_3

OH O

C

H_3C CH_3

Chapter 18 Practice Test

1. Indicate whether each of the following reactions is an oxidation, a reduction, or neither.

a. $CH_3CH_2C(=O)CH_3 \longrightarrow CH_3CH_2CH(OH)CH_3$

b. $CH_3CH_2C(=O)H \longrightarrow CH_3CH_2C(=O)OH$

2. Give the product of each of the following reactions, or if no reaction will occur, so state.

a. $CH_3CH_2CH_2CH_2OH \xrightarrow[H_2SO_4]{Na_2Cr_2O_7}$

b. $CH_3CH_2CH_2C(=O)OCH_3 \xrightarrow[Pt]{H_2}$

c. $CH_3CH_2C(=O)NHCH_2CH_3 \xrightarrow[2.\ H^+,\ H_2O]{1.\ LiAlH_4}$

d. 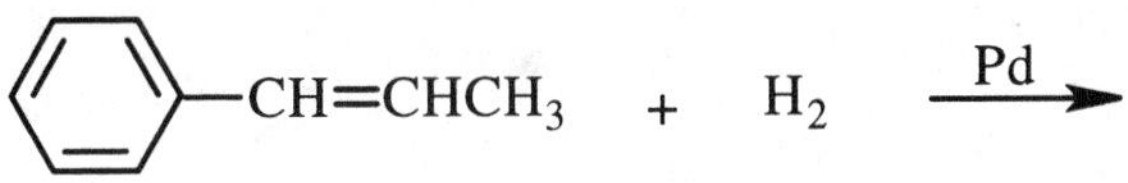 $C_6H_5{-}CH{=}CHCH_3 + H_2 \xrightarrow{Pd}$

e. $CH_3CH_2CH_2CH_2C(=O)H + RC(=O)OOH \longrightarrow$

f. 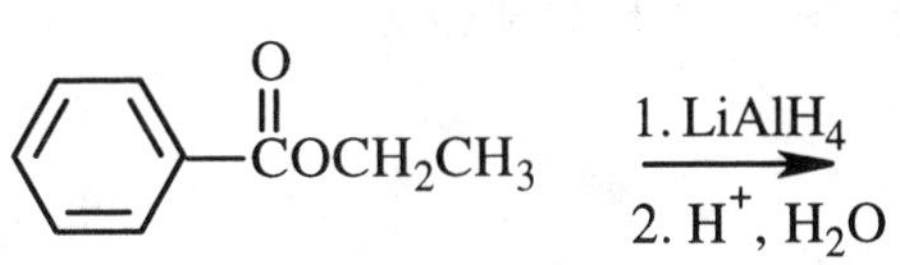 $C_6H_5{-}C(=O)OCH_2CH_3 \xrightarrow[2.\ H^+,\ H_2O]{1.\ LiAlH_4}$

g. $CH_3CH_2CH_2C(=O)CH_2CH_3 \xrightarrow[2.\ H^+,\ H_2O]{1.\ NaBH_4}$

h. $CH_3CH_2CH_2C(CH_3){=}CHCH_2CH_3 \xrightarrow[2.\ H_2O_2]{1.\ O_3}$

3. Which of the following reagents **cannot** be used to convert an aldehyde into an alcohol?

H_2/Pt	1. $NaBH_4$ 2. H^+, H_2O	1. $LiAlH_4$ 2. H^+, H_2O	1. Ag_2O/NH_3 2. H^+, H_2O	H_2/Raney Nickel

4. Two alkenes, when treated with ozone and then with $(CH_3)_2S$, both form only the ketone shown below. Identify the alkenes.

$$CH_3CH_2\overset{O}{\overset{\|}{C}}CH_3$$

5. Indicate whether each of the following is true or false.

a. $NaBH_4$ is a weaker reducing agent than $LiAlH_4$. T F

b. Esters are easier to reduce than ketones. T F

c. In an oxidation-reduction reaction the oxidizing agent is oxidized. T F

d. Ketones are reduced to primary alcohols. T F

e. Aldehydes are oxidized to carboxylic acids. T F

f. Acyl halides are oxidized to aldehydes. T F

g. Alkenes cannot be reduced with $NaBH_4$. T F

6. Describe how the following compound can be prepared from the given starting material.

$$CH_3CH_2CH_2CH_3 \longrightarrow CH_3CH_2\overset{O}{\overset{\|}{C}}CH_3$$

7. Which pair of reagents could be used to carry out the following reaction?

$$CH_3\overset{CH_3}{\overset{|}{C}}{=}CHCH_3 \longrightarrow CH_3\overset{O}{\overset{\|}{C}}CH_3 + CH_3\overset{O}{\overset{\|}{C}}H$$

a. 1. O_3 2. Zn, H_2O and 1. O_3 2. H_2O_2

b. 1. OsO_4 2. $NaHSO_3$ and $KMnO_4/H^+$

c. 1. O_3 2. Zn, H_2O and 1. O_3 2. $(CH_3)_2S$

CHAPTER 19
Carbonyl Compounds III: Reactions at the α-Carbon

Important Terms

acetoacetic ester synthesis synthesis of a methyl ketone using ethyl acetoacetate as the starting material.

aldol addition a reaction between two molecules of an aldehyde (or two molecules of a ketone) that connects the α-carbon of one with the carbonyl carbon of the other.

aldol condensation an aldol addition followed by elimination of water.

ambident nucleophile a nucleophile with two nucleophilic sites.

annulation reaction a ring-forming reaction.

α-carbon a carbon adjacent to a carbonyl carbon.

carbon acid a compound that contains a carbon bonded to a relatively acidic hydrogen.

Claisen condensation a reaction between two molecules of an ester that connects the α-carbon of one with the carbonyl carbon of the other and eliminates an alkoxide ion.

condensation reaction a reaction combining two molecules while removing a small molecule (usually water or an alcohol).

crossed aldol addition (mixed aldol addition) an aldol addition in which two different carbonyl compounds are used.

decarboxylation loss of carbon dioxide.

Dieckmann condensation an intramolecular Claisen condensation.

β-diketone a ketone with a second ketone carbonyl group at the β-position.

enolization keto-enol interconversion.

gluconeogenesis the synthesis of D-glucose from pyruvate.

glycolysis the breakdown of D-glucose into two molecules of pyruvate.

haloform reaction the conversion of a methyl ketone to a carboxylic acid and haloform.

Hell-Volhard-Zelinski (HVZ) reaction conversion of a carboxylic acid into an α-bromocarboxylic acid using Br_2 + P.

Hundsdiecker reaction conversion of a carboxylic acid into an alkyl halide by heating a heavy metal salt of the carboxylic acid with bromine or iodine.

α-hydrogen a hydrogen bonded to the carbon adjacent to a carbonyl carbon.

keto-enol tautomerism (keto-enol interconversion) interconversion of keto and enol tautomers.

β-keto ester an ester with a ketone carbonyl group at the β-position.

Kolbe-Schmidt carboxylation reaction reaction of a phenolate ion with carbon dioxide under pressure.

malonic ester synthesis synthesis of a carboxylic acid using diethyl malonate as the starting material.

Michael reaction the addition of an α-carbanion to the β-carbon of an α,β-unsaturated carbonyl compound.

mixed aldol addition (crossed aldol addition) an aldol addition in which two different carbonyl compounds are used.

mixed Claisen condensation a Claisen condensation in which two different esters are used.

Robinson annulation a Michael reaction followed by an intramolecular aldol condensation.

Stork enamine reaction uses an enamine as a nucleophile in a Michael reaction.

α-substitution reaction a reaction that puts a substituent on an α-carbon in place of an α-hydrogen.

tautomers isomers that differ in the location of a double bond and a hydrogen.

Solutions to Problems

1. The electrons left behind when a proton is removed from propene are delocalized—they are shared by two carbon atoms. Therefore, propene is more acidic than an alkane, because the electrons left behind when a proton is removed from an alkane are localized—they belong to a single (carbon) atom.

$$CH_2{=}CHCH_3 \xrightarrow{-H^+} CH_2{=}CH\ddot{C}H_2^- \longleftrightarrow {}^-\ddot{C}H_2CH{=}CH_2$$

$$CH_3CH_2CH_3 \xrightarrow{-H^+} CH_3\ddot{C}HCH_3^-$$

But propene is not as acidic as a carbonyl compound, because the electrons left behind when a proton is removed from an α-carbon are shared by a carbon and an electronegative oxygen atom.

$$RCH_2{-}\overset{\overset{:\ddot{O}}{\|}}{C}{-}R \xrightarrow{-H^+} R\overset{-}{\ddot{C}}H{-}\overset{\overset{:\ddot{O}}{\|}}{C}{-}R \longleftrightarrow RCH{=}\overset{\overset{:\ddot{O}:^-}{|}}{C}{-}R$$

2.

a. $CH_3\overset{O}{\overset{\|}{C}}CH_2C{\equiv}N$
a β-keto nitrile

b. $CH_3O\overset{O}{\overset{\|}{C}}CH_2\overset{O}{\overset{\|}{C}}OCH_3$
a β-diester

3.

a. $CH_3\overset{O}{\overset{\|}{C}}H > HC{\equiv}CH > CH_2{=}CH_2 > CH_3CH_3$

b. $CH_3\overset{O}{\overset{\|}{C}}CH_2\overset{O}{\overset{\|}{C}}CH_3 > CH_3\overset{O}{\overset{\|}{C}}CH_2\overset{O}{\overset{\|}{C}}OCH_3 > CH_3O\overset{O}{\overset{\|}{C}}CH_2\overset{O}{\overset{\|}{C}}OCH_3 > CH_3\overset{O}{\overset{\|}{C}}CH_3$

c. cyclohexanone > δ-valerolactone (O) > N-methyl δ-lactam (NCH_3)

The ketone is the strongest acid because there is no competition for delocalization of the electrons that are left behind when the α-hydrogen is removed. The lactam is the weakest acid because nitrogen is better than oxygen at delocalizing its nonbonding electrons onto the carbonyl oxygen. Therefore, nitrogen is better at competing with the electrons left behind when an α-hydrogen is removed for delocalization onto the carbonyl oxygen.

4. Both the keto and enol tautomers of 2,4-pentanedione can form hydrogen bonds with water. Neither the keto nor the enol tautomer can form hydrogen bonds with hexane. However, the enol tautomer can form an intramolecular hydrogen bond. This intramolecular hydrogen bonding stabilizes the enol tautomer. Thus, the enol tautomer is more stable relative to the keto tautomer in hexane than in water.

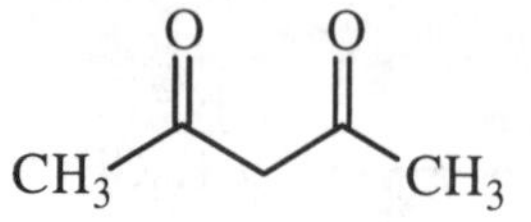

keto tautomer 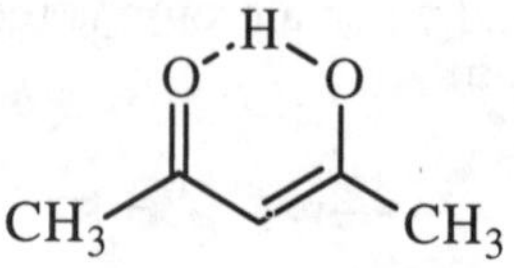 enol tautomer

2,4-pentanedione

5.

a. $CH_3CH{=}C(OH)CH_2CH_3$

b. 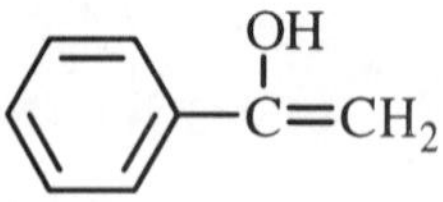 $C_6H_5C(OH){=}CH_2$

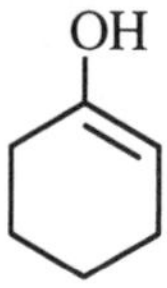

c. 1-cyclohexenol (OH)

d. 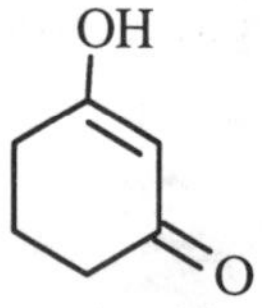 (two enol forms of 1,3-cyclohexanedione) and

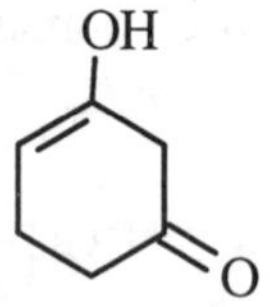

more stable because the double bonds are conjugated

e. $CH_3CH_2C(OH){=}CHC(O)CH_2CH_3$ and $CH_3CH{=}C(OH)CH_2C(O)CH_2CH_3$

more stable because the double bonds are conjugated

f. 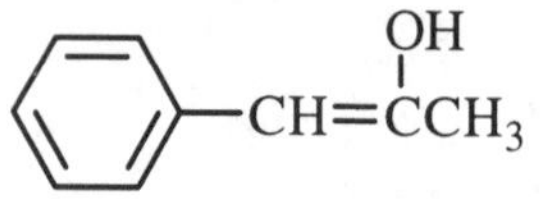 $C_6H_5CH{=}C(OH)CH_3$ and $C_6H_5CH_2C(OH){=}CH_2$

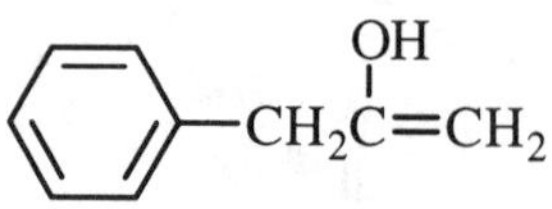

more stable because the double bond is conjugated with the benzene ring

6. The aldehyde hydrogen cannot be removed by ^-OD. The aldehyde hydrogen is not acidic, because the electrons left behind if it were to be removed cannot be delocalized.

7. A bromine-bromine bond is weaker and easier to break than a chlorine-chlorine bond. Because the rates of bromination and chlorination are the same, you know that breaking the bromine-bromine (or chlorine-chlorine) bond takes place after the rate-determining step. Therefore, the rate-determining step must be removal of the proton from the α-carbon of the ketone.

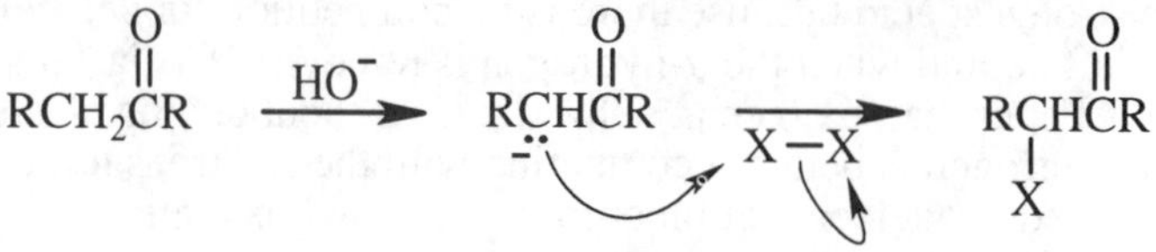

$$RCH_2\overset{O}{\overset{\|}{C}}R \xrightarrow{HO^-} R\ddot{C}^-H\overset{O}{\overset{\|}{C}}R \xrightarrow{X-X} RCH(X)\overset{O}{\overset{\|}{C}}R$$

8.

a. $CH_3CH_2\overset{O}{\overset{\|}{C}}H + Br_2 \xrightarrow[H_2O]{H^+} CH_3\underset{Br}{\underset{|}{C}}H\overset{O}{\overset{\|}{C}}H \xrightarrow{CH_3NHCH_3} CH_3\underset{N(CH_3)_2}{\underset{|}{C}}H\overset{O}{\overset{\|}{C}}H$

b. $CH_3CH_2\overset{O}{\overset{\|}{C}}H + Br_2 \xrightarrow[H_2O]{H^+} CH_3\underset{Br}{\underset{|}{C}}H\overset{O}{\overset{\|}{C}}H \xrightarrow{HO^-} CH_3\underset{OH}{\underset{|}{C}}H\overset{O}{\overset{\|}{C}}H$

c. $\text{cyclohexanone} + Br_2 \xrightarrow[H_2O]{H^+} \text{2-bromocyclohexanone (Br)} \xrightarrow{CH_3O^-} \text{2-methoxycyclohexanone } (OCH_3)$

d. $\text{cyclopentanone} + Br_2 \xrightarrow[H_2O]{H^+} \text{2-bromocyclopentanone (Br)} \xrightarrow{C_6H_5O^-} \text{2-phenoxycyclopentanone } (O\text{–}C_6H_5)$

9.

a. $CH_3CH_2CH_2\overset{O}{\overset{\|}{C}}CH_2CH_2CH_3 \xrightarrow[H^+,\ H_2O]{Br_2} CH_3CH_2\underset{Br}{\underset{|}{C}}H\overset{O}{\overset{\|}{C}}CH_2CH_2CH_3$

$\downarrow$ *tert*-BuO$^-$, *tert*-BuOH

$CH_3CH{=}CH\overset{O}{\overset{\|}{C}}CH_2CH_2CH_3$

b. $\text{1-methylcyclohexyl–}\overset{O}{\overset{\|}{C}}CH_2CH_3 \xrightarrow[H^+,\ H_2O]{Br_2} \text{1-methylcyclohexyl–}\overset{O}{\overset{\|}{C}}\underset{Br}{\underset{|}{C}}HCH_3 \xrightarrow[tert\text{-}BuOH]{tert\text{-}BuO^-} \text{1-methylcyclohexyl–}\overset{O}{\overset{\|}{C}}CH{=}CH_2$

10.

a. cyclohexanone $\xrightarrow[H^+,\ H_2O]{Br_2}$ 2-bromocyclohexanone (Br) $\xrightarrow{^-C\equiv N}$ 2-cyanocyclohexanone (C≡N)

b. cyclohexanone $\xrightarrow[H^+,\ H_2O]{Br_2}$ 2-bromocyclohexanone (Br) $\xrightarrow[tert\text{-}BuOH]{tert\text{-}BuO^-}$ cyclohex-2-enone $\xrightarrow[HCl]{^-C\equiv N}$ 3-cyanocyclohexanone (C≡N)

c. cyclohexanone $\xrightarrow[H^+,\ H_2O]{Br_2}$ 2-bromocyclohexanone (Br) $\xrightarrow[tert\text{-}BuOH]{tert\text{-}BuO^-}$ cyclohex-2-enone $\xrightarrow{CH_3SH}$ 3-(methylthio)cyclohexanone (SCH_3)

d. cyclohexanone $\xrightarrow[H^+,\ H_2O]{Br_2}$ 2-bromocyclohexanone (Br) $\xrightarrow[tert\text{-}BuOH]{tert\text{-}BuO^-}$ cyclohex-2-enone $\xrightarrow[2.\ H^+,\ H_2O]{1.(CH_3)_2CuLi}$ 3-methylcyclohexanone (CH_3)

11.

cyclohexanone $\xrightarrow[D_2O]{^-OD}$ 2,2,6,6-tetradeuteriocyclohexanone (D, D, D, D)

12. Alkylation of an alpha carbon is an S_N2 reaction. S_N2 reactions work best with primary alkyl halides because there is less steric hindrance in a primary alkyl halide than in a secondary alkyl halide. S_N2 reactions don't work at all with tertiary alkyl halides because they are the most sterically hindered of the alkyl halides. Therefore, the substitution reaction cannot compete with the elimination reaction.

13.

a. Cyclohexanone $\xrightarrow[2.\ ICH_2CH{=}CH_2]{1.\ LDA/THF}$ 2-($CH_2CH{=}CH_2$)cyclohexanone

b. $CH_3CH_2\overset{O}{\overset{\|}{C}}CH_2CH_3$ $\xrightarrow[2.\ C_6H_5CH_2Br]{1.\ LDA/THF}$ C_6H_5–$CH_2\underset{CH_3}{\underset{|}{C}H}\overset{O}{\overset{\|}{C}}CH_2CH_3$

14.

a. Cyclohexanone + pyrrolidine (N–H) ⟶ 1-pyrrolidinylcyclohexene (enamine) $\xrightarrow{CH_3CH_2CH_2Br}$ iminium ion (N⁺=C) with 2-$CH_2CH_2CH_3$, Br^-

$\downarrow$ H^+, H_2O

pyrrolidinium ion ($\overset{+}{N}H_2$) Br^- + 2-($CH_2CH_2CH_3$)cyclohexanone

b. Cyclohexanone + pyrrolidine (N–H) ⟶ 1-pyrrolidinylcyclohexene (enamine) $\xrightarrow{CH_3CH_2\overset{O}{\overset{\|}{C}}Cl}$ iminium ion (N⁺=C) with 2-$\overset{O}{\overset{\|}{C}}CH_2CH_3$, Cl^-

$\downarrow$ H^+, H_2O

pyrrolidinium ion ($\overset{+}{N}H_2$) Cl^- + 2-($\overset{O}{\overset{\|}{C}}CH_2CH_3$)cyclohexanone

15.

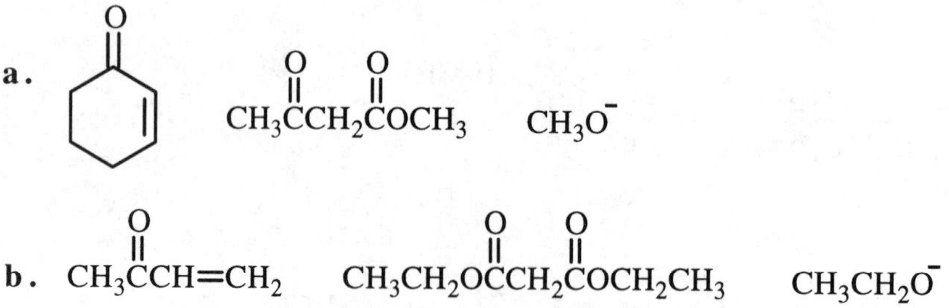

a. (2-cyclohexenone) $CH_3\overset{O}{\overset{\|}{C}}CH_2\overset{O}{\overset{\|}{C}}OCH_3$ CH_3O^-

b. $CH_3\overset{O}{\overset{\|}{C}}CH{=}CH_2$ $CH_3CH_2O\overset{O}{\overset{\|}{C}}CH_2\overset{O}{\overset{\|}{C}}OCH_2CH_3$ $CH_3CH_2O^-$

16.

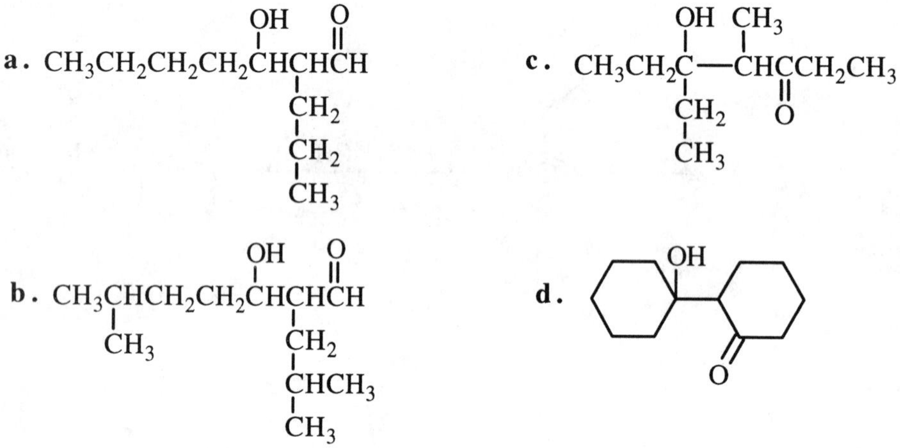

a. $CH_3CH_2CH_2CH_2CH(OH)CH(CH_2CH_2CH_3)CHO$

b. $CH_3CH(CH_3)CH_2CH_2CH(OH)CH(CH_2CH(CH_3)CH_3)CHO$

c. $CH_3CH_2C(OH)(CH_2CH_3){-}CH(CH_3)\overset{O}{\overset{\|}{C}}CH_2CH_3$

d. OH (2-(1-hydroxycyclohexyl)cyclohexanone)

17.

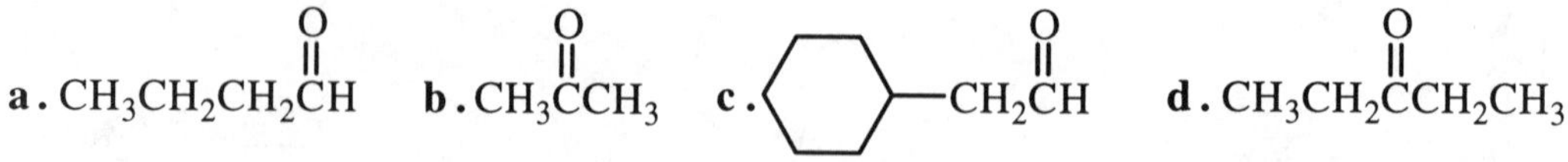

a. $CH_3CH_2CH_2\overset{O}{\overset{\|}{C}}H$ b. $CH_3\overset{O}{\overset{\|}{C}}CH_3$ c. (cyclohexyl)$-CH_2\overset{O}{\overset{\|}{C}}H$ d. $CH_3CH_2\overset{O}{\overset{\|}{C}}CH_2CH_3$

18.

$CH_3CH_2\overset{O}{\overset{\|}{C}}H \underset{-H^+}{\overset{H^+}{\rightleftharpoons}} CH_3CH(H){-}CH{=}\overset{+}{O}H$ ($H_2\ddot{O}:$) $\rightleftharpoons CH_3CH{=}CH\ddot{O}H$

$CH_3CH{=}CH\ddot{O}H + CH_3CH_2CH{=}\overset{+}{O}H \rightarrow$

$CH_3CH_2CH(OH)CH(CH_3)CH{=}\overset{+}{O}H \underset{H^+}{\overset{-H^+}{\rightleftharpoons}} CH_3CH_2CH(OH)CH(CH_3)\overset{O}{\overset{\|}{C}}H$

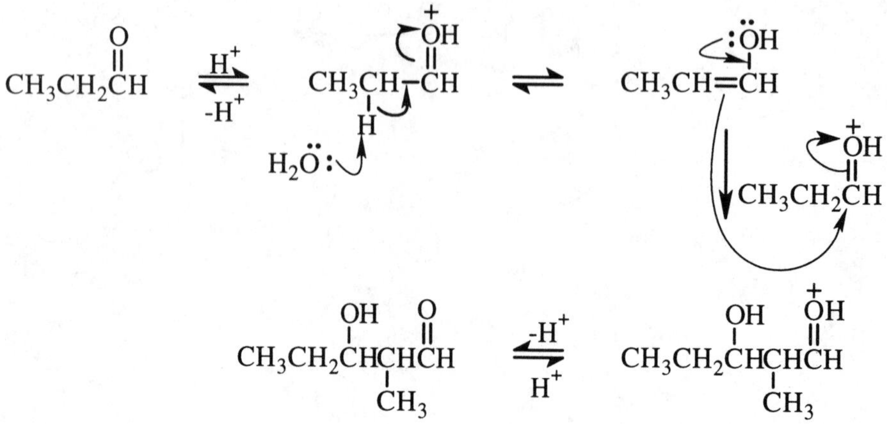

19.

a. Solved in the text.

b. $2\ CH_3CH_2\overset{O}{\overset{\|}{C}}H \xrightleftharpoons{HO^-} CH_3CH_2\overset{OH}{\overset{|}{C}}H\underset{CH_3}{\underset{|}{C}}H\overset{O}{\overset{\|}{C}}H \xrightarrow[\Delta]{H^+} CH_3CH_2CH{=}\underset{CH_3}{\underset{|}{C}}\overset{O}{\overset{\|}{C}}H$

$\xrightarrow{\text{excess } H_2,\ Pt} CH_3CH_2CH_2\underset{CH_3}{\underset{|}{C}}HCH_2OH$

c. $CH_3\overset{O}{\overset{\|}{C}}CH_3 \xrightleftharpoons{HO^-} CH_3\underset{CH_3}{\underset{|}{\overset{OH}{\overset{|}{C}}}}CH_2\overset{O}{\overset{\|}{C}}CH_3 \xrightarrow[\Delta]{H^+} CH_3\underset{CH_3}{\underset{|}{C}}{=}CH\overset{O}{\overset{\|}{C}}CH_3$

$\xrightarrow{H_2,\ Pt} CH_3\underset{CH_3}{\underset{|}{C}}HCH_2\overset{OH}{\overset{|}{C}}HCH_3 + CH_3\underset{CH_3}{\underset{|}{C}}HCH_2\overset{O}{\overset{\|}{C}}CH_3$

$CH_3\underset{CH_3}{\underset{|}{C}}HCH_2\overset{O}{\overset{\|}{C}}CH_3 \xleftarrow[H_2SO_4]{KMnO_4} CH_3\underset{CH_3}{\underset{|}{C}}HCH_2\overset{OH}{\overset{|}{C}}HCH_3$

Because it is easier to reduce a carbon-carbon double bond than a carbonyl group by catalytic hydrogenation, with careful reduction the desired ketone may be obtained directly from the α,β-unsaturated ketone. If, however, both the alkene and carbonyl groups are reduced, the ketone can be obtained by a subsequent oxidation.

20.

a. $CH_3CH_2CH_2\overset{OH}{\overset{|}{C}}H\underset{CH_2CH_3}{\underset{|}{C}}H\overset{O}{\overset{\|}{C}}H$ $\quad CH_3CH_2CH_2CH_2\overset{OH}{\overset{|}{C}}H\underset{CH_2CH_2CH_3}{\underset{|}{C}}H\overset{O}{\overset{\|}{C}}H$ $\quad CH_3CH_2CH_2\overset{OH}{\overset{|}{C}}H\underset{CH_2CH_2CH_3}{\underset{|}{C}}H\overset{O}{\overset{\|}{C}}H$

$CH_3CH_2CH_2CH_2\overset{OH}{\overset{|}{C}}H\underset{CH_2CH_3}{\underset{|}{C}}H\overset{O}{\overset{\|}{C}}H$

b. $CH_3C(OH)(CH_3)CH_2C(=O)CH_3$ $CH_3CH_2C(OH)(CH_2CH_3)—CH(CH_3)C(=O)CH_2CH_3$ $CH_3C(OH)(CH_3)—CH(CH_3)C(=O)CH_2CH_3$

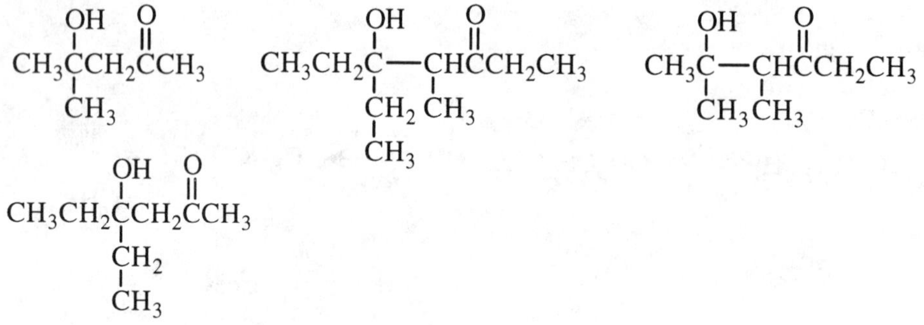

$CH_3CH_2C(OH)(CH_2CH_3)CH_2C(=O)CH_3$

c. 2-(1-hydroxycyclohexyl)cyclohexanone $CH_3CH_2CH(OH)CH(CH_3)CH=O$ (1-hydroxycyclohexyl)$CH(CH_3)CH=O$ $CH_3CH_2CH(OH)$–(2-oxocyclohexyl)

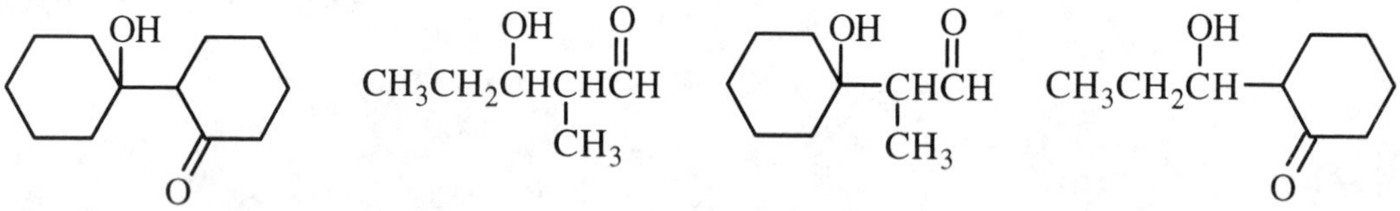

d. $HOCH_2CH(CH_3)CH=O$ $CH_3CH_2CH(OH)CH(CH_3)CH=O$

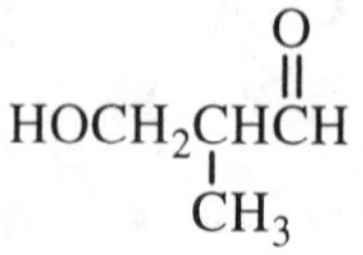

major product 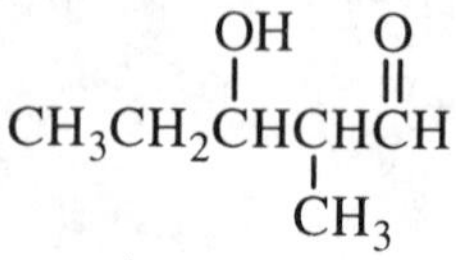 minor product

21.

a. $HC(=O)H + CH_3CH_2C(=O)CH_2CH_3 \xrightarrow{HO^-} CH_3CH_2C(=O)CH(CH_3)CH_2OH$

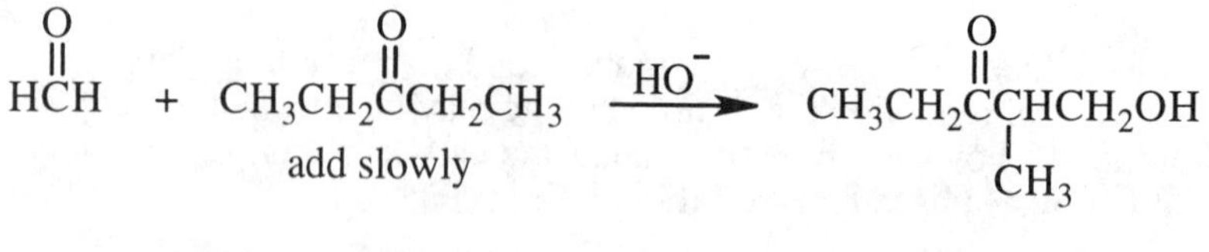

add slowly (under $CH_3CH_2C(=O)CH_2CH_3$)

b. $C_6H_5C(=O)H$ (excess) $+ CH_3C(=O)CH_3$ (add slowly) $\xrightarrow{HO^-} C_6H_5CH=CHC(=O)CH=CHC_6H_5 + H_2O$

c. $C_6H_5C(=O)H$ + cyclohexanone (add slowly) $\xrightarrow{HO^-}$ 2-benzylidenecyclohexanone $+ H_2O$

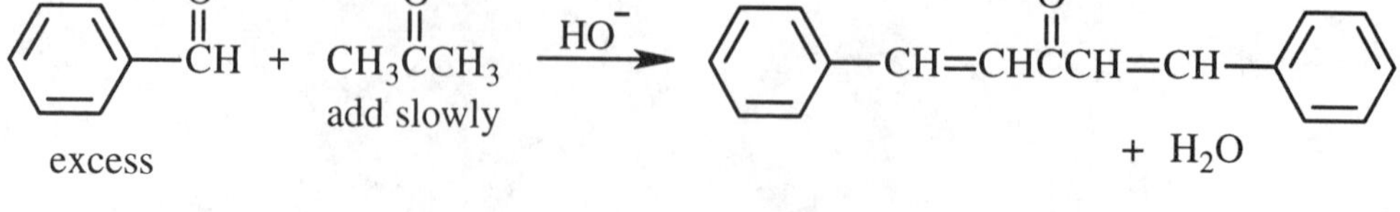

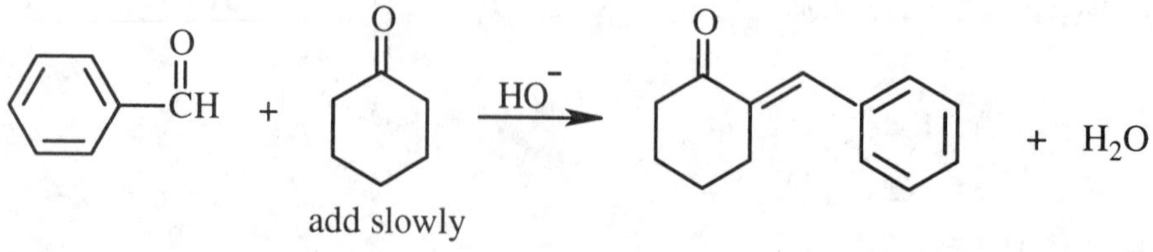

22.

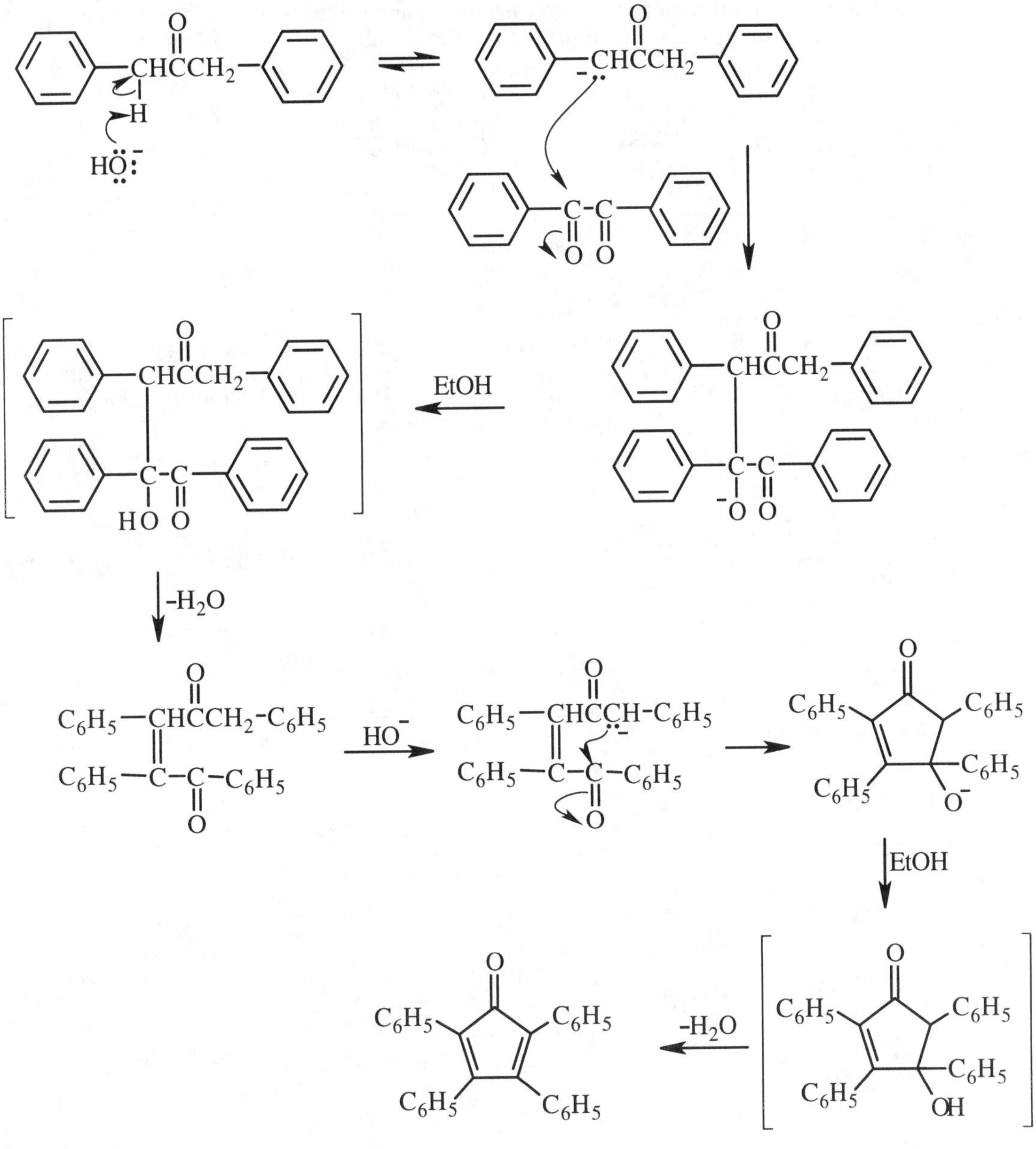

CHCCH$_2$ ⇌ CHCCH$_2$

HÖ:$^-$ H

C-C
O O

EtOH

CHCCH$_2$
C-C
HO O

CHCCH$_2$
C-C
$^-$O O

-H$_2$O

C_6H_5—CHCCH$_2$-C_6H_5
C_6H_5—C-C-C_6H_5
O

HO$^-$ →

C_6H_5—CHCCH-C_6H_5
C_6H_5—C-C-C_6H_5
O

→

O
C_6H_5 C_6H_5
C_6H_5 C_6H_5
O$^-$

EtOH

O
C_6H_5 C_6H_5
C_6H_5 C_6H_5
OH

-H$_2$O

O
C_6H_5 C_6H_5
C_6H_5 C_6H_5

23.

a. 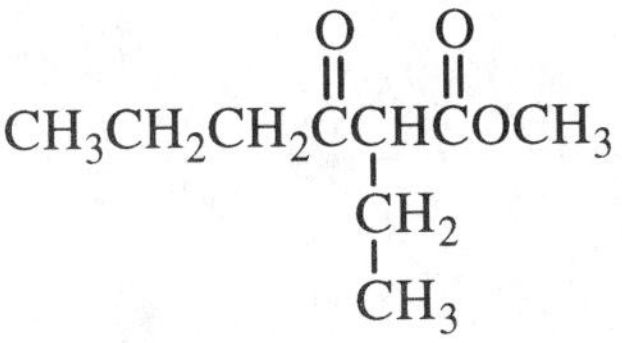 $CH_3CH_2CH_2\overset{O}{\overset{\|}{C}}\underset{CH_2CH_3}{CH}\overset{O}{\overset{\|}{C}}OCH_3$

b. 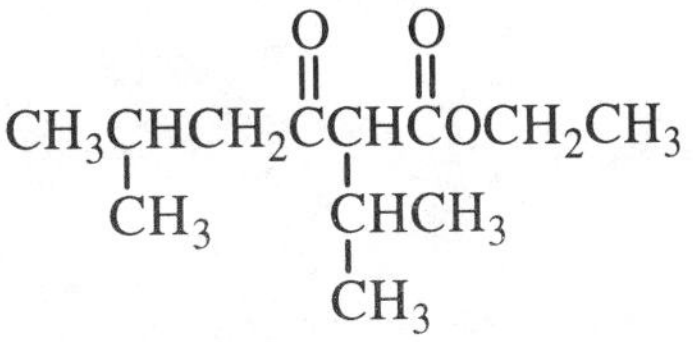 $CH_3\underset{CH_3}{CH}CH_2\overset{O}{\overset{\|}{C}}\underset{CH(CH_3)CH_3}{CH}\overset{O}{\overset{\|}{C}}OCH_2CH_3$

24. An α-hydrogen has to be removed from the condensation product in order to drive the reaction toward completion. If the original ester had only one α-hydrogen, the condensation product would not have an α-hydrogen, so the reaction cannot be driven to completion.

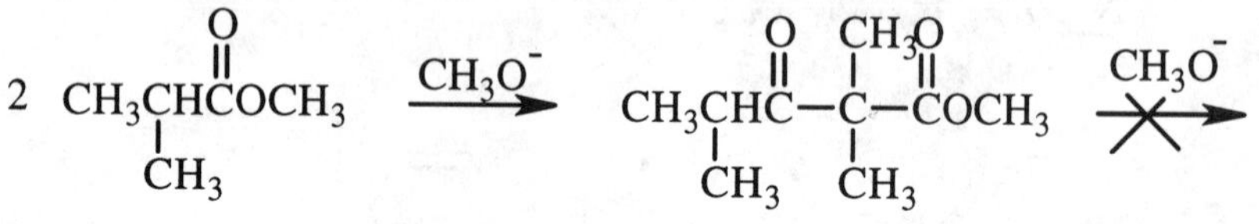

$$2\ CH_3CH(CH_3)COCH_3\ (\text{ester, C=O}) \xrightarrow{CH_3O^-} CH_3CH(CH_3)C(=O)-C(CH_3)_2-C(=O)OCH_3 \overset{CH_3O^-}{\not\longrightarrow}$$

25.

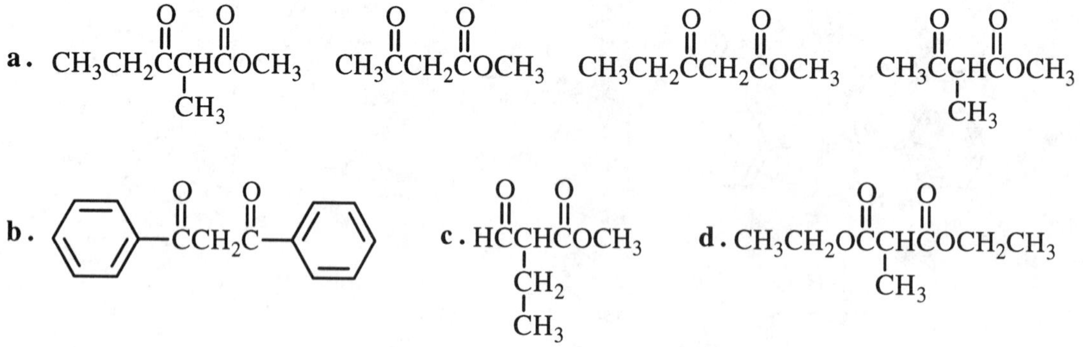

a. $CH_3CH_2C(=O)CH(CH_3)C(=O)OCH_3$ $CH_3C(=O)CH_2C(=O)OCH_3$ $CH_3CH_2C(=O)CH_2C(=O)OCH_3$ $CH_3C(=O)CH(CH_3)C(=O)OCH_3$

b. $C_6H_5C(=O)CH_2C(=O)C_6H_5$

c. $HC(=O)CH(CH_2CH_3)C(=O)OCH_3$

d. $CH_3CH_2OC(=O)CH(CH_3)C(=O)OCH_2CH_3$

26. Solved in the text.

27.

$CH_3OC(=O)CH_2CH_2CH_2CH_2CH_2C(=O)OCH_3 \underset{}{\overset{CH_3O^-}{\rightleftharpoons}}$ $CH_3OC(=O)\bar{C}HCH_2CH_2CH_2C(=O)OCH_3$ + CH_3OH $\rightleftharpoons$ tetrahedral intermediate (cyclohexane ring bearing $COCH_3$ (C=O), OCH_3 and O^-) $\rightleftharpoons$ CH_3O^- + 2-(methoxycarbonyl)cyclohexanone ($COCH_3$ with C=O, ring C=O)

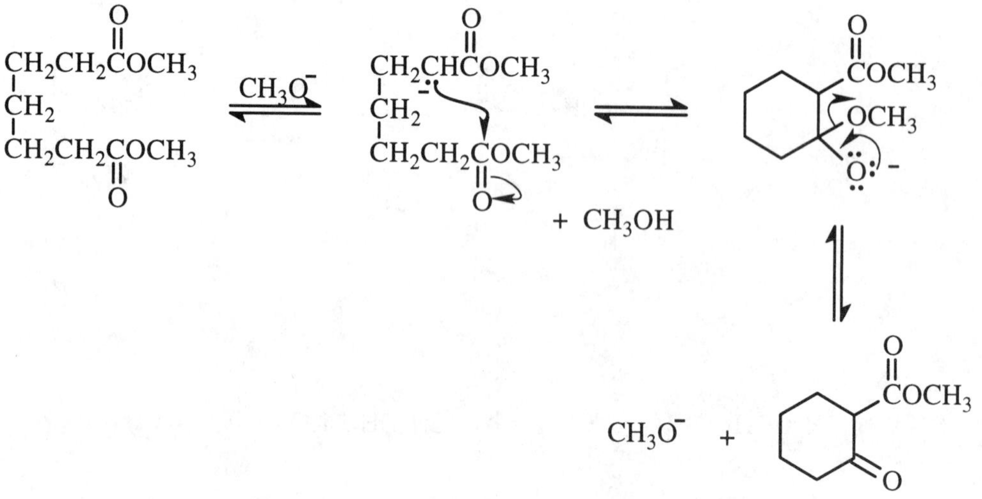

28.

a. 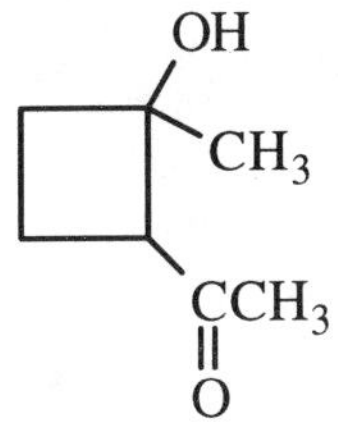 [structure: 1-hydroxy-1-methyl-2-acetylcyclobutane; labels OH, CH_3, $\overset{\displaystyle O}{\overset{\|}{C}}CH_3$]

b. 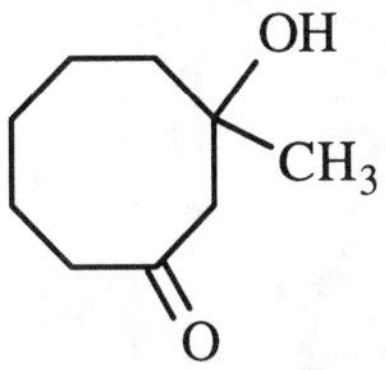 [structure: cyclooctanone with OH and CH_3 substituents; labels OH, CH_3, O]

29. No, because an intramolecular reaction would lead to a strained four-membered ring. Therefore, the intermolecular reaction will be preferred.

$$CH_3\overset{\overset{\displaystyle O}{\|}}{C}CH_2\overset{\overset{\displaystyle O}{\|}}{C}CH_3$$

30. Solved in the text.

31.

a. 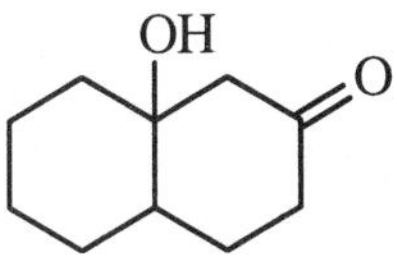 [structure: bicyclic hydroxy ketone; labels OH, O]

b. 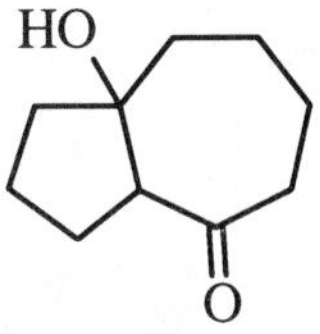 [structure: bicyclic hydroxy ketone; labels HO, O]

c. 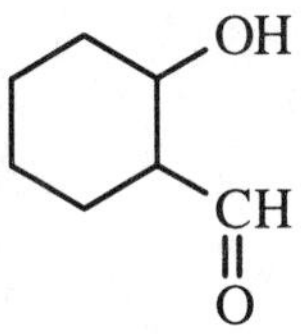 [structure: cyclohexane with OH and CHO; labels OH, $\overset{\displaystyle}{C}H$, O]

d. 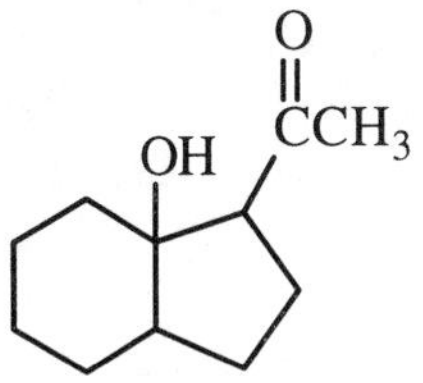 [structure: bicyclic with OH and acetyl; labels OH, $\overset{\overset{\displaystyle O}{\|}}{C}CH_3$]

32.

a. 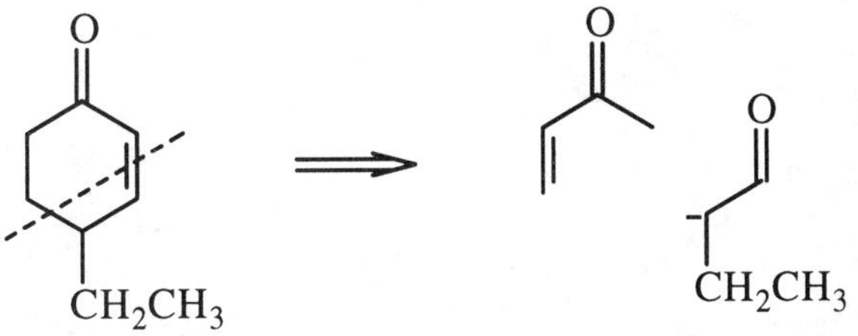 [retrosynthesis: 4-ethylcyclohex-2-enone ⟹ methyl vinyl ketone + enolate of butanal; labels O, CH_2CH_3]

$$CH_2{=}CH\overset{\overset{\displaystyle O}{\|}}{C}CH_3 \text{ and } CH_3CH_2CH_2\overset{\overset{\displaystyle O}{\|}}{C}H \xrightarrow[\Delta]{HO^-}$$ [4-ethylcyclohex-2-enone; labels O, CH_2CH_3]

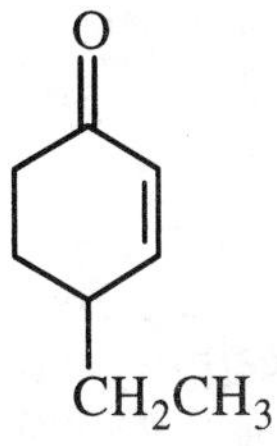

b.

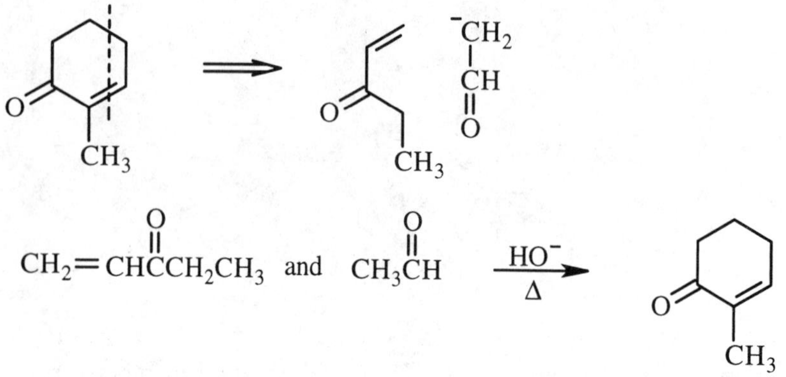

$$CH_2{=}CH\overset{O}{\overset{\|}{C}}CH_2CH_3 \text{ and } CH_3\overset{O}{\overset{\|}{C}}H \xrightarrow[\Delta]{HO^-}$$

c.

$$CH_2{=}\underset{CH_3}{\underset{|}{C}}\overset{O}{\overset{\|}{C}}CH_3 \text{ and } CH_3CH_2\overset{O}{\overset{\|}{C}}C_6H_5 \xrightarrow[\Delta]{HO^-}$$

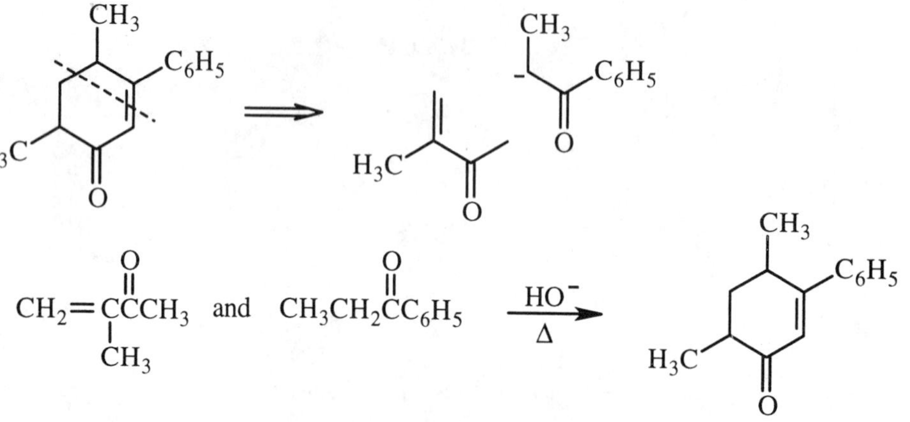

d.

$$CH_2{=}CH\overset{O}{\overset{\|}{C}}CH_2CH_3 + \xrightarrow[\Delta]{HO^-}$$

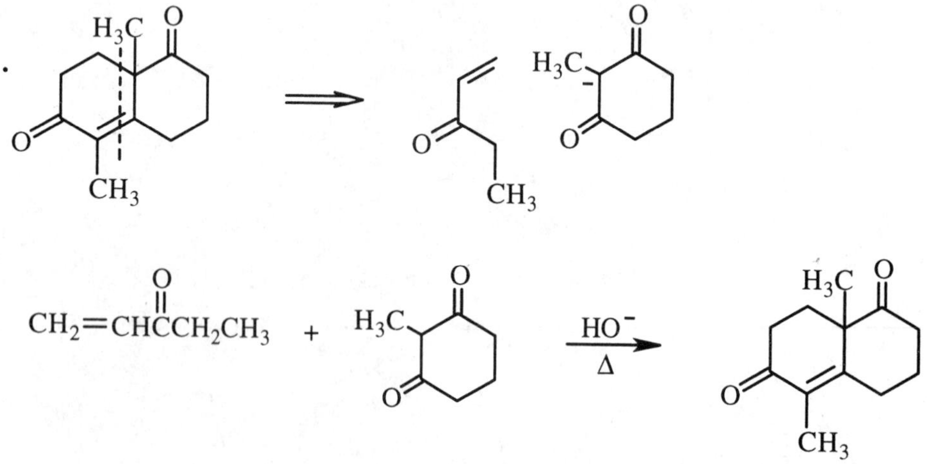

33. **a** and **d** can be decarboxylated.
b can't be decarboxylated, because it doesn't have a carboxyl group.
The electrons left behind if **c** were decarboxylated cannot be delocalized onto an oxygen.

34.

a. methyl bromide
b. methyl bromide (twice)
c. benzyl bromide
d. propyl bromide
e. isobutyl bromide
f. propyl bromide and methyl bromide

35.

a. An S_N2 reaction cannot be done on bromobenzene. (Section 9.8 of the text.)

b. An S_N2 reaction cannot be done on vinyl bromide. (Section 9.8 of the text.)

c. An S_N2 reaction cannot be done on a tertiary alkyl halide. (Only elimination occurs; Section 10.8 of the text.)

36.

a. $CH_3CH_2O\overset{O}{\overset{\|}{C}}CH_2\overset{O}{\overset{\|}{C}}OCH_2CH_3 \xrightarrow[2.\ Br(CH_2)_5Br]{1.\ CH_3CH_2O^-} CH_3CH_2O\overset{O}{\overset{\|}{C}}\underset{CH_2CH_2CH_2CH_2CH_2Br}{\underset{|}{C}H}\overset{O}{\overset{\|}{C}}OCH_2CH_3$

$\xrightarrow{CH_3CH_2O^-}$ 1,1-cyclohexanedicarboxylate diethyl ester (cyclohexane ring carrying two $\overset{O}{\overset{\|}{C}}OCH_2CH_3$ groups on the same carbon) $\xrightarrow[\Delta]{H^+,\ H_2O}$ cyclohexane ring carrying $\overset{O}{\overset{\|}{C}}OH$

b. $CH_3CH_2O\overset{O}{\overset{\|}{C}}CH_2\overset{O}{\overset{\|}{C}}OCH_2CH_3 \xrightarrow[2.\ Br(CH_2)_5Br]{1.\ CH_3CH_2O^-} CH_3CH_2O\overset{O}{\overset{\|}{C}}\underset{CH_2CH_2CH_2CH_2CH_2Br}{\underset{|}{C}H}\overset{O}{\overset{\|}{C}}OCH_2CH_3$

$\xrightarrow{\text{malonic ester},\ CH_3CH_2O^-}$ $CH_3CH_2O\overset{O}{\overset{\|}{C}}\underset{CH_2CH_2CH_2CH_2CH_2\underset{\underset{\overset{\|}{O}}{COCH_2CH_3}}{\underset{|}{C}H}\overset{O}{\overset{\|}{C}}OCH_2CH_3}{\underset{|}{C}H}\overset{O}{\overset{\|}{C}}OCH_2CH_3$

$\xrightarrow[\Delta]{H^+,\ H_2O}$ $HO\overset{O}{\overset{\|}{C}}CH_2CH_2CH_2CH_2CH_2CH_2CH_2\overset{O}{\overset{\|}{C}}OH$

37.

a. ethyl bromide **b.** pentyl bromide **c.** benzyl bromide

38.

a. $C_6H_5C(=O)CH_3$ + pyrrolidine (N–H) → $C_6H_5C(=CH_2)$–N(pyrrolidinyl) $\xrightarrow{CH_3CH_2Br}$ $C_6H_5C(=N^+\text{pyrrolidinyl})CH_2CH_2CH_3$ Br^- $\xrightarrow{H^+, H_2O}$ $C_6H_5CCH_2CH_2CH_3$ (C=O)

b. cyclohexene $\xrightarrow{RCOOH\ (C=O)}$ cyclohexene oxide $\xrightarrow{1.\ CH_3CH_2MgBr;\ 2.\ H^+}$ trans-2-ethylcyclohexanol (OH, CH_2CH_3)

c. C_6H_5CH (C=O) + $\underset{-}{C}H_2CH$ (C=O) → $C_6H_5CHCH_2CH$ (O^-, C=O) $\xrightarrow{H^+}$ $C_6H_5CH{=}CHCH$ (C=O) $\xrightarrow{1.\ Ag_2O,\ NH_3;\ 2.\ H_3O^+}$ $C_6H_5CH{=}CHCOH$ (C=O)

CH_3CH (C=O) $\xrightarrow{LDA}$ $\underset{-}{C}H_2CH$ (C=O)

d. 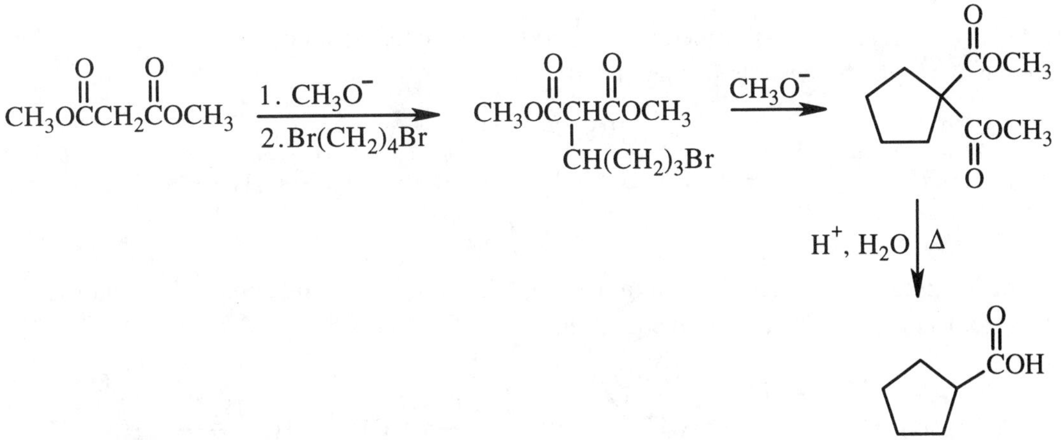 $CH_3O\overset{O}{\overset{\|}{C}}CH_2\overset{O}{\overset{\|}{C}}OCH_3 \xrightarrow[2.\,Br(CH_2)_4Br]{1.\,CH_3O^-} CH_3O\overset{O}{\overset{\|}{C}}\underset{CH(CH_2)_3Br}{\underset{|}{C}H}\overset{O}{\overset{\|}{C}}OCH_3 \xrightarrow{CH_3O^-}$ (cyclopentane-1,1-diyl)bis($\overset{O}{\overset{\|}{C}}OCH_3$) $\xrightarrow{H^+,\ H_2O,\ \Delta}$ cyclopentyl–$\overset{O}{\overset{\|}{C}}OH$

39. Because the catalyst is hydroxide ion rather than an enzyme, four stereoisomers will be formed since two chirality centers are created in the product. One of the four is fructose.

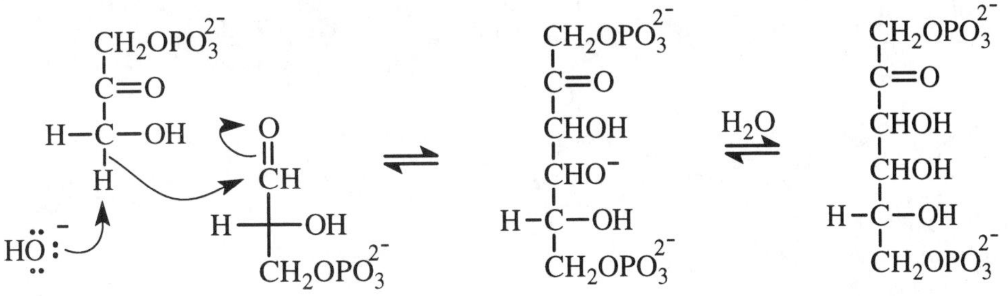

$CH_2OPO_3^{2-}$–C=O–H–C–OH–H (with HO^- removing H) + O=CH–H–C–OH–$CH_2OPO_3^{2-}$ $\rightleftharpoons$ $CH_2OPO_3^{2-}$–C=O–CHOH–CHO^-–H–C–OH–$CH_2OPO_3^{2-}$ $\underset{}{\overset{H_2O}{\rightleftharpoons}}$ $CH_2OPO_3^{2-}$–C=O–CHOH–CHOH–H–C–OH–$CH_2OPO_3^{2-}$

40. Seven moles. The first two carbons in the fatty acid come from acetyl CoA. Each subsequent two-unit piece comes from malonyl CoA. Because this amounts to fourteen carbons for the synthesis of the 16-carbon fatty acid, seven moles of malonyl CoA are required.

41. a. **Three** deuteriums would be incorporated into palmitic acid because only one CD_3COSR is used in the synthesis.

b. Seven deuteriums would be incorporated into palmitic acid because seven $^-OOCCD_2COSR$ are used in the synthesis (for a total of 14 D's), and each $^-OOCCD_2COSR$ loses one deuterium in the dehydration step (14 D's - 7 D's = 7 D's).

42. It tells you that an imine is formed as an intermediate. Because an imine is formed, the only source of oxygen for acetone formation is $H_2{}^{18}O$.

$$CH_3\overset{\overset{+}{N}HR}{\overset{\|}{C}}-CH_2-\overset{O}{\overset{\|}{C}}-O^- \longrightarrow CH_3\overset{:NHR}{\overset{|}{C}}=CH_2 + CO_2 \xrightarrow{H_2{}^{18}O} CH_3\overset{\overset{+}{N}HR}{\overset{\|}{C}}-CH_3 + H^{18}O^- \xrightarrow{H_2{}^{18}O} CH_3\overset{^{18}O}{\overset{\|}{C}}CH_3 + H_2NR$$

If decarboxylation occurred without imine formation, acetone would contain ^{16}O and some ^{18}O from a hydration-dehydration equilibrium.

$$CH_3\overset{O}{\overset{\|}{C}}-CH_2-\overset{O}{\overset{\|}{C}}-O^- \longrightarrow CH_3\overset{O^-}{\overset{|}{C}}=CH_2 + CO_2 \xrightarrow{H_2O^{18}} CH_3\overset{O}{\overset{\|}{C}}CH_3 \xrightarrow{H_2O^{18}} CH_3\overset{O^{18}}{\overset{\|}{C}}CH_3$$

43.

a. $CH_3\overset{O}{\overset{\|}{C}}CH_2\overset{O}{\overset{\|}{C}}OCH_2CH_3$

b. $HO\overset{O}{\overset{\|}{C}}\underset{CH_3}{\underset{|}{C}H}\overset{O}{\overset{\|}{C}}OH$

c. $R\overset{O}{\overset{\|}{C}}CH_2\overset{O}{\overset{\|}{C}}OR$

d. 1-hydroxycyclopentene (cyclopentene ring with OH on the double-bond carbon)

e. $CH_3CH_2CH_2CH_2\overset{O}{\overset{\|}{C}}OH$

44.

a. 2-oxocyclohexane with $\overset{O}{\overset{\|}{C}}OCH_2CH_3$ on the adjacent carbon

b. $CH_3CH_2CH_2\underset{Br}{\underset{|}{C}H}\overset{O}{\overset{\|}{C}}OH$

c. $CH_3\overset{O}{\overset{\|}{C}}CH_2\overset{O}{\overset{\|}{C}}CH_3$

d. CH_3CH_2– cyclopentanone –$\overset{O}{\overset{\|}{C}}OCH_2CH_3$ (substituents on both carbons adjacent to C=O)

e. $CH_3\underset{CH_3}{\underset{|}{C}H}CH_2CH_2\overset{O}{\overset{\|}{C}}OH + CO_2$

f. C_6H_5–$\overset{O}{\overset{\|}{C}}CH_2\overset{O}{\overset{\|}{C}}OCH_2CH_3$

g. cyclohexane-1,3-dione with $CH_2CH{=}CH_2$ on C-2

i. 2-ethylcyclopentanone (CH_2CH_3) + pyrrolidinium ion ($\overset{+}{N}$ with H, H)

h. 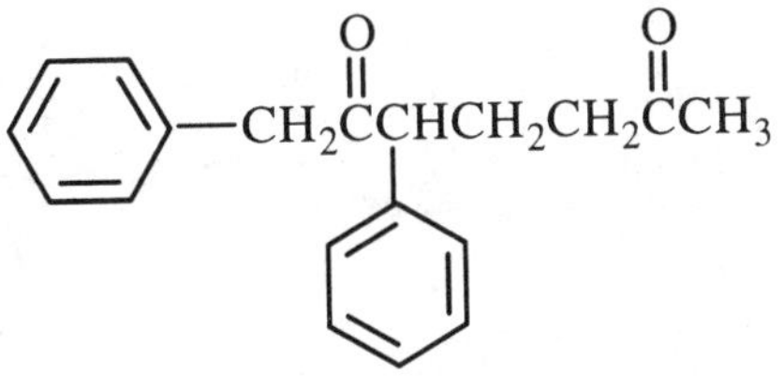 $C_6H_5-CH_2\overset{O}{\overset{\|}{C}}\underset{C_6H_5}{CH}CH_2CH_2\overset{O}{\overset{\|}{C}}CH_3$

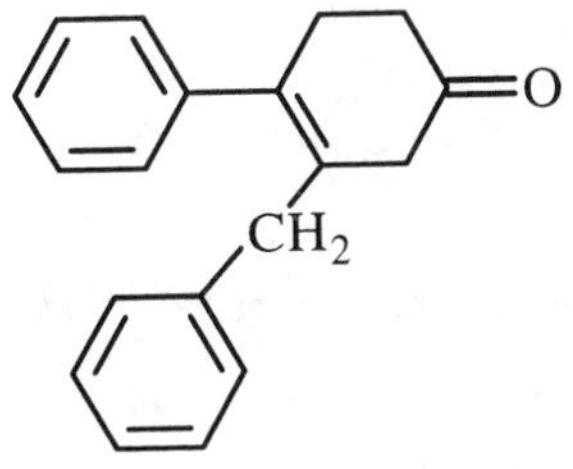

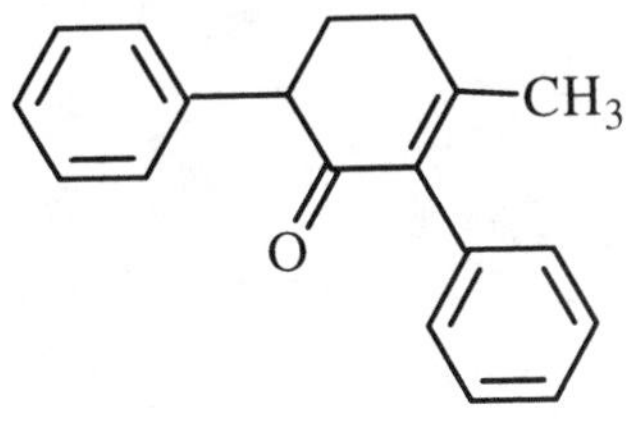

j. 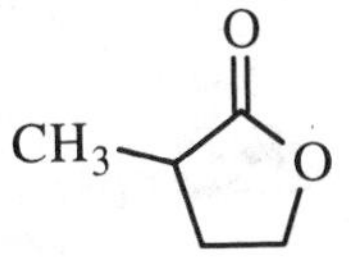 CH_3 (lactone, ring O)

k. 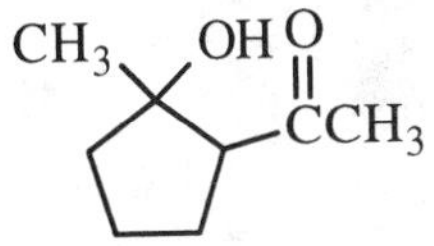 CH_3 OH $\overset{O}{\overset{\|}{C}}CH_3$

l. 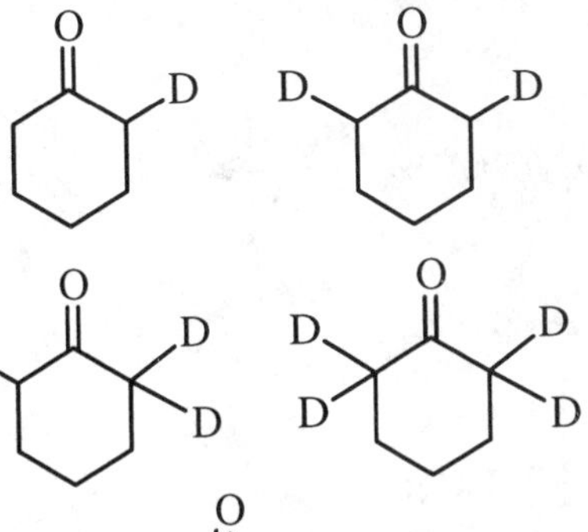 D D D D D D D D D D D

m. 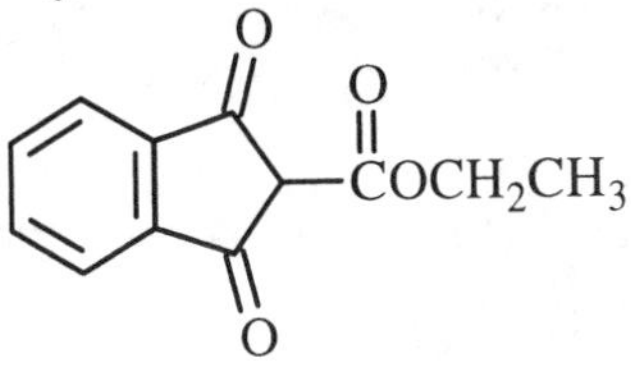 $\overset{O}{\overset{\|}{C}}OCH_2CH_3$

45. **a.** When the ketone enolizes, the chirality center is lost.

When the enol reforms the ketone, it can form the *R* and *S* enantiomer equally as easily, so a racemic mixture is obtained.

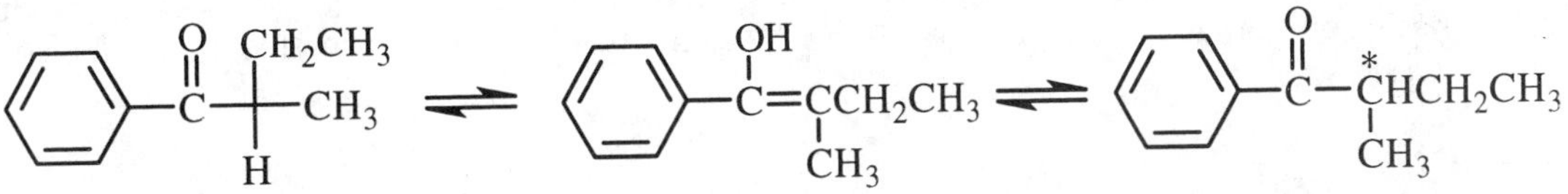

$$C_6H_5-\overset{O}{\overset{\|}{C}}-\underset{H}{\overset{CH_2CH_3}{C}}-CH_3 \rightleftharpoons C_6H_5-\overset{OH}{\overset{|}{C}}=\underset{CH_3}{C}CH_2CH_3 \rightleftharpoons C_6H_5-\overset{O}{\overset{\|}{C}}-\underset{CH_3}{\overset{*}{C}}HCH_2CH_3$$

(*R*)-2-methyl-1-phenyl-1-butanone

b. You want a ketone that has an α-carbon that is a chirality center. Racemization will occur when an α-hydrogen is removed from the chirality center. If you don't want racemization to have to compete with removal of a hydrogen from a second α-carbon, it is best to chose a compound that can form only one enol.

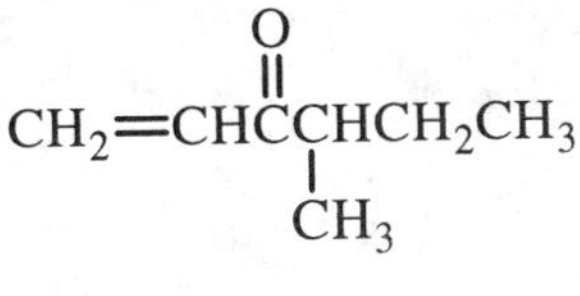

$$CH_2=CH\overset{O}{\overset{\|}{C}}\underset{CH_3}{C}HCH_2CH_3$$

46.

$$\underset{\substack{\textbf{A}\\ C_5H_8O_3}}{CH_3\overset{O}{\overset{\|}{C}}CH_2\overset{O}{\overset{\|}{C}}OCH_3} \xrightarrow[\Delta]{H^+,\ H_2O} \underset{\textbf{B}}{CH_3\overset{O}{\overset{\|}{C}}CH_3} \xrightarrow{HO^-} \underset{\textbf{C}}{CH_3\underset{CH_3}{\overset{O^-}{C}}CH_2\overset{O}{\overset{\|}{C}}CH_3} \xrightarrow[\Delta]{H^+} \underset{\textbf{D}}{CH_3\underset{CH_3}{C}{=}CH\overset{O}{\overset{\|}{C}}CH_3}$$

$$\textbf{A} \xrightarrow{1.\ CH_3O^-\quad 2.\ CH_3Br} \underset{\textbf{E}}{CH_3\overset{O}{\overset{\|}{C}}\underset{CH_3}{C}H\overset{O}{\overset{\|}{C}}OCH_3} \xrightarrow[\Delta]{H^+,\ H_2O} \underset{\textbf{H}}{CH_3\overset{O}{\overset{\|}{C}}CH_2CH_3} \xrightarrow[I_2\ \text{excess}]{HO^-\ \text{excess}} \underset{\textbf{I}}{CH_3CH_2\overset{O}{\overset{\|}{C}}OH} \xrightarrow{SOCl_2} \underset{\textbf{J}}{CH_3CH_2\overset{O}{\overset{\|}{C}}Cl}$$

$$\textbf{E} \xrightarrow{1.\ CH_3O^-\quad 2.\ CH_3Br} \underset{\textbf{F}}{CH_3\overset{O}{\overset{\|}{C}}-\underset{CH_3}{\overset{CH_3}{C}}-\overset{O}{\overset{\|}{C}}OCH_3} \xrightarrow[\Delta]{H^+,\ H_2O} \underset{\textbf{G}}{CH_3\overset{O}{\overset{\|}{C}}\underset{CH_3}{C}HCH_3}$$

$$\textbf{J} \xrightarrow{CH_3OH} \underset{\textbf{K}}{CH_3CH_2\overset{O}{\overset{\|}{C}}OCH_3} \xrightarrow{1.\ CH_3O^-\quad 2.\ H^+} \underset{\textbf{L}}{CH_3CH_2\overset{O}{\overset{\|}{C}}\underset{CH_3}{C}H\overset{O}{\overset{\|}{C}}OCH_3}$$

47. Remember that there are no positively charged organic reactants, intermediates, or products in a basic solution, and no negatively charged organic reactants, intermediates, or products in an acidic solution.

a. 3-cyclohexenone + HO^- (removes H) $\rightarrow$ enolate carbanion $\longleftrightarrow$ resonance form $\xrightarrow{H_2O}$ 2-cyclohexenone + HO^-

b. 3-cyclohexenone $\underset{-H^+}{\overset{H^+}{\rightleftharpoons}}$ protonated ketone (^+OH) $\xrightarrow{H_2\ddot{O}:}$ dienol ($:\ddot{O}H$) $\xrightarrow{H_3O^+}$ protonated enone (^+OH) $\underset{H^+}{\overset{-H^+}{\rightleftharpoons}}$ 2-cyclohexenone + H_2O

48.

a. $C_6H_5CHO + {}^-CH_2COCCH_3$ (from CH_3COCCH_3 + CH_3CO^-; O O over the carbonyls) $\longrightarrow C_6H_5CH(OH)CH_2COCCH_3 \xrightarrow{-H_2O} C_6H_5CH{=}CHCOCCH_3$

b. $C_6H_5CH{=}CHCOCCH_3 \xrightarrow{H_2O} C_6H_5CH{=}CHCOH + CH_3COH$

c. $C_6H_5CHO + C_2H_5OC\overset{-}{C}HCOC_2H_5$ (from $C_2H_5OCCH_2COC_2H_5$ + $CH_3CH_2O^-$) $\longrightarrow C_6H_5CH(OH)CH(COC_2H_5)COC_2H_5 \xrightarrow{-H_2O} C_6H_5CH{=}C(COC_2H_5)COC_2H_5$

d. $C_6H_5CH{=}C(COC_2H_5)COC_2H_5 \xrightarrow[\Delta]{H_3O^+} C_6H_5CH{=}CHCOH + CO_2 + 2\ CH_3CH_2OH$

49.

a. $CH_3CH_2CH_2\overset{O}{\overset{\|}{C}}H + BrCH_2\overset{O}{\overset{\|}{C}}OCH_3 \xrightarrow[2.\ H_2O]{1.\ Zn} CH_3CH_2CH_2\overset{OH}{\overset{|}{C}}HCH_2\overset{O}{\overset{\|}{C}}OCH_3$

b. $CH_3CH_2\overset{O}{\overset{\|}{C}}H + CH_3CH_2\underset{Br}{\underset{|}{C}}H\overset{O}{\overset{\|}{C}}OCH_3 \xrightarrow[2.\ H_2O]{1.\ Zn} CH_3CH_2\overset{OH}{\overset{|}{C}}H\underset{CH_2CH_3}{\underset{|}{C}}H\overset{O}{\overset{\|}{C}}OCH_3$

$\xrightarrow{H^+,\ H_2O \text{ excess}} CH_3CH_2\overset{OH}{\overset{|}{C}}H\underset{CH_2CH_3}{\underset{|}{C}}H\overset{O}{\overset{\|}{C}}OH$

c. $CH_3CH_2\overset{O}{\overset{\|}{C}}H + CH_3\underset{Br}{\underset{|}{C}}H\overset{O}{\overset{\|}{C}}OCH_3 \xrightarrow[2.\ H_2O]{1.\ Zn} CH_3CH_2\overset{OH}{\overset{|}{C}}H\underset{CH_3}{\underset{|}{C}}H\overset{O}{\overset{\|}{C}}OCH_3$

$\xrightarrow{H^+,\ H_2O \text{ excess}} CH_3CH_2\overset{OH}{\overset{|}{C}}H\underset{CH_3}{\underset{|}{C}}H\overset{O}{\overset{\|}{C}}OH \xrightarrow[\Delta]{H_2SO_4} CH_3CH_2CH{=}\underset{CH_3}{\underset{|}{C}}\overset{O}{\overset{\|}{C}}OH$

d. $CH_3CH_2\overset{O}{\overset{\|}{C}}CH_2CH_3 + BrCH_2\overset{O}{\overset{\|}{C}}OCH_3 \xrightarrow[2.\ H_2O]{1.\ Zn} CH_3CH_2\underset{CH_2CH_3}{\underset{|}{\overset{OH}{\overset{|}{C}}}}CH_2\overset{O}{\overset{\|}{C}}OCH_3$

50. The ketone is 2-hexanone. Therefore, the alkyl halide is a propyl halide (propyl bromide, propyl chloride, or propyl iodide).

$$\underbrace{CH_3\overset{O}{\overset{\|}{C}}CH_2}_{\text{This part comes from acetoacetic ester.}}CH_2CH_2CH_3$$

51.

a. Cyclohexanone $\xrightarrow[\text{2. } CH_3CH_2CH_2CH_2Br]{\text{1. LDA/THF}}$ 2-($CH_2CH_2CH_2CH_3$)cyclohexanone

b. 1. Cyclohexanone $\xrightarrow[\text{2. } CH_3CH_2CH_2\overset{O}{\overset{\|}{C}}Cl \quad \text{3. } H^+, H_2O]{\text{1. pyrrolidine (N–H)}}$ 2-($\overset{O}{\overset{\|}{C}}CH_2CH_2CH_3$)cyclohexanone

2. Cyclohexanone $\xrightarrow[\text{2. } CH_3CH_2CH_2\overset{O}{\overset{\|}{C}}Cl]{\text{1. LDA}}$ 2-($\overset{O}{\overset{\|}{C}}CH_2CH_2CH_3$)cyclohexanone

c. Cyclohexanone $\xrightarrow[\text{2. } CH_2{=}CH\overset{O}{\overset{\|}{C}}CH_3 \quad \text{3. } H_2O]{\text{1. LDA}}$ 2-($CH_2CH_2\overset{O}{\overset{\|}{C}}CH_3$)cyclohexanone

d. Cyclohexanone $\xrightarrow[\text{2. } CH_3CH_2O\overset{O}{\overset{\|}{C}}OCH_2CH_3 \text{ (excess)} \quad \text{3. } H^+]{\text{1. LDA}}$ 2-($\overset{O}{\overset{\|}{C}}OCH_2CH_3$)cyclohexanone

e. Cyclohexanone $\xrightarrow[\text{2. } H\overset{O}{\overset{\|}{C}}OCH_2CH_3 \text{ (excess)} \quad \text{3. } H^+]{\text{1. LDA}}$ 2-($\overset{O}{\overset{\|}{C}}H$)cyclohexanone

f. cyclohexanone + $CH_2{=}CH\overset{O}{\overset{\|}{C}}CH_3$ $\xrightarrow[\text{a Robinson annulation}]{HO^-}$ (bicyclic hydroxy ketone, OH) $\xrightarrow{H_2SO_4}$ (bicyclic enone) $\xrightarrow{H_2,\ Pt}$ (bicyclic alcohol, HO) + (bicyclic ketone)

(bicyclic alcohol) $\xrightarrow[H_2SO_4]{Na_2Cr_2O_7}$ (bicyclic ketone)

The carbon-carbon double bond is more easily reduced than the carbonyl group by hydrogenation. Any alcohol that forms as a result of the reduction reaction can be oxidized back to the ketone.

52. The positive iodoform test indicates that the compound with two singlets with a ratio of 3 : 1 is a methyl ketone.

3,3-Dimethyl-1-butyne will show two singlets with a ratio of 9 : 1, and when it adds water, it will form a methyl ketone with two singlets with a ratio of 3 : 1.

$$(CH_3)_3CC{\equiv}CH \xrightarrow[H_2O]{H^+} (CH_3)_3C\overset{O}{\overset{\|}{C}}CH_3$$

2 singlets (9 : 1) 2 singlets (3 : 1)

53.

a. $CH_3\overset{O}{\overset{\|}{C}}CH_3$ $\xrightarrow[\text{2. } H\overset{O}{\overset{\|}{C}}OCH_3 \text{ excess; 3. } H_2O,\ H^+]{\text{1. } CH_3O^-}$ $CH_3\overset{O}{\overset{\|}{C}}CH_2\overset{O}{\overset{\|}{C}}H$

b. Ph–C(=O)CH$_2$CH$_3$ $\xrightarrow[2.\ CH_3CH_2O^-]{1.\ Br_2,\ H^+}$ Ph–C(=O)CH=CH$_2$ $\xrightarrow[CH_3CH_2O^-]{CH_3CH_2OC(=O)CH_2C(=O)OCH_2CH_3}$ Ph–C(=O)CH$_2$CH$_2$C$^-$(C(=O)OCH$_2$CH$_3$)C(=O)OCH$_2$CH$_3$ $\xrightarrow[\Delta]{H^+,\ H_2O}$ Ph–C(=O)CH$_2$CH$_2$CH$_2$C(=O)OH

c. cyclopentanone $\xrightarrow[2.\ CH_3O^-]{1. Br_2,\ H^+}$ cyclopentenone $\xrightarrow[CH_3O^-]{CH_3C(=O)CH_2C\equiv N}$ 3-[CH$^-$(C≡N)C(=O)CH$_3$]cyclopentanone $\xrightarrow[\Delta]{H^+,\ H_2O}$ 3-(CH$_2$C(=O)CH$_3$)cyclopentanone + CO_2

d. CH$_3$C(=O)CH$_2$C(=O)OCH$_2$CH$_3$ $\xrightarrow[2.\ \text{cyclopentyl–Br}]{1. CH_3CH_2O^-}$ CH$_3$C(=O)CH(cyclopentyl)C(=O)OCH$_2$CH$_3$ $\xrightarrow[\Delta]{H^+,\ H_2O}$ CO_2 + CH$_3$CH$_2$OH + CH$_3$C(=O)CH$_2$–cyclopentyl

e. CH$_3$C(=O)CH$_2$CH$_2$CH$_2$C(=O)OCH$_3$ $\xrightarrow[\text{Claisen condensation}]{CH_3O^-}$ cyclohexane-1,3-dione $\xrightarrow[CH_3I\ (2\ equiv.)]{CH_3O^-}$ 2,2-dimethylcyclohexane-1,3-dione (CH$_3$, CH$_3$)

f. $CH_3CH_2O\overset{O}{\overset{\|}{C}}CH_2CH_2CH_2CH_2\overset{O}{\overset{\|}{C}}OCH_2CH_3$ $\xrightarrow[\text{Claisen condensation}]{CH_3CH_2O^-}$ (cyclopentanone enolate with $\overset{O}{\overset{\|}{C}}OCH_2CH_3$) $\xrightarrow[\Delta]{H^+, H_2O}$ cyclopentanone $+ CO_2$ $\xrightarrow[\text{2. } CH_3CH_2CH_2Br]{\text{1. LDA/THF}}$ 2-($CH_2CH_2CH_3$)cyclopentanone

54.

benzene $\xrightarrow[\text{2. } H_2O]{\text{1. } CH_3CH_2\overset{O}{\overset{\|}{C}}Cl,\ AlCl_3}$ $C_6H_5\overset{O}{\overset{\|}{C}}CH_2CH_3$ $\xrightarrow[FeCl_3]{Cl_2}$ Cl-$C_6H_4\overset{O}{\overset{\|}{C}}CH_2CH_3$ (meta) $\xrightarrow{Cl_2,\ H^+, H_2O}$ Cl-$C_6H_4\overset{O}{\overset{\|}{C}}CHCH_3$ (with Cl on CH) $\xrightarrow{(CH_3)_3CNH_2}$ Cl-$C_6H_4\overset{O}{\overset{\|}{C}}CHCH_3$ with $+NH_2C(CH_3)_3$ Cl^-

55. Synthesizing benzaldehyde from benzene would be easy if a one carbon acyl chloride could be used, but there is no such compound. The only way a single carbon can be placed on a benzene ring is via a Friedel-Crafts alkylation.

benzene $\xrightarrow{\text{1. } CH_3Cl/AlCl_3;\ \text{2. NBS}/\Delta;\ \text{3. } HO^-;\ \text{4. PCC}}$ $C_6H_5\overset{O}{\overset{\|}{C}}H$ $\xrightarrow{^-CH_2\overset{O}{\overset{\|}{C}}CH_3}$ (from $CH_3\overset{O}{\overset{\|}{C}}CH_3$ + LDA) $C_6H_5\overset{O^-}{\overset{|}{C}}HCH_2\overset{O}{\overset{\|}{C}}CH_3$ $\xrightarrow[-H_2O]{H^+}$ $C_6H_5CH{=}CH\overset{O}{\overset{\|}{C}}CH_3$ $\xrightarrow{HO^-\text{ excess},\ I_2\text{ excess}}$ $C_6H_5CH{=}CH\overset{O}{\overset{\|}{C}}O^-$ $\xrightarrow[\text{2. } CH_3CH_2OH]{\text{1. } SOCl_2}$ $C_6H_5CH{=}CH\overset{O}{\overset{\|}{C}}OCH_2CH_3$

56.

Benzene $\xrightarrow[\text{2. } H_2O]{\text{1. } CH_3CCl(=O),\ AlCl_3}$ A (acetophenone, $C_6H_5CCH_3$ with C=O) $\xrightarrow{HO^-}$ B ($C_6H_5C(CH_3){=}CHC(=O){-}C_6H_5$)

B $\xrightarrow[\text{2. } H^+, H_2O]{\text{1. } CH_3MgBr}$ C ($C_6H_5C(CH_3){=}CHC(OH)(CH_3){-}C_6H_5$)

C $\xrightarrow{\Delta}$ D ($C_6H_5C(CH_3){=}CHC(=CH_2){-}C_6H_5$)

D $\xrightarrow[\text{2. } H_2O_2]{\text{1. } O_3}$ E ($C_6H_5C(CH_3){=}O$) + F ($HOC(=O){-}C(=O){-}C_6H_5$) + G (CO_2)

57. **a.** The first reaction is an intermolecular Claisen condensation. The base then removes a hydrogen from the α-carbon that has two α-hydrogens, and an intramolecular Claisen condensation occurs.

$EtOCCH_2\overset{-}{C}HCOEt$ (both C=O) attacks $EtOCCH_2CH_2COEt$ (both C=O) $\longrightarrow$ EtO(C=O)–CH₂CH₂–C(=O)–CH(COEt)(CH₂C(=O)OEt) + EtO^- $\longrightarrow$ EtO(C=O)–$\overset{-}{C}$HCH₂–C(=O)–CH(COEt)(CH₂C(=O)OEt) $\longrightarrow$ EtO^- + cyclohexane-1,4-dione bearing two COEt (C=O) groups (2,5-bis(ethoxycarbonyl)cyclohexane-1,4-dione)

b. The reaction involves two successive aldol condensations.

$$\text{phthalaldehyde} + \text{cyclohexane-1,4-dione} \longrightarrow \text{(monocondensation product)} + H_2O \longrightarrow \text{(tricyclic dione)} + H_2O$$

58.

a. $CH_3CH_2\overset{O}{\overset{\|}{C}}OH \xrightarrow[2.\,H_3O^+]{1.\,Br_2,\,PBr_3} CH_3\underset{Br}{\underset{|}{C}}H\overset{O}{\overset{\|}{C}}OH \xrightarrow[2.\,H_3O^+]{\text{excess } 1.\,NH_3} CH_3\underset{+NH_3}{\underset{|}{C}}H\overset{O}{\overset{\|}{C}}OH$

b. phthalimide $\xrightarrow[2.\,CH_3CH_2O\overset{O}{\overset{\|}{C}}\underset{Br}{\underset{|}{C}}H\overset{O}{\overset{\|}{C}}OCH_2CH_3]{1.\,HO^-}$ N-phthalimido-$CH(\overset{O}{\overset{\|}{C}}OCH_2CH_3)_2$

$\xrightarrow{H^+,\ H_2O,\ \Delta}$ phthalic acid (benzene-1,2-di-COOH) + $\underset{+NH_3}{\underset{|}{C}}H_2\overset{O}{\overset{\|}{C}}OH$ + CO_2

59. **a** can be prepared by removing an α-hydrogen from acetone with LDA and then adding formaldehyde.

$$CH_3\overset{O}{\overset{\|}{C}}CH_3 \xrightarrow{LDA} \underline{C}H_2\overset{O}{\overset{\|}{C}}CH_3 \xrightarrow{H\overset{O}{\overset{\|}{C}}H} \overset{OH}{\overset{|}{C}}H_2CH_2\overset{O}{\overset{\|}{C}}CH_3 \xrightarrow[\Delta]{H_2SO_4} CH_2{=}CH\overset{O}{\overset{\|}{C}}CH_3$$

b can be prepared by removing an α-hydrogen from propionaldehyde with LDA and then slowly adding acetaldehyde.

$$CH_3CH_2\overset{O}{\overset{\|}{C}}H \xrightarrow{LDA} CH_3\underline{C}H\overset{O}{\overset{\|}{C}}H \xrightarrow[\text{dropwise}]{CH_3\overset{O}{\overset{\|}{C}}H} CH_3\overset{OH}{\overset{|}{C}}H\underset{CH_3}{\underset{|}{C}}H\overset{O}{\overset{\|}{C}}H \xrightarrow[\Delta]{H_2SO_4} CH_3CH{=}\underset{CH_3}{\underset{|}{C}}\overset{O}{\overset{\|}{C}}H$$

c can be prepared by removing an α-hydrogen from 3-hexanone with LDA and then adding formaldehyde. The yield is poor because 3-hexanone is an asymmetrical ketone therefore, two different α-carbanions are formed.

$$CH_3CH_2\overset{O}{\overset{\|}{C}}CH_2CH_2CH_3 \xrightarrow{LDA} CH_3\underline{C}H\overset{O}{\overset{\|}{C}}CH_2CH_2CH_3 \xrightarrow{HCH(=O)} \overset{OH}{CH_2}\underset{CH_3}{CH}\overset{O}{\overset{\|}{C}}CH_2CH_2CH_3$$

$$\xrightarrow{H_2SO_4,\ \Delta} CH_2{=}\underset{CH_3}{C}\overset{O}{\overset{\|}{C}}CH_2CH_2CH_3$$

d can be prepared by removing an α-hydrogen from acetaldehyde with LDA and then slowly adding acetone.

$$CH_3\overset{O}{\overset{\|}{C}}H \xrightarrow{LDA} \bar{C}H_2\overset{O}{\overset{\|}{C}}H \xrightarrow[\text{dropwise}]{CH_3\overset{O}{\overset{\|}{C}}CH_3} CH_3\underset{CH_3}{\overset{OH}{C}}CH_2\overset{O}{\overset{\|}{C}}H \xrightarrow[\Delta]{H_2SO_4} CH_3\underset{CH_3}{C}{=}CH\overset{O}{\overset{\|}{C}}H$$

e cannot be prepared by a mixed aldol condensation. It can be prepared via an intramolecular aldol condensation using 5-oxooctanal.

$$CH_3CH_2CH_2\overset{O}{\overset{\|}{C}}CH_2CH_2CH_2\overset{O}{\overset{\|}{C}}H \xrightarrow{LDA} CH_3CH_2\underline{C}H\overset{O}{\overset{\|}{C}}CH_2CH_2CH_2\overset{O}{\overset{\|}{C}}H \longrightarrow \text{2-ethyl-3-hydroxycyclohexanone} \longrightarrow \text{2-ethyl-2-cyclohexenone}$$

f cannot be prepared by a mixed aldol condensation. It can be prepared via an intramolecular aldol condensation using 1,7-heptanedial.

$$H\overset{O}{\overset{\|}{C}}CH_2CH_2CH_2CH_2CH_2\overset{O}{\overset{\|}{C}}H \xrightarrow{LDA} H\overset{O}{\overset{\|}{C}}CH_2CH_2CH_2CH_2\underline{C}H\overset{O}{\overset{\|}{C}}H \longrightarrow \text{2-hydroxycyclohexane-CH=O} \longrightarrow \text{cyclohexene-1-CH=O}$$

g cannot be prepared by a mixed aldol condensation. A small amount of **g** can be formed via an intramolecular aldol condensation using 7-oxooctanal, but this compound can form two different α-carbanions that can react with a carbonyl group to form a six-membered ring. Because an aldehyde is more reactive than a ketone, the desired compound will be formed only as a minor product.

$$\text{minor product} \longleftarrow CH_3\overset{O}{\overset{\|}{C}}CH_2CH_2CH_2CH_2CH_2\overset{O}{\overset{\|}{C}}H \longrightarrow \text{major product}$$

minor product: 2-methylcyclohex-1-ene-1-carbaldehyde ($\overset{O}{\overset{\|}{C}}H$, CH_3) — major product: 1-acetylcyclohexene ($\overset{O}{\overset{\|}{C}}CH_3$)

minor product major product

Theoretically **h** can be prepared by a mixed aldol condensation. However, because of steric hindrance, the reaction of propanal with dimethylpropanal to form the desired product would be very slow.

$$CH_3CH_2\overset{O}{\overset{\|}{C}}H \xrightarrow{\text{LDA}} CH_3\underset{-}{C}H\overset{O}{\overset{\|}{C}}H \xrightarrow{(CH_3)_3C\overset{O}{\overset{\|}{C}}H} CH_3\underset{CH_3}{\overset{CH_3}{C}}{-}\overset{OH}{C}H\underset{CH_3}{C}H\overset{O}{\overset{\|}{C}}H \xrightarrow[\Delta]{H_2SO_4} CH_3\underset{CH_3}{\overset{CH_3}{C}}CH{=}\underset{CH_3}{C}\overset{O}{\overset{\|}{C}}H$$

i can be prepared was using propanal and 3,3-dimethylbutanal.

$$CH_3CH_2\overset{O}{\overset{\|}{C}}H \xrightarrow{\text{LDA}} CH_3\overset{-}{C}H\overset{O}{\overset{\|}{C}}H \xrightarrow[\text{dropwise}]{CH_3\underset{CH_3}{\overset{CH_3}{C}}CH_2\overset{O}{\overset{\|}{C}}H} CH_3\underset{CH_3}{\overset{CH_3}{C}}CH_2\overset{OH}{C}H\underset{CH_3}{C}H\overset{O}{\overset{\|}{C}}H \xrightarrow[\Delta]{H_2SO_4} CH_3\underset{CH_3}{\overset{CH_3}{C}}CH_2CH{=}\underset{CH_3}{C}\overset{O}{\overset{\|}{C}}H$$

60. At low temperatures (- 78°), the proton will be more apt to be removed from the methyl group because its hydrogens are the most accessible and are slightly more acidic.

$$\text{cyclopentyl}{-}\overset{O}{\overset{\|}{C}}CH_3 \text{ (with } CH_2CH_2CH_2CH_2Br \text{ chain)} \xrightarrow[-78°]{\text{LDA}} \overset{O}{\overset{\|}{C}}\overset{-}{C}H_2 \text{ (attacks C–Br)} \longrightarrow \text{bicyclic ketone}$$

At higher temperatures (25°), the proton will be more apt to be removed from the more substituted α-carbon because in that way the more stable enolate is formed (the one with the more stable double bond).

LDA
25°

CH$_3$ Br CH$_3$ Br CH$_3$ O$^-$ CH$_3$ Br

61. The single equivalent of base removes the most acidic proton (from a carbon flanked by two carbonyl groups), and alkylation takes place at that position.

$$CH_3\overset{O}{\overset{\|}{C}}CH_2\overset{O}{\overset{\|}{C}}OCH_2CH_3 \xrightarrow{CH_3O^-} CH_3\overset{O}{\overset{\|}{C}}\ddot{\underset{-}{C}}H\overset{O}{\overset{\|}{C}}OCH_2CH_3 \xrightarrow[CH_3Br]{} CH_3\overset{O}{\overset{\|}{C}}\underset{CH_3}{\underset{|}{C}H}\overset{O}{\overset{\|}{C}}OCH_2CH_3$$

The second equivalent of base removes a proton from the methyl group. Since the methyl group is a weaker acid than the methylene group, the conjugate base of the methyl group is the stronger base and alkylation takes place at that position.

$$CH_3\overset{O}{\overset{\|}{C}}CH_2\overset{O}{\overset{\|}{C}}OCH_2CH_3 \xrightarrow{2\ CH_3O^-} \ddot{\underset{-}{C}}H_2\overset{O}{\overset{\|}{C}}\ddot{\underset{-}{C}}H\overset{O}{\overset{\|}{C}}OCH_2CH_3 \xrightarrow[CH_3Br]{} \underset{CH_3}{\underset{|}{C}H_2}\overset{O}{\overset{\|}{C}}\ddot{\underset{-}{C}}H\overset{O}{\overset{\|}{C}}OCH_2CH_3$$

$$\xrightarrow{CH_3OH} \underset{CH_3}{\underset{|}{C}H_2}\overset{O}{\overset{\|}{C}}CH_2\overset{O}{\overset{\|}{C}}OCH_2CH_3$$

62. The compound that gives the 1H NMR spectrum is 4-phenyl-3-propen-2-one. The singlet at 2.3 ppm is the methyl group, the doublet at ~ 6.7 ppm and the doublet buried under the benzene ring protons are the hydrogens bonded to the *sp*2 carbons. The compounds that would form this compound (via an aldol condensation) are benzaldehyde and acetone.

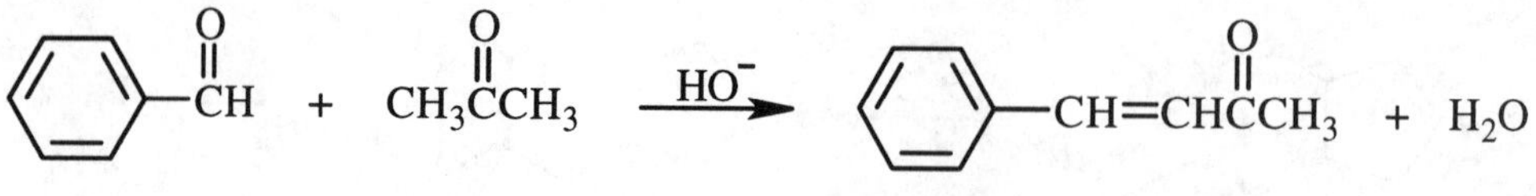

$$C_6H_5CH(=O) + CH_3C(=O)CH_3 \xrightarrow{HO^-} C_6H_5CH{=}CHC(=O)CH_3 + H_2O$$

4-phenyl-3-buten-2-one

63.

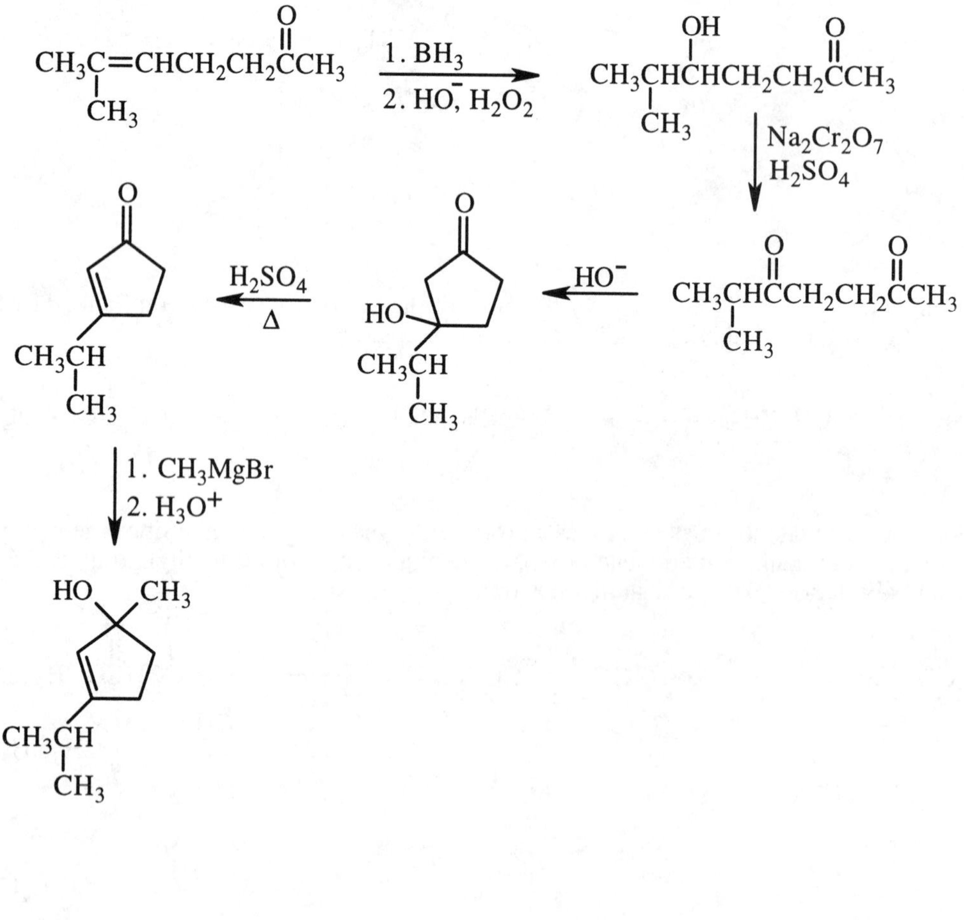

$$CH_3C(CH_3){=}CHCH_2CH_2C(=O)CH_3 \xrightarrow[2.\ HO^-,\ H_2O_2]{1.\ BH_3} CH_3CH(CH_3)CH(OH)CH_2CH_2C(=O)CH_3$$

$$\xrightarrow[H_2SO_4]{Na_2Cr_2O_7} CH_3CH(CH_3)C(=O)CH_2CH_2C(=O)CH_3$$

$$\xrightarrow{HO^-} \text{3-hydroxy-3-(}CH_3CHCH_3\text{)cyclopentanone} \xrightarrow[\Delta]{H_2SO_4} \text{3-(}CH_3CHCH_3\text{)-2-cyclopentenone}$$

$$\xrightarrow[2.\ H_3O^+]{1.\ CH_3MgBr} \text{1-methyl-3-(}CH_3CHCH_3\text{)-2-cyclopenten-1-ol (HO, }CH_3\text{)}$$

64. The middle carbonyl group is hydrated because it is stabilized by the electron-withdrawing carbonyl groups on either side of it.

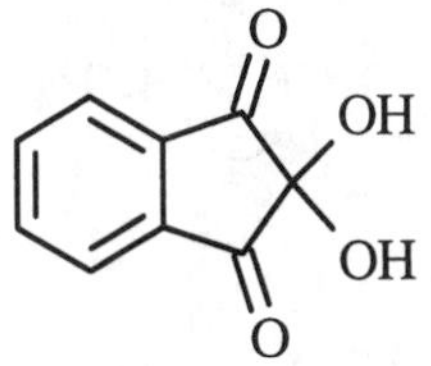

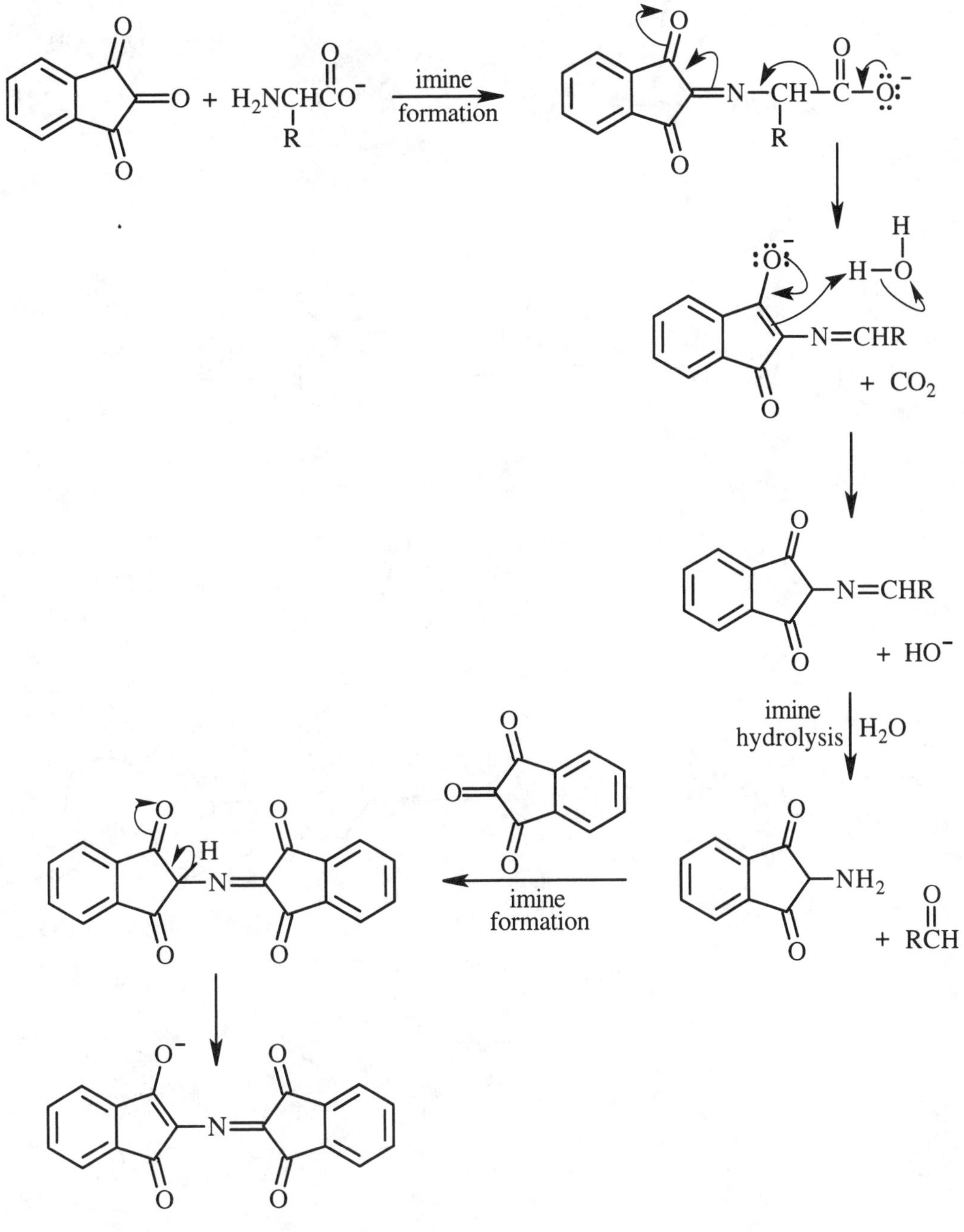

OH
OH

H_2NCHCO^- (R)

imine formation

N=CHR

$+ CO_2$

$+ HO^-$

imine hydrolysis H_2O

NH_2

$+ RCH$ (=O)

imine formation

65.

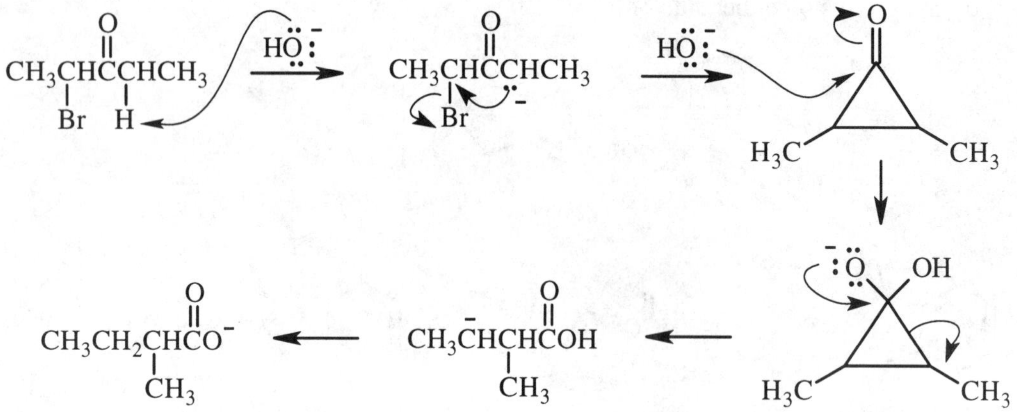

$CH_3CHCCHCH_3$ (Br, H) $\xrightarrow{HO^-}$ $CH_3CHCCHCH_3$ (Br) $\xrightarrow{HO^-}$ (cyclopropanone with H_3C, CH_3) $\rightarrow$ (O⁻, OH cyclopropane with H_3C, CH_3) $\rightarrow$ $CH_3\bar{C}HCHCOH$ (CH_3) $\rightarrow$ $CH_3CH_2CHCO^-$ (CH_3)

66.

a. 2-bromocyclohexanone $\xrightarrow{HO^-}$ enolate $\rightarrow$ bicyclic cyclopropanone $\xrightarrow{HO^-}$ (O⁻, OH adduct) $\rightarrow$ cyclopentyl anion with COOH $\rightarrow$ cyclopentane-COO^-

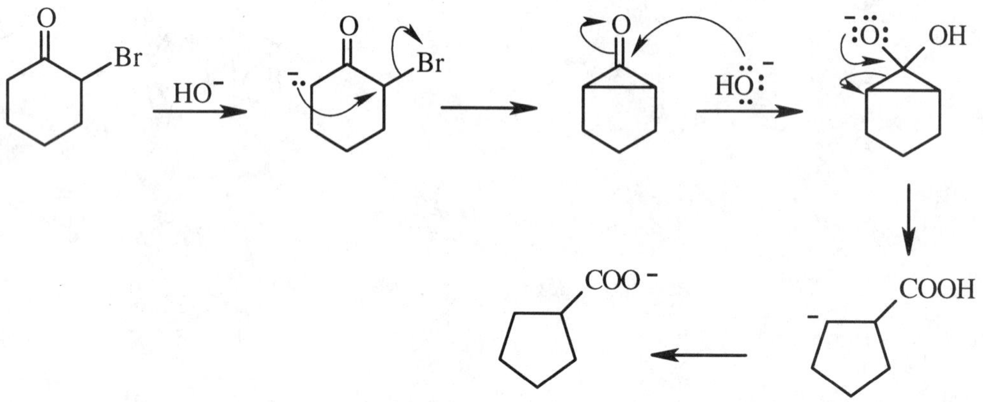

b. (bicyclic ketone with $OCCH_3$) $\xrightarrow{HO^-}$ (carbanion with $OCCH_3$) + H_2O $\rightarrow$ (bicyclic ketone anion with $OCCH_3$) $\xrightarrow{H_2O}$ (bicyclic ketone with $OCCH_3$) $\xrightarrow{HO^-}$ (COO^- / $OCCH_3$ cyclohexane) + (COO^- / $OCCH_3$ cyclohexane) $\xrightarrow[H_2O]{HO^-}$ CH_3CO^- + (COO^- / OH cyclohexane) + (COO^- / OH cyclohexane)

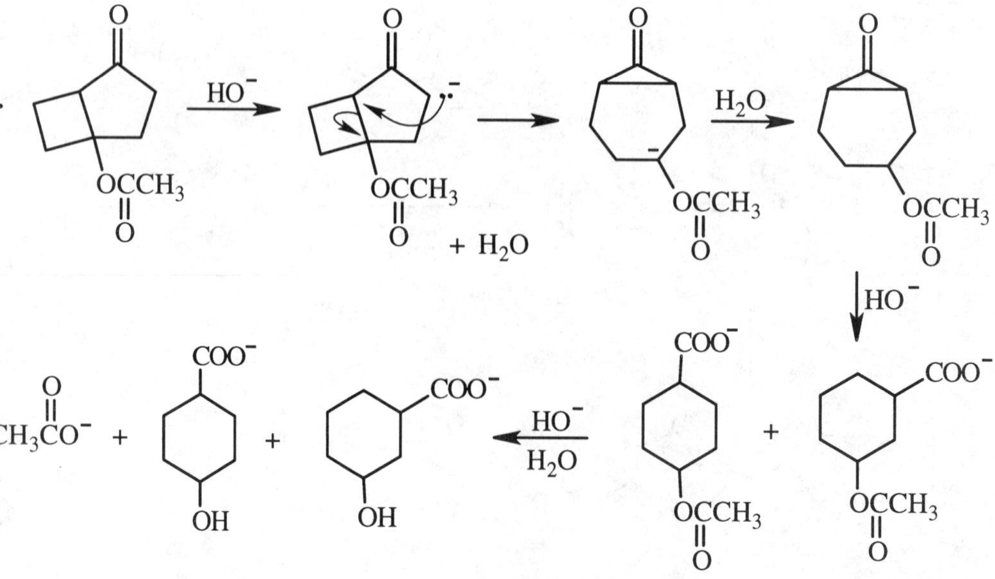

67.

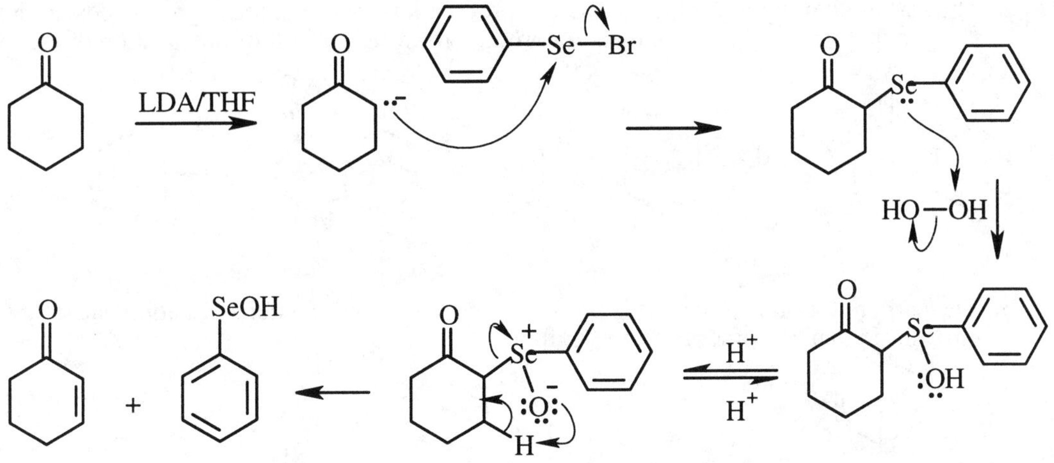

LDA/THF

Se—Br

HO—OH

H^+

H^+

SeOH

+

68.

HÖ:⁻

CH_2OH $\xleftarrow{H_2O}$ CH_2O^- + CO^-

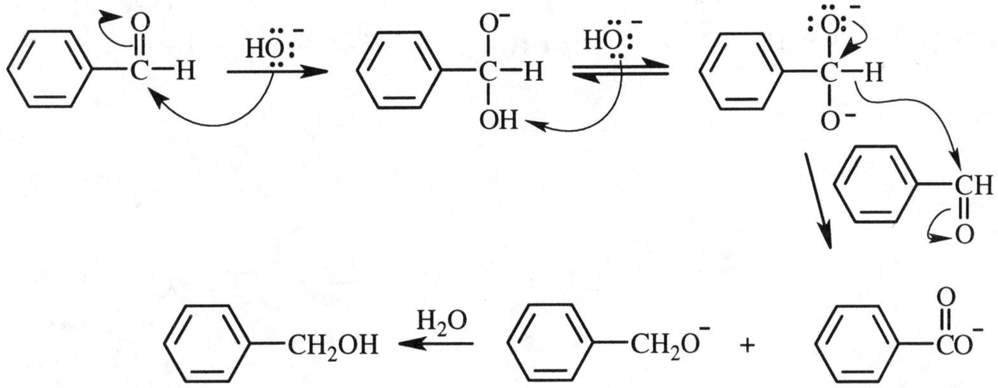

69. **a.** The carbonyl group is the most basic but protonation of the carbonyl group does not lead to a productive reaction. The reaction product forms as a result of protonation of the alkene.

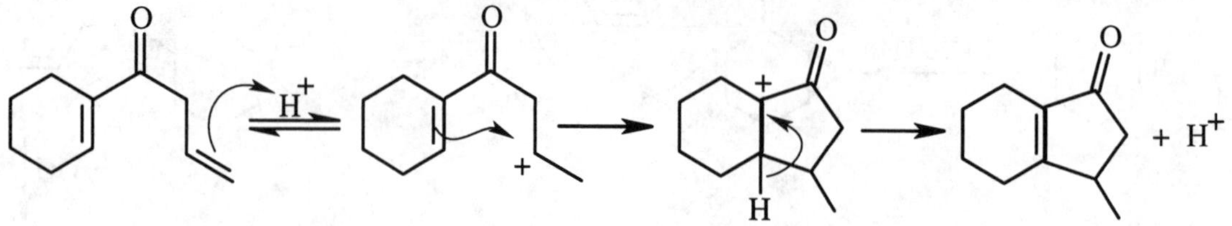

$\xrightarrow{H^+}$ + H^+

b. The carbonyl group on the left is protonated because electron delocalization causes it to be more basic than the other carbonyl group.

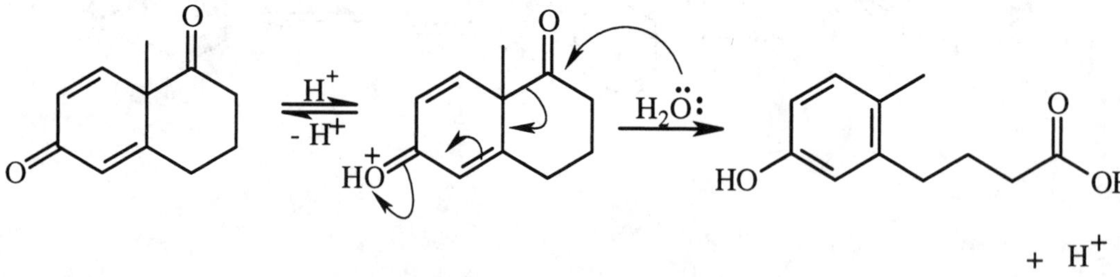

$\underset{-H^+}{\overset{H^+}{\rightleftharpoons}}$ $\xrightarrow{H_2\ddot{O}:}$ HO ... OH + H^+

70.

a. $CH_3CH_2CH_2OH \xrightarrow{PCC} CH_3CH_2CH{=}O \xrightarrow{HO^-} CH_3CH_2CH(OH)CH(CH_3)CH{=}O$

$\xrightarrow{\Delta,\ H_2SO_4} CH_3CH_2CH{=}C(CH_3)CH{=}O \xrightarrow[Pd/C]{H_2} CH_3CH_2CH_2CH(CH_3)CH_2OH$

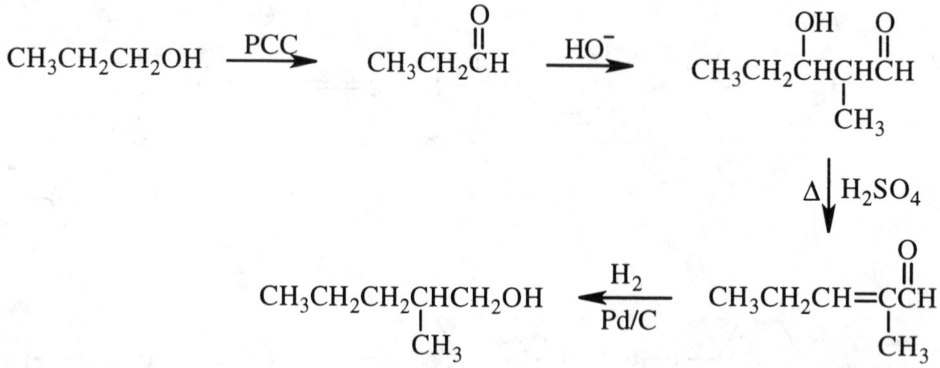

b. $CH_3CH(CH_3)OH \xrightarrow[H_2SO_4]{Na_2Cr_2O_7} CH_3C(=O)CH_3 \xrightarrow{HO^-} CH_3C(OH)(CH_3)CH_2C(=O)CH_3$

$\xrightarrow[2.\ H_3O^+]{1.\ NaBH_4} CH_3C(OH)(CH_3)CH_2CH(OH)CH_3 \xrightarrow[\Delta]{H_2SO_4} \left\{ \begin{array}{l} CH_3C(CH_3){=}CHCH{=}CH_2 \\ CH_2{=}C(CH_3)CH{=}CHCH_3 \end{array} \right. \xrightarrow[Pd/C]{H_2} CH_3CH(CH_3)CH_2CH_2CH_3$

c. $CH_3CH_2OH \xrightarrow{PCC} CH_3CH{=}O \xrightarrow[2.\ CH_3CH{=}O]{1.\ LDA/THF} CH_3CH(OH)CH_2CH{=}O \xrightarrow[2.\ H_3O^+]{1.\ NaBH_4} CH_3CH(OH)CH_2CH_2OH$

$CH_3CH(CH_3)OH \xrightarrow[H_2SO_4]{Na_2Cr_2O_7} CH_3C(=O)CH_3$; with H^+ → 2,2,4-trimethyl-1,3-dioxane (cyclic acetal: ring O–C(CH_3)(CH_3)–O, ring carbon bearing CH_3)

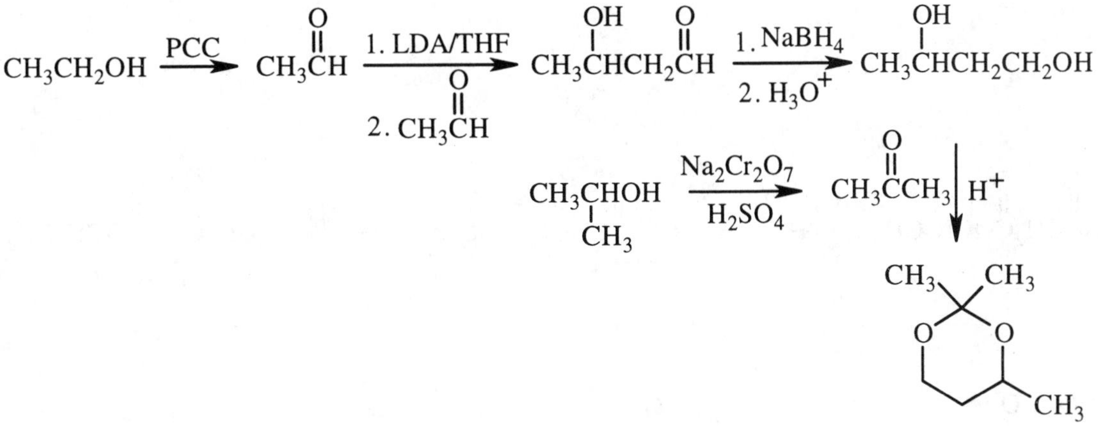

71. Hydroxide ion cannot be used in place of cyanide ion because the aldehydic hydrogen cannot be removed unless the electrons left behind can be delocalized onto an electronegative atom. The nitrogen of the cyano group serves that purpose.

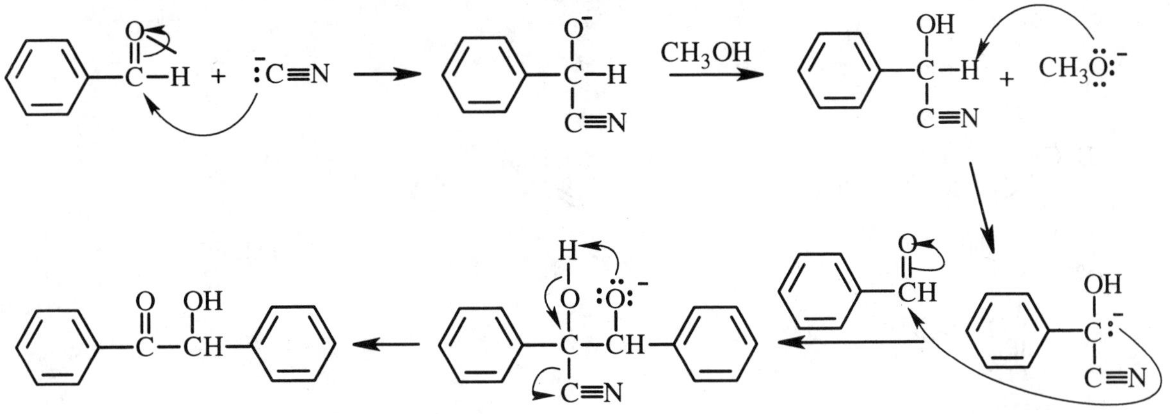

$C_6H_5CH{=}O + {}^-{:}C{\equiv}N \longrightarrow C_6H_5CH(O^-)C{\equiv}N \xrightarrow{CH_3OH} C_6H_5CH(OH)C{\equiv}N + CH_3\ddot{O}{:}^-$

$\longrightarrow C_6H_5\ddot{C}^-(OH)C{\equiv}N + C_6H_5CH{=}O \longrightarrow C_6H_5C(OH)(C{\equiv}N)CH(\ddot{O}{:}^-)C_6H_5 \longrightarrow C_6H_5C(=O)CH(OH)C_6H_5$

72. To arrive at the final product, three equivalents of malonyl thioester are needed. (See page 865 in the text.)

$$CH_3\overset{O}{\overset{\|}{C}}SR\ (*) + 3\ {}^-O\overset{O}{\overset{\|}{C}}CH_2\overset{O}{\overset{\|}{C}}SR \longrightarrow CH_3\overset{O}{\overset{\|}{C}}CH_2\overset{O}{\overset{\|}{C}}CH_2\overset{O}{\overset{\|}{C}}CH_2\overset{O}{\overset{\|}{C}}SR \longrightarrow$$

(aromatic product: benzene ring bearing CH_3, COOH, OH, and HO; the labeled carbon * is the ring carbon bearing CH_3)

In Chapter 23, you will see that malonyl thioester is synthesized from acetyl thioester. So if the lichen is not fed unlabeled malonyl thioester, its malonyl thioester will also be labeled, and the product will have labels in the indicated positions.

$$CH_3\overset{O}{\overset{\|}{C}}CH_2\overset{O}{\overset{\|}{C}}CH_2\overset{O}{\overset{\|}{C}}CH_2\overset{O}{\overset{\|}{C}}SR \ (\text{each carbonyl carbon } *) \longrightarrow$$

(aromatic product: benzene ring bearing CH_3, COOH, OH, and HO; labels * on the ring carbon bearing CH_3, the COOH carbon, and the ring carbons bearing HO and OH)

73.

$$CH_3\overset{O}{\overset{\|}{C}}CH_2\overset{O}{\overset{\|}{C}}OCH_2CH_3 \xrightarrow{CH_3CH_2O^-} CH_3\overset{O}{\overset{\|}{C}}\underset{-}{C}H\overset{O}{\overset{\|}{C}}OCH_2CH_3 \xrightarrow{H_2C=O} CH_3\overset{O}{\overset{\|}{C}}\underset{CH_2OH}{\underset{|}{C}H}\overset{O}{\overset{\|}{C}}OCH_2CH_3$$

$$\downarrow$$

$$CH_3\overset{O}{\overset{\|}{C}}\underset{CH_2}{\underset{\|}{C}}\overset{O}{\overset{\|}{C}}OCH_2CH_3 \xrightarrow{CH_3\overset{O}{\overset{\|}{C}}\underset{-}{C}H\overset{O}{\overset{\|}{C}}OCH_2CH_3} CH_3\overset{O}{\overset{\|}{C}}CH\overset{O}{\overset{\|}{C}}OCH_2CH_3 \text{—}CH_2\text{—}CH_3\overset{O}{\overset{\|}{C}}CH\overset{O}{\overset{\|}{C}}OCH_2CH_3$$

$$\xrightarrow{CH_3CH_2O^-} CH_3\overset{O}{\overset{\|}{C}}CH\overset{O}{\overset{\|}{C}}OCH_2CH_3 \text{—}CH_2\text{—}\ddot{C}H_2\overset{O}{\overset{\|}{C}}CH\overset{O}{\overset{\|}{C}}OCH_2CH_3$$

$$\downarrow$$

(cyclohexanone ring bearing H_3C, OH, two $\overset{O}{\overset{\|}{C}}OCH_2CH_3$ groups, and O=) $\xrightarrow[\Delta]{H^+}$ (cyclohexenone ring bearing CH_3 and $\overset{O}{\overset{\|}{C}}OCH_2CH_3$) $+\ H_2O + CO_2 + CH_3CH_2OH$

74.

$$CH_3CH_2O\overset{O}{\overset{\|}{C}}CH_2\overset{O}{\overset{\|}{C}}OCH_2CH_3 \xrightarrow[2.\ CH_3CH_2Br]{1.\ CH_3CH_2O^-} CH_3CH_2O\overset{O}{\overset{\|}{C}}\underset{CH_2CH_3}{\underset{|}{C}H}\overset{O}{\overset{\|}{C}}OCH_2CH_3$$

1. $CH_3CH_2O^-$
2. $CH_3\underset{CH_3}{\underset{|}{C}H}CH_2CH_2Br$

$CH_3CH_2O\overset{O}{\overset{\|}{C}}C(CH_2CH_3)(CH_2CH_2\underset{CH_3}{\underset{|}{C}H}CH_3)\overset{O}{\overset{\|}{C}}OCH_2CH_3$ $\xrightarrow[CH_3CH_2O^-,\ \Delta]{H_2N\overset{O}{\overset{\|}{C}}NH_2}$ 5-ethyl-5-($CH_2CH_2\underset{CH_3}{\underset{|}{C}H}CH_3$) barbiturate (HN, N-H, three C=O)

75. a.

$$CH_3\overset{O}{\overset{\|}{C}}CH_2\overset{O}{\overset{\|}{C}}OCH_2CH_3 \xrightarrow{CH_3CH_2O^-} CH_3\overset{O}{\overset{\|}{C}}\ddot{C}^-H\overset{O}{\overset{\|}{C}}OCH_2CH_3$$

+ epoxide (CH_2CH_3-substituted) $\longrightarrow$ $CH_3\overset{O}{\overset{\|}{C}}CH(CH_2\underset{CH_2CH_3}{CH}-\ddot{O}:^-)\overset{O}{\overset{\|}{C}}OCH_2CH_3$

$\downarrow$

tetrahedral intermediate (ring with CH_3CH_2, $\overset{O}{\overset{\|}{C}}CH_3$, $:\ddot{O}:^-$, OCH_2CH_3) $\longrightarrow$ $CH_3CH_2O^-$ + lactone (CH_3CH_2, $\overset{O}{\overset{\|}{C}}CH_3$, O, =O)

b.

dihydropyran + BrCH_2-aryl (HO, NO_2) $\longrightarrow$ oxocarbenium ($+\ddot{O}$, $H\ddot{O}$, NO_2) + Br^-

$\downarrow$

protonated cyclic acetal ($\ddot{O}$, $\ddot{O}^+$H, NO_2) $\xrightarrow{-H^+}$ cyclic acetal ($\ddot{O}$, $\ddot{O}$, NO_2)

76.

a.

$C_6H_5CH_2CH_2Cl \xrightarrow[2.\ H_2/Pt]{1.\ {}^-N_3} C_6H_5CH_2CH_2NH_2$

$C_6H_5CH_2CH_2Cl \xrightarrow[2.\ H^+,\ H_2O,\ \Delta \quad 3.\ HO^-]{1.\ \text{phthalimide anion}} C_6H_5CH_2CH_2NH_2$

b.

$C_6H_5CH_2Cl \xrightarrow{{}^-C{\equiv}N} C_6H_5CH_2C{\equiv}N \xrightarrow[Pt]{H_2} C_6H_5CH_2CH_2NH_2$

c.

$C_6H_5CH{=}O \xrightarrow{{}^-C{\equiv}N} C_6H_5CH(OH)C{\equiv}N \xrightarrow[Pt]{H_2} C_6H_5CH(OH)CH_2NH_2 \xrightarrow[\Delta]{H_2SO_4} C_6H_5CH{=}CH\overset{+}{N}H_3$

$C_6H_5CH{=}CH\overset{+}{N}H_3 \xrightarrow{H_2,\ Pt} C_6H_5CH_2CH_2\overset{+}{N}H_3 \xrightarrow{HO^-} C_6H_5CH_2CH_2NH_2$

or

$C_6H_5CH{=}O \xrightarrow[2.\ H_3O^+]{1.\ NaBH_4} C_6H_5CH_2OH \xrightarrow{SOCl_2} C_6H_5CH_2Cl \xrightarrow{{}^-C{\equiv}N} C_6H_5CH_2C{\equiv}N \xrightarrow[Pt]{H_2} C_6H_5CH_2CH_2NH_2$

d.

$C_6H_5CH_2CH_2NH_2$ $\xrightarrow{CH_3CCl\ (\text{C=O})}$ $C_6H_5CH_2CH_2NHCCH_3$ (C=O) $\xrightarrow[H_2SO_4]{HNO_3}$ $p\text{-}O_2NC_6H_4CH_2CH_2NHCCH_3$ (C=O)

$\xrightarrow[2.\ HO^-]{1.\ Sn/HCl}$ $p\text{-}H_2NC_6H_4CH_2CH_2NHCCH_3$ (C=O) $\xrightarrow[0°]{HCl,\ NaNO_2}$ $p\text{-}\overset{+}{N}{\equiv}NC_6H_4CH_2CH_2NHCCH_3$ (C=O) $\xrightarrow[\Delta]{H_2O}$ $p\text{-}HOC_6H_4CH_2CH_2NHCCH_3$ (C=O)

$\xrightarrow[2.\ HO^-]{1.\ H_3O^+,\ \Delta}$ $p\text{-}HOC_6H_4CH_2CH_2NH_2$

e.

$p\text{-}HOC_6H_4CH_2CH_2NH_2$ $\xrightarrow[H_2SO_4]{HNO_3}$ 4-OH, 3-NO_2 $C_6H_3CH_2CH_2\overset{+}{N}H_3$ $\xrightarrow[2.\ HO^-]{1.\ Sn/HCl}$ 4-OH, 3-NH_2 $C_6H_3CH_2CH_2NH_2$

$\xrightarrow[0°]{HCl,\ NaNO_2}$ 4-OH, 3-$\overset{+}{N}{\equiv}N$ $C_6H_3CH_2CH_2\overset{+}{N}{\equiv}N$ $\xrightarrow{H_2O}$ 3,4-(OH)$_2C_6H_3CH_2CH_2OH$ $\xrightarrow[2.\ ^-N_3\ \ 3.\ H_2/Pt]{1.\ PCl_3}$ 3,4-(OH)$_2C_6H_3CH_2CH_2NH_2$

77.

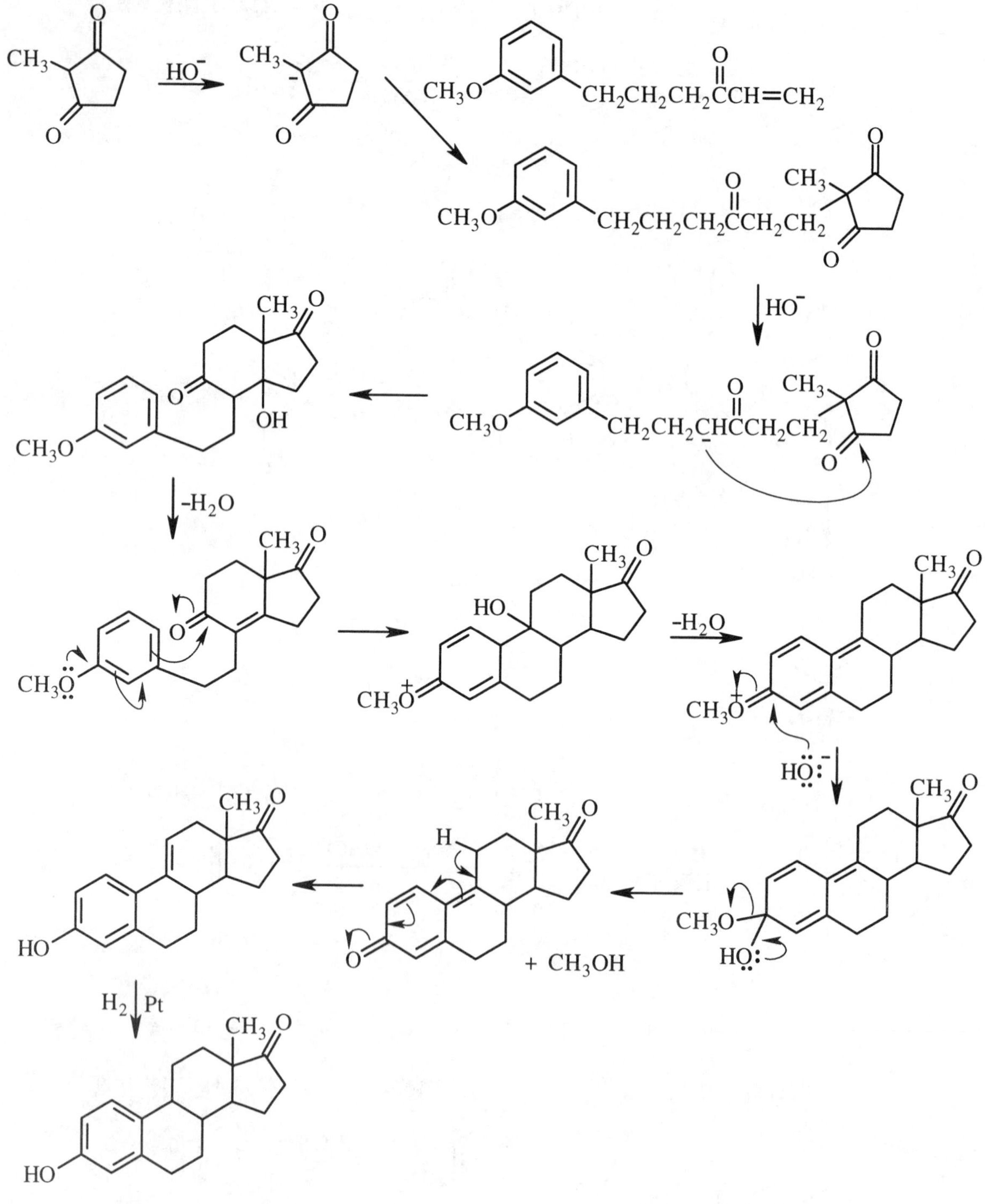

CH_3O $CH_2CH_2CH_2\overset{O}{\overset{\|}{C}}CH{=}CH_2$

HO^-

CH_3O $CH_2CH_2CH_2\overset{O}{\overset{\|}{C}}CH_2CH_2$ CH_3

HO^-

CH_3O $CH_2CH_2\overset{-}{C}H\overset{O}{\overset{\|}{C}}CH_2CH_2$ CH_3

CH_3O CH_3 OH

$-H_2O$

CH_3O CH_3

HO CH_3 $CH_3\overset{+}{O}$

$-H_2O$

CH_3 $CH_3\overset{+}{O}$

HO^-

CH_3 CH_3O HO

H CH_3 O + CH_3OH

CH_3 HO

H_2 Pt

CH_3 HO

78. The substituent in ibuprofen is placed on the ring by a Friedel-Crafts acylation reaction. A Friedel-Crafts reaction cannot be done in the synthesis of ketoprofen because the benzene rings are deactivated and deactivated rings cannot undergo Friedel-Crafts reactions.

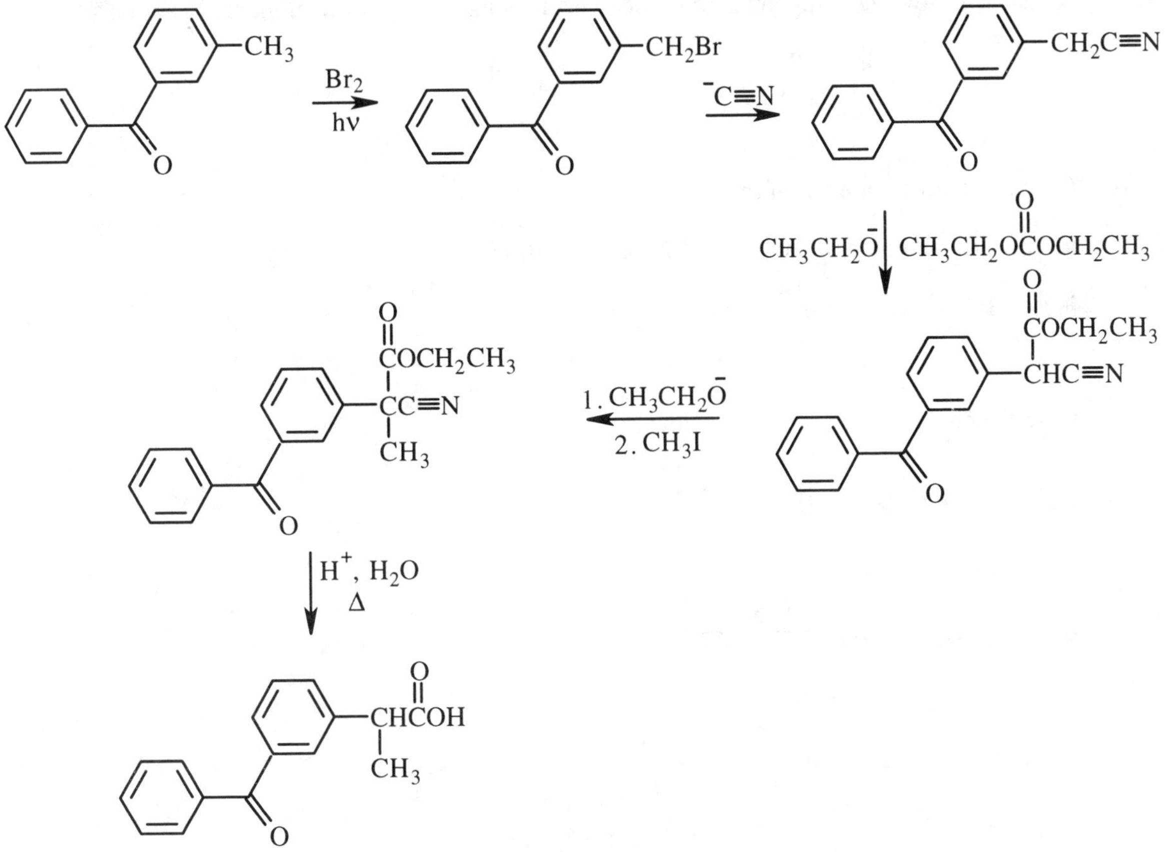

Br_2, $h\nu$

$-CH_2Br$

$^-C{\equiv}N$

$-CH_2C{\equiv}N$

$CH_3CH_2O^-$ | $CH_3CH_2OCOCH_2CH_3$ (with C=O)

$-CHC{\equiv}N$ with $COCH_2CH_3$ (C=O)

1. $CH_3CH_2O^-$
2. CH_3I

$-CC{\equiv}N$ with $COCH_2CH_3$ (C=O) and CH_3

H^+, H_2O, Δ

$-CHCOH$ (C=O) with CH_3

Chapter 19 Practice Test

1. Rank the following compounds in order of decreasing acidity. (Label the most acidic #1.)

$$CH_3\overset{\overset{O}{\|}}{C}CH_2\overset{\overset{O}{\|}}{C}OCH_3 \qquad CH_3\overset{\overset{O}{\|}}{C}CH_2\overset{\overset{O}{\|}}{C}CH_3 \qquad CH_3\overset{\overset{O}{\|}}{C}CH_3$$

2. Give a structure for each of the following:

a. the most stable enol tautomer of 2,4-pentanedione

b. a β-keto ester

3. Give the product of each of the following reactions.

a. $CH_3CH_2\overset{\overset{O}{\|}}{C}CH_2\overset{\overset{O}{\|}}{C}OH \xrightarrow{\Delta}$

b. $CH_3CH_2CH_2\overset{\overset{O}{\|}}{C}H \xrightarrow{Br_2,\ HO^-}$

c. $2\ CH_3CH_2CH_2\overset{\overset{O}{\|}}{C}OCH_3 \xrightarrow[2.\ H^+]{1.\ CH_3O^-}$

d. $CH_2{=}CH\overset{\overset{O}{\|}}{C}CH_3 + CH_3\overset{\overset{O}{\|}}{C}CH_2\overset{\overset{O}{\|}}{C}CH_3 \xrightarrow{HO^-}$

e. cyclopentanone $\xrightarrow[2.\ tert\text{-}BuO^-]{1.\ Br_2,\ H^+}$

f. cyclohexanone $+\ Cl_2 \xrightarrow{H^+,\ H_2O}$

g. $CH_3CH_2O\overset{\overset{O}{\|}}{C}CH_2\overset{\overset{O}{\|}}{C}OCH_2CH_3 \xrightarrow[2.\ CH_3CH_2CH_2Br]{1.\ CH_3CH_2O^-}$
3. H^+, H_2O, Δ

4. Give an example of each of the following:

a. an aldol addition

b. an aldol condensation

c. a Claisen condensation

d. a Dieckmann condensation

e. a malonic ester synthesis

f. an acetoacetic ester synthesis

5. Give the products of the following crossed aldol addition.

$$CH_3\underset{\displaystyle CH_3}{\underset{|}{C}}HCH_2CH_2\overset{\displaystyle O}{\overset{\|}{C}}H \quad + \quad CH_3CH_2CH_2CH_2\overset{\displaystyle O}{\overset{\|}{C}}H \xrightarrow{HO^-}$$

CHAPTER 20
Carbohydrates

Important Terms

aldaric acid a dicarboxylic acid with an OH group bonded to each carbon. Obtained by oxidizing the aldehyde and primary alcohol groups of an aldose.

alditol a compound with an OH group bonded to each carbon. Obtained by reducing an aldose or a ketose.

aldonic acid a carboxylic acid with an OH group bonded to each carbon. Obtained by oxidizing the aldehyde group of an aldose.

aldose a polyhydroxyaldehyde.

amino sugar a sugar in which one of the OH groups is replaced by an NH_2 group.

anomeric carbon the carbon in a cyclic sugar that is the carbonyl carbon in the straight-chain form.

anomeric effect preference for the axial position by certain substituents bonded to the anomeric carbon

anomers two cyclic sugars that differ in configuration only at the carbon that is the carbonyl carbon in the straight-chain form.

antibody a compound that recognizes foreign particles in the body.

antigen a compound that can generate a response from the immune system.

bioorganic compound an organic compound that is found in a biological system.

carbohydrate a sugar, a saccharide. Naturally occurring carbohydrates have the D-configuration.

complex carbohydrate contains two or more sugar molecules linked together; it can be hydrolyzed to simple sugars.

deoxy sugar a sugar in which one of the OH groups has been replaced by a hydrogen.

disaccharide a compound containing two sugar molecules linked together.

epimers monosaccharides that differ in configuration at only one carbon.

furanose a five-membered ring sugar.

furanoside a five-membered ring glycoside.

glycoprotein a protein that is covalently bonded to a polysaccharide.

glycoside the acetal of a sugar.

***N*-glycoside** a glycoside with a nitrogen instead of an oxygen at the glycosidic linkage.

glycosidic bond — the bond between the anomeric carbon and the alcohol residue in a glycoside.

α-1,4'-glycosidic linkage — a glycosidic linkage between the C-1 of one sugar and the C-4 of a second sugar with the oxygen atom at C-1 in the axial position.

β-1,4'-glycosidic linkage — a glycosidic linkage between the C-1 of one sugar and the C-4 of a second sugar with the oxygen atom at C-1 in the equatorial position.

Haworth projection — a way to show the structure of a sugar in which the five- and six-membered rings are represented as being flat.

heptose — a monosaccharide with seven carbons.

hexose — a monosaccharide with six carbons.

ketose — a polyhydroxyketone.

Kiliani-Fischer synthesis — a method used to increase the number of carbons in an aldose by one, resulting in the formation of a pair of C-2 epimers.

molecular recognition — the ability of molecules to recognize one another.

monosaccharide — a single sugar molecule.

mutarotation — a slow change in optical rotation to an equilibrium value.

nonreducing sugar — a sugar that cannot be oxidized by reagents such as Ag^+ and Cu^+. Nonreducing sugars are not in equilibrium with the open-chain aldose or ketose.

oligosaccharide — three to ten sugar molecules linked by glycosidic bonds.

osazone — the product obtained by reacting an aldose or a ketose with excess phenylhydrazine. An osazone contains two imine bonds.

oxocarbenium ion — an ion in which the positive charge is shared by a carbon and an oxygen.

pentose — a monosaccharide with five carbons.

photosynthesis — the synthesis of glucose and O_2 from CO_2 and H_2O.

polysaccharide — a compound containing ten or more sugar molecules linked together.

pyranose — a six-membered ring sugar.

pyranoside — a six-membered ring glycoside.

reducing sugar — a sugar that can be oxidized by reagents such as Ag^+ and Cu^+. Reducing sugars are in equilibrium with the open-chain aldose or ketose.

Ruff degradation — a method used to shorten an aldose by one carbon.

simple carbohydrate	a single sugar molecule.
tetrose	a monosaccharide with four carbons.
triose	a monosaccharide with three carbons.

Solutions to Problems

1. D-Ribose is an aldopentose.
D-Sedoheptulose is a ketoheptose.
D-Mannose is an aldohexose.

2.

```
      HC=O                 CH2OH
  HO──┼──H                  C=O
   H──┼──OH              H──┼──OH
  HO──┼──H              HO──┼──H
  HO──┼──H              HO──┼──H
      CH2OH                 CH2OH
```

L-glucose L-fructose

3.

```
         HC=O              H                  CH2OH
HOCH2────┼────OH    HO─────┼─────CH2OH    HO──┼──H
         H                HC=O                HC=O
```

L-glyceraldehyde L-glyceraldehyde D-glyceraldehyde

4. **a.** enantiomers

b. diastereomers

5. **a.** D-ribose **b.** L-talose **c.** L-allose

6. **a.** (2*R*,3*S*,4*R*,5*R*)-2,3,4,5,6-pentahydroxyhexanal

b. (2*S*,3*R*,4*S*,5*S*)-2,3,4,5,6-pentahydroxyhexanal

7. D-psicose

8. **a.** A ketoheptose has four chirality centers ($2^4 = 16$ stereoisomers).

b. An aldoheptose has five chirality centers ($2^5 = 32$ stereoisomers).

c. A ketotriose has no chirality centers; therefore, it has no stereoisomers.

9. **a.** When D-idose is reduced, D-iditol is formed.

```
       CH=O                              CH2OH
  HO ──┼── H                        HO ──┼── H
   H ──┼── OH        H2              H ──┼── OH
  HO ──┼── H      ───────►          HO ──┼── H
   H ──┼── OH        Pd/C            H ──┼── OH
       CH2OH                             CH2OH
      D-idose                           D-iditol
```

b. When D-sorbose is reduced, C-2 becomes a chirality center, so both D-iditol and the C-2 epimer of D-iditol (D-gulitol) are formed.

```
      CH2OH                        CH2OH                 CH2OH
      C=O                     HO ──┼── H            H ──┼── OH
  H ──┼── OH       H2          H ──┼── OH           H ──┼── OH
 HO ──┼── H     ──────►       HO ──┼── H      +    HO ──┼── H
  H ──┼── OH       Pd/C        H ──┼── OH           H ──┼── OH
      CH2OH                        CH2OH                 CH2OH
    D-sorbose                    D-iditol              D-gulitol
```

10. **a.** **1.** D-Altrose is reduced to the same alditol as D-talose.
The easiest way to answer this question is to draw D-talose and its alditol. Then draw the monosaccharide with the same configuration at C-2, C-3, C-4, and C-5 as D-talose, reversing the functional groups at C-1 and C-6. (Put the alcohol group at the top and the aldehyde group at the bottom.) The resulting Fischer projection can be rotated 180° in the plane of the paper, and the monosaccharide can then be identified.

```
      CH=O                    CH2OH                  CH2OH                               CH=O
 HO ──┼── H              HO ──┼── H            HO ──┼── H                       HO ──┼── H
 HO ──┼── H     H2       HO ──┼── H     H2     HO ──┼── H      rotate 180°       H ──┼── OH
 HO ──┼── H   ─────►     HO ──┼── H   ◄─────   HO ──┼── H     ◄- - - - - ►       H ──┼── OH
  H ──┼── OH    Pd/C      H ──┼── OH    Pd/C    H ──┼── OH                       H ──┼── OH
      CH2OH                   CH2OH                  CH=O                                CH2OH
    D-talose                                                                           D-altrose
```

2. L-Galactose is reduced to the same alditol as D-galactose.

D-galactose $\xrightarrow[\text{Pd/C}]{H_2}$ alditol $\xleftarrow[\text{Pd/C}]{H_2}$ (rotate 180°) L-galactose

CH=O		CH$_2$OH		CH$_2$OH		CH=O
H—OH		H—OH		H—OH		HO—H
HO—H	H_2 / Pd/C →	HO—H	← H_2 / Pd/C	HO—H	← rotate 180° →	H—OH
HO—H		HO—H		HO—H		H—OH
H—OH		H—OH		H—OH		HO—H
CH$_2$OH		CH$_2$OH		CH=O		CH$_2$OH
D-galactose						L-galactose

b. 1. The ketohexose (D-tagatose) with the same configuration at C-3, C-4, and C-5 as D-talose will give the same alditol as D-talose. The other alditol is the one with the opposite cofiguration at C-2.

CH$_2$OH		CH$_2$OH		CH$_2$OH
C=O		HO—H		H—OH
HO—H	H_2 / Pd/C →	HO—H	+	HO—H
HO—H		HO—H		HO—H
H—OH		H—OH		H—OH
CH$_2$OH		CH$_2$OH		CH$_2$OH
D-tagatose		D-talitol		D-galactitol

2. D-fructose

CH$_2$OH		CH$_2$OH		CH$_2$OH
C=O		HO—H		H—OH
HO—H	H_2 / Pd/C →	HO—H	+	HO—H
H—OH		H—OH		H—OH
H—OH		H—OH		H—OH
CH$_2$OH		CH$_2$OH		CH$_2$OH
D-fructose		D-mannitol		D-glucitol

11. Removal of an α-hydrogen creates an enol that can enolize back to the ketone (using the OH at C-2) or can enolize to an aldehyde (using the OH at C-1). The aldehyde has a new chirality center; one of the epimers is D-glucose, and the other is D-mannose.

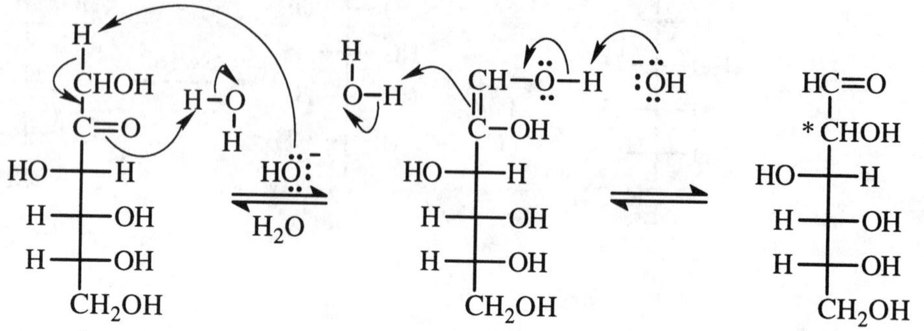

D-fructose

D-glucose
D-mannose

12. D-tagatose, D-galactose, and D-talose

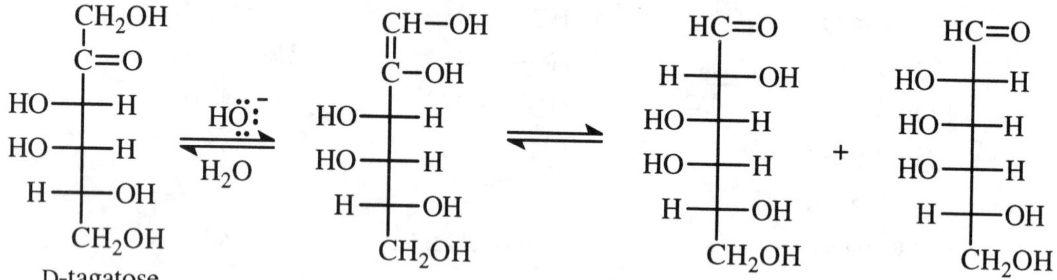

D-tagatose

D-galactose

D-talose

13. **a.** L-gulose

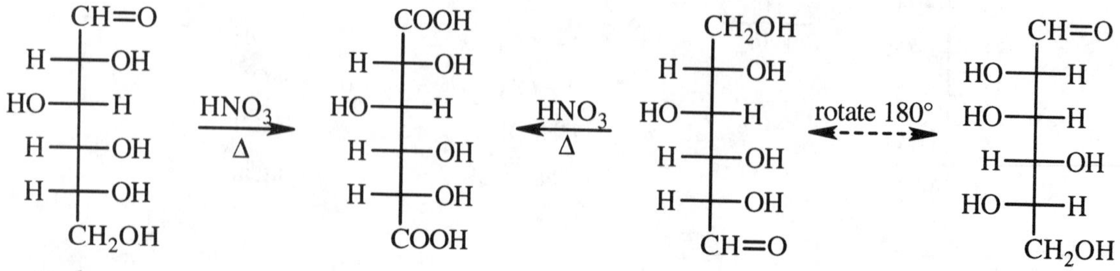

D-glucose

D-glucaric acid

rotate 180°

L-gulose

b. L-Gularic acid, because it is also the oxidation product of L-gulose.

c. D-allose and L-allose, D-altrose and D-talose, L-altrose and L-talose, D-galactose and L-galactose

14. **a.** D-arabinose and D-ribulose

b. D-allose and D-psicose

c. L-gulose and L-sorbose

d. D-talose and D-tagatose

15. D-gulose and D-idose

16. **a.** D-gulose and D-idose

b. L-xylose and L-lyxose

17. **a.** D-glucose and D-mannose

b. D-erythrose and D-threose

c. L-allose and L-altrose

18. Solved in the text.

19.

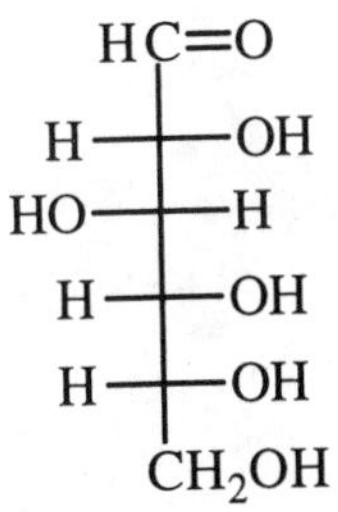

A	B	C	D
HC=O	HC=O	HC=O	HC=O
H—OH	HO—H	HO—H	H—OH
HO—H	HO—H	H—OH	H—OH
H—OH	H—OH	H—OH	CH_2OH
H—OH	H—OH	CH_2OH	
CH_2OH	CH_2OH		
D-glucose	D-mannose	D-arabinose	D-erythrose

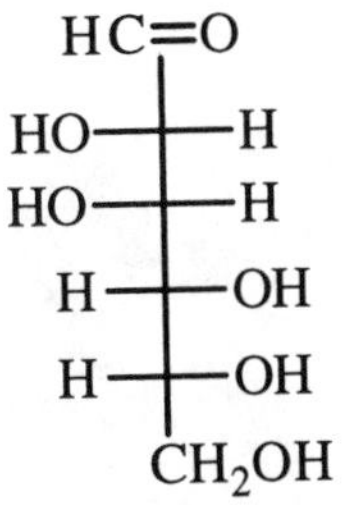

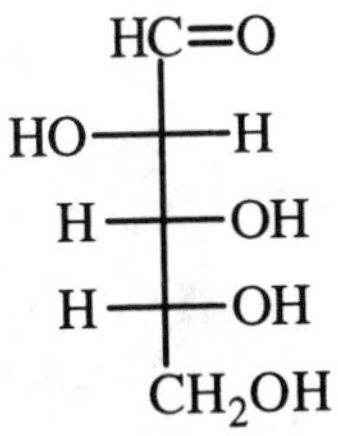

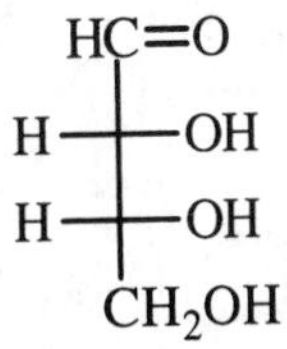

20. The hemiacetals in **a** and **b** have one chirality center; therefore, each has two stereoisomers. The hemiacetals in **c** and **d** have two chirality centers; therefore, each has four stereoisomers.

c.

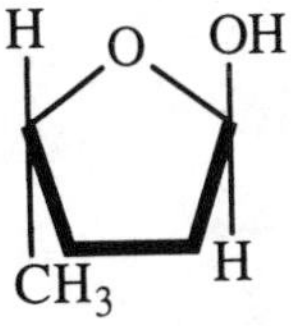

CH_3 O H, H OH CH_3 O OH, H H   H O H, CH_3 OH   H O OH, CH_3 H

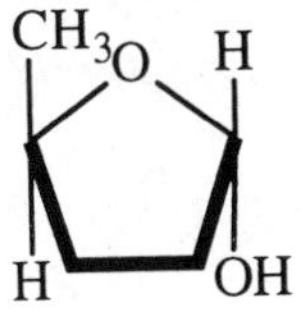

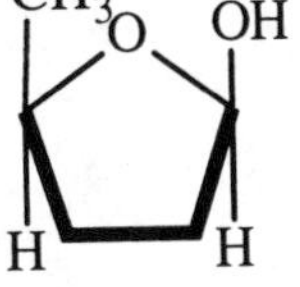

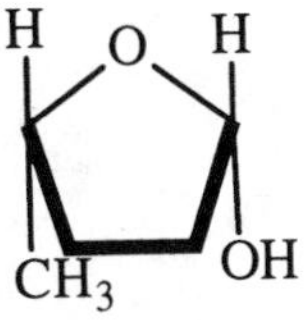

d.

$CH_3CH_2CH_2$ O H, H OH

$CH_3CH_2CH_2$ O OH, H H

H O H, $CH_3CH_2CH_2$ OH

H O OH, $CH_3CH_2CH_2$ H

21.

a.

CH_2OH, HO, O, OH, H, OH, H, H, H, H, OH

β-D-galactose

b.

$HOCH_2$ O CH_2OH, OH HO, H OH, H H

α-D-tagatose
furanose form

H, H O CH_2OH, H, OH HO, HO OH, H H

α-D-tagatose
pyranose form

c.

CH_2OH, H O H, H, OH H, HO OH, H OH

α-D-glucose

CH_2OH, H O H, H, H HO, HO OH, OH H

α-L-glucose
the mirror image of α-D-glucose

22.

CH_2OH, HO—H, O H, OH H, H OH, H OH

23. First recall that in the more stable chair conformation: an α-anomer has the anomeric carbon in the axial position, while a β-anomer has the anomeric carbon in the equatorial position. Then recall that glucose has all its OH groups in equatorial positions. Now this question can be answered easily.

a. Mannose is a C-2 epimer of glucose. Therefore, the OH group at C-2 in β-D-mannose is in the axial position.

b. Idose differs in configuration from glucose at C-2, C-3, and C-4. Therefore, the OH groups at C-2, C-3, and C-4 in β-D-idose are in the axial position.

c. Allose is a C-3 epimer of glucose. Therefore, the OH group at C-3 is in the axial position, and since it is the α-anomer, the OH group at C-1 (the anomeric carbon) is in the axial position.

24. The contributing resonance structure on the left is more stable since the other contributing resonance structure has an atom (the C-1 carbon) with an incomplete octet.

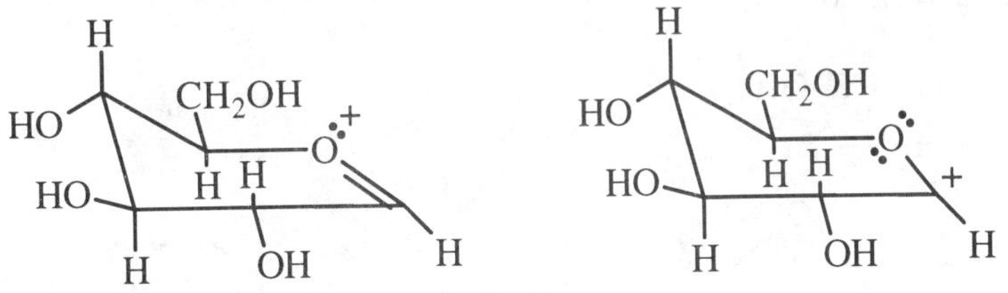

more stable

25. If more than a trace amount of acid is used, the amine that acts as a nucleophile when it forms the *N*-glycoside becomes protonated, and a protonated amine is not nucleophilic.

26. **a.** propyl β-D-allopyranoside, a nonreducing sugar. (It does not have a hemiacetal linkage.)
b. methyl α-D-galactopyranoside, a nonreducing sugar. (It does not have a hemiacetal linkage.)
c. α-D-talopyranose, a reducing sugar. (It has a hemiacetal linkage.)
d. ethyl β-D-psicofuranoside, a nonreducing sugar. (It does not have a hemiacetal linkage.)

27.

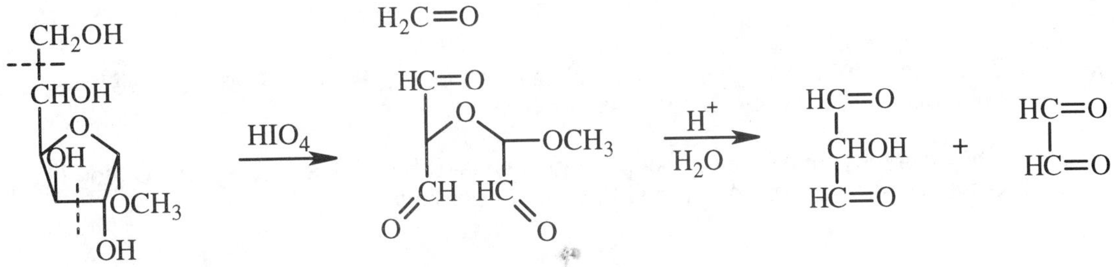

28. **a.** Amylose has α-1,4'-glycosidic linkages, whereas cellulose has β-1,4'-glycosidic linkages.

b. Amylose has α-1,4'-glycosidic linkages, whereas amylopectin has both α-1,4'-glycosidic linkages and α-1,6'-glycosidic linkages.

c. Glycogen and amylopectin have the same kind of linkages, but glycogen has a higher frequency of α-1,6'-glycosidic linkages.

d. Cellulose has a hydroxyl at C-2, whereas chitin has an *N*-acetylamino group at that position.

29. A proton is more easily lost from the C-3 OH group because the anion that is formed when the proton is removed is more stable than the anion that is formed when a proton is removed from the C-2 OH group. When a proton is removed from the C-3 OH group, the electrons that are left behind can be delocalized onto another oxygen atom. When a proton is removed from the C-2 OH group, the electrons that are left behind are delocalized onto a carbon atom. Because oxygen is more electronegative than carbon, a negatively charged oxygen is more stable than a negatively charged carbon.

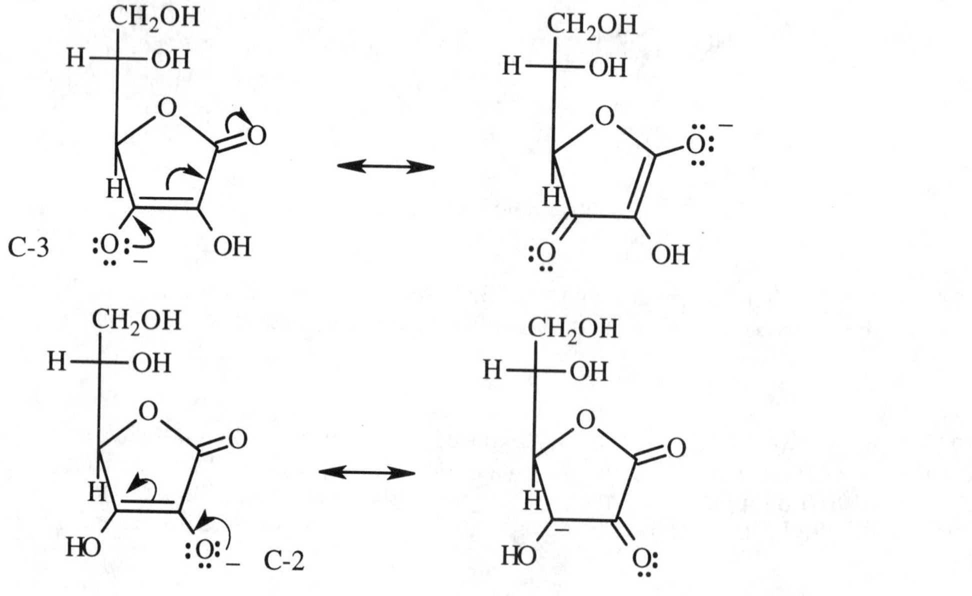

30. **a.** They can only receive Type O blood, because Type A, B, and AB blood have sugar components that Type O blood does not have.

b. They can only give blood to people with Type AB blood, because Type AB blood has sugar components that Type A, B, or O blood does not have.

31.

a. 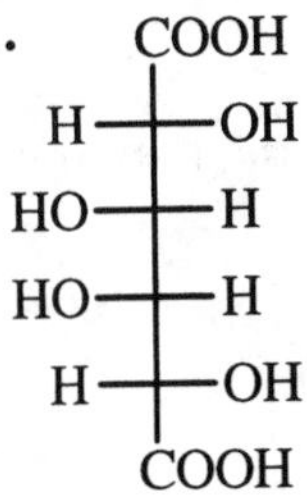

b. 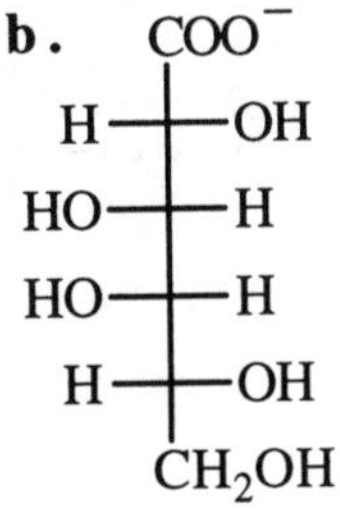

c. 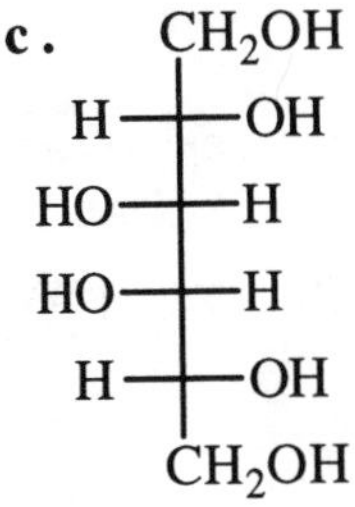

d. 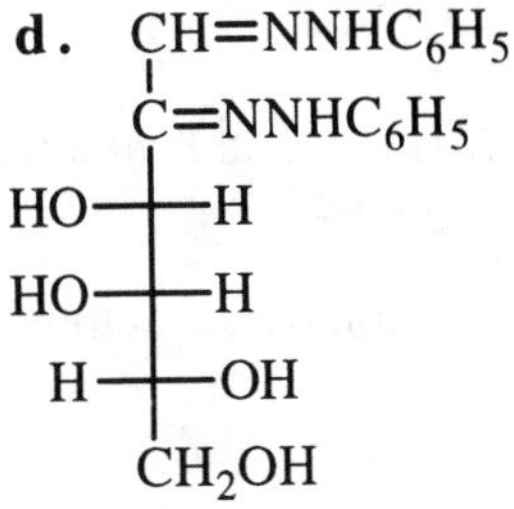

e. 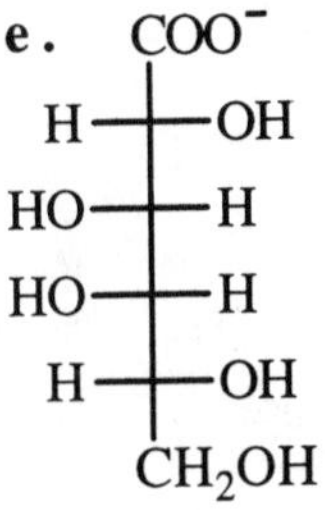

f. 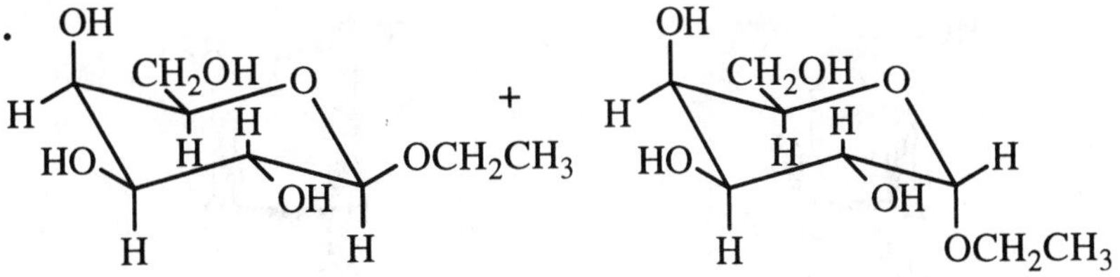

g. 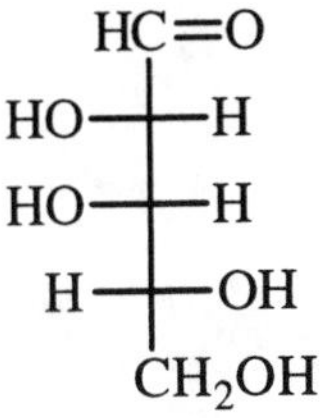

32. **a.** D-lyxose **b.** D-tagatose **c.** D-talose **d.** D-psicose

33.

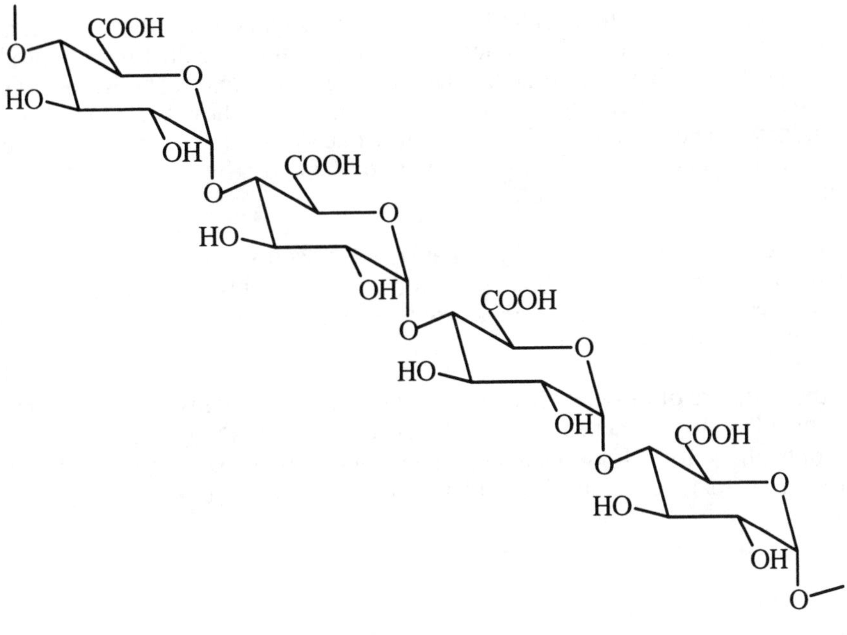

34. **a.** D-ribose and L-ribose, D-arabinose and L-arabinose, D-xylose and L-xylose, D-lyxose and L-lyxose

b. D-ribose and D-arabinose, L-ribose and L-arabinose, D-xylose and D-lyxose, L-xylose and L-lyxose

c. D-arabinose, L-arabinose, D-lyxose, and L-lyxose

35.

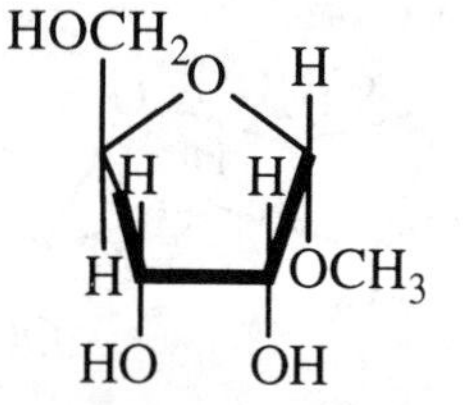

methyl α-D-ribofuranose

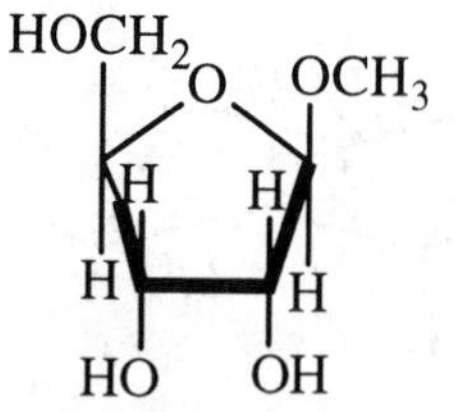

methyl β-D-ribofuranose

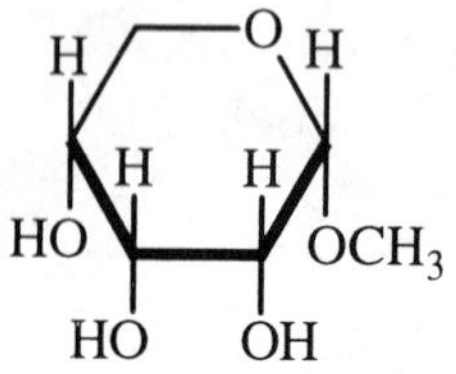

methyl α-D-ribopyranose

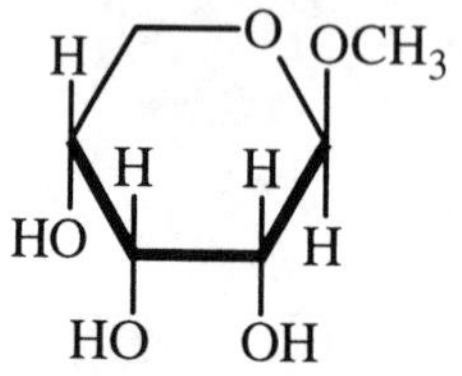

methyl β-D-ribopyranose

36. **1.** As Fischer did, we can narrow our search to eight 6-carbon sugars, since there are eight pairs of enantiomers. First we need to find an aldopentose that forms (+)-galactose as a product of a Kiliani-Fischer synthesis. That sugar is the one known as (-)-lyxose. The Kiliani-Fischer synthesis on (-)-lyxose yields two sugars with melting points that show them to be the sugars known as (+)-galactose and (+)-talose. Now we know that (+)-galactose and (+)-talose are C-2 epimers. They are sugars 1 and 2, 3 and 4, 5 and 6, or 7 and 8. (See Table 20.1 on page 882 of the text.)

2. When (+)-galactose and (+)-talose react with HNO_3, (+)-galactose forms an optically inactive product and (+)-talose forms an optically active product. Thus, (+)-galactose and (+)-talose are sugars 1 and 2 or 7 and 8. Since (+)-galactose is the one that forms the optically inactive oxidation product, it is either sugar 1 or 7.

3. To determine the structure of (+)-galactose, we can go back to (-)-lyxose, the sugar that forms sugars 7 and 8 by a Kiliani-Fischer synthesis, and oxidize it with HNO_3. Finding that the aldaric acid is optically active allows us to conclude that (+)-galactose is sugar 7, because the aldopentose that leads to sugars 1 and 2 would give an optically inactive aldaric acid.

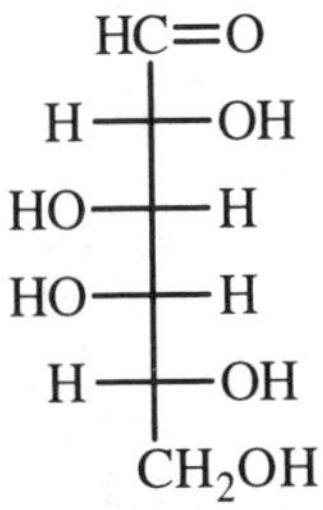

HC=O
H—OH
HO—H
HO—H
H—OH
CH_2OH
7
D-galactose

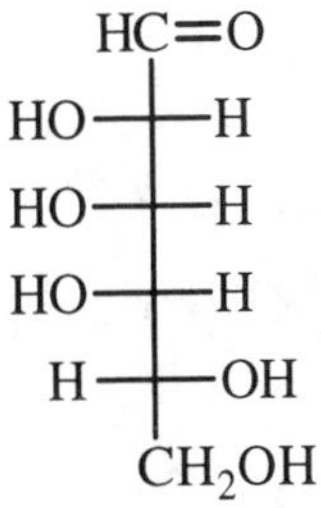

HC=O
HO—H
HO—H
HO—H
H—OH
CH_2OH
8
D-talose

37. A monosaccharide with a molecular weight of 150 must have five carbons (five C's = 60, five O's = 80, and 10 H's = 10 for a total of 150). All aldopentoses are optically active. Therefore, the compound must be a ketopentose. The following is the only ketopentose that would not be optically active.

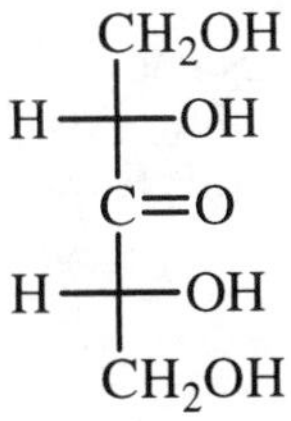

CH_2OH
H—OH
C=O
H—OH
CH_2OH

38. The hydrogen that is bonded to the anomeric carbon will be the hydrogen that is farthest downfield, because it is the only hydrogen that is bonded to a carbon that is bonded to two electronegative oxygen atoms. So the two anomeric hydrogens, one on the α-anomer and one on the β-anomer, are responsible for the two low-field doublets.

39.

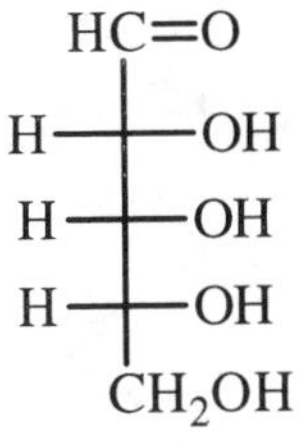

HC=O
H—OH
H—OH
H—OH
CH_2OH
D-ribose
A

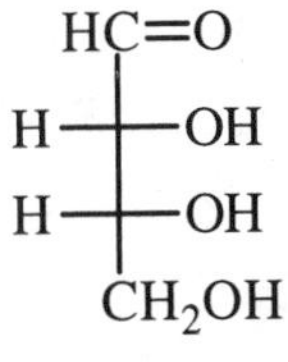

HC=O
H—OH
H—OH
CH_2OH
D-erythrose
B

40.

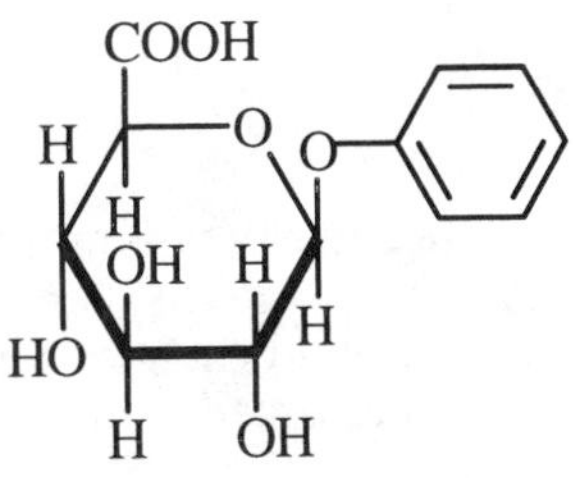

the β-D-glucuronide

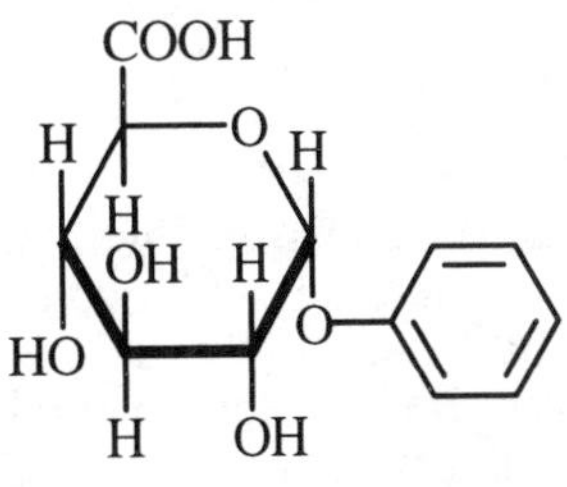

the α-D-glucuronide

41.

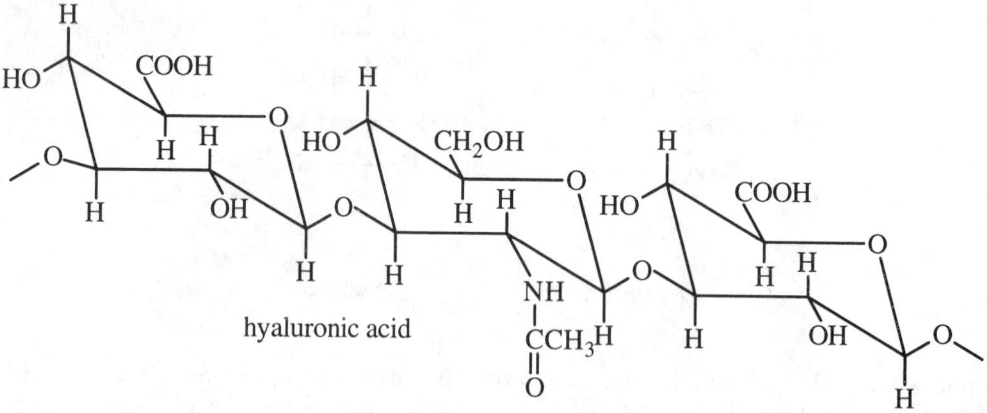

hyaluronic acid

42. She can take a sample of one of the sugars and oxidize it with nitric acid to an aldaric acid or reduce it with sodium borohydride to an alditol. If the product is optically active, the sugar was D-lyxose. If the product is not optically active, the sugar was D-xylose; if the product is optically active, the sugar was D-lyxose.

D-lyxose $\xrightarrow{HNO_3}$ **optically active**

D-xylose $\xrightarrow{HNO_3}$ **optically inactive**

D-lyxose $\xrightarrow{1. NaBH_4 \quad 2. H^+, H_2O}$ **optically active**

D-xylose $\xrightarrow{1. NaBH_4 \quad 2. H^+, H_2O}$ **optically inactive**

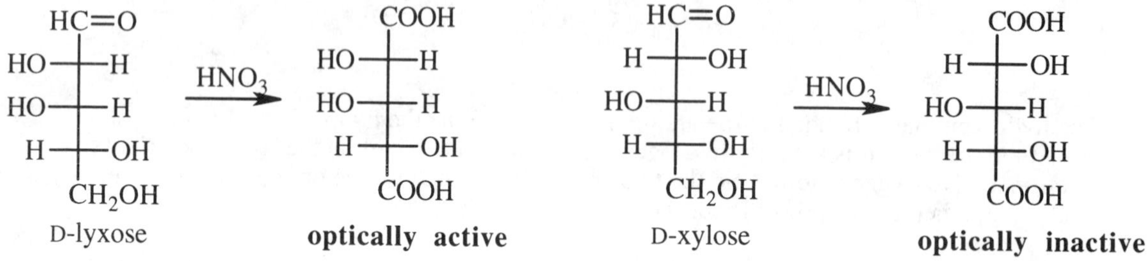

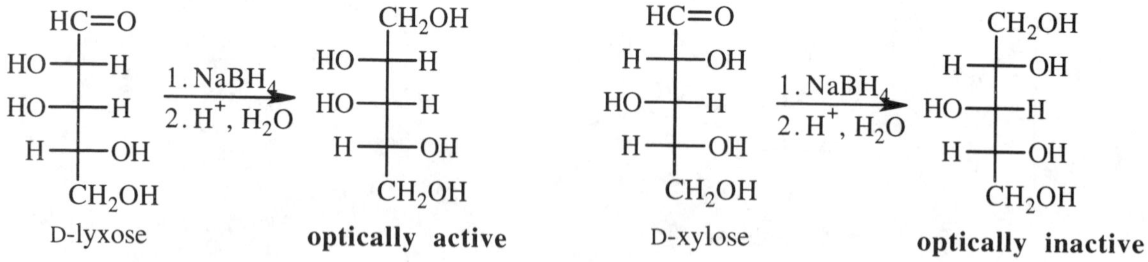

43. One pathway leads to a compound with one deuterium atom attached to carbon, and a second pathway leads to a compound with two deuterium atoms attached to carbon. Since both pathways can occur, when the reaction reaches equilibrium there will be a mixture of the two deuterated products. So the average number of deuterium atoms attached to carbon per molecule is greater than 1, but less than 2.

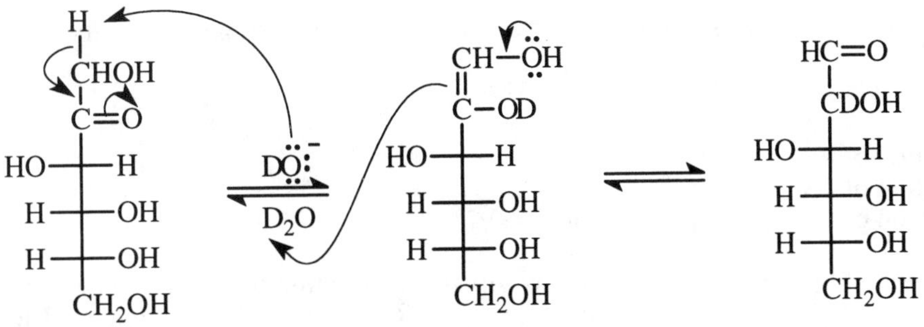

and

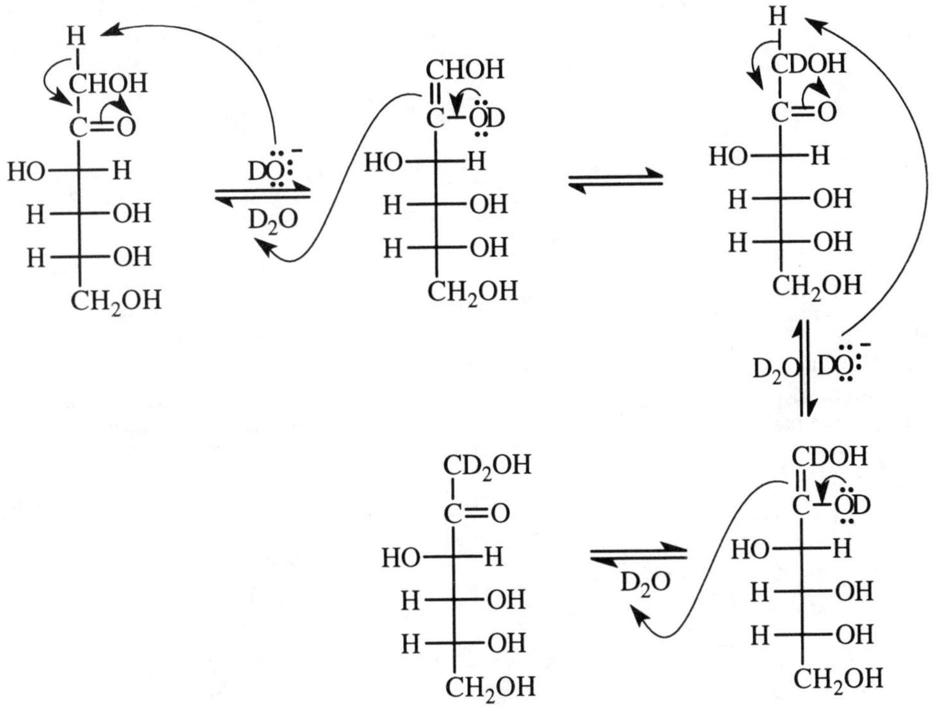

44. **D-arabinose**. The only D-aldopentoses that are oxidized to optically active aldaric acids are D-arabinose and D-lyxose.

A Ruff degradation of D-arabinose forms D-erythrose, while a Ruff degradation of D-lyxose forms D-threose. Since D-erythrose forms an optically inactive aldaric acid but D-threose does not, the D-aldopentose is D-arabinose.

45. **10** aldaric acids

Each of the following pairs forms the same aldaric acid:

D-allose and L-allose	L-altrose and L-talose
D-galactose and L-galactose	D-glucose and L-gulose
D-altrose and D-talose	L-glucose and D-gulose

Thus twelve aldohexoses form six aldaric acids. The other four aldohexoses each form a distinctive aldaric acid, and (6 + 4 = 10).

46. Let A = the fraction of glucose in the α-form and B = the fraction of glucose in the β-form.

$$A + B = 1$$
$$B = 1 - A$$

specific rotation of A = 112.2°
specific rotation of B = 18.7°
specific rotation of the equilibrium mixture = 52.7°

specific rotation of the mixture = specific rotation of A x fraction of glucose in the α-form + specific rotation of B x fraction of glucose in the β-form

$$52.7 = 112.2\,A + (1 - A)\,18.7$$
$$52.7 = 112.2\,A + 18.7 - 18.7\,A$$
$$34.0 = 93.5\,A$$
$$A = 0.36$$
$$B = 0.64$$

This calculation shows that 36% is in the α-form and 64% is in the β-form.

47. D-Altrose will most likely exist as a furanose because
(1) the furanose is particularly stable, since all the large substituents are trans to each other, and
(2) the pyranose has two of its OH groups in the unstable axial position.

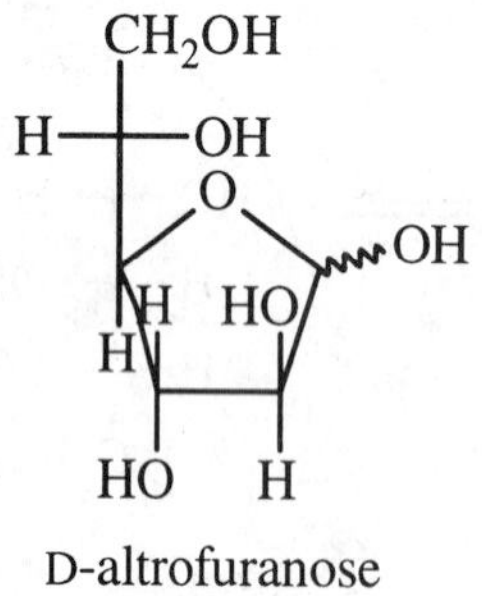

D-altrofuranose

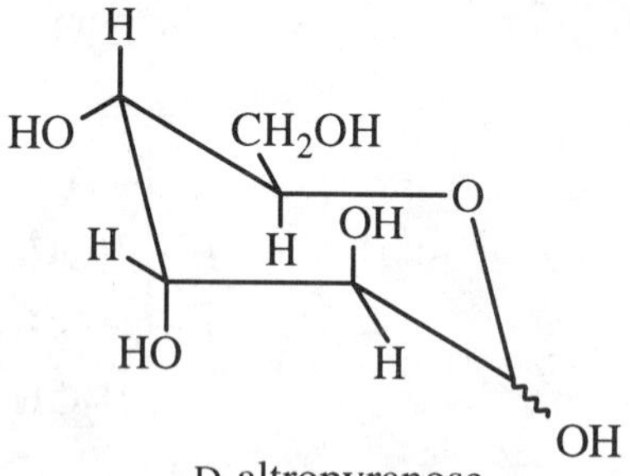

D-altropyranose

48.

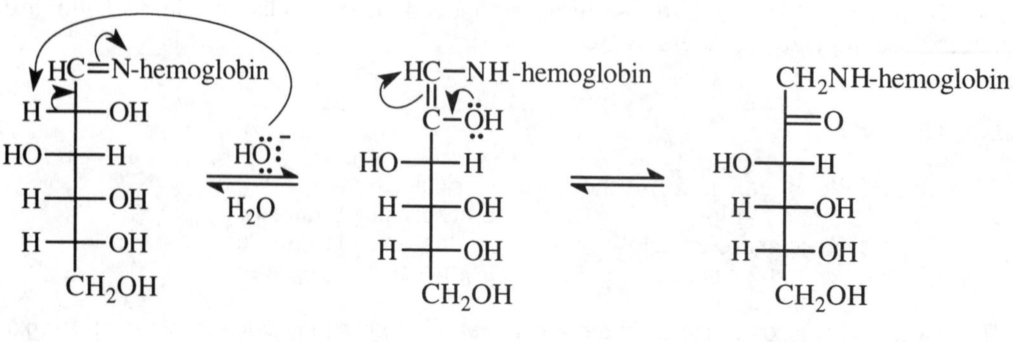

HC=N-hemoglobin
H—OH
HO—H
H—OH
H—OH
CH2OH
HO:⁻
H2O
HC—NH-hemoglobin
C—OH
HO—H
H—OH
H—OH
CH2OH
CH2NH-hemoglobin
=O
HO—H
H—OH
H—OH
CH2OH

49. **a.** Because mannose is missing the methyl substituent on the oxygen at C-6, the disaccharide must be formed using the C-6 OH group of mannose and the anomeric carbon of galactose.

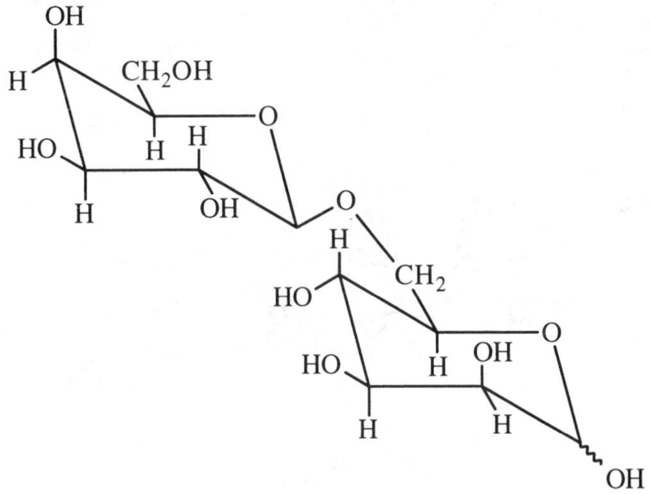

b. Silver oxide increases the leaving tendency of the iodide ion from methyl iodide, thereby allowing the nucleophilic substitution reaction to take place with the weakly nucleophilic alcohol groups.

50. The sugar unit of the disaccharide that possesses the acetal linkage can be determined by treating the disaccharide with excess methyl iodide in the presence of Ag_2O, followed by hydrolysis under acidic conditions. The two acetal linkages are hydrolyzed but the ether linkages are untouched. The products will be 2,3,4,6-tetra-*O*-methylgalactose and 2,3,6-tri-*O*-methyl-glucose. This tells us that galactose contains the acetal since it was able to react with methyl iodide at C-4.

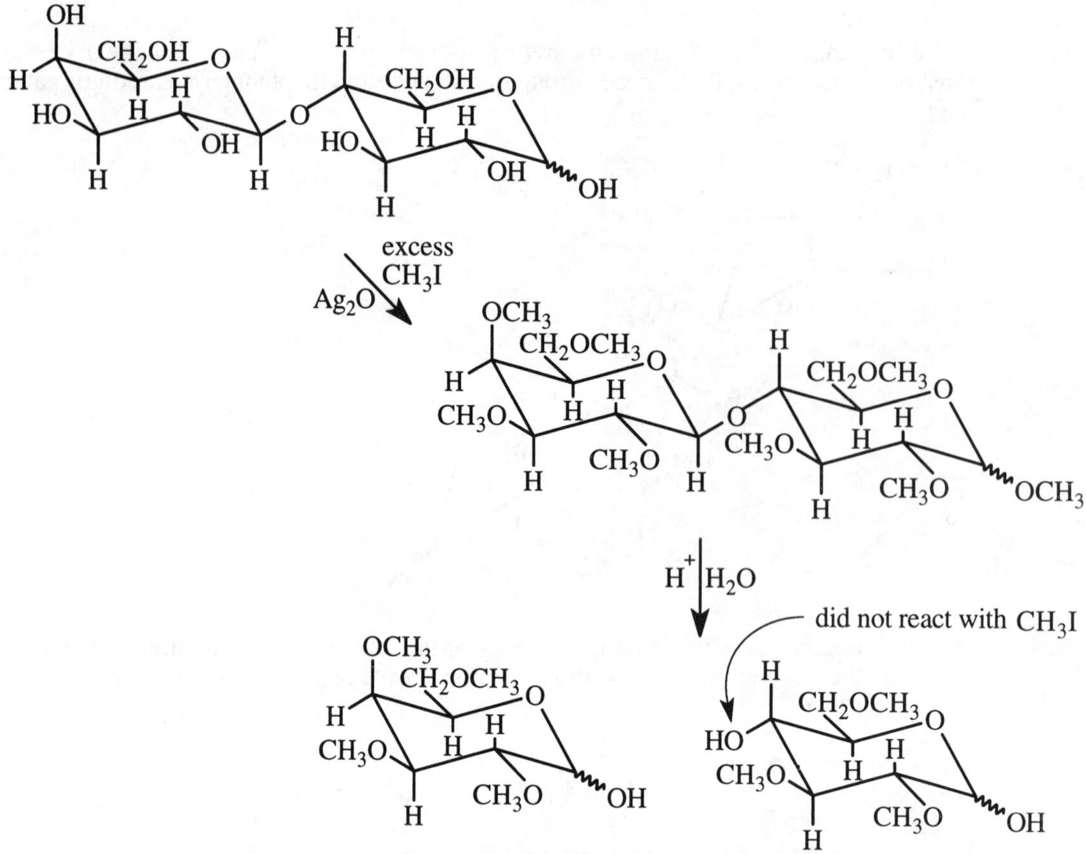

excess CH_3I

Ag_2O

H^+ H_2O

did not react with CH_3I

2,3,4,6-tetra-*O*-methylgalactose

2,3,6-tri-*O*-methylglucose

51.

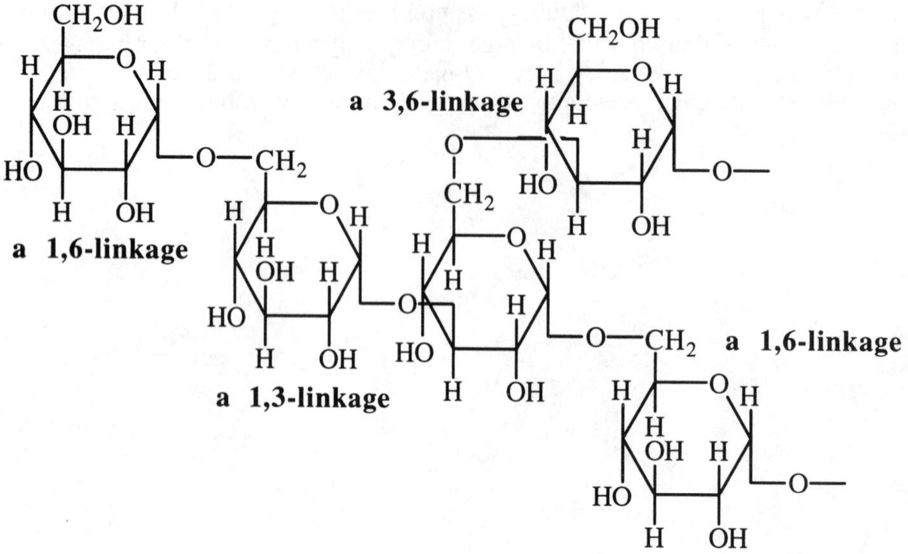

a 1,6-linkage

a 3,6-linkage

a 1,3-linkage

a 1,6-linkage

52. In the case of D-idose, the chair conformer with the OH substituents at C-1 and C-5 in the axial position (which is necessary for formation of the anhydro form) has the OH substituents at C-2, C-3, and C-4 all in equatorial positions. Thus, this is a preferred conformer since, three of the five large substituents are in the more stable equatorial position.

In the case of D-glucose, the chair conformer with the OH substituents at C-1 and C-5 in the axial position has the OH substituents at C-2, C-3, and C-4 all in axial positions. This is a relatively unstable conformer, since all the large substituents are in less stable axial positions.

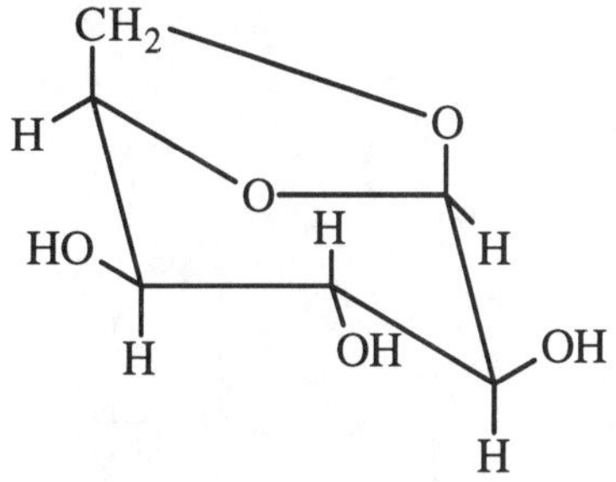

anhydro form of D-idose

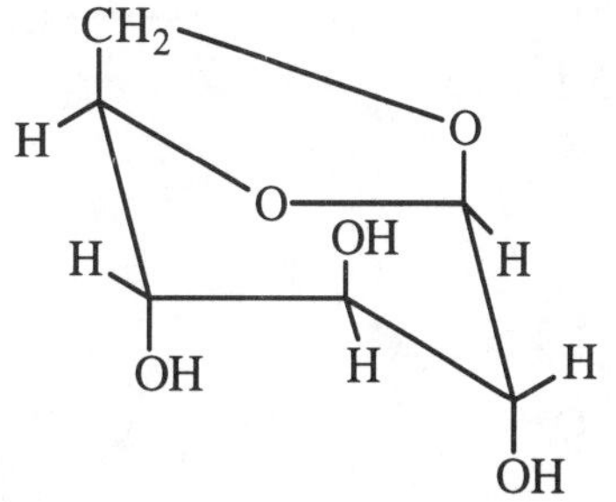

anhydro form of D-glucose

53. Periodate oxidatively cleaves at positions where there are adjacent OH groups. The sites of periodate cleavage are shown by wiggly lines in the furanose and pyranose.

Analyzing the oxidation products will indicate whether methyl-β-D-glucose is a furanose or a pyranose: if it is a furanose, formaldehyde will be formed; if it is a pyranose, formic acid will be formed.

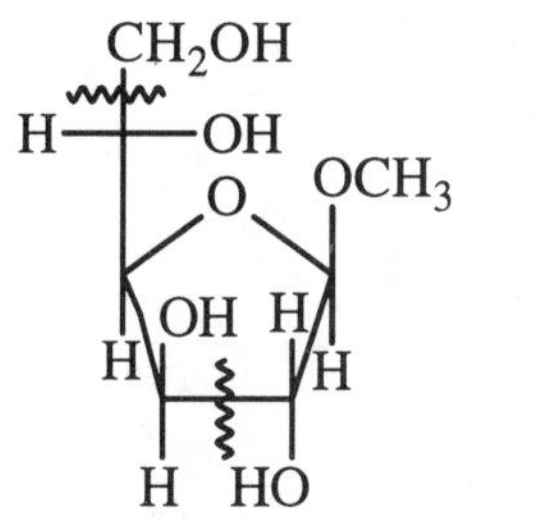

methyl-β-D-glucofuranose

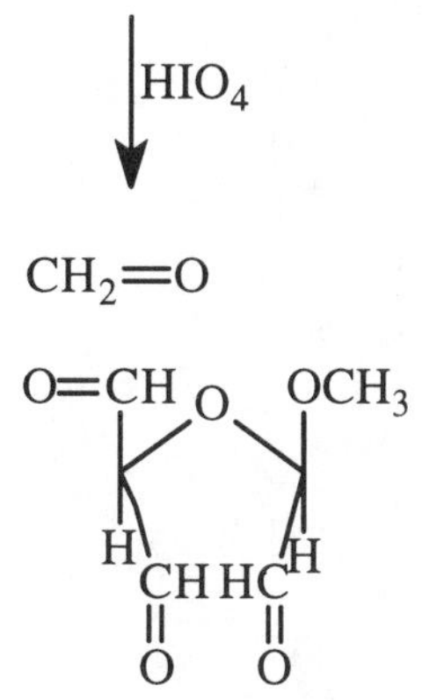

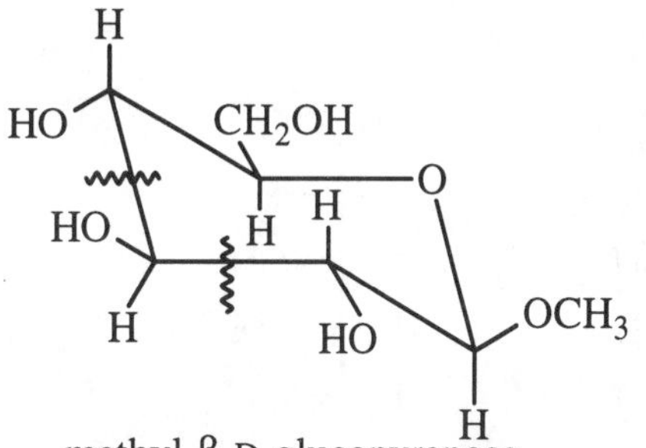

methyl-β-D-glucopyranose

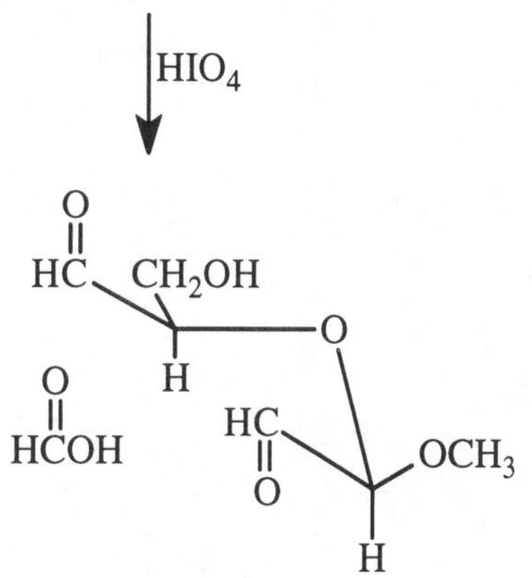

HIO_4

$CH_2{=}O$

$HCOH$ (formic acid, $HC(=O)OH$)

54.

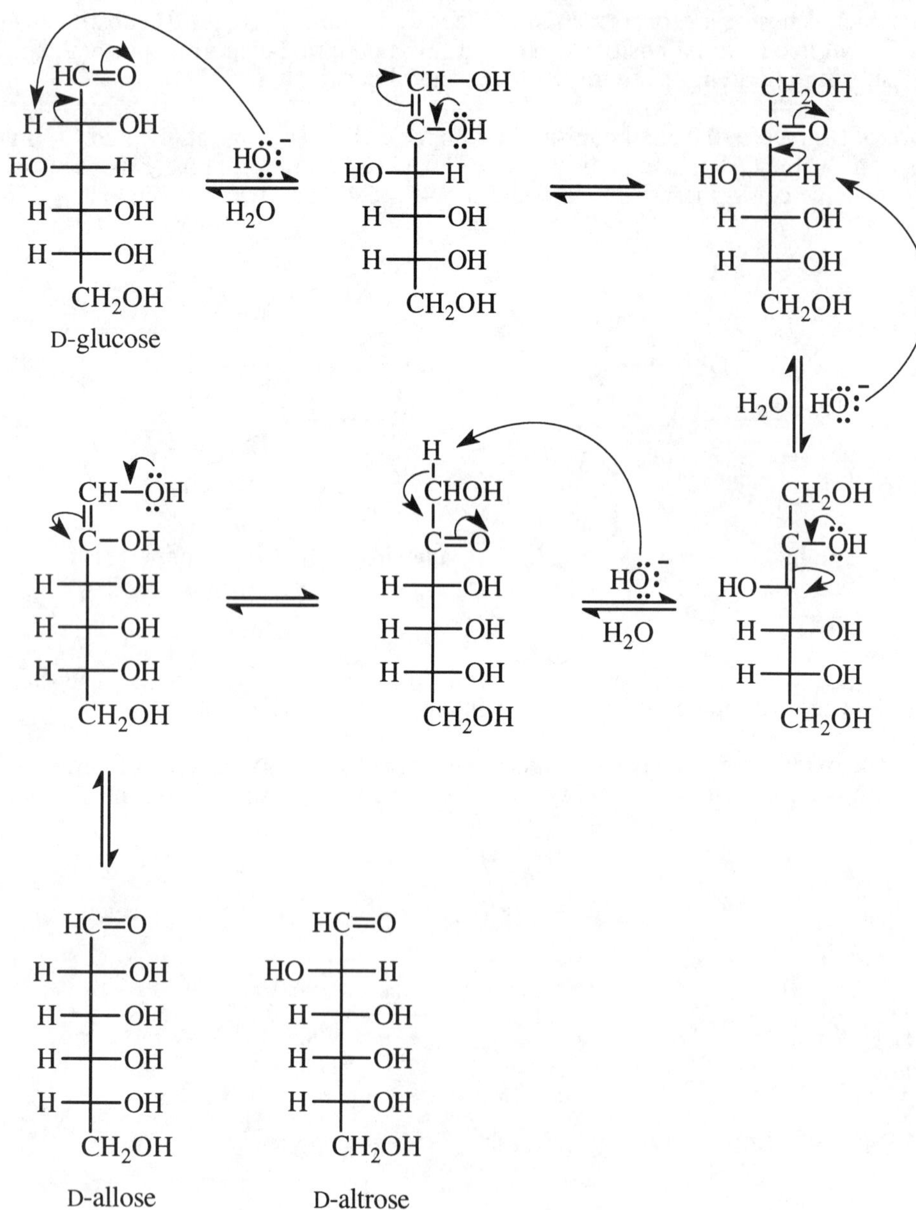
HC=O
H
OH
HO
H
H
OH
H
OH
CH_2OH
D-glucose
HO:⁻
H_2O
CH—OH
C—OH
CH_2OH
C=O
H
CHOH
C—OH
D-allose
D-altrose

Chapter 20 Practice Test

1. Give the product(s) of each of the following reactions.

a. Fischer projection: $CH{=}O$; H–C–OH; HO–C–H; H–C–OH; H–C–OH; CH_2OH $\xrightarrow[\Delta]{HNO_3}$

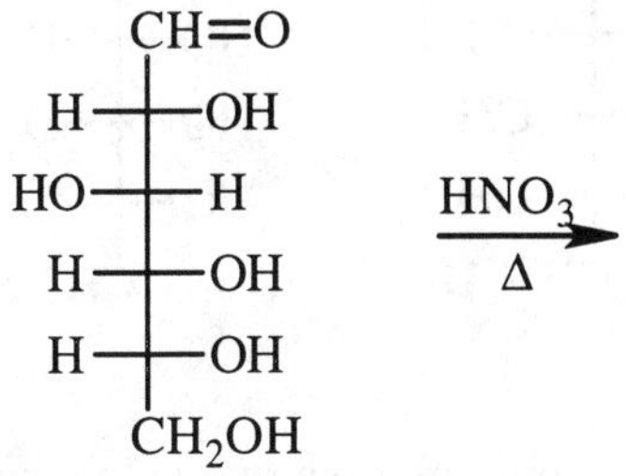

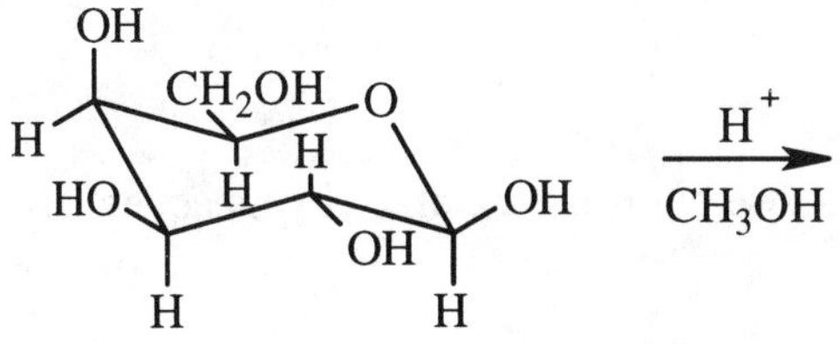

b. Pyranose ring (chair) with substituents OH, H, CH_2OH, O, H, HO, H, OH, OH, H, H $\xrightarrow[CH_3OH]{H^+}$

c. Fischer projection: $HC{=}O$; HO–C–H; HO–C–H; H–C–OH; CH_2OH

1. $HC{\equiv}N$
2. $H_2/Pd/BaSO_4$
3. H^+, H_2O

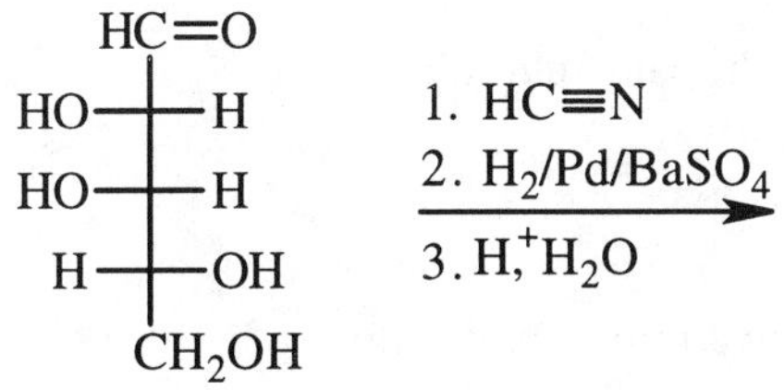

d. Fischer projection: $HC{=}O$; H–C–OH; H–C–OH; H–C–OH; CH_2OH + Br_2 $\xrightarrow{H_2O}$

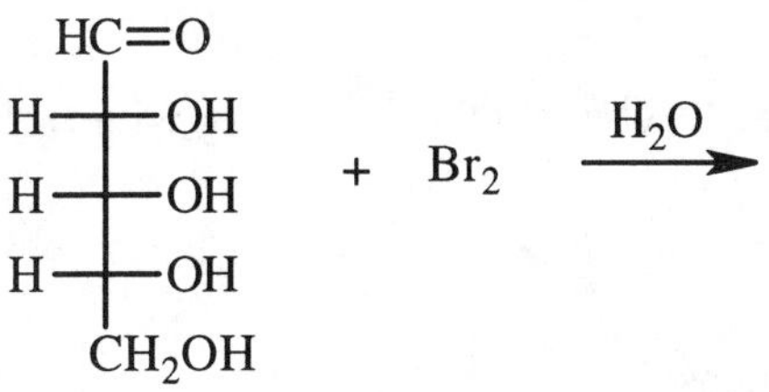

2. Indicate whether the following statements are true or false.

a. Glycogen contains α-1,4'- and β-1,6'-glycosidic linkages. T F

b. D-Mannose is a C-1 epimer of D-glucose. T F

c. D-Glucose and L-glucose are anomers. T F

d. D-Erythrose and D-threose are diastereomers. T F

e. Ruff degradations of D-glucose and D-gulose form the same aldopentose. T F

3. Which of the following sugars will form an optically active aldaric acid?

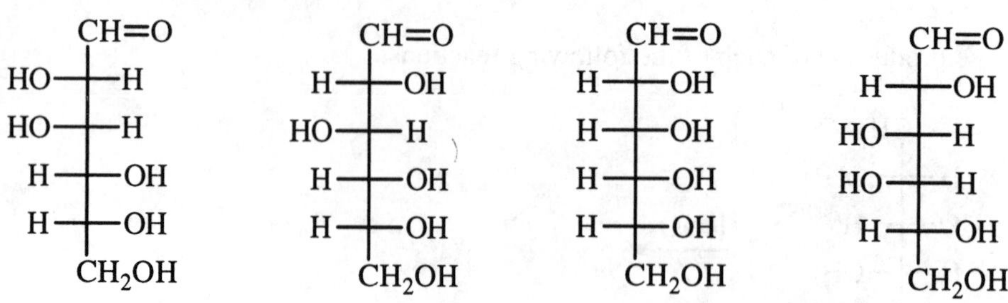

4. When crystals of D-fructose are dissolved in a basic aqueous solution, two aldohexoses are obtained. Identify the aldohexoses.

5. A sugar forms the same osazone as D-galactose but is not oxidized by an aqueous solution of Br_2. Identify the sugar.

6. D-Talose and _____________ are reduced to the same alditol.

7. What is the main structural difference between amylose and cellulose?

8. What aldohexoses are formed from a Kiliani-Fischer synthesis starting with D-xylose?

9. What aldohexose is the C-3 epimer of D-glucose?

10. Draw the most stable chair conformer of β-D-allose, a C-3 epimer of β-D-glucose.

CHAPTER 21
Amino Acids, Peptides, and Proteins

Important Terms

amino acid — an α-aminocarboxylic acid. Naturally occurring amino acids have the L-configuration.

D- amino acid — the configuration of an amino acid drawn in a Fischer projection such that the carboxyl group is on top, the hydrogen is on the left, and the amino group is on the right.

L-amino acid — the configuration of an amino acid drawn in a Fischer projection such that the carboxyl group is on top, the hydrogen is on the right, the amino group is on the left.

amino acid analyzer — an instrument that automates the ion-exchange separation of amino acids.

amino acid residue — a monomeric unit of a peptide or protein.

anion-exchange resin — a resin that binds anions.

antiparallel β-pleated sheet — the adjacent hydrogen-bonded peptide chains in a β-pleated sheet run in opposite directions.

automated solid-phase peptide synthesis — an automated technique that synthesizes a peptide while its C-terminal amino acid is attached to a solid support.

cation-exchange resin — a resin that binds cations.

coil conformation (loop conformation) — that part of a protein that is highly ordered but not in an α-helix or a β-sheet.

denaturation — destruction of the highly organized tertiary structure of a protein.

dipeptide — two amino acids linked together by an amide bond.

disulfide — a compound with an -S-S- bond.

disulfide bridge — a disulfide (-S-S-) bond in a peptide or protein.

Edman's reagent — phenyl isothiocyanate. A reagent used to determine the N-terminal amino acid of a polypeptide.

electrophoresis — a technique that separates amino acids on the basis of their pI values.

endopeptidase — an enzyme that hydrolyzes a peptide bond that is not at the end of a peptide chain.

enkephalin — a pentapeptide synthesized by the body to control pain.

enzyme — a protein that is a catalyst.

essential amino acid — an amino acid that humans must obtain from their diet because they either cannot synthesize it at all or they cannot synthesize it in adequate amounts.

exopeptidase	an enzyme that hydrolyzes a peptide bond at the end of a peptide chain.
fibrous protein	a water insoluble protein that has its polypeptide chains arranged in bundles.
globular protein	a water soluble protein that tends to have a roughly spherical shape.
α-helix	the backbone of a polypeptide coiled in a right-handed spiral with hydrogen bonding occurring within the helix.
hydrophobic interactions	interactions between nonpolar groups. They increase stability by decreasing the amount of structured water (increasing entropy).
interchain disulfide bridge	a disulfide bridge between two cysteine residues in different peptide chains.
intrachain disulfide bridge	a disulfide bridge between two cysteine residues in the same peptide chain.
ion-exchange chromatography	a technique that uses a column packed with an insoluble resin to separate compounds on the basis of their charge and polarity.
isoelectric point (p*I*)	the pH at which there is no net charge on an amino acid.
kinetic resolution	separating enantiomers based on the difference in their rate of reaction with an enzyme.
oligomer	a protein with more than one peptide chain.
oligopeptide	three to ten amino acids linked by amide bonds.
paper chromatography	a technique that separates amino acids based on polarity.
parallel β-pleated sheet	the adjacent hydrogen-bonded peptide chains in a β-pleated sheet run in the same direction.
partial hydrolysis	a technique that hydrolyzes only some of the peptide bonds in a polypeptide.
peptide	polymer of amino acids linked together by amide bonds.
peptide bond	the amide bond that links the amino acids in a peptide or protein.
β-pleated sheet	the backbone of a polypeptide is extended in a zigzag structure with hydrogen bonding between neighboring chains.
polypeptide	many amino acids linked by amide bonds.
primary structure	the sequence of amino acids in a protein.
protein	polymer of amino acids linked together by amide bonds.
quaternary structure	a description of the way the individual polypeptide chains of a protein are

arranged with respect to one other.

random coil the conformation of a totally denatured protein.

secondary structure a description of the conformation of the backbone of a protein.

structural protein a protein that gives strength to a biological structure.

subunit an individual chain of an oligomer.

C-terminal amino acid the terminal amino acid of a peptide (or protein) that has a free carboxyl group.

N-terminal amino acid the terminal amino acid of a peptide (or protein) that has a free amino group.

tertiary structure a description of the three-dimensional arrangement of all the atoms in a protein.

thin-layer chromatography a technique that separates compounds on the basis of their polarity.

tripeptide three amino acids linked by amide bonds.

zwitterion a compound with a negative charge and a positive charge on nonadjacent atoms.

Solutions to Problems

1. **a.** The nonbonding pair of electrons on the doubly-bonded nitrogen are localized, while the nonbonding electrons on the other ring nitrogen are delocalized. The localized electrons are protonated, since the delocalized electrons are part of the π cloud that causes the imidazole ring to be aromatic.

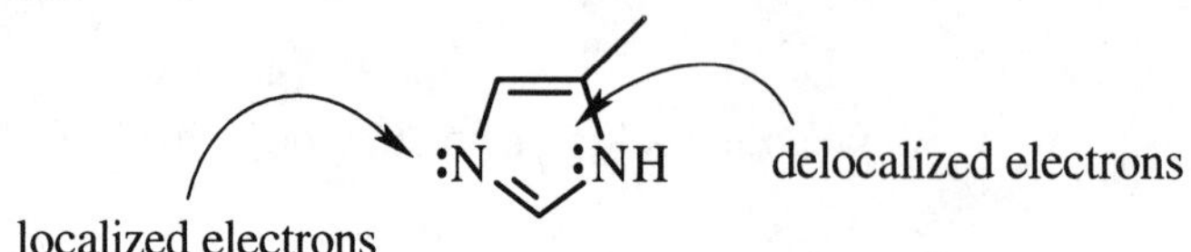

b. The nonbonding pair of electrons on the sp^2 hybridized nitrogen are localized, while the nonbonding electrons on the other nitrogens are delocalized. The localized electrons are protonated since the others are involved in resonance.

localized electrons

delocalized electrons

delocalized electrons

$$H_2\ddot{N}-\underset{}{\overset{\overset{\ddot{N}H}{\|}}{C}}-\ddot{N}H-$$

guanidino group

2. **a.** (R)-alanine

b. (R)-aspartic acid

c. The α-carbons of all the D-amino acids except cysteine have the R-configuration. Similarly, the α-carbon of all the L-amino acids except cysteine has the S-configuration. (In all the amino acids except cysteine, the amino group has the highest priority and the carboxyl group has the second-highest priority. In cysteine, the thiomethyl group has a higher priority than the carboxyl group because sulfur has a greater atomic number than oxygen.)

3. isoleucine and threonine

4. The electron-withdrawing $^+NH_3$ substituent on the α-carbon increases the acidity of the carboxyl group.

5.

a. $^{-}OC(=O)CH_2CH(^{+}NH_3)C(=O)O^{-}$

b. (imidazol-4-yl)$CH_2CH(^{+}NH_3)C(=O)O^{-}$

c. $H_2NC(=O)CH_2CH_2CH(^{+}NH_3)C(=O)O^{-}$

d. $H_3\overset{+}{N}CH_2CH_2CH_2CH_2CH(^{+}NH_3)C(=O)O^{-}$

e. $NH_2C(=^{+}NH_2)NHCH_2CH_2CH_2CH(^{+}NH_3)C(=O)O^{-}$

f. HO–C_6H_4–$CH_2CH(^{+}NH_3)C(=O)O^{-}$

6.

a. $HOC(=O)CH_2CH_2CH(^{+}NH_3)C(=O)OH$

b. $HOC(=O)CH_2CH_2CH(^{+}NH_3)C(=O)O^{-}$

c. $^{-}OC(=O)CH_2CH_2CH(^{+}NH_3)C(=O)O^{-}$

d. $^{-}OC(=O)CH_2CH_2CH(NH_2)C(=O)O^{-}$

7. **a.** The carboxyl group of the aspartic acid side chain is a stronger acid than the carboxyl group of the glutamic acid side chain because the aspartic acid side chain is closer to the electron-withdrawing protonated amino group.

b. The lysine side chain is a stronger acid that the arginine side chain. The arginine side chain has less of a tendency to lose a proton because the positive charge is delocalized over three nitrogen atoms.

8. In order for the amino acid to have no net charge, the two amino groups must have a +1 charge between them in order to cancel out the -1 charge of the carboxylate group. Because they are positively charged in their acidic forms and neutral in their basic forms, the sum of their charges will be +1 at the midpoint between their pK_a's.

9.

a. asparagine $pI = \frac{2.02 + 8.84}{2} = \frac{10.86}{2} = 5.43$

b. arginine $pI = \frac{9.04 + 12.48}{2} = \frac{21.52}{2} = 10.76$

c. serine $pI = \frac{2.21 + 9.15}{2} = \frac{11.36}{2} = 5.68$

10. **a.** Asp ($pI = 2.98$)

b. Arg ($pI = 10.76$)

c. Asp

d. Met, because at pH = 6.20 Met is farther away from the pH at which it has no net charge (pI of Met = 5.75, pI of Gly = 5.97).

11. Tyrosine and cysteine each have two groups that are neutral in their acidic forms and negatively charged in their basic forms. Unlike other amino acids that have similarly ionizing groups, the two groups in tyrosine and cysteine do not have pK_a's that are close in value and quite different from the pK_a of the group that ionizes differently. Therefore, the third group cannot be ignored in calculating the pI.

12. Leucine and isoleucine both have butyl side chains and, therefore, have the same polarity. Consequently, the spots for both amino acids appear at the same place on the chromatographic plate.

13. Because the amino acid analyzer contains a cation-exchange resin (it binds cations), the less positively charged the amino acid, the less tightly it is bound to the column. Using buffer solutions of increasingly higher pH to elute the column causes the amino acids bound to the column to become increasingly less positively charged, so they can be released from the column.

14. Cation-exchange chromatography releases amino acids in order of their pI's. The amino acid with the lowest pI is released first because at a given pH it will be the amino acid with the highest concentration of negative charge, and negatively charged molecules are not bound by the negatively charged resin. The relatively nonpolar resin will release polar amino acids before nonpolar amino acids.

a. Asp ($pI = 2.98$) is more negative than Ser ($pI = 5.68$).
b. Gly is more polar than Ala.
c. Val is more polar than Leu.
d. Tyr is more polar than Phe.

15. A column containing an anion-exchange resin releases amino acids in order of their p*I* 's. The amino acid with the highest p*I* is released first, because at a given pH it will be the amino acid with the highest concentration of positive charge.

His > Val > Ser > Asp

16. The first equivalent of ammonia will react with the acidic proton of the carboxylic acid to form ammonium ion, which is not nucleophilic and, therefore, cannot substitute for Br. A second equivalent of ammonia therefore is needed for the desired nucleophilic substitution reaction.

17. Convert the amino acids into esters using $SOCl_2$ followed by ethanol. The products obtained after treatment with pig liver esterase will be an L-amino acid and unreacted ester of the D-amino acid. These compounds can be readily separated, and the D-amino acid can be obtained by acid-catalyzed hydrolysis of the ester. This technique is called a *kinetic resolution*. If the difference in the rate of the enzyme-catalyzed reaction with the two enantiomers is greater than about 3 kcal/mol, a kinetic resolution will be successful.

18. **a.** The reactions below show that pyruvic acid forms alanine, oxaloacetic acid forms aspartic acid, and α-ketoglutarate forms glutamic acid. If reductive amination is carried out in the cell, only the L-isomer of each amino acid will be formed.

b. If reductive amination is carried out in the laboratory, both the D- and L-isomer of each amino acid will be formed.

$$CH_3C(=O)COOH \xrightarrow[\text{Raney Ni}]{H_2} CH_3CH(\overset{+}{N}H_3)COOH$$

pyruvic acid → alanine

$$HOOCCH_2C(=O)COOH \xrightarrow[\text{Raney Ni}]{H_2} HOOCCH_2CH(\overset{+}{N}H_3)COOH$$

oxaloacetic acid → aspartic acid

$$HOOCCH_2CH_2C(=O)COOH \xrightarrow[\text{Raney Ni}]{H_2} HOOCCH_2CH_2CH(\overset{+}{N}H_3)COOH$$

α-ketoglutaric acid → glutamic acid

19.

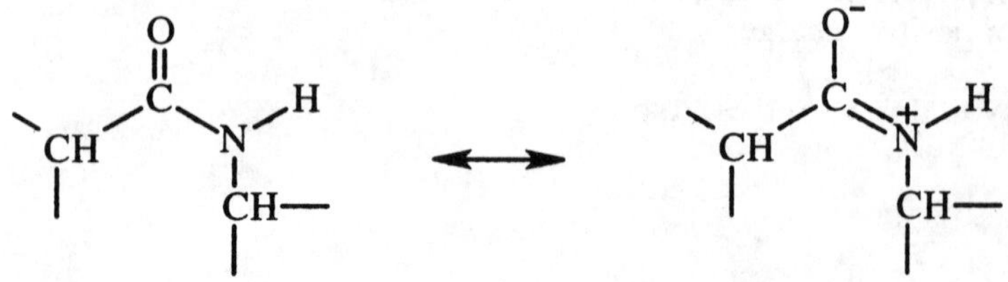

[Resonance structures: amide group —CH—C(=O)—N(H)—CH— ⟷ —CH—C(O⁻)=N⁺(H)—CH—]

20. **a.** $20^8 = 25{,}600{,}000{,}000$ **b.** 20^{100}

21. The bonds on either side of the α-carbon can freely rotate. In other words, the bond between the α-carbon and the carbonyl carbon and the bond between the α-carbon and the nitrogen (the bonds indicated by arrows) can freely rotate.

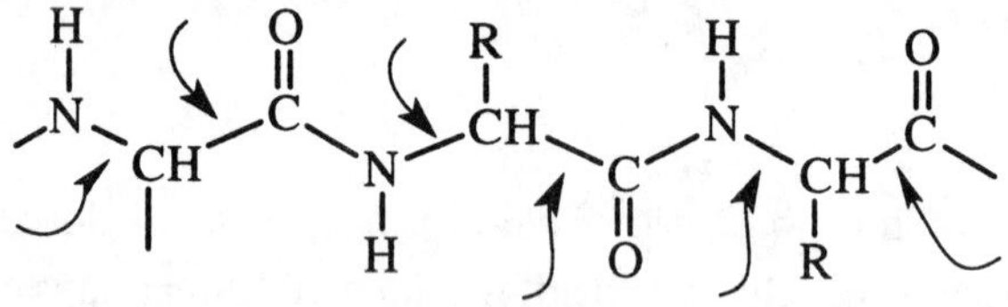

[Structure: peptide backbone —N(H)—CH—C(=O)—N(H)—CH(R)—C(=O)—N(H)—CH(R)—C(=O)— with arrows on the bonds on either side of each α-carbon]

22. In forming the amide linkage, the amino group of cysteine reacts with the γ-carboxyl group rather than with the α-carboxyl group of glutamic acid.

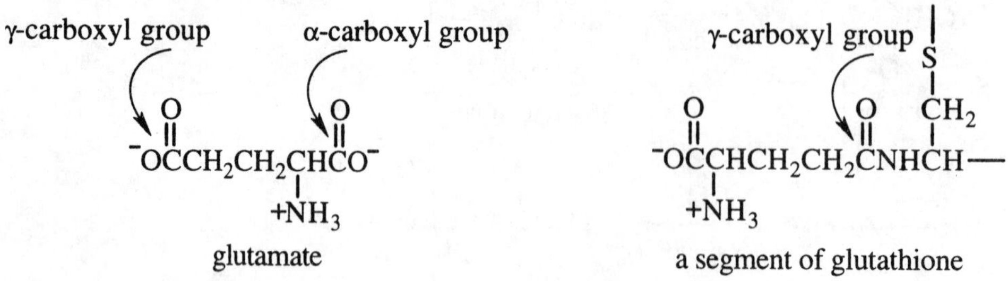

γ-carboxyl group α-carboxyl group

$^{-}OOCCH_2CH_2CH(^{+}NH_3)COO^{-}$

glutamate

γ-carboxyl group

$^{-}OOCCH(^{+}NH_3)CH_2CH_2C(=O)NHCH(CH_2S-)$—

a segment of glutathione

23. **Leu-Val** and **Val-Val** will be formed because the amino group of leucine is not reactive (so leucine could not be the C-terminal amino acid) but the amino group of valine would react equally easily with the carboxyl group of leucine and the carboxyl group of valine.

$$(CH_3)_3C{-}O{-}\overset{O}{\overset{\|}{C}}{-}NH\underset{CH_2CH(CH_3)_2}{CH}\overset{O}{\overset{\|}{C}}O^- + H_3\overset{+}{N}\underset{CH(CH_3)_2}{CH}\overset{O}{\overset{\|}{C}}O^- \longrightarrow H_3\overset{+}{N}\underset{CH_2CH(CH_3)_2}{CH}\overset{O}{\overset{\|}{C}}{-}NH\underset{CH(CH_3)_2}{CH}\overset{O}{\overset{\|}{C}}O^-$$

N-protected leucine **valine** **Leu-Val**

$$H_3\overset{+}{N}\underset{CH(CH_3)_2}{CH}\overset{O}{\overset{\|}{C}}O^- + H_3\overset{+}{N}\underset{CH(CH_3)_2}{CH}\overset{O}{\overset{\|}{C}}O^- \longrightarrow H_3\overset{+}{N}\underset{CH(CH_3)_2}{CH}\overset{O}{\overset{\|}{C}}{-}NH\underset{CH(CH_3)_2}{CH}\overset{O}{\overset{\|}{C}}O^-$$

valine **valine** **Val-Val**

24. If valine's carboxyl group is activated with thionyl chloride, the OH group of serine, as well as the NH_2 group of serine, would react readily with the very reactive acyl halide, forming both an ester and an amide.

$$H_2N\underset{CH(CH_3)_2}{CH}\overset{O}{\overset{\|}{C}}O^- \xrightarrow{SOCl_2} H_2N\underset{CH(CH_3)_2}{CH}\overset{O}{\overset{\|}{C}}Cl \xrightarrow{H_2N\underset{CH_2OH}{CH}\overset{O}{\overset{\|}{C}}O^-} \underset{\text{ester}}{R\overset{O}{\overset{\|}{C}}OCH_2\underset{NH_2}{CH}\overset{O}{\overset{\|}{C}}O^-} + \underset{\text{amide}}{R\overset{O}{\overset{\|}{C}}NH\underset{CH_2OH}{CH}\overset{O}{\overset{\|}{C}}O^-}$$

If valine's carboxyl group is activated with DCC, an imidate will be formed. Because an imidate is less reactive than an acyl chloride, the imidate will react with the more reactive NH_2 group in preference to the OH group.

$$H_2N\underset{(CH_3)_2CH}{CH}\overset{O}{\overset{\|}{C}}O^- \xrightarrow{DCC} \underset{\text{an imidate}}{H_2N\underset{(CH_3)_2CH}{CH}\overset{O}{\overset{\|}{C}}O{-}\underset{NH{-}C_6H_{11}}{\overset{N{-}C_6H_{11}}{\overset{\|}{C}}}} \xrightarrow{H_2N\underset{CH_2OH}{CH}\overset{O}{\overset{\|}{C}}O^-} H_2N\underset{(CH_3)_2CH}{CH}\overset{O}{\overset{\|}{C}}{-}NH\underset{CH_2OH}{CH}\overset{O}{\overset{\|}{C}}O^- + O{=}\underset{NH{-}C_6H_{11}}{\overset{NH{-}C_6H_{11}}{C}}$$

25.

$$\underset{CH_3}{\overset{CH_3}{CH_3\overset{|}{\underset{|}{C}}}}-O\overset{O}{\overset{\|}{C}}O\overset{O}{\overset{\|}{C}}O-\underset{CH_3}{\overset{CH_3}{\overset{|}{\underset{|}{C}}}}CH_3 + H_2N\underset{\underset{(CH_3)_2CH}{|}}{\underset{CH_2}{\underset{|}{C}}}H\overset{O}{\overset{\|}{C}}O^- \longrightarrow \underset{CH_3}{\overset{CH_3}{CH_3\overset{|}{\underset{|}{C}}}}-O\overset{O}{\overset{\|}{C}}NH\underset{\underset{(CH_3)_2CH}{|}}{\underset{CH_2}{\underset{|}{C}}}H\overset{O}{\overset{\|}{C}}O^-$$

1. DCC
2. $H_2N\underset{CH_2C_6H_5}{\underset{|}{C}}H\overset{O}{\overset{\|}{C}}O^-$

$$\underset{CH_3}{\overset{CH_3}{CH_3\overset{|}{\underset{|}{C}}}}-O\overset{O}{\overset{\|}{C}}NH\underset{\underset{(CH_3)_2CH}{|}}{\underset{CH_2}{\underset{|}{C}}}H\overset{O}{\overset{\|}{C}}NH\underset{\underset{C_6H_5}{|}}{\underset{CH_2}{\underset{|}{C}}}H\overset{O}{\overset{\|}{C}}O^-$$

1. DCC
2. $H_2N\underset{(CH_2)_4NH_2}{\underset{|}{C}}H\overset{O}{\overset{\|}{C}}O^-$

$$\underset{CH_3}{\overset{CH_3}{CH_3\overset{|}{\underset{|}{C}}}}-O\overset{O}{\overset{\|}{C}}NH\underset{\underset{(CH_3)_2CH}{|}}{\underset{CH_2}{\underset{|}{C}}}H\overset{O}{\overset{\|}{C}}NH\underset{\underset{C_6H_5}{|}}{\underset{CH_2}{\underset{|}{C}}}H\overset{O}{\overset{\|}{C}}NH\underset{\underset{NH_2}{|}}{\underset{(CH_2)_4}{\underset{|}{C}}}H\overset{O}{\overset{\|}{C}}O^-$$

1. DCC
2. $H_2N\underset{CH(CH_3)_2}{\underset{|}{C}}H\overset{O}{\overset{\|}{C}}O^-$

$$\underset{CH_3}{\overset{CH_3}{CH_3\overset{|}{\underset{|}{C}}}}-O\overset{O}{\overset{\|}{C}}NH\underset{\underset{(CH_3)_2CH}{|}}{\underset{CH_2}{\underset{|}{C}}}H\overset{O}{\overset{\|}{C}}NH\underset{\underset{C_6H_5}{|}}{\underset{CH_2}{\underset{|}{C}}}H\overset{O}{\overset{\|}{C}}NH\underset{\underset{NH_2}{|}}{\underset{(CH_2)_4}{\underset{|}{C}}}H\overset{O}{\overset{\|}{C}}NH\underset{CH(CH_3)_2}{\underset{|}{C}}H\overset{O}{\overset{\|}{C}}O^-$$

CF_3COOH
CH_2Cl_2

$$\overset{CH_3}{\overset{|}{CH_3\underset{\underset{CH_2}{\|}}{C}}} + CO_2 + H_3\overset{+}{N}\underset{\underset{(CH_3)_2CH}{|}}{\underset{CH_2}{\underset{|}{C}}}H\overset{O}{\overset{\|}{C}}NH\underset{\underset{C_6H_5}{|}}{\underset{CH_2}{\underset{|}{C}}}H\overset{O}{\overset{\|}{C}}NH\underset{\underset{NH_2}{|}}{\underset{(CH_2)_4}{\underset{|}{C}}}H\overset{O}{\overset{\|}{C}}NH\underset{CH(CH_3)_2}{\underset{|}{C}}H\overset{O}{\overset{\|}{C}}OH$$

26. **a.** 5.8%

2	**3**	**4**	**5**	**6**	**7**	**8**	**9**
70%	49%	34%	24%	17%	12%	8.2%	5.8%

b. 4.4%

2	**3**	**4**	**5**	**6**	**7**	**8**	**9**	**10**	**11**	**12**	**13**	**14**	**15**
80%	64%	51%	41%	33%	26%	21%	17%	13%	11%	8.6%	6.9%	5.5%	4.4%

27.

$$(CH_3)_3CO\text{-}\overset{O}{\overset{\|}{C}}Cl + H_2NCH(CH(CH_3)_2)\overset{O}{\overset{\|}{C}}O^- \longrightarrow (CH_3)_3CO\text{-}\overset{O}{\overset{\|}{C}}NHCH(CH(CH_3)_2)\overset{O}{\overset{\|}{C}}O^-$$

$$\xrightarrow{ClCH_2\text{-}C_6H_4\text{-}\bullet} (CH_3)_3CO\text{-}\overset{O}{\overset{\|}{C}}NHCH(CH(CH_3)_2)\overset{O}{\overset{\|}{C}}OCH_2\text{-}C_6H_4\text{-}\bullet$$

$$\xrightarrow[CH_2Cl_2]{CF_3COOH} CH_3C(CH_3){=}CH_2 + CO_2 + H_2NCH(CH(CH_3)_2)\overset{O}{\overset{\|}{C}}OCH_2\text{-}C_6H_4\text{-}\bullet$$

$$(CH_3)_3CO\text{-}\overset{O}{\overset{\|}{C}}NHCH((CH_2)_4{}^+NH_3)\overset{O}{\overset{\|}{C}}O^- \xrightarrow{DCC} (CH_3)_3CO\text{-}\overset{O}{\overset{\|}{C}}NHCH((CH_2)_4{}^+NH_3)\overset{O}{\overset{\|}{C}}ODCC$$

$$\longrightarrow (CH_3)_3CO\text{-}\overset{O}{\overset{\|}{C}}NHCH((CH_2)_4{}^+NH_3)\overset{O}{\overset{\|}{C}}NHCH(CH(CH_3)_2)\overset{O}{\overset{\|}{C}}OCH_2\text{-}C_6H_4\text{-}\bullet$$

$$\xrightarrow[CH_2Cl_2]{CF_3COOH} CH_3C(CH_3){=}CH_2 + CO_2 + H_2NCH((CH_2)_4{}^+NH_3)\overset{O}{\overset{\|}{C}}NHCH(CH(CH_3)_2)\overset{O}{\overset{\|}{C}}OCH_2\text{-}C_6H_4\text{-}\bullet$$

$$(CH_3)_3CO\text{-}\overset{O}{\overset{\|}{C}}NHCH(CH_2C_6H_5)\overset{O}{\overset{\|}{C}}O^- \xrightarrow{DCC} (CH_3)_3CO\text{-}\overset{O}{\overset{\|}{C}}NHCH(CH_2C_6H_5)\overset{O}{\overset{\|}{C}}ODCC \longrightarrow$$

$$(CH_3)_3CO{-}\overset{O}{\overset{\|}{C}}NH\underset{CH_2C_6H_5}{CH}\overset{O}{\overset{\|}{C}}NH\underset{(CH_2)_4\overset{+}{N}H_3}{CH}\overset{O}{\overset{\|}{C}}NH\underset{CH(CH_3)_2}{CH}\overset{O}{\overset{\|}{C}}OCH_2{-}C_6H_4{-}\text{(resin)}$$

$$\xrightarrow[CH_2Cl_2]{CF_3COOH}$$

$$CH_3\underset{CH_2}{\overset{CH_3}{C}} + CO_2 + H_2N\underset{CH_2C_6H_5}{CH}\overset{O}{\overset{\|}{C}}NH\underset{(CH_2)_4\overset{+}{N}H_3}{CH}\overset{O}{\overset{\|}{C}}NH\underset{CH(CH_3)_2}{CH}\overset{O}{\overset{\|}{C}}OCH_2{-}C_6H_4{-}\text{(resin)}$$

$$(CH_3)_3CO{-}\overset{O}{\overset{\|}{C}}NH\underset{CH_2CH(CH_3)_2}{CH}\overset{O}{\overset{\|}{C}}O^- \xrightarrow{DCC} (CH_3)_3CO{-}\overset{O}{\overset{\|}{C}}NH\underset{CH_2CH(CH_3)_2}{CH}\overset{O}{\overset{\|}{C}}ODCC$$

$$(CH_3)_3CO{-}\overset{O}{\overset{\|}{C}}NH\underset{CH_2CH(CH_3)_2}{CH}\overset{O}{\overset{\|}{C}}NH\underset{CH_2C_6H_5}{CH}\overset{O}{\overset{\|}{C}}NH\underset{(CH_2)_4\overset{+}{N}H_3}{CH}\overset{O}{\overset{\|}{C}}NH\underset{CH(CH_3)_2}{CH}\overset{O}{\overset{\|}{C}}OCH_2{-}C_6H_4{-}\text{(resin)}$$

$$\xrightarrow[CH_2Cl_2]{CF_3COOH}$$

$$CH_3\underset{CH_2}{\overset{CH_3}{C}} + CO_2 + H_2N\underset{CH_2CH(CH_3)_2}{CH}\overset{O}{\overset{\|}{C}}NH\underset{CH_2C_6H_5}{CH}\overset{O}{\overset{\|}{C}}NH\underset{(CH_2)_4\overset{+}{N}H_3}{CH}\overset{O}{\overset{\|}{C}}NH\underset{CH(CH_3)_2}{CH}\overset{O}{\overset{\|}{C}}OCH_2{-}C_6H_4{-}\text{(resin)}$$

$$\xrightarrow{HF,\ H_2O}$$

$$H_3\overset{+}{N}\underset{CH_2CH(CH_3)_2}{CH}\overset{O}{\overset{\|}{C}}NH\underset{CH_2C_6H_5}{CH}\overset{O}{\overset{\|}{C}}NH\underset{(CH_2)_4\overset{+}{N}H_3}{CH}\overset{O}{\overset{\|}{C}}NH\underset{CH(CH_3)_2}{CH}\overset{O}{\overset{\|}{C}}OH + HOCH_2{-}C_6H_4{-}\text{(resin)}$$

28.

$$-NHCH(C=O)- \text{ with side chain } CH_2-\ddot{S}H + I-CH_2COOH \longrightarrow -NHCH(C=O)- \text{ with side chain } CH_2-\overset{+}{S}H-CH_2COOH + I^- \underset{+H^+}{\overset{-H^+}{\rightleftharpoons}} -NHCH(C=O)- \text{ with side chain } CH_2-\ddot{S}-CH_2COOH$$

29. Knowing that the N-terminal amino acid is Gly, look for a peptide fragment that contains Gly.

"**f**" tells you that the 2nd amino acid is Arg

"**e**" tells you the next two are Ala-Trp or Trp-Ala. "**d**" tells you that Glu is next to Ala, so 3 and 4 must be Trp-Ala and the 5th is Glu

"**g**" tells you the 6th amino acid is Leu

"**h**" tells you the next two are Met-Pro or Pro-Met. "**c**" tells you that Pro is next to Val, so 7 and 8 must be Met-Pro and the 9th is Val

"**b**" tells you the last amino acid is Asp

Gly-Arg-Trp-Ala-Glu-Leu-Met-Pro-Val-Asp

30. Cysteine can react with cyanogen bromide, but the lactone will not be formed, because it would be a strained four-membered-ring lactone and the sulfur would not be positively charged, causing it to be a poor leaving group. Without lactone formation, cleavage cannot occur.

31. Treatment with Edman's reagent would release two PTH-amino acids in approximately equal amounts.

32. Solved in the text.

33. The data from treatment with Edman's reagent and carboxypeptidase A identify the first and last amino acids.

Leu ___ ___ ___ ___ ___ ___ Ser

The data from cleavage with cyanogen bromide identify the position of Met and identify the other amino acids in the pentapeptide and tripeptide but not their order.

cleavage with cyanogen bromide

Arg, Lys, Tyr | Arg, Phe

Leu ___ ___ ___ Met | ___ ___ Ser

The data from treatment with trypsin put the remaining amino acids in the correct position.

Leu Tyr Lys | Arg | Met Phe Arg | Ser

34. 74 amino acids/ 3.6 amino acids per turn = 20.6 turns of the helix
20.6 x 5.4 Å = **111 Å** in an α-helix

74 amino acids x 3.5 Å = **259 Å** in a straight chain

35. It would fold so that its nonpolar residues are on the outside of the protein in contact with the nonpolar membrane and its polar residues are on the inside of the protein.

36. **a.** A cigar-shaped protein has the greatest surface area to volume ratio, so it has the highest percentage of polar amino acids.

b. A subunit of a hexamer would have the smallest percentage of polar amino acids because part of the surface of the subunit can be on the inside of the hexamer and, therefore, have nonpolar amino acids on its surface.

37. An amino acid is insoluble in diethyl ether (a relatively nonpolar solvent) because an amino acid exits as a zwitterion at neutral pH. In contrast, carboxylic acids and amines have a single charge at neutral pH and, therefore, are less polar.

38.

a. His-Lys Leu-Val-Glu-Pro-Arg Ala-Gly-Ala

b. Leu-Gly-Ser-Met-Phe-Pro-Tyr Gly-Val

c. Val-Arg-Gly-Met-Arg-Ala Ser

d. Ser-Phe-Lys-Met Pro-Ser-Ala-Asp

e. Arg Ser-Pro-Lys Lys Ser-Glu-Gly

39.

$$\overset{+}{H_3N}\underset{\underset{COO^-}{|}\atop{CH_2}}{CH}\overset{O}{\overset{\|}{C}}-NH\underset{CH_2C_6H_5}{CH}\overset{O}{\overset{\|}{C}}OCH_3$$

40.

a. $HO\overset{O}{\overset{\|}{C}}CH_2\underset{+NH_3}{CH}\overset{O}{\overset{\|}{C}}OH$ **b.** $HO\overset{O}{\overset{\|}{C}}CH_2\underset{+NH_3}{CH}\overset{O}{\overset{\|}{C}}O^-$ **c.** $^-O\overset{O}{\overset{\|}{C}}CH_2\underset{+NH_3}{CH}\overset{O}{\overset{\|}{C}}O^-$ **d.** $^-O\overset{O}{\overset{\|}{C}}CH_2\underset{NH_2}{CH}\overset{O}{\overset{\|}{C}}O^-$

41. The student is correct. At the pI the total of the positive charges on the tripeptide's amino groups must be 1 to balance the one negative charge of the carboxyl group. When the pH of the solution is equal to the pK_a of a lysine residue, the three lysine groups each have one-half a positive charge for a total of one and one-half positive charges. Thus, the solution must be more basic than this in order to have just one positive charge.

42. Since the mixture of amino acids is in a solution of pH = 5, **His** will have an overall positive charge and **Glu** will have an overall negative charge. **His**, therefore, will migrate to the cathode, and **Glu** will migrate to the anode. **Ser** is more polar than **Thr** (both have OH groups, but **Thr** has an additional carbon); **Thr** is more polar than **Met** (**Met** has S instead of O and an additional carbon); and **Met** is more polar than **Leu**.

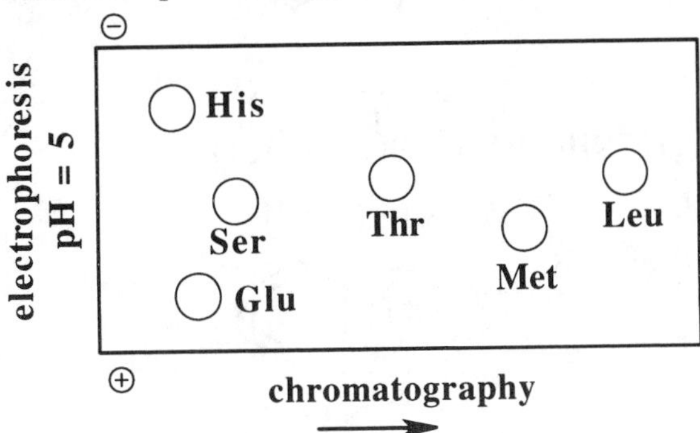

43. You would expect serine and cysteine to have lower pK_a's than alanine since a hydroxymethyl and a thiomethyl group are more electron withdrawing than a methyl group. Because oxygen is more electronegative than sulfur, one would probably expect serine to have a lower pK_a than cysteine. The fact that cysteine has the lower pK_a can be explained by stabilization of serine's carboxyl proton by hydrogen bonding to the β-OH group of serine, which causes it to have less of a tendency to be removed by base.

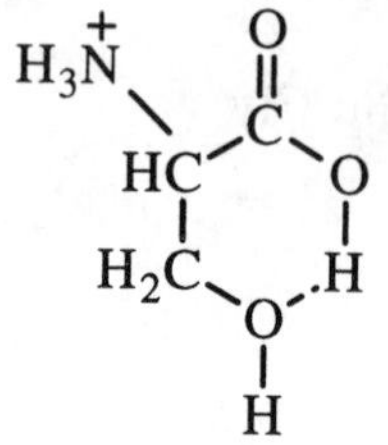

$H_3\overset{+}{N}$ O
HC C O
H_2C O H
H

44. Each compound has two groups that can act as a buffer, one amino group and one carboxyl group. Thus, the compound in higher concentration (0.2 M glycine) will be a more effective buffer.

45.

a. 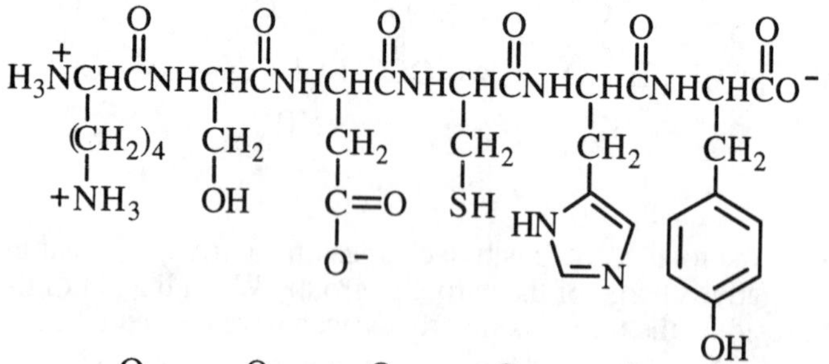 $H_3\overset{+}{N}CHCNHCHCNHCHCNHCHCNHCHCNHCHCO^-$ (each C=O); side chains: $(CH_2)_4$–$^+NH_3$, CH_2–OH, CH_2–C=O–O^-, CH_2–SH, CH_2–(imidazole, HN ... N), CH_2–(phenyl)–OH

b. $H_3\overset{+}{N}CHCNHCHCNHCHCNHCHCNHCHCNHCHCO^-$ (each C=O); side chains: $(CH_2)_4$–$^+NH_3$, CH_2–OH, CH_2–C=O–O^-, CH_2–SH, CH_2–(imidazole, HN ... $\overset{+}{N}H$), CH_2–(phenyl)–OH

c. $H_3\overset{+}{N}CHCNHCHCNHCHCNHCHCNHCHCNHCHCO^-$ (each C=O); side chains: $(CH_2)_4$–$^+NH_3$, CH_2–OH, CH_2–C=O–O^-, CH_2–S^-, CH_2–(imidazole, HN ... N), CH_2–(phenyl)–OH

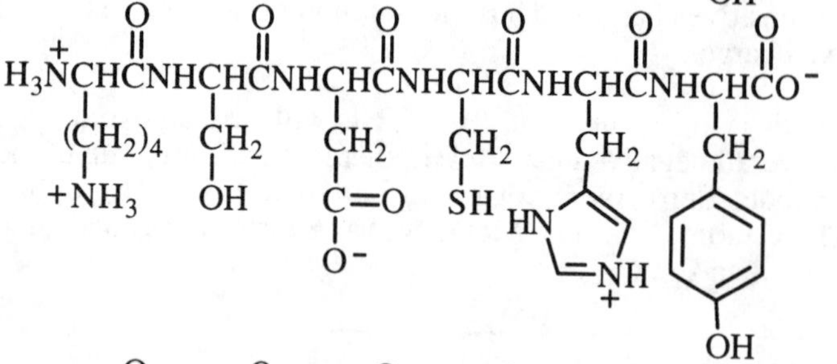

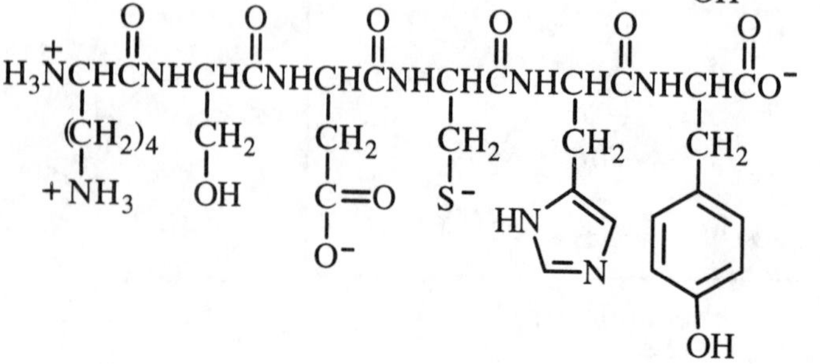

46. **a.** When the polypeptide is treated with maleic anhydride, lysine reacts with maleic anhydride, but the amino group of arginine is not sufficiently nucleophilic to react. Therefore, trypsin will cleave only at arginine residues, since it will no longer recognize lysine residues. Trypsin also will not cleave the Arg-Pro bond.

$$-NHCHC(=O)-(CH_2)_4NH_2 \text{ (lysine)} + \text{maleic anhydride} \longrightarrow -NHCHC(=O)-(CH_2)_4NH-\text{(tetrahedral intermediate, } O^-) \longrightarrow -NHCHC(=O)-(CH_2)_4-NH-C(=O)-CH=CH-C(=O)-O^-$$

lysine

maleic anhydride

b. The N-terminal end of each fragment will be positively charged because of the $^+NH_3$ group, the C-terminal end will be negatively charged because of the COO^- group, arginine residues will be positively charged, aspartic acid and glutamic acid residues will be negatively charged, and lysine residues will be negatively charged because they are attached to the maleic acid group.

	Fragment	Charges	
A	Gly-Ser-Asp-Ala-Leu-Pro-Gly-Ile-Thr-Ser-Arg	+ - + -	overall charge = 0
B	Asp-Val-Ser-Lys-Val-Glu-Tyr-Phe-Glu-Ala-Gly-Arg	+ - - - - + -	overall charge = -3
C	Ser-Glu-Phe-Lys-Glu-Pro-Arg	+ - - - + -	overall charge = -2
D	Leu-Tyr-Met-Lys-Val-Glu-Gly-Arg-Pro-Val-Ser-Ala-Gly-Leu-Trp	+ - - + -	overall charge = -1

c. Elution order: **A** > **D** > **C** > **B**

47. First mark off where the chains would have been cleaved by chymotrypsin (C-side of Phe, Trp, Tyr).

Val-Met-Tyr | Ala-Cys-Ser-Phe | Ala-Glu-Ser

Ser-Cys-Phe | Lys-Cys-Trp | Lys-Tyr | Cys-Phe | Arg-Cys-Ser

Then, from the pieces given, you can determine where the disulfide brides are.

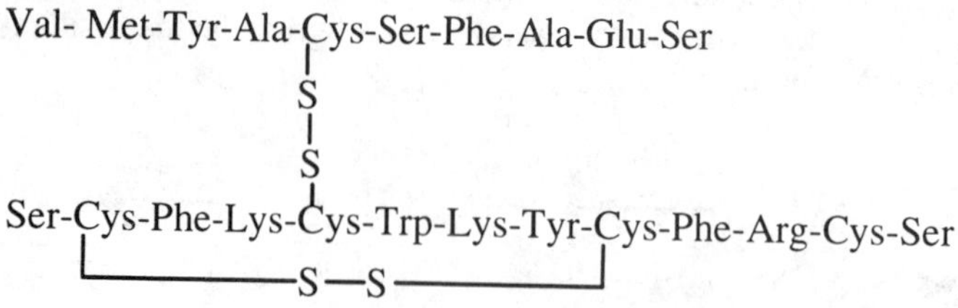

Val- Met-Tyr-Ala-Cys-Ser-Phe-Ala-Glu-Ser

S

S

Ser-Cys-Phe-Lys-Cys-Trp-Lys-Tyr-Cys-Phe-Arg-Cys-Ser

S—S

48. The methyl ester of phenylalanine rather than phenylalanine itself should be added in the peptide bond-forming step because if esterification is done after amide bond formation, both carboxyl groups could be esterified. Some product will be obtained in which the amide bond is formed with the γ-carboxyl group of aspartate rather than with the α-carboxyl group.

$H_2NCH(CH_2COO^-)C(=O)O^-$ + $(CH_3)_3C{-}OC(=O){-}O{-}C(=O)O{-}C(CH_3)_3$ ⟶ $(CH_3)_3C{-}OC(=O)NHCH(CH_2COO^-)C(=O)O^-$

DCC ↓

$(CH_3)_3C{-}OC(=O)NHCH(CH_2COO^-)C(=O)O{-}C(=N{-}C_6H_{11})NH{-}C_6H_{11}$

+ $H_2NCH(CH_2C_6H_5)C(=O)OCH_3$ ⟶

$(CH_3)_3C{-}OC(=O)NHCH(CH_2COO^-)C(=O)NHCH(CH_2C_6H_5)C(=O)OCH_3$

CF_3COOH | CH_2Cl_2 ↓

$CH_3C(CH_3){=}CH_2$ + CO_2 + $H_3\overset{+}{N}CH(CH_2COO^-)C(=O){-}NHCH(CH_2C_6H_5)C(=O)OCH_3$

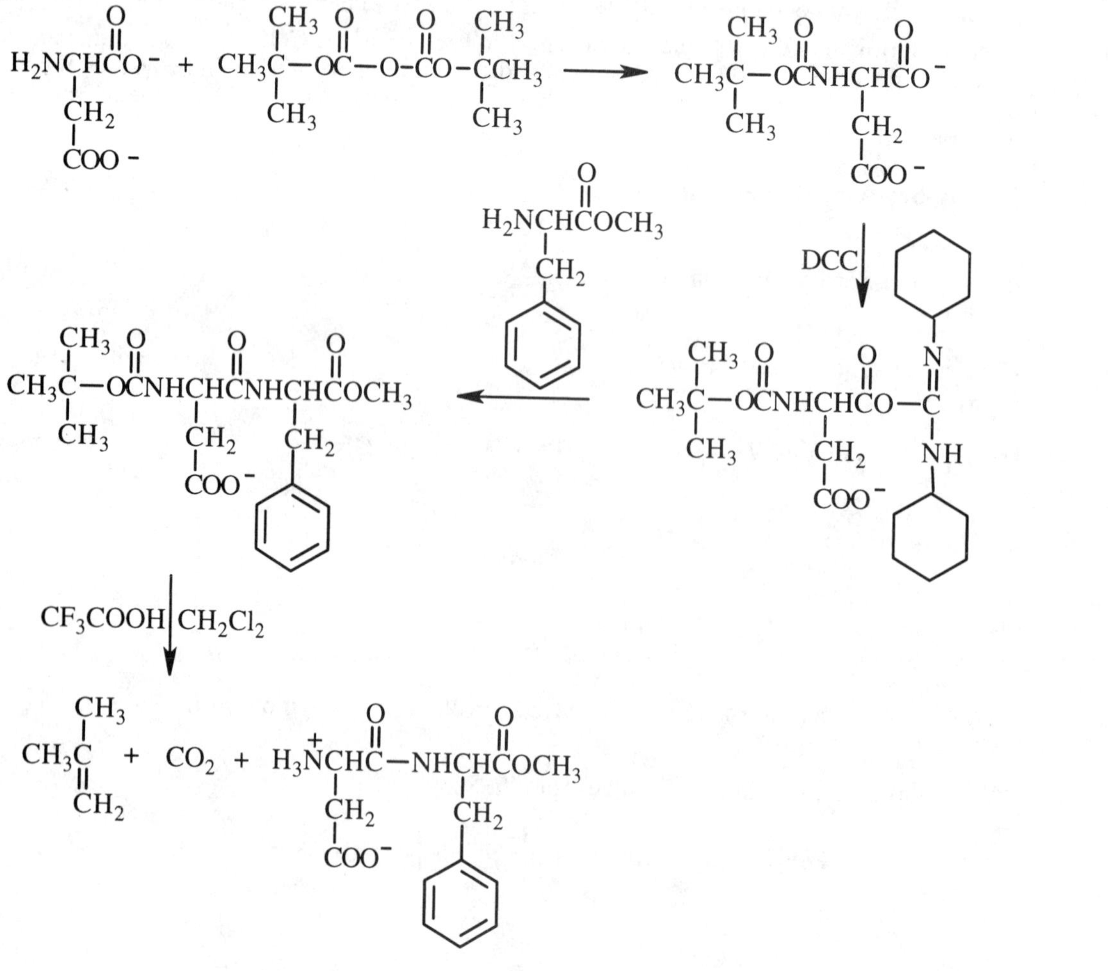

49. Ser-Glu-Leu-Trp-Lys-Ser-Val-Glu-His-Gly-Ala-Met

From the experiment with carboxypeptidase, we know the C-terminal amino acid is Met.

“l” tells us the amino acid adjacent to Met is Ala

“e” tells us the next amino acid is Gly

“b” tells us the next amino acid is His

“g” tells us the next amino acid is Glu

“j” tells us the next amino acid is Val

“c” tells us the next amino acid is Ser

“i” tells us the next amino acid is Lys

“a” tells us the next amino acid is Trp

“h” tells us the next amino acid is Leu

“k” tells us the next amino acid is Glu

“k” tells us the next (first) amino acid is Ser

50. The p*K*a of the carboxylic acid of the dipeptide is higher than the p*K*a of the carboxylic acid of the amino acid because the positively charged amino group of the amino acid is more strongly electron withdrawing than the amide group of the peptide. This causes the amino acid to be a stronger acid and have a lower p*K*a.

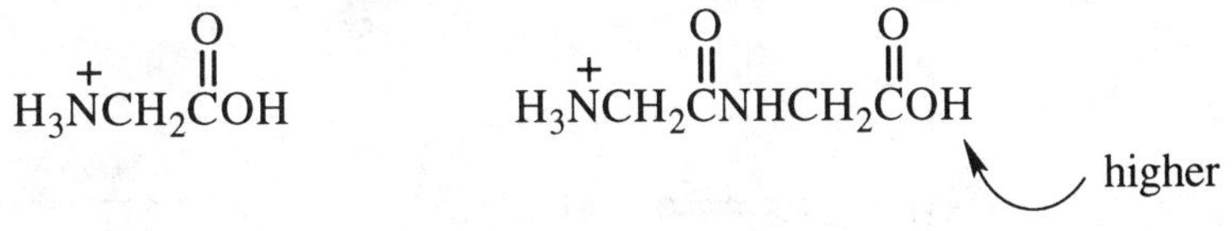

$$H_3\overset{+}{N}CH_2\overset{\overset{O}{\|}}{C}OH \qquad H_3\overset{+}{N}CH_2\overset{\overset{O}{\|}}{C}NHCH_2\overset{\overset{O}{\|}}{C}OH$$

higher

The p*K*a of the amino group of the peptide is lower than the p*K*a of the amino group of the amino acid because the amide group of the peptide is more strongly electron withdrawing than the carboxylate group of the amino acid.

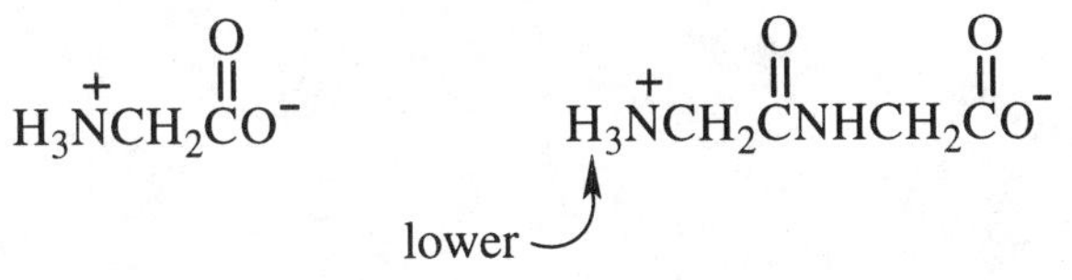

$$H_3\overset{+}{N}CH_2\overset{\overset{O}{\|}}{C}O^- \qquad H_3\overset{+}{N}CH_2\overset{\overset{O}{\|}}{C}NHCH_2\overset{\overset{O}{\|}}{C}O^-$$

lower

51. Finding that there is one less spot than the number of amino acids tells you that the spots for two of the amino acids superimpose. Since leucine and isoleucine have identical polarities, they are good candidates for being the amino acids that migrate to the same location.

Val (pI = 5.97), Trp (pI = 5.89), and Met (pI = 5.75) can be ordered based on their pI 's since the one with the greatest pI will be the one with the greatest amount of positive charge at pH = 5.

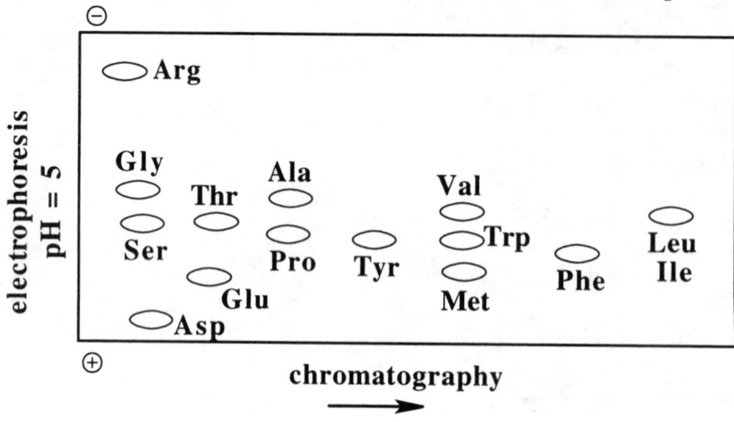

52. Oxidation of dithiothreitol is an intramolecular reaction so it occurs with a larger rate constant than the oxidation of 2-mercaptoethanol, which is an intermolecular reaction. The reverse reduction reaction should occur with about the same rate constant in both cases. Increasing the rate of the oxidation reaction while keeping the rate of the reduction reaction constant is responsible for the greater equilibrium constant, since $K_{eq} = k_1/k_{-1}$.

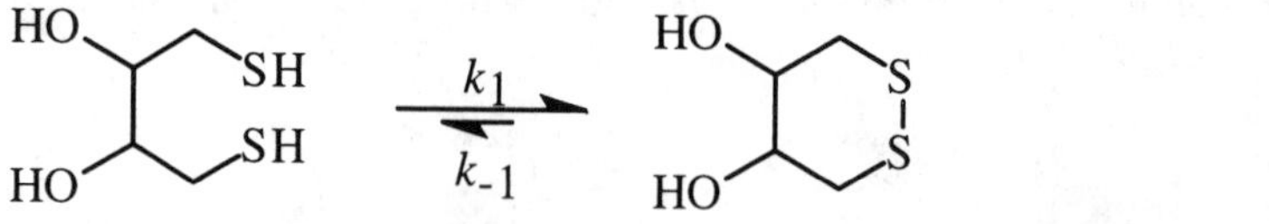

$$2\ HOCH_2CH_2SH \underset{k_{-1}}{\overset{k_1}{\rightleftharpoons}} HOCH_2CH_2S{-}SCH_2CH_2OH$$

53.

a. $\overset{\displaystyle O}{\overset{\|}{RCH}} \xrightarrow{NH_3} \overset{\displaystyle NH_2}{\overset{\|}{RCH}} \xrightarrow{^{-}C\equiv N} \overset{\displaystyle NH_2}{\overset{|}{RCHC\equiv N}} \xrightarrow[H_2O]{H^+} \overset{\displaystyle +NH_3}{\overset{|}{RCHCOOH}}$

intermediate I (the imine, $RCH{=}NH_2$ as drawn); intermediate II ($RCH(NH_2)C\equiv N$)

b. $\overset{\displaystyle CH_3\quad O}{CH_3\overset{|}{C}HCH_2\overset{\|}{C}H}$ (3-methylbutanal) $\longrightarrow$ $CH_3CH(CH_3)CH_2CH(^{+}NH_3)COOH$ (leucine)

c. $\overset{\displaystyle CH_3\;O}{CH_3\overset{|}{C}H{-}\overset{\|}{C}H}$ (2-methylpropanal) $\longrightarrow$ $CH_3CH(CH_3)CH(^{+}NH_3)COOH$ (valine)

54.

amino acid	Absorbance at 280 nm	molar absorptivity = absorbance/concentration
tryptophan	0.61	610
tyrosine	0.15	150
phenylalanine	0.01	1

55. The spot marked with an **X** is the peptide that is different in the normal and mutant polypeptide. The spot is closer to the anode, indicating that the substituted amino acid in the mutant has a higher pI and is less polar.

The fingerprints are those of hemoglobin (normal) and sickle-cell hemoglobin (mutant). In sickle-cell hemoglobin, a glutamic acid in the normal polypeptide is substituted with a valine. This agrees with our observation that the substituted amino acid is less negative and more nonpolar. (See page 1089 of the text.)

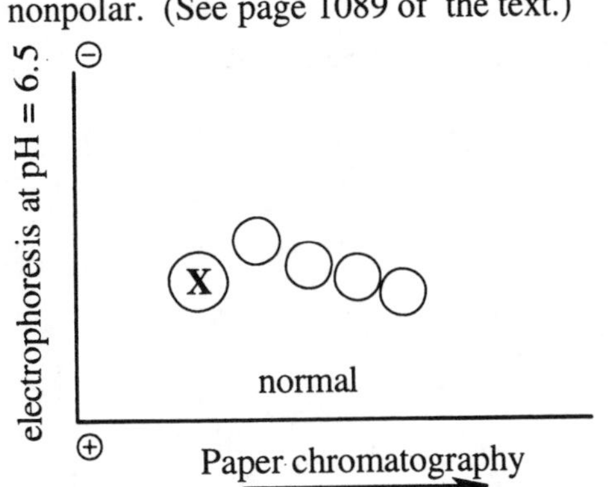

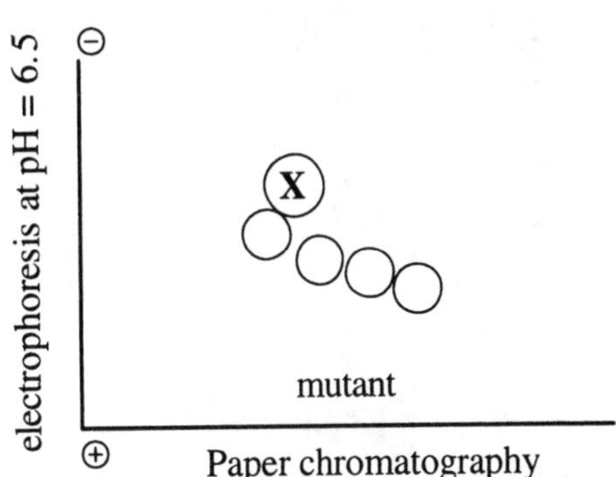

56.

a. Acid-catalyzed hydrolysis indicates the peptide contains 12 amino acids.

___ ___ ___ ___ ___ ___ ___ ___ ___ ___ ___ ___

b. Treatment with Edman's reagent indicates that Val is the N-terminal amino acid.

Val ___ ___ ___ ___ ___ ___ ___ ___ ___ ___ ___

c. Treatment with carboxypeptidase A indicates that Ala is the C-terminal amino acid.

Val ___ ___ ___ ___ ___ ___ ___ ___ ___ ___ Ala

d. Treatment with cyanogen bromide indicates that Met is the 5th amino acid with Arg, Gly, Ser in an unknown order in positions 2, 3, and 4.

Val ___ ___ ___ Met ___ ___ ___ ___ ___ ___ Ala
Arg, Gly, Ser

e. Treatment with chymotrypsin indicates that Tyr is the 6th amino acid, Phe is the 10th, and Ser the 11th .

Val ___ ___ ___ Met Tyr ___ ___ ___ Phe Ser Ala
Arg, Gly, Ser

The tetrapeptide that contains Phe, also contains Pro and 2 Lys. Cleavage would not have occurred at Tyr if it had been adjacent to Pro. Therefore, the 7th amino acid is Lys.

Val ___ ___ ___ Met Tyr Lys ___ ___ Phe Ser Ala
Arg, Gly, Ser Pro, Lys

f. Treatment with trypsin indicates that Arg is the 3rd amino acid, and Ser is 2nd and Gly 4th.

Val Ser Arg Gly Met Tyr Lys ___ ___ Phe Ser Ala
Pro, Lys

Since trypsin catalyzed cleavage between the 7th and 8th amino acids, Lys must be the 8th amino acid, since cleavage would not have occurred at Pro.

Val Ser Arg Gly Met Tyr Lys Lys Pro Phe Ser Ala

57. **a.** The C-terminal end of the protein contains 3 nonpolar amino acids and 4 polar amino acids. Each of the seven amino acids has two atoms that can form hydrogen bonds for a total of 14 atoms that can form hydrogen bonds. Gln and Asp each have two additional groups, and each serine one additional group that can form hydrogen bonds. Thus, the terminal end of the protein has 3 hydrophobic groups and can form 20 hydrogen bonds. So there are 20 hydrogen bonds formed between protein groups and water that must be broken before the protein groups can hydrogen bond to each other. So 20 hydrogen bonds are broken and 20 hydrogen bonds are formed (10 hydrogen bonds from 20 atoms involved in intramolecular hydrogen bonds, and 10 hydrogen bonds from 20 liberated water molecules forming hydrogen bonds with each other). So ΔG° comes only from removing the three hydrophobic groups from water (3 x -4) = **-12 kcal/mol.**

b. If two of the polar groups do not form intramolecular bonds in the interior of the protein, 20 hydrogen bonds will be broken, but only 19 will be formed, so ΔG° = -12 + 3 = **-9 kcal/mol**.

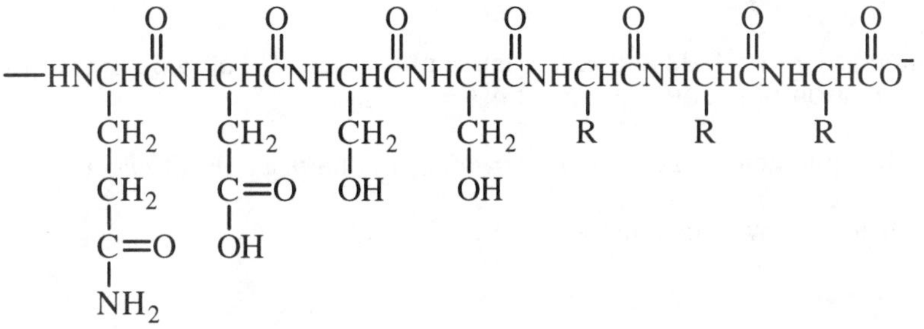

```
          O         O         O         O         O         O         O
          ‖         ‖         ‖         ‖         ‖         ‖         ‖
—HNCHCNHCHCNHCHCNHCHCNHCHCNHCHCNHCHCO⁻
     |         |         |         |         |         |         |
    CH2       CH2       CH2       CH2        R         R         R
     |         |         |         |
    CH2       C=O       OH        OH
     |         |
    C=O       OH
     |
    NH2
```

58. Because the native enzyme has four disulfide bridges, the denatured enzyme has eight cysteine residues. The first cysteine has a 1 in 7 chance of forming a disulfide bridge with the correct cysteine. The first cysteine of the next pair has a 1 in 5 chance, and the first cysteine of the third pair has a 1 in 3 chance.

$$\frac{1}{7} \times \frac{1}{5} \times \frac{1}{3} = 0.0095$$

If disulfide bridge formation were entirely random, the recovered enzyme should have 0.95% of its original activity. The fact that the enzyme Professor Thorpe recovered had 80% of its original activity proves her hypothesis that disulfide bridges form after the minimum energy conformation of the protein has been achieved. In other words, disulfide bridge formation is not random but is determined by the tertiary structure of the protein.

Chapter 20 Practice Test

1. Give the structure of the following amino acids at pH = 7.

   a. glutamic acid **b.** lysine **c.** isoleucine **d.** arginine **e.** asparagine

2. Draw the form of histidine that predominates at:

   a. pH = 1 **b.** pH = 4 **c.** pH = 8 **d.** pH = 11

3. Answer the following:

   a. Alanine has a pI = 6.02 and serine has a pI = 5.68. Which would have the highest concentration of positive charge at pH = 5.50?

   b. Which amino acid is the only one that does not have a chirality center?

   c. Which are the two most nonpolar amino acids?

   d. Which amino acid has the lowest pI?

4. Why does the carboxyl group of alanine have a lower pK_a than the carboxyl group of propanoic acid?

$$\underset{\text{alanine}}{CH_3\underset{\underset{+NH_3}{|}}{C}H\overset{\overset{O}{\|}}{C}O^-} \quad pK_a = 2.2 \qquad \underset{\text{propanoic acid}}{CH_3CH_2\overset{\overset{O}{\|}}{C}O^-} \quad pK_a = 4.7$$

5. Indicate whether each of the following is true or false.

   a. A cigar-shaped protein has a greater percentage of polar residues than a spherical protein. T F

   b. Naturally occurring amino acids have the L-configuration. T F

   c. There is free rotation about a peptide bond. T F

6. Give the compound obtained from mild oxidation of cysteine.

7. Define the following:

a. the primary structure of a protein

b. the tertiary structure of a protein

c. the quaternary structure of a protein

8. Identify the spots.

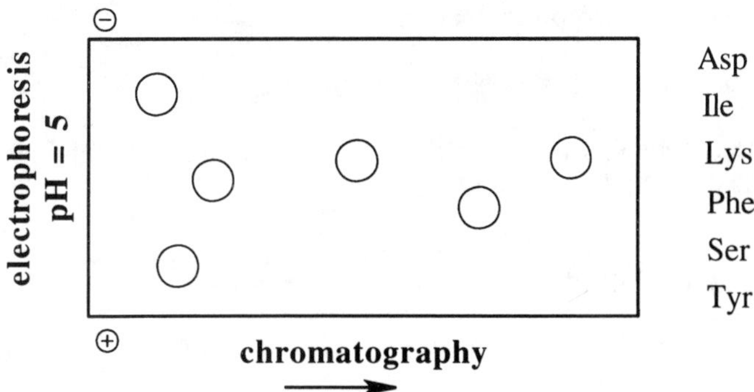

9. Calculate the pI of each of the following amino acids.

a. phenylalanine (pK_a's = 1.83, 9.13) **b.** arginine (pK_a's = 2.17, 9.04, 12.48)

10. From the following information, determine the primary sequence of the decapeptide.

1. Acid hydrolysis gives: Ala, 2 Arg, Gly, His, Ile, Lys, Met, Phe, Ser
2. Reaction with Edman's reagent liberated Ala
3. Reaction with carboxypeptidase A liberated Ile

4. Reaction with cyanogen bromide (cleaves on the C-side of Met)
a. Gly, 2 Arg, Ala, Met, Ser
b. Lys, Phe, Ile, His

5. Reaction with trypsin (cleaves on the C-side of Arg and Lys)
a. Arg, Gly
b. Ile
c. Phe, Lys, Met, His
d. Arg, Ser, Ala

6. Reaction with thermolysin (cleaves on the N-side of Leu, Ile, Phe, Trp, Tyr)
a. Lys, Phe
b. 2 Arg, Ser, His, Gly, Ala, Met
c. Ile

CHAPTER 22
Catalysis

Important Terms

acid catalyst	a catalyst that increases the rate of a reaction by donating a proton.
active site	a pocket or cleft in an enzyme where the substrate is bound.
acyl-enzyme intermediate	an amino acid residue of an enzyme that has been acylated while catalyzing a reaction.
anchimeric assistance (intramolecular catalysis)	catalysis in which the catalyst that facilitates the reaction is part of the molecule undergoing reaction.
base catalyst	a catalyst that increases the rate of a reaction by removing a proton.
catalyst	a substance that increases the rate of a reaction without itself being consumed in the overall reaction.
catalytic antibody	a compound that facilitates a reaction by forcing the conformation of the substrate in the direction of the transition state.
covalent catalysis (nucleophilic catalysis)	catalysis that occurs as a result of a nucleophile forming a covalent bond with one of the reactants.
effective molarity	the concentration of the reagent that would be required in an intermolecular reaction for it to have the same rate as an intramolecular reaction.
electrophilic catalyst	an electrophile that facilitates a reaction.
electrostatic catalysis	stabilization of a charge by an opposite charge.
enzyme	a protein that is a catalyst.
***gem*-dialkyl effect**	two alkyl groups on a carbon whose effect is to increase the probability that the molecule will be in the proper conformation for ring closure.
general-acid catalysis	catalysis in which a proton is transferred to the reactant during the slow step of the reaction.
general-base catalysis	catalysis in which a proton is removed from the reactant during the slow step of the reaction.
induced fit model	a model that describes the specificity of an enzyme for its substrate: the shape of the active site does not become completely complementary to the shape of the substrate until after the enzyme has bound the substrate.
intramolecular catalysis (anchimeric assistance)	catalysis in which the catalyst that facilitates the reaction is part of the molecule undergoing reaction.
lock-and-key model	a model that describes the specificity of an enzyme for its substrate: the substrate fits the enzyme like a key fits into a lock.
metal-ion catalysis	catalysis in which the species that facilitates the reaction is a metal ion.

molecular recognition the recognition of one molecule by another as a result of specific interactions; for example, the specificity of an enzyme for its substrate.

nucleophilic catalysis (covalent catalysis) catalysis that occurs as a result of a nucleophile forming a covalent bond with one of the reactants.

nucleophilic catalyst a catalyst that increases the rate of a reaction by acting as a nucleophile.

oxocarbenium ion an ion in which the positive charge is shared by a carbon and an oxygen.

pH-activity profile or pH-rate profile a plot of the activity of an enzyme as a function of the pH of the reaction mixture.

relative rate obtained by dividing the actual rate constant by the rate constant of the slowest reaction in the group being compared.

site-specific mutagenesis a technique that substitutes one amino acid of a protein for another.

specific-acid catalysis catalysis in which the proton is fully transferred to the reactant before the slow step of the reaction.

specific-base catalysis catalysis in which the proton is completely removed from the reactant before the slow step of the reaction.

substrate the reactant of an enzyme-catalyzed reaction.

transition-state analog a compound that is structurally similar to the transition state of an enzyme-catalyzed reaction.

Solutions to Problems

1. $\Delta H^{\dagger}$, E_a, $\Delta S^{\dagger}$, $\Delta G^{\dagger}$, k_{rate} (These are the parameters that measure the difference between the reactant and the transition state.)

2. Note that **(1)** and **(2)** have only the first phase (because the final product of the reaction is a tetrahedral intermediate), **(3)** has only the second phase (because the initial reactant is a tetrahedral intermediate and **(4)** has both the first and second phases.

First Phase

$$R-\overset{O}{\overset{\|}{C}}-OCH_3 \xrightarrow{H^+} R-\overset{\overset{+}{O}H}{\overset{\|}{C}}-OCH_3 \xrightarrow[H_2\ddot{O}:]{} R-\underset{\underset{H}{+\ddot{O}H}}{\overset{OH}{C}}-OCH_3 \xrightarrow{-H^+} R-\underset{OH}{\overset{OH}{C}}-OCH_3$$

acid-catalyzed ester hydrolysis

$$(1)\quad R-\overset{O}{\overset{\|}{C}}-R \xrightarrow{H^+} R-\overset{\overset{+}{O}H}{\overset{\|}{C}}-R \xrightarrow[H_2\ddot{O}:]{} R-\underset{\underset{H}{+\ddot{O}H}}{\overset{OH}{C}}-R \xrightarrow{-H^+} R-\underset{OH}{\overset{OH}{C}}-R$$

$$(2)\quad R-\overset{O}{\overset{\|}{C}}-H \xrightarrow{H^+} R-\overset{\overset{+}{O}H}{\overset{\|}{C}}-H \xrightarrow[R\ddot{O}H]{} R-\underset{\underset{H}{+\ddot{O}R}}{\overset{OH}{C}}-H \xrightarrow{-H^+} R-\underset{OR}{\overset{OH}{C}}-H$$

$$(4)\quad R-\overset{O}{\overset{\|}{C}}-NHCH_3 \xrightarrow{H^+} R-\overset{\overset{+}{O}H}{\overset{\|}{C}}-NHCH_3 \xrightarrow[H_2\ddot{O}:]{} R-\underset{\underset{H}{+\ddot{O}H}}{\overset{OH}{C}}-NHCH_3 \xrightarrow{-H^+} R-\underset{OH}{\overset{OH}{C}}-NHCH_3$$

a. **similarities:** the first step is protonation of the carbonyl compound, and the second step is attack of a nucleophile on the protonated carbonyl compound, and the third step is loss of a proton.

b. **differences:** the carbonyl compound that is used as the starting material, and the nucleophile that is used in **(2)** is an alcohol rather than water.

Second Phase

$$R-C(OH)_2-OCH_3 \xrightarrow{H^+} R-C(OH)_2-\overset{+}{O}HCH_3 \longrightarrow R-C(=\overset{+}{O}H)-OCH_3 \xrightarrow{-H^+} R-C(=O)-OCH_3$$

acid-catalyzed ester hydrolysis

$$(3)\quad R-CH(OH)-OCH_3 \xrightarrow{H^+} R-CH(\overset{+}{O}H_2)-OCH_3 \longrightarrow R-CH=\overset{+}{O}CH_3 \xrightarrow{CH_3OH} R-CH(OCH_3)-\overset{+}{O}HCH_3 \xrightarrow{-H^+} R-CH(OCH_3)_2$$

$$(4)\quad R-C(OH)_2-NHCH_3 \xrightarrow{H^+} R-C(OH)_2-\overset{+}{N}H_2CH_3 \longrightarrow R-C(=\overset{+}{O}H)-OH \xrightarrow{-H^+} R-C(=O)-OH$$

a. similarities: the first step is protonation of the tetrahedral intermediate, and the second step is elimination of a group from the tetrahedral intermediate. In two of the three reactions, the third step is loss of a proton.

b. differences: the nature of the group that is eliminated from the tetrahedral intermediate. In acetal formation, the third step is not loss of a proton because the intermediate does not have a proton to lose. Instead, the third step is attack of a nucleophile, and the fourth step is loss of a proton.

3.

CH₂ H–A → CH₃ A⁻ (HO⁺ ring-closed intermediate) → CH₃ HA (2-methylchromane, O)

4. Solved in the text.

5. Hydroxide ion catalyzes formation of the tetrahedral intermediate by acting as a nucleophilic catalyst. (It is a better nucleophile than water.) Hydroxide ion catalyzes collapse of the tetrahedral intermediate by acting as a specific-base catalyst. It removes a proton from the neutral tetrahedral intermediate, creating a negatively charged oxygen. It is easier for a negatively charged oxygen to expel a leaving group because the transition state is more stable than the transition state formed from a neutral oxygen, which would have a partial positive charge on the oxygen.

6.

$C_6H_4(COOCH_2CH_3)(CH_2O-H)$ + $:\bar{B}$ (base) → tetrahedral intermediate ($:\ddot{O}:^-$, OCH_2CH_3, ring O) + HB → lactone + CH_3CH_2OH + $:\bar{B}$

7. A general-base catalyst catalyzes formation of the tetrahedral intermediate by removing a proton from water, thereby creating a stronger nucleophile than water. It also can catalyze collapse of the neutral tetrahedral intermediate by removing a proton, thereby making it easier to expel the leaving group.

$CH_3-C(=O)-OCH_3$ + $H-\ddot{O}:-H$ + $\bar{B}$ (general base) ⇌ $CH_3-C(O^-)(OH)-OCH_3$ ⇌ $CH_3-C(=O)-OH$ + CH_3O^-

HO^- / H_2O

$\bar{B}$ + $CH_3-C(OH)(OH)-OCH_3$ ⇌ $CH_3-C(=O)-O^-$ + CH_3OH

8.

$$\frac{1.5 \times 10^6\ s^{-1}\ M^{-1}}{0.6\ s^{-1}\ M^{-1}} = 2.5 \times 10^6$$

9. The metal ion catalyzes the decarboxylation reaction by complexing with the negatively charged oxygen of the carboxyl group and carbonyl oxygen of the β-keto group, thereby making it easier for the carbonyl oxygen to accept the electrons that are left behind when CO_2 is eliminated.

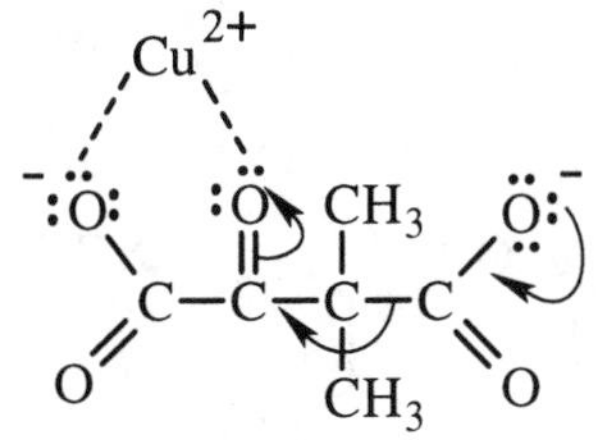

Because acetoacetate and the monoethyl ester of dimethyloxaloacetate do not have a negatively charged oxygen on one carbon and a carbonyl group on an adjacent carbon with which to form a complex, a metal ion does not catalyze decarboxylation of these compounds.

10. Co^{2+} can catalyze the reaction in three different ways. It can complex with the reactant, increasing the susceptibility of the carbonyl group to nucleophilic attack. It can also complex with water, increasing the tendency of water to lose a proton, resulting in a stronger nucleophile for hydrolysis. And it can complex with the leaving group, decreasing its basicity and thereby making it a better leaving group.

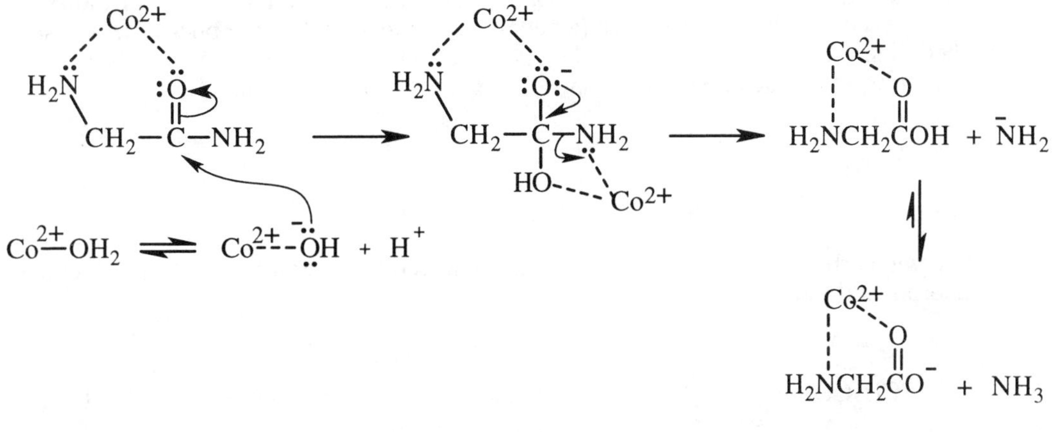

11. Because the reacting groups in the trans isomer are pointed in opposite directions, they cannot react in an intramolecular reaction. Because they can only react via an intermolecular pathway, they will have approximately the same rate of reaction as they would have if the reacting groups were in separate molecules. Consequently, the relative rate would be expected to be close to one.

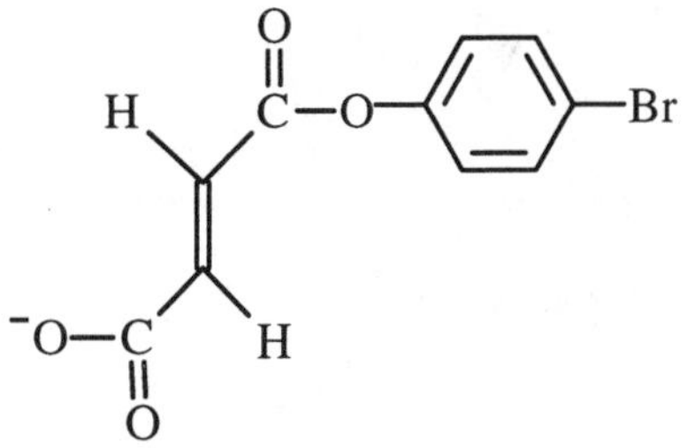

12. **a.** The nucleophile can attack the back side of either of the two ring carbons to which the sulfur is bonded, thereby forming two trans products. There are two nucleophiles (water and ethanol), so a total of four products will be formed.

b.

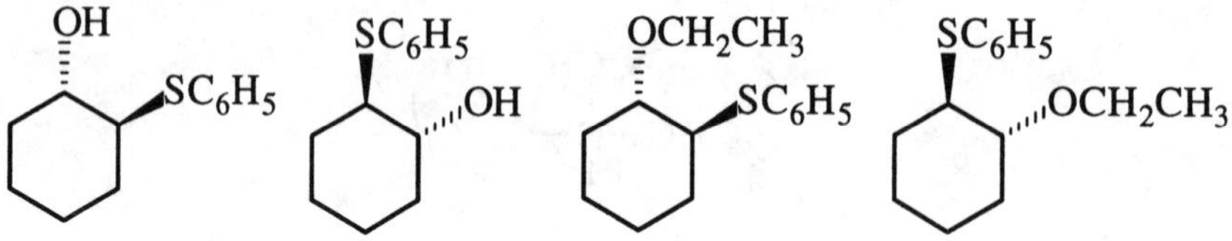

OH
SC_6H_5
SC_6H_5
OH
OCH_2CH_3
SC_6H_5
SC_6H_5
OCH_2CH_3

13. Solved in the text.

14. The tetrahedral intermediate has two leaving groups, a carboxylate ion and a phenolate ion. The carboxylate ion is a weaker base (a better leaving group) than the phenolate ion, so the tetrahedral intermediate re-forms **A**. The 2,4-dinitrophenoxide ion is a weaker base (better leaving group) than the carboxylate ion, so the tetrahedral intermediate forms **B**.

15. If the *ortho*-carboxyl substituent acts as an intramolecular general-base catalyst, ^{18}O would not be incorporated into salicylic acid.

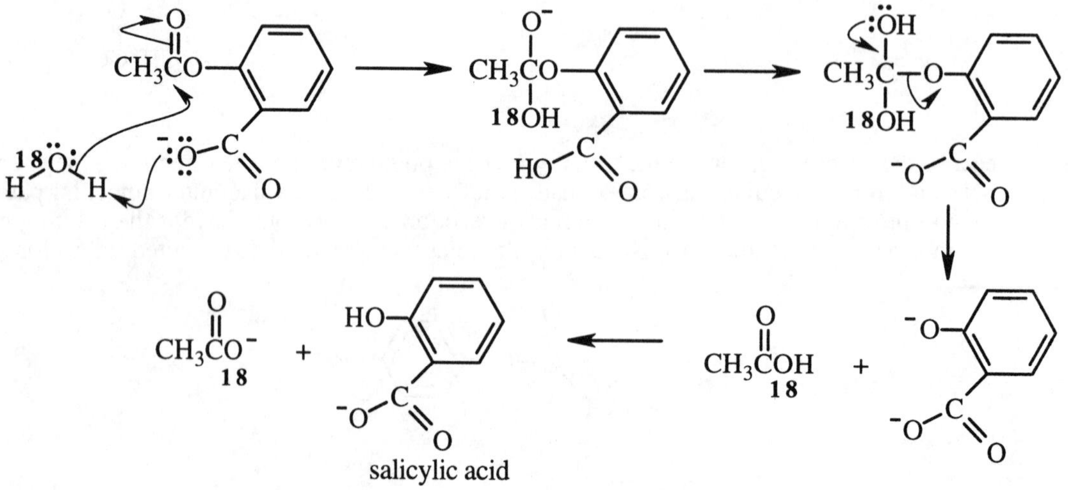

CH_3CO
CH_3CO
^{18}OH
HO
CH_3C
^{18}OH
CH_3CO^-
18
HO
salicylic acid
CH_3COH
18

If the *ortho*-carboxyl substituent acts as an intramolecular nucleophilic catalyst, ^{18}O would be incorporated into salicylic acid.

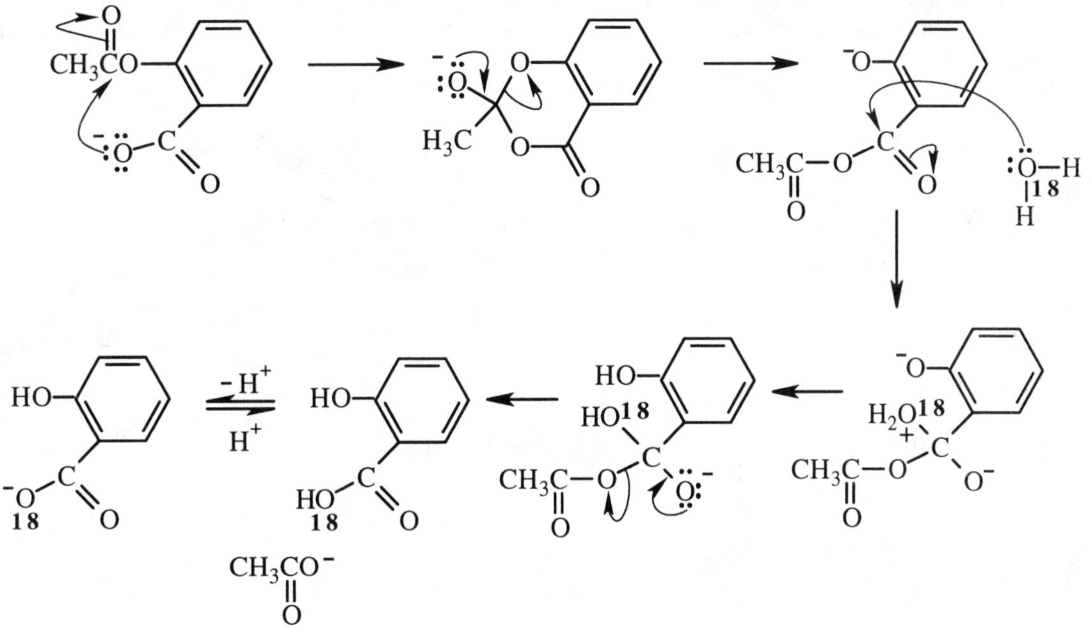

Not all the salicylic acid would contain ^{18}O, because the anhydride intermediate can be hydrolyzed in two different ways.

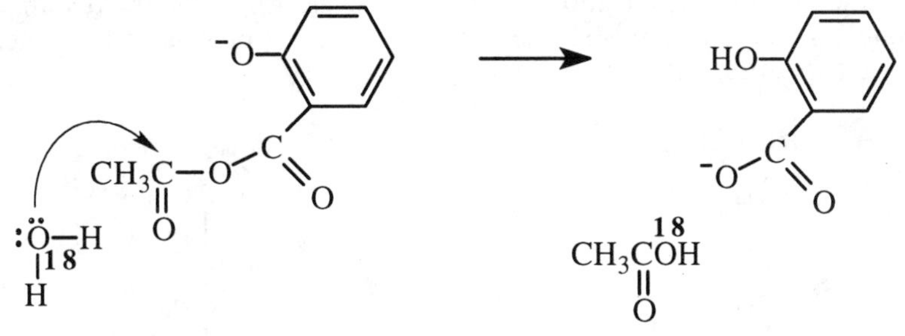

16. Ser-Ala-Phe would be more readily cleaved by carboxypeptidase A because the phenyl substituent of phenylalanine would be more attracted to the hydrophobic pocket of the enzyme than would the negatively charged substituent of aspartic acid.

17. Glu 270 attacks the carbonyl group of the ester, forming a tetrahedral intermediate. Collapse of the tetrahedral intermediate is most likely catalyzed by a general-acid group of the enzyme in order to increase the leaving ability of the RO group. (Perhaps the HO substituent of tyrosine is close enough in the esterase to act as the catalyst.) The group that donates the proton can then act as a general-base catalyst to remove a proton from water as it hydrolyzes the anhydride.

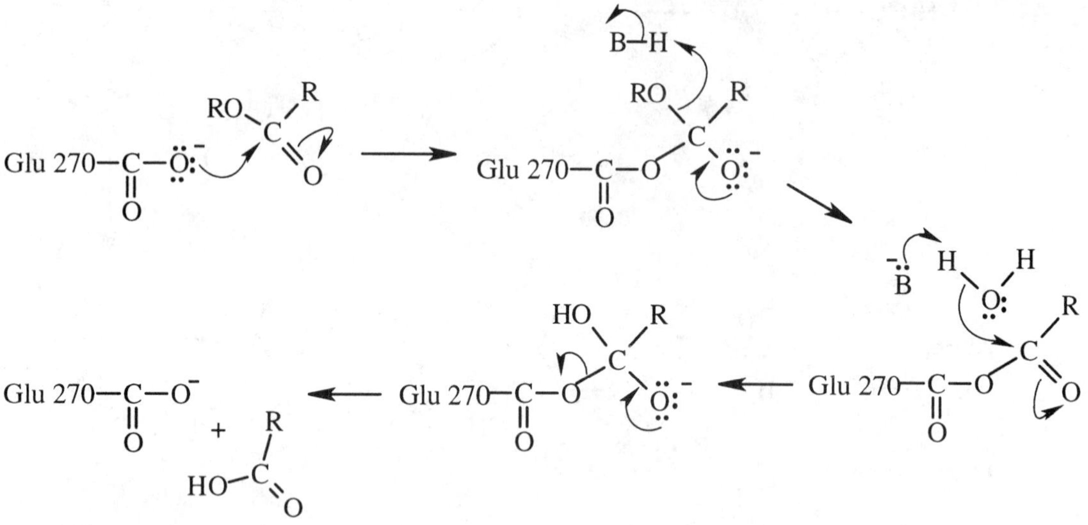

18. Because arginine extends farther into the binding pocket, it must be the one that forms direct hydrogen bonds. Lysine, which is shorter, needs the mediation of a water molecule in order to engage in bond formation with aspartic acid.

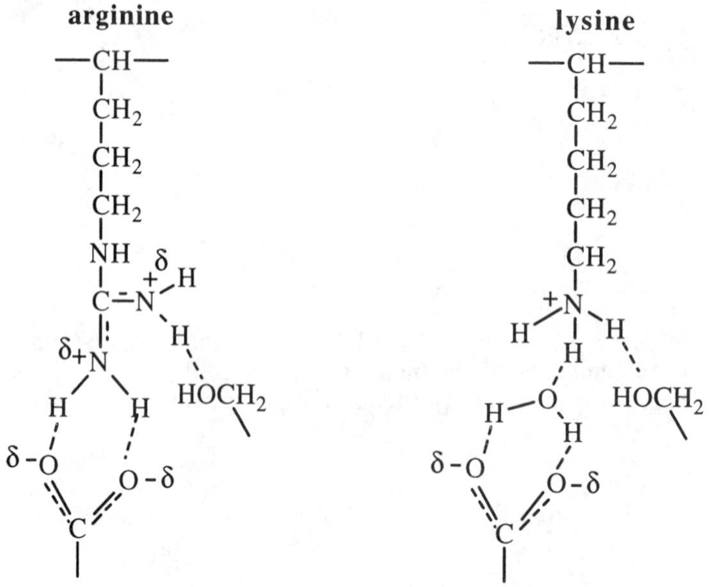

arginine **lysine**

19. The side chains of D-Arg and D-Lys are not positioned to bind correctly at the active site.

20. NAM would contain ^{18}O because it is the ring that would undergo nucleophilic attack by $H_2{}^{18}O$.

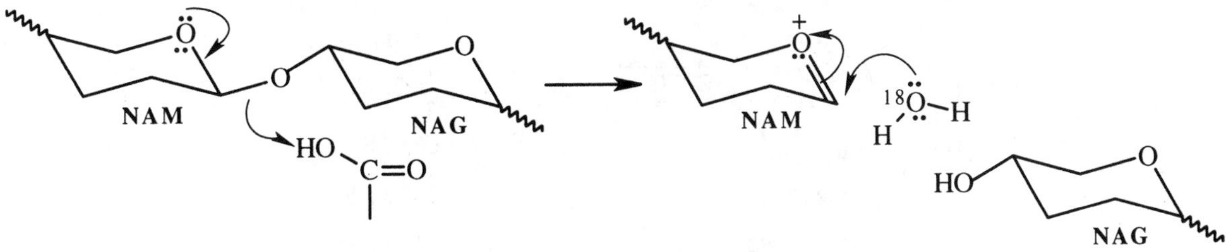

NAM NAG HO C=O → NAM ^{18}O–H H HO NAG

21. Lemon juice contains citric acid. Some of the side chains of an enzyme will become protonated in an acidic solution. This will change the charge of the group (for example a negatively charged aspartic acid, when protonated, becomes neutral), and because the shape of an enzyme is determined by the interaction of the side chains, changing the charges of the side chains will cause the enzyme to undergo a conformational change that leads to denaturation. When the enzyme is denatured, it loses its ability to catalyze the reaction that causes apples to turn brown.

22. The positively charged nitrogen atom of the Schiff base serves as an electron sink to accept the electrons that are left behind when the C3-C4 bond breaks.

In the absence of the Schiff base, the electrons would be delocalized onto a neutral oxygen. The neutral oxygen is not as electron withdrawing as the positively charged nitrogen. In other words, the electron sink that is present as a result of imine formation makes it easier to break the carbon-carbon bond.

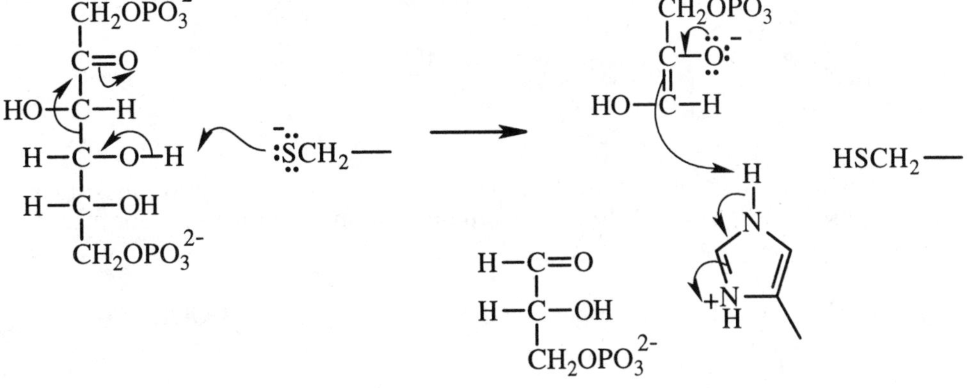

$CH_2OPO_3^{2-}$ C=O HO–C–H H–C–O–H H–C–OH $CH_2OPO_3^{2-}$:$\ddot{S}CH_2$— → $CH_2OPO_3^{2-}$ C–O:$^-$ HO–C–H H N +N H HSCH$_2$— H–C=O H–C–OH $CH_2OPO_3^{2-}$

23. In order to break the C3-C4 bond, the carbonyl group has to be at the #2 position as it is in fructose, so it can accept the electrons; the carbonyl group at the #1 position in glucose cannot serve as an electron sink. Therefore, glucose must isomerize to a ketose, so the carbonyl group will be at the #2 position.

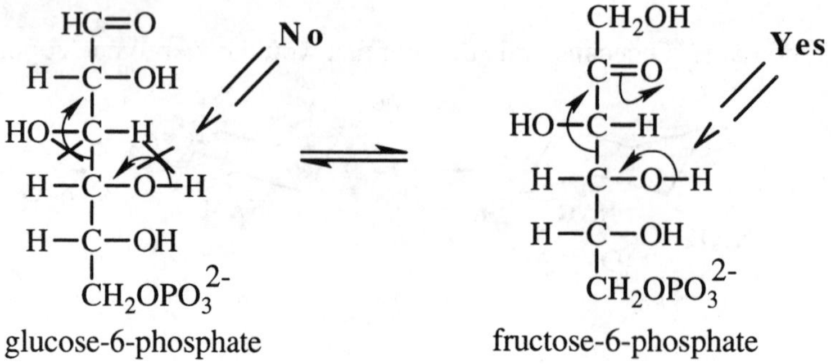

HC=O | H—C—OH | HO—C—H | H—C—O—H | H—C—OH | $CH_2OPO_3^{2-}$ — No

glucose-6-phosphate

CH_2OH | C=O | HO—C—H | H—C—O—H | H—C—OH | $CH_2OPO_3^{2-}$ — Yes

fructose-6-phosphate

24. Cysteine residues are known to react with iodoacetic acid. If a cysteine residue is at the active site of the enzyme, adding a substituent to the sulfur atom could interfere with the enzyme's being able to bind the substrate or it could interfere with positioning the tyrosine residue that is involved in catalyzing the reaction. Adding a substituent to cysteine might also cause a conformational change in the enzyme that could destroy its activity

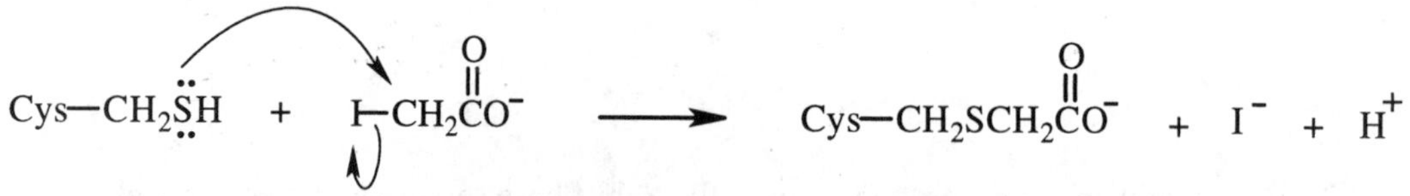

$$\text{Cys—CH}_2\text{SH} + \text{I—CH}_2\text{CO}^- (\text{C=O}) \longrightarrow \text{Cys—CH}_2\text{SCH}_2\text{CO}^- + \text{I}^- + \text{H}^+$$

25. The following compound will lose HBr more rapidly because the negatively charged oxygen is in position to act as an intramolecular general-base catalyst.

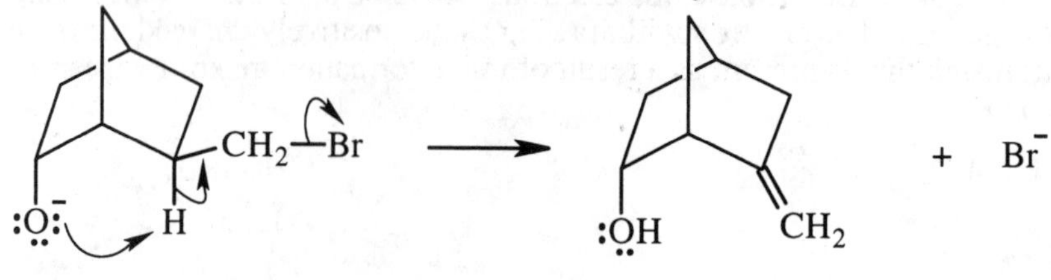

:O:⁻ … H … CH_2—Br ⟶ :OH … CH_2 + Br^-

26. The compounds shown below will be the more reactive because the methyl substituent causes the reacting groups (COOH and OH) to stay in a more favorable conformation for reaction.

a. H_3C, COOH, OH

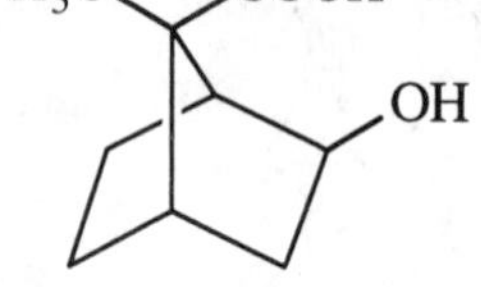

b. OH, COOH, CH_3

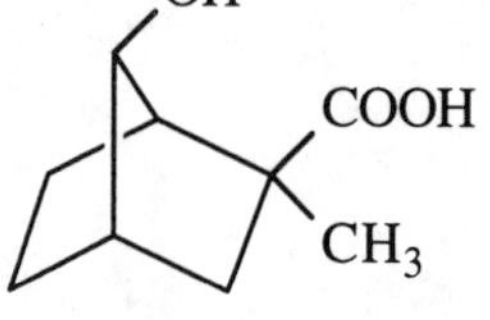

27. The compound shown below will form a lactone more rapidly because it forms a five-membered-ring lactone, which is less strained and than the seven-membered-ring lactone formed by the other compound. The greater stability of the five-membered-ring product will cause the transition state leading to its formation to be more stable than the transition state leading to the seven-membered-ring product.

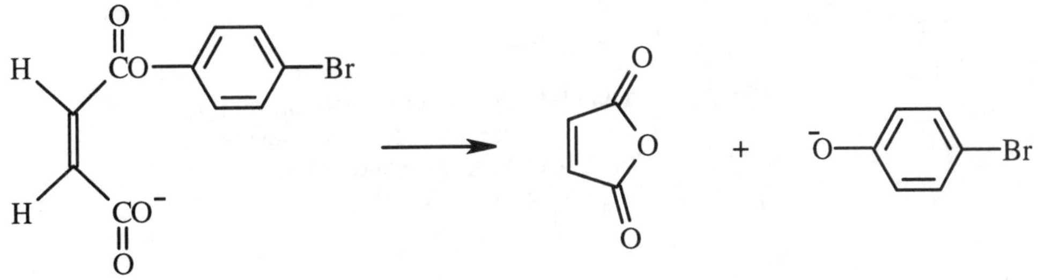

28. In order to hydrolyze an amide, the NH_2 group in the tetrahedral intermediate has to leave in preference to the less basic OH group. This can happen if the NH_2 group is protonated because $^{+}NH_3$ is a weaker base and, therefore, easier to eliminate than OH. Of the four compounds, two have substituents that can protonate the NH_2 by acting as general-acid catalysts, *ortho*-carboxybenzamide and *ortho*-hydroxybenzamide.

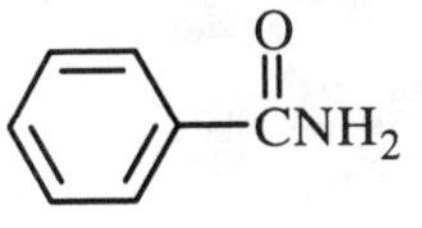

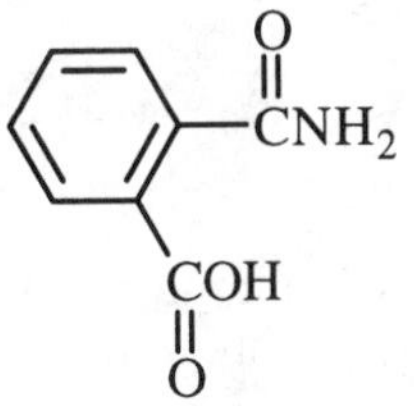

ortho-carboxybenzamide

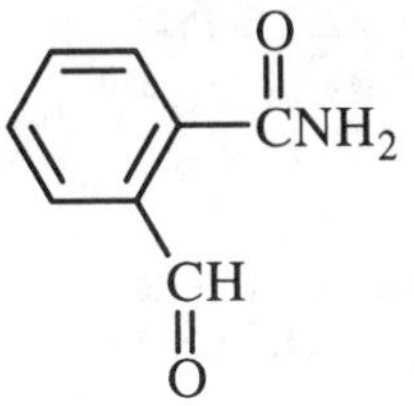

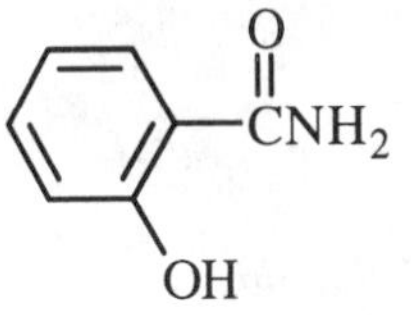

ortho-hydroxybenzamide

Because the carboxy group is electron-withdrawing and the phenolic OH group is electron-donating, formation of the tetrahedral intermediate will be faster for *o*-carboxybenzamide. Because the carboxy group is a stronger acid than the phenolic OH group, the tetrahedral intermediate of *o*-carboxybenzamide will collapse to products faster. Therefore, the *o*-carboxybenzamide has the faster rate of hydrolysis.

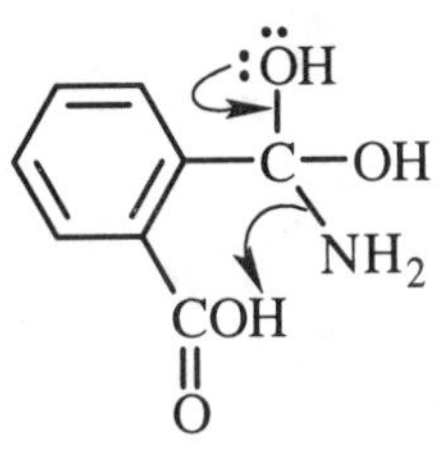

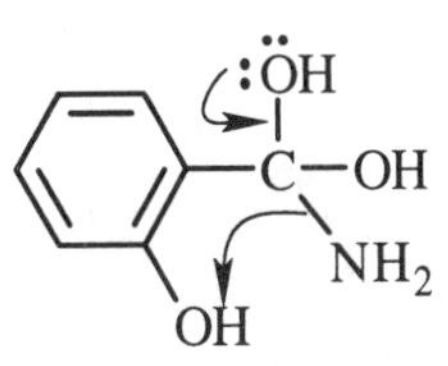

29.

a.  $CH_3CH_2\ddot{\underset{\cdot\cdot}{S}}CH_2CH_2$—Cl intramolecular nucleophilic catalysis

b. 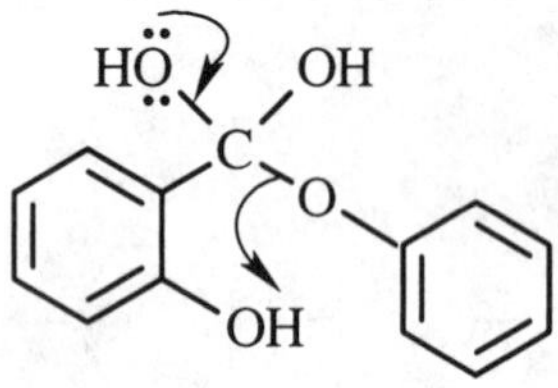 intramolecular general-acid catalysis

HO OH C O OH

The OH substituent is protonating the leaving group as it departs, causing it to be a weaker base and, thus, a better leaving group.

30. If the *ortho*-carboxy substituent is acting as a general-base catalyst, the kinetic isotope effect will be greater than 1.0 because an OH (or OD) bond is broken in the slow step of the reaction.

O CH_3CO— D O D O C O

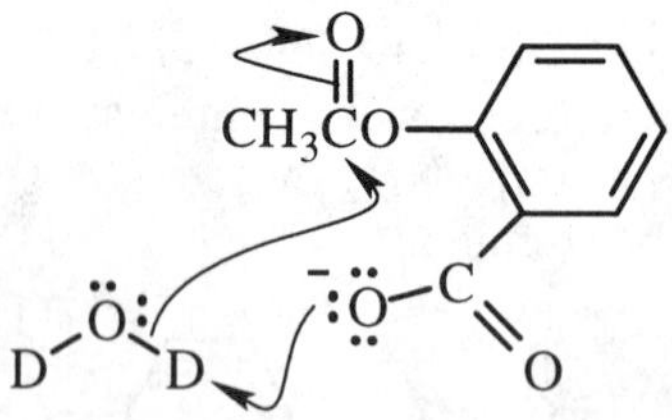

If the *ortho*-carboxyl substituent is acting as a nucleophilic catalyst, the kinetic isotope effect will be about 1.0 because an OH (or OD) bond is not broken in the slow step of the reaction.

O CH_3CO— O C O

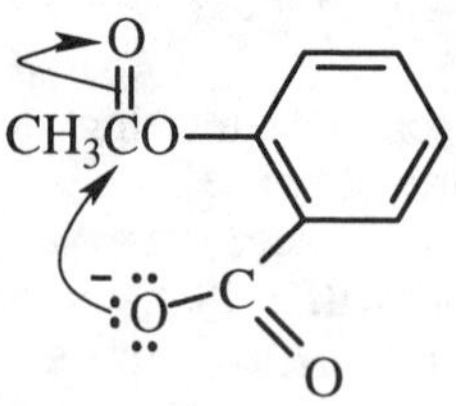

31. Because the catalytic group is a general-acid catalyst, it will be active in its acidic form and inactive in its basic form. The pH at the midpoint of the curve corresponds to the pK_a of the group responsible for the catalysis.

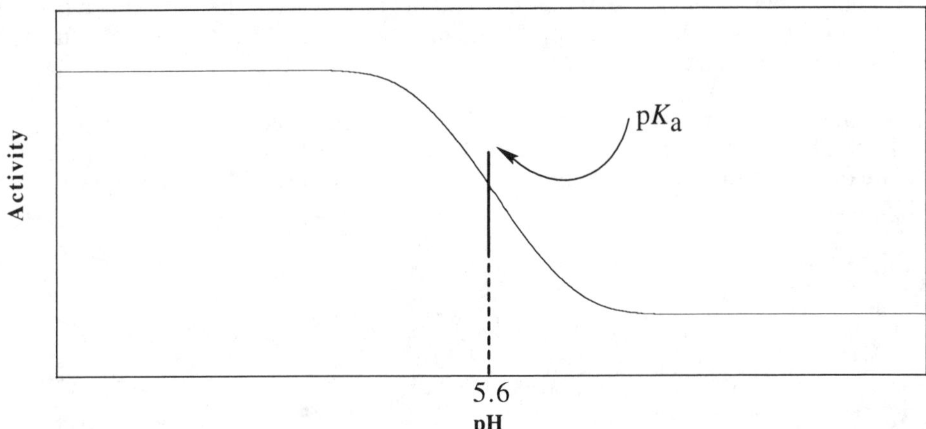

32. Co^{2+} can catalyze the hydrolysis reaction by complexing with three nitrogen atoms in the substrate as well as with water. Complexation increases the acidity of water, thereby providing a stronger nucleophile for the hydrolysis reaction. Complexation with the substrate locks the nucleophile into the correct position for attack on the carbonyl carbon.

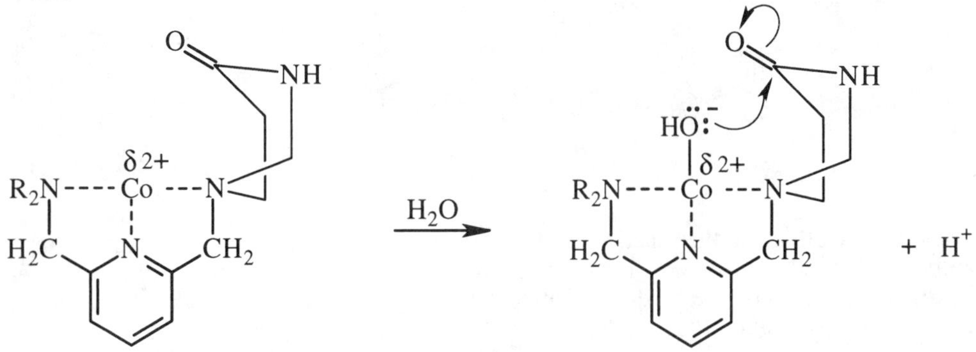

33.

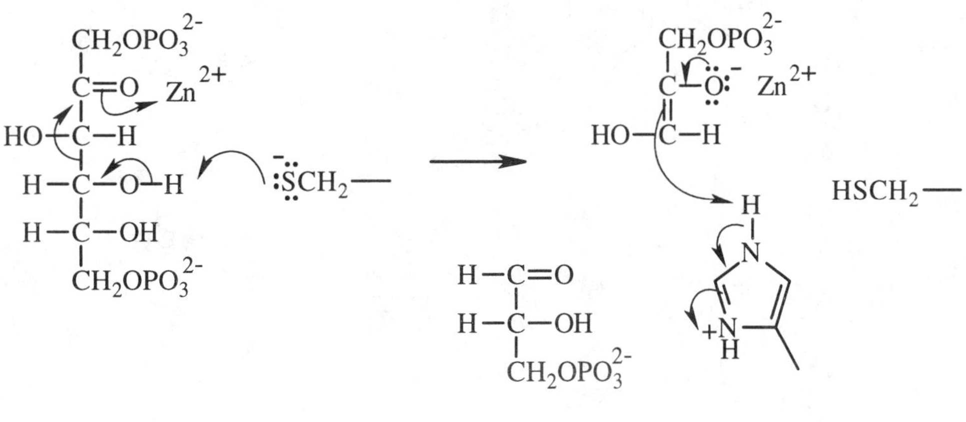

34. There are two possible mechanisms that involve morpholine as a nucleophilic catalyst that can account for the observed catalysis by morpholine. In the absence of morpholine, the first step in the reaction is attack of water on the ester. In the presence of morpholine, the first step in the reaction (in both mechanisms) is attack of morpholine on the aldehyde, which is a faster reaction because morpholine is a stronger nucleophile than water and an aldehyde is more susceptible to nucleophilic attack than an ester.

In one mechanism, the negatively charged aldehyde oxygen (which is a much stronger nucleophile than water) is then the nucleophile that attacks the ester. The tetrahedral intermediate collapses to form a lactone. Imine formation followed by imine hydrolysis gives the final product.

CH HN O COCH3 O O OCH3 O N O + CH_3OH O N O H C=N+ O CO^- O H_2O O CH HN O CO^- O

In the other mechanism, the reaction of morpholine with the aldehyde forms an imine. The positively charged imine makes it easier for water to attack the ester by stabilizing the negative charge that develops on the oxygen. Collapse of the tetrahedral intermediate and hydrolysis of the imine form the final product.

O CH HN O O C OCH_3 H C=N+ O O C $H_2\ddot{O}$: OCH_3 H C=N+ O :Ö:⁻ C HO OCH_3 H C=N+ O O C OH + CH_3OH H_2O HN O + O CH O C O^-

35.

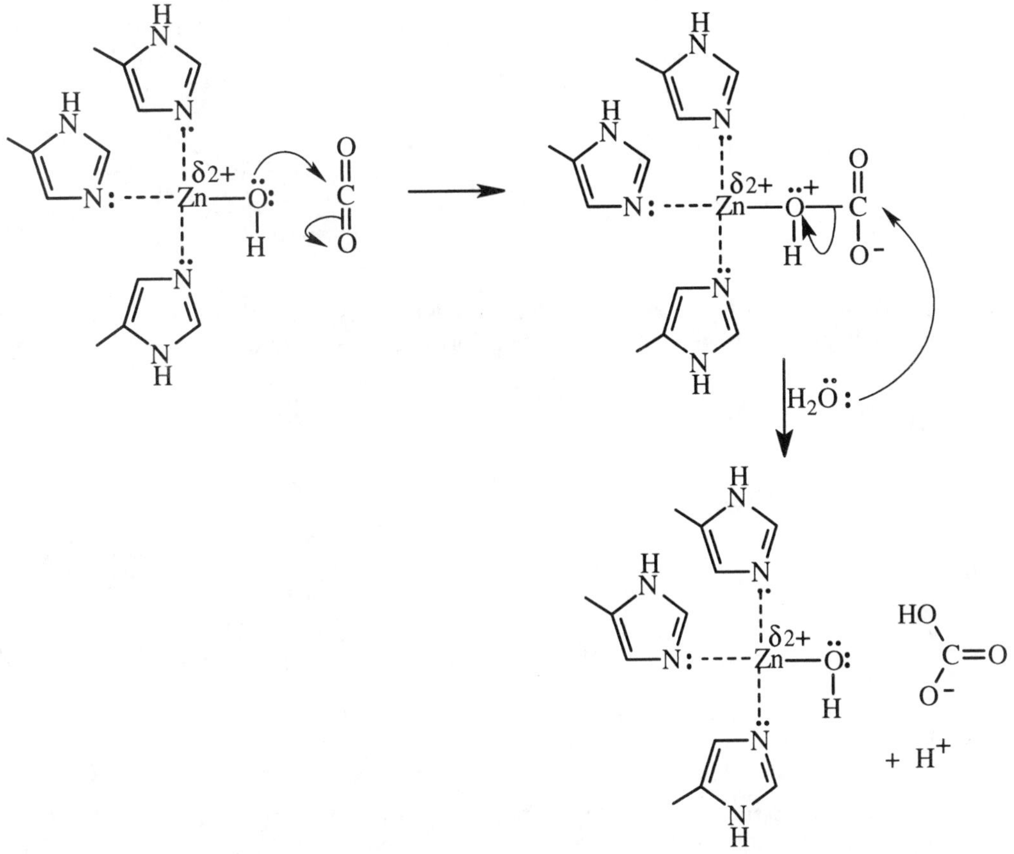

36. At pH = 12, the nucleophile is hydroxide ion. Attack of hydroxide ion on the carbonyl group is faster in **A** because the negative charge that is created in the tetrahedral intermediate is stabilized by the positively charged nitrogen.

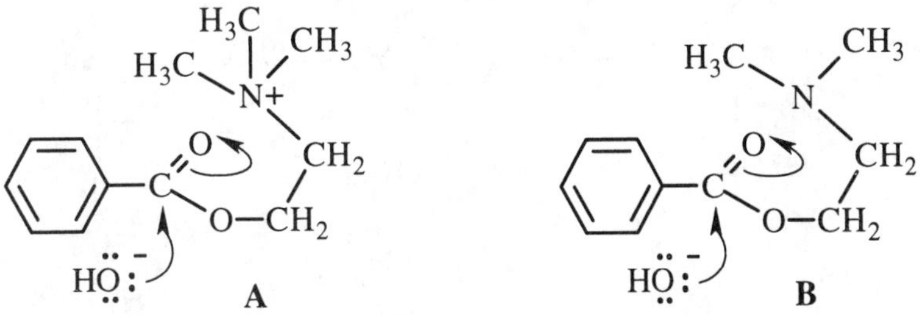

At pH = 8, the nucleophile is water. Attack of water on the carbonyl group is faster in **B** because the amino group can act as a general-base catalyst to make water a better nucleophile.

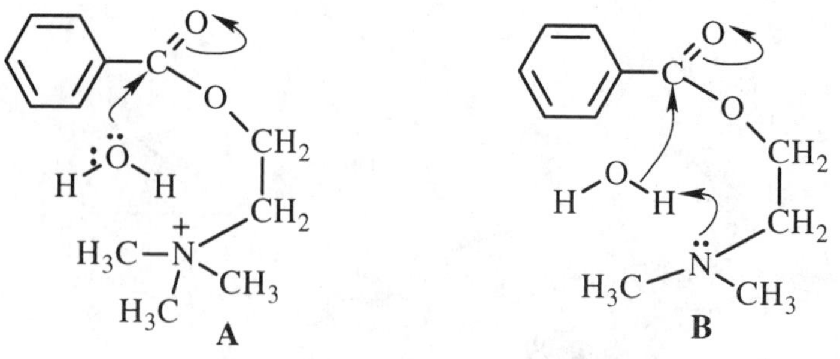

37. **a.** The cis reactants each undergo a direct S_N2 reaction. Because the acetate displaces the tosyl group by backside attack, each cis reactant forms a trans product.

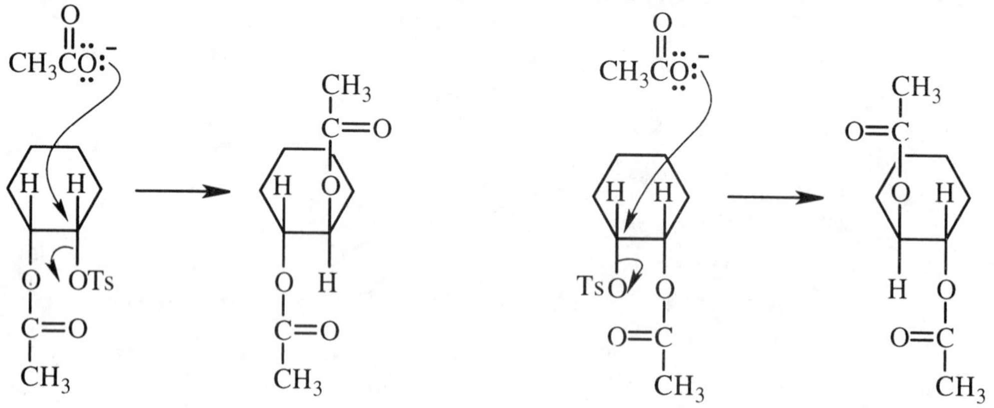

b. The acetate group in a trans reactant is positioned to be able to displace the tosyl leaving group by an intramolecular S_N2 reaction. Acetate ion then attacks in a second S_N2 reaction from the backside of the group it displaces, so trans products are formed. Because both trans reactants form the same intermediate, they both will form the same product. Because the acetate ion can attack either of the carbons in the intermediate equally as easily, a racemic mixture will be formed.

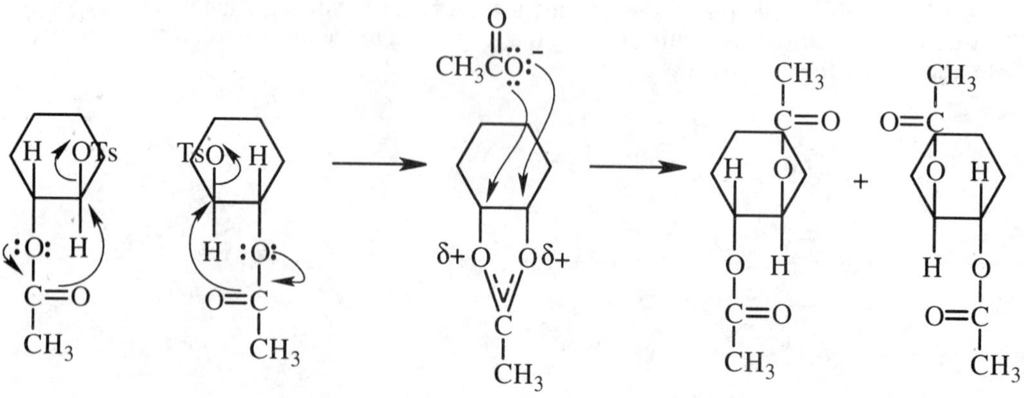

c. The trans reactant is more reactive because the tosyl leaving group is displaced in an intramolecular reaction, forming a positively charged cis intermediate that is considerably more reactive than the neutral cis isomer.

38. Hydrolysis of the phosphate group of DNA occurs by an in-line displacement mechanism; the phosphorous-oxygen double bond is not broken when DNA is cleaved. (See page 1068 of the text). A possible mechanism for the hydrolysis reaction is for Arg to neutralize the negative charge on the phosphate group so a negatively charged nucleophile can approach it. (See Figure 25.2 on page 1074 of the text.) Ca^{2+} increases the acidity of water, forming metal-bound hydroxide ion that is a stronger nucleophile than water. Glu can function as a general-acid catalyst, donating a proton to the leaving group, thereby decreasing its basicity and making it a better leaving group.

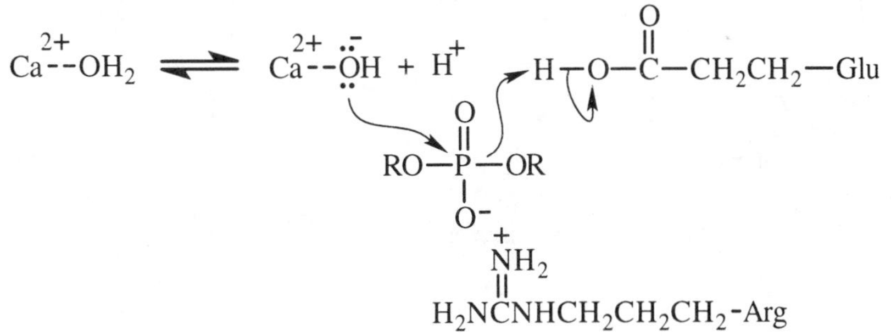

$$Ca^{2+}\text{--}OH_2 \rightleftharpoons Ca^{2+}\text{--}\ddot{O}H^- + H^+$$

H–O–C(=O)–CH_2CH_2–Glu

RO–P(=O)(O$^-$)–OR

$H_2N\overset{+}{C}(=NH_2)NHCH_2CH_2CH_2$-Arg

39. Reduction of the imine linkage with sodium borohydride causes fructose to become permanently attached to the enzyme because the hydrolyzable imine bond has been lost. Acid-catalyzed hydrolysis removes the phosphate groups and hydrolyzes the peptide bonds, so the radioactive fragment that is isolated after hydrolysis is the lysine residue (covalently attached to fructose) of the enzyme that originally formed the Schiff base.

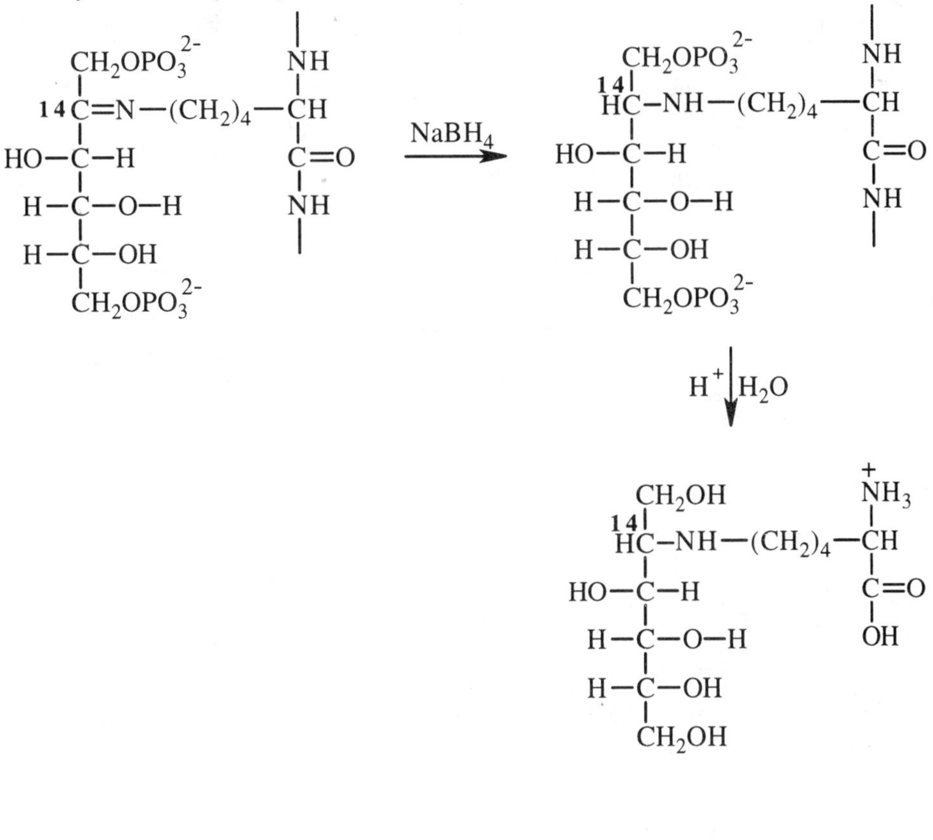

$CH_2OPO_3^{2-}$–^{14}C(=N–$(CH_2)_4$–CH(NH)–C=O–NH)–(HO–C–H)–(H–C–O–H)–(H–C–OH)–$CH_2OPO_3^{2-}$

$\xrightarrow{NaBH_4}$

$CH_2OPO_3^{2-}$–^{14}HC–NH–$(CH_2)_4$–CH(NH)–C=O–NH; HO–C–H; H–C–O–H; H–C–OH; $CH_2OPO_3^{2-}$

$\downarrow H^+,\ H_2O$

CH_2OH–^{14}HC–NH–$(CH_2)_4$–CH($\overset{+}{N}H_3$)–C(=O)–OH; HO–C–H; H–C–O–H; H–C–OH; CH_2OH

40. **a.** 3-Amino-2-oxindole catalyzes the decarboxylation of an α-keto acid by first forming an imine that is followed by a prototropic shift.

H NH$_2$ =O + RC—COH (α-keto acid) —imine formation→ H N=CR C=O O H =O; H$^+$

↓

N=CHR —OH ←decarboxylation— N—CHR C=O O H =O

↓ enolization

H N=CHR =O —imine hydrolysis→ H NH$_2$ =O O=CHR

b. 3-Aminoindole would not be as effective a catalyst, because the electrons left behind when CO_2 is eliminated cannot be delocalized onto an electronegative atom.

NH$_2$ (3-aminoindole) + RC—COH (α-keto acid) —imine formation→ R N=C C=O O H

41. **a.** Intramolecular nucleophilic attack on an alkyl halide occurs more rapidly than intermolecular attack on an alkyl halide, because the reacting groups are tethered together in the former. The intramolecular reaction is followed by another relatively rapid reaction, because the strain in the three-membered ring causes it to break easily.

$CH_3CH_2CH_2CH_2$—Cl + :Nu ⟶ $CH_3CH_2CH_2CH_2$—Nu

$ClCH_2CH_2\ddot{N}(CH_3)CH_2CH_2$—Cl ⟶ $ClCH_2CH_2(CH_3)\overset{+}{N}$ (aziridinium) Cl^- + :Nu ⟶ Nu—$CH_2CH_2N(CH_3)CH_2CH_2Cl$

b.

$ClCH_2CH_2\ddot{N}(CH_3)CH_2CH_2$—Cl ⟶ $ClCH_2CH_2(CH_3)\overset{+}{N}$ (aziridinium) Cl^- + guanine (HN, H_2N, N, R) ⟶ N-7-alkylated guanine ($CH_2CH_2\ddot{N}(CH_3)CH_2CH_2$-Cl) Cl^-

↓

Cl^- CH_3 / CH_2CH_2—$\overset{+}{N}$ (aziridinium) attached to guanine Cl^- + guanine (NH, NH_2, N, R)

↓

Guanine—CH_2CH_2—N(CH_3)—CH_2CH_2—guanine, both N+ with Cl^- (HN, H_2N, N, R; NH, NH_2, N, R)

42.

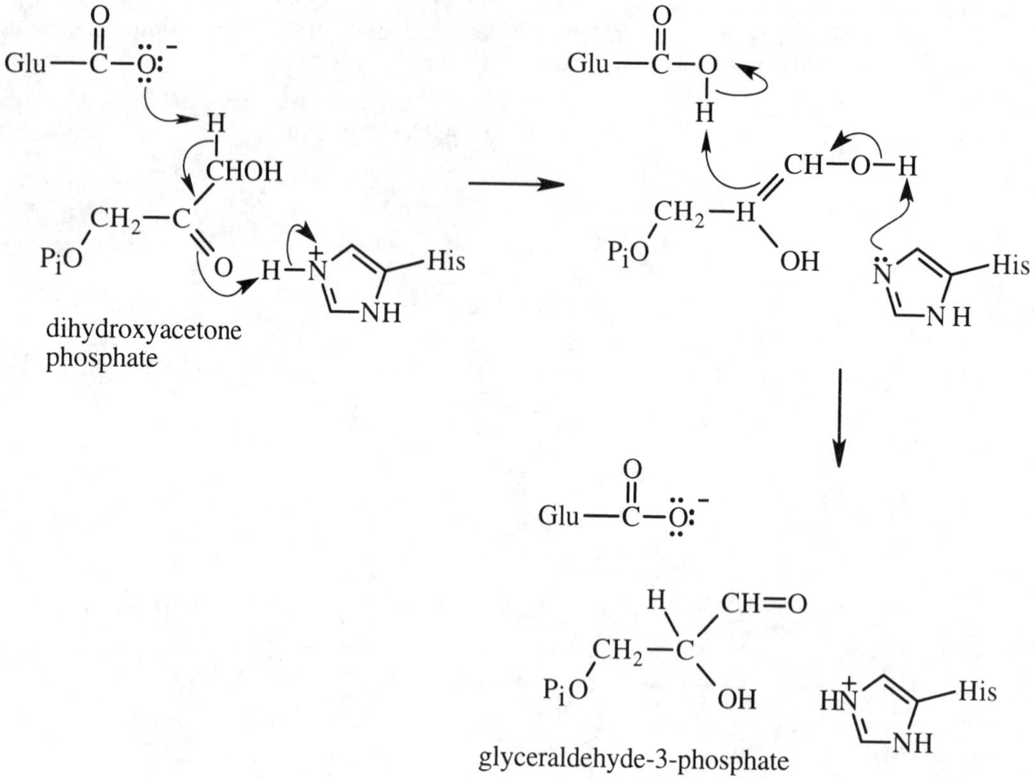

Glu—C(=O)—Ö:⁻
H
CHOH
CH_2—C
P_iO
O H—N⁺ His
NH
dihydroxyacetone phosphate

Glu—C(=O)—O
H
CH—O—H
CH_2—H
P_iO
OH
N His
NH

Glu—C(=O)—Ö:⁻
H CH=O
CH_2—C
P_iO OH
HN⁺ His
NH
glyceraldehyde-3-phosphate

CHAPTER 23
The Organic Mechanisms of the Coenzymes. Metabolism

Important Terms

anabolism — the reactions living organisms carry out that result in the synthesis of complex biomolecules from simple precursor molecules.

apoenzyme — an enzyme without its cofactor.

biotin — the coenzyme required by enzymes that catalyze carboxylation of a carbon adjacent to an ester or a keto group.

catabolism — the reactions living organisms carry out to provide energy and simple precursor molecules for synthesis.

coenzyme — a cofactor that is an organic molecule.

coenzyme A — a thiol used by biological organisms to form thioesters.

coenzyme B_{12} — the coenzyme required by enzymes that catalyze certain rearrangement reactions.

cofactor — an organic molecule or a metal ion that an enzyme needs in order to catalyze a reaction.

competitive inhibitor — a compound that inhibits an enzyme by competing with the substrate for binding at the active site.

dehydrogenase — an enzyme that carries out an oxidation reaction by removing hydrogen from the substrate.

electron sink — a site to which electrons can be delocalized.

flavin adenine dinucleotide (FAD) — a coenzyme required in certain oxidation reactions. It is reduced to $FADH_2$, which is a coenzyme required in certain reduction reactions.

flavin mononucleotide (FMN) — a coenzyme required in certain oxidation reactions. It is reduced to $FMNH_2$, which is a coenzyme required in certain reduction reactions.

heterocycle — a cyclic compound in which one or more of the ring atoms is an atom other than carbon.

holoenzyme — an enzyme plus its cofactor.

lipoate — a coenzyme required in certain oxidation reactions.

mechanism-based inhibitor — an inhibitor that inactivates an enzyme by undergoing part of the normal catalytic mechanism.

metabolism — reactions living organisms carry out in order to obtain the energy they need and to synthesize the compounds they require.

metalloenzyme — an enzyme that has a tightly bound metal ion.

nicotinamide adenine dinucleotide (NAD^+)	a coenzyme required in certain oxidation reactions. It is reduced to NADH, which is a coenzyme required in certain reduction reactions.
nicotinamide adenine dinucleotide phosphate ($NADP^+$)	a coenzyme required in certain oxidation reactions. It is reduced to NADPH, which is a coenzyme required in certain reduction reactions.
nucleotide	a heterocycle attached in the β-position to a phosphorylated ribose.
pyridoxal phosphate	the coenzyme required by enzymes that catalyze certain transformations of amino acids.
suicide inhibitor (mechanism-based inhibitor)	a compound that inactivates an enzyme by undergoing part of its normal catalytic mechanism.
tetrahydrofolate (THF)	the coenzyme required by enzymes that catalyze a reaction that donates a group containing a single carbon to its substrate.
thiamine pyrophosphate (TPP)	the coenzyme required by enzymes that catalyze a reaction that transfers a two-carbon fragment to its substrate.
transamination	a reaction in which an amino group is transferred from one compound to another.
transimination	the reaction of a primary amine with an imine to form a new imine and a primary amine derived from the original imine.
vitamin	a substance needed in small amounts for normal body function that the body cannot synthesize or cannot synthesize in adequate amounts.
vitamin KH_2	the coenzyme required by the enzyme that catalyzes the carboxylation of glutamate side chains.

Solutions to Problems

1. The metal ion (Zn^{2+}) makes the carbonyl carbon more susceptible to nucleophilic attack, increases the nucleophilicity of water by making it more like hydroxide ion, and stabilizes the negative charge on the transition state.

2. FAD contains a diphosphate linkage. One of the phosphate groups comes from FMN and the other comes from ATP. Since ATP uses only one of its three phosphate groups in forming FAD, the other product of the reaction must be pyrophosphate to account for loss of the two phosphate groups. In other words, the phosphate group of FMN must attack the α-phosphorus of ATP, eliminating pyrophosphate. (See pages 719 and 1071 of the text.)

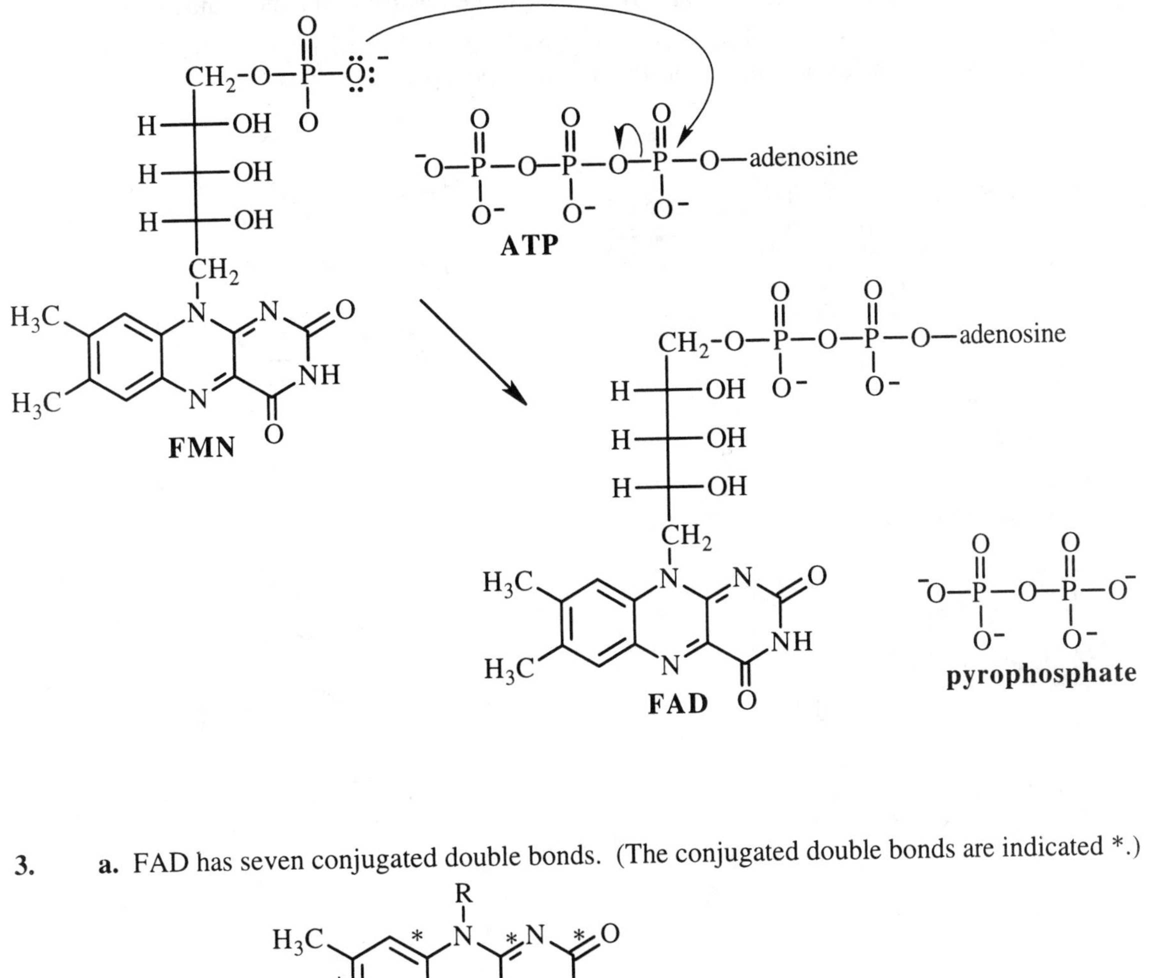

FMN

ATP

FAD

pyrophosphate

3. **a.** FAD has seven conjugated double bonds. (The conjugated double bonds are indicated *.)

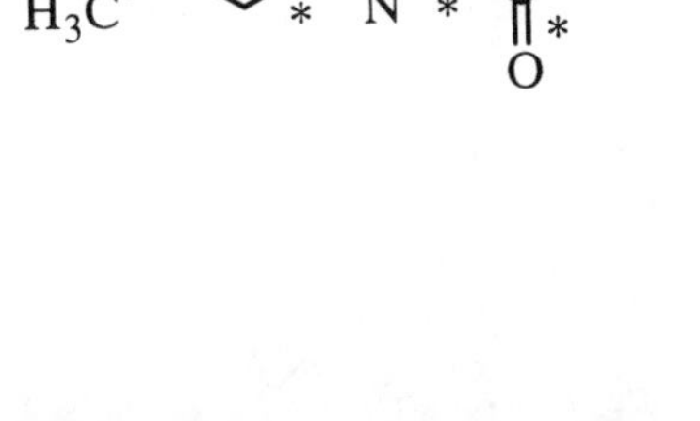

b. $FADH_2$ has three conjugated double bonds. It also has two conjugated double bonds that are isolated from the other three.

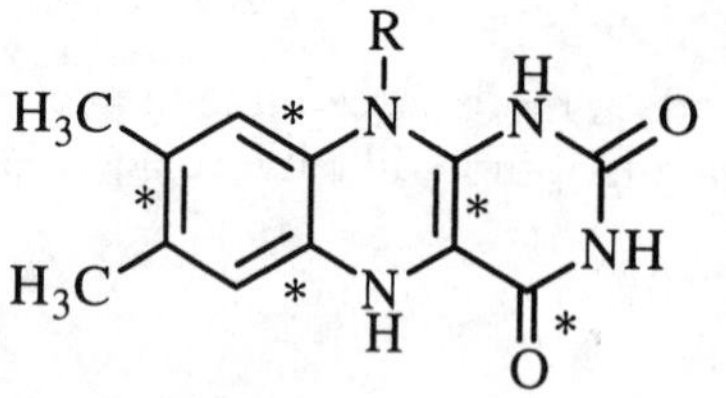

4. The mechanism for reduction of lipoate by $FADH_2$ is the reverse of the mechanism for oxidation of dihydrolipoate by FAD. Because the pK_a of the N-1 hydrogen is 7, $FADH_2$ is present in both the acidic and basic forms at physiological pH (pH = 7.3).

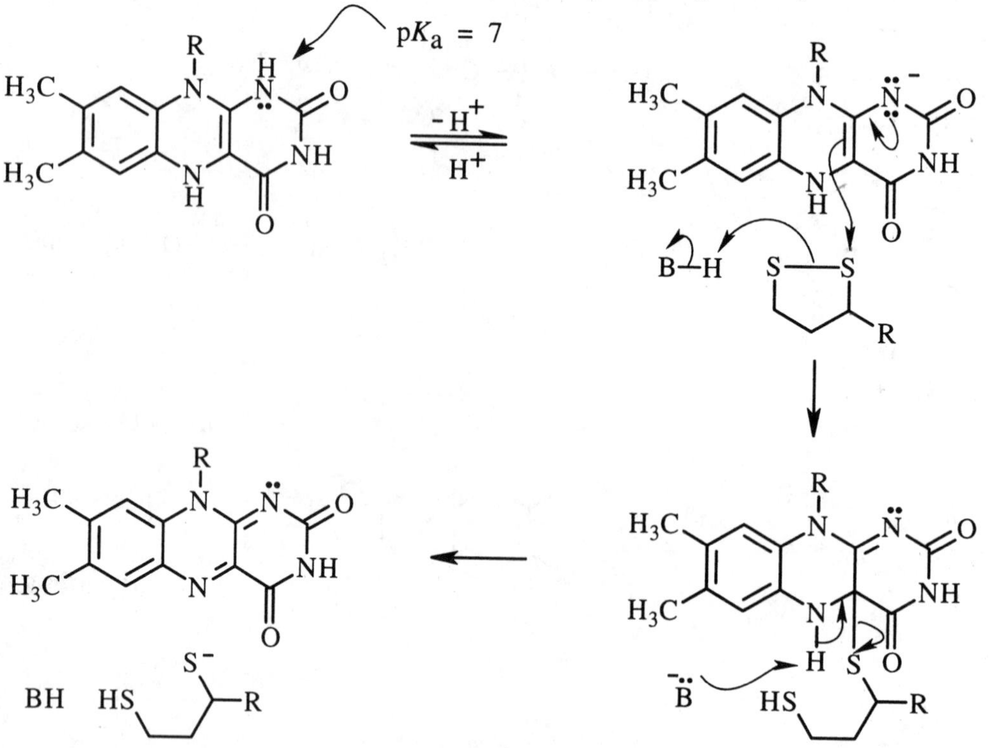

$pK_a = 7$

$-H^+$

H^+

5. Solved in the text.

6. When a proton is removed from the methyl group at C-8, the electrons that are left behind can be delocalized onto the oxygen at position #2 or onto the oxygen at position #4.

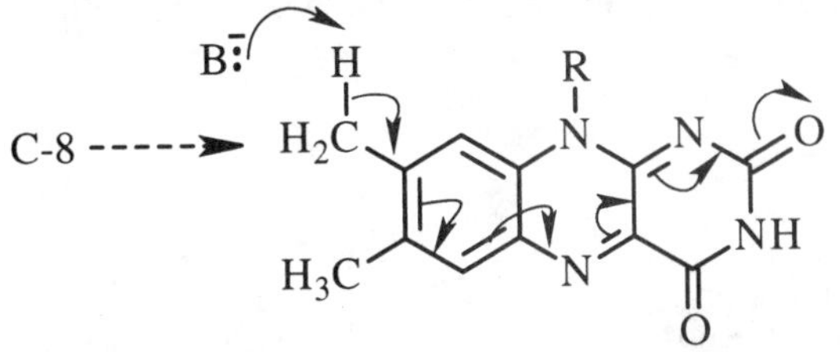

When a proton is removed from the methyl group at C-7, the electrons that are left behind can be delocalized only onto carbon atoms that, being less electronegative than oxygen atoms, are less able to accommodate the electrons. (Try pushing electrons to prove to yourself that this is so.)

7.

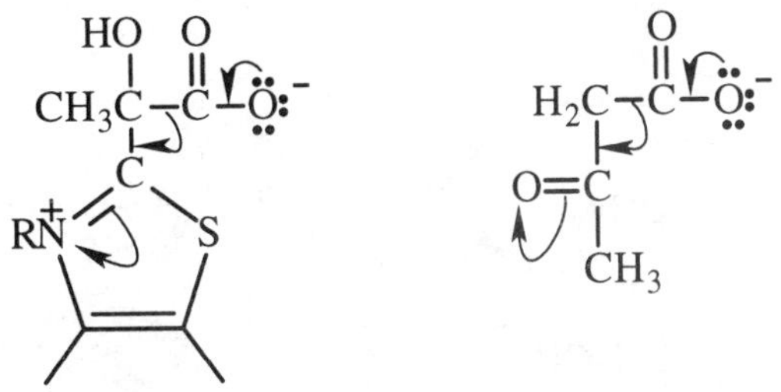

8. The only difference in the mechanisms of pyruvate decarboxylase and acetolactate synthase is the species the two-carbon fragment (the carbanion) is transferred to: a proton in the case of pyruvate carboxylase and pyruvate in the case of acetolactate synthase.

acetolactate

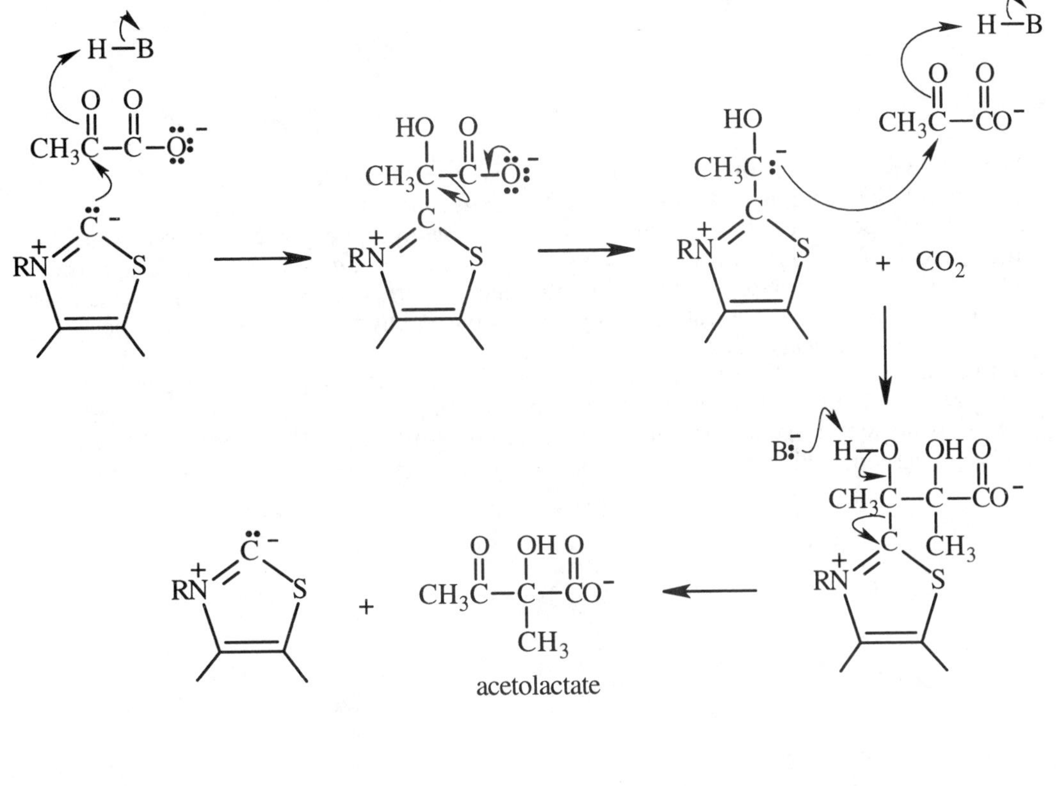

9. Notice that the only difference in this reaction and that in Problem 8 is that the species to which the two-carbon fragment is transferred to has an ethyl group in place of the methyl group.

H—B
O O
CH3C—C—O:⁻
C:⁻
RN⁺ S
→
HO O
CH3C—C—O:⁻
C
RN⁺ S
→
HO
CH3C:⁻
C
RN⁺ S
+ CO2
H—B
O O
CH3CH2C—CO⁻
↓
B:⁻ H—O OH O
CH3C—C—CO⁻
C CH2CH3
RN⁺ S
←
C:⁻
RN⁺ S
+
O OH O
CH3C—C—CO⁻
CH2CH3

10. Solved in the text.

11. Acetaldehyde is responsible for the physiological effects known as a hangover. A way to "cure" a hangover, therefore, is to get rid of the acetaldehyde that is formed from the oxidation of ethanol. The coenzyme form of vitamin B_1 (thiamine pyrophosphate) can convert acetaldehyde into acetyl-CoA. It does this by attacking the carbonyl carbon of acetaldehyde to form an intermediate. Loss of a proton from the intermediate forms the carbanion that is the reactive intermediate in the reaction catalyzed by the pyruvate dehydrogenase system. The final product of this pathway is acetyl-CoA.
Notice that except for the second step of the reaction, the reaction is the same as the reaction catalyzed by the pyruvate dehydrogneanse system. In the second step of reaction catalyzed by the pyruvate dehydrogenase system, CO_2 is removed from the carbon that formerly was the carbonyl carbon; in this reaction it is a proton rather than CO_2 that is removed in that step.

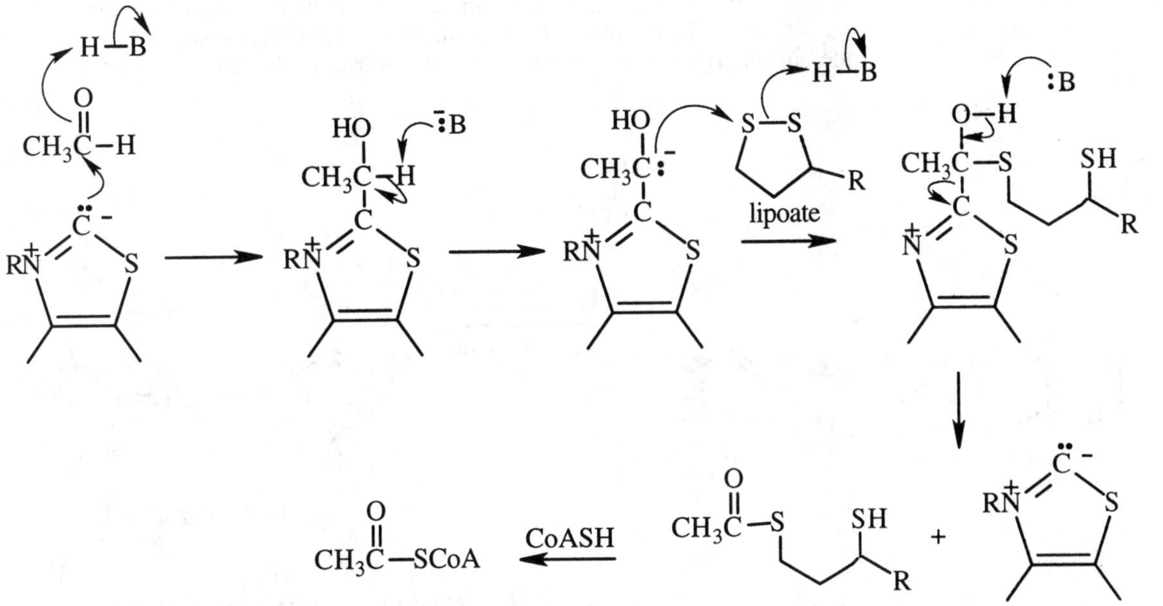

H–B :B CH$_3$C–H RN S HO CH$_3$C–H C lipoate S–S R O–H CH$_3$C–S SH R

$$CH_3\overset{O}{\overset{\|}{C}}-SCoA \xleftarrow{CoASH} CH_3\overset{O}{\overset{\|}{C}}-S\text{(lipoate, SH, R)} + \text{thiazolium ylide}$$

12. **a.** coenzyme = pyridoxal phosphate; other organic compound = α-ketoglutarate

b. This is the bond that breaks.

OH, CH_2COO^-, HO

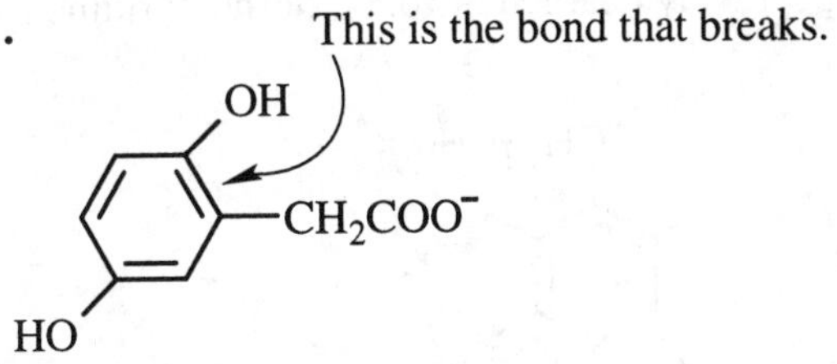

c. *S*-adenosylmethionine

13. The α-keto group that accepts the amino group from pyridoxamine is converted into an amino group.

a. $CH_3\overset{O}{\overset{\|}{C}}-\overset{O}{\overset{\|}{C}}O^- \longrightarrow CH_3\underset{+NH_3}{CH}\overset{O}{\overset{\|}{C}}O^-$

pyruvate → alanine

b. $^-O\overset{O}{\overset{\|}{C}}CH_2\overset{O}{\overset{\|}{C}}-\overset{O}{\overset{\|}{C}}O^- \longrightarrow {}^-O\overset{O}{\overset{\|}{C}}CH_2\underset{+NH_3}{CH}\overset{O}{\overset{\|}{C}}O^-$

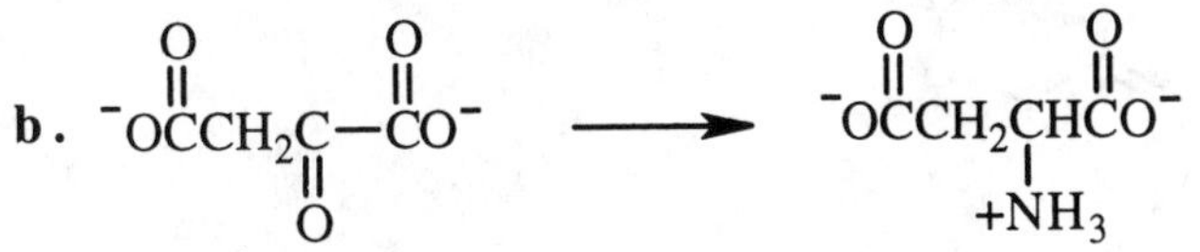

oxaloacetate → aspartate

14. In the first step, a proton is removed from the α-carbon, and the electrons left behind expel the leaving group from the β-carbon. Hydrolysis of the imine regenerates pyridoxal and produces an enamine. The enamine tautomerizes to an imine that is hydrolyzed to the final product.

$$X{-}CH_2{-}C(H)(N{=}CH{-}Py^+)\text{-}CO^- \xrightarrow{:B^-} X^- + CH_2{=}C(N{=}CH{-}Py^+){-}CO^- \xrightarrow[\text{transimination}]{E{-}(CH_2)_4NH_2} E{-}(CH_2)_4{-}N{=}CH{-}Py^+ + CH_2{=}C(NH_2){-}CO^-$$

$$CH_2{=}C(NH_2){-}CO^- \xrightleftharpoons{\text{tautomerization}} CH_3{-}C({=}\overset{+}{N}H_2){-}CO^- \xrightarrow{H_2O} \overset{+}{N}H_4 + CH_3{-}C({=}O){-}CO^-$$

15. The compound on the right is more easily decarboxylated because the electrons left behind when CO_2 is eliminated are delocalized onto the positively charged nitrogen of the pyridine ring.

$$CH_2CH_2{-}C({=}O){-}\ddot{O}:^- \text{ (on pyridinium ring)} \qquad CH_2{-}C({=}O){-}\ddot{O}:^- \text{ (on pyridinium ring)}$$

16. The first step in all amino acid transformations is removal of a substituent from the α-carbon of the amino acid. The electrons left behind when the substituent is removed are delocalized onto the positively charged nitrogen of the pyridine ring. If the ring nitrogen were not protonated, it would be less attractive to the electrons. In other words, it would be a less effective electron sink.

$$R{-}C(H)(X){-}N{=}CH{-}Py^+ \quad (:B^- \text{ removes } H)$$

17. The hydrogen of the OH substituent forms a hydrogen bond with the nitrogen of the imine linkage (see above structure). This puts a partial positive charge on the nitrogen, which makes it easier for the amino acid to attack the imine carbon in the transimination reaction that attaches the amino acid to the coenzyme. It also makes it easier to remove a substituent from the α-carbon of the amino acid. If the OH substituent is replaced by an OCH_3 substituent, there is no longer a proton available to form the hydrogen bond.

18. The mechanism is the same as that shown in the text for dioldehydrase. The tetrahedral intermediate that is formed as a result of the coenzyme B_{12}-catalyzed isomerization is unstable and loses ammonia to give acetaldehyde, the final product of the reaction.

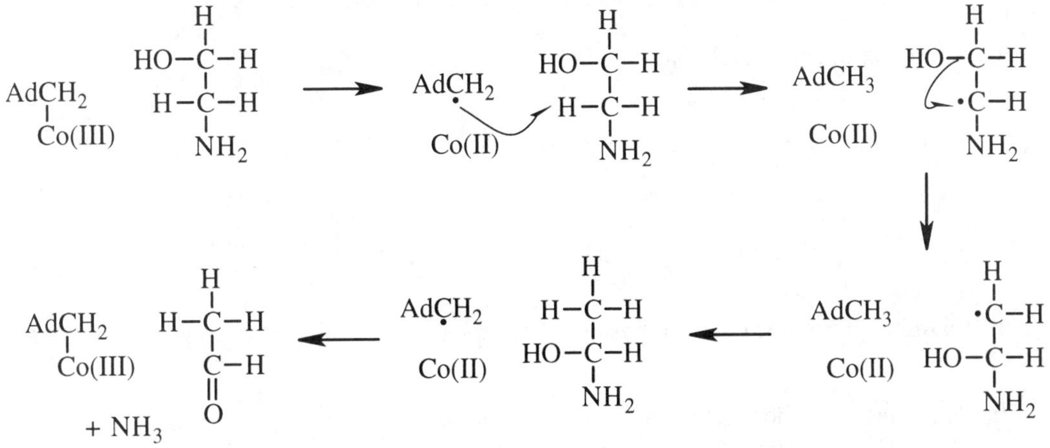

$AdCH_2$–Co(III) + $H(HO)C(H)–C(H)(H)–NH_2$ ⟶ $AdCH_2\cdot$ Co(II) + $H(HO)C(H)–C(H)(H)–NH_2$ ⟶ $AdCH_3$ Co(II) + $H(HO)C(H)–\cdot C(H)–NH_2$
↓
$AdCH_3$ Co(II) + $H–\cdot C(H)–C(HO)(H)–NH_2$ ⟶ $AdCH_2\cdot$ Co(II) + $H–C(H)(H)–C(HO)(H)–NH_2$ ⟶ $AdCH_2$–Co(III) + $H–C(H)(H)–C(H)=O$ + NH_3

19.

$$CH_3CH_2\overset{O}{\overset{\|}{C}}SCoA \xrightarrow[\text{biotin}\; HCO_3^-,\, ATP,\, Mg^{2+}]{\text{propionyl-CoA carboxylase}} CH_3\underset{COO^-}{\underset{|}{C}H}\overset{O}{\overset{\|}{C}}SCoA \xrightarrow[\text{coenzyme } B_{12}]{\text{methylmalonyl-CoA mutase}} \underset{COO^-}{\underset{|}{C}H_2}CH_2\overset{O}{\overset{\|}{C}}SCoA$$

propionyl-CoA ... succinyl-CoA

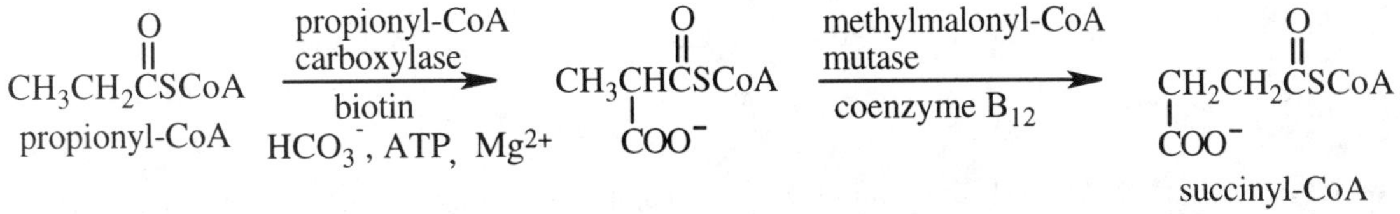

20. The methyl group in thymidine comes from the methylene group of N^5, N^{10}-methylene-THF. The methylene group of N^5, N^{10}-methylene-THF comes from the CH_2OH group of serine by means of a PLP-catalyzed C_α-C_β cleavage.

21. Two thiol groups are necessary for each of the two oxidations in the conversion of vitamin K epoxide to vitamin KH_2. Since lipoate has two thiol groups, each oxidation involves an intramolecular reaction. When thiols such as ethanethiol or propanethiol are used, the oxidation involves an intermolecular reaction. These thiols react more slowly than lipoate because the two thiol groups are not in the same molecule. Therefore, they have to find each other in order to react.

22. **a.** NAD^+, $NADP^+$, FAD, FMN, lipoate, vitamin K epoxide

b. N^5-methyl-THF, N^5, N^{10}-methylene-THF, N^5, N^{10}-methenyl-THF, N^5-formyl-THF, N^{10}-formyl-THF, N^5-formimino-THF, *S*-adenosylmethionine
One could include biotin and vitamin KH_2; they accept a one carbon group (from HCO_3^- and CO_2, respectively) and transfer it to the substrate.

c. $-CH_3$ $\quad$ $-CH_2-$ $\quad$ $-\overset{O}{\overset{\|}{C}}H$

d. FAD oxidizes dihydrolipoate back to lipoate.

e. NAD^+ oxidizes $FADH_2$ back to FAD.

f. $-NH-CH(CH_2CH_2COO^-)-\overset{O}{\overset{\|}{C}}- \longrightarrow -NH-CH(CH_2CH(COO^-)_2)-\overset{O}{\overset{\|}{C}}-$

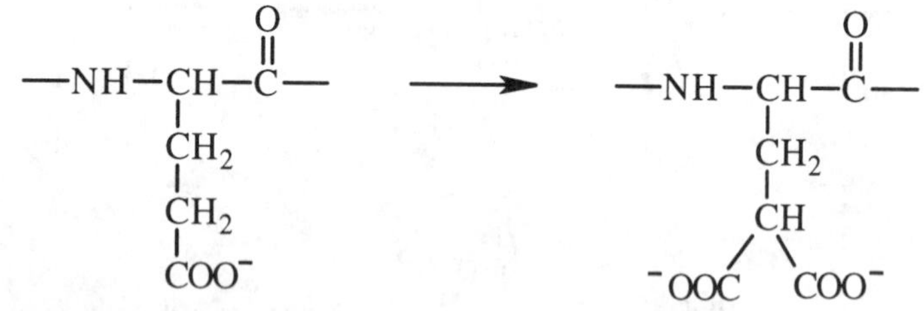

g. thiamine pyrophosphate and pyridoxal phosphate

h. Thiamine pyrophosphate is used for the decarboxylation of α-keto acids. Pyridoxal phosphate is used for the decarboxylation of amino acids.

i. biotin and vitamin KH_2

j. Biotin carboxylates a carbon adjacent to a carbonyl group. Vitamin KH_2 carboxylates the γ-carbon of a glutamate.

23. **a.** NAD^+, $NADP^+$, FAD, FMN, thiamine pyrophosphate, carboxybiotin, pyridoxal phosphate

b. All coenzymes that bind the substrate do so to activate it for some kind of further reaction. So all the coenzymes can be considered to activate substrates for further reaction except those that directly deliver an atom or group to the substrate or directly receive an atom or group from the substrate without binding it, such as NAD^+, $NADP^+$, NADH, NADPH, and the tetrahydrofolate coenzymes that deliver a one-carbon fragment in one step such as N^5-methyl-THF. In other words, the answers are: FAD, FMN, $FADH_2$, $FMNH_2$, thiamine pyrophosphate, biotin, pyridoxal phosphate, coenzyme B_{12}, some tetrahydrofolate coenzymes, and vitamin KH_2.

c. thiamine pyrophosphate

d. vitamin KH_2

24. **a.** acetyl-CoA carboxylase, biotin

b. dihydrolipoyl dehydrogenase, FAD

c. methylmalonyl-CoA mutase, coenzyme B_{12}

d. lactate dehydrogenase, NADH

e. aspartate transaminase, pyridoxal phosphate

f. propionyl-CoA carboxylase, biotin

25.

$$^{-}OCCH(\overset{+}{N}H_3)CH_2CH_2\ddot{S}CH_3 \text{ (with C=O)} + \text{ATP (triphosphate–}CH_2\text{–ribose–adenine)}$$

$$\downarrow$$

$$\text{triphosphate } (^{-}O-P(O)(O^-)-O-P(O)(O^-)-O-P(O)(O^-)-O^-) + {}^{-}OCCH(\overset{+}{N}H_3)CH_2CH_2\overset{+}{S}(CH_3)CH_2\text{–ribose(OH, OH)–adenine}$$

26. **a.** thiamine pyrophosphate, lipoate, coenzyme A, FAD, NAD+

b.

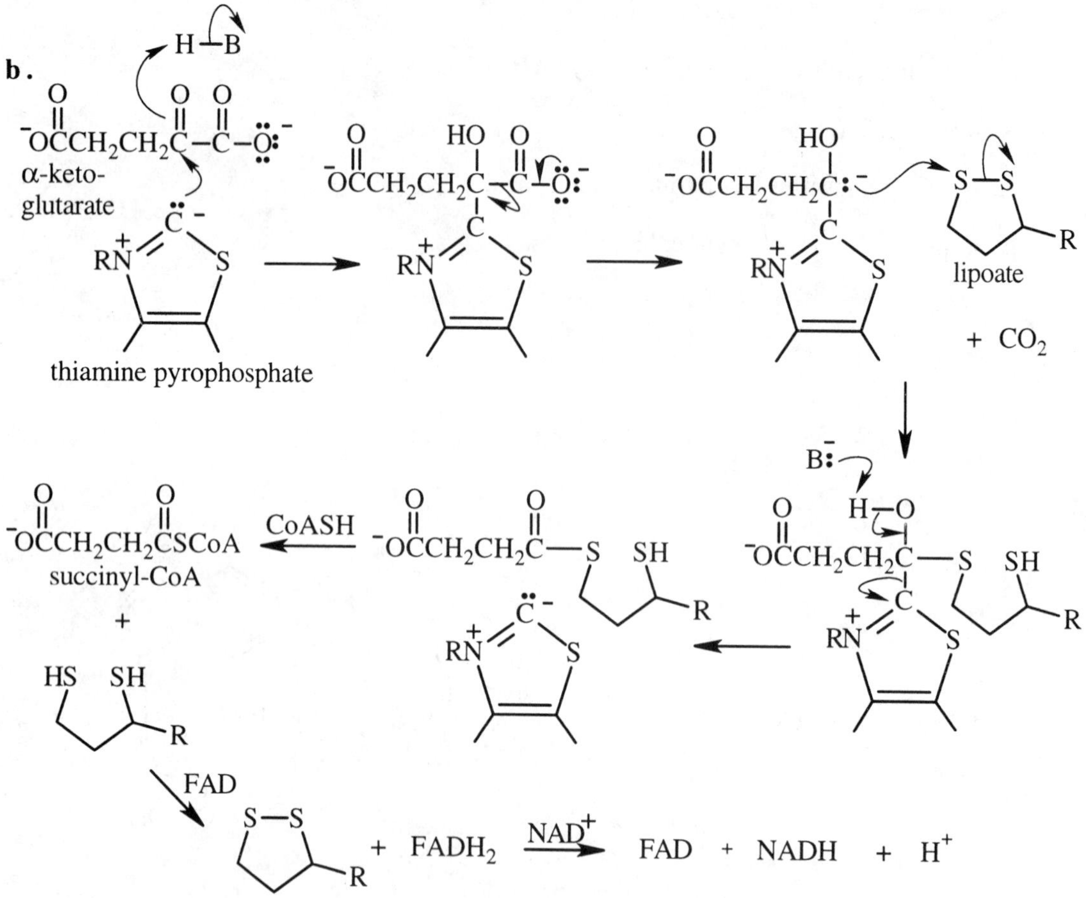

α-keto-glutarate

thiamine pyrophosphate

lipoate

$+ CO_2$

CoASH

succinyl-CoA

+

FAD

$+ FADH_2 \xrightarrow{NAD^+} FAD + NADH + H^+$

27. The products of the reaction are shown below.

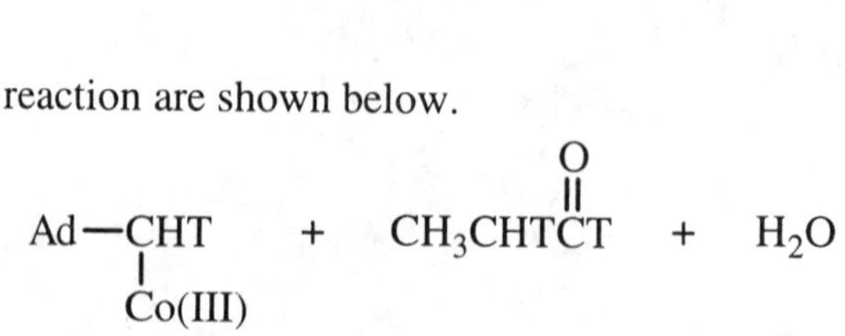

$$\text{Ad—CHT(—Co(III))} + CH_3CHTC(=O)T + H_2O$$

The following mechanism explains the above products.

$$AdCH_2\text{–}Co(III) + HO\text{–}C(CH_3)(T)\text{–}C(T)(T)\text{–}OH \longrightarrow Ad\dot{C}H_2\ Co(II) + HO\text{–}C(CH_3)(T)\text{–}C(T)(T)\text{–}OH \longrightarrow AdCH_2T\ Co(II) + HO\text{–}C(CH_3)(T)\text{–}\dot{C}(T)\text{–}OH$$

$$\downarrow$$

$$AdCHT\text{–}Co(III) + H\text{–}C(CH_3)(T)\text{–}C(T){=}O + H_2O \longleftarrow Ad\dot{C}HT\ Co(II) + H\text{–}C(CH_3)(T)\text{–}C(T)(OH)\text{–}OH \longleftarrow H\text{–}AdCHT\ Co(II) + \dot{C}(CH_3)(T)\text{–}C(T)(OH)\text{–}OH$$

If there is only a limited amount of coenzyme, the second H of the coenzyme will be replaced by T in a subsequent round of catalysis, which means that the final product will be:

$$Ad\text{–}CT_2\text{–}Co(III) \quad + \quad CH_3CHTCT(=O) \quad + \quad H_2O$$

28.

$$AdCH_2\text{–}Co(III) + {}^-OOC\text{–}CH(CSCoA{=}O)\text{–}CH_2\text{–}H \longrightarrow Ad\dot{C}H_2\ Co(II) + {}^-OOC\text{–}CH(CSCoA{=}O)\text{–}CH_2\text{–}H \longrightarrow AdCH_3\ Co(II) + {}^-OOC\text{–}CH(CSCoA{=}O)\text{–}\dot{C}H_2$$

$$\downarrow$$

$$AdCH_2\text{–}Co(III) + {}^-OOC\text{–}CH_2\text{–}CH_2\text{–}CSCoA(=O) \longleftarrow Ad\dot{C}H_2\ Co(II) + {}^-OOC\text{–}CH_2\text{–}CH_2\text{–}CSCoA(=O) \longleftarrow H\text{–}AdCH_2\ Co(II) + {}^-OOC\text{–}CH_2\text{–}\dot{C}H\text{–}CSCoA(=O)$$

29.

a. 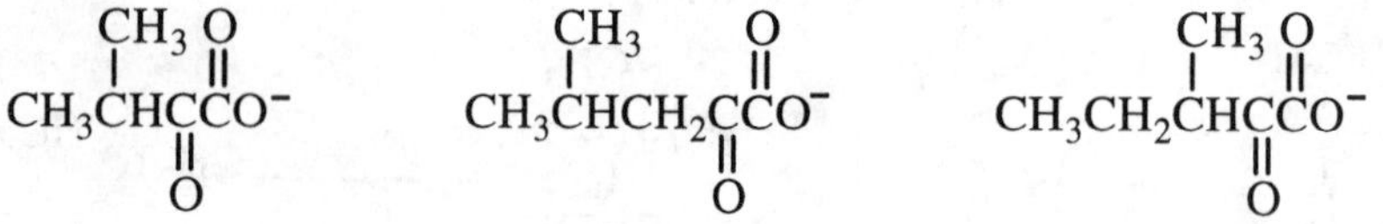 $CH_3CH(CH_3)C(=O)CO^-$ $\quad$ $CH_3CH(CH_3)CH_2C(=O)CO^-$ $\quad$ $CH_3CH_2CH(CH_3)C(=O)CO^-$

b. 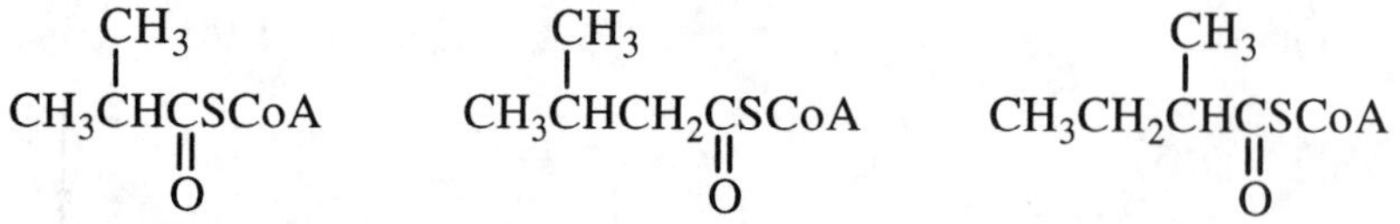 $CH_3CH(CH_3)C(=O)SCoA$ $\quad$ $CH_3CH(CH_3)CH_2C(=O)SCoA$ $\quad$ $CH_3CH_2CH(CH_3)C(=O)SCoA$

c. thiamine pyrophosphate, lipoate, coenzyme A, FAD, NAD^+

d. The disease can be treated by a diet low in branched-chain amino acids.

30.

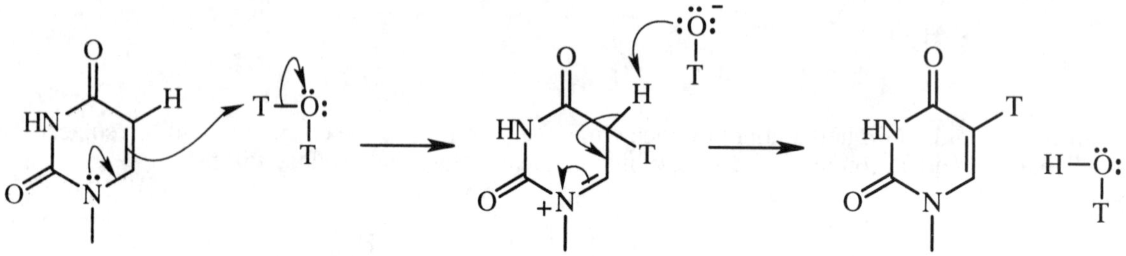

HN, O, H, T–Ö:, T → HN, O, +N, H, T, :Ö:⁻, T → HN, O, N, T + H–Ö:, T

31.

:B⁻ ; $HO-CH_2-C(H)(N=CH-\text{pyridinium})-CO^-$ → HB ; $HO^- + CH_2=C(N=CH-\text{pyridinium})-CO^-$ $\xrightarrow[\text{transimination}]{E-(CH_2)_4-NH_2}$ $E-(CH_2)_4-N=CH-\text{pyridinium}$ + $CH_2=C(NH_2)-CO^-$

$CH_2=C(NH_2)-CO^-$ $\xrightleftharpoons{\text{tautomerization}}$ $CH_3-C(=NH)-CO^-$ $\xrightarrow[\text{imine hydrolysis}]{H_2O}$ $\overset{+}{N}H_4 + CH_3-C(=O)-CO^-$

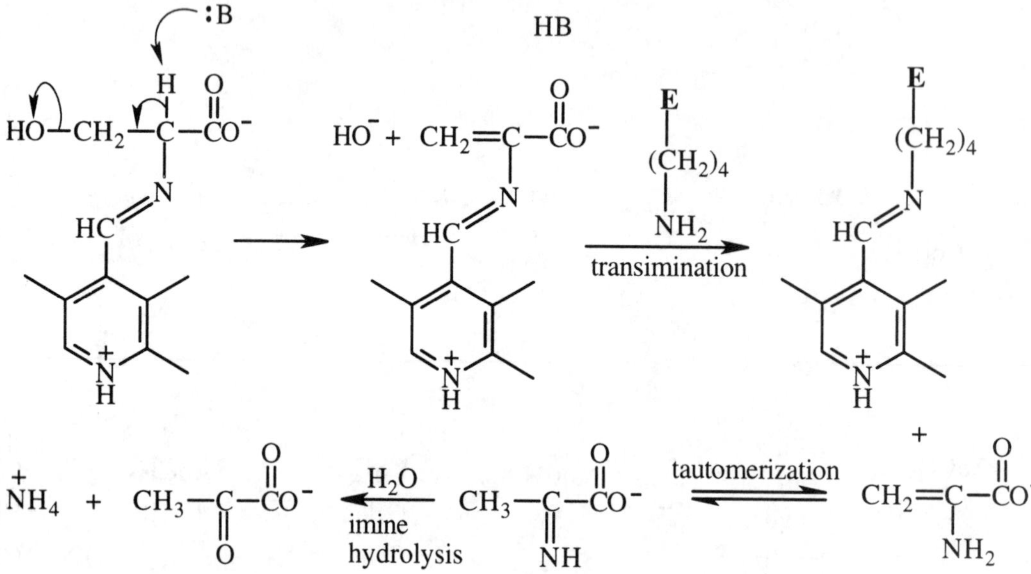

32.

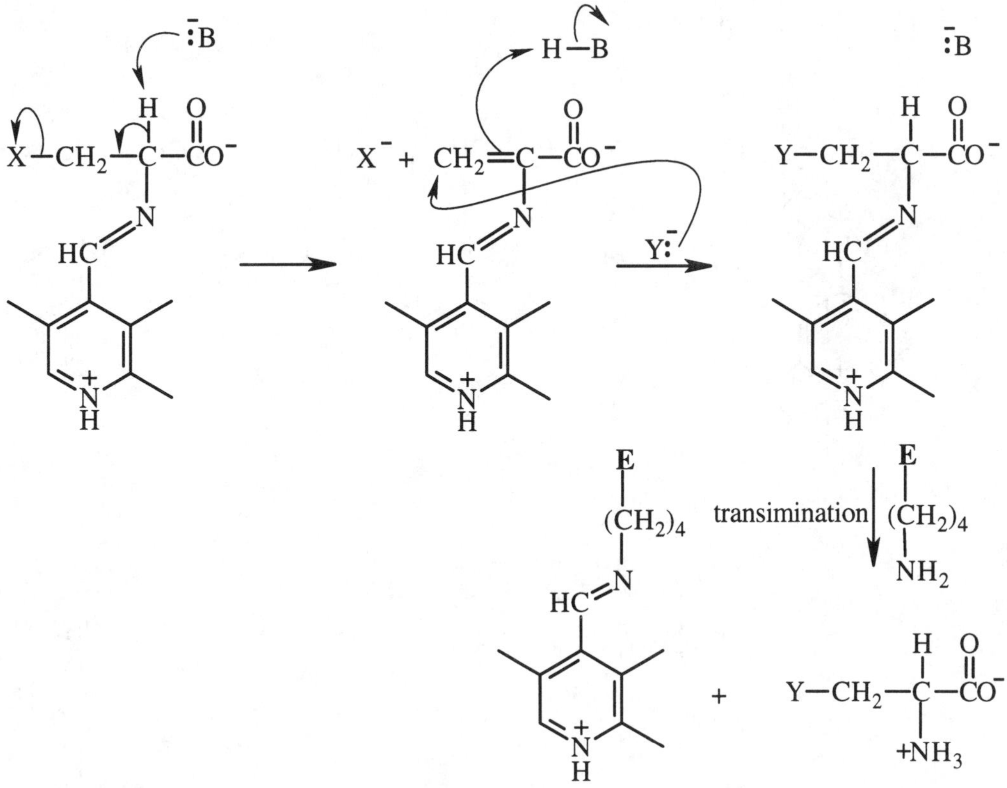

$$X{-}CH_2{-}\underset{N=CH\text{-PLP}}{\overset{H}{C}}{-}CO_2^- \xrightarrow{:B^-} X^- + CH_2{=}\underset{N=CH\text{-PLP}}{C}{-}CO_2^- \xrightarrow{Y:^-} Y{-}CH_2{-}\underset{N=CH\text{-PLP}}{\overset{H}{C}}{-}CO_2^-$$

transimination (E–$(CH_2)_4$–NH_2)

$$\text{E}{-}(CH_2)_4{-}N{=}CH\text{-PLP} + Y{-}CH_2{-}\underset{^+NH_3}{\overset{H}{C}}{-}CO_2^-$$

33.

$X-CH_2-CH_2-CH(-N=CH-Ar)-CO^-$ (pyridinium) → $X-CH_2-CH_2-C(=N-CH=Ar)-CO^-$ + HB → $CH_2=CH-C(=N-CH=Ar)-CO^-$ + X^- + HB → $CH_3CH=C(-N=CH-Ar)-CO^-$ (pyridinium)

B:, :B, H–B, HB

E–$(CH_2)_4$–NH_2, transimination → E–$(CH_2)_4$–N=CH–Ar (pyridinium) + $CH_3CH=C(-{}^+NH_3)-CO^-$

tautomerization

$CH_3CH_2-C(={}^+NH_2)-CO^-$ $\xrightarrow[\text{imine hydrolysis}]{H_2O}$ $CH_3CH_2-C(=O)-CO^-$ + $\overset{+}{N}H_4$

34.

$CH_2-\overset{O}{\overset{\|}{C}}O^-$ with $+NH_3$ $\xrightarrow[\textbf{PLP}]{\textbf{E}_1}$ PLP imine ($CH_2-C(=O)-O^-$ attached through N to HC= of the pyridinium ring, N–H) attacking S–S of the lipoate ring $\xrightarrow{\textbf{E}_2}$ CH_2-S SH (imine-bound to PLP)

$\downarrow$ $E_3-(CH_2)_4-NH_2$, **THF**

N^5, N^{10}-methylene-THF + $E_3-(CH_2)_4-N=CH$-(pyridinium, N–H) + $\overset{+}{N}H_4$ + HS SH

$\downarrow$ **E_4 FAD**

FAD + NADH + H^+ $\xleftarrow{NAD^+}$ $FADH_2$ + S—S (ring)

35. Nonenzyme-bound FAD has a greater oxidation potential than NAD^+. Consequently, FAD oxidizes NADH to NAD^+. When the enzyme binds FAD, it changes its oxidation potential, causing it to become less than that of NAD^+ . Consequently, NAD^+ oxidizes enzyme-bound $FADH_2$ to FAD. In addition, there will much more NAD^+ relative to enzyme-bound $FADH_2$ in the mitochondria (where the oxidation of $FADH_2$ by NAD^+ takes place). Therefore, the equilibrium favors the oxidation of $FADH_2$ by NAD^+.

$$\text{E-FADH}_2 + \underset{\text{excess}}{NAD^+} \rightleftharpoons \text{E-FAD} + \text{NADH} + H^+$$

36.

FADH$_2$ ⟷ (resonance form with N^+–H and O^-) + $RCH{=}CH\overset{O}{\overset{\|}{C}}SR$

$\downarrow$ - e$^-$, - H$^+$ (loss of an electron and a proton)

FADH• + $R\dot{C}HCH{=}\overset{OH}{\overset{|}{C}}SR$ ⇌ $R\dot{C}HCH_2\overset{O}{\overset{\|}{C}}SR$

$\swarrow$ - e$^-$, - H$^+$

$RCH_2CH_2\overset{O}{\overset{\|}{C}}SR$ + FAD

CHAPTER 24
Lipids

Important Terms

adrenal cortical steroids	steroids synthesized in the adrenal cortex.
anabolic steroids	steroids that aid in the development of muscle.
androgens	steroids responsible for the development of male secondary sex characteristics.
angular methyl group	a methyl substituent at the 10- or 13-position of a steroid ring system.
carotenoid	a class of compound (a tetraterpene) responsible for the red and orange colors of fruits, vegetables, and fall leaves.
cephalin	a phosphoacylglycerol in which the second OH group of phosphate has formed an ester with ethanolamine.
cerebroside	a sphingolipid in which the primary OH group of sphingosine is bonded to a sugar residue.
cholesterol	a steroid that is the precursor of all other steroids.
cis fused	two rings fused together such that if one ring is considered to be two substituents of the other ring, the substituents would be on the same side of the first ring.
essential oil	fragrances and flavorings isolated from plants that do not leave a residue when they evaporate. Most are terpenes.
estrogens	steroids responsible for the development of female secondary sex characteristics.
fat	a triester of glycerol that exists as a solid at room temperature.
fatty acid	a carboxylic acid with a long hydrocarbon side chain.
hormone	an organic compound synthesized in a gland and delivered by the bloodstream to its target tissue.
isoprene rule	head-to-tail linkage of isoprene units.
lecithin	a phosphoacylglycerol in which the second OH group of phosphate has formed an ester with choline.
leukotriene	compounds that induce contraction of the muscle that lines the airways to the lungs.
lipid	a water-insoluble compound found in a living system.
lipid bilayer	two layers of phosphoacylglycerols arranged so that their polar heads are on the outside and their nonpolar fatty acid chains are on the inside.

membrane the material the surrounds the cell in order to isolate its contents.

mixed triacylglycerol a triacylglycerol in which the fatty acid components are different.

monoterpene a terpene that contains 10 carbons.

oil a triester of glycerol that exists as a liquid at room temperature.

phosphatidic acid a phosphoacylglycerol in which only one of the OH groups of phosphate is in an ester linkage.

phosphoacylglycerol (phosphoglyceride) formed when two OH groups of glycerol form esters with fatty acids and the terminal OH group forms a phosphate ester.

phospholipid a lipid that contains a phosphate group.

polyunsaturated fatty acid a fatty acid with more than one double bond.

progestins a class of steroid hormones.

prostacyclin a compound that dilates blood vessels and inhibits platelet aggregation.

prostaglandin a carboxylic acid, derived from arachidonic acid, that is responsible for a variety of physiological functions.

sesquiterpene a terpene that contains 15 carbons.

simple triacylglycerol a triacylglycerol in which the fatty acid components are the same.

sphingolipid a lipid that contains sphingosine.

sphingomyelin a sphingolipid in which the primary OH group of sphingosine is bonded to a phosphocholine or to a phosphoethanolamine.

squalene a triterpene that is a precursor of steroid molecules.

steroid a class of compounds that contains a steroid ring system.

steroid ring system

α-substituent a substituent on the opposite side of a steroid ring system as the angular methyl groups.

β-substituent a substituent on the same side of a steroid ring system as the angular methyl groups.

terpene a lipid isolated from a plant that contains carbon atoms in multiples of five.

terpenoid a terpene that contains oxygen.

tetraterpene a terpene that contains 40 carbons.

thromboxane a compound that constricts blood vessels and stimulates platelet aggregation.

trans fused two rings fused together such that if one ring is considered to be two substituents of the other ring, the substituents would be on opposite sides of the first ring.

triacylglycerol the compound formed when the three OH groups of glycerol are esterified with fatty acids.

triterpene a terpene that contains 30 carbons.

wax an ester formed from a long-chain carboxylic acid and a long-chain alcohol.

Solutions to Problems

1. **a.** Stearic acid has the higher melting point because it has two more methylene groups (giving it a greater surface area) than palmitic acid.

   b. Palmitic acid has the higher melting point because it does not have any carbon-carbon double bonds, while palmitoleic acid has a cis double bond that prevents the molecules from packing closely together.

   c. Oleic acid has the higher melting point because it has one double bond, while linoleic acid has two double bonds, which give greater interference to close packing of the molecules.

2.

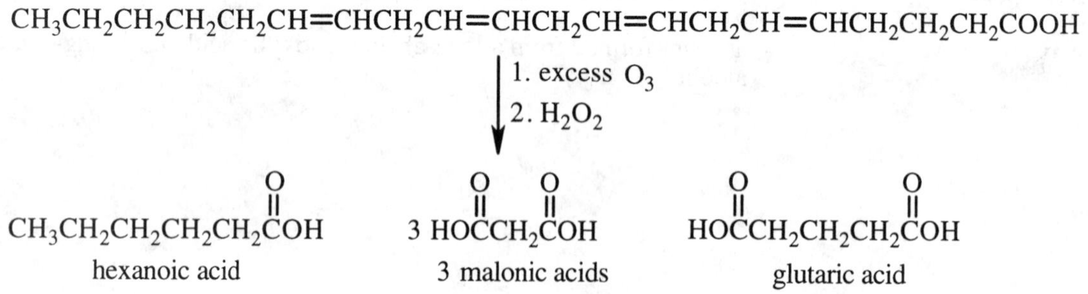

$$CH_3CH_2CH_2CH_2CH_2CH{=}CHCH_2CH{=}CHCH_2CH{=}CHCH_2CH{=}CHCH_2CH_2CH_2COOH$$

$$\downarrow \text{1. excess } O_3 \quad \text{2. } H_2O_2$$

$$\underset{\text{hexanoic acid}}{CH_3CH_2CH_2CH_2CH_2\overset{O}{\overset{\|}{C}}OH} \qquad \underset{\text{3 malonic acids}}{3\ HO\overset{O}{\overset{\|}{C}}CH_2\overset{O}{\overset{\|}{C}}OH} \qquad \underset{\text{glutaric acid}}{HO\overset{O}{\overset{\|}{C}}CH_2CH_2CH_2\overset{O}{\overset{\|}{C}}OH}$$

3. All triacylglycerols do not have the same number of chirality centers. If the carboxylic acid components at C-1 and C-3 of glycerol are not identical, the triacylglycerol has one chirality center (C-2). If the carboxylic acid components at C-1 and C-3 of glycerol are identical, the triacylglycerol has no chirality centers.

4. Glyceryl tripalmitate has a higher melting point because the carboxylic acid components are saturated and can, therefore, pack more closely together than the unsaturated carboxylic acid components of glyceryl tripalmitolate.

5.

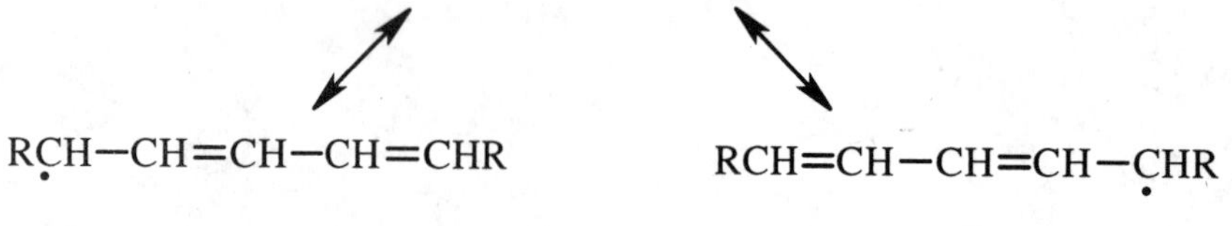

$$RCH{=}CH{-}\dot{C}H{-}CH{=}CHR$$

$$\swarrow\!\!\nearrow \qquad \nwarrow\!\!\searrow$$

$$R\dot{C}H{-}CH{=}CH{-}CH{=}CHR \qquad RCH{=}CH{-}CH{=}CH{-}\dot{C}HR$$

6. Because the interior of a membrane is nonpolar and the surface of a membrane is polar, integral proteins will have a higher percentage of nonpolar amino acids.

7. The bacteria could synthesize phosphoacylglycerols with more saturated fatty acids because these triacylglycerols would pack more tightly in the lipid bilayer and, therefore, would have higher melting points and be less fluid.

8. The sphingomyelins can differ in the fatty acid component of the amide and have either choline or ethanolamine attached to the phosphate group.

a.

$CH{=}CH(CH_2)_{12}CH_3$
$CH{-}OH$
$CH{-}NH{-}C(=O)(CH_2)_{12}CH_3$
$CH_2{-}O{-}P(=O)(O^-){-}OCH_2CH_2\overset{+}{N}(CH_3)_3$

$CH{=}CH(CH_2)_{12}CH_3$
$CH{-}OH$
$CH{-}NH{-}C(=O)(CH_2)_{14}CH_3$
$CH_2{-}O{-}P(=O)(O^-){-}OCH_2CH_2\overset{+}{N}(CH_3)_3$

$CH{=}CH(CH_2)_{12}CH_3$
$CH{-}OH$
$CH{-}NH{-}C(=O)(CH_2)_{14}CH_3$
$CH_2{-}O{-}P(=O)(O^-){-}OCH_2CH_2NH_2$

b.

$CH{=}CH(CH_2)_{12}CH_3$
$CH{-}OH$
$CH{-}NH{-}C(=O)(CH_2)_{14}CH_3$
CH_2
O — (sugar ring: HO, CH_2OH, O, HO, OH)

9. Membranes must be kept in a semifluid state in order to allow transport across them. Cells closer to the hoof of an animal are going to be in a colder average environment than cells closer to the body. Therefore, the cells closer to the hoof have a higher degree of unsaturation to give them a lower melting point so the membranes will not solidify at the colder temperature.

10.

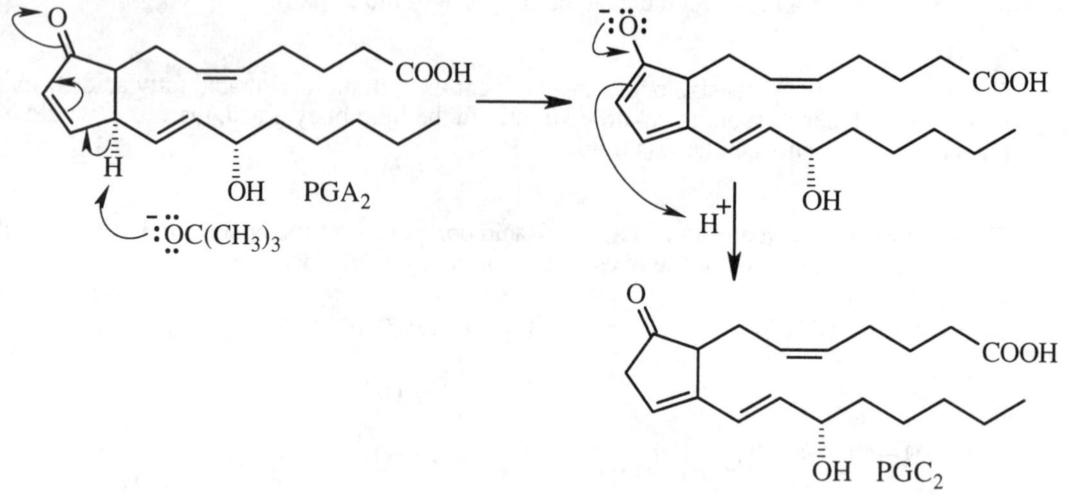

COOH

H

OH PGA_2

$^{-}:\ddot{O}C(CH_3)_3$

$^{-}:\ddot{O}:$

COOH

OH

H^+

O

COOH

OH PGC_2

11.

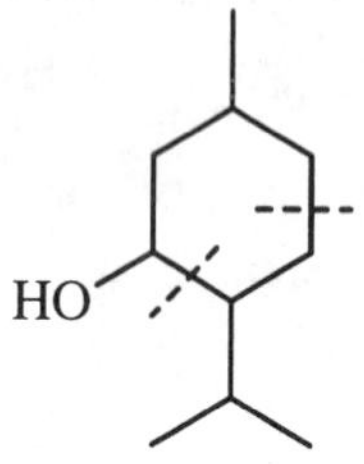

HO

menthol

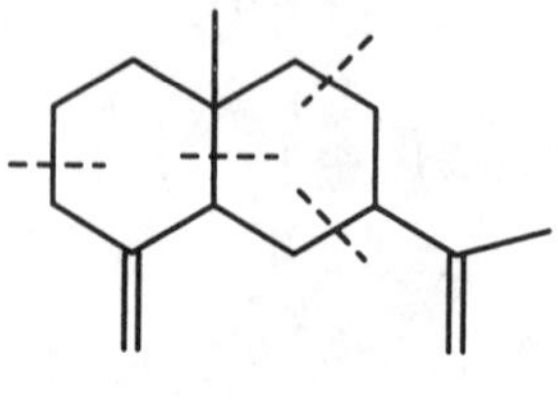

β-selinene

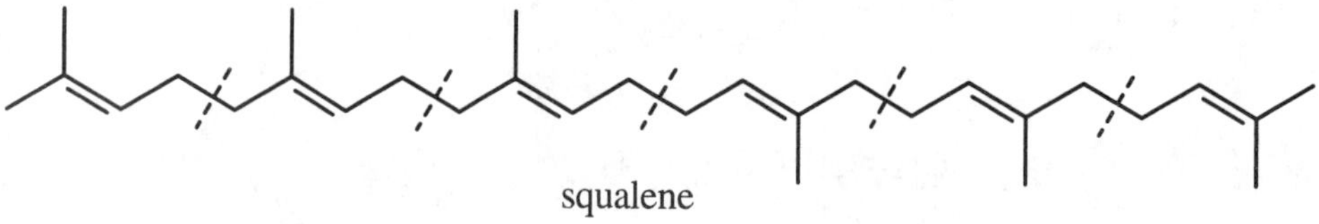

squalene

12. The fact that the tail-to-tail linkage occurs in the exact center of the molecule indicates that the two halves are synthesized (in a head-to-tail fashion) and then joined together in a tail-to-tail linkage.

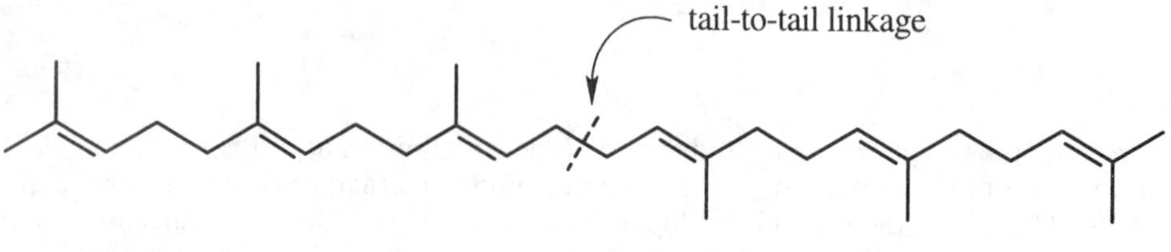

tail-to-tail linkage

13. Squalene, lycopene, and β-carotene are all synthesized in the same way. In each case, two halves are synthesized (in a head-to-tail fashion) and then joined together in a tail-to-tail linkage.

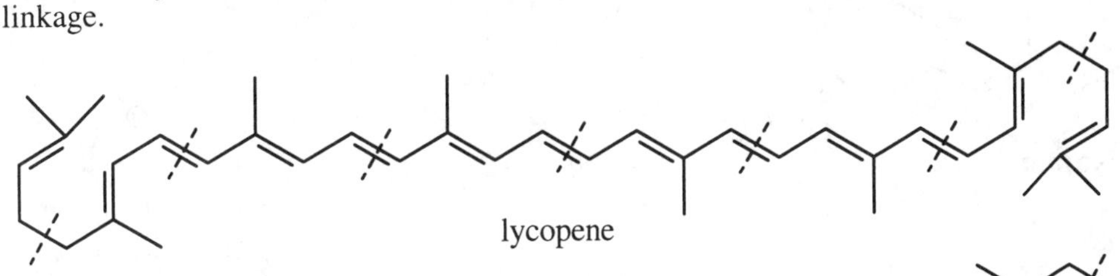

lycopene

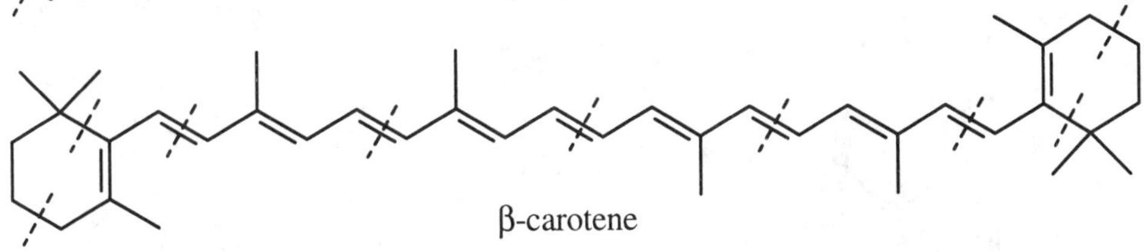

β-carotene

14. Solved in the text.

15.

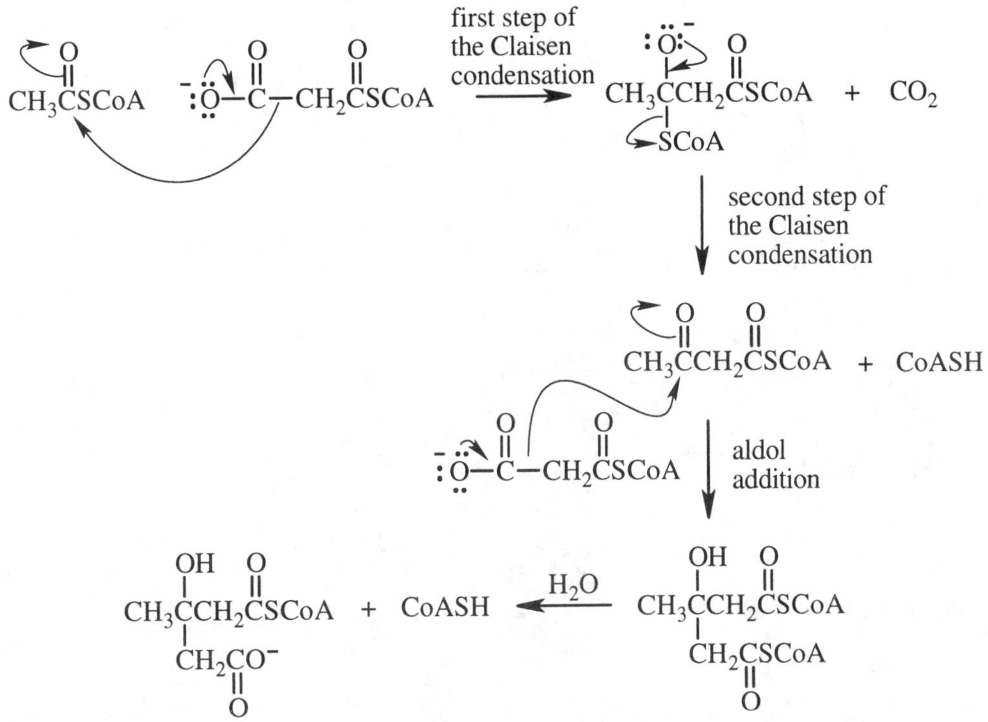

CH_3CSCoA (C=O) + $^{-}O-C(=O)-CH_2C(=O)SCoA$ —first step of the Claisen condensation→ $CH_3C(O^-)(SCoA)CH_2C(=O)SCoA$ + CO_2

—second step of the Claisen condensation→ $CH_3C(=O)CH_2C(=O)SCoA$ + CoASH

+ $^{-}O-C(=O)-CH_2C(=O)SCoA$ —aldol addition→ $CH_3C(OH)(CH_2C(=O)SCoA)CH_2C(=O)SCoA$

—H_2O→ $CH_3C(OH)(CH_2CO^-(=O))CH_2C(=O)SCoA$ + CoASH

16.

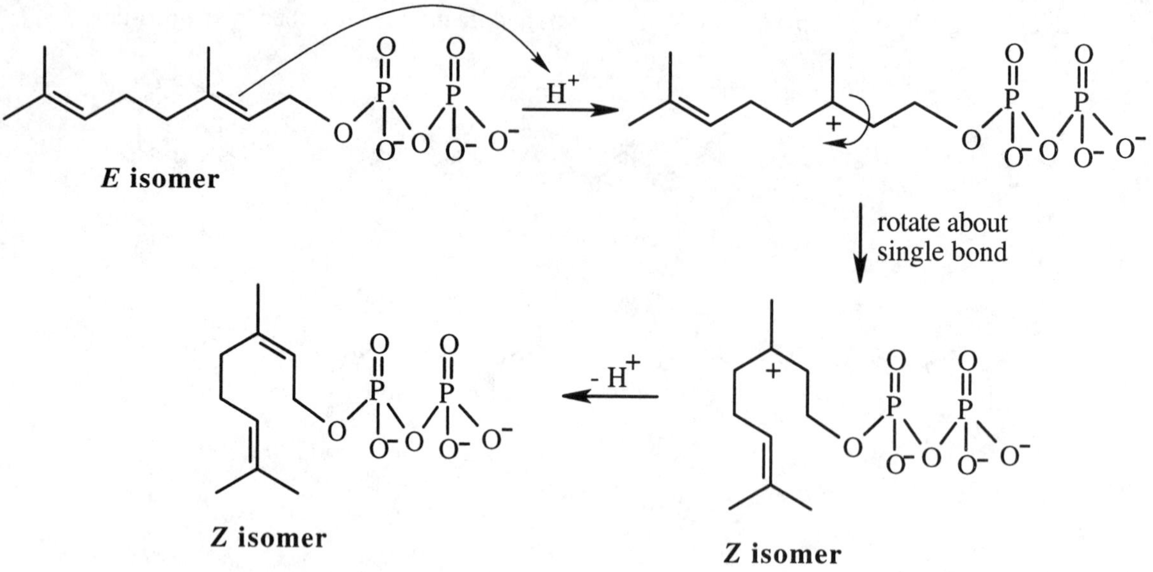

H^+

***E* isomer**

rotate about single bond

$-H^+$

***Z* isomer**

***Z* isomer**

17.

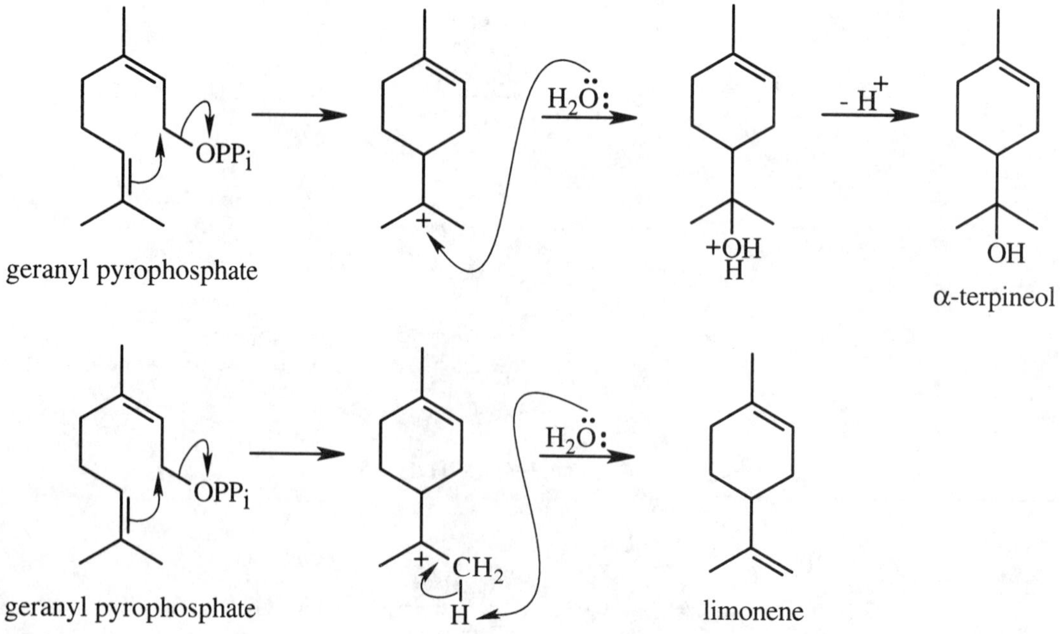

$H_2\ddot{O}:$

$-H^+$

OPP$_i$

geranyl pyrophosphate

+OH H

OH

α-terpineol

OPP$_i$

$H_2\ddot{O}:$

CH$_2$ H

geranyl pyrophosphate

limonene

18.

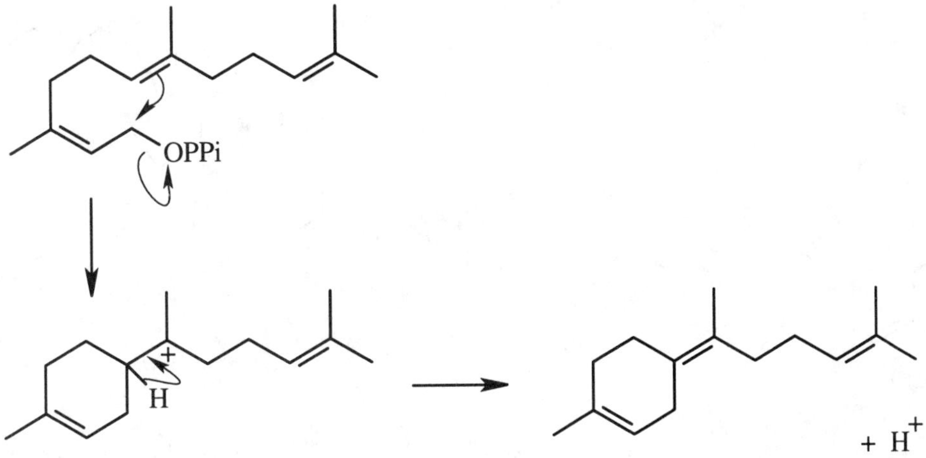

19. Solved in the text.

20. A β-**hydrogen** at C-5 means that the A and B rings are **cis** fused, while an α-**hydrogen** at C-5 means that the A and B rings are **trans** fused.

21. Because the OH substituent is on the **same side** of the steroid ring system as the angular methyl groups, it is a β-**substituent**.

22. The hemiacetal is formed by reaction of the primary alcohol with the aldehyde.

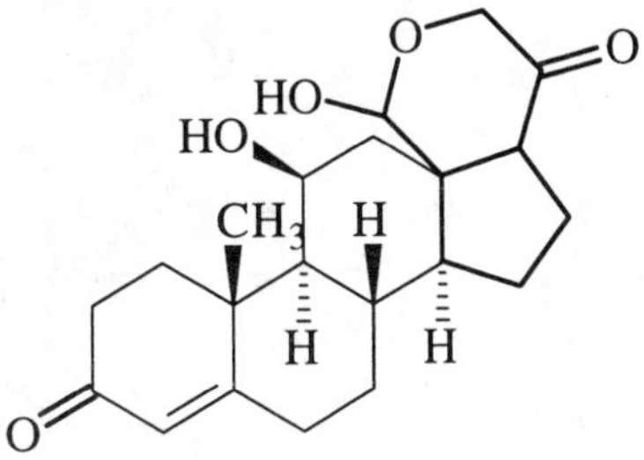

23.

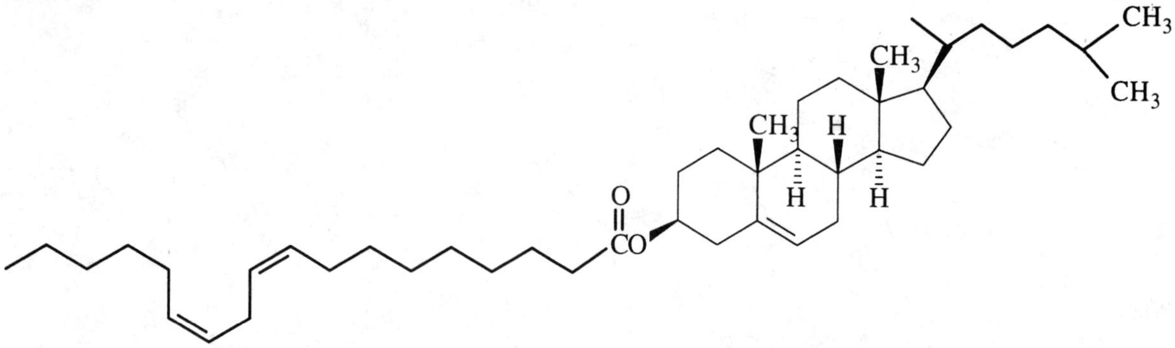

$$CH_3(CH_2)_4CH=CHCH_2CH=CH(CH_2)_7\overset{O}{\overset{\|}{C}}O\text{-cholesterol}$$

24. Because the three OH substituents of cholic acid are on the **opposite side** of the steroid ring system as the angular methyl groups, they are all α-**substituents**. Two of the OH substituents are axial substituents and one is an equatorial substituent.

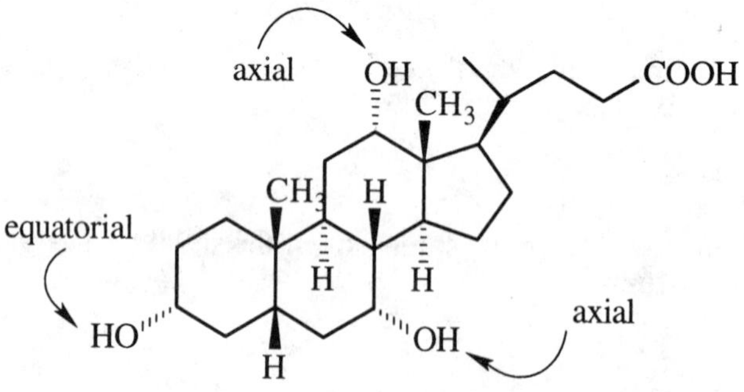

axial OH; equatorial HO; axial OH; COOH; CH_3; H

25. There are two hydride shifts and two methyl shifts. The last step is elimination of a proton.

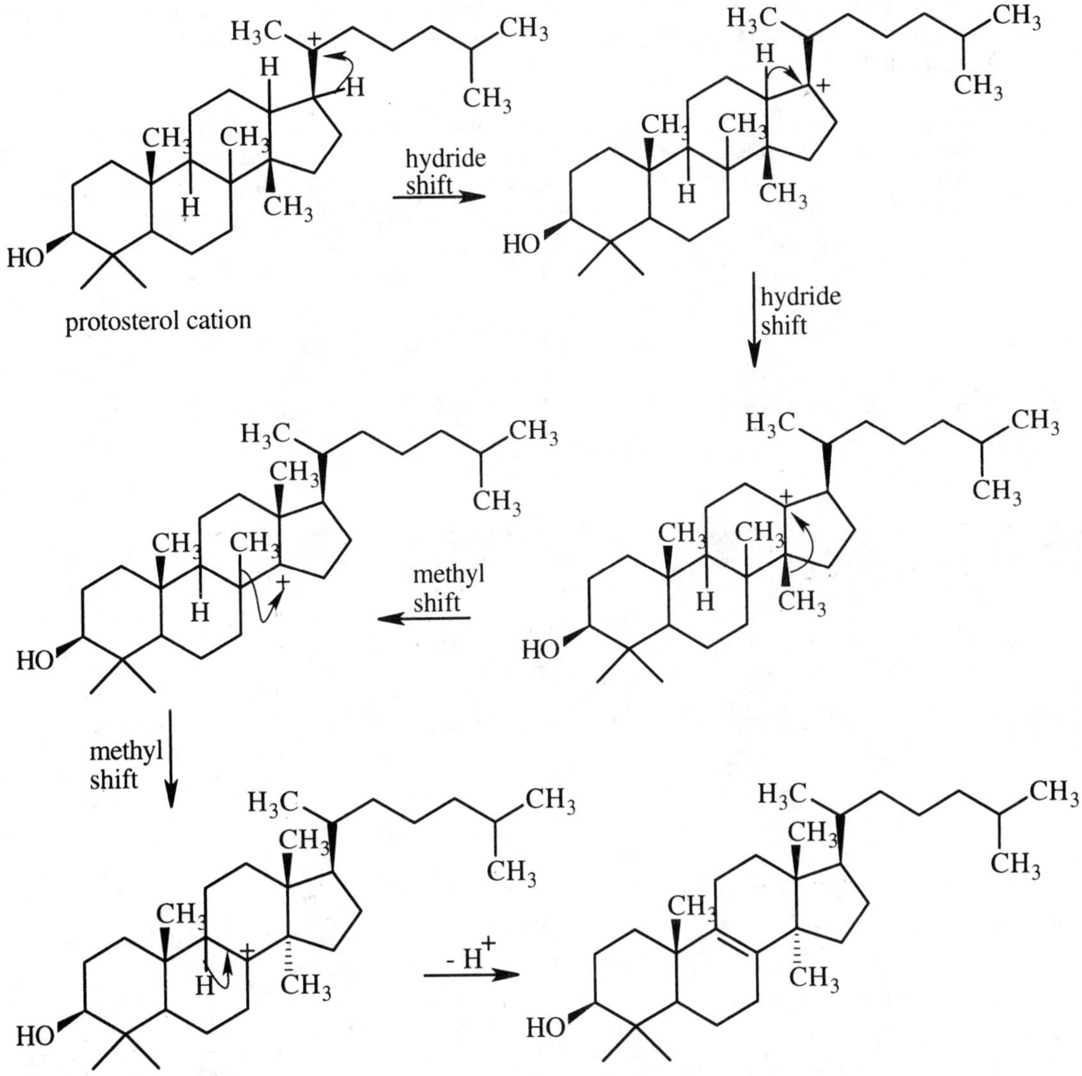

protosterol cation

hydride shift

hydride shift

methyl shift

methyl shift

$- H^+$

26. If stearic acid were at C-1 and C-3, the fat would not be optically active. Therefore, stearic acid must be at C-1 and C-2 (or C-2 and C-3).

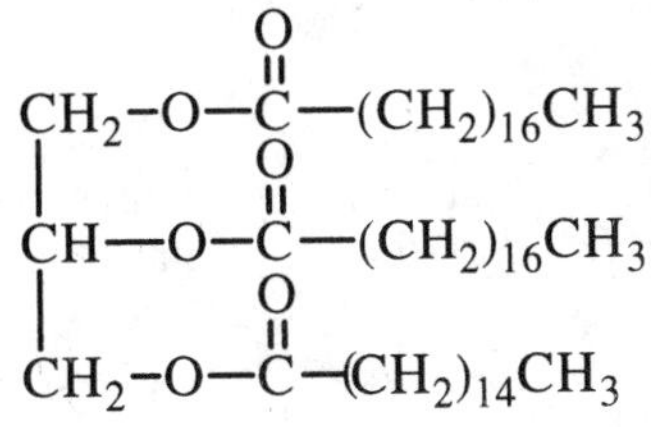

$$
\begin{array}{l}
CH_2-O-\overset{O}{\overset{\|}{C}}-(CH_2)_{16}CH_3 \\
| \\
CH-O-\overset{O}{\overset{\|}{C}}-(CH_2)_{16}CH_3 \\
| \\
CH_2-O-\overset{O}{\overset{\|}{C}}-(CH_2)_{14}CH_3
\end{array}
$$

27. **a.** There are three triacylglycerols in which one of the fatty acid components is lauric acid and two are myristic acid. Myristic acid can be at C-1 and C-3 of glycerol, in which case the triacylglycerol does not have any chirality centers. If myristic acid is at C-1 and C-2 of glycerol, C-2 is a chirality center, and consequently, two enantiomers are possible for the compound.

b. There are six triacylglycerols in which one of the fatty acid components is lauric acid, one is myristic acid, and one is palmitic acid. The three possible arrangements are shown below (with the fatty acid components abbreviated as L, M, and P). Since each has a chirality center, each can exist as a pair of enantiomers for a total of six triacylglycerols.

CH_2–O–L	CH_2–O–L	CH_2–O–M
*CH–O–M	*CH–O–P	*CH–O–L
CH_2–O–P	CH_2–O–M	CH_2–O–P

28.

R^1CO^- (C=O), R^2CO^- (C=O), R^3CO^- (C=O), R^4CO^- (C=O)

3 CH_2–OH / CH–OH / CH_2–OH

2 $HO–P(=O)(O^-)–O^-$

29.

CH_2–O–C(=O)–CH_3
CH–O–C(=O)–CH_3
CH_2–O–C(=O)–CH_3

The structure at left has a molecular formula = $C_9H_{14}O_6$ and a molecular weight = 218.

Subtracting 218 from the total molecular weight gives the molecular weight of the methylene (CH_2) groups in the triacylglycerol.

$$722 - 218 = 504$$

Dividing 504 by the molecular weight of a methylene group (14) will give the number of methylene groups.

$$\frac{504}{14} = 36$$

Since there are 36 methylene groups, each fatty acid in the triacylglycerol has 12 methylene groups.

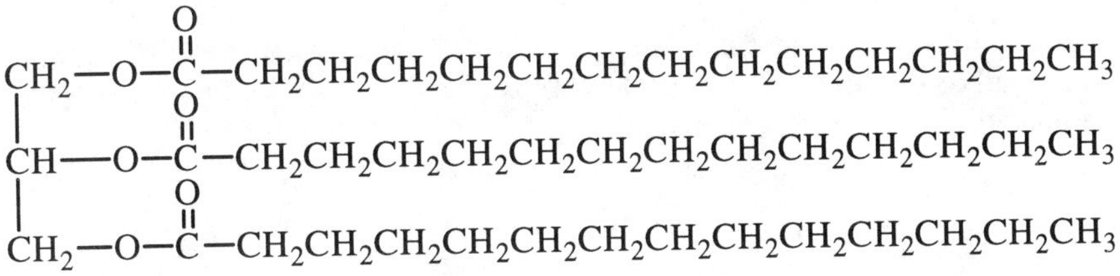

$$
\begin{array}{l}
CH_2-O-\overset{O}{\overset{\|}{C}}-CH_2CH_2CH_2CH_2CH_2CH_2CH_2CH_2CH_2CH_2CH_2CH_2CH_3 \\
| \\
CH-O-\overset{O}{\overset{\|}{C}}-CH_2CH_2CH_2CH_2CH_2CH_2CH_2CH_2CH_2CH_2CH_2CH_2CH_3 \\
| \\
CH_2-O-\overset{O}{\overset{\|}{C}}-CH_2CH_2CH_2CH_2CH_2CH_2CH_2CH_2CH_2CH_2CH_2CH_2CH_3
\end{array}
$$

nutmeg

30.

a. [steroid structure: H_3C, CH_3, CH_3, CH_3, CH_3, HO, OH]

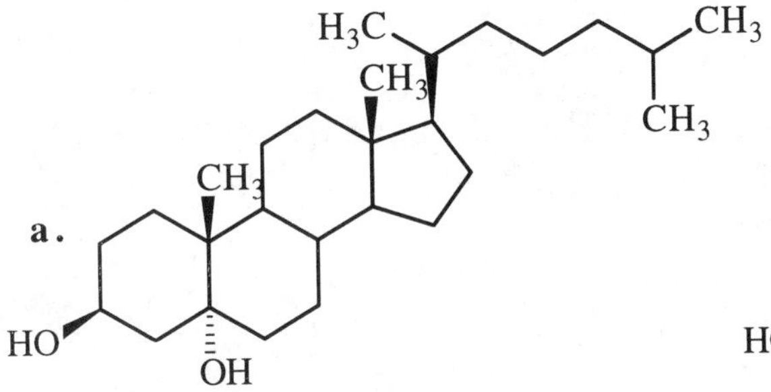

d. [steroid structure: H_3C, CH_3, CH_3, CH_3, CH_3, HO, OH, Br]

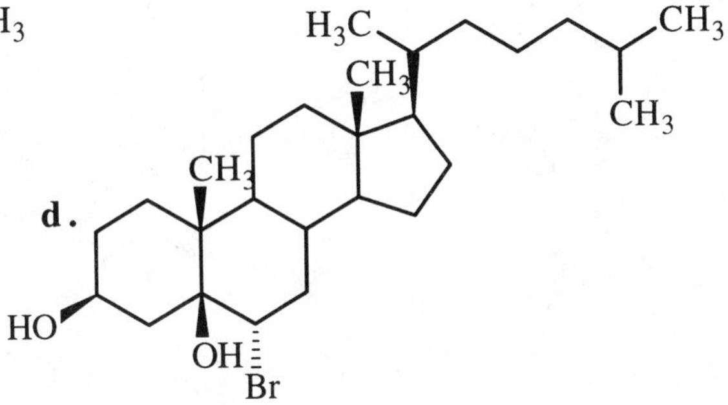

b. [steroid structure: H_3C, CH_3, CH_3, CH_3, CH_3, HO, H, OH]

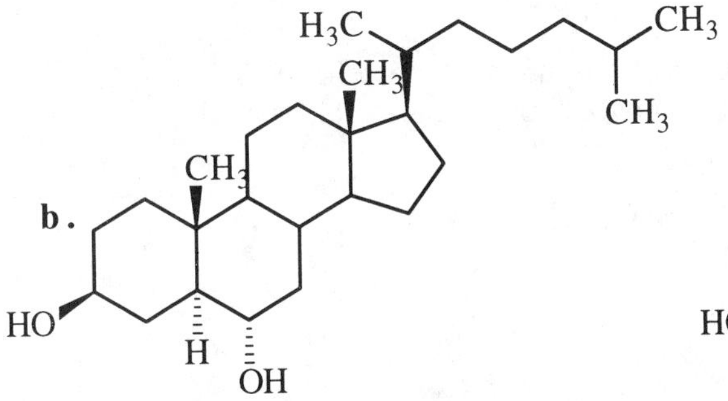

e. [steroid structure: H_3C, CH_3, CH_3, CH_3, CH_3, HO, O]

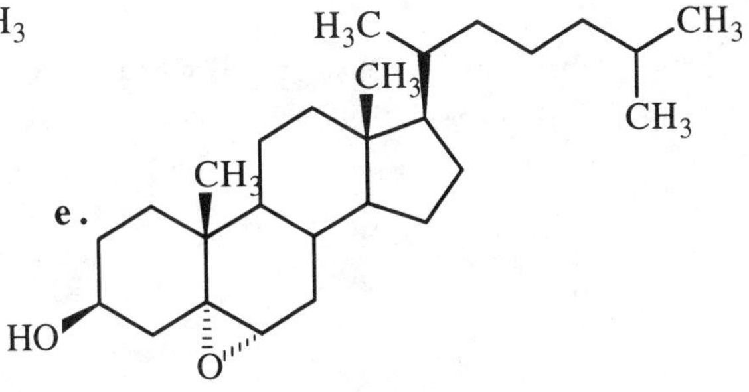

c. [steroid structure: H_3C, CH_3, CH_3, CH_3, CH_3, HO, H]

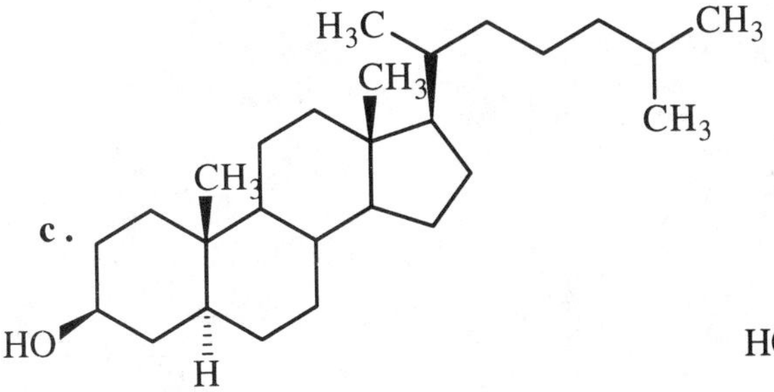

f. [steroid structure: H_3C, CH_3, CH_3, CH_3, CH_3, HO, OH, OCH_3]

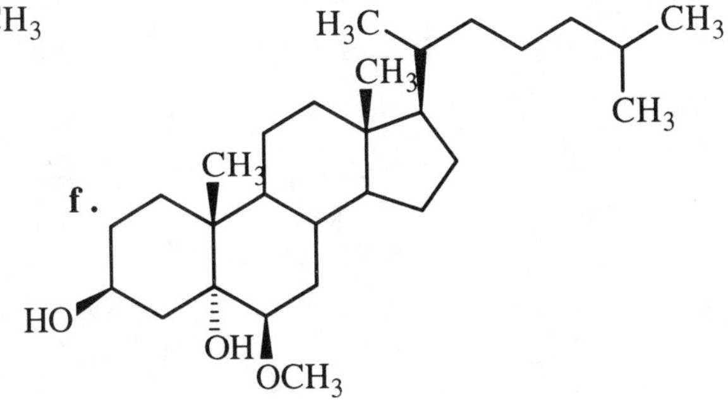

31. **a.** Starting with mevalonyl pyrophosphate, you can trace the location of the label in the compounds that lead to the formation of geranyl pyrophosphate, and geranyl pyrophosphate is converted to citronellal.

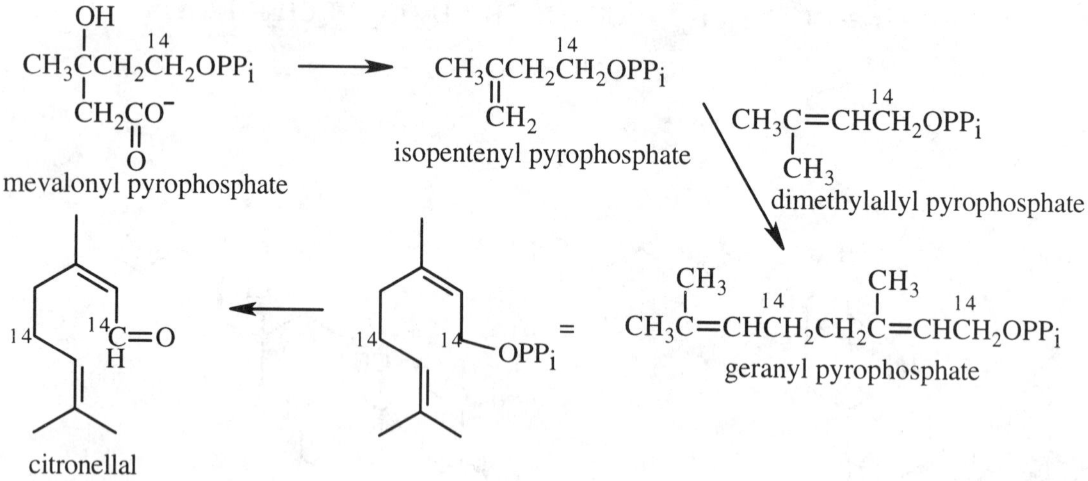

$$CH_3C(OH)(CH_2CO^-{=}O)CH_2\overset{14}{C}H_2OPP_i \longrightarrow CH_3C(=CH_2)CH_2\overset{14}{C}H_2OPP_i$$

mevalonyl pyrophosphate → isopentenyl pyrophosphate

$$CH_3C(CH_3)=CH\overset{14}{C}H_2OPP_i$$

dimethylallyl pyrophosphate

$$CH_3C(CH_3)=CH\overset{14}{C}H_2CH_2C(CH_3)=CH\overset{14}{C}H_2OPP_i$$

geranyl pyrophosphate

= (skeletal structure with two 14-labeled carbons, –OPP$_i$) ⟶ citronellal (skeletal structure, 14 label on the CH$_2$ carbon and on $^{14}C{=}O$, H)

citronellal

b. The label is lost from sample B when mevalonic pyrophosphate loses CO_2 to form isopentenyl pyrophosphate, so none of the carbons will be labeled in citronellal.

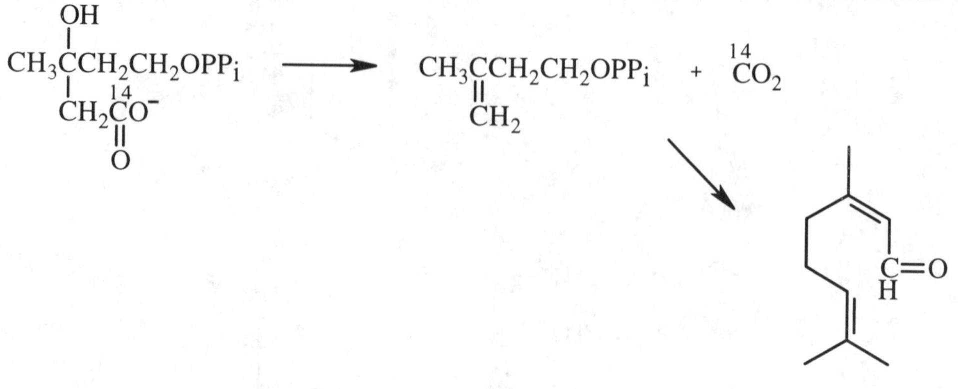

$$CH_3C(OH)(CH_2\overset{14}{C}O^-{=}O)CH_2CH_2OPP_i \longrightarrow CH_3C(=CH_2)CH_2CH_2OPP_i + \overset{14}{C}O_2$$

↘ citronellal (skeletal structure, unlabeled, C=O, H)

c. Because the methyl groups are equivalent in the carbocation that is formed as an intermediate when isopentenyl pyrophosphate is converted to dimethylallyl pyrophosphate, either of the methyl groups can be labeled in dimethylallyl pyrophosphate. This means that either of the two methyl groups can be labeled in geranyl pyrophosphate and in citronellal.

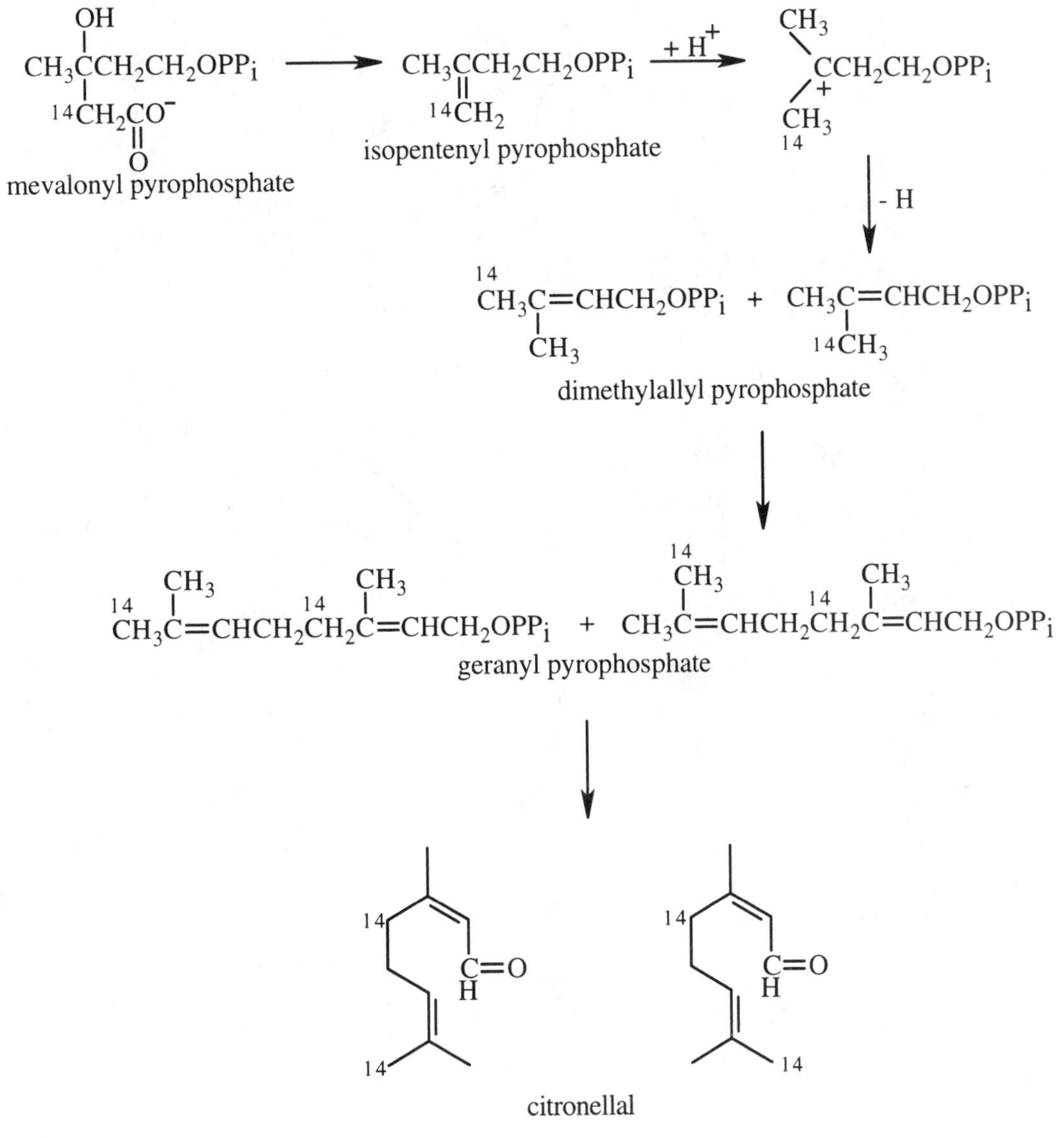
OH
CH3CCH2CH2OPPi
14CH2CO⁻
O
mevalonyl pyrophosphate
CH3CCH2CH2OPPi
14CH2
isopentenyl pyrophosphate
+ H+
CH3
CCH2CH2OPPi
CH3
14
- H
14
CH3C=CHCH2OPPi + CH3C=CHCH2OPPi
CH3
14CH3
dimethylallyl pyrophosphate
CH3 CH3
14 14
CH3C=CHCH2CH2C=CHCH2OPPi + CH3C=CHCH2CH2C=CHCH2OPPi
geranyl pyrophosphate
C=O
H
14
citronellal

32. There are five possible structures for compound A, and two possible structures for compound B.

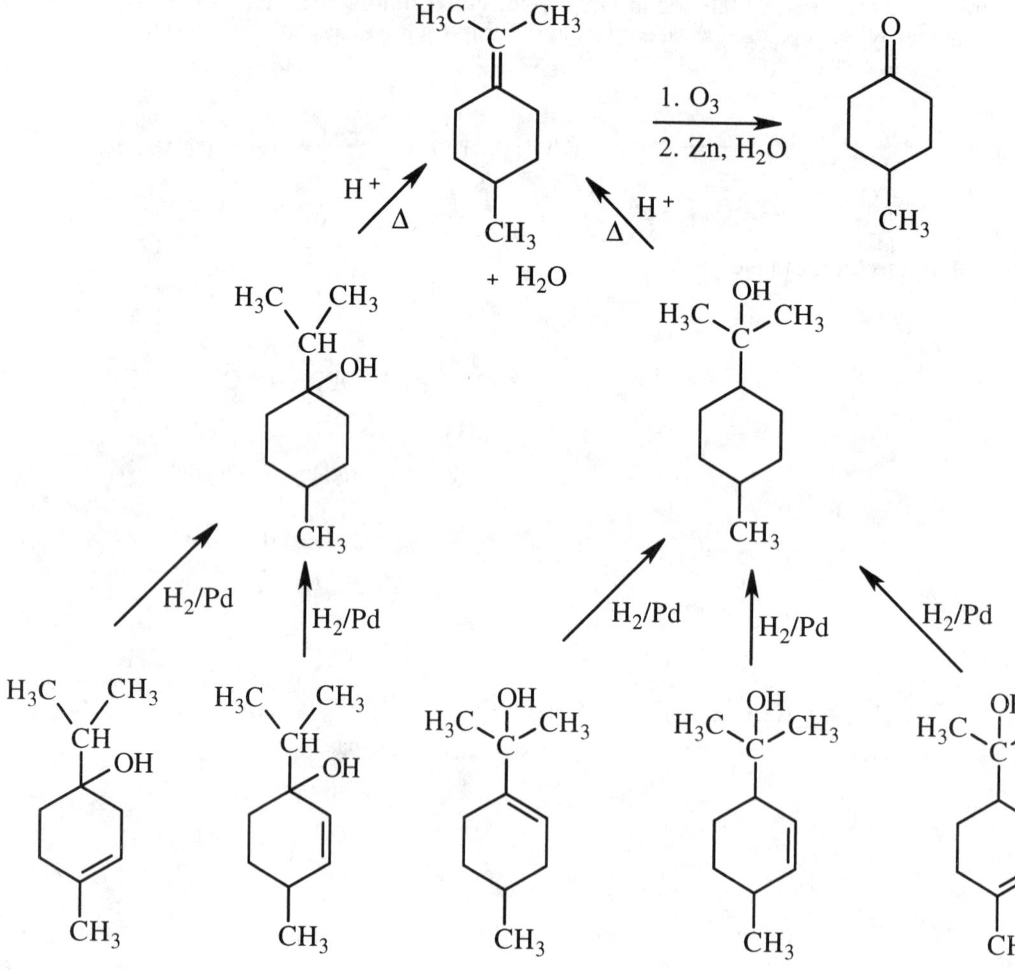

H_3C CH_3 C

1. O_3
2. Zn, H_2O

O

H^+ Δ

CH_3

H^+ Δ

CH_3

+ H_2O

H_3C CH_3 CH OH

OH H_3C CH_3 C

CH_3

CH_3

H_2/Pd H_2/Pd H_2/Pd H_2/Pd H_2/Pd

H_3C CH_3 CH OH

H_3C CH_3 CH OH

OH H_3C CH_3 C

OH H_3C CH_3 C

OH H_3C CH_3 C

CH_3 CH_3 CH_3 CH_3 CH_3

33. Because acetyl-CoA is converted into malonyl-CoA (see Section 23.5), mevalonyl pyrophosphate will contain three labeled carbons, which means that juniper oil will contain six labeled carbons.

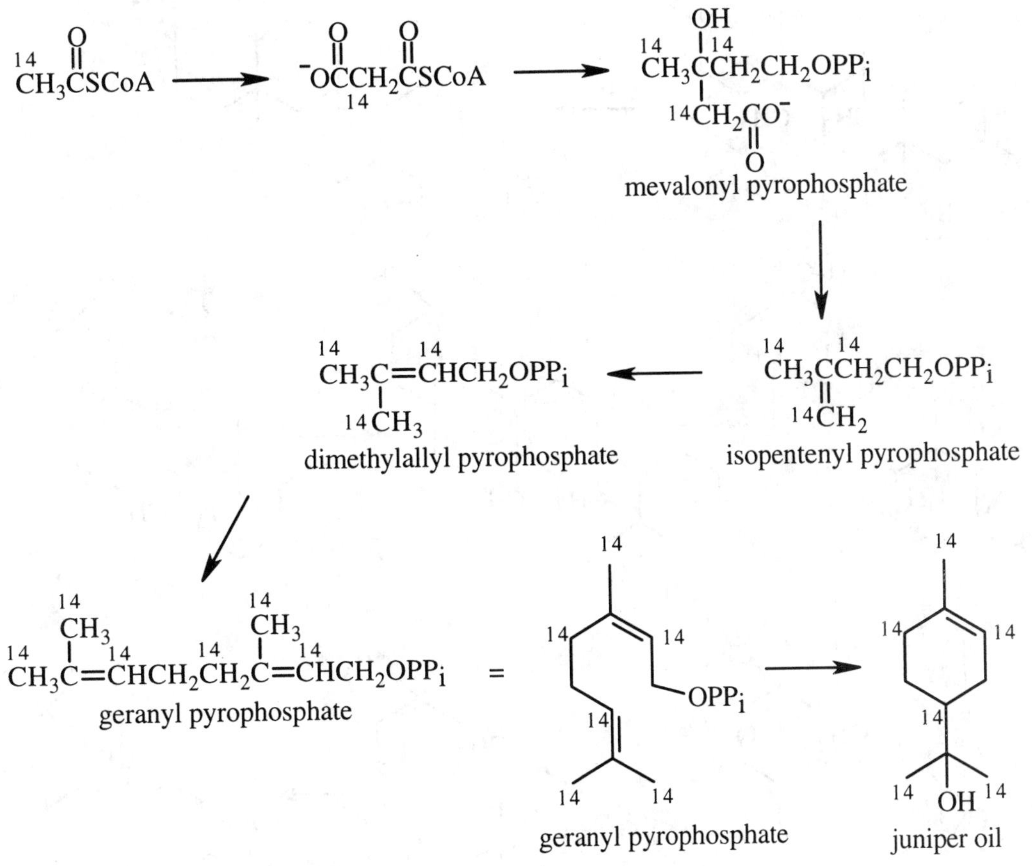

$$^{14}CH_3C(=O)SCoA \longrightarrow {}^{-}OC(=O)\,{}^{14}CH_2C(=O)SCoA \longrightarrow {}^{14}CH_3\,{}^{14}C(OH)(CH_2CH_2OPP_i)\,{}^{14}CH_2CO^{-}(=O)$$

mevalonyl pyrophosphate

$$\downarrow$$

$$^{14}CH_3\,{}^{14}C(={}^{14}CH_2)CH_2CH_2OPP_i$$

isopentenyl pyrophosphate

$$\downarrow$$

$$^{14}CH_3\,{}^{14}C(^{14}CH_3)=CHCH_2OPP_i$$

dimethylallyl pyrophosphate

$$\downarrow$$

$$^{14}CH_3\,{}^{14}C(^{14}CH_3)={}^{14}CHCH_2CH_2\,{}^{14}C(^{14}CH_3)={}^{14}CHCH_2OPP_i$$

geranyl pyrophosphate

= geranyl pyrophosphate (skeletal structure, labeled carbons marked 14) $\longrightarrow$ juniper oil (skeletal structure, six labeled carbons marked 14, with OH)

geranyl pyrophosphate juniper oil

34. **a.**

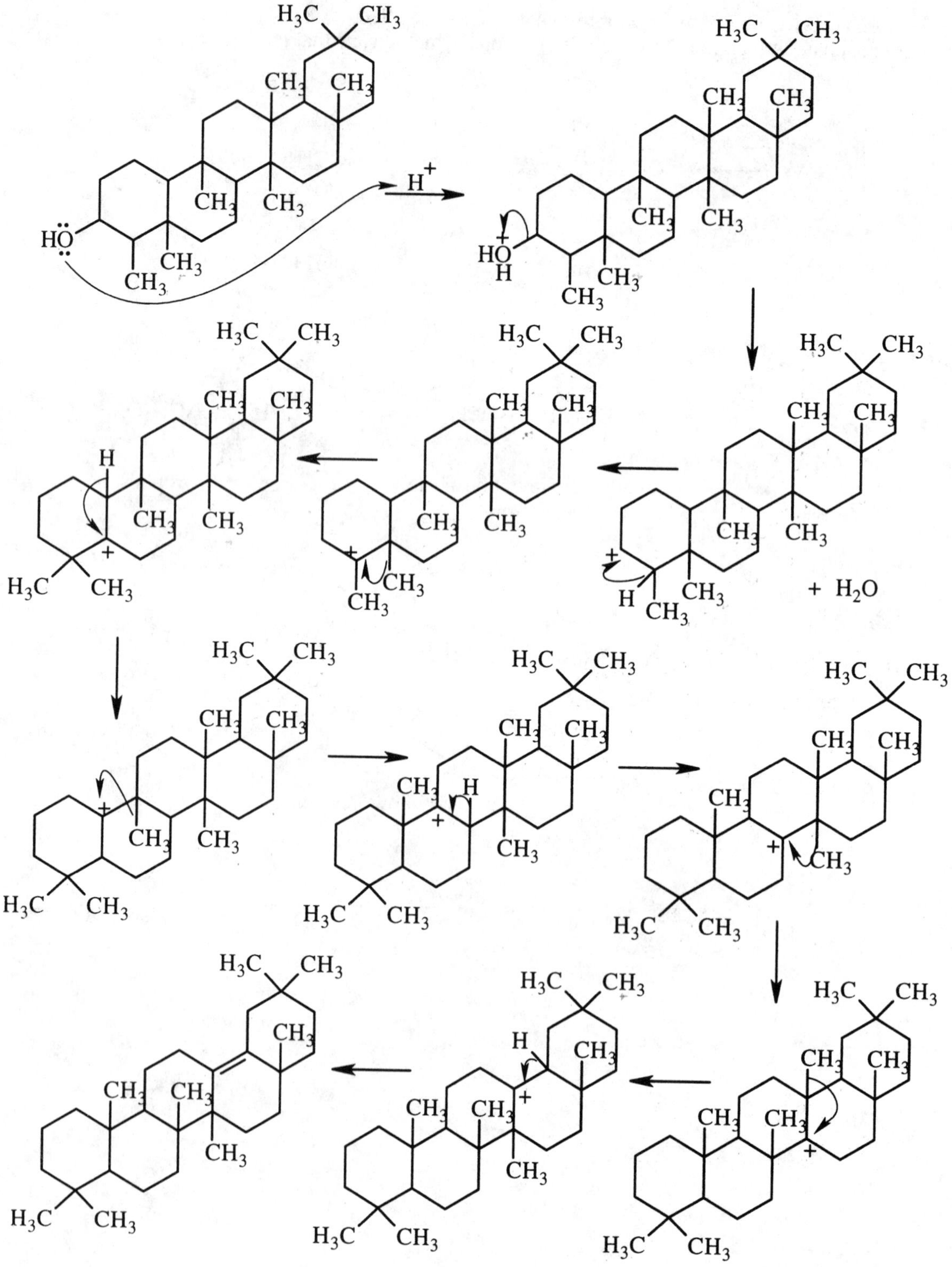

$\xrightarrow{H^+}$

$+ H_2O$

b. It has 30 carbons, so it is a triterpene.

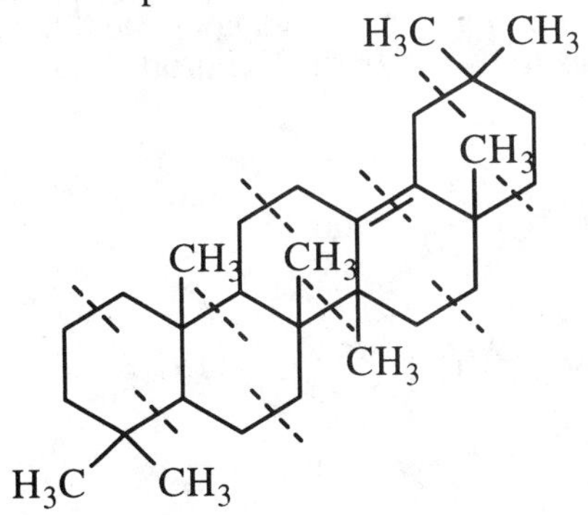

H_3C CH_3 CH_3 CH_3 CH_3 CH_3 H_3C CH_3

35.

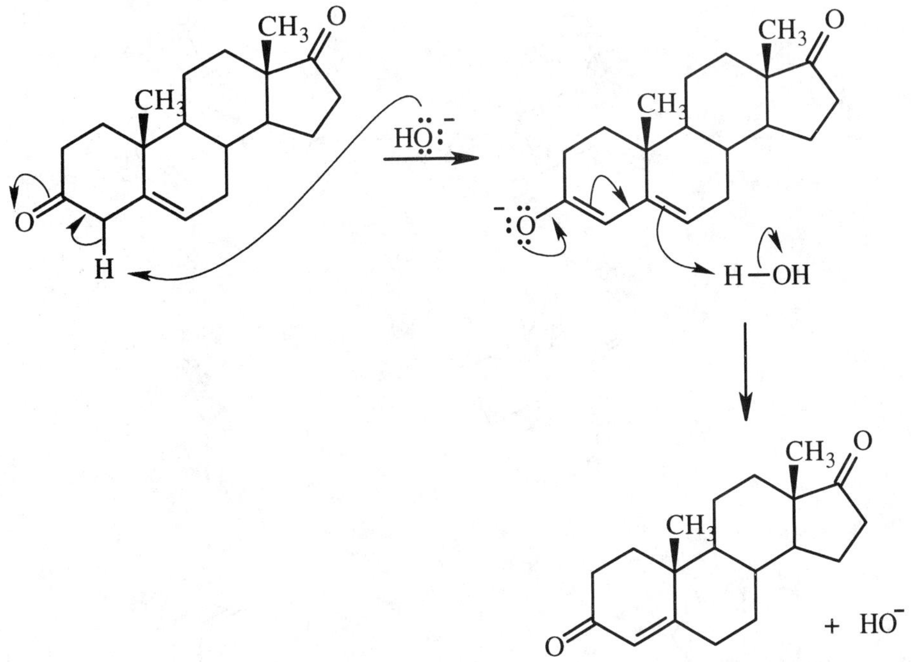

CH_3 O CH_3 HO^- CH_3 O CH_3 $H-OH$ CH_3 O CH_3 O + HO^-

36. The OH groups will react only if they are in equatorial positions, because introduction of bulky axial substituents would decrease the stability of the molecule.

In the case of 5α-cholestane-3β,7β-diol, the two OH groups are on the same side of the ring system as the angular methyl group, which means that they are in equatorial positions. Both OH groups react with ethyl chloroformate.

In the case of 5α-cholestane-3β,7α-diol, only one of the OH groups is on the same side of the ring system as the angular methyl group. The other is on the opposite side of the ring, which means that it is in an axial position. Only the OH group that is in the equatorial position reacts with ethyl chloroformate.

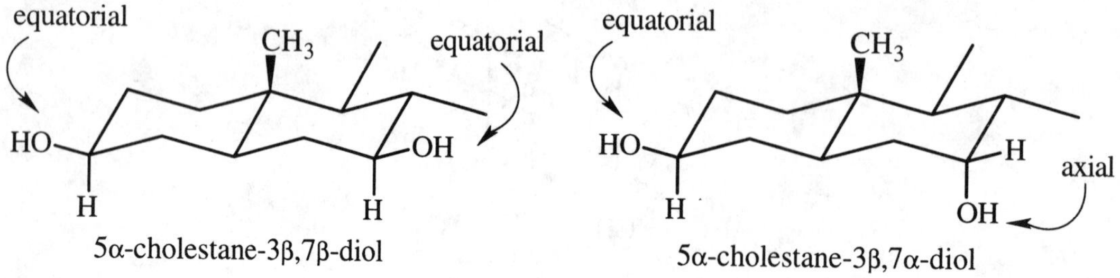

equatorial CH_3 equatorial HO OH H H
5α-cholestane-3β,7β-diol

equatorial CH_3 HO H H axial OH
5α-cholestane-3β,7α-diol

37.

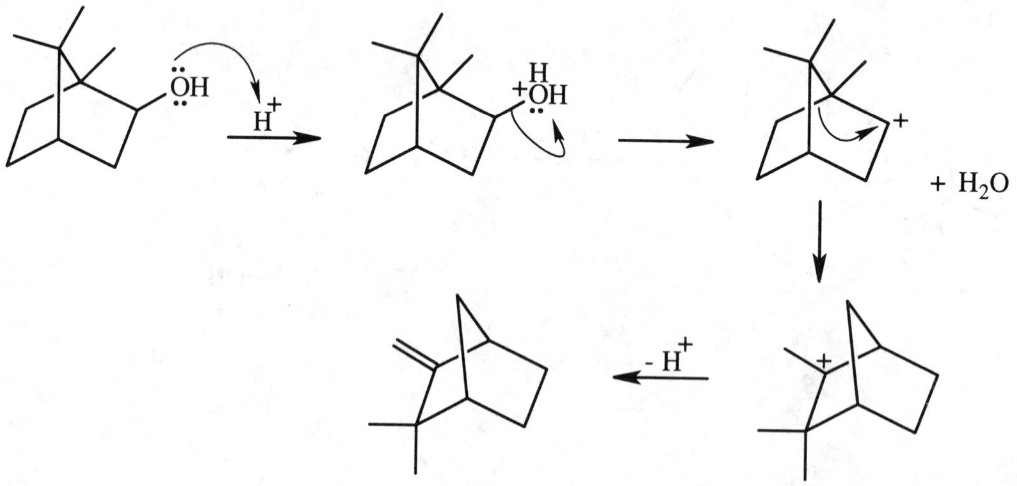

OH H^+ ^+OH H + H_2O $- H^+$

38.

CH_3 OH H H H HO
estradiol

OH HO
DES

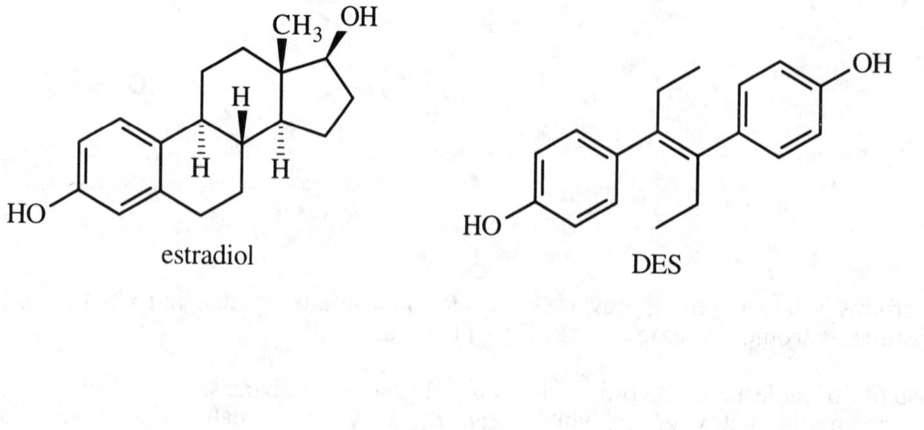

CHAPTER 25
Nucleosides, Nucleotides, and Nucleic Acids

Important Terms

acyl adenylate

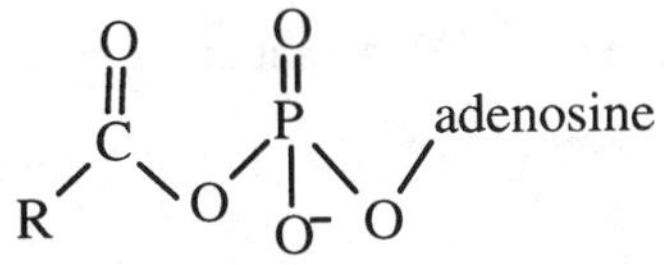

$$R-C(=O)-O-P(=O)(O^-)-O-\text{adenosine}$$

acyl phosphate

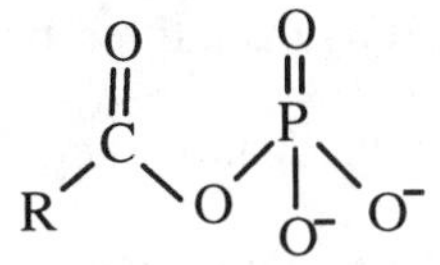

$$R-C(=O)-O-P(=O)(O^-)-O^-$$

acyl pyrophosphate

$$R-C(=O)-O-P(=O)(O^-)-O-P(=O)(O^-)-O^-$$

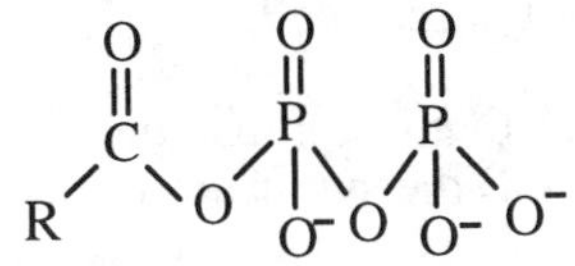

anticodon — the three bases at the bottom of the middle loop in a tRNA.

antigene agent — a polymer designed to bind to DNA at a particular site.

antisense agent — a polymer designed to bind to mRNA at a particular site.

antisense strand (template strand) — the strand in DNA that is read during transcription.

autoradiograph (autorad) — the exposed photographic plate obtained in autoradiography.

autoradiography — a technique used to determine the base sequence of DNA.

base — a heterocyclic compound (a purine or a pyrimidine) in DNA and RNA.

codon — a sequence of three bases of mRNA that specifies the amino acid to be incorporated into a protein.

deamination — a hydrolysis reaction that results in removal of ammonia.

deoxyribonucleic acid (DNA) — a polymer of deoxyribonucleotides.

deoxyribonucleotide — a nucleotide where the sugar component is D-2-deoxyribose.

dideoxy method — a method used to determine the sequence of bases in DNA.

dinucleotide two nucleotides linked by phosphodiester bonds.

double helix the term used to describe the secondary structure of DNA.

eukaryotic organism as organism with cells that contains a nucleus.

exon a stretch of bases in DNA that are a portion of a gene.

gene a segment of DNA.

gene therapy a technique that inserts a synthetic gene into the DNA of an organism defective in that gene.

genetic code the amino acid specified by each three-base sequence of mRNA.

high-energy bond a bond that releases a great deal of energy when it is broken.

Hoogsteen base pairing the pairing between a base in a synthetic strand of DNA with a base pair in double-stranded DNA.

human genome the total DNA of a human cell.

informational strand (sense strand) the strand in DNA that is not read during transcription; it has the same sequence of bases as the synthesized mRNA strand (with a U, T difference).

in-line displacement mechanism nucleophilic attack on a phosphorus concerted with breaking a phosphoanhydride bond.

intron a stretch of bases in DNA that contain no genetic information.

major groove the wider and deeper of the two alternating grooves in DNA.

minor groove the narrower and more shallow of the two alternating grooves in DNA.

nucleic acid the two kinds of nucleic acid are DNA and RNA.

nucleoside a heterocyclic base (purine or pyrimidine) bonded to the anomeric carbon of a sugar (D-ribose or D-2-deoxyribose).

nucleotide a nucleoside with one of its OH groups bonded to phosphoric acid in an ester linkage.

oligonucleotide three to ten nucleotides linked by phosphodiester bonds.

phosphoanhydride bond the bond holding two phosphoric acid molecules together.

phosphoryl transfer reaction the transfer of a phosphate group from one compound to another.

polynucleotide many nucleotides linked by phosphodiester bonds.

primary structure the sequence of bases in the nucleic acid.

prokaryotic organism a unicellular organism without a nucleus.

promoter site a short sequence of bases at the beginning of a gene.

rational drug design designing drugs with a particular structure to achieve a specific purpose.

replication the synthesis of identical copies of DNA.

replication fork the position on DNA where replication begins.

restriction endonuclease an enzyme that cleaves DNA at a specific base sequence.

restriction fragment a fragment that is formed when DNA is cleaved by a restriction endonuclease.

retrovirus a virus whose genetic information is stored in its RNA.

ribonucleic acid (RNA) a polymer of ribonucleotides.

ribonucleotide a nucleotide where the sugar component is D-ribose.

ribosome a particle composed of about 40% protein and 60% RNA on which protein biosynthesis takes place.

ribozyme an RNA molecule that acts as a catalyst.

RNA splicing the step in RNA processing that cuts out nonsense bases and splices informational pieces together.

sedimentation constant designates where a species sediments in an ultracentrifuge.

semiconservative replication the mode of replication that results in a daughter molecule of DNA having one of the original DNA strands plus a newly synthesized strand.

sense strand (informational strand) the strand in DNA that is not read during transcription; it has the same sequence of bases as the synthesized mRNA strand (with a U, T difference).

site-specific recognition recognition by a molecule of a specific site on another molecule.

stacking interactions van der Waals interactions between the mutually induced dipoles of adjacent pairs of bases in DNA.

stop codon a codon that says "stop protein synthesis here."

template strand the strand in DNA that is read during transcription.

transcription the synthesis of mRNA from a DNA blueprint.

translation the synthesis of a protein from a mRNA blueprint.

Solutions to Problems

1. The ring is protonated at its most basic position. In the case of a purine, this is the #7 position. In the next step, the bond between the heterocyclic base and the sugar breaks, with the anomeric carbocation being stabilized by the ring oxygen's nonbonding electrons.

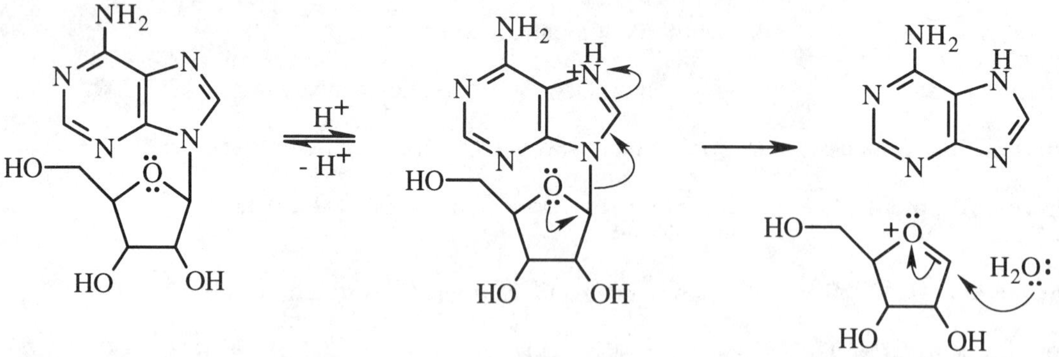

NH_2 N N N N HO– O HO OH $\underset{-\ H^+}{\overset{H^+}{\rightleftharpoons}}$ NH_2 H +N N N N HO– O HO OH → NH_2 H N N N N HO– +O HO OH H_2O:

The mechanism is exactly the same for pyrimidines except that the initial protonation takes place at the #1 position.

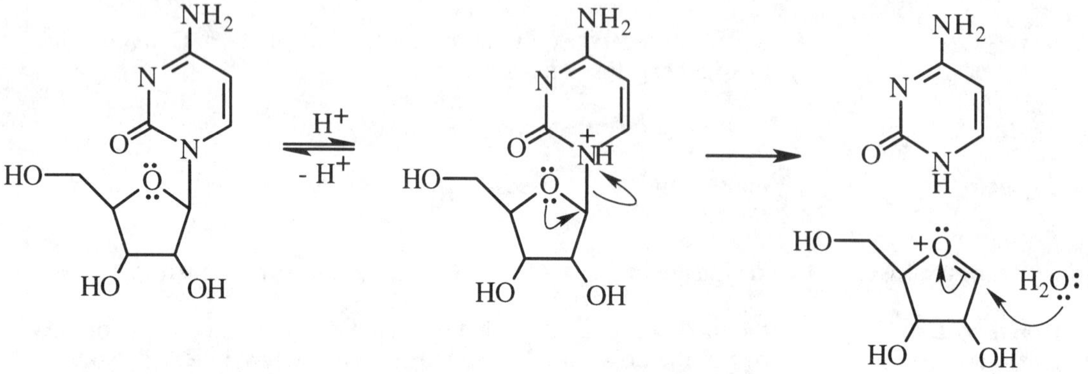

NH_2 N O N HO– O HO OH $\underset{-\ H^+}{\overset{H^+}{\rightleftharpoons}}$ NH_2 N O +NH HO– O HO OH → NH_2 N O N H HO– +O HO OH H_2O:

2.

a. NH_2 N O N O O P P ^-O O$^-$ O O$^-$ O O HO

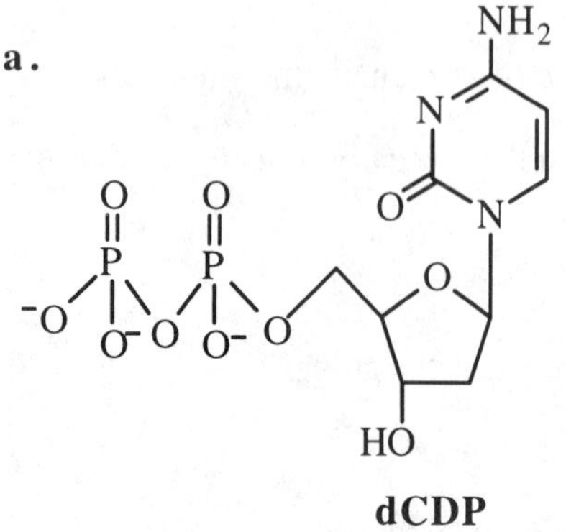

dCDP

c. O HN O N O P ^-O O$^-$ O O HO

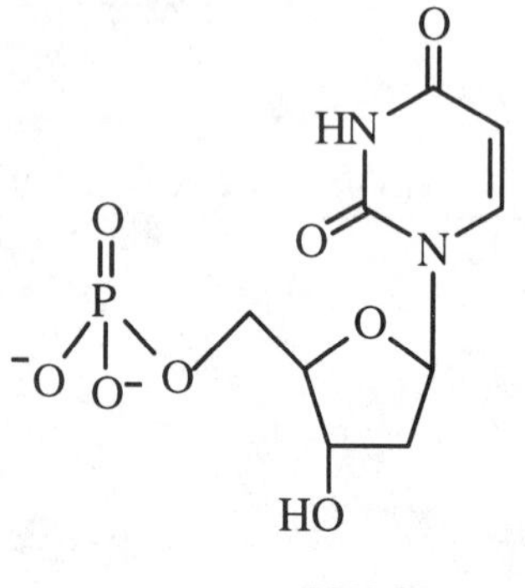

dUMP

b.

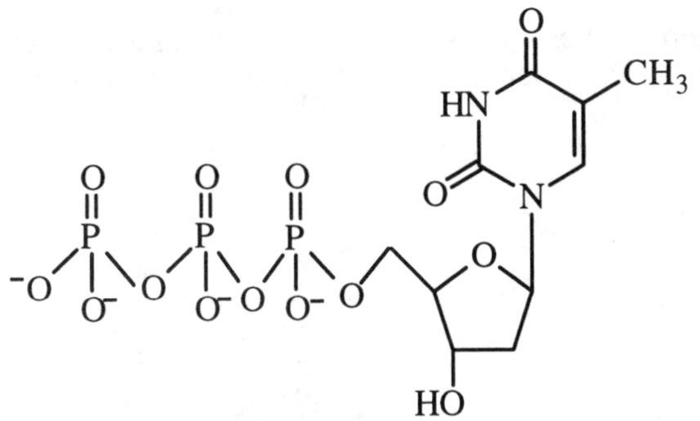

dTTP

d.

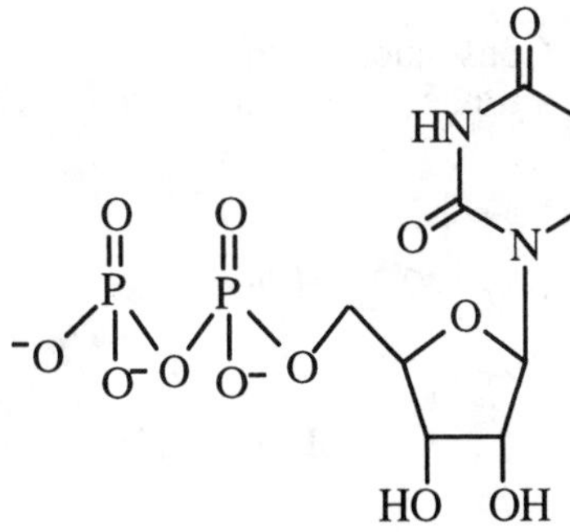

UDP

e.

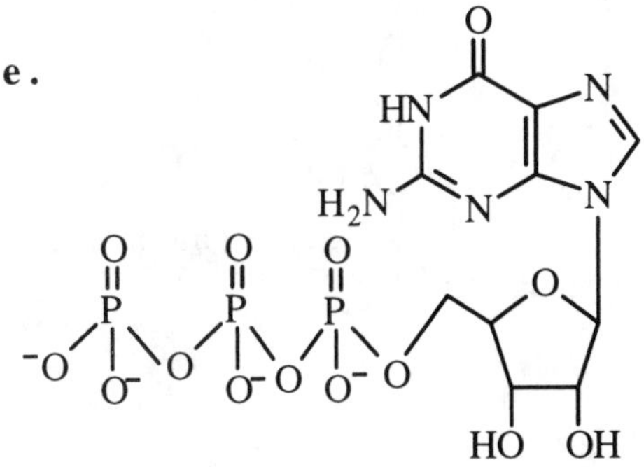

guanosine 5'-triphosphate
GTP

f.

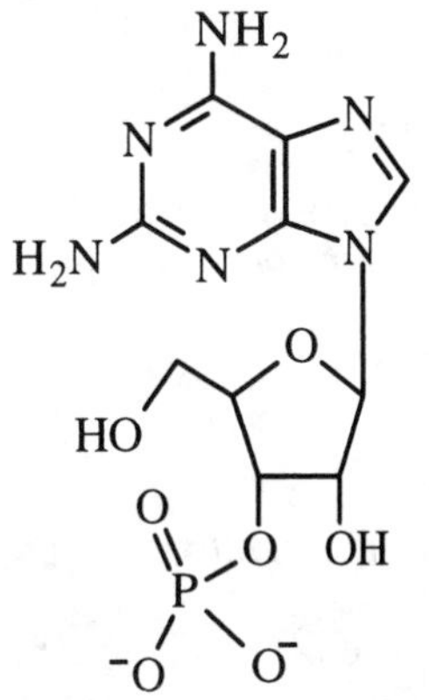

adenosine 3'-monophosphate

3. Solved in the text.

4. The $\Delta G°$ for formation of ATP is + 7.3 kcal/mol. This means that for a compound to drive the formation of ATP, it must hydrolyze with a $\Delta G°$ that is more negative than - 7.3 kcal/mol. Phosphocreatine is the only one of the four compounds that hydrolyzes with sufficient energy.

5. Phosphoric acid has three pK_a's: the first OH group to lose a proton has a pK_a of 1.5, the second OH group has a pK_a of 7.0, and the third OH group a pK_a of 12.2.

When the bond linking the β-phosphorus to the α-phosphorus breaks, the phosphorus leaving group attains its second negative charge which means that the pK_a of the conjugate acid of the leaving group is ~ 7.0.

When the bond linking the β-phosphorus to the γ-phosphorus breaks, the phosphorus leaving group attains its third negative charge, which means that the pK_a of the conjugate acid of the leaving group is ~ 12.2.

Consequently, the bond to the γ-phosphorus never breaks because the leaving group is a much stronger base than the leaving group that leaves when the bond to the α-phosphorus breaks.

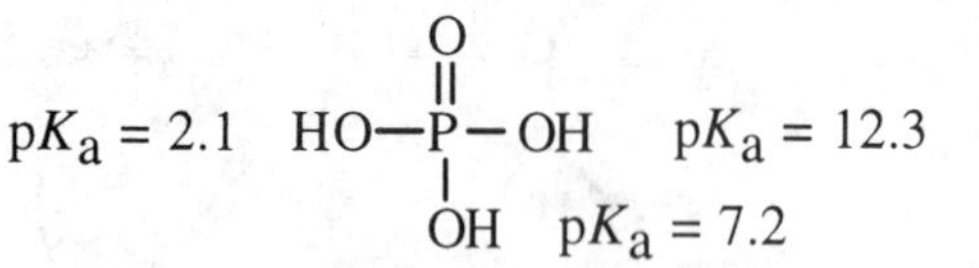

$pK_a = 2.1$ HO—P(=O)(OH)—OH $pK_a = 12.3$

OH $pK_a = 7.2$

phosphoric acid

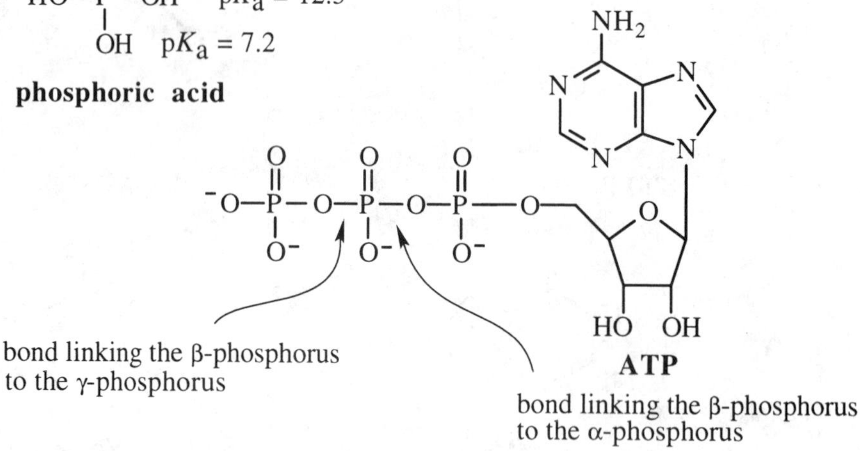

ATP

bond linking the β-phosphorus to the γ-phosphorus

bond linking the β-phosphorus to the α-phosphorus

6. **a.** Solved in the text.

b. Because pH 7.3 is much more basic than the pK_a values of the first two ionizations of ADP, these two groups will be in their basic forms at that pH, giving ADP two negative charges. We can determine the fraction of the group with pK_a of 6.8 that will be in its basic form at pH 7.3 using the method shown in part **a**.

$$\frac{K_a}{K_a + [H^+]} = \frac{1.6 \times 10^{-7}}{1.6 \times 10^{-7} + 5.0 \times 10^{-8}}$$

$$= \frac{1.6 \times 10^{-7}}{1.6 \times 10^{-7} + 0.5 \times 10^{-7}} = 0.8$$

total negative charge on ADP = 2.0 + 0.8 = 2.8

c. At pH 7.3 the OH group with a pK_a of 2.1 will account for one negative charge and the OH group with a pK_a of 12.3 will have no negative charge. We need to calculate the fraction of the group with a pK_a value of 7.2 that will be negatively charged at pH 7.3.

$$\frac{K_a}{K_a + [H^+]} = \frac{6.3 \times 10^{-8}}{6.3 \times 10^{-8} + 5.0 \times 10^{-8}} = 0.6$$

total negative charge on phosphate = 1.0 + 0.6 = 1.6

7.

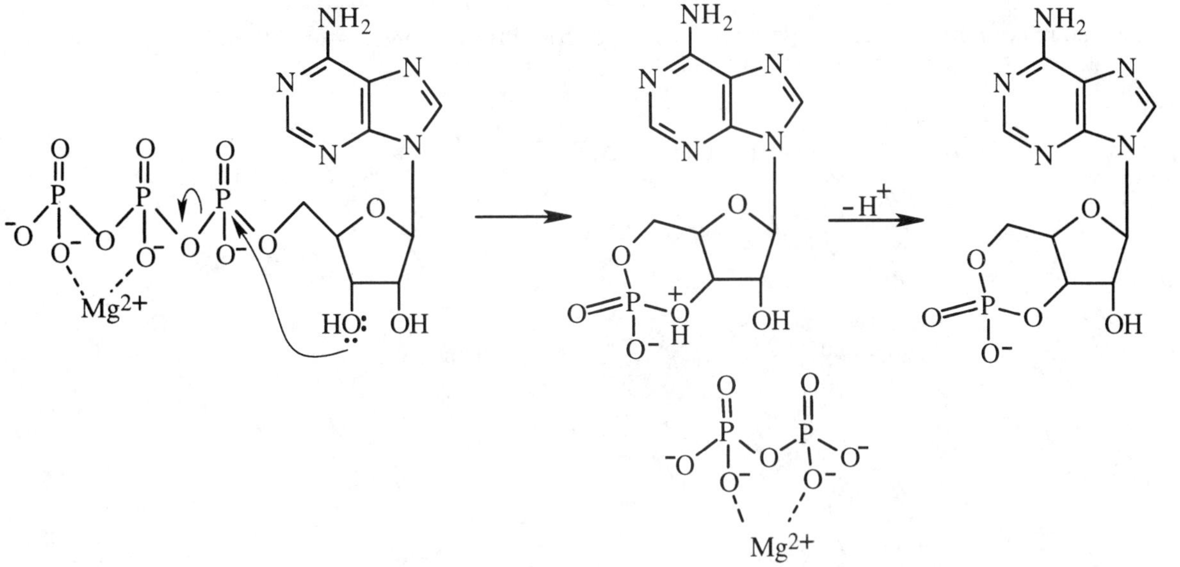

NH_2 N N N N O O O O P P P $^-$O O O O O^- O^- O^- O O Mg^{2+} HO: OH → O O P O O^- H OH $-H^+$ → O P O O O^- OH

O O P P $^-$O O O^- O^- O^- Mg^{2+}

8. The NH_2 groups could serve either as hydrogen bond acceptors, using the nonbonding electrons on nitrogen, or as hydrogen bond donors, using the hydrogen bonded to the nitrogen.

The **A**, **D**, and **A/D** designations show that the maximum number of hydrogen bonds that can form are two between thymine and adenine and three between cytosine and guanine. Notice that uracil and thymine have the same designations.

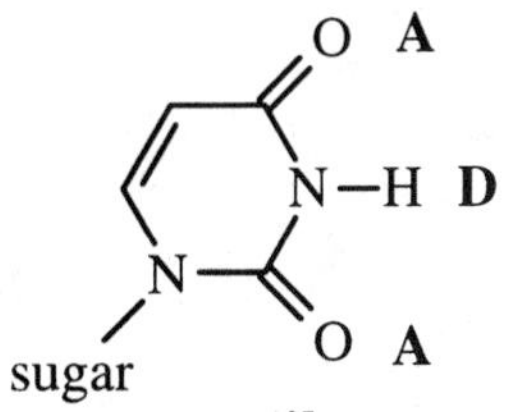

O **A** N–H **D** N O **A** sugar **uracil**

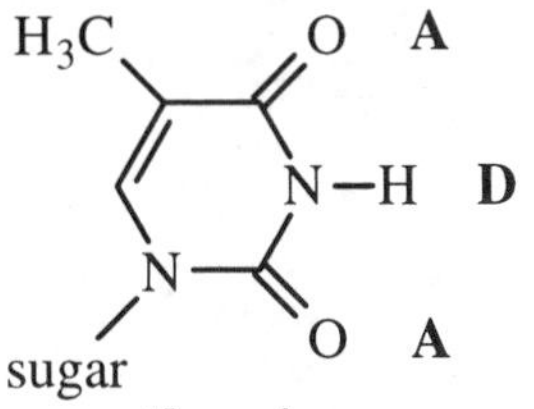

H_3C O **A** N–H **D** N O **A** sugar **thymine**

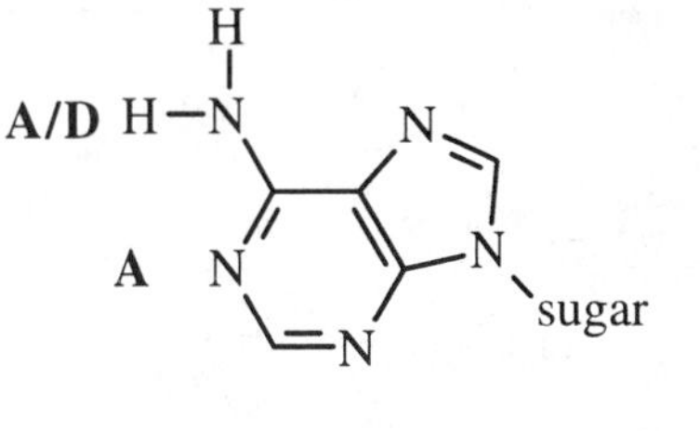

H **A/D** H–N N **A** N N N sugar **adenine**

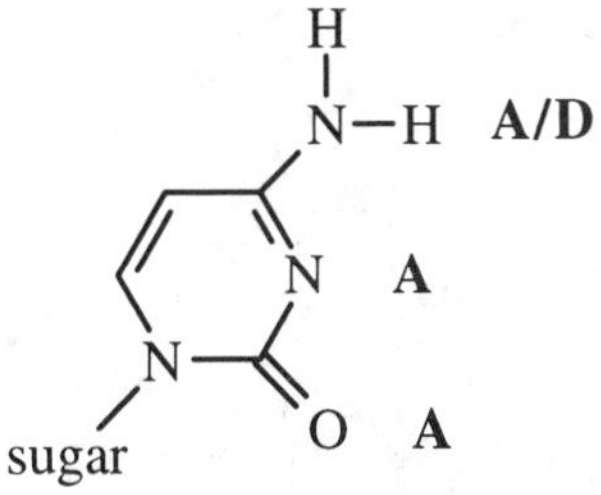

H N–H **A/D** N **A** N O **A** sugar **cytosine**

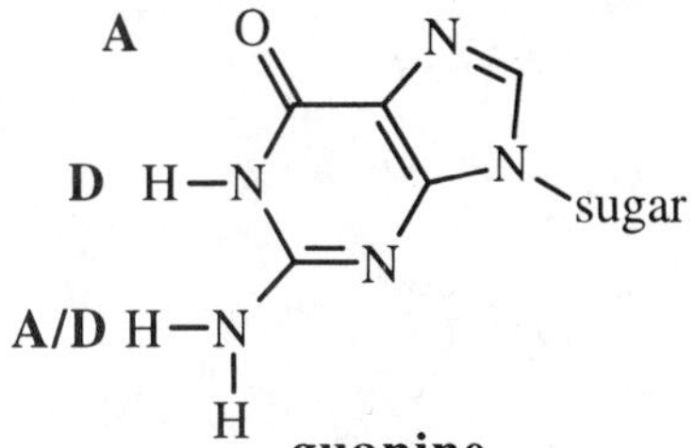

A O N **D** H–N N N sugar **A/D** H–N H **guanine**

9. If the bases existed in the enol form, the OH groups and NH_2 groups could act either as hydrogen bond acceptors or hydrogen bond donors.

The maximum number of hydrogen bonds that could form is one between thymine and adenine and two between cytosine and guanine.

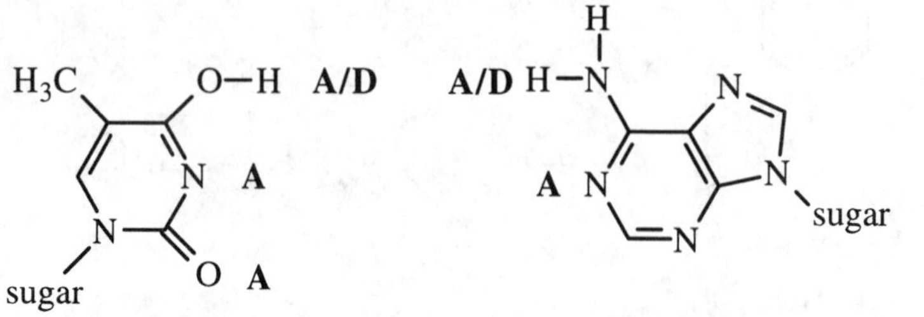

thymine

adenine

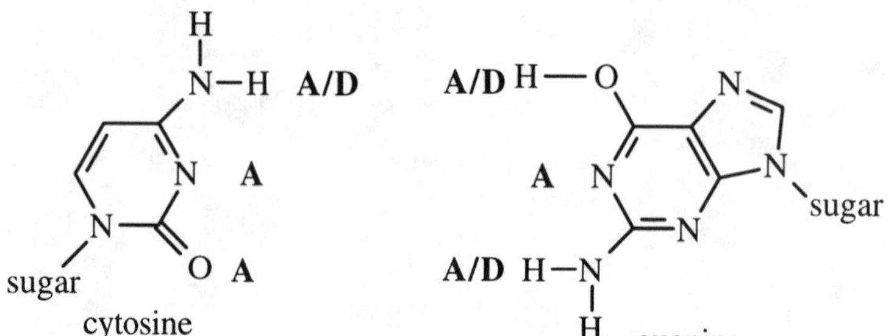

cytosine

guanine

10.

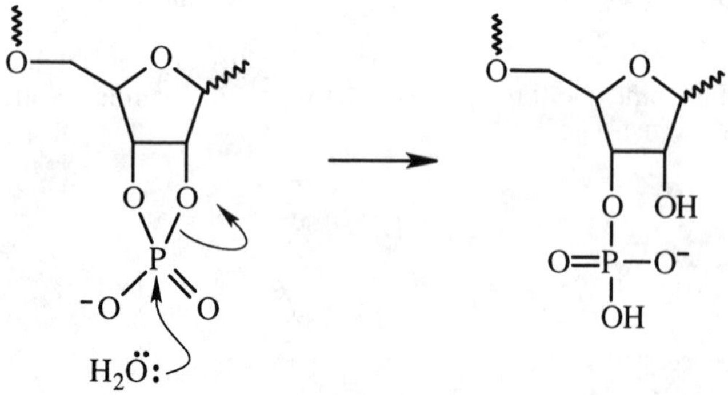

or

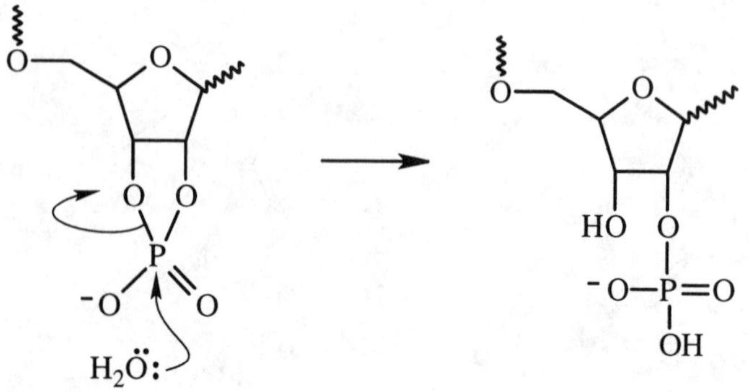

11.

a. 3'—C—C—T—G—T—T—A—G—A—C—G— 5'

b. guanine

12. **a.** Eleven base pairs per turn: (11 x 2.3 Å = 25 Å).

b. Ten base pairs per turn: (10 x 3.4 Å = 34 Å).

c. Twelve base pairs per turn: (12 x 3.8 Å = 46 Å).

13.

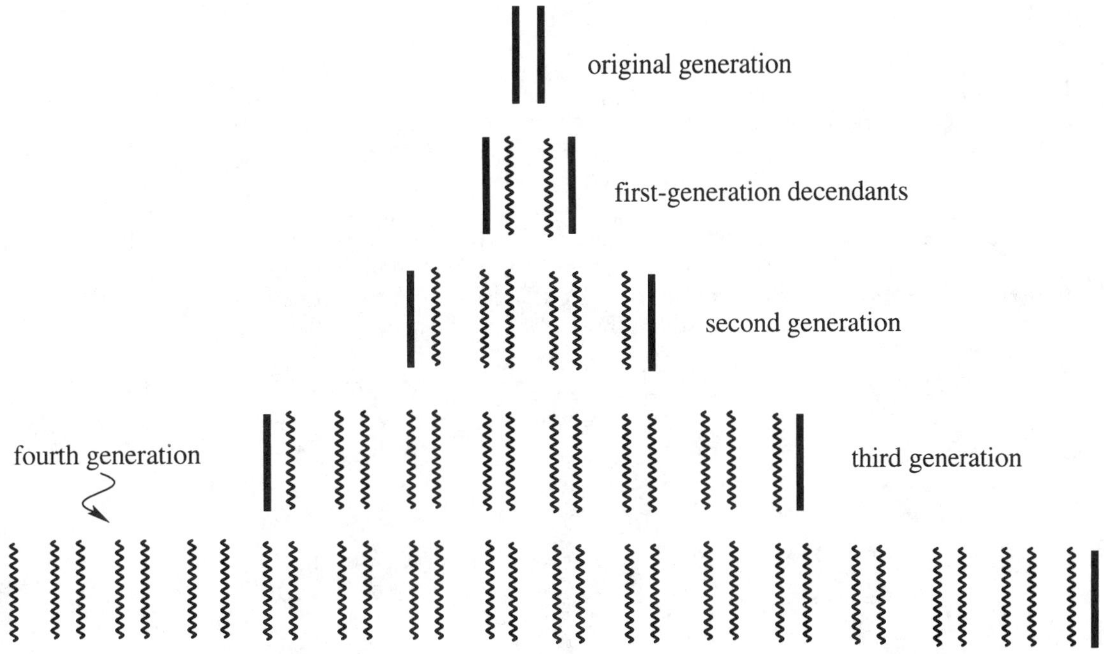

14. The rise per turn in B DNA (the form found in living organisms) is 34 Å, and there are 10 base pairs per turn:

$$3{,}100{,}000{,}000 \times 3.4 \text{ Å} = 3.1 \times 10^{9} \times 3.4 \text{ Å} = 10^{10} \text{ Å}$$

15. It requires energy to break the hydrogen bonds that hold the two chains together, so an enormous amount of energy would be required to unravel the chain completely. However, as the new nucleotides that are incorporated into the growing chain form hydrogen bonds with the parent chain, energy is released, and this energy can be used to unwind the next part of the double helix.

16. Thymine and uracil differ only in that thymine has a methyl substituent that uracil does not have. (Thymine is 5-methyluracil.) Because they both have the same groups in the same positions that can participate in hydrogen bonding, they will both call for the incorporation of the same purine. Because thymine and uracil form one hydrogen bond with guanine and two with adenine, they will incorporate adenine in order to maximize hydrogen bonding.

17. Because methionine is known to be the first base incorporated into the heptapeptide, the mRNA sequence is read beginning at AUG, since that is the only codon that codes for methionine.

Met-Asp-Pro-Val-Ile-Lys-His

18. Met-Asp-Pro-Leu-Leu-Asn

19. It does not cause protein synthesis to stop, because the sequence UAA does not occur within a triplet. The reading frame causes the triplets to be AU**U** and **AA**A.

20. The sequence of bases in the template strand of DNA specify the sequence of bases in mRNA, so the bases in the template strand and the bases in mRNA are complementary. Therefore, the sequence of bases in the sense strand of DNA are identical to the sequence of bases in mRNA, except wherever there is a U in mRNA, there is a T in the sense strand of DNA.

5'—— G-C-A-T-G-G-A-C-C-C-C-G-T-T-A-T-T-A-A-A-C-A-C —— 3'

21.

	Met	Asp	Pro	Val	Ile	Lys	His
codons	AUG	GAU	CCU	GUU	AUU	AAA	CAU
		GAC	CCC	GUC	AUC	AAG	CAC
			CCA	GUA	AUA		
			CCG	GUG			

anticodons
Note that the anticodons are stated in the 5'→ 3' direction. For example, the anticodon of AUG is stated as CAU.

codon 5' AUG 3'
anticodon 3' UAC 5'

CAU	AUC	AGG	AAC	AAU	UUU	AUG
	GUC	GGG	GAC	GAU	CUU	GUG
		UGG	UAC	UAU		
		CGG	CAC			

22.

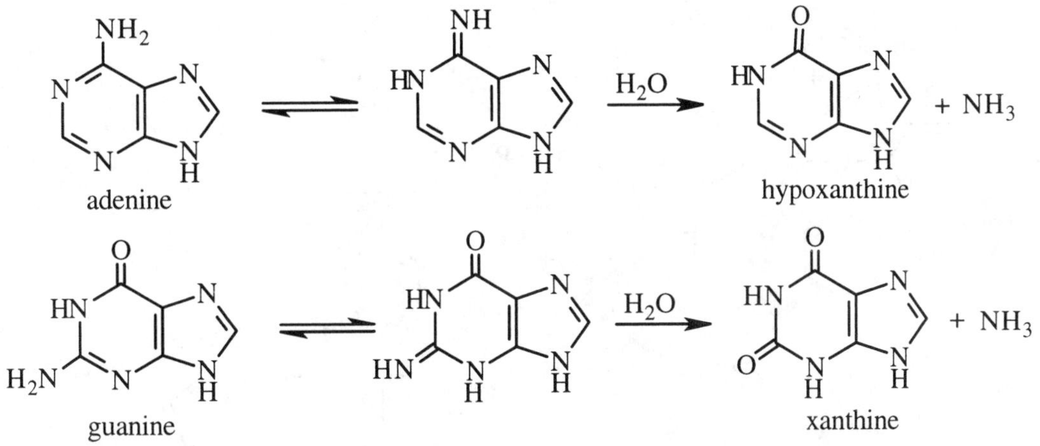

adenine $\rightleftharpoons$ $\xrightarrow{H_2O}$ hypoxanthine + NH_3

guanine $\rightleftharpoons$ $\xrightarrow{H_2O}$ xanthine + NH_3

23. Thymine does not have an amino substituent on the ring, which means that it cannot form an imine. Deamination involves hydrolyzing an imine linkage to a carbonyl group and ammonia.

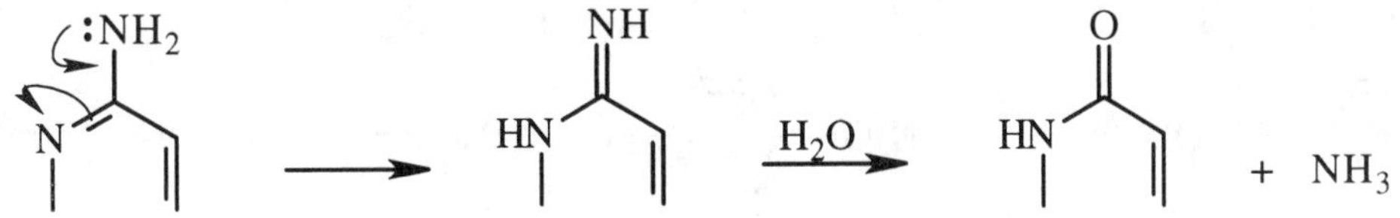

$\longrightarrow$ $\xrightarrow{H_2O}$ + NH_3

24. **a** is the only sequence that has a chance of being recognized by a restriction endonuclease because it is the only one that has the same sequence of bases in the 5' to 3' direction that the complementary strand has in the 5' to 3' direction.

ACGCGT

ACGCGT

25. All the fragments will end in "G".

^{32}P—TCCGAGGTCACTAGG

^{32}P—TCCGAGGTCACTAG

^{32}P—TCCGAGG

^{32}P—TCCGAG

^{32}P—TCCG

26.

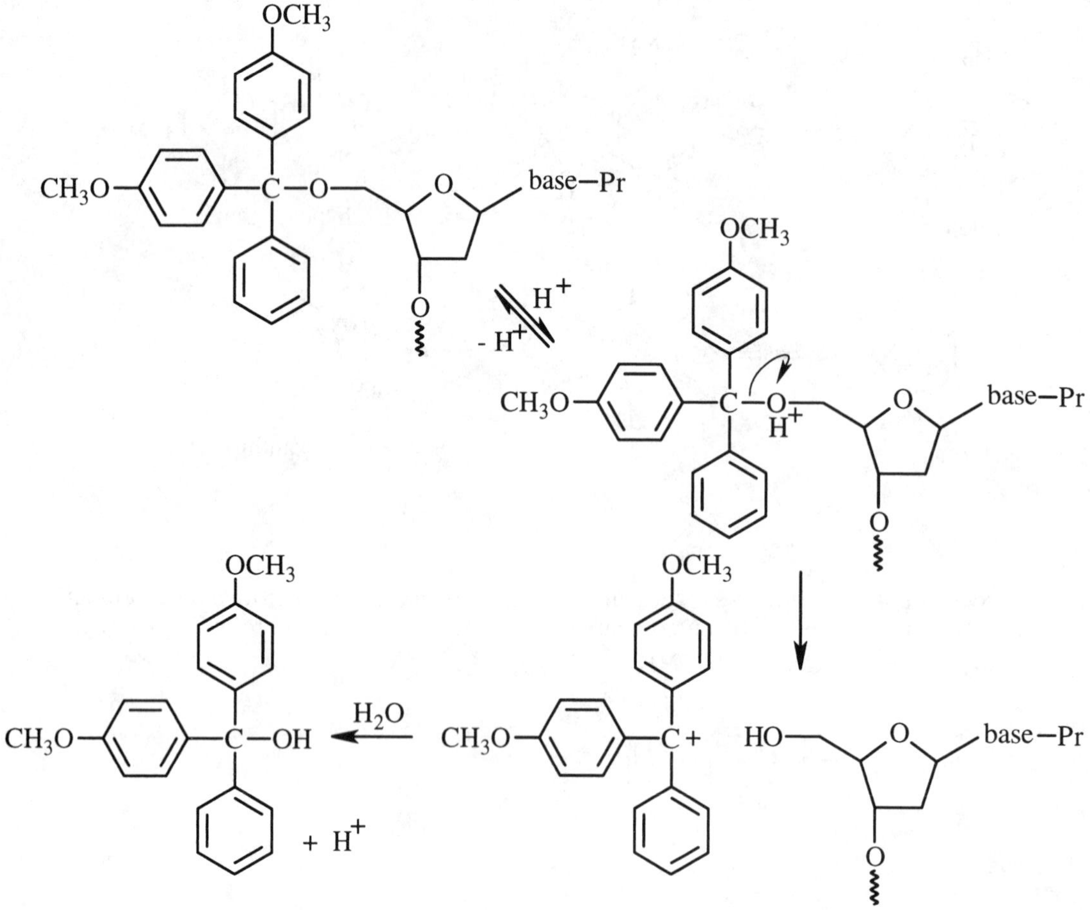

$$CH_3O-C_6H_4-C(C_6H_4OCH_3)(C_6H_5)-O-CH_2-\text{(deoxyribose)}-\text{base}-Pr \underset{-H^+}{\overset{H^+}{\rightleftharpoons}} CH_3O-C_6H_4-C(C_6H_4OCH_3)(C_6H_5)-\overset{+}{O}H-CH_2-\text{(deoxyribose)}-\text{base}-Pr$$

$$\longrightarrow CH_3O-C_6H_4-\overset{+}{C}(C_6H_4OCH_3)(C_6H_5) + HO-CH_2-\text{(deoxyribose)}-\text{base}-Pr$$

$$CH_3O-C_6H_4-\overset{+}{C}(C_6H_4OCH_3)(C_6H_5) \xrightarrow{H_2O} CH_3O-C_6H_4-C(C_6H_4OCH_3)(C_6H_5)-OH + H^+$$

27. 5-Bromouracil is incorporated into DNA in place of thymine because of their similar size. Thymine pairs with adenine via two hydrogen bonds. 5-Bromouracil exists primarily in the enol form. The enol can form only one hydrogen bond with adenine, but it can form three hydrogen bonds with guanine. Therefore, 5-bromouracil pairs with guanine. Because 5-bromouracil causes guanine to be incorporated instead of adenine into newly synthesized DNA strands, it causes mutations.

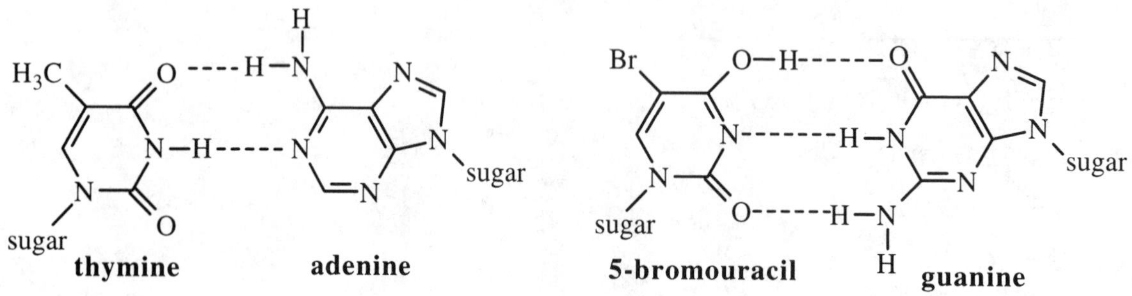

28. **a.** guanosine 3'-monophosphate

b. cytidine 5'-diphosphate

c. 2'-deoxyadenosine 5'-monophosphate

d. 2'-deoxythymidine

29. Lys-Val-Gly-Tyr-Pro-Gly-Met-Val-Val

30. The third base in each codon has some variability.

mRNA 5'-GG(UCA or G)UC(UCA or G)CG(UCA or G)GU(UCA or G)CA(U or C)GA(A or G)-3'
or AG(U or C) AG(A or G)

DNA
template 3'-CC(AGT or C)AG(AGT or C)GC(AGT or C)CA(AGT or C)GT(A or G)CT(T or C)-5'
or TC(A or G) TC(T or C)
sense 5'-GG(TCA or G)TC(TCA or G)CG(TCA or G)GT(TCA or G)CA(T or C)GA(A or G)-3'
or AG(T or C) AG(A or G)

Notice that Ser and Arg are two of three amino acids that can be specified by six different codons.

31.

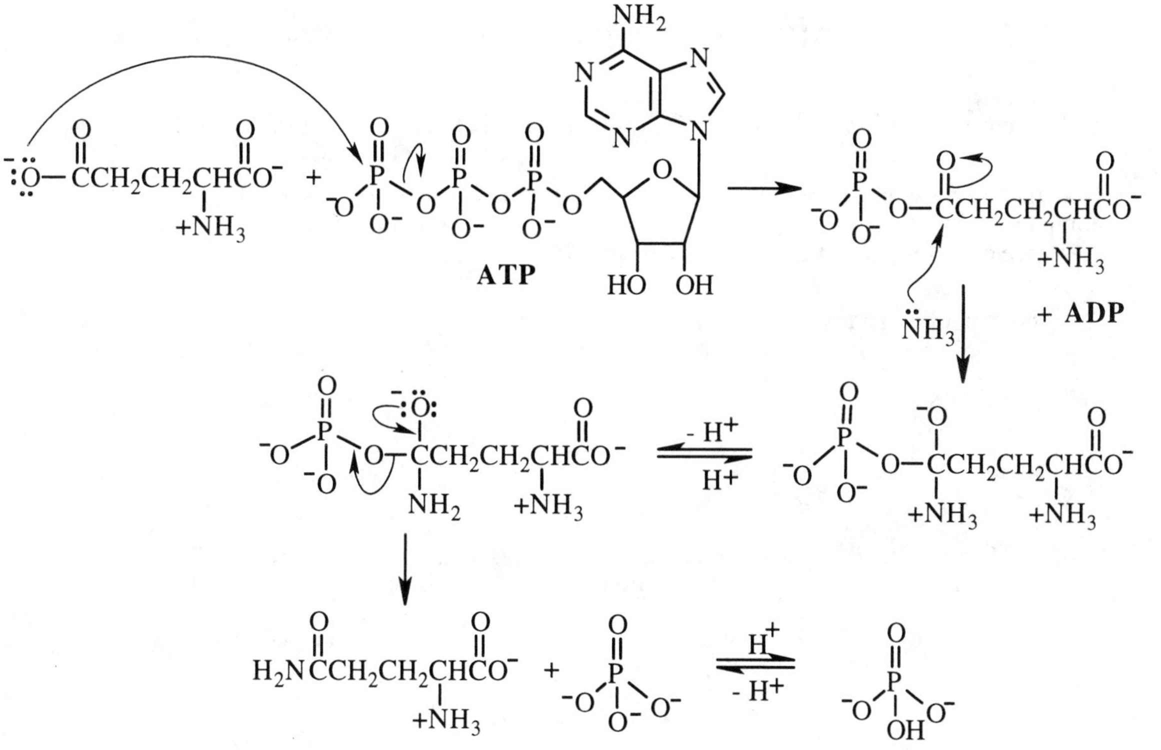

ATP

+ ADP

32.

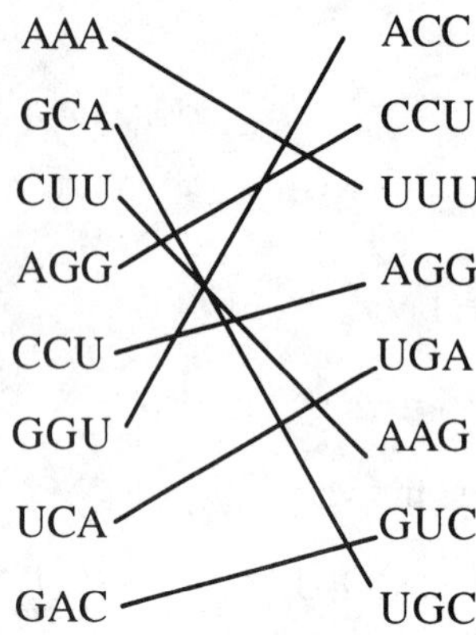

33.

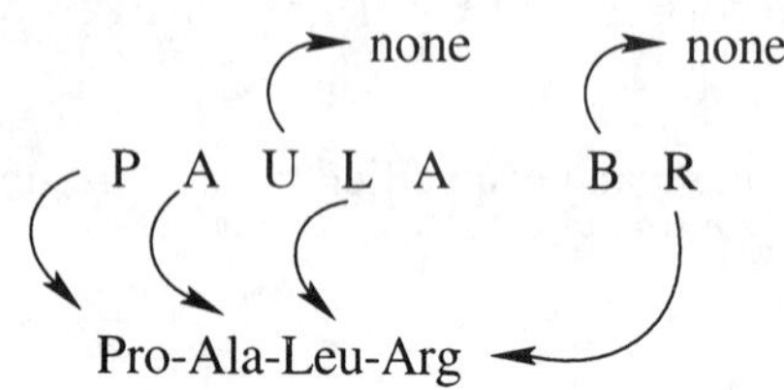

mRNA CC(UCA or G)GC(UCA or G)CU(UCA or G)CG(UCA or G)
UU(A or G) AG(A or G)

DNA (sense strand) CC(TCA or G)GC(TCA or G)CT(TCA or G)CG(TCA or G)
TT(A or G) AG(A or G)

Note that because mRNA is complementary to the template strand of DNA, which is complementary to the sense strand, the sense strand of DNA and mRNA have the same sequence of bases (except DNA has a T where RNA has a U).

Also note that Leu and Arg are specified by six codons.

34. **a.** CC and GG **c.** CA and TG

CA and TG are formed in equal amounts, since A pairs with T and C pairs with G. (Remember that the dinucleotides are written in the 5' to 3' direction.)

```
5' CA 3'
   ¦¦
3' GT 5'
```

35. The number of different possible codons using four nucleotides is $(4)^n$ where n is the number of letters (nucleotides) in the code.

for a two-letter code: $(4)^2 = 16$
for a three-letter code: $(4)^3 = 64$
for a four-letter code: $(4)^4 = 256$

Since there are 20 amino acids that must be specified, a two-letter code would not provide an adequate number of codons. A three-letter code provides enough codes for all the amino acids and also provides the necessary stop codons. A four-letter code provides many more codes than would be needed.

36. In the first step of the reaction, the imidazole ring of one histidine acts as a general-base catalyst, removing a proton from the 2-OH group to make it a better nucleophile. The imidazole ring of the other histidine acts as a general-acid catalyst, protonating the leaving group to make it a weaker base and therefore a better leaving group. In the second step of the reaction the roles of the two imidazole rings are reversed.

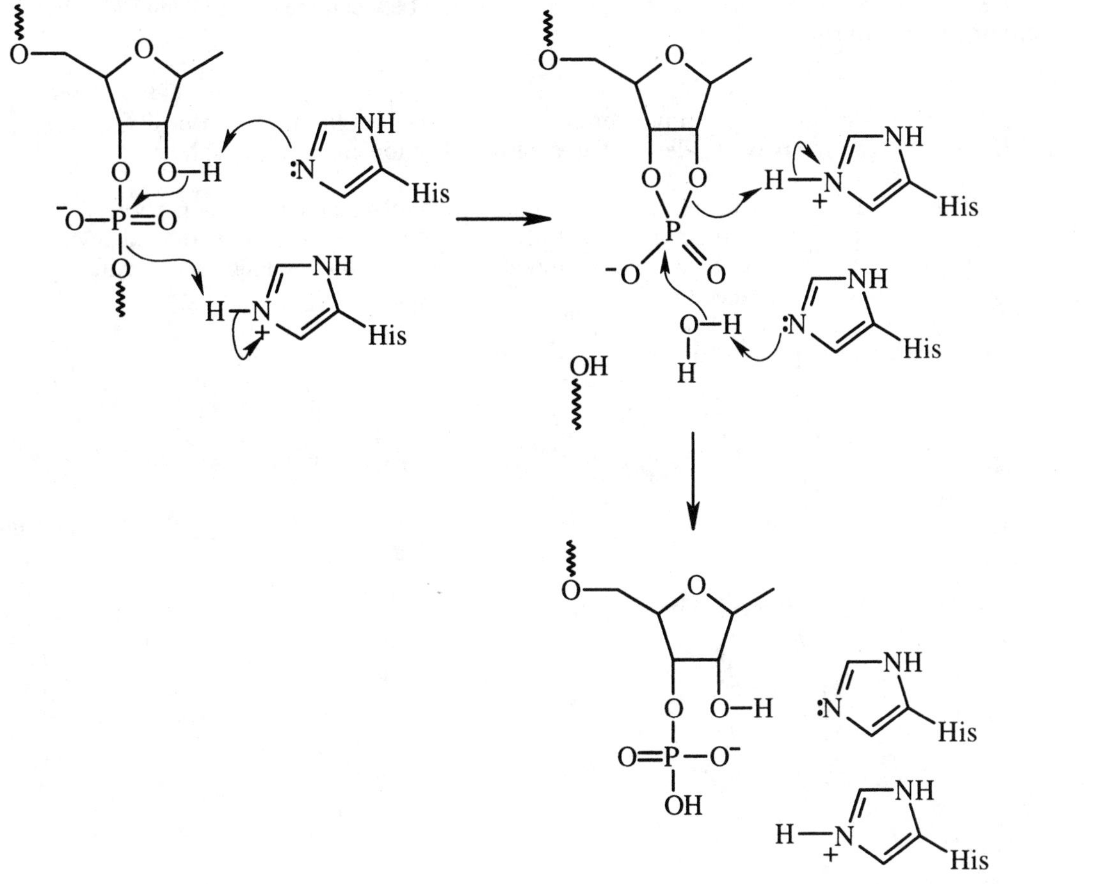

37. The normal and mutant peptides would have the following base sequence in their mRNA.

normal: CA(AG) UA(UC) GG(UCAG) AC(UCAG) CG(UCAG) UA(UC) GU(UCAG)

mutant: CA(AG) UC(UCAG) GA(AG) CC(UCGA) GG(UCGA) AC(UCAG)

a. The middle nucleotide (A) in the second triplet was deleted. This means that an A was deleted in the sense strand of DNA or a T was deleted in the template strand of DNA.

b. The mRNA for the mutant peptide has an unused 3'-terminal two-letter code, U(UCGA). The last amino acid in the octapeptide of the normal fragment is leucine, so its last triplet is UU(AG) or CU(UCAG).

This means that the triplet for the last amino acid in the mutant is U(UCGA)(UC) and that the last amino acid in the mutant is one of the following: Phe, Ser, Tyr, or Cys.

38. Because the compound that will react in the second step with the activated carboxylic acid group is excluded from the incubation mixture, the reaction between the carboxylate ion and ATP will come to equilibrium.

If radioactively labeled pyrophosphate is put into the incubation mixture, ATP will become radioactive if the mechanism involves attack on the α-phosphorus, because ATP is formed from the reverse reaction between the activated carboxylate ion and pyrophosphate.

ATP will not become radioactive if the mechanism involves attack on the β-phosphorus because ATP is formed from reaction between the activated carboxylate ion and AMP. (In other words, because pyrophosphate is not a product of the reaction, it cannot become incorporated into ATP in the reverse reaction.)

attack on the α-phosphorus

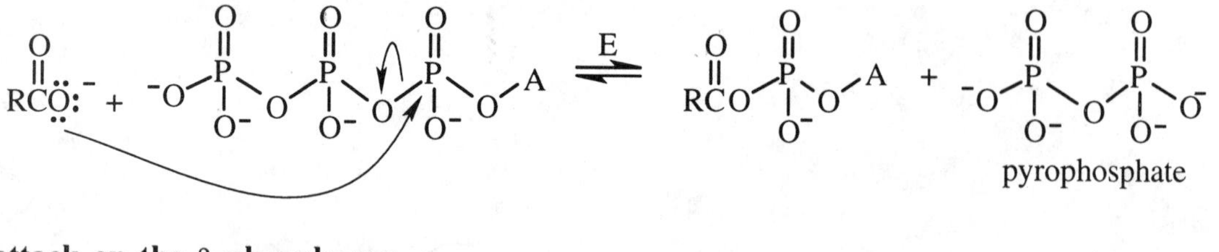

pyrophosphate

attack on the β-phosphorus

AMP

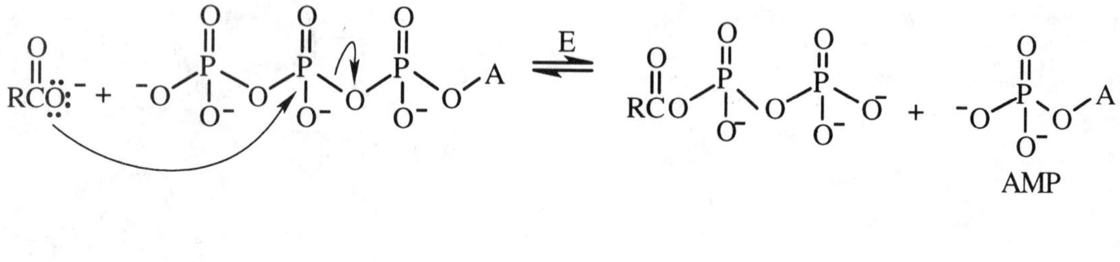

39. If radioactive AMP is added to the reaction mixture, the results will be opposite. If the mechanism involves attack on the α-phosphorus, ATP will not become radioactive. If the mechanism involves attack on the β-phosphorus, ATP will become radioactive.

40. If deamination does not occur, the mRNA sequence will be:

AUG-UCG-CUA-AUC which will code for the following tetrapeptide
Met - Ser - Leu - Ile

Deamination of a cytosine results in a uracil.
If the cytosines are deaminated, the mRNA sequence will be:

AUG-UUG-UUA-AUU which will code for the following tetrapeptide
Met - Leu - Leu - Ile

The only cytosine that will change the particular amino acid that is incorporated into the peptide is the first one. Therefore, this is the cytosine that could cause the most damage to an organism if it were deaminated.

41. In an acidic environment, nitrite ion is protonated to nitrous acid. We have seen that the nitrosonium ion is formed from nitrous acid (Section 15.10).

$$Na^+\ {}^-O{-}N{=}O \underset{-H^+}{\overset{H^+}{\rightleftharpoons}} HO{-}N{=}O \underset{-H^+}{\overset{H^+}{\rightleftharpoons}} H\overset{+}{O}(H){-}N{=}O \rightleftharpoons H_2O + \overset{+}{N}{=}O$$

sodium nitrite; nitrosonium ion

The nitrosonium ion reacts with a primary amino group to form a diazonium ion, which can be displaced by water (Section 15.10).

cytosine $\xrightarrow{\overset{+}{N}=O}$ (diazonium, $\overset{+}{N}{\equiv}N$) $\xrightarrow{H_2O}$ (OH) $\xrightarrow{\text{ketonization}}$ uracil

42. The smaller subunit of a prokaryotic ribosome (30S) contains a binding site for the growing peptide chain and a binding site for the next amino acid to be incorporated into the chain.

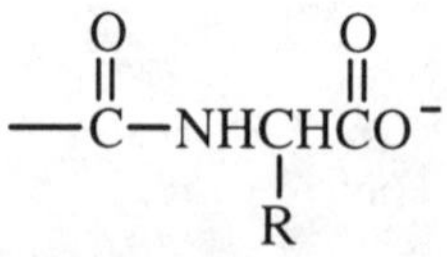

$$-\overset{O}{\overset{\|}{C}}-NH\underset{R}{\underset{|}{C}}H\overset{O}{\overset{\|}{C}}O^- \qquad H_2N\underset{R}{\underset{|}{C}}H\overset{O}{\overset{\|}{C}}O^-$$

peptide binding site 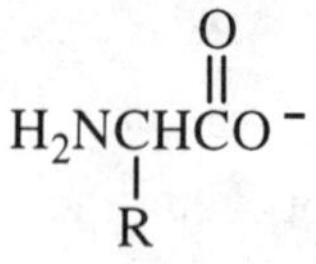 amino acid binding site

All peptide bonds are formed by the reaction of an amino acid with a peptide, except formation of the first peptide bond, which has to be formed by the reaction of two amino acids. Therefore, in the synthesis of the first peptide bond, one of the amino acids has to be a peptide that will fit into the peptide binding site.

The formyl group of *N*-formylmethionine will provide the peptide group that will be recognized by the peptide binding site, and the second amino acid will be bound in the amino acid binding site.

$$H\overset{O}{\overset{\|}{C}}-NH\underset{CH_2CH_2SCH_3}{\underset{|}{C}}H\overset{O}{\overset{\|}{C}}O^-$$

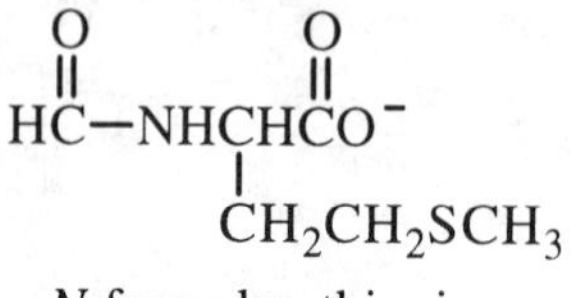

N-formylmethionine

CHAPTER 26
Synthetic Polymers

Important Terms

addition polymer (chain-growth polymer) — made by adding monomers to the growing end of a chain.

alpha olefin — a monosubstituted olefin.

alternating copolymer — a copolymer in which two monomers alternate.

anionic polymerization — chain-growth polymerization where the initiator is a nucleophile; the propagation site, therefore, is an anion.

aramide — an aromatic polyamide.

atactic polymer — a polymer in which the substituents are randomly oriented on the extended carbon chain.

biodegradable polymer — a polymer that can be broken into small segments by an enzyme-catalyzed reaction.

biopolymer — a polymer that is synthesized in nature.

block copolymer — a copolymer in which there are blocks of each kind of monomer.

cationic polymerization — chain-growth polymerization where the initiator is an electrophile; the propagation site, therefore, is a cation.

chain-growth polymer (addition polymer) — made by adding monomers to the growing end of a chain.

chain transfer — a growing polymer chain reacts with a molecule XY in a manner that allows X to terminate the chain, leaving behind Y· to initiate a new chain.

condensation polymer (step-growth polymer) — made by combining two molecules while removing a small molecule (usually water or an alcohol).

conducting polymer — a polymer that can conduct electricity down its backbone.

copolymer — a polymer formed using two or more different monomers.

cross-linking — connecting polymer chains by intermolecular bond formation.

crystallites — regions of a polymer in which the chains are highly ordered.

elastomer — a polymer that can stretch and then revert back to its original shape.

epoxy resin — formed by mixing a low molecular weight prepolymer with a compound that forms a cross-linked polymer.

graft copolymer — a copolymer that contains branches of a polymer of one monomer grafted onto the backbone of a polymer made from another monomer.

head-to-tail addition	the head of one molecule is added to the tail of another molecule.
homopolymer	a polymer that contains only one kind of monomer.
isotactic polymer	a polymer in which all the substituents are on the same side of the fully extended carbon chain.
living polymer	a nonterminated chain-growth polymer that remains active. Therefore, the polymerization reaction can continue upon addition of more monomer.
materials science	the science of creating new materials to take the place of known materials such as metal, glass, wood, cardboard, and paper.
monomer	a repeating unit in a polymer.
oriented polymer	a polymer obtained by stretching out polymer chains and putting them back together in a parallel fashion.
plasticizer	an organic molecule that dissolves in a polymer and allows the polymer chains to slide by each other.
polyamide	a polymer with many amide groups.
polycarbonate	a step-growth polymer in which the dicarboxylic acid is carbonic acid.
polyester	a polymer with many ester groups.
polymer	a large molecule made by linking monomers together.
polymer chemistry	the field of chemistry that deals with synthetic polymers; part of the larger discipline known as materials science.
polymerization	the process of linking up monomers to form a polymer.
polyurethane	a polymer with many urethane groups.
propagating site	the reactive end of a chain-growth polymer.
radical polymerization	chain-growth polymerization where the initiator is a radical; the propagation site, therefore, is a radical.
random copolymer	a copolymer with a random distribution of monomers.
ring-opening polymerization	a chain-growth polymerization that involves opening the ring of the monomer.
step-growth polymer (condensation polymer)	made by combining two molecules while removing a small molecule (usually water or an alcohol).
syndiotactic polymer	a polymer in which the substituents regularly alternate on both sides of the fully extended carbon chain.
synthetic polymer	a polymer that is not synthesized in nature.

thermoplastic polymer a polymer that has both ordered crystalline regions and amorphous non-crystalline regions.

thermosetting polymer cross-linked polymers that, after they are hardened, cannot be re-melted by heating.

urethane a compound with a carbonyl group that is both an amide and an ester.

vinyl polymer a polymer in which the monomer is ethylene or a substituted ethylene.

vulcanization increasing the flexibility of rubber by heating it with sulfur.

Ziegler-Natta catalyst an aluminum-titanium initiator that controls the stereochemistry of a polymer.

Solutions to Problems

1.

a. $CH_2{=}CHCl$

b. 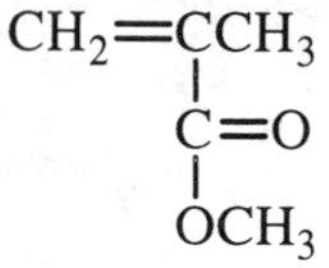 $CH_2{=}C(CH_3)C({=}O)OCH_3$

c. $CF_2{=}CF_2$

2. Poly(vinyl chloride) would be more apt to contain head-to-head linkages because a chloro substituent is less able (compared with a phenyl substituent) to stabilize the growing end of the polymer chain by resonance.

3.

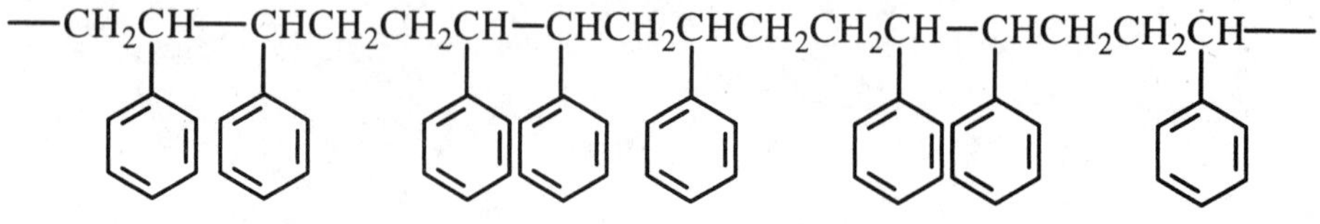

$$-CH_2CH(C_6H_5)-CH(C_6H_5)CH_2CH_2CH(C_6H_5)-CH(C_6H_5)CH_2CH(C_6H_5)CH_2CH_2CH(C_6H_5)-CH(C_6H_5)CH_2CH_2CH(C_6H_5)-$$

4.

$$HO{-}OH \longrightarrow 2\,HO\cdot$$

$$HO\cdot + CH_2{=}CHCl \longrightarrow HO{-}CH_2CH(Cl)\cdot$$

$$HO{-}CH_2CH(Cl)\cdot + CH_2{=}CHCl \longrightarrow HO{-}CH_2CH(Cl)CH_2CH(Cl)\cdot$$

$$HO{-}CH_2CH(Cl)CH_2CH(Cl)\cdot + CH_2{=}CHCl \longrightarrow HO{-}CH_2CH(Cl)CH_2CH(Cl)CH_2CH(Cl)\cdot$$

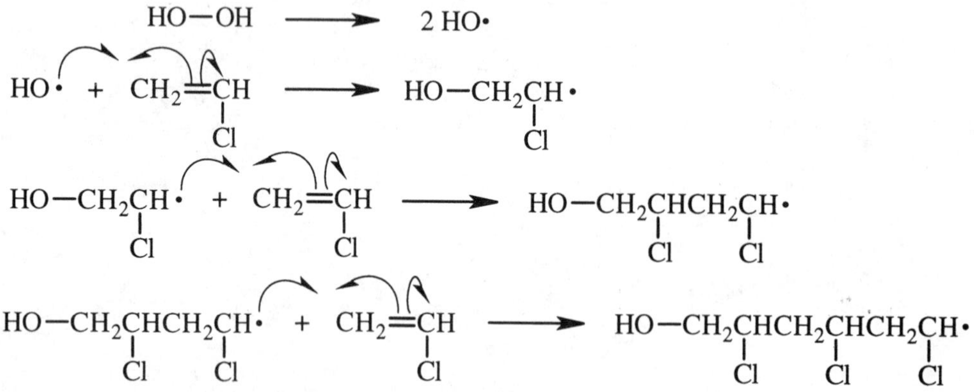

5. Since branching increases the flexibility of the polymer, beach balls are made from more highly branched polyethylene.

6.

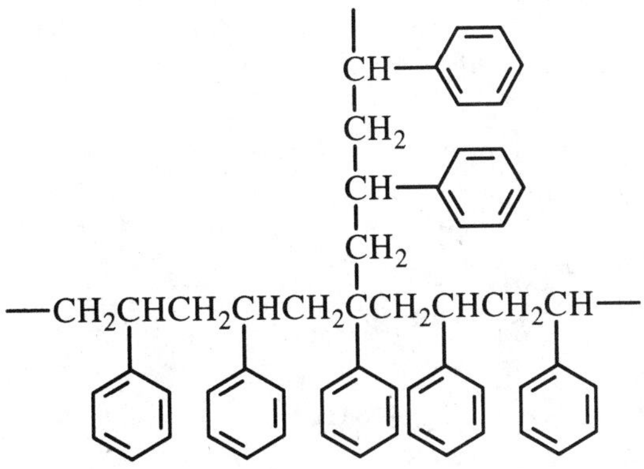

$$-CH_2CHCH_2CHCH_2CCH_2CHCH_2CH-$$

(the central C bears a branch chain $-CH_2-CH(C_6H_5)-CH_2-CH(C_6H_5)-$; each CH in the main chain and the central C bear phenyl rings)

7. Decreasing ability to undergo cationic polymerization is in the same order as decreasing stability of the carbocation intermediate. (Electron donation increases the stability of the carbocation.)

a. $CH_2{=}CH-C_6H_4-OCH_3$ > $CH_2{=}CH-C_6H_4-CH_3$ > $CH_2{=}CH-C_6H_4-NO_2$

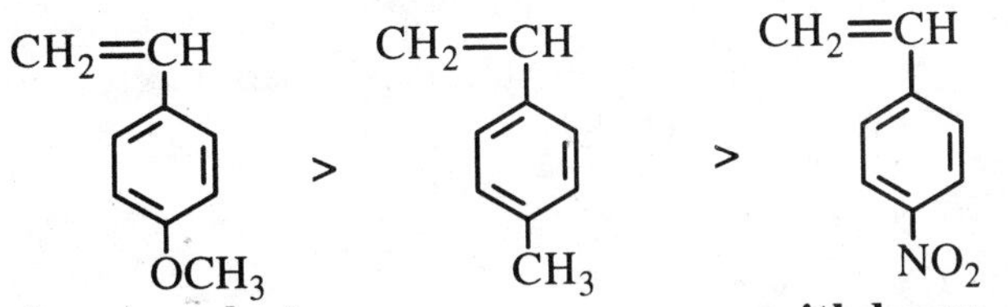

donates electrons by resonance (OCH$_3$) … **withdraws electrons by resonance** (NO$_2$)

b. $CH_2{=}CHOCCH_3$ (with C=O) > $CH_2{=}CHCH_3$ > $CH_2{=}CHCOCH_3$ (with C=O)

donates electrons by resonance … **withdraws electrons by resonance**

c. $CH_2{=}CCH_3$ (with phenyl on C) > $CH_2{=}CH$ (with phenyl on CH)

A tertiary benzylic carbocation is more stable than a secondary benzylic carbocation.

8. Decreasing ability to undergo anionic polymerization is in the same order as decreasing stability of the carbanion intermediate. (Electron withdrawal increases the stability of the carbanion.)

a. $CH_2{=}CH$–C_6H_4–NO_2 > $CH_2{=}CH$–C_6H_4–CH_3 > $CH_2{=}CH$–C_6H_4–OCH_3

withdraws electrons by resonance (NO_2)

donates electrons by resonance (OCH_3)

b. $CH_2{=}CHC{\equiv}N$ > $CH_2{=}CHCl$ > $CH_2{=}CHCH_3$

withdraws electrons by resonance ($CH_2{=}CHC{\equiv}N$)

9. In anionic polymerization, nucleophilic attack occurs at the less substituted carbon because it is the less sterically hindered. In cationic polymerization, nucleophilic attack occurs at the more substituted carbon, because the ring opens to give the more stable partial carbocation.

(Propylene oxide, CH_3-substituted epoxide)

position of nucleophilic attack in cationic polymerization (more substituted carbon)

position of nucleophilic attack in anionic polymerization (less substituted carbon)

10.

a. $R\ddot{O}:^-$ + (2,2-dimethyloxirane) ⟶ $RO{-}CH_2C(CH_3)_2O^-$

$RO{-}CH_2C(CH_3)_2O:^-$ + (2,2-dimethyloxirane) ⟶ $RO{-}CH_2C(CH_3)_2OCH_2C(CH_3)_2O^-$

$RO{-}CH_2C(CH_3)_2OCH_2C(CH_3)_2O:^-$ + (2,2-dimethyloxirane) ⟶ $RO{-}CH_2C(CH_3)_2OCH_2C(CH_3)_2OCH_2C(CH_3)_2O^-$

b. [reaction scheme: protonation of 2,2-dimethyloxirane by H^+; ring-opening attack by a second epoxide gives $HOCH_2C(CH_3)_2$–$\overset{+}{O}$ oxiranium; attack by a third epoxide gives $HOCH_2C(CH_3)_2OCH_2C(CH_3)_2$–$\overset{+}{O}$ oxiranium]

11.

a. $CH_2{=}C(CH_3)CH_3$ + BF_3

b. $CH_2{=}CH$–N(pyrrolidine) + BF_3

c. ethylene oxide + CH_3O^-

d. $CH_2{=}CHCOCH_3$ (C=O) + BuLi

12. 3,3-Dimethyloxacyclobutane undergoes cationic polymerization by the following mechanism.

[mechanism: protonated 3,3-dimethyloxetane attacked by 3,3-dimethyloxetane gives oxetanium–$CH_2C(CH_3)_2CH_2OH$; further attack gives oxetanium–$CH_2C(CH_3)_2CH_2OCH_2C(CH_3)_2CH_2OH$; further attack gives oxetanium–$CH_2C(CH_3)_2CH_2OCH_2C(CH_3)_2CH_2OCH_2C(CH_3)_2CH_2OH$]

13.

14. Notice that a branch occurs as a result of nucleophilic attack on C-3 instead of on C-1.

$R\ddot{O}:^- \quad CH_2{=}CH{-}CH{=}CH_2 \longrightarrow RO{-}CH_2CH{=}CH\ddot{C}H_2^-$

$RO{-}CH_2CH{=}CH\ddot{C}H_2^- \quad CH{=}CH_2(-CH{=}CH_2) \longrightarrow RO{-}CH_2CH{=}CHCH_2\ddot{C}H(CH{=}CH_2)\ddot{C}H_2^-$

$RO{-}CH_2CH{=}CHCH_2CH(CH{=}CH_2)\ddot{C}H_2^- \quad CH_2{=}CH{-}CH{=}CH_2 \longrightarrow RO{-}CH_2CH{=}CHCH_2CH(CH{=}CH_2)CH_2CH_2CH{=}CH\ddot{C}H_2^-$

This polymerization also could take place by a radical mechanism or by a cationic mechanism.

15.

a. $-NHCH_2CH_2CH_2\overset{O}{\overset{\|}{C}}NHCH_2CH_2CH_2\overset{O}{\overset{\|}{C}}-$

b. (2-pyrrolidinone: five-membered ring with C=O and NH)

c. $-NH(CH_2)_4NH\overset{O}{\overset{\|}{C}}CH_2CH_2\overset{O}{\overset{\|}{C}}NH(CH_2)_4NH\overset{O}{\overset{\|}{C}}CH_2CH_2\overset{O}{\overset{\|}{C}}-$

16.

$$-\overset{O}{\overset{\|}{C}}(CH_2)_4\overset{O}{\overset{\|}{C}}\left[NH(CH_2)_6NH-\overset{O}{\overset{\|}{C}}(CH_2)_4\overset{O}{\overset{\|}{C}}\right]_n NH(CH_2)_6NH-$$

$$\downarrow H_2SO_4$$

$$n\ HO\overset{O}{\overset{\|}{C}}(CH_2)_4\overset{O}{\overset{\|}{C}}OH + n\ H_3\overset{+}{N}(CH_2)_6\overset{+}{N}H_3$$

17. They hydrolyze to give monomers of dicarboxylic acids and diols.

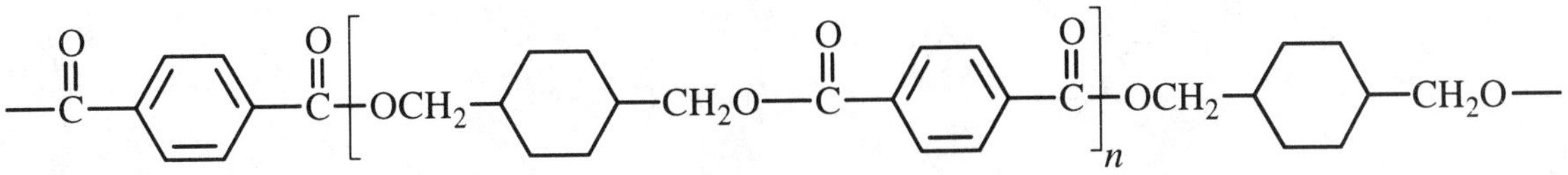

$$-\overset{O}{\overset{\|}{C}}-C_6H_4-\overset{O}{\overset{\|}{C}}\left[-OCH_2-C_6H_{10}-CH_2O-\overset{O}{\overset{\|}{C}}-C_6H_4-\overset{O}{\overset{\|}{C}}\right]_n-OCH_2-C_6H_{10}-CH_2O-$$

Kodel

$\xrightarrow{NaOH}$

$$n\ {}^{-}O\overset{O}{\overset{\|}{C}}-C_6H_4-\overset{O}{\overset{\|}{C}}O^{-} + n\ HOCH_2-C_6H_{10}-CH_2OH$$

$$-\overset{O}{\overset{\|}{C}}-C_6H_4-\overset{O}{\overset{\|}{C}}\left[-OCH_2CH_2O-\overset{O}{\overset{\|}{C}}-C_6H_4-\overset{O}{\overset{\|}{C}}\right]_n-OCH_2CH_2O-$$

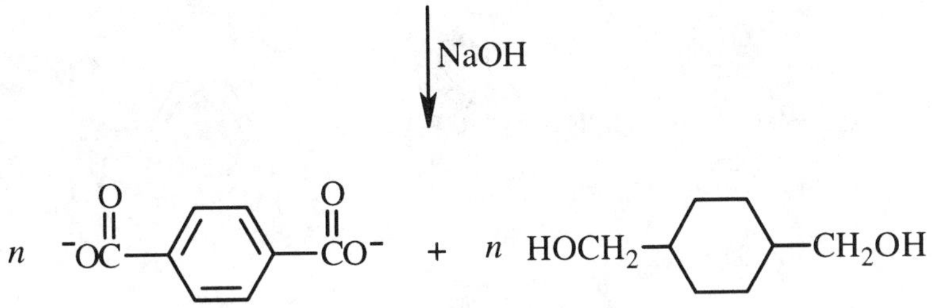

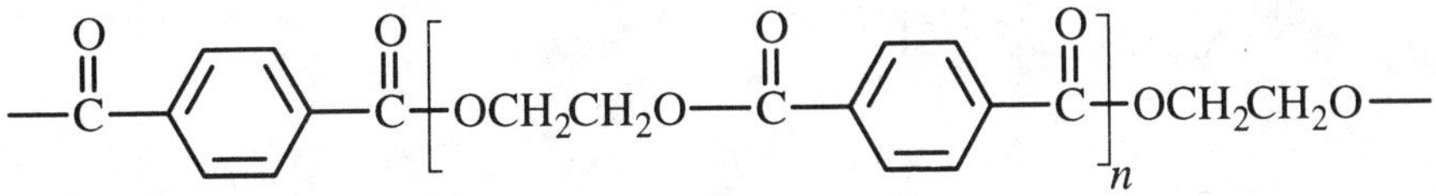

Dacron

$\xrightarrow{NaOH}$

$$n\ {}^{-}O\overset{O}{\overset{\|}{C}}-C_6H_4-\overset{O}{\overset{\|}{C}}O^{-} + n\ HOCH_2CH_2OH$$

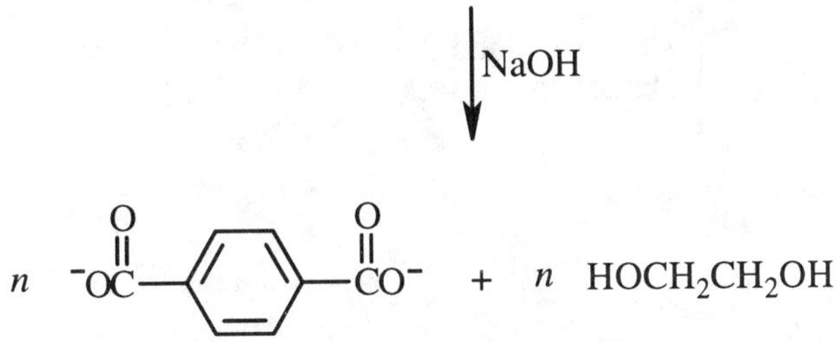

18.

a.

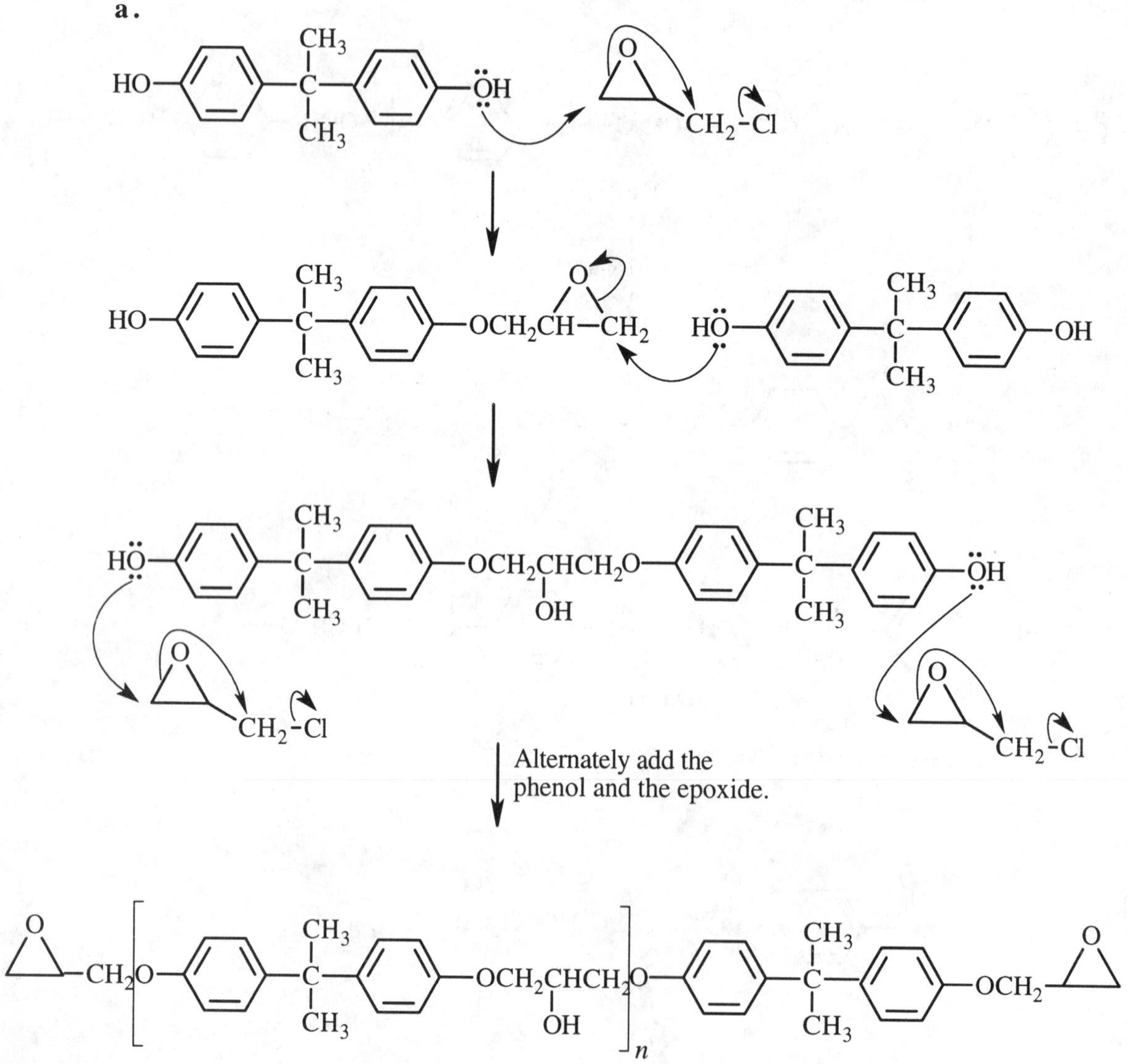

Alternately add the phenol and the epoxide.

b.

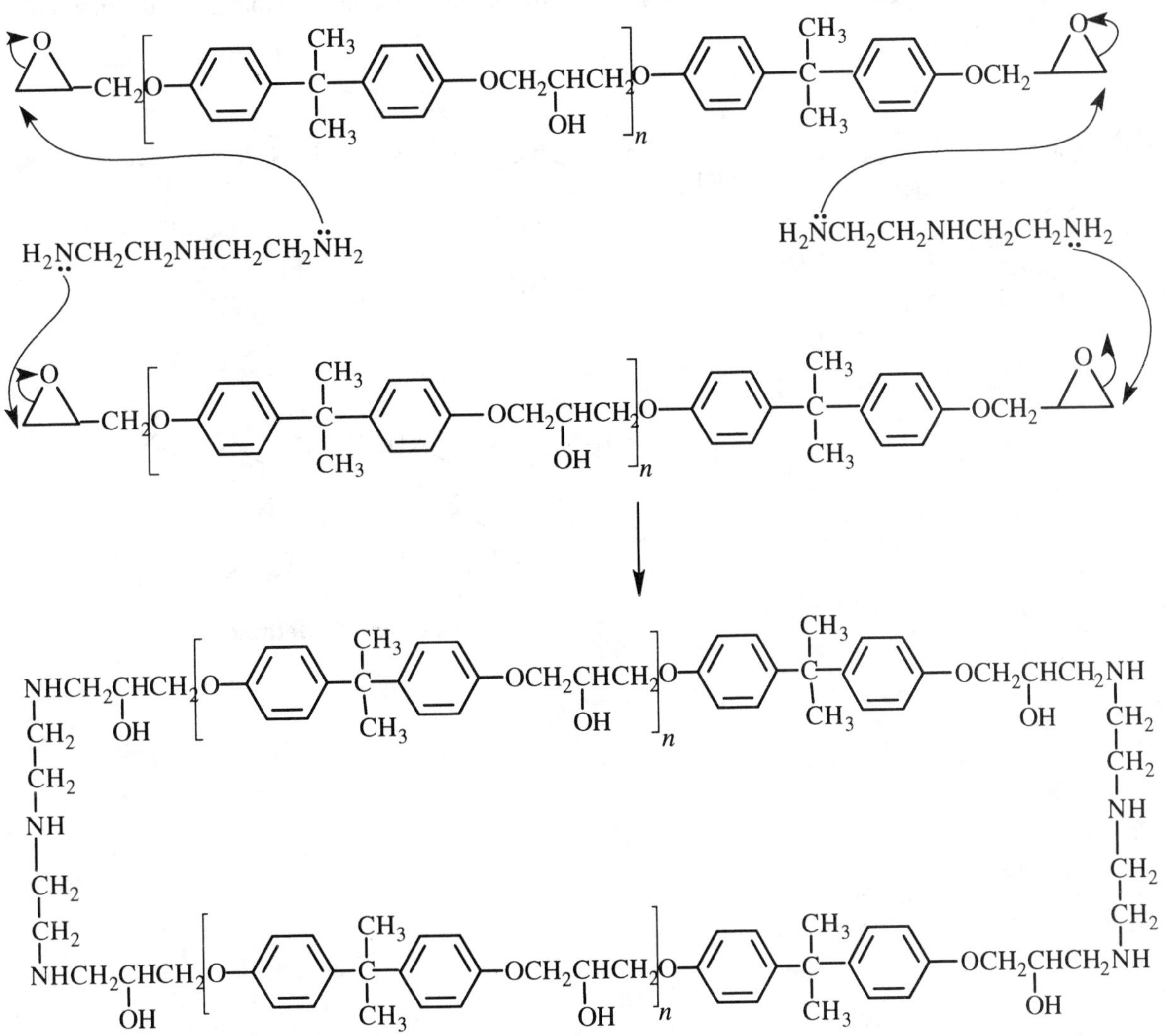

$H_2NCH_2CH_2NHCH_2CH_2NH_2$ $H_2NCH_2CH_2NHCH_2CH_2NH_2$

19.

Glycerol cross-links the polymer chains.

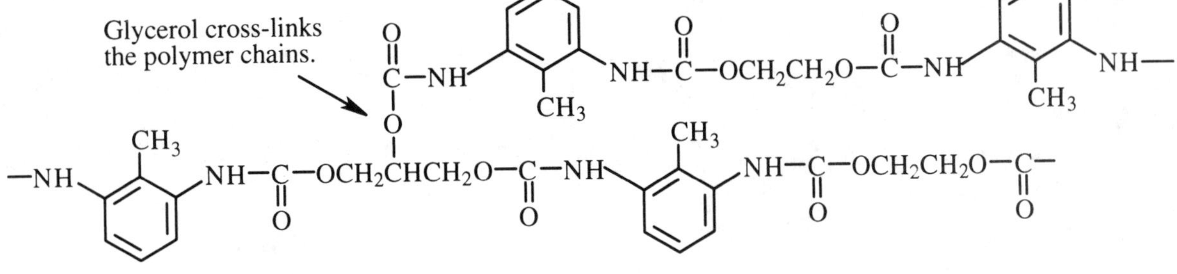

20. Formation of an imine between formaldehyde and one amino group, followed by reaction of the imine with a second amino group, accounts for formation of the linkage that holds the monomers together.

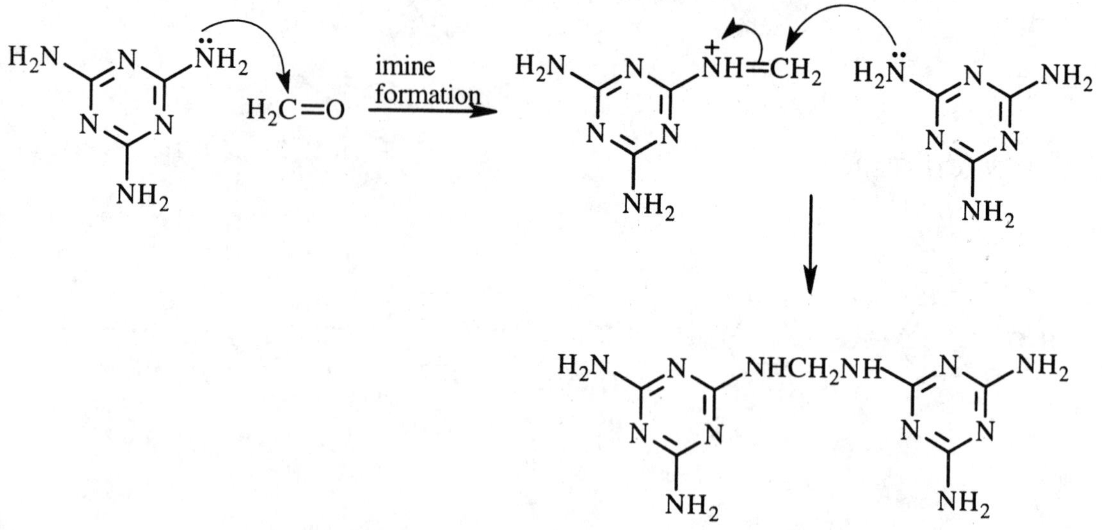

H₂N, N, NH₂, N, N, NH₂ + $H_2C{=}O$ —imine formation→ H₂N, N, $\overset{+}{N}H{=}CH_2$, N, N, NH₂ + H₂N, N, NH₂, N, N, NH₂

↓

H₂N, N, NHCH₂NH, N, NH₂, N, N, NH₂, N, N, NH₂

a dimer of Melmac

21.

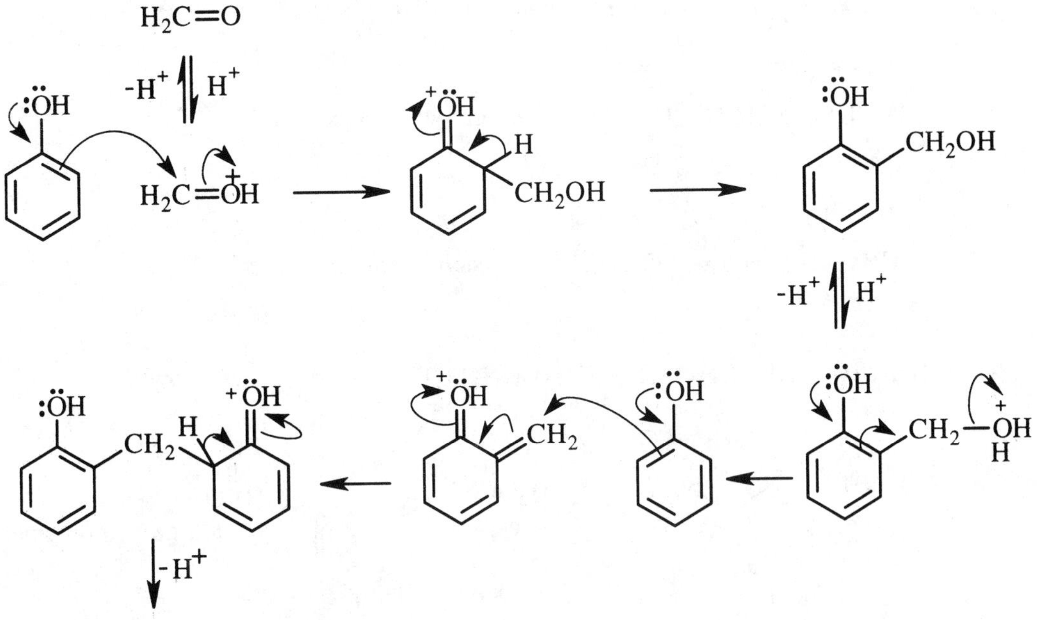

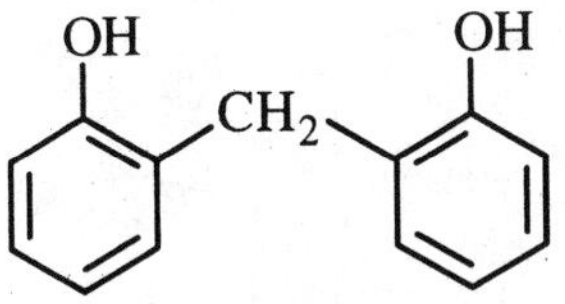

$H_2C{=}O$

$-H^+$ ⇌ H^+

:ÖH

$H_2C{=}\overset{+}{O}H$

CH_2OH

CH_2OH

$-H^+$ ⇌ H^+

:ÖH

$CH_2{-}\overset{+}{O}H$

H

CH_2

CH_2

$-H^+$

OH OH

CH_2

monomer used in the polymerization

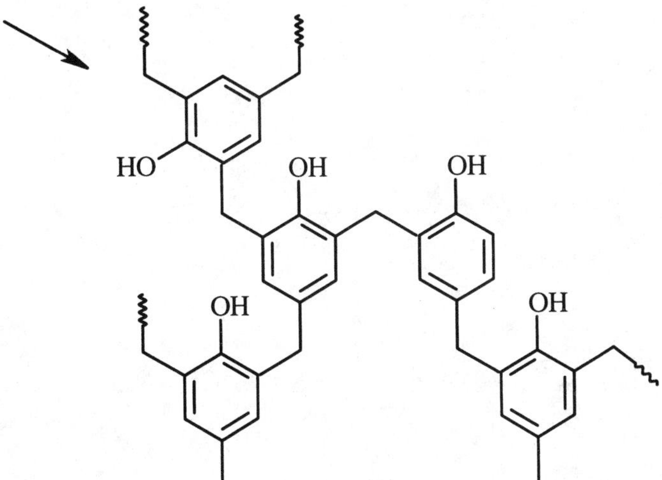

HO OH OH OH OH

Bakelite

22.

a. $-CH_2CHCH_2CHCH_2CH-$ (each CH bearing F) chain-growth polymer

b. $-CH_2CHCH_2CHCH_2CH-$ (each CH bearing CO_2H) chain-growth polymer

c. $-O(CH_2)_5\overset{O}{\overset{\|}{C}}O(CH_2)_5\overset{O}{\overset{\|}{C}}O(CH_2)_5\overset{O}{\overset{\|}{C}}-$ step-growth polymer

d. $-NH(CH_2)_5NH\overset{O}{\overset{\|}{C}}(CH_2)_5\overset{O}{\overset{\|}{C}}NH(CH_2)_5NH\overset{O}{\overset{\|}{C}}(CH_2)_5\overset{O}{\overset{\|}{C}}-$ step-growth polymer

e. $-O\overset{O}{\overset{\|}{C}}NH-C_6H_3(CH_3)-NH\overset{O}{\overset{\|}{C}}OCH_2CH_2O\overset{O}{\overset{\|}{C}}NH-C_6H_3(CH_3)-NH\overset{O}{\overset{\|}{C}}O-$ step-growth polymer

23.

a. $-CH_2CH_2OCH_2CH_2N\langle piperazine \rangle N-$

b. $-C(CH_3)_2-C_6H_4-C(CH_3)_2O-C_6H_4-CH_2-C_6H_4-O-$

c. $-C_6H_4-OCH_2CH_2CH_2O-C_6H_4-N{=}CHCH{=}N-$

d. $-CH{=}C_6H_{10}{=}CH-C_6H_4-$

24.

a. $CH_2{=}CHCH_2CH_3$

b. 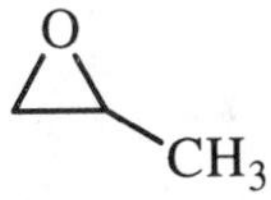 (methyloxirane: three-membered ring containing O, with CH_3 substituent)

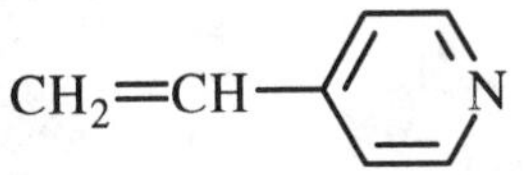

c. 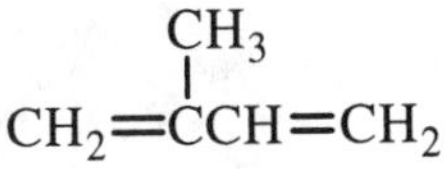 $CH_2{=}C(CH_3)CH{=}CH_2$

d. 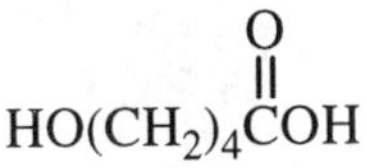 $HO(CH_2)_4COH$ (with C=O)

e. 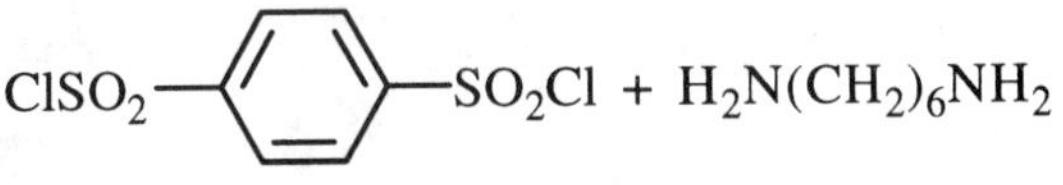 $ClSO_2$–C_6H_4–SO_2Cl + $H_2N(CH_2)_6NH_2$

f. $CH_2{=}CH$–(pyridyl, N)

g. $CH_2{=}CCH_3$ (with phenyl on C)

h. $HOC(=O)$–C_6H_4–$C(=O)OH$ + $HOCH_2CH_2OH$

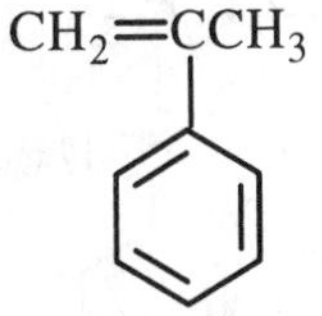

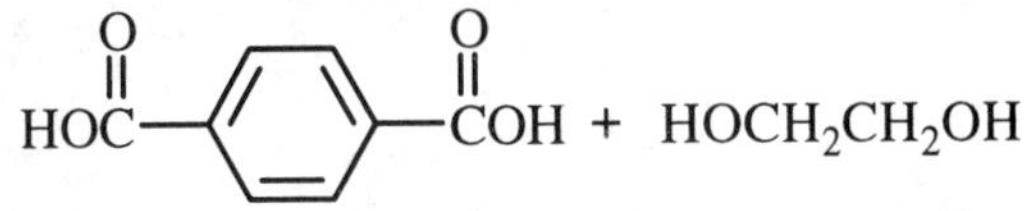

a, **b**, **c**, **f**, and **g** are chain-growth polymers.

d, **e**, and **h** are step-growth polymers.

25. Whether a polymer is isotactic, syndiotactic, or atactic depends on whether the substituents are all on one side of the carbon chain, alternate on both sides of the chain, or are random with respect to the chain. Because a polymer of isobutylene has two identical substituents at each carbon in the chain, different configurations are not possible.

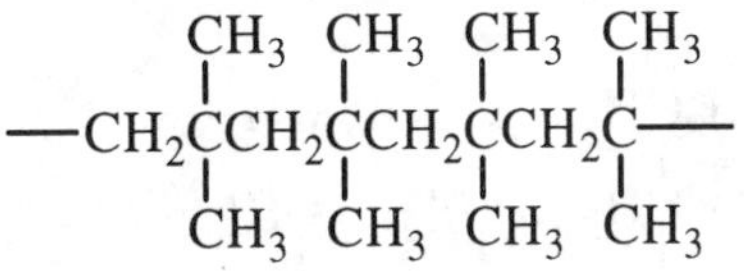

$$—CH_2C(CH_3)_2CH_2C(CH_3)_2CH_2C(CH_3)_2CH_2C(CH_3)_2—$$

polymer of isobutylene

26.

a. 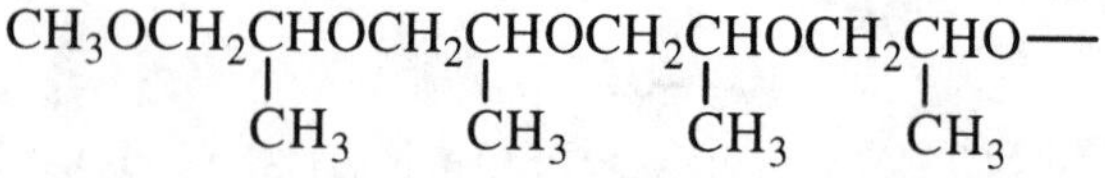 $CH_3OCH_2CHOCH_2CHOCH_2CHOCH_2CHO—$
$\quad\quad\quad\; | \quad\quad\quad | \quad\quad\quad | \quad\quad\quad |$
$\quad\quad\; CH_3 \quad CH_3 \quad CH_3 \quad CH_3$

b. 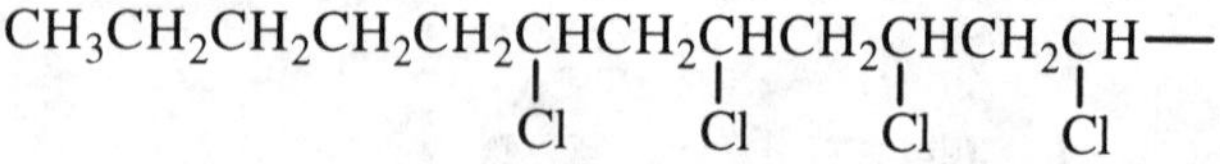 $CH_3CH_2CH_2CH_2CH_2CHCH_2CHCH_2CHCH_2CH—$
$\quad\quad\quad\quad\quad\quad\quad\quad | \quad\quad\;\; | \quad\quad\;\; | \quad\quad\;\; |$
$\quad\quad\quad\quad\quad\quad\quad\; Cl \quad\;\; Cl \quad\;\; Cl \quad\;\; Cl$

c. 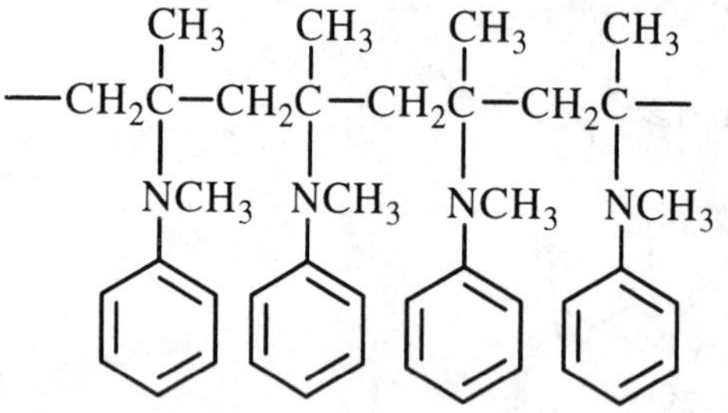 $—CH_2C(CH_3)(N(CH_3)C_6H_5)—CH_2C(CH_3)(N(CH_3)C_6H_5)—CH_2C(CH_3)(N(CH_3)C_6H_5)—CH_2C(CH_3)(N(CH_3)C_6H_5)—$

(each quaternary carbon bears CH_3 above and NCH_3 below, with a phenyl ring attached to each N)

d. 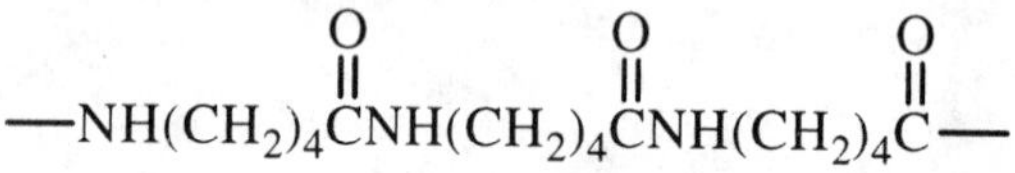 $—NH(CH_2)_4\overset{O}{\overset{\|}{C}}NH(CH_2)_4\overset{O}{\overset{\|}{C}}NH(CH_2)_4\overset{O}{\overset{\|}{C}}—$

e. Depending on the particular Ziegler-Natta catalyst used, the double bonds can be either cis or trans.

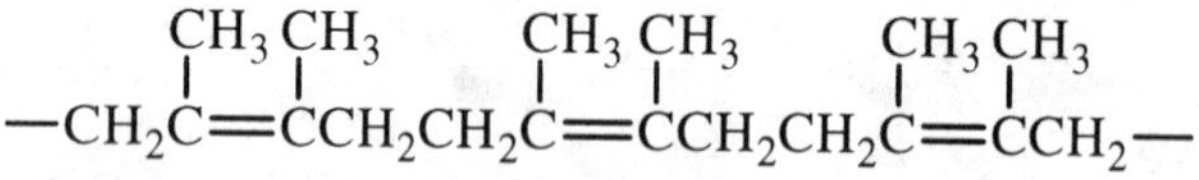

$—CH_2C(CH_3){=}C(CH_3)CH_2CH_2C(CH_3){=}C(CH_3)CH_2CH_2C(CH_3){=}C(CH_3)CH_2—$ or

$—CH_2C(CH_3){=}C(CH_3)CH_2CH_2C(CH_3){=}C(CH_3)CH_2CH_2C(CH_3){=}C(CH_3)CH_2—$

(in the first structure both CH_3 groups of each double bond are drawn above the chain; in the second one CH_3 is drawn above and one below)

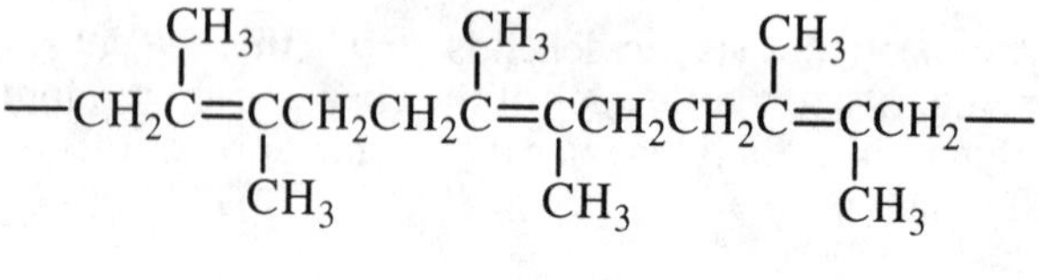

f. 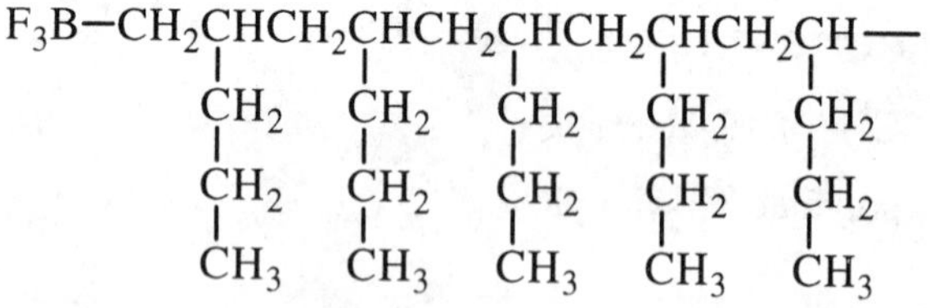 $F_3\bar{B}—CH_2CHCH_2CHCH_2CHCH_2CHCH_2CH—$

(each CH bears a $CH_2CH_2CH_3$ group: CH_2 / CH_2 / CH_3 drawn below)

27.

a. 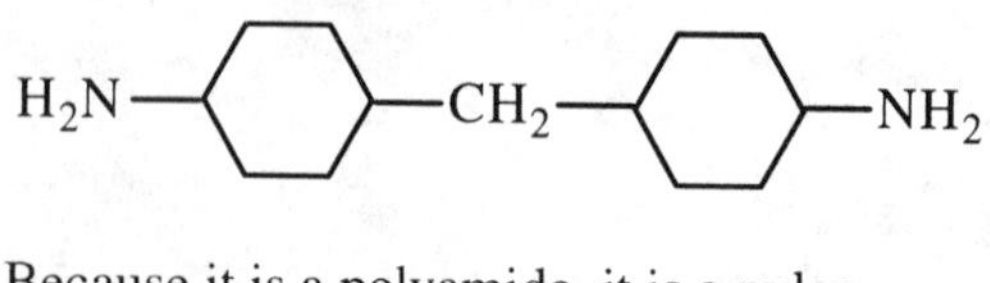 $H_2N—C_6H_{10}—CH_2—C_6H_{10}—NH_2$ (cyclohexane rings) and $HO\overset{O}{\overset{\|}{C}}(CH_2)_6\overset{O}{\overset{\|}{C}}OH$

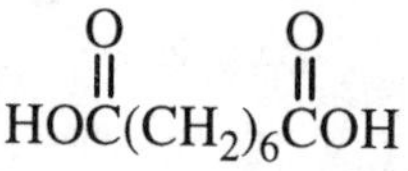

b. Because it is a polyamide, it is a nylon.

28. A copolymer is a polymer composed of more than one kind of monomer. Because the initially formed carbocation can rearrange, two different monomers are involved in formation of the polymer.

$$F_3\bar{B}{-}CH_2\overset{+}{C}H\underset{CH_3}{\overset{CH_3}{C}}CH_3 \xrightarrow{\text{1,2-methyl shift}} F_3\bar{B}{-}CH_2\underset{}{\overset{CH_3}{C}H}\underset{CH_3}{\overset{+}{C}}CH_3$$

unrearranged monomer — rearranged monomer

29. The polymer in the flask that contained a high molecular weight polymer and little material of intermediate molecular weight was formed by a chain-growth mechanism, while the polymer in the flask that contained mainly material of intermediate molecular weight was formed by a step-growth mechanism.

In a chain-growth mechanism, monomers are added to the growing end of a chain. This means that at any one time there will be polymeric chains and monomers.

Step-growth polymerization is not a chain reaction; any two monomers can react. Therefore, high molecular weight material will be formed by the reaction of two pieces of intermediate molecular weight.

30. **a.** Vinyl alcohol is unstable, it tautomerizes to acetaldehyde.

$$CH_2{=}\overset{OH}{\overset{|}{C}H} \longrightarrow CH_3\overset{O}{\overset{\|}{C}}H$$

vinyl alcohol — acetaldehyde

b. It is not a true polyester. It has ester groups as substituents **on** the backbone of the chain so it does have "polyester" groups, but it does not have ester groups **within** the backbone of the polymer chain. A true polyester has ester groups in the backbone of the polymer chain.

31. Each of the five carbocations shown below can add the growing end of the polymer chain.

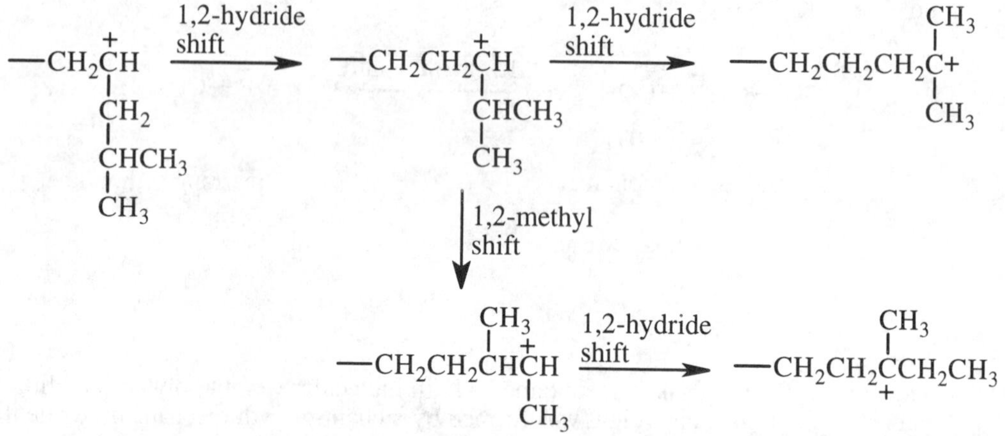

$$-CH_2\overset{+}{C}H(CH_2CH(CH_3)CH_3) \xrightarrow[\text{shift}]{\text{1,2-hydride}} -CH_2CH_2\overset{+}{C}H(CH(CH_3)CH_3) \xrightarrow[\text{shift}]{\text{1,2-hydride}} -CH_2CH_2CH_2\overset{+}{C}(CH_3)_2$$

$$-CH_2CH_2\overset{+}{C}H(CH(CH_3)CH_3) \xrightarrow[\text{shift}]{\text{1,2-methyl}} -CH_2CH_2CH(CH_3)\overset{+}{C}HCH_3 \xrightarrow[\text{shift}]{\text{1,2-hydride}} -CH_2CH_2\overset{+}{C}(CH_3)CH_2CH_3$$

32. Because 1,4-divinylbenzene has substituents on both ends of the benzene ring that can engage in polymerization, the polymer chains can become cross-linked, which increases the rigidity of the polymer.

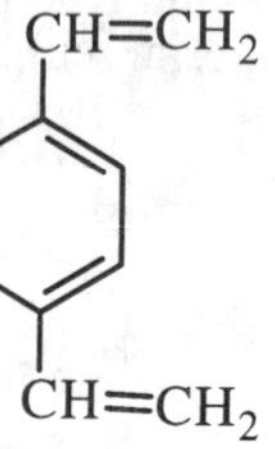

$CH{=}CH_2$ — C_6H_4 — $CH{=}CH_2$

1,4-divinylbenzene

$$-CH-CH_2-CH-CH_2-CH-CH_2-CH-CH_2-CH-$$

$$-CH-CH_2-CH-CH_2-CH-CH_2-CH-CH_2-CH-CH_2-CH-$$

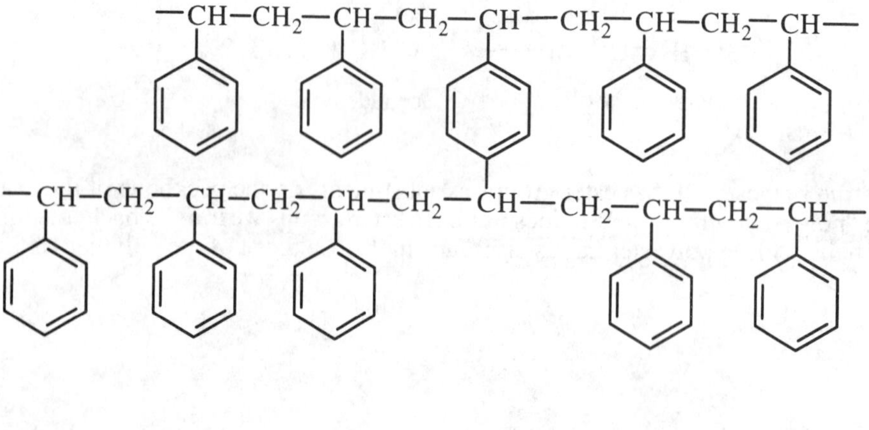

33. Glyptal gets its strength from cross-linking.

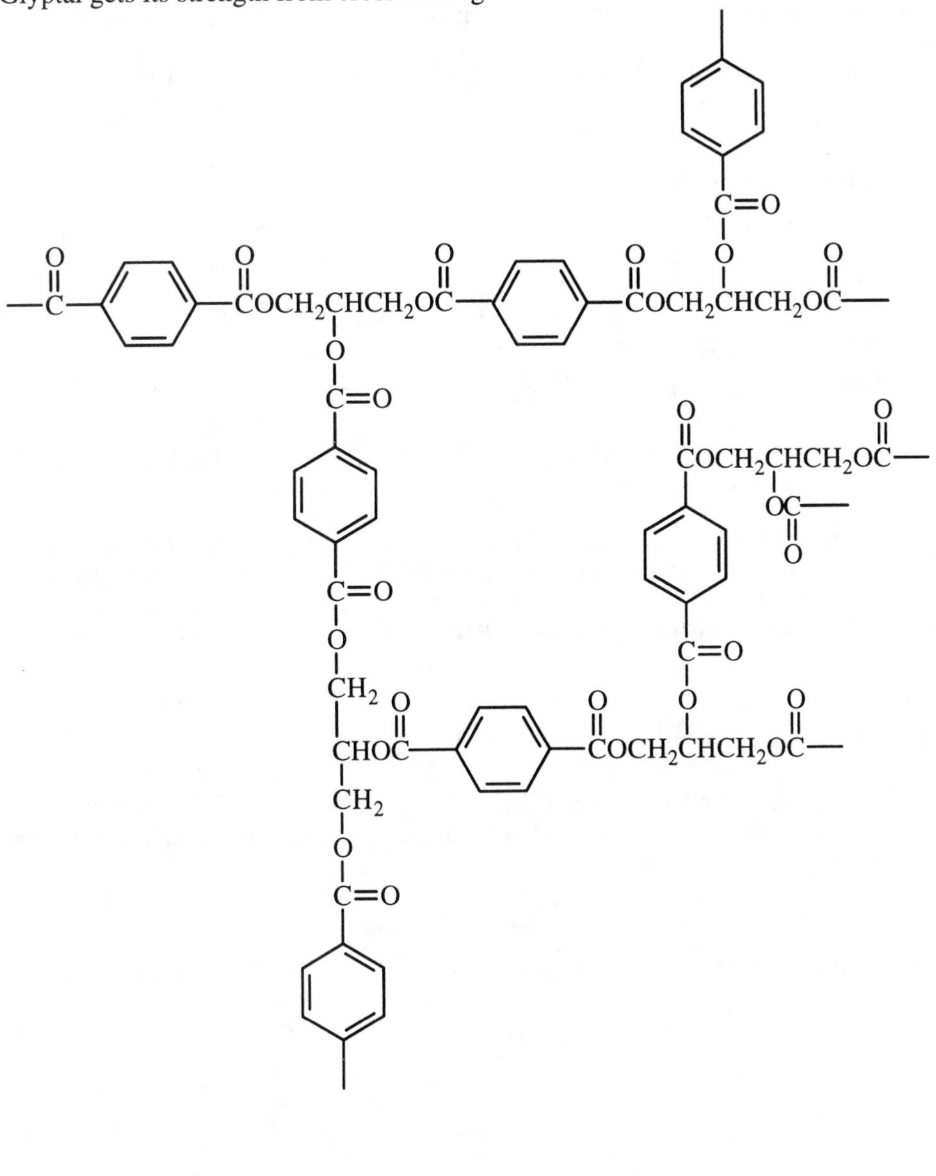

34.

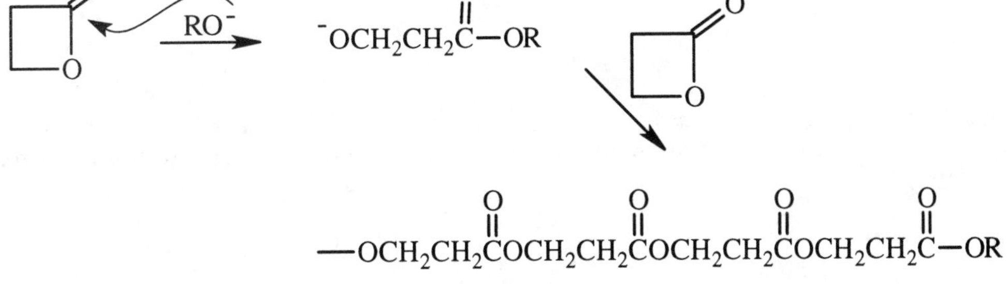

$$\text{RO}^- + \text{(β-propiolactone)} \longrightarrow {}^-OCH_2CH_2\overset{O}{\overset{\|}{C}}-OR \xrightarrow{\text{β-propiolactone}} -OCH_2CH_2\overset{O}{\overset{\|}{C}}OCH_2CH_2\overset{O}{\overset{\|}{C}}OCH_2CH_2\overset{O}{\overset{\|}{C}}OCH_2CH_2\overset{O}{\overset{\|}{C}}-OR$$

35. Both compounds can form esters via intramolecular or intermolecular reactions. The product of the intramolecular reaction is a lactone; the intermolecular reaction leads to a polymer.

5-Hydroxypentanoic acid reacts intramolecularly to form a six-membered-ring lactone, while 6-hydroxyhexanoic acid reacts intramolecularly to form a seven-membered-ring lactone.

$HOCH_2CH_2CH_2CH_2\overset{O}{\overset{\|}{C}}OH$ ⟶ (six-membered-ring lactone)

5-hydroxypentanoic acid

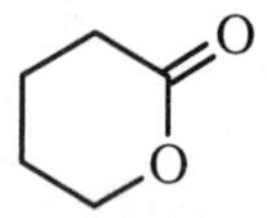

$HOCH_2CH_2CH_2CH_2CH_2\overset{O}{\overset{\|}{C}}OH$ ⟶ (seven-membered-ring lactone)

6-hydroxypentanoic acid

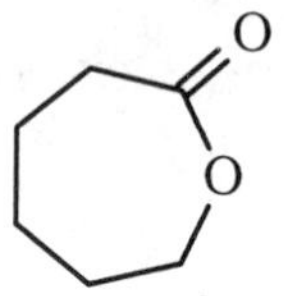

The compound that forms the most polymer will be the one that forms the least lactone, because the two reactions compete with one another.

The six-membered-ring lactone is more stable and so is the transition state for its formation, compared to a seven-membered ring lactone. Since it is easier for 5-hydroxypentanoic acid to form the six-membered ring lactone, than for 6-hydroxyhexanoic acid to form the seven-membered ring lactone, **6-hydroxypentanoic acid** will form the most polymer.

36. Rubber contains cis double bonds. Ozone, which is present in the air, oxidizes double bonds to carbonyl groups, destroying the polymer chain. Polyethylene does not contain double bonds, so it is not air-oxidized.

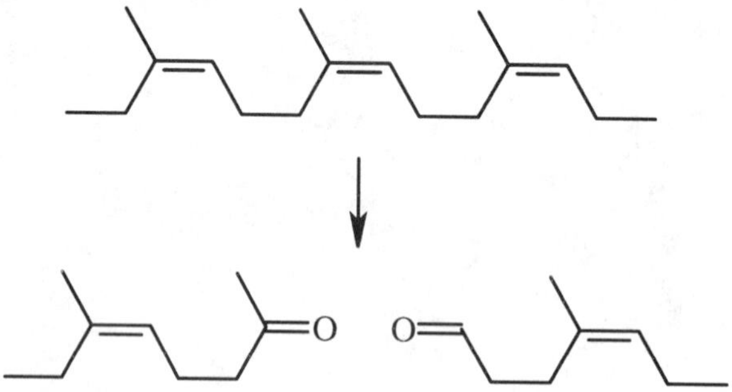

37. The plasticizer that keeps vinyl soft and pliable can vaporize over time, causing the polymer to become brittle. For this reason, high boiling materials are preferred over low boiling materials as plasticizers.

38. **a.** Because the negative charge on the propagation site can be delocalized onto the carbonyl oxygen, the polymer is best prepared by anionic polymerization.

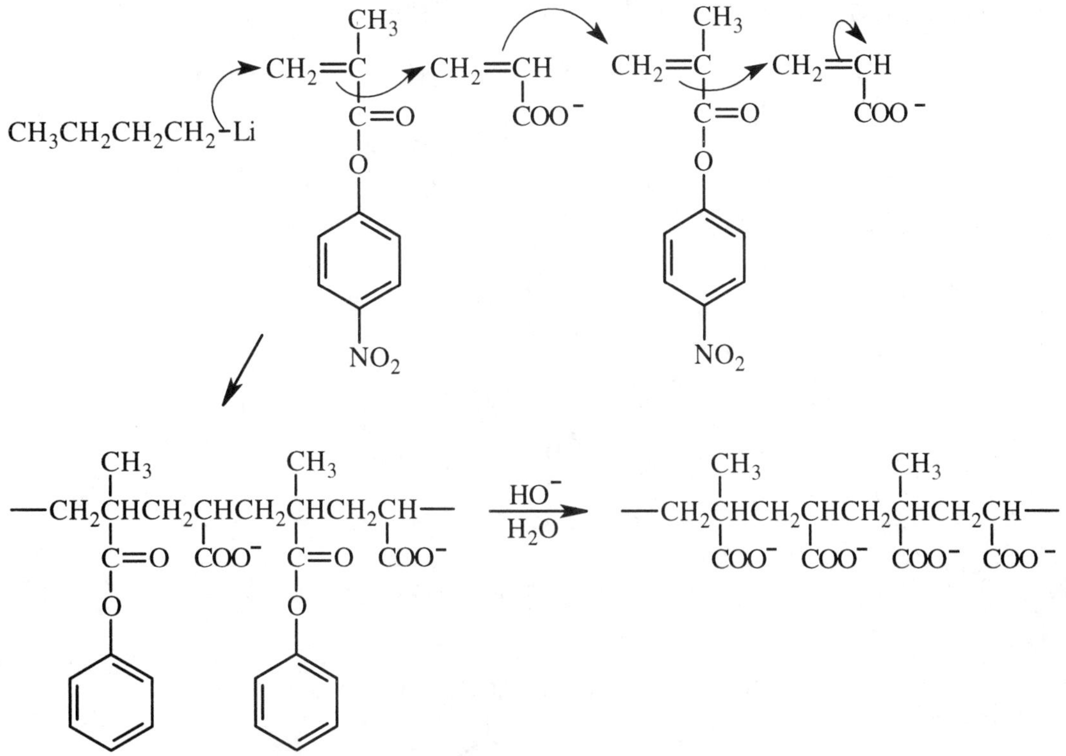

$CH_3CH_2CH_2CH_2$–Li + $CH_2=C(CH_3)C(=O)O$–C_6H_4–NO_2 + $CH_2=CHCOO^-$ + $CH_2=C(CH_3)C(=O)O$–C_6H_4–NO_2 + $CH_2=CHCOO^-$

$$\longrightarrow \;—CH_2C(CH_3)(C{=}O\text{–}O\text{–}C_6H_5)CH_2CH(COO^-)CH_2C(CH_3)(C{=}O\text{–}O\text{–}C_6H_5)CH_2CH(COO^-)— \xrightarrow[H_2O]{HO^-} —CH_2C(CH_3)(COO^-)CH_2CH(COO^-)CH_2C(CH_3)(COO^-)CH_2CH(COO^-)—$$

b. The carboxyl substituent is in position to remove a proton from water, making water a stronger nucleophile. (See page 971 of the text.)

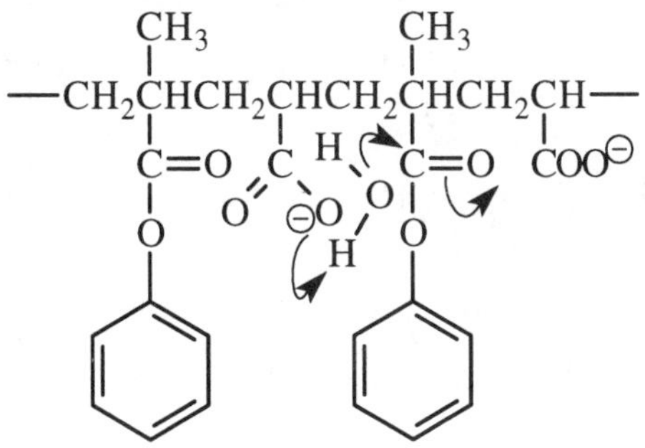

$$—CH_2C(CH_3)(C{=}O\text{–}O\text{–}C_6H_5)CH_2CH(COO^-)CH_2C(CH_3)(C{=}O\text{–}O\text{–}C_6H_5)CH_2CH(COO^-)—$$

39. Hydrolysis converts the ester substituents into alcohol substituents, which can react with ethylene oxide to graft a polymer of ethylene oxide onto the backbone of the random polymer of styrene and vinyl acetate.

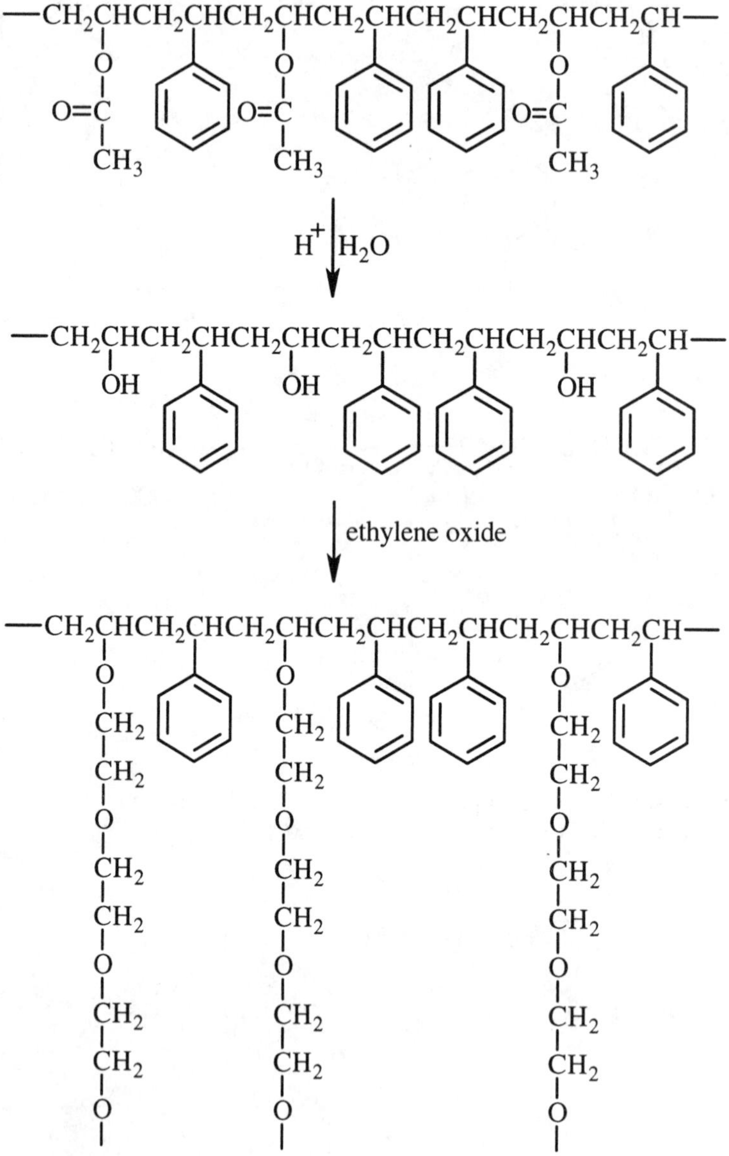

$$-CH_2CHCH_2CHCH_2CHCH_2CHCH_2CHCH_2CHCH_2CH-$$

H^+ / H_2O

$$-CH_2CHCH_2CHCH_2CHCH_2CHCH_2CHCH_2CHCH_2CH-$$

ethylene oxide

$$-CH_2CHCH_2CHCH_2CHCH_2CHCH_2CHCH_2CHCH_2CH-$$

40. The desired polymer is an alternating copolymer of ethylene and 1,2-dibromoethylene.

$CH_2{=}CH_2$ $\quad$ $BrCH{=}CHBr$

CHAPTER 27
Heterocyclic Compounds

Important Terms

alkaloid	a natural product with a nitrogen heteroatom found in the leaves, bark, or seeds of plants.
corrin ring system	the ring system found in vitamin B_{12}. It is similar to the porphyrin ring system, but one of the methine bridges of the porphyrin ring system is missing.
deamination	loss of ammonia.
furan	a five-membered ring aromatic compound containing an oxygen heteroatom.
heteroatom	an atom other than a carbon atom or a hydrogen atom.
heterocyclic compound (heterocycle)	a cyclic compound in which one or more of the atoms of the ring are heteroatoms.
imidazole	a five-membered ring compound with two nitrogen heteroatoms and two double bonds.
iron protoporphyrin IX	the porphyrin ring system of heme ligated to an iron atom.
ligation	sharing of nonbonded electrons with a metal.
natural product	a product synthesized in nature.
porphyrin ring system	consists of four pyrrole rings joined by one carbon bridge.
protoporphyrin IX	the porphyrin ring system of heme.
purine	a pyrimidine ring fused to an imidazole ring.
pyrimidine	a benzene ring with nitrogens at the #1 and #3 positions.
pyrrole	a five-membered ring aromatic compound containing a nitrogen heteroatom.
saturated heterocycle	a heterocyclic compound with no double bonds.
thiophene	a five-membered ring aromatic compound containing a sulfur heteroatom.

Solutions to Problems

1. **a.** 2,2-dimethylaziridine
   b. 4-ethylpiperidine
   c. 3-methylazacyclobutane
   d. 2-methylthiacyclopropane
   e. 2,3-dimethyltetrahydrofuran
   f. 2-ethyloxacyclobutane

2. The electron-withdrawing oxygen atom of morpholine decreases the strength of the nitrogen-hydrogen bond, making it easier for the proton to dissociate from the compound.

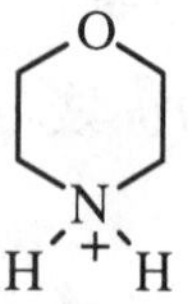

conjugate acid of morpholine
$pK_a = 9.28$

conjugate acid of piperidine
$pK_a = 11.12$

3. **a.**

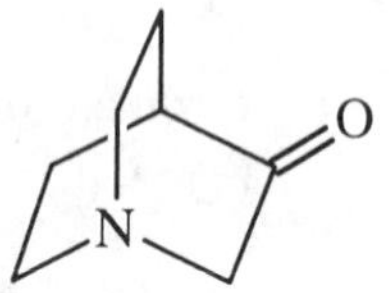

b. The conjugate acid of 3-quinuclidinone has a lower pK_a than the conjugate acid of morpholine ($pK_a = 9.28$) because the electron-withdrawing oxygen atom in 3-quinuclidinone is closer to the site of the acidic proton.

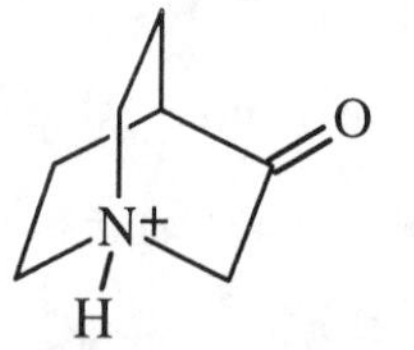

conjugate acid of 3-quinuclidinone

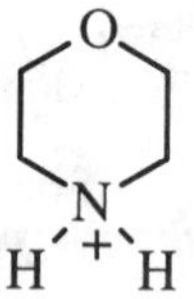

conjugate acid of morpholine

c. The conjugate acid of 3-chloroquinuclidine has a lower pK_a than the conjugate acid of 3-bromoquinuclidine because chlorine is more electronegative than bromine. This means that 3-bromoquinuclidine is a stronger base than 3-chloroquinuclidine.

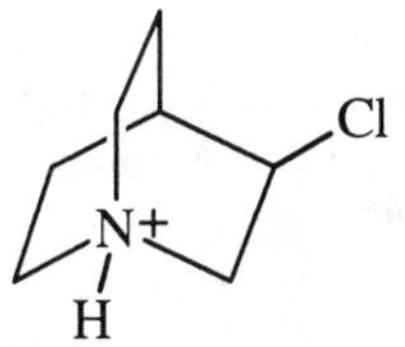

conjugate acid of 3-chloroquinuclidine

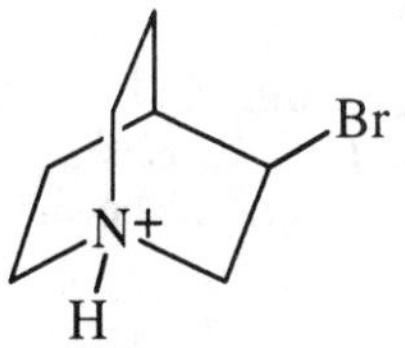

conjugate acid of 3-bromoquinuclidine

4.

a. 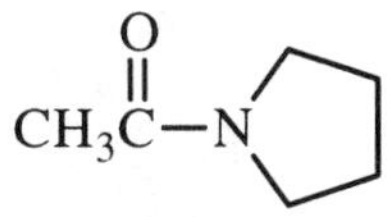 $CH_3C(=O)-N$ (pyrrolidine ring)

b. $HOCH_2CH_2CH_2CH_2CH_2I$

if excess HI is used

$ICH_2CH_2CH_2CH_2CH_2I$

c. $HOCH_2CH_2CH_2CH_2\overset{O}{\overset{\|}{C}}OH$

d. 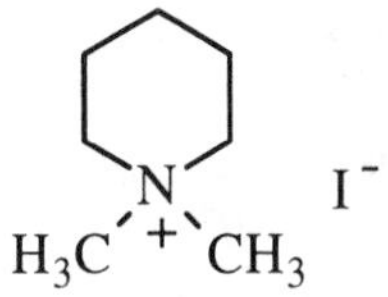 (piperidinium ring) H_3C $\overset{+}{N}$ CH_3 I^-

5.

D^+

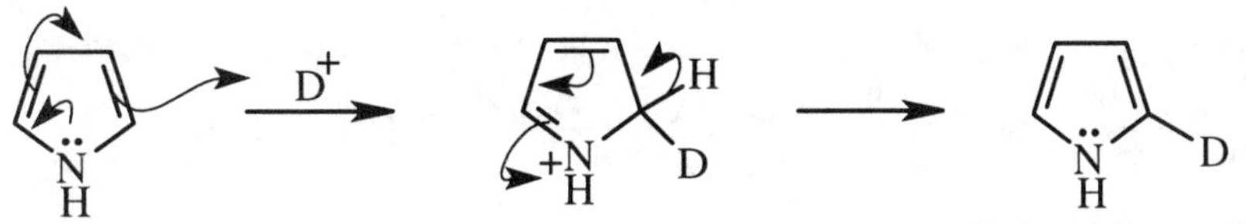

2-deuteriopyrrole

6. Protonation on C-2 leads to a cation with three resonance contributors. Protonation on nitrogen leads to a cation with no resonance contributors. (The positively charged nitrogen cannot accept electrons by resonance, because that would put ten electrons around the nitrogen.)

protonation on C-2

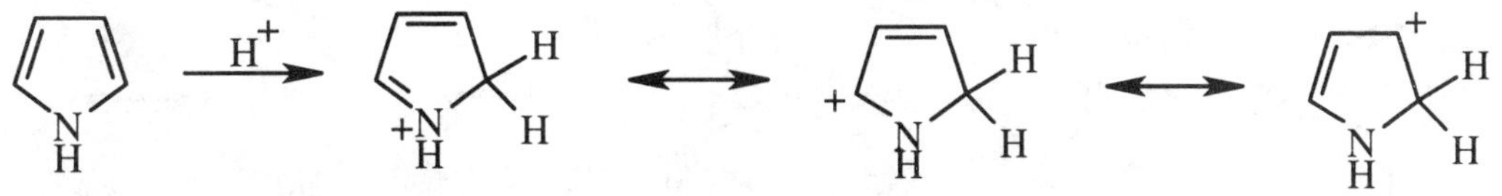

H^+

protonation on nitrogen

H^+

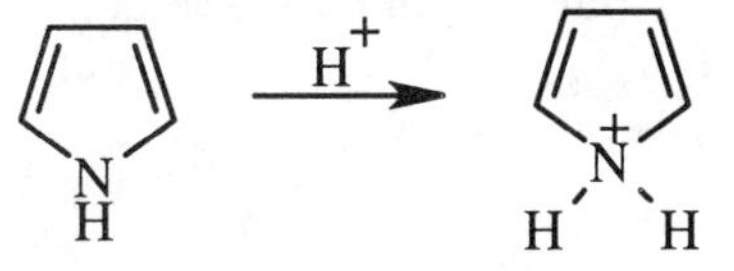

7. Pyrrole is aromatic in both its acidic and basic forms. Cyclopentadiene does not become aromatic until it loses a proton. It is the drive to become a stable aromatic compound that causes cyclopentadiene to be more acidic than pyrrole.

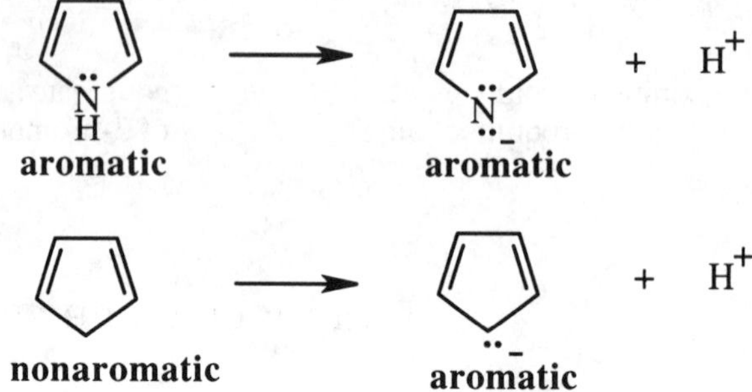

8. Solved in the text.

9. Pyridine will act as an amine with the acid chloride. However, the amide that is formed is very reactive because of its positively charged nitrogen atom. Therefore, it will undergo a nucleophilic acyl substitution reaction with methanol, forming an ester as the final product of the reaction.

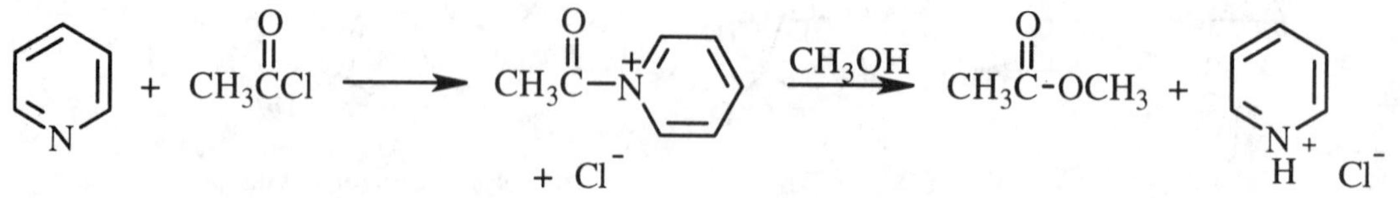

pyridine + $CH_3\overset{O}{\overset{\|}{C}}Cl$ ⟶ $CH_3\overset{O}{\overset{\|}{C}}{-}\overset{+}{N}$(pyridinium) + Cl^- $\xrightarrow{CH_3OH}$ $CH_3\overset{O}{\overset{\|}{C}}{-}OCH_3$ + pyridinium ($\overset{+}{N}$–H) Cl^-

10. The increase in the electron density of the ring as a result of resonance donation of electrons by oxygen causes pyridine-*N*-oxide to be more reactive toward electrophilic substitution than pyridine.

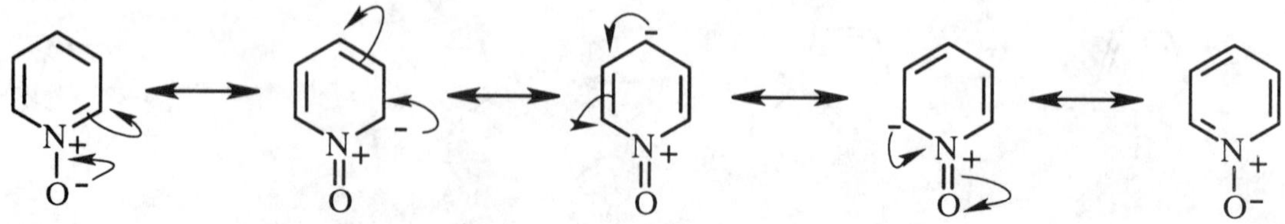

From the resonance structures you can see that the increased electron density is at the #2 and #4 positions. Because the #2 position is somewhat sterically hindered, pyridine-*N*-oxide undergoes electrophilic substitution primarily at the #4 position.

11. The first reaction is a two-step nucleophilic aromatic substitution reaction (S_NAr). The second reaction is a one-step nucleophilic substitution reaction (S_N2).

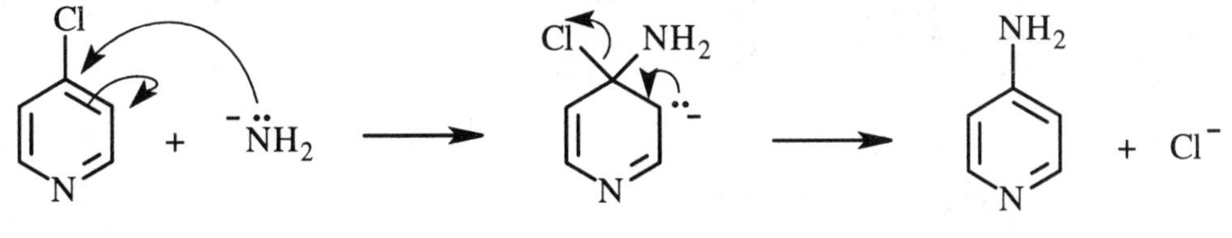

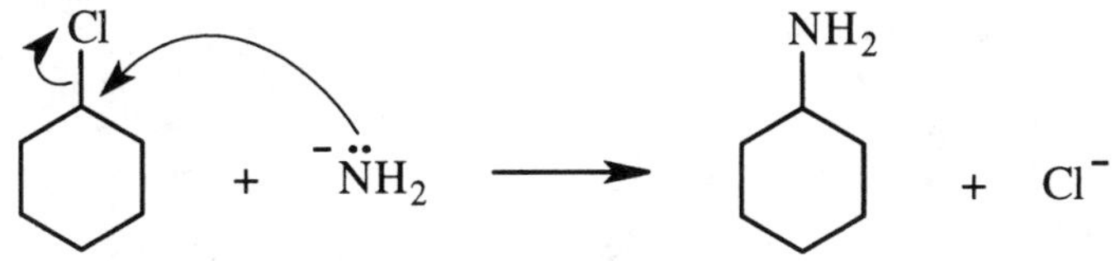

Cl, + $^-\ddot{N}H_2$ ⟶ Cl NH$_2$ ⟶ NH$_2$ + Cl$^-$

Cl + $^-\ddot{N}H_2$ ⟶ NH$_2$ + Cl$^-$

12.

a. + HÖ:$^-$ ⟶ H, ÖH; $\underset{H_2O}{\overset{HO^-}{\rightleftharpoons}}$ H, Ö:$^-$ + H$^-$ + H_2O ⟶ 2 HO$^-$

HO$^-$ + [N–H, O] $\xleftarrow{H_2O}$ [N$^-$, O] ⟵ [N, Ö:$^-$]

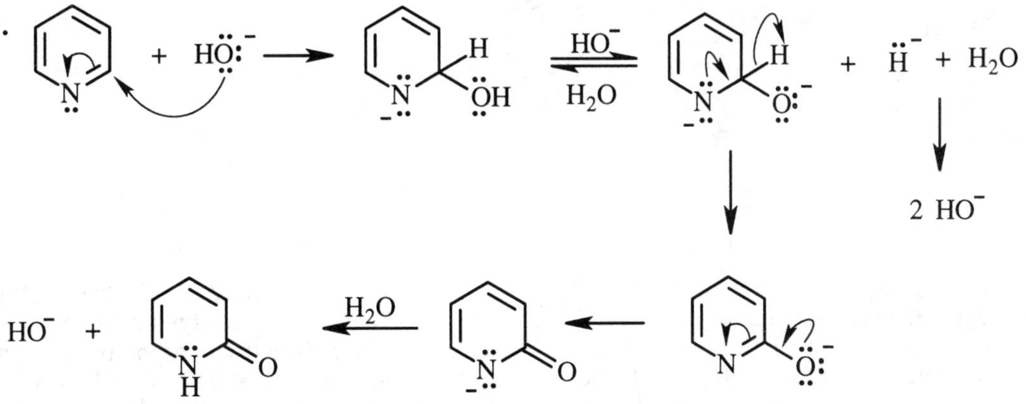

2-pyridinone

b. 4-Pyridinone is also formed because nucleophilic attack by hydroxide ion can take place at the #4 position as well as at the #2 position.

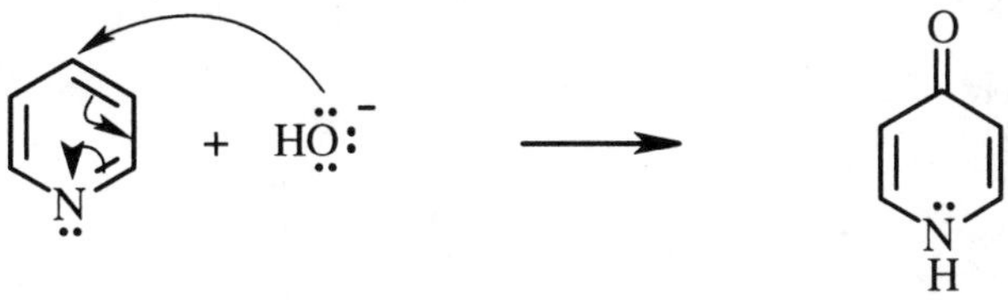

+ HÖ:$^-$ ⟶ O, N, H

4-pyridinone

13.

a. 1. $\bar{N}HCH_3$, Δ 2. H_2O → NHCH$_3$

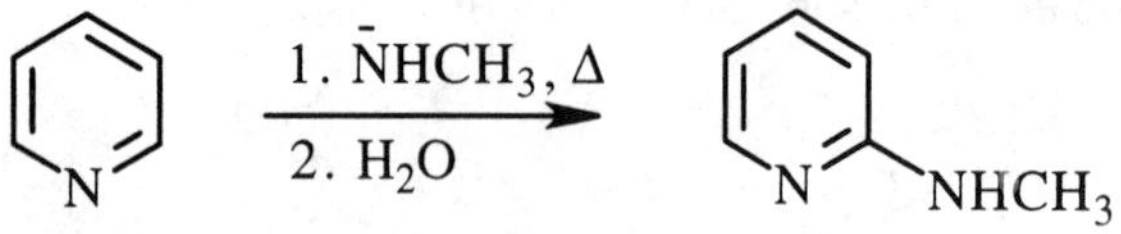

b. Cl_2, $AlCl_3$, Δ → Cl

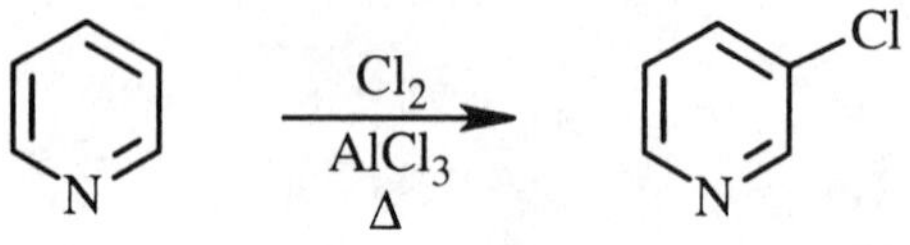

c. H_2O_2 → N^+–O^-; Cl_2, $AlCl_3$, Δ → Cl, N^+–O^-; PCl_3 → Cl

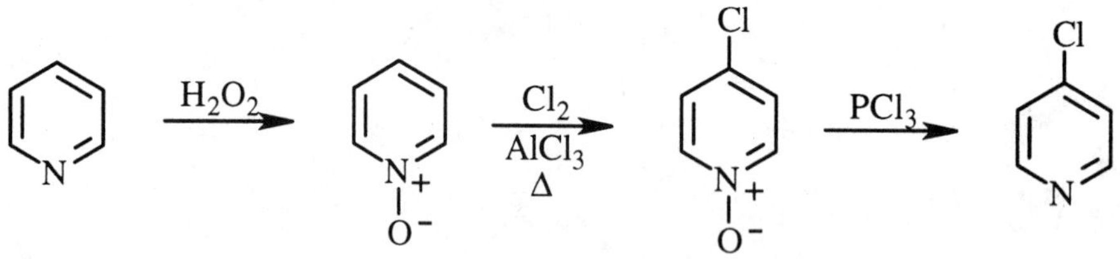

d. 1. $\bar{N}HCH_3$, Δ 2. H_2O → NHCH$_3$

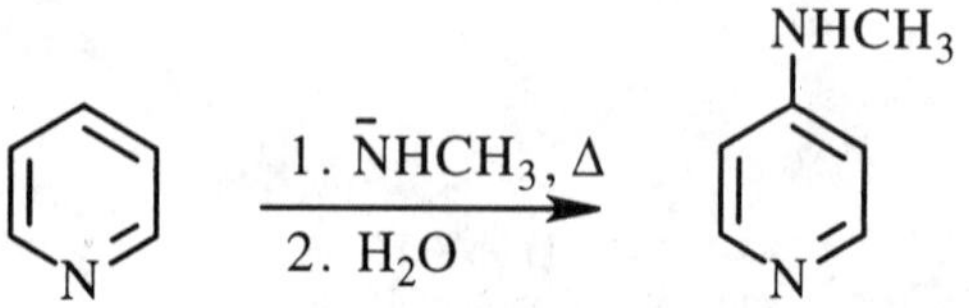

14. It is easiest to remove a proton from the *N*-alkylated pyridine because the electrons left behind when the proton is removed can be delocalized onto the positively charged nitrogen atom. It is easier to remove a proton from 4-methylpyridine than from 3-methylpyridine because in the former, the electrons left behind when the proton is removed can be delocalized onto the electronegative nitrogen atom. In 3-methylpyridine, the electrons can be delocalized only onto carbon atoms.

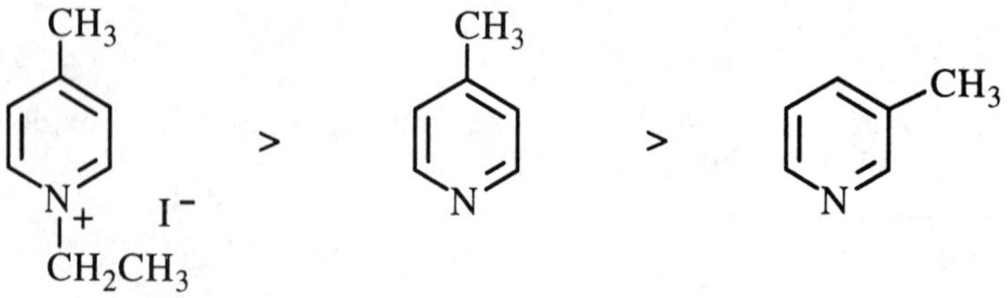

CH_3, N^+, I^-, CH_2CH_3 > CH_3 > CH_3

15.

a. 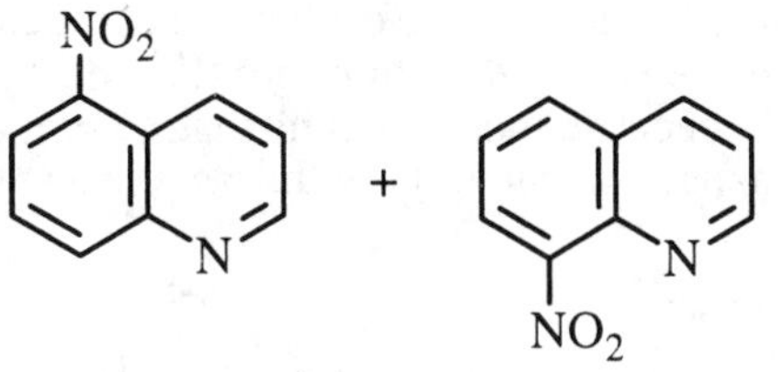 5-nitroquinoline + 8-nitroquinoline

NO_2 ... N + N ... NO_2

b. 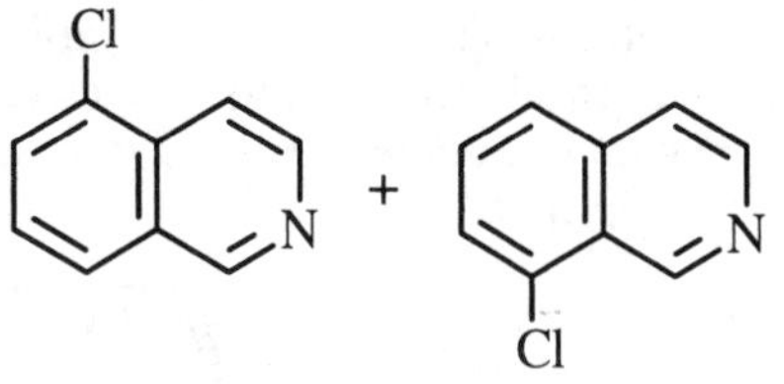 Cl ... N + N ... Cl

c. 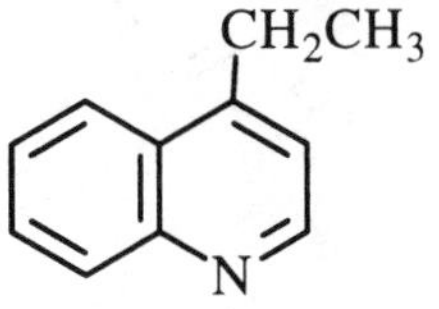 CH_2CH_3 ... N

d. 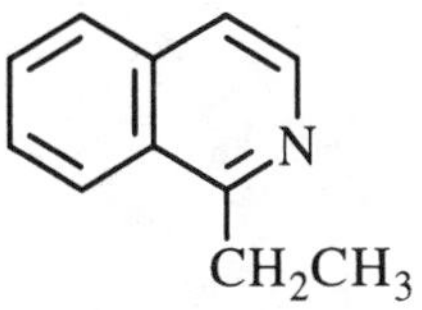 N ... CH_2CH_3

16. **a.** There are three possible sites for electrophilic substitution: C-2, C-4, and C-5. To determine the major product, compare the relative stabilities of the carbocations formed in the first step of the reaction.

Substitution at C-2 leads to a carbocation with three contributing resonance structures. One of them is particularly unstable, because the positive charge is on a nitrogen atom that does not have a complete octet.

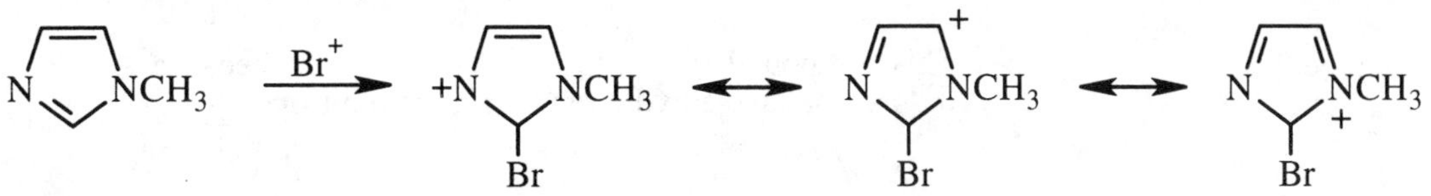

N NCH_3 $\xrightarrow{Br^+}$ +N NCH_3 (Br) ⟷ N NCH_3 (Br, +) ⟷ N NCH_3 (+, Br)

Substitution at C-4 leads to a carbocation with only two contributing resonance structures.

N NCH_3 $\xrightarrow{Br^+}$ N NCH_3 (Br, +) ⟷ N NCH_3 (Br, +)

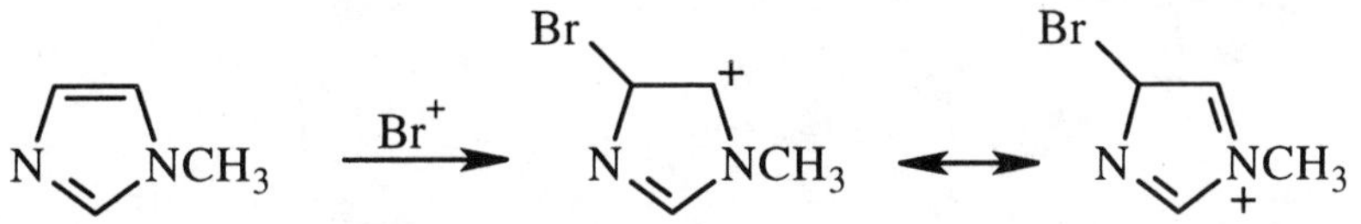

Substitution at C-5 leads to a carbocation with three contributing resonance structures. Because none of them are as unstable as one of the contributing resonance structures obtained from substitution at C-2, it is the most stable of the three possible carbocation intermediates.

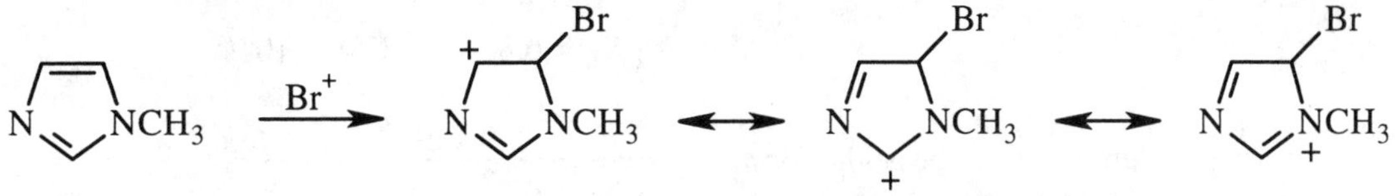

N NCH_3 $\xrightarrow{Br^+}$ N NCH_3 (Br, +) ⟷ N NCH_3 (Br, +) ⟷ N NCH_3 (Br, +)

Therefore, the major product of the reaction is 5-bromo-*N*-methylimidazole.

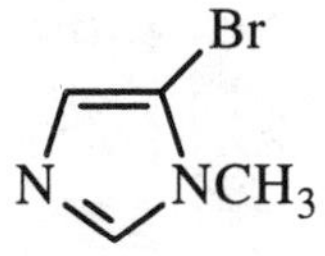

Br

N NCH_3

5-bromo-*N*-methylimidazole

b. Pyrrole and imidazole are more reactive than benzene because each has a carbocation intermediate that can be stabilized by resonance electron donation into the ring by a nitrogen atom. Pyrrole is more reactive than imidazole because the second nitrogen atom of imidazole cannot donate electrons into the ring by resonance but can only withdraw electrons from the ring inductively.

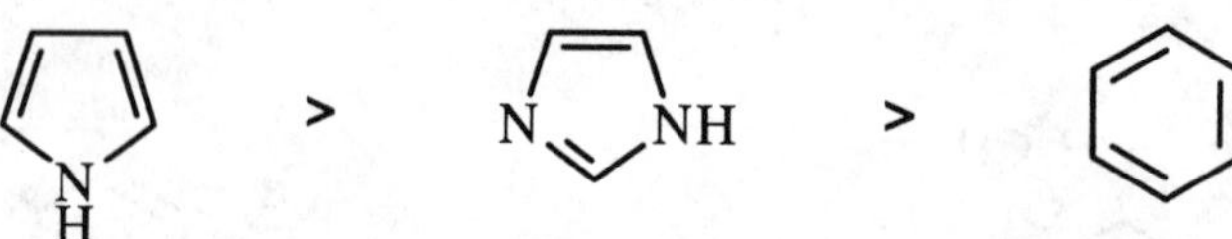

17. Imidazole forms intermolecular hydrogen bonds, whereas *N*-methylimidazole cannot form hydrogen bonds. Because the hydrogen bonds have to be broken in order for the compound to boil, imidazole has a higher boiling point.

This compound cannot form hydrogen bonds, because it does not have a hydrogen bonded to a nitrogen.

18.

$$\text{fraction of imidazole in the acidic form} = \frac{[H^+]}{K_a + [H^+]}$$

pH = 7.3; $[H^+] = 5.0 \times 10^{-8}$

$pK_a = 6.8$; $K_a = 1.58 \times 10^{-7}$

$$\text{fraction of imidazole in the acidic form} = \frac{5.0 \times 10^{-8}}{1.58 \times 10^{-7} + 5.0 \times 10^{-8}}$$

$$= \frac{5.0 \times 10^{-8}}{2.08 \times 10^{-7}}$$

$$= 0.24$$

percent of imidazole in the acidic form = 24%

19. Yes, porphyrin is aromatic, because it fulfills the two requirements for aromaticity. It has an uninterrupted ring of *p* orbital-bearing atoms (it is cyclic and planar), and the π cloud contains an odd number of pairs (thirteen pairs) of π electrons.

20.

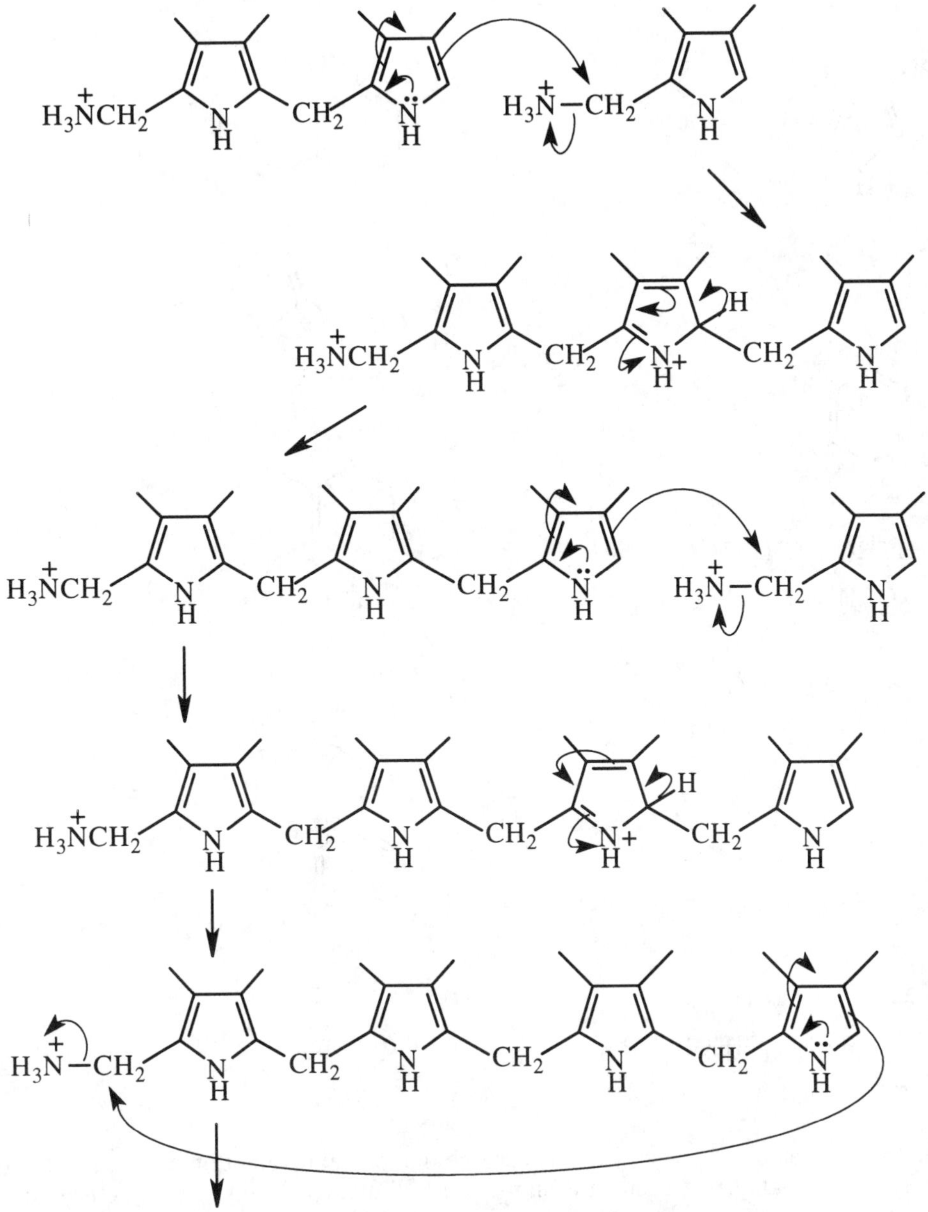

A proton is lost in the next step, which restores the aromaticity of the right-hand ring. Subsequent oxidation results in the formation of porphyrin.

21. **a.** 2-methylazacyclobutane
b. 4-chloropyrimidine
c. 4-bromo-8-chloroquinoline
d. 2-methylimidazole
e. 3,5-dimethylindole
f. 2-ethyl-5-methylpiperidine

22.

a. 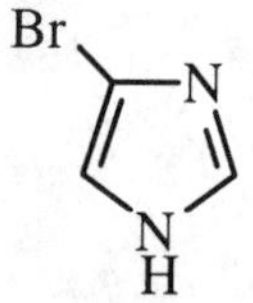 Br, N, N, H

b. 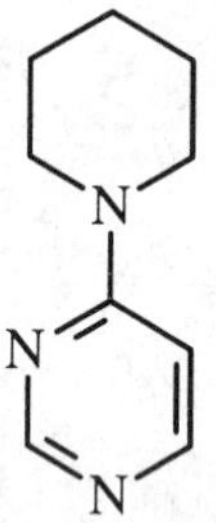 N, N, N

c. $CH_3\underset{|}{C}HCH_2CH_2\overset{+}{N}H_3$ with OCH_2CH_3 on the second carbon

d. 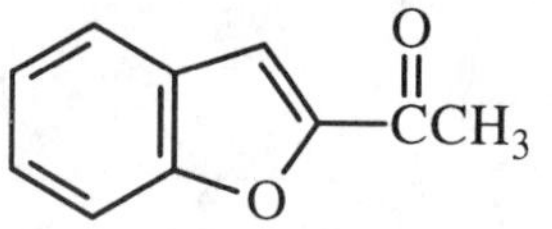 O, $\overset{O}{\overset{||}{C}}CH_3$

e. 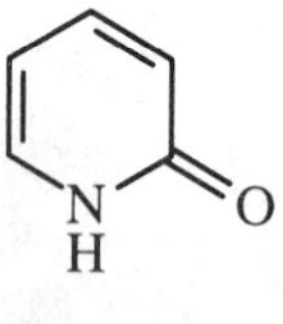 N, H, O

f. $\overset{+}{N}H_2CH_2CH_2CH_3$ Br^-

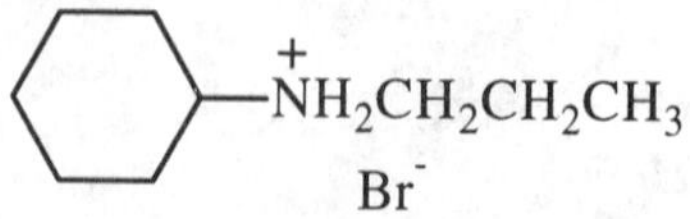

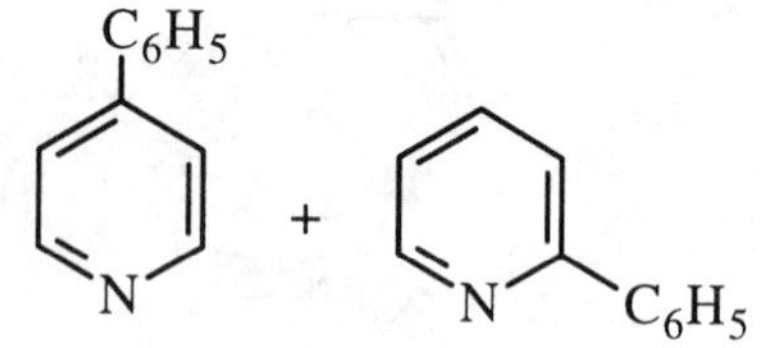

g. C_6H_5, N + N, C_6H_5

h. $CH_3\underset{\underset{O}{||}}{C}$, O, CH_3

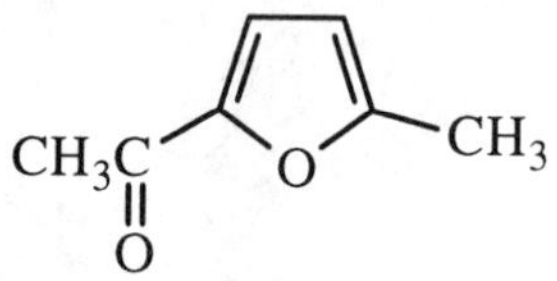

i. 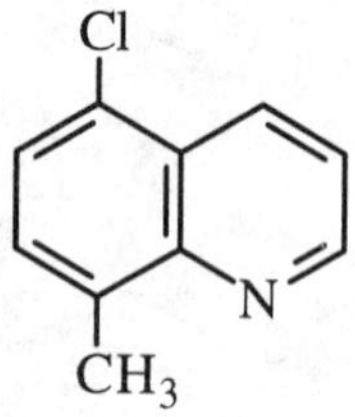 Cl, N, CH_3

j. 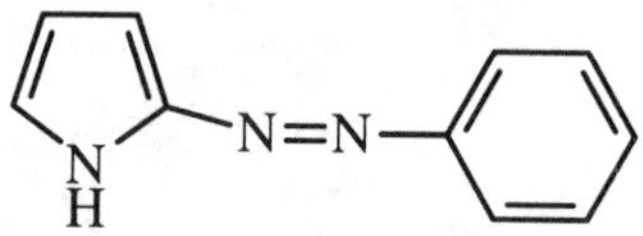 N, H, N=N

k. 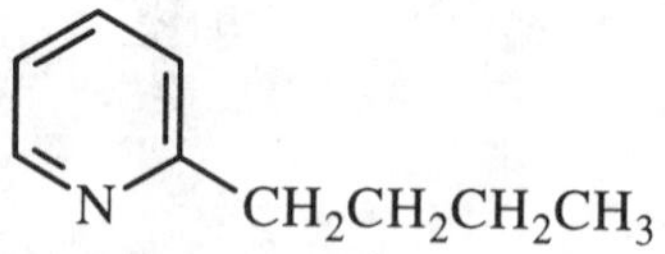 N, $CH_2CH_2CH_2CH_3$

l. 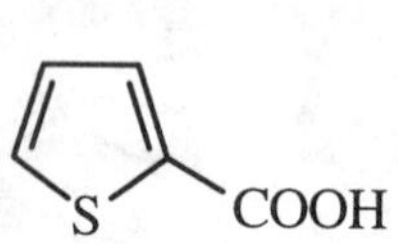 S, COOH

Because it is not stated otherwise, you can assume that one equivalent of each reagent is used. Therefore, the monoalkylated product is shown for "**f.**" If excess alkyl halide is used, dialkylated and trialkylated products will be obtained.

23. **A** is the most acidic because it becomes aromatic when it loses a proton.

B, **C**, and **D** are the next most acidic because in all three, the proton is bonded to a positively charged nitrogen atom.

B and **C** are more acidic than **D** because in **B** and **C** the proton to be lost is bonded to an sp^2 hybridized nitrogen atom, which is more electronegative than the sp^3 hybridized nitrogen atom in **D**.

B is more acidic than **C** because the uncharged nitrogen atom in **C** can donate electrons by resonance to the positively charged nitrogen atom.

Neutral compounds **E**, **F**, and **G** are the least acidic, with **E** and **F** more acidic than**G**, because **E** and **F** lose a proton from an sp^2 nitrogen atom, while a proton is lost from a less electronegative sp^3 nitrogen atom in **G**.

E is more acidic than **F** because of the electron-withdrawing second nitrogen in **E**.

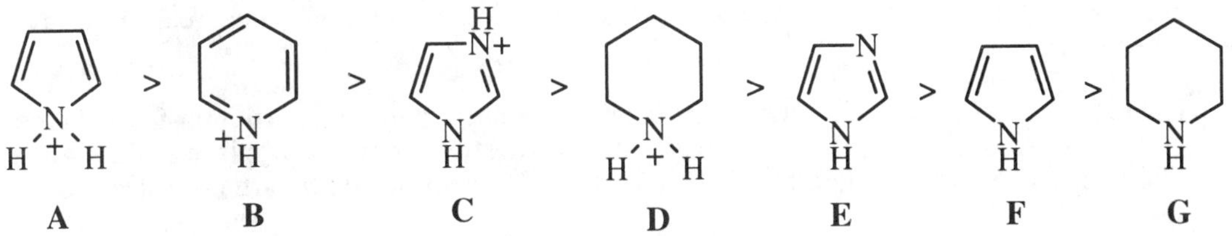

A > B > C > D > E > F > G

24. The compound on the right (shown below) is easier to decarboxylate because the electrons left behind when CO_2 is removed can be delocalized onto nitrogen and stabilized by accepting a proton from oxygen. The electrons left behind when the other compound loses CO_2 cannot be delocalized.

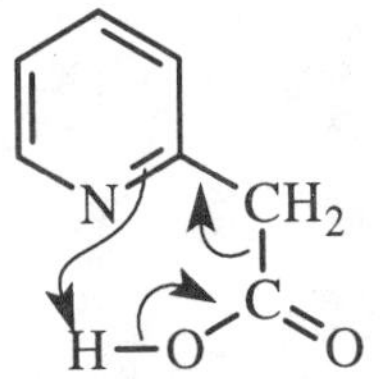

25. Note that the *N*-, *O*-, and *S*-substituted benzenes have the same relative reactivity toward electrophilic substitution as the *N*-, *O*-, and *S*-containing five-membered heterocyclic rings and for the same reason.

The *N*-substituted benzene is more reactive than the *O*-substituted benzene, because nitrogen is more effective than oxygen at donating electrons into the benzene ring, since it is less electronegative than oxygen. (Recall that electrophilic substitution is aided by electron donation into the ring.)

The *S*-substituted benzene is the least reactive because the nonbonding electrons of sulfur are in a 3*p* orbital, whereas the nonbonding electrons of nitrogen and oxygen are in a 2*p* orbital. The overlap of a 3*p* orbital with the 2*p* orbitals of carbon is less effective than the overlap of a 2*p* orbital with the 2*p* orbitals of carbon.

$NHCH_3$ > OCH_3 > SCH_3

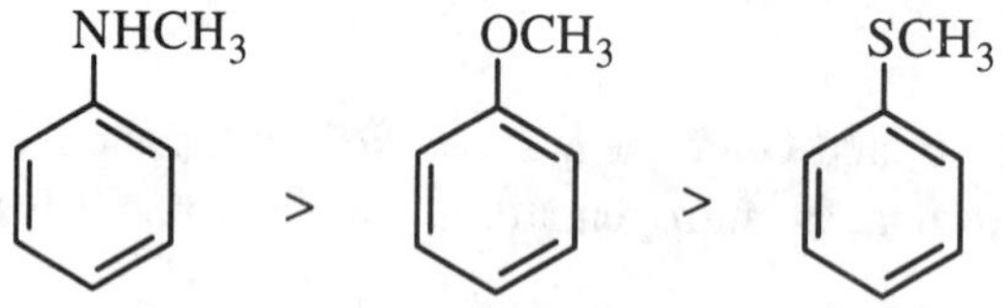

26. Because the slow step in aromatic electrophilic substitution is getting the electrophile onto the ring, electrophilic substitution will take place at the ring position with the greatest electron density.

The ketone substituent withdraws electrons by resonance, putting a partial positive charge on the #1, #3, and #5 positions. Therefore, electrophilic substitution will take place at the #4 and #6 positions, since they are more electron dense than the positions with partial positive charges.

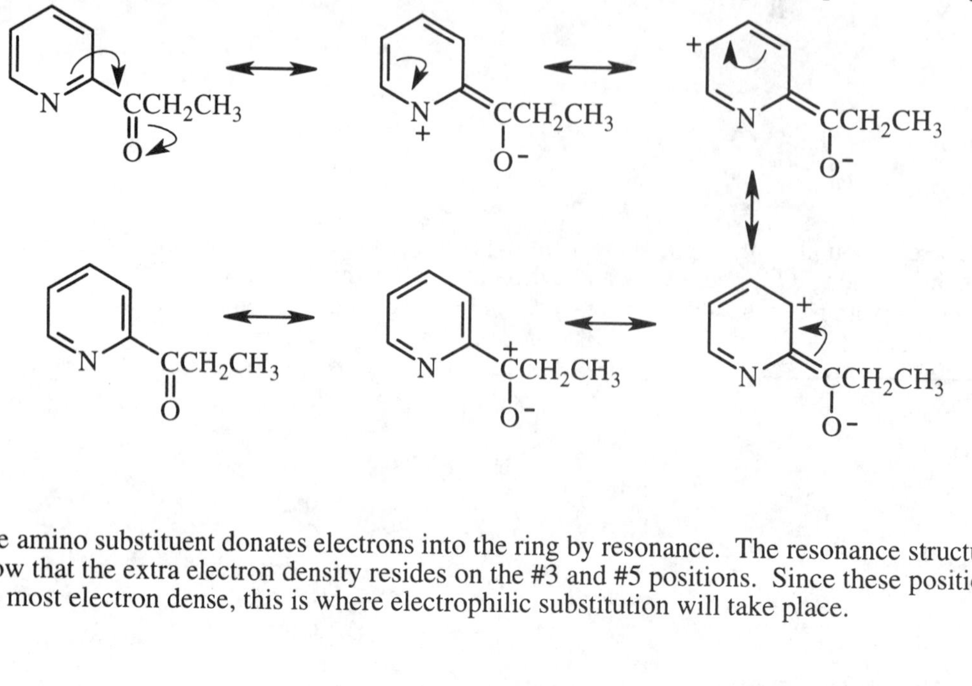

N, CCH_2CH_3, O ⟷ N+, CCH_2CH_3, O- ⟷ +, N, CCH_2CH_3, O-

N, CCH_2CH_3, O ⟷ N, $\overset{+}{C}CH_2CH_3$, O- ⟷ +, N, CCH_2CH_3, O-

The amino substituent donates electrons into the ring by resonance. The resonance structures show that the extra electron density resides on the #3 and #5 positions. Since these positions are the most electron dense, this is where electrophilic substitution will take place.

NHCH$_2$CH$_3$ ⟷ $\overset{+}{N}$HCH$_2$CH$_3$ ⟷ $\overset{+}{N}$HCH$_2$CH$_3$ ⟷ $\overset{+}{N}$HCH$_2$CH$_3$ ⟷ NHCH$_2$CH$_3$

27. **a.** The Lewis acid, $AlCl_3$, complexes with nitrogen, causing the aziridine ring to open when it is attacked by the nucleophilic benzene ring. The ring will open in the direction that puts the partial positive charge on the more substituted carbon (more stable carbocation).

CH$_3$ δ− δ+N---AlCl$_3$ δ+ CH$_3$

The major and minor products are those shown.

major	minor
C$_6$H$_5$CH(CH$_3$)CH$_2$NHCH$_3$	C$_6$H$_5$CH$_2$CH(CH$_3$)NHCH$_3$

b. Yes, epoxides can undergo similar reactions.

28. Oxygen is the negative end of the dipole in both compounds. Tetrahydrofuran has the greater dipole moment because in furan the effect of the electron-withdrawing oxygen is mitigated by the ability of oxygen to donate electrons by resonance into the ring.

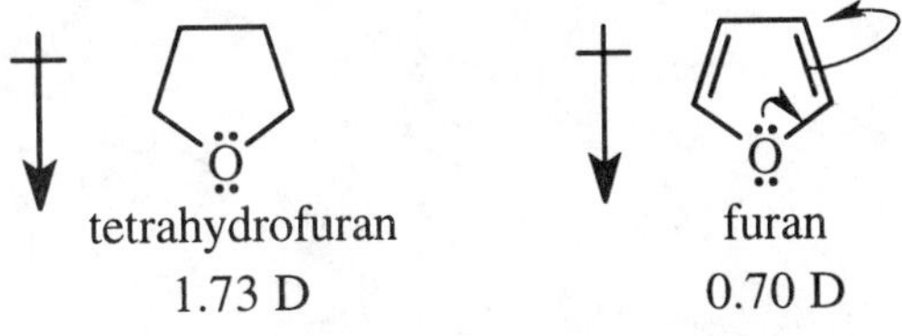

29. We saw in Section 15.2 of the text that an alkyl substituent bonded to a benzene ring can be oxidized to a carboxylic acid substituent. Similarly, an alkyl substituent bonded to a pyridine ring can be oxidized to a carboxylic acid substituent. Nicotine is an alkaloid found in tobacco leaves. It is used in agriculture as an insecticide.

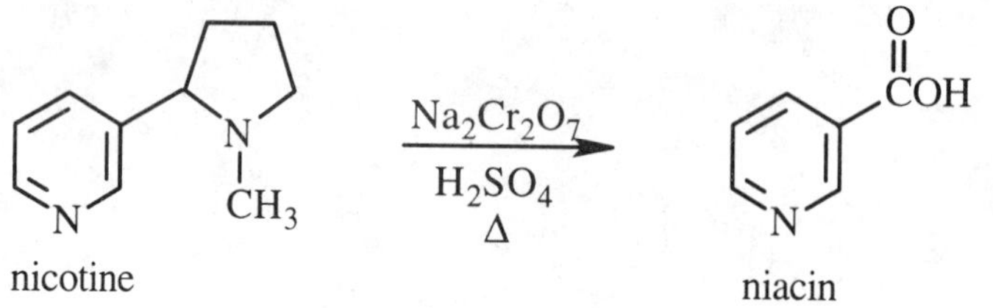

nicotine $\xrightarrow[\text{H}_2\text{SO}_4,\ \Delta]{\text{Na}_2\text{Cr}_2\text{O}_7}$ niacin

30.

$H_2C{=}\overset{+}{O}H$ (from $H_2C{=}O$, H^+ / $-H^+$) → pyrrole attacks → CH_2OH adduct ($\overset{+}{N}H$) → 2-(hydroxymethyl)pyrrole (CH_2OH)

$-H^+$ / H^+ ⇌ $CH_2{-}\overset{+}{O}H_2$ → loss of water ($+ H_2O$) → $\overset{+}{N}H{=}CH_2$ intermediate, attacked by pyrrole → CH_2-bridged cation → dipyrrolylmethane (pyrrole–CH_2–pyrrole)

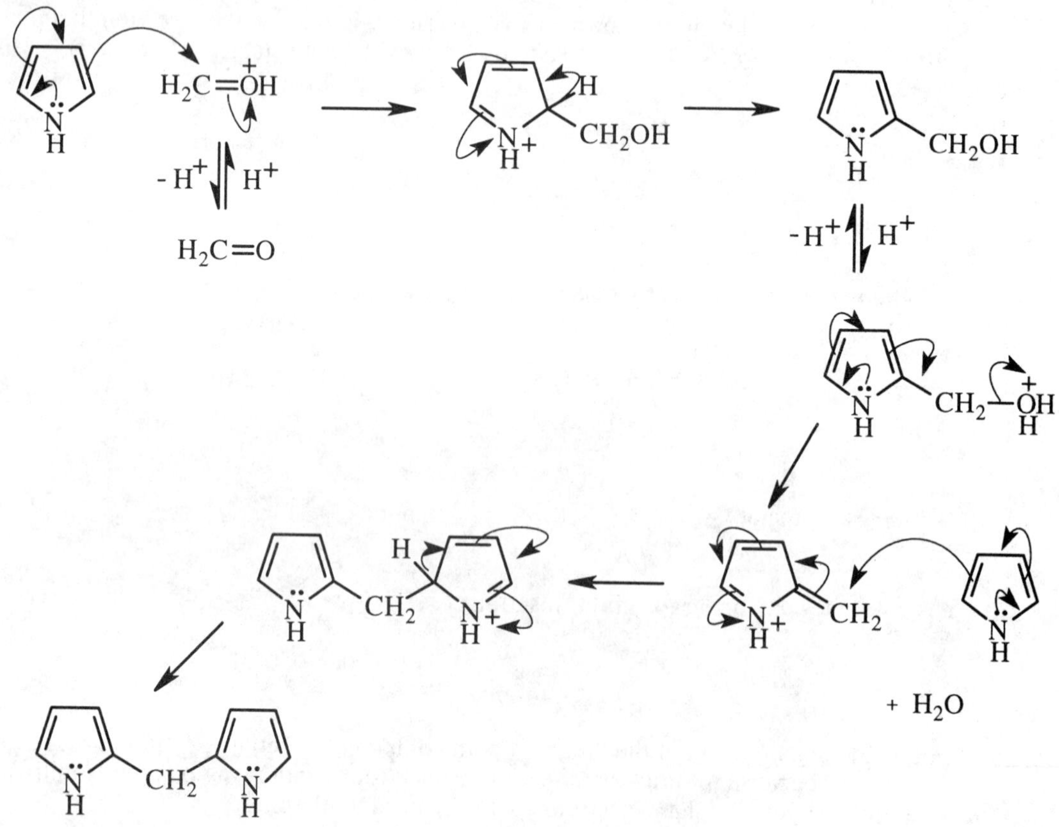

31. Pyrrolidine is a saturated nonaromatic compound, whereas pyrrole and pyridine are unsaturated aromatic compounds. The C-2 hydrogens of pyrrolidine are at δ 2.82, about where one would expect the signal for hydrogens bonded to an sp^3 carbon adjacent to an electron-withdrawing amino group.

The C-2 hydrogens of pyrrole and pyridine are expected to be farther downfield because of magnetic anisotropy. Because the nitrogen of pyrrole donates electrons into the ring and the nitrogen of pyridine withdraws electrons from the ring, the C-2 hydrogens of pyrrole are in an environment with a greater electron density, so they should show a signal upfield relative to the C-2 hydrogens of pyridine. Thus, the C-2 hydrogens of pyrrole are at δ 6.42, and the C-2 hydrogens of pyridine are at δ 8.50.

32. A UV spectrum results from the π electron system. The lone pair of electrons on the nitrogen atom in aniline are delocalized into the benzene ring and, thus, are part of the π system. Protonation of aniline removes two electrons from the π system, which has a significant effect on its UV spectrum.

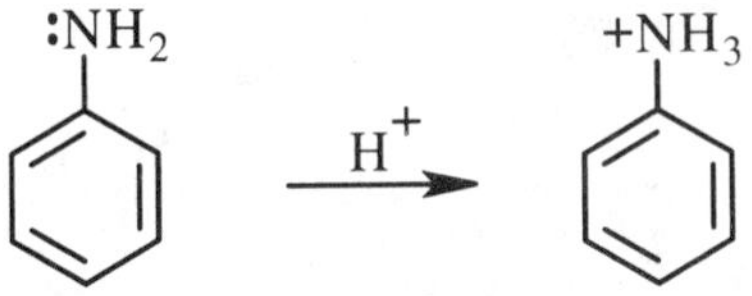

:NH_2 $\xrightarrow{H^+}$ $+NH_3$

The lone pair of electrons on the nitrogen atom in pyridine are sp^2 electrons and thus are not part of the π system. Protonation of pyridine, therefore, does not remove any electrons from the π system and has only a minor effect on the UV spectrum.

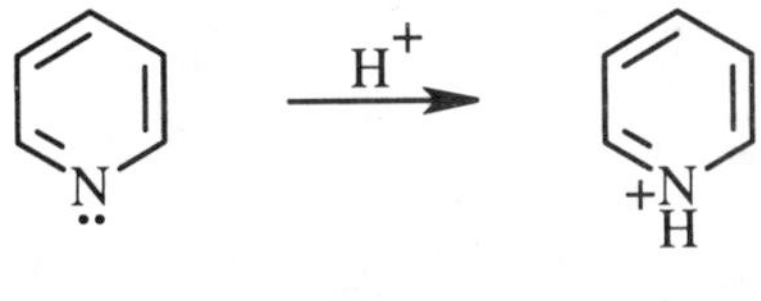

N: $\xrightarrow{H^+}$ +N H

33. When ammonia loses a proton, the electrons left behind remain on nitrogen.

$\ddot{N}H_3 \xrightarrow{-H^+} {}^{-}\!:\ddot{N}H_2$

When pyrrole loses a proton, the electrons left behind can be delocalized onto the four ring carbons. Electron delocalization stabilizes the anion and makes it easier to form.

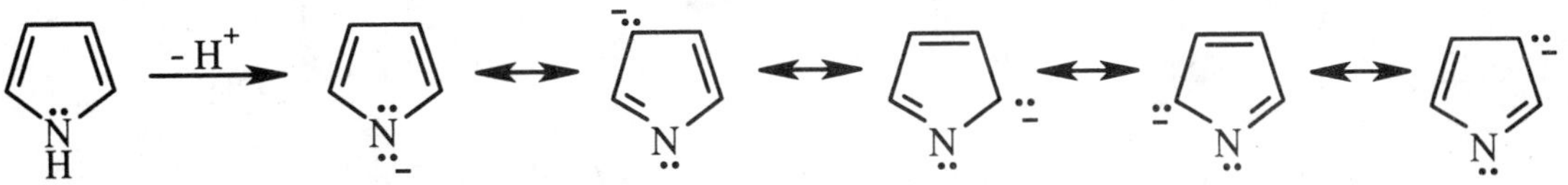

N H $\xrightarrow{-H^+}$ N ⟷ N ⟷ N ⟷ N ⟷ N

34. Before we can answer the questions, we must figure out the mechanism of the reaction. Once the mechanism is known, it will be easy to determine how a change in a reactant will affect the product. The mechanism is shown below:

Acrolein, an α,β-unsaturated aldehyde, undergoes a conjugate addition reaction with aniline. This is followed by an intramolecular electrophilic aromatic substitution reaction. Dehydration of the alcohol results in 1,2-dihydroquinoline, which is oxidized to quinoline by nitrobenzene.

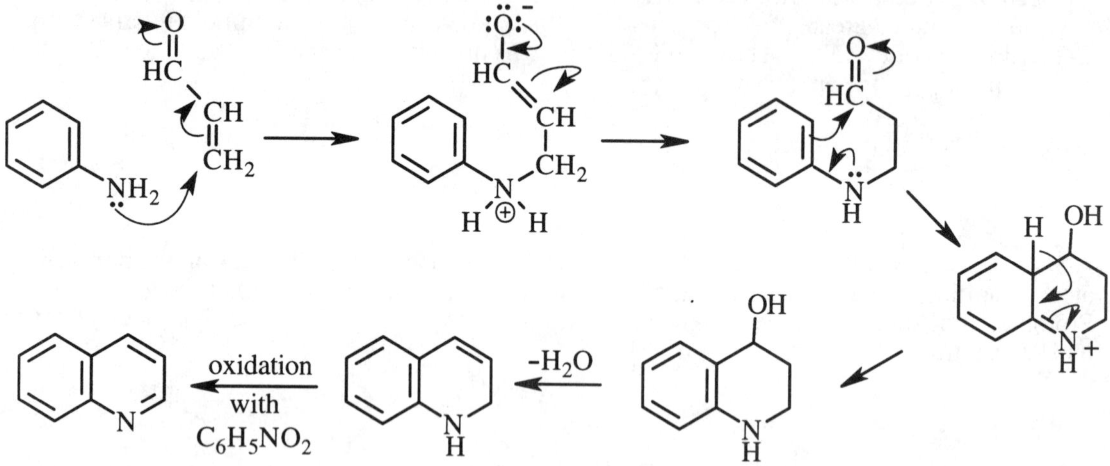

oxidation with $C_6H_5NO_2$

$-H_2O$

a. CH_3CH_2–C₆H₄–NH_2 + $HC(=O)CH=CH_2$ → 6-ethylquinoline

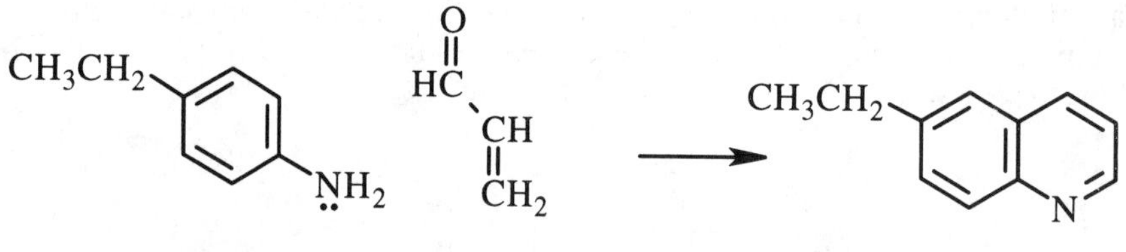

b. $C_6H_5NH_2$ + $CH_3C(=O)CH=CHCH_2CH_3$ → 4-methyl-2-ethylquinoline

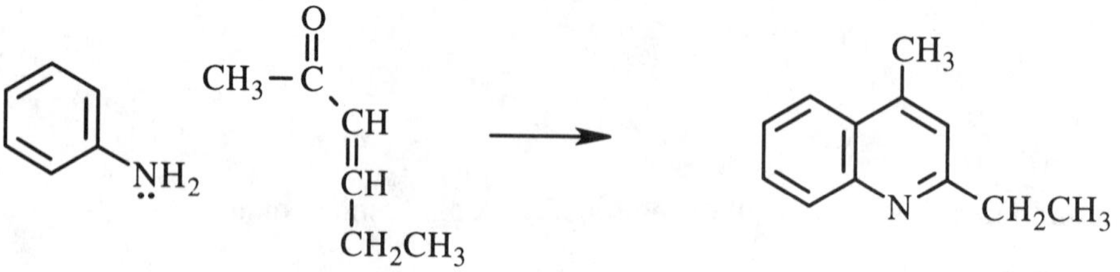

c. 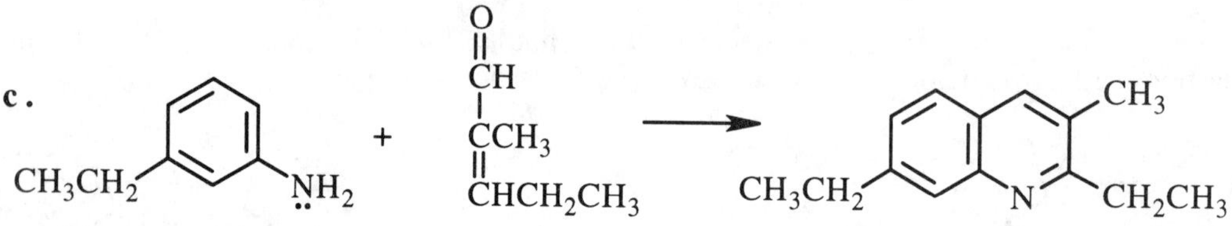 *meta*-ethylaniline + 2-methyl-2-pentenal → 2,7-diethyl-3-methylquinoline

meta-ethylaniline   2-methyl-2-pentenal   2,7-diethyl-3-methylquinoline

35.

a.

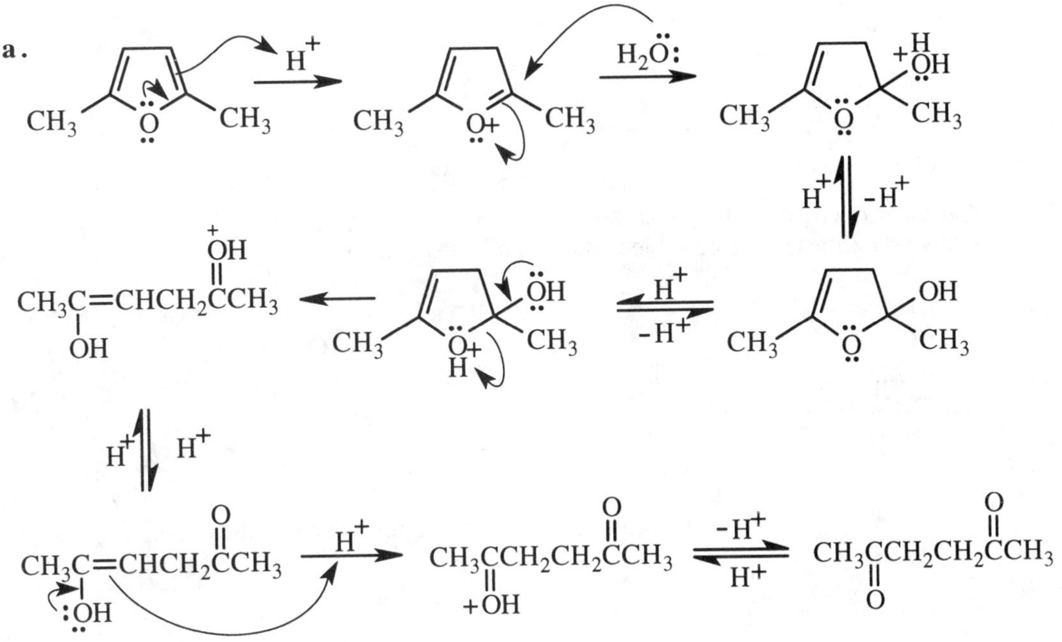

$$CH_3C=CHCH_2\overset{O}{\overset{\|}{C}}CH_3 \xrightarrow{H^+} CH_3\underset{+OH}{\underset{\|}{C}}CH_2CH_2\overset{O}{\overset{\|}{C}}CH_3 \underset{H^+}{\overset{-H^+}{\rightleftharpoons}} CH_3\underset{O}{\underset{\|}{C}}CH_2CH_2\overset{O}{\overset{\|}{C}}CH_3$$

b.

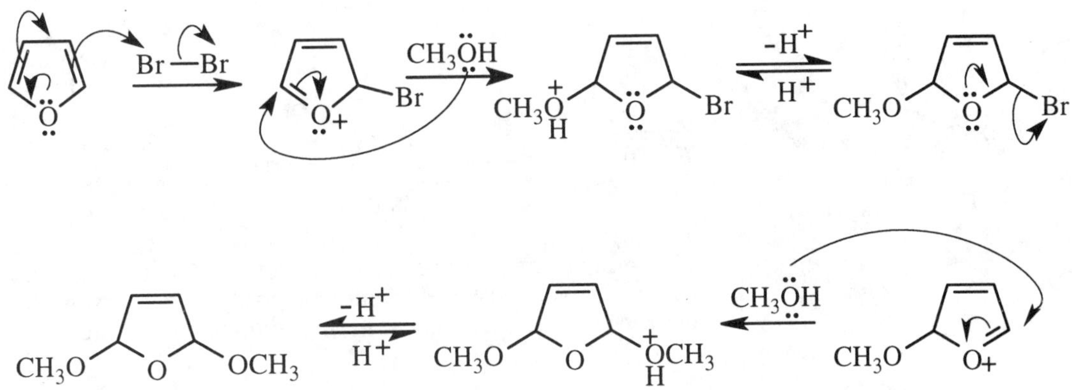

36.

a. 3-acetylfuran $\xrightarrow{HNO_3}$ 2-nitro-4-acetylfuran (O_2N, $\overset{O}{\overset{\|}{C}}CH_3$)

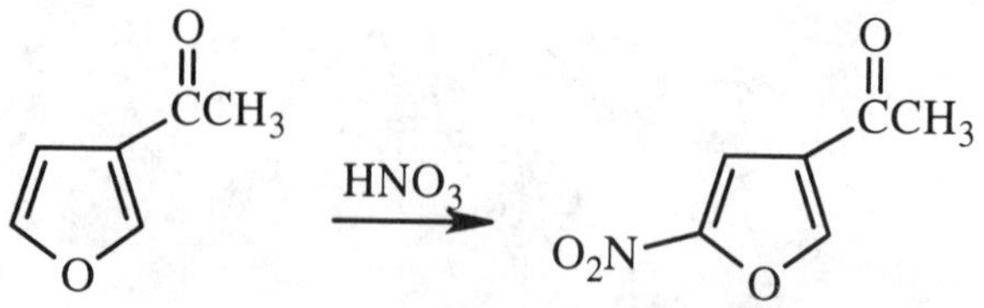

You can see why the nitro substituent goes to this position by examining the relative stabilities of the possible carbocation intermediates.

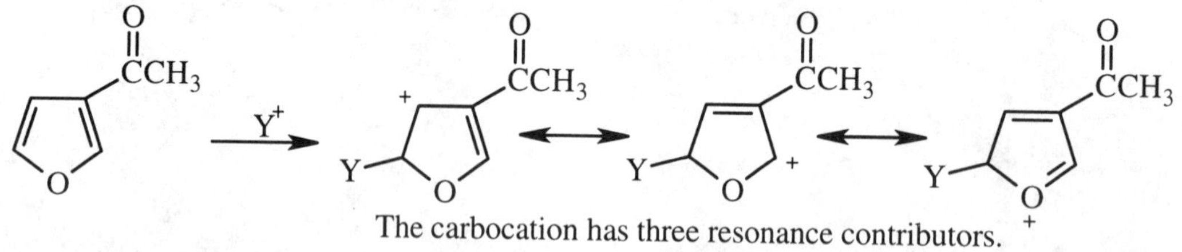

$\overset{O}{\overset{\|}{C}}CH_3$, Y^+, Y, $+$

The carbocation has three resonance contributors.

$\overset{O}{\overset{\|}{C}}CH_3$, Y^+, Y, $+$

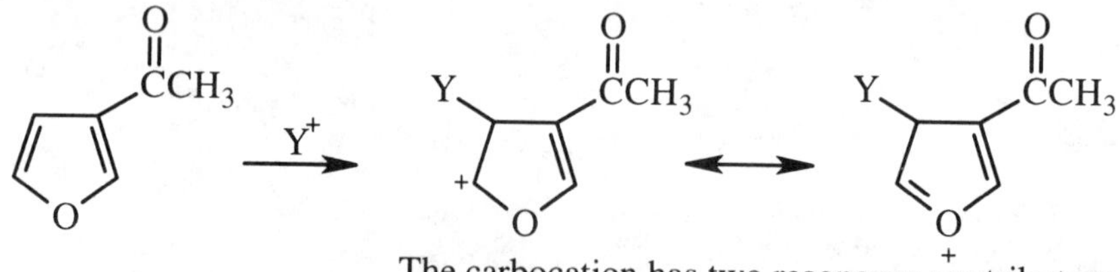

The carbocation has two resonance contributors.

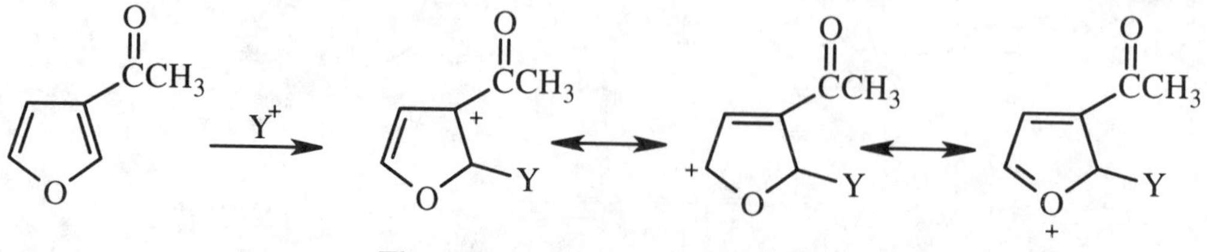

$\overset{O}{\overset{\|}{C}}CH_3$, Y^+, Y, $+$

The carbocation has three resonance contributors but the first one is relatively unstable because the positive charge is on the carbon attached to the electron-withdrawing substituent.

b.

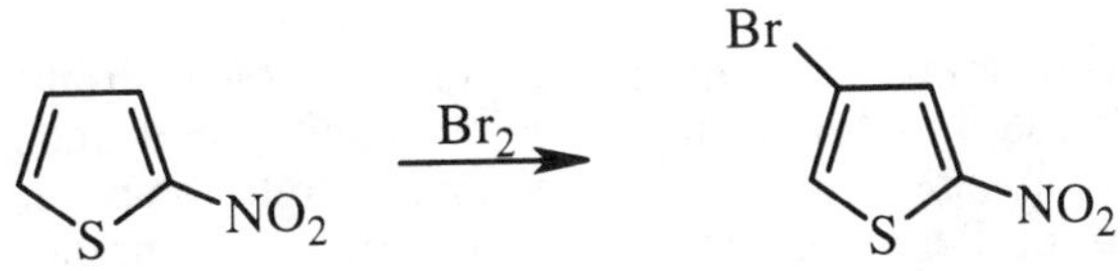

$\xrightarrow{Br_2}$ (2-nitrothiophene → 4-bromo-2-nitrothiophene)

You can see why the bromo substituent goes to this position by examining the relative stabilities of the possible carbocation intermediates.

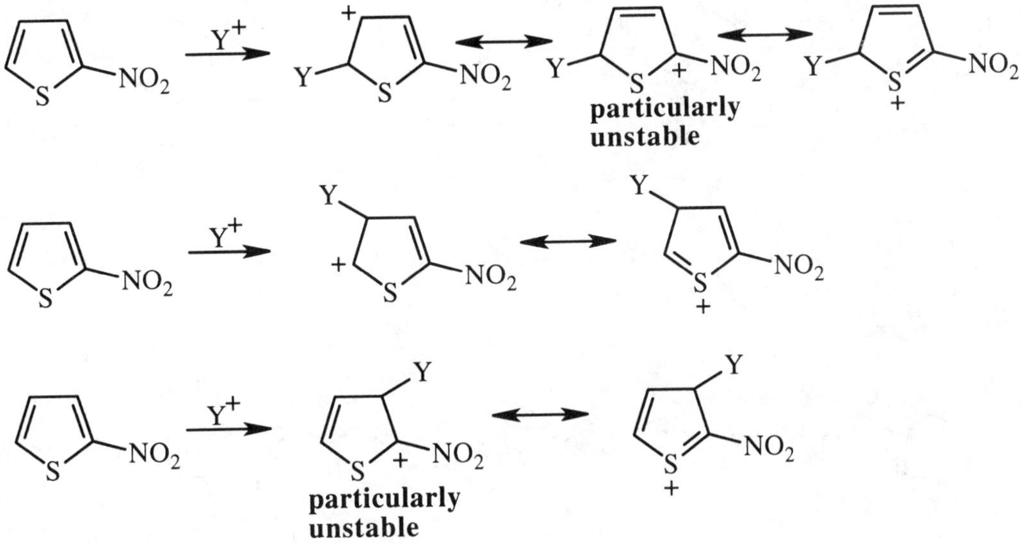

$\xrightarrow{Y^+}$ Y, NO_2 ↔ Y, NO_2 ↔ Y, NO_2

particularly unstable

$\xrightarrow{Y^+}$ Y, NO_2 ↔ Y, NO_2

$\xrightarrow{Y^+}$ Y, NO_2 ↔ Y, NO_2

particularly unstable

c.

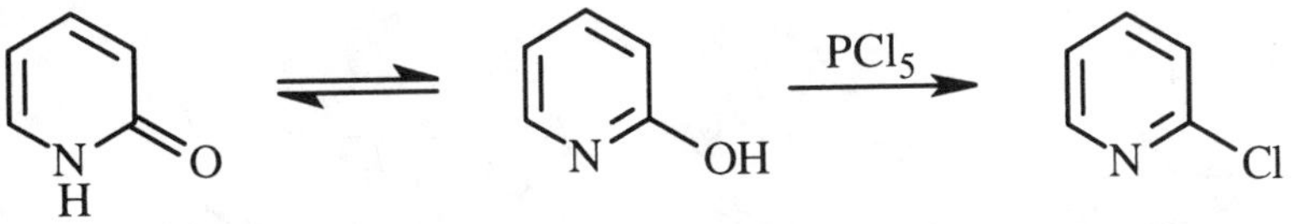

$\rightleftharpoons$ N–H, O ; N, OH $\xrightarrow{PCl_5}$ N, Cl

PCl_5 substitutes a Cl for an OH, as it does in alcohols and carboxylic acids.

d.

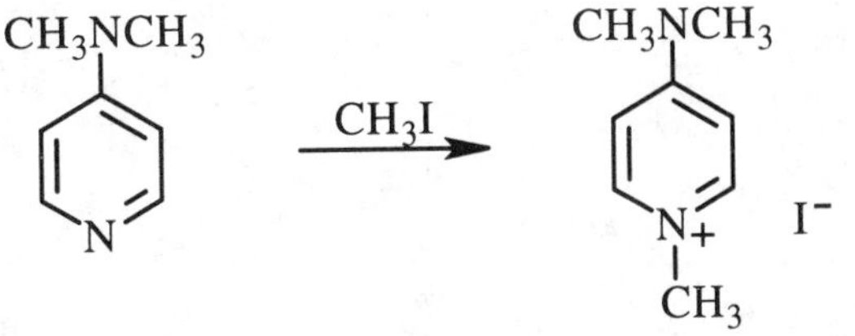

CH_3NCH_3 $\xrightarrow{CH_3I}$ CH_3NCH_3, N^+–CH_3 I^-

The other nitrogen is not alkylated, because its nonbonding electrons are delocalized into the pyridine ring, so they are not available to react with the alkyl halide.

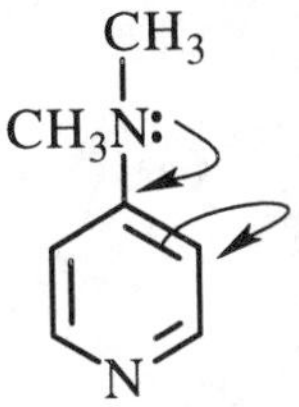

CH_3, $CH_3\ddot{N}$, N

e. In the first step, hydroxide ion removes the most acidic hydrogen from the compound. A hydrogen bonded to the C-4 methyl group is the most acidic hydrogen because the electrons left behind when the proton is removed can be delocalized into the pyridinium ring. In contrast, the electrons left behind when a proton is removed from the N-methyl group cannot be delocalized.

CH_3 N+ CH_3 $\xrightarrow{HO^-}$ $\ddot{C}H_2^-$ N+ CH_3 $\xrightarrow{H_2C{=}O}$ $CH_2CH_2O^-$ N+ CH_3 $\xrightarrow{H^+}$ CH_2CH_2OH N+ CH_3

f. $\ddot{N}$H + $CH_3CH_2{-}MgBr$ ⟶ $\ddot{N}^-$ + CH_3CH_3 + Mg^{2+} + Br^-

g. N $\xrightarrow{H_2O_2}$ N+ O^- $\xrightarrow[\Delta]{Br_2, FeBr_3}$ Br N+ O^- $\xrightarrow{PCl_3}$ Br N

37.

N+ CH_3 $:\ddot{O}:^-$ + $CH_3\overset{O}{\overset{\|}{C}}O\overset{O}{\overset{\|}{C}}CH_3$ ⟶ N+ CH_3 $:\ddot{O}{-}\overset{CH_3}{C}(\ddot{O}:^-){-}O\overset{O}{\overset{\|}{C}}CH_3$ ⟶ N+ $CH_2{-}H$ $:\ddot{O}{-}\overset{O}{\overset{\|}{C}}CH_3$ + $^-O\overset{O}{\overset{\|}{C}}CH_3$

⇌

$\ddot{N}$ $=CH_2$ $:\ddot{O}{-}\overset{O}{\overset{\|}{C}}CH_3$ ⟶ N $CH_2O\overset{O}{\overset{\|}{C}}CH_3$

38.

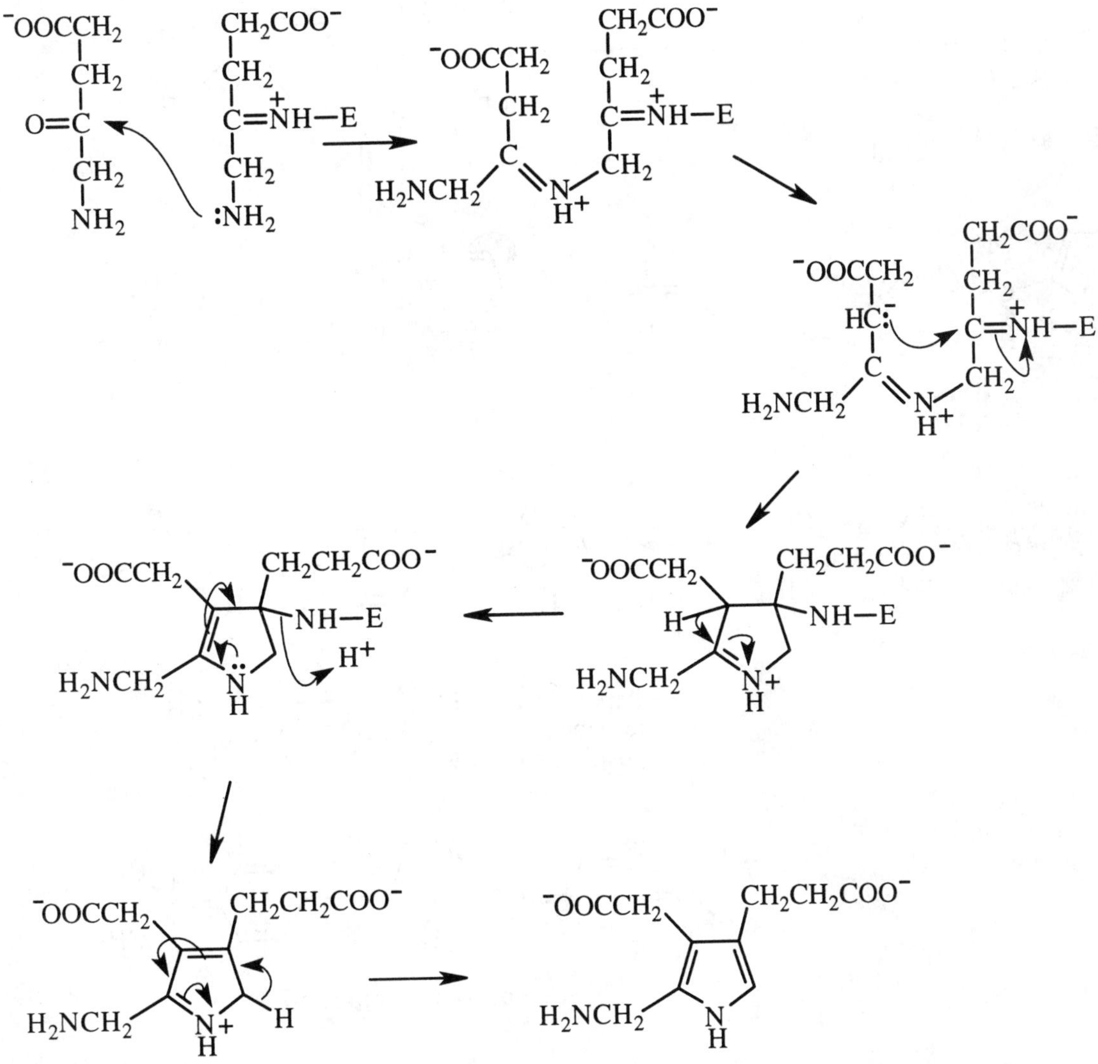

39.

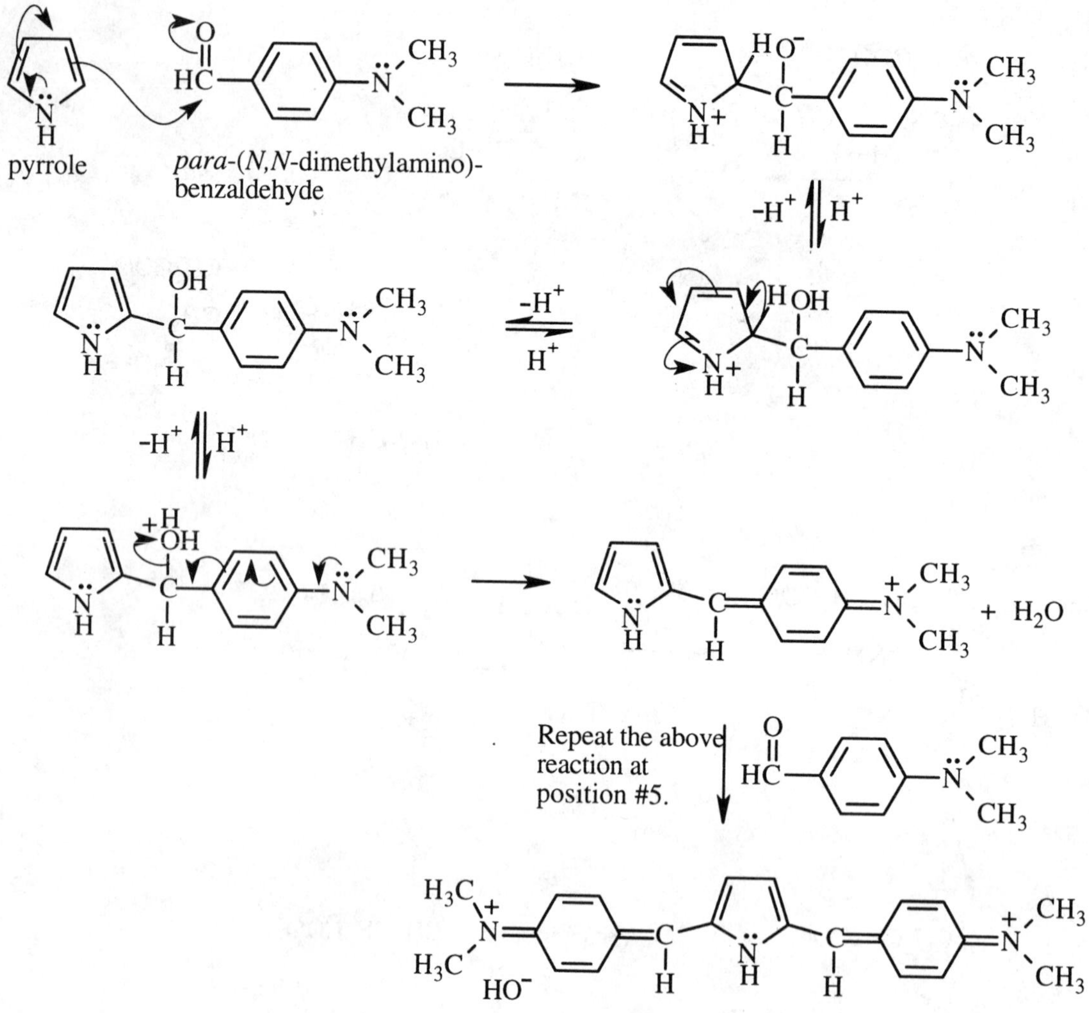

pyrrole

para-(*N,N*-dimethylamino)-benzaldehyde

$-H^+$ / H^+

$-H^+$ / H^+

$-H^+$ / H^+

+ H_2O

Repeat the above reaction at position #5.

colored compound

40.

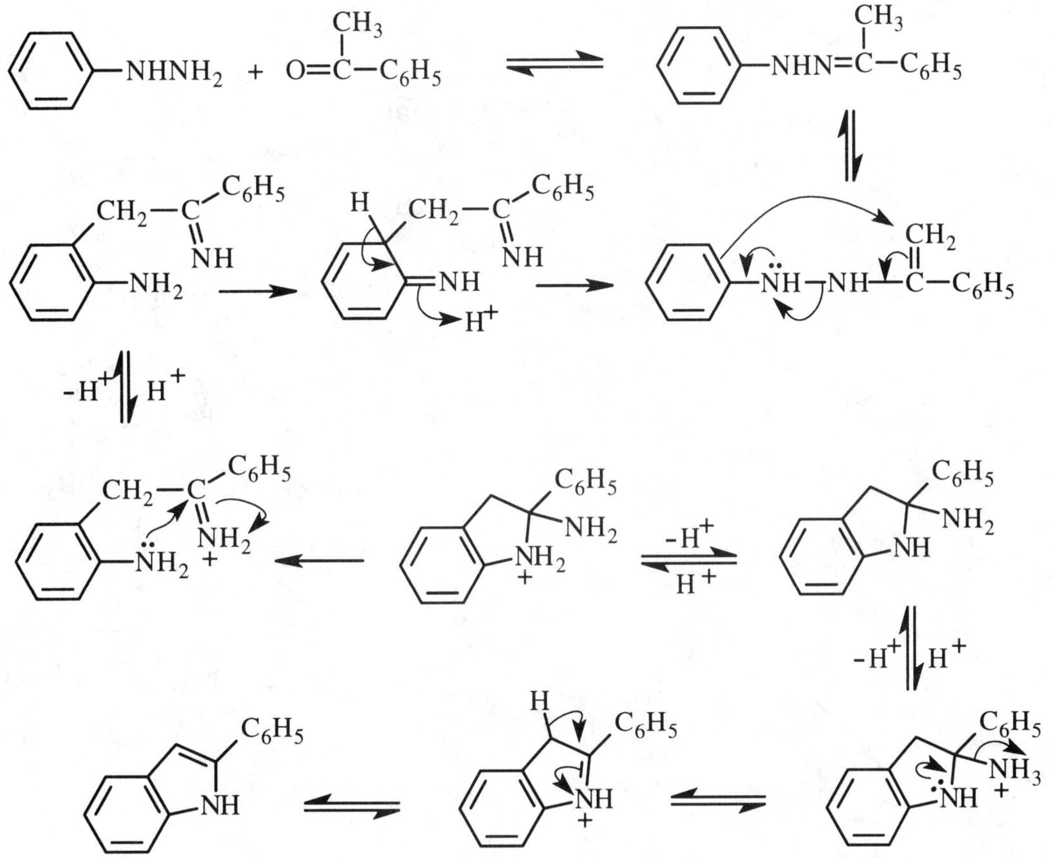

$C_6H_5NHNH_2 + O{=}C(CH_3){-}C_6H_5 \rightleftharpoons C_6H_5NHN{=}C(CH_3){-}C_6H_5$

$\rightleftharpoons C_6H_5\ddot{N}H{-}NH{-}C(=CH_2){-}C_6H_5 \longrightarrow$ (H, $CH_2{-}C(=NH){-}C_6H_5$, $=NH$, H^+) $\longrightarrow$ $CH_2{-}C(=NH){-}C_6H_5$, $-NH_2$

$-H^+ \rightleftharpoons H^+$

$CH_2{-}C(=\overset{+}{N}H_2){-}C_6H_5$, $-\ddot{N}H_2$ $\longrightarrow$ C_6H_5, NH_2, $\overset{+}{N}H_2$ $\underset{H^+}{\overset{-H^+}{\rightleftharpoons}}$ C_6H_5, NH_2, NH

$-H^+ \rightleftharpoons H^+$

C_6H_5, $\overset{+}{N}H_3$, $\ddot{N}H$ $\rightleftharpoons$ H, C_6H_5, $\overset{+}{N}H$ $\rightleftharpoons$ C_6H_5, NH

41.

a. 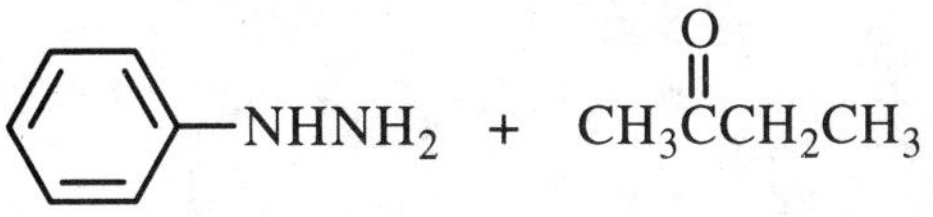 $C_6H_5NHNH_2 + CH_3\overset{O}{\overset{\|}{C}}CH_2CH_3$

b. 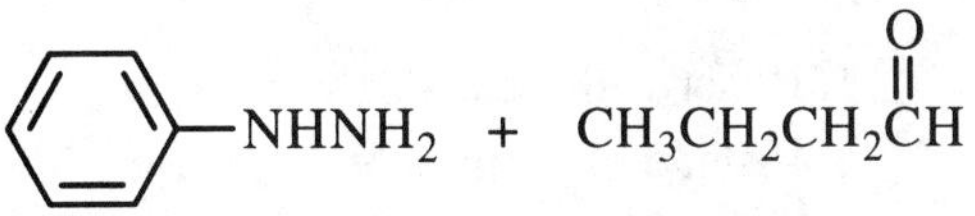 $C_6H_5NHNH_2 + CH_3CH_2CH_2\overset{O}{\overset{\|}{C}}H$

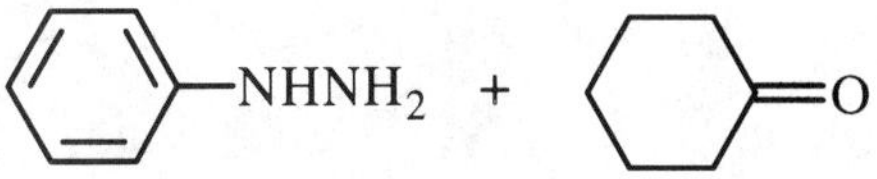

c. $C_6H_5NHNH_2$ + cyclohexanone (=O)

42.

HC=O: BF$_3$ → HC=$\overset{+}{O}\overset{-}{B}F_3$ → H $O\bar{B}F_3$ CH N+ H C_6H_5 → -H$^+$ → $O\bar{B}F_3$ CH N H C_6H_5

-H$^+$ ⇌ H$^+$

H +$O\bar{B}F_3$ CH N H C_6H_5

N H

CH N+ H C_6H_5 N H

N H CH H N+ H C_6H_5

N H CH N H C_6H_5

Repeat two more times.

N H CH N H CH N H CH N H C_6H_5 C_6H_5 C_6H_5

1. React with benzaldehyde.
2. Instead of using a new pyrrole, do an intramolecular reaction with the pyrrole at the end of the chain.

desired product

CHAPTER 28
Pericyclic Reactions

Important Terms

antarafacial bond formation	formation of two σ bonds from opposite sides of the π system.
antarafacial rearrangement	a rearrangement where the migrating group moves to the opposite face of the π system.
antibonding π molecular orbital	a molecular orbital that results when two atomic orbitals with opposite signs interact. Electrons in an antibonding orbital decrease bond strength.
asymmetric molecular orbital	a molecular orbital in which the left half is not a mirror image of the right half.
bonding π molecular orbital	a molecular orbital that results when two atomic orbitals with the same sign interact. Electrons in a bonding orbital increase bond strength.
Claisen rearrangement	a [3,3] sigmatropic rearrangement of an allyl vinyl ether.
conrotatory ring closure	achieves head-to-head overlap of π orbitals by rotating the orbitals in the same direction.
conservation of orbital symmetry theory	a theory that explains the relationship between the structure and stereochemistry of the reactant, the conditions under which a pericyclic reaction takes place, and the stereochemistry of the product.
Cope rearrangement	a [3,3] sigmatropic rearrangement of a 1,5-diene.
cycloaddition reaction	a reaction in which two π-bond-containing molecules react to form a cyclic compound.
disrotatory ring closure	achieves head-to-head overlap of π orbitals by rotating the orbitals in opposite directions.
electrocyclic reaction	a reaction in which a π bond in the reactant is lost so that a cyclic compound with a new σ bond can be formed.
excited state	a description of which orbitals the electrons of an atom or molecule occupy when an electron in the ground state has been moved to a higher energy orbital.
frontier orbital analysis	determining the outcome of a pericyclic reaction using frontier molecular orbitals.
frontier orbitals	the HOMO and the LUMO.
frontier orbital theory	a theory that, like the conservation of orbital symmetry, explains the relationship between reactant, product, and reaction conditions in a pericyclic reaction.

ground state	a description of which orbitals the electrons of an atom or molecule occupy when they are all in their lowest energy orbitals.
highest occupied molecular orbital (HOMO)	the molecular orbital of highest energy that contains an electron.
linear combination of atomic orbitals (LCAO)	the combination of atomic orbitals to produce a molecular orbital.
lowest unoccupied molecular orbital (LUMO)	the molecular orbital of lowest energy that does not contain an electron.
molecular orbital (MO) theory	describes a model in which the electrons occupy orbitals as they do in atoms but with the orbitals extending over the entire molecule.
pericyclic reaction	a reaction that occurs as a result of a cyclic reorganization of electrons.
photochemical reaction	a reaction that takes place when a reactant absorbs light.
polar reaction	the reaction between a nucleophile and an electrophile.
radical reaction	a reaction in which a new bond is formed using one electron from one reagent and one electron from another reagent.
selection rules	the rules that determine the outcome of a pericyclic reaction.
sigmatropic rearrangement	a reaction in which a σ bond is broken in the reactant, a new σ bond is formed in the product, and the π bonds rearrange.
suprafacial bond formation	formation of two σ bonds from the same side of the π system.
suprafacial rearrangement	a rearrangement where the migrating group remains on the same face of the π system.
symmetric molecular orbital	a molecular orbital in which the left half is a mirror image of the right half.
symmetry-allowed pathway	a pathway that leads to overlap of in-phase orbitals.
symmetry-forbidden pathway	a pathway that leads to overlap of out-of-phase orbitals.
thermal reaction	a reaction that takes place without the reactant having to absorb light.
Woodward-Hoffmann rules	a series of selection rules for pericyclic reactions.

Solutions to Problems

1. **a.** electrocyclic reaction **c.** cycloaddition reaction

   b. sigmatropic rearrangement **d.** cycloaddition reaction

2. **a.** bonding orbitals = ψ_1, ψ_2, ψ_3; antibonding orbitals = ψ_4, ψ_5, ψ_6

   b. ground-state HOMO = ψ_3; ground-state LUMO = ψ_4

   c. excited-state HOMO = ψ_4; excited-state LUMO = ψ_5

   d. symmetric orbitals = ψ_1, ψ_3, ψ_5; asymmetric orbitals = ψ_2, ψ_4, ψ_6

   e. The HOMO and LUMO have opposite symmetries.

3. **a.** 8 π molecular orbitals **b.** ψ_4 **c.** 8 nodes (includes the node that passes through the centers of the *p* orbitals).

4. **a.** 1,3-Pentadiene has two conjugated π bonds, so it has the same π molecular orbital description as 1,3-butadiene, a compound that also has two conjugated π bonds. (See Figure 28.2 on page 1168 of the text.)

   b. The π bonds in 1,4-pentadiene are isolated, so its π molecular orbital description is the same as ethene, a compound with an isolated π bond. (See Figure 28.1 on page 1167 of the text.)

   c. 1,3,5-Heptatriene has three conjugated π bonds, so it has the same π molecular orbital description as 1,3,5-hexatriene, a compound that also has three conjugated π bonds. (See Figure 28.3 on page 1169 of the text.)

   d. 1,3,5,8-Nonatetraene has three conjugated π bonds and an isolated π bond. The three conjugated π bonds are described by Figure 28.3 and the isolated π bond by Figure 28.1.

5. **a.**

2, 4, or 6 conjugated double bonds

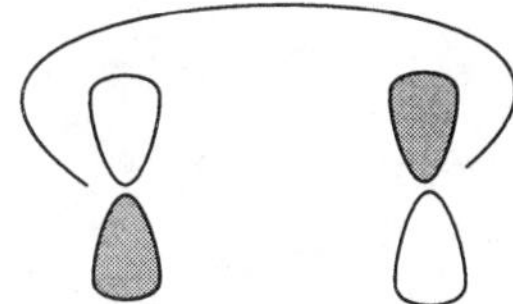

3, 5, or 7 conjugated double bonds

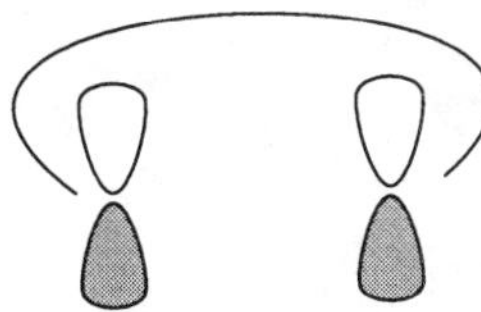

6. **a.** (2*E*,4Z,6Z,8*E*)-Decatetraene has an even number of conjugated π bonds (4). Therefore, under thermal conditions, ring closure will be conrotatory.

b. The substituents point in opposite directions, and conrotatory ring closure of such substituents will cause them to be trans in the ring-closed product.

c. Under photochemical conditions, a compound with four conjugated π bonds will undergo disrotatory ring closure.

d. Because ring closure is disrotatory and the substituents point in opposite directions, the product will have the cis configuration.

7. **a.** correct **b.** correct **c.** correct

8. **1.** **a.** conrotatory **b.** trans

2. **a.** disrotatory **b.** cis

9. The reaction of maleic anhydride with 1,3-butadiene involves three π bonds in the reacting system, and such a reaction under thermal conditions involves suprafacial ring closure.

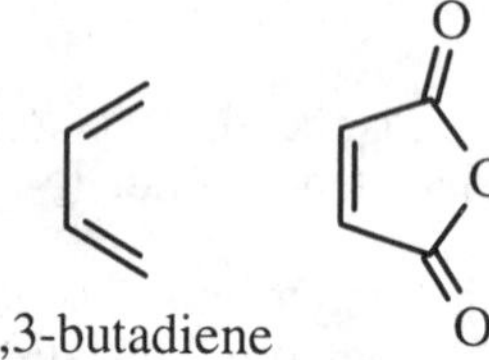

The reaction of maleic anhydride with ethylene involves two π bonds in the reacting system, and such a reaction under thermal conditions involves antarafacial ring closure, which cannot occur with a four-membered ring.

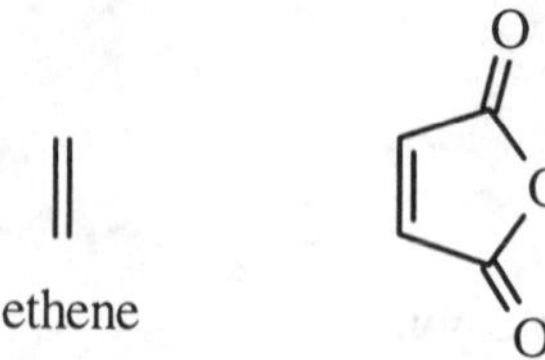

10. Solved in the text.

11. Because we are dealing with formation of a small ring, ring closure will have to be suprafacial. Under photochemical conditions, suprafacial ring closure requires an even number of π bonds in the reacting system. Thus, a concerted reaction will occur but it will use only one of the π bonds of 1,3-butadiene.

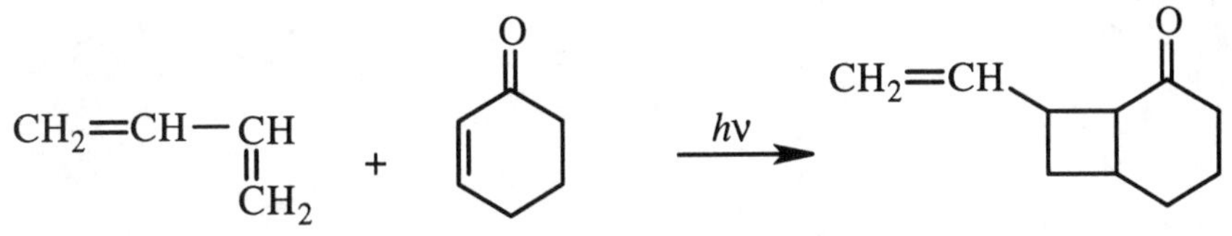

$CH_2{=}CH{-}CH({=}CH_2)$ + (cyclohexenone) $\xrightarrow{h\nu}$ $CH_2{=}CH$–(bicyclic cyclobutane-fused cyclohexanone)

12. a. 1. [1,7] sigmatropic rearrangement

2. [1,5] sigmatropic rearrangement

3. [5,5] sigmatropic rearrangement

4. [3,3] sigmatropic rearrangement

b. 1. $CH_2{-}H$ ... CH_2 $\xrightarrow{\Delta}$ CH_2 ... CH_3

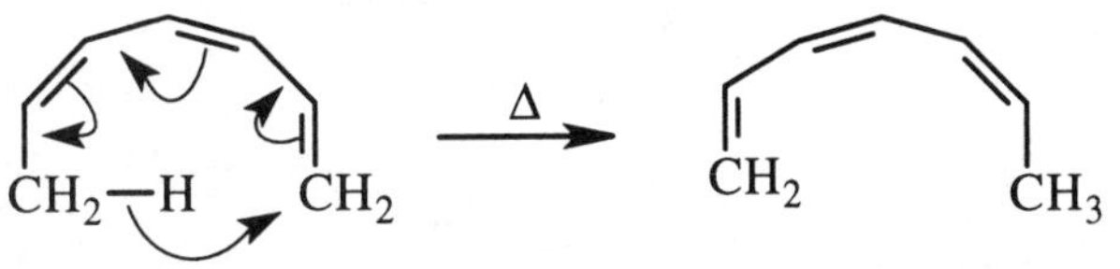

2. CH_2, H, CH_2 $\xrightarrow{\Delta}$ CH_3, CH_2

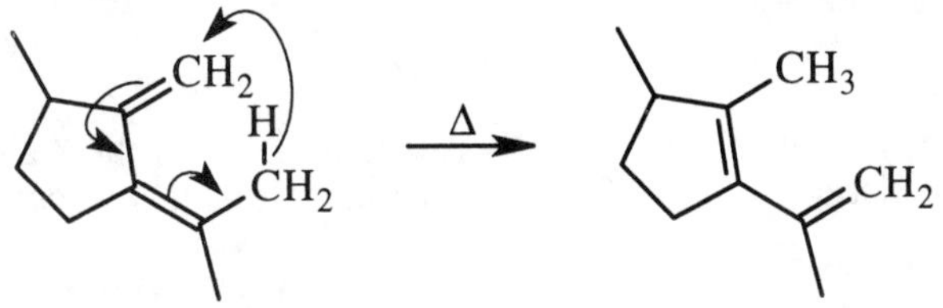

3. $\xrightarrow{\Delta}$

4. O ... CH_3 $\xrightarrow{\Delta}$ O ... H ... CH_3 $\longrightarrow$ OH ... CH_3

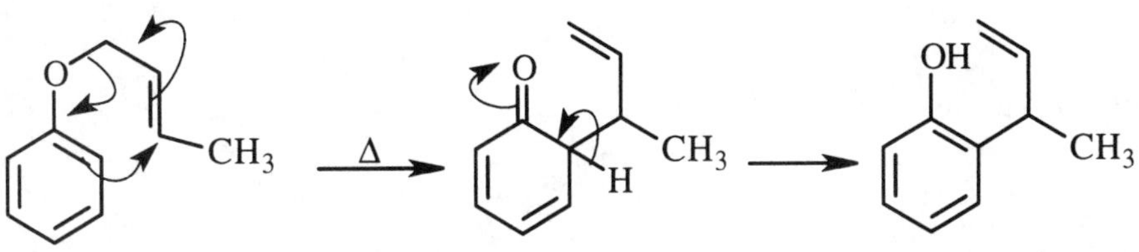

13.

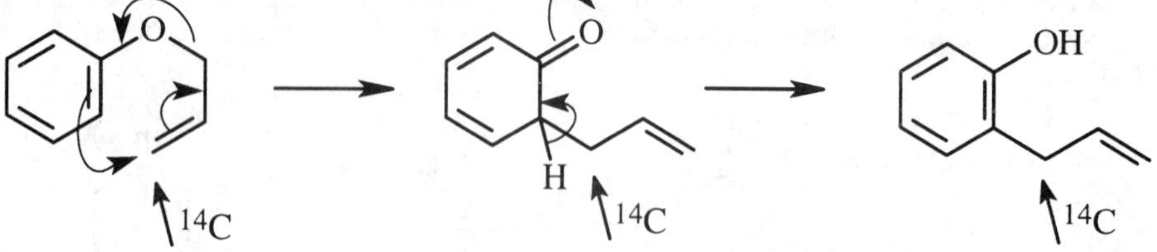

OH

^{14}C ^{14}C ^{14}C

H

14. If a nondeuterated reactant had been used, the product would have been identical to the reactant. Therefore, the rearrangement would not have been detectable.

15. A suprafacial rearrangement can take place under photochemical conditions if there are an even number of electrons in the reacting system. Therefore, a 1,3-hydrogen shift occurs.

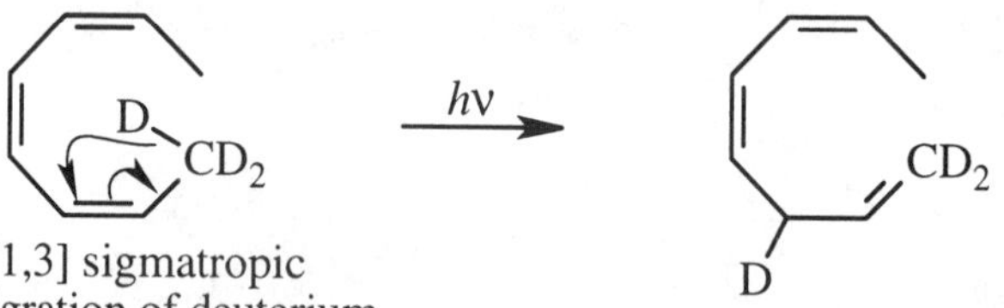

D CD_2 $h\nu$ CD_2 D

a [1,3] sigmatropic migration of deuterium

A suprafacial rearrangement can take place under thermal conditions if there are an odd number of electrons in the reacting system. Therefore, a 1,5-hydrogen shift occurs.

D CD_2 Δ D CD_2

a [1,5] sigmatropic migration of deuterium

16. Solved in the text.

17. [1,3] Sigmatropic migrations of hydrogen cannot occur under thermal conditions, because the four-membered transition state does not allow the required antarafacial rearrangement.

[1,3] Sigmatropic migrations of carbon can occur under thermal conditions because carbon can achieve the required antarafacial rearrangement by using both of its lobes when it migrates.

18. **a.** Because 1,3-migration of carbon requires carbon to migrate using both of its lobes (it involves an even number of pairs of electrons, so it takes place by an antarafacial pathway), migration will occur with inversion of configuration.

b. Because 1,5-migration of carbon requires carbon to migrate using only one of its lobes (it involves an odd number of pairs of electrons, so it takes place by a suprafacial pathway), migration will occur with retention of configuration.

19. Because the [1,7] sigmatropic rearrangement takes place under thermal conditions and involves an even number (4) of pairs of electrons, migration of hydrogen involves antarafacial rearrangement.

20. Because the reactant (provitamin D_3) has an odd number (3) of conjugated π bonds and reacts under photochemical conditions, ring closure is conrotatory. The methyl and hydrogen substituents point in opposite directions in provitamin D_3. Conrotatory ring closure will cause substituents that point in opposite directions in the reactant to be trans in the product.

21.

a. 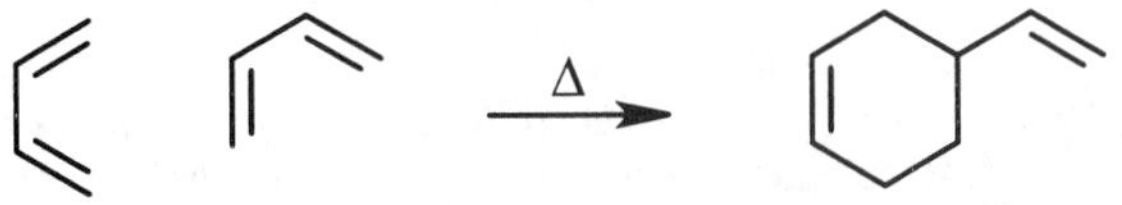 Δ

b. 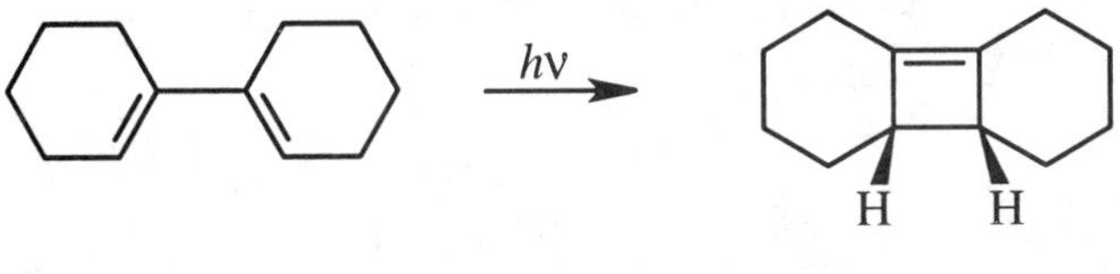 hν

H H

c. 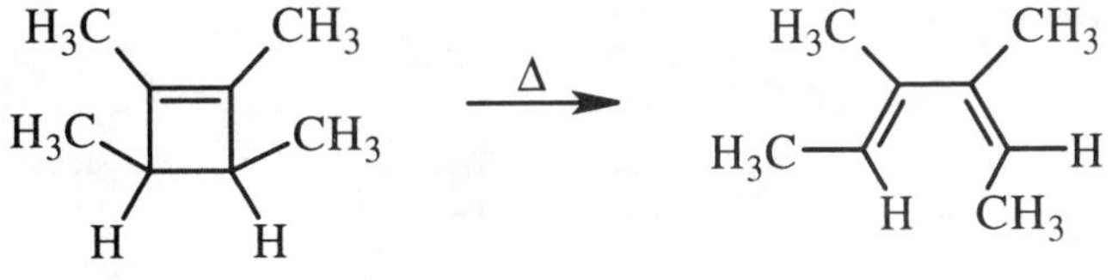 H_3C CH_3 H_3C CH_3 H H Δ H_3C CH_3 H_3C H H CH_3

d.  Δ

e. 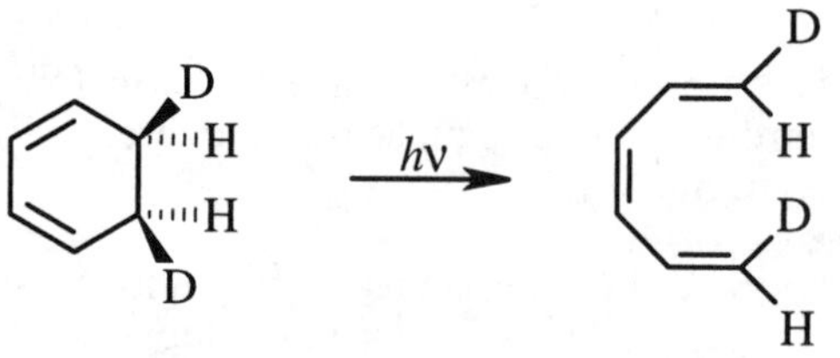 D, H, H, D — $h\nu$ → D, H, D, H

f. 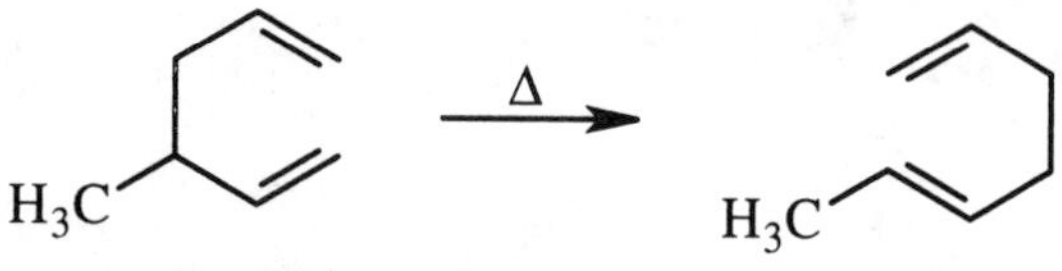 H_3C — Δ → H_3C

g.  H, CH_3, CH_3, H — Δ → CH_3, CH_3

h. Δ → H, H

22. 1. Because the compound has an even number of π bonds, it will undergo conrotatory ring closure under thermal conditions and disrotatory ring closure under photochemical conditions. Because the two methyl substituents point in opposite directions, they will be trans in the ring-closed product when ring closure is conrotatory and cis in the ring-closed product when ring closure is disrotatory.

a. 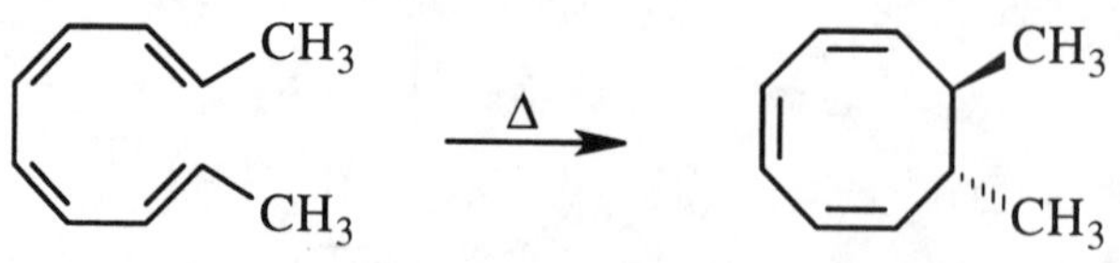 CH_3, CH_3 — Δ → CH_3, CH_3

b. 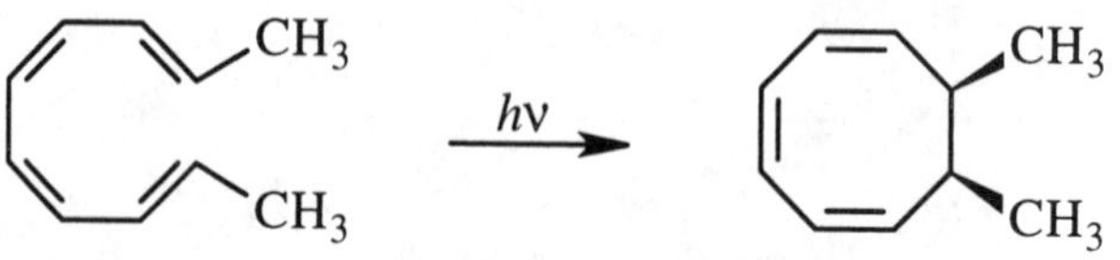 CH_3, CH_3 — $h\nu$ → CH_3, CH_3

2. Because the compound has an even number of π bonds, it will undergo conrotatory ring closure under thermal conditions and disrotatory ring closure under photochemical conditions. Because the two methyl substituents point in the same direction, they will be cis in the ring-closed product when ring closure is conrotatory and trans in the ring-closed product when ring closure is disrotatory.

a.

$\xrightarrow{\Delta}$

b.

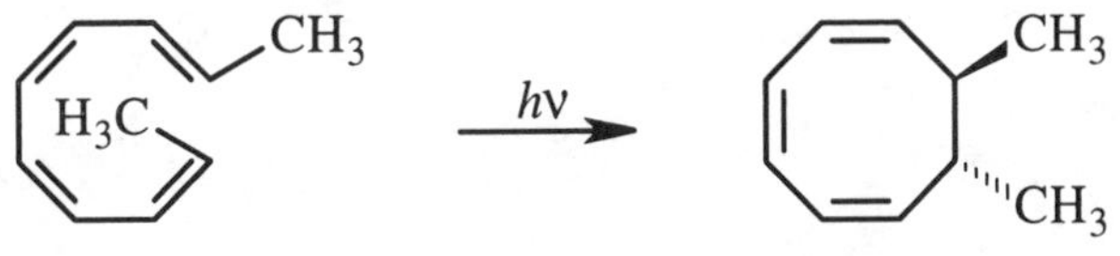

$\xrightarrow{h\nu}$

23. Chorismate mutase catalyzes a [3,3] sigmatropic rearrangement.

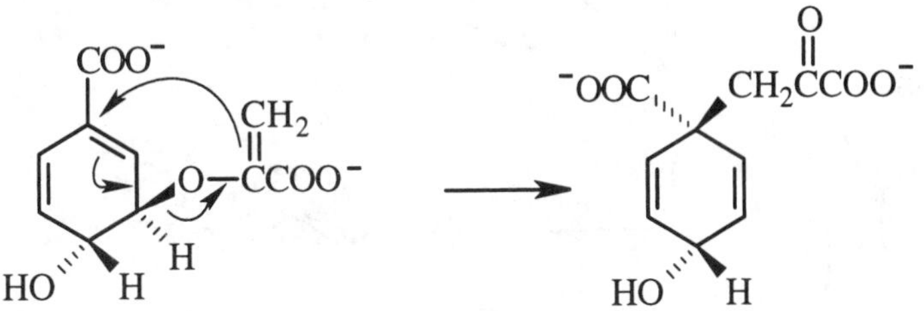

24. The hydrogens that end up at the ring juncture in the first reaction point in opposite directions in the reactant. Because ring closure is disrotatory (odd number of π bonds, thermal conditions), the hydrogens in the ring-closed product are cis. (See Table 28.2 on page 1168 of the text.)

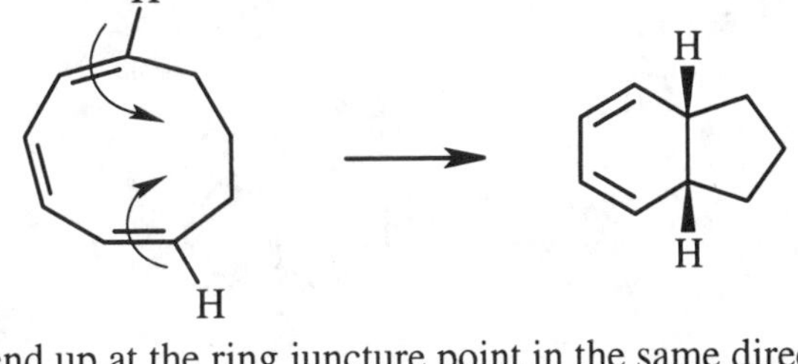

The hydrogens that end up at the ring juncture point in the same directions in the reactant. Ring closure is still disrotatory, so the hydrogens in the ring-closed product are trans.

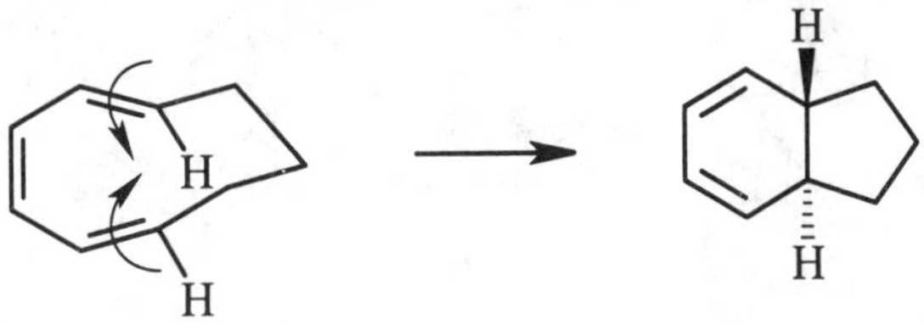

25.

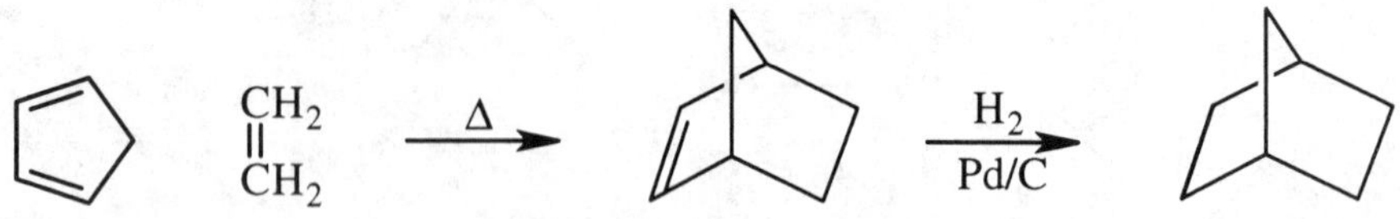

$CH_2=CH_2$ Δ → H₂ Pd/C →

26.

a. hν →

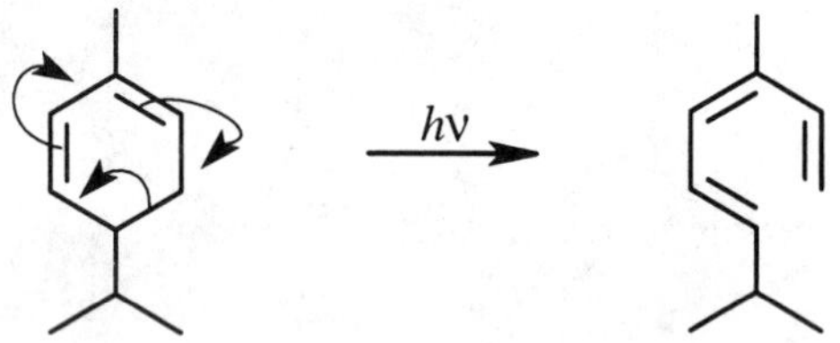

b. OH Δ → OH

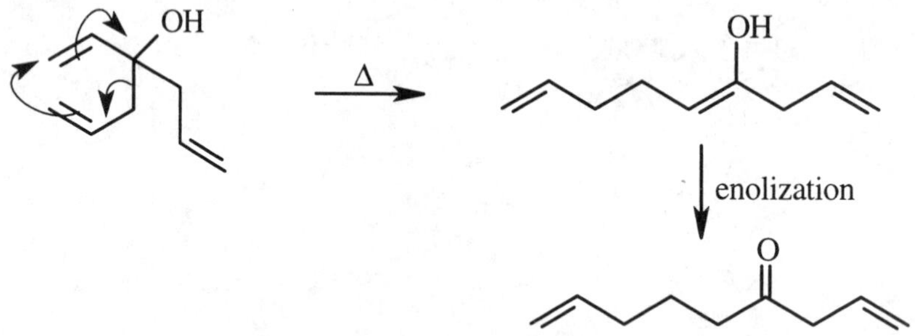

enolization

O

c. CH_3 CH_3 Δ → CH_3 CH_3

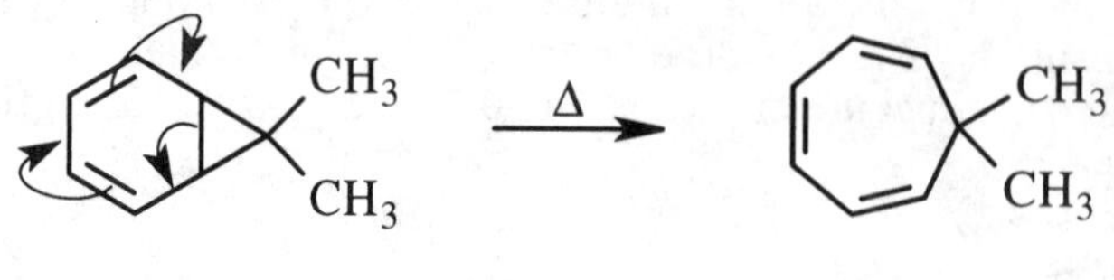

d. O Δ → H O H

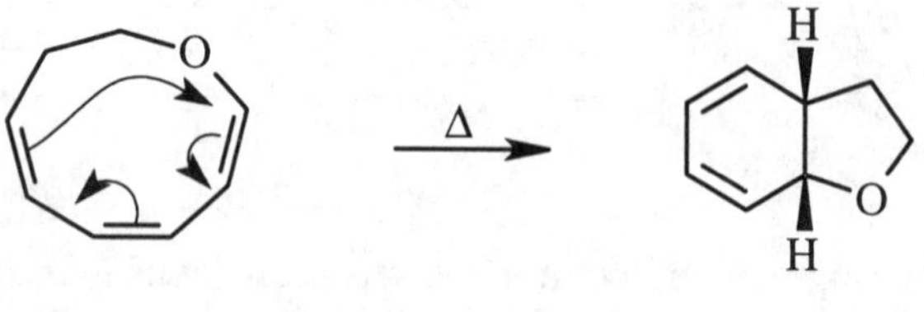

e. CH_3 O Δ → CH_3 O H H_2C → CH_3 OH H_2C

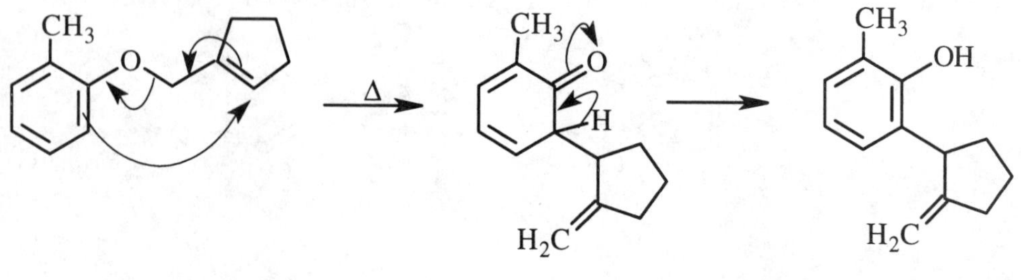

27. **B** is the product. Because the reaction is a [1,3] sigmatropic rearrangement, antarafacial ring closure is required. Carbon, therefore, must migrate using both of its lobes. This means that the configuration of the migrating carbon will undergo inversion. The configuration of the migrating carbon has been inverted in **B** and retained in **A**.

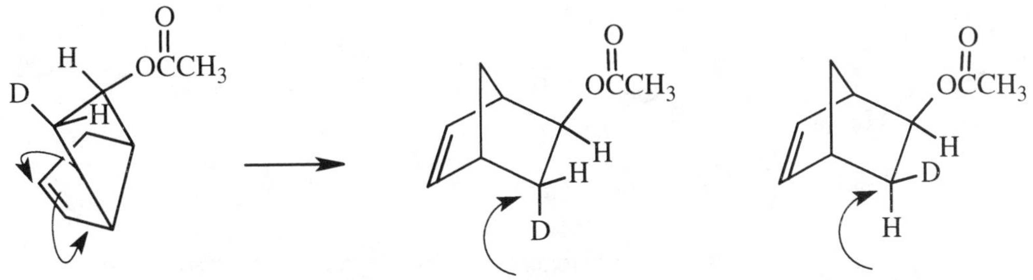

retention
A

inversion
B

28. At first glance it is surprising that the isomerization of Dewar benzene (a highly strained and unstable molecule) to benzene (a stable aromatic compound) is so slow. However, the isomerization requires conrotatory ring-opening, which is symmetry-forbidden under thermal conditions. The reaction, therefore, cannot take place by a concerted pathway and has to take place by a much slower stepwise process.

29. Hydrogen cannot undergo a [1,3] sigmatropic rearrangement, because it cannot migrate by a suprafacial pathway, since its HOMO is asymmetric. Carbon can undergo a [1,3] sigmatropic rearrangement because it can migrate by a suprafacial pathway if it uses both of its lobes. So, the first compound can undergo only a 1,3-methyl group migration.

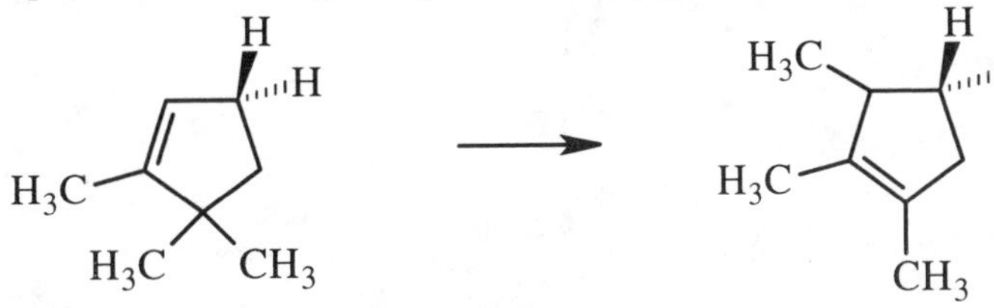

The second compound can under the 1,3-methyl group migration that the first compound undergoes, and the *sec*-butyl group can also undergo a [1,3] sigmatropic rearrangement. The migrating *sec*-butyl group will have its configuration inverted.

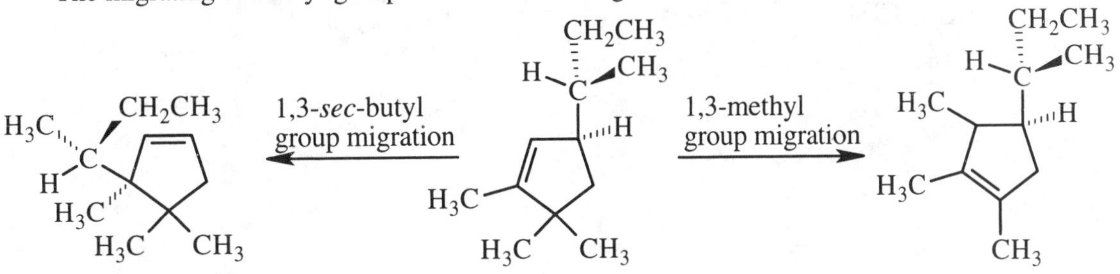

1,3-*sec*-butyl group migration

1,3-methyl group migration

30. An infrared absorption band is indicative of a carbonyl group. A [3,3] sigmatropic rearrangement of the reactant leads to a compound with two enolic groups. Tautomerization of the enols results in keto groups. The ketone carbonyl groups are what give the absorbance at 1715 cm^{-1}.

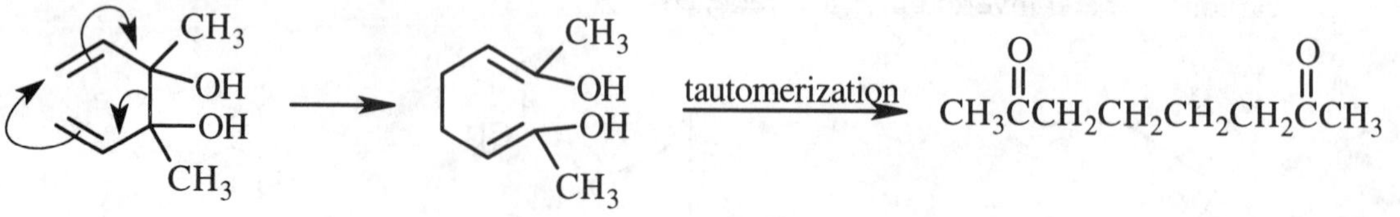

CH_3, OH, OH, CH_3 → CH_3, OH, OH, CH_3 —tautomerization→ $CH_3\overset{O}{\overset{||}{C}}CH_2CH_2CH_2CH_2\overset{O}{\overset{||}{C}}CH_3$

31. The reaction is a [1,7] sigmatropic rearrangement. Since the reaction involves four pairs of electrons, antarafacial rearrangement occurs. Thus, when H migrates, because it is above the plane of the reactant molecule, it ends up below the plane of the product molecule. When D migrates, because it is below the plane of the reactant molecule, it ends up above the plane of the product molecule.

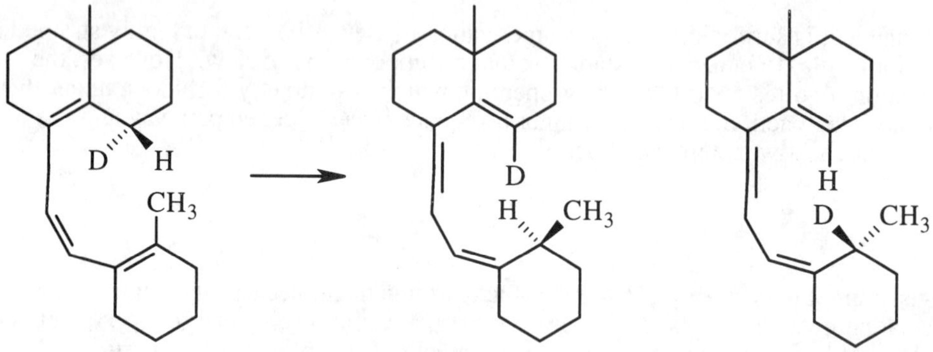

D, H, CH_3 → D, H, CH_3; H, D, CH_3

32.

a. 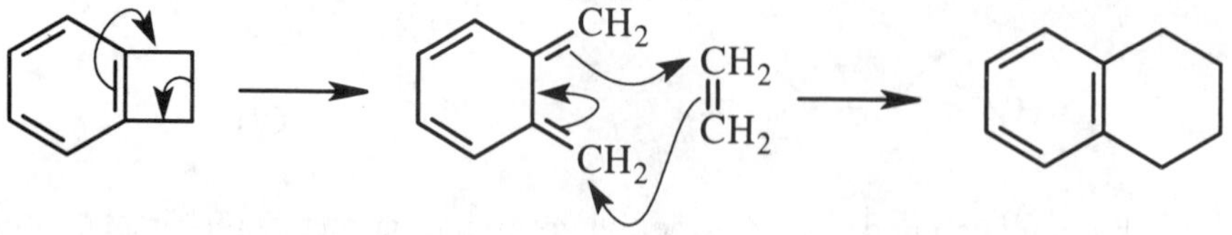 → CH_2, CH_2 + CH_2=CH_2 →

b. 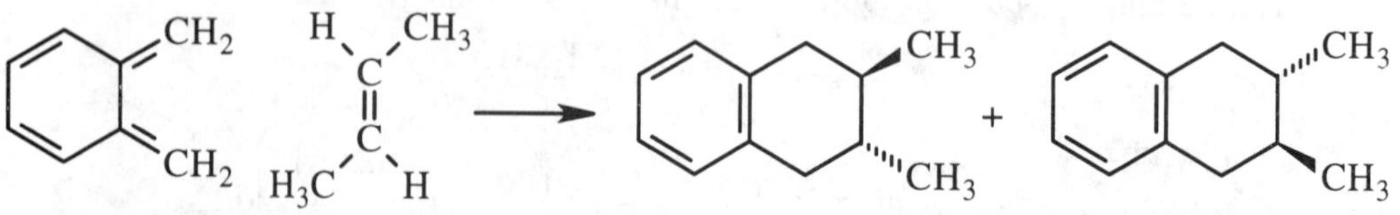 CH_2, CH_2 + H, CH_3, C=C, H_3C, H → CH_3, CH_3 + CH_3, CH_3

33. Disrotatory ring closure of (2*E*,4*Z*,6*Z*)-octatriene leads to the trans isomer, which can exist as a pair of enantiomers. One enantiomer is formed if the "top lobes" of the *p* orbitals rotate toward one other , and the other enantiomer is formed if the "bottom lobes" of the *p* orbitals rotate toward one other.

CH_3 H CH_3 H $\xrightarrow{\Delta}$ CH_3 CH_3 + CH_3 CH_3

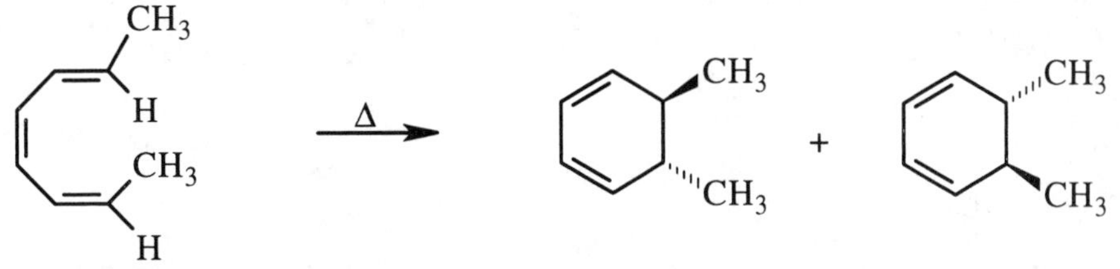

(2*E*,4*Z*,6*Z*)-octatriene

In contrast, disrotatory ring closure of (2*E*,4*Z*,6*E*)-octatriene leads to the cis isomer, which is a meso compound and, consequently, does not have a nonsuperimposable mirror image. Therefore, the same compound is formed from the "top lobes" of the *p* orbitals rotating toward one other and from the "bottom lobes" of the *p* orbitals rotating toward one other.

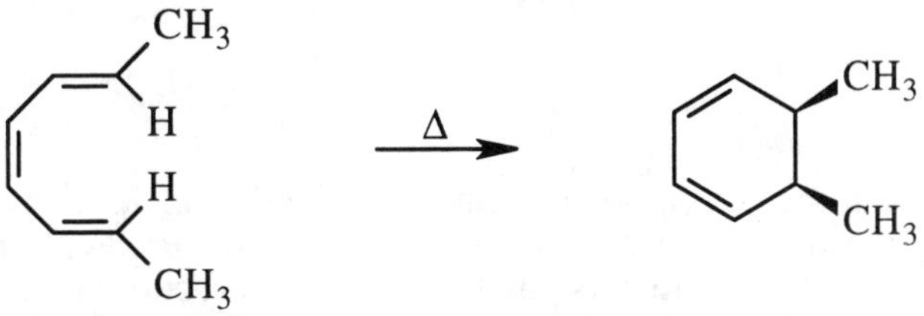

CH_3 H H CH_3 $\xrightarrow{\Delta}$ CH_3 CH_3

(2*E*,4*Z*,6*E*)-octatriene

34.

a. O $\xrightarrow{[3,3]}$ O H $\longrightarrow$ OH

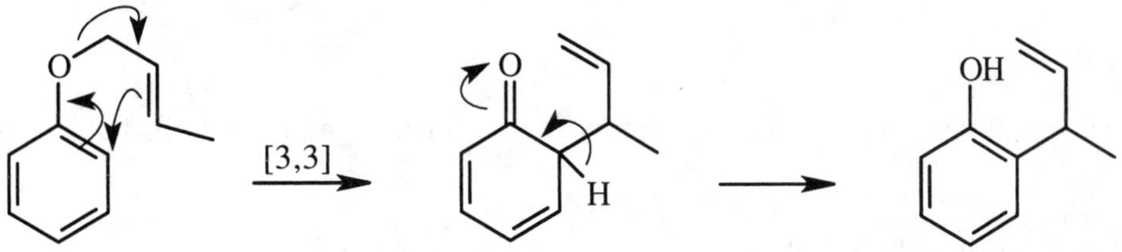

b. O $\xrightarrow{[3,3]}$ O H $\longrightarrow$ OH

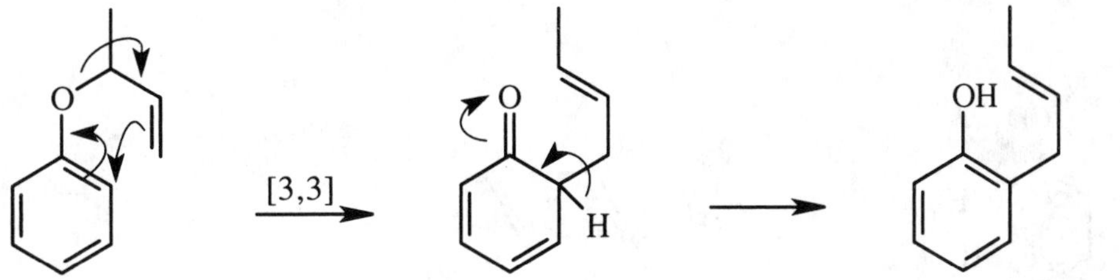

c. [5,5]

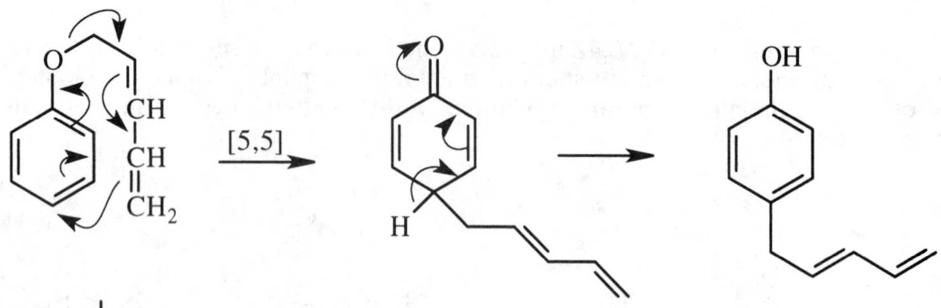

d. [5,5]

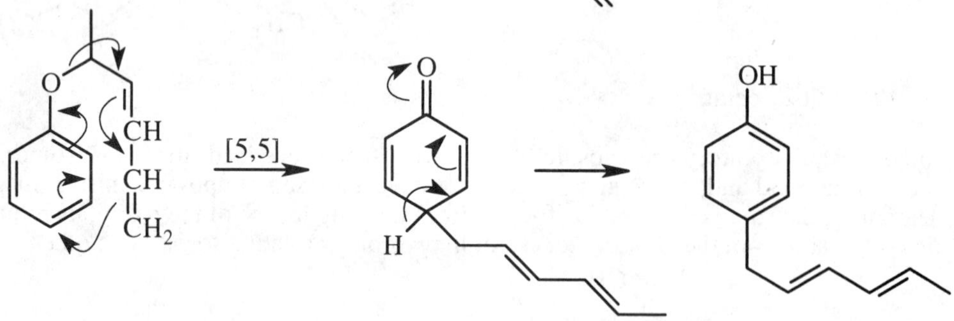

35. Under thermal conditions a compound with two π bonds undergoes conrotatory ring closure. Conrotatory ring closure that results in a ring-closed compound with the substituents cis to one another requires that the substituents point in the same direction in the reactant. Therefore, the product with the methyl substituents pointing in the same direction is obtained in 99% yield.

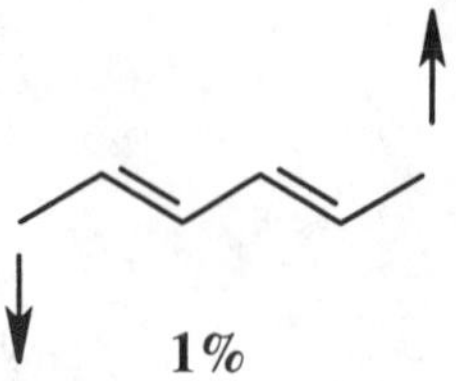

1%
methyl groups point in opposite directions

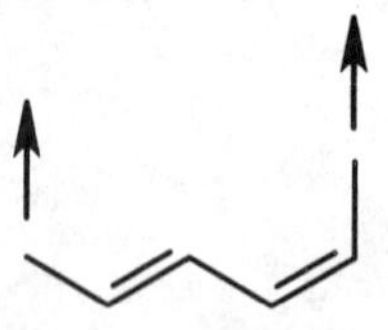

99%
methyl groups point in the same direction

36.

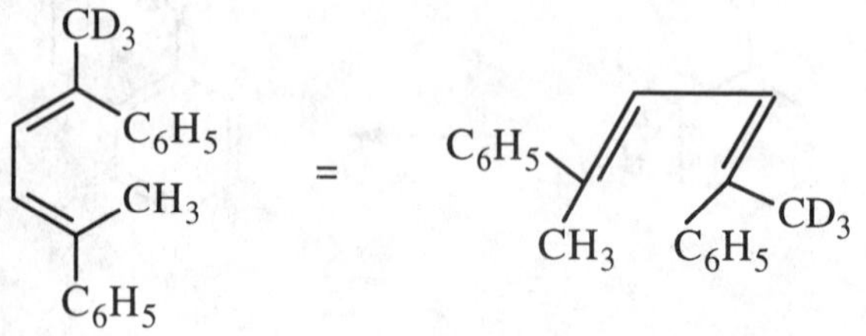

Because the reactant has two π bonds, ring closure is conrotatory.
Two different compounds, **X** and **Y**, can be formed because conrotatory ring closure can occur in either a clockwise or a counterclockwise direction.

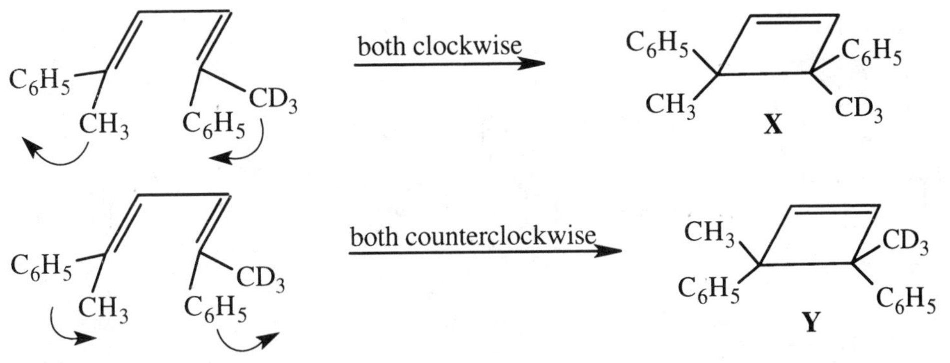

both clockwise → X

both counterclockwise → Y

Each of the compounds (**X** and **Y**) can undergo a conrotatory ring-opening reaction in either a clockwise or a counterclockwise direction to form either **A** or **B**.
A and **B** are the only isomers that can be formed.

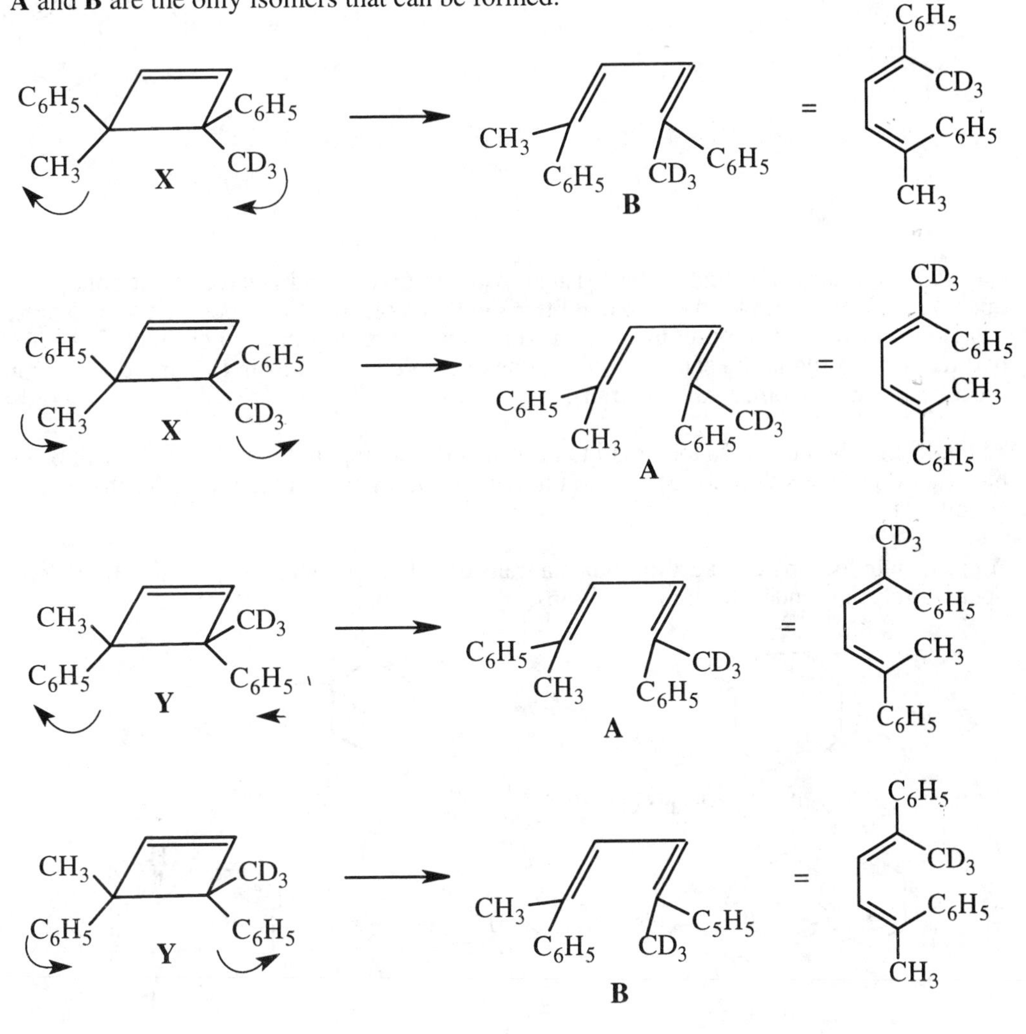

X → B

X → A

Y → A

Y → B

37.

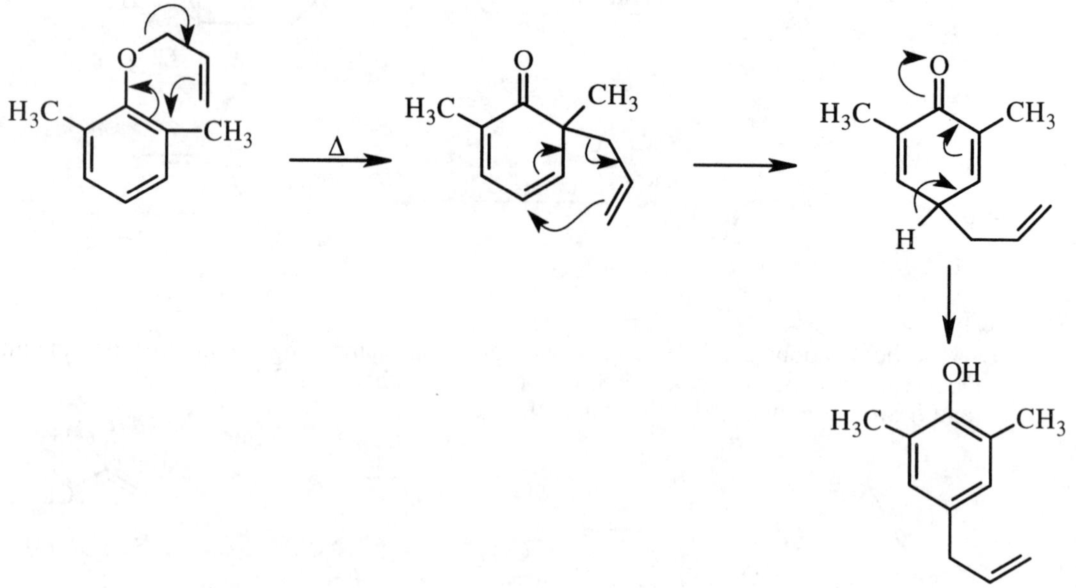

H_3C CH_3 O $\xrightarrow{\Delta}$ H_3C O CH_3 $\longrightarrow$ H_3C O CH_3 H $\longrightarrow$ OH H_3C CH_3

38. Because the compounds that undergo ring closure to give **A** and **B** have two π bonds, ring opening of **A** and **B** under thermal conditions will be conrotatory. Because the hydrogens in **A** and **B** are cis, they must point in the same direction in the ring-opened product. To have the two hydrogens pointing in the same direction, one of the double bonds in the ring-opened compound must be cis and the other must be trans.

An eight-membered ring is too small to accommodate cojugated double bonds with one cis and the other trans, so **A** will not be able to undergo a ring-opening reaction under thermal conditions.

A ten-membered ring can accommodate a trans double bond, so **B** will be able to undergo a ring-opening reaction under thermal conditions.

Both double bonds are cis. H H **A**

H H cis double bond trans double bond H H **B**

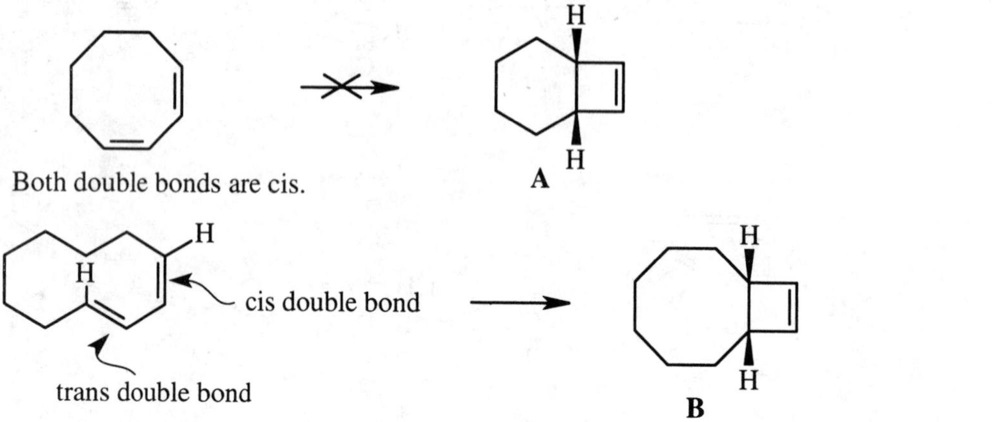

39. The compound undergoes a 1,5-hydrogen shift of D or a 1,5-hydrogen shift of H. In each case, an unstable nonaromatic intermediate is formed that undergoes a subsequent 1,5-hydrogen shift to form an aromatic product.

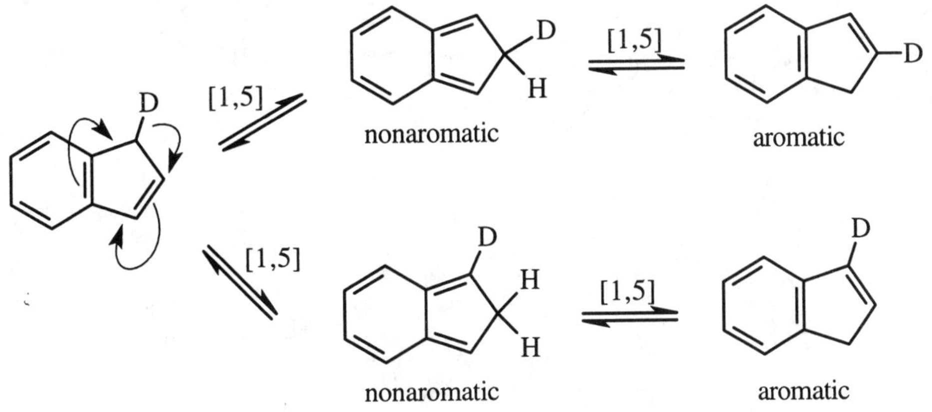

D [1,5] D H [1,5] D
nonaromatic aromatic
[1,5] D H H [1,5] D
nonaromatic aromatic

40. Because the reacting system of the ring-opened compound has three conjugated π bonds involved in the reaction, conrotatory ring closure will occur under photochemical conditions, and trans hydrogens require that the hydrogens point in the opposite direction in the ring-opened compound. Thermal ring closure of a three π bond system is disrotatory, and disrotatory ring closure of a compound with hydrogens that point in opposite directions will cause those hydrogens to be cis in the ring-closed product.

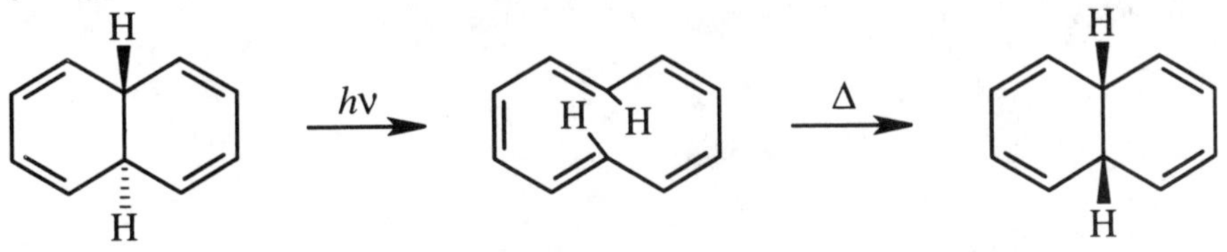

H H $\xrightarrow{h\nu}$ H H $\xrightarrow{\Delta}$ H H

41. A Diels-Alder reaction is followed by an extrusion reaction that eliminates CO_2. (See page 1208 in the text)

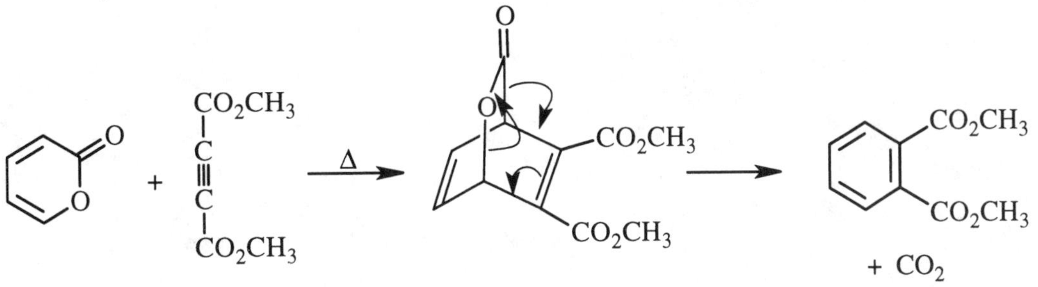

O O + CO_2CH_3 C ≡ C CO_2CH_3 $\xrightarrow{\Delta}$ O O CO_2CH_3 CO_2CH_3 ⟶ CO_2CH_3 CO_2CH_3 + CO_2

42.

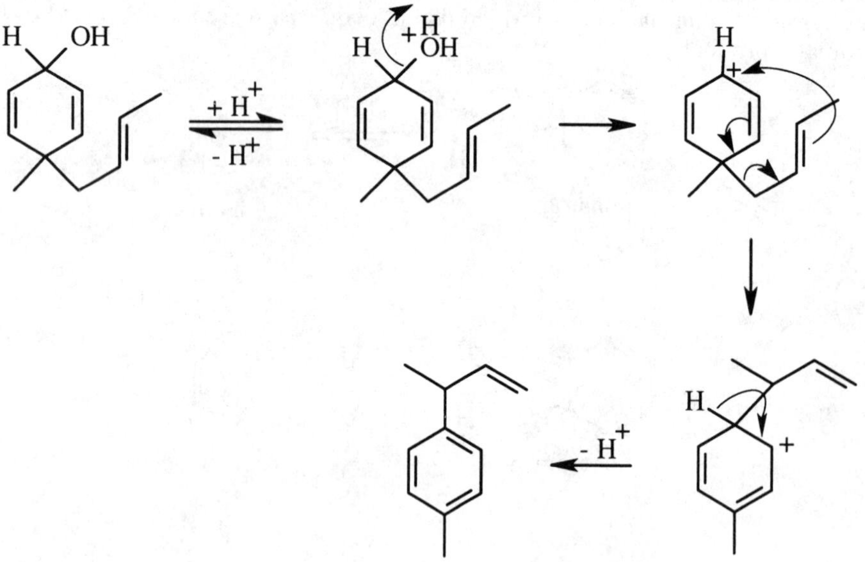

CHAPTER 29
More About Multistep Organic Synthesis

Important Terms

chiral auxiliary an enantiomerically pure compound that, when attached to a reactant, causes a product with particular stereochemistry to be formed.

combinatorial library a group of structurally related compounds.

combinatorial organic synthesis the synthesis of a library of compounds by covalently connecting sets of building blocks of varying structure.

convergent synthesis a synthesis in which pieces of the target compound are individually prepared and then assembled.

Darzen's condensation the reaction of the carbanion of an α-bromosubstituted ester with an aldehyde or ketone to form an epoxide.

disconnection breaking a bond to carbon, in a retrosynthetic analysis, to give a simpler molecule.

enantioselective reaction a reaction that forms an excess of one enantiomer.

epimerization changing the configuration of a chirality center by removing a proton from a carbon and then reprotonating the same carbon.

extrusion reaction a reaction in which a neutral molecule (for example CO_2, CO, N_2) is eliminated from a molecule.

Favorskii ring contraction the reaction of an α-bromosubstituted cyclic ketone with base to form a carboxy substituted cycloalkane with one fewer carbon in the ring than the starting material.

functional group interconversion converting one functional group into another functional group.

iterative synthesis a synthesis in which a reaction sequence is carried out more than once.

linear synthesis a synthesis that builds a molecule step by step from starting materials.

multistep synthesis preparation of a compound via a route that requires several steps.

natural product a compound synthesized in nature.

organic synthesis preparation of organic compounds from other organic compounds.

protecting group a reagent that protects a functional group from a synthetic operation that it otherwise would not survive.

retrosynthesis or retrosynthetic analysis working backward (on paper) from the target molecule to available starting materials.

synthetic equivalent the reagent actually used as the source of a synthon.

synthetic tree an outline of the available routes to get to the desired product from available starting materials.

synthon a fragment of a disconnection.

target molecule the desired end product of a synthesis.

umpolung reversing the normal polarity of a functional group.

vinylogy transmission of reactivity through double bonds.

Solutions to Problems

1. The six-step linear synthesis results in a 26% yield, and the five-step convergent synthesis results in a 51% yield.

a linear synthesis

$$A \longrightarrow B \longrightarrow C \longrightarrow D \longrightarrow E \longrightarrow F \longrightarrow G$$

B	C	D	E	F	G
80%	0.80 x 0.80	0.80 x 0.64	0.80 x 0.51	0.80 x 0.41	0.80 x 0.33
	64%	51%	41%	33%	26%

a convergent synthesis

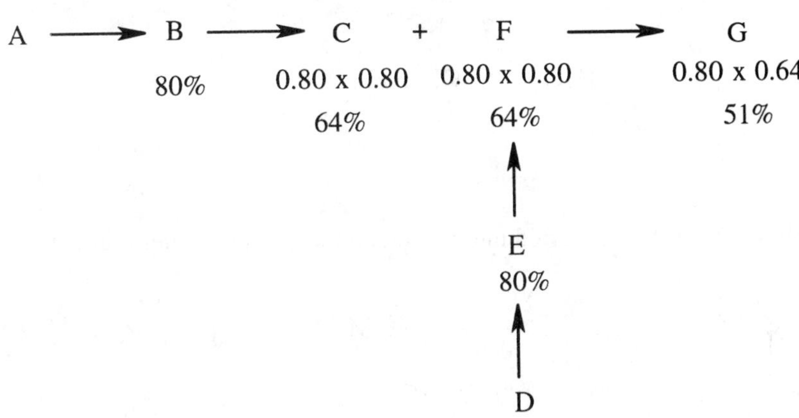

2. The reaction of a Grignard reagent with ethylene oxide cannot be used for the synthesis of isobutyl alcohol, because this reaction leads to an alcohol with a $-CH_2CH_2OH$ group, and isobutyl alcohol does not have such a group.

$$RMgBr \xrightarrow[\text{2. } H^+]{\text{1. ethylene oxide}} RCH_2CH_2OH \qquad CH_3CH(CH_3)CH_2OH$$

isobutyl alcohol

3. A reaction that changes the carbon skeleton of the starting material is one that forms (or breaks) a carbon-carbon bond. The only reaction in Figure 28.1 on page 1209 of the text that changes the carbon skeleton of the starting materials is the reaction of a Grignard reagent with ethylene oxide.

4. There are nine methods for the synthesis of a carboxylic acid listed in Appendix IV on page A-14 of the text. Compare your answer with the nine methods listed in the Appendix.

5. Methods for the synthesis of primary amines are discussed in Section 19.4 of the text.

1. $CH_3CH_2CH{=}CH_2 \xrightarrow[\text{peroxide}]{\text{HBr}} CH_3CH_2CH_2CH_2Br$

$CH_3CH_2CH_2CH_2Br \xrightarrow{N_3^-} CH_3CH_2CH_2CH_2N{=}\overset{+}{N}{=}\overset{-}{N}$

$CH_3CH_2CH_2CH_2N{=}\overset{+}{N}{=}\overset{-}{N} \xrightarrow[\text{Pt}]{H_2} CH_3CH_2CH_2CH_2NH_2 + N_2$

2. $CH_3CH_2CH{=}CH_2 \xrightarrow[\text{peroxide}]{\text{HBr}} CH_3CH_2CH_2CH_2Br$

$CH_3CH_2CH_2CH_2Br \xrightarrow{\text{1. phthalimide anion; 2. } H^+, H_2O, \Delta\text{; 3. } HO^-} CH_3CH_2CH_2CH_2NH_2$

6. Electrophilic bromocyclohexane can be converted into a nucleophile by converting it into a Grignard reagent.

$$\text{cyclohexyl}\overset{\delta+}{C}\text{–Br} \xrightarrow[Et_2O]{Mg} \text{cyclohexyl}\overset{\delta-}{C}\text{–MgBr}$$

7. **a.** *umpolung*
b. normal (An α-hydrogen can be removed by a strong base.)

c. normal (The β-carbon of an α,β-unsaturated carbonyl compound is electrophilic.)
d. *umpolung*

e. normal (A carbonyl carbon is electrophilic.)
f. *umpolung*

g. *umpolung*
h. normal (An α-hydrogen can be removed by a strong base.)

8. **a.** Use 1-bromohexane instead of 1-bromopropane.

b. Use 2-methyl-1,3-dithiane instead of 1,3-dithiane.

9.

a. 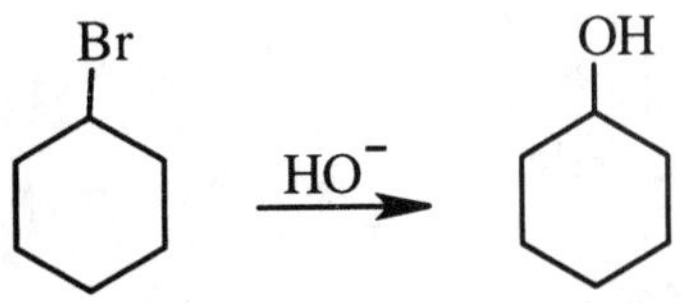 Bromocyclohexane $\xrightarrow{HO^-}$ cyclohexanol (cyclohexane–OH)

b. 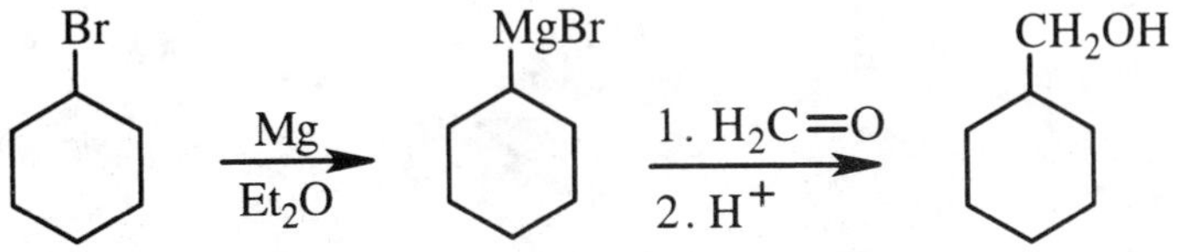 Bromocyclohexane $\xrightarrow[Et_2O]{Mg}$ cyclohexyl–MgBr $\xrightarrow[2.\ H^+]{1.\ H_2C{=}O}$ cyclohexyl–CH_2OH

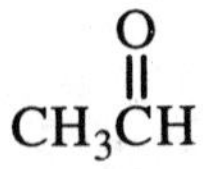

c.  Bromocyclohexane $\xrightarrow{^-C\equiv N}$ cyclohexyl–$C\equiv N$

d. 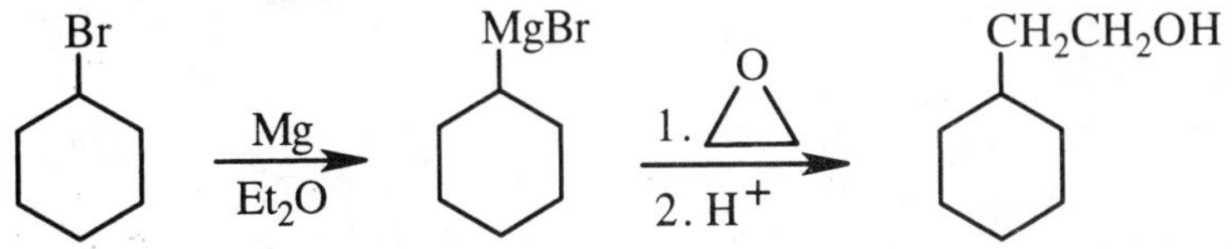 Bromocyclohexane $\xrightarrow[Et_2O]{Mg}$ cyclohexyl–MgBr $\xrightarrow[2.\ H^+]{1.\ \text{ethylene oxide}}$ cyclohexyl–CH_2CH_2OH

10.

a. 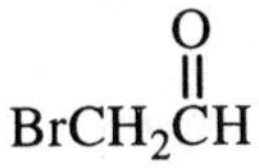 $BrCH_2CH{=}O$ ($BrCH_2\overset{O}{\overset{\|}{C}}H$)

b. $CH_3\overset{O}{\overset{\|}{C}}H$

c. $CH_3\overset{O}{\overset{\|}{C}}Cl$

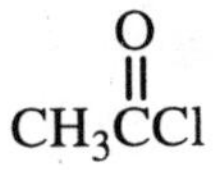

d. 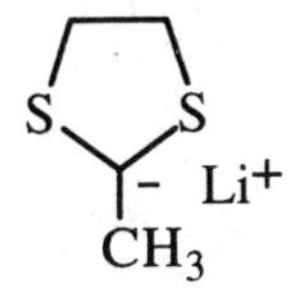 1,3-dithiolane carbanion (S, S ring) with CH_3, $^-$ Li^+

e. 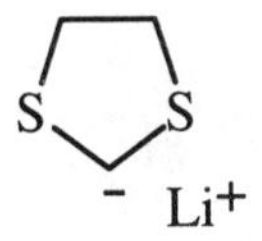 1,3-dithiolane carbanion (S, S ring), $^-$ Li^+

f. 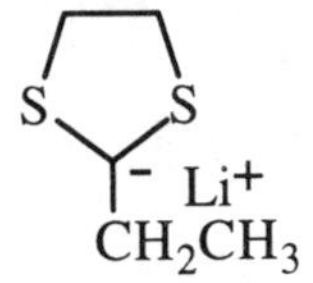 1,3-dithiolane carbanion (S, S ring) with CH_2CH_3, $^-$ Li^+

g. $CH_3CH_2CH_2Br$

h. $CH_3CH_2CH_2MgBr$

i.  ethylene oxide (three-membered ring with O)

j. $BrCH_2CH_2CH_2OH$

11. **a.** Solved in the text.

b.

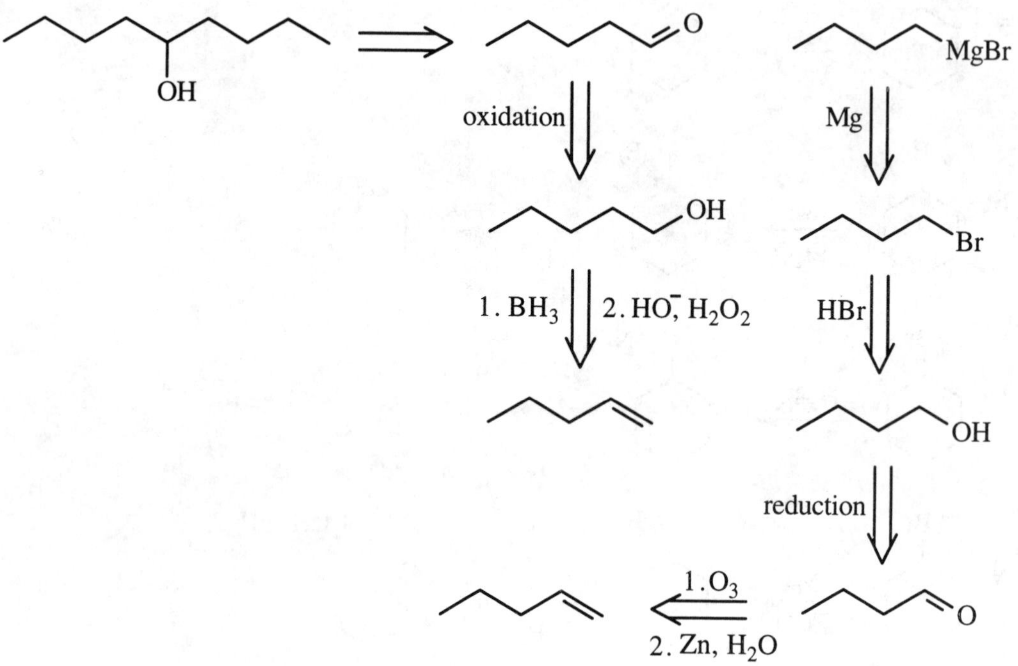

OH ⟹ O MgBr

oxidation ⇓ Mg ⇓

OH Br

1. BH_3 2. HO^-, H_2O_2 ⇓ HBr ⇓

OH

reduction ⇓

⟸ 1. O_3 2. Zn, H_2O O

12.

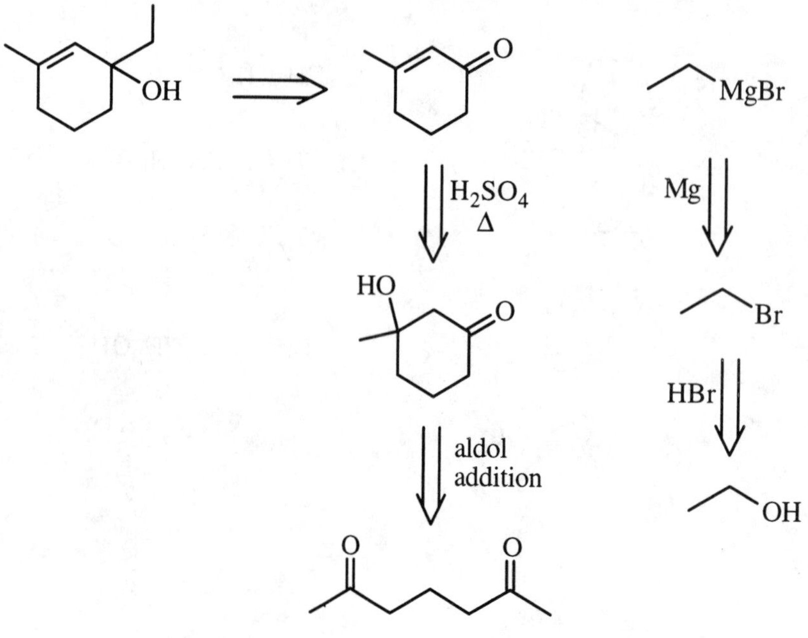

OH ⟹ O MgBr

⇓ H_2SO_4 Δ Mg ⇓

HO O Br

⇓ aldol addition HBr ⇓

O O OH

13.

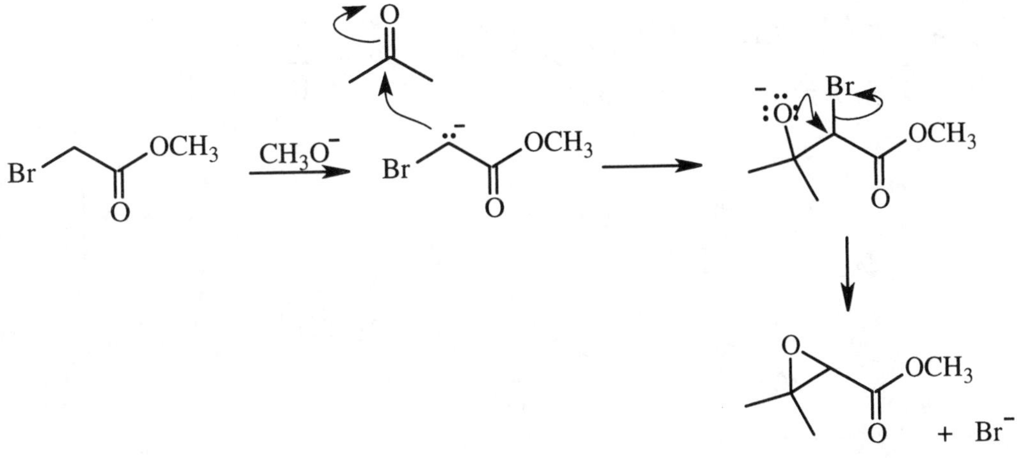

Br–CH₂–C(=O)OCH₃ $\xrightarrow{CH_3O^-}$ Br–CH⁻–C(=O)OCH₃ (adds to acetone) ⟶ alkoxide intermediate ⟶ epoxide–C(=O)OCH₃ + Br⁻

14. Putting a bromo substituent on the α-carbon increases the acidity of the other α-hydrogens. Thus, it will be easier for hydroxide ion to remove an α-hydrogen from the monobrominated compound than from the original nonbrominated reactant, which would lead to a dibrominated product. Consequently, the monobrominated compound cannot be prepared under basic conditions.

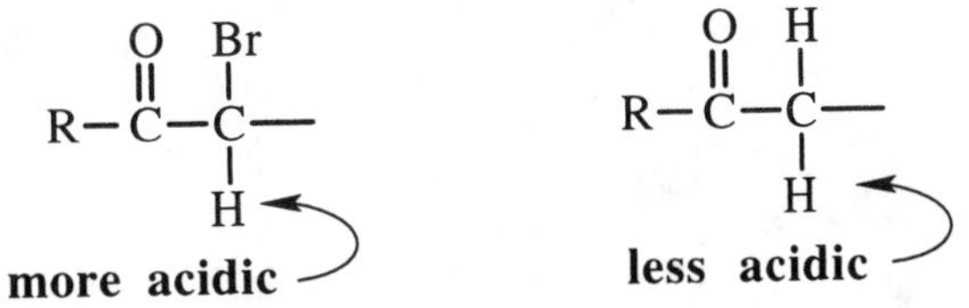

R–C(=O)–C(Br)(H)– **more acidic**

R–C(=O)–C(H)(H)– **less acidic**

15.

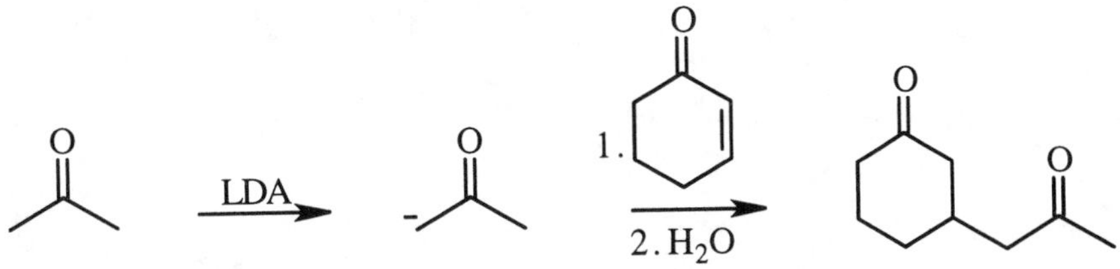

acetone $\xrightarrow{LDA}$ enolate $\xrightarrow[2.\ H_2O]{1.\ \text{cyclohexenone}}$ 3-(2-oxopropyl)cyclohexanone

16.

a. 2,4-pentanedione ⟹ 4-hydroxy-2-pentanone ⟹ acetaldehyde + ⁻CH₂C(=O)CH₃

⟸ LDA

acetone

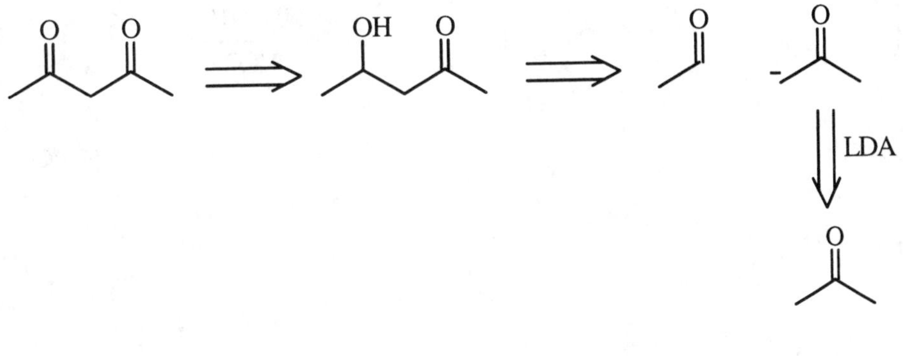

b. OCH$_3$ $\xRightarrow{\text{1. } SOCl_2 \text{ 2. } CH_3OH}$ OH $\xRightarrow{H_2SO_4,\ \Delta}$ OH OH

reduction

H^+ H_2O

OCH$_3$ OH

Br OCH$_3$

N H

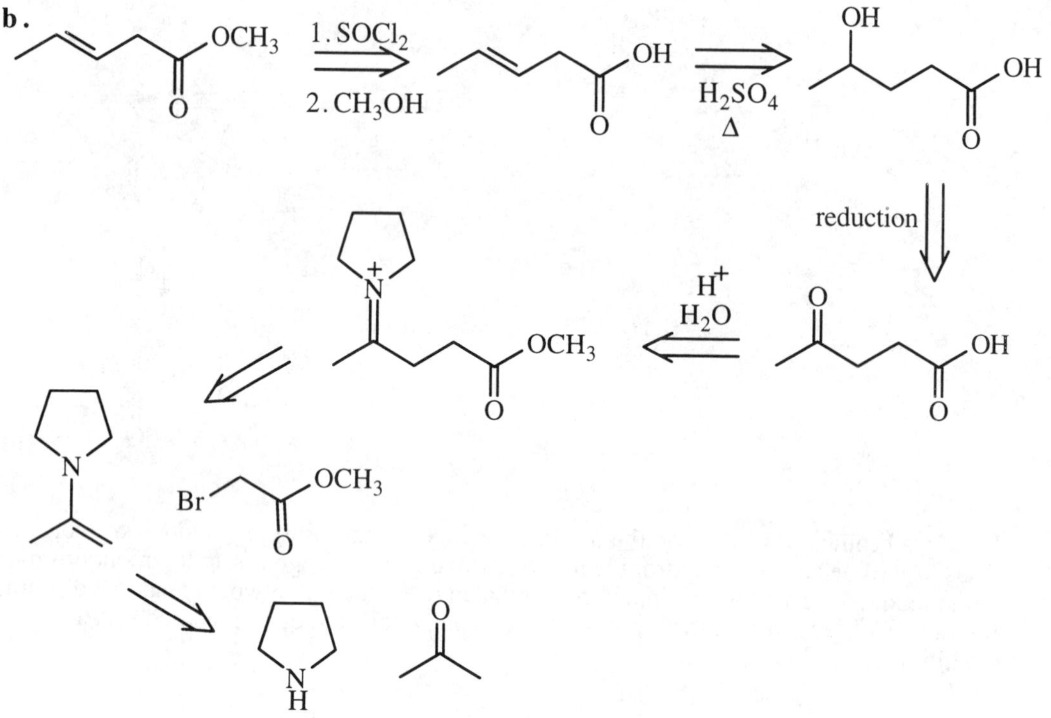

c. OH oxidation OH OH H^+, H_2O Δ C≡N OH H $^-$C≡N

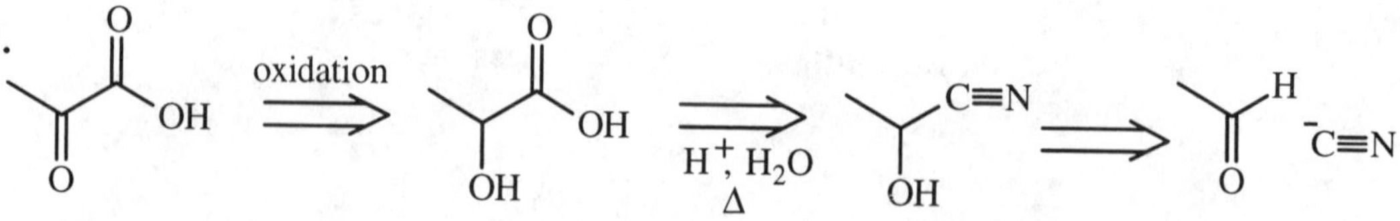

17. **a.** Solved in the text.

b. Δ NH_2NH_2 HO^- Δ H_2 Pd/C

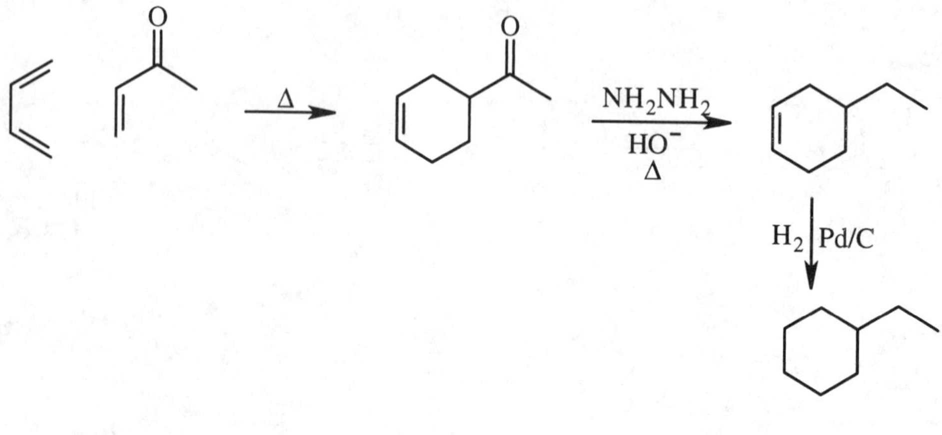

c. [diene + ethene] $\xrightarrow{\Delta}$ cyclohexene $\xrightarrow[\text{2. Zn, } H_2O]{\text{1. } O_3}$ H(C=O)CH$_2$CH$_2$CH$_2$CH$_2$(C=O)H

$\downarrow$ 1. CH_3MgBr 2. H_3O^+

$CH_3CH(OH)CH_2CH_2CH_2CH_2CH(OH)CH_3$

d. [diene + ethene] $\xrightarrow{\Delta}$ cyclohexene $\xrightarrow[\Delta]{\text{NBS}}$ 3-bromocyclohexene

18.

a. $CH_3CH_2CH(CH_2CH_3)CH{=}O$ + $CH_2{=}P(C_6H_5)_3$ ⟶ $CH_3CH_2CH(CH_2CH_3)CH{=}CH_2$
3-ethyl-1-pentene

b. The target molecule is the only product of the Wittig reaction. When an alkyl halide and an acetylide ion are used to make the target molecule, two products are formed, because in the first step of the reaction, an elimination product is formed in addition to the desired substitution product.

$CH_3CH_2CH(CH_2CH_3)Br$ + $^{-}C{\equiv}CH$ ⟶ $CH_3CH_2CH(CH_2CH_3)C{\equiv}CH$ (substitution product) + $CH_3CH{=}CHCH_2CH_3$ (elimination product)

$\downarrow$ H_2, Lindlar's catalyst

$CH_3CH_2CH(CH_2CH_3)CH{=}CH_2$
3-ethyl-1-pentene

19.

retrosynthetic analysis

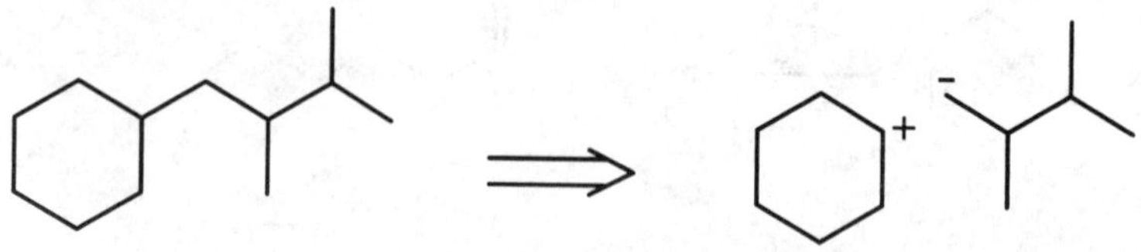

(cyclohexylmethyl-substituted branched alkane) $\Longrightarrow$ (cyclohexyl cation)$^{+}$ + $^{-}$(branched alkyl anion)

synthesis

$(C_6H_5)_3P$ + $CH_3CH(CH_3)CH(CH_3)CH_2Br$ $\longrightarrow$ $(C_6H_5)_3\overset{+}{P}{-}CH_2CH(CH_3)CH(CH_3)CH_3$

triphenyl-
phosphine

$\downarrow$ $CH_3(CH_2)_2CH_2Li$

$(C_6H_5)_3\overset{+}{P}{-}\overset{-}{C}HCH(CH_3)CH(CH_3)CH_3$

$\updownarrow$

$(C_6H_5)_3P{=}CHCH(CH_3)CH(CH_3)CH_3$

+ cyclohexanone (O) $\longrightarrow$ (alkene product)

$\xrightarrow{H_2,\ Pd/C}$ (alkane product)

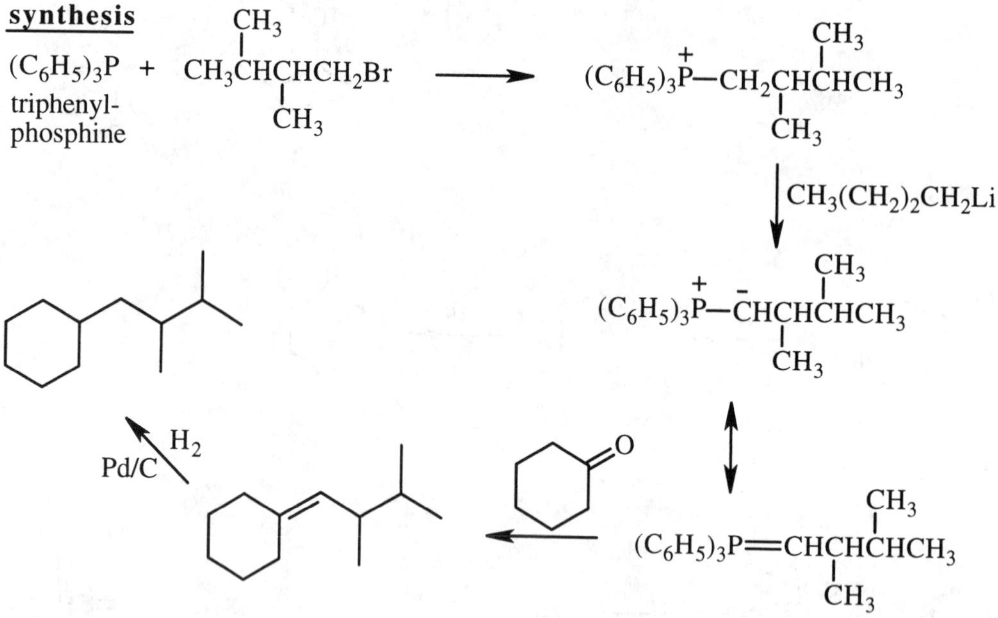

20. The Grignard reagent removes a proton from the OH group, forming an alkane and an alkoxide ion. Therefore, a second equivalent of the Grignard reagent is needed to carry out the desired reaction.

$$RCH_2MgBr \ + \ ROH \longrightarrow RCH_3 \ + \ RO^- \ + \ Mg^{2+} \ + \ Br^-$$

21.

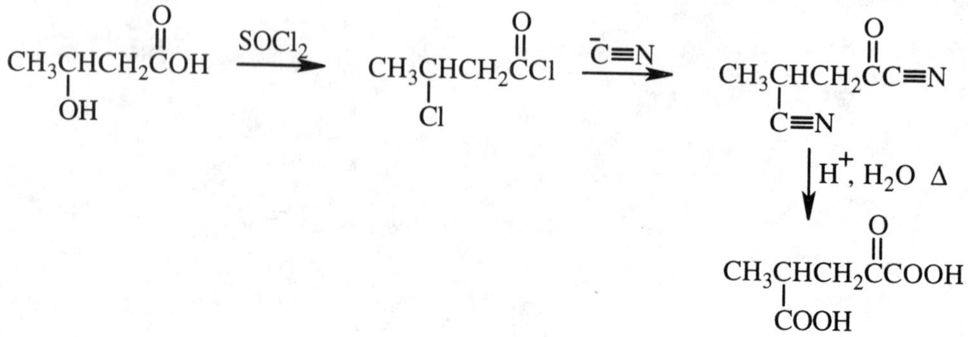

$CH_3CH(OH)CH_2\overset{O}{\overset{\|}{C}}OH$ $\xrightarrow{SOCl_2}$ $CH_3CH(Cl)CH_2\overset{O}{\overset{\|}{C}}Cl$ $\xrightarrow{^{-}C\equiv N}$ $CH_3CH(C{\equiv}N)CH_2\overset{O}{\overset{\|}{C}}C{\equiv}N$

$\downarrow$ H^+, H_2O Δ

$CH_3CH(COOH)CH_2\overset{O}{\overset{\|}{C}}COOH$

22. Acetyl chloride would protect the amino group of the amino acid by reacting with it to form an amide. Removing the protecting group would require the amide to be hydrolyzed. The peptide amide groups would also be hydrolyzed.

23. If alanine's amino group were not protected, both glycine and alanine could react with the acyl halide of alanine, forming Ala-Ala as well as the desired Ala-Gly.

24. In order to convert (*S*)-2-octanol into (*R*)-2-bromooctane, one S_N2 reaction must take place. An alcohol reacts with PBr_3 by first converting the OH group into a good leaving group that is subsequently displaced by Br^- in an S_N2 reaction. (Note that you cannot convert the alcohol into an alkyl bromide by reacting it directly with HBr, because secondary alcohols react by an S_N1 mechanism that will result in a racemic mixture.)

$$CH_3CH_2CH_2CH_2CH_2CH_2-C(OH)(H)-CH_3 \xrightarrow{PBr_3} CH_3CH_2CH_2CH_2CH_2CH_2-C(H)(Br)-CH_3$$

(*S*)-2-octanol → (*R*)-2-bromooctane

In order to convert (*S*)-2-octanol into (*S*)-2-bromooctane, two successive S_N2 reactions must occur.

$$CH_3CH_2CH_2CH_2CH_2CH_2-C(OH)(H)-CH_3 \xrightarrow{TsCl} CH_3CH_2CH_2CH_2CH_2CH_2-C(OTs)(H)-CH_3$$

(*S*)-2-octanol

$$\xrightarrow{I^-} CH_3CH_2CH_2CH_2CH_2CH_2-C(H)(I)-CH_3$$

(*R*)-2-iodooctane

$$\xrightarrow{Br^-} CH_3CH_2CH_2CH_2CH_2CH_2-C(Br)(H)-CH_3$$

(*S*)-2-bromooctane

25.

a. $CH_3CH_2C{\equiv}CH \xrightarrow{^-NH_2} CH_3CH_2C{\equiv}C^- \xrightarrow{CH_3CH_2Br} CH_3CH_2C{\equiv}CCH_2CH_3 \xrightarrow[\text{Lindlar's catalyst}]{H_2}$ (H)(CH_3CH_2)C=C(H)(CH_2CH_3)

cis-3-hexene

b. $CH_3CH_2C{\equiv}CH \xrightarrow{^-NH_2} CH_3CH_2C{\equiv}C^- \xrightarrow{CH_3CH_2CH_2Br} CH_3CH_2C{\equiv}CCH_2CH_2CH_3 \xrightarrow{Na/NH_3}$ (H)(CH_3CH_2)C=C(H)($CH_2CH_2CH_3$)

trans-3-heptene

26.

(H)(H_3C)C=C(CH_3)(H) *trans*-2-butene $\xrightarrow[CH_2Cl_2]{Br_2}$ $CH_3CHBrCHBrCH_3$ $\xrightarrow{\text{excess } ^-NH_2}$ $CH_3C{\equiv}CCH_3$ $\xrightarrow{H_2,\ \text{Lindlar's catalyst}}$ (H)(H_3C)C=C(H)(CH_3)

cis-2-butene

27. Each epoxide reacts with methylmagnesium bromide to form two products. The major product results from attack of the Grignard reagent on the less sterically hindered carbon of the epoxide (the one bonded to the methyl substituent), and the minor product results from attack of the Grignard reagent on the other carbon of the epoxide. It will be easier to answer this question if you use molecular models.

Epoxide (H_3C, CH_3CH_2 on one carbon; CH_3, H on the other; O hashed) $\xrightarrow[2.\ H^+,\ H_2O]{1.\ CH_3MgBr}$

major: Fischer projection — top $CH(CH_3)_2$; left HO; right CH_3; bottom CH_2CH_3 — *S*

minor: Fischer projection — top CH_3; left H; right OH; bottom $CH(CH_3)_2$ then CH_2CH_3 — *S*

Epoxide (H_3C, CH_3CH_2 on one carbon; CH_3, H on the other; O wedged) $\xrightarrow[2.\ H^+,\ H_2O]{1.\ CH_3MgBr}$

major: Fischer projection — top $CH(CH_3)_2$; left CH_3; right OH; bottom CH_2CH_3 — *R*

minor: Fischer projection — top CH_3; left HO; right H; bottom $CH(CH_3)_2$ then CH_2CH_3 — *R*

28. The addition of Br_2 to an alkene is a stereoselective reaction because not all possible stereoisomers are formed: the trans-alkene forms the erythro pair of enantiomers, while the cis-alkene forms the threo pair of enantiomers. It is a stereospecific reaction because a cis-alkene leads to a different set of products than a trans-alkene. It is not enantioselective, because if a chirality center is created, both enantiomers are obtained in equal amounts, and if two chirality centers are created, two pairs of enantiomers (or a meso form and a pair of enantiomers) are obtained.

29.

1. CH_3MgBr
2. H^+, H_2O

H_2SO_4, Δ

RCOOH (O, ||)

Mg, Et_2O

Br

MgBr

H_3C OH

H_3C O^-

H^+

30.

H^+ / $-H^+$

$H_2\ddot{O}$:

enedione

$-H^+$ / H^+

$H_2\ddot{O}$:

diol

31.

benzyne

HO^-

Diels-Alder reaction

triptycene

32. The intermediates shown are those involved in removing each of the two carboxyl substituents.

$SOCl_2$

t-BuOOH

solvent

RH + HOC$(CH_3)_3$ ← R• CO_2 •OC$(CH_3)_3$

33. 2-Bromo-2,4-cyclopentadienone readily undergoes a [4+2] cycloaddition reaction, forming a dimer. Therefore, it cannot be isolated at room temperature.

34.

a. $(CH_3CH_2)_2NH$ $(CH_3CH_2)_2NH$

H Br Br Br → H Br Br → Br

2,3,4-tribromocyclopentanone

2-bromo-2,4-cyclopentadienone

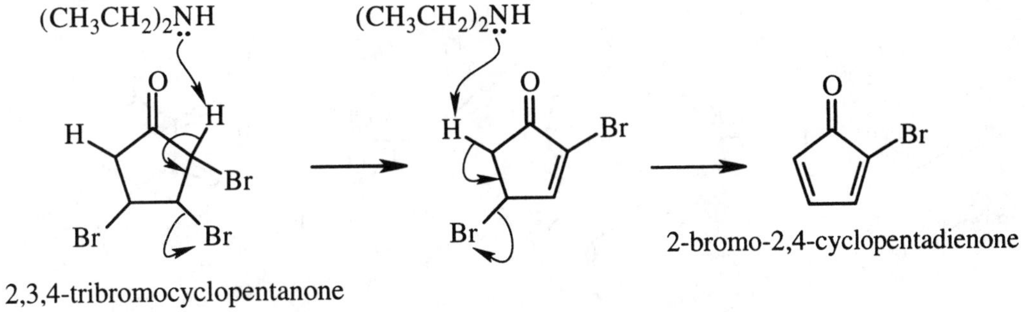

b. O → NBS / Δ → O Br → Br_2 / CH_2Cl_2 → O Br Br Br

2-cyclopentenone

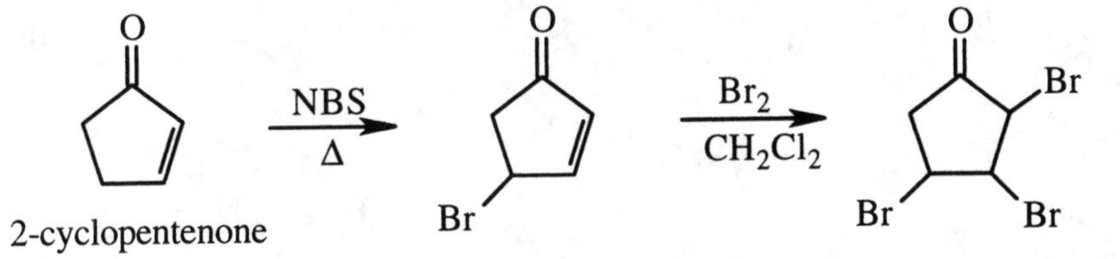

35. The first step in a Favorskii ring contraction is removal of an α-hydrogen by hydroxide ion. This is followed by an intramolecular reaction that expels a bromide ion and forms a cyclopropanone ring. Attack of hydroxide ion on the carbonyl group forms a tetrahedral intermediate that, when it collapses, breaks a carbon-carbon bond, resulting in a ring with one fewer atom than the original ring.

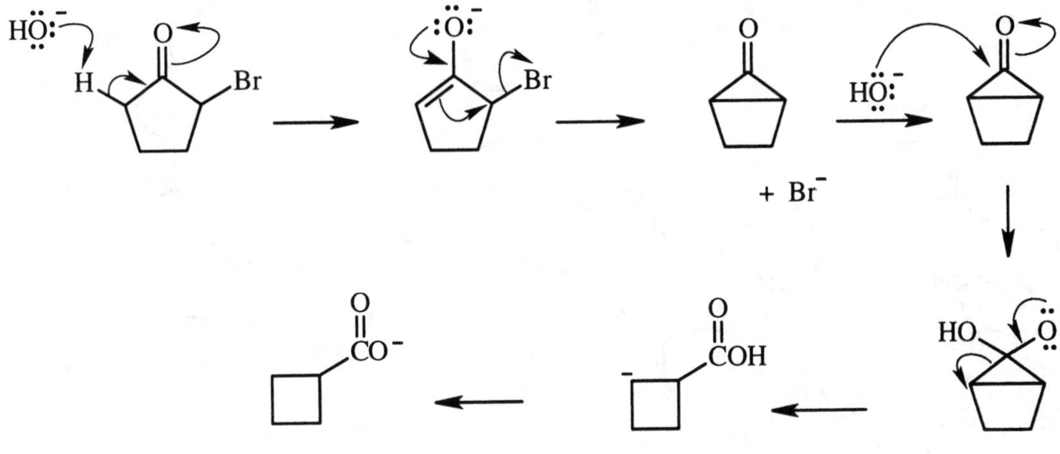

HO^- H O Br → O^- Br → O + Br^- → HO^- → HO O^- → COH → CO^-

36.

$SOCl_2$

1. $AlCl_3$
2. H_2O

H^+, H_2O, Br_2

$HNCH_3$

HCl

$NaOCH_3$

$NaBH_4$

$SOCl_2$

NaCN

H_2O, HO^-, Δ

Raney Ni

37. Note: FGI = functional group interconversion

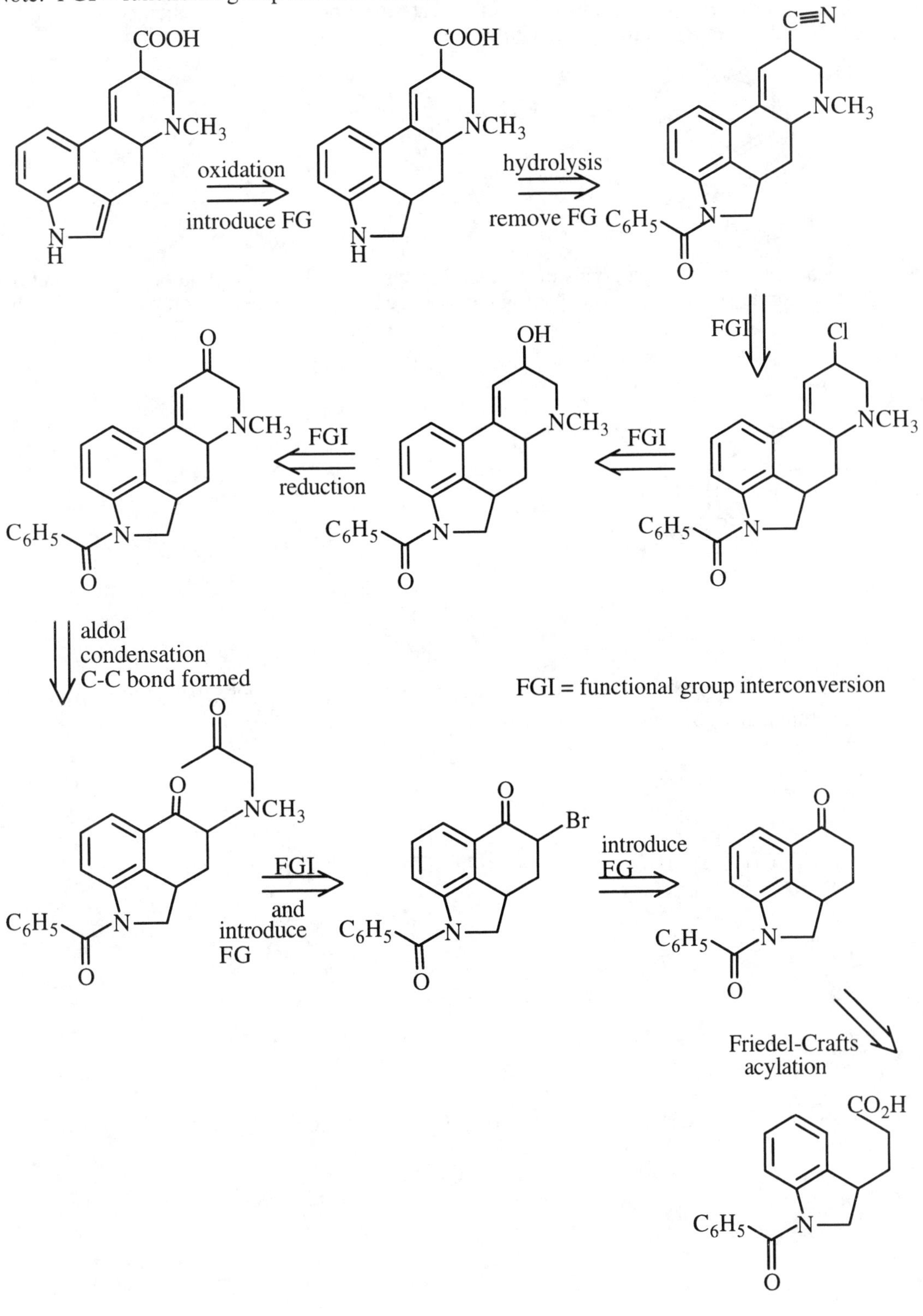

COOH NCH$_3$ N H — oxidation / introduce FG ⟹ COOH NCH$_3$ N H — hydrolysis / remove FG ⟹ C≡N NCH$_3$ C_6H_5 N O

FGI ⟹ Cl NCH$_3$ C_6H_5 N O — FGI ⟹ OH NCH$_3$ C_6H_5 N O — FGI / reduction ⟹ O NCH$_3$ C_6H_5 N O

aldol condensation C-C bond formed ⟹

FGI = functional group interconversion

O O NCH$_3$ C_6H_5 N O — FGI and introduce FG ⟹ O Br C_6H_5 N O — introduce FG ⟹ O C_6H_5 N O

Friedel-Crafts acylation ⟹

CO_2H C_6H_5 N O

38. In the presence of acid, indole is protonated at the #3 position. It can then be attacked by a nucleophile (a second indole molecule or another nucleophile).

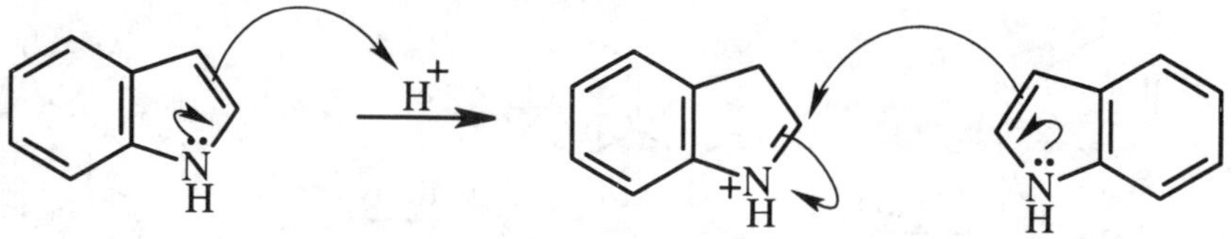

$\xrightarrow{H^+}$

N H +N H N H

39. Because there are two π bonds in the reacting system, only under photochemical conditions can the required suprafacial mode of ring closure occur. (See Table 28.3 on page 1169 of the text.)

40.

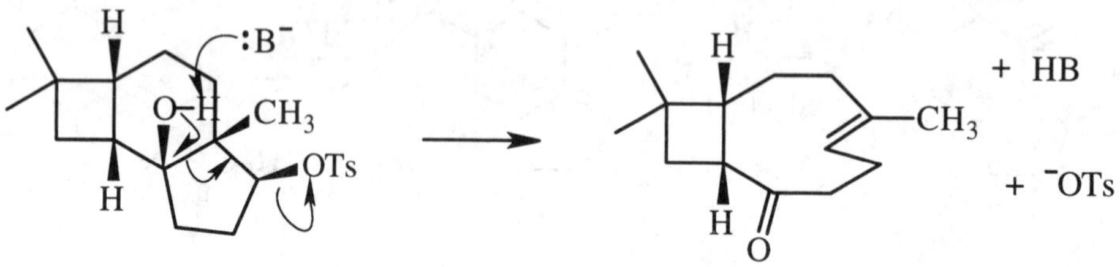

H :B⁻ O–H CH₃ OTs H → H CH₃ H O + HB + ⁻OTs

41. The electrons on the other α-carbon cannot be delocalized, because the carbon at the ring juncture cannot become sp^2 hybridized due to excessive angle strain.

42.

a. $CH_3CH_2CH_2CH_2Br \xrightarrow[Et_2O]{Mg} CH_3CH_2CH_2CH_2MgBr$

1.  ethylene oxide
2. H^+

$CH_3CH_2CH_2CH_2CH_2CH_2OH$

b. $CH_3CH_2CH_2CH_2CH_2Br \xrightarrow[Et_2O]{Mg} CH_3CH_2CH_2CH_2CH_2MgBr$

1.  ethylene oxide
2. H^+

$CH_3CH_2CH_2CH_2CH_2CH_2CH_2OH$

c. $CH_2{=}CHCH_2Br \xrightarrow[Et_2O]{Li} CH_2{=}CHCH_2Li \xrightarrow[Et_2O]{CuI} (CH_2{=}CHCH_2)_2CuLi$

$\downarrow CH_3CH_2CH_2CH_2CH_2Br$

$CH_3CH_2CH_2CH_2CH_2CH_2CH_2CH_2OH \xleftarrow[2.\ H_2O_2,\ HO^-]{1.\ BH_3} CH_3CH_2CH_2CH_2CH_2CH_2CH{=}CH_2$

d. $CH_3CH_2CH_2CH_2Br \xrightarrow[Et_2O]{Mg} CH_3CH_2CH_2CH_2MgBr \xrightarrow[2.\ H^+]{1.\ CH_3CH(=O)} CH_3CH_2CH_2CH_2CH(OH)CH_3$

e. $CH_3CH_2CH_2CH_2Br \xrightarrow[Et_2O]{Mg} CH_3CH_2CH_2CH_2MgBr \xrightarrow[2.\ H^+]{1.\ CH_3C(=O)CH_3} CH_3CH_2CH_2CH_2C(CH_3)(OH)CH_3$

f. $CH_3CH_2CH_2Br \xrightarrow[Et_2O]{Mg} CH_3CH_2CH_2MgBr \xrightarrow[2.\ H^+]{1.\ CH_3C(=O)CH_2CH_2CH_3} CH_3CH_2CH_2C(CH_3)(OH)CH_2CH_2CH_3$

or

$2\ CH_3CH_2CH_2Br \xrightarrow[Et_2O]{Mg} 2\ CH_3CH_2CH_2MgBr \xrightarrow[2.\ H^+]{1.\ CH_3C(=O)OCH_3} CH_3CH_2CH_2C(CH_3)(OH)CH_2CH_2CH_3$

43.

a. 1-(HO)-1-($CH_2CH{=}CH_2$)cyclohexane $\Longrightarrow$ 1-hydroxycyclohexyl cation (HO, +) + $^-CH_2CH{=}CH_2$

1-hydroxycyclohexyl cation $\Downarrow$ cyclohexanone (O)

$^-CH_2CH{=}CH_2 \Downarrow BrMgCH_2CH{=}CH_2 \Downarrow BrCH_2CH{=}CH_2$

b. HO $CH_2CH{=}CH_2$ (cyclohexane) ⟹ HO $\overset{+}{C}H_2$ (cyclohexane) $\overset{-}{C}H{=}CH_2$

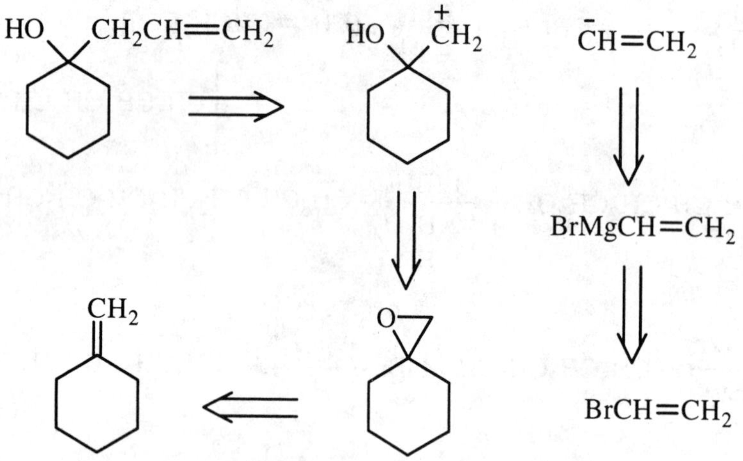

⟹ $BrMgCH{=}CH_2$ ⟹ $BrCH{=}CH_2$

CH_2 (methylenecyclohexane) ⟸ O (epoxide)

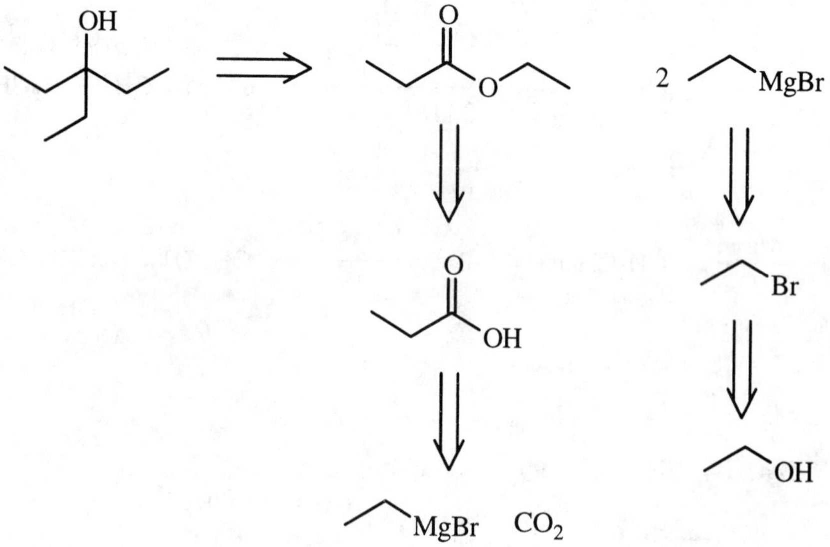

44.

a. OH ⟹ O, O 2 MgBr

⟹ O, OH ⟹ Br

⟹ MgBr CO_2 ⟹ OH

b. O, H, H, O ⟹ ⟹ OH

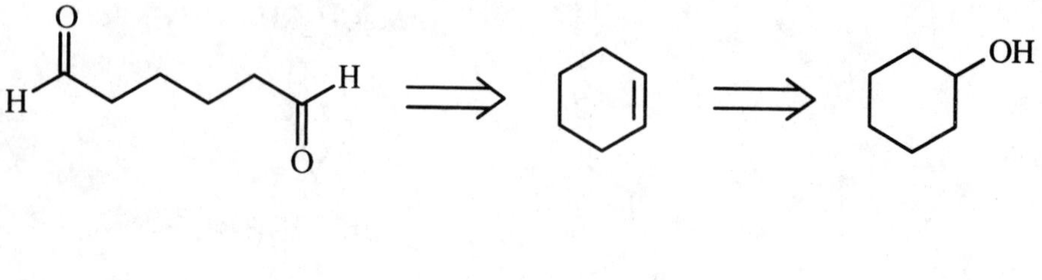

c. [retrosynthesis scheme: 3-cyclohexyl-3-pentanol (OH) ⟹ ethyl cyclohexanecarboxylate + 2 CH_3CH_2MgBr; ester ⟹ cyclohexanecarboxylic acid (OH) ⟹ cyclohexyl-MgBr + CO_2 ⟹ cyclohexyl-Br ⟹ cyclohexanol (OH); CH_3CH_2MgBr ⟹ CH_3CH_2Br ⟹ CH_3CH_2OH]

OH

O

2 MgBr

O

OH

Br

OH

OH Br MgBr

CO_2

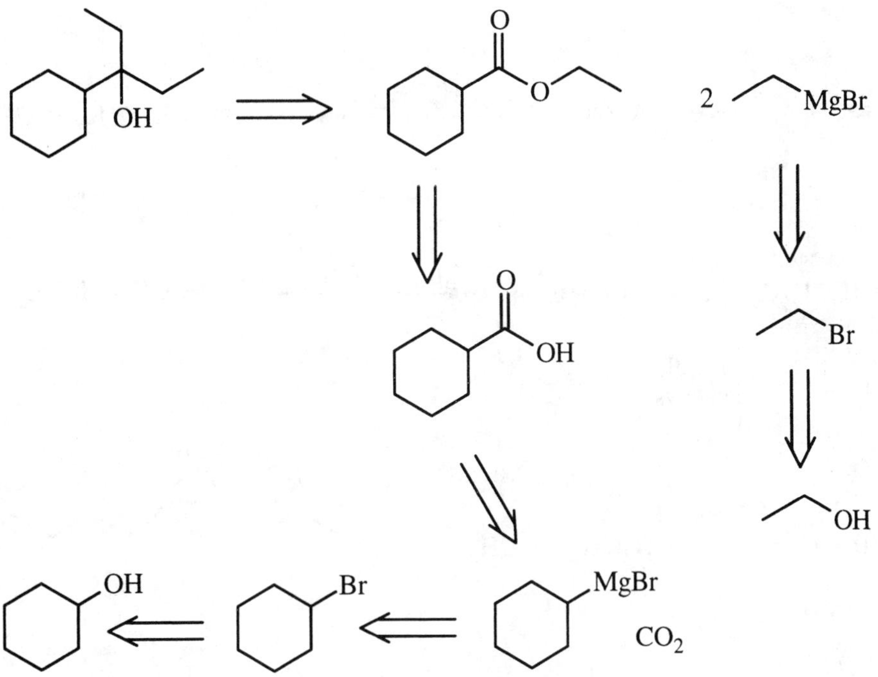

d. [retrosynthesis scheme: bicyclohexyl ⟹ cyclohexyl-MgBr + cyclohexyl-Br; cyclohexyl-MgBr ⟹ cyclohexyl-Br ⟹ cyclohexanol; cyclohexyl-Br ⟹ cyclohexanol]

MgBr Br

Br OH

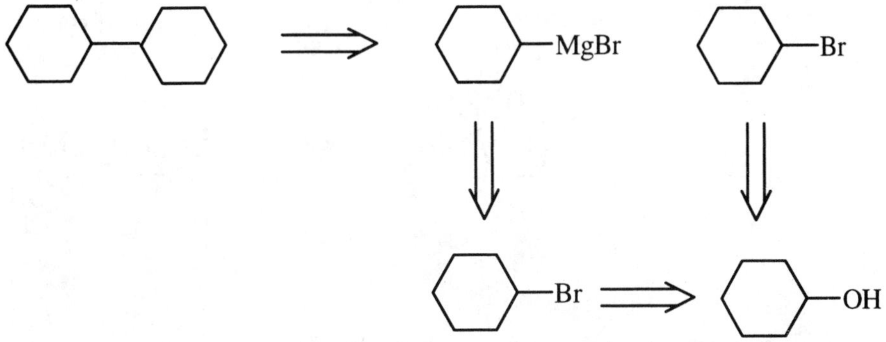

45.

CH_3

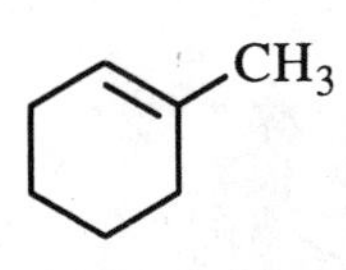

1-methylcyclohexene

a. HBr

b. H_2O, H_2SO_4

c. 1. BH_3
2. H_2O_2, HO^-

d. $RCOOH$ (with C=O: $R\overset{O}{\overset{\|}{C}}OOH$)

e. 1. OsO_4
2. $NaHSO_3$, H_2O

f. 1. $R\overset{O}{\overset{\|}{C}}OOH$
2. HO^-

g. D_2, Pd/C

h. Br_2, CH_2Cl_2

46.

a. $HC{\equiv}CH \xrightarrow{^{-}NH_2} HC{\equiv}C^- \xrightarrow{CH_3CH_2CH_2CH_2Br} HC{\equiv}CCH_2CH_2CH_2CH_3$

$\downarrow {^{-}NH_2}$

$CH_3CH_2CH_2C{\equiv}CCH_2CH_2CH_2CH_3 \xleftarrow{CH_3CH_2CH_2Br} {^{-}C}{\equiv}CCH_2CH_2CH_2CH_3$

$\downarrow$ H_2 Lindlar's catalyst

cis-$CH_3CH_2CH_2(H)C{=}C(H)CH_2CH_2CH_2CH_3 \xrightarrow{MMPP}$ cis epoxide

b. $CH_3C(=CH_2)CH_2Br \xrightarrow[Et_2O]{Mg} CH_3C(=CH_2)CH_2MgBr \xrightarrow[2.\ H^+]{1.\ \text{2-methyloxirane}} CH_3C(=CH_2)CH_2CH_2CH(OH)CH_3$

$\downarrow H^+$

2,2,5-trimethyltetrahydrofuran $\leftarrow CH_3\overset{+}{C}(CH_3)CH_2CH_2CH(OH)CH_3$

47. Sodium hydride is a better base than a nucleophile. Therefore, instead of attacking the carbonyl carbon, it will remove a proton from an α-carbon, generating an enolate.

48. Three possible methods are shown for the synthesis of the target molecule.

1. $C_6H_5CCH_3$ (C=O) $\xrightarrow{LDA} C_6H_5C\bar{C}H_2$ (C=O) $\xrightarrow{CH_3CHO} C_6H_5CCH{=}CHCH_3$ (C=O)

2. Benzene + $CH_3CH{=}CHCCl$ (C=O) $\xrightarrow{AlCl_3}$ $C_6H_5CCH{=}CHCH_3$ (C=O)

$CH_3CH{=}CHCCl$ (C=O) is prepared from $CH_3CH{=}CHCOH$ (C=O) with $SOCl_2$.

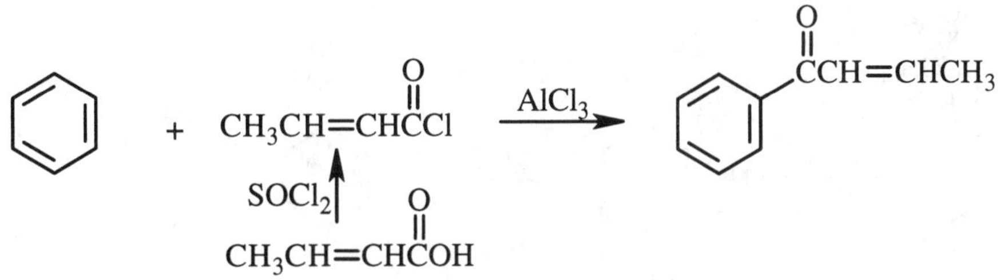

3. 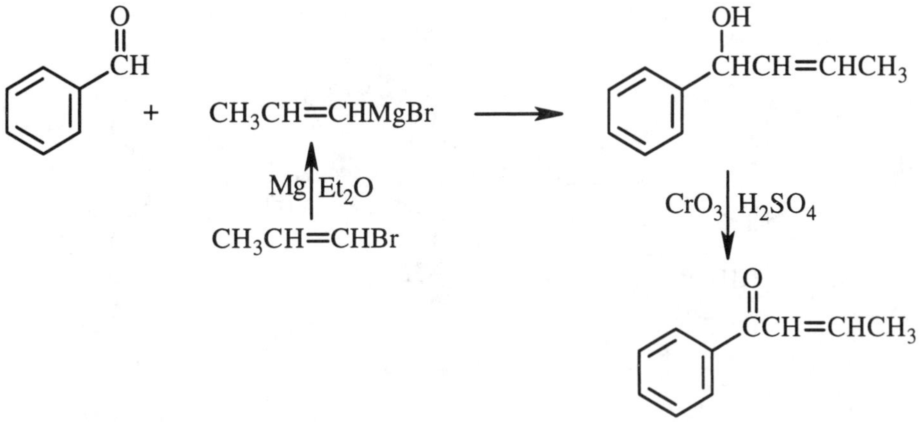 C_6H_5CH (C=O) + $CH_3CH{=}CHMgBr$ $\longrightarrow$ $C_6H_5CHCH{=}CHCH_3$ (OH)

$CH_3CH{=}CHMgBr$ is prepared from $CH_3CH{=}CHBr$ with Mg, Et_2O.

$C_6H_5CHCH{=}CHCH_3$ (OH) $\xrightarrow{CrO_3,\ H_2SO_4}$ $C_6H_5CCH{=}CHCH_3$ (C=O)

49.

a. 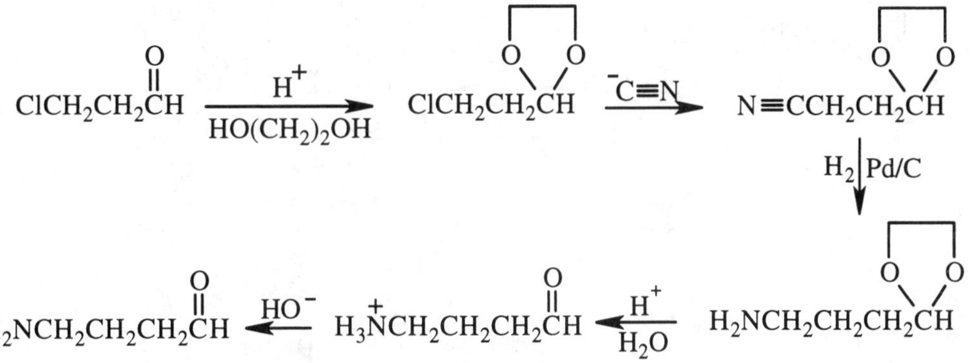 $ClCH_2CH_2CH$ (C=O) $\xrightarrow[HO(CH_2)_2OH]{H^+}$ $ClCH_2CH_2CH$ (1,3-dioxolane acetal) $\xrightarrow{^-C{\equiv}N}$ $N{\equiv}CCH_2CH_2CH$ (1,3-dioxolane acetal) $\xrightarrow{H_2,\ Pd/C}$ $H_2NCH_2CH_2CH_2CH$ (1,3-dioxolane acetal) $\xrightarrow[H_2O]{H^+}$ $H_3\overset{+}{N}CH_2CH_2CH_2CH$ (C=O) $\xrightarrow{HO^-}$ $H_2NCH_2CH_2CH_2CH$ (C=O)

b. 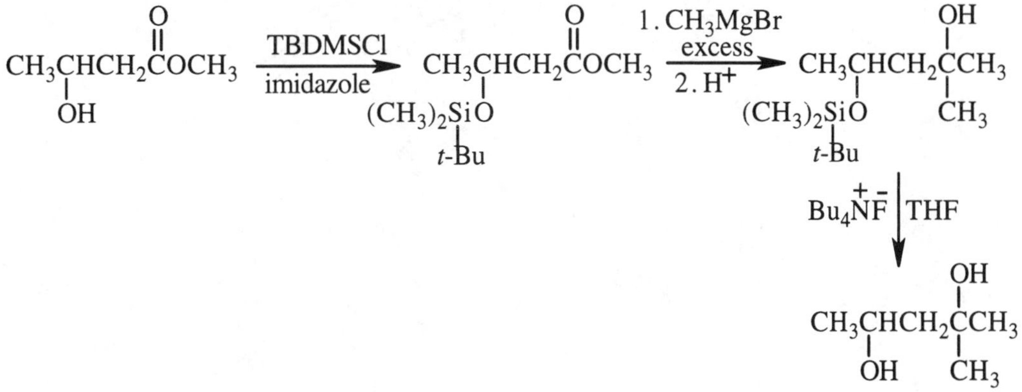 $CH_3CHCH_2COCH_3$ (C=O; OH on C-3) $\xrightarrow[imidazole]{TBDMSCl}$ $CH_3CHCH_2COCH_3$ (C=O; C-3 bears $OSi(CH_3)_2$-*t*-Bu) $\xrightarrow[2.\ H^+]{1.\ CH_3MgBr\ (excess)}$ $CH_3CHCH_2CCH_3$ (OH and CH_3 on C-4; C-2 bears $OSi(CH_3)_2$-*t*-Bu) $\xrightarrow{Bu_4\overset{+}{N}\overset{-}{F},\ THF}$ $CH_3CHCH_2CCH_3$ (OH on C-2; OH and CH_3 on C-4)

c.

4-hydroxycyclohexanecarboxylic acid (COH=O, OH) $\xrightarrow[\text{excess}]{H^+,\ CH_3CH_2OH}$ ethyl 4-hydroxycyclohexanecarboxylate ($COCH_2CH_3$, OH) $\xrightarrow{SOCl_2}$ ethyl 4-chlorocyclohexanecarboxylate ($COCH_2CH_3$, Cl) $\xrightarrow{^-N_3}$ ethyl 4-azidocyclohexanecarboxylate ($COCH_2CH_3$, $N{=}\overset{+}{N}{=}\overset{-}{N}$) $\xrightarrow[Pt]{H_2}$ ethyl 4-aminocyclohexanecarboxylate ($COCH_2CH_3$, NH_2) $\xrightarrow{H^+,\ H_2O}$ 4-ammoniocyclohexanecarboxylic acid (COH=O, $+NH_3$)

d.

2-oxocyclohexanecarboxylic acid (O, COOH) + $HOCH_2C(CH_3)_2NH_2$ $\xrightarrow{\Delta}$ 2-(4,4-dimethyl-oxazolinyl)cyclohexanone (O; N, C, O; CH_3, CH_3) $\xrightarrow{(C_6H_5)_3P{=}CHCH_3}$ ($CH_3CH{=}$; oxazoline, CH_3, CH_3) $\xrightarrow[H_2O]{H^+}$ ($CHCH_3$, COOH) $\xrightarrow[Pt]{H_2}$ (CH_2CH_3, COOH)

50.

a.

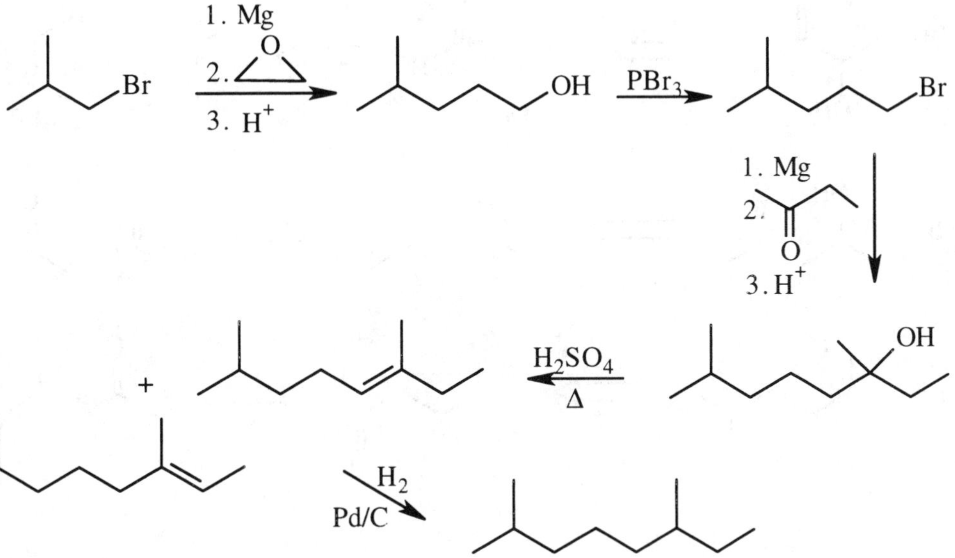
Br
1. Mg
2.
O
3. H+
OH
PBr3
Br
1. Mg
2.
O
3. H+
OH
H2SO4
Δ
+
H2
Pd/C

b.

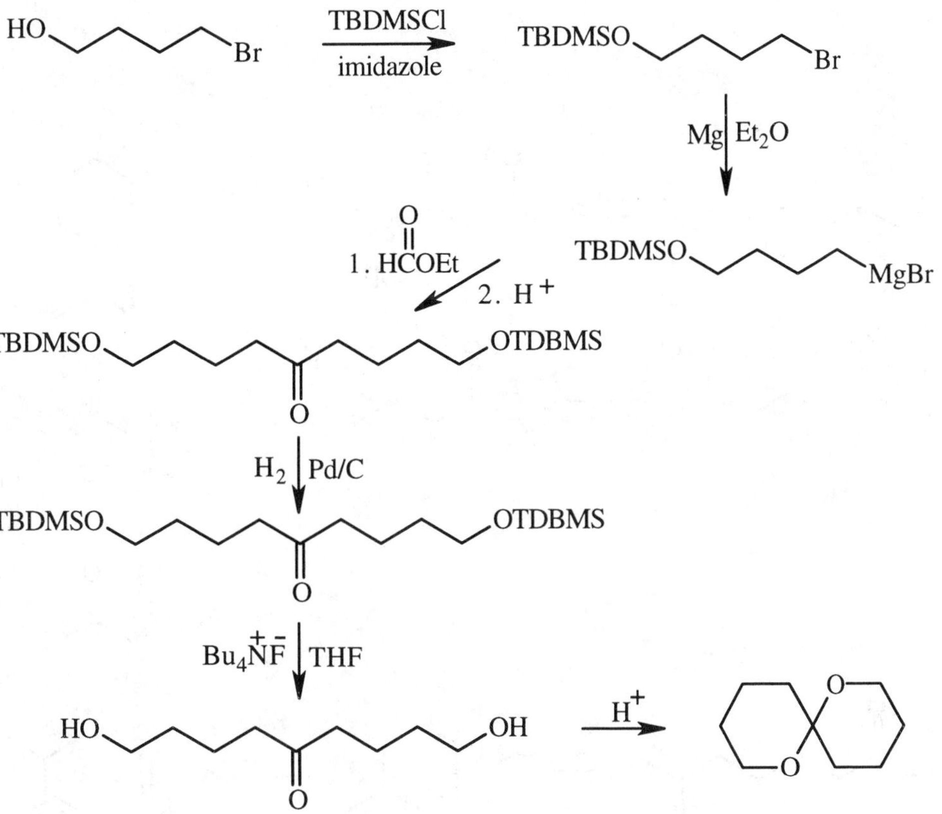
HO
Br
TBDMSCl
imidazole
TBDMSO
Br
Mg Et2O
TBDMSO
MgBr
O
1. HCOEt
2. H+
TBDMSO
OTDBMS
O
H2 Pd/C
TBDMSO
OTDBMS
O
Bu4N+F− THF
HO
OH
O
H+
O
O

51.

a. 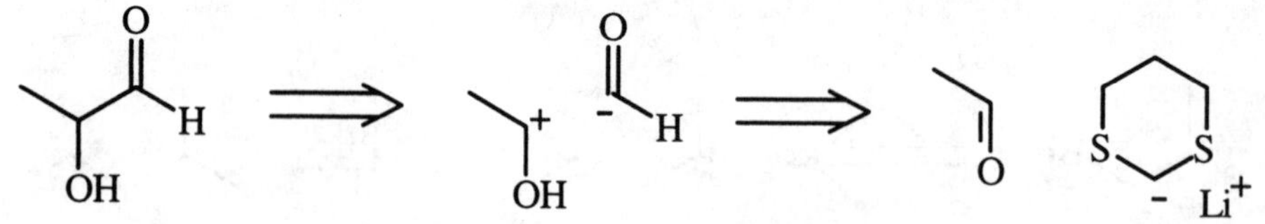 [structure] ⟹ [structure] ⟹ [structure] Li+

b. 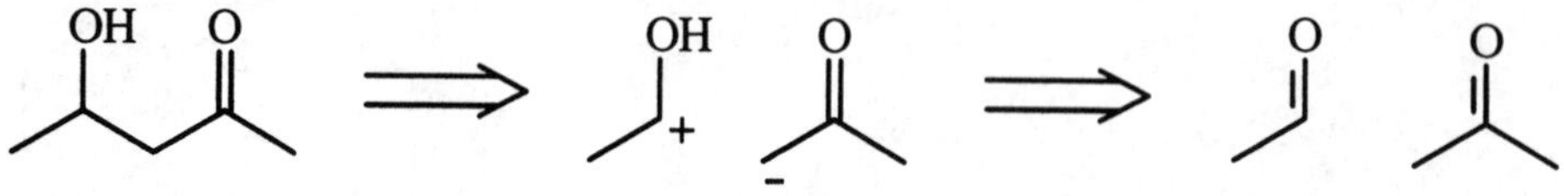 [structure] ⟹ [structure] ⟹ [structure]

c. 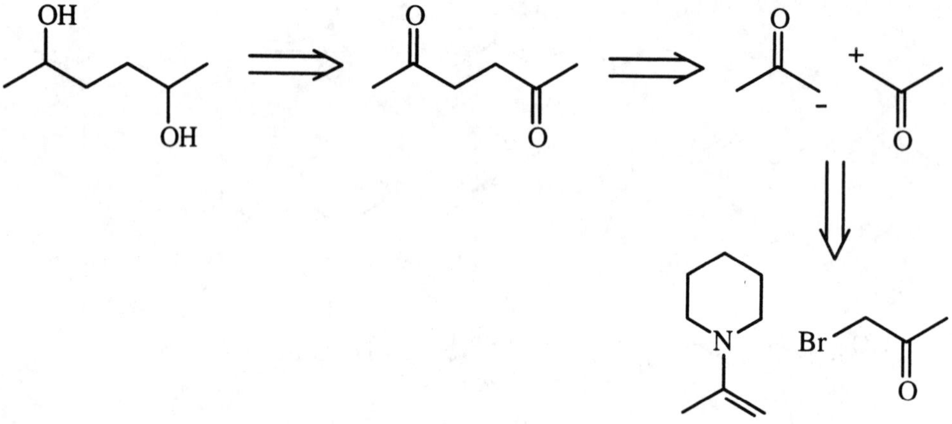 [structure] ⟹ [structure] ⟹ [structure] ⟹ [structure] Br

d. 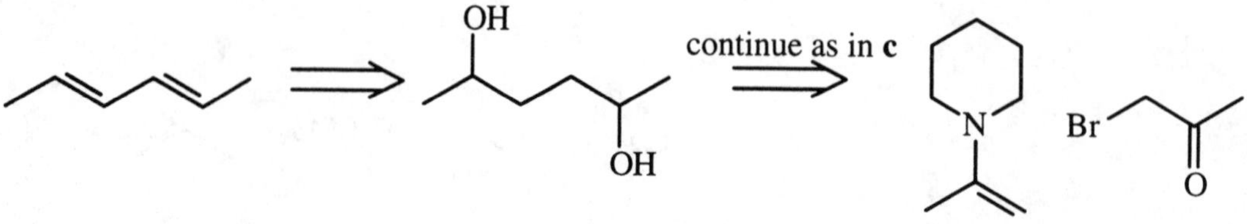 [structure] ⟹ [structure] continue as in **c** ⟹ [structure] Br

e. 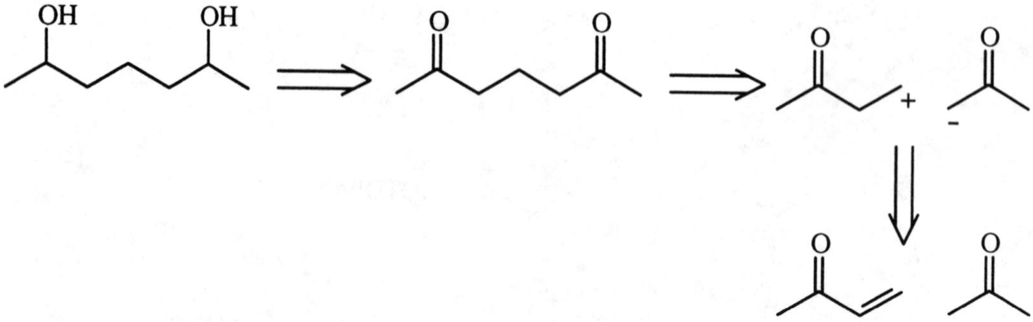 [structure] ⟹ [structure] ⟹ [structure] ⟹ [structure]

f. 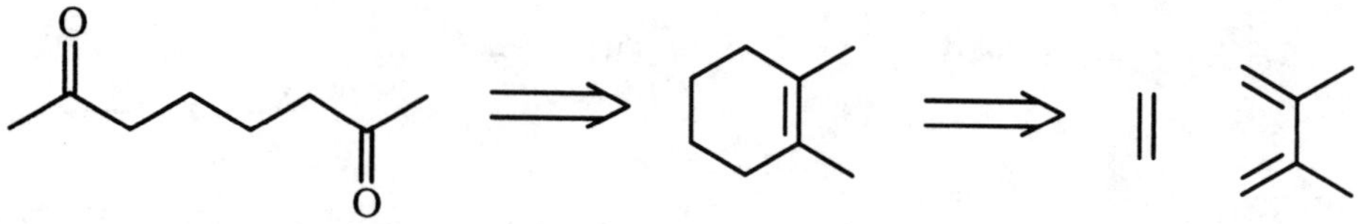 [structure] ⟹ [structure] ⟹ [structure]

52. **a.** Recognizing that the target molecule is an acetal gets you started. The tertiary alcohol can be formed by the reaction of a Grignard reagent with a ketone. The required ß-hydroxyketone can be prepared by an aldol addition.

retrosynthetic analysis

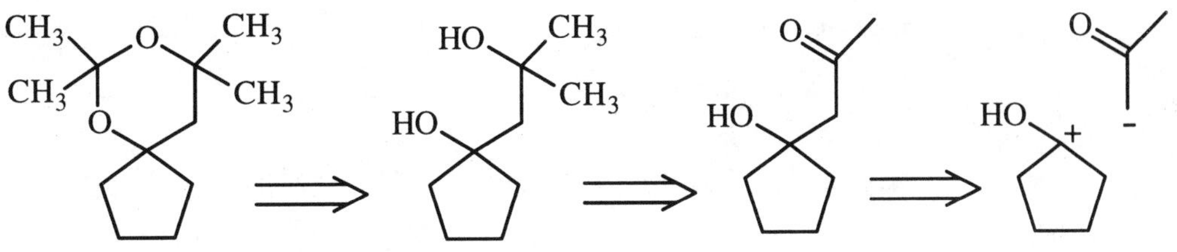

CH_3 O CH_3 / CH_3 O CH_3 ⟹ HO CH_3 / HO CH_3 ⟹ O / HO ⟹ O / HO + -

synthesis

O ⟶ (1. LDA; 2. O=cyclopentanone) ⟶ ^{-}O / O ⟶ (H^+, Δ) ⟶ O ⟶ (CH_3MgBr) ⟶ ^{-}O

H^+, H_2O, Δ ↓

H_3C O CH_3 / H_3C O CH_3 ⟵ ($CH_3\overset{O}{\overset{||}{C}}CH_3$, H^+) ⟵ HO / HO

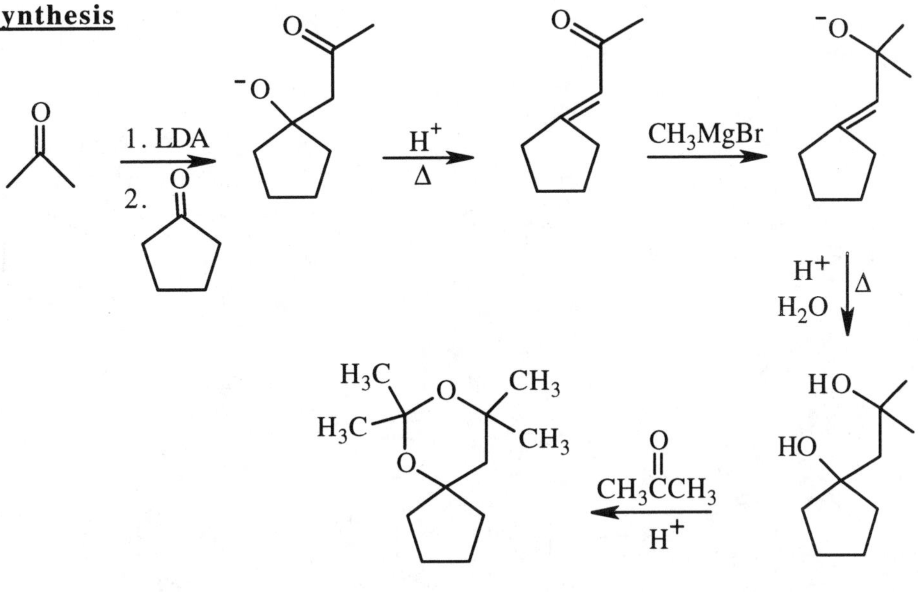

Notice that the product of the aldol addtion is dehydrated before reaction with the Grignard reagent because of the incompatibility of the Grignard reagent with the OH group. Addition of water to the double bond after the tertiary alcohol has been formed, reforms the alcohol.

b. $CH_3CH_2CH_2CH_2Br$ → (1. Mg, Et_2O; 2. ethylene oxide; 3. H^+) → $CH_3CH_2CH_2CH_2CH_2CH_2OH$

$CH_3CH_2CH_2CH_2CH_2CH_2OH$ → (PBr_3) → $CH_3CH_2CH_2CH_2CH_2CH_2Br$

$CH_3CH_2CH_2CH_2CH_2CH_2Br$ → (Mg, Et_2O) → $CH_3CH_2CH_2CH_2CH_2CH_2MgBr$

$CH_3CH_2CH_2CH_2CH_2CH_2MgBr$ → (1. cyclopentanone; 2. H^+) → 1-($CH_2CH_2CH_2CH_2CH_2CH_3$)-cyclopentanol (HO)

53.

a. benzoquinone + butadiene → (Diels–Alder adduct) → (+ butadiene) → (bis-adduct)

b. benzoquinone + cyclopentadiene → (adduct, O, H, H) → base → (OH, H, O) → base → (hydroquinone product, OH, OH)

c.

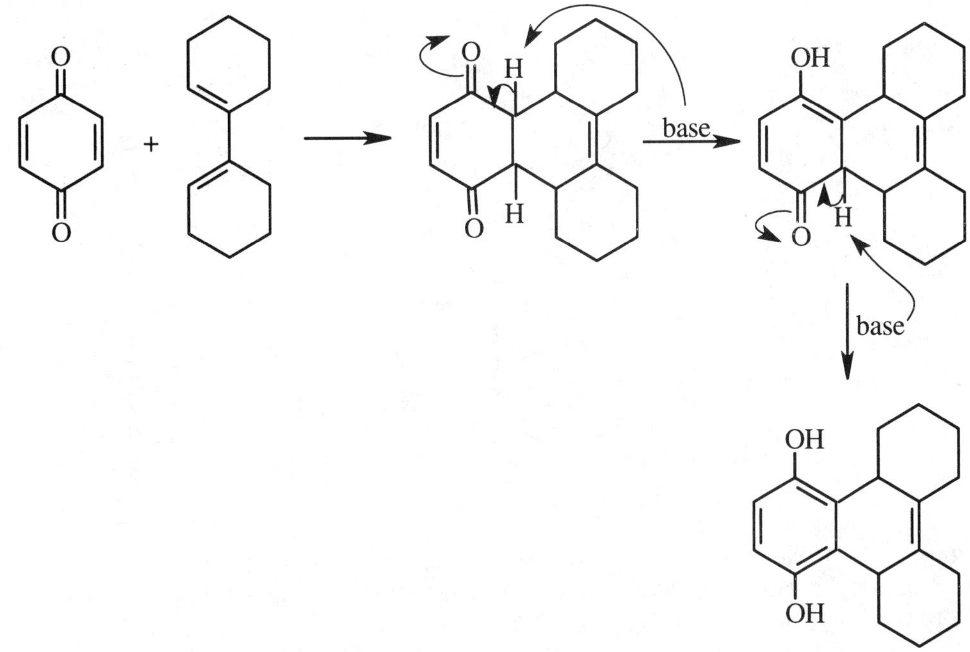

base

base

54. **a.** The first step is nucleophilic attack by cyanide ion on the carbonyl group. This creates an alkoxide ion that eliminates the bromide ion, thereby forming the epoxide ring. Because the alkoxide ion must attack the carbon bonded to the bromine from its backside, cyanide ion must attack the carbonyl group from the side of the ring where the bromine is.

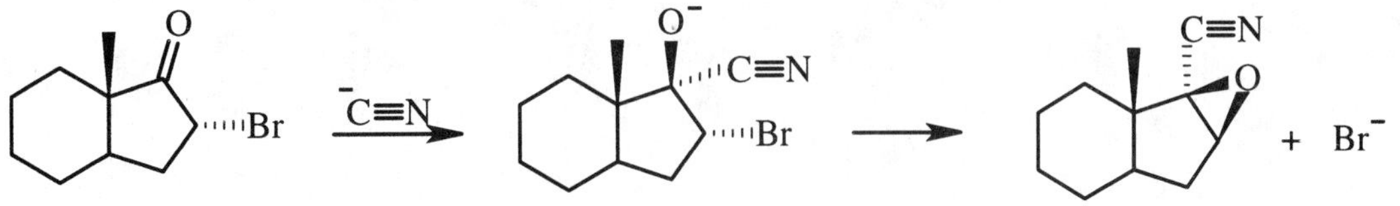

$^{-}C{\equiv}N$

+ Br^-

b. The desired isomer can be obtained if the stereochemistry of the starting material is changed before cyanide ion is added to the reaction mixture. This can be done by using a base such as pyridine to remove a proton from the α-carbon and then reprotonating it.

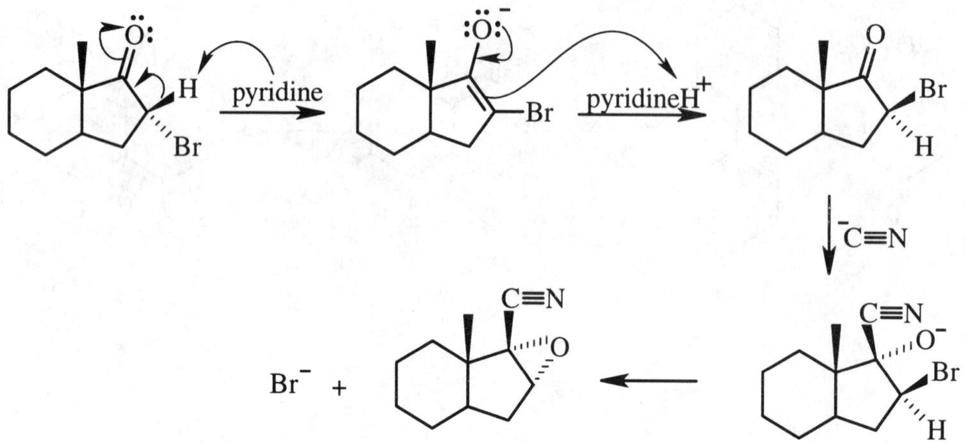

pyridine

pyridineH$^+$

$^-$C≡N

Br$^-$ +

55. Reduction of the quinone to a hydroquinone places an OH group on the ring that reacts with the carbonyl group in an intramolecular reaction, eliminating the group that was protected by forming a six-membered ring lactone.

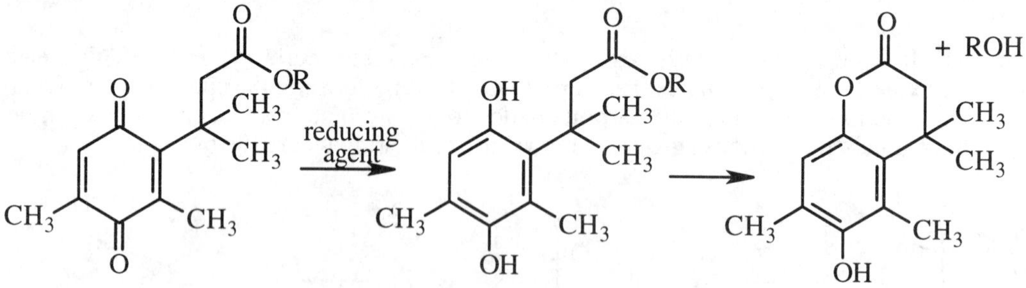

reducing agent

+ ROH

56. **a.** Bromination of the alkane is followed by an S_N2 reaction resulting in an alcohol. Oxidation is followed by bromination of the α-carbon. Addition of methoxide ion results in a Favorskii ring contraction.

Br | OH

Br_2 / $h\nu$ → ; HO^- → ; $Na_2Cr_2O_7$ / H_2SO_4 ↓

O | O | Br | O

CH_3O^- ← ; Br_2 / H^+, H_2O ←

^-O OCH_3 | $COOCH_3$ | $COOCH_3$

→ ; H^+ →

b. Converting the OH group into a good leaving group is followed by formation of an organolithium compound that undergoes an elimination reaction to form a benzyne intermediate, which reacts with furan in a Diels-Alder reaction. Rearrangement of the Diels-Alder product results in the target molecule.

OCH_3 I OH | OCH_3 I OSO_2CF_3 | OCH_3 Li OSO_2CF_3

CF_3SO_2Cl → ; Li →

OCH_3 OH | OCH_3 O H | OCH_3 O

CHAPTER 30
The Organic Chemistry of Drugs: Discovery and Design

Important Terms

antiviral drug — a drug that interferes with DNA or RNA synthesis in order to prevent a virus from replicating.

bactericidal drug — a drug that kills bacteria

bacteriostatic drug — a drug that inhibits the further growth of bacteria

blind screen (random screen) — the search for a pharmacologically active compound without any information about what chemical structures might show activity.

brand name (proprietary name, trade name) — identifies a commercial product and distinguishes it from other products. It can be used only by the owner of the registered trademark.

combinatorial organic synthesis — the synthesis of a library of compounds by covalently connecting sets of building blocks of varying structure.

distribution coefficient — the ratio of the amount of a compound dissolving in each of two solvents.

drug — a compound that reacts with a biological molecule, triggering a physiological effect.

drug resistance — resistance to a particular drug.

drug synergism — when the effect of two drugs used in combination is greater than the sum of the effects obtained when administered individually.

generic name — a name for a drug that anyone can manufacture.

lead compound — the prototype in a search for other biologically active compounds.

molecular modeling — computer-assisted design of a compound with a structure similar to that of a compound with the desired activity.

molecular modification — changing the structure of a lead compound.

orphan drug — a drug for a disease or condition that affects fewer than 200,000 people.

proprietary name — identifies a commercial product and distinguishes it from other products. It can be used only by the owner of the registered trademark.

quantitative structure-reactivity relationship (QSAR) — the relation between a particular property of a series of compounds and their biological activity.

random screen (blind screen) — the search for a pharmacologically active compound without any information about what chemical structures might show activity.

receptor — the site where a drug binds in order to exert its physiological effect.

suicide inhibitor a compound that inactivates an enzyme by undergoing part of the normal catalytic mechanism.

therapeutic index the ratio of the lethal dose of a drug to the therapeutic dose.

trademark a registered name, symbol, or picture.

trade name identifies a commercial product and distinguishes it from other products. It can be used only by the owner of the registered trademark.

Solutions to Problems

1. **a.** ethyl *para*-aminobenzoate **b.** 2-(*N,N*-diethyl)ethyl *para*-aminobenzoate

2.

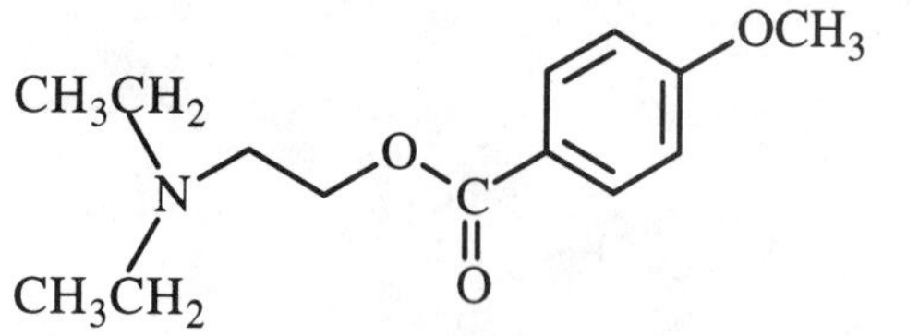

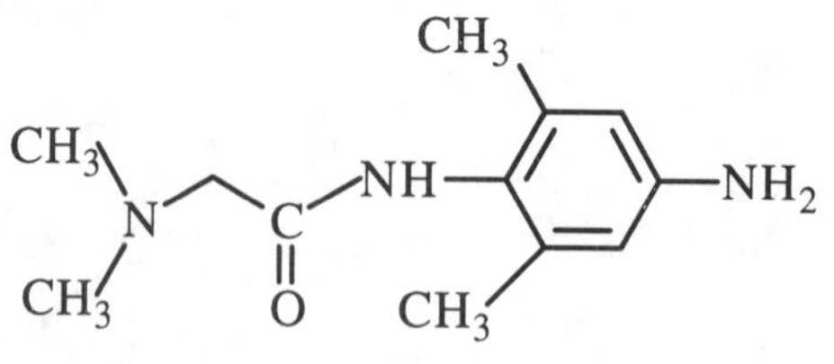

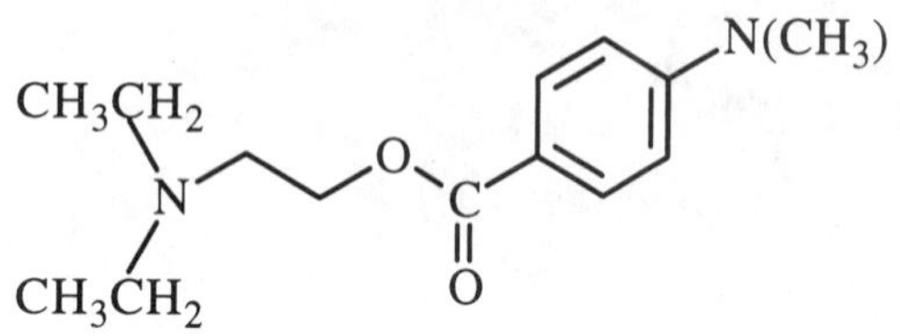

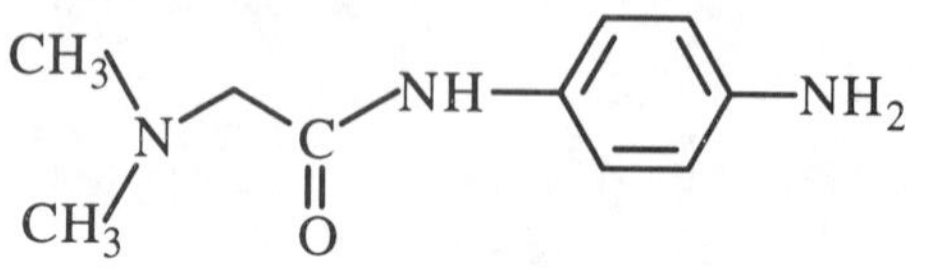

3. The compound on the left has an electron-donating substituent, while the compound on the right has an electron-withdrawing substituent in the same position. Because the compounds known to be effective tranquilizers have an electron-withdrawing substituent in that position (Cl or NO_2), the compound on the right is more likely to show activity as a tranquilizer.

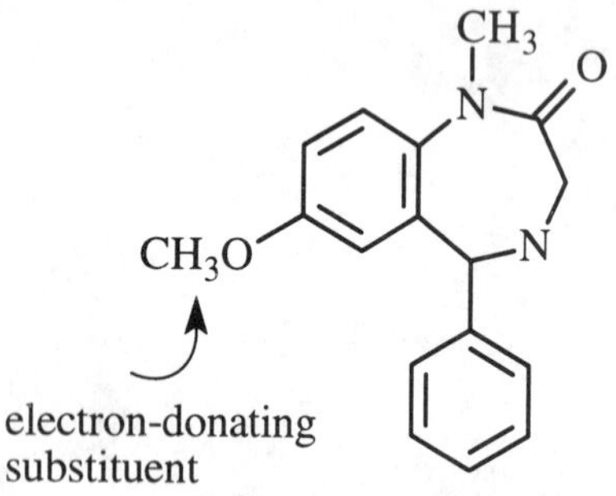

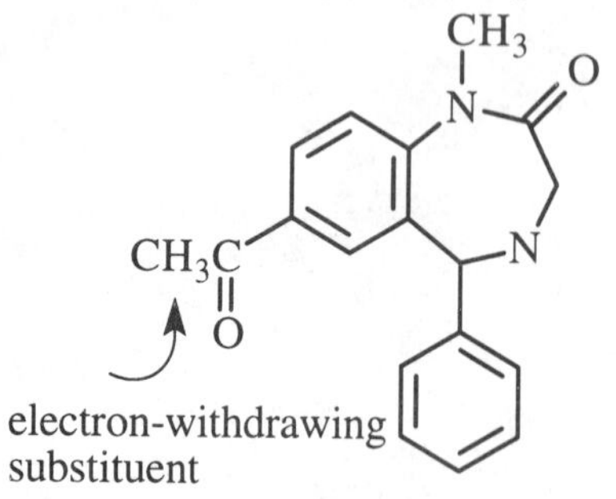

electron-donating substituent

electron-withdrawing substituent

4. Anesthetics have been found to have similar distribution coefficients, which means that they have similar polarities. Since diethyl ether is a known anesthetic, ethyl methyl ether with a polarity similar to that of diethyl ether is more apt to be a general anesthetic than propanol, which is considerably more polar.

5. The electron-withdrawing SO_2 group is weakly basic and, therefore, an excellent leaving group. The weak carbon-sulfur bond aids imine formation.

6. The chloro-substituted compound would be expected to be a more potent inhibitor because potency increases with the ability of the substituents to donate electrons.

a.

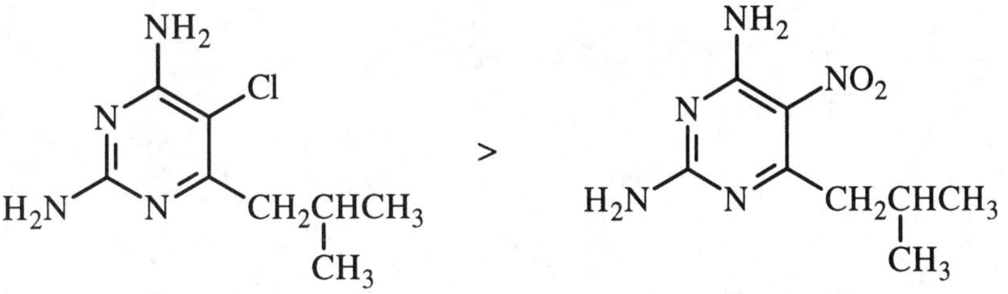

The compound with the pentyl substituent would be expected to be a more potent inhibitor because potency increases with increasing hydrophobicity of one of the substituents.

b.

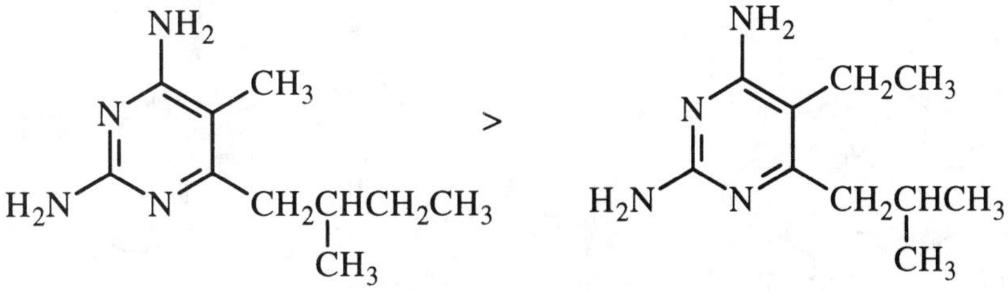

7. Tetrahydrocannabinol is a safer drug because it has a higher therapeutic index.

$$\text{therapeutic index} = \frac{\text{lethal dose}}{\text{therapeutic dose}}$$

$$\text{tetrahydrocannabinol} = \frac{2.0\ \text{g/kg}}{20\ \text{mg/kg}} = \frac{2000\ \text{mg/kg}}{20\ \text{mg/kg}} = 100$$

$$\text{sodium pentothal} = \frac{100\ \text{mg/kg}}{30\ \text{mg/kg}} = 3.3$$

8. Hydrolysis frees the coenzyme and forms a reactive α,β-unsaturated amino acid that can react with the coenzyme bound to its enzyme. This reaction causes the imine linkage to be converted into an amine linkage that cannot be hydrolyzed to release the enzyme. In this way the enzyme is deactivated.

:B̄

H O

HC≡C—C—CO⁻ $\qquad$ CH_2=C=C—CO⁻ $\qquad$ HC=O $\qquad$ + CH_2=C=C—CO⁻

N $\qquad$ N $\qquad$ NH_2

HC $\qquad$ HC

N⁺ H $\qquad$ N⁺ H $\qquad$ N⁺ H

H_2O

E—$(CH_2)_4NH_2$

O NH CH_2 $NH(CH_2)_4$—E $\qquad$ ⁺$NH(CH_2)_4$—E

⁻OC—C—C—CH $\qquad$ CH_2=C=C—CO⁻ $\qquad$ HC

:NH_2

N⁺ H $\qquad$ N⁺ H

H_2O

O O CH_2 $NH(CH_2)_4$—E

⁻OC—C—C—CH

N⁺ H $\qquad$ + NH_3

9. Cytosar differs from cytidine in that it has the 2'-OH group in the β-position.

Vira-A differs from adenosine in that it has the 2'-OH group in the β-position.

Herplex differs from 2'-deoxyuridine in that it has an iodo substituent in the 5-position.

Aclovir differs from guanosine in that it has a hydroxy-ether substituent rather than a ribose.

Viramid has an unusual heterocyclic base attached to ribose.

10.

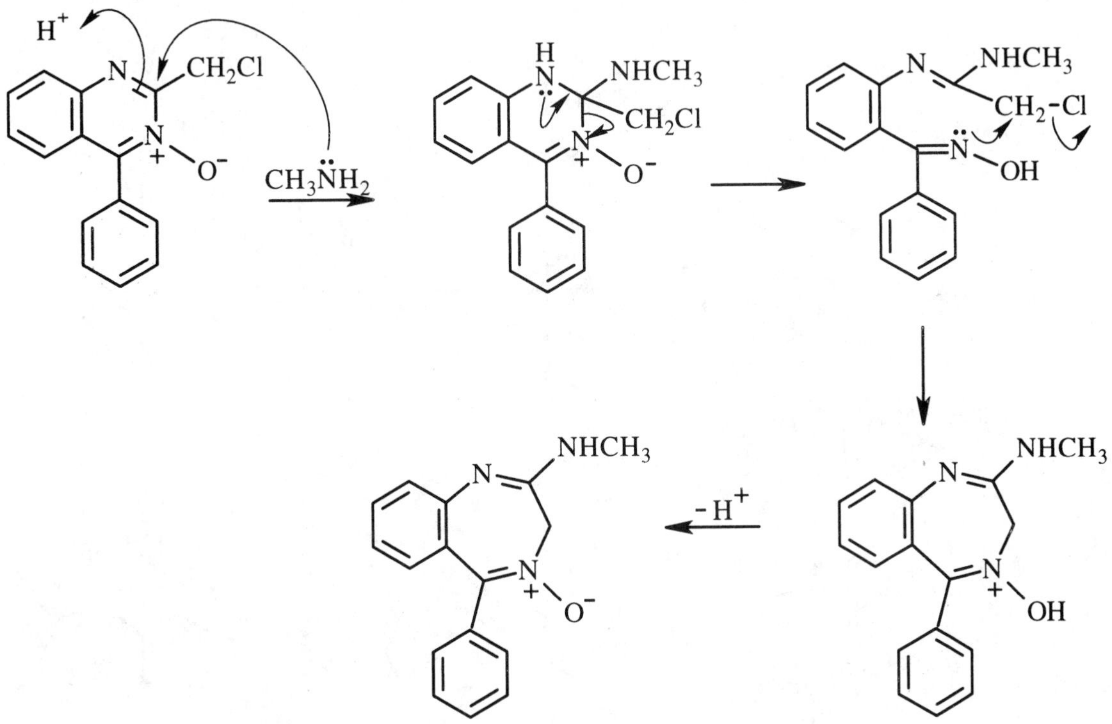

H^+ CH_2Cl $CH_3\ddot{N}H_2$ $NHCH_3$ CH_2Cl $NHCH_3$ CH_2-Cl OH

$NHCH_3$ $-H^+$ $NHCH_3$ OH

11.

CH_3CCl (acetyl chloride); C_6H_5CCl (benzoyl chloride); H^+, H_2O, Δ; $(CH_3)_3COCNHCH_2CF$; 1. CH_3COOH, CH_2Cl_2 2. Δ; HO^-; CH_3I

NH₂ → NHCCH₃ → NHCCH₃ / C=O → NH₂ / C=O → NH / NHCOC(CH₃)₃ / C=O → H N / O / N → N / O⁻ / N → CH₃ N / O / N

Answers to Chapter 1 Practice Test

1. **a.** HBr **b.** NH_3 **c.** a carbon-fluorine bond **d.** CCl_4

2. $\overset{+}{C}H_3$ sp^2 $\overset{-}{C}H_3$ sp^3 $\overset{\bullet}{C}H_3$ sp^2

3. 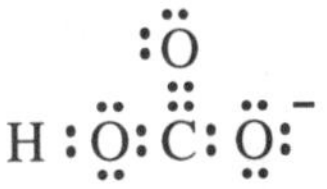 H:Ö:C:Ö:⁻ (with :O above C, double-bonded)

4. CH_3COO^- CH_3CH_2OH CH_3OH $CH_3CH_2\overset{+}{N}H_3$

5. $^{+}NH_4$

6. $^{-}NH_2$

7. $CH_3CH_2CH_2CH{=}CH_2$ $CH_3CH_2CH{=}CHCH_3$ $CH_3CHCH{=}CH_2$ with CH_3 on C2

8. **a.** $CH_3OH + \overset{+}{N}H_4 \rightleftharpoons CH_3\overset{+}{O}H(H) + NH_3$ **b.** reactants

9. **a.** $1s^2\,2s^2\,2p_x\,2p_y$ **b.** $1s^2\,2s\,2p_x\,2p_y\,2p_z$

10. $CH_3CH_2CHCOOH$ with Cl on C2

11. O=C=O (C: *sp*) HCOH with C=O (C: sp^2) HC≡N (C: *sp*) CH_3OCH_3 (C: sp^3) $CH_3CH{=}CH_2$ (CH_3: sp^3)

12.
a. A pi bond is stronger than a sigma bond. F
b. A triple bond is shorter than a double bond. T
c. The oxygen-hydrogen bonds in water are formed by the overlap of an sp^2 orbital of oxygen with an *s* orbital of hydrogen. F
d. HO^- is a stronger base than $^{-}NH_2$. F
e. A double bond is stronger than a single bond. T
f. A tetrahedral carbon has bond angles of 107.5°. F
g. A Lewis acid is a compound that accepts a share in a pair of electrons. T

Answers to Chapter 2 Practice Test

1. 3-methyloctane

2. **a.** 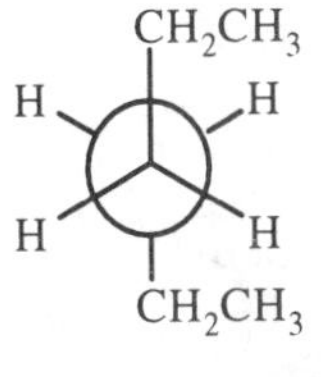 Newman projection: front CH_2CH_3 (up), H, H; back CH_2CH_3 (down), H, H
b. 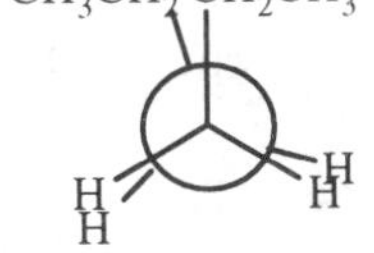 Newman projection: CH_3CH_2 CH_2CH_3 eclipsed (up); H, H, H, H
c. 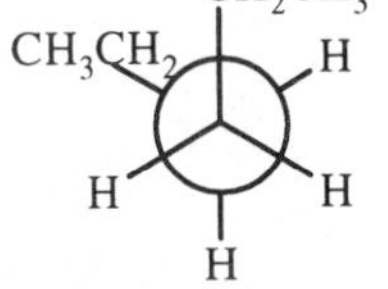 Newman projection: CH_2CH_3, CH_3CH_2, H; H, H, H

3. **a.** *sec*-butyl chloride, 2-chlorobutane **b.** isohexyl alcohol, 4-methyl-1-pentanol
c. cyclopentyl bromide, bromocyclopentane

4. a. $CH_3CH_2CH_2CH_2CH_2Br$ **1** $CH_3CH_2CH_2Br$ **3** $CH_3CH_2CH_2CH_2Br$ **2**

b. $CH_3CH_2CH_2CH_2CH_3$ **3** $CH_3CH_2CH_2CH_2OH$ **1** $CH_3CH_2CH_2CH_2Cl$ **2**

c.
CH_3 CH_3
$CH_3C—CCH_3$
CH_3 CH_3
3

$CH_3CH_2CH_2CH_2CH_2CH_2CH_2CH_3$ **1**

$CH_3CHCH_2CH_2CH_2CH_2CH_3$
CH_3
2

5. **a.** 6-methyl-3-heptanol **b.** 3-ethoxyheptane **c.** 1-bromo-3-methylcyclopentane **d.** 1,4-dichloro-5-methylheptane

6. 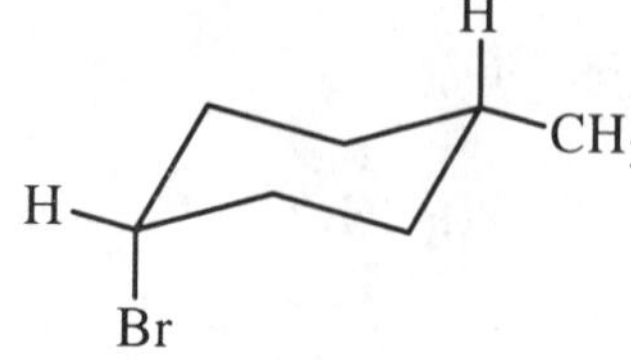

7. H CH_3 $CHCH_3$ CH_3 H

8. **a.** butyl alcohol **b.** 1-butanol **c.** hexane **d.** pentylamine **e.** ethyl alcohol

9. **a.** 1-bromo-3-methylbutane isopentyl bromide **b.** 3-methyl-1-butanol isopentyl alcohol **c.** 3-methyl-1-butanamine isopentylamine

10. **a.** a staggered conformer
b. the chair conformer of methylcyclohexane with the methyl group in the equatorial position
c. cyclohexane

11. a.
CH_3CHCH_3
Br

d.
CH_3CHCH_3
CH_3

e. $CH_3CH_2CH_2OH$
CH_3CHOH
CH_3
$CH_3CH_2OCH_3$

b. $CH_3CH_2NHCH_3$

c.
CH_3CHCH_3
CH_3

CH_3
CH_3CCH_3
CH_3

CH_4

CH_3 CH_3
$CH_3C—CCH_3$
CH_3 CH_3

Answers to Chapter 3 Practice Test

1. a.
CH_3
$CH_3\overset{}{C}CH_3$
$+$

b. $CH_3CH_2\dot{C}HCH_3$

c. $CH_3CH_2\overset{+}{C}H_2$

d. CH_3

e. H_3C H
C=C
H CH_3

2. $CH_3CH_2CH_2CH{=}CH_2$

3. **B**

4. 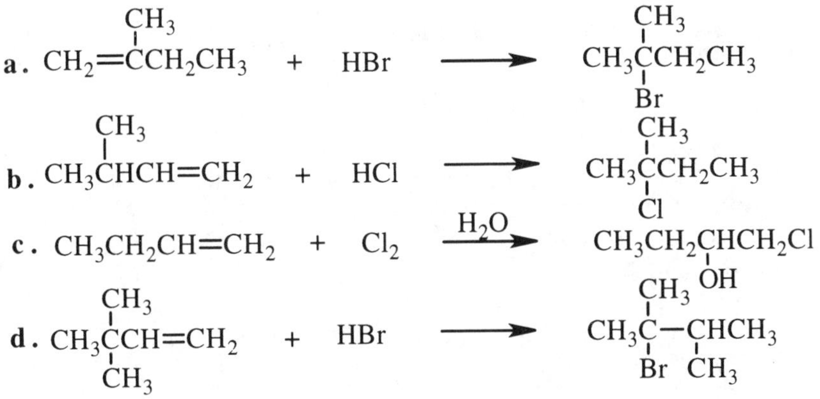 a. $CH_2{=}C(CH_3)CH_2CH_3$ + HBr ⟶ $CH_3C(CH_3)(Br)CH_2CH_3$

b. $CH_3CH(CH_3)CH{=}CH_2$ + HCl ⟶ $CH_3C(CH_3)(Cl)CH_2CH_3$

c. $CH_3CH_2CH{=}CH_2$ + Cl_2 $\xrightarrow{H_2O}$ $CH_3CH_2CH(OH)CH_2Cl$

d. $CH_3C(CH_3)_2CH{=}CH_2$ + HBr ⟶ $CH_3C(CH_3)(Br){-}CH(CH_3)CH_3$

5. 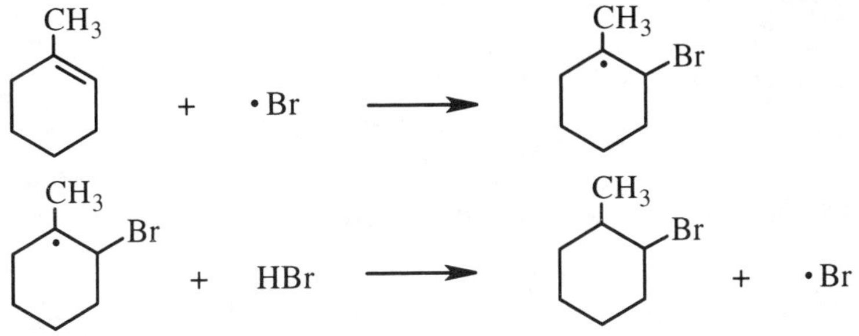 1-methylcyclohexene (CH_3) + ·Br ⟶ radical (·, at the CH_3-bearing carbon) with Br on adjacent carbon

radical (CH_3, ·, Br) + HBr ⟶ 1-bromo-2-methylcyclohexane (CH_3, Br) + ·Br

6.

$-\overset{O}{\overset{\|}{C}}CH_3$	$-CH{=}CH_2$	$-Cl$	$-C{\equiv}N$
2	**4**	**1**	**3**

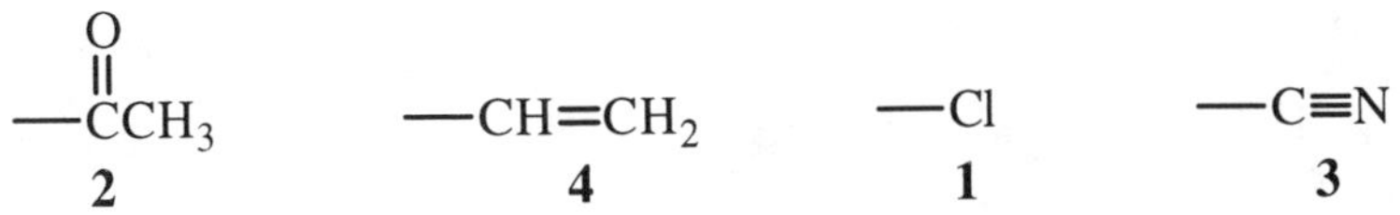

7. **a.** 2-pentene
b. 3-methyl-1-hexene
c. 3-methyl-2-pentene
d. 1-methylcyclohexene

8. 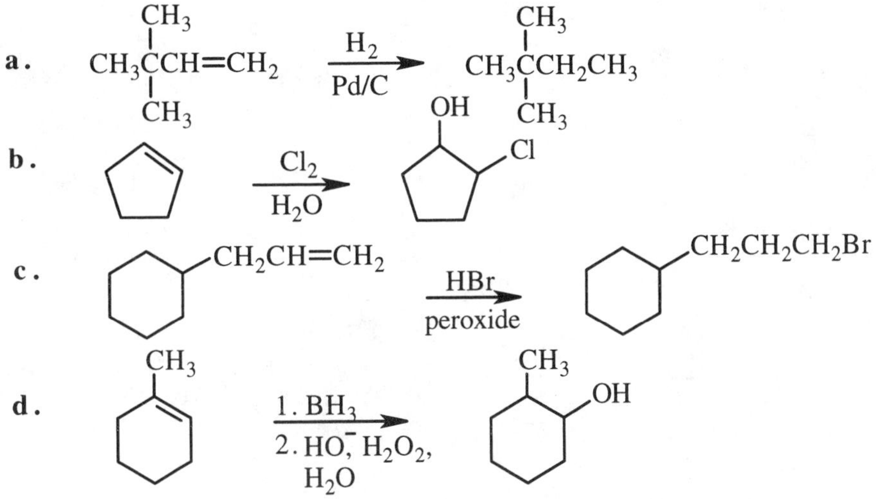 a. $CH_3C(CH_3)_2CH{=}CH_2$ $\xrightarrow[Pd/C]{H_2}$ $CH_3C(CH_3)_2CH_2CH_3$

b. cyclopentene $\xrightarrow[H_2O]{Cl_2}$ 2-chlorocyclopentanol (OH, Cl)

c. cyclohexyl–$CH_2CH{=}CH_2$ $\xrightarrow[\text{peroxide}]{HBr}$ cyclohexyl–$CH_2CH_2CH_2Br$

d. 1-methylcyclohexene (CH_3) $\xrightarrow[2.\ HO^-,\ H_2O_2,\ H_2O]{1.\ BH_3}$ 2-methylcyclohexanol (CH_3, OH)

9. $CH_3CH_2\overset{+}{C}H\overset{CH_3}{\overset{|}{C}}HCH_3$ CH_3 (cyclohexane ring) +

10. a. $CH_3\overset{CH_3}{\overset{|}{C}}H\underset{OH}{\underset{|}{C}}HCH_3$ b. CH_2CH_3 (cyclopentane ring) OH

11.
a. The addition of Br_2 to 2-butene to form 1,2-dibromobutane is a concerted reaction. F
b. Decreasing the entropy of the products compared to the entropy of the reactants makes the equilibrium constant more favorable. F
c. An exergonic reaction is one with a $-\Delta G°$. T
d. An alkene is an electrophile. F
e. The higher the energy of activation, the more slowly the reaction will take place. T
f. Another name for *trans*-2-butene is *Z*-2-butene. F
g. The reaction of 1-butene with HCl will form 1-chlorobutane as the major product if hydrogen peroxide is added to the reaction mixture. F
h. 2,3-Dimethyl-2-pentene is more stable than 3,4-dimethyl-2-pentene. T

Answers to Chapter 4 Practice Test

1. a pair of enantiomers 2. -3.0°

3. Fischer projection: CH_3 (top); H—C—Cl; H—C—Cl; CH_3 (bottom)

4. $CH_3CH_2CH_2CH_2Cl$ $CH_3CH_2\underset{Cl}{\underset{|}{C}}HCH_3$ $CH_3\underset{CH_3}{\underset{|}{C}}HCH_2Cl$ $CH_3\overset{CH_3}{\underset{Cl}{C}}CH_3$

5. a. CH_3CH_2—C(Cl)(H)—CH_3 + CH_3—C(Cl)(H)—CH_2CH_3

b. $CH_3CH_2CH_2$—C(Br)(H)—CH_3 + CH_3—C(Br)(H)—$CH_2CH_2CH_3$ + $CH_3CH_2\underset{Br}{\underset{|}{C}}HCH_2CH_3$

c. Fischer projection: CH_2CH_3 (top); H—C—Br; H—C—Br; CH_2CH_3 (bottom)

d. Fischer projection: CH_2CH_3 (top); H—C—Br; H—C—Br; $CH_2CH_2CH_3$ (bottom) + Fischer projection: CH_2CH_3 (top); Br—C—H; Br—C—H; $CH_2CH_2CH_3$ (bottom)

6. (-)-2-Methylbutanoic acid has the *R* configuration.

7. a. CH_3 Br / Br CH_3 + Br CH_3 / CH_3 Br

c. H H / CH_3 CH_3

b. CH_3 Br / Br CH_2CH_3 + Br CH_2CH_3 / CH_3 Br

d. H H / CH_3 CH_2CH_3 + H H / CH_3CH_2 CH_3

8. CH_3CH_2—C(H)(Br)—CH_3 and CH_3CH_2—C(CH_2CH_2Br)(Br)—CH_3

9. H Cl / Cl H or Cl H / H Cl

10. a. Diastereomers have the same melting points. F
b. The addition of HBr to 3-methyl-2-pentene is stereospecific. F
c. The addition of HBr to to 3-methyl-2-pentene is stereoselective. F
d. The addition of HBr to to 3-methyl-2-pentene is regioselective. T
e. Meso compounds do not rotate polarized light. T
f. 2,3-Dichloropentane has a stereoisomer that is a meso compound. F
g. In the most stable conformation of *cis*-1-ethyl-2-methylcyclohexane, both substituents are in the equatorial position. F
h. A compound with three chirality centers can have a maximum of nine stereoisomers. F

Answers to Chapter 5 Practice Test

1. a. 1. disiamylborane 2. HO^-, H_2O_2, H_2O
b. H_2/Lindlar's catalyst

2. (cyclohexenyl)—OH

3. $CH_3CH_2CHC{\equiv}CCH_2CHCH_3$ (with CH_3 on each CH)

4. a. A terminal alkyne is more stable than an internal alkyne. F
b. Propyne is more reactive than propene toward reaction with HBr. F
c. 1-Butyne is more acidic than 1-butene. T
d. An sp^2 hybridized carbon is more electronegative than an sp^3 hybridized carbon. T

5. 1-bromo-5-methyl-3-hexyne

6. $CH_3CH_2CH_2C{\equiv}CH$

7. NH_3 (3) $CH_3C{\equiv}CH$ (2) CH_3CH_3 (5) H_2O (1) $CH_3CH{=}CH_2$ (4)

8. $CH_3CH_2CH_2\overset{O}{\overset{\|}{C}}CH_2CH_3$ 3-methylcyclohexanone (O= ring with CH_3)

9. a. $CH_3CH_2C{\equiv}CH \xrightarrow{\bar{N}H_2} CH_3CH_2C{\equiv}C^- \xrightarrow{CH_3CH_2Br} CH_3CH_2C{\equiv}CCH_2CH_3 \xrightarrow{H_2/Pt} CH_3CH_2CH_2CH_2CH_2CH_3$

b. $CH_3CH_2C{\equiv}CH \xrightarrow{H_2/\text{Lindlar's catalyst or Na/NH}_3\text{ liq}} CH_3CH_2CH{=}CH_2 \xrightarrow[\text{peroxide}]{HBr} CH_3CH_2CH_2CH_2Br$

c. $CH_3CH_2C{\equiv}CH \xrightarrow{\bar{N}H_2} CH_3CH_2C{\equiv}C^- \xrightarrow{CH_3CH_2Br} CH_3CH_2C{\equiv}CCH_2CH_3 \xrightarrow{H_2O,\ H_2SO_4} CH_3CH_2\overset{O}{\overset{\|}{C}}CH_2CH_2CH_3$

Answers to Chapter 6 Practice Test

1. a. $C_6H_5\overset{+}{C}H_2$

b. $CH_3\bar{C}HC{\equiv}CH$

c. $CH_3\bar{C}H\overset{O}{\overset{\|}{C}}CH_3$

d. $CH_2{=}CH\overset{+}{C}H_2$

e. cyclohexenyl–$\overset{+}{C}HCH_3$

2. a. $CH_3CH{=}CH{-}\ddot{O}CH_3 \longleftrightarrow CH_3\bar{C}H{-}CH{=}\overset{+}{O}CH_3$

b. $CH_3CH{=}CH{-}CH{=}CH{-}\overset{+}{C}H_2 \longleftrightarrow CH_3CH{=}CH{-}\overset{+}{C}H{-}CH{=}CH_2 \longleftrightarrow CH_3\overset{+}{C}H{-}CH{=}CH{-}CH{=}CH_2$

c. $\bar{C}H_2{-}CH{=}CH{-}\overset{O}{\overset{\|}{C}}H \longleftrightarrow CH_2{=}CH{-}\bar{C}H{-}\overset{O}{\overset{\|}{C}}H \longleftrightarrow CH_2{=}CH{-}CH{=}\overset{O^-}{\overset{|}{C}}H$

3. $CH_3\overset{+}{C}(CH_3)CH_2CH{=}CH_2$ $CH_2{=}CHCH_2CH{=}CH_2$ $CH_3CH_2NHCH_2CH{=}CHCH_3$

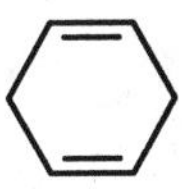

4. c. $CH_3\overset{O}{\overset{\|}{C}}OH$ and $CH_3\overset{O^-}{\overset{|}{C}}{=}\overset{+}{O}H$

5. 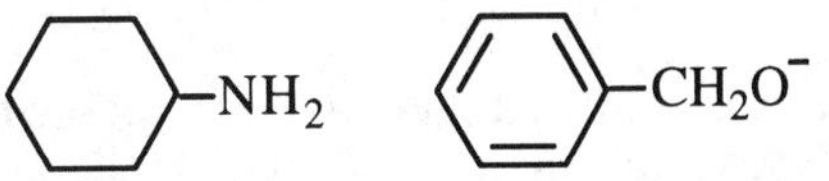 —NH_2 —CH_2O^-

6. a. 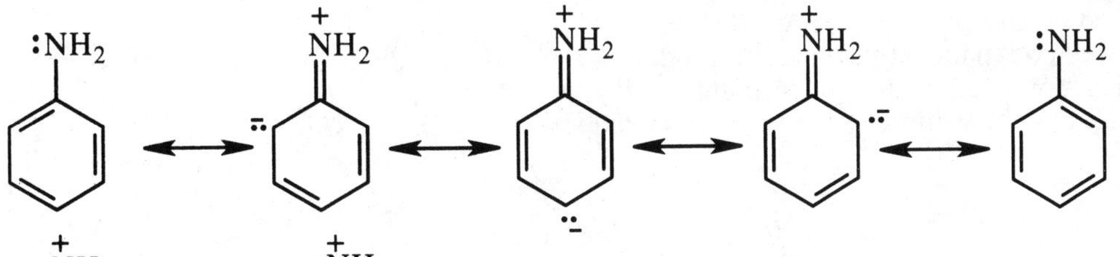 $:NH_2$ $\overset{+}{N}H_2$ $\overset{+}{N}H_2$ $\overset{+}{N}H_2$ $:NH_2$

b. 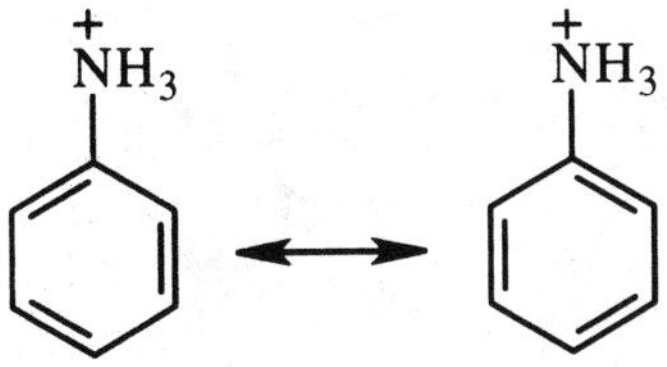 $\overset{+}{N}H_3$ $\overset{+}{N}H_3$

c. 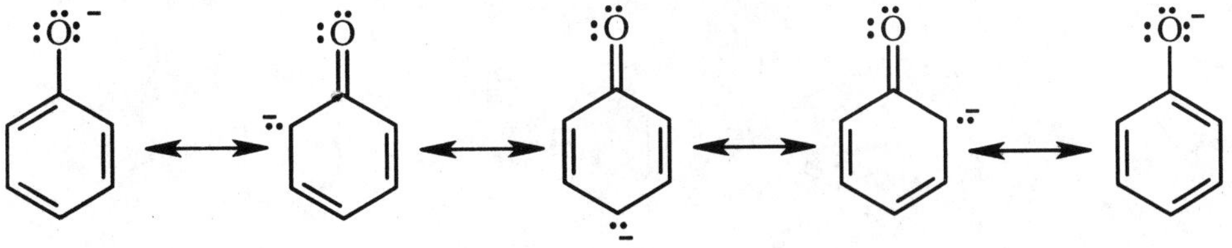 $:\ddot{O}:^-$ $:\ddot{O}$ $:\ddot{O}$ $:\ddot{O}$ $:\ddot{O}:^-$

6.

7. a.  H H b.  c. 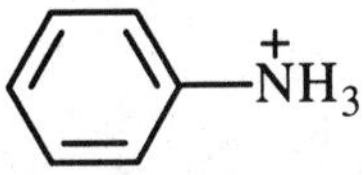 $\overset{+}{N}H_3$

8. a. 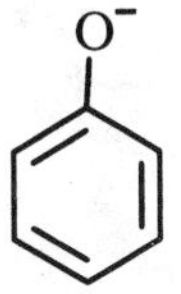 O^- b. 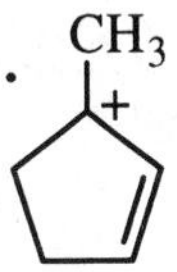 CH_3 +

Answers to Chapter 7 Practice Test

1. a. 5-chloro-1,3-cyclohexadiene
b. 1-octen-6-yne
c. 4-methyl-2-cyclohexen-1-ol

2. a. 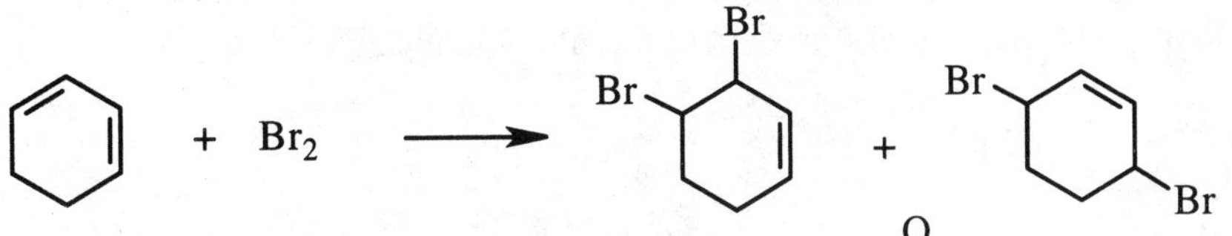 + Br_2 ⟶ Br Br + Br Br

b. 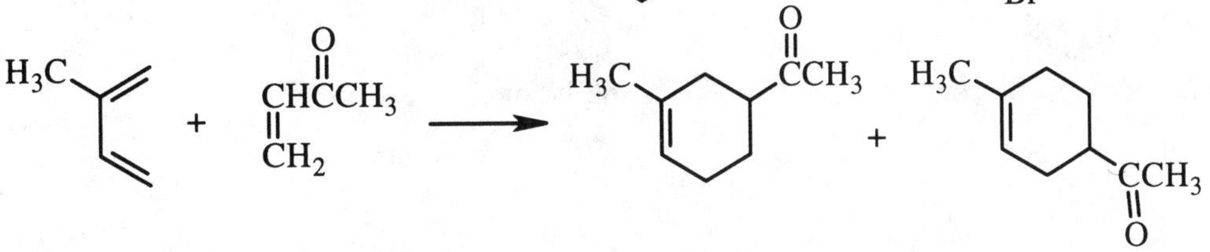 H_3C + $CH\overset{O}{\overset{\|}{C}}CH_3$, $\|$ CH_2 ⟶ H_3C $\overset{O}{\overset{\|}{C}}CH_3$ + H_3C $\underset{O}{\underset{\|}{C}}CH_3$

3\. CH_2 CH_2 CH_2

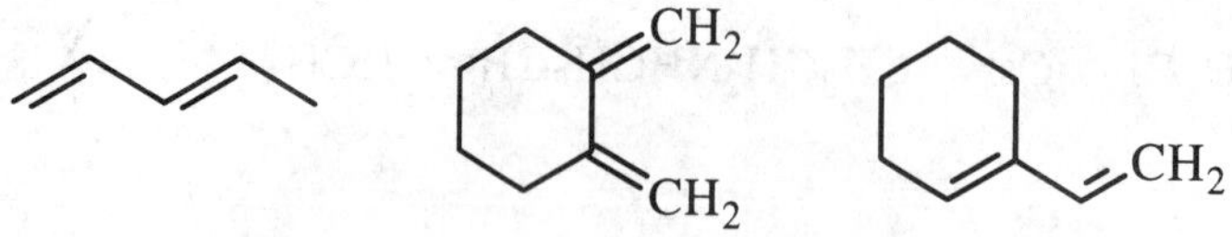

4\.
a. A conjugated diene is more stable than an isomeric isolated diene. T
b. A single bond formed by an sp^2—sp^2 overlap is longer than a single bond formed by an sp^2—sp^3 overlap. F
c. The thermodynamically controlled product is the major product obtained when the reaction is carried out under mild conditions. F
d. 1,3-Hexadiene is more stable than 1,4-hexadiene. T

5\. 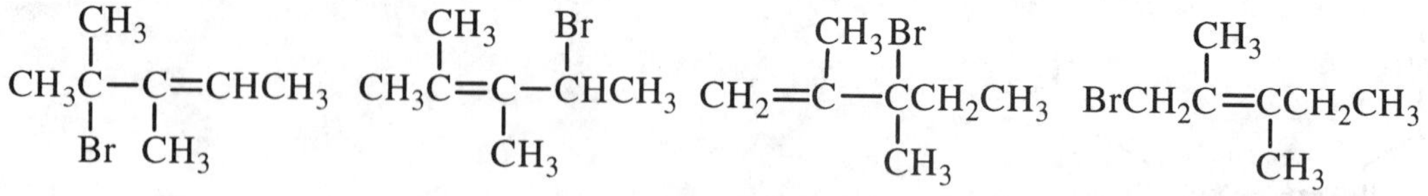 $CH_3C(CH_3)(Br){-}C(CH_3){=}CHCH_3$ $CH_3C(CH_3){=}C(CH_3){-}CH(Br)CH_3$ $CH_2{=}C(CH_3){-}C(Br)(CH_3)CH_2CH_3$ $BrCH_2C(CH_3){=}C(CH_3)CH_2CH_3$

6\. CH_3 CH_3 + O

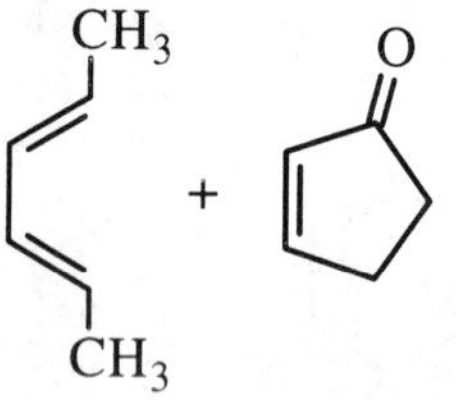

7\. a. 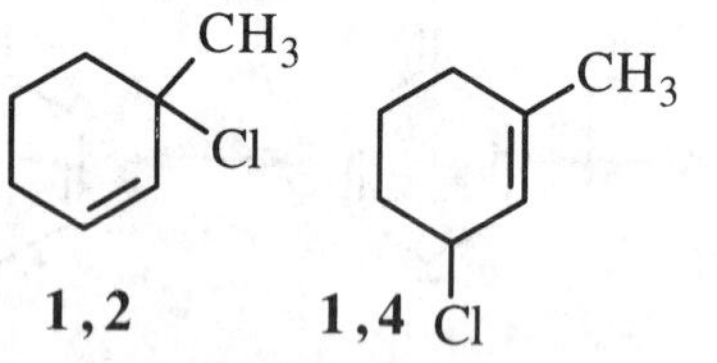 CH_3 Cl CH_3 Cl
1,2 1,4

b. 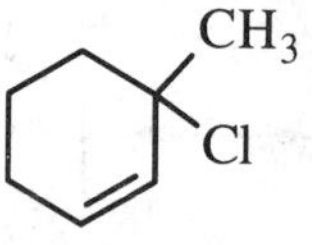 CH_3 Cl

8\. $\xrightarrow[H_2O]{Br_2}$ HO— —Br

9\. CH_3 CH_3

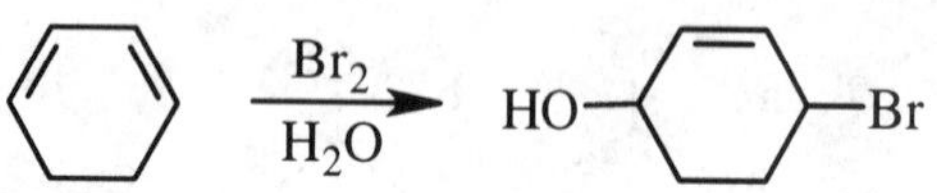

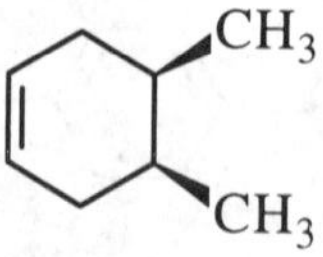

10\. a. 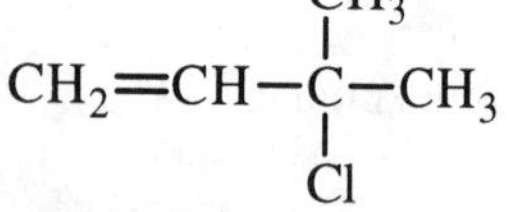 $CH_2{=}CH{-}C(CH_3)(Cl){-}CH_3$
product of kinetic control

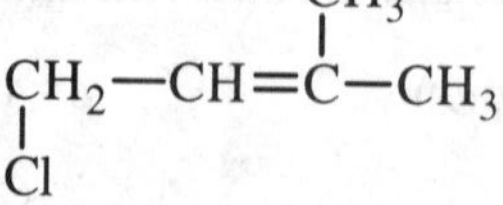

$CH_2(Cl){-}CH{=}C(CH_3){-}CH_3$
product of thermodynamic control

b. 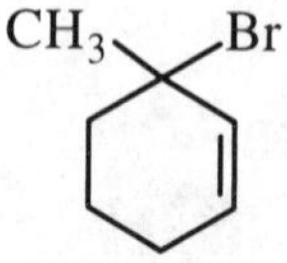 CH_3 Br
product of kinetic control

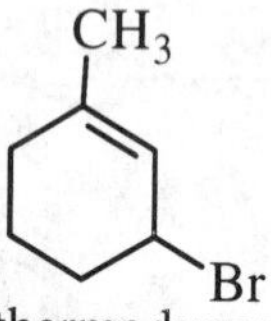

CH_3 Br
product of thermodynamic control

Answers to Chapter 8 Practice Test

1. 3

2. CH_3CH_3 + $\cdot Cl$ ⟶ $CH_3\dot{C}H_2$ + HCl

3. Fischer projections: (1) CH_3 top; Br–C–H; Br–C–H; CH_2CH_3 bottom. (2) CH_3 top; Br–C–H; H–C–Br; CH_2CH_3 bottom.

4. CH_3CH_3 + $Cl\cdot$ ⟶ $CH_3\dot{C}H_2$ + HCl ΔH° = 101 - 103 = -2

$CH_3\dot{C}H_2$ + Cl_2 ⟶ CH_3CH_2Cl + $Cl\cdot$ ΔH° = 58 - 69 = -11 **4.**

5. $\dot{C}H_2CH_2CH_2CH{=}CH_2$ **3** $CH_3CH_2\dot{C}HCH{=}CH_2$ **1**

$CH_3CH_2CH_2CH{=}\dot{C}H$ **4** $CH_3\dot{C}HCH_2CH{=}CH_2$ **2**

6. a. methylenecyclopentane ($=CH_2$) $\xrightarrow[\Delta]{NBS}$ 2-bromo-1-methylenecyclopentane ($=CH_2$, Br) + 1-(bromomethyl)cyclopentene (CH_2Br)

b. $CH_3CH_2CH{=}CH_2$ $\xrightarrow[\Delta]{NBS}$ $CH_3CH(Br)CH{=}CH_2$ + $CH_3CH{=}CHCH_2Br$

7. a. $CH_3C(CH_3)(CH_2Cl)CH_2CH_2CH(CH_3)CH_3$ $CH_3C(CH_3)_2$—$CH(Cl)CH_2CH(CH_3)CH_3$ $CH_3C(CH_3)_2CH_2CH(Cl)CH(CH_3)CH_3$

$CH_3C(CH_3)_2CH_2CH_2C(CH_3)(Cl)CH_3$ $CH_3C(CH_3)_2CH_2CH_2CH(CH_3)CH_2Cl$

b. $CH_3C(CH_3)(CH_2Cl)CH_2CH_2CH(CH_3)CH_3$ c. $CH_3C(CH_3)_2CH_2CH_2C(CH_3)(Br)CH_3$

8. 23%

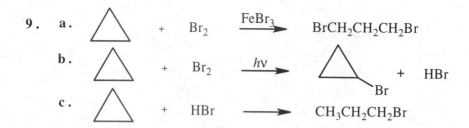

9. a. (cyclopropane) + Br_2 $\xrightarrow{FeBr_3}$ $BrCH_2CH_2CH_2Br$

b. (cyclopropane) + Br_2 $\xrightarrow{h\nu}$ (bromocyclopropane) + HBr

c. (cyclopropane) + HBr ⟶ $CH_3CH_2CH_2Br$

Answers to Chapter 9 Practice Test

1. a. $CH_3CH_2CH_2CH(CH_3)CH_2Br$

b. $CH_3CH{=}CHCH(Br)CH_3$

2. a. $CH_3CH_2CH(CH_3)Br$

b. C_6H_5–CH_2Br

3. $CH_3CH(CH_3)CH(Br)CH_3$

(1-bromo-2-methylcyclopentane: cyclopentane ring with CH_3 and Br on adjacent carbons)

4.
a. Increasing the concentration of the nucleophile favors an S_N1 reaction over an S_N2 reaction. F
b. Ethyl iodide is more reactive than ethyl chloride in an S_N2 reaction. T
c. In an S_N1 reaction, the product with the retained configuration is obtained in greater yield. F
d. The rate of a substitution reaction in which none of the reactants is charged will increase if the polarity of the solvent is increased. T
e. An S_N2 reaction is a two-step reaction. F
f. The pK_a of a carboxylic acid is greater in water than it is in a mixture of dioxane and water. F

5. a. CH_3O^- b. CH_3S^-

6. a. $CH_3CH_2CH_2Cl$ + HO^-

b. $CH_3CH_2CH_2I$ + HO^-

c. $CH_3CH_2CH_2Br$ + HO^-

d. $CH_3CH(Br)CH_3$ $\xrightarrow[CH_3OH]{CH_3O^-}$

e. $BrCH_2CH_2CH_2CH_2NHCH_3$

7. All are aprotic solvents except ethanol.

8. a. The rate would increase.

b. The rate would increase.

c. The rate would decrease.

d. The pK_a would increase.

e. The pK_a would increase.

Answers to Chapter 10 Practice Test

1. a. $C_6H_5CH_2CHCH_3$ (Br on C-2) b. $CH_2{=}CHCH_2CHCH_3$ (Br on C-4) 2. CH_3CHBr (CH₃ on the CHBr carbon)

3. CH_3CH_2–C(OCH₃)(H)–CH_3 CH_3–C(OCH₃)(H)–CH_2CH_3

H₃C(H)C=C(H)CH₃ (trans) major H₃C(H)C=C(H)CH₃ (cis) minor

4. a. $CH_3CH_2C(CH_3)_2O^-$ + $CH_3CH_2CH_2Br$ b. cyclohexyl–Br + $C_6H_5O^-$ c. cyclohexyl–O^- + CH_3Br

5. a. $CH_3CHClCH_3$ + HO^-
b. $CH_3CH_2CH_2I$ + HO^-
c. $CH_3CH_2CH_2Br$ + HO^-
d. $CH_3CHBrCH_3$ $\xrightarrow[CH_3OH]{CH_3O^-}$
e. $CH_3CBr(CH_3)CH_3$ + HO^-

6. a. C_6H_5(H)C=C(H)CHCH₃(CH₃) b. cyclopentadiene c. $CH_3CH{=}C(CH_3)CH_3$ d. $CH_2{=}CHCH(CH_3)CH_3$

7. a. *cis*-1-bromo-2-methylcyclohexane
b. The cis and trans isomers react at about the same rate.

Answers to Chapter 11 Practice Test

1. HBr 2. $SOCl_2$

3. a. $H_2C{=}C(CH_3)Br$ $\xrightarrow[Et_2O]{(CH_3CH_2)_2CuLi}$ $H_2C{=}C(CH_3)CH_2CH_3$

b. cyclopentyl–CH_2OH $\xrightarrow{PBr_3}$ cyclopentyl–CH_2Br $\xrightarrow[Et_2O]{Mg}$ cyclopentyl–CH_2MgBr $\xrightarrow[2.\ H^+]{1.\ H_2C\text{–}CH_2\ (\text{ethylene oxide})}$ cyclopentyl–$(CH_2)_3OH$

4. 2,2-diethyloxirane or 1,2-epoxy-2-ethylbutane

5. a. 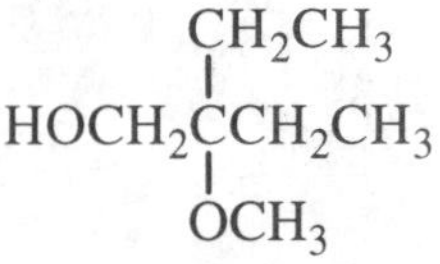 $HOCH_2C(CH_2CH_3)(OCH_3)CH_2CH_3$ b. 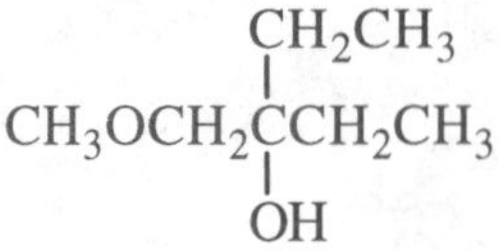 $CH_3OCH_2C(CH_2CH_3)(OH)CH_2CH_3$

6. a. 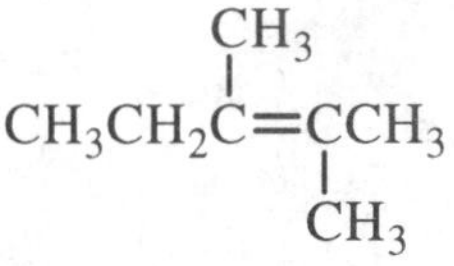 $CH_3CH_2C(CH_3){=}C(CH_3)CH_3$ b. 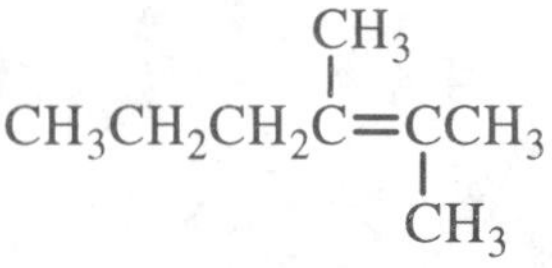 $CH_3CH_2CH_2C(CH_3){=}C(CH_3)CH_3$ c. $CH_3CH_2CH{=}CHCH_3$ d. 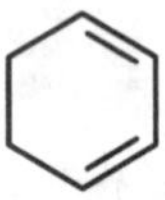 (1,3-cyclohexadiene)

7.

a. Tertiary alcohols are easier to dehydrate than secondary alcohols. T
b. Alcohols are more acidic than thiols. F
c. Alcohols have higher boiling points than thiols. T

8. a. 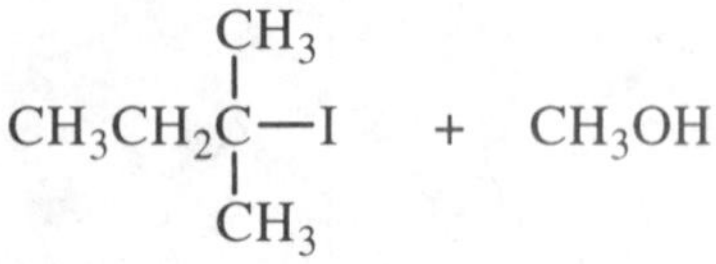 $CH_3CH_2C(CH_3)_2{-}I$ + CH_3OH b. 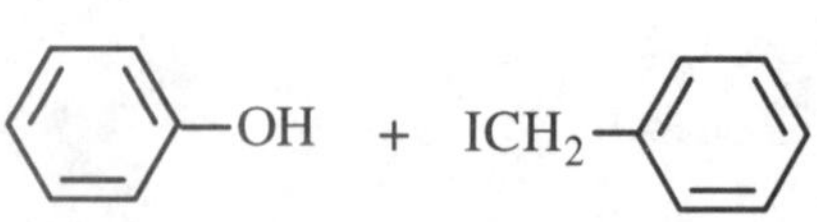 $C_6H_5{-}OH$ + $ICH_2{-}C_6H_5$

9. a. $CH_3CH_2CH_2CH_2CH_2OH$ b. 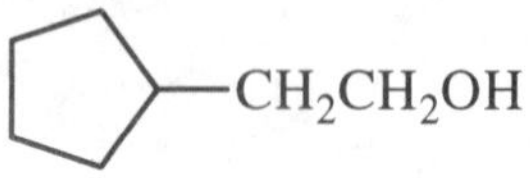 $C_5H_9{-}CH_2CH_2OH$ (cyclopentyl)

10. a. 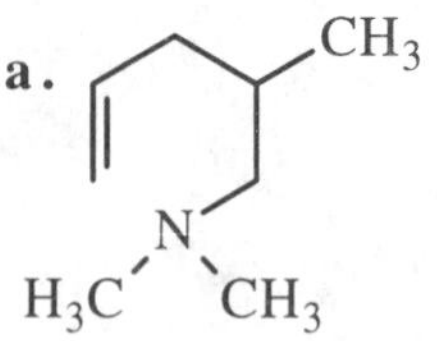 (structure: CH_3, N, H_3C, CH_3) b. 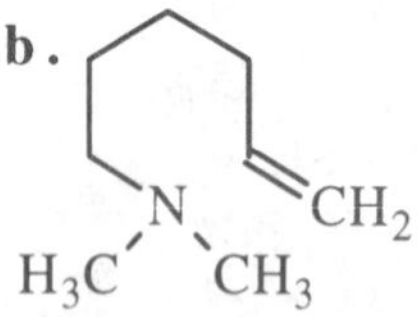 (structure: N, CH_2, H_3C, CH_3) c. 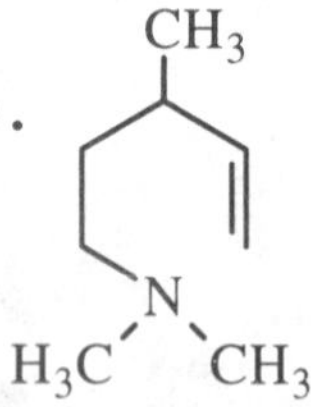 (structure: CH_3, N, H_3C, CH_3)

Answers to Chapter 12 Practice Test

1.

a.  $CH_3CH_2CH_2CH{=}O$ ~2700 cm^{-1}

b. 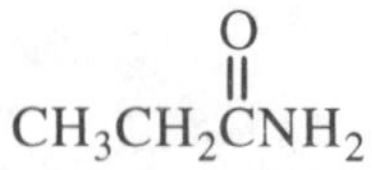 $CH_3CH_2C({=}O)NH_2$ ~3300 cm^{-1}

c. $CH_3CH_2CH_2CH_2OH$ ~3600-3200 cm^{-1}

d. 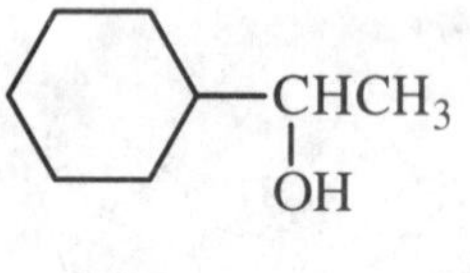 $C_6H_{11}{-}CH(OH)CH_3$ ~1380 cm^{-1}

e. 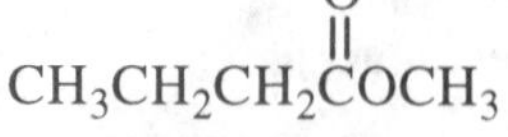 $CH_3CH_2CH_2C({=}O)OCH_3$ ~1050 or ~1250 cm^{-1}

f. $CH_3CH_2CH{=}CHCH_3$ ~1600 cm^{-1} ~3100 cm^{-1} $CH_3CH_2C{\equiv}CCH_3$ ~2100 cm^{-1}

g. $CH_3CH_2C{\equiv}CH$ ~3300 cm^{-1}

2. a.  (cyclohexadiene structure) b. 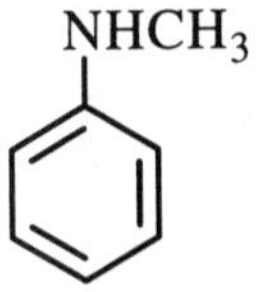 $NHCH_3$ (benzene ring) c. 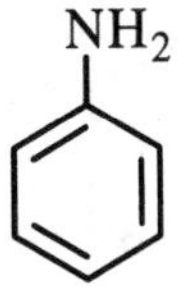 NH_2 (benzene ring)

3. molar absorbtivity = 19.7

4. $CH_3CH_2C(CH_3)(OH)CH_2CH_2CH_3$

5. a. 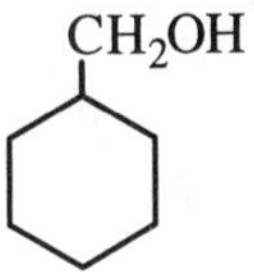 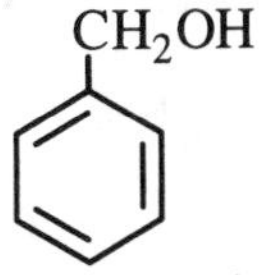 CH_2OH (cyclohexane) ; CH_2OH (benzene): ~ 3100 cm^{-1}, ~ 1500 cm^{-1}

c. 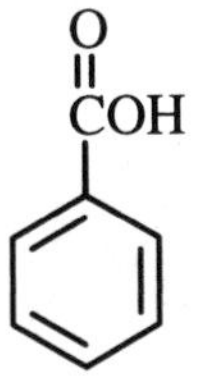 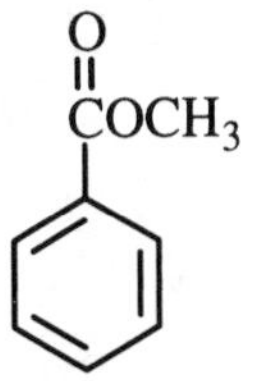 $COOH$ (benzene): ~ 3300-2500 cm^{-1} ; $COOCH_3$ (benzene): ~ 2900 cm^{-1}, ~ 1380 cm^{-1}, ~ 1050 cm^{-1}

d. 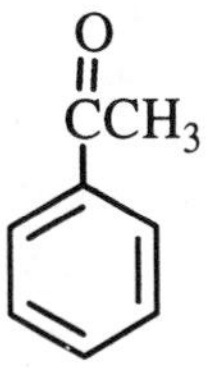 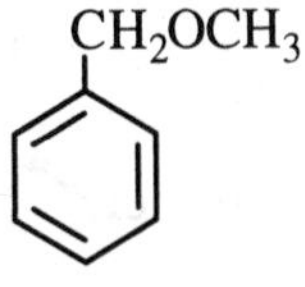 $COCH_3$ (benzene): ~ 1700 cm^{-1} ; CH_2OCH_3 (benzene): ~ 1050 cm^{-1}

b. 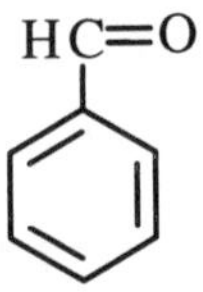 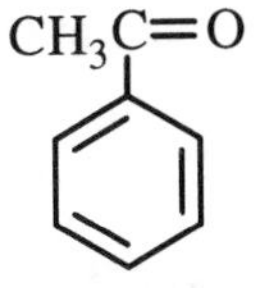 $HC{=}O$ (benzene): ~ 2700 cm^{-1} ; $CH_3C{=}O$ (benzene): ~ 2900 cm^{-1}, ~ 1380 cm^{-1}

e. 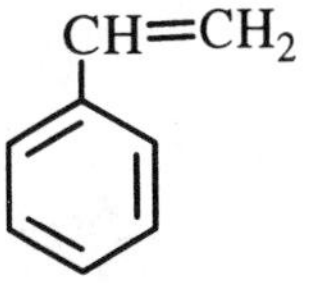 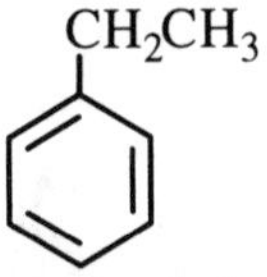 $CH{=}CH_2$ (benzene): ~ 3100 cm^{-1} ; CH_2CH_3 (benzene): ~ 2900 cm^{-1}, ~ 1380 cm^{-1}

6.
a. The O-H stretch of a concentrated solution of an alcohol occurs at a higher frequency than the O-H stretch of a dilute solution. F
b. Light of 2 μm is of higher energy than light of 3 μm. T
c. It takes more energy for a bending vibration than for a stretching vibration. F
d. Propyne will not have an absorption band at 3100 cm^{-1} because there is no change in the dipole moment. F
e. Light of 8 μm has the same energy as light of 1250 cm^{-1}. T
f. Ultraviolet radiation is higher in energy than infrared radiation. T
g. The M + 2 peak of an alkyl chloride is half the height of the M peak. F
h. A chromophore exhibiting both $n \rightarrow \pi^*$ and $\pi \rightarrow \pi^*$ transitions will have the $n \rightarrow \pi^*$ transition at a longer wavelength. T

Answers to Chapter 13 Practice Test

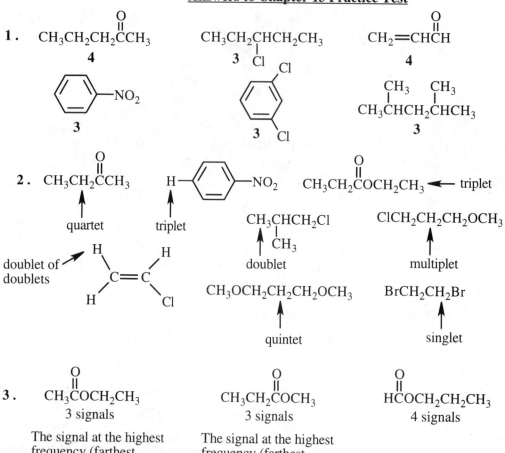

1. $CH_3CH_2CH_2C(=O)CH_3$ 4 $CH_3CH_2CH(Cl)CH_2CH_3$ 3 $CH_2{=}CHC(=O)H$ 4

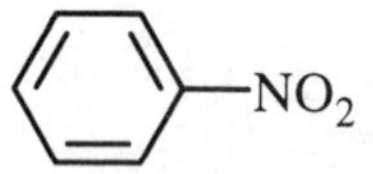

nitrobenzene ($C_6H_5NO_2$) 3 1,3-dichlorobenzene 3 $CH_3CH(CH_3)CH_2CH(CH_3)CH_3$ 3

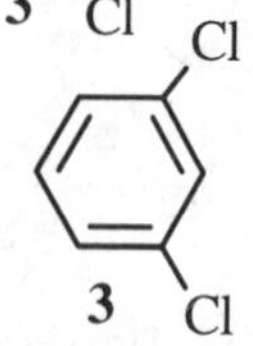

2. $CH_3CH_2C(=O)CH_3$ — quartet (arrow to CH_2)

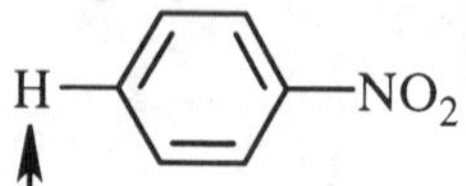

$H{-}C_6H_4{-}NO_2$ — triplet (arrow to H)

$CH_3CH_2C(=O)OCH_2CH_3$ ← triplet

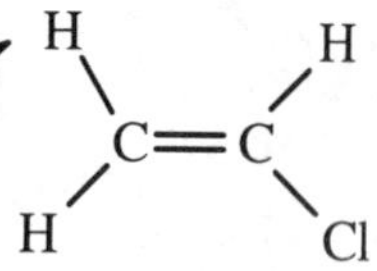

$CH_3CH(CH_3)CH_2Cl$ — doublet (arrow to CH_3)

$ClCH_2CH_2CH_2OCH_3$ — multiplet (arrow to central CH_2)

$H_2C{=}CHCl$ — doublet of doublets (arrow to H)

$CH_3OCH_2CH_2CH_2OCH_3$ — quintet (arrow to central CH_2)

$BrCH_2CH_2Br$ — singlet

3. $CH_3C(=O)OCH_2CH_3$ 3 signals

The signal at the highest frequency (farthest downfield) is a quartet.

$CH_3CH_2C(=O)OCH_3$ 3 signals

The signal at the highest frequency (farthest downfield) is a signal.

$HC(=O)OCH_2CH_2CH_3$ 4 signals

4. a. The peaks on the right of an NMR spectrum are deshielded compared to the peaks on the left. F

b. Dimethyl ketone has the same number of signals in its ^{1}H NMR spectrum as in its ^{13}C NMR spectrum. F

c. In the ^{1}H NMR spectrum of the compound shown below, the signal farthest upfield is a singlet and the signal farthest downfield is a doublet. T

$O_2N{-}C_6H_4{-}CH_3$

5. a. $CH_3CH_2CH_2Cl$ — triplet (arrow to CH_2Cl); 3 signals

b. $CH_3CH_2C(=O)OCH_3$ — singlet (arrow to OCH_3); 4 signals

c. $CH_3CH(Br)CH_3$ — doublet (arrow to CH_3); 2 signals

Answers to Chapter 14 Practice Test

1\.

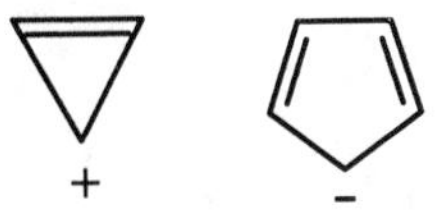

2\.

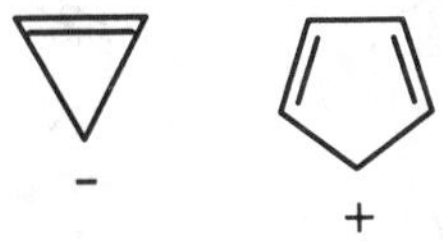

3\.

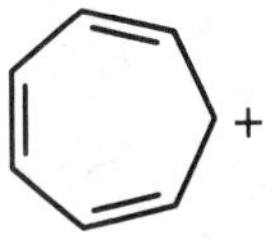

4\. 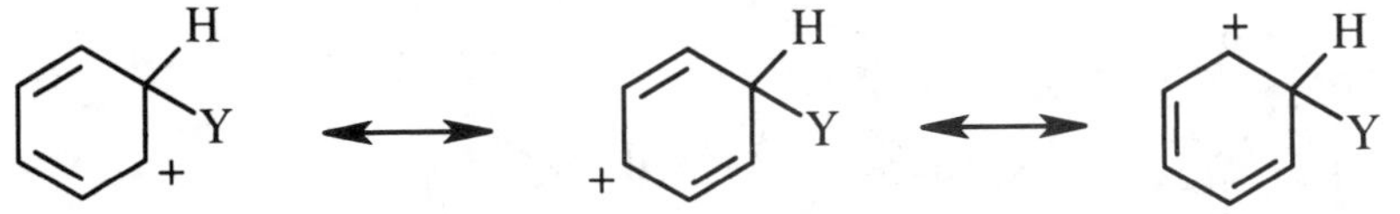 H Y + ⟷ H Y + ⟷ + H Y

5\. O SOH O H H

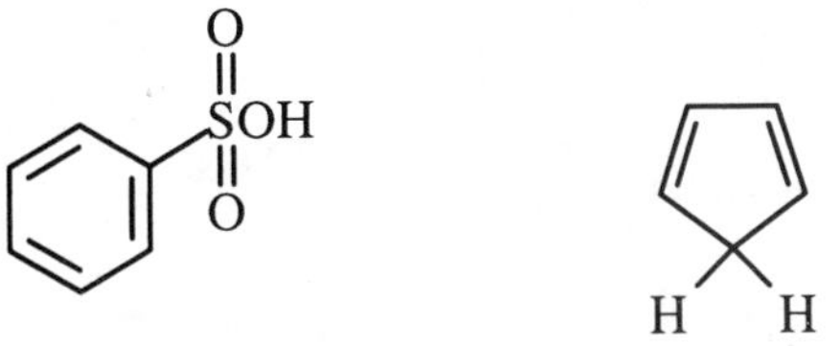

6\. $CH_3CH_2CH_2CH_2Cl$ $CH_3CH{=}CHCH_3$ $CH_3CH_2CH_2CH_2OH$

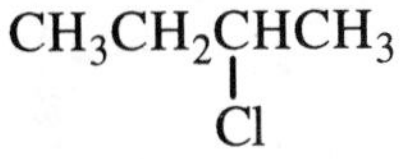

$CH_3CH_2CHCH_3$ (Cl) $CH_2{=}CHCH_2CH_3$ $CH_3CH_2CHCH_3$ (OH)

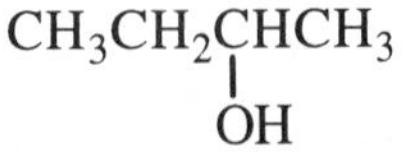

7\. $CH_3CH_2\overset{O}{\overset{\|}{C}}O\overset{O}{\overset{\|}{C}}CH_2CH_3$

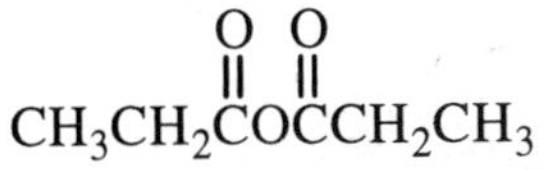

8\. a. 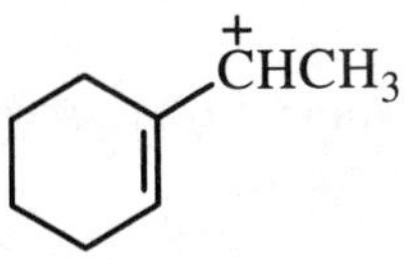 $\overset{+}{C}HCH_3$ b. $CH_3\overset{+}{C}HCH{=}CH_2$ c. 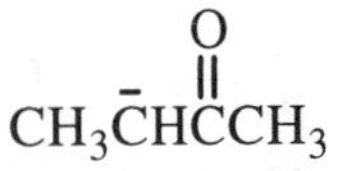 $CH_3\overset{-}{C}H\overset{O}{\overset{\|}{C}}CH_3$

9\. $H\ddot{O}{-}NO_2 + H{-}OSO_3H \rightleftharpoons HO(H)^{+}{-}NO_2 \rightleftharpoons H_2O + {}^{+}NO_2$

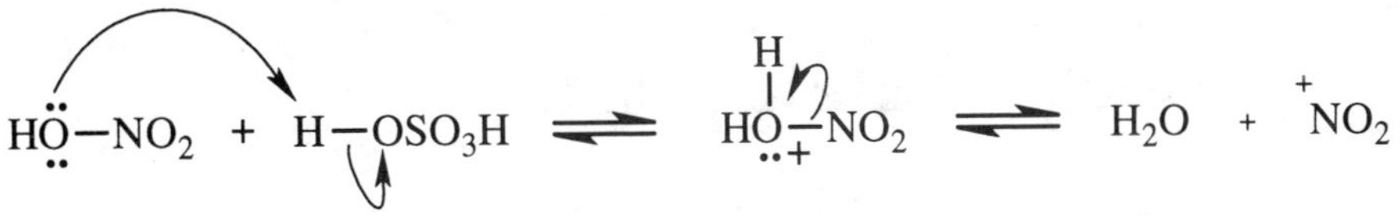

Answers to Chapter 15 Practice Test

1\. a. *meta*-nitrotoluene
b. 1,2,4-tribromobenzene
c. *ortho*-ethylbenzoic acid
d. *para*-chlorophenol

2. 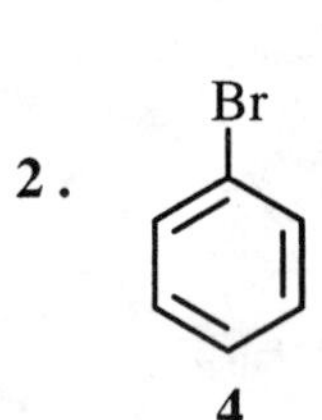

Br (4); $NHCCH_3$ with C=O (1); CH_2CH_3 (2); $CNHCH_3$ with C=O (5); benzene (3)

3. a. COOH, Cl b. OH, NO_2 c. $\overset{+}{N}H_3$, CH_3 d. COOH

4. **a.** *para*-bromonitrobenzene **b.** *para*-bromoethylbenzene

5. a. NO_2, SO_3H b. OCH_3, CH_3 + OCH_3, CH_3 c. COOH

d. OCH_3, NO_2 e. Cl, NO_2, OCH_3 f. CCH_3 with C=O, Cl

6. **a.** Benzoic acid is more reactive than benzene towards electrophilic substitution. F
b. *para*-Chlorobenzoic acid is more acidic than *para*-methoxybenzoic acid. T
c. A -CH=CH_2 group is a meta director. F
d. *para*-Nitroaniline is more basic than *para*-chloroaniline. F

Answers to Chapter 16 Practice Test

1. a. $CH_3\overset{O}{\overset{\|}{C}}OCH_3$ b. $CH_3\overset{O}{\overset{\|}{C}}O$–$C_6H_5$ c. $CH_3\overset{O}{\overset{\|}{C}}O$–$C_6H_4$–$NO_2$ d. $CH_3\overset{O}{\overset{\|}{C}}Cl$

2. **a.** *N*-ethylpentanamide
b. 3-methylpentanoic acid
c. methyl 4-phenylbutanoate
d. ethanoic propanoic anhydride

3\. a. $CH_3CH_2\overset{O}{\overset{\|}{C}}O\overset{O}{\overset{\|}{C}}CH_2CH_3$

b. $CH_3\overset{O}{\overset{\|}{C}}Cl + H_2O \longrightarrow CH_3\overset{O}{\overset{\|}{C}}OH + HCl$

Any reaction in which one of the reactants is cleaved as a result of reaction with water.

c. $CH_3\overset{O}{\overset{\|}{C}}OCH_3 + CH_3CH_2OH \overset{H^+}{\rightleftharpoons} CH_3\overset{O}{\overset{\|}{C}}OCH_2CH_3 + CH_3OH$

4\. a. $CH_3-\overset{O}{\overset{\|}{C}}-OH$ b. $CH_3-\overset{O}{\overset{\|}{C}}-OCH_3$ c. $CH_3-\overset{O}{\overset{\|}{C}}-NH_2$ d. $CH_3-\overset{O}{\overset{\|}{C}}-OH$

5\. a. $CH_3CH_2\overset{O}{\overset{\|}{C}}OH + \overset{+}{N}H_4$

b. $CH_3CH_2\overset{O}{\overset{\|}{C}}O^- + NH_3$

c. $CH_3CH_2CH_2\overset{O}{\overset{\|}{C}}OH + CH_3CH_2\overset{O}{\overset{\|}{C}}OH$

d. $CH_3CH_2\overset{O}{\overset{\|}{C}}NHCH_2CH_3 + Cl^-$

e. 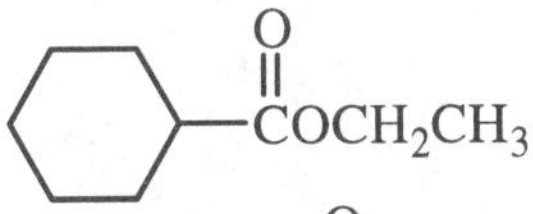 $C_6H_{11}-\overset{O}{\overset{\|}{C}}OCH_2CH_3 + CH_3OH$ (cyclohexyl)

f. $CH_3CH_2CH_2\overset{O}{\overset{\|}{C}}OCH_2CH_2CH_3$

g. $CH_3CH_2\overset{O}{\overset{\|}{C}}OH + CH_3\overset{+}{N}H_3$

h. $CH_3CH_2\overset{O}{\overset{\|}{C}}OH + HO-C_6H_5$ (phenol)

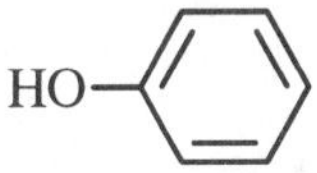

Answers to Chapter 17 Practice Test

1\. a. 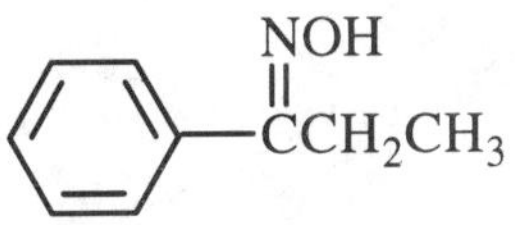 $C_6H_5-\overset{NOH}{\overset{\|}{C}}CH_2CH_3 + H_2O$

b. 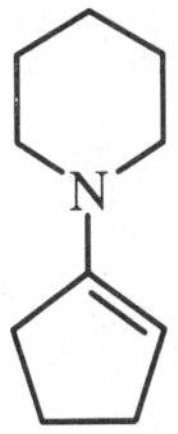 N-(1-cyclopentenyl)piperidine $+ H_2O$

c. $CH_3CH_2CH_2\overset{O}{\overset{\|}{C}}OH$

d. 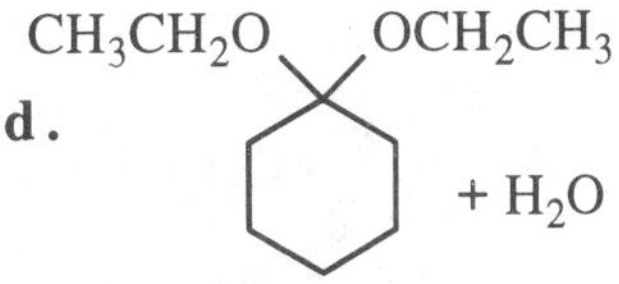 1,1-diethoxycyclohexane (CH_3CH_2O, OCH_2CH_3 on the same ring carbon) $+ H_2O$

e. $CH_3CH_2\overset{OH}{\overset{|}{C}}CH_2CH_3$ with CH_3 on the central carbon

f.  cyclohexyl$-NH_2$

g. $CH_3CH_2\overset{OH}{\overset{|}{C}}CH_2CH_3$ with $C{\equiv}N$ on the central carbon

h. $CH_3CH{=}CH\overset{OH}{\overset{|}{C}}HCH_3$

i. 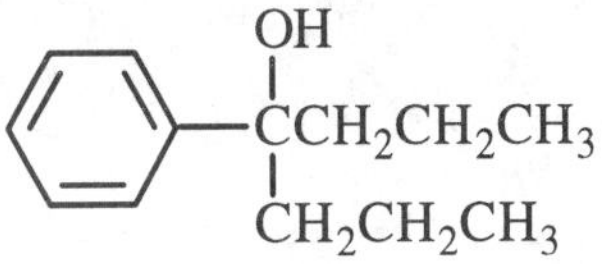 $C_6H_5-\overset{OH}{\overset{|}{C}}CH_2CH_2CH_3$ with $CH_2CH_2CH_3$ on the central carbon

2. $CH_3CH_2C(OH)(CH_3)CH_2CH_2CH_3$

3. 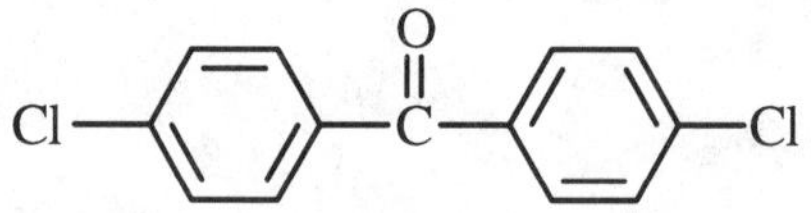 Cl–C₆H₄–C(=O)–C₆H₄–Cl

4. a. 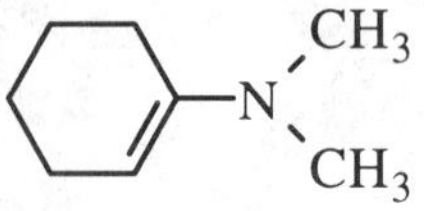 1-cyclohexenyl–$N(CH_3)_2$

b. $CH_3CH_2CH(OCH_3)_2$

c. 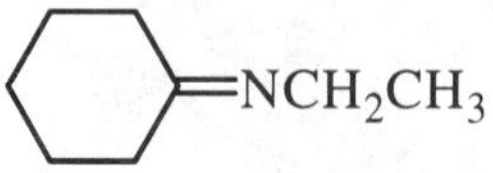 cyclohexylidene=NCH_2CH_3

d. $CH_3CH_2CH(OH)OCH_3$

e. 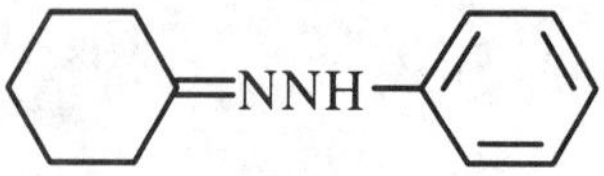 cyclohexylidene=NNH–C_6H_5

5. **a.** butanal **b.** 3-pentanone

6. $CH_3CH_2CH_2Br \xrightarrow{Mg} CH_3CH_2CH_2MgBr \xrightarrow{CO_2} CH_3CH_2CH_2CO_2^-$

$\xrightarrow{SOCl_2} CH_3CH_2CH_2CCl$ (C=O)

$\xrightarrow{CH_3CH_2OH} CH_3CH_2CH_2COCH_2CH_3$ (C=O)

Answers to Chapter 18 Practice Test

1. **a.** reduction **b.** oxidation

2. a. $CH_3CH_2CH_2COH$ (C=O)

b. no reaction

c. $CH_3CH_2CH_2NHCH_2CH_3$

d. 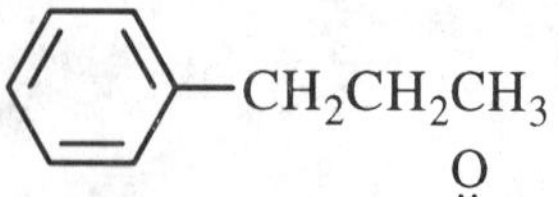 C_6H_5–$CH_2CH_2CH_3$

e. $CH_3CH_2CH_2CH_2COH$ (C=O)

f. 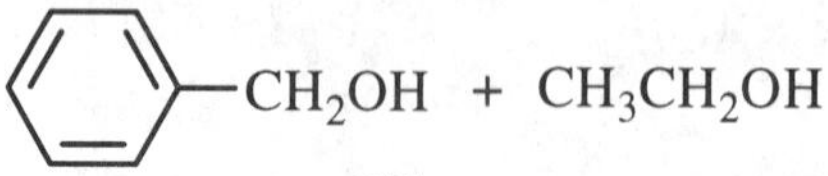 C_6H_5–CH_2OH + CH_3CH_2OH

g. $CH_3CH_2CH_2CH(OH)CH_2CH_3$

h. $CH_3CH_2CH_2C(CH_3){=}O$ + CH_3CH_2COH (C=O)

3. 1. Ag_2O/NH_3
2. H_3O^+

4. 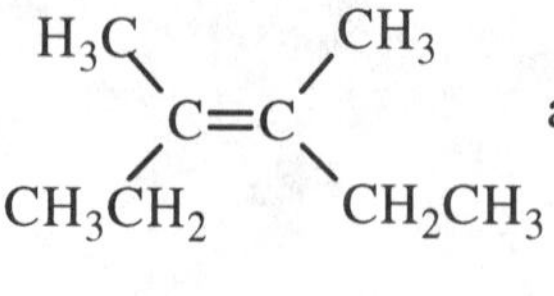 $(H_3C)(CH_3CH_2)C{=}C(CH_3)(CH_2CH_3)$ and $(H_3C)(CH_3CH_2)C{=}C(CH_2CH_3)(CH_3)$

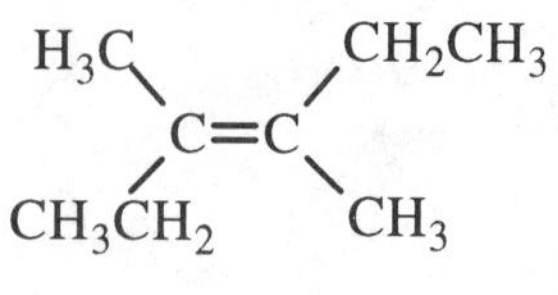

5. **a.** NaBH4 is a weaker reducing agent than LiAlH4. T
b. Esters are easier to reduce than ketones. F
c. In an oxidation-reduction reaction the oxidizing agent is oxidized. F
d. Ketones are reduced to primary alcohols. F
e. Aldehydes are oxidized to carboxylic acids. T
f. Acyl halides are oxidized to aldehydes. F
g. Alkenes cannot be reduced with NaBH4. T

6. $CH_3CH_2CH_2CH_3 \xrightarrow[h\nu]{Br_2} CH_3CH_2\underset{Br}{\underset{|}{C}}HCH_3 \xrightarrow{HO^-} CH_3CH_2\overset{OH}{\overset{|}{C}}HCH_3 \xrightarrow{H_2CrO_4} CH_3CH_2\overset{\overset{O}{\|}}{C}CH_3$

7. **c.** 1. O_3 2. Zn, H_2O or 1. O_3 2. $(CH_3)_2S$

Answers to Chapter 19 Practice Test

1. $CH_3\overset{\overset{O}{\|}}{C}CH_2\overset{\overset{O}{\|}}{C}OCH_3$ **2** $CH_3\overset{\overset{O}{\|}}{C}CH_2\overset{\overset{O}{\|}}{C}CH_3$ **1** $CH_3\overset{\overset{O}{\|}}{C}CH_3$ **3**

1.

$CH_3\overset{\overset{O}{\|}}{C}CH_2\overset{\overset{O}{\|}}{C}OCH_3$ **2** $CH_3\overset{\overset{O}{\|}}{C}CH_2\overset{\overset{O}{\|}}{C}CH_3$ **1** $CH_3\overset{\overset{O}{\|}}{C}CH_3$ **3**

2. a. $CH_3\overset{OH}{\overset{|}{C}}{=}CH\overset{\overset{O}{\|}}{C}CH_3$ b. $CH_3\overset{\overset{O}{\|}}{C}CH_2\overset{\overset{O}{\|}}{C}OCH_3$

3. a. $CH_3CH_2\overset{\overset{O}{\|}}{C}CH_3 + CO_2$

b. $CH_3CH_2\underset{Br}{\underset{|}{\overset{Br}{\overset{|}{C}}}}-\overset{\overset{O}{\|}}{C}H + 2\ Br^-$

c. $CH_3CH_2CH_2\overset{\overset{O}{\|}}{C}\underset{CH_2CH_3}{\underset{|}{C}}H\overset{\overset{O}{\|}}{C}OCH_3$

d. $CH_3\underset{\underset{O}{\|}}{C}\overset{CH_2CH_2\overset{\overset{O}{\|}}{C}CH_3}{\overset{|}{C}}H\underset{\underset{O}{\|}}{C}CH_3$

e. 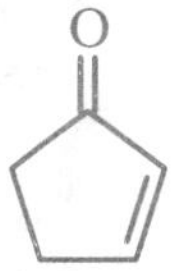 O

f. 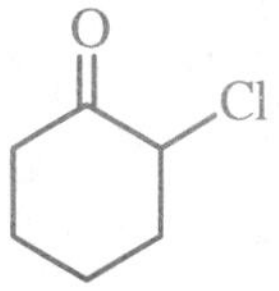 O Cl

g. $CH_3CH_2CH_2CH_2\overset{\overset{O}{\|}}{C}OH$

4\. a. $2\ CH_3CH_2\overset{O}{\overset{\|}{C}}H \xrightarrow{HO^-} CH_3CH_2\overset{OH}{\overset{|}{C}}H\underset{CH_3}{\underset{|}{C}}H\overset{O}{\overset{\|}{C}}H$

b. $2\ CH_3CH_2\overset{O}{\overset{\|}{C}}H \xrightarrow{HO^-} CH_3CH_2\overset{OH}{\overset{|}{C}}H\underset{CH_3}{\underset{|}{C}}H\overset{O}{\overset{\|}{C}}H \xrightarrow[\Delta]{H_2SO_4} CH_3CH_2CH{=}$

c. $2\ CH_3CH_2\overset{O}{\overset{\|}{C}}OCH_3 \xrightarrow{CH_3O^-} CH_3CH_2\overset{O}{\overset{\|}{C}}\underset{CH_3}{\underset{|}{C}}H\overset{O}{\overset{\|}{C}}OCH_3 + CH_3OH$

d. 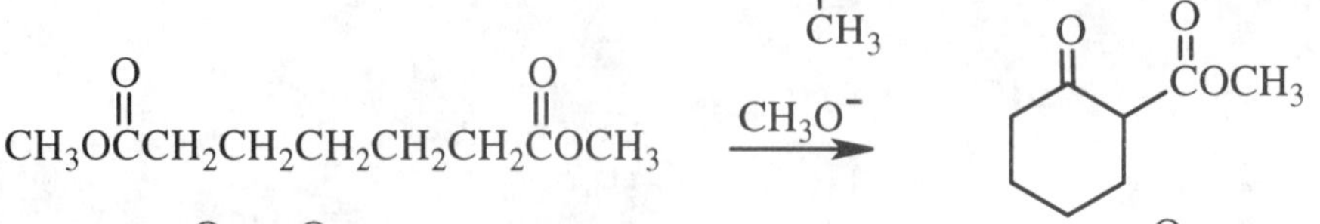 $CH_3O\overset{O}{\overset{\|}{C}}CH_2CH_2CH_2CH_2CH_2\overset{O}{\overset{\|}{C}}OCH_3 \xrightarrow{CH_3O^-}$ 2-($\overset{O}{\overset{\|}{C}}OCH_3$)cyclohexanone

e. $CH_3CH_2O\overset{O}{\overset{\|}{C}}CH_2\overset{O}{\overset{\|}{C}}OCH_2CH_3 \xrightarrow[\text{2. } CH_3CH_2Br \quad \text{3. } H^+,\ H_2O,\ \Delta]{\text{1. } CH_3CH_2O^-} CH_3CH_2CH_2\overset{O}{\overset{\|}{C}}OH$

f. $CH_3\overset{O}{\overset{\|}{C}}CH_2\overset{O}{\overset{\|}{C}}OCH_2CH_3 \xrightarrow[\text{2. } CH_3CH_2Br \quad \text{3. } H^+,\ H_2O,\ \Delta]{\text{1. } CH_3CH_2O^-} CH_3CH_2CH_2\overset{O}{\overset{\|}{C}}CH_3$

5\. $CH_3\underset{CH_3}{\underset{|}{C}}HCH_2CH_2\overset{OH}{\overset{|}{C}}H\underset{CH_2CH_2CH_3}{\underset{|}{C}}H\overset{O}{\overset{\|}{C}}H$ $\qquad CH_3\underset{CH_3}{\underset{|}{C}}HCH_2CH_2\overset{OH}{\overset{|}{C}}H\underset{CH_2CH(CH_3)CH_3}{\underset{|}{C}}H\overset{O}{\overset{\|}{C}}H$

$CH_3CH_2CH_2CH_2\overset{OH}{\overset{|}{C}}H\underset{CH_2CH_2CH_3}{\underset{|}{C}}H\overset{O}{\overset{\|}{C}}H$ $\qquad CH_3CH_2CH_2CH_2\overset{OH}{\overset{|}{C}}H\underset{CH_2CH(CH_3)CH_3}{\underset{|}{C}}H\overset{O}{\overset{\|}{C}}H$

Answers to Chapter 20 Practice Test

1\. a.

	COOH	
H	—	OH
HO	—	H
H	—	OH
H	—	OH
	COOH	

b. 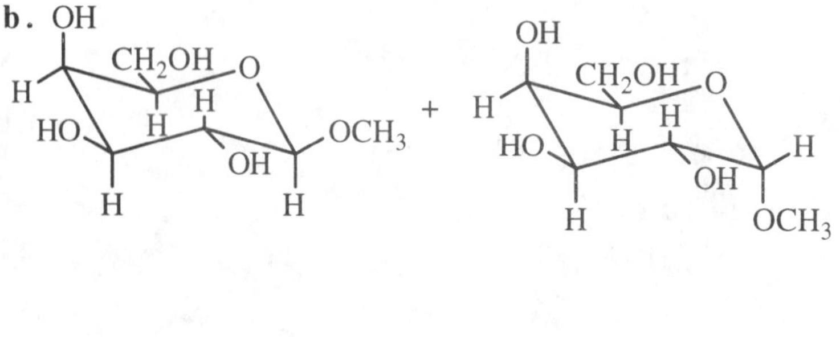 (chair structures) OH, CH_2OH, O, H, H, H, HO, OCH_3, OH, H, H + OH, CH_2OH, O, H, H, H, HO, H, OH, H, OCH_3

c. D-idose + D-gulose Fischer projections:

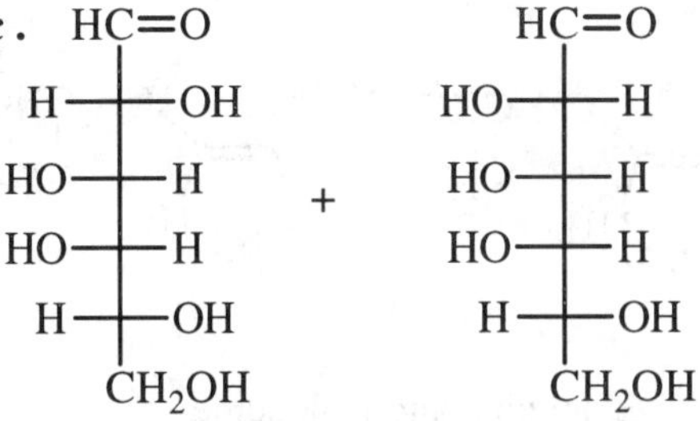

HC=O		HC=O
H—OH		HO—H
HO—H	+	HO—H
HO—H		HO—H
H—OH		H—OH
CH_2OH		CH_2OH

d.

COOH
H—OH
H—OH
H—OH
CH_2OH

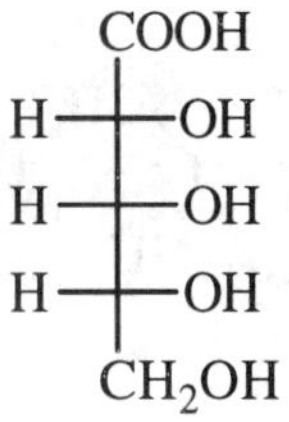

2. a. Glycogen contains α-1,4' and β-1,6'-glycosidic linkages. F
b. D-Mannose is a C-1 epimer of D-glucose. F
c. D-Glucose and L-glucose are anomers. F
d. D-Erythrose and D-threose are diastereomers. T
e. Ruff degradations of D-glucose and D-gulose form the same aldotetrose. F

3.

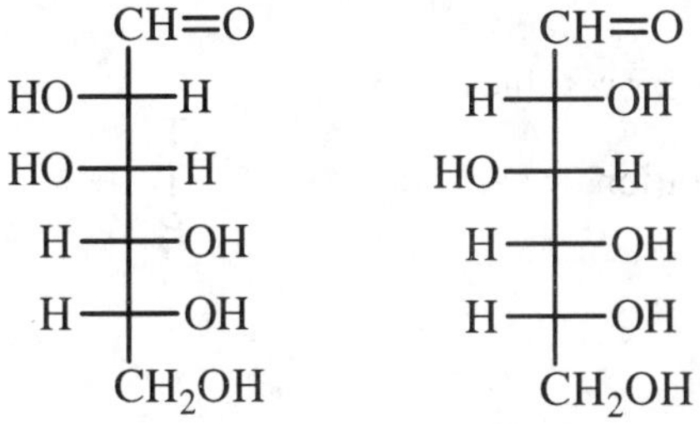

CH=O	CH=O
HO—H	H—OH
HO—H	HO—H
H—OH	H—OH
H—OH	H—OH
CH_2OH	CH_2OH

4. D-mannose and D-glucose 5. D-tagatose 6. D-altrose

7. Amylose has α-1,4'-glycosidic linkages, while cellulose has β-1,4'-glycosidic linkages.

8. D-gulose and D-idose 9. D-allose

10. 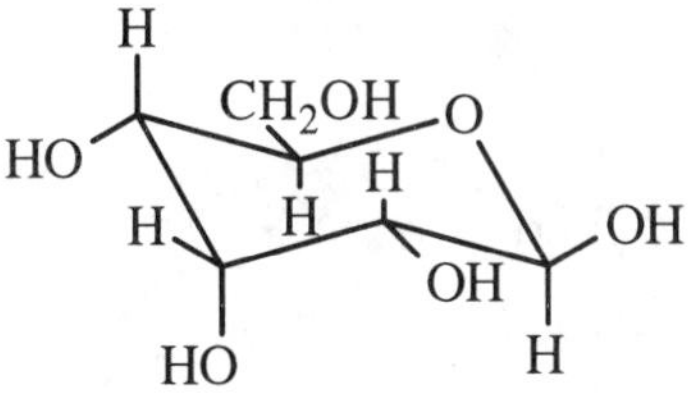 (chair structure; labels: H, CH_2OH, O, HO, H, H, H, OH, OH, HO, H)

Answers to Chapter 21 Practice Test

1. a. 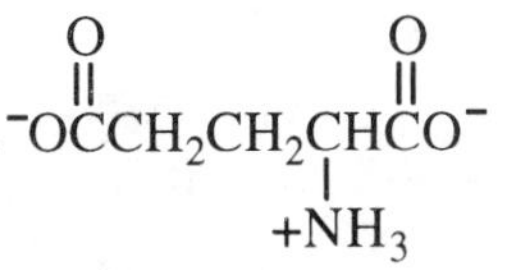 $^{-}OOCCH_2CH_2CH(\overset{+}{N}H_3)COO^{-}$

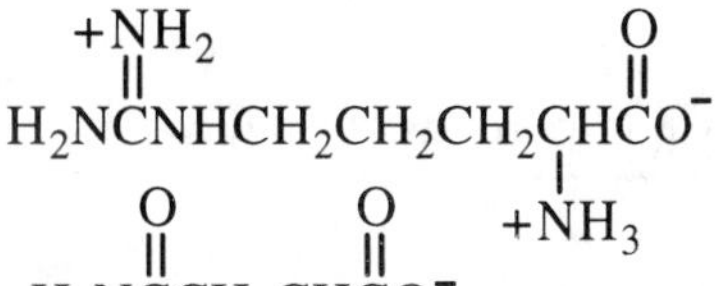

b. 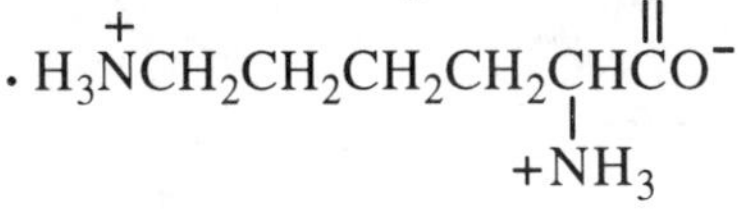 $H_3\overset{+}{N}CH_2CH_2CH_2CH_2CH(\overset{+}{N}H_3)COO^{-}$

c. $CH_3CH_2CH(CH_3)CH(\overset{+}{N}H_3)COO^{-}$

d. $H_2NC(=\overset{+}{N}H_2)NHCH_2CH_2CH_2CH(\overset{+}{N}H_3)COO^{-}$

e. 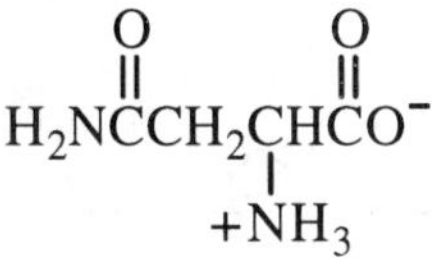 $H_2NC(=O)CH_2CH(\overset{+}{N}H_3)COO^{-}$

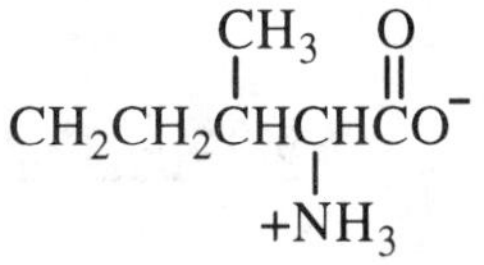

2. a. Histidine with imidazolium ring (HN+ … NH) and side chain CH_2CHCOH (C=O), with $+NH_3$

b. CH_2CHCO^- (C=O), $+NH_3$, ring HN+ … NH

c. CH_2CHCO^- (C=O), $+NH_3$, ring N … NH

d. CH_2CHCO^- (C=O), NH_2, ring N … NH

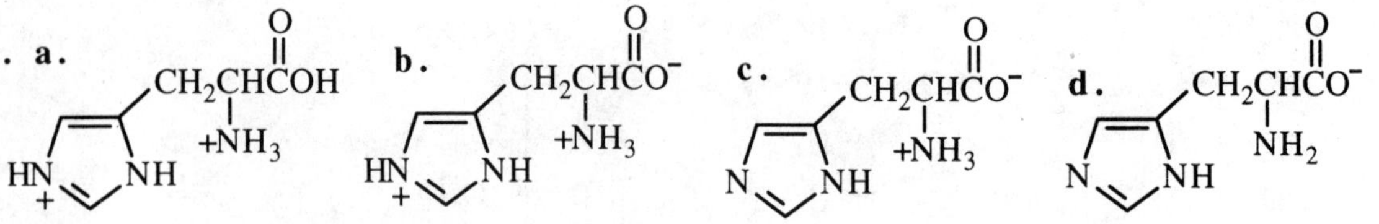

3. **a.** Alanine, because it is farther away from its pI.
   b. glycine
   c. leucine and isoleucine
   d. aspartic acid

4. The electron-withdrawing protonated amino group causes the carboxyl group of alanine to have a lower pK_a.

5. **a.** A cigar shaped protein has a greater percentage of polar residues than a spherical protein. T
   b. Naturally occurring amino acids have the L-configuration. T
   c. There is free rotation about a peptide bond. F

6. $^-OCCHCH_2S{-}SCH_2CHCO^-$ (each C=O), with $+NH_3$ on each α-carbon

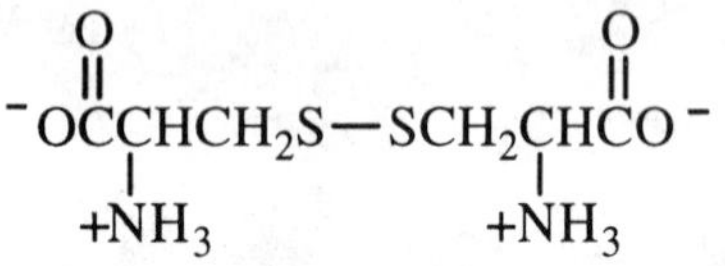

7. **a.** The sequence of the amino acids in the protein chain.
   b. The three-dimensional arrangement of all the atoms in the protein.
   c. A description of the way the subunits of an oligomer are arranged in space.

8.

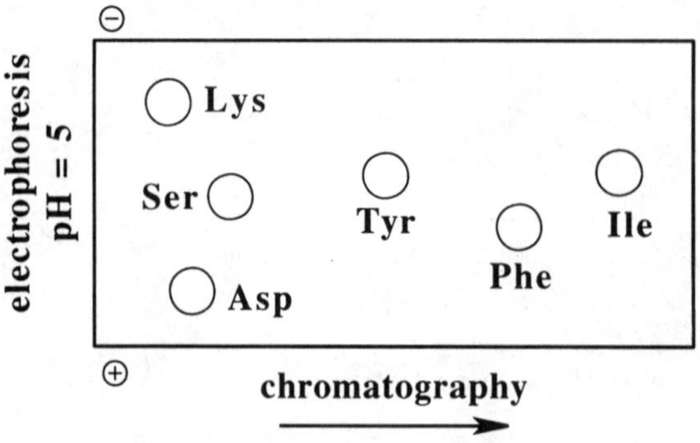

⊖

Lys

Ser Tyr Ile

Phe

Asp

⊕

electrophoresis pH = 5

chromatography →

9. a. $\frac{1.83 + 9.13}{2} = \frac{10.96}{2} = 5.48$ b. $\frac{9.04 + 12.48}{2} = \frac{21.52}{2} = 10.76$

10. Ala Ser Arg Gly Arg Met His Phe Lys Ile